THERMAL RADIATION
HEAT TRANSFER

THERMAL RADIATION HEAT TRANSFER

Third Edition

Robert Siegel
Lewis Research Academy
NASA Lewis Research Center

John R. Howell
Baker-Hughes Professor of Mechanical Engineering
University of Texas at Austin

⊙ HEMISPHERE PUBLISHING CORPORATION
A member of the Taylor & Francis Group
Washington   Philadelphia   London

USA	Publishing Office:	Taylor & Francis
		1101 Vermont Avenue, N.W., Suite 200
		Washington, DC 20005-3521
		Tel: (202) 289-2174
		Fax: (202) 289-3665
	Distribution Center:	Taylor & Francis Inc.
		1900 Frost Road, Suite 101
		Bristol PA 19007-1598
		Tel: (215) 785-5800
		Fax: (215) 785-5515
UK		Taylor & Francis Ltd.
		4 John St.
		London WC1N 2ET
		Tel: 071 405 2237
		Fax: 071 831 2035

THERMAL RADIATION HEAT TRANSFER, Third Edition

1 2 3 4 5 6 7 8 9 0   B R B R   9 8 7 6 5 4 3 2

This book was set in Times Roman by Edwards Brothers. The editors were Mary Prescott and Carolyn V. Ormes. Cover design by Michelle Fleitz. Printing and binding by Braun-Brumfield, Inc.

A CIP catalog record for this book is available from the British Library.

∞ The paper in this publication meets the requirements of the ANSI Standard Z39.48-1984 (Permanence of Paper)

Library of Congress Cataloging-in-Publication Data
Siegel, Robert, date.
    Thermal radiation heat transfer / Robert Siegel, John R. Howell. — 3rd ed.
      p. cm.
    Includes bibliographical references and index.

    1. Heat—Transmission and absorption. 2. Heat—Transmission.
    3. Materials—Thermal properties. I. Howell, John R. II. Title.
QC331.S55   1992
621.402′2—dc20   92-10974
                                                              CIP

ISBN 0-89116-271-2

CONTENTS

*Topics preceded by an asterisk can form the basis of one-semester course as discussed in the Preface to the First Edition.

PREFACE

It has been eleven years since publication of the second edition of *Thermal Radiation Heat Transfer*. During that time, developments have continued as prompted by many current interests and applications involving thermal radiation. These include thermal control in space technology, efforts to improve combustion, use of high temperature ceramics and glass forming technology, solar energy utilization, hypersonic flight, high temperature engines, furnace technology, cryogenic insulation, and studies of the earth's energy balance. This third edition is a comprehensive revision that incorporates new general information, advances in analytical and computational techniques, and new reference material. Because the organization of material in the previous editions has been found successful in the classroom, the general order of subjects has been retained, although some chapters have been combined and rearranged. Briefly, this order is to present basic background information, blackbody properties, fundamental definitions, and material properties, then continue with configuration factors, enclosure theory, and the inclusion of conduction and convection at enclosure boundaries. The final seven chapters deal with media, such as gases, ceramics, and glass, that absorb, emit, and scatter radiation throughout their volume.

The original objective, which was to provide detailed fundamental derivations so the text can be used for self-study as well as in the classroom, has been retained. This is aided by providing numerical examples in each chapter. The number of examples has been increased in this edition. Additional homework problems are also provided, many with answers. The reference material is presented in sufficient detail so the book can also be used for reference purposes.

A significant development in the past eleven years has been the availability of computers with increased capacity and speed. Many problem areas that previously were treated only by approximate analytical techniques, or that could not be solved, can now be investigated and solved by numerical methods. For this reason information on numerical methods has been added to this edition. Finite-difference, finite-element, Monte Carlo, and other methods are presented as they apply to various radiation problems involving conduction and convection, with or without radiatively participating media. The previous edition contained two chapters on Monte Carlo methods. These chapters have been changed to incorporate the additional numerical methods. The chapter on approximate methods in radiative transfer, such as the P-N method, has been considerably updated to reflect the large amount of literature using these methods that was published during the past eleven years. Finite-difference and finite-element methods are included because their use is increasing in radiation problems.

To improve the organization of the previous material and provide space for new material, some of the chapters have been combined and rearranged. The material on enclosure theory has been unified by including black and diffuse-gray surfaces in the same chapter. The various approaches for enclosure solutions are compared to show how they are related, and alternative forms of the equations are included. The material on gas properties and scattering has been combined into a single chapter to eliminate some previous repetition of fundamental definitions. The engineering approach to gas radiation problems has been presented earlier to tie in more closely with the enclosure methods. The information on soot radiation has been updated.

In the sections on emitting, absorbing, and scattering materials, the previous editions treated mainly plane layers. The treatment has now been generalized to include multidimensional configurations, and results are given for rectangular regions. Techniques for transient solutions are also presented. The information on radiating materials with nonunity refractive indices has been expanded to discuss boundary conditions and multiple layers. The catalog of available configuration factors has been revised to include new factors that have become available.

The basic radiation and physical constants, such as the Stefan-Boltzmann constant, have been revised to the values currently in use as a result of the latest international revision. The numerical examples have all been reworked using these new values. The text is now primarily in SI units.

ROBERT SIEGEL
JOHN R. HOWELL

PREFACE TO THE SECOND EDITION

In this second edition of *Thermal Radiation Heat Transfer,* the authors have tried to retain the features that made the first edition unique as a radiation textbook and reference. These features include detailed derivation of the fundamentals of radiative transfer, careful exposition of the assumptions underlying the radiation laws, and sufficient explanatory material so that the student or practicing engineer can use the book as both a self-study text and a reference.

In the several years since the first edition, there has been considerable development in the field of solar-energy utilization. For this reason a new chapter was added on the transmission behavior of windows and multiple windows (as used in solar collectors), coated surfaces, and thin films. The material on glass properties has been expanded.

To improve the organization, two chapters were eliminated by consolidation with material in other chapters. The material on soot radiation was expanded and then combined with the information on radiation of gases in furnaces and combustion chambers. The treatment of scattering was combined with that on absorption to yield a more unified development of radiative behavior in attenuating media.

Many other additions have been made. The catalog of available configuration factors has been brought up to date. There has been considerable recent work on band radiation correlations, so this section was expanded. There is an updated section on the differential approximation, and a number of new homework problems and examples. The use of SI units has been increased but they are not used exclusively, as English units are also commonly used in engineering heat transfer calculations in the United States.

ROBERT SIEGEL
JOHN R. HOWELL

PREFACE TO THE FIRST EDITION

Several years ago it was realized that thermal radiation was becoming of increasing importance in aerospace research and design. This importance arose from several areas: high temperatures associated with increased engine efficiencies, high-velocity flight accompanied by elevated temperatures from frictional heating, and the operation of devices beyond the earth's atmosphere where convection vanishes and radiation becomes the only external mode of heat transfer. As a result, a course in thermal radiation at about a first-year graduate level was initiated at the NASA Lewis Research Center as part of an internal advanced study program.

The course was divided into three main sections. The first dealt with the radiation properties of opaque materials, including a discussion of the blackbody, electromagnetic theory, and measured properties. The second discussed radiation exchange in enclosures both with and without convection and wall conduction. The third section treated radiation in partially transmitting materials—chiefly gases.

When the course was originated, a single radiation textbook that covered the desired span was not available. As a result we began writing a set of notes; the present publication is an outgrowth of these notes. During the past few years, a few radiation textbooks have appeared in the literature; hence the need for a single reference has been partially satisfied.

The objectives of this volume are more extensive than providing the content of a standard textbook intended for a one-semester course. Many parts of the present discussion have been made quite detailed so that they will serve as a source of reference for some of the more subtle points in radiation theory. The detailed treatment has resulted in some rather long sections, but the intent in these instances was to be thorough rather than to try to conserve space. The sections have been subdivided so that specific portions can be located for easy reference.

This volume is divided into 21 chapters. The first five deal with some fundamentals of radiative transfer, the blackbody, electromagnetic theory, and the properties of solid materials. Chapters 6 through 12 treat energy exchange between surfaces and in enclosures when no attenuating medium is present. The final nine chapters concern radiative transfer in the presence of an attenuating medium.

Topics felt to be of primary interest for a one-semester course at a first-year graduate level have been marked with an asterisk in the contents. Experience in teaching the course has indicated that these topics can be covered provided that some of the subsections are considered in less detail according to the emphasis desired by the instructor. The authors have tried in most of the work to use a conversational style and, at the risk of being wordy, explain things in fundamental detail. As a teaching philosophy, the material usually proceeds from specific to more general (and hence usually more complex) situations. For these reasons the text is also intended for self-study by engineers having little previous knowledge of thermal radiation.

Each chapter contains numerical examples to acquaint the reader with the use of the analytical relations. It is hoped that these examples will help bridge the gap between theory and practical application. Homework exercises are also included at the end of each chapter.

As a closing note, let it be stated that this book comes as close as possible to being a 50:50 effort by the authors. This paragraph is written prior to flipping a coin to determine first authorship. Therefore, whoever appears as first author does not claim major credit for the work, but only a hot gambling hand.

The authors would like to acknowledge the help from many of their associates at NASA who reviewed the material and provided valuable detailed comments. Special thanks are extended to Curt H. Liebert who carefully reviewed the entire manuscript.

ROBERT SIEGEL
JOHN R. HOWELL

ONE

INTRODUCTION

All substances continuously emit electromagnetic radiation by virtue of the molecular and atomic agitation associated with the internal energy of the material. In the equilibrium state, this internal energy is proportional to the temperature of the substance. The emitted radiant energy can range from radio waves, which can have wavelengths of tens of meters, to cosmic rays with wavelengths less than 10^{-14} meter. This book will consider only radiation that is detected as heat or light. Such radiation is termed *thermal radiation,* and it occupies an intermediate wavelength range (defined explicitly in Sec. 1-5).

Although radiant energy constantly surrounds us, we are not very aware of it because our bodies are able to detect only portions of it directly. Other portions require detection by using instruments. Our eyes are sensitive detectors of light, being able to form images of objects, but are relatively insensitive to heat (infrared) radiation. Our skin is a direct detector for heat radiation but not a good one. The skin is not aware of images of warm or cool surfaces around us unless the heat radiation is large. We require indirect means, such as an infrared-imaging video camera [1], to form images using heat radiation.

Before discussing the nature of thermal radiation in detail, it is well to consider why thermal radiation is important in current technology.

1-1 IMPORTANCE OF THERMAL RADIATION

One factor that accounts for the importance of thermal radiation in some applications is the manner in which radiant emission depends on temperature. For conduction and convection the transfer of energy between two locations depends on the temperature difference of the locations to approximately the first power.* The

---

*For free convection, or when variable property effects are included, the power of the temperature difference may become larger than 1 but usually in convection and conduction does not approach 2.

transfer of energy by thermal radiation between two bodies, however, depends on the difference between the individual absolute temperatures of the bodies each raised to a power in the range of about 4 or 5.

From this basic difference between radiation and the convection and conduction energy-exchange mechanisms, it is evident that the importance of radiation becomes intensified at high absolute-temperature levels. Consequently, radiation contributes substantially to heat transfer in furnaces and combustion chambers and to the energy emission from a nuclear explosion. The laws of radiation govern the temperature distribution within the sun and the radiant emission from the sun or from a radiation source duplicating the sun in a solar simulator. The nature of radiation from the sun is of obvious importance in the technology for solar-energy utilization. Some devices for space applications are designed to operate at high temperature levels to achieve high thermal efficiency. Hence radiation must often be considered in calculating thermal effects in devices such as a rocket nozzle, a nuclear power plant for space applications, or a gaseous-core nuclear rocket.

A second distinguishing feature of radiative transfer is that *no* medium need be present between two locations for radiant exchange to occur. Radiative energy passes perfectly through a vacuum. This is in contrast to convection and conduction, where a physical medium must be present to carry energy with the convective flow or to transport it by thermal conduction. When no medium is present, radiation becomes the only significant mode of heat transfer. Some examples are heat leakage through the evacuated space in the wall of a Dewar flask or thermos bottle and heat dissipation from the heated filament of a vacuum tube. Many important examples are in devices operating in outer space or in earth orbit. On earth, coolants such as air or water can be readily employed to dissipate waste heat from devices such as power plants, automobile engines, and computers and other electronic devices. For devices operating in space, waste energy must ultimately be rejected by radiation as no surrounding coolants are available. For large space power plants, such as for future space stations, the radiating devices to reject heat may become components of major size and weight. Another device that depends on radiation transfer is the temperature control of a satellite. The equilibrium temperature depends on the heat balance of absorbed solar energy, emitted and reflected energy from earth (for near-earth orbits), radiation from the satellite surfaces, and sources of internal energy production such as on-board electronics.

Radiation can be of importance in some instances even though the temperature level is not elevated and other modes of heat transfer are present. The following example is from a Cleveland newspaper published in the spring of 1964. A florist "noted the recurrence of a phenomenon he has observed for two seasons since using plastic coverings over [flower] flats. Water collecting in the plastic has formed ice a quarter-inch thick [at night] when the official [temperature] reading was well above freezing. 'I'd like an answer to that, I supposed you couldn't get ice without freezing temperatures.'" The florist's oversight was in considering only the convection to the air and omitting the nighttime radiation loss occurring between the water-covered surface and the cold heat sink of the night sky.

A similar illustration is the discomfort a person experiences in a room where cold surfaces are present. Cold window surfaces can have a chilling effect as the body radiates to them without receiving compensating energy from them. Covering cold windows with a shade or drape will decrease bodily discomfort. Another example is the warming effect of sunshine felt by a person outside in cold weather. In northern climates we often hear of the "wind-chill factor," referring to wind-augmented convective heat losses from the human body. Perhaps there should be a compensating "sun-warm factor" for evaluating human comfort.

Radiation is important in some instances because its action from a distance provides local heat sources that modify temperature distributions, thereby influencing conduction, free convection, or forced convection. In fiberglass insulation radiation can penetrate deeply to augment the heat flow and the usual conduction heat transfer. Radiation can heat the walls of an enclosure, producing free convection where it would not ordinarily occur. In boundary-layer flows of gases that absorb radiation, the presence of external radiation can alter the convective heat transfer.

An important application of thermal radiation is in the practical utilization of the sun's radiation as an energy source on earth. Solar energy is transferred to a solar collector on earth through the vacuum of space and the earth's atmosphere. The collector converts solar radiation into internal energy. If the arriving energy is not concentrated by a curved mirror or lens, the collector normally functions at temperatures near to, or at most a few hundred degrees Celsius above, ambient. The balance between the available solar energy, the useful energy transferred to a working fluid, and the various convective, conductive, and radiative losses is very sensitive, and it makes the design of an efficient collector quite challenging.

Finally, we note that the thermal radiation we shall examine is in the wavelength region that gives humans heat, light, photosynthesis, and all their attendant benefits. This is strong justification for studying thermal radiation. Our existence depends on the solar radiant energy incident upon the earth. Understanding the interaction of this radiation with the atmosphere and surface of the earth is important and can provide additional benefits in its use.

1-2 SYMBOLS

A	surface area
c	speed of electromagnetic radiation propagation in *medium* other than a vacuum
c_0	speed of electromagnetic radiation propagation in *vacuum*
k	thermal conductivity
n	index of refraction, c_0/c
q_c	energy per unit area per unit time resulting from heat conduction
q_r	radiant energy per unit area per unit time arriving at surface element
q_s	radiant energy per unit area per unit time arriving from unit surface element

q_v	radiant energy per unit area per unit time arriving from unit volume element
S	surface area
T	temperature
V	volume
x, y, z	coordinates in cartesian system
ζ	arbitrary direction
η	wavenumber in vacuum, $1/\lambda$; η_m in a medium
λ	wavelength in vacuum; λ_m in a medium
ν	frequency, $c_0/\lambda - c/\lambda_m$

1-3 COMPLEXITIES INHERENT IN RADIATION PROBLEMS

First consider some of the mathematical complexities that arise from the basic nature of radiation exchange. In conduction and convection, energy is transported by means of a physical medium. The energy transferred into and from an infinitesimal volume element of solid or fluid depends on the temperature gradients and physical properties in the *immediate vicinity* of the element. For example, for the relatively simple case of heat conduction in a material (no convection) with temperature distribution $T(x,y,z)$ and constant thermal conductivity k, the energy equation is derived by locally applying the Fourier conduction law:

$$q_c|_{\text{in } \zeta \text{ direction}} = -k\frac{\partial T}{\partial \zeta} \tag{1-1}$$

For an elemental cube within a solid as shown in Fig. 1-1a, consideration of the net heat flow in and out of all the faces, using the terms given in the figure, yields the Laplace equation governing heat conduction within the material:

$$\frac{\partial^2 T}{\partial x^2} + \frac{\partial^2 T}{\partial y^2} + \frac{\partial^2 T}{\partial z^2} = 0 \tag{1-2}$$

The terms in this energy-balance equation depend only on *local* temperature derivatives in the material.

A similar although more complex analysis can be made for the convection process, again demonstrating that the heat balance depends only on conditions in the immediate vicinity of the location being considered.

In radiation, energy is transmitted between separated elements *without* the need of a medium between the elements. Consider a heated enclosure of surface S and volume V filled with radiating material (such as hot gas, a cloud of particles, or glass) as shown in Fig. 1-1b. If $q_s\,dS$ is the radiant energy flux (energy per unit area and per unit time) arriving at dA from an element of the surface dS of the enclosure, and $q_v\,dV$ arrives at dA from an element of the gas dV, then the total radiation arriving per unit area at dA is

$$q_r = \int_S q_s\,dS + \int_V q_v\,dV \tag{1-3}$$

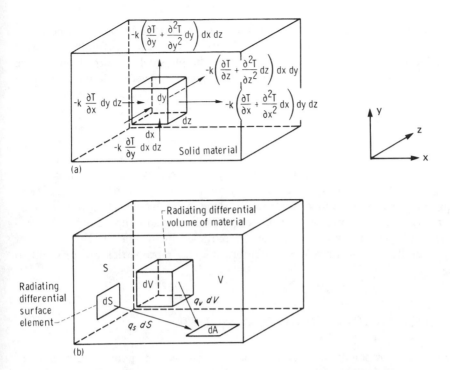

Figure 1–1 Comparison of types of terms for conduction and radiation heat balances (*a*) Heat conduction terms for volume element in solid; (*b*) radiation terms for enclosure filled with radiating material.

These types of terms lead to heat balances in the form of integral equations, which are generally not as familiar to the engineer as differential equations. When radiation is combined with conduction and/or convection, the presence of both integral and differential terms having different powers of temperature can lead to nonlinear integrodifferential equations that are difficult to solve. A complication for numerical solutions is that the relatively large grid sizes reasonable for integrating the radiation terms (especially in three dimensions) are often not adequate to give good accuracy for the conduction and/or convection terms. This can cause difficulty in the convergence of numerical solutions.

Equation (1-3) emphasizes the geometric aspects of radiative transfer that require integrations over surfaces and volumes. There is another integration that can be very important; this is over a spectral variable such as wavelength or frequency. Radiative properties are often quite wavelength dependent; this is true of both surface properties and volume properties such as for radiating gases. To account for property variations, the solution is carried out in individual spectral regions and then the heat flows are integrated over all wavelengths to obtain total energy quantities. The boundary conditions may be specified in terms of total energies, as in an electrical heating, so the spectral heat flows at the boundaries are unknown. This can require an iterative solution to satisfy the boundary conditions.

In addition to the mathematical complexities there is the difficulty, inherent in radiation problems, of accurately specifying the physical property values to be inserted into the equations. This will limit engineering accuracy even if exact mathematical solutions are available. The difficulty in specifying accurate property values arises because the properties for solids depend on many variables, such as surface roughness and degree of polish, purity of material, thickness of a coating such as paint on a surface (for a thin coating the underlying material may have an effect), temperature, wavelength of radiation, and angle at which radiation leaves the surface. Unfortunately, many measurements have been reported where the pertinent surface conditions have not all been precisely defined. For radiating gases the properties usually depend on wavelength in a very irregular way. The properties are a function of pressure and temperature plus composition for a gas mixture. There can be particles in the gas, such as soot or ash, that emit, absorb, or scatter radiation, and their properties are difficult to define.

1-4 WAVE AGAINST QUANTUM MODEL

Radiant energy propagation can be considered from two viewpoints: classical electromagnetic wave theory and quantum mechanics. The classical view of the interaction of radiation and matter yields, in most cases, equations that are similar to the quantum-mechanical results. With a few exceptions, thermal radiation may therefore be viewed as based on the classical concept of energy transport by electromagnetic waves. The exceptions, however, include some of the most important effects common to radiative-transfer studies, such as the spectral distribution of the energy emitted from a body and the behavior of the radiative properties of gases. These can be explained and derived only on the basis of quantum effects in which energy is carried by discrete particles (photons). The "true" nature of electromagnetic energy (that is, waves or quanta) is not generally important to the engineer. Throughout the present work, the wave theory will usually be applied because it has the greatest utility in engineering calculations and generally produces the same formal equations as the quantum theory. Some reference will be made to phenomena for which quantum arguments must be invoked, such as transitions between energy levels that account for the irregular behavior of gas properties as a function of wavelength.

1-5 ELECTROMAGNETIC SPECTRUM

Within the framework of wave theory, electromagnetic radiation follows the laws for transverse waves oscillating perpendicular to the direction of travel. The speed of propagation for electromagnetic radiation in vacuum is the same as for light;

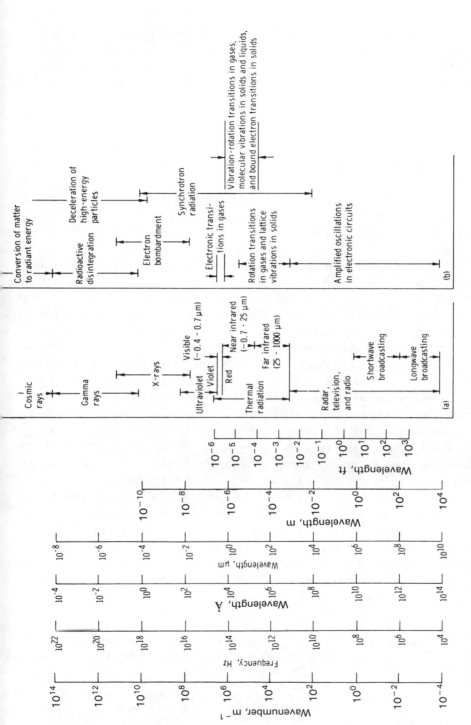

Figure 1-2 Spectrum of electromagnetic radiation (wavelength and wavenumber in vacuum). (*a*) Type of radiation; (*b*) production mechanism.

7

light, after all, is simply the special case of electromagnetic radiation in a small region of the spectrum. In vacuum the speed of propagation $c_0 = 2.9979 \times 10^8$ m/s (186,000 miles per second). The speed c in a medium is less than c_0 and is commonly given in terms of the index of refraction $n = c_0/c$, where n is greater than unity.* For glass, n is about 1.5, while for gases n is very close to 1.

The types of electromagnetic radiation can be classified according to their wavelength λ in vacuum, frequency ν where $c_0 = \lambda\nu$, or wavenumber $\eta = 1/\lambda$. Frequency has an advantage for some calculations as it does not change when a wave goes from one material into another with a different refractive index. Common units for wavelength are the micrometer† (μm), where 1 μm = 10^{-6} m or 10^{-4} cm, and the angstrom (Å), where 1 Å = 10^{-10} m. Hence 10^4 Å = 1 μm. A chart of the radiation spectrum is in Fig. 1-2. A set of conversion factors for units in radiative transfer is given in Tables A-2 and A-3 in Appendix A.

The wavelength region of interest here includes a portion of the long-wave fringe of the ultraviolet, the visible-light region that extends from approximately $\lambda = 0.4$ to 0.7 μm, and the infrared region that extends from beyond the red end of the visible spectrum to about $\lambda = 1000$ μm. The infrared region is sometimes divided into the near infrared, extending from the visible region to about $\lambda = 25$ μm, and the far infrared, composed of the longer-wavelength portion of the infrared spectrum. In the visible region the colors are at the following approximate wavelengths in vacuum: violet, 0.43 μm; green, 0.53 μm; red, 0.64 μm.

The column at the far right in Fig. 1-2 indicates the various mechanisms by which electromagnetic radiation is produced. Some of the descriptions reflect a quantum-mechanical viewpoint in which electrons or molecules in a state of agitation undergo transitions from one energy state to another at lower energy. These transitions result in radiative energy release. The transitions may occur spontaneously or may be initiated by the presence of a radiation field.

In this chapter we have discussed the importance of thermal radiation, the difficulties inherent in radiation problems, and the wavelength region occupied by thermal radiation within the electromagnetic spectrum. In the next chapter, we examine the radiative behavior of the *ideal* radiating body, termed a *blackbody*. With the behavior of this ideal as a standard for comparison, the behavior of radiative energy for conditions of interest to the engineer will be discussed in succeeding chapters.

---

*For attenuating media such as metals the index of refraction is a complex quantity of which n is only the real part. In some instances, such as in the region of anomalous dispersion, n can be less than unity, which might convey the impression that the propagation speed can be greater than c_0. This is not the case; in this instance the waves propagating in the medium can be complicated in form. The c is then the phase velocity of the wave, which has no physical significance when it exceeds c_0. For further discussion see section 1.3 of Born and Wolf [2] and Sec. 4-5.2 of this text.

†The term *micron* (μ) has also been used to represent 10^{-6} m; micrometer (μm), however, is now preferred.

REFERENCES

1. Silverman, J., J. M. Mooney, and F. D. Shepherd: Infrared Video Cameras, *Sci. Am.*, pp. 78–83, March 1992.
2. Born, Max, and Emil Wolf: "Principles of Optics," 2d rev. ed., The Macmillan Company, New York, 1964.

PROBLEMS

1-1 What are the wavenumber range in vacuum and the frequency range for the visible spectrum (0.4 to 0.7 μm)? What are the wavenumber and frequency values at the boundary between the near- and far-infrared regions?

Answer: 2.5×10^6 to 1.4286×10^6 m^{-1}, 4.283×10^{14} to 7.495×10^{14} s^{-1}, 4×10^4 m^{-1}, 1.1992×10^{13} s^{-1}

1-2 Radiant energy at a wavelength of 2 μm is traveling through a vacuum. It then enters a medium with a refractive index of 1.5.

(a) Find the following quantities for the radiation in the vacuum: speed, frequency, wavenumber.

(b) Find the following quantities for the radiation in the medium: speed, frequency, wavenumber, wavelength.

Answer: (a) 2.998×10^8 m/s, 1.4990×10^{14} s^{-1}, 5×10^5 m^{-1}; (b) 1.9986×10^8 m/s, 1.4990×10^{14} s^{-1}, 7.5×10^5 m^{-1}, 1.3333×10^{-6} m

1-3 Radiation within a transparent medium is found to have a wavelength of 2.7 μm and a speed of 2.29×10^8 m/s.

(a) What is the refractive index of the medium?

(b) What is the wavelength of the radiation in a vacuum?

Answer: (a) 1.3092; (b) 3.535 μm

1-4 What range of wavelengths will be present within a glass sheet that has a wavelength-independent refractive index of 1.52 when the sheet is exposed in vacuum to incident radiation in the visible range (0.4 to 0.7 μm)?

Answer: 0.2631 to 0.4605 μm

1-5 A material has an index of refraction $n(x)$ that varies with position x within its thickness. Obtain an expression in terms of c_0 and $n(x)$ for the transit time for radiation to pass through a thickness L. If $n(x) = n_i(1 + kx)$, where n_i and k are constants, what is the relation for transit time? How does wavenumber (relative to that in a vacuum) vary with position within the medium?

Answer: $\dfrac{n_i}{c_0}\left(L + \dfrac{kL^2}{2}\right)$, $n_i(1 + kx)$

TWO

RADIATION FROM A BLACKBODY

2-1 INTRODUCTION

Before discussing the idealized concept of the blackbody, let us consider a few aspects of the interaction of incident radiant energy with matter. The idea we are concerned with is that the interaction at the surface of a body is not the result of only a surface property but depends as well on the bulk material beneath the surface.

When radiation is incident on a homogeneous body, some of the radiation is reflected and the remainder penetrates into the body. The radiation may then be absorbed as it travels through the medium. If the material thickness required to substantially absorb the radiation is large compared with the thickness dimension of the body, then most of the radiation will be transmitted entirely through the body and will emerge with its nature unchanged. If, on the other hand, the material is a strong internal absorber, the radiation that is not reflected from the body will be converted into internal energy within a thin layer adjacent to the surface.

A distinction must be made between the ability of a material to let radiation pass through its surface and its ability to internally absorb the radiation after it has passed into the body. For example, a highly polished metal will generally reflect all but a small portion of the incident radiation, but the radiation passing into the body will be strongly absorbed and converted to internal energy within a very short distance within the material. Thus the metal has strong *internal* absorption ability, although it is a poor absorber for the incident beam because most of the incident beam is reflected. Nonmetals may exhibit the opposite tendency. Nonmetals may allow a substantial portion of the incident beam to pass into the material (small surface reflection), but a larger thickness is usually required than

11

in the case of a metal to absorb the radiation internally and convert it into internal energy. A glass window readily allows radiation to pass through its surface but is a poor absorber for visible radiation, hence that radiation is transmitted. When *all* the radiation that passes into the body is absorbed internally, the body is *opaque*.

To be a good absorber for incident energy, a body must have a low surface reflectivity and internal absorption sufficiently high to prevent the radiation from passing through. If metals in the form of very fine particles are deposited on a subsurface, the result is a surface of low reflectivity. This effect, combined with the high internal absorption of the metal, causes this type of coating to be a good absorber. This is the basis for formation of the metallic "blacks," such as platinum or gold black [1]. These materials can have less than 1% reflection for the solar spectrum and provide complete internal absorption within a few micrometers of thickness. A *blackbody* must have *zero surface reflection* and *complete internal absorption*.

2-2 SYMBOLS

A	surface area
c	speed of electromagnetic radiation propagation in a *medium* other than vacuum
c_0	speed of electromagnetic radiation propagation in *vacuum*
C_1, C_2	constants in Planck's spectral energy distribution (See Table A-4 of Appendix A)
C_3	constant in Wien's displacement law (see Table A-4 of Appendix A)
e	emissive power (usually with subscript); with superscript, exponential function
E	energy emitted per unit time
$F_{0-\lambda}$	fraction of total blackbody intensity or emissive power lying in spectral region 0 to λ
h	Planck's constant
i	radiant intensity
k	Boltzmann constant
n	refractive index
Q	energy per unit time
R	radius
T	absolute temperature
ζ	the quantity $C_2/\lambda T$
η	wavenumber
θ	polar or cone angle (measured from normal of surface)
κ	extinction coefficient for electromagnetic radiation
λ	wavelength in vacuum
λ_m	wavelength in a *medium* other than vacuum
ν	frequency
σ	Stefan-Boltzmann constant [Eq. (2-27)]

φ circumferential, or azimuthal, angle

ω solid angle

Subscripts

b blackbody

max   corresponding to maximum energy

n normal direction

p projected

s sphere

η wavenumber dependent

λ spectrally (wavelength) dependent

$\lambda_1-\lambda_2$ in wavelength span from λ_1 to λ_2

λT evaluated at λT

ν frequency dependent

Superscript

'   directional quantity

2-3 DEFINITION OF A BLACKBODY

A *blackbody* is defined as an ideal body that allows *all* the incident radiation to pass into it (no reflected energy) and internally absorbs *all* the incident radiation (no transmitted energy). This is true of radiation for all wavelengths and for all angles of incidence. Hence *the blackbody is a perfect absorber for all incident radiation*. All other qualitative aspects of blackbody behavior can be derived from this definition.

The concept of a blackbody is basic to the study of radiative energy transfer. As a perfect absorber, it serves as a standard with which real absorbers can be compared. As will be seen, the blackbody also emits the maximum radiant energy and hence serves as an ideal standard of comparison with a real body emitting radiation. The quantitative radiative properties of the ideal blackbody have been well established by quantum theory and have been verified by experiment.

Only a few materials, such as carbon black, carborundum, platinum black, gold black, and some specially formulated black paints on absorbing substrates, approach the blackbody in their ability to absorb radiant energy. The blackbody derives its name from the observation that good absorbers of incident visible light do indeed appear black to the eye. However, except for the visible region, the eye is not a good indicator of absorbing ability over the wavelength range of thermal radiation. For example, a surface coated with white oil-base paint is a very good absorber for infrared radiation emitted at room temperature, although it is a poor absorber for the shorter-wavelength region characteristic of visible light as evidenced by its white appearance.

2-4 PROPERTIES OF A BLACKBODY

Aside from being a perfect absorber for all incident radiation, the blackbody has other important properties, now to be discussed.

2-4.1 Perfect Emitter

Consider a blackbody at a uniform temperature placed in vacuum within a perfectly insulated enclosure of arbitrary shape whose walls are blackbodies that are at a uniform temperature initially different from that of the enclosed blackbody (Fig. 2-1). After a period of time, the blackbody and the enclosure will attain a common uniform equilibrium temperature. In this equilibrium condition, the blackbody must radiate away exactly as much energy as it absorbs. To prove this, consider what would happen if the incoming and outgoing amounts of radiation were not equal with the system at a uniform temperature. Then the enclosed blackbody would either increase or decrease in temperature. This would involve a net amount of heat transferred between two bodies at the same temperature, which violates the second law of thermodynamics. It follows that, because the blackbody is by definition absorbing the maximum possible radiation incident from the enclosure at each wavelength and from each direction, it must also be emitting the maximum total amount of radiation. This is made clear by considering any less-than-perfect absorber, which must consequently emit less energy than the blackbody to remain in equilibrium. The fact that a body must continue to emit radia-

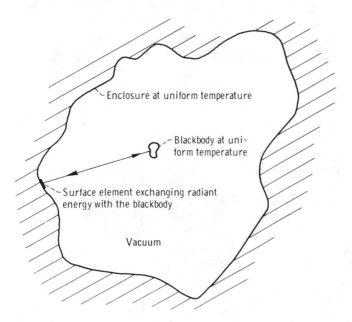

Figure 2-1 Enclosure geometry for derivation of blackbody properties.

tion even when in thermal equilibrium with its surroundings is called *Prevost's law*.

2-4.2 Radiation Isotropy in a Black Enclosure

Now, consider the isothermal enclosure with black walls and arbitrary shape shown in Fig. 2-1, move the blackbody to another position, and rotate it to another orientation. The blackbody must still be at the same temperature because the whole enclosure remains isothermal. Consequently the blackbody must be emitting the same amount of radiation as before. To be in equilibrium, the body must still be receiving the same amount of radiation from the enclosure walls. Thus, the total radiation received by the blackbody is independent of body orientation or position throughout the enclosure. Hence, the radiation traveling through any point within the enclosure is independent of position or direction. This means that the black radiation filling the enclosure is *isotropic*.

In addition to emitting the maximum possible total radiation, the blackbody emits the maximum possible energy in each direction and at each wavelength. This is shown by the following arguments.

2-4.3 Perfect Emitter in Each Direction

Consider an area element on the surface of the black isothermal enclosure and an elemental blackbody within the enclosure. Some of the radiation from the surface element strikes the elemental body and is at an angle to the body surface. All this radiation, by definition, is absorbed. To maintain thermal equilibrium and isotropic radiation throughout the enclosure, the radiation emitted back into the incident direction must equal that received. Since the body is absorbing the maximum radiation from any direction, it must be emitting the maximum in any direction. Because the black radiation filling the enclosure is isotropic, the radiation received and hence emitted in *any* direction by the enclosed black surface, per unit projected area normal to that direction, must be the same as that in any other direction.

2-4.4 Perfect Emitter at Every Wavelength

Consider a blackbody inside an evacuated enclosure with the whole system in thermal equilibrium. The enclosure boundary is specified as being of a special type—it emits and absorbs radiation only in the small wavelength interval $d\lambda_1$ around λ_1. The blackbody, being a perfect absorber, absorbs all the incident radiation in this wavelength interval. To maintain the thermal equilibrium of the enclosure, the blackbody must re-emit radiation in this same wavelength interval; the radiation can then be absorbed by the enclosure boundary, which absorbs only in this particular wavelength interval. Since the blackbody is absorbing a maximum of the radiation in $d\lambda_1$, it must be emitting a maximum in $d\lambda_1$. A second enclosure can now be specified that emits and absorbs only the interval $d\lambda_2$ around

λ_2. The blackbody must then emit a maximum in $d\lambda_2$. In this manner the blackbody is shown to be a perfect emitter at each wavelength. The special nature of the enclosure assumed in this discussion is of no significance relative to the blackbody, because the emissive properties of a body depend only on the nature of the body and are independent of the evacuated enclosure.

2-4.5 Total Radiation into Vacuum Is a Function Only of Temperature

If the enclosure temperature is altered, the enclosed blackbody temperature must adjust and become equal to the new enclosure temperature. The system is again isothermal and evacuated, and the absorbed energy and emitted energy of the blackbody will again be equal to each other, although different in magnitude from the value for the previous enclosure temperature. Since by definition the body absorbs (and hence emits) the maximum amount corresponding to this temperature, the characteristics of the evacuated surroundings do not affect the emissive behavior of the blackbody. Hence, the *total radiant energy emitted by a blackbody in vacuum is a function only of its temperature*.

The second law of thermodynamics forbids net energy transfer from a cooler to a hotter surface without doing work on the system. If the radiant energy emitted by a blackbody increased with decreasing temperature, we could build a device to violate this law. Consider, for example, the infinite parallel black plates shown in Fig. 2-2. The upper plate is held at temperature T_1, which is higher than the temperature T_2 of the lower plate. If the emission of energy decreases with increasing temperature, then the energy E_2 emitted per unit time by plate 2 is larger than E_1, that emitted by plate 1. Because the plates are black, each absorbs all energy emitted by the other. To maintain the temperature of the plates, an amount of energy $Q_1 = E_2 - E_1$ must be extracted from plate 1 per unit time and an equal amount added to plate 2. Thus we are transferring net energy from the colder to the warmer plate without doing external work, a violation of the second law of thermodynamics. Therefore, the radiant energy emitted by a blackbody must in-

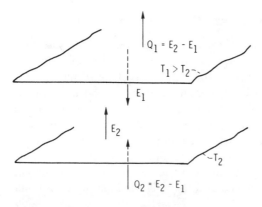

Figure 2-2 Device violating second law of thermodynamics

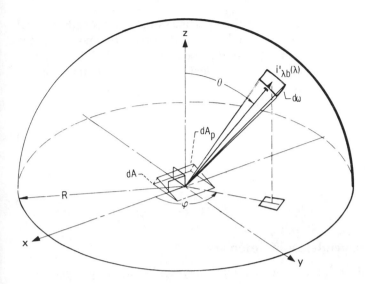

Figure 2-3 Spectral emission intensity from black surface.

crease with temperature. The total radiant energy emitted by a blackbody is thus expected to be proportional only to a monotonically increasing function of temperature.

2-5 EMISSIVE CHARACTERISTICS OF A BLACKBODY

2-5.1 Definition of Blackbody Radiation Intensity

Consider an elemental surface area dA surrounded by a hemisphere of radius R as shown in Fig. 2-3. A hemisphere has a surface area of $2\pi R^2$ and subtends a solid angle of 2π steradians (sr) about a point at the center of its base. Hence by considering a hemisphere of unit radius, the solid angle about the center of the base can be regarded directly as the area on the unit hemisphere. Direction is measured by the angles θ and φ as shown in Fig. 2-3, where the angle θ is measured from the direction *normal* to the surface. The angular position for $\varphi = 0$ is arbitrary, but is usually measured from the x axis.

The radiation emitted in any direction will be defined in terms of the *intensity*. There are two types of intensities: the *spectral intensity* refers to radiation in an interval $d\lambda$ around a *single* wavelength, while the *total intensity* refers to the combined radiation including *all* wavelengths. The spectral intensity of a blackbody will be given by $i'_{\lambda b}(\lambda)$. The subscripts denote, respectively, that a *spectral* quantity is being considered and that the properties are for a *blackbody*. The prime denotes that radiation per unit solid angle in a *single direction* is being considered. The notation is explained in detail in Sec. 3-1.2. The emitted spectral intensity is defined as the energy emitted per unit time per unit small wavelength interval around the wavelength λ, per unit elemental *projected surface area* normal to the (θ, φ) direction and into a unit elemental solid angle centered around the direction (θ, φ). As will be shown in Sec. 2-5.2, the blackbody intensity defined in this way (that is, on the basis of projected area) is *independent* of direction; hence, the symbol for blackbody intensity is not modified by any (θ, φ) designation. The *total* intensity i_b' is defined analogously to $i_{\lambda b}'$ except that it includes the radiation for all wavelengths; hence, the subscript λ and the functional dependence (λ) do not appear. The total intensity is the integral of the spectral intensity over all wavelengths

$$i_b' = \int_{\lambda=0}^{\infty} i_{\lambda b}'(\lambda) \, d\lambda \tag{2-1}$$

2-5.2 Angular Independence of Intensity

The angular independence of the blackbody intensity can be shown by considering a spherical isothermal blackbody enclosure of radius R with a blackbody element dA at its center, as shown in Fig. 2-4a. Once again, the enclosure and the central elemental body are in thermal equilibrium. Thus all radiation in transit throughout the enclosure must be isotropic. Consider radiation in a wavelength interval $d\lambda$ about λ that is emitted by an element dA_s on the enclosure surface and travels toward the central element dA (Fig. 2-4b). The emitted energy in this direction per unit solid angle and time is $i_{\lambda b,n}'(\lambda) \, dA_s \, d\lambda$. The normal spectral intensity of a blackbody is used because the energy is emitted normal to the black wall element dA_s of the spherical enclosure. The amount of energy per unit time that impinges

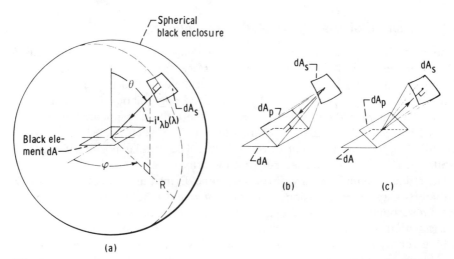

(a)

Figure 2-4 Energy exchange between element of enclosure surface and element within enclosure. (a) Black element dA within black spherical enclosure; (b) energy transfer from dA_s to dA_p; (c) energy transfer from dA_p to dA_s.

upon dA depends on the solid angle that dA occupies when viewed from the location of dA_s. This solid angle is the projected area of dA normal to the (θ, φ) direction divided by R^2. The projected area of dA is

$$dA_p = dA \cos \theta \tag{2-2}$$

Then the energy absorbed by dA is

$$d^3 Q'_{\lambda b}(\lambda,\theta,\varphi) = i'_{\lambda b,n}(\lambda)\, dA_s\, d\lambda \frac{dA \cos \theta}{R^2} \tag{2-3}$$

The energy emitted by dA in the (θ, φ) direction and incident on dA_s (Fig. 2-4c) must equal that absorbed from dA_s, or equilibrium would be disturbed; hence,

$$i'_{\lambda b}(\lambda,\theta,\varphi)\, dA_p\, \frac{dA_s}{R^2}\, d\lambda = d^3 Q'_{\lambda b}(\lambda,\theta,\varphi) = i'_{\lambda b,n}(\lambda)\, dA_s\, \frac{dA \cos \theta}{R^2}\, d\lambda$$

Then, by virtue of (2-2),

$$i'_{\lambda b}(\lambda,\theta,\varphi) = i'_{\lambda b,n}(\lambda) \neq \text{function of } \theta,\varphi \tag{2-4}$$

Equation (2-4) shows that the *intensity of radiation from a blackbody*, as defined on the basis of projected area, *is independent of the direction of emission*. Neither the subscript n nor the (θ, φ) notation is really needed for complete description of the black intensity. Since the blackbody is always a perfect absorber and emitter, these properties of the blackbody are independent of its surroundings. Hence, these results are independent of both the assumptions used in the derivation of a spherical enclosure and thermodynamic equilibrium with the surroundings.*

2-5.3 Blackbody Emissive Power—Definition and Cosine-Law Dependence

The intensity is defined on the basis of projected area. It is also useful to have a quantity for the energy emitted in a given direction per unit of actual (unprojected) surface area. This is $e'_{\lambda b}(\lambda, \theta, \varphi)$, which is the energy emitted by a black surface per unit time within a unit small wavelength interval centered around the wavelength λ, per unit elemental surface area, and into a unit elemental solid angle centered around the direction (θ, φ). The energy in the wavelength interval $d\lambda$

---

*Some exceptions exist for most of the blackbody "laws" presented in this chapter. The exceptions are of minor importance in almost any practical engineering situation but need to be considered when extremely rapid transients are present in a radiative-transfer process. If the transient period is of the order of the time scale of whatever process is governing the emission of radiation from the body in question, then the emission properties of the body may lag behind the absorption properties. In such a case, the concepts of temperature used in the derivation of the blackbody laws no longer hold rigorously. The treatment of such problems is outside the scope of this work (but see Sec. 12-8.2 for further discussion).

centered about λ emitted per unit time in any direction $d^3Q'_{\lambda b}(\lambda, \theta, \varphi)$ can then be expressed in the two forms

$$d^3Q'_{\lambda b}(\lambda,\theta,\varphi) = e'_{\lambda b}(\lambda,\theta,\varphi)\, dA\, d\omega\, d\lambda = i'_{\lambda b}(\lambda)\, dA\, \cos\theta\, d\omega\, d\lambda$$

Consequently, there exists the relation

$$e'_{\lambda b}(\lambda,\theta,\varphi) = i'_{\lambda b}(\lambda)\cos\theta = e'_{\lambda b}(\lambda,\theta) \tag{2-5}$$

This is illustrated in Fig. 2-5. It is evident from the $i'_{\lambda b}(\lambda)\cos\theta$ term in Eq. (2-5) that $e'_{\lambda b}(\lambda, \theta, \varphi)$ does not depend on φ and hence can be written as $e'_{\lambda b}(\lambda, \theta)$. The $e'_{\lambda b}(\lambda, \theta)$ is the *directional spectral emissive power* for a black surface. In the case of nonblack surfaces, there can be a dependence of e'_λ on φ.

Equation (2-5) is known as *Lambert's cosine law,* and surfaces having a directional emissive power that follows this relation are known as *diffuse* or *cosine-law* surfaces. A blackbody, because it is always a diffuse surface, serves as a standard for comparison with the directional properties of real surfaces that do not, in general, follow the cosine law.

2-5.4 Hemispherical Spectral Emissive Power of a Blackbody

In calculations of radiant energy rejection by a surface, one needs the directional spectral emissive power integrated over all solid angles of a hemispherical envelope placed over a black surface. This yields the *hemispherical spectral emissive power,* $e_{\lambda b}(\lambda)$. It is the energy leaving a black surface per unit time per unit area and per unit wavelength interval around λ. Figure 2-6 shows the elemental area dA at the center of the base of a unit hemisphere. By definition, a solid angle anywhere above dA is equal to the intercepted area on the unit hemisphere. As shown in Fig. 2-6, an element of this hemispherical area is given by $(\sin\theta\, d\varphi)(d\theta)$ so that

$$d\omega = \sin\theta\, d\theta\, d\varphi \tag{2-6}$$

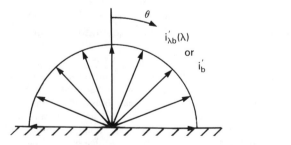
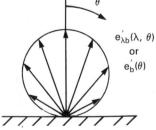

Figure 2-5 The angular distribution of blackbody intensity (independent of θ) and blackbody directional emissive power ($\cos\theta$ function).

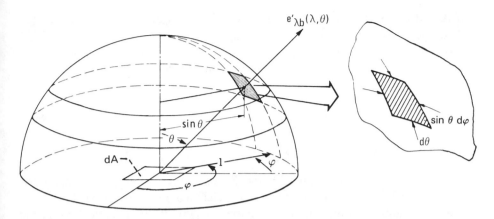

Figure 2-6 Unit hemisphere used to obtain relation between blackbody hemispherical emissive power and intensity.

Hence, the spectral emission from dA per unit time and unit surface area passing through the element on the hemisphere is given by use of (2-5) as

$$e'_{\lambda b}(\lambda, \theta)\, d\omega = e'_{\lambda b}(\lambda, \theta) \sin \theta\, d\theta\, d\varphi = i'_{\lambda b}(\lambda) \cos \theta \sin \theta\, d\theta\, d\varphi \qquad (2\text{-}7)$$

To obtain the blackbody emission passing through the entire hemisphere in terms of the blackbody intensity, Eq. (2-7) is integrated over all solid angles to give

$$e_{\lambda b}(\lambda) = i'_{\lambda b}(\lambda) \int_{\varphi=0}^{2\pi} \int_{\theta=0}^{\pi/2} \cos \theta \sin \theta\, d\theta\, d\varphi \qquad (2\text{-}8a)$$

The integration is carried out to yield

$$e_{\lambda b}(\lambda) = 2\pi i'_{\lambda b}(\lambda) \int_{0}^{1} \sin \theta\, d(\sin \theta) = \pi i'_{\lambda b}(\lambda) \qquad (2\text{-}8b)$$

where the prime notation is absent in the designation of the hemispherical quantity. From (2-5), for emission normal to the surface ($\theta = 0$), $i'_{\lambda b}(\lambda) = e'_{\lambda b,n}(\lambda)$ so (2-8b) can also be written as

$$e_{\lambda b}(\lambda) = \pi i'_{\lambda b}(\lambda) = \pi e'_{\lambda b,n}(\lambda) \qquad (2\text{-}9)$$

Hence, purely from the geometry involved, this simple relation is found: *The blackbody hemispherical emissive power is π times the intensity, or π times the directional emissive power normal to the surface.* This relation for a blackbody will be very useful in relating directional and hemispherical quantities in the following chapters.

2-5.5 Spectral Emissive Power through a Finite Solid Angle

The emission may be desired through only part of the hemispherical solid angle enclosing an area element. The emission within the solid angle extending from

θ_1 to θ_2 and φ_1 to φ_2 is found by modifying the limits of integration in Eq. (2-8a) and integrating to obtain

$$e_{\lambda b}(\lambda, \theta_1 - \theta_2, \varphi_1 - \varphi_2) = i'_{\lambda b}(\lambda) \int_{\varphi_1}^{\varphi_2} \int_{\theta_1}^{\theta_2} \cos\theta \sin\theta \, d\theta \, d\varphi$$

$$= i'_{\lambda b}(\lambda) \frac{\sin^2\theta_2 - \sin^2\theta_1}{2} (\varphi_2 - \varphi_1) \tag{2-10}$$

2-5.6 Spectral Distribution of Emissive Power; Planck's Law

We have discussed some of the blackbody characteristics: the blackbody is defined as a perfect absorber and also a perfect emitter; its total intensity and total emissive power in vacuum are functions only of the temperature of the blackbody; the emitted blackbody energy follows Lambert's cosine law. These properties have been demonstrated by thermodynamic arguments. However, a very important fundamental property of the blackbody remains to be presented. This is the formula that gives the magnitude of the emitted intensity at each of the wavelengths that, taken together, constitute the radiation spectrum. This relation cannot be obtained from purely thermodynamic arguments. Indeed, the search for this formula led Planck to an investigation and hypothesis that became the foundation of quantum theory. The derivation of the spectral distribution is not within the scope of the present discussion, and the results will be presented here without derivation. The interested reader may consult various standard physics texts for the complete development.

It has been shown by the quantum arguments of Planck [2] and verified experimentally that for a blackbody the spectral distributions of hemispherical emissive power and radiant intensity in *vacuum* are given as a function of wavelength and the blackbody's absolute temperature by

$$e_{\lambda b}(\lambda, T) = \pi i'_{\lambda b}(\lambda, T) = \frac{2\pi C_1}{\lambda^5 (e^{C_2/\lambda T} - 1)} \tag{2-11}$$

This is known as *Planck's spectral distribution of emissive power*. As will be shown, for radiation into a *medium* within which the speed of light is not close to c_0, Eq. (2-11) must be modified by including an index-of-refraction multiplying factor (see Sec. 2-5.12). For engineering work the radiant emission is usually into air or other gases with an index of refraction $n = c_0/c$ close to unity, so (2-11) is applicable. The values of the constants C_1 and C_2 are given in Table A-4 in two common systems of units. These constants are equal to $C_1 = hc_0^2$ and $C_2 = hc_0/k$, where h is Planck's constant and k is the Boltzmann constant.* Equation (2-11) is of great importance, as it provides *quantitative* results for the radiation from a blackbody.

---

*$h = 6.6260755 \times 10^{-34}$ J · s and $k = 1.380658 \times 10^{-23}$ J/K. In some literature, the constant C_1 is defined as $2\pi hc_0^2$.

EXAMPLE 2-1 A black surface element at temperature 1000° C is radiating into vacuum. What are the intensity and the directional spectral emissive power of the blackbody at an angle of 60° from the normal and at a wavelength of 6 μm?

From Eq. (2-11),

$$i'_{\lambda b}(6 \; \mu m) = \frac{2 \times 0.59552 \times 10^8 \; W \cdot \mu m^4/m^2}{6^5 \mu m^5(e^{14,388/(6\times 1273)} - 1) \; sr} = 2746 \; W/(m^2 \cdot \mu m \cdot sr)$$

From Eq. (2-5) the directional spectral emissive power is

$$e'_{\lambda b}(6 \; \mu m, \; 60°) = 2746 \cos 60° = 1373 \; W/(m^2 \cdot \mu m \cdot sr)$$

EXAMPLE 2-2 The sun emits like a blackbody at 5780 K. What is the sun's intensity at the center of the visible spectrum?

From Fig. 1-2 the wavelength of interest is 0.55 μm. Then, from (2-11),

$$i'_{\lambda b}(0.55 \; \mu m) = \frac{2 \times 0.59552 \times 10^8 \; W \cdot \mu m^4/m^2}{0.55^5 \; \mu m^5(e^{14388/(0.55\times 5780)} - 1)sr}$$

$$= 0.256 \times 10^8 \; W/(m^2 \cdot \mu m \cdot sr)$$

Alternative forms of Eq. (2-11) are employed where frequency or wavenumber is used rather than wavelength. The use of frequency has an advantage when spectral radiation travels from one medium into another, since in that situation the *frequency remains constant* while the wavelength changes because of the change in propagation velocity. To make the transformation of (2-11) to frequency, note that in vacuum $\lambda = c_0/\nu$, and hence $d\lambda = -(c_0/\nu^2) \; d\nu$. Then the hemispherical emissive power in the wavelength interval $d\lambda$ becomes

$$e_{\lambda b}(\lambda) \, d\lambda = \frac{2\pi C_1 \, d\lambda}{\lambda^5(e^{C_2/\lambda T} - 1)} = \frac{-2\pi C_1 \nu^3 \, d\nu}{c_0^4(e^{C_2\nu/c_0 T} - 1)} = e_{\nu b}(\nu) \, d\nu \qquad (2\text{-}12)$$

The quantity $e_{\nu b}(\nu)$ is the emissive power in vacuum *per unit frequency interval* about ν. The intensity is $i'_{\nu b}(\nu) = e_{\nu b}(\nu)/\pi$ so that

$$i'_{\nu b}(\nu) = \frac{2C_1\nu^3}{c_0^4(e^{C_2\nu/c_0 T} - 1)} = \frac{2h\nu^3}{c_0^2(e^{h\nu/kT} - 1)} \qquad (2\text{-}13)$$

The wavenumber $\eta = 1/\lambda$ is the number of waves per unit length. Then $d\lambda = -(1/\eta^2) \; d\eta$ and

$$e_{\lambda b}(\lambda) \, d\lambda = -\frac{2\pi C_1 \eta^3 \, d\eta}{e^{C_2\eta/T} - 1} = -e_{\eta b}(\eta) \, d\eta \qquad (2\text{-}14)$$

The quantity $e_{\eta b}(\eta)$ is the emissive power *per unit wavenumber interval* about η. The intensity is

$$i'_{\eta b}(\eta) = \frac{e_{\eta b}(\eta)}{\pi} = \frac{2C_1 \eta^3}{e^{C_2 \eta / T} - 1} \tag{2-15}$$

For better understanding of the implications of Eq. (2-11), it is plotted in Fig. 2-7. The hemispherical spectral emissive power is given as a function of wavelength for several different values of the absolute temperature. It is evident that the energy emitted at all wavelengths increases as the temperature increases. It was shown in Sec. 2-4.5, and is known from common experience, that the total (that is, including all wavelengths) radiated energy must increase with tempera-

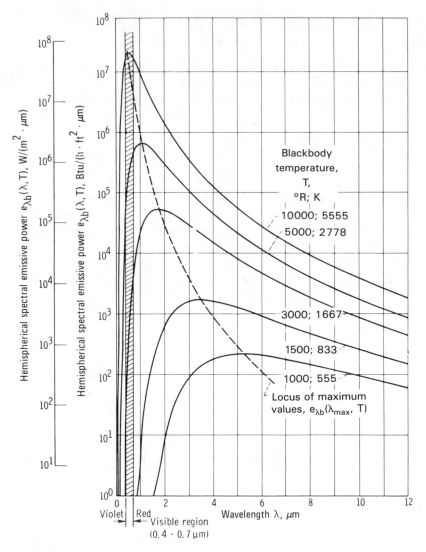

Figure 2-7 Hemispherical spectral emissive power of blackbody for several different temperatures.

ture; the curves show that this is also true for the energy at each wavelength. Another characteristic is that the peak spectral emissive power shifts toward a smaller wavelength as the temperature is increased. A cross plot of Fig. 2-7 giving energy as a function of temperature for fixed wavelengths shows that the energy emitted at the shorter-wavelength end of the spectrum increases more rapidly with temperature than the energy at the long wavelengths.

The location of the visible spectrum is included in Fig. 2-7. For a body at 555 K (1000° R) only a very small amount of energy is in the visible region, and it is insufficient to be detected by eye. Since the curves at the lower temperatures slope downward from the red toward the violet end of the spectrum, the red light becomes visible first as the temperature is raised.* Higher temperatures make visible additional wavelengths of the visible light range, and at a sufficiently high temperature the light emitted becomes white, representing emission composed of a mixture of all the visible wavelengths. The visible brightness also increases substantially as the temperature is raised.

For the filament of an incandescent lamp to operate efficiently, the temperature must be high; otherwise too much of the electrical energy is dissipated as radiation in the infrared region rather than in the visible range. Most tungsten filament lamps operate at about 3000 K (5400° R), and thus do give off a large fraction of their energy in the infrared, but their filament vaporization rate limits the temperature to near this value. The sun emits a spectrum quite similar to that of a blackbody at 5780 K (10,400° R), and an appreciable amount of the energy is in what we sense as the visible region. This may be because evolution has caused the human eye to be most sensitive in the spectral region of greatest energy received.[†] If the eye were sensitive in other regions (for example, the infrared, so that we could see thermal images in the "dark"), then our definition of the "visible region of the spectrum" would change. If we find life in other solar systems having a sun with an effective temperature different from ours, it will be interesting to discover what wavelength range encompasses the "visible spectrum" if the beings there possess sight.

Equation (2-11) can be placed in a convenient form that eliminates the need to provide a separate curve for each value of T. This is done by dividing by the fifth power of temperature to obtain

$$\frac{e_{\lambda b}(\lambda,T)}{T^5} = \frac{\pi i'_{\lambda b}(\lambda,T)}{T^5} = \frac{2\pi C_1}{(\lambda T)^5(e^{C_2/\lambda T}-1)} \tag{2-16}$$

This equation gives the quantity $e_{\lambda b}(\lambda,T)/T^5$ in terms of the single variable λT. A plot is in Fig. 2-8 and it replaces the multiple curves in Fig. 2-7. Numerical values are in Table A-5.

---

*This occurs at the so-called *Draper point* of 798 K (977° F) [3], at which red light first becomes visible from a heated object in darkened surroundings.

[†]Although human vision is well-tuned to the sun, this is not true of all beings on earth. Turtles have eyes sensitive to infrared but not blue; bees have eyes sensitive to ultraviolet but not red.

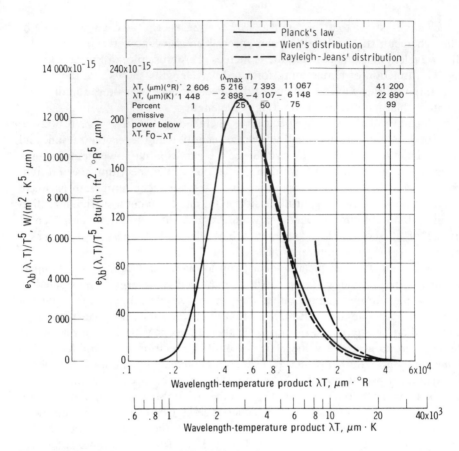

Figure 2-8 Spectral distribution of blackbody hemispherical emissive power as a function of λT.

Equation (2-14) can also be placed in a more universal form in terms of the variable η/T ($= 1/\lambda T$),

$$\frac{i'_{\eta b}(\eta,T)}{T^3} = \frac{2C_1(\eta/T)^3}{e^{C_2(\eta/T)} - 1} \qquad (2\text{-}17)$$

This function is shown in Fig. 2-9.

EXAMPLE 2-3 For a blackbody at 1200 K, what is the hemispherical spectral emissive power at a wavelength of 2 μm? Use Table A-5.

The value of λT is 2400 μm · K. From Table A-5, at this λT, $e_{\lambda b}/T^5 = 117.37 \times 10^{-13}$ W/(m² · μm · K⁵). Then $e_{\lambda b}(2\ \mu\text{m}) = 117.37 \times 10^{-13}(1200)^5 = 29.21$ kW/(m² · μm).

EXAMPLE 2-4 What is the blackbody intensity at $\lambda = 2$ μm for $T = 2500$ K?

From Table A-5 at $\lambda T = 5000 \ \mu m \cdot K$, $e_{\lambda b}/T^5 = 713.97 \times 10^{-14} \ W/ (m^2 \cdot \mu m \cdot K^5)$. Then since $i'_{\lambda b} = e_{\lambda b}/\pi$, $i'_{\lambda b}(2 \ \mu m) = 713.97 \times 10^{-14}(1/\pi)(2500^5) = 222 \ kW/(m^2 \cdot \mu m \cdot sr)$.

Planck's spectral distribution gives the maximum (blackbody) intensity of radiation that any body can emit in vacuum at a given wavelength and temperature. This serves as an optimum with which real surface performance can be compared. In Chap. 3, the methods of comparison will be defined. Planck's distribution also provides a means of evaluating the maximum emissive performance that can be attained for any radiating device.

2-5.7 Approximations for Spectral Distribution

Some approximate forms of Planck's distribution are occasionally useful in derivations because of their simplicity. Care must be taken to use them only in the range for which their accuracy is acceptable.

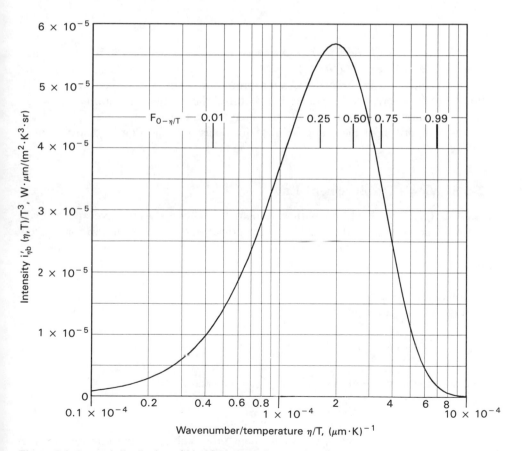

Figure 2-9 Spectral distribution of blackbody intensity as a function of η/T.

Wien's formula If the term $e^{C_2/\lambda T}$ is much larger than 1, Eq. (2-16) reduces to

$$\frac{i'_{\lambda b}(\lambda,T)}{T^5} = \frac{2C_1}{(\lambda T)^5 e^{C_2/\lambda T}} \tag{2-18}$$

which is known as *Wien's formula*. It is accurate to within 1% for λT less than 3000 μm $\cdot$ K (5400 μm $\cdot$ °R).

Rayleigh-Jeans formula Another approximation is found by expanding the denominator of (2-16) in a series to obtain

$$e^{C_2/\lambda T} - 1 = 1 + \frac{C_2}{\lambda T} + \frac{1}{2!}\left(\frac{C_2}{\lambda T}\right)^2 + \frac{1}{3!}\left(\frac{C_2}{\lambda T}\right)^3 + \cdots - 1 \tag{2-19}$$

For λT much larger than C_2, this series can be approximated by the single term $C_2/\lambda T$, and Eq. (2-16) becomes

$$\frac{i'_{\lambda b}(\lambda,T)}{T^5} = \frac{2C_1}{C_2}\frac{1}{(\lambda T)^4} \tag{2-20}$$

This is the *Rayleigh-Jeans formula* and is accurate to within 1% for λT greater than 7.78×10^5 μm $\cdot$ K (14×10^5 μm $\cdot$ °R). This is well outside the range generally encountered in thermal radiation problems, since a blackbody emits over 99.9% of its energy at λT values below this. The formula has utility for long-wave radiation of other classifications, such as radio waves.

A comparison of these approximate formulas with the Planck distribution is in Fig. 2-8.

2-5.8 Wien's Displacement Law

Another quantity of interest with regard to the blackbody emissive spectrum is the wavelength λ_{max} at which the emissive power $e_{\lambda b}(\lambda)$ is a maximum for a given temperature. This maximum shifts toward shorter wavelengths as the temperature is increased, as shown by the dashed line in Fig. 2-7. The value of $\lambda_{max}T$ can be found at the peak of the distribution curve given in Fig. 2-8. Alternatively, it can be found analytically by differentiating Planck's distribution from Eq. (2-16) and setting the left side equal to zero. This gives the transcendental equation

$$\lambda_{max}T = \frac{C_2}{5}\frac{1}{1 - e^{-C_2/\lambda_{max}T}} \tag{2-21}$$

The solution to this equation for $\lambda_{max}T$ is a constant,

$$\lambda_{max}T = C_3 \tag{2-22}$$

which is one form of *Wien's displacement law*. Values of the constant C_3 are given in Table A-4. Equation (2-22) indicates that at a higher temperature the peak emissive power and intensity shift to a shorter wavelength that is in inverse proportion to T.

EXAMPLE 2-5 For a blackbody to radiate its maximum emissive power $e_{\lambda b}$ at the center of the visible spectrum, what would its temperature need to be?

Figure 1-2 shows that the visible spectrum spans the range 0.4–0.7 μm, and the center of the range is at 0.55 μm. From Eq. (2-22)

$$T = \frac{C_3}{\lambda_{max}} = \frac{2897.8 \ \mu m \cdot K}{0.55 \ \mu m} \ 5269 \ K \ (9484°R)$$

This is close to the effective radiating surface temperature of the sun, which is 5780 K (10,400°R).

EXAMPLE 2-6 At what wavelength is the maximum emission from a blackbody at room temperature?

Using a room temperature of 21° C (69.8°F), the Wien displacement law gives

$$\lambda_{max} = \frac{C_3}{T} = \frac{2897.8 \ \mu m \cdot K}{294.2 \ K} = 9.85 \ \mu m$$

which is in the middle of the near-infrared region.

2-5.9 Total Intensity and Emissive Power

The previous discussion has provided the energy per unit wavelength interval that a blackbody radiates into vacuum at each wavelength. Now the *total intensity* will be determined, which includes the radiation for all wavelengths. The result is a surprisingly simple relation.

The intensity emitted over the small wavelength interval $d\lambda$ is $i'_{\lambda b}(\lambda) \ d\lambda$. Integrating over all wavelengths $\lambda = 0-\infty$ gives the *total intensity*

$$i'_b = \int_0^\infty i'_{\lambda b}(\lambda) \ d\lambda \tag{2-23}$$

This integral is evaluated by substitution of Planck's distribution from (2-16) and a transformation of variables using $\zeta = C_2/\lambda T$. Equation (2-23) then becomes

$$i'_b - \int_0^\infty \frac{2C_1}{\lambda^5 (e^{C_2/\lambda T} - 1)} \ d\lambda$$

$$= \int_\infty^0 \left(\frac{C_2}{\lambda T}\right)^5 \left(\frac{T}{C_2}\right)^5 \frac{2C_1}{e^{C_2/\lambda T} - 1} \left(\frac{\lambda T}{C_2}\right)^2 \left(\frac{-C_2}{T}\right) d\left(\frac{C_2}{\lambda T}\right)$$

$$= \frac{2C_1 T^4}{C_2^4} \int_0^\infty \frac{\zeta^3}{e^\zeta - 1} \ d\zeta \tag{2-24}$$

From a table of definite integrals [4] this is evaluated as

$$i_b' = \frac{2C_1 T^4}{C_2^4} \frac{\pi^4}{15}$$

(2-25)

Defining a new constant σ results in

$$i_b' = \frac{\sigma}{\pi} T^4$$

(2-26)

where the constant is

$$\sigma = \frac{2C_1 \pi^5}{15 C_2^4} = 5.67051 \times 10^{-8} \text{ W}/(\text{m}^2 \cdot \text{K}^4)$$

$$= 0.17123 \times 10^{-8} \text{ Btu}/(\text{h} \cdot \text{ft}^2 \cdot {}^\circ\text{R}^4)$$

(2-27)

The *hemispherical total emissive power* of a blackbody radiating into vacuum is then

$$e_b = \int_0^\infty e_{\lambda b}(\lambda) \, d\lambda = \int_0^\infty \pi i_{\lambda b}'(\lambda) \, d\lambda = \sigma T^4$$

(2-28)

This is the *Stefan-Boltzmann law,* and σ is the *Stefan-Boltzmann constant.*

EXAMPLE 2-7 A beam emitted normally by a blackbody surface is found to have a total radiation per unit solid angle and per unit surface area of 10,000 W/(m² · sr). What is the surface temperature?

The hemispherical total emissive power is related to the total emissive power in the normal direction by $e_b = \pi e_{b,n}'$. Hence, from Eq. (2-28) $T = (\pi e_{b,n}'/\sigma)^{1/4} = (10,000\pi/5.67051 \times 10^{-8})^{1/4} = 862.7$ K.

EXAMPLE 2-8 A black surface is radiating with a hemispherical total emissive power of 20 kW/m². What is its surface temperature? At what wavelength is its maximum spectral emissive power?

From the Stefan-Boltzmann law, the temperature of the blackbody is $T = (e_b/\sigma)^{1/4} = (20,000/5.67051 \times 10^{-8})^{1/4} = 770.6$ K. Then from Wien's displacement law, $\lambda_{max} = C_3/T = 2898/770.6 = 3.76$ μm.

EXAMPLE 2-9 An electric flat-plate heater that is square with a 0.1 m edge length is radiating 10^2 W from each side. If the heater can be considered black, what is its temperature?

Using the Stefan-Boltzmann law,

$$T = \left(\frac{Q}{A\sigma}\right)^{1/4} = \left[\frac{10^2 \text{ W}}{(0.1 \text{ m})^2 \, 5.67051 \times 10^{-8} \text{ W}/(\text{m}^2 \cdot \text{K}^4)}\right]^{1/4} = 648 \text{ K}$$

The spectral intensity of a black surface $i_{\lambda b}'(\lambda)$ was shown in Eq. (2-4) to be independent of the angle of emission. Integrating over all wavelengths of course did not change this angular independence. The intensity of a surface is what the eye interprets as "brightness." A black surface will exhibit the same brightness when viewed from any angle.

2-5.10 Behavior of Maximum Intensity with Temperature

The intensity at a given wavelength is found from Planck's spectral distribution. Substituting Wien's displacement law [Eq. (2-22)] into Eq. (2-16) gives the maximum intensity as

$$i'_{\lambda_{max}b} = T^5 \frac{2C_1}{C_3^5(e^{C_2/C_3} - 1)} = 4.09579 \times 10^{-12} \text{ W}/(\text{m}^2 \cdot \mu\text{m} \cdot \text{K}^5)T^5 = C_4T^5 \quad (2\text{-}29)$$

This shows that the maximum blackbody intensity increases as temperature to the *fifth power*. Indeed, because $i'_{\lambda b}/T^5$ is a function only of λT as shown by Eq. (2-16), it is evident that if the blackbody temperature is changed from T_1 to T_2 and at the same time the wavelengths λ_1 and λ_2 are chosen such that $\lambda_1 T_1 = \lambda_2 T_2$, the value of $i'_{\lambda b}/T^5$ remains unchanged. Therefore, the intensity at λ_2 for temperature T_2 changes as $(T_2/T_1)^5$ from the intensity at λ_1 for temperature T_1. This is the general statement of Wien's law.

2-5.11 Blackbody Radiation in a Wavelength Interval

The Stefan-Boltzmann law shows that the hemispherical total emissive power of a blackbody radiating into vacuum is given by $e_b = \pi i'_b = \int_0^\infty e_{\lambda b}(\lambda) \, d\lambda = \pi \int_0^\infty i'_{\lambda b}(\lambda) \, d\lambda = \sigma T^4$. It is often necessary in calculations of radiative exchange to determine the fraction of the total emissive power that is emitted in a given wavelength band as illustrated by Fig. 2-10. This fraction is designated by $F_{\lambda_1-\lambda_2}$ and is given by the ratio (the dash in the subscript means "to")

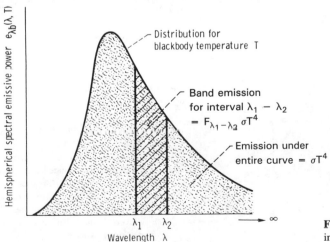

Figure caption text within image:
- Distribution for blackbody temperature T
- Band emission for interval $\lambda_1 - \lambda_2$ = $F_{\lambda_1-\lambda_2} \sigma T^4$
- Emission under entire curve = σT^4
- λ_1 λ_2
- Wavelength λ
- Hemispherical spectral emissive power $e_{\lambda b}(\lambda, T)$

Figure 2-10 Emitted energy in wavelength band.

$$F_{\lambda_1-\lambda_2} = \frac{\int_{\lambda_1}^{\lambda_2} e_{\lambda b}(\lambda)\,d\lambda}{\int_0^\infty e_{\lambda b}(\lambda)\,d\lambda} = \frac{1}{\sigma T^4}\int_{\lambda_1}^{\lambda_2} e_{\lambda b}(\lambda)\,d\lambda \tag{2-30}$$

The last integral in Eq. (2-30) can be expressed by two integrals each with a lower limit at $\lambda = 0$:

$$F_{\lambda_1-\lambda_2} = \frac{1}{\sigma T^4}\left[\int_0^{\lambda_2} e_{\lambda b}(\lambda)\,d\lambda - \int_0^{\lambda_1} e_{\lambda b}(\lambda)\,d\lambda\right] = F_{0-\lambda_2} - F_{0-\lambda_1} \tag{2-31}$$

The fraction of the emissive power for any wavelength band can therefore be found by having values of $F_{0-\lambda}$ as a function λ. The $F_{0-\lambda_1}$ function is illustrated by Fig. 2-11a, where it equals the crosshatched area divided by the total area (shaded) under the curve.

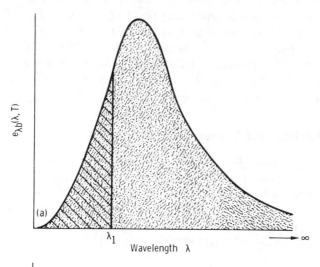

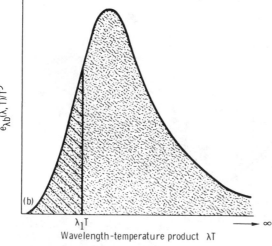

Figure 2-11 Physical representation of F factor, where $F_{0-\lambda_1}$ or $F_{0-\lambda_1 T}$ is ratio of crosshatched to total shaded area. (a) In terms of curve for specific temperature. Entire area under curve is σT^4. (b) In terms of universal curve. Entire area under curve is σ.

For a blackbody, because of the simple manner in which hemispherical emissive power is related to intensity [Eq. (2-8b)], the $F_{\lambda_1-\lambda_2}$ function also gives the fraction of the intensity within the wavelength interval λ_1 to λ_2. Since $e_{\lambda b}$ depends on T, the application of (2-31) would require that $F_{0-\lambda}$ be tabulated for each T. There is no need to have this complexity, however, as it is possible to arrange the F function in terms of only the single variable λT (Fig. 2-11b). In this way a universal set of F values is obtained that applies for all temperatures and wavelengths. The universal form is found by rewriting (2-31) as

$$F_{\lambda_1-\lambda_2} = F_{\lambda_1 T-\lambda_2 T} = \frac{1}{\sigma}\left[\int_0^{\lambda_2 T}\frac{e_{\lambda b}(\lambda)}{T^5}d(\lambda T) - \int_0^{\lambda_1 T}\frac{e_{\lambda b}(\lambda)}{T^5}d(\lambda T)\right]$$

$$= F_{0-\lambda_2 T} - F_{0-\lambda_1 T} \tag{2-32}$$

As shown by Eq. (2-16), $e_{\lambda b}/T^5$ is a function only of λT so that the integrands in Eq. (2-32) depend only on the λT variable.

A convenient series form for $F_{0-\lambda T}$ can be found by using the substitution $\zeta = C_2/\lambda T$ to obtain from (2-24)

$$F_{0-\lambda T} = \frac{2\pi C_1}{\sigma T^4}\int_0^\lambda \frac{d\lambda}{\lambda^5(e^{C_2/\lambda T} - 1)} = \frac{2\pi C_1}{\sigma C_2^4}\int_\zeta^\infty \frac{\zeta^3}{e^\zeta - 1}d\zeta$$

Using the definition of σ and the value of the definite integral in Eqs. (2-24) and (2-25), this reduces to

$$F_{0-\lambda T} = 1 - \frac{15}{\pi^4}\int_0^\zeta \frac{\zeta^3}{e^\zeta - 1}d\zeta$$

By using the series expansion $(1 - e^\zeta)^{-1} = 1 + e^\zeta + e^{2\zeta} + \cdots$ and then integrating by parts, the $F_{0-\lambda T}$ becomes

$$F_{0-\lambda T} = \frac{15}{\pi^4}\sum_{n=1}^\infty \left[\frac{e^{-n\zeta}}{n}\left(\zeta^3 + \frac{3\zeta^2}{n} + \frac{6\zeta}{n^2} + \frac{6}{n^3}\right)\right] \tag{2-33}$$

where $\zeta = C_2/\lambda T$. This form is given in [5]; it was used to evaluate the blackbody tables in [6]. The infinite series in (2-33) converges very rapidly, and the first three terms give good results over most of the range of $F_{0-\lambda T}$. As T becomes large (small ζ), a larger number of terms is required. The series is useful in computer solutions, rather than tabulated results. Some additional expressions that have been used for $F_{0-\lambda T}$ are referenced in Appendix A. For computer solutions, direct numerical integration can also be used. The $F_{0-\lambda T}$ values generated from Eq. (2-33) are given in Table A-5. In generating the tables, the series was continued until no change of magnitude greater than 10^{-9} occurred between successive terms. A plot of $F_{0-\lambda T}$ as a function of λT is in Fig. 2-12.

When working with wavenumber or frequency, the fraction $F_{0-\eta/T} = F_{0-\nu/c_0 T} = F_{0-1/\lambda T}$ is needed. Because of the inverse relation between wavenumber and wavelength, there is the relation

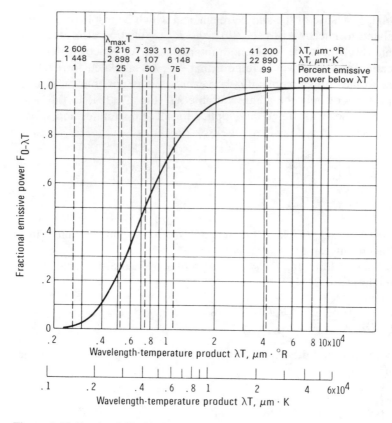

Figure 2-12 Fractional blackbody emissive power in range 0 to λT.

$$F_{0-\eta/T} = 1 - F_{0-\lambda T} \qquad (2\text{-}34)$$

Some values of $F_{0-\eta/T}$ are shown in Fig. 2-9.

The blackbody functions are easily computed and tabulated by using the analytical expressions given in this chapter. Extensive tables of $F_{0-\lambda T}$ values have been published. The tables of Pivovonsky and Nagel [7], for example, give values for every λT interval of 10 μm $\cdot$ K over a very wide range of λT. These tables use earlier values of the fundamental constants c_0, h, and k in their computation. The $F_{0-\lambda T}$ function has a number of uses as illustrated in the following examples.

EXAMPLE 2-10 A blackbody is radiating at a temperature of 3000 K. An experimenter wishes to measure the total radiant emission by use of a radiation detector. This detector absorbs all radiation in the λ range 0.8–5 μm but detects no energy outside that range. What percentage correction will the experimenter have to apply to the energy measurement? If the sensitivity of the detector could be extended in range by 0.5 μm at only one end of the sensitive range, which end should be extended?

Taking $\lambda_1 T = 0.8 \times 3000 = 2400$ μm $\cdot$ K and $\lambda_2 T = 5 \times 3000 = 15,000$ μm $\cdot$ K results in the fraction of energy outside the sensitive range being $F_{0-\lambda_1 T} + F_{\lambda_2 T-\infty} = F_{0-\lambda_1 T} + (1 - F_{0-\lambda_2 T}) = (0.14026 + 1 - 0.96893) = 0.17133$, or a correction of 17.1% of the total incident energy. Extending the sensitive range to the longer-wavelength side of the measurement interval adds little accuracy because of the small slope of the curve of F against λT in that region, so extending to shorter wavelengths would provide the greatest increase in detected energy. The quantitative results to demonstrate this are found from the factors $F_{0-0.3\times3000} = 0.00009$ and $F_{0-5.5\times3000} = 0.97581$. The correction factors after extending the short- and long-wavelength boundaries of the interval are then $0.00009 + 1 - 0.96893 = 0.03116$ and $0.14026 + 1 - 0.97581 = 0.16445$.

EXAMPLE 2-11 The experimenter of the previous example has designed a radiant energy detector that can be made sensitive over only any 1-μm range of wavelength. The experimenter wants to measure the total emissive power of two blackbodies, one at 2500 K and the other at 5000 K, and plans to adjust the 1-μm interval to give a 0.5-μm sensitive band on each side of the peak blackbody emissive power. For which blackbody would you expect to detect the greatest percentage of the total emissive power? What will the percentage be in each case?

Wien's displacement law tells us that the peak emissive power will occur at $\lambda_{max} = 2897.8/T$ μm in each case. For the higher temperature a wavelength interval of 1 μm will give a wider spread of λT values around the peak of $\lambda_{max} T$ on the normalized blackbody curve (Fig. 2-8), so the measurement should be more accurate for the 5000 K case. For the 5000 K blackbody, $\lambda_{max} = 0.5796$ μm, and $\lambda_1 T = (0.5796 - 0.5000) \times 5000 = 398$ μm $\cdot$ K. Similarly, $\lambda_2 T = 1.0796 \times 5000 = 5398$ μm $\cdot$ K. The percentage of detected emissive power is then

$$100(F_{0-5398} - F_{0-398}) = 100 \times (0.6801 - 0) = 68.0\%$$

A similar calculation for the 2500 K blackbody shows that 48.3% of the emissive power is detected.

EXAMPLE 2-12 A light-bulb filament is at 3000 K. Assuming the filament radiates a spectrum like a blackbody, what fraction of its energy is emitted in the visible region?

The visible region is within $\lambda = 0.4$–0.7 μm. The desired fraction is then

$$F_{0-2100\ \mu m\cdot K} - F_{0-1200\ \mu m\cdot K} = 0.08306 - 0.00213 = 0.0809$$

Some commonly used values of $F_{0-\lambda T}$ and $F_{0-\eta/T}$ are given in Table 2-1. It is interesting to note that exactly one-fourth of the total emissive power lies in the wavelength range below the peak of the Planck spectral distribution at any temperature. This relation appears to have no simple physical explanation and is grouped

Table 2-1 Fraction of blackbody emission contained in the range 0 to λT or 0 to η/T

λT			η/T	
$\mu m \cdot °R$	$\mu m \cdot K$	$F_{0-\lambda T}$	$(\mu m \cdot K)^{-1}$	$F_{0-\eta/T}$
2,606	1,448	0.01	0.4369×10^{-4}	0.01
$5,216 = \lambda_{max}T$	$2,898 = \lambda_{max}T$	0.25	1.627×10^{-4}	0.25
7,393	4,107	0.50	2.435×10^{-4}	0.50
11,067	6,148	0.75	3.451×10^{-4}	0.75
41,200	22,890	0.99	6.906×10^{-4}	0.99

with other phenomena, such as gravitational attraction and the Stefan-Boltzmann fourth-power law, for which nature provides a simple law to describe an apparently complex event.

2-5.12 Blackbody Emission into a Medium Other than Vacuum

The previous expressions have been for blackbody emission into vacuum or a medium where the refractive index $n \approx 1$. When emission is considered from a blackbody into a large volume of medium other than vacuum, the quantities C_1 and C_2 appearing in Planck's energy distribution equation [Eq. (2-11)] are replaced by $C_1' = hc^2$ and $C_2' = hc/k$ so that

$$e_{\lambda mb}(\lambda_m)\,d\lambda_m = \frac{2\pi C_1'}{\lambda_m{}^5(e^{C_2'/\lambda_m T} - 1)}\,d\lambda_m \tag{2-35}$$

where k is the Boltzmann constant, h is Planck's constant, c is the speed of electromagnetic propagation *in the medium*, and λ_m is the wavelength *in the medium*.

Since the speed c depends on the medium, it is better to define C_1 and C_2 in terms of c_0, the speed in vacuum, so that C_1 and C_2 are then truly constants. The speed in the medium is given by $c = c_0/n$, where n is the index of refraction. Planck's distribution for the energy in a wavelength interval $d\lambda_m$ *in the medium* then becomes

$$e_{\lambda mb}(\lambda_m)\,d\lambda_m = \frac{2\pi c^2 h}{\lambda_m^5(e^{ch/k\lambda_m T} - 1)}\,d\lambda_m = \frac{2\pi c_0^2 h}{n^2\lambda_m^5(e^{c_0 h/nk\lambda_m T} - 1)}\,d\lambda_m$$

$$e_{\lambda mb}(\lambda_m)\,d\lambda_m = \frac{2\pi C_1}{n^2\lambda_m^5(e^{C_2/n\lambda_m T} - 1)}\,d\lambda_m \tag{2-36}$$

In Eq. (2-36), $C_1 = hc_0^2$ and $C_2 = hc_0/k$, which are the values of C_1 and C_2 in Table A-4. The $\lambda_m = \lambda/n$ where λ is the value in vacuum. The energy in the medium within $d\lambda$ measured in vacuum is

$$e_{\lambda b}(\lambda)\,d\lambda = \frac{2\pi n^2 C_1}{\lambda^5(e^{C_2/\lambda T} - 1)}\,d\lambda \tag{2-37}$$

In frequency and wavenumber form, Eq. (2-37) becomes

$$e_{\nu b}(\nu)\, d\nu = \frac{2\pi n^2 C_1 \nu^3}{c_0^4(e^{C_2\nu/c_0 T} - 1)}\, d\nu \qquad e_{\eta b}(\eta)\, d\eta = \frac{2\pi n^2 C_1 \eta^3}{e^{C_2\eta/T} - 1}\, d\eta \qquad (2\text{-}38a,b)$$

where ν in the medium is the *same* as in vacuum, which makes the frequency form especially convenient. The η is in vacuum; in a medium, $\eta_m = n\eta$.

When n is constant (independent of λ), the integration of (2-37) over all wavelengths follows as in (2-24). This yields the Stefan-Boltzmann law for hemispherical total emissive power into a medium of constant refractive index n:

$$e_{b,m} = n^2 \sigma T^4 \qquad (2\text{-}39)$$

Blackbody emission *within* glass ($n \approx 1.5$) is thus 2.25 times that of a blackbody in vacuum (see Sec. 18-6.2).

Finally, by similar arguments Wien's displacement law becomes

$$\lambda_{\max,m} T = \frac{C_3}{n} \qquad (2\text{-}40)$$

where $\lambda_{\max,m}$ is the wavelength at peak emission *into a medium*.

The n factors are included in a few succeeding sections that consider special topics where $n > 1$. A notable application is the work of Gardon and others [8–10] dealing with radiation effects in molten glass. These types of problems are discussed in Sec. 18-6 and in detail by Viskanta and Anderson [11]. Section 18-6 discusses how a portion of the blackbody radiation within a medium with $n > 1$ cannot pass through the interface into an adjacent medium with a smaller n. When entering into vacuum, there is a $1/n^2$ interface-reflection factor that removes the n^2 in Eq. (2-39).

2-6 EXPERIMENTAL PRODUCTION OF A BLACKBODY

When making experimental measurements of radiative properties of materials, it is desirable to have a blackbody source for reference so that a direct comparison can be made with the material surface. Since perfectly black surfaces do not exist in nature, a special technique is utilized to provide a very close approximation to a black area. Figure 2-13 shows a metal cylinder that has been hollowed out to form a cavity with a small opening. If an incident beam passes into the cavity as shown, it strikes the cavity wall, and part is absorbed with the remainder being reflected. The reflected portion strikes other parts of the wall and is again partially absorbed. It is evident that, if the opening to the cavity is very small, very little of the original incident beam will escape through the opening. Thus, if the opening is made sufficiently small, the opening area approaches the behavior of a black surface because essentially all the radiation passing in through it is absorbed. To help keep the cavity at a uniform temperature so that the internal radiation will

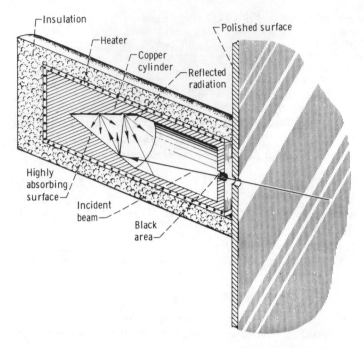

Figure 2-13 Cavity used to produce blackbody area.

all be in thermal equilibrium, the cavity shown in Fig. 2-13 is machined from a copper cylinder and surrounded by insulation. By heating the cavity, a source of black radiation is obtained at the opening since, as previously discussed in Sec. 2-4.1, a perfectly absorbing surface is also perfectly emitting. The polished surface at the front of the cavity aids in shielding the radiation leaving the opening from the effects of stray radiation from the surroundings. A good discussion of various cavity designs is in [12].

The attainment of isothermal conditions in such a cavity (often referred to as a *hohlraum*) is a difficult but necessary condition in the accurate experimental determination of radiative properties. Another difficulty in the use of a blackbody cavity is the upper temperature limit imposed by the heater and wall materials. This limits the energy output, especially in the short-wavelength portion of the spectrum. A high-temperature (3500 K) blackbody cavity made from graphite is described in [13]. Grooves were used on the inside of the cavity to increase its emissivity. At the aperture an emissivity of 0.996–0.999 was obtained.

If the cavity is assumed perfectly isothermal and perfectly insulated, and if the exit hole is infinitesimally small so that it does not disturb the radiative equilibrium in the enclosure, then the radiation from the hole is blackbody radiation and hence the cavity is completely filled with blackbody radiation. An interesting feature is that under these conditions, the local radiation leaving the cavity wall will be blackbody radiation even though the wall is not a perfect emitter. When the blackbody radiation within the cavity strikes the wall, part of it will be absorbed and the remainder will be reflected. Since the wall is perfectly insulated on the outside, the absorbed energy must be re-emitted on the inside. The combination of reflected and emitted energy leaving the wall must equal the incident blackbody radiation.

If a body that is small, so that it does not disturb conditions in the cavity, is placed in the cavity, it will come to equilibrium at the cavity temperature. Then, since the net energy exchange with the body must be zero, the radiation leaving the body by combined reflection and emission must be blackbody radiation. Hence the radiation leaving the body will be the same as that from all the surroundings, and the body will not be visible inside the cavity.

2-7 SUMMARY OF BLACKBODY PROPERTIES

It has been shown in this chapter that the blackbody possesses certain ideal fundamental properties that make it a standard with which real radiating bodies can be compared. These properties, listed here for convenience, are the following:

1. The blackbody is the best possible absorber and emitter of radiant energy at any wavelength and in any direction.
2. The total radiant intensity and hemispherical total emissive power of a blackbody into a medium with constant index of refraction n are given by the Stefan-Boltzmann law:

$$\pi i_b' = e_b = n^2 \sigma T^4$$

3. The blackbody spectral and total intensities are *independent* of direction. The blackbody directional-spectral and directional-total emissive powers follow Lambert's cosine law:

$$e_{\lambda b}'(\lambda, \theta) = e_{\lambda b, n}'(\lambda) \cos \theta$$

$$e_b'(\theta) = e_{b,n}' \cos \theta$$

4. The blackbody spectral distribution of intensity is given by Planck's distribution:

$$i_{\lambda b}'(\lambda) = \frac{e_{\lambda b}'}{\pi} = \frac{2C_1}{\lambda^5(e^{C_2/\lambda T} - 1)} \qquad \text{emission into vacuum}$$

$$i_{\lambda_m b}'(\lambda_m) = \frac{2C_1}{n^2 \lambda_m^5(e^{C_2/n\lambda_m T} - 1)} \qquad \begin{array}{l}\text{emission into a medium} \\ (\lambda_m \text{ is } \textit{inside} \text{ the medium})\end{array}$$

5. The wavelength at which the maximum spectral intensity of radiation for a blackbody occurs is given by Wien's displacement law:

Table 2-2 Blackbody radiation quantities ($n \approx 1$)

Symbol	Name	Definition	Geometry	Formula
$i'_{\lambda b}(\lambda, T)$	Spectral intensity	Emission in any direction per unit of projected area normal to that direction, and per unit time, wavelength interval about λ, and solid angle		$\dfrac{2C_1}{\lambda^5(e^{C_2/\lambda T}-1)}$
$i'_b(T)$	Total intensity	Emission, including all wavelengths, in any direction per unit of projected area normal to that direction, and per unit time and solid angle		$\dfrac{\sigma T^4}{\pi}$
$e'_{\lambda b}(\lambda, \theta, T)$	Directional spectral emissive power	Emission per unit solid angle in direction θ per unit surface area, wavelength interval, and time		$i'_{\lambda b}(\lambda, T)\cos\theta$
$e'_b(\theta, T)$	Directional total emissive power	Emission, including all wavelengths, in direction θ per unit surface area, solid angle, and time		$\dfrac{\sigma T^4}{\pi}\cos\theta$
$e_{\lambda b}(\lambda, \theta_1-\theta_2, \varphi_1-\varphi_2, T)$	Finite solid-angle spectral emissive power	Emission in solid angle $\theta_1 \le \theta \le \theta_2$, $\varphi_1 \le \varphi \le \varphi_2$ per unit surface area, wavelength interval, and time		$i'_{\lambda b}(\lambda, T)\dfrac{\sin^2\theta_2 - \sin^2\theta_1}{2}(\varphi_2 - \varphi_1)$

Symbol	Name	Description		Formula
$e_b(\theta_1 - \theta_2, \varphi_1 - \varphi_2, T)$	Finite solid-angle total emissive power	Emission, including all wavelengths, in solid angle $\theta_1 \le \theta \le \theta_2, \varphi_1 \le \varphi \le \varphi_2$ per unit surface area and time		$\dfrac{\sigma T^4}{\pi}\dfrac{\sin^2\theta_2 - \sin^2\theta_1}{2}(\varphi_2 - \varphi_1)$
$e_{\lambda b}(\lambda_1 - \lambda_2, \theta_1 - \theta_2, \varphi_1 - \varphi_2, T)$	Finite solid-angle band emissive power	Emission in solid angle $\theta_1 \le \theta \le \theta_2, \varphi_1 \le \varphi \le \varphi_2$ and wavelength band $\lambda_1 - \lambda_2$ per unit surface area and time		$\dfrac{\sigma T^4}{\pi}\dfrac{\sin^2\theta_2 - \sin^2\theta_1}{2}(\varphi_2 - \varphi_2)$ $\times (F_{0-\lambda_2} - F_{0-\lambda_1})$
$e_{\lambda b}(\lambda, T)$	Hemispherical spectral emissive power	Emission into hemispherical solid angle per unit surface area, wavelength interval, and time		$\pi i'_{\lambda b}(\lambda, T)$
$e_{\lambda b}(\lambda_1 - \lambda_2, T)$	Hemispherical band emissive power	Emission in wavelength band $\lambda_1 - \lambda_2$ into hemispherical solid angle per unit surface area and time		$(F_{0-\lambda_2} - F_{0-\lambda_1})\sigma T^4$
$e_b(T)$	Hemispherical total emissive power	Emission, including all wavelengths, into hemispherical solid angle per unit surface area and time		σT^4

$$\lambda_{max} = \frac{C_3}{T} \qquad \text{emission into vacuum}$$

$$\lambda_{max,m} = \frac{\lambda_{max}}{n} = \frac{C_3}{nT} \qquad \text{emission into a medium}$$

The many definitions introduced in this chapter are conveniently summarized in Table 2-2. The formulas for the quantities are given in terms of either the spectral intensity $i'_{\lambda b}(\lambda)$, which is computed from Planck's law, or the surface temperature T.

2-8 HISTORICAL DEVELOPMENT

The derivation of the approximate spectral distributions of Wien and of Rayleigh and Jeans, the Stefan-Boltzmann law, and Wien's displacement law are all logical consequences of the spectral distribution of intensity derived by Max Planck. However, it is interesting to note that all these relations were formulated *prior* to publication of Planck's work in 1901 and were originally derived through fairly complex thermodynamic arguments.

Joseph Stefan [14] proposed in 1879, after study of some experimental results, that emissive power was related to the fourth power of the absolute temperature of a radiating body. Ludwig Edward Boltzmann [15] was able to derive the same relation in 1884 by analyzing a Carnot cycle in which radiation pressure was assumed to act as the pressure of the working fluid.

Wilhelm Carl Werner Otto Fritz Franz (Willy) Wien [16] derived the displacement law in 1891 by consideration of a piston moving within a mirrored cylinder. He found that the spectral energy density in an isothermal enclosure and the spectral emissive power of a blackbody are both directly proportional to the fifth power of the absolute temperature when "corresponding wavelengths" are chosen. The relation presented in Sec. 2-5.8 [Eq. (2-22)] is more often cited as Wien's displacement law, but is actually a consequence of the previous sentence.

Wien [17] also derived his spectral distribution of intensity through thermodynamic argument plus assumptions concerning the absorption and emission processes. Lord Rayleigh (1900) and Sir James Jeans (1905) based their spectral distribution on the assumption that the classical idea of equipartition of energy was valid [18,19].

The fact that measurements and some theoretical considerations indicated Wien's expression for the spectral distribution to be invalid at high temperatures* and/ or large wavelengths led Planck to an investigation of harmonic oscillators that were assumed to be the emitters and absorbers of radiant energy. Various further

---

*It was felt that as temperature approaches large values, the intensity of a blackbody should not approach a finite limit. Examination of Wien's formula [Eq. (2-18)] shows that this condition is not met. Planck's distribution law [Eq. (2-11)], however, does satisfy the condition.

assumptions as to the average energy of the oscillators led Planck to derive both the Wien and the Rayleigh-Jeans distributions. Planck finally found an empirical equation that fit the measured energy distributions over the entire spectrum. In determining what modifications to the theory would allow derivation of this empirical equation, he was led to the assumptions that form the basis of quantum theory. His equation leads directly to all the results derived previously by Wien, Stefan, Boltzmann, Rayleigh, and Jeans.

For an interesting and informative comprehensive review of the history of the field of thermal radiation, the article by Barr [20] is recommended. Lewis [21] discusses the derivation of Planck's law.

REFERENCES

1. O'Neill, P., A. Ignatiev, and C. Doland: The Dependence of Optical Properties on the Structural Composition of Solar Absorbers: Gold Black, *Sol. Energy,* vol. 21, no. 6, pp. 465–468, 1978.
2. Planck, Max: Distribution of Energy in the Spectrum, *Ann. Phys.,* vol. 4, no. 3, pp. 553–563, 1901.
3. Draper, John W.: On the Production of Light by Heat, *Phil. Mag.,* ser. 3, vol. 30, pp. 345–360, 1847.
4. Dwight, Herbert B.: "Tables of Integrals and Other Mathematical Data," 4th ed., p. 231, The Macmillan Company, New York, 1961.
5. Chang, S. L., and K. T. Rhee: Blackbody Radiation Functions, *Int. Comm. Heat Mass Transfer,* vol. 11, pp. 451–455, 1984.
6. Lowan, Arnold N., Technical Director, "Planck's Radiation Functions and Electronic Functions," Federal Works Agency Work Projects Administration for the City of New York, under the sponsorship of the U.S. National Bureau of Standards Computation Laboratory, 1941.
7. Pivovonsky, Mark, and Max R. Nagel: "Tables of Blackbody Radiation Functions," The Macmillan Company, New York, 1961.
8. Gardon, Robert: The Emissivity of Transparent Materials, *J. Am. Ceram. Soc.,* vol. 39, no. 8, pp. 278–287, 1956.
9. Gardon, Robert: A Review of Radiant Heat Transfer in Glass, *J. Am. Ceram. Soc.,* vol. 44, no. 7, pp. 305–312, 1961.
10. Kellett, B. S.: The Steady Flow of Heat through Hot Glass, *J. Opt. Soc. Am.,* vol. 42, no. 5, pp. 339–343, 1952.
11. Viskanta, R., and E. E. Anderson: Heat Transfer in Semitransparent Solids, in J. P. Hartnett and T. Irvine Jr. (eds.) "Advances in Heat Transfer," vol. 11, pp. 317–458, Academic Press, New York, 1975.
12. Bedford, R. E.: Calculation of Effective Emissivities of Cavity Sources of Thermal Radiation, in D. P. DeWitt and G. D. Nutter (eds.) "Theory and Practice of Radiation Thermometry," chapter 12, John Wiley and Sons, Inc., New York, 1988.
13. Gu, S., G. Fu, and Q. Zhang: 3500-K High-Frequency Induction-Heated Blackbody Source, *J. Thermophys. Heat Transfer,* vol. 3, no. 1, pp. 83–85, 1989.
14. Stefan, Joseph: Über die Beziehung zwischen der Wärmestrahlung und der Temperatur, *Sitzber. Akad. Wiss. Wien,* vol. 79, pt. 2, pp. 391–428, 1879.
15. Boltzmann, Ludwig: Ableitung des Stefan'schen Gesetzes, betreffend die Abhängigkeit der Wärmestrahlung von der Temperatur aus der electromagnetischen Lichttheorie, *Ann. Phys.,* ser. 2, vol. 22, pp. 291–294, 1884.
16. Wien, Willy: Temperatur und Entropie der Strahlung, *Ann. Phys.,* ser. 2, vol. 52, pp. 132–165, 1894.

17. Wien, Willy: Über die Energievertheilung im Emissionsspectrum eines schwarzen Körpers, *Ann. Phys.*, ser. 3, vol. 58, pp. 662–669, 1896.
18. Lord Rayleigh: The Law of Complete Radiation, *Phil. Mag.*, vol. 49, pp. 539–540, 1900.
19. Jeans, Sir James: On the Partition of Energy between Matter and the Ether, *Phil. Mag.*, vol. 10, pp. 91–97, 1905.
20. Barr, E. Scott: Historical Survey of the Early Development of the Infrared Spectral Region, *Am. J. Phys.*, vol. 28, no. 1, pp. 42–54, 1960.
21. Lewis, Henry R.: Einstein's Derivation of Planck's Radiation Law, *Am. J. Phys.*, vol. 41, no. 1, pp. 38–44, 1973.

PROBLEMS

2-1 A blackbody in air is at a temperature of 1025 K.

(a) What is the spectral intensity emitted in a direction normal to the black surface at $\lambda = 3 \ \mu m$?

(b) What is the spectral intensity at $\theta = 60°$ with respect to the normal of the black surface at $\lambda = 3 \ \mu m$?

(c) What is the directional spectral emissive power from the surface at $\theta = 60°$ and $\lambda = 3 \ \mu m$?

(d) At what λ is the maximum spectral intensity emitted from this blackbody, and what is the value of this intensity?

(e) What is the hemispherical total emissive power of the blackbody?

Answer: (a) 4596 W/(m² · μm · sr); (b) 4596 W/(m² · μm · sr); (c) 2298 W/(m² · μm · sr); (d) 2.8271 μm, 4634 W/(m² · μm · sr); (e) 62592 W/m²

2-2 Plot the hemispherical spectral emissive power $e_{\lambda b}$ for a blackbody in air [W/(m² · μm)] as a function of wavelength (μm) for surface temperatures of 1000 and 5000 K.

2-3 For a blackbody at 1500 K, find:

(a) the maximum emitted spectral intensity (kW/m² · μm · sr).

(b) the hemispherical total emissive power (kW/m²).

(c) the emissive power in the spectral range between 2 and 5 μm.

(d) the ratio of spectral intensity at 2 μm to that at 5 μm.

Answer: (a) 31.1 kW/(m² · μm · sr); (b) 287 kW/m²; (c) 161.1 kW/m²; (d) 4.7277

2-4 A blackbody at 1150 K is radiating to space.

(a) What is the ratio of the spectral intensity of the blackbody at $\lambda = 1 \ \mu m$ to the spectral intensity at $\lambda = 5 \ \mu m$?

(b) What fraction of the blackbody emissive power lies between the wavelengths $\lambda = 1 \ \mu m$ and $\lambda = 5 \ \mu m$?

(c) At what wavelength does the peak in the spectrum occur for this blackbody?

(d) How much energy is emitted by the blackbody in the range $1 \le \lambda \le 5 \ \mu m$?

Answer: (a) 0.12911; (b) 0.71408; (c) 2.5198 μm; (d) 70,821 W/m²

2-5 The surface of the sun has an effective blackbody temperature of 5780 K.

(a) What percentage of the radiant emission of the sun lies in the visible range (0.4–0.7 μm)?

(b) What percentage is in the ultraviolet?

(c) At what wavelength and frequency is the maximum energy per unit wavelength emitted?

(d) What is the maximum value of the hemispherical spectral emissive power?

Answer: (a) 36.7%; (b) 12.2%; (c) 0.5013 μm, 5.98×10^{14} Hz; (d) 8.301×10^7 W/(m² · μm)

2-6 A blackbody radiates such that the wavelength at maximum emissive power is 1.5 μm. What fraction of the total emissive power from this blackbody is in the range $\lambda = 1$ to $\lambda = 4 \ \mu m$?

Answer: 0.788

2-7 A blackbody has hemispherical spectral emissive power of 0.012923 W/(m² · μm) at a wavelength of 96.67 μm. What is the wavelength for the maximum emissive power of this black-body?

Answer: 28.97 μm

2-8 A radiometer is sensitive to radiation only in the interval $4 \leq \lambda \leq 10$ μm. The radiometer is used to calibrate a blackbody source at 750 K. The radiometer records that the emitted energy is 9500 W/m^2. What percentage of the expected blackbody energy in the prescribed wavelength range is the source emitting?

Answer: 94.36%

2-9 What temperature must a blackbody have for one-quarter of the emitted energy to be in the visible region?

Answer: 4343 K, 12461 K (Note: two solutions are possible!)

2-10 Show that $i'_{\lambda b}$ increases with T at any fixed value of λ.

2-11 Blackbody radiation is leaving a small hole in a furnace at 1400 K (see the figure). What fraction of the radiation is intercepted by the annular disk? What fraction passes through the hole in the disk?

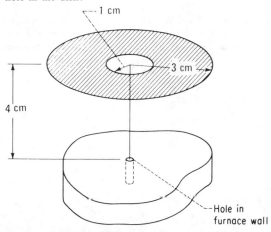

Answer: 0.3012, 0.0588

2-12 A sheet of silica glass transmits 92% of the incident radiation in the wavelength range between 0.35 and 2.7 μm and is essentially opaque to radiation at longer and shorter wavelengths. Estimate the percent of solar radiation that the glass will transmit. (Consider the sun as a black-body at 5780 K.)

If the garden in a greenhouse radiates as a black surface and is at 38° C, what percent of this radiation will be transmited through the glass?

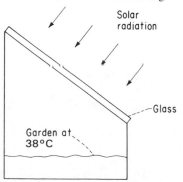

Answer: 82.9%, 0.003%.

2-13 Derive Wien's displacement law by differentiation of Planck's spectral distribution in terms of wavenumber, and show that $T/\eta_{max} = 5099.4 \ \mu m \cdot K$.

2-14 A student notes that the peak emission of the sun according to Wien's displacement law is at a wavelength of about $\lambda_{max} = C_3/5780 \ K = 2897.8/5780 = 0.501 \ \mu m$. Using $\eta_{max} = 1/0.501 \ \mu m$, the student solves again for the solar temperature using the result derived in Problem 2-13. Does this computed temperature agree with the solar temperature? Why? (This is not trivial—put some thought into *why*.)

2-15 Derive the relation between the wavenumber and the wavelength at the peak of the blackbody spectrum. (You may use the result of Prob. 2-13).

Answer: $\eta_{max} \ (cm^{-1}) = 5682.6/\lambda_{max} \ (\mu m)$

2-16 A solid copper sphere 3.5 cm in diameter has a thin black coating. Initially, the sphere is at 925 K, and it is then placed in vacuum with very cold surroundings. How long will it take for the sphere to cool to 460 K? Because of the high thermal conductivity of copper, it is assumed that the temperature within the sphere is uniform at any instant during the cooling process. (Properties of copper: density, 8950 kg/m³; specific heat, 383 J/kg·K.)

Answer: 0.294 h

2-17 A black rectangular sheet of metal, 5 cm by 8 cm in size, is heated uniformly with 775 W. One face of the rectangle is well insulated. The other face is exposed to vacuum and very cold surroundings. At thermal equilibrium, what fraction of the emitted energy is in the wavenumber range from 0.25 to 0.5 $(\mu m)^{-1}$?

Answer: 0.4748

THREE

DEFINITIONS OF PROPERTIES FOR NONBLACK OPAQUE SURFACES

3-1 INTRODUCTION

The radiative behavior of a blackbody was presented in Chap. 2. The ideal behavior of the blackbody serves as a standard with which the performance of real radiating bodies can be compared. The radiative behavior of a real body depends on many factors such as composition, surface finish, temperature, wavelength of the radiation, angle at which radiation is either emitted or intercepted, spectral distribution of the incident radiation, and whether the body is opaque. Various emissive, absorptive, and reflective properties, both unaveraged and averaged, are used to describe the radiative behavior of real materials relative to blackbody behavior.

The definitions of radiative properties of opaque materials are given in this chapter. To make them of greatest value, the definitions are presented rigorously and in detail. Since the definitions are numerous, the reader should not expect to read the chapter in as much detail as the blackbody information in Chap. 2. Rather, some of the alternative ways of defining the same quantity can be briefly scanned to obtain an overview of the information available, and the chapter can then be used for reference. The sections have been subdivided and made fairly independent to facilitate reference use.

The rigorous examination of radiative-property definitions arises from the need to interpret properly and use available property data in heat transfer computations. Only limited amounts of data in the literature provide detailed directional and spectral measurements. Because of the complexities in making these measurements (see [1] for example) most tabulated property values are *averaged* quantities. An averaged radiative performance has been measured that includes all directions, all wavelengths, or both. A clear understanding of the averages can be obtained from the definitions in this chapter. The definitions reveal relations among various averaged properties in the form of equalities or reciprocity relations. This enables maximum use of available property information; for example, absorptivity data can be obtained from emissivity data *if certain restrictions are observed*.

Examination of the derivations of property definitions reveals the restrictions on the property relations. As an aid to understanding these definitions, Fig. 3-1 provides a representation of the types of directional properties. To provide a physical interpretation, the various parts of this figure will be referred to as each definition is introduced. Table 3-1 lists each property, its symbolic notation, and the equation number of its definition. The notation is described in Sec. 3-1.2. Measured properties of materials are given in Chap. 5; these demonstrate the use of the relations derived here.

Quantities such as emissivity ϵ and absorptivity α are usually thought of as surface properties. As discussed at the beginning of Chap. 2, for an opaque material the portion of incident radiation that is not reflected is absorbed within a layer that extends below the surface. This layer can be quite thin (small fraction of a millimeter) for a material with high internal absorptance, such as a metal, or can be a few millimeters or more for a less strongly internally absorbing dielectric material. Emission also arises from within the material extending below the surface. The amount of energy leaving the surface depends on how much of the energy emitted by each internal volume element can reach and then be transmitted through the surface. As will be discussed in Sec. 18-6, some of the energy reaching the surface can be internally reflected from the surface back into the body, depending on the refractive index of the body relative to that of the external medium. Thus, the emissive ability of a body depends on the refractive index of its surrounding medium. With rare exceptions, emissivity measurements are made for emission into air or vacuum, so the tabulated values of spectral or total emissivity (ϵ_λ or ϵ) are for an external medium with $n \approx 1$. It should be noted that if the medium adjacent to the surface has $n > 1$, then ϵ_λ and α_λ, and hence ϵ and α, can be different from the tabulated values. Layers of material that are *not* opaque, and therefore transmit a portion of the incident radiation, are discussed beginning with Chap. 14.

3-1.1 Nomenclature for Properties

A number of suggestions have been made in an effort to standardize the radiation nomenclature. One issue is the ending *-ivity* for various radiative properties. The National Institute of Standards and Technology (NIST) [formerly the National Bureau of Standards (NBS)] is attempting to standardize nomenclature and reserves this ending for the properties of an optically smooth substance with an uncontaminated surface (emissivity, reflectivity, etc.), while assigning the *-ance* ending (emittance, reflectance, etc.) to measured properties where there is a need

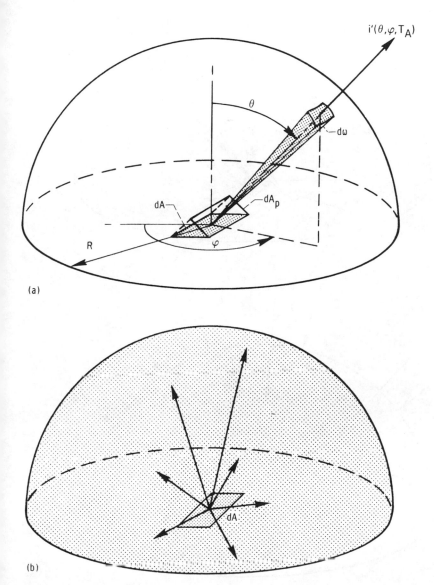

Figure 3-1 Pictorial description of directional and hemispherical radiation properties. (*a*) Directional emissivity $\epsilon'(\theta, \varphi, T_A)$; (*b*) hemispherical emissivity $\epsilon(T_A)$.

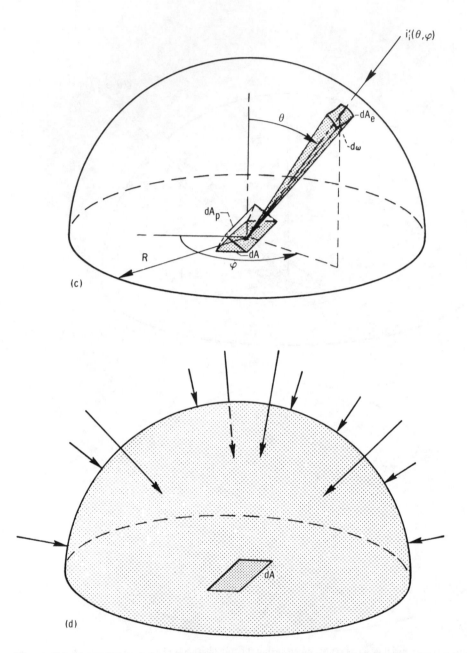

Figure 3-1 Pictorial description of directional and hemispherical radiation properties (*Continued*). (*c*) directional absorptivity $\alpha'(\theta, \varphi, T_A)$; (*d*) hemispherical absorptivity $\alpha(T_A)$.

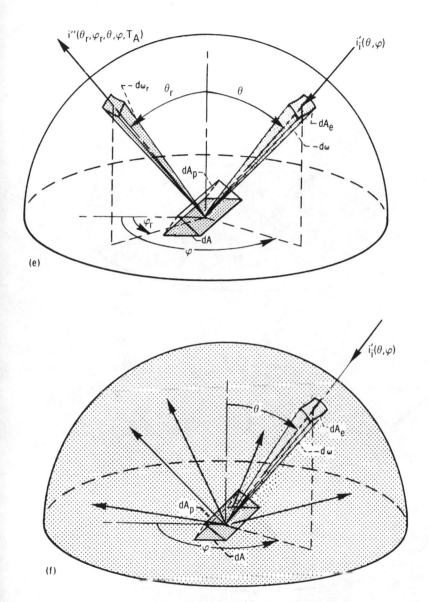

Figure 3-1 Pictorial description of directional and hemispherical radiation properties (*Continued*). (*e*) bidirectional reflectivity $\rho''(\theta_r, \varphi_r, \theta, \varphi, T_A)$; (*f*) directional-hemispherical reflectivity $\rho'(\theta, \varphi, T_A)$.

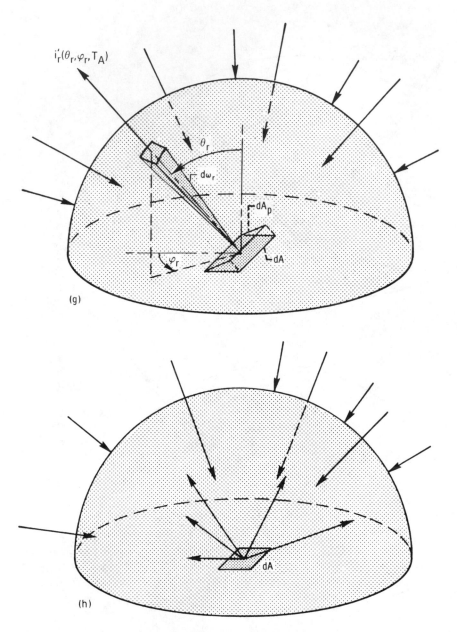

Figure 3-1 Pictorial description of directional and hemispherical radiation properties (*Continued*). (*g*) hemispherical-directional reflectivity $\rho'(\theta_r, \varphi_r, T_A)$; (*h*) hemispherical reflectivity $\rho(T_A)$.

Table 3-1 Summary of surface-property definitions

Quantity	Symbol	Defining equation	Descriptive figure
	Emissivity		
Directional spectral	$\epsilon'_\lambda(\lambda, \theta, \varphi)$	(3-2)	3-1a
Directional total	$\epsilon'(\theta, \varphi)$	(3-3)	3-1a
Hemispherical spectral	$\epsilon_\lambda(\lambda)$	(3-5)	3-1b
Hemispherical total	ϵ	(3-6)	3-1b
	Absorptivity		
Directional spectral	$\alpha'_\lambda(\lambda, \theta, \varphi)$	(3-10a)	3-1c
Directional total	$\alpha'(\theta, \varphi)$	(3-14)	3-1c
Hemispherical spectral	$\alpha_\lambda(\lambda)$	(3-16)	3-1d
Hemispherical total	α	(3-18)	3-1d
	Reflectivity		
Bidirectional spectral	$\rho''_\lambda(\lambda, \theta_r, \varphi_r, \theta, \varphi)$	(3-20)	3-1e
Directional-hemispherical spectral	$\rho_\lambda(\lambda, \theta, \varphi)$	(3-24)	3-1f
Hemispherical-directional spectral	$\rho'_\lambda(\lambda, \theta_r, \varphi_r)$	(3-26)	3-1g
Hemispherical spectral	$\rho_\lambda(\lambda)$	(3-29)	3-1h
Bidirectional total	$\rho''(\theta_r, \varphi_r, \theta, \varphi)$	(3-39)	3-1e
Directional-hemispherical total	$\rho'(\theta, \varphi)$	(3-41a)	3-1f
Hemispherical-directional total	$\rho'(\theta_r, \varphi_r)$	(3-41b)	3-1g
Hemispherical total	ρ	(3-43)	3-1h

to specify surface conditions. In addition, NIST has published nomenclature for reflectance useful in the fields of illumination and measurement [2]. The nomenclature is quite close to that adopted here.

It is the practice in most fields of science to assign the *-ivity* ending to intensive properties, such as electrical resistivity, thermal conductivity, or diffusivity. The *-ance* ending is reserved for extensive properties as in electrical resistance or conductance. Use of the term *emittance* as defined in the previous paragraph does not follow this convention, since the emittance of an opaque material would not vary with thickness. Further, it seems cumbersome to define two terms for the same concept, using one term to differentiate the one special case of a perfectly prepared pure substance.

For these reasons, the *-ivity* ending is used throughout this book for radiative properties of opaque materials, whether for ideal uncontaminated surfaces or for properties with a given surface condition. The *-ance* ending is reserved for an extensive property such as the emittance of a partially transmitting isothermal layer of water or glass, where the emittance would vary with layer thickness. The derived relations of course apply regardless of the nomenclature adopted. It should be noted that the *-ance* ending is often found in the literature dealing with the experimental determination of surface properties. The term emittance is also used in some references to describe what we have called emissive power.

3-1.2 Notation

Because of the many independent variables that must be specified for radiative properties, a concise but accurate notation is required. The notation used here is an extension of that in the preceding chapter. A functional notation is used to give explicitly the variables upon which a quantity depends. For example, $\epsilon'_\lambda(\lambda,\ \theta,\ \varphi,\ T_A)$ shows that ϵ'_λ depends on the four variables noted. The prime denotes a *directional quantity*, and the λ subscript specifies that the quantity is *spectral*. Certain quantities depend upon *two* directions (four angles); these quantities have a *double prime*. A hemispherical quantity will not have a prime, and a total quantity will not have a λ subscript. A hemispherical-directional or directional-hemispherical quantity will have a single prime. A quantity that is directional in nature, that is, must be evaluated on a per unit solid angle basis, will always have a prime even if in a specific case its numerical value is independent of direction; the independence of direction is denoted by the absence of $(\theta,\ \varphi)$ in the functional notation. Similarly, a spectral quantity will have a λ subscript even when in specific cases the numerical value does not vary with wavelength; such a specific case would not have a λ in the functional notation. In later chapters, the functional notation may be abbreviated or omitted to simplify the equation format.

Additional notation is needed for Q, the energy rate for a *finite area*, to keep consistent mathematical forms for energy balances. Thus, $d^2Q'_\lambda$ denotes a directional-spectral quantity, but the second differential is added to specify that the energy is of differential order in both wavelength and solid angle. The dQ' and dQ_λ are differential quantities with respect to solid angle and wavelength, respectively. If the energy is for a *differential area*, the order of the derivative is correspondingly increased.

This notation may appear somewhat redundant, but its usefulness will become clear in this chapter when dealing with certain special cases, such as gray and diffuse bodies. In addition, the convenience of referring to ϵ'_λ rather than writing out the term *directional spectral emissivity* is apparent. A study of Table 3-1 will help clarify the notation system.

The three main sections of this chapter each deal with a different property: emissivity, absorptivity, and reflectivity. In each section the most basic unaveraged property is presented first; for example, the first section begins with the directional spectral emissivity. Then averaged quantities are obtained by integration. The section on absorptivity also contains forms of Kirchhoff's law relating absorptivity to emissivity. The section on reflectivity includes the reciprocity relations.

3-2 SYMBOLS

A	surface area
C	a coefficient

e	radiative emissive power
F	fraction of blackbody total emissive power
i	radiation intensity
q	energy flux; energy per unit area per unit time
Q	energy rate; energy per unit time
S	distance between emitting and absorbing elements
T	absolute temperature
α	absorptivity
ϵ	emissivity
θ	polar angle measured from normal of surface
λ	wavelength
ρ	reflectivity
σ	Stefan-Boltzmann constant, Table A-4
φ	circumferential angle
ω	solid angle
$\int_\cap$	integration over solid angle of entire enclosing hemisphere

Subscripts

a	absorbed
A	of surface A
b	blackbody
d	diffuse
e	emitted or emitting
i	incident
p	projected
r	reflected
s	specular
λ	spectrally dependent

Superscripts

$'$	directional
$''$	bidirectional

3-3 EMISSIVITY

The *emissivity* specifies how well a real body radiates energy as compared with a blackbody. The emissivity can depend on factors such as body temperature, wavelength of the emitted energy, and angle of emission. The emissivity is usually measured experimentally in a direction normal to the surface and as a function of wavelength. In calculating the entire energy loss by a body, an emissivity is needed that includes all directions and wavelengths. For radiant interchange between surfaces, emissivities averaged over wavelength but not direction might be needed;

in other cases, when spectral effects become large, spectral values averaged only over direction may be used. Thus, various averaged emissivities may be required, and they must often be obtained from available measured values.

In this section, the basic definition of the directional spectral emissivity is given. This emissivity is then averaged with respect to wavelength, direction, and wavelength and direction simultaneously. Values averaged with respect to all wavelengths are termed *total* quantities; averages with respect to all directions are termed *hemispherical quantities*.

3-3.1 Directional Spectral Emissivity $\epsilon'_\lambda(\lambda, \theta, \varphi, T_A)$

Consider the geometry for emitted radiation in Fig. 3-1a. As discussed in Chap. 2, the radiation intensity is the energy per unit time emitted in direction (θ, φ) per unit of *projected* area dA_p normal to this direction, per unit solid angle and per unit wavelength interval. By basing the intensity on the projected area there is the advantage that for a black surface the intensity has the same value for all directions. Unlike the intensity from a blackbody, the emitted intensity from a real body *does* depend on direction, and hence the (θ, φ) designation is included in the notation. The energy leaving a real surface dA of temperature T_A per unit time in the wavelength interval $d\lambda$ and within the solid angle $d\omega$ is then given by

$$d^3Q'_\lambda(\lambda,\theta,\varphi,T_A) = i'_\lambda(\lambda,\theta,\varphi,T_A)\, dA \cos\theta\, d\lambda\, d\omega = e'_\lambda(\lambda,\theta,\varphi,T_A)\, dA\, d\lambda\, d\omega \qquad (3\text{-}1a)$$

For a blackbody the intensity is independent of direction and was designated in Chap. 2 by $i'_{\lambda b}(\lambda)$. The T_A notation is introduced here to clarify when quantities are temperature dependent, so that the blackbody intensity is $i'_{\lambda b}(\lambda, T_A)$. The energy leaving a black area element per unit time within $d\lambda$ and $d\omega$ is

$$d^3Q'_{\lambda b}(\lambda, \theta, T_A) = i'_{\lambda b}(\lambda, T_A)\, dA \cos\theta\, d\lambda\, d\omega = e'_{\lambda b}(\lambda, \theta, T_A)\, dA\, d\lambda\, d\omega \qquad (3\text{-}1b)$$

The emissivity is then defined as the ratio of the emissive ability of the real surface to that of a blackbody,

$$\textit{Directional spectral emissivity} \equiv \epsilon'_\lambda(\lambda,\theta,\varphi,T_A) = \frac{d^3Q'_\lambda(\lambda,\theta,\varphi,T_A)}{d^3Q'_{\lambda b}(\lambda,\theta,T_A)}$$

$$= \frac{i'_\lambda(\lambda,\theta,\varphi,T_A)}{i'_{\lambda b}(\lambda,T_A)} = \frac{e'_\lambda(\lambda,\theta,\varphi,T_A)}{e'_{\lambda b}(\lambda,\theta,T_A)} \qquad (3\text{-}2)$$

This is the most fundamental emissivity, because it includes the dependence on wavelength, direction, and surface temperature.

EXAMPLE 3-1 At 60° from the normal, a surface heated to 1000 K has a directional spectral emissivity of 0.70 at a wavelength of 5 μm. The emissivity is isotropic with respect to the angle φ. What is the spectral intensity in this direction?

From Table A-5, for a blackbody at λT_A of 5000 μm $\cdot$ K, $e_{\lambda b}(\lambda, T_A)/T_A^5$ = 713.97 $\times$ 10^{-14} W/(m$^2 \cdot \mu$m $\cdot$ K^5). Then using $i'_{\lambda b}(\lambda, T_A) = e_{\lambda b}(\lambda, T_A)/\pi$,

$$i'_\lambda(5\ \mu m, 60°, 1000\ K) = \epsilon'_\lambda(5\ \mu m, 60°, 1000\ K)i'_{\lambda b}(5\ \mu m, 1000\ K)$$

$$= 0.70 \times \frac{713.97 \times 10^{-14}}{\pi}\ (1000)^5$$

$$= 1591\ W/(m^2 \cdot \mu m \cdot sr)$$

From the directional spectral emissivity in Eq. (3-2), an averaged emissivity can be derived by proceeding along each of two approaches: averaging over all wavelengths or averaging over all directions.

3-3.2 Directional Total Emissivity $\epsilon'(\theta, \varphi, T_A)$

To obtain an average over all wavelengths, the radiation emitted into direction (θ, φ), including contributions from all wavelengths, is found by integrating the directional spectral emissive power to give the *directional total emissive power* (the term *total* denotes that radiation from all wavelengths is included):

$$e'(\theta,\varphi,T_A) = \int_0^\infty e'_\lambda(\lambda,\theta,\varphi,T_A)\,d\lambda$$

Similarly, from Table 2-2 the directional total emissive power for a *blackbody* is

$$e'_b(\theta, T_A) = \int_0^\infty e'_{\lambda b}(\lambda, \theta, T_A)\,d\lambda = \frac{\sigma T_A^4}{\pi} \cos\theta$$

The directional total emissivity is the ratio of $e'(\theta, \varphi, T_A)$ for the real surface to $e'_b(\theta, T_A)$ emitted by a blackbody at the same temperature; that is,

$$Directional\ total\ emissivity \equiv \epsilon'(\theta, \varphi, T_A) = \frac{e'(\theta, \varphi, T_A)}{e'_b(\theta, T_A)}$$

$$= \frac{\pi \int_0^\infty e'_\lambda(\lambda, \theta, \varphi, T_A)\,d\lambda}{\sigma T_A^4 \cos\theta} = \frac{\pi \int_0^\infty i'_\lambda(\lambda, \theta, \varphi, T_A)\,d\lambda}{\sigma T_A^4}$$

$$(3\text{-}3a)$$

The $e'_\lambda(\lambda, \theta, \varphi, T_A)$ or $i'_\lambda(\lambda, \theta, \varphi, T_A)$ in the numerator can be replaced in terms of $\epsilon'_\lambda(\lambda, \theta, \varphi, T_A)$ by using Eq. (3-2) to give

Directional total emissivity (in terms of directional spectral emissivity)

$$\equiv \epsilon'(\theta, \varphi, T_A) = \frac{\pi \int_0^\infty \epsilon'_\lambda(\lambda, \theta, \varphi, T_A)e'_{\lambda b}(\lambda, \theta, T_A)\,d\lambda}{\sigma T_A^4 \cos\theta}$$

$$= \frac{\pi \int_0^\infty \epsilon_\lambda'(\lambda, \theta, \varphi, T_A) i_{\lambda b}'(\lambda, T_A) \, d\lambda}{\sigma T_A^4}$$

$$= \frac{\int_0^\infty \epsilon_\lambda'(\lambda, \theta, \varphi, T_A) e_{\lambda b}(\lambda, T_a) d\lambda}{\sigma T_A^4} \tag{3-3b}$$

Thus if the wavelength dependence of $\epsilon_\lambda'(\lambda, \theta, \varphi, T_A)$ is known, the $\epsilon'(\theta, \varphi, T_A)$ is obtained as an integrated average weighted by the blackbody emissive power or intensity. The $\epsilon_\lambda'(\lambda, \theta, \varphi, T_A)$ must be known accurately in the region where $i_{\lambda b}'(\lambda, T_A)$ is large, so that the integrand of Eq. (3-3b) will be accurate where it has large values.

EXAMPLE 3-2 At $T_A = 700$ K the $\epsilon_\lambda'(\lambda, \theta, \varphi, T_A)$ can be approximated by 0.8 for $\lambda = 0$–5 μm, and 0.4 for $\lambda > 5$ μm. What is the value of $\epsilon'(\theta, \varphi, T_A)$?
From Eq. (3-3b),

$$\epsilon'(\theta, \varphi, T_A) = \int_0^{5T_A} 0.8 \frac{e_{\lambda b}(\lambda, T_A)}{\sigma T_A^5} \, d(\lambda T_A) + \int_{5T_A}^\infty 0.4 \frac{e_{\lambda b}(\lambda, T_A)}{\sigma T_A^5} \, d(\lambda T_A)$$

From Eq. (2-32),

$$\epsilon'(\theta, \varphi, T_A) = 0.8 F_{0-3500} + 0.4 F_{3500-\infty}$$
$$= 0.8(0.38291) + 0.4(0.61709) = 0.553$$

Since 61.7% of the emitted blackbody energy at 700 K is in the region for $\lambda > 5$ μm, the result is weighted toward the 0.4 emissivity value.

3-3.3 Hemispherical Spectral Emissivity $\epsilon_\lambda(\lambda, T_A)$

Now return to Eq. (3-2) and consider the average obtained by integrating the directional spectral quantities over all directions of a hemispherical envelope centered over dA (Fig. 3-1b). The spectral radiation emitted by a unit surface area into all directions of the hemisphere is the *hemispherical spectral emissive power* found by integrating the spectral energy per unit solid angle over all solid angles. This is analogous to Eq. (2-8a) for a blackbody and is given by

$$e_\lambda(\lambda, T_A) = \int_{\varphi=0}^{2\pi} \int_{\theta=0}^{\pi/2} i_\lambda'(\lambda, \theta, \varphi, T_a) \cos \theta \sin dv \, d\varphi$$

$$= \int_\Omega i_\lambda'(\lambda, \theta, \varphi, T_A) \cos \theta \, d\omega$$

The notation $\int_\cap d\omega$ signifies integration over the hemispherical solid angle and $d\omega = \sin\theta\, d\theta\, d\varphi$. Here, $i'_\lambda(\lambda, \theta, \varphi, T_A)$ cannot in general be removed from under the integral sign as was done for a blackbody. By using Eq. (3-2) this can be written as

$$e_\lambda(\lambda,T_A) = i'_{\lambda b}(\lambda,T_A) \int_\cap \epsilon'_\lambda(\lambda,\theta,\varphi,T_A)\cos\theta\, d\omega \tag{3-4a}$$

For a blackbody the hemispherical spectral emissive power is, from (2-8b),

$$e_{\lambda b}(\lambda,T_A) = \pi i'_{\lambda b}(\lambda,T_A) \tag{3-4b}$$

The ratio of actual to blackbody emission from the surface, (3-4a) divided by (3-4b), provides the following definition:

Hemispherical spectral emissivity (in terms of directional spectral emissivity)

$$\equiv \epsilon_\lambda(\lambda,T_A) = \frac{e_\lambda(\lambda,T_A)}{e_{\lambda b}(\lambda,T_A)} = \frac{1}{\pi}\int_\cap \epsilon'_\lambda(\lambda,\theta,\varphi,T_A)\cos\theta\, d\omega \tag{3-5}$$

3-3.4 Hemispherical Total Emissivity $\epsilon(T_A)$

To derive the hemispherical total emissivity, consider that from a unit area the spectral emissive power in any direction is derived from Eq. (3-2) as $\epsilon'_\lambda(\lambda, \theta, \varphi, T_A)i'_{\lambda b}(\lambda, T_A)\cos\theta$. This is integrated over all λ and ω to give the *hemispherical total emissive power*. Dividing by σT_A^4, which is the hemispherical total emissive power for a blackbody, results in the emissivity:

Hemispherical total emissivity (in terms of directional spectral emissivity)

$$\equiv \epsilon(T_A) = \frac{e(T_A)}{e_b(T_A)} = \frac{\int_\cap [\int_0^\infty e'_\lambda(\lambda,\theta,\varphi,T_A)\, d\lambda]\, d\omega}{\sigma T_A^4}$$

$$= \frac{\int_\cap [\int_0^\infty \epsilon'_\lambda(\lambda,\theta,\varphi,T_A)\, i'_{\lambda b}(\lambda,T_A)\, d\lambda]\cos\theta\, d\omega}{\sigma T_A^4} \tag{3-6a}$$

By using Eq. (3-3b) this can be placed in a second form,

Hemispherical total emissivity (in terms of directional total emissivity)

$$\equiv \epsilon(T_A) = \frac{1}{\pi}\int_{\varphi=0}^{2\pi}\int_{\theta=0}^{\pi/2} \epsilon'(\theta,\varphi,T_A)\cos\theta\sin\theta\, d\theta\, d\varphi$$

$$= \frac{1}{\pi}\int_\cap \epsilon'(\theta,\varphi,T_A)\cos\theta\, d\omega \tag{3-6b}$$

If the order of the integrations is interchanged in Eq. (3-6a), there results

$$\epsilon(T_A) = \frac{\int_0^\infty i'_{\lambda b}(\lambda,T_A)[\int_\cap \epsilon'_\lambda(\lambda,\theta,\varphi,T_A)\cos\theta\, d\omega]\, d\lambda}{\sigma T_A^4}$$

Equation (3-5) is then utilized to obtain a third form:

Hemispherical total emissivity (in terms of hemispherical spectral emissivity)

$$\equiv \epsilon(T_A) = \frac{\pi \int_0^\infty \epsilon_\lambda(\lambda, T_A) \, i'_{\lambda b}(\lambda, T_A) \, d\lambda}{\sigma T_A^4} \tag{3-6c}$$

Substituting Eq. (3-4b) gives

$$\epsilon(T_A) = \frac{\int_0^\infty \epsilon_\lambda(\lambda, T_A) \, e_{\lambda b}(\lambda, T_A) \, d\lambda}{\sigma T_A^4} \tag{3-6d}$$

To interpret Eq. (3-6d) physically, look at Fig. 3-2. In Fig. 3-2a is shown the emissivity ϵ_λ for a surface temperature T_A. The solid curve in Fig. 3-2b is the

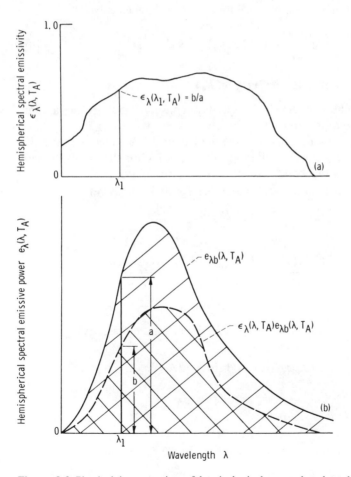

Figure 3-2 Physical interpretation of hemispherical spectral and total emissivities. (*a*) Measured emissivity values; (*b*) interpretation of emissivity as ratio of actual emissive power to blackbody emissive power.

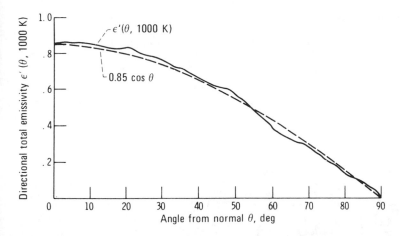

Figure 3-3 Directional total emissivity at 1000 K for Example 3-3.

hemispherical spectral emissive power for a blackbody at T_A. The area under the solid curve is σT_A^4, which is the denominator of Eq. (3-6d) and is equal to the radiation emitted per unit area by a black surface including all wavelengths and directions. The dashed curve in Fig. 3-2b is the product $\epsilon_\lambda(\lambda, T_A)e_{\lambda b}(\lambda, T_A)$, and the area under this curve is the integral in the numerator of Eq. (3-6d), which is the emission from the real surface. Hence $\epsilon(T_A)$ is the ratio of the area under the dashed curve to that under the solid curve. At each λ the quantity ϵ_λ is the ordinate of the dashed curve divided by the ordinate of the solid curve. In Fig. 3-2, the hemispherical spectral emissivity for λ_1 is $\epsilon_\lambda(\lambda_1, T_A) = b/u$.

EXAMPLE 3-3 A surface at 1000 K is isotropic in the sense that ϵ' is independent of φ, but depends on θ, as shown in Fig. 3-3. What are the hemispherical total emissivity and the hemispherical total emissive power?

The $\epsilon'(\theta, 1000 \text{ K})$ is approximated by the function $0.85 \cos \theta$. Then, from Eq. (3-6b), the hemispherical total emissivity is

$$\epsilon(T_A) = \frac{1}{\pi} \int_{\varphi=0}^{2\pi} \int_{\theta=0}^{\pi/2} 0.85 \sin \theta \cos^2 \theta \, d\theta \, d\varphi = -1.70 \left. \frac{\cos^3 \theta}{3} \right|_0^{\pi/2} = 0.567$$

The hemispherical total emissive power is then

$$e(T_A) = \epsilon(T_A)\sigma T_A^4 = 0.567 \times 5.67051 \times 10^{-8} \times 1000^4 = 32,150 \text{ W/m}^2$$

Generally the $\epsilon'(\theta, T_A)$ will not be well approximated by a convenient analytical function, and numerical integration is used.

EXAMPLE 3-4 The $\epsilon_\lambda(\lambda, T_A)$ for a surface at $T_A = 1000$ K can be approximated as shown in Fig. 3-4. What are the hemispherical total emissivity and the hemispherical total emissive power of the surface?

From Eq. (3-6d),

$$\epsilon(T_A) = \frac{1}{\sigma T_A^4} \int_0^\infty \epsilon_\lambda(\lambda, T_A)\, e_{\lambda b}(\lambda, T_A)\, d\lambda = \frac{1}{\sigma} \int_0^2 0.1\, \frac{e_{\lambda b}(\lambda, T_A)}{T_A^5}\, T_A\, d\lambda$$

$$+ \frac{1}{\sigma} \int_2^6 0.4\, \frac{e_{\lambda b}(\lambda, T_A)}{T_A^5}\, T_A\, d\lambda + \frac{1}{\sigma} \int_6^\infty 0.2\, \frac{e_{\lambda b}(\lambda, T_A)}{T_A^5}\, T_A\, d\lambda$$

This yields

$$\epsilon(T_A) = \frac{0.1}{\sigma} \int_0^{2000} \frac{e_{\lambda b}}{T_A^5}\, d(\lambda T_A) + \frac{0.4}{\sigma} \int_{2000}^{6000} \frac{e_{\lambda b}}{T_A^5}\, d(\lambda T_A)$$

$$+ \frac{0.2}{\sigma} \int_{6000}^\infty \frac{e_{\lambda b}}{T_A^5}\, d(\lambda T_A)$$

where the quantity $e_{\lambda b}/T_A^5$ is a function of λT_A. From Eq. (2-32) this is written as

$$\epsilon(T_A) = 0.1 F_{0-2000} + 0.4(F_{0-6000} - F_{0-2000}) + 0.2(1 - F_{0-6000})$$

$$= -0.3 F_{0-2000} + 0.2 F_{0-6000} + 0.2$$

$$= -0.3(0.06673) + 0.2(0.73779) + 0.2 = 0.3275$$

The hemispherical total emissive power is

$$e(T_A) = \epsilon(T_A)\sigma T_A^4 = 0.3275 \times 5.67051 \times 10^{-8} \times 1000^4 = 18.57\ \text{kW/m}^2$$

EXAMPLE 3-5 The $\epsilon_\lambda(\lambda, T_A)$ for a surface at $T_A = 650$ K is approximated as shown in Fig. 3-5. What is the hemispherical total emissivity?

In this instance numerical integration will be used for the portion where $3.5 \leq \lambda \leq 9.5\ \mu m$. This part of the spectral emissivity is given by the relation $\epsilon_\lambda(\lambda, T_A) = 1.27917 - 0.10833\lambda$. The λT bounds are $3.5 \times 650 = 2275$ $\mu m \cdot K$ and $9.5 \times 650 = 6175\ \mu m \cdot K$. Then as in Example 3-4, and by use of Eq. (2-11),

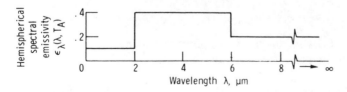

Figure 3-4 Hemispherical spectral emissivity for Example 3-4. Surface temperature $T_A = 1000$ K.

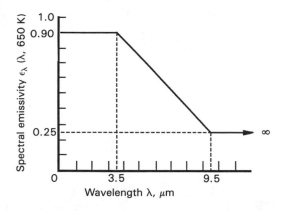

Figure 3-5 Spectral emissivity for Example 3-5.

$$\epsilon(650 \text{ K}) = 0.90 F_{0-2275}$$
$$+ \frac{1}{\sigma 650^4} \int_{3.5}^{9.5} (1.27917 - 0.10833\lambda) \frac{2\pi C_1}{\lambda^5 (e^{C_2/650\lambda} - 1)} \, d\lambda$$
$$+ 0.25(1 - F_{0-6175})$$

Evaluating the integral numerically by use of a computer subroutine yields

$$\epsilon(650 \text{ K}) = 0.90 \times 0.11517 + 0.40140 + 0.25(1 - 0.75213) = 0.5670$$

3-4 ABSORPTIVITY

The *absorptivity* is defined as the fraction of the energy incident on a body that is absorbed by the body. The incident radiation is the result of the radiative conditions at the *source* of the incident energy. The spectral distribution of the incident radiation is independent of the temperature or physical nature of the absorbing surface (unless radiation emitted from the surface is partially reflected back to the surface). Compared with emissivity, additional complexities are introduced into the absorptivity because the directional and spectral characteristics of the incident radiation must be accounted for.

Experimentally it is often easier to measure emissivity than absorptivity; hence, it is desirable to have relations between these quantities so that measured values of one will allow the other to be calculated. Such relations are developed in this section along with the absorptivity definitions.

3-4.1 Directional Spectral Absorptivity $\alpha'_\lambda(\lambda, \theta, \varphi, T_A)$

Figure 3-6a illustrates the energy incident on a surface element dA from the (θ, φ) direction. The line from dA in the direction (θ, φ) passes normally through an area element dA_e on the surface of a hemisphere of radius R placed over dA. The incident spectral intensity passing through dA_e is $i'_{\lambda, i}(\lambda, \theta, \varphi)$. This is the

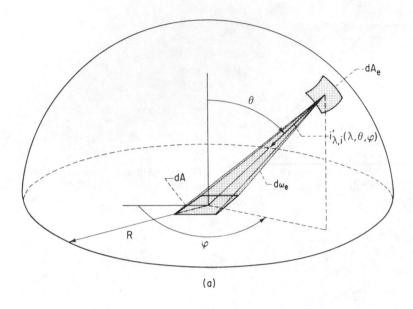

(a)

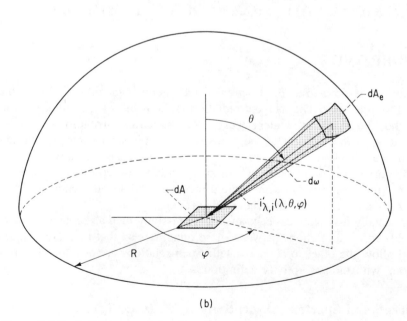

(b)

Figure 3-6 Equivalent ways of showing energy from dA_e that is incident upon dA. (a) Incidence within solid angle $d\omega_e$ having origin at dA_e; (b) incidence within solid angle $d\omega$ having origin at dA.

energy per unit area of the hemisphere, per unit solid angle $d\omega_e$, per unit time, and per unit wavelength interval. The energy within the incident solid angle $d\omega_e$ strikes the area dA of the absorbing surface. Hence, the energy per unit time incident from the direction (θ, φ) in the wavelength interval $d\lambda$ is

$$d^3Q'_{\lambda,i}(\lambda,\theta,\varphi) = i'_{\lambda,i}(\lambda,\theta,\varphi)\,dA_e\,d\omega_e\,d\lambda = i'_{\lambda,i}(\lambda,\theta,\varphi)\,dA_e\,\frac{dA\cos\theta}{R^2}\,d\lambda \qquad (3\text{-}7)$$

where $dA\cos\theta/R^2$ is the solid angle $d\omega_e$ subtended by dA when viewed from dA_e.

Equation (3-7) will be expressed in terms of the solid angle $d\omega$ shown in Fig. 3-6b. This is the solid angle subtended by dA_e when viewed from dA. The $d\omega$ has its vertex at dA and hence is the convenient solid angle to use when integrating to obtain energy incident from more than one direction. The use of this solid angle is also consistent with that used in an absorbing, emitting, and scattering medium as discussed in Chap. 12. For a *nonabsorbing medium* in the region above the surface, as is being considered here, the incident intensity *does not change along the path* from dA_e to dA (this is proved in Sec. 12-3). For these reasons, in the figures that follow, the energy incident on dA from dA_e will be pictured as that arriving in $d\omega$ as shown in Fig. 3-6b rather than as the energy leaving dA_e in $d\omega_e$ as in Fig. 3-6a. To place (3-7) in terms of $d\omega$, note that

$$\frac{dA\cos\theta}{R^2}\,dA_e = \frac{dA_e}{R^2}\,dA\cos\theta = d\omega\,dA\cos\theta \qquad (3\text{-}8)$$

Equation (3-7) can then be written as

$$d^3Q'_{\lambda,i}(\lambda, \theta, \varphi) = i'_{\lambda,i}(\lambda, \theta, \varphi)\,d\omega\,dA\cos\theta\,d\lambda \qquad (3\text{-}9)$$

The fraction of the incident energy $d^3Q'_{\lambda,i}(\lambda, \theta, \varphi)$ that is absorbed is defined as the *directional spectral absorptivity* $\alpha'_\lambda(\lambda, \theta, \varphi, T_A)$. In addition to depending on the wavelength and direction of the incident radiation, the spectral absorptivity is a function of the absorbing surface temperature. The amount of the incident energy that is absorbed is designated as $d^3Q'_{\lambda,a}(\lambda, \theta, \varphi, T_A)$. Then the ratio is formed

$$\textit{Directional spectral absorptivity} \equiv \alpha'_\lambda(\lambda,\theta,\varphi,T_A) = \frac{d^3Q'_{\lambda,a}(\lambda,\theta,\varphi,T_A)}{d^3Q'_{\lambda,i}(\lambda,\theta,\varphi)}$$

$$= \frac{d^3Q'_{\lambda,a}(\lambda,\theta,\varphi,T_A)}{i'_{\lambda,i}(\lambda,\theta,\varphi)\,dA\cos\theta\,d\omega\,d\lambda} \qquad (3\text{-}10a)$$

If the incident energy is from *black surroundings* at uniform temperature T_b, then there is the special case

$$\alpha'_\lambda(\lambda,\theta,\varphi,T_A) = \frac{d^3Q'_{\lambda,a}(\lambda,\theta,\varphi,T_A)}{i'_{\lambda b,i}(\lambda,T_b)\,dA\cos\theta\,d\omega\,d\lambda} \qquad (3\text{-}10b)$$

3-4.2 Kirchhoff's Law

This law is concerned with relating the emitting and absorbing abilities of a body. The law can have various necessary conditions imposed on it, depending on whether spectral, total, directional, or hemispherical quantities are being considered. From Eqs. (3-1) and (3-2) the energy emitted per unit time by an element dA in a wavelength interval $d\lambda$ and solid angle $d\omega$ is

$$d^3Q'_{\lambda,e} = i'_\lambda(\lambda,\theta,\varphi,T_A)\, dA \cos\theta\, d\omega\, d\lambda$$

$$= \epsilon'_\lambda(\lambda,\theta,\varphi,T_A)\, i'_{\lambda b}(\lambda,T_A)\, dA \cos\theta\, d\omega\, d\lambda \tag{3-11}$$

If the element dA at temperature T_A is assumed to be placed in an isothermal black enclosure also at temperature T_A, then the intensity of the energy incident on dA from the direction $(\theta,\ \varphi)$ (recalling the isotropy of intensity in a black enclosure) will be $i'_{\lambda b}(\lambda,\ T_A)$. To maintain the isotropy of the radiation within the black enclosure (that is, to maintain an energy balance with the element of the black enclosure that is emitting the intensity that is incident on dA at $\lambda,\ \theta,\ \varphi$), the absorbed and emitted energies given by Eqs. (3-10b) and (3-11) must be equal. Equating these gives

$$\epsilon'_\lambda(\lambda,\theta,\varphi,T_A) = \alpha'_\lambda(\lambda,\theta,\varphi,T_A) \tag{3-12}$$

This relation between the properties of the material holds *without restriction*. This is the most *general form of Kirchhoff's law.**

3-4.3 Directional Total Absorptivity $\alpha'(\theta,\ \varphi,\ T_A)$

The directional total absorptivity is the ratio of the energy, including all wavelengths, that is absorbed from a given direction to the energy incident from that direction. The total energy incident from the given direction is obtained by integrating the spectral incident energy [Eq. (3-9)] over all wavelengths to obtain

---

*As will be discussed in Chap. 4, radiation is polarized in the sense of having two wave components vibrating at right angles to each other and to the propagation direction. For the special case of black radiation the two components of polarization are equal. To be strictly accurate, Eq. (3-12) holds only for each component of polarization, and for (3-12) to be valid as written for all incident energy the incident radiation must have equal components of polarization.

Kirchhoff's law was proved for thermodynamic equilibrium in an isothermal enclosure and hence is strictly true only when there is no *net* heat transfer to or from the surface. In an actual application there is usually a net heat transfer, so it is an approximation that Eq. (3-12) will still apply. The validity of this approximation is based on experimental evidence that in most applications α'_λ and ϵ'_λ are not significantly influenced by the surrounding radiation field. Another way of stating this is that the material is able to maintain itself in a local thermodynamic equilibrium in which the populations of energy states that take part in the absorption and emission processes are given to a very close approximation by their equilibrium distributions. Thus the extension of Kirchhoff's law to nonequilibrium systems is not a result of simple thermodynamic considerations. Rather it results from the physics of materials which allows them in most instances to maintain themselves in local thermodynamic equilibrium and thus have their properties not depend on the surrounding radiation field.

$$d^2Q_i'(\theta, \varphi) = dA \cos \theta \, d\omega \int_0^\infty i_{\lambda,i}'(\lambda, \theta, \varphi) \, d\lambda \tag{3-13a}$$

The radiation absorbed is determined by integrating Eq. (3-10a) over all wavelengths,

$$d^2Q_a'(\theta, \varphi, T_A) = dA \cos \theta \, d\omega \int_0^\infty \alpha_\lambda'(\lambda, \theta, \varphi, T_A) i_{\lambda,i}'(\lambda, \theta, \varphi) \, d\lambda \tag{3-13b}$$

The following ratio is then formed:

$$\textit{Directional total absorptivity} \equiv \alpha'(\theta,\varphi,T_A) = \frac{d^2Q_a'(\theta,\varphi,T_A)}{d^2Q_i'(\theta,\varphi)}$$

$$= \frac{\int_0^\infty \alpha_\lambda'(\lambda,\theta,\varphi,T_A)\, i_{\lambda,i}'(\lambda,\theta,\varphi)\, d\lambda}{\int_0^\infty i_{\lambda,i}'(\lambda,\theta,\varphi)\, d\lambda} \tag{3-14a}$$

By use of Kirchhoff's law (3-12), an alternative form of (3-14a) is

$$\alpha'(\theta,\varphi,T_A) = \frac{\int_0^\infty \epsilon_\lambda'(\lambda,\theta,\varphi,T_A)\, i_{\lambda,i}'(\lambda,\theta,\varphi)\, d\lambda}{\int_0^\infty i_{\lambda,i}'(\lambda,\theta,\varphi)\, d\lambda} \tag{3-14b}$$

3-4.4 Kirchhoff's Law for Directional Total Properties

The general form of Kirchhoff's law [Eq. (3-12)] shows that the spectral properties ϵ_λ' and α_λ' are equal. It is now of interest to examine this equality for the *directional total* quantities. This can be accomplished by comparing a special case of Eq. (3-14b) with (3-3b). If in Eq. (3-14b) the incident radiation has a spectral distribution *proportional to that of a blackbody at* T_A, then $i_{\lambda,i}'(\lambda, \theta, \varphi) = C(\theta, \varphi)i_{\lambda b}'(\lambda, T_A)$, and Eq. (3-14b) becomes

$$\alpha'(\theta,\varphi,T_A) = \frac{\int_0^\infty \epsilon_\lambda'(\lambda,\theta,\varphi,T_A)\, i_{\lambda b}'(\lambda,T_A)\, d\lambda}{\int_0^\infty i_{\lambda b}'(\lambda,T_A)\, d\lambda(=\sigma T_A^4/\pi)} = \epsilon'(\theta,\varphi,T_A)$$

Hence when ϵ_λ' and α_λ' are dependent on wavelength, $\alpha'(\theta, \varphi, T_A) = \epsilon'(\theta, \varphi, T_A)$ *only when the incident radiation meets the restriction* $i_{\lambda,i}'(\lambda, \theta, \varphi) = C(\theta, \varphi)i_{\lambda b}'$ (λ, T_A), *where C is independent of wavelength.*

There is another important case when the relation $\alpha'(\theta, \varphi, T_A) = \epsilon'(\theta, \varphi, T_A)$ is valid. If the directional emission from a surface has the same wavelength dependence as the emission from a blackbody, $i_\lambda'(\lambda, \theta, \varphi, T_A) = C(\theta, \varphi)i_{\lambda b}'(\lambda, T_A)$, then ϵ_λ' is independent of λ. From Eqs. (3-3b) and (3-14b), if $\epsilon_\lambda'(\theta, \varphi, T_A)$ and hence $\alpha_\lambda'(\theta, \varphi, T_A)$ do not depend on λ, then, *for the direction* (θ, φ), ϵ_λ', α_λ', ϵ', and α' are all equal. A surface exhibiting such behavior is termed a *directional gray* surface.

3-4.5 Hemispherical Spectral Absorptivity $\alpha_\lambda(\lambda, T_A)$

Now consider energy in a wavelength interval $d\lambda$. The hemispherical spectral absorptivity is the fraction of the spectral energy that is absorbed from the spectral energy that is incident from all directions of a surrounding hemisphere (Fig. 3-1d). The spectral energy from an element dA_e on the hemisphere that is intercepted by a surface element dA is given by Eq. (3-9). The incident energy on dA from all directions of the hemisphere is then given by the integral

$$d^2Q_{\lambda,i} = dA \, d\lambda \int_{\varphi=0}^{2\pi} \int_{\theta=0}^{\pi/2} i'_{\lambda,i}(\lambda, \theta, \varphi) \cos \theta \sin \theta \, d\theta \, d\varphi$$

$$= dA \, d\lambda \int_{\cap} i'_{\lambda,i}(\lambda, \theta, \varphi) \cos \theta \, d\omega \tag{3-15a}$$

The amount absorbed is found by integrating Eq. (3-10a) over the hemisphere:

$$d^2Q_{\lambda,a} = dA \, d\lambda \int_{\cap} \alpha'_\lambda(\lambda,\theta,\varphi,T_A) i'_{\lambda,i}(\lambda,\theta,\varphi) \cos \theta \, d\omega \tag{3-15b}$$

The ratio of these quantities gives

$$\textit{Hemispherical spectral absorptivity} \equiv \alpha_\lambda(\lambda,T_A) = \frac{d^2Q_{\lambda,a}}{d^2Q_{\lambda,i}}$$

$$= \frac{\int_{\cap} \alpha'_\lambda(\lambda,\theta,\varphi,T_A) i'_{\lambda,i}(\lambda,\theta,\varphi) \cos \theta \, d\omega}{\int_{\cap} i'_{\lambda,i}(\lambda,\theta,\varphi) \cos \theta \, d\omega} \tag{3-16a}$$

or by using Kirchhoff's law

$$\alpha_\lambda(\lambda,T_A) = \frac{\int_{\cap} \epsilon'_\lambda(\lambda,\theta,\varphi,T_A) i'_{\lambda,i}(\lambda,\theta,\varphi) \cos \theta \, d\omega}{\int_{\cap} i'_{\lambda,i}(\lambda,\theta,\varphi) \cos \theta \, d\omega} \tag{3-16b}$$

The hemispherical spectral absorptivity and emissivity can now be compared by looking at Eqs. (3-16b) and (3-5). It is found that for the general case, where α'_λ and ϵ'_λ are functions of λ, θ, φ, and T_A, $\alpha_\lambda(\lambda, T_A) = \epsilon_\lambda(\lambda, T_A)$ only if $i'_{\lambda,i}(\lambda)$ is independent of θ and φ, that is, if the incident spectral intensity is uniform over all directions. If this is so, the $i'_{\lambda,i}$ can be canceled in (3-16b) and the denominator becomes π, which then compares with (3-5).

For the case $\alpha'_\lambda(\lambda, T_A) = \epsilon'_\lambda(\lambda, T_A)$, that is, the directional spectral properties are independent of angle, the hemispherical spectral properties are related by $\alpha_\lambda(\lambda, T_A) = \epsilon_\lambda(\lambda, T_A)$ for any angular variation of incident intensity. Such a surface is termed a *diffuse spectral* surface.

3-4.6 Hemispherical Total Absorptivity $\alpha(T_A)$

The hemispherical total absorptivity represents the fraction of energy absorbed that is incident from all directions of the enclosing hemisphere and for all wavelengths, as shown in Fig. 3-1d. The total incident energy that is intercepted by a surface element dA is determined by integrating Eq. (3-9) over all λ and all (θ, φ) of the hemisphere, which results in

$$dQ_i = dA \int_{\varphi=0}^{2\pi} \int_{\theta=0}^{\pi/2} \left[\int_0^\infty i'_{\lambda,i}(\lambda, \theta, \varphi) \, d\lambda \right] \cos \theta \sin \theta \, d\theta \, d\varphi$$

$$= dA \int_{\cap} \left[\int_0^\infty i'_{\lambda,i}(\lambda, \theta, \varphi) \, d\lambda \right] \cos \theta \, d\omega \tag{3-17a}$$

Similarly, by integrating (3-10a), the total amount of energy absorbed is equal to

$$dQ_a(T_A) = dA \int_{\cap} \left[\int_0^\infty \alpha'_\lambda(\lambda,\theta,\varphi,T_A) i'_{\lambda,i}(\lambda,\theta,\varphi) \, d\lambda \right] \cos \theta \, d\omega \tag{3-17b}$$

The ratio of absorbed to incident energy provides the definition

Hemispherical total absorptivity (in terms of directional spectral absorptivity or emissivity)

$$\equiv \alpha(T_A) = \frac{dQ_a(T_A)}{dQ_i} = \frac{\int_{\cap}[\int_0^\infty \alpha'_\lambda(\lambda,\theta,\varphi,T_A) i'_{\lambda,i}(\lambda,\theta,\varphi) \, d\lambda] \cos \theta \, d\omega}{\int_{\cap}[\int_0^\infty i'_{\lambda,i}(\lambda,\theta,\varphi) \, d\lambda] \cos \theta \, d\omega} \tag{3-18a}$$

or from Kirchhoff's law

$$\alpha(T_A) = \frac{\int_{\cap}[\int_0^\infty \epsilon'_\lambda(\lambda,\theta,\varphi,T_A) i'_{\lambda,i}(\lambda,\theta,\varphi) \, d\lambda] \cos \theta \, d\omega}{\int_{\cap}[\int_0^\infty i'_{\lambda,i}(\lambda,\theta,\varphi) \, d\lambda] \cos \theta \, d\omega} \tag{3-18b}$$

Equation (3-18b) can be compared with (3-6a) to determine under what conditions the hemispherical total absorptivity and emissivity are equal. It is recalled in (3-6a) that $\int_{\cap}[\int_0^\infty i'_{\lambda b}(\lambda, T_A) \, d\lambda] \cos \theta \, d\omega = \sigma T_A^4$. The comparison reveals that for the general case when ϵ'_λ and α'_λ vary with both wavelength and angle, $\alpha(T_A) = \epsilon(T_A)$ *only when the incident intensity is independent of the incident angle and has the same spectral form as that emitted by a blackbody with temperature equal to the surface temperature* T_A, that is, only when $i'_{\lambda,i}(\lambda, \theta, \varphi) = C i'_{\lambda b}(\lambda, T_A)$, where C is a constant. Some more restrictive cases are listed in Table 3-2.

Substituting Eq. (3-14a) into (3-18a) gives the following alternative forms:

Hemispherical total absorptivity (in terms of directional total absorptivity)

$$\equiv \alpha(T_A) = \frac{\int_{\cap} \alpha'(\theta, \varphi, T_A) \left[\int_0^\infty i'_{\lambda,i}(\lambda, \theta, \varphi) \, d\lambda \right] \cos \theta \, d\omega}{\int_{\cap} \left[\int_0^\infty i'_{\lambda,i}(\lambda, \theta, \varphi) \, d\lambda \right] \cos \theta \, d\omega} \tag{3-18c}$$

or

$$\alpha(T_A) = \frac{\int_{\cap} \alpha'(\theta,\varphi,T_A) i'_i(\theta,\varphi) \cos \theta \, d\omega}{\int_{\cap} i'_i(\theta,\varphi) \cos \theta \, d\omega} \tag{3-18d}$$

Table 3-2 Summary of Kirchhoff's-law relations between absorptivity and emissivity

Type of quantity	Equality	Restrictions
Directional spectral	$\alpha'_\lambda(\lambda, \theta, \varphi, T_A) = \epsilon'_\lambda(\lambda, \theta, \varphi, T_A)$	None
Directional total	$\alpha'(\theta, \varphi, T_A) = \epsilon'(\theta, \varphi, T_A)$	Incident radiation must have a spectral distribution proportional to that of a blackbody at T_A, $i'_{\lambda,i}(\lambda, \theta, \varphi) = C(\theta, \varphi)i'_{\lambda b}(\lambda, T_A)$; or $\alpha'_\lambda(\theta, \varphi, T_A) = \epsilon'_\lambda(\theta, \varphi, T_A)$ are independent of wavelength (directional-gray surface)
Hemispherical spectral	$\alpha_\lambda(\lambda, T_A) = \epsilon_\lambda(\lambda, T_A)$	Incident radiation must be independent of angle, $i'_{\lambda,i}(\lambda) = C(\lambda)$; or $\alpha'_\lambda(\lambda, T_A) = \epsilon'_\lambda(\lambda, T_A)$ do not depend on angle (diffuse-spectral surface)
Hemispherical total	$\alpha(T_A) = \epsilon(T_A)$	Incident radiation must be independent of angle and have a spectral distribution proportional to that of a blackbody at T_A, $i'_{\lambda,i}(\lambda) = Ci'_{\lambda b}(\lambda, T_A)$; or incident radiation independent of angle and $\alpha'_\lambda(\theta, \varphi, T_A) = \epsilon'_\lambda(\theta, \varphi, T_A)$ are independent of λ (directional-gray surface); or incident radiation from each direction has spectral distribution proportional to that of a blackbody at T_A and $\alpha'_\lambda(\lambda, T_A) = \epsilon'_\lambda(\lambda, T_A)$ are independent of angle (diffuse-spectral surface); or $\alpha'_\lambda(T_A) = \epsilon'_\lambda(T_A)$ are independent of wavelength and angle (diffuse-gray surface)

where $i'_i(\theta, \varphi)$ is the incident *total* intensity from direction (θ, φ). Changing the order of integration in Eq. (3-18a) and substituting Eq. (3-16a) give

Hemispherical total absorptivity (in terms of hemispherical spectral absorptivity)

$$\equiv \alpha(T_A) = \frac{\int_0^\infty [\alpha_\lambda(\lambda,T_A) \int_\omega i'_{\lambda,i}(\lambda,\theta,\varphi) \cos\theta \, d\omega] \, d\lambda}{\int_0^\infty [\int_\omega i'_{\lambda,i}(\lambda,\theta,\varphi) \cos\theta \, d\omega] \, d\lambda} \tag{3-18e}$$

or

$$\alpha(T_A) = \frac{\int_0^\infty \alpha_\lambda(\lambda,T_A) \, d^2Q_{\lambda,i}}{\int_0^\infty d^2Q_{\lambda,i}} \tag{3-18f}$$

where $d^2Q_{\lambda,i}$ is the spectral energy incident *from all directions* that is intercepted by the surface element dA.

A special case is in Prob. 3-1 at the end of this chapter. If there is a uniform incident intensity from a gray source at T_i, and if $\epsilon_\lambda(\lambda, T_A)$ is independent of T_A, then the hemispherical total absorptivity for the incident radiation is equal to the hemispherical total emissivity of the material evaluated at the source temperature T_i, $\alpha(T_A) = \epsilon(T_i)$.

EXAMPLE 3-6 The hemispherical spectral emissivity of a surface at 300 K is 0.8 for $0 \leqslant \lambda \leqslant 3$ μm, and 0.2 for $\lambda > 3$ μm. What is the hemispherical total absorptivity for diffuse incident radiation from a black source at $T_i = 1000$ K? What is it for diffuse incident solar radiation?

For diffuse incident radiation $\alpha_\lambda(\lambda, T_A) = \epsilon_\lambda(\lambda, T_A)$, and for a black source $i'_{\lambda,i}(\lambda, \theta, \varphi) = i'_{\lambda b,i}(\lambda)$. Then (3-18e) becomes

$$\alpha(T_A) = \frac{\int_0^\infty \epsilon_\lambda(\lambda,T_A) i'_{\lambda b, i}(\lambda, T_i)\, d\lambda}{\int_0^\infty i'_{\lambda b, i}(\lambda, T_i)\, d\lambda} = \frac{\int_0^\infty \epsilon_\lambda(\lambda,T_A) e_{\lambda b, i}(\lambda, T_i)\, d\lambda}{\sigma T_i^4}$$

Expressing $\alpha(T_A)$ in terms of the two wavelength regions over which $\epsilon_\lambda(\lambda, T_A)$ is constant gives $\alpha(T_A) = 0.8 F_{0-3T_i} + 0.2(1 - F_{0-3T_i})$. Using Table A-5, a source at 1000 K gives $F_{0-3000} = 0.27323$ so that

$$\alpha(T_A) = 0.8 F_{0-3000} + 0.2(1 - F_{0-3000}) = 0.364$$

For incident solar radiation, $T_i = 5780$ K, $F_{0-17,340} = 0.97880$ so that

$$\alpha(T_A) = 0.8 F_{0-17,340} + 0.2(1 - F_{0-17,340}) = 0.787$$

Hence, there is a considerable increase in $\alpha(T_A)$ as a result of the shift of the incident-energy spectrum toward shorter wavelengths as the source temperature is raised.

3-4.7 Diffuse-Gray Surface

As will be discussed in Chap. 7, a common assumption in enclosure calculations is that surfaces are *diffuse-gray*. *Diffuse* signifies that the directional emissivity and directional absorptivity do not depend on direction. Hence in the case of emission, the emitted intensity is uniform over all directions as for a blackbody. The term *gray* signifies that the spectral emissivity and absorptivity do not depend on wavelength. They can, however, depend on temperature. Thus at each surface temperature, the emitted spectral radiation is a fixed fraction of blackbody spectral radiation for all wavelengths.

The diffuse-gray surface therefore *absorbs a fixed fraction of incident radiation from any direction and at any wavelength*. It emits radiation that is a *fixed fraction of blackbody radiation for all directions and all wavelengths* (this is the motivation for the term gray). The directional-spectral absorptivity and emissivity then become

$$\alpha'_\lambda(\lambda, \theta, \varphi, T_A) = \alpha'_\lambda(T_A) \qquad \text{and} \qquad \epsilon'_\lambda(\lambda, \theta, \varphi, T_A) = \epsilon'_\lambda(T_A)$$

From Kirchhoff's law, Eq. (3-12), it follows that $\alpha'_\lambda(T_A) = \epsilon'_\lambda(T_A)$.

From Eqs. (3-18a), (3-18b), and (3-6a), since the $\alpha'_\lambda(T_A)$ and $\epsilon'_\lambda(T_A)$ are not functions of either direction or wavelength, they can be taken out of the integrals and the equations reduce to $\alpha(T_A) = \alpha'_\lambda(T_A) = \epsilon'_\lambda(T_A) = \epsilon(T_A)$. Thus for a diffuse-gray surface the directional-spectral and the hemispherical-total values of absorptivity and emissivity are *all equal*. The hemispherical total absorptivity is *independent* of the nature of the incident radiation.

3-4.8 Summary of Kirchhoff's-Law Relations

The restrictions on application of Kirchhoff's law are summarized in Table 3-2.

EXAMPLE 3-7 The surface in Example 3-5 at $T_A = 650$ K is subject to incident radiation from a diffuse-gray source at $T_i = 925$ K. What is the hemispherical-total absorptivity?

For a diffuse-gray source at T_i, $i'_{\lambda,\,i}(\lambda,\,\theta,\,\varphi) = Ci'_{\lambda b}(\lambda,\,T_i)$ where C is a constant. From Table 3-2, since there is no angular dependence, $\alpha_\lambda(\lambda,\,T_A) = \epsilon_\lambda(\lambda,\,T_A)$. Then from Eq. (3-18e),

$$
\alpha(T_A,\,T_i) = \frac{\displaystyle\int_0^\infty \left[\epsilon_\lambda(\lambda,\,T_A) \int_\cap Ci'_{\lambda b}(\lambda,\,T_i) \cos\theta\,d\omega \right] d\lambda}{\displaystyle\int_0^\infty \left[\int_\cap Ci'_{\lambda b}(\lambda,\,T_i) \cos\theta\,d\omega \right] d\lambda}
$$

$$
= \frac{1}{\sigma T_i^4} \int_0^\infty \epsilon_\lambda(\lambda,\,T_A) e_{\lambda b}(\lambda,\,T_i)\,d\lambda
$$

As in Example 3-5 (using λT_i values of $3.5 \times 925 = 3238$ $\mu\mathrm{m}\cdot\mathrm{K}$ and $9.5 \times 925 = 8788$ $\mu\mathrm{m}\cdot\mathrm{K}$)

$$
\alpha = 0.90 F_{0-3238} + \frac{1}{\sigma 925^4} \int_{3.5}^{9.5} (1.27917 - 0.10833\lambda) \frac{2\pi C_1}{\lambda^5 (e^{C_2/925\lambda} - 1)}\,d\lambda
$$

$$
+ 0.25(1 - F_{0-8788})
$$

Using numerical integration yields,

$$
\alpha = 0.90 \times 0.32650 + 0.37871 + 0.25(1 - 0.88377) = 0.7016
$$

3-5 REFLECTIVITY

The reflective properties of a surface are more complicated to specify than the emissivity or absorptivity. This is because the reflected energy depends not only on the angle at which the incident energy impinges on the surface but also on the direction being considered for the reflected energy. The important reflectivity quantities are now defined.

3-5.1 Spectral Reflectivities

Bidirectional spectral reflectivity $\rho_\lambda''(\lambda, \theta_r, \varphi_r, \theta, \varphi)$ Consider spectral radiation incident on a surface from direction (θ, φ) as shown in Fig. 3-1e. Part of this energy is reflected into the (θ_r, φ_r) direction and provides part of the reflected intensity in the (θ_r, φ_r) direction. The subscript r denotes quantities evaluated at the reflection angle. The entire $i_{\lambda,r}'(\lambda, \theta_r, \varphi_r)$ is the result of summing the reflected intensities produced by the incident intensities $i_{\lambda,i}'(\lambda, \theta, \varphi)$ from all incident directions (θ, φ) of the hemisphere surrounding the surface element. The contribution to $i_{\lambda,r}'(\lambda, \theta_r, \varphi_r)$ produced by the incident energy from only one (θ, φ) is designated as $i_{\lambda,r}''(\lambda, \theta_r, \varphi_r, \theta, \varphi)$ and it depends on both the incidence and reflection angles.

The energy from direction (θ, φ) intercepted by dA per unit area and wavelength is, from Eq. (3-9),

$$\frac{d^3 Q_{\lambda,i}'(\lambda,\theta,\varphi)}{dA\, d\lambda} = i_{\lambda,i}'(\lambda,\theta,\varphi) \cos\theta\, d\omega \tag{3-19}$$

The *bidirectional spectral reflectivity* is a ratio expressing the contribution that $i_{\lambda,i}'(\lambda, \theta, \varphi) \cos\theta\, d\omega$ makes to the reflected spectral intensity in the (θ_r, φ_r) direction:

$$Bidirectional\ spectral\ reflectivity \equiv \rho_\lambda''(\lambda,\theta_r,\varphi_r,\theta,\varphi) = \frac{i_{\lambda,r}''(\lambda,\theta_r,\varphi_r,\theta,\varphi)}{i_{\lambda,i}'(\lambda,\theta,\varphi) \cos\theta\, d\omega} \tag{3-20}$$

Although the reflectivity is a function of surface temperature, the T_A notation modifying ρ is omitted for simplicity. The ratio in Eq. (3-20) is a reflected intensity divided by the intercepted energy arriving within solid angle $d\omega$. Having $\cos\theta\, d\omega$ in the denominator means that when $\rho_\lambda''(\lambda, \theta_r, \varphi_r, \theta, \varphi)i_{\lambda,i}'(\lambda, \theta, \varphi) \cos\theta\, d\omega$ is integrated over all incidence angles to provide the reflected intensity $i_{\lambda,r}'(\lambda, \theta_r, \varphi_r)$, this reflected intensity will be properly weighted by the amount of energy intercepted from each direction. Since $i_{\lambda,r}''$ is generally one differential order smaller than $i_{\lambda,i}'$,* the $d\omega$ in the denominator prevents $\rho_\lambda''(\lambda, \theta_r, \varphi_r, \theta, \varphi)$ from being a differential quantity. For a *diffuse reflection* the incident energy from (θ, φ) contributes equally to the reflected intensity for all (θ_r, φ_r). It will also be shown that the form of Eq. (3-20) leads to some convenient reciprocity relations.

Reciprocity for bidirectional spectral reflectivity The $\rho_\lambda''(\lambda, \theta_r, \varphi_r, \theta, \varphi)$ is symmetric with regard to reflection and incidence angles; that is, ρ_λ'' for energy incident at (θ, φ) and reflected at (θ_r, φ_r) is equal to ρ_λ'' for energy incident at

---

*For a mirrorlike (specular) reflection, the $i_{\lambda,r}''$ is of the same order as $i_{\lambda,i}'$ and ρ_λ'' can become quite large. Thus, in contrast to the other radiative properties, ρ_λ'' can be larger than unity. In the notation for ρ'' the outgoing angles are given first, followed by the incoming angles.

(θ_r, φ_r) and reflected at (θ, φ). This is demonstrated by considering a nonblack element dA_2 located within an isothermal black enclosure as shown in Fig. 3-7. For the isothermal condition, the net energy exchange between black elements dA_1 and dA_3 must be zero. This energy is exchanged by two possible paths. The first is the direct exchange along the dashed line. This direct exchange between black elements is uninfluenced by the presence of dA_2 and hence is zero as it would be in a black isothermal enclosure without dA_2. If the net exchange along this path is zero and net exchange including all paths between dA_1 and dA_3 is zero, then net exchange along the remaining path having reflection from dA_2 must also be zero. We can now write the following for the energy traveling along the reflected path:

$$d^4Q''_{\lambda,\,1-2-3} = d^4Q''_{\lambda,\,3-2-1} \tag{3-21a}$$

The energy reflected from dA_2 that reaches dA_3 is $d^4Q''_{\lambda,1-2-3} = i''_{\lambda,r}(\lambda,\,\theta_r,\,\varphi_r,\,\theta,\,\varphi)$ $\cos\,\theta_r\,dA_2(dA_3\cos\,\theta_3/S_2^2)\,d\lambda$ or, using Eq. (3-20),

$$d^4Q''_{\lambda,\,1-2-3} = \rho''_\lambda(\lambda,\theta_r,\varphi_r,\theta,\varphi)i'_{\lambda,1}(\lambda,T)\cos\,\theta\,\frac{dA_1\cos\,\theta_1}{S_1^2}\cos\,\theta_r\,dA_2\,\frac{dA_3\cos\,\theta_3}{S_2^2}\,d\lambda$$

$$\tag{3-21b}$$

Similarly,

$$d^4Q''_{\lambda,\,3-2-1} = \rho''_\lambda(\lambda,\theta,\varphi,\theta_r,\varphi_r)i'_{\lambda,3}(\lambda,T)\cos\,\theta_r\,\frac{dA_3\cos\,\theta_3}{S_2^2}\cos\,\theta\,dA_2\,\frac{dA_1\cos\,\theta_1}{S_1^2}\,d\lambda$$

$$\tag{3-21c}$$

Substituting Eqs. (3-21b) and (3-21c) into (3-21a) gives $\rho''_\lambda(\lambda,\,\theta_r,\,\varphi_r,\,\theta,\,\varphi)i'_{\lambda,1}$ $(\lambda,\,T) = \rho''_\lambda(\lambda,\,\theta,\,\varphi,\,\theta_r,\,\varphi_r)i'_{\lambda,3}(\lambda,\,T)$ or, because $i'_{\lambda,1}(\lambda,\,T) = i'_{\lambda,3}(\lambda,\,T) = i'_{\lambda b}$ $(\lambda,\,T)$, we find the following reciprocity relation for ρ''_λ:

$$\rho''_\lambda(\lambda,\theta_r,\varphi_r,\theta,\varphi) = \rho''_\lambda(\lambda,\theta,\varphi,\theta_r,\varphi_r) \tag{3-22}$$

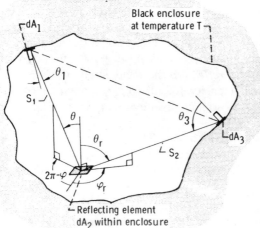

Figure 3-7 Enclosure used to prove reciprocity of bidirectional spectral reflectivity.

Directional spectral reflectivities Multiplying $i''_{\lambda,r}$ by $d\lambda \cos \theta_r \, dA \, d\omega_r$ and integrating it over the hemisphere for all θ_r and φ_r gives the energy per unit time that is reflected into the entire hemisphere as the result of an incident intensity from one direction:

$$d^3 Q'_{\lambda,r}(\lambda,\theta,\varphi) = d\lambda \, dA \int_{\cap} i''_{\lambda,r}(\lambda,\theta_r,\varphi_r,\theta,\varphi) \cos \theta_r \, d\omega_r$$

where $d\omega_r = \sin \theta_r \, d\theta_r \, d\varphi_r$. By use of Eq. (3-20) this is equal to

$$d^3 Q'_{\lambda,r}(\lambda,\theta,\varphi) = i'_{\lambda,i}(\lambda,\theta,\varphi) \cos \theta \, d\omega \, d\lambda \, dA \int_{\cap} \rho''_\lambda(\lambda,\theta_r,\varphi_r,\theta,\varphi) \cos \theta_r \, d\omega_r \qquad (3\text{-}23)$$

The directional-hemispherical spectral reflectivity is defined as the energy reflected into all solid angles divided by the energy incident from one direction (Fig. 3-1f). This gives Eq. (3-23) divided by the incident energy from Eq. (3-19):

Directional-hemispherical spectral reflectivity (in terms of bidirectional spectral reflectivity)

$$\equiv \rho'_\lambda(\lambda,\theta,\varphi) = \frac{d^3 Q'_{\lambda,r}(\lambda,\theta,\varphi)}{d^3 Q'_{\lambda,i}(\lambda,\theta,\varphi)} = \int_{\cap} \rho''_\lambda(\lambda,\theta_r,\varphi_r,\theta,\varphi) \cos \theta_r \, d\omega_r \qquad (3\text{-}24)$$

Equation (3-24) defines how much of the radiant energy incident from one direction is reflected into all directions.

Another directional reflectivity is useful when one is concerned with the reflected intensity into one direction resulting from incident radiation coming from all directions. It is the *hemispherical-directional spectral reflectivity* (Fig. 3-1g). The reflected intensity into the (θ_r, φ_r) direction is found by integrating Eq. (3-20) over all incident directions:

$$i'_{\lambda,r}(\lambda,\theta_r,\varphi_r) = \int_{\cap} \rho''_\lambda(\lambda,\theta_r,\varphi_r,\theta,\varphi) i'_{\lambda,i}(\lambda,\theta,\varphi) \cos \theta \, d\omega \qquad (3\text{-}25)$$

where $d\omega = \sin \theta \, d\theta \, d\varphi$. The hemispherical-directional spectral reflectivity is defined as the reflected intensity into the (θ_r, φ_r) direction divided by the integrated average incident intensity:

Hemispherical-directional spectral reflectivity (in terms of bidirectional spectral reflectivity)

$$\equiv \rho'_\lambda(\lambda,\theta_r,\varphi_r) = \frac{\int_{\cap} \rho''_\lambda(\lambda,\theta_r,\varphi_r,\theta,\varphi) i'_{\lambda,i}(\lambda,\theta,\varphi) \cos \theta \, d\omega}{(1/\pi) \int_{\cap} i'_{\lambda,i}(\lambda,\theta,\varphi) \cos \theta \, d\omega} \qquad (3\text{-}26)$$

Multiplying both the numerator and denominator of Eq. (3-26) by π shows that $\rho'_\lambda(\lambda, \theta_r, \varphi_r)$ may also be interpreted as π times the reflected intensity into a given direction, divided by the energy incident from all directions.

Reciprocity for directional spectral reflectivity A reciprocity relation can also be found for ρ'_λ. When the incident intensity is *uniform* over all incident directions, (3-26) reduces to

Hemispherical-directional spectral reflectivity (for *uniform* incident intensity)

$$\equiv \rho_\lambda'(\lambda,\theta_r,\varphi_r) = \int_\frown \rho_\lambda''(\lambda,\theta_r,\varphi_r,\theta,\varphi) \cos\theta \, d\omega \qquad (3\text{-}27)$$

By comparing Eqs. (3-24) and (3-27) and noting (3-22), the *reciprocal relation for* ρ_λ' is obtained (restricted to *uniform incident intensity*):

$$\rho_\lambda'(\lambda,\theta,\varphi) = \rho_\lambda'(\lambda,\theta_r,\varphi_r) \qquad (3\text{-}28)$$

where (θ_r, φ_r) and (θ, φ) are the *same* angles. This means that the reflectivity of a material irradiated at a given angle of incidence (θ, φ) as measured by the energy collected over the entire hemisphere of reflection is equal to the reflectivity for *uniform* irradiation from the hemisphere as measured by collecting the energy at a single angle of reflection (θ_r, φ_r) when (θ_r, φ_r) are the *same* angles as (θ, φ). This relation is employed in the design of "hemispherical reflectometers" for measuring radiative properties [3].

Hemispherical spectral reflectivity $\rho_\lambda(\lambda)$ If the incident spectral radiation arrives from all angles of the hemisphere (Fig. 3-1*h*), then all the radiation intercepted by the area element dA of the surface is given by Eq. (3-15*a*) as

$$d^2Q_{\lambda,i}(\lambda) = d\lambda \, dA \int_{\varphi=0}^{2\pi} \int_{\theta=0}^{\pi/2} i_{\lambda,i}'(\lambda, \theta, \varphi) \cos\theta \sin\theta \, d\theta \, d\varphi$$

$$= d\lambda \, dA \int_\frown i_{\lambda,i}'(\lambda, \theta, \varphi) \cos\theta \, d\omega$$

The amount of $d^2Q_{\lambda,i}$ that is reflected is, by integration of Eq. (3-24),

$$d^2Q_{\lambda,r}(\lambda) = \int_\frown \rho_\lambda'(\lambda,\theta,\varphi) \, d^3Q_{\lambda,i}'(\lambda,\theta,\varphi)$$

$$= d\lambda \, dA \int_\frown \rho_\lambda'(\lambda,\theta,\varphi) i_{\lambda,i}'(\lambda,\theta,\varphi) \cos\theta \, d\omega$$

The fraction of $d^2Q_{\lambda,i}(\lambda)$ that is reflected provides the definition

Hemispherical spectral reflectivity (in terms of directional-hemispherical spectral reflectivity)

$$\equiv \rho_\lambda(\lambda) = \frac{d^2Q_{\lambda,r}(\lambda)}{d^2Q_{\lambda,i}(\lambda)} = \frac{\displaystyle\int_\frown \rho_\lambda'(\lambda, \theta, \varphi)i_{\lambda,i}'(\lambda, \theta, \varphi) \cos\theta \, d\omega}{\displaystyle\int_\frown i_{\lambda,i}'(\lambda, \theta, \varphi) \cos\theta \, d\omega} \qquad (3\text{-}29)$$

Limiting cases for spectral surfaces Two important limiting cases of spectrally reflecting surfaces will be discussed in this section.

Diffuse surfaces For a diffuse surface the incident energy from the direction (θ, φ) that is reflected produces a reflected intensity that is uniform over all (θ_r, φ_r) directions. When a diffuse surface element irradiated by an incident beam is viewed, the element appears equally bright from all viewing directions. The bidirectional spectral reflectivity is independent of (θ_r, φ_r), and Eq. (3-24) simplifies to $\rho'_{\lambda,d}(\lambda, \theta, \varphi) = \rho''_{\lambda,d}(\lambda, \theta, \varphi) \int_{\cap} \cos \theta_r \, d\omega_r = \pi \rho''_{\lambda,d}(\lambda, \theta, \varphi)$. Note, however, that $\rho'_{\lambda,d}(\lambda, \theta, \varphi) = 1 - \alpha'_{\lambda,d}(\lambda)$, which is *not* a function of (θ, φ). The directional hemispherical reflectivity must be *independent of angle of incidence* because the absorptivity of a diffuse surface is, by definition, independent of angle of incidence. Then for a *diffuse* surface

$$\rho'_{\lambda,d}(\lambda) = \pi \rho''_{\lambda,d}(\lambda) \tag{3-30}$$

so that for any incidence angle the directional-hemispherical spectral reflectivity is equal to π times the bidirectional spectral reflectivity. This is because $\rho'_{\lambda,d}$ accounts for the energy reflected into all (θ_r, φ_r) directions, while $\rho''_{\lambda,d}$ accounts for the reflected intensity into only one direction. This is analogous to the relation between blackbody hemispherical emissive power and intensity, $e_{\lambda b}(\lambda) = \pi i'_{\lambda b}(\lambda)$.

Equation (3-25) provides the intensity in the (θ_r, φ_r) direction when the incident radiation is distributed over (θ, φ) values. If the surface is *diffuse, and if the incident intensity is uniform for all incident angles,* Eq. (3-25) reduces to

$$i'_{\lambda,r}(\lambda) = \rho''_{\lambda,d}(\lambda) i'_{\lambda,i}(\lambda) \int_{\cap} \cos \theta \, d\omega = \pi \rho''_{\lambda,d}(\lambda) i'_{\lambda,i}(\lambda) \tag{3-31u}$$

By using Eqs. (3-30) and (3-29) with $\rho'_\lambda(\lambda, \theta, \varphi) = \rho'_{\lambda,d}(\lambda)$, one has

$$i'_{\lambda,r}(\lambda) = \rho'_{\lambda,d}(\lambda) \, i'_{\lambda,i}(\lambda) = \rho_{\lambda,d}(\lambda) \, i'_{\lambda,i}(\lambda) \tag{3-31b}$$

so that the reflected intensity in any direction for a diffuse surface is the incident intensity times either the hemispherical-directional reflectivity or the hemispherical reflectivity. For the assumed uniform irradiation, the incident spectral intensity is related to the spectral energy per unit time intercepted by the surface element dA from all angular directions of the hemisphere by $d^2 Q_{\lambda,i}(\lambda) = \pi i'_{\lambda,i}(\lambda) \, dA \, d\lambda$.

Specular surfaces Mirrorlike, or *specular,* surfaces obey well-known laws of reflection. The specular reflector and the diffuse surface provide two relatively simple special cases that can be used for calculating heat exchange in enclosures. For an incident beam from a single direction, a specular reflector, by definition, provides a definite relation between incident and reflected angles. The reflected beam is at the same angle away from the surface normal as the incident beam, and is in the same plane as that formed by the incident beam and surface normal. Hence,

$$\theta_r = \theta \qquad \varphi_r = \varphi + \pi \tag{3-32}$$

At all other angles the bidirectional reflectivity of a specular surface is zero. We can write

$$\rho_\lambda''(\lambda,\theta,\varphi,\theta_r,\varphi_r)_{\text{specular}} = \rho_\lambda''(\lambda,\theta,\varphi,\theta_r = \theta,\varphi_r = \varphi + \pi) \equiv \rho_{\lambda,s}''(\lambda,\theta,\varphi) \tag{3-33}$$

and the $\rho_{\lambda,s}''$ is a function of only the incident direction.

For the intensity of radiation reflected from a specular surface into the solid angle around (θ_r, φ_r), Eq. (3-25) gives, for an *arbitrary* directional distribution of incident intensity, $i_{\lambda,r}'(\lambda, \theta_r, \varphi_r) = \int_\cap \rho_{\lambda,s}''(\lambda, \theta, \varphi)i_{\lambda,i}'(\lambda, \theta, \varphi) \cos\theta \, d\omega$. The integrand has a nonzero value only in the small solid angle around the direction (θ, φ) because of the properties of $\rho_{\lambda,s}''(\lambda, \theta, \varphi)$. Hence,

$$i_{\lambda,r}'(\lambda, \theta_r, \varphi_r)/i_{\lambda,i}'(\lambda, \theta, \varphi) = \rho_{\lambda,s}''(\lambda, \theta, \varphi) \cos\theta \, d\omega \tag{3-34}$$

Now consider Eq. (3-20) for bidirectional spectral reflectivity. For a specular surface, it becomes

$$i_{\lambda,r}''(\lambda,\theta_r = \theta, \varphi_r = \varphi + \pi) = \rho_{\lambda,s}''(\lambda,\theta,\varphi) i_{\lambda,i}'(\lambda,\theta,\varphi) \cos\theta \, d\omega \tag{3-35}$$

This result is the intensity reflected into a solid angle around (θ_r, φ_r) from a single beam incident at $(\theta = \theta_r, \varphi = \varphi_r - \pi)$. The right side of Eq. (3-35) is identical to the right side of (3-34), which gives the intensity reflected into the solid angle around (θ_r, φ_r) from *distributed* incident radiation. The point of this reasoning is the rather obvious fact that when examining the radiation reflected from a specular surface into a given direction, only that radiation incident at the (θ, φ) defined by (3-32) need be considered as contributing to the reflected intensity.

If a surface tends to be specular, the use of the bidirectional reflectivity can be of less practical value than in a situation with more diffuse reflection characteristics. For a specular reflection $i_{\lambda,r}''$ can be of the same order as $i_{\lambda,i}'$; hence, the $\rho_{\lambda,s}''$ becomes very large because of the $d\omega$ on the right side of (3-35). The reflectivities discussed next, which equal the *well-behaved quantity* $\rho_{\lambda,s}'' \cos\theta \, d\omega$, are more useful. From (3-34), it is this entire quantity that equals the ratio of reflected to incident intensities.

From Eq. (3-26) the hemispherical-directional spectral reflectivity for uniform irradiation of a specular surface is given by

$$\rho_{\lambda,s}'(\lambda,\theta_r,\varphi_r) = \frac{\int_\cap \rho_{\lambda,s}''(\lambda,\theta,\varphi) i_{\lambda,i}'(\lambda) \cos\theta \, d\omega}{(1/\pi)\int_\cap i_{\lambda,i}'(\lambda) \cos\theta \, d\omega} = \frac{i_{\lambda,r}'(\lambda,\theta_r,\varphi_r)}{i_{\lambda,i}'(\lambda)} \tag{3-36a}$$

Then from Eq. (3-34)

$$\rho_{\lambda,s}'(\lambda, \theta_r, \varphi_r) = \rho_{\lambda,s}''(\lambda, \theta, \varphi) \cos\theta \, d\omega = i_{\lambda,r}'(\lambda, \theta_r, \varphi_r)/i_{\lambda,i}'(\lambda, \theta, \varphi) \tag{3-36b}$$

Use of reciprocity, Eq. (3-28), shows that the directional-hemispherical reflectivity $\rho_{\lambda,s}'(\lambda, \theta, \varphi)$ for a single incident beam is equal to

$$i'_{\lambda,r}(\lambda, \theta_r, \varphi_r)/i'_{\lambda,i}(\lambda, \theta, \varphi) = \rho'_{\lambda,s}(\lambda, \theta, \varphi) = \rho'_{\lambda,s}(\lambda, \theta_r, \varphi_r) \tag{3-36c}$$

The angular relations of Eq. (3-32) apply so that all the reflected radiation is in one direction.

The hemispherical spectral reflectivity of a specular reflector irradiated with a uniform intensity is, from (3-29),

$$\rho_{\lambda,s}(\lambda) = \frac{1}{\pi} \int_{\circ} \rho'_{\lambda,s}(\lambda,\theta,\varphi) \cos \theta \, d\omega \tag{3-37}$$

If $\rho'_{\lambda,s}$ is independent of incident angle, Eq. (3-37) becomes

$$\rho_{\lambda,s}(\lambda) = \rho'_{\lambda,s}(\lambda) \tag{3-38}$$

3-5.2 Total Reflectivities

The previous reflectivity definitions have been for spectral radiation, but the expressions can be readily generalized to include all wavelengths.

Bidirectional total reflectivity $\rho''(\theta_r, \varphi_r, \theta, \varphi)$ This gives the contribution made by the total *energy* incident from direction (θ, φ) to the reflected total *intensity* into the direction (θ_r, φ_r). By analogy with Eq. (3-20)

$$Bidirectional\ total\ reflectivity \equiv \rho''(\theta_r,\varphi_r,\theta,\varphi) = \frac{\int_0^\infty i''_{\lambda,r}(\lambda,\theta_r,\varphi_r,\theta,\varphi)\,d\lambda}{\cos \theta \, d\omega \int_0^\infty i'_{\lambda,i}(\lambda,\theta,\varphi)\,d\lambda}$$

$$= \frac{i''_r(\theta_r,\varphi_r,\theta,\varphi)}{i'_i(\theta,\varphi)\cos \theta \, d\omega} \tag{3-39a}$$

As an alternative form, the reflected intensity is obtained by integrating (3-20) over all wavelengths $i''_r(\theta_r, \varphi_r, \theta, \varphi) = \cos \theta \, d\omega \int_0^\infty \rho''_\lambda(\lambda, \theta_r, \varphi_r, \theta, \varphi)i'_{\lambda,i}(\lambda, \theta, \varphi) \, d\lambda$ so that Eq. (3-39a) can be written as

Bidirectional total reflectivity (in terms of bidirectional spectral reflectivity)

$$\equiv \rho''(\theta_r, \varphi_r, \theta, \varphi) = \frac{\displaystyle\int_0^\infty \rho''_\lambda(\lambda, \theta_r, \varphi_r, \theta, \varphi)i'_{\lambda,i}(\lambda, \theta, \varphi) \, d\lambda}{\displaystyle\int_0^\infty i'_{\lambda,i}(\lambda, \theta, \varphi) \, d\lambda} \tag{3-39b}$$

Reciprocity Rewriting Eq. (3-39b) for the case of energy incident from direction (θ_r, φ_r) and reflected into direction (θ, φ) gives

$$\rho''(\theta, \varphi, \theta_r, \varphi_r) = \frac{\displaystyle\int_0^\infty \rho_\lambda''(\lambda, \theta, \varphi, \theta_r, \varphi_r) i_{\lambda,i}'(\lambda, \theta_r, \varphi_r)\, d\lambda}{\displaystyle\int_0^\infty i_{\lambda,i}'(\lambda, \theta_r, \varphi_r)\, d\lambda} \tag{3-39c}$$

Comparing Eqs. (3-39b) and (3-39c) shows that

$$\rho''(\theta,\varphi,\theta_r,\varphi_r) = \rho''(\theta_r,\varphi_r,\theta,\varphi) \tag{3-40}$$

if the *spectral distribution of incident intensity is the same for all directions or, in a less restrictive sense, if* $i_{\lambda,i}'(\lambda, \theta, \varphi) = Ci_{\lambda,i}'(\lambda, \theta_r, \varphi_r)$.

Directional total reflectivity ρ' The *directional-hemispherical total reflectivity* is the fraction of the total energy incident from a single direction that is reflected into all angular directions. The spectral energy from a given direction that is intercepted by the surface is $i_{\lambda,i}'(\lambda, \theta, \varphi) \cos\theta\, d\omega\, d\lambda\, dA$. The portion that is reflected is $\rho_\lambda'(\lambda, \theta, \varphi) i_{\lambda,i}'(\lambda, \theta, \varphi) \cos\theta\, d\omega\, d\lambda\, dA$. By integrating over all wavelengths to provide total values, the following definition is formed:

Directional-hemispherical total reflectivity (in terms of directional-hemispherical spectral reflectivity)

$$\equiv \rho'(\theta,\varphi) = \frac{d^2 Q_r'(\theta,\varphi)}{d^2 Q_i'(\theta,\varphi)} = \frac{\int_0^\infty \rho_\lambda'(\lambda,\theta,\varphi)\, i_{\lambda,i}'(\lambda,\theta,\varphi)\, d\lambda}{\int_0^\infty i_{\lambda,i}'(\lambda,\theta,\varphi)\, d\lambda} \tag{3-41a}$$

Another directional total reflectivity specifies the fraction of radiation reflected into a given (θ_r, φ_r) direction when there is diffuse irradiation. The total radiation intensity reflected into the (θ_r, φ_r) direction when the incident intensity is uniform for all directions is

$$i_r'(\theta_r, \varphi_r) = \int_0^\infty i_{\lambda,r}'(\lambda, \theta_r, \varphi_r)\, d\lambda = \int_0^\infty \rho_\lambda'(\lambda, \theta_r, \varphi_r) i_{\lambda,i}'(\lambda)\, d\lambda$$

where $\rho_\lambda'(\lambda, \theta_r, \varphi_r)$ is in Eq. (3-27). The reflectivity is defined as the reflected intensity divided by the incident intensity

Hemispherical-directional total reflectivity (for diffuse irradiation)

$$\equiv \rho'(\theta_r,\varphi_r) = \frac{\int_0^\infty \rho_\lambda'(\lambda,\theta_r,\varphi_r)\, i_{\lambda,i}'(\lambda)\, d\lambda}{\int_0^\infty i_{\lambda,i}'(\lambda)\, d\lambda} \tag{3-41b}$$

Reciprocity   Equations (3-41a) and (3-41b) are now compared bearing in mind that the latter is restricted to *uniform* incident intensity. With this restriction, from Eq. (3-28) it is found that

$$\rho'(\theta_r,\varphi_r) = \rho'(\theta, \varphi) \tag{3-42}$$

where (θ_r, φ_r) and (θ, φ) are the *same* angles, when there is a *fixed spectral distribution of the incident radiation* such that the $i'_{\lambda,i}$ in (3-41a) is related to that in (3-41b) by

$$i'_{\lambda,i}(\lambda,\theta,\varphi) = Ci'_{\lambda,i}(\lambda)$$

Hemispherical total reflectivity If the incident total radiation arrives from all angles of the hemisphere, the total radiation intercepted by a unit area at the surface is given by Eq. (3-17a). The amount of this radiation that is reflected is $dQ_r = dA \int_{\varphi=0}^{2\pi} \int_{\theta=0}^{\pi/2} \rho'(\theta, \varphi)i'_i(\theta, \varphi) \cos \theta \sin \theta \, d\theta \, d\varphi$. The ratio of these two quantities is the hemispherical total reflectivity, the fraction of all the incident energy that is reflected including all directions of reflection,

Hemispherical total reflectivity (in terms of directional-hemispherical total reflectivity)

$$\equiv \rho = \frac{dQ_r}{dQ_i} = \frac{dA}{dQ_i} \int_{\circ} \rho'(\theta,\varphi)\, i'_i(\theta,\varphi) \cos \theta \, d\omega \tag{3-43a}$$

Another form is found by using $d^2Q_{\lambda,i}(\lambda)$, the incident hemispherical spectral energy intercepted by the surface. The amount that is reflected is $\rho_\lambda(\lambda)d^2Q_{\lambda,i}$, where $\rho_\lambda(\lambda)$ is the hemispherical spectral reflectivity from Eq. (3-29). Integrating yields

Hemispherical total reflectivity (in terms of hemispherical spectral reflectivity)

$$\equiv \rho = \frac{\int_0^\infty \rho_\lambda(\lambda)\, d^2Q_{\lambda,i}(\lambda)}{dQ_i} \tag{3-43b}$$

3-5.3 Summary of Restrictions on Reflectivity Reciprocity Relations

Table 3-3 presents a summary of the restrictive conditions necessary for application of the various reflectivity reciprocity relations.

3-6 RELATIONS AMONG REFLECTIVITY, ABSORPTIVITY, AND EMISSIVITY

From the definitions of absorptivity and reflectivity as fractions of incident energy absorbed or reflected, it is evident that for an opaque body (no radiation transmitted through the body) some simple relations exist between these surface properties. By using Kirchhoff's law (Sec. 3-4.8) and taking note of the restrictions involved, further relations between the emissivity and the reflectivity can be found in certain cases.

Because the *spectral* energy per unit time $d^3Q'_{\lambda,i}$ incident upon dA of an opaque body from a solid angle $d\omega$ is either absorbed or reflected, it is evident that $d^3Q'_{\lambda,i}(\lambda, \theta, \varphi) = d^3Q'_{\lambda,a}(\lambda, \theta, \varphi, T_A) + d^3Q'_{\lambda,r}(\lambda, \theta, \varphi, T_A)$, or

$$\frac{d^3Q'_{\lambda,a}(\lambda,\theta,\varphi,T_A)}{d^3Q'_{\lambda,i}(\lambda,\theta,\varphi)} + \frac{d^3Q'_{\lambda,r}(\lambda,\theta,\varphi,T_A)}{d^3Q'_{\lambda,i}(\lambda,\theta,\varphi)} = 1 \tag{3-44}$$

Table 3-3 Summary of reciprocity relations between reflectivities

Type of quantity	Equality	Restrictions
A. Bidirectional spectral [Eq. (3-22)]	$\rho''_\lambda(\lambda, \theta, \varphi, \theta_r, \varphi_r) = \rho''_\lambda(\lambda, \theta_r, \varphi_r, \theta, \varphi)$	None
B. Directional spectral [Eq. (3-28)]	$\rho'_\lambda(\lambda, \theta, \varphi) = \rho'_\lambda(\lambda, \theta_r, \varphi_r)$ where $\theta = \theta_r$, $\varphi = \varphi_r$	$\rho'_\lambda(\lambda, \theta_r, \varphi_r)$ is for uniform incident intensity or $\rho''_\lambda(\lambda)$ independent of θ, φ, θ_r, and φ_r
C. Bidirectional total [Eq. (3-40)]	$\rho''(\theta, \varphi, \theta_r, \varphi_r) = \rho''(\theta_r, \varphi_r, \theta, \varphi)$	$i'_{\lambda,i}(\lambda, \theta, \varphi) = Ci'_{\lambda,i}(\lambda, \theta_r, \varphi_r)$ or $\rho''_\lambda(\theta, \varphi, \theta_r, \varphi_r)$ independent of wavelength
D. Directional total [Eq. (3-42)]	$\rho'(\theta, \varphi) = \rho'(\theta_r, \varphi_r)$ where $\theta = \theta_r$, $\varphi = \varphi_r$	One restriction from both B and C

Since the energy is incident from the direction (θ, φ), the two energy ratios of Eq. (3-44) are the directional spectral absorptivity [Eq. (3-10a)] and the directional-hemispherical spectral reflectivity [Eq. (3-24)]. Substituting gives

$$\alpha'_\lambda(\lambda,\theta,\varphi,T_A) + \rho'_\lambda(\lambda,\theta,\varphi,T_A) = 1 \tag{3-45}$$

Kirchhoff's law [Eq. (3-12)] can then be applied without restriction to yield

$$\epsilon'_\lambda(\lambda,\theta,\varphi,T_A) + \rho'_\lambda(\lambda,\theta,\varphi,T_A) = 1 \tag{3-46}$$

When the *total* energy arriving at dA from a given direction is considered, (3-44) becomes

$$\frac{d^2Q'_a(\theta,\varphi,T_A)}{d^2Q'_i(\theta,\varphi)} + \frac{d^2Q'_r(\theta,\varphi,T_A)}{d^2Q'_i(\theta,\varphi)} = 1 \tag{3-47}$$

Substituting Eqs. (3-14a) and (3-41a) for the energy ratios results in

$$\alpha'(\theta,\varphi,T_A) + \rho'(\theta,\varphi,T_A) = 1 \tag{3-48}$$

The absorptivity is the directional total value, and the reflectivity is the directional-hemispherical total value. Kirchhoff's law for directional total properties (Sec. 3-4.4) can then be applied to give

$$\epsilon'(\theta,\varphi,T_A) + \rho'(\theta,\varphi,T_A) = 1 \tag{3-49}$$

under the restriction that the incident radiation obeys the relation $i'_{\lambda,i}(\lambda, \theta, \varphi) = C(\theta, \varphi)i'_{\lambda b}(\lambda, T_A)$ *or* the surface is *directional-gray*.

If the incident spectral energy is arriving at dA from all directions over the hemisphere, then (3-44) gives

$$\frac{d^2Q_{\lambda,a}(\lambda,T_A)}{d^2Q_{\lambda,i}(\lambda)} + \frac{d^2Q_{\lambda,r}(\lambda,T_A)}{d^2Q_{\lambda,i}(\lambda)} = 1 \tag{3-50}$$

Equation (3-50) can then be written as

$$\alpha_\lambda(\lambda,T_A) + \rho_\lambda(\lambda,T_A) = 1 \tag{3-51}$$

where the radiative properties are hemispherical spectral values from Eqs. (3-16) and (3-29). Substitution of the hemispherical spectral emissivity $\epsilon_\lambda(\lambda, T_A)$ for $\alpha_\lambda(\lambda, T_A)$ in this relation is valid only if the intensity of incident radiation is independent of incident angle, that is, it is uniform over all incident directions (diffuse irradiation), or if the α_λ and ϵ_λ do not depend on angle (see Sec. 3-4.8). Under these restrictions, Eq. (3-51) becomes

$$\epsilon_\lambda(\lambda,T_A) + \rho_\lambda(\lambda,T_A) = 1 \tag{3-52}$$

If the energy incident on dA is integrated over all wavelengths and directions, Eq. (3-44) becomes

$$\frac{dQ_a(T_A)}{dQ_i} + \frac{dQ_r(T_A)}{dQ_i} = 1 \tag{3-53}$$

The energy ratios are now the hemispherical total values of absorptivity and reflectivity [Eqs. (3-18) and (3-43a), respectively], and Eq. (3-53) becomes

$$\alpha(T_A) + \rho(T_A) = 1 \tag{3-54}$$

Again, certain restrictions apply if $\epsilon(T_A)$ is substituted for $\alpha(T_A)$ to obtain

$$\epsilon(T_A) + \rho(T_A) = 1 \tag{3-55}$$

The principal restrictions are that the incident spectral intensity is proportional to the emitted spectral intensity of a blackbody at T_A *and* the incident intensity is uniform over all incident angles; that is, $i'_{\lambda,i}(\lambda) = C i'_{\lambda b}(\lambda, T_A)$. Other special cases where the substitution $\alpha(T_A) = \epsilon(T_A)$ can be made are listed in Sec. 3-4.8.

When the body is not opaque, such as a window or layer of water or ice, so that some radiation passes entirely through it, a transmitted fraction is introduced. The fractions of incident radiation that are reflected, absorbed, and transmitted must sum to unity. This is discussed later in connection with radiation in absorbing and transmitting media (see Sec. 12-7 and Chap. 18). Transmission (and hence absorption) properties of semitransparent materials are often quite spectrally dependent.

EXAMPLE 3-8 Radiation from the sun is incident on a surface in orbit above the earth's atmosphere. The surface is at 1000 K, and the directional total emissivity is given in Fig. 3-3. If the solar energy is incident at an angle of 25° from the normal to the surface, what is the reflected energy flux?

From Fig. 3-3, $\epsilon'(25°, 1000 \text{ K}) = 0.8$. Section 3-4.8 shows that for directional total properties, $\alpha'(25°, 1000 \text{ K}) = \epsilon'(25°, 1000 \text{ K})$ only when the incident spectrum is proportional to that emitted by a blackbody at 1000 K. This is not the case here, since the solar spectrum is like that of a blackbody at 5780 K. Hence $\alpha' \neq 0.8$, and without α' we cannot determine ρ'; the emissivity data given are insufficient to work the problem.

EXAMPLE 3-9 A surface at $T_A = 500$ K has a spectral emissivity in the normal direction that can be approximated as shown in Fig. 3-8. The surface is maintained at 500 K by cooling water and is enclosed by a black hemisphere heated to $T_i = 1500$ K. What is the reflected radiant intensity in the direction normal to the surface?

From Eq. (3-46), $\rho'_\lambda(\lambda, \theta = 0°, T_A) = 1 - \epsilon'_\lambda(\lambda, \theta = 0°, T_A)$, which is the reflectivity into the hemisphere for radiation arriving from the normal direction. From reciprocity, for uniform incident intensity over the hemisphere, $\rho'_\lambda(\lambda, \theta_r = 0°, T_A) = \rho'_\lambda(\lambda, \theta = 0°, T_A)$. Hence the reflectivity into the normal direction resulting from the incident radiation from the hemisphere is (by use of Fig. 3-8): $\rho'_\lambda(0 \leqslant \lambda < 2, \theta_r = 0°, T_A) = 0.7$; $\rho'_\lambda(2 \leqslant \lambda < 5, \theta_r = 0°, T_A) = 0.2$; $\rho'_\lambda(5 \leqslant \lambda \leqslant \infty, \theta_r = 0°, T_A) = 0.5$. The incident intensity is $i'_{\lambda,i}(\lambda, T_i) = i'_{\lambda b}(\lambda, 1500 \text{ K})$. From the relation preceding Eq. (3-41b), the reflected intensity is

$$i'_r(\theta_r = 0°) = \int_0^\infty i'_{\lambda b}(\lambda, T_i) \rho'_\lambda(\lambda, \theta_r = 0°, T_A)\, d\lambda$$

$$= \frac{\sigma T_i^4}{\pi} \int_0^\infty \frac{e_{\lambda b}(\lambda, T_i)}{\sigma T_i^5} \rho'_\lambda(\lambda, \theta_r = 0°, T_A)\, d(\lambda T_i)$$

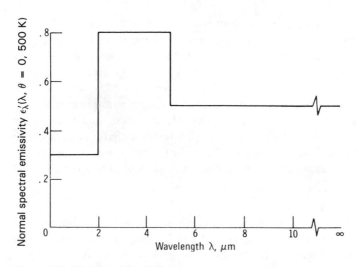

Figure 3-8 Directional spectral emissivity in normal direction for Example 3-9.

From Eq. (2-32) this becomes,

$$i_r'(\theta_r = 0°) = \frac{\sigma T_i^4}{\pi} (0.7F_{0-2T_i} + 0.2F_{2T_i-5T_i} + 0.5F_{5T_i-\infty})$$

$$= \frac{5.67051 \times 10^{-8}}{\pi} (1500)^4 [0.7(0.27323)$$

$$+ 0.2(0.83437 - 0.27323) + 0.5(1 - 0.83437)]$$

$$= 35.3 \text{ kW/(m}^2 \cdot \text{sr)}$$

EXAMPLE 3-10 A large flat plate at T_A = 850 K has a hemispherical spectral emissivity that can be approximated by a straight line decreasing from 0.85 to 0 as λ increases from 0 to 10.5 μm. The flat plate faces a second large plate that has a reflectivity of 0.35 for $0 < \lambda < 4.5$ μm and 0.82 for $\lambda > 4.5$ μm. Both plates are diffuse. Consider the emitted energy from the first plate that arrives at the second plate and is then reflected back to the first plate. What is the energy flux being reabsorbed by the first plate as a result of this single reflection?

The spectral energy emitted in $d\lambda$ is $\epsilon_\lambda(\lambda, T_A)e_{\lambda b}(\lambda, T_A) d\lambda$. For the geometry of large plates all of this energy will reach the second plate and the fractions 0.35 and 0.82 will be reflected back for the respective ranges $\lambda < 4.5$ μm and $\lambda > 4.5$ μm. From Kirchhoff's law the spectral absorptivity is equal to $\epsilon_\lambda(\lambda, T_A)$. Hence the reabsorbed energy is

$$q_{abs} = 0.35 \int_0^{4.5} \epsilon_\lambda^2(\lambda, T_A)e_{\lambda b}(\lambda, T_A) d\lambda + 0.82 \int_{4.5}^{\infty} \epsilon_\lambda^2(\lambda, T_A)e_{\lambda b}(\lambda, T_A) d\lambda$$

The $\epsilon_\lambda(\lambda, T_A)$ is given by $0.85[1 - (\lambda/10.5)]$ for $\lambda \leq 10.5$ μm and is zero for $\lambda > 10.5$ μm. Then

$$q_{abs} = 0.35 \times 0.85^2 \int_0^{4.5} (1 - \lambda/10.5)^2 e_{\lambda b}(\lambda, T_A) d\lambda$$

$$+ 0.82 \times 0.85^2 \int_{4.5}^{10.5} (1 - \lambda/10.5)^2 e_{\lambda b}(\lambda, T_A) d\lambda$$

where $e_{\lambda b}(\lambda, T_A) = 2\pi C_1/\lambda^5(e^{C_2/\lambda T_A} - 1)$. Numerical integration yields

$$q_{abs} = 0.35 \times 0.85^2 \times 10,541 + 0.82 \times 0.85^2 \times 2745 = 4292 \text{ W/m}^2$$

3-7 CONCLUDING REMARKS

In this chapter, a precise system of nomenclature has been introduced and careful definitions of the radiative properties have been given. The defining equations are summarized in Table 3-1 for convenience, along with the symbols used here.

By using these definitions it was possible to examine the restrictions on the various forms of Kirchhoff's law relating emissivity to absorptivity. These restrictions are sometimes a source of confusion, and it is hoped that the summary (Table 3-2) will make clear the conditions when α can be set equal to ϵ. These restrictions are also invoked when deriving the relation $\epsilon + \rho = 1$ from the general relation $\alpha + \rho = 1$ for opaque bodies.

The detailed definitions made it possible to derive the reciprocal relations for reflectivities and examine the restrictions involved. These restrictions are listed in a convenient summary in Table 3-3.

REFERENCES

1. Shafey, H. M., Y. Tsuboi, M. Fujita, T. Makino, and T. Kunitomo: Experimental Study on Spectral Reflective Properties of a Painted Layer, *AIAA J.*, vol. 20, no. 12, pp. 1747–1753, 1982.
2. Nicodemus, F. E., et al.: Geometrical Considerations and Nomenclature for Reflectance, NBS monograph 160, National Bureau of Standards, United States Department of Commerce, 1977.
3. Brandenberg, W. M.: The Reflectivity of Solids at Grazing Angles, in Joseph C. Richmond (ed.), "Measurement of Thermal Radiation Properties of Solids," NASA SP-31, pp. 75–82, 1963.

PROBLEMS

3-1 A material has a hemispherical spectral emissivity that varies considerably with wavelength but is fairly independent of surface temperature (see, for example, the behavior of tungsten in Fig. 5-16). Radiation from a gray source at T_i is incident on the surface uniformly from all directions. Show that the total absorptivity for the incident radiation is equal to the total emissivity of the material evaluated at the source temperature T_i.

3-2 Using Fig. 5-16, estimate the hemispherical total emissivity of tungsten at 2800 K.

Answer: 0.308

3-3 Suppose that ϵ_λ is independent of λ (gray-body radiation). Show that $F_{0-\lambda T}$ represents the fraction of the total radiant output of the gray body in the range from 0 to λT.

3-4 For a surface with hemispherical spectral emissivity ϵ_λ, does the maximum of the e_λ distribution occur at the same λ as the maximum of the $e_{\lambda b}$ distribution at the same temperature? (*Hint:* Examine the behavior of $de_\lambda/d\lambda$.) Plot the distributions of e_λ as a function of λ for the data of Fig. 3-8 at 500 K and for the data in Prob. 3-8a at 475 K. At what λ is the maximum of e_λ? How does this compare with the maximum of $e_{\lambda b}$?

3-5 The hemispherical spectral absorptivity of a surface is measured when it is exposed to isotropic incident blackbody intensity, and the results are shown below. What is the total hemispherical emissivity of this surface when it is at a temperature of 2500 K?

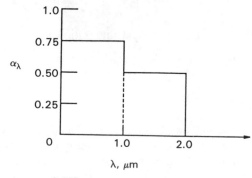

Answer: 0.357

3-6 A white ceramic surface has a hemispherical spectral emissivity distribution at 1525 K as shown. What is the hemispherical total emissivity of the surface?

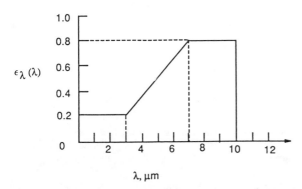

$\epsilon_\lambda (\lambda)$

$\lambda, \mu m$

Answer: 0.289

3-7 A flat plate in earth orbit is insulated on one side, and the other side is oriented normal to the solar intensity. The incident solar flux is 1350 W/m². A coating on the plate surface facing the sun has a total hemispherical emissivity of 0.1 over a broad range of plate temperatures. Surroundings above the plate are at a very low temperature. Telemetry signals to earth indicate that the plate temperature is 650 K.

(a) What is the normal solar absorptivity of the plate?

(b) If α_{solar} is independent of angle, what is the plate temperature if the plate is tilted so that its normal is 25° away from the solar direction?

Answer: (a) 0.750; (b) 634.2 K

3-8 A surface has the following values of hemispherical spectral emissivity at a temperature of 475 K.

$\lambda, \mu m$	$\epsilon_\lambda (\lambda, 475 \text{ K})$
<1	0
1	0
1.5	0.2
2	0.4
2.5	0.6
3	0.8
3.5	0.8
4	0.8
4.5	0.7
5	0.6
6	0.4
7	0.2
8	0
>8	0

(a) What is the hemispherical total emissivity of the surface at 475 K?

(b) What is the hemispherical total absorptivity of the surface at 475 K if the incident radiation is from a gray source at 1200 K that has an emissivity of 0.76? The incident radiation is uniform over all incident angles.

Answer: (a) 0.192; (b) 0.510

3-9 A diffuse surface at 750 K has a hemispherical spectral emissivity that can be approximated by the solid line shown:

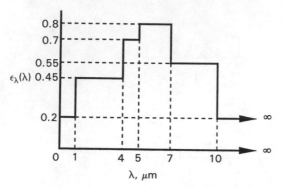

(a) What is the hemispherical total emissive power of the surface? What is the total intensity emitted in a direction 45° from the normal to the surface?

(b) What percentage of the total emitted energy is in the wavelength range $4 < \lambda < 7$ μm? How does this compare with the percentage emitted in this wavelength range by a gray body at 750 K with $\epsilon = 0.35$?

Answer: (a) 9801 W/m², 3120 W/(m²·sr); (b) 54.25%, 39.05%

3-10 The $\epsilon_\lambda(\lambda)$ for a metal at 1200 K is as shown, and it does not vary significantly with the metal temperature.

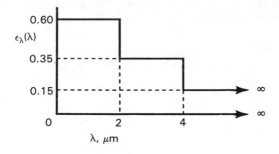

(a) What is α for incident radiation from a gray source at 1500 K with $\epsilon_{source} = 0.75$?

(b) What is α for incident radiation from a source at 1500 K made from the same metal as the receiving plate? Neglect any directional effects.

Answer: (a) 0.3659; (b) 0.4405

3-11 The directional total absorptivity of a gray surface is given by the expression:

$\alpha'(\theta) = 0.5 \cos^2 \theta$

(a) What is the hemispherical total emissivity of the surface?

(b) What is the hemispherical-hemispherical total reflectivity of this surface for diffuse incident radiation (uniform incident intensity)?

(c) What is the hemispherical-directional total reflectivity for diffuse incident radiation reflected into a direction 50° from the normal?

Answer: (a) 0.25; (b) 0.75; (c) 0.793

3-12 A diffuse spectral coating has the characteristics shown below. The coating is placed on one face of a thin sheet of metal. The sheet is placed in an orbit around the sun where the solar flux is 1350 W/m^2. The other face of the sheet is coated with a diffuse-gray coating of hemispherical total emissivity $\epsilon = 0.8$.

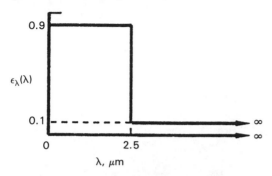

What is the temperature (K) of the sheet if
(a) the side with the spectral coating is normal to the sun?
(b) the gray side is normal to the sun?
(c) What is the normal-hemispherical total reflectivity of the diffuse-spectral coating when exposed to solar radiation? Take the effective solar radiating temperature to be 5780 K. Note any necessary assumptions.

Answer: (a) 389.7 K; (b) 381 K; (c) 0.127

3-13 Using Fig. 5-10, estimate the total absorptivity of typewriter paper for normally incident radiation from a blackbody source at 1178 K.

3-14 The spectral absorptivity of an SiO-Al selective surface can be approximated as shown below. The surface is in earth orbit around the sun and has incident on it in the normal direction the solar flux 1353 W/m^2. What will the equilibrium temperature of the surface be if the surroundings are very cold?

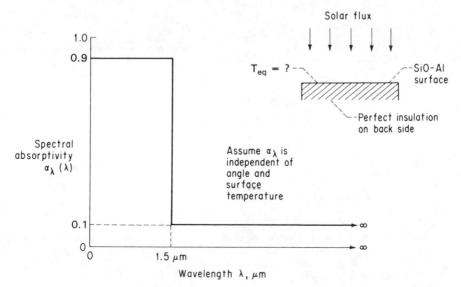

Answer: 661 K

3-15 A gray surface has directional emissivity as shown. The properties are isotropic with respect to circumferential angle φ.

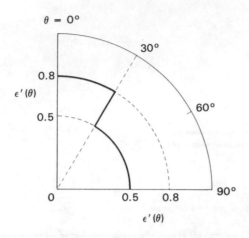

(a) What is the value of the hemispherical emissivity of this surface?
(b) If the energy from a blackbody source at 350 K is incident uniformly from all directions, what fraction of the incident energy will be absorbed by this surface?
(c) If the surface is placed in an environment at absolute zero temperature, at what rate must energy be added per unit area to maintain the surface temperature at 550 K?

Answer. (a) 0.575; (b) 0.575; (c) 2984 W/m²

3-16 A thin plate has a directional gray surface on one side with the emissivity shown below on the left. On the other side of the plate is a coating with diffuse-spectral emissivity shown below on the right. The surroundings are at very low temperature.

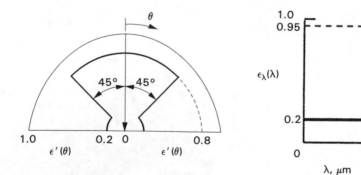

Find the equilibrium temperature of the plate if it is exposed in vacuum to a normal solar flux of 1353 W/m² with a solar spectrum equivalent to a blackbody at 5780 K when
(a) the directional-gray side is normal to the sun.
(b) the diffuse-spectral side is normal to the sun.

Answer: (a) 339 K; (b) 263 K

3-17 A gray surface has a directional total emissivity that depends on angle of incidence as $\epsilon(\theta) = 0.82 \cos \theta$. Uniform radiant energy from a single direction normal to the cylinder axis is incident on a long cylinder of radius R. What fraction of energy striking the cylinder is reflected? What is the result if the body is a sphere rather than a cylinder?

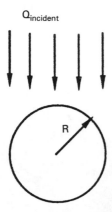

$Q_{incident}$

R

Answer: 0.356, 0.453

3-18 A flat metal plate 0.1 m wide by 1.0 m long has a temperature that varies only along the long direction. The temperature is 1050 K at one end and decreases linearly over the 1-m length to 450 K. The hemispherical spectral emissivity of the plate does not change significantly with temperature but is a function of wavelength. The wavelength dependence is approximated by a linear function decreasing from $\epsilon_\lambda(\lambda) = 0.65$ at $\lambda = 0$ to $\epsilon_\lambda(\lambda) = 0$ at $\lambda = 9$ μm. What is the rate of radiative energy loss from one side of the plate? The surroundings are at a very low temperature.

Answer: 647 W

FOUR

PREDICTION OF RADIATIVE PROPERTIES BY CLASSICAL ELECTROMAGNETIC THEORY

4-1 INTRODUCTION

James Clerk Maxwell published in 1864 a crowning achievement of classical physics, the relation between electric and magnetic fields and the realization that electromagnetic waves propagate with the speed of light, indicating that light itself is in the form of an electromagnetic wave [1]. Although quantum effects have since been shown to be the controlling phenomena in electromagnetic energy propagation, it is possible and indeed necessary to describe many of the properties of light and radiant heat by the classical wave approach. It will be shown in this chapter that the reflectivity, emissivity, and absorptivity of materials can in certain cases be calculated from the optical and electrical properties of the materials.

The relations between radiative, optical, and electrical properties are obtained by considering the interaction when an electromagnetic wave traveling through one medium is incident on the surface of another medium. The analysis is based on the assumption that there is an ideal interaction between the incident wave and the surface. Physically this means that the results are for optically smooth, clean surfaces that reflect in a specular fashion. The wave surface interaction will be investigated in a somewhat simplified fashion by using Maxwell's fundamental equations relating electric and magnetic fields. For ideal surface conditions it is possible to perform more accurate property computations by using theory that is more rigorous than the wave analysis. However, the labor involved is generally not justified because neither the simplified nor the more sophisticated approach can account for the effects of surface condition. The departures of real materials from the ideal materials assumed in the theory are often responsible for large variations of measured property values from theoretical predictions. These departures are caused by factors such as surface roughness, surface contamination, impurities, and crystal-structure modification by surface working.

Although there can be large effects of surface condition, the theory does serve a number of useful purposes. It provides an understanding of why there are basic differences in the radiative properties of insulators and electrical conductors and reveals general trends that help unify the presentation of experimental data. These trends are useful when it is required for engineering calculations to extrapolate limited experimental data into another range. The theory has utility in understanding the angular behavior of the directional reflectivity, absorptivity, and emissivity. Since the theory applies for pure substances with ideally smooth surfaces, it provides a means for computing one limit of attainable properties, such as the maximum reflectivity or minimum emissivity of a metallic surface.

The derivation of radiative property relations from classical theory is in Secs. 4-3 to 4-5. The results are applied to radiative property predictions in Sec. 4-6. Readers interested only in the use of the results for property predictions can pass over the derivation portions to Sec. 4-6.

4-2 SYMBOLS

a	absorption coefficient
c	speed of electromagnetic wave in a medium other than vacuum
c_0	speed of electromagnetic wave in vacuum
C_1, C_2	constants in Planck spectral energy distribution
e	emissive power
E	amplitude of electric intensity wave
$\mathbf{E}$	electric intensity vector
H	amplitude of magnetic intensity wave
$\mathbf{H}$	magnetic intensity vector
K	dielectric constant
n	refractive index; ratio n_2/n_1 in a few equations
$\bar{n}$	complex refractive index, $n - i\kappa$
r_e	electrical resistivity
S	instantaneous rate of energy transport per unit area
$\mathbf{S}$	Poynting vector, Eq. (4-22)
t	time
T	absolute temperature
x, y, z	coordinates in cartesian system
x', y', z'	
α	constant in propagation velocity of electromagnetic wave
β	extinction coefficient in x direction
γ	permittivity
δ	propagation angle in medium
ϵ	emissivity

θ	angle measured from normal of surface, polar angle
κ	extinction coefficient
λ	wavelength
μ	magnetic permeability
ν	frequency
ρ	reflectivity
φ	circumferential angle
χ	angle of refraction
ω	angular frequency
$\int_{\cap}$	integration over solid angle of entire enclosing hemisphere

Subscripts

A	property of body or surface A
b	black
i	incident
m	in a medium
M	maximum value
n	normal
r	reflected
s	specular
t	transmitted
x, y, z	components in x, y, z directions
x', y', z'	components in x', y', z' directions
λ	spectral
0	in vacuum
1, 2	medium 1 or 2
$\perp$	perpendicular component
$\parallel$	parallel component

Superscript

$'$	directional quantity (except in x', y', z')

4-3 FUNDAMENTAL EQUATIONS OF ELECTROMAGNETIC THEORY

Maxwell's equations can be used to describe the interaction of electric and magnetic fields within any isotropic medium, including vacuum, under the condition of no accumulation of static charge. With these restrictions the equations are, in SI (mks) units,

$$\nabla \times \mathbf{H} = \gamma \frac{\partial \mathbf{E}}{\partial t} + \frac{\mathbf{E}}{r_e} \qquad (4\text{-}1)$$

$$\nabla \times \mathbf{E} = -\mu \frac{\partial \mathbf{H}}{\partial t} \tag{4-2}$$

$$\nabla \cdot \mathbf{E} = 0 \tag{4-3}$$

$$\nabla \cdot \mathbf{H} = 0 \tag{4-4}$$

where **H** and **E** are the magnetic and electric intensities, γ is the permittivity, r_e is the electrical resistivity, and μ is the magnetic permeability of the medium. The SI units for these quantities are in Table 4-1. Zero subscripts denote quantities in vacuum.

The solutions to these equations reveal how radiation waves travel within a material and what the interaction is between the electric and magnetic fields. By knowing how waves move in each of two adjacent media and applying coupling relations at the interface, the relations governing absorption and reflection are obtained.

4-4 RADIATIVE WAVE PROPAGATION WITHIN A MEDIUM

Propagation is first considered within an infinite, homogeneous, isotropic medium. The derivation of wave propagation within a *perfect dielectric* will be considered in Sec. 4-4.1; it is found that a wave is not attenuated in such a material. Media of finite electrical conductivity are then analyzed in Sec. 4-4.2; these media can be *imperfect dielectrics* (poor conductors) or *metals* (good conductors). Waves do attenuate in these materials because of absorption of energy.

Table 4-1 Quantities for use in electromagnetic equations in SI units

Symbol	Quantity	Units	Value
c	Speed of electromagnetic wave propagation	m/s	
c_0	Speed of electromagnetic wave propagation in vacuum	m/s	2.997925×10^8
E	Electric intensity	N/C [newtons/coulomb]	
H	Magnetic intensity	C/(m · s)	
K	Dielectric constant, γ/γ_0		
r_e	Electrical resistivity	$\Omega \cdot$ m; N · m^2 · s/C^2	
S	Instantaneous rate of energy transport per unit area	N · m/(s · m^2); W/m^2	
x, y, z x', y', z'	Cartesian coordinate position	m	
γ	Electrical permittivity	C^2/(N · m^2)	
γ_0	Electrical permittivity of vacuum	C^2/(N · m^2)	$\dfrac{1}{\mu_0 c_0^2} = \dfrac{10^{-9}}{4\pi \times 8.98755}$
μ	Magnetic permeability	N · s^2/C^2	
μ_0	Magnetic permeability of vacuum	N · s^2/C^2	$4\pi \times 10^{-7}$

4-4.1 Propagation in Perfect Dielectric Media

For simplicity we first consider the situation in which the medium is a vacuum or other insulator having an electrical resistivity so large that the term E/r_e can be neglected in Eq. (4-1). Equations (4-1) and (4-2) can then be written in cartesian coordinates to provide two sets of three equations relating the x, y, and z components of the electric and magnetic intensities,

$$\frac{\partial H_z}{\partial y} - \frac{\partial H_y}{\partial z} = \gamma \frac{\partial E_x}{\partial t} \qquad \frac{\partial H_x}{\partial z} - \frac{\partial H_z}{\partial x} = \gamma \frac{\partial E_y}{\partial t} \qquad \text{(4-5a,b)}$$

$$\frac{\partial H_y}{\partial x} - \frac{\partial H_x}{\partial y} = \gamma \frac{\partial E_z}{\partial t} \qquad \text{(4-5c)}$$

$$\frac{\partial E_z}{\partial y} - \frac{\partial E_y}{\partial z} = -\mu \frac{\partial H_x}{\partial t} \qquad \frac{\partial E_x}{\partial z} - \frac{\partial E_z}{\partial x} = -\mu \frac{\partial H_y}{\partial t} \qquad \text{(4-6a,b)}$$

$$\frac{\partial E_y}{\partial x} - \frac{\partial E_x}{\partial y} = -\mu \frac{\partial H_z}{\partial t} \qquad \text{(4-6c)}$$

From Eqs. (4-3) and (4-4), we get

$$\frac{\partial E_x}{\partial x} + \frac{\partial E_y}{\partial y} + \frac{\partial E_z}{\partial z} = 0 \qquad \frac{\partial H_x}{\partial x} + \frac{\partial H_y}{\partial y} + \frac{\partial H_z}{\partial z} = 0 \qquad \text{(4-7,8)}$$

The coordinate system x, y, z is fixed to the path of a wave that is propagating in the x direction (Fig. 4-1). For simplicity, a plane wave is considered; that is, all the quantities concerned with the wave are constant over any yz plane at any time. Hence $\partial/\partial y = \partial/\partial z = 0$. For these conditions, (4-5) to (4-8) reduce to

$$0 = \gamma \frac{\partial E_x}{\partial t} \qquad -\frac{\partial H_z}{\partial x} = \gamma \frac{\partial E_y}{\partial t} \qquad \frac{\partial H_y}{\partial x} = \gamma \frac{\partial E_z}{\partial t} \qquad \text{(4-9a,b,c)}$$

$$0 = -\mu \frac{\partial H_x}{\partial t} \qquad -\frac{\partial E_z}{\partial x} = -\mu \frac{\partial H_y}{\partial t} \qquad \frac{\partial E_y}{\partial x} = -\mu \frac{\partial H_z}{\partial t} \qquad \text{(4-10a,b,c)}$$

$$\frac{\partial E_x}{\partial x} = 0 \qquad \frac{\partial H_x}{\partial x} = 0 \qquad \text{(4-11,12)}$$

The H components are then eliminated. By differentiating Eq. (4-9b) with respect to t and (4-10c) with respect to x, the results may be combined to eliminate H_z. Similarly, by differentiating (4-9c) with respect to t and (4-10b) with respect to x, the results combine to eliminate H_y. This yields

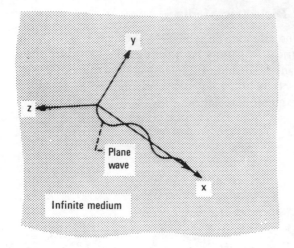

Figure 4-1 Wave propagation in homogeneous, isotropic material.

$$\mu\gamma\,\frac{\partial^2 E_y}{\partial t^2} = \frac{\partial^2 E_y}{\partial x^2} \qquad \mu\gamma\,\frac{\partial^2 E_z}{\partial t^2} = \frac{\partial^2 E_z}{\partial x^2} \tag{4-13a,b}$$

These wave equations govern the propagation of E_y and E_z in the x direction. For simplicity in the remainder of the derivation, it is assumed that the electromagnetic waves are polarized such that the vector **E** is contained only within the xy plane (Fig. 4-2). Then E_z and its derivatives are zero and Eq. (4-13b) need not be considered. The vector **E** will have only x and y components.

With regard to the x components of **E** and **H**, from (4-9a), (4-10a), (4-11), and (4-12), $\partial E_x/\partial t = \partial E_x/\partial x = \partial H_x/\partial t = \partial H_x/\partial x = 0$. Hence, the electric and magnetic intensity components in the direction of propagation are steady and independent of the propagation direction x. Consequently, the only time-varying component of **E** is E_y as governed by (4-13a). Since this component is normal to the x direction of propagation, the wave is a *transverse* wave. Equation (4-13a) is the *wave equation* for propagation of E_y in the x direction. The general solution is

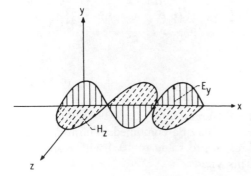

Figure 4-2 Electric field wave polarized in xy plane, traveling in x direction with companion magnetic field wave.

$$E_y = f\left(x - \frac{t}{\sqrt{\mu\gamma}}\right) + g\left(x + \frac{t}{\sqrt{\mu\gamma}}\right) \tag{4-14a}$$

where f and g are *any* differentiable functions. The f provides propagation in the positive x direction, while g accounts for propagation in the negative x direction. Since the present discussion deals with a wave moving in the positive direction, only the f function will be present.

To obtain the wave propagation speed, consider an observer moving with the wave; the observer will always be at a fixed value of E_y. The x location of the observer must then vary with time such that the argument of f, $x - t/\sqrt{\mu\gamma}$, is also fixed. Hence, $dx/dt = 1/\sqrt{\mu\gamma}$. The relation

$$E_y = f\left(x - \frac{t}{\sqrt{\mu\gamma}}\right) \tag{4-14b}$$

thus represents a wave with y component E_y, propagating in the positive x direction with speed $1/\sqrt{\mu\gamma}$. In free space, the propagation speed is c_0, *the speed of electromagnetic radiation in vacuum,* so there is the relation $c_0 = 1/\sqrt{\mu_0\gamma_0}$. Independent measurements of μ_0, γ_0, and c_0 validate this result. The fact that Maxwell's equations predict that all electromagnetic radiation propagates in vacuum with speed c_0 was considered convincing evidence that light is a form of electromagnetic radiation and was one of the early triumphs of the electromagnetic theory.

Accompanying the E_y wave component is a companion wave component of the magnetic field. If (4-9b) is differentiated with respect to x and (4-10c) with respect to t, the results can be combined to yield

$$\mu\gamma \frac{\partial^2 H_z}{\partial t^2} = \frac{\partial^2 H_z}{\partial x^2} \tag{4-15}$$

Equation (4-15) is the same wave equation as (4-13a). Hence, the H_z component of the magnetic field propagates along with E_y as shown in Fig. 4-2.

Any propagating waveform as designated by the f function in Eq. (4-14b) can be represented using Fourier series as a superposition of waves, each wave having a different fixed wavelength. Let us consider only one such spectral wave and note that any waveform could be built up from a number of spectral components. For convenience in later portions of the analysis, the wave component will be expressed using complex algebra. At the origin ($x = 0$) let the waveform variation with time be

$$E_y = E_{yM} \exp(i\omega t) = E_{yM} (\cos \omega t + i \sin \omega t)$$

A position on the wave that leaves the origin at time t_1 arrives at location x after a time interval x/c, where c is the wave speed in the medium. Hence the time of arrival is $t = t_1 + (x/c)$, so that the time of leaving the origin was $t_1 = t - (x/c)$. A wave traveling in the positive x direction is then given by

$$E_y = E_{yM} \exp\left[i\omega\left(t - \frac{x}{c}\right)\right] = E_{yM} \exp[i\omega(t - \sqrt{\mu\gamma}\,x)] \tag{4-16a}$$

This is a solution to the governing wave equation (4-13a), as is shown by comparison with (4-14b). If desired, other forms of the solution can be obtained by using the relations $\omega = 2\pi\nu = 2\pi c/\lambda = 2\pi c_0/\lambda_0$, where λ and λ_0 are the wavelengths in the medium and in vacuum.

The simple refractive index n is defined as the ratio of the wave speed in vacuum c_0 to the speed in the medium $c = 1/\sqrt{\mu\gamma}$. Hence, $n = c_0/c = c_0\sqrt{\mu\gamma} = \sqrt{\mu\gamma/\mu_0\gamma_0}$, and (4-16a) can be written as

$$E_y = E_{yM} \exp\left[i\omega\left(t - \frac{n}{c_0}x\right)\right] \tag{4-16b}$$

As shown by (4-16), the wave propagates with undiminished amplitude in the medium. This is a consequence of the medium being a *perfect dielectric*, that is, having zero electrical conductivity. In many real materials the conductivity is significant and the last term on the right in (4-1) cannot be neglected. This will cause the propagating wave to attenuate.

4-4.2 Propagation in Isotropic Media of Finite Conductivity

This section includes both *imperfect dielectrics* that have low electrical conductivity and *metals*. For simplicity a single plane wave is again considered. If an exponential attenuation with distance is introduced [it will be shown by Eqs. (4-19) to (4-21) that this obeys Maxwell's equations], the wave takes the form

$$E_y = E_{yM} \exp\left[i\omega\left(t - \frac{n}{c_0}x\right)\right] \exp\left(-\frac{\omega}{c_0}\kappa x\right) \tag{4-17a}$$

where κ is the *extinction coefficient* in the medium. The attenuation term indicates an absorption of energy of the wave as it travels through the medium; such a wave is called *evanescent*. The specific form of the attenuation exponent was chosen so the exponential terms would combine into the relation

$$E_y = E_{yM} \exp\left\{i\omega\left[t - (n - i\kappa)\frac{x}{c_0}\right]\right\} \tag{4-17b}$$

A comparison of (4-17b) with (4-16b) shows that the simple refractive index n has been replaced by a complex quantity called the *complex refractive index* $\bar{n}$,

$$\bar{n} = n - i\kappa \tag{4-18}$$

It remains to be shown that (4-17b) is a solution of the governing equations with the last term on the right of (4-1) included. With this term retained, (4-13a) takes the form

$$\mu\gamma \frac{\partial^2 E_y}{\partial t^2} = \frac{\partial^2 E_y}{\partial x^2} - \frac{\mu}{r_e}\frac{\partial E_y}{\partial t} \tag{4-19}$$

The wave given by (4-17b) is substituted into (4-19) and the following equality results:

$$c_0^2 \mu \gamma = (n - i\kappa)^2 + \frac{i\mu\lambda_0 c_0}{2\pi r_e} \qquad (4\text{-}20a)$$

Equation (4-20a) is a relation necessary for the wave to satisfy Maxwell's equations. Equating the real and imaginary parts of (4-20a) yields

$$n^2 - \kappa^2 = \mu\gamma c_0^2 \quad \text{and} \quad n\kappa = \frac{\mu\lambda_0 c_0}{4\pi r_e} \qquad (4\text{-}20b,c)$$

These equations are solved for the components of the complex refractive index,

$$n^2 = \frac{\mu\gamma c_0^2}{2} \left\{ 1 + \left[1 + \left(\frac{\lambda_0}{2\pi c_0 r_e \gamma} \right)^2 \right]^{1/2} \right\} \qquad (4\text{-}21a)$$

$$\kappa^2 = \frac{\mu\gamma c_0^2}{2} \left\{ -1 + \left[1 + \left(\frac{\lambda_0}{2\pi c_0 r_e \gamma} \right)^2 \right]^{1/2} \right\} \qquad (4\text{-}21b)$$

Comparison of (4-16b), the solution for perfect dielectric media, with (4-17b), the solution for conducting media, shows the solutions to be identical with one exception: the simple refractive index n appearing in the perfect dielectric solution is replaced for conductors by the complex refractive index $n - i\kappa$. This important observation means that some of the relations for perfect dielectrics will also hold for conductors, providing the complex index $n - i\kappa$ is substituted for the simple refractive index n. Use will be made of this analogy in succeeding sections, but there are some instances in which the analogy does not apply.

4-4.3 Energy of an Electromagnetic Wave

The instantaneous energy carried per unit time and per unit area by an electromagnetic wave is given by the cross product of the electric and magnetic intensity vectors. This is called the *Poynting* vector, $\mathbf{S} = \mathbf{E} \times \mathbf{H}$, and according to the properties of the cross product, $\mathbf{S}$ is a vector propagating at right angles to the $\mathbf{E}$ and $\mathbf{H}$ vectors in a direction defined by the right-hand rule. For the plane wave under consideration in Fig. 4-2, the propagation is in the positive x direction, and the magnitude of $\mathbf{S}$ is

$$|\mathbf{S}| = E_y H_z \qquad (4\text{-}22)$$

If E_y is given by (4-17b), then (4-10c), which holds for conductors as well as perfect dielectrics, can be used to find H_z as follows:

$$-\mu \frac{\partial H_z}{\partial t} = \frac{\partial E_y}{\partial x} = \frac{-i\omega}{c_0}(n - i\kappa)E_y = -\frac{i\omega\bar{n}}{c_0}E_y$$

Then noting the t dependence of E_y in (4-17b) and integrating yield the following relation between electric and magnetic intensities:

$$H_z = \frac{\bar{n}}{\mu c_0} E_y \tag{4-23}$$

The constant of integration has been taken to be zero. The constant would correspond to the presence of a steady magnetic intensity in addition to that induced by E_y and is zero for the conditions of the present discussion.

When H_z is substituted in (4-22), the magnitude of the Poynting vector becomes

$$|\mathbf{S}| = \frac{\bar{n}}{\mu c_0} E_y{}^2 \tag{4-24a}$$

Thus, the instantaneous energy per unit time and area carried by the wave is proportional to the square of the amplitude of the electric intensity. Because $|\mathbf{S}|$ is a spectral quantity, it is seen by examination of its definition to be proportional to the quantity we have called spectral intensity. For radiation passing through a medium, the exponential decay factor in the spectral intensity must be, by virtue of (4-24a), equal to the square of the decay term in E_y. Thus, from (4-17a), the intensity decay factor is $\exp(-2\omega\kappa x/c_0)$ or $\exp(-4\pi\kappa x/\lambda_0)$. Later in this book, starting with Chap. 12, an absorption coefficient a will be used to describe exponential attenuation caused by absorption in a medium. As a result of absorption the intensity decays as $\exp(-ax)$. Thus there is the *very useful relation between a and κ* given by $a = 4\pi\kappa/\lambda_0$. This provides a means for obtaining the spectral absorption coefficient $a(\lambda_0)$ from optical data for κ as a function of wavelength. The more general vector form of (4-24a) is

$$|\mathbf{S}| = \frac{\bar{n}}{\mu c_0} |\mathbf{E}|^2 \tag{4-24b}$$

4-5 LAWS OF REFLECTION AND REFRACTION

In the previous derivations the wave nature of the propagating radiation was revealed, and the characteristics of movement through *infinite* homogeneous isotropic media were found. Now the interaction of an electromagnetic wave with the interface between two media will be considered. This will provide laws of reflection and refraction in terms of the indices of refraction and extinction coefficients, which are in turn related to the electric and magnetic properties of the two media by means of Eq. (4-21).

4-5.1 Reflection and refraction at the interface between two perfect dielectrics ($\kappa \to 0$)

The interaction at the optically smooth interface between two nonattenuating (*perfect dielectric*) materials will now be considered. For simplicity, a simple cosine wave is utilized as obtained by retaining only the real part of (4-17b). An x', y', z' coordinate system is fixed to the path of the incident wave, and the wave is moving in the x' direction. The wave strikes the interface between two media as shown in Fig. 4-3, where the interface is in the yz plane of the x, y, z coordinate system attached to the media. The plane containing both the normal to the interface and the incident direction x' is defined as the *plane of incidence*. In Fig. 4-3 the coordinate system has been drawn so that the y' direction is in the plane of incidence. The interaction of the wave with the interface depends on the wave orientation relative to the plane of incidence. For example, if the amplitude vector of the incident wave is in the plane of incidence (amplitude vector in y' direction), then the amplitude vector is at an angle to the interface. If the amplitude vector is normal to the plane of incidence (amplitude vector in z' direction), then the incident wave vector is parallel to the interface.

Figure 4-3 shows a plane, transverse wave front propagating in the x' direction. Although the wave will bend as it moves across the interface because of the difference in propagation velocity in the two media, the wave is continuous, so the velocity component tangent to the interface (y component) is the same in both media at the interface. This continuity relation is used in deriving the laws of reflection.

Consider an incident wave $E_{\parallel,i}$ polarized so that it has amplitude only in the $x'y'$ plane (Fig. 4-4) and hence is parallel to the plane of incidence. From Eq.

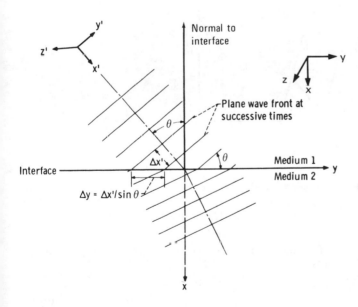

Figure 4-3 Plane wave incident upon interface between two media.

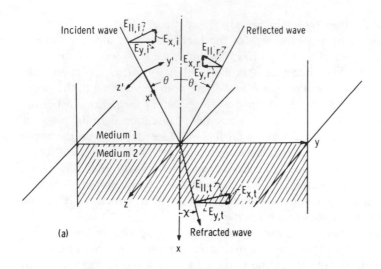

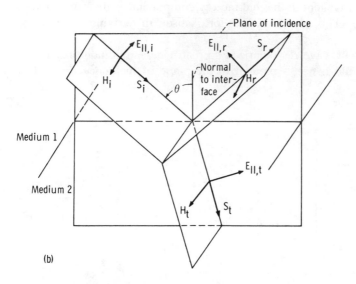

Figure 4-4 Interaction of electromagnetic wave with boundary between two media. (*a*) Plane electric field wave polarized in *xy* plane striking intersection of two media; (*b*) electric intensity, magnetic intensity, and Poynting vectors for incident wave polarized in plane of incidence.

(4-16*b*), retaining only the real part (cosine term), the wave propagating in the x' direction is characterized by

$$E_{\parallel,i} = E_{M\parallel,i} \cos\left[\omega\left(t - \frac{n_1 x'}{c_0}\right)\right]$$

(4-25)

From Fig. 4-4*a*, the components of the incident wave in the *x*, *y*, *z* coordinate system are (components are taken to be positive in the positive coordinate directions)

$$E_{x,i} = -E_{\|,i} \sin \theta \qquad E_{y,i} = E_{\|,i} \cos \theta \qquad E_z = 0 \qquad (4\text{-}26a,b,c)$$

Substituting (4-25) into (4-26) and noting that x', the distance the wave front travels in a given time, is related to the y distance the front travels along the interface by (see Fig. 4-3)

$$x' = y \sin \theta \qquad (4\text{-}27)$$

we obtain for the incident components

$$E_{x,i} = -E_{M\|,i} \sin \theta \cos \left[\omega \left(t - \frac{n_1 y \sin \theta}{c_0} \right) \right] \qquad (4\text{-}28a)$$

$$E_{y,i} = E_{M\|,i} \cos \theta \cos \left[\omega \left(t - \frac{n_1 y \sin \theta}{c_0} \right) \right] \qquad E_{z,i} = 0 \qquad (4\text{-}28b,c)$$

Upon striking the interface between media 1 and 2, the incident wave separates into $E_{\|,r}$ reflected at angle θ_r and $E_{\|,t}$ refracted at angle χ and transmitted into medium 2. From the geometry in Fig. 4-4 the components in the positive coordinate directions of the reflected ray evaluated at the interface are

$$E_{x,r} = -E_{M\|,r} \sin \theta_r \cos \left[\omega \left(t - \frac{n_1 y \sin \theta_r}{c_0} \right) \right] \qquad (4\text{-}29a)$$

$$E_{y,r} = -E_{M\|,r} \cos \theta_r \cos \left[\omega \left(t - \frac{n_1 y \sin \theta_r}{c_0} \right) \right] \qquad E_{z,r} = 0 \qquad (4\text{-}29b,c)$$

The direction of $E_{\|,r}$ was drawn such that $E_{\|,r}$, H_r, and S_r would be consistent with the right-hand rule connecting the Poynting vector with the E and H fields. In a similar fashion, from Fig. 4-4, the components of the refracted portion of the wave are

$$E_{x,t} = -E_{M\|,t} \sin \chi \cos \left[\omega \left(t - \frac{n_2 y \sin \chi}{c_0} \right) \right] \qquad (4\text{-}30a)$$

$$E_{y,t} = E_{M\|,t} \cos \chi \cos \left[\omega \left(t - \frac{n_2 y \sin \chi}{c_0} \right) \right] \qquad E_{z,t} = 0 \qquad (4\text{-}30b,c)$$

Certain boundary conditions must be followed by the waves at the interface of the two media. The sum of the components, parallel to the interface, of the electric intensities of the reflected and incident waves must equal the intensity of the refracted wave in the same plane. This is because the intensity in medium 1 is the superposition of the incident and reflected intensities. For the polarized wave considered here, this condition gives the following equality for the y components (parallel to interface) in the two media:

$$\left\{ E_{M\parallel,i} \cos\theta \cos\left[\omega\left(t - \frac{n_1 y \sin\theta}{c_0}\right)\right] - E_{M\parallel,r} \cos\theta_r \cos\left[\omega\left(t - \frac{n_1 y \sin\theta_r}{c_0}\right)\right] \right.$$

$$\left. = E_{M\parallel,t} \cos\chi \cos\left[\omega\left(t - \frac{n_2 y \sin\chi}{c_0}\right)\right] \right\}_{x=0} \tag{4-31}$$

Since Eq. (4-31) must hold for arbitrary t and y and the angles θ, θ_r, and χ are independent of t and y, the cosine terms involving time must be equal. This can be true only if

$$n_1 \sin\theta = n_1 \sin\theta_r = n_2 \sin\chi \tag{4-32}$$

which provides that

$$\theta = \theta_r \tag{4-33}$$

The angle of reflection of an electromagnetic wave from an ideal interface is thus equal to its angle of incidence rotated about the normal to the interface through a circumferential angle of $\varphi = \pi$. These are the relations that define mirrorlike or specular reflections.

Equation (4-32) yields the following relation between θ and χ:

$$\frac{\sin\chi}{\sin\theta} = \frac{n_1}{n_2} \quad \text{(Snell's law)} \tag{4-34}$$

This relates the angle of refraction to the angle of incidence by means of the ratio of refractive indices and is known as *Snell's law*.

With the cosine terms involving time equal, and with the use of Eq. (4-33), there also follows from (4-31)

$$(E_{M\parallel,i}\cos\theta - E_{M\parallel,r}\cos\theta = E_{M\parallel,t}\cos\chi)_{x=0} \tag{4-35}$$

This can be used to find how the reflected electric intensity is related to the incident value. The refracted component $E_{M\parallel,t}$ must be eliminated; to accomplish this the magnetic intensities must be considered.

The magnetic intensity parallel to the boundary must be continuous at the boundary plane. The magnetic intensity vector is perpendicular to the electric intensity; since the electric intensity being considered is in the plane of incidence, the magnetic intensity will be parallel to the boundary. Continuity at the boundary provides that

$$(H_i + H_r = H_t)_{x=0} \tag{4-36}$$

The relation between electric and magnetic components was shown in (4-23). Although for simplicity this relation was derived for only the specific components H_z and E_y, it is true more generally so that the magnitudes of the **E** and **H** vectors are related by

$$|\mathbf{H}| = \frac{\bar{n}}{\mu c_0} |\mathbf{E}| \tag{4-37}$$

For both dielectrics and metals the magnetic permeability is very close to that of a vacuum so that $\mu \approx \mu_0$. Then (4-36) can be written as

$$(\bar{n}_1 E_{M\parallel,i} + \bar{n}_1 E_{M\parallel,r} = \bar{n}_2 E_{M\parallel,t})_{x=0} \tag{4-38}$$

Equations (4-35) and (4-38) are combined to eliminate $E_{M\parallel,t}$ and give the reflected electric intensity in terms of the incident intensity for nonattenuating materials ($\bar{n} \to n$) as

$$\frac{E_{M\parallel,r}}{E_{M\parallel,i}} = \frac{\cos\theta/\cos\chi - n_1/n_2}{\cos\theta/\cos\chi + n_1/n_2} \tag{4-39}$$

If the preceding derivation is repeated for an incident plane electric wave polarized perpendicular to the incident plane, the relation between reflected and incident components is

$$\frac{E_{M\perp,r}}{E_{M\perp,i}} = -\frac{\cos\chi/\cos\theta - n_1/n_2}{\cos\chi/\cos\theta + n_1/n_2} \tag{4-40}$$

Equation (4-34) can be used in Eqs. (4-39) and (4-40) to eliminate n_1/n_2 in terms of $\sin\chi/\sin\theta$. After some manipulation, the expressions become

$$\frac{E_{M\parallel,r}}{E_{M\parallel,i}} = \frac{\tan(\theta - \chi)}{\tan(\theta + \chi)} \qquad \frac{E_{M\perp,r}}{E_{M\perp,i}} = -\frac{\sin(\theta - \chi)}{\sin(\theta + \chi)} \tag{4-41a,b}$$

The energy carried by a wave is proportional to the square of the amplitude of the wave as shown by (4-24). Squaring the ratio $E_{M,r}/E_{M,i}$ gives the ratio of the energy reflected from a surface to the energy incident upon the surface from a given direction. Because electromagnetic radiation for the ideal conditions examined here was shown by (4-33) to reflect specularly and because the electromagnetic theory relations are for spectral waves, the energy ratio is the directional-hemispherical spectral specular reflectivity (Sec. 3-5.1). The spectral dependence arises from the variation of optical constants with wavelength. The values of $\rho'_{\lambda,s}(\lambda, \theta, \varphi)$ for incident parallel and perpendicular polarized components are then obtained as

$$\rho'_{\lambda\parallel,s}(\lambda, \theta, \varphi) = \left(\frac{E_{M\parallel,r}}{E_{M\parallel,i}}\right)^2 \qquad \rho'_{\lambda\perp,s}(\lambda, \theta, \varphi) = \left(\frac{E_{M\perp,r}}{E_{M\perp,i}}\right)^2 \tag{4-42a,b}$$

The subscript s denotes a specular reflectivity. Because all reflectivities predicted by electromagnetic theory are specular, the subscript s will not be carried from this point on, to simplify an already complicated notation. For the ideal surfaces considered, there is no dependence on the angle φ; hence, notation showing dependence on this variable will not be retained.

For unpolarized incident radiation the electric field has no definite orientation relative to the incident plane and can be resolved into equal parallel and perpendicular components. Then the directional-hemispherical spectral specular reflectivity is the average of $\rho'_{\lambda\|}(\lambda, \theta)$ and $\rho'_{\lambda\perp}(\lambda, \theta)$. Equations (4-41) to (4-42) result in

$$\rho'_\lambda(\lambda,\theta) = \frac{\rho'_{\lambda\|}(\lambda,\theta) + \rho'_{\lambda\perp}(\lambda,\theta)}{2} = \frac{1}{2}\left[\frac{\tan^2(\theta-\chi)}{\tan^2(\theta+\chi)} + \frac{\sin^2(\theta-\chi)}{\sin^2(\theta+\chi)}\right]$$

$$= \frac{1}{2}\frac{\sin^2(\theta-\chi)}{\sin^2(\theta+\chi)}\left[1 + \frac{\cos^2(\theta+\chi)}{\cos^2(\theta-\chi)}\right] \tag{4-43}$$

Equation (4-43) is known as *Fresnel's equation,* and it gives the directional-hemispherical spectral reflectivity for an unpolarized ray incident upon an interface between two perfect (nonattenuating) dielectric media. The relation between χ and θ is given by (4-34).

In the special case when the incident radiation is normal to the interface, $\cos \theta = \cos \chi = 1$, and Eqs. (4-39) and (4-40) yield

$$\frac{E_{M\|,r}}{E_{M\|,i}} = -\frac{E_{M\perp,r}}{E_{M\perp,i}} = \frac{1 - n_1/n_2}{1 + n_1/n_2} = \frac{n_2 - n_1}{n_2 + n_1} \tag{4-44}$$

The normal directional-hemispherical spectral specular reflectivity is then

$$\rho'_{\lambda,n}(\lambda) = \rho'_\lambda(\lambda, \theta = \theta_r = 0) = \left(\frac{n_2 - n_1}{n_2 + n_1}\right)^2 = \left[\frac{(n_2/n_1) - 1}{(n_2/n_1) + 1}\right]^2 \tag{4-45}$$

The foregoing reflectivities are spectral quantities because n_1 and n_2 are functions of λ.

4-5.2 Incidence on an Absorbing Medium ($\kappa \neq 0$)

As was shown by Eq. (4-17b), the propagation of a wave in an infinite attenuating medium is governed by the same relations as in a nonattenuating medium if the index of refraction n for the latter case is replaced by $\bar{n} = n - i\kappa$. When the interaction of a wave with a boundary is considered, the theoretical expressions for reflected wave amplitudes as derived previously for $\kappa = 0$ also apply if $\bar{n}$ is used instead of n, but this leads to some complexities in interpretation. For example, Snell's law becomes

$$\frac{\sin \chi}{\sin \theta} = \frac{\bar{n}_1}{\bar{n}_2} = \frac{n_1 - i\kappa_1}{n_2 - i\kappa_2} \tag{4-46}$$

Because this relation is complex, $\sin \chi$ is complex, and the angle χ can no longer be interpreted physically as a simple angle of refraction for propagation into the material. Except in the special case of normal incidence, n is no longer directly related to the propagation velocity. Some discussion will now be given by considering oblique incidence on an attenuating medium. This discussion is of background interest and is not needed for the application of the derived reflection laws, which will be expressed in terms of n, κ, and the incidence angle.

Figure 4-5 shows a plane wave incident from vacuum on an absorbing material with $\bar{n} = n - i\kappa$. After refraction, the planes of equal phase are still normal to the direction of propagation and these planes move with the *phase velocity,* which will be called c_0/α and is related to n and κ. In the case of a nonattenuating medium the phase velocity is simply c_0/n. The attenuation of the wave must depend on the distance traveled within the medium, and hence the planes of constant amplitude must be parallel to the interface. The wave in the material is called an *inhomogeneous wave* as the planes of constant amplitude and constant phase are not along the same direction. Only for normal incidence are the two sets of planes parallel, so the waves in the medium are homogeneous for normal incidence. The extinction coefficient in the x direction will be called β, and only for normal incidence will $\beta = \kappa$. By analogy with (4-17a) the wave in the medium can be described as

$$E_{y'} = E_{y'M} \exp\left[i\omega\left(t - \frac{\alpha}{c_0} x' \right) \right] \exp\left(-\frac{\omega}{c_0} \beta x \right) \qquad (4\text{-}47)$$

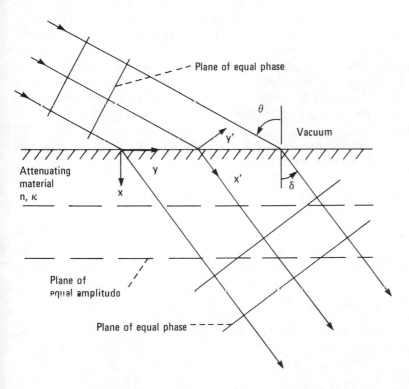

Figure 4-5 Planes of equal phase and amplitude for propagation into an attenuating material.

where propagation is in the x' direction and extinction depends on the x direction. The x coordinate can be written as $x = x' \cos \delta - y' \sin \delta$ where, from the phase velocity, $\delta = \sin^{-1}[(1/\alpha) \sin \theta]$. The propagation angle δ is not equal to the χ in (4-46) as χ is complex in this instance. Then $E_{y'}$ becomes

$$E_{y'} = E_{y'M} \exp(i\omega t) \exp\left[- x' \left(\frac{i\omega\alpha}{c_0} + \frac{\omega}{c_0} \beta \cos \delta\right)\right] \exp\left(y' \frac{\omega}{c_0} \beta \sin \delta\right) \quad (4\text{-}48)$$

Since $E_{y'}$ is a function of t, x', and y', the wave equation (4-19) is written in two space dimensions as

$$\mu\gamma \frac{\partial^2 E_{y'}}{\partial t^2} + \frac{\mu}{r_e} \frac{\partial E_{y'}}{\partial t} = \frac{\partial^2 E_{y'}}{\partial x'^2} + \frac{\partial^2 E_{y'}}{\partial y'^2}$$

Substituting (4-48) yields

$$-\mu\gamma\omega^2 + \frac{\mu}{r_e} i\omega = \left(\frac{i\omega\alpha}{c_0} + \frac{\omega}{c_0} \beta \cos \delta\right)^2 + \left(\frac{\omega}{c_0} \beta \sin \delta\right)^2$$

Equating real and imaginary parts and using Eqs. (4-20b) and (4-20c) yields the relations between α and β and n and κ: $\alpha^2 - \beta^2 = n^2 - \kappa^2$ and $\alpha\beta \cos \delta = n\kappa$, and, as given earlier, $\delta = \sin^{-1}[(1/\alpha) \sin \theta]$. This provides three simultaneous equations from which β, α, and δ can be calculated from θ, n, and κ. This yields the attenuation, propagation velocity (phase velocity), and direction of propagation within the material. It is evident that the propagation velocity c_0/α depends on δ; that is, the velocity depends on direction within the material even though the material is isotropic. In the case of normal incidence, $\delta = 0$, and then $\alpha = n$ and $\beta = \kappa$. Hence, only in this instance are n directly related to the propagation velocity by $c = c_0/n$ and κ a direct measure of the rate of extinction with depth within the material.

Returning now to the reflection laws, we first consider the case of normal incidence. From (4-44) with $\bar{n}$ replacing n,

$$\frac{E_{M\|,r}}{E_{M\|,i}} = -\frac{E_{m\perp,r}}{E_{M\perp,i}} = \frac{\bar{n}_2 - \bar{n}_1}{\bar{n}_2 + \bar{n}_1} = \frac{n_2 - i\kappa_2 - (n_1 - i\kappa_1)}{n_2 - i\kappa_2 + n_1 - i\kappa_1}$$

It was noted in (4-24) that the energy in the wave depends on $|\mathbf{E}|^2$. For a complex quantity z, $|z|^2 = zz^*$, where z^* is the complex conjugate. Then because the relations for $\|$ and $\perp$ polarization are the same for normal incidence, the reflectivity is

$$\rho'_{\lambda,n}(\lambda) = \left[\frac{n_2 - i\kappa_2 - (n_1 - i\kappa_1)}{n_2 - i\kappa_2 + n_1 - i\kappa_1}\right]\left[\frac{n_2 + i\kappa_2 - (n_1 + i\kappa_1)}{n_2 + i\kappa_2 + n_1 + i\kappa_1}\right]$$

$$= \frac{(n_2 - n_1)^2 + (\kappa_2 - \kappa_1)^2}{(n_2 + n_1)^2 + (\kappa_2 + \kappa_1)^2} \quad (4\text{-}49)$$

For an incident ray in air, $n_1 \approx 1$ and $\kappa_1 \approx 0$. When the material is also nonattenuating (completely transparent), $\kappa_2 \to 0$.

For oblique incidence the directional-hemispherical reflectivity can be derived by starting from (4-39) and (4-40) and using the complex index of refraction. For incident rays polarized parallel or perpendicular to the plane of incidence, (4-39) and (4-40) give the complex ratios

$$\frac{E_{M\|,r}}{E_{M\|,i}} = \frac{\cos\theta/\cos\chi - (n_1 - i\kappa_1)/(n_2 - i\kappa_2)}{\cos\theta/\cos\chi + (n_1 - i\kappa_1)/(n_2 - i\kappa_2)} \tag{4-50a}$$

$$\frac{E_{M\perp,r}}{E_{M\perp,i}} = -\frac{\cos\chi/\cos\theta - (n_1 - i\kappa_1)/(n_2 - i\kappa_2)}{\cos\chi/\cos\theta + (n_1 - i\kappa_1)/(n_2 - i\kappa_2)} \tag{4-50b}$$

The real and imaginary parts of (4-50a,b) correspond respectively to the changes in amplitude and phase. The reflectivity for the parallel or perpendicular component is found by multiplying by its complex conjugate. This requires considerable manipulation, as $\cos\chi$ is complex as shown by (4-46).

To provide some specific results, the important case will be considered of radiation incident in air or vacuum on a material with properties n and κ. Then from (4-50) and (4-46) with $\bar{n} = n - i\kappa$,

$$\frac{E_{M\|,r}}{E_{M\|,i}} = \frac{\bar{n}\cos\theta - \cos\chi}{\bar{n}\cos\theta + \cos\chi} \qquad \frac{E_{M\perp,r}}{E_{M\perp,i}} = -\frac{\bar{n}\cos\chi - \cos\theta}{\bar{n}\cos\chi + \cos\theta}$$

$$\frac{\sin\chi}{\sin\theta} = \frac{1}{n - i\kappa} = \frac{1}{\bar{n}} \tag{4-51a,b}$$

The $\bar{n}\cos\chi$ in (4-51) can be found as

$$\bar{n}\cos\chi = \bar{n}(1 - \sin^2\chi)^{1/2} = (\bar{n}^2 - \sin^2\theta)^{1/2} \tag{4-52}$$

The results can be presented more conveniently by letting $a - ib = (\bar{n}^2 - \sin^2\theta)^{1/2}$. By squaring and equating real and imaginary parts, the resulting simultaneous equations can be solved for a and b to obtain

$$2a^2 = [(n^2 - \kappa^2 - \sin^2\theta)^2 + 4n^2\kappa^2]^{1/2} + n^2 - \kappa^2 - \sin^2\theta \tag{4-53a}$$

$$2b^2 = [(n^2 - \kappa^2 - \sin^2\theta)^2 + 4n^2\kappa^2]^{1/2} - (n^2 - \kappa^2 - \sin^2\theta) \tag{4-53b}$$

The quantity $a - ib$ is substituted for $\bar{n}\cos\chi$ in (4-51), and the resulting equations are multiplied through by their complex conjugates to yield the reflectivities [note that $\bar{n}^2 = (a - ib)^2 + \sin^2\theta$]:

$$\rho'_{\lambda\|}(\lambda,\theta) = \frac{a^2 + b^2 - 2a\sin\theta\tan\theta + \sin^2\theta\tan^2\theta}{a^2 + b^2 + 2a\sin\theta\tan\theta + \sin^2\theta\tan^2\theta}\rho'_{\lambda\perp}(\lambda,\theta) \tag{4-54a}$$

$$\rho'_{\lambda\perp}(\lambda,\theta) = \frac{a^2 + b^2 - 2a\cos\theta + \cos^2\theta}{a^2 + b^2 + 2a\cos\theta + \cos^2\theta} \tag{4-54b}$$

If the incident beam has no specific polarization the reflectivity is an average of the parallel and perpendicular components, as in Eq. (4-43). Another form for these reflectivity relations is given by Isard [2]. Some results are given in [3] for the more complex situation of oblique incidence from an absorbing material rather than from vacuum or air as given here.

To this point in this chapter, the wave nature of radiation has been shown from a consideration of Maxwell's equations. Then the interactions of waves with nonabsorbing and absorbing media have been discussed in terms of the refractive index n and the complex refractive index $\bar{n}$. Now the results will be applied more specifically for radiative properties.

4-6 APPLICATION OF ELECTROMAGNETIC-THEORY RELATIONS TO RADIATIVE-PROPERTY PREDICTIONS

The electromagnetic theory applied here to radiative-property prediction has a number of limitations for practical calculations. Aside from the assumptions used in the derivations, the theory becomes invalid when the frequencies being considered become of the order of molecular vibrational frequencies. This restricts the equations used here to wavelengths longer than the visible spectrum. The theory neglects the effects of surface conditions on radiative properties. This is its most serious limitation, since perfectly clean, optically smooth interfaces are not often encountered in practice. The greatest usefulness of the theory is probably in providing help in intelligent extrapolation when only limited experimental data are available. In the following sections, electromagnetic theory results that are useful for the prediction of properties will be examined.

4-6.1 Radiative Properties of Ideal Dielectrics ($\kappa \to 0$)

The measured index of refraction of the medium is in general a function of wavelength, and thus the calculated radiative properties will be wavelength dependent. If, however, the refractive index is calculated from the permittivity γ or the *dielectric constant K* (where $K = \gamma/\gamma_0$), which are not generally given as functions of wavelength, the spectral dependence is lost. Because of these considerations, no notation is used in the following equations to signify spectral dependence, but such dependence can be included if the optical or electromagnetic properties are known as wavelength functions.

The surfaces are assumed to be "optically smooth," that is, smooth in comparison with the wavelength of the incident radiation so that specular reflections result.

Reflectivity The directional-hemispherical specular reflectivity of a wave incident on a surface at angle θ and polarized parallel or perpendicular to the plane of incidence may be obtained from Eqs. (4-42) and (4-41) as

$$\rho'_\parallel(\theta) = \left[\frac{\tan(\theta - \chi)}{\tan(\theta + \chi)}\right]^2 \qquad \rho'_\perp(\theta) = \left[\frac{\sin(\theta - \chi)}{\sin(\theta + \chi)}\right]^2 \qquad (4\text{-}55a,b)$$

where χ is the angle of refraction in the medium on which the ray impinges. For a given incident angle θ, the angle χ can be determined from (4-34) as

$$\frac{\sin \chi}{\sin \theta} = \frac{n_1}{n_2} = \frac{\sqrt{\gamma_1}}{\sqrt{\gamma_2}} = \frac{\sqrt{K_1}}{\sqrt{K_2}} \tag{4-56}$$

The n, γ, and K are assumed not to have any angular dependence.

Alternative forms containing only θ are obtained by eliminating χ in (4-55a) and (4-55b) by using (4-56):

$$\rho'_{\parallel}(\theta) = \left\{ \frac{(n_2/n_1)^2 \cos \theta - [(n_2/n_1)^2 - \sin^2 \theta]^{1/2}}{(n_2/n_1)^2 \cos \theta + [(n_2/n_1)^2 - \sin^2 \theta]^{1/2}} \right\}^2 \tag{4-57a}$$

$$\rho'_{\perp}(\theta) = \left\{ \frac{[(n_2/n_1)^2 - \sin^2 \theta]^{1/2} - \cos \theta}{[(n_2/n_1)^2 - \sin^2 \theta]^{1/2} + \cos \theta} \right\}^2 \tag{4-57b}$$

The $\rho'_{\parallel}(\theta) = 0$ when $\theta = \tan^{-1}(n_2/n_1)$; this θ is called *Brewster's angle*. Radiation reflected from this incidence angle will all be perpendicularly polarized.

The reflectivity for unpolarized incident radiation was shown in Fresnel's equation (4-45) to be given by

$$\rho'(\theta) = \frac{1}{2} \frac{\sin^2(\theta - \chi)}{\sin^2(\theta + \chi)} \left[1 + \frac{\cos^2(\theta + \chi)}{\cos^2(\theta - \chi)} \right] \tag{4-58a}$$

or the average of (4-57a) and (4-57b) can be conveniently used to be in terms of θ. For normal incidence ($\theta = 0$)

$$\rho'_n = \left(\frac{n_2 - n_1}{n_2 + n_1} \right)^2 = \left(\frac{(n_2/n_1) - 1}{(n_2/n_1) + 1} \right)^2 \tag{4-58b}$$

EXAMPLE 4-1 An unpolarized beam of radiation is incident at angle $\theta = 30°$ from the normal on a dielectric surface (medium 2) in air (medium 1). The dielectric has $\kappa_2 \approx 0$ and index of refraction $n_2 = 3.0$. Find the directional-hemispherical reflectivity of each polarized component and of the unpolarized beam.

Because the incident beam is in air, $n_1 - i\kappa_1 \approx 1$, and from Eq. (4-56), $n_1/n_2 = 1/3.0 = \sin \chi/\sin 30°$; therefore, $\chi = 9.6°$. The reflectivity of the parallel component is, from (4-55a), $\rho'_{\parallel}(\theta = 30°) = (\tan 20.4°/\tan 39.6°)^2 = 0.202$, and that of the perpendicular component is, from (4-55b), $\rho'_{\perp}(\theta = 30°) = (\sin 20.4°/\sin 39.6°)^2 = 0.301$. The reflectivity for the unpolarized beam obtained from (4-55a) or, more simply here, from the average of the components, is $\rho'(\theta = 30°) = (0.202 + 0.301)/2 = 0.252$.

EXAMPLE 4-2 What fraction of light is reflected given normal incidence from air on a glass surface or the surface of water?

For glass $n \approx 1.55$, and for water $n \approx 1.33$. Then from (4-58b)

$$\rho'_n(\text{glass}) = \left(\frac{n-1}{n+1}\right)^2 = \left(\frac{0.55}{2.55}\right)^2 = 0.047$$

$$\rho'_n(\text{water}) = \left(\frac{0.33}{2.33}\right)^2 = 0.020$$

The reflectivity, as in Example 4-1, is a directional-hemispherical reflectivity in that it provides all the reflected energy resulting from an incident beam from one direction. It is a spectral quantity since the indices of refraction can correspond to a particular wavelength if the details of the wavelength dependence are available. Finally, it is a specular quantity that obeys the constraints of (4-33).

Emissivity After the reflectivity has been evaluated, the directional spectral emissivity can be found from (3-46) as $\epsilon'(\theta) = 1 - \rho'(\theta)$ when the body is *opaque* since this relation is valid when the transmitted radiation through the body is zero (if κ is very small, the body must be thick).

A graph of the directional emissivity is shown in Fig. 4-6 for various ratios n_2/n_1 where $n_2 \geq n_1$ for the discussion here. When $n_2 < n_1$ there is a limiting angle such that radiation at larger incidence angles is totally reflected, giving $\epsilon'(\theta) = 0$ in that region. This special situation is discussed in Sec. 18-6.2. When $\rho'(\theta)$ is computed for an incident beam in air ($n_1 \approx 1$), the ratio n_2/n_1 reduces to the refractive index for the material on which the beam is incident. Figure 4-6 can thus be regarded as giving the emissivity of a dielectric into air when the value of n_2/n_1 is set equal to the simple refractive index n of the emitting dielectric.

For $n_2/n_1 = 1$, the emissivity is unity (blackbody case), and the curve in Fig. 4-6 is circular with a radius of unity. As n_2/n_1 increases, the curves remain circular up to about $\theta = 70°$ and then decrease rapidly to zero at $\theta = 90°$. Thus, dielectric materials emit poorly at large angles away from the normal direction. For angles less than 70°, the emissivities are quite high, so that, in a hemispherical sense, dielectrics are good emitters. The assumptions used for the present interpretation of Maxwell's equations restrict these findings to wavelengths longer than the visible spectrum, as borne out by comparisons with experimental measurements.

From the directional spectral emissivity, the hemispherical spectral emissivity can be computed from Eq. (3-5) as $\epsilon_\lambda(\lambda, T_A) = (1/\pi) \int_\cap \epsilon'_\lambda(\lambda, \theta, \varphi, T_A) \cos \theta \, d\omega$. Then an integration can be performed over all wavelengths to obtain the hemispherical total emissivity as given by (3-6a). Since the optical properties are generally not known in sufficient detail that a wavelength integration of ϵ_λ can be made, in the theory spectal ϵ_λ values are used for total ϵ values for lack of anything better.

The integration of $\epsilon'(\theta)$ to evaluate ϵ is complicated, as indicated by the forms of (4-57a) and (4-57b), but the integration has been carried out to yield

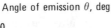

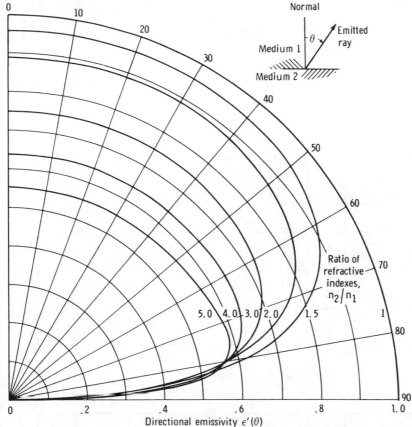

Figure 4-6 Directional emissivity predicted from electromagnetic theory.

$$\epsilon = \frac{1}{2} - \frac{(3n + 1)(n - 1)}{6(n + 1)^2} - \frac{n^2(n^2 - 1)^2}{(n^2 + 1)^3} \ln\left(\frac{n - 1}{n + 1}\right) + \frac{2n^3(n^2 + 2n - 1)}{(n^2 + 1)(n^4 - 1)}$$

$$- \frac{8n^4(n^4 + 1)}{(n^2 + 1)(n^4 - 1)^2} \ln n \qquad (n = n_2/n_1) \tag{4-59}$$

The normal emissivity provides a convenient value to which the hemispherical value may be referenced. The normal emissivity is computed from Eq. (4-45) as

$$\epsilon_n' = 1 - \left(\frac{n - 1}{n + 1}\right)^2 = \frac{4n}{(n + 1)^2} \qquad (n = n_2/n_1) \tag{4-60}$$

The ϵ_n' is shown as a function of n_2/n_1 in Fig. 4-7a. Note that normal emissivities less than about 0.50 correspond to $n_2/n_1 > 6$. Such large n_2/n_1 values are not common for dielectrics, so that the curve is not extended to smaller ϵ_n'. The ratio of hemispherical to normal emissivity for dielectrics is provided as a function of normal emissivity in Fig. 4-7b.

EXAMPLE 4-3 A dielectric has a refractive index of 1.41. What is its hemispherical emissivity into air at the wavelength at which the refractive index was measured?

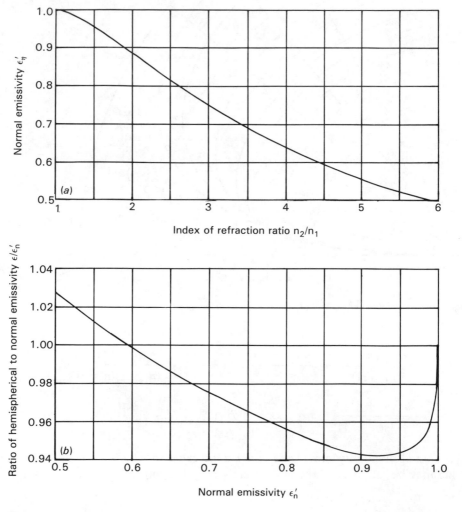

Figure 4-7 Predicted emissivities of dielectric materials for emission into medium with refractive index n_1 from medium with refractive index n_2 where $n_2 \geqslant n_1$. (a) Normal emissivity as function of refractive index ratio; (b) relation between hemispherical and normal emissivity.

From Eq. (4-60) the normal emissivity is $\epsilon'_n = 1 - (0.41/2.41)^2 = 0.97$. From Eq. (4-59), $\epsilon = 0.92$, or from Fig. 4-7b, $\epsilon/\epsilon'_n = 0.95$ and the hemispherical emissivity is $\epsilon = 0.97 \times 0.95 = 0.92$.

For large n_2/n_1 the ϵ'_n values are relatively low, and with increasing n_2/n_1 the curves in Fig. 4-6 depart more from the circular form corresponding to $n_2/n_1 = 1$. Figure 4-7b reveals that the flattening of the curves of Fig. 4-6 in the region near the normal causes the hemispherical emissivity to exceed the normal value at large n_2/n_1. For n_2/n_1 near unity (ϵ'_n near 1), the hemispherical value is lower than the normal value because of the poor emission at large θ, as shown in Fig. 4-6.

4-6.2 Absorbing Materials ($\kappa \neq 0$)

The previous section dealt with perfect dielectric materials ($\kappa = 0$). In this instance there is no attenuation of radiation as it travels within the material (the material is perfectly transparent except for interface reflections). For real dielectrics there is attenuation so κ is finite. Glass, for example, has small attenuation in the visible region but has high attenuation in the infrared. Metals are usually highly absorbing and the κ must be included in the theoretical relations. With κ included, Eq. (4-54) can be used with the a and b from Eq. (4-53). In [4] these results have been given in the following alternative form (these relations are for incidence from air or vacuum),

$$\rho'_\perp(\theta) = \frac{(n\beta - \cos\theta)^2 + (n^2 + \kappa^2)\alpha - n^2\beta^2}{(n\beta + \cos\theta)^2 + (n^2 + \kappa^2)\alpha - n^2\beta^2} \tag{4-61a}$$

$$\rho'_\parallel(\theta) = \frac{(n\gamma - \alpha/\cos\theta)^2 + (n^2 + \kappa^2)\alpha - n^2\gamma^2}{(n\gamma + \alpha/\cos\theta)^2 + (n^2 + \kappa^2)\alpha - n^2\gamma^2} \tag{4-61b}$$

where

$$\alpha^2 = \left(1 + \frac{\sin^2\theta}{n^2 + \kappa^2}\right)^2 - \frac{4n^2}{n^2 + \kappa^2}\left(\frac{\sin^2\theta}{n^2 + \kappa^2}\right)$$

$$\beta^2 = \frac{n^2 + \kappa^2}{2n^2}\left(\frac{n^2 - \kappa^2}{n^2 + \kappa^2} - \frac{\sin^2\theta}{n^2 + \kappa^2} + \alpha\right)$$

$$\gamma = \frac{n^2 - \kappa^2}{n^2 + \kappa^2}\beta + \frac{2n\kappa}{n^2 + \kappa^2}\left(\frac{n^2 + \kappa^2}{n^2}\alpha - \beta^2\right)^{1/2}$$

For nonpolarized incident radiation, $\epsilon'(\theta) = 1 - [\rho'_\perp(\theta) + \rho'_\parallel(\theta)]/2$ was evaluated and then integrated over the hemisphere to obtain the hemispherical emissivity, $\epsilon = \int_0^1 \epsilon'(\theta)\, d(\sin^2\theta)$, as a function of n and κ. If $\kappa = 0$, Eq. (4-61) reduces to (4-57). Equation (4-61) was evaluated and presented graphically in detail in [4]. One set of results, for $n \geq 1$, is in Fig. 4-8 as a function of n and κ/n. This shows how the hemispherical emittance decreases as both n and κ increase.

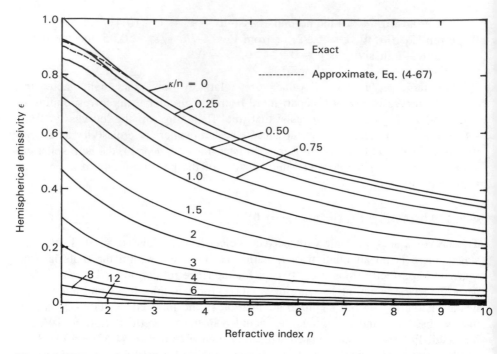

Figure 4-8 Exact and approximate hemispherical emissivity results ($n \geqslant 1.0$), from [4].

4-6.3 Radiative Properties of Metals (Large κ)

Reflectivity and emissivity relations using optical constants Since metals are usually highly absorbing, the large extinction coefficient κ provides simplifying assumptions that lead to more convenient equations than the preceding general results. The main difficulty in applying the theoretical results is that the optical properties for use in these equations are difficult to obtain; when measured values are available, they can be inaccurate because of experimental measurement problems. For large κ the $\sin^2 \theta$ terms in (4-61) can be neglected relative to $n^2 + \kappa^2$. Then $\alpha = \beta = \gamma = 1$, and Eq. (4-61) yields for incidence from air or vacuum to a medium with $\bar{n} = n - i\kappa$

$$\rho'_\parallel(\theta) = \frac{(n \cos \theta - 1)^2 + (\kappa \cos \theta)^2}{(n \cos \theta + 1)^2 + (\kappa \cos \theta)^2} \qquad \rho'_\perp(\theta) = \frac{(n - \cos \theta)^2 + \kappa^2}{(n + \cos \theta)^2 + \kappa^2} \qquad \text{(4-62a,b)}$$

For an unpolarized beam,

$$\rho'(\theta) = \frac{\rho'_\parallel(\theta) + \rho'_\perp(\theta)}{2} \qquad \text{(4-63)}$$

For the normal direction ($\theta = 0$), $\rho'_n = [(n - 1)^2 + \kappa^2]/[(n + 1)^2 + \kappa^2]$, which is the same as the exact relation (4-49).

The corresponding emissivity values are found from $\epsilon'(\theta) = 1 - \rho'(\theta)$, and these simplify to

$$\epsilon'_\parallel(\theta) = \frac{4n \cos\theta}{(n^2 + \kappa^2)\cos^2\theta + 2n \cos\theta + 1} \tag{4-64a}$$

$$\epsilon'_\perp(\theta) = \frac{4n \cos\theta}{\cos^2\theta + 2n \cos\theta + n^2 + \kappa^2} \tag{4-64b}$$

For emission, which is unpolarized,

$$\epsilon'(\theta) = \frac{\epsilon'_\parallel(\theta) + \epsilon'_\perp(\theta)}{2} \tag{4-65}$$

In the normal direction ($\theta = 0$) this becomes

$$\epsilon'_n = \frac{4n}{(n+1)^2 + \kappa^2} \tag{4-66}$$

The use of these emissivity relations is demonstrated in Fig. 4-9 for a pure smooth platinum surface at a wavelength of 2 μm, and it is evident by comparison with the experimental data that, although the general shape of the curve predicted by (4-65) is correct, the magnitude is in error, probably because of uncertainty in the optical constants. (The data for n and κ for platinum, taken from [5], are $n = 5.7$ and $\kappa = 9.7$.*) For metals, as illustrated by Fig. 4-9, the emissivity is

*The complex refractive index can be defined in other ways than $\bar{n} = n - i\kappa$ as used here. It is also commonly given as $\bar{n} = n - in\kappa$, and occasionally with a positive sign in front of the extinction factor. When consulting data references, care should be taken in determining what definition is used so that conversion to the system used in this text can be carried out if necessary.

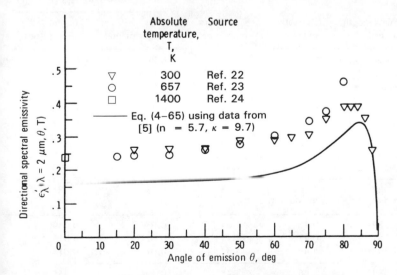

Figure 4-9 Directional spectral emissivity of platinum at wavelength $\lambda = 2$ μm.

essentially constant for about 50° away from the normal, and then increases to a maximum within a few degrees of the tangent to the surface. This angular dependence for emission by metals is in contrast to the behavior for dielectrics, for which emission decreases substantially as the angle from the normal becomes larger than about 70°.

In Table 4-2 the prediction of normal spectral emissivity by using Eq. (4-66) is compared with measured values. A wavelength of $\lambda = 0.589$ μm is used for some of the comparisons because of the wealth of data available. This is because of the ease with which a sodium-vapor lamp, which emits at this wavelength, has been employed as an intense spectral energy source in the laboratory. Since this wavelength is in the visible range, it is in the borderline short-wavelength region where the electromagnetic theory becomes inaccurate.

Comparison of the values in Table 4-2 shows the agreement between predicted and measured $\epsilon'_{\lambda,n}$ to be good, for example, for nickel and tungsten, but a factor of 4 in error for magnesium. For the cases of poor agreement, it is difficult to ascribe the error specifically to the optical constants, to the measured emissivity, or to the theory itself. Most probably the optical constants are somewhat in error, and the experimental samples do not meet the standards of perfection in surface preparation demanded by the theory.

A good recent source of property values for n and κ is [6]. This contains extensive information for metals such as copper, gold, silver, nickel, aluminum,

Table 4-2 Comparison of spectral normal emissivity predictions from electromagnetic theory with experiment (Data from [5])

Metal	Wavelength λ, μm	Refractive index n	Extinction coefficient κ	Spectral normal emissivity $\epsilon'_{\lambda,n}(\lambda)$ Experimental	Calculated from (4-66)
Copper	0.650	0.44	3.26	0.20	0.140
	2.25	1.03	11.7	0.041	0.029
	4.00	1.87	21.3	0.027	0.014
Gold	0.589	0.47	2.83	0.176	0.184
	2.00	0.47	12.5	0.032	0.012
Iron	0.589	1.51	1.63	0.43	0.674
Magnesium	0.589	0.37	4.42	0.27	0.070
Nickel	0.589	1.79	3.33	0.355	0.381
	2.25	3.95	9.20	0.152	0.145
Silver	0.589	0.18	3.64	0.074	0.049
	2.25	0.77	15.4	0.021	0.013
	4.50	4.49	33.3	0.015	0.014
Tungsten	0.589	3.46	3.25	0.49	0.455

and tungsten. Semiconductors such as germanium, indium arsenide, silicon, and lead telluride are included, as well as some insulating materials such as lithium fluoride, glass, and salt (sodium chloride). The information on κ is also useful for obtaining the spectral absorption coefficient a used later in this book to describe exponential attenuation by absorption in a medium according to the relation $\exp(-ax)$. The relation between a and κ is $a = 4\pi\kappa/\lambda_0$, where both a and κ are functions of wavelength.

Within the approximation of neglecting $\sin^2 \theta$ relative to $n^2 + \kappa^2$, the hemispherical emissivity for a metal in air or vacuum is found by substituting (4-65) into (3-5). After carrying out the integration, this yields

$$\epsilon = 4n - 4n^2 \ln \frac{1 + 2n + n^2 + \kappa^2}{n^2 + \kappa^2} + \frac{4n(n^2 - \kappa^2)}{\kappa} \tan^{-1} \frac{\kappa}{n + n^2 + \kappa^2}$$

$$+ \frac{4n}{n^2 + \kappa^2} - \frac{4n^2}{(n^2 + \kappa^2)^2} \ln(1 + 2n + n^2 + \kappa^2) - \frac{4n(\kappa^2 - n^2)}{\kappa(n^2 + \kappa^2)^2} \tan^{-1} \frac{\kappa}{1 + n} \quad (4\text{-}67)$$

Evaluation of (4-67) may involve small differences of large numbers, and sufficient significant figures should be carried in the calculations.

Without neglecting $\sin^2 \theta$, the hemispherical ϵ was calculated by numerical integration by Hering and Smith [4] as shown in Fig. 4-8, and the results were compared with those from (4-67) for various n and κ values. They found that for an accuracy of (4-67) within 1, 2, 5, or 10%, the value of $n^2 + \kappa^2$ should be larger than 40, 3.25, 1.75, and 1.25, respectively. For most metals in view of the optical constants as indicated by Table 4-2, Eq. (4-67) should usually be accurate within a few percent.

From (4-66) the normal emissivity from a metal into air can be computed as a function of n and κ, and this is shown in Fig. 4-10a; more complete results are in [4]. It is of interest to compare the hemispherical emissivity with the normal value. The practical use for this is that the normal emissivity is often measured experimentally because of the relative simplicity of placing a radiation detector in this orientation. With regard to the total amount of heat dissipation, however, it is the hemispherical emissivity that is desired. Figure 4-10b shows the ratio of hemispherical to normal emissivity as a function of the normal value. Equation (4-67) divided by ϵ_n' has been plotted for the case where $\kappa = n$. This is valid at large wavelengths for many metals [see Eq. (4-68)]. The curve is seen to be close to that presented by Jakob [6] for metals as derived from approximate equations and to lie somewhat below the curve for insulators (as taken from Fig. 4-7b) at high normal emissivities.

For polished metals when ϵ_n' is less than about 0.5, the hemispherical emissivity is larger than the normal value because of the increase in emissivity in the direction near tangency to the surface, as was pointed out in Fig. 4-9. Hence, in a table listing emissivity values for polished metals, if ϵ_n' is given, it should be multiplied by a factor larger than unity such as obtained from Fig. 4-10b to estimate the hemispherical value. Real surfaces that have roughness or may be slightly

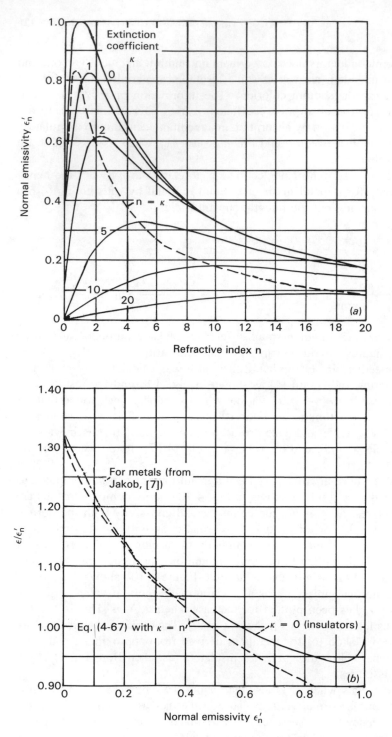

Figure 4-10 Emissivity of metals as computed from electromagnetic theory. (*a*) Normal emissivity of attenuating media emitting into air; (*b*) ratio of hemispherical to normal emissivity for emission into air.

oxidized often tend to have a directional emissivity that is more diffuse than for polished specimens. For a practical case, therefore, the emissivity ratio may be closer to unity than indicated by Fig. 4-10b.

Relation between emissive and electrical properties The wave solutions to Maxwell's equations provide a means for determining n and κ from the electric and magnetic properties of a material. The relations for n and κ are given by (4-21). For metals where r_e is small, and for relatively long wavelengths, say $\lambda_0 > \sim 5$ μm, the term $\lambda_0/2\pi c_0 r_e \gamma$ becomes dominating, and Eqs. (4-21a,b) then reduce to (the magnetic permeability is taken equal to μ_0)

$$n = \kappa = \sqrt{\frac{\lambda_0 \mu_0 c_0}{4\pi r_e}} = \sqrt{\frac{0.003\lambda_0}{r_e}} \qquad (\lambda_0 \text{ in } \mu\text{m, } r_e \text{ in ohm-cm}) \qquad (4\text{-}68)$$

This is the *Hagen-Rubens equation* [8]. Predictions of n and κ from this equation can be greatly in error, as shown in Table 4-3. Nevertheless, some useful results are eventually obtained.

With the simplification that $n = \kappa$, Eq. (4-66) reduces to the following expression for a material with refractive index n radiating in the normal direction into air or vacuum:

$$\epsilon'_{\lambda,n}(\lambda) = 1 - \rho'_{\lambda,n}(\lambda) = \frac{4n}{2n^2 + 2n + 1} \qquad (4\text{-}69a)$$

where n can be inserted from (4-68). Although there is no difficulty in evaluating (4-69a), a simplification is often made by expanding in a series to obtain

$$\epsilon'_{\lambda,n}(\lambda) = \frac{2}{n} - \frac{2}{n^2} + \frac{1}{n^3} - \frac{1}{2n^5} + \frac{1}{2n^6} - \cdots \qquad (4\text{-}69b)$$

Because the index of refraction of metals as predicted from (4-68) is generally large at the long wavelengths being considered here, $\lambda_0 > \sim 5$ μm (see Table 4-3, column 6), only one or two terms of the series are often retained, and the normal spectral emissivity is than obtained by substituting (4-68) to obtain the *Hagen-Rubens emissivity relation* (λ_0 in μm, r_e in ohm-cm),

$$\epsilon'_{\lambda,n}(\lambda) = \frac{2}{\sqrt{0.003}} \left(\frac{r_e}{\lambda_0}\right)^{1/2} - \frac{2}{0.003}\frac{r_e}{\lambda_0} + \cdots \qquad (4\text{-}70a)$$

A two-term approximation is made by adjusting the coefficient of the second term to account for the remaining terms in the series; this yields

$$\epsilon'_{\lambda,n}(\lambda) = 36.5 \left(\frac{r_e}{\lambda_0}\right)^{1/2} - 464\frac{r_e}{\lambda_0} \qquad (4\text{-}70b)$$

Data for polished nickel are shown in Fig. 4-11, and the extrapolation to long wavelengths by Eq. (4-70b) appears reasonable. The predictions of normal spec-

Table 4-3 Comparison of measured optical constants with electromagnetic theory predictions

Metal	Wavelength λ_0, μm	Measured values Electrical resistivity (at 293 K) r_e, Ω-cm [a]	Refractive index n	Extinction coefficient κ	Calculated from (4-68) $n = \kappa$	Spectral normal emissivity $\epsilon_{\lambda,n}(\lambda)$ Measured	Calculated from (4-70b)
Aluminum	12	2.82×10^{-6} [b]	33.6 [b]	76.4 [b]	113	0.02 [a]	0.018
	4.20	1.72×10^{-6} [b]	1.92 [b]	22.8 [b]	86	0.027 [a,c]	0.023
	4.20	1.72×10^{-6} [b]	1.92 [b]	22.8 [b]	86	0.015 [d]	0.023
Copper	5.50	1.72×10^{-6} [b]	3.16 [a]	28.4 [a]	98	0.012 [d]	0.020
Gold	5.00	2.44×10^{-6} [b]	1.81 [a]	32.8 [a]	78	0.031 [a,c]	0.025
Platinum	5.00	10×10^{-6} [b]	11.5 [a]	15.7 [a]	39	0.050 [d]	0.051
	4.50	1.63×10^{-6} [b]	4.49 [a]	33.3 [a]	91	0.015 [a,c]	0.022
Silver	4.37	1.63×10^{-6} [b]	4.34 [a]	32.6 [b]	90	0.015 [a,c]	0.022

[a]Data from [5]. [b]Data from [20]. [c]Measured at 4 μm. [d]Data from [21].

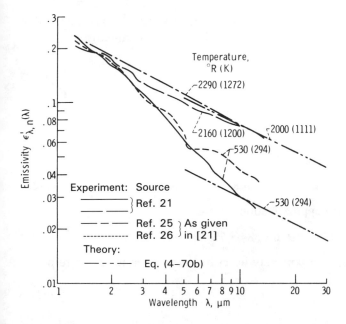

Figure 4-11 Comparison of measured values with theoretical predictions for normal spectral emissivity of polished nickel.

tral emissivity at long wavelengths as presented in Table 4-3 are much better than the prediction of the optical constants.

The angular behavior of the spectral emissivity can be obtained by substituting (4-68) into (4-64) to yield for the two components of polarization

$$\epsilon'_{\lambda\parallel}(\lambda,\ \theta) = \frac{4(0.003\lambda_0/r_e)^{1/2}\cos\theta}{0.006(\lambda_0/r_e)\cos^2\theta + 2(0.003\lambda_0/r_e)^{1/2}\cos\theta + 1} \qquad (4\text{-}71a)$$

$$\epsilon'_{\lambda\perp}(\lambda,\ \theta) = \frac{4(0.003\lambda_0/r_e)^{1/2}\cos\theta}{\cos^2\theta + 2(0.003\lambda_0/r_e)^{1/2}\cos\theta + 0.006\lambda_0/r_e} \qquad (4\text{-}71b)$$

The normal spectral emissivity given in (4-70) can be integrated with respect to wavelength to yield a normal total emissivity, $\epsilon'_n(T) = \pi \int_0^\infty \epsilon'_{\lambda,n}(\lambda,\ T)i'_{\lambda b}(\lambda,\ T)\ d\lambda/\sigma T^4$. Equation (4-70) is valid only for $\lambda_0 > \sim 5\ \mu m$, so in performing the integration starting from $\lambda = 0$ the condition is being imposed that the metal temperature is such that the energy radiated from $\lambda_0 = 0$ to $5\ \mu m$ is small compared with that at wavelengths longer than $5\ \mu m$. Then substituting into the integral the first term of (4-70) and (2-11) for $i'_{\lambda b}$ provides

$$\epsilon'_n(T) \approx \frac{\pi \int_0^\infty 2(r_e/0.003\lambda_0)^{1/2}2C_1/[\lambda_0^5(e^{C_2/\lambda_0 T}-1)]\ d\lambda_0}{\sigma T^4}$$

$$= \frac{4\pi C_1(Tr_e)^{1/2}}{(0.003)^{1/2}\sigma C_2^{4.5}}\int_0^\infty \frac{\zeta^{3.5}}{e^\zeta - 1}\ d\zeta \qquad (4\text{-}72)$$

where $\zeta = C_2/\lambda_0 T$ as was used in conjunction with Eq. (2-24). The integration is carried out by use of Γ functions to yield (T in K, r_e in ohm-cm)

$$\epsilon'_n(T) \approx \frac{4\pi C_1(Tr_e)^{1/2}}{(0.003)^{1/2}\sigma C_2^{4.5}}(12.27) = 0.575(r_eT)^{1/2} \tag{4-73}$$

If additional terms in the series (4-70a) are retained, then

$$\epsilon'_n(T) = 0.575(r_eT)^{1/2} - 0.177r_eT + 0.058(r_eT)^{3/2} - \cdots \tag{4-74a}$$

The recommended formula from [7] is

$$\epsilon'_n(T) = 0.576(r_eT)^{1/2} - 0.124r_eT \qquad (T \text{ in K, } r_e \text{ in ohm-cm}) \tag{4-74b}$$

and from [9] the three-term approximation

$$\epsilon'_n(T) = 0.578(r_eT)^{1/2} - 0.178r_eT + 0.0584(r_eT)^{3/2} \tag{4-74c}$$

For pure metals, r_e is approximately described near room temperature by

$$r_e \approx r_{e,273}\frac{T}{273} \tag{4-75}$$

where $r_{e,273}$ is the electrical resistivity in ohm-cm at 273 K. Substituting Eq. (4-75) into (4-73) gives the approximate result

$$\epsilon'_n(T) \approx 0.0348\sqrt{r_{e,273}}\,T \tag{4-76}$$

This indicates that the total emissivity of pure metals should be directly proportional to temperature. This result was originally derived by Aschkinass [10] in 1905. In some cases it holds to unexpectedly high temperatures where considerable radiation is in the short-wavelength region (for platinum, to near 1800 K), but in general applies only below about 550 K. This is illustrated in Fig. 4-12 for platinum and tungsten (data from [5]).

In Fig. 4-13, a comparison is made at 373 K of the normal total emissivity from experiment and from (4-76) for various polished surfaces of pure metals. Agreement is generally satisfactory. The experimental values are the minimum values of results available in three standard compilations [5,11,12].

Using the angular dependence from (4-71) combined with (4-65), an integration was made over all directions to provide hemispherical quantities. The following approximate equations for the hemispherical total emissive power fit the results in two ranges:

$$e(T) = (0.751\sqrt{r_eT} - 0.396r_eT)\sigma T^4 \qquad 0 < r_eT < 0.2 \tag{4-77a}$$

and

$$e(T) = (0.698\sqrt{r_eT} - 0.266r_eT)\sigma T^4 \qquad 0.2 < r_eT < 0.5 \tag{4-77b}$$

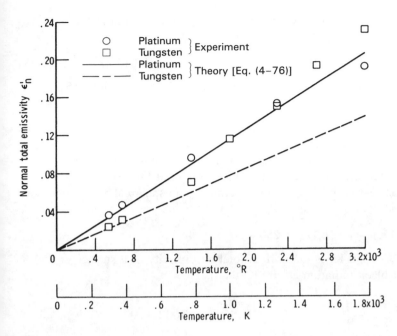

Figure 4-12 Temperature dependence of normal total emissivity of polished metals.

where the numerical factors in the parentheses and those used in specifying the ranges of validity apply for T in K and r_e in ohm-cm. The resistivity r_e depends on T to the first power so that the first term inside the parentheses of each of these equations provides a T^5 dependence, indicating that the temperature dependence for energy emission by metals is higher than the fourth power that is valid

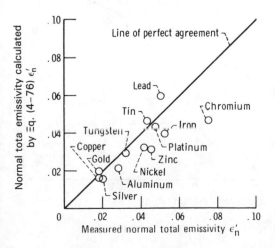

Figure 4-13 Comparison of data with calculated normal total emissivity for polished metals at 373 K.

for a blackbody. The hemispherical total emissivity expression of Parker and Abbott [9] is

$$\epsilon(T) = 0.766(r_eT)^{1/2} - (0.309 - 0.0889 \ln r_eT)r_eT - 0.0175(r_eT)^{3/2} \quad (4\text{-}78)$$

Comparisons of (4-74c) and (4-78) with data are in Figs. 4-14 and 4-15.

EXAMPLE 4-4 A polished platinum surface is maintained at temperature T_A = 250 K. Energy is incident upon the surface from a black enclosure, at temperature T_i = 500 K, that encloses the surface. What is the hemispherical-directional total reflectivity into the direction normal to the surface?

Equation (3-48) shows that the directional-hemispherical total reflectivity can be found from

$$\rho'_n(T_A = 250 \text{ K}) = 1 - \alpha'_n(T_A = 250 \text{ K})$$

where $\alpha'_n(T_A = 250 \text{ K})$ is the normal total absorptivity of a surface at 250 K for incident black radiation at 500 K; that is,

$$\alpha'_n(T_A = 250 \text{ K}) = \frac{\int_0^\infty \alpha'_{\lambda,n}(\lambda, T_A = 250 \text{ K})i'_{\lambda b}(\lambda, 500 \text{ K}) \, d\lambda}{\int_0^\infty i'_{\lambda b}(\lambda, 500 \text{ K}) \, d\lambda}$$

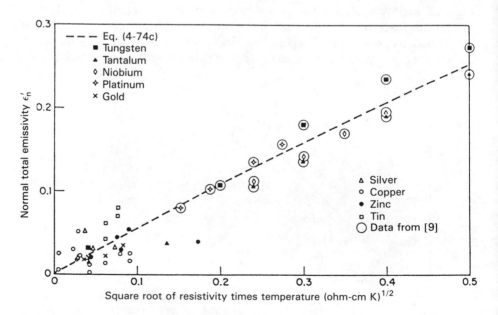

Figure 4-14 Normal total emissivity of various metals compared with theory [9].

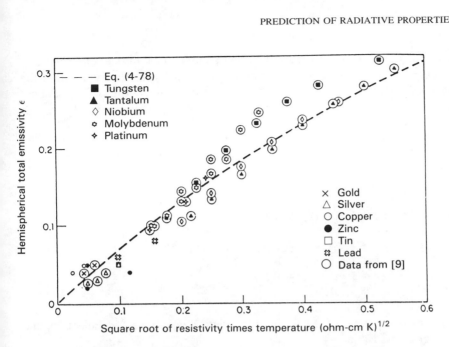

Figure 4-15 Hemispherical total emissivity of various metals compared with theory [9].

For spectral quantities, $\alpha'_{\lambda,n}(\lambda, T_A = 250 \text{ K}) = \epsilon'_{\lambda,n}(\lambda, T_A = 250 \text{ K})$. From Eq. (4-70) the variation of r_e with temperature provides the approximate emissivity variation $\epsilon'_{\lambda,n}(\lambda, T_A) \propto T_A^{1/2}$. Then $\epsilon'_{\lambda,n}(\lambda, T_A = 250 \text{ K}) = \epsilon'_{\lambda,n}(\lambda, T_A = 500 \text{ K})(250/500)^{1/2}$, and we obtain

$$\alpha'_n(T_A = 250 \text{ K}) = \frac{\sqrt{\dfrac{1}{2}} \displaystyle\int_0^\infty \epsilon'_{\lambda,n}(\lambda, T_A = 500 \text{ K}) i'_{\lambda b}(\lambda, 500 \text{ K}) \, d\lambda}{\displaystyle\int_0^\infty i'_{\lambda b}(\lambda, 500 \text{ K}) \, d\lambda}$$

$$= \frac{\epsilon'_n(T_A = 500 \text{ K})}{\sqrt{2}}$$

where the last equality is obtained by examination of the emissivity definition, Eq. (3-3b). The normal total emissivity of platinum at 500 K is given by (4-76) as plotted in Fig. 4-12 as

$$\epsilon'_n(T_A = 500 \text{ K}) = 0.0348\sqrt{r_{e,273}} \, T = 0.0348 \sqrt{r_{e,293}} \, \sqrt{\frac{273}{293}} \, T$$

$$= 0.348\sqrt{10 \times 10^{-6}} \, \sqrt{\frac{273}{293}} \times 500 = 0.053$$

Note that (4-76) is to be used only when temperatures are such that most of the energy involved is at wavelengths greater than 5 μm. Examination of the blackbody functions, Table A-5, shows that for a temperature of 500 K about 10% of the energy is at less than 5 μm so that possibly a small error is introduced.

The reciprocity relation of (3-28) for uniform incident intensity can now be employed to give the final result for the hemispherical-directional total reflectivity:

$$\rho_n'(T_A = 250 \text{ K}) = 1 - \alpha_n'(T_A = 250 \text{ K}) \approx 1 - \frac{\epsilon_n'}{\sqrt{2}} (T_A = 500 \text{ K})$$

$$= 1 - \frac{0.053}{\sqrt{2}} = 0.963$$

A summary of some of the property prediction equations for dielectrics and metals is given in Table 4-4.

Table 4-4 Summary of some equations for property prediction by electromagnetic theory

Property	Equation	Conditions
	Ideal dielectrics ($\kappa = 0$)	
Directional reflectivity	(4-57a)	Polarized in plane parallel to plane of incidence
Directional reflectivity	(4-57b)	Polarized in plane perpendicular to plane of incidence
Directional reflectivity	(4-58)	Unpolarized
	(4-56)	
Normal reflectivity	(4-45)	Polarized or unpolarized
Hemispherical emissivity	(4-59)	Emission into medium having $n = 1$
Normal emissivity	(4-60)	Emission into medium having $n = 1$
	Metals (in contact with transparent medium of unity refractive index)	
Directional reflectivity	(4-54a), (4-62a)	Parallel polarized component
Directional reflectivity	(4-54b), (4-62b)	Perpendicular polarized component
Directional reflectivity	(4-63)	Unpolarized
Directional emissivity	(4-65)	Unpolarized
Hemispherical emissivity	(4-67), (4-78)	Unpolarized
Normal spectral emissivity	(4-66)	Unpolarized
	(4-69a)	Unpolarized $\lambda > \sim 5$ μm
	(4-70)	
Normal total emissivity	(4-74), (4-76)	$T < \sim 550$ K ($1000°$R)

4-7 EXTENSIONS OF THE THEORY OF RADIATIVE PROPERTIES

Much work has been expended in improving the theory for radiative properties of materials, using both classical wave theory and quantum theory. A number of authors have removed some restrictions that are present in the classical development presented here. The contributions of Davisson and Weeks [13], Foote [14], Schmidt and Eckert [15], and Parker and Abbott [9] extended the emissivity relations for metals to shorter wavelengths and higher temperatures, and Mott and Zener [16] derived predictions for metal emissivity at very short wavelengths on the basis of quantum relations. Kunitomo [17] gives accurate predictions of high- and low-temperature properties of metals and alloys by including the effect of bound electrons on optical properties. Edwards [18] and Kunitomo [17] review the advances in predictions of surface properties, and Sievers [19] and Kunitomo [17] give additional theory and results.

None of these treatments, however, can account for surface conditions. Because of the difficulty of specifying surface conditions and controlling surface preparation, it is found that comparison of theory with experiment is not always adequate for even the refined theories. In fact, comparison to the less exact and simpler relations given here is often better. For even the purest materials given the most meticulous preparation, the elementary relations are often more accurate because the errors in the simpler theory are in the direction that provides compensation for surface working.

Makino [28] modeled the radiative properties of real surfaces by combining electromagnetic theory for pure substances with models for surface roughness and microgeometry and for surface films. Encouraging agreement of the theory predictions with measured properties that show the effect of oxidation layer growth [29] has been reported [30].

REFERENCES

1. Maxwell, James Clerk: A Dynamical Theory of the Electromagnetic Field, in W. D. Niven (ed.), "The Scientific Papers of James Clerk Maxwell," vol. 1, Cambridge University Press, London, 1890.
2. Isard, J. O.: Surface Reflectivity of Strongly Absorbing Media and Calculation of the Infrared Emissivity of Glasses, *Infrared Phys.*, vol. 20, pp. 249–256, 1980.
3. Harpole, G. M.: Radiative Absorption by Evaporating Droplets, *Int. J. Heat Mass Transfer*, vol. 23, pp. 17–26, 1980.
4. Hering, R. G., and T. F. Smith: Surface Radiation Properties from Electromagnetic Theory, *Int. J. Heat Mass Transfer*, vol. 11, pp. 1567–1571, 1968.
5. Weast, Robert C. (ed.): "Handbook of Chemistry and Physics," 44th ed., Chemical Rubber Company, Cleveland, 1962.
6. Palik, E. D. (ed.): "Handbook of Optical Constants of Solids," Academic Press, New York, 1985.
7. Jakob, Max: "Heat Transfer," vol. 1, Wiley, New York, 1949.

8. Hagen, E., and H. Rubens: Metallic Reflection, *Ann. Phys.*, vol. 1, no. 2, pp. 352–375, 1900. (See also E. Hagen and H. Rubens: Emissivity and Electrical Conductivity of Alloys, *Deutsch. Phys. Ges. Verhandl.*, vol. 6, no. 4, pp. 128–136, 1904.)

9. Parker, W. J., and G. L. Abbott: Theoretical and Experimental Studies of the Total Emittance of Metals, *Symp. Therm. Radiat. Solids*, NASA SP-55, pp. 11–28, 1964.

10. Aschkinass, E.: Heat Radiation of Metals, *Ann. Phys.*, vol. 17, no. 5, pp. 960–976, 1905.

11. Hottel, H. C.: Radiant Heat Transmission, in William H. McAdams (ed.), "Heat Transmission," 3d ed., pp. 55–125, McGraw-Hill, New York, 1954.

12. Eckert, E. R. G., and R. M. Drake, Jr.: "Heat and Mass Transfer," 2d ed., McGraw-Hill, New York, 1959.

13. Davisson, C., and J. R. Weeks, Jr.: The Relation between the Total Thermal Emissive Power of a Metal and Its Electrical Resistivity, *J. Opt. Soc. Am.*, vol. 8, no. 5, pp. 581–605, 1924.

14. Foote, Paul D.: The Emissivity of Metals and Oxides, III. The Total Emissivity of Platinum and the Relation between Total Emissivity and Resistivity, *NBS Bull.*, vol. 11, no. 4, pp. 607–612, 1915.

15. Schmidt, E., and E. R. G. Eckert: Über die Richtungsverteilung der Wärmestrahlung von Oberflächen, *Forsch. Geb. Ingenieurwes.*, vol. 6, no. 4, pp. 175–183, 1935.

16. Mott, N. F., and C. Zener: The Optical Properties of Metals, *Cambridge Philos. Soc. Proc.*, vol. 30, pt. 2, pp. 249–270, 1934.

17. Kunitomo, T.: Present Status of Research on Radiative Properties of Materials, *Int. J. Thermophys.*, vol. 5, no. 1, pp. 73–90, 1984.

18. Edwards, D. K.: Radiative Transfer Characteristics of Materials, *J. Heat Transfer*, vol. 91, pp. 1–15, 1969.

19. Sievers, A. J.: Thermal Radiation from Metal Surfaces, *J. Opt. Soc. Am.*, vol. 68, no. 11, pp. 1505–1516, 1978.

20. Garbuny, M.: "Optical Physics," Academic Press, New York, 1965.

21. Seban, R. A.: Thermal Radiation Properties of Materials, pt. III, WADD-TR-60-370, University of California, Berkeley, August 1963.

22. Brandenberg, W. M.: The Reflectivity of Solids at Grazing Angles, in Joseph C. Richmond (ed.), "Measurement of Thermal Radiation Properties of Solids," NASA SP-31, pp. 75–82, 1963.

23. Brandenberg, W. M., and O. W. Clausen: The Directional Spectral Emittance of Surfaces between 200° and 600° C, *Symp. Therm. Radiat. Solids*, NASA SP-55, pp. 313–320, 1964.

24. Price, Derek J.: The Emissivity of Hot Metals in the Infra-Red, *Proc. Phys. Soc. London*, ser. A, vol. 59, pt. 1, pp. 118–131, 1947.

25. Hurst, C.: The Emission Constants of Metals in the Near Infra-Red, *Proc. R. Soc. London*, ser. A, vol. 142, no. 847, pp. 466–490, 1933.

26. Pepperhoff, W.: "Temperaturstrahlung," D. Steinkopf, Darmstadt, 1956.

27. Toscano, W. M., and E. G. Cravalho: Thermal Radiative Properties of the Noble Metals at Cryogenic Temperatures, *J. Heat Transfer*, vol. 98, no. 3, pp. 438–445, 1976.

28. Makino, T., and K. Kaga: Scattering of Radiation at a Rough Surface Modeled by a Three-Dimensional Super-Imposition Technique, *Proc. 3rd ASME/JSME Thermal Engineering Joint Conf.*, Reno, vol. 4, pp. 27–33, 1991.

29. Makino, T., O. Sotokawa, and Y. Iwata: Transient Behaviors in Thermal Radiation Characteristics of Heat-Resisting Metals and Alloys in Oxidation Processes, *Int. J. Thermophysics*, vol. 9, no. 6, pp. 1121–1130, 1988.

30. Makino, T., K. Kaga, and H. Murata: Numerical Experiment on Transient Behavior in Reflection Characteristics of a Real Surface of a Metal, *Proc. 28th Natl. Symp. Heat Transfer, Japan*, Fukuoka, vol. 2, pp. 568–570, 1991.

PROBLEMS

4-1 An electrical insulator has a refractive index of $n = 1.8$ and is radiating into air. What is the directional emissivity for the direction normal to the surface? What is it for the direction 85° from the normal?

Answer: 0.9184, 0.3707

4-2 A hot ceramic sphere with index of refraction $n = 1.7$ is photographed with an infrared camera. Calculate how bright the image is at locations B and C relative to that at A. (Camera is distant from sphere.)

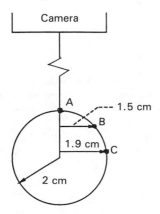

Answer: 0.982, 0.829

4-3 A particular dielectric material has a refractive index near 2. Estimate:

(a) the hemispherical emissivity of the material for emission into air.
(b) the directional emissivity at $\theta = 70°$ into air.
(c) the directional hemispherical reflectivity in air for both components of polarized reflectivity. Plot both components for $n = 2$ on a graph similar to Fig. 4-9. Let θ be the angle of incidence.

Answer: (a) 0.839; (b) 0.764

4-4 A dielectric material has a normal spectral emissivity at wavelength of 10 μm of $\epsilon'_{\lambda,n} = 0.95$. Find or estimate values for:

(a) the hemispherical spectral emissivity ϵ_λ at the same wavelength.
(b) the perpendicular component of the directional hemispherical spectral reflectivity $\rho'_{\lambda,\perp}(\theta)$ at the same wavelength and $\theta = 30°$.

Answer: (a) 0.897; (b) 0.0709

4-5 An inventor wants to use a light source and some Polaroid glasses to determine when the wax finish is worn from his favorite bowling alley. He reasons that the wax will reflect as a dielectric (with $n = 1.450$) and that the parallel component of light from the source will be preferentially absorbed and the perpendicular component strongly reflected by the wax. When the wax is worn away, the wood will reflect diffusely. At what height should the light source be placed to maximize the ratio of perpendicular to parallel polarization from the wax as seen by the viewer?

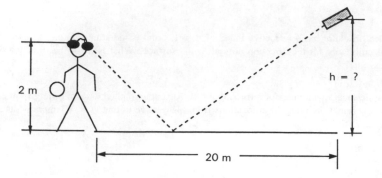

Answer: 11.8 m

4-6 A smooth ceramic dielectric has an index of refraction $n = 1.65$, which is independent of wavelength. If a flat ceramic disk is at 675 K, how much emitted energy per unit time is received by the detector when it is placed at $\theta = 0°$ and $\theta = 80°$? Use relations from electromagnetic theory.

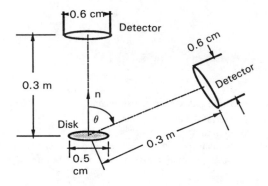

Answer: 21.7×10^{-6} W at $\theta = 0°$, 2.38×10^{-6} W at $\theta = 80°$.

4-7 At a temperature of 300 K these metals have the resistivities

Silver	1.65×10^{-6} Ω-cm
Platinum	11.0×10^{-6} Ω-cm
Lead	20.8×10^{-6} Ω-cm

What are the theoretical hemispherical total emissivities of these materials, and how do they compare with tabulated values for clean, unoxidized, polished surfaces?
Answer: 0.017, 0.042, 0.057

4-8 A highly polished metal disk is found to have a measured normal spectral emissivity of 0.2 at a wavelength of 15 μm. What is

(a) the hemispherical spectral emissivity of the disk at this wavelength?
(b) the electrical resistivity of the metal (Ω-cm)?
(c) the normal spectral emissivity of the metal at $\lambda = 10$ μm?
(d) the refractive index of the metal at $\lambda = 10$ μm?
(Note any assumptions that you make in obtaining your answers.)

Answer: (a) 0.228; (b) 5.267×10^{-4} Ω-cm; (c) 0.241; (d) 7.249

4-9 For a metal with normal spectral emissivity of $\epsilon'_{\lambda,n} = 0.15$, find the value of the electrical resistivity at a wavelength of 10 μm.

Answer: 1.892×10^{-4} ohm-cm

4-10 Evaluate the normal spectral reflectivity of aluminum at 365 K when $\lambda_0 = 5$ μm, 10 μm, 20 μm. For aluminum, the temperature coefficient of resistivity is 0.0039.

Answer: 0.969, 0.978, 0.985

4-11 Polished gold at 300 K is irradiated normally by a gray-body source at 800 K. Evaluate the absorptivity α''_n. (Use the method of Example 4-4.)

Answer: 0.0257

4-12 A smooth polished gold surface must radiate 225 W/m². What is its surface temperature as calculated from the Hagen-Rubens relation? At this temperature, what is the normal total emissivity? What is the ratio of hemispherical total emissivity to normal total emissivity, and how does this value compare with that from Fig. 4-10?

Answer: 568.8 K, 0.0296, 1.283.

4-13 The hemispherical total emissive power emitted by a polished metallic surface is 1900 W/m² at temperature T_S. What would you expect the emissive power to be if the temperature were doubled? What assumptions are involved in your answer?

Answer: 60,800 W/m²

4-14 The figure below gives some experimental data for the hemispherical spectral reflectivity of polished aluminum at room temperature. Extrapolate the data to $l = 12$ μm. Use whatever method you want, but list your assumptions. Discuss the probably accuracy of your extrapolation. (*Hint:* The electrical resistivity of pure aluminum is about 2.82×10^{-6} Ω-cm at 293 K. At 12 μm, $\bar{n} = 33.6 - 76.4i$. You may use any, all, or none of these data as you wish.)

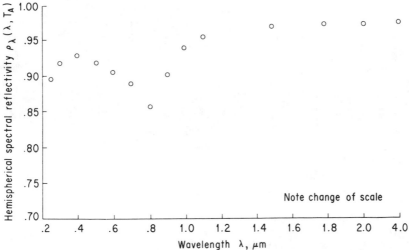

(Figure from G. Hass, Filmed Surfaces for Reflecting Optics, *J. Opt. Soc. Am.*, vol. 45, p. 945, 1955.)

4-15 An unoxidized titanium sphere is heated until it is glowing red. From a distance, it appears as a red disk. From electromagnetic theory how would you expect the brightness to vary across the disk? What would you expect after looking at Fig. 5-14?

4-16 Using the Hagen-Rubens emissivity relation, plot the normal spectral emissivity as a function of wavelength for a polished aluminum surface used in a cryogenic application at 30 K. What is the normal total emissivity? (*Note:* Do not use any relations valid only near room temperature.)

4-17 A sample of highly polished platinum has a value of normal spectral emissivity of 0.050 at a wavelength of 5.0 μm at 293 K. What value of normal spectral absorptivity do you expect the sample to have at

(a) a wavelength of 10 μm and at 293 K?
(b) a wavelength of 10 μm and at 570 K?

Answer: (a) 0.0355; (b) 0.0493

4-18 Metals cooled to very low temperatures approaching absolute zero become superconducting; i.e., the value of $r_e(T \rightarrow 0) \rightarrow 0$. Based on electromagnetic theory predictions, what is your estimate of the values of the simple refractive index n and the extinction coefficient κ and of the normal spectral and normal total emissivities at such conditions? What assumptions are implicit in your estimate? (The results predicted by the Hagen-Rubens relation, and other results from classical electromagnetic theory, become inaccurate at $T < 100$ K. Predictions of radiative properties at low absolute temperatures using more exact theoretical approaches are reviewed by Toscano and Cravalho [27].)

4-19 A smooth polished stainless steel surface must emit a total intensity in the normal direction of 33.1 W/(m$^2 \cdot$ sr). What is its surface temperature as calculated from equations derived from the Hagen-Rubens relation? The r_e of the steel is 11.9×10^{-6} ohm-cm at 293 K.

Answer: 438 K

FIVE

RADIATIVE PROPERTIES OF REAL MATERIALS

5-1 INTRODUCTION

In this chapter, the general characteristics of radiative properties of *real* materials are examined. These properties can differ considerably from the results for ideal, optically smooth materials as predicted by electromagnetic theory in Chap. 4. The analytical predictions yield useful trends and provide a unifying basis to help explain various radiation phenomena. However, the analyses are inadequate in the sense that the engineer is generally dealing with surfaces that are not perfectly clean but have varying amounts of contaminants or oxides on them. They are usually not optically smooth and have a surface roughness that is difficult to specify completely. Examples of typical variations of radiative properties as a function of these and other parameters are presented to illustrate the types of variations that can occur. This will provide an appreciation of how sensitive the radiative performance is to the surface condition. In addition to the typical properties presented, a number of atypical examples are given to demonstrate that a careful examination of individual properties must be made to properly select the property values to be used in radiative exchange calculations.

Most of the discussion is for opaque solids, where opaque means that no transmission of radiant energy occurs through the entire thickness of the body. A composite body such as a thin coating on a substrate of a different material can have partial transmission through the coating, but for an opaque body, none of the transmitted radiation will pass entirely through the substrate. Information is also included on the transmission ability of glass and water. This illustrates how transmission can differ considerably in short- and long-wavelength regions. In Chap. 18 the effects of thin films on various substrates are discussed in some detail.

No attempt is made to compile comprehensive property data. Extensive tabulations and graphs of radiative properties have been gathered in [1–8]. A limited tabulation is in Appendix D for the convenience of the reader.

As discussed in Chap. 4, basic differences in the radiative behavior of dielectrics and metals are evident from electromagnetic theory. For this reason the first two sections in this chapter deal with these two classes of materials. Then some special surfaces are discussed that have specific desirable variations of properties with wavelength and direction.

5-2 SYMBOLS

a_λ	spectral absorption coefficient
A	area
c_0	speed of electromagnetic wave propagation in vacuum
d	thickness of plate; dimension in Fig. 5-36
e	emissive power
E	overall emittance of plate
$F_{0-\lambda}$	fraction of blackbody energy in spectral range 0 to λ
n	index of refraction
p	probability function
q	energy flux; energy per unit time per unit area
Q	energy rate; energy per unit time
r_e	electrical resistivity
R	overall reflectance of plate
T	absolute temperature; overall transmittance of plate
z	height of surface roughness
α	absorptivity
γ	electrical permittivity
ϵ	emissivity
θ	angle measured from normal of surface; polar angle
λ	wavelength
μ	magnetic permeability
ρ	reflectivity
σ	Stefan-Boltzmann constant (Table A-4)
σ_0	root-mean-square height of surface roughness
τ	transmittance along a path
φ	circumferential or azimuthal angle

Subscripts

a	absorber
A	of surface A

b	blackbody condition
c	evaluated at cutoff wavelength
e	emitted
eq	at thermal equilibrium
i	incident
max	maximum value
n	normal direction
r	reflected
R	radiator
s	specular
λ	spectrally dependent
$0-\lambda$	in wavelength range 0 to λ

Superscripts

$'$	directional
$''$	bidirectional

5-3 RADIATIVE PROPERTIES OF OPAQUE NONMETALS

Nonmetals are usually characterized by large values of hemispherical total emissivity and absorptivity at moderate temperatures and therefore have small values of reflectivity in comparison with metals (see tabulated values in Appendix D). For a clean, optically smooth surface, several results were obtained by use of electromagnetic theory in Chap. 4. These provide the following generalizations (bearing in mind the rather stringent assumptions of the theory): the directional emissivity will decrease with increasing angle from the surface normal; wavelength dependence may be weak in the infrared region as it enters the predicted properties through the refractive index, which can vary slowly with wavelength for some nonmetals; finally, the temperature dependence of the spectral properties of nonmetals will also be small, since temperature also enters the spectral property prediction only through the refractive index, which is often a weak function of temperature.

The difficulty with these generalizations is that most nonmetals cannot be polished to the degree necessary for their surfaces to be considered ideal, although some common exceptions exist, such as glass, large crystals of various types, gemstones, and some plastics (some of these are not opaque materials like those being discussed here). As a result of having nonideal surface finishes, many nonmetals have behavior that deviates radically from that predicted by electromagnetic theory. Available property measurements for nonmetals are less detailed than for metals. Specifications of surface composition, texture, and so forth are often lacking.

An effect that complicates the interpretation of the measured properties of nonmetals is that radiation passing into such a material may penetrate quite far before being absorbed (this is evident for visible wavelengths in glass). To be opaque, a specimen must be of sufficient thickness to absorb essentially all the radiation that enters; otherwise, transmitted radiation must be accounted for. Transmission is not considered in the present discussion. Often, samples of non-metals such as paints are sprayed onto a metallic or other opaque base (substrate), and the properties of the composite are measured. If it is desired to have the surface behave completely as the coating material, the thickness of the coating must be large enough to ensure that no significant radiation is transmitted through it. Otherwise, in reflectivity measurements, some of the incident radiation can be reflected from the substrate and then transmitted again through the coating to reappear as measured energy. The data will then be a function of both the coating material and the substrate.

For emissivity measurements of a coating material, the coating must be thick enough so that no emitted energy from the substrate penetrates the coating. An illustration is given by Liebert [9], who examined the spectral emissivity of zinc oxide on various substrates, using various oxide thicknesses. The effect of coating thickness on the emissivity of the composite formed by the zinc oxide coating and a substrate of approximately constant normal spectral emissivity is shown in Fig. 5-1. The effect of increasing the coating thickness becomes small in the thickness range 0.2 to 0.4 mm, indicating that the emissivity of zinc oxide alone is being approached.

Another example is in Fig. 5-2, where the effects of different substrates were measured. A white paint film was placed on a dense thick white paint substrate, on a black substrate, and on specularly reflecting aluminum [10]. The paint film thickness was 14.4 μm and the wavelength of the incident radiation varied from 0.5 to 10 μm. The reflectivity is shown, and the nature of the substrate has a considerable effect. Consider the black substrate: as the wavelength decreases and

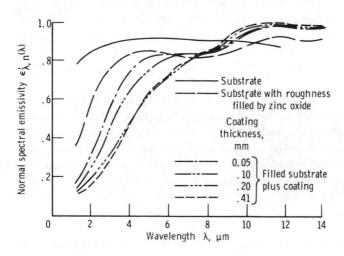

Figure 5-1 Emissivity of zinc oxide coatings on oxidized stainless steel substrate. Surface temperature, 880 ± 8 K. Data from [9].

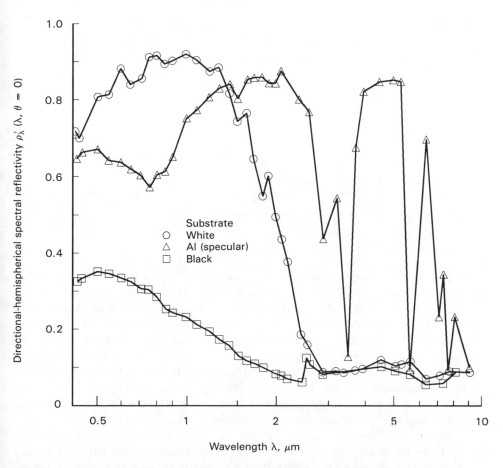

Figure 5-2 Effect of substrate reflection characteristics on the hemispherical reflection characteristics of a TiO₂ paint film for normal incidence: Film thickness, 14.4 μm; volume concentration of pigment 0.017. Data from [10].

becomes small relative to the film thickness, the reflectivity increases as it tends to become more like that of the white paint film and less like that of the black substrate.

We shall now examine the effects of wavelength, temperature, and surface roughness on the radiative properties of dielectrics and then briefly examine the radiative properties of semiconductors.

5-3.1 Spectral Variations of Properties

Figure 5-3 shows the hemispherical normal spectral reflectivity of three paint coatings on steel. From Kirchhoff's law and the reflectivity reciprocity relations, the difference between unity and these reflectivity values is the normal spectral emis-

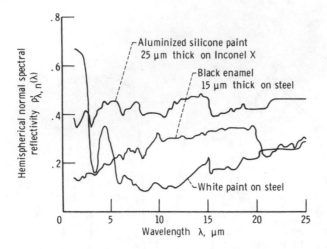

Figure 5-3 Spectral reflectivity of paint coatings. Specimens at room temperature. Data from [58].

sivity. The three paints shown exhibit somewhat different characteristics. White paint has a high reflectivity (low emissivity) at short wavelengths, and the reflectivity decreases at longer wavelengths. Black paint, on the other hand, has a relatively low reflectivity over the entire wavelength region shown. By using aluminum powder in a silicone base as a paint, the reflectivity is increased, as would be expected for the more metallic coating. This particular specimen of aluminized paint acts approximately as a "gray" surface since the properties are reasonably independent of wavelength. Because of the large variation in spectral emissivity at short wavelengths, the gray approximation would be poor for the white paint unless very little of the participating radiation was at the shorter wavelengths.

Figure 5-4 illustrates that at the short wavelengths in the visible range the reflectivity for some nonmetals may decrease substantially. This behavior is very important in considering the suitability of a specific nonmetallic coating for reflecting radiation from a high-temperature source, where much of the energy will be at short wavelengths.

The normal spectral emissivities of several nonmetals are shown in Fig. 5-5. The $\epsilon'_{\lambda,n}(\lambda)$ values for the nitrides vary over a range of about 0.2, but the other three materials have strong variations with λ. They exhibit large changes in $\epsilon'_{\lambda,n}(\lambda)$ in some wavelength regions.

5-3.2 Variation of Total Properties with Temperature

The effect of surface temperature on the total emissivity of several nonmetallic materials is shown in Figs. 5-6 to 5-8. Both increasing and decreasing trends with temperature are observed. Some of these effects may be caused by the dielectric coating being rather thin; hence, the properties are influenced by the temperature and spectral characteristics of the underlying material (substrate). For example, as shown in Fig. 5-6, magnesium oxide refractory exhibits a significant emissivity decrease with increasing temperature. For a silicon carbide coating on graphite,

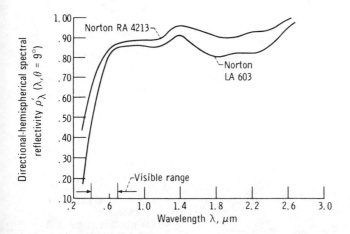

Figure 5-4 Directional-hemispherical spectral reflectivity of aluminum oxide. Incident angle, 9°; specimens at room temperature. Data from [5].

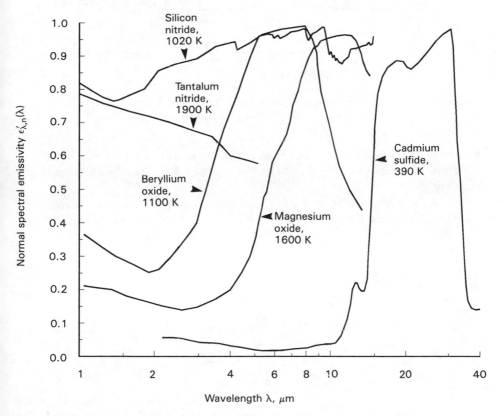

Figure 5-5 Normal spectral emissivity of some nonmetal samples at the temperatures shown. Typical data replotted from [7].

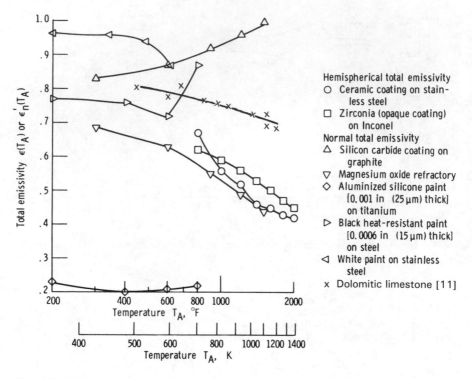

Figure 5-6 Effect of surface temperature on emissivity of dielectrics. Data from [1] and [11].

however, the emissivity increases with temperature; this may be partly caused by the emissive behavior of the graphite substrate, which increases with temperature (see Fig. 5-17).

White and black paint both have high emissivities for the temperature range shown, as is typical for ordinary oil-base paint. Aluminized paint is considerably

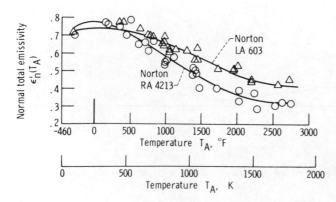

Figure 5-7 Effect of surface temperature on normal total emissivity of aluminum oxide. Data from [5].

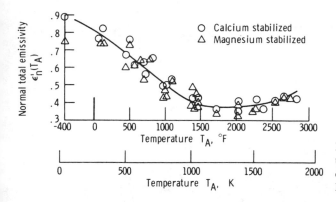

Figure 5-8 Effect of surface temperature on normal total emissivity of zirconium oxide. Data from [5].

lower in emissive ability since it behaves partly like a metal. Note that the emissivity for aluminized paint in Fig. 5-6 is about one-half that in Fig. 5-3. This illustrates the variation in properties that can be found for samples having the same general description. For applications where property values are critical, it is advisable to make radiation measurements for the specific samples or materials being used. For example, to study fluidized-bed combustion of high-sulfur coal, the emissivity of dolomitic limestone was needed at high temperatures. The normal total emissivity measured in [11] was found to decrease linearly as T increased from 539 to 1223 K.

As stated earlier, the *spectral* properties of a dielectric often vary slowly with temperature since the index of refraction is not a strong function of temperature. Then Eq. (3-6d) for the total emissivity has the form $\epsilon(T_A) = \int_0^\infty \epsilon_\lambda(\lambda) e_{\lambda b}(\lambda, T_A)\, d\lambda / \sigma T_A^4$, where the spectral variation $\epsilon_\lambda(\lambda)$ is approximately independent of temperature. This demonstrates that the most significant factor in the variation of ϵ with T_A is the wavelength shift in the blackbody function $e_{\lambda b}$ as T_A changes. Spectral data for ϵ_λ obtained at one temperature can be used in the integral to calculate accurate $\epsilon(T_A)$ values over a range of nearby temperatures.

Figure 5-9 gives the normal total absorptivity of a few materials for blackbody radiation incident from sources at various temperatures. White paper is a good absorber for the spectrum emitted at low temperatures but a poor absorber for the spectrum emitted at temperatures of a few thousand degrees K. It is thus a reasonably good reflector for energy incident from the sun. An asphalt pavement or a gray slate roof, on the other hand, absorbs energy from the sun very well. Absorptivities for incident solar radiation are given in Appendix D as needed for space vehicle, satellite, and solar collector design.

5-3.3 Effect of Surface Roughness

An important parameter in characterizing surface roughness effects is the *optical roughness*. This is the ratio σ_0/λ of a characteristic roughness height, usually the root-mean-square (rms) roughness σ_0, to the wavelength of the radiation. If the

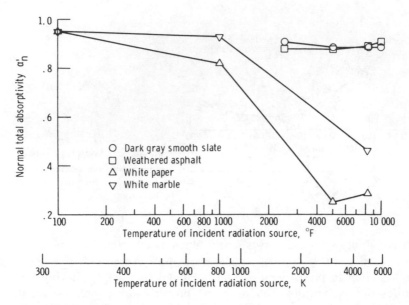

Figure 5-9 Normal total absorptivity of nonmetals at room temperature for incident black radiation from the sources at temperatures indicated. Data from [1].

surface imperfections on a material are much smaller than the radiation wavelength, the material is said to be *optically smooth*. A surface that is optically smooth for long wavelengths may be quite optically rough at short wavelengths.

In Fig. 5-10 the bidirectional total reflectivity of typewriter paper is shown for three angles of incidence. For an ideal (polished, smooth) surface, a specular peak would be expected with the angles of reflection and incidence symmetric about the normal; obviously, the surface finish of typewriter paper is not ideal, since the reflected intensity occupies a rather large angular envelope around the direction of specular reflection. Reference [12] provides detailed bidirectional reflectivity measurements for magnesium oxide ceramic with optical roughness σ_0/λ varying from 0.46 to 11.6. As the roughness and the incidence angle of the incoming radiation are increased, off-specular peaks are obtained, as will be discussed in connection with Fig. 5-21. Bidirectional spectral reflectivity measurements were made in [13] for six materials used for solar energy absorption. Peaks in the specular direction were found.

Dielectrics generally show a slight increase in emissivity with roughness. However, Cox [14] has shown that for materials with a ratio of roughness to radiation mean penetration distance* in the dielectric of about 0.05, the emissivity may be *less* than for a smooth surface. This result was predicted by analysis and confirmed by experiment. Because of the porosity of many dielectrics, it is difficult to study roughness parameters below a certain limiting smoothness.

---

*See Sec. 12-4.2 for definition.

The type of curves shown in Fig. 5-10 has suggested characterizing reflected energy as a combination of a purely diffuse plus a purely specular component. This approximation has merit in some cases and results in a simplification of radiant interchange calculations in comparison with the use of exact directional properties [15, 16]; in other cases, however, the approximation would fail completely. An example is in Fig. 5-11, which shows the observed bidirectional total reflectivity for visible light reflected from the moon. These particular curves are for the mountainous regions, but similar curves have been obtained for other areas. The interesting feature is that the peak of the reflected radiation is back into the direction of the incident radiation. This peak is located at a circumferential angle φ of 180° away from where a specular peak would occur.

Observations confirm that curves of this type characterize lunar reflectivity. At full moon, which occurs when the sun, earth, and moon are almost in a straight line (Fig. 5-12), the moon appears equally bright across its face. For this to be true, it follows that an observer on earth sees equal intensities from all points on the moon. However, the solar energy incident upon a unit area of the lunar surface varies as the cosine of the angle θ between the sun and the normal to the lunar surface. The angle θ varies from 0 to 90°, as the position of the incident energy varies from the center to the edge of the lunar disk. To reflect a constant intensity to an observer on earth from all observable points on the lunar surface therefore requires that the product $\rho''(\theta, \theta_r) \cos \theta$ be constant. Consequently, the value of the bidirectional reflectivity in the direction of incidence must increase approximately in proportion to $1/\cos \theta$ (shown by the dashed line in Fig. 5-11) as the angle of incidence increases. This change in reflectivity with incidence angle will compensate for the reduced energy incident per unit area on the moon at the large

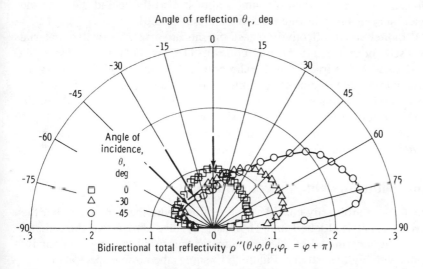

Figure 5-10 Bidirectional total reflectivity of typewriter paper in plane of incidence. Source temperature, 1178 K (2120° R). Replotted data from [59].

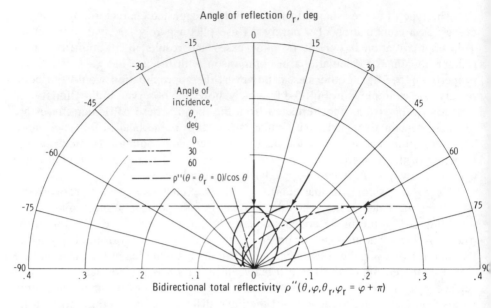

Figure 5-11 Bidirectional total reflectivity in plane of incidence ($\varphi_r = \varphi + \pi$) for mountainous regions of lunar surface. (After Orlova [60].)

angles. The reflectivity behavior is confirmed by the curves in Fig. 5-11. Hence a uniformly bright moon does not imply that the moon is a diffuse reflector. If the moon were diffuse, it would appear bright at the center and dark at the edges. The strong backscattering of the lunar surface causes the moon's brightness to peak strongly at full moon when the sun is almost directly behind the observer. Further discussion of lunar infrared behavior is in [17–20].

In [21], bidirectional reflectivity measurements are given for twelve materials commonly used for spacecraft thermal control. The "specularity" of these materials was defined as the fraction of the directional-hemispherical reflectivity that is contained within the specular solid angle, and it is also presented. Specularity values can be used in surface property models that assume a combination of diffuse and specular reflectivities (Sec. 9-5). Most measurements are at a wavelength of $\lambda = 0.488$ μm, but some white paints were measured at four discrete wavelengths covering the range $0.488 \le \lambda \le 10.63$ μm.

5-3.4 Semiconductors and Superconductors

Semiconductors are arbitrarily considered here along with the nonmetals, but they behave partly as metals. Liebert and Thomas [22] have shown that the radiative properties can be determined by electromagnetic theory by treating semiconductors as metals with high resistivity. Figure 5-13 shows the normal spectral emissivity of a silicon semiconductor. The Hagen-Rubens relation shown for comparison is based on the dc resistivity measured for the same sample, one of the few cases

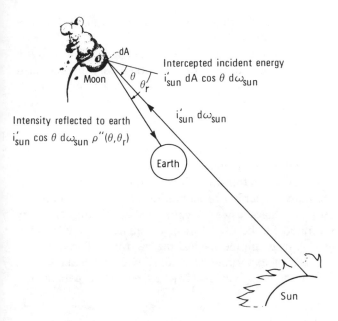

Figure 5-12 Reflected energy at full moon.

where such comparable emissive and electrical data are available. Agreement does not become good until wavelengths are reached that are much greater than those giving agreement for metals. This difference in range of agreement can be traced to the assumption used in deriving the Hagen-Rubens equation (Sec. 4 6.3): $(\lambda / 2\pi c_0 r_e \gamma)^2 \gg 1$ where c_0 and λ are in vacuum. For semiconductors, in which the resistivity is larger than it is for metals, this inequality cannot hold until a range of wavelengths larger than those for metals is reached. The shape of the curve measured for silicon (Fig. 5-13) resembles what would be expected for a polished metal (see, for example, the tungsten data in Fig. 5-16). The emissivity increases with decreasing wavelength over much of the measured spectrum, with a peak occurring at shorter wavelengths. However, most of the features of the semicon-

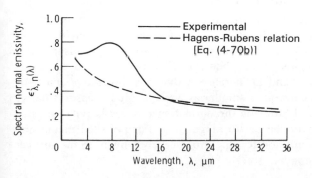

Figure 5-13 Normal spectral emissivity of a highly doped silicon semiconductor at room temperature. Data from [22].

ductor curve occur at longer wavelengths than for a metal; the peak emissivity, for example, is well outside the visible region.

Liebert and Thomas [22] were also able to show excellent agreement between the measured emissivity and predictions from electromagnetic theory that included the effects of free electrons and were more sophisticated than those in Chap. 4. The theoretical equations were evaluated by using physical properties that were measured from the specific samples on which the emissivity measurements were made.

The theory for spectral reflectivity of high-temperature superconductors is examined in [23] and compared with experimental data. The material is a ceramic, Y-Ba-Cu-O, and is used in the form of a thin film. It is operated at liquid nitrogen temperature. The material has almost perfect reflection in the far-infrared region, and the reflectivity decreases to 0.85 at a wavelength of about 6 μm. The very high reflectivity in the far infrared can be useful in space applications as a radiation shield for liquid nitrogen. A quantum-mechanical theory (Mattis-Bardeen [24]) provides predictions within 12% of experimental results, with better agreement as wavelength increases. Relations are given in [23] that provide excellent predictions for engineering use.

5-4 RADIATIVE PROPERTIES OF METALS

Pure, smooth metals are often characterized by low values of emissivity and absorptivity and therefore high values of reflectivity. Figure 4-13 demonstrates that the emissivity in the direction normal to the surface is quite low for various polished metals. However, low emissivity values are not an absolute rule for metals; in some of the examples that will be given the spectral emissivity rises to 0.5 or larger as the wavelength becomes short, and the total emissivity becomes large as the surface temperature is elevated.

5-4.1 Directional Variations

A behavior typical of polished metals is that the directional emissivity tends to increase with increasing angle θ measured away from the surface normal, except near $\theta = 90°$. This is predicted by electromagnetic theory and was shown for platinum in Fig. 4-9. At wavelengths shorter than the range for which the simple electromagnetic theory of Chap. 4 applies, a deviation from this behavior might be expected as illustrated by the directional spectral emissivity of polished titanium in Fig. 5-14. At wavelengths greater than about 1 μm, the directional spectral emissivity of titanium does tend to increase with increasing θ over most of the θ range. The increase with θ becomes smaller as wavelength decreases; finally, at wavelengths less than about 1 μm, the directional spectral emissivity actually decreases with increasing θ over the entire range of θ. For polished metals the typical behavior of increased emission for directions nearly tangent to the surface may not occur at short wavelengths.

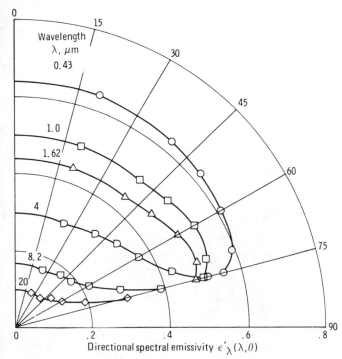

Figure 5-14 Effect of wavelength on directional spectral emissivity of pure titanium. Surface ground to 16 μin (0.4 μm) rms. Data from [6].

5-4.2 Effect of Wavelength

In the infrared region, it was shown in Chap. 4 that the spectral emissivity of metals tends to increase as wavelength decreases. This trend remains true over a large span of wavelength as illustrated for several metals in Fig. 5-15 for the spectral emissivity in the normal direction. For other directions, the same effect is illustrated in Fig. 5-14 except at large angles from the normal, where curves for various wavelengths may cross. The curve for the copper sample in Fig. 5-15 is an exception as the emissivity remains relatively constant with wavelength.

At very short wavelengths, the assumptions on which the simplified electromagnetic theory of Chap. 4 are based become invalid. Indeed, most metals exhibit a peak emissivity somewhere near the visible region, and the emissivity then decreases rapidly with further decrease in wavelength. This is illustrated by the behavior of tungsten in Fig. 5-16.

5-4.3 Effect of Surface Temperature

The Hagen-Rubens relation [Eq. (4-70)] shows that, for wavelengths that are not too short ($\lambda > \sim 5$ μm), the spectral emissivity of a metal tends to be proportional

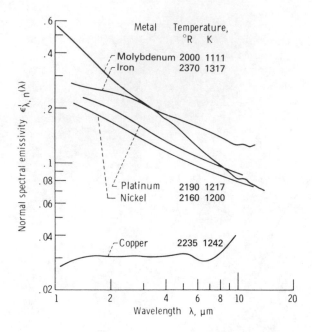

Figure 5-15 Variation with wavelength of normal spectral emissivity for polished metals. Data from [61].

to the resistivity of the metal to the one-half power. Hence, we can expect the spectral emissivity of pure metals to increase with temperature as does the resistivity, and this is found to be the case in most instances. Figure 5-16 is an example for the hemispherical spectral emissivity of tungsten. The expected trend is observed for $\lambda > 1.27$ μm. Figure 5-16 also illustrates a phenomenon characteristic

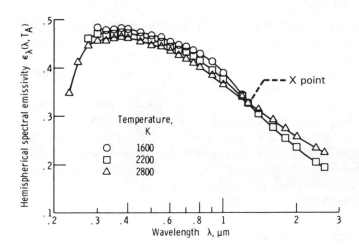

Figure 5-16 Effect of wavelength and surface temperature on hemispherical spectral emissivity of tungsten [62].

of many metals as discussed in [25]. At short wavelengths (in the case of tungsten, $\lambda < 1.27\ \mu m$), the temperature effect is reversed and the spectral emissivity decreases as temperature is increased. The emissivity curves all cross at the same point, which has been called the "X point." Some other X point values are: iron, 1.0 μm; nickel, 1.5 μm; copper, 1.7 μm; and platinum, 0.7 μm.

The increase of spectral emissivity with decreasing wavelength for metals in the infrared region (wavelengths longer than visible region), as discussed in Sec. 5-4.2, accounts for the increase in total emissivity with temperature. With increased temperature the peak of the blackbody radiation curve (Fig. 2-7) moves toward shorter wavelengths. Consequently, as the surface temperature is increased, proportionately more radiation is emitted in the region of higher spectral emissivity, which results in an increased total emissivity. Some examples are shown in Fig. 5-17. Here the behavior of metals is contrasted with that of a dielectric, magnesium oxide, for which the emissivity decreases with increasing temperature.

In insulation systems consisting of a series of radiation shields in vacuum, an important heat loss is by radiation. At the low temperatures involved in the insulation of cryogenic systems, there is a lack of experimental property data, especially at the long wavelengths characteristic of emission at these temperatures. The Hagen-Rubens result indicates that the normal spectral emissivity is proportional to $(r_e/\lambda)^{1/2}$, and the normal total emissivity is proportional to $(r_eT)^{1/2}$. If electrical resistivity is directly proportional to temperature, then $\epsilon'_{\lambda,n}(\lambda, T) \propto (T/\lambda)^{1/2}$ and $\epsilon'_n(T) \propto T$. This indicates that the emissivities should become quite small at low T and large λ. Experimental measurements summarized in [26] for copper, silver, and gold indicate that the ϵ values do not decrease to such small values. As refinements in the theory, the Drude free-electron theory and anomalous skin-effect theory, described in [26], include additional electron and quantum-me-

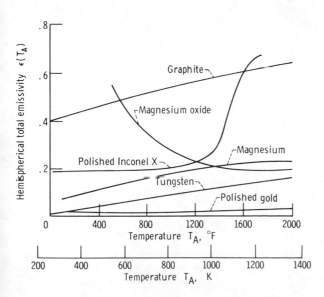

Figure 5-17 Effect of temperature on hemispherical total emissivity of several metals and one dielectric. Data from [1].

chanical interactions. The Drude theory, which reduces to the Hagen-Rubens relation at long wavelengths, predicts ϵ values that decrease much more rapidly with temperature than has been observed. The anomalous skin-effect model with diffuse electron reflections was found to predict the emissivity most accurately. Figure 5-18 shows the results of the two theoretical models. The limited data at low temperatures lie somewhat above the anomalous skin-effect model. Similar results for gold films are in [27].

5-4.4 Effect of Surface Roughness

The radiative properties of optically smooth materials can be predicted within the limitations of electromagnetic theory as discussed in Chap. 4. When the surface optical roughness σ_0/λ is greater than about 1, there are multiple reflections in the cavities between roughness elements. As a result the roughness increases the trapping of incident radiation, thereby increasing the surface absorptivity and consequently its emissivity.

When the roughness is large, it has a considerable effect on the directional emission and reflection characteristics. For $\sigma_0/\lambda > 1$ the concepts of geometric optics can be used to trace the radiation paths reflected within the cavities or from the roughness elements. If the roughness geometry is completely specified, it is

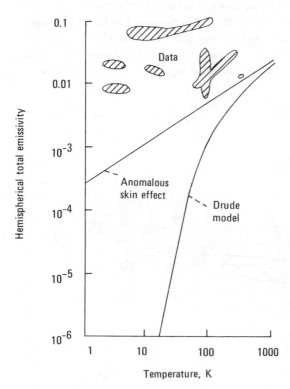

Figure 5-18 Effect of low temperature on hemispherical total emissivity of copper. Results from [26].

possible in certain cases to predict the directional behavior. An example is the directional emissivity of a parallel-grooved surface in Sec. 5-5.3. Ordinarily the roughness is very irregular, and a statistical model must be assumed. For example, the roughness can be represented as randomly oriented facets that each reflect in a specular manner. A roughness model composed of spherical cap indentations was used in [28].

When the optical roughness is small ($\sigma_0/\lambda < 1$), multiple reflection effects in the roughness cavities usually become small, and the hemispherical properties approach those for optically smooth surfaces. However, as a result of diffraction effects, the directional properties (especially the bidirectional reflectivity) can be significantly influenced by the roughness.

Several analyses have been made with the intent of predicting the effect of surface roughness on the radiative properties of materials. They have provided a basis for understanding many of the observed roughness effects. A significant difficulty in the prediction of radiative properties is in the precise definition of the surface characteristics for use in the analytical relations. Perhaps the most common way to characterize surface roughness is by the method of preparation (lapping, grinding, etching, etc.) plus a specification of rms roughness. The latter is usually obtained with a profilometer, which traverses a sharp stylus over the surface and reads out the vertical perturbations in terms of an rms value. It does not account for the horizontal spacing of the roughness and gives no indication of the *distribution* of the size of roughness around the rms value. It does not give any information on the average slope of the sides of the roughness peaks, which influences the behavior of the cavities between them. At present, there is no generally accepted method of accurately specifying surface characteristics, and none of those mentioned in this paragraph is completely adequate for prediction of radiative properties.

Some of the analytical approaches are now discussed briefly and some comparisons given with experimental data. Davies [29] used diffraction theory to examine the reflecting properties of a surface with roughness assumed to be distributed according to a gaussian (normal) probability distribution, specified as a probability $p(z)$ of having a roughness of height z given by $p(z) = (1/\sigma_0\sqrt{2\pi})$ exp $(-z^2/2\sigma_0^2)$. The individual surface irregularities were assumed to be of sufficiently small slope that shadowing could be neglected, and σ_0 was assumed very much smaller than the wavelength of incident radiation. The material was assumed to be a perfect electrical conductor so that from (4-21b) the extinction coefficient is infinite. From (4-49) this provides perfect reflection, and consequently the theory is concerned with the directional distribution of the energy that is reflected rather than with the amount that is reflected. The reflected distribution was found to consist of a specular component and a distribution about the specular peak.

A similar derivation, with σ_0 assumed much larger than λ, again yielded a distribution of reflected intensity about the specular peak, this time of larger angular spread than for $\sigma_0 \ll \lambda$. This would be expected since the surface should behave increasingly like an ideal specular reflector as the roughness becomes very small compared with the wavelength of the incident radiation. Davies' treatment is found to be very inaccurate at near-grazing angles because of the neglect of shadowing by the roughness elements.

Porteus [30] extended Davies' approach by removing the restrictions on the relation between σ_0 and λ and including more parameters for specification of the surface roughness characteristics. Some success in predicting the roughness characteristics of prepared samples from measured reflectivity data was obtained, but certain types of surface roughness led to poor agreement. Measurements were mainly at normal incidence, and the neglect of shadowing makes the results of doubtful value at near-grazing angles.

A more satisfactory treatment is by Beckmann and Spizzichino [31]. Their method includes the autocorrelation distance of the roughness in the prescription of the surface. This is a measure of the spacing of the characteristic roughness peaks on the surface and hence is related to the rms slope of the roughness elements. The method provides better data correlation than the earlier analyses. A critical evaluation and comparison of the Davies and Beckmann analyses is given by Houchens and Hering [32].

Some observed effects of surface roughness for small σ_0/λ are shown in Figs. 5-19 and 5-20. The former shows the directional emissivity of titanium [6] at a

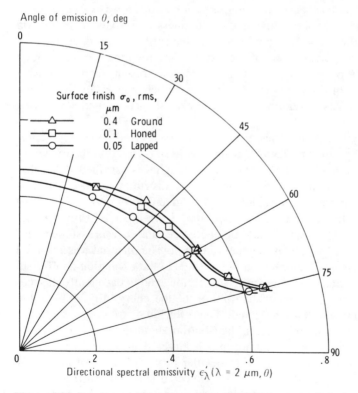

Figure 5-19 Roughness effects for small optical roughness, $\sigma_0/\lambda < 1$; effect of surface finish on directional spectral emissivity of pure titanium. Wavelength, 2 μm. Data from [6].

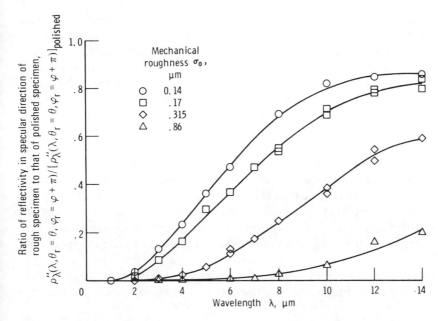

Figure 5-20 Effect of roughness on bidirectional reflectivity in specular direction for ground nickel specimens. Mechanical roughness for polished specimen, 0.015 μm. Data from [63].

wavelength of 2 μm for three surface roughnesses as obtained by grinding, honing, and lapping. The maximum roughness is 0.4 μm so, relative to the 2 μm wavelength, the specimens are smooth. As a result, the emissivity changes only a small amount as the roughness varies from 0.05 to 0.4 μm. Also there is very little effect on the directional variation of the emissivity. Reference [6] also gives data for sandblasted surfaces that produced larger increases in emissivity.

Figure 5-20 provides the reflectivity of nickel for energy reflected *into the specular direction* from a beam incident at an angle 10° from the normal. In this figure, the reflectivities of the rough specimens are expressed as a ratio to the reflectivity of a polished surface to exhibit the effect of roughness on the directional characteristics rather than on the magnitude of the reflectivity. The polished surface used for comparison had a roughness about 10 times less than any of the rough specimens. A high value of the ordinate means that the specimen is behaving more like a polished surface. Data are shown for ground nickel specimens with four different roughnesses with $\sigma_0/\lambda < 1$. The reflectivity rises as wavelength is increased (that is, optical roughness is decreased) because for a given roughness the surface is smoother relative to the incident radiation. As expected, for a fixed wavelength the reflectivity for the specular direction decreases as the roughness is increased. Data exhibiting the same trends for aluminum have been well correlated in [33] by use of the Beckmann theory.

For an optical roughness $\sigma_0/\lambda > 1$ some detailed experimental measurements of the bidirectional reflectivity, along with an analysis using geometric optics, are given in [12, 34]. The analysis substantiated the important trends of the data. Some typical results for the bidirectional reflectivity of aluminum are in Fig. 5-21 for an optical roughness of 2.6. This gives the bidirectional reflectivity in the plane of incidence (plane formed by incident path and normal to surface) as a function of reflection angle θ_r. The reflectivity is ratioed to the value in the specular direction. The various curves correspond to different angles of incidence. For a diffuse surface the reflected intensity is independent of θ_r so the ratio given in the figure is unity as shown by the dashed line. A specular reflection would appear as a sharp high peak at $\theta_r = \theta$. For incidence at 30° the reflected intensity peaks in the specular direction ($\theta_r = 30°$). However, at larger incidence angles the maximum in ρ_λ'' shifts to angles larger than the specular angle; for example, when $\theta = 60°$ the reflectivity peak is at $\theta_r = 85°$. This is in contrast to the behavior for a smooth surface, where the peak would be at $\theta_r = \theta$. The theory indicates that this off-specular reflection, which occurs for large optical roughness and large incidence angles, is the result of mutual shadowing effects by the roughness elements.

References [35] and [36] analyze the angular distribution of the emissivity from a rough surface by considering the emission from regular V-grooves in a

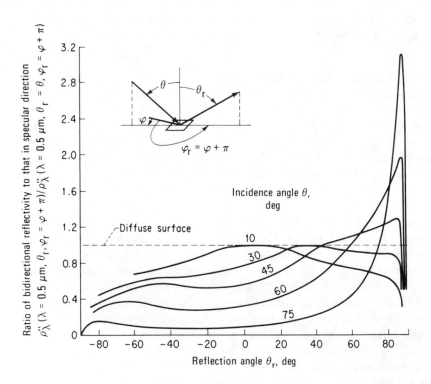

Figure 5-21 Bidirectional reflectivity in plane of incidence for various incidence angles; material, aluminum (2024-T4), aluminum coated; rms roughness, $\sigma_0 = 1.3$ μm; wavelength of incident radiation, $\lambda = 0.5$ μm. Data from [34].

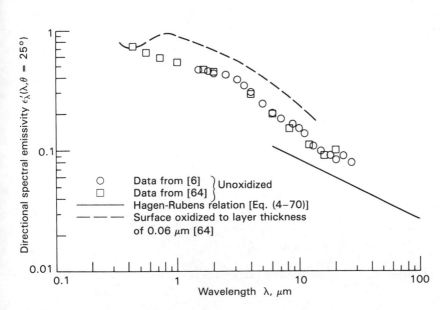

Figure 5-22 Effect of oxide layer on directional spectral emissivity of titanium. Emission angle, θ = 25°; surface lapped to 0.05 μm rms; temperature, 294 K.

parallel or circular pattern. A more realistic model is the randomly roughened metal surface in [37, 38]. The distributions of heights and slopes of the roughnesses were assumed gaussian, and random blockage of radiation by adjacent roughness elements in the path of observation was included. Multiple reflections between surface elements were neglected. Calculations carried out for gold and chromium indicated that for directions less than about 60° from the surface normal, the emissivity increases with surface roughness. At larger angles the emissivity was less than for an ideal smooth surface; this was a result of both roughness effects and the behavior of smooth metallic surfaces for which the emissivity becomes large at angles nearly tangent to the surface. The roughness models have been used with ray tracing methods to simulate realistic visual reflection from objects [39].

5-4.5 Effect of Surface Impurities

Impurities in this context include contaminants of any type that cause deviations of the surface properties from those of an optically smooth pure metal. The most common contaminants are thin layers deposited either by adsorption, such as water vapor, or by chemical reaction, such as a thin layer of oxide. Because dielectrics generally have high values of emissivity, an oxide or other nonmetallic contaminant layer will usually increase the emissivity of an otherwise ideal metallic body.

Figure 5-22 shows the directional spectral emissivity of titanium at an angle of 25° to the surface normal. The data points are for the unoxidized metal, and

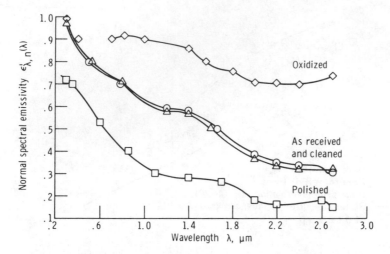

Figure 5-23 Effect of oxidation on normal spectral emissivity of Inconel X. Data from [5].

the solid line is the ideal emissivity predicted from electromagnetic theory. The dashed curve shown above the data points is the observed emissivity when an oxide layer only 0.06 μm thick is present. The emissivity is increased by a factor of almost 2 from that of the pure material over much of the wavelength range. Figure 5-23 shows a similar large increase in the normal spectral emissivity of Inconel X for an oxidized surface as compared with that for the polished metal.

Figures 5-24 and 5-25a illustrate the effect of an oxide coating on the normal total emissivity of stainless steel and on the hemispherical total emissivity of copper. The details of the oxide coatings are not specified, but the large effect of

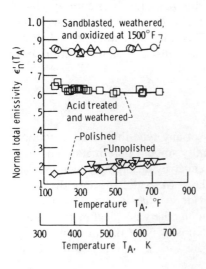

Figure 5-24 Effect of surface condition and oxidation on normal total emissivity of stainless steel type 18-8. Data from [5].

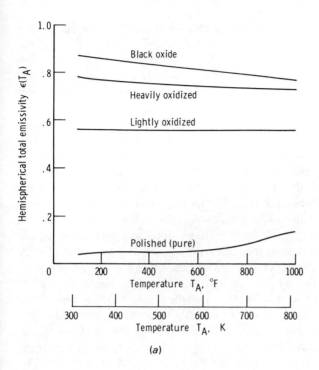

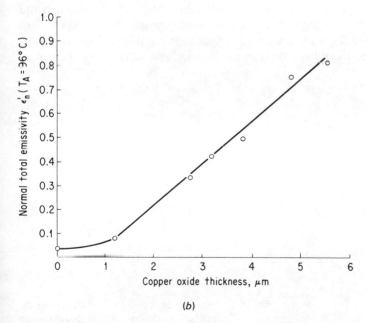

Figure 5-25 Effect of oxide coating on emissive properties of copper. (*a*) Effect of oxide coating on hemispherical total emissivity of copper. Data from [1]. (*b*) Effect of oxide thickness on normal total emissivity of copper at 369 K. Data from [40].

surface oxidation is apparent. More precise indications of oxide coating effect are shown in Figs. 5-25b and 5-26 for the normal total emissivity of copper and the hemispherical total emissivity of aluminum. An oxide thickness of even a few micrometers provides a very substantial emissivity increase. For oxidized aluminum some results of a nature similar to Fig. 5-26 are given by Brannon and Goldstein [40].

Makino et al. [40a] carried out measurements of the spectral near-normal bidirectional reflectivity of a polished chromium surface as a function of time during the growth of a surface oxide layer. The measurements covered the wavelength range of 0.35 to 10 μm. For a high-temperature surface (T = 1000 K) oxidizing in air over a period of 1.5 h, the reflectivity was found to follow very definite trends that were explained through diffraction/interference effects within the oxide film and a model of the microgeometry of the surface.

Figure 5-27 shows approximately the directional total absorptivity of an anodized aluminum surface for radiation incident from various θ directions and originating from sources at various temperatures. The quantity $\rho_s'(\theta)$ is the fraction of the incident energy that is reflected into the specular direction; hence, $1 - \rho_s'(\theta)$ is the fraction of the incident energy that is absorbed plus the fraction of the incident energy reflected into directions other than the specular direction. For the specimens tested, only a small percentage of the energy was reflected into directions other than the specular direction. Thus, in Fig. 5-27, the quantity $1 - \rho_s'(\theta)$ can be regarded as a good approximation to the directional total absorptivity. The curves have all been normalized to pass through unity at $\theta = 0$; hence, it is the shapes of the curves that are significant. At low source temperatures, the incident radiation is predominantly in the long-wavelength region. This incident radiation is barely influenced by the thin oxide film on the anodized surface; consequently,

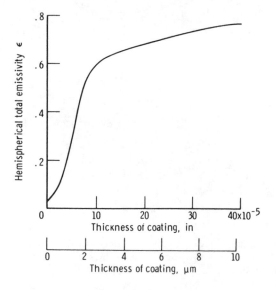

Figure 5-26 Typical curve illustrating effect of electrolytically produced oxide thickness on hemispherical total emissivity of aluminum. Temperature, 311 K. Data from [1].

Angle of incidence θ, deg

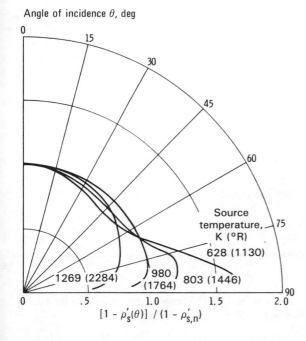

$[1 - \rho'_s(\theta)] / (1 - \rho'_{s,n})$

Figure 5-27 Approximate directional total absorptivity of anodized aluminum at room temperature relative to value for normal incidence. Redrawn data from [59].

the specimen acts like a bare metal and has large absorptivities at large angles from the normal. At high source temperatures where the incident radiation is predominantly at shorter wavelengths, the thin oxide film has a significant effect, and the surface behaves as a nonmetal where the absorptivity decreases with increasing θ.

The structure of the surface coating can also have a substantial effect on the radiative behavior. Figure 5-28 shows the hemispherical spectral reflectivity of aluminum coated with lead sulfide. The mass of the coating per unit area of surface is the same for both sets of data shown. The difference in crystal structure and size causes the reflectivity of the coated specimens to differ by a factor of 2 at wavelengths longer than 3 μm.

5-4.6 Molten Metals

In [41] the spectral and total emissivities of liquid sodium were measured in the normal direction. The liquid metal was kept in an argon atmosphere to avoid oxidation. Unlike the behavior for many solid metals, the normal total emissivity did not have a significant temperature dependence. For temperatures increasing from 450 to 715 K the ϵ_n increased from 0.045 to 0.057 and then decreased to 0.044 as the temperature further increased to 795 K. The spectral emissivity in the range of λ from 3 to 14 μm did not vary with wavelength for temperatures below 700 K. Above 700 K the $\epsilon'_{\lambda,n}(\lambda)$ had an increasing trend with λ.

The measurements in [42] for three rare earth liquid metals showed a similar lack of temperature dependence as given in the first three lines of Table 5-1. These

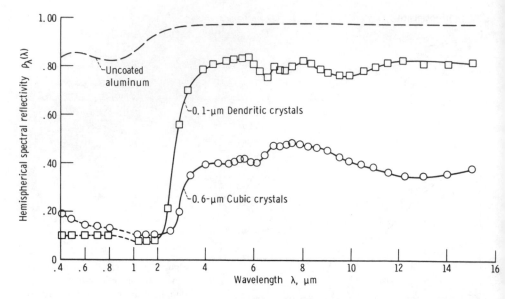

Figure 5-28 Hemispherical spectral reflectivity for normal incident beam on aluminum coated with lead sulfide. Coating mass per unit surface area, 0.68 mg/cm². Data from [65].

are normal spectral values at $\lambda = 0.645$ μm. Normal spectral emissivities at this same λ were measured for the liquid state at the melting temperatures of several metals [43]. The values (Table 5-1) are generally comparable to those for solid metals at high temperatures near their melting points. In [43a] the reflectivity of a thin film of silicon was measured as it was melted by incident laser radiation.

Table 5-1 Normal spectral emissivities of liquid metals at $\lambda = 0.645$ μm [43]

Metal	Melting point, K	Spectral emissivity	Temperature range, K
Lanthanum	1193	0.282	1193–1493
Cerium	1071	0.309	1123–1498
Praseodymium	1204	0.294	1203–1498
Cobalt	1767	0.335	1767
Chromium	2133	0.262	2133
Copper	1358	0.147	1358
Iron	1811	0.357	1811
Molybdenum	2895	0.306	2895
Niobium	2744	0.317	2744
Nickel	1728	0.346	1728
Palladium	1827	0.354	1827
Tantalum	3256	0.309	3256
Titanium	1946	0.434	1946
Vanadium	2178	0.343	2178
Zirconium	2128	0.318	2128

An approximate increase in reflectivity of 0.1 was noted, possibly because of the elimination of the silicon polycrystalline structure in the molten state.

5-5 SELECTIVE AND DIRECTIONAL OPAQUE SURFACES, AND SELECTIVE TRANSMISSION

For engineering purposes, it is often desirable to tailor the radiative properties of surfaces to increase or decrease their natural ability to absorb, emit, or reflect radiant energy. This can be done to provide a desired spectral or directional performance. Information is also given here on the radiative transmission through glass and water. These materials have wavelength-dependent behaviors somewhat like the special surfaces being discussed.

5-5.1 Modification of Surface Spectral Characteristics

For surfaces used in the collection of radiant energy, such as in solar distillation units, solar furnaces, or solar collectors for energy conversion, it is desirable to maximize the energy absorbed by a surface while minimizing the loss by emission. In solar thermionic or thermoelectric devices, it is desirable to maintain the highest possible equilibrium temperature of the surface exposed to the sun. Here again a maximum-collection, minimum-loss performance is needed. For situations in which it is desirable to keep a surface exposed to the sun cool, it is desirable to have maximum reflection of solar energy accompanied by maximum radiative emission from the surface.

For solar-energy collection, a black surface will of course maximize the absorption of incident solar energy; unhappily, it also maximizes the emissive losses. However, if a surface could be manufactured that had an absorptivity large in the spectral region of short wavelengths about the peak solar energy, yet small in the spectral region of longer wavelengths where the peak surface emission would occur, it might be possible to absorb nearly as well as a blackbody while emitting very little energy. Such surfaces are called "spectrally selective." One method of manufacture is to coat a thin, nonmetallic layer onto a metallic substrate. For radiation with large wavelengths, the thin coating is essentially transparent, and the surface behaves as a metal yielding low values for spectral absorptivity and emissivity. At short wavelengths, however, the radiation characteristics approach those of the nonmetallic coating, so the spectral emissivity and absorptivity are relatively large. Some examples of this type are in Fig. 5-29.

An *ideal* solar selective surface would absorb a maximum of solar energy while emitting a minimum amount of energy. The surface would thus have an absorptivity of unity over the range of short wavelengths where the incident solar energy has a large intensity. At longer wavelengths, the absorptivity should drop sharply to zero. The wavelength λ_c at which this sharp drop occurs, as in Fig. 5-29, is termed the *cutoff wavelength*.

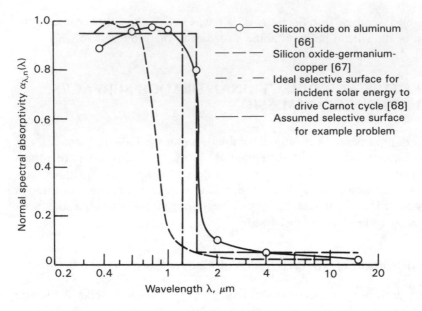

Figure 5-29 Characteristics of some spectrally selective surfaces.

EXAMPLE 5-1 An ideal selective surface is exposed to a normally incident flux of radiation corresponding to the average solar constant $q_i = 1353$ W/m^2. The only means of heat transfer to or from the exposed surface is by radiation. Determine the maximum equilibrium temperature T_{eq} corresponding to a cutoff wavelength of $\lambda_c = 1$ μm. (The energy arriving from the sun can be assumed to have a spectral distribution proportional to that of a blackbody at 5780 K.)

Since the only means of heat transfer is by radiation, the radiant energy absorbed must equal that emitted from the exposed side. Since we have specified an ideal selective absorber, the hemispherical emissivity and absorptivity are given by

$$\epsilon_\lambda(\lambda) = \alpha_\lambda(\lambda) = 1 \quad 0 \le \lambda < \lambda_c \qquad \epsilon_\lambda(\lambda) = \alpha_\lambda(\lambda) = 0 \quad \lambda_c \le \lambda < \infty$$

The energy absorbed by the surface per unit time is

$$Q_a = \alpha_\lambda F_{0-\lambda_c}(T_R)q_iA = (1)\, F_{0-\lambda_c}(T_R)q_iA$$

where $F_{0-\lambda_c}(T_R)$ is the fraction of blackbody energy in the range of wavelengths between zero and the cutoff value, for a radiating source at temperature T_R. In this case, T_R is the effective solar radiating temperature of 5780 K. Similarly, the energy emitted by the selective surface is

$$Q_e = \epsilon_\lambda F_{0-\lambda_c}(T_{eq})\sigma T_{eq}^4A = (1)\, F_{0-\lambda_c}(T_{eq})\sigma T_{eq}^4A$$

Equating Q_e and Q_a, we obtain

$$T_{eq}^4 F_{0-\lambda_c}(T_{eq}) = \frac{q_i F_{0-\lambda_c}(T_R)}{\sigma}$$

For the chosen value for λ_c, all terms on the right are known, and we can solve for T_{eq} by trial and error. The equilibrium temperature for $\lambda_c = 1\ \mu m$, as specified in the problem, is 1334 K. Values of T_{eq} corresponding to other values of λ_c are given in the following table:

Cutoff wavelength λ_c, μm	Equilibrium temperature T_{eq}, K	(°R)
0.6	1811	(3261)
0.8	1523	(2741)
1.0	1334	(2402)
1.2	1210	(2178)
1.5	1041	(1873)
→ ∞	393	(707.4)

For a blackbody surface ($\lambda_c \to \infty$), the equilibrium temperature is 393 K (707° R); this is the equilibrium temperature of the surface of a black object in space at the earth's orbit when exposed to solar radiation and with all other surfaces of the object perfectly insulated. The same equilibrium temperature is reached by a gray body, since a gray emissivity would cancel out of the energy-balance equation.

As smaller values of λ_c are taken, T_{eq} continues to increase even though less energy is absorbed, because it becomes relatively more difficult to emit energy as λ_c is decreased.

A common measure of the performance of a given selective surface is the ratio of the directional total absorptivity $\alpha'(\theta, \varphi, T_A)$ of the surface for incident solar energy to the hemispherical total emissivity of the surface $\epsilon(T_A)$. The ratio α'/ϵ for the condition of incident solar energy is a measure of the theoretical maximum temperature that an otherwise insulated surface can attain when exposed to solar radiation. The significance of α'/ϵ is shown as follows.

The energy absorbed per unit time by any surface when exposed to a directional energy flux $q_i(\theta, \varphi)$ incident from direction (θ, φ) on a surface element dA can be written as

$$dQ_a'(\theta, \varphi, T_A) = \alpha'(\theta, \varphi, T_A)\, q_i(\theta, \varphi)\, dA \cos \theta \qquad (5\text{-}1)$$

The total energy emitted per unit time by the surface element is

$$dQ_e = \epsilon(T_A)\sigma T_A^4\, dA \qquad (5\text{-}2)$$

If the only energy absorbed by the surface is that given by (5-1) and the surface loses energy only by radiation, then the emitted and absorbed energies given by (5-2) and (5-1) are equated to give

$$\frac{\alpha'(\theta,\varphi,T_{eq})}{\epsilon(T_{eq})} = \frac{\sigma T_{eq}^4}{q_i \cos \theta} \tag{5-3}$$

where T_{eq} is the equilibrium temperature that is achieved. Thus the ratio $\alpha'(\theta, \varphi, T_A)/\epsilon(T_A)$ is a measure of the equilibrium temperature of the element. Note also that the temperature at which the properties α' and ϵ are selected should be the equilibrium temperature that the body attains. A common situation is for normal incidence $\cos \theta = 1$, so that α' becomes α'_n.

Equation (5-3) shows that the smaller the value of α'_n/ϵ that can be reached, the smaller will be the equilibrium temperature. For a cryogenic storage tank in space exposed to solar flux, α'_n/ϵ should be as small as possible in order to reduce losses of the stored fluid. In practice, values of α'_n/ϵ in the range 0.20 to 0.25 can be obtained.

For the collection and utilization of solar energy on earth or in outer-space applications, a high α'_n/ϵ is desired. For the relatively low temperatures of solar collection in ground-based systems without solar concentrators, selective paints on an aluminum substrate have yielded $\alpha' = 0.92$ and $\epsilon = 0.10$ [44]. Thickness-insensitive paints that do not require a special substrate are also under development. Low-emittance metallic flakes have been mixed in a binder with high-absorptance metallic oxides to yield coatings with $\alpha' = 0.88$ and $\epsilon = 0.40$. Various data are in [44]. It is hoped that economical paints can be developed for application to large solar collection areas.

To attain high equilibrium temperatures for space power systems, α'_n/ϵ should be as large as possible. Polished metals attain α'_n/ϵ values of 5 to 7, and specially manufactured surfaces have values of α'_n/ϵ approaching 20. Coatings with $\alpha'_n/\epsilon \approx 13$ and stability at temperatures up to about 900 K in air were reported in [45]. Space power systems usually have a concentrator such as a parabolic mirror. This increases the collection area relative to the area for emission and thus effectively increases the absorption-to-emission ratio.

EXAMPLE 5-2 The properties of a real SiO-Al selective surface are approximated by the long-dashed curve in Fig. 5-29 (it is assumed that the long-dashed curve can be extrapolated toward $\lambda = 0$ and $\lambda = \infty$). What is the equilibrium temperature of the surface for normally incident solar radiation at earth orbit when heat transfer is only by radiation? What is α'_n/ϵ for the surface? Describe the spectra of the absorbed and emitted energy at the surface. (Assume normal and hemispherical emissivities are equal.)

As in the derivation of Eq. (5-3), we equate the absorbed and emitted energies. The emissivity has nonzero values on both sides of the cutoff wavelength, so that

$$Q_a = \epsilon_{0-\lambda_c} F_{0-\lambda_c}(T_R) q_i A + \epsilon_{\lambda_c-\infty} F_{\lambda_c-\infty}(T_R) q_i A = \alpha'_n q_i A$$

$$Q_e = \epsilon_{0-\lambda_c} F_{0-\lambda_c}(T_{eq}) \sigma T_{eq}^4 A + \epsilon_{\lambda_c-\infty} F_{\lambda_c-\infty}(T_{eq}) \sigma T_{eq}^4 A = \epsilon \sigma T_{eq}^4 A$$

Equating Q_e and Q_a gives

$$\{0.95 F_{0-\lambda_c}(T_R) + 0.05[1 - F_{0-\lambda_c}(T_R)]\} q_i$$
$$= \{0.95 F_{0-\lambda_c}(T_{eq}) + 0.05[1 - F_{0-\lambda_c}(T_{eq})]\} \sigma T_{eq}^4$$

Solving by trial and error we obtain, for $\lambda_c = 1.5$ μm, $T_{eq} = 789$ K. For q_i = 1353 W/m^2, Eq. (5-3) gives $\alpha_n'/\epsilon = \sigma(789)^4/1353 = 16.28$. The small difference in properties in this example from the properties of an ideal selective surface produces a significant change in T_{eq}, which in the previous example was 1041 K for an ideal selective surface with the same λ_c. The spectral curve of incident solar energy is given by $e_{\lambda,i}(\lambda, T_R) \propto e_{\lambda b}(\lambda, T_R)$. It has the shape of the blackbody curve at the solar temperature, but it is reduced in magnitude so that the integral of $e_{\lambda,i}$ over all λ is equal to q_i, the total incident solar energy per unit area at earth orbit. Multiplying this curve by the spectral absorptivity of the selective surface gives the spectrum of the absorbed energies. The spectrum of emitted energy is that of a blackbody at 789 K multiplied by the spectral emissivity of the selective surface. The integrated energies under the spectral curves of absorbed and emitted energy are equal.

The energy equation solved in Example 5-2 is a two-spectral-band approximation to the following more general energy equation for a *diffuse* surface: $\int_{\lambda=0}^{\infty} \alpha_\lambda(\lambda, T_{eq}) dq_{\lambda,i}(\lambda) = \int_{\lambda=0}^{\infty} \epsilon_\lambda(\lambda, T_{eq}) e_{\lambda b}(\lambda, T_{eq}) d\lambda$. The $dq_{\lambda,i}$ can have any spectral distribution, and by Kirchhoff's law $\alpha_\lambda(\lambda, T_{eq}) = \epsilon_\lambda(\lambda, T_{eq})$. A more general situation is in Fig. 5-30. The absorption of incident energy from the θ direction depends on the directional spectral absorptivity $\alpha_\lambda'(\lambda, \theta, \varphi, T_{eq})$. The emission from the surface depends on its hemispherical spectral emissivity $\epsilon_\lambda(\lambda, T_{eq})$. The q_e is the heat flux supplied to the

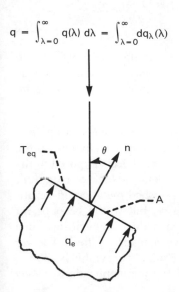

$$q = \int_{\lambda=0}^{\infty} q(\lambda) \, d\lambda = \int_{\lambda=0}^{\infty} dq_\lambda(\lambda)$$

Figure 5-30 Energy incident on a selective surface.

surface by any other means, such as convection, electrical heating, or radiation to its lower side. The heat balance then gives

$$\cos \theta \int_{\lambda=0}^{\infty} \alpha_\lambda'(\lambda, \theta, \varphi, T_{eq}) q(\lambda) \, d\lambda + q_e = \int_{\lambda=0}^{\infty} \epsilon_\lambda(\lambda, T_{eq}) e_{\lambda b}(\lambda, T_{eq}) \, d\lambda \qquad (5\text{-}4)$$

or

$$\cos \theta \, \alpha'(\theta, \varphi, T_{eq}) q + q_e = \epsilon(T_{eq}) \sigma T_{eq}^4 \qquad (5\text{-}5)$$

EXAMPLE 5-3 A selective surface having spectral characteristics as given in the previous example is to be used as a solar-energy absorber. The surface is to be maintained at a temperature of $T_A = 393$ K by extracting energy for use in a power-generating cycle. If the absorber is placed in orbit around the sun at the same radius as the earth, how much net energy will a square meter of the surface provide? How does this energy compare with that provided by a black surface at the same temperature? Reflected solar energy and emitted energy by earth are neglected.

The net energy extracted from the surface is the difference between that absorbed and that emitted. The absorbed energy flux is, as in Example 5-2,

$$q_a = \{0.95 \, F_{0-\lambda_c}(T_R) + 0.05[1 - F_{0-\lambda_c}(T_R)]\} q_i$$

$$= [0.95(0.880) + 0.05(1 - 0.880)] \, 1353 = 1139 \text{ W/m}^2$$

The emitted flux is

$$q_e = \{0.95 \, F_{0-\lambda_c}(T_A) + 0.05[1 - F_{0-\lambda_c}(T_A)]\} \sigma T_A^4$$

$$= \{0.95 \times (\sim 0) + 0.05[1 - (\sim 0)]\} 5.6705 \times 10^{-8} \times (393)^4$$

$$= 67.63 \text{ W/m}^2$$

and the net energy that can be used for power generation is $1139 - 68 = 1071$ W/m². For a blackbody or gray body, the equilibrium temperature was found in Example 5-1 as 393 K, so that the net useful energy that could be removed from such a surface would be zero.

Spectrally selective surfaces can also be employed to advantage in applications where it is desirable to cool an object exposed to incident radiation from a high-temperature source. The most common situations are objects exposed to the sun, such as a gasoline storage tank, a cryogenic fuel tank in space, or the roof of a building. A highly reflecting coating such as a polished metal could be utilized. This would reflect much of the incident energy, but would be poor for radiating away energy that was absorbed or generated within the enclosure such as by electronic equipment. Also, some metals have a tendency toward lower reflectivity at the shorter wavelengths characteristic of solar energy; this is shown, for example, for uncoated aluminum in Fig. 5-28. For some applications it may be advantageous to use a material that is spectrally selective; white paint as shown in Fig. 5-31 is an example. This not only reflects the incident radiation predominant at short wavelengths but also radiates well at the longer wavelengths characteristic of the relatively low temperature of the body. The optical solar reflector (OSR) is a mirror composed of a glass layer silvered on the back side. The glass, being transparent in the short-wavelength region, $\lambda < {\sim}2.5~\mu m$, which includes the visible range, lets the silver reflect incident radiation in this spectral region. The small fraction of short-wavelength energy that is absorbed by the silver and the energy absorbed by the glass at longer wavelengths are radiated away by the glass in the longer-wavelength infrared region where glass emits well. Commonly used thin plastic sheets for solar reflection are Kapton, Mylar, and Teflon with silver or aluminum coated on the back side.

Radiative dissipation is vital in outer space applications as there are no other means for elimination of waste energy except for dissipating small quantities by using expendable coolants. For a device on the ground or in the earth's atmosphere, convection and conduction to the surrounding environment are available. The significance of each of the three heat transfer modes depends on the particular conditions in the heat balance. An interesting example of such a heat balance is discussed in [46] to show that radiative cooling might be useful for cooling buildings to help with air conditioning. Objects exposed to the night sky can cool by radiation below the ambient air temperature. This cooling effect can also be utilized during the day if the solar reflectivity of a surface is high (greater than about 0.95) and its emissivity is large in the infrared. For this purpose titanium dioxide white paint was found somewhat superior, as an external solar-selective coating,

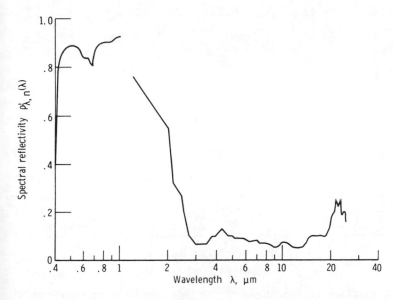

Figure 5-31 Reflectivity of white paint coating on aluminum. Replotted from [69].

to polyvinyl fluoride film with an aluminized coating 12 μm thick on the underside. These are the same types of materials, as well as many others, that are used for spacecraft thermal control (properties are given in [8]).

5-5.2 Selective Transmission through Glass and Water

The previous section discussed wavelength-selective opaque surfaces. It is also important to point out the selective transmission and absorption behavior of two common nonopaque materials, glass and water. Figure 5-32 shows the overall spectral transmittance of glass plates for normally incident radiation. As shown in Sec. 18-3.1 the overall transmittance includes the effect of absorption within the glass and multiple surface reflections; it is given by Eq. (18-2) as $T_\lambda = \tau_\lambda (1 - \rho_\lambda)^2/(1 - \rho_\lambda^2 \tau_\lambda^2)$ where $\tau_\lambda = \exp(-a_\lambda d)$ and $\rho_\lambda = [(n - 1)/(n + 1)]^2$. For small $a_\lambda d$ this reduces to $T_\lambda = (1 - \rho_\lambda)/(1 + \rho_\lambda)$. Typically for glass $n \approx 1.5$ so $\rho_\lambda = (0.5/2.5)^2 = 0.04$. Then including only reflection losses gives $T_\lambda = (1 - 0.04)/(1 + 0.04) = 0.92$. In Fig. 5-32 the fused silica has very low absorption in the range $\lambda = 0.2$ to 2 μm, and $T_\lambda \approx 0.9$ in this region as a result of surface reflections. Ordinary glasses typically have two strong cutoff wavelengths beyond which the glass becomes highly absorbing and T_λ decreases rapidly to near zero except for very thin plates. The measured curve for fused silica in Fig.

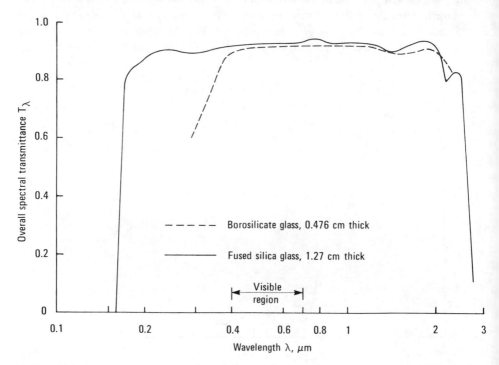

Figure 5-32 Normal overall spectral transmittance of glass plate (includes surface reflections) at 298 K. Replotted from [7].

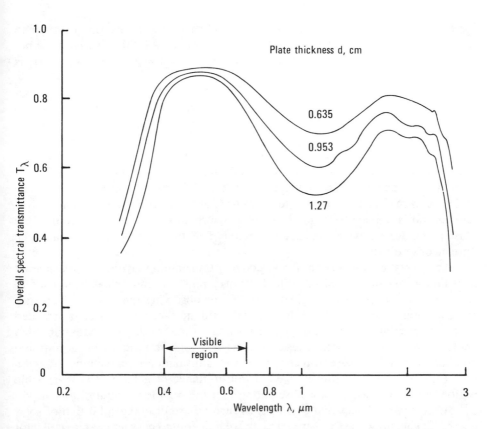

Figure 5 33 Effect of plate thickness on normal overall spectral transmittance of soda-lime glass (includes surface reflections) at 298 K. Replotted from [7].

5-32 shows this clearly. There are strong cutoffs in the far ultraviolet at $\lambda \approx 0.17$ μm and in the near infrared at $\lambda \approx 2.5$ μm. The glass will therefore be a strong absorber or emitter for $\lambda < 0.17$ μm and $\lambda > 2.5$ μm. Figure 5-33 shows the overall transmittance for various thicknesses of soda-lime glass, which is more absorbing than fused silica. The effect of absorption is illustrated quite well as the thickness increases. Typical optical constants for glass are given in [47].

For windows in high-temperature devices, such as furnaces or solar-cavity receivers, emission from the windows can be significant. From Kirchhoff's law the overall spectral emittance, including the effect of surface reflections, is equal to the spectral absorptance. Hence, from (18-3) for an isothermal window, $E_\lambda = (1 - \rho_\lambda)(1 - \tau_\lambda)/(1 - \rho_\lambda\tau_\lambda)$. For a thick plate, beyond the cutoff wavelength, $\tau_\lambda \to 0$ and $E_\lambda \approx 1 - \rho_\lambda$. In this instance reflection from only one surface is significant as all the radiation is absorbed before it can be transmitted through to the second surface of the plate. If the n for glass is 1.5, then $\rho_\lambda = 0.04$ for incidence from the normal direction; hence, $E_\lambda = 0.96$ in the normal direction for the highly absorbing wavelength regions of the glass. In a fashion similar to Example 4-3, the hemispherical value is found as $E_\lambda = 0.90$. This is the upper value in Fig. 5-34, which shows the emittance of window-glass sheets of various thicknesses.

The transmission behavior shown in Figs. 5-32 and 5-33 provides glass windows with the important ability to trap solar energy. The sun radiates a spectral energy distribution very much like a blackbody at 5780 K (10,400°R). Considering the range $0.3 < \lambda < 2.7$ μm as being between the cutoff wavelengths in Fig. 5-33, the blackbody tables in Appendix A show that 95% of the solar energy will be in this range. This means that glass has a low absorptance for solar radiation; consequently solar radiation passes readily through a glass window. The emission from objects at ambient temperature inside the enclosure is at long wavelengths and is trapped because of the high absorptance (poor transmission) of the glass in this long-wavelength spectral region. This trapping behavior is called the "greenhouse effect."

An interesting application of a selectively transmitting layer is the transparent heat mirror. As mentioned in [48], the transparent heat mirror has been used to construct transparent metallurgical furnaces for observing the growth of crystals at temperatures up to 1300 K. The thermal insulation for this furnace is provided by a gold film about 0.02 μm thick deposited on the inside of a Pyrex tube that encloses the heated region. These films have a high reflectance in the infrared and are equivalent to several inches of asbestos insulation in preventing radiation heat loss. The films, however, have a transmittance of about 0.2 in the visible region, which is adequate for observation into the high-temperature furnace.

A selectively transparent coating may also be useful in the collection of solar energy. It can allow the short-wavelength solar energy to pass into a solar collector and prevent the escape of long-wavelength radiation re-emitted by the receiver. Some possible coating materials are indium trioxide (In_2O_3), magnesium oxide (MgO), tin dioxide (SnO_2), and zinc oxide. Thin films of these materials are

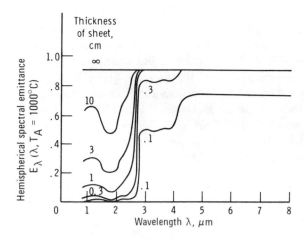

Figure 5-34 Emittance of sheets of window glass at 1000° C. Data from [70].

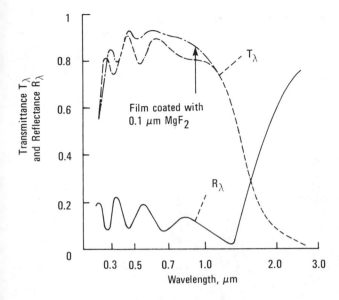

Figure 5-35 Transmittance and reflectance of 0.35-μm-thick film of Sn-doped In_2O_3 film on Corning 7059 glass. Also shown is the effect on T_λ of an antireflection coating of MgF_2. From [48].

transparent in the solar part of the spectrum and have a rapid increase of reflectance in the infrared. The measured transmittance and reflectance of a 0.35-μm-thick layer of Sn-doped In_2O_3 deposited on Corning 7059 glass is in Fig. 5-35. The application to a solar-energy receiver is discussed in [49].

Water is also a selective absorber for the transmission of radiation. Table 5-2 gives the spectral absorption coefficient of water as taken from [50] (see [51–53] for additional information). The absorption coefficient determines the exponential attenuation of the intensity as given by Eq. (12-21). In the visible region ($\lambda = 0.4$ to 0.7 μm) the absorption coefficient is quite small. In the vicinity of 1 μm, a_λ begins to increase, and at longer wavelengths in the near-infrared region the absorption becomes quite large. In the visible region a_λ is especially low in the blue-green region (0.5 to 0.55 μm). This accounts for the green appearance of sunlight penetrating underwater to large depths.

Application of the values of the absorption coefficient in Table 5-2 to the solar spectrum results in the energy penetrations given in Table 5-3 (from [54]). The second column shows the portions of the solar energy spectrum that are in various wavelength intervals. The successive columns demonstrate the high transparency for visible radiation as compared with very strong energy absorption in most of the near-infrared region. A similar table is given in [53] for the capture of solar radiation by shallow bodies of water ("solar ponds").

Ice also has a low absorption coefficient in the visible range, and the absorption coefficient increases by a factor of the order of 10^3 as the radiation wavelength increases from about 0.55 to 1.2 μm. Radiation in the visible and near-

Table 5-2 Absorption coefficient of water (from [50])

λ, μm	a_λ, cm^{-1}	λ, μm	a_λ, cm^{-1}
0.20	0.0691	2.4	50.1
0.25	0.0168	2.6	153
0.30	0.0067	2.8	5160
0.35	0.0023	3.0	11,400
0.40	0.00058	3.2	3630
0.45	0.00029	3.4	721
0.50	0.00025	3.6	180
0.55	0.000045	3.8	112
0.60	0.0023	4.0	145
0.65	0.0032	4.2	206
0.70	0.0060	4.4	294
0.75	0.0261	4.6	402
0.80	0.0196	4.8	393
0.85	0.0433	5.0	312
0.90	0.0679	5.5	265
0.95	0.388	6.0	2240
1.0	0.363	6.5	758
1.2	1.04	7.0	574
1.4	12.4	7.5	546
1.6	6.72	8.0	539
1.8	8.03	8.5	543
2.0	69.1	9.0	557
2.2	16.5	9.5	587
		10.0	638

visible range can therefore be passed through ice that is not cloudy due to impurities or air bubbles. If an ice layer has adhered to a surface, it may be possible to pass visible and near-infrared radiation through the ice, thereby heating the surface and providing a means for ice removal [55]. Information on melting of semitransparent materials is given in [56].

Table 5-3 Fractions of solar radiation spectrum transmitted through various thicknesses of water [54]

Spectral interval λ, μm	Incident solar-energy distribution	Transmitted energy distribution for water-layer thickness, cm							
		0.001	0.01	0.1	1	10	100	1000	10000
0.3–0.6	0.237	0.237	0.237	0.237	0.237	0.236	0.229	0.173	0.014
0.6–0.9	0.360	0.360	0.360	0.359	0.353	0.305	0.129	0.010	
0.9–1.2	0.179	0.179	0.178	0.172	0.123	0.008			
1.2–1.5	0.087	0.086	0.082	0.063	0.017				
1.5–1.8	0.080	0.078	0.064	0.027					
1.8–2.1	0.025	0.023	0.011						
2.1–2.4	0.025	0.025	0.019	0.001					
2.4–2.7	0.007	0.006	0.002						
Totals	1.000	0.994	0.953	0.859	0.730	0.549	0.358	0.183	0.014

5-5.3 Modification of Surface Directional Characteristics

As discussed previously, surface roughness can have profound effects on radiative properties and can become a controlling factor when roughness is large compared with the wavelength of the energy being considered. This leads to the concept of controlling roughness to tailor the directional characteristics of a surface.

A surface used as an emitter might be roughened or designed to emit strongly in preferred directions, while reducing emission in unwanted directions. Commercial radiant area-heating equipment would operate more efficiently by using such surfaces to direct energy where it is most needed. The most common device for controlling the directional distribution of electromagnetic radiation in the visible region is called a *lamp shade*.

If the directional surface were used primarily as an absorber, then, using a solar absorber as an example, we would make it strongly absorbing in the direction of incident solar radiation but as close as possible to nonabsorbing in other directions. The surface would, because of Kirchhoff's law for directional properties, emit strongly toward the sun but weakly in other directions. The surface would absorb the same energy as a nondirectional absorber since the incident energy is only from the direction of the sun, but would emit less than a surface that emits well in all directions.

The characteristics of one such surface are in Fig. 5-36. The surface has long grooves of angle 18.2° running parallel to each other. For each groove, a highly reflecting specular coating is on the side walls, and a black surface is at the base. The solid line gives the behavior predicted by analysis of such an ideal surface, while the data points show experimental results at 8 μm for an actual surface. It is seen that the directional emissivity is very high for angles of emission less than about 30° from the surface normal. It then drops rapidly as the angle becomes larger. Many other such surface configurations exhibit similar characteristics. Results for various vee, rectangular, and other grooves are in [28].

EXAMPLE 5-4 Suppose that a directional surface has a directional total emissivity given for all φ by

$$\epsilon(\theta) = 1 \quad 0 \leqslant \theta \leqslant 30° \qquad \epsilon(\theta) = 0 \quad \theta > 30°$$

For solar radiation incident normally on such a surface in earth orbit with no other heat exchange except by radiation from the directional surface, what is the equilibrium temperature of the surface? How does this temperature compare with that achieved by a black surface?

The absorptivity of this surface for normal incident radiation is unity. Therefore, the absorbed energy per unit time is $Q_a = (1)q_i A$ where $q_i = 1353$ W/m^2 is the solar constant for an object at a distance from the sun equal to the mean radius of the earth's orbit.

The energy emitted by the body when it is at thermal equilibrium is $Q_e = \epsilon \sigma T_{eq}^4 A$, where ϵ is the hemispherical total emissivity given by Eq. (3-6b) as

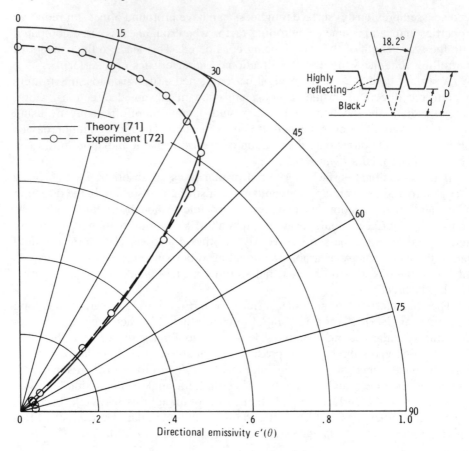

Angle of emission θ, deg

Directional emissivity $\epsilon'(\theta)$

Theory [71]
Experiment [72]

Figure 5-36 Directional emissivity of grooved surface with highly reflecting specular side walls and highly absorbing base; $d/D = 0.649$. Results in plane perpendicular to groove direction; data at 8 μm.

$$\epsilon(T_{eq}) = \frac{1}{\pi} \int_{\cap} \epsilon'(\theta, \varphi, T_{eq}) \cos \theta \, d\omega$$

For this problem ϵ becomes

$$\epsilon = \frac{2\pi}{\pi} \int_{\theta=0}^{30^\circ} \sin \theta \cos \theta \, d\theta = 0.25$$

Equating Q_a and Q_e for radiative thermal equilibrium gives

$$T_{eq} = \left(\frac{q_i}{\epsilon\sigma}\right)^{1/4} = \left(\frac{1353}{0.25 \times 5.6705 \times 10^{-8}}\right)^{1/4} = 556 \text{ K } (1000^\circ \text{ R})$$

This is larger than the equilibrium temperature of a black or diffuse gray body of 393 K (707° R) as shown in Example 5-1.

Note that Eq. (5-3) can be used for the α'_n/ϵ of directional as well as spectrally selective surfaces. For the surface in this example, $\alpha'_n/\epsilon = 4.0$. Combining selective and directional effects would be a way of obtaining considerably increased values of α'_n/ϵ for a given surface.

It should not be inferred that the directional distribution of emissivity assumed in this example corresponds to that of the parallel grooved surface in Fig. 5-36. In the case of Fig. 5-36, there is a strong dependence on the angle φ, which has been ignored in this example. The situation here is somewhat like the surface covered with indented spherical caps in [28].

5-6 CONCLUDING REMARKS

The radiative-property examples in this chapter have illustrated a number of features that may be encountered when dealing with real surfaces. Certain broad generalizations could be attempted. For example, the total emissivities of dielectrics at moderate temperatures are larger than those for metals, and the spectral emissivity of metals increases with temperature over a broad range of wavelengths. However, these types of rules can be misleading because of the large property variations that can occur as a result of surface roughness, contamination, oxide coating, grain structure, and so forth. The presently available analytical procedures cannot account for all these factors, so it is usually not possible to predict radiative property values except for surfaces that approach ideal conditions of composition and finish. By coupling analytical trends with observations of experimental trends, it is possible to gain insight into what classes of surfaces would be expected to be suitable for specific applications and how surfaces may be fabricated to obtain certain types of radiative behavior. The latter include spectrally selective surfaces that are of great value in practical applications such as collection of solar energy and spacecraft temperature control. With regard to temperature control, Furukawa [56a] considers recent information for solar and earth radiation fluxes on area elements in orbit.

Some other factors affecting radiative properties are outside the scope of this work. It is well known that exposure to ultraviolet radiation, cosmic rays, neutron, gamma, and proton bombardment, atomic oxygen, and the solar wind can cause significant changes in radiative properties. For the design of spacecraft, these effects are of major concern and are under investigation by satellite experiments [57].

Finally, some comment on the measurement of radiative properties should be given. We note that relatively few detailed measurements of directional spectral properties have been made. The reason for this lies in one of the many practical difficulties involved: For a directional measurement, the energy available for detection at a small solid angle centered about a given direction is itself small. If only the portion of this small energy that lies within a wavelength band is measured to obtain directional spectral values, an even smaller energy is available for detection. Minor absolute errors in the measurement of the energy can then lead to large percentage errors in the directional properties being determined. Further, the sheer magnitude of the amount of data generated for such combined directional spectral properties precludes its gathering unless a very specific problem requires it. These and similar practical problems make the field of thermal-radiation property measurement a most exacting and difficult one.

REFERENCES

1. Gubareff, G. G., J. E. Janssen, and R. H. Torberg: "Thermal Radiation Properties Survey," 2d ed., Honeywell Research Center, Minneapolis, 1960.
2. Svet, Darii Ia.: "Thermal Radiation; Metals, Semiconductors, Ceramics, Partly Transparent Bodies, and Films," Consultants Bureau, Plenum Publishing, New York, 1965.
3. Goldsmith, Alexander, and Thomas E. Waterman: Thermophysical Properties of Solid Materials, WADC TR 58-476, Armour Research Foundation, January 1959.
4. Hottel, H. C.: Radiant Heat Transmission, in William H. McAdams (ed.), "Heat Transmission," 3d ed., pp. 472–479, McGraw-Hill, New York, 1954.
5. Wood, W. D., H. W. Deem, and C. F. Lucks: "Thermal Radiative Properties," Plenum Press, New York, 1964.
6. Edwards, D. K., and Ivan Catton: Radiation Characteristics of Rough and Oxidized Metals, in Serge Gratch (ed.), "Adv. Thermophys. Properties Extreme Temp. Pressures," pp. 189–199, ASME, New York, 1964.
7. Touloukian, Y. S., et al.: *Thermal Radiative Properties*, vol. 7, "Metallic Elements and Alloys," vol. 8, "Nonmetallic Solids," vol. 9, "Coatings," Thermophysical Properties Research Center of Purdue University, Data Series, Plenum Publishing, New York, 1970.
8. Henninger, J. H.: "Solar Absorptance and Thermal Emittance of Some Common Spacecraft Thermal-Control Coatings," NASA Reference Publication 1121, 1984.
9. Liebert, Curt H.: Spectral Emittance of Aluminum Oxide and Zinc Oxide on Opaque Substrates, NASA TN D-3115, 1965.
10. Shafey, H. M., Y. Tsuboi, M. Fuijita, T. Makino, and T. Kunitomo: Experimental Study on Spectral Reflective Properties of a Painted Layer, *AIAA J.*, vol. 20, no. 12, pp. 1747–1753, 1982.
11. Grewal, N. S., and M. Kestoras: Total Normal Emittance of Dolomitic Limestone, *Int. J. Heat Mass Transfer*, vol. 31, no. 1, pp. 207–209, 1988.
12. Torrance, K. E., and E. M. Sparrow: Off-Specular Peaks in the Directional Distribution of Reflected Thermal Radiation, *J. Heat Transfer*, vol. 88, no. 2, pp. 223–230, 1966.
13. DeSilva, A. A., and B. W. Jones: Bidirectional Spectral Reflectance and Directional-Hemispherical Spectral Reflectance of Six Materials Used as Absorbers of Solar Energy, *Solar Energy Mater.*, vol. 15, pp. 391–401, 1987.
14. Cox, R. L.: Radiant Emission from Cavities in Scattering and Absorbing Media, *SMU Research Rep.* 68-2, Southern Methodist University Institute of Technology, Dallas, October 1968.
15. Sarofim, A. F., and H. C. Hottel: Radiative Exchange among Non-Lambert Surfaces, *J. Heat Transfer*, vol. 88, no. 1, pp. 37–44, 1966.
16. Sparrow, E. M., and S. L. Lin: Radiation Heat Transfer at a Surface Having Both Specular and Diffuse Reflectance Components, *Int. J. Heat Mass Transfer*, vol. 8, pp. 769–779, 1965.
17. Saari, J. M., and R. W. Shorthill: Review of Lunar Infrared Observations, in S. F. Singer (ed.), "Physics of the Moon," vol. 13, AAS Science and Technology Series, 1967.
18. Harrison, James K.: Non-Diffuse Infrared Emission from the Lunar Surface, *Int. J. Heat Mass Transfer*, vol. 12, pp. 689–697, 1969.

19. Birkebak, Richard C.: Thermophysical Properties of Lunar Materials: Part I, Thermal Radiation Properties of Lunar Materials from the Apollo Missions, in J. P. Hartnett and T. F. Irvine, Jr. (eds.), "Advances in Heat Transfer," vol. 10, Academic Press, New York, 1974.
20. Birkebak, Richard C.: Spectral Emittance of Apollo-12 Lunar Fines, *J. Heat Transfer*, vol. 94, no. 3, pp. 323–324, 1972.
21. Drolen, B. L.: Bidirectional Reflectance and Surface Specularity Results for a Variety of Spacecraft Thermal Control Surfaces, AIAA Paper 91-1326, June 1991.
22. Liebert, C. H., and R. D. Thomas: Spectral Emissivity of Highly Doped Silicon, in G. B. Heller (ed.), *Thermophysics of Spacecraft and Planetary Bodies*, vol. 20, AIAA Progress in Astronautics and Aeronautics, pp. 17–40, Academic Press, New York, 1967. (Also, NASA TN D-4303, April 1968.)
23. Phelan, P. E., M. I. Flik, and C. L. Tien: Radiative Properties of Superconducting Y-Ba-Cu-O Thin Films, *J. Heat Transfer*, vol. 113, no. 2, pp. 487–493, 1991; also, Phelan, P. E., G. Chen, and C. L. Tien: Thickness-Dependent Radiative Properties of Y-Ba-Cu-O Thin Films, *J. Heat Transfer,* vol. 114, no. 1, pp. 227–233, 1992.
24. Mattis, D. C., and J. Bardeen: Theory of the Anomalous Skin Effect in Normal and Superconducting Metals, *Physical Review*, vol. 111, pp. 412–417, 1958.
25. Sadykov, B. S.: Temperature Dependence of the Radiating Power of Metals, *High Temp.*, vol. 3, no. 3, pp. 352–356, 1965.
26. Toscano, W. M., and E. G. Cravalho: Thermal Radiative Properties of the Noble Metals at Cryogenic Temperatures, *J. Heat Transfer*, vol. 98, no. 3, pp. 438–445, 1976.
27. Tien, C. L., and G. R. Cunnington: Cryogenic Insulation Heat Transfer, in T. F. Irvine, Jr., and J. P. Hartnett (eds.), *Advances in Heat Transfer*, vol. 9, pp. 349–417, Academic Press, New York, 1973.
28. Demont, P., M. Huetz-Aubert, and H. Tran N'Guyen: Experimental and Theoretical Studies of the Influence of Surface Conditions on Radiative Properties of Opaque Materials, *Int. J. Thermophys.*, vol. 3, no. 4, pp. 335–364, 1982.
29. Davies, H.: The Reflection of Electromagnetic Waves from a Rough Surface, *Proc. Inst. Elec. Eng. London*, vol. 101, pp. 209–214, 1954.
30. Porteus, J. O.: Relation between the Height Distribution of a Rough Surface and the Reflectance at Normal Incidence, *J. Opt. Soc. Am.*, vol. 53, no. 12, pp. 1394–1402, 1963.
31. Beckmann, P., and A. Spizzichino: "The Scattering of Electromagnetic Waves from Rough Surfaces," Macmillan, New York, 1963.
32. Houchens, A. F., and R. G. Hering: Bidirectional Reflectance of Rough Metal Surfaces, *Progr. Astronautics and Aeronautics: Thermophys. Spacecraft Planetary Bodies*, vol. 20, pp. 65–89, 1967.
33. Smith, T. F., and R. G. Hering: Comparison of Bidirectional Measurements and Model for Rough Metallic Surfaces, *Fifth Symp. Thermophys. Properties, ASME*, Boston, 1970.
34. Torrance, K. E., and E. M. Sparrow: Theory for Off-Specular Reflection from Roughened Surfaces, *J. Opt. Soc. Am.*, vol. 57, no. 9, pp. 1105–1114, 1967.
35. Ody-Sacadura, J. F.: Influence de la rugosité sur le rayonnement thermique émis par les surfaces opaques: Essai de modèle (Influence of Surface Roughness on the Radiative Heat Emitted by Opaque Surfaces: A Test Model), *Int. J. Heat Mass Transfer*, vol. 15, no. 8, pp. 1451–1465, 1972.
36. Kanayama, K.: Apparent Directional Emittance of V-Groove and Circular-Groove Rough Surfaces, *Heat Transfer Jpn. Res.*, vol. 1, no. 1, pp. 11–22, 1972.
37. Birkebak, R. C., and A. Abdulkadir: Random Rough Surface Model for Spectral Directional Emittance of Rough Metal Surfaces, *Int. J. Heat Mass Transfer*, vol. 19, no. 9, pp. 1039–1043, 1976.
38. Abdulkadir, A., and R. C. Birkebak: Spectral Directional Emittance of Rough Metal Surfaces: Comparison Between Semi-Random and Pyramidal Surface Approximations. AIAA paper 78-848, *2d AIAA/ASME Thermophys. Heat Transfer Conf.*, Palo Alto, 1978.
39. Wolff, L. B., and D. J. Kurlander: Ray Tracing with Polarization Parameters, *IEEE Computer Graphics and Applications*, vol. 10, no. 6, pp. 44–55, November 1990.

40. Brannon, R. R., Jr., and R. J. Goldstein: Emittance of Oxide Layers on a Metal Substrate, *J. Heat Transfer*, vol. 92, no. 2, pp. 257–263, 1970.

40*a*. Makino, T., O. Sotokawa, and Y. Iwata: Transient Behaviors in Thermal Radiation Characteristics of Heat-Resisting Metals and Alloys in Oxidation Processes, *Int. J. Thermophysics*, vol. 9, no. 6, pp. 1121–1130, 1988.

41. Hattori, N., H. Takasu, and T. Iguchi: Emissivity of Liquid Sodium, *Heat Transfer Jpn. Res.*, vol. 13, no. 1, pp. 30–40, 1984.

42. Moscowitz, C. M., L. A. Stretz, and R. G. Bautista: The Spectral Emissivities of Lanthanum, Cerium, and Praseodymium, *High Temp. Sci.*, vol. 4, no. 5, pp. 372–378, 1972.

43. Bonnell, D. W., J. A. Treverton, A. J. Valerga, and J. L. Margrave: The Emissivities of Liquid Metals at Their Fusion Temperatures, *Fifth Symp. Temperature, June 1971*. In "Temperature, Its Measurement and Control in Science and Industry," vol. 4, pp. 483–487, American Institute of Physics, 1972.

43*a*. Grigoropoulos, C. P., W. E. Dutcher, Jr., and K. E. Barclay: Radiative Phenomena in CW Laser Annealing, *J. Heat Transfer*, vol. 113, no. 3, pp. 657–662, 1991.

44. Moore, S. W.: Solar Absorber Selective Paint Research, *Solar Energy Mater.*, vol. 12, pp. 435–447, 1985; Progress on Solar Absorber Selective Paint Research, *Solar Energy Mater.*, vol. 12, pp. 449–460, 1985.

45. Craighead, H. G., et al.: Metal/Insulator Composite Selective Absorbers, *Solar Energy Mater.*, vol. 1, nos. 1 and 2, pp. 105–124, 1979.

46. Berdahl, P., M. Martin, and F. Sakkal: Thermal Performance of Radiative Cooling Panels, *Int. J. Heat Mass Transfer*, vol. 26, no. 6, pp. 871–880, 1983.

47. Hsieh, C. K., and K. C. Su: Thermal Radiative Properties of Glass From 0.32 to 206 μm, *Sol. Energy*, vol. 22, no. 1, pp. 37–43, 1979.

48. Fan, John C. C., and Frank J. Bachner: Transparent Heat Mirrors for Solar-Energy Applications, *Appl. Opt.*, vol. 15, no. 4, pp. 1012–1017, 1976.

49. Jarvinen, P. O.: Heat Mirrored Solar Energy Receivers, paper no. 77-728, *AIAA 12th Thermosphys. Conf.*, Albuquerque, 1977.

50. Hale, George M., and Marvin R. Querry: Optical Constants of Water in the 200-nm to 200-μm Wavelength Region, *Appl. Opt.*, vol. 12, no. 3, pp. 555–563, 1973.

51. Irvine, William M., and James B. Pollack: Infrared Optical Properties of Water and Ice Spheres, *Icarus*, vol. 8, pp. 324–360, 1968.

52. Pinkley, L. W., P. P. Sethna, and D. Williams: Optical Constants of Water in the Infrared: Influence of Temperature, *J. Opt. Soc. Am.*, vol. 67, no. 4, pp. 494–499, 1977.

53. Rabl, A., and C. E. Nielsen: Solar Ponds for Space Heating, *Solar Energy*, vol. 17, pp. 1–12, 1975.

54. Kondratyev, Ya K.: "Radiation in the Atmosphere," Academic Press, New York, 1969.

55. Seki, N., M. Sugawara, and S. Fukusako: Back-Melting of a Horizontal Cloudy Ice Layer with Radiative Heating, *J. Heat Transfer*, vol. 101, no. 1, pp. 90–95, 1979.

56. Diaz, L., and R. Viskanta: Experiments and Analysis on the Melting of a Semitransparent Material by Radiation, *Warme Stoffubertrag.*, vol. 20, pp. 311–321, 1986.

56*a*. Furukawa, M.: Practical Method for Calculating Radiation Incident upon a Panel in Orbit, *J. Thermophys. Heat Transfer*, vol. 6, no. 1, pp. 173–177, 1992.

57. Stevens, N. J.: Method for Estimating Atomic Oxygen Surface Erosion in Space Environments, *J. Spacecr. Rockets*, vol. 27, no. 1, pp. 93–95, 1990.

58. Ohlsen, P. E., and G. A. Etamad: Spectral and Total Radiation Data of Various Aircraft Materials, Rep. NA57-330, North American Aviation, July 23, 1957.

59. Munch, B.: "Directional Distribution in the Reflection of Heat Radiation and Its Effect in Heat Transfer," Ph.D. thesis, Swiss Technical College of Zurich, 1955.

60. Orlova, N. S.: Photometric Relief of the Lunar Surface, *Astron. Z.*, vol. 33, no. 1, pp. 93–100, 1956.

61. Seban, R. A.: Thermal Radiation Properties of Materials, pt. III, WADD TR-60-370, University of California, Berkeley, 1963.

62. De Vos, J. C.: A New Determination of the Emissivity of Tungsten Ribbon, *Physica*, vol. 20, pp. 690–714, 1954.

63. Birkebak, R. C., and E. R. G. Eckert: Effects of Roughness of Metal Surfaces on Angular Distribution of Monochromatic Reflected Radiation, *J. Heat Transfer*, vol. 87, no. 1, pp. 85–94, 1965.

64. Edwards, D. K., and N. Bayard de Volo: Useful Approximations for the Spectral and Total Emissivity of Smooth Bare Metals, in Serge Gratch (ed.), "Adv. Thermophys. Properties Extreme Temp. Pressures," pp. 174–188, ASME, New York, 1965.

65. Williams, D. A., T. A. Lappin, and J. A. Duffie: Selective Radiation Properties of Particulate Coatings, *J. Eng. Power*, vol. 85, no. 3, pp. 213–220, 1963.

66. Long, R. L.: A Review of Recent Air Force Research on Selective Solar Absorbers, *J. Eng. Power*, vol. 87, no. 3, pp. 277–280, 1965.

67. Hibbard, R. R.: Equilibrium Temperatures of Ideal Spectrally Selective Surfaces, *Sol. Energy*, vol. 5, no. 4, pp. 129–132, 1961.

68. Shaffer, L. H.: Wavelength-Dependent (Selective) Processes for the Utilization of Solar Energy, *J. Sol. Energy Sci. Eng.*, vol. 2, nos. 3 and 4, pp. 21–26, 1958.

69. Dunkle, R. V.: Thermal Radiation Characteristics of Surfaces, in J. A. Clark (ed.), "Theory and Fundamental Research in Heat Transfer," pp. 1–31, Pergamon Press, New York, 1963.

70. Gardon, Robert: The Emissivity of Transparent Materials, *J. Am. Ceram. Soc.*, vol. 39, no. 8, pp. 278–287, 1956.

71. Perlmutter, Morris, and John R. Howell: A Strongly Directional Emitting and Absorbing Surface, *J. Heat Transfer*, vol. 85, no. 3, pp. 282–283, 1963.

72. Brandenberg, W. M., and O. W. Clausen: The Directional Spectral Emittance of Surfaces between 200° and 600° C, in S. Katzoff (ed.), "Symp. Thermal Radiation Solids," NASA SP-55 (AFML-TDR-64-159), 1965.

PROBLEMS

5-1 The normal spectral absorptivity of an SiO-Al selective surface can be approximated as shown by the long-dashed line in Fig. 5-29. The surface receives a flux q from the normal direction. The equilibrium temperature of the surface is 1200 K. Assume $\epsilon_\lambda = \alpha'_\lambda(\theta - 0)$. What is the value of q if it comes from a gray body source at 3500 K?

Answer: 1.551×10^4 W/m^2

5-2 A directionally selective gray surface has properties as shown below. The α' is isotropic with respect to the azimuthal angle φ.

(a) What is the ratio $\alpha'(\theta = 0)/\epsilon$ (the directional absorptivity over the hemispherical emissivity) for this surface?

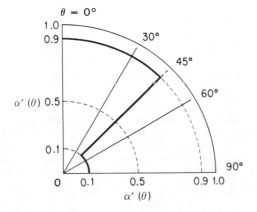

(b) If a thin plate with the above properties is in earth orbit around the sun with incident solar flux of 1350 W/m², what equilibrium temperature will it reach? Assume that the plate is oriented normal to the sun's rays and is perfectly insulated on the side away from the sun.

(c) What is the equilibrium temperature if the plate is oriented at 60° to the sun's rays?

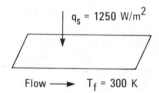

(d) What is the equilibrium temperature if the plate is normal to the sun's rays but is not insulated? Assume the plate is very thin and has the same directional properties on both sides. Neglect radiation emitted by or reflected from the earth.

Answer: (a) 1.8; (b) 455 K; (c) 221 K; (d) 383 K

5-3 A gray surface has directional total absorptivity given by $\alpha' = 0.75 \cos^3 \theta$. This flat surface is exposed to normally incident sunlight of flux 1250 W/m². A fluid flows past the back of the thin collector at $T_{\text{fluid}} = 300$ K at a velocity that gives a heat transfer coefficient of $h = 60$ W/(m²·K). What is the equilibrium temperature of this collector?

$$q_s = 1250 \text{ W/m}^2$$

Flow ⟶ $T_f = 300$ K

Answer: 313 K

5-4 Using Fig. 5-28, estimate the ratio of normal total solar absorptivity to hemispherical total emissivity at a surface temperature of 1000 K for aluminum with a coating of 0.1-μm dendritic lead sulfide crystals. Assume the surface is diffuse. (The solar temperature can be taken as 5780 K.)

Answer: 2.262

5-5 A plate containing material combining directional and spectral selectivity has a normal solar absorptivity of 0.97 and an infrared hemispherical total emissivity of 0.02. When placed in sunlight normal to the sun's rays, what temperature will the plate reach (neglecting conduction and convection and with no heat losses from the unexposed side of the plate)? What assumptions did you make in reaching your answer? Take the incident solar flux ("insolation") as 1150 W/m².

Answer: 681 K

5-6 A water heater consists of a sheet of glass 1 cm thick over a black surface that is assumed to be in perfect contact with the water below it. Estimate the water temperature for normally incident solar radiation. (Assume that Fig. 5-34 can be used for the glass properties and that the glass is perfectly transparent for wavelengths shorter than those shown. Take into account approximately the reflections at the glass surfaces; this will be treated in more detail in Chap. 18.)

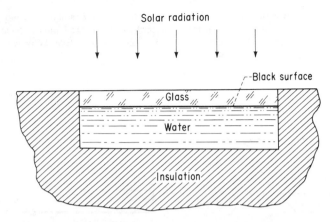

Answer: 389 K

5-7 A gasoline storage tank is in direct sun with the incident radiation normal to the top of the tank at 1150 W/m². The tank is painted with white paint having the reflectivity in Fig. 5-31.

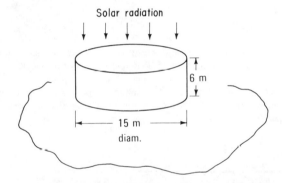

(a) Estimate the equilibrium temperature that the tank can achieve. (Neglect emitted and re-flected radiation from the ground. Do not account for free or forced convection to the air, although this will be appreciable. Take the ambient radiating temperature of the surround-ings to be 300 K.)

(b) What will the tank temperature be if the top is painted white as before but the sides have a gray coating with an emissivity of 0.9?

(c) What if the entire tank were painted with the gray coating?

Answer: (a) 311 K; (b) 311 K; (c) 355 K

5-8 A spinning sphere 0.30 m in diameter is in earth orbit and is receiving solar radiation. As a consequence of the rotation, the surface temperature of the sphere is assumed uniform. The sphere exterior is an SiO-Al selective surface as in Example 5-2, with a cutoff wavelength of 1.5 μm. The surface properties do not depend on angle. Neglect energy emitted by the earth and solar energy reflected from the earth.

(a) What is the equilibrium surface temperature for heat transfer only by radiation? How does the temperature depend on the sphere diameter?

(b) It is desired to maintain the sphere surface at 675 K. At what rate must energy be supplied to the entire sphere to accomplish this?

Answer: (a) 563 K; (b) 87.0 W

5-9 A flat-plate radiator in space in earth orbit is oriented normal to the solar radiation. It is receiving direct solar radiation, radiation emission from the earth, and solar radiation reflected from the earth. What must the radiator temperature be to dissipate a total of 1050 W of waste heat from both sides of each 1 m² of the radiator?

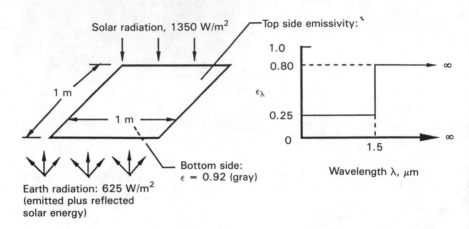

Answer: 380.8 K

5-10 A thin ceramic plate, insulated on one side, is radiating energy from its exposed side into vacuum at very low temperature. The plate is initially at 1550 K and is to cool to 450 K. At any instant, the plate is assumed to be at uniform temperature across its thickness and over its exposed area. The plate is 0.25 cm thick, and the surface hemispherical-spectral emissivity is as shown and is independent of temperature. What is the cooling time? The density of the ceramic is 3200 kg/m³, and its specific heat is 710 J/(kg · K).

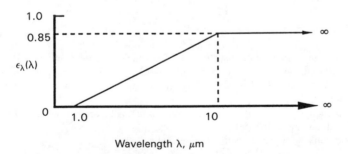

Answer: $\tau = 11.2$ min

5-11 A cylindrical concentrator (reflector) is long in the direction normal to the cross section shown, so that end effects may be neglected. The concentrator is gray and reflects 95% of the incident solar energy onto the central tube. The central tube receiving the energy is assumed to be at uniform temperature and is coated with a material that has the spectral properties shown. The properties are independent of direction. Emitted energy from the tube that is reflected by the concentrator may be neglected, as may emission from the concentrator. The surrounding environment is at low temperature.

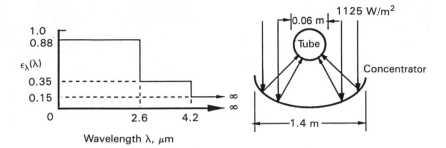

(a) If the heat exchange is only by radiation, compute the temperature of the central tube.
(b) If the tube is cooled to 550 K by passing a coolant through its interior, how much energy must be removed by the coolant per meter of tube length?

Answer: (a) 817.8 K; (b) 1117.2 W/meter of length

SIX

CONFIGURATION FACTORS FOR SURFACES TRANSFERRING UNIFORM DIFFUSE RADIATION

6-1 INTRODUCTION TO ENCLOSURE THEORY AND USE OF CONFIGURATION FACTORS

The determination of radiation interchange between surface elements is required in heat transfer, illumination engineering, and applied optics. Studies have been conducted for many years, as evidenced by the publication dates of [1, 2]. Since 1960 the study of radiant interchange has been given impetus by technological advances that provided systems in which thermal radiation is often very significant. Examples are satellite temperature control, energy leakage into cryogenic vacuum systems, high-temperature phenomena in hypersonic flight, heat transfer in nuclear propulsion systems, devices for collection and utilization of solar energy, and space station power systems.

The configuration factors to be derived here are an important aspect of radiation exchange. Before considering configuration factors, some introductory comments are made about "enclosure" theory. This will motivate the need for and use of the factors. In Chaps. 6–11 the theory will be developed for radiation exchange within enclosures that are *evacuated* or contain radiatively *nonparticipating* media. First it must be understood what is meant by an *enclosure*. Any surface can be considered as completely surrounded by an envelope of other solid surfaces or open areas. This envelope is the *enclosure* for the surface; thus, an enclosure accounts for *all directions surrounding a surface*. By considering the radiation going from the surface to all parts of the enclosure and the radiation arriving at the surface from all parts of the enclosure, it is assured that all radiative contributions are accounted for. A convenient enclosure is usually evident from the physical configuration. An open area can be considered as a plane of zero reflectivity. It will also act as a radiation source when radiation is entering the enclosure from the surrounding environment.

Chapters 1–5 discussed radiative properties of solid opaque surfaces. For some materials the properties vary substantially with wavelength, surface temperature, and direction. For radiation computations within enclosures, the geometric exchange effects are a complication in addition to the surface property variations. For simple geometries it may be possible to account in detail for property variations without the problem becoming complex. As the geometry becomes more involved, it is often necessary to invoke more idealizations of the surface properties so that the problem can be solved with reasonable effort. The treatment here could begin with the most general situation where properties vary with wavelength, temperature, and direction, and radiation fluxes vary arbitrarily over the enclosure surfaces. All other situations would then be simplified cases. However, this would plunge the uninitiated reader into the most complex treatment. Hence, the development will begin with simple situations; successive complexities will then be added to build more comprehensive treatments.

6-1.1 Enclosures with Diffuse Surfaces

In the simplest enclosure all surfaces are black and each is isothermal. For black surfaces there is no reflected radiation to be accounted for and all emitted energy is diffuse (the *intensity* leaving a surface is independent of direction). The local heat balance at a surface involves the enclosure geometry, which governs how much radiation leaving a surface will reach another surface. For a black enclosure the geometric effects are expressed in terms of the diffuse configuration factors obtained in this chapter. A diffuse factor is the *fraction of uniform diffuse radiation leaving a surface that directly reaches another surface*.

The computation of configuration factors involves integration over the solid angles by which surfaces can view each other. Because these integrations can be tedious and require numerical evaluation, it is desirable to use relations that exist between configuration factors. This may make it possible to obtain the desired factor from factors that are already known. These relations, along with various shortcut methods that can be used to obtain configuration factors, are presented in this chapter. Appendix B provides references where configuration factors can be found for approximately 230 different geometric configurations. This includes reference to computer programs for numerical evaluation of the factors required for complex radiative analyses. Appendix C provides a catalog of analytical expressions for some useful factors.

The next step in enclosure complexity is to have *gray* surfaces that emit and reflect *diffusely*. It is assumed that both emitted and reflected energies are *uniform* over each surface. The diffuse configuration factors then apply for radiation leaving a surface by both emission and reflection. For gray surfaces, reflections among surfaces must be accounted for; this is done in Chap. 7.

6-1.2 Enclosures with Nondiffuse and Nongray Surfaces

In some instances the approximations of black or diffuse-gray surfaces are in-adequate, and directional and/or spectral effects must be considered. The neces-sity of treating spectral effects was noticed quite early in the field of radiative transfer. In the remarkable paper [3] published in 1800 by Sir William Herschel entitled "Investigation of the Powers of the Prismatic Colours to Heat and Illu-minate Objects; with Remarks, that prove the Different Refrangibility of Radiant Heat, to which is added, an Inquiry into the Method of Viewing the Sun Advan-tageously, with Telescopes of Large Apertures and High Magnifying Powers" appears the following statement: "In a variety of experiments I have occasionally made, relating to the method of viewing the sun, with large telescopes, to the best advantage, I used various combinations of differently coloured darkening glasses. What appeared remarkable was, that when I used some of them, I felt a sensation of heat, though I had but little light; while others gave me much light, with scarce any sensation of heat. Now, as in these different combinations, the sun's image was also differently coloured, it occurred to me, that the prismatic rays might have the power of heating bodies very unequally distributed among them." This paper was the first in which what is now called the infrared region of the spectrum was defined and the energy radiated as "heat" shown to be of wavelengths different than those for "light." The quotation shows an awareness that in some instances spectral effects must be included in a radiative analysis. The performance of spectrally selective surfaces, such as those for satellite tem-perature control and solar collectors, can be understood only by considering wave-length variations of the surface properties. If the spectral surfaces are diffuse, the configuration factors developed here are applicable for enclosure analysis.

In some instances the surface properties have significant directional charac-teristics. In Chap. 5 a number of directionally dependent surface properties were examined, and some were found to differ considerably from the diffuse approx-imation. The diffuse configuration factors cannot be used when surfaces emit or reflect in a significantly nondiffuse manner. Other methods are required, and these will be given later.

A special directional surface treated later is a mirror reflector. *Emission* from this type of surface is often approximated as being *diffuse*; hence, the emitted energy is treated by using diffuse configuration factors. Reflected energy, how-ever, is followed within the enclosure by using the mirror characteristic that the angles of reflection and incidence are equal in magnitude.

6-1.3 Enclosures Involving Energy Transfer by Combined Modes

When conduction and/or convection is combined with radiative heat transfer at an opaque surface, the radiative interaction is usually considered to occur only at the surface. Thus, in the absence of significant radiation penetration below the surface, the radiation and convection combine into a boundary condition that governs heat conduction within the solid. When there are thermal transients, the radiative terms are applied at the opaque boundary at each instant when solving the energy equation for the transient temperature distribution within the body.

Heat conduction depends on local temperature derivatives, while convection depends on local differences between fluid and surface temperatures. Radiative exchange, however, depends approximately on differences of fourth powers of the surface temperatures and on the sum of the radiation incident from all the surroundings. The fourth-power terms cause the energy relations for a combined convection, conduction, and radiation problem to be a nonlinear system of equations. Numerical methods are usually required for multimode problems. When the enclosure surfaces can be considered to be diffuse and have uniformly distributed outgoing radiative energy fluxes, the configuration factors can be used for the radiative portion of the analysis. This provides a simplification for formulating the radiative contributions.

6-1.4 Notation

The notation is now briefly reviewed. A *prime* denotes a *directional* quantity, while a λ subscript specifies a *spectral* quantity; for example, ϵ_λ' is the directional-spectral emissivity. Quantities such as bidirectional reflectivities can depend on two directions, that is, for incoming and outgoing radiation. Bidirectional quantities are denoted by a double prime. A hemispherical quantity will not have a prime, and a total quantity will not have a λ subscript; thus, ϵ is the hemispherical-total emissivity. In addition, notation such as $\epsilon_\lambda'(\lambda, \theta, \varphi, T)$ can be used to emphasize the functional dependences or to state more specifically at what wavelength, angle, and surface temperature the quantity is being evaluated.

Additional notation is needed for the energy rate Q for a finite area to keep consistent mathematical forms for energy balances. The quantity d^2Q_λ' is directional-spectral, and the second order differential indicates that the energy is of differential order in both solid angle and wavelength. The quantities dQ' and dQ_λ are of differential order with respect to solid angle and to wavelength, respectively. If a differential area is involved, the order of the derivative is correspondingly increased.

6-2 SYMBOLS

A	area
e	emissive power
f	function defined by Eq. (6-53)
F	configuration factor
i	intensity
l, m, n	direction cosines, Eq. (6-52)
N	number of surfaces in an enclosure
P, Q, R	functions in contour integration used in Sec. 6-4.3

q	energy per unit time and per unit area
Q	energy per unit time
r	radius
S	distance between two differential elements
S, T, U, V	distances used in Eq. (6-61)
T	temperature
U	number of unknowns in equations describing an N-sided enclosure
v	coordinate used in Eq. (6-61)
x, y, z	cartesian coordinate positions
α, γ, δ	angles in direction cosines
β	angle in yz plane
θ	angle from normal
λ	wavelength
μ	angle used in Eq. (6-61)
σ	Stefan-Boltzmann constant
ω	solid angle

Subscripts

$d1, d2$	evaluated at differential element $d1$ or $d2$
i	inner
j, k	jth or kth surface
N	Nth surface
ring	ring area
s	sun
strip	elemental strip
λ	quantity or parameter is wavelength dependent
1, 2	at area 1 or 2

Superscript

denotes quantity is per unit solid angle in one direction

6-3 RADIATIVE GEOMETRIC CONFIGURATION FACTORS BETWEEN TWO SURFACES

One of the complexities in calculating radiative transfer between surfaces is the geometric relations for how the surfaces view each other. This results in integrations of the radiative interchange over the finite areas involved in the exchange process. In this section, a quantity is developed to account for the geometry; this is the *geometric configuration factor*. Such factors allow computation of radiative transfer in many systems by referring to formulas or tabulated values obtained for the geometric relations between any two surfaces. This simplifies a time-consuming and often error-prone portion of the analysis. As described in the introduction, these factors are for *uniform diffuse energy* leaving a surface.

6-3.1 Configuration Factor for Energy Exchange Between Diffuse Differential Elements

The radiative transfer from a diffuse differential element to another element is considered first, as it will be used to derive relations for transfer between finite areas. Consider the two differential area elements in Fig. 6-1. The elements dA_1 and dA_2 are at temperatures T_1 and T_2, are arbitrarily oriented, and have their normals at angles θ_1 and θ_2 to the line of length S joining them.

If i_1' is the total intensity leaving dA_1, the total energy per unit time leaving dA_1 and incident on dA_2 is

$$d^2Q_{d1-d2}' = i_1' \, dA_1 \cos\theta_1 \, d\omega_1 \tag{6-1}$$

where $d\omega_1$ is the solid angle subtended by dA_2 when viewed from dA_1. Equation (6-1) follows directly from the definition of i_1' as the total energy leaving surface 1 per unit time, per unit area projected normal to S, and per unit solid angle. Since i_1' is diffuse, it is independent of the angle at which it leaves the surface. It may consist of both diffusely emitted and diffusely reflected portions. The prime indicates a quantity in a single direction. The d^2Q' is a second differential to denote its dependence on two differential quantities, dA_1 and $d\omega_1$.

Equation (6-1) can also be written for radiation in a wavelength interval $d\lambda$,

$$d^3Q_{\lambda,d1-d2}' = i_{\lambda,1}'(\lambda) \, d\lambda \, dA_1 \cos\theta_1 \, d\omega_1$$

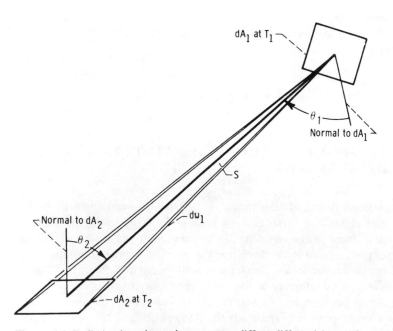

Figure 6-1 Radiative interchange between two diffuse differential area elements.

Total radiation quantities are then found by integrating over all wavelengths:

$$d^2Q'_{d1-d2} = \int_{\lambda=0}^{\infty} d^3Q'_{\lambda,d1-d2} = dA_1 \cos \theta_1 \, d\omega_1 \int_0^{\infty} i'_{\lambda,1}(\lambda) \, d\lambda$$

For a diffuse surface $i'_\lambda(\lambda)$ does not depend on direction; hence all the geometric factors can be removed from under the integral sign, and the integration over wavelength is independent of geometry. Thus the results that follow for diffuse geometric configuration factors involving finite areas apply for *both spectral* and *total* quantities. For simplicity in notation, the development is carried out for total quantities.

The solid angle $d\omega_1$ is related to the projected area of dA_2 and the distance between the differential elements by

$$d\omega_1 = \frac{dA_2 \cos \theta_2}{S^2} \tag{6-2}$$

Substituting this into Eq. (6-1) gives the following equation for the total energy per unit time leaving dA_1 that is incident upon dA_2:

$$d^2Q'_{d1-d2} = \frac{i'_1 \, dA_1 \cos \theta_1 \, dA_2 \cos \theta_2}{S^2} \tag{6-3}$$

An analogous derivation for the radiation leaving a diffuse dA_2 that arrives at dA_1 results in

$$d^2Q'_{d2-d1} = \frac{i'_2 \, dA_2 \cos \theta_2 \, dA_1 \cos \theta_1}{S^2} \tag{6-4}$$

Note that (6-3) and (6-4) both contain the same geometric factors.

For later use, d^2Q' has been defined in Eqs. (6-3) and (6-4) as the energy leaving one element that is *incident* upon the second element. For the special case of a black receiving element, all incident energy is absorbed so that in this case Eqs. (6-3) and (6-4) also give the energy from one element that is *absorbed* by the second.

The *fraction* of energy leaving diffuse surface element dA_1 that arrives at element dA_2 is defined as the *geometric configuration factor* dF_{d1-d2}. The dash in the subscript means "to." [Either total or spectral energy can be considered, as discussed with regard to Eq. (6-1), and the same results are obtained for dF. The total energy is used here for convenience in not carrying the λ notation] Using (6-3), the definition of dF gives

$$dF_{d1-d2} = \frac{d^2Q'_{d1-d2}}{\pi i'_1 \, dA_1} = \frac{i'_1 \cos \theta_1 \cos \theta_2 \, dA_1 \, dA_2/S^2}{\pi i'_1 \, dA_1} = \frac{\cos \theta_1 \cos \theta_2}{\pi S^2} \, dA_2 \tag{6-5}$$

where $\pi i'_1 \, dA_1$ is the total diffuse energy leaving dA_1 within the entire hemispherical solid angle over dA_1. Equation (6-5) shows that dF_{d1-d2} *depends only on the* *size of dA_2 and its orientation with respect to dA_1.* By substituting (6-2), Eq. (6-5) can also be written in the form

$$dF_{d1-d2} = \frac{\cos \theta_1 \, d\omega_1}{\pi} \tag{6-6}$$

Consequently, elements dA_2 have the same configuration factor if they subtend the same solid angle $d\omega_1$ when viewed from dA_1 and are positioned along a path at angle θ_1 with respect to the normal of dA_1.

The factor dF_{d1-d2} has various names, being called the view, angle, shape, interchange, exchange, or configuration factor. The last seems most specific, implying a dependence on both orientation and shape, the latter variable entering when finite areas are involved.

The notation used here for configuration factors is based on subscript designation for the types of areas involved in the energy exchange and a derivative notation consistent with the mathematical meaning of the configuration factor. For the subscript notation, $d1$, $d2$, etc. indicate differential area elements, while 1, 2, etc. indicate areas of finite size. Thus dF_{d1-d2} is a factor between two differential elements, as in Eq. (6-5). The dF_{1-d2} is a factor from finite area A_1 to differential area dA_2. The derivative notation dF indicates that the configuration factor is for energy transfer *to a differential* element, as in (6-5). This is redundant with the subscript notation but keeps the mathematical form of equations such as (6-5) consistent in that a differential quantity appears on both sides. The quantity F denotes a factor *to a finite* area. Thus F_{d1-2} is the configuration factor from differential element dA_1 to finite area A_2.

Reciprocity for differential-element configuration factors By a derivation similar to that for Eq. (6-5), the configuration factor for calculating energy from element dA_2 to dA_1 is

$$dF_{d2-d1} = \frac{\cos \theta_1 \cos \theta_2}{\pi S^2} \, dA_1 \tag{6-7}$$

Multiplying (6-5) by dA_1 and (6-7) by dA_2 gives the *reciprocity relation*

$$dF_{d1-d2} \, dA_1 = dF_{d2-d1} \, dA_2 = \frac{\cos \theta_1 \cos \theta_2}{\pi S^2} \, dA_1 \, dA_2 \tag{6-8}$$

Some sample configuration factors between differential elements The derivation of configuration factors in terms of the system geometry will now be illustrated by considering some sample cases.

EXAMPLE 6-1 The two elemental areas in Fig. 6-2 are located on strips that have parallel generating lines. Derive an expression for the configuration factor between dA_1 and dA_2.

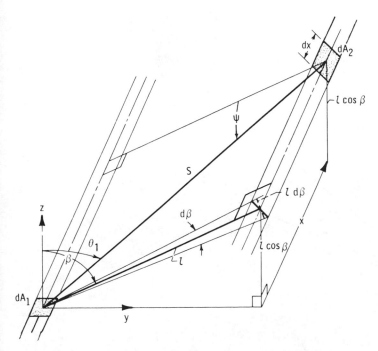

Figure 6-2 Geometry for configuration factor between elements on strips formed by parallel generating lines.

The yz plane is drawn normal to the generating lines. The distance S can be expressed as $S^2 = l^2 + x^2$, and $\cos \theta_1 = (l \cos \beta)/S = (l \cos \beta)/(l^2 + x^2)^{1/2}$. The angle β is in the cross section (yz plane) normal to the two strips. The solid angle subtended by dA_2, when viewed from dA_1, is

$$d\omega_1 = \frac{\text{projected area of } dA_2}{S^2} = \frac{(\text{projected width of } dA_2)(\text{projected length of } dA_2)}{S^2}$$

$$= \frac{(l\, d\beta)(dx \cos \psi)}{S^2} = \frac{l\, d\beta\, dx}{S^2} \frac{l}{S}$$

Substituting into (6-6) gives

$$dF_{a1\text{-}a2} = \frac{\cos \theta_1\, d\omega_1}{\pi} = \frac{l \cos \beta}{(l^2 + x^2)^{1/2}} \frac{1}{\pi} \frac{l^2\, d\beta\, dx}{(l^2 + x^2)^{3/2}} = \frac{l^3 \cos \beta\, d\beta\, dx}{\pi(l^2 + x^2)^2}$$

which is the desired configuration factor between dA_1 and dA_2.

EXAMPLE 6-2 Find the configuration factor between an elemental area and an infinitely long strip of differential width oriented as in Fig. 6-3, so that the generating lines of dA_1 and $dA_{\text{strip},2}$ are parallel.

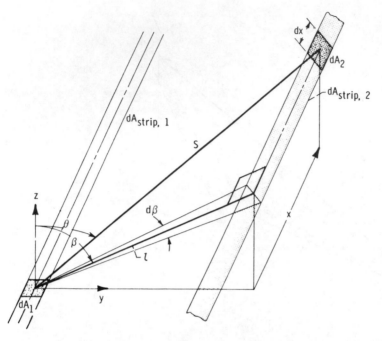

Figure 6-3 Geometry for configuration factor between elemental area and infinitely long strip of differential width; area and strip are on parallel generating lines.

Example 6-1 gave the configuration factor between differential element dA_1 and area element dA_2 of length dx as $dF_{d1-d2} = l^3 \cos \beta \, d\beta \, dx/\pi(l^2 + x^2)^2$. To find the factor when dA_2 becomes an infinite strip as in Fig. 6-3, integrate over all x to obtain

$$
\begin{aligned}
dF_{d1-\text{strip}, 2} &= \frac{l^3 \cos \beta \, d\beta}{\pi} \int_{-\infty}^{\infty} \frac{dx}{(l^2 + x^2)^2} \\
&= \frac{l^3 \cos \beta \, d\beta}{\pi} \left[\frac{x}{2l^2(l^2 + x^2)} + \frac{1}{2l^3} \tan^{-1} \frac{x}{l} \right]_{-\infty}^{\infty} \\
&= \frac{\cos \beta \, d\beta}{2} = \tfrac{1}{2} d(\sin \beta)
\end{aligned}
$$

where the angle β is in the yz plane, which is normal to the strips. This useful relation will be used in later examples.

Figure 6-3 also shows that, since dA_1 lies on an *infinite* strip $dA_{\text{strip},1}$ with elements parallel to $dA_{\text{strip},2}$, the configuration factor $dF_{d1-\text{strip},2} = \tfrac{1}{2}d(\sin \beta)$ will be valid for dA_1 regardless of where dA_1 is along $dA_{\text{strip},1}$. Then, since any element dA_1 on $dA_{\text{strip},1}$ has the same fraction of its energy reaching $dA_{\text{strip},2}$, it follows that the fraction of energy from the *entire* $dA_{\text{strip},1}$ that reaches $dA_{\text{strip},2}$ is the same as the fraction for each element dA_1. Thus, the configuration factor between *two infinitely long strips* of differential width and having parallel generating lines must also be the same as for element dA_1 to $dA_{\text{strip},2}$, or $\frac{1}{2}d(\sin \beta)$. The angle β is always in a plane normal to the generating lines of both strips.

EXAMPLE 6-3 Consider an infinitely long wedge-shaped groove as shown in cross section in Fig. 6-4. Determine the configuration factor between the differential strips dx and $d\xi$ in terms of x, ξ, and α.

As discussed in Example 6-1, the configuration factor is given by

$$dF_{dx-d\xi} = \tfrac{1}{2}d(\sin \beta) = \tfrac{1}{2} \cos \beta \, d\beta$$

From the construction in Fig. 6-4b, $\cos \beta = (\xi \sin \alpha)/L$. The quantity $d\beta$ is the angle subtended by the projection of $d\xi$ normal to L, that is,

$$d\beta = \frac{d\xi \cos (\alpha + \beta)}{L} = \frac{d\xi}{L} \frac{x \sin \alpha}{L}$$

From the law of cosines, $L^2 = x^2 + \xi^2 - 2x\xi \cos \alpha$. Then

$$dF_{dx-d\xi} = \frac{1}{2} \cos \beta \, d\beta = \frac{1}{2} \frac{x\xi \sin^2 \alpha}{L^3} \, d\xi = \frac{1}{2} \frac{x\xi \sin^2 \alpha}{(x^2 + \xi^2 - 2x\xi \cos \alpha)^{3/2}} \, d\xi$$

6-3.2 Configuration Factor between a Differential Element and a Finite Area

Consider now an isothermal diffuse element dA_1 with uniform emissivity at temperature T_1 exchanging energy with a finite area A_2 that is isothermal at temperature T_2 as shown in Fig. 6-5. The relations for exchange between two differential elements must be extended to permit A_2 to be finite. Figure 6-5 shows (compare the solid and dashed cases) that the angle θ_2 will be different for different positions

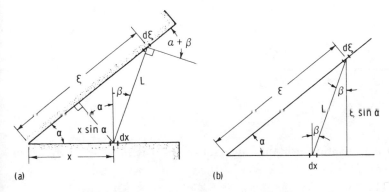

(a) (b)

Figure 6-4 Configuration factor between two strips on sides of wedge groove. (*a*) Wedge-shaped groove geometry; (*b*) auxiliary construction.

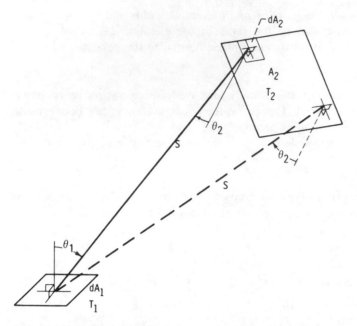

Figure 6-5 Radiant interchange between differential element and finite area.

on A_2 and that θ_1 and S will also vary as different differential elements on A_2 are viewed from dA_1.

There are two configuration factors to be considered. The factor F_{d1-2} is from the differential area dA_1 to the finite area A_2, and dF_{2-d1} is from A_2 to dA_1. Each of these will be considered by using as the definition of configuration factor, the fraction of energy leaving one diffusely emitting and reflecting surface that reaches the second surface. To derive F_{d1-2}, note that the total radiation leaving element dA_1 is $dQ_1 = \pi i_1' \, dA_1$. The energy reaching dA_2 on A_2 is $d^2Q_{d1-d2}' = \pi i_1' \, [(\cos \theta_1 \cos \theta_2)/\pi S^2] \, dA_1 \, dA_2$. Then, integrating over A_2 to obtain the energy reaching all of A_2 and dividing the total energy leaving dA_1 result in

$$F_{d1-2} = \frac{\int_{A_2} d^2Q_{d1-d2}'}{dQ_1} = \frac{\int_{A_2} i_1' \dfrac{\cos \theta_1 \cos \theta_2 \, dA_1}{S^2} \, dA_2}{\pi i_1' \, dA_1}$$

$$= \int_{A_2} \frac{\cos \theta_1 \cos \theta_2}{\pi S^2} \, dA_2 \tag{6-9}$$

From Eq. (6-5) the quantity inside the integral of (6-9) is dF_{d1-d2}, so F_{d1-2} can also be written as

$$F_{d1-2} = \int_{A_2} dF_{d1-d2} \tag{6-10}$$

This expresses the fact that the fraction of the energy reaching A_2 is the sum of the fractions that reach all the parts of A_2.

Now consider the configuration factor from the finite area A_2 to the elemental area dA_1. The energy reaching dA_1 from A_2 is, by integrating (6-4) over A_2,

$$dQ_{2-d1} = dA_1 \int_{A_2} i_2' \frac{\cos \theta_1 \cos \theta_2}{S^2} dA_2 \tag{6-11}$$

The total hemispherical energy leaving A_2 is

$$Q_2 = \int_{A_2} \pi i_2' \, dA_2 \tag{6-12}$$

The configuration factor dF_{2-d1} is then the ratio of dQ_{2-d1} to Q_2

$$dF_{2-d1} = \frac{dA_1 \int_{A_2} i_2' \frac{\cos \theta_1 \cos \theta_2}{S^2} dA_2}{\int_{A_2} \pi i_2' \, dA_2} = \frac{dA_1}{A_2} \int_{A_2} \frac{\cos \theta_1 \cos \theta_2}{\pi S^2} dA_2 \tag{6-13}$$

The last integral on the right was obtained subject to the imposed condition that A_2 has uniform emitted plus reflected intensity over its entire area. From (6-5) the quantity under the integral sign in Eq. (6-13) is dF_{d1-d2}, so the alternative form is obtained:

$$dF_{2-d1} = \frac{dA_1}{A_2} \int_{A_2} dF_{d1-d2} \tag{6-14}$$

Reciprocity for configuration factor between differential and finite areas By use of (6-10) the factor dF_{2-d1} given by (6-14) can be written as $dF_{2-d1} = (dA_1/A_2)F_{d1-2}$ or

$$A_2 dF_{2-d1} = dA_1 F_{d1-2} \tag{6-15}$$

which is a useful reciprocity relation.

Some sample configuration factors involving a differential and a finite area Certain geometries have configuration factors in the form of closed-form algebraic expressions (see Appendix C), while others require numerical integration of Eq. (6-9). Configuration factors can be tabulated for common geometries so that they need not be re-evaluated. A list of references for available factors is in Appendix B, and a comprehensive catalog is in [4]. Two geometries having closed-form configuration factors are in the next examples, which illustrate how the factors are obtained.

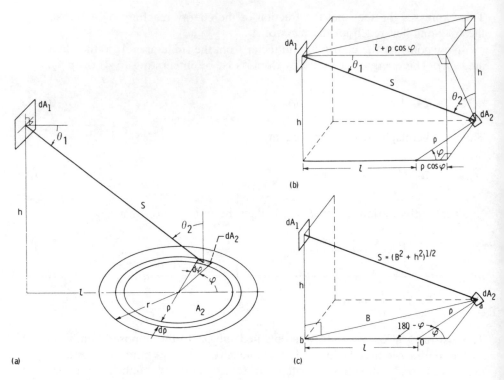

Figure 6-6 Geometry for radiative exchange between differential area and circular disk. (*a*) Geometry of problem; (*b*) auxiliary construction for determining $\cos \theta_1$ and $\cos \theta_2$; (*c*) auxiliary construction for determining S.

EXAMPLE 6-4 An elemental area dA_1 is perpendicular to a circular disk of finite area A_2 and outer radius r as shown in Fig. 6-6a. Find an equation describing the configuration factor F_{d1-2} in terms of the appropriate parameters h, l, and r.

The first step is to find expressions for the quantities inside the integral of (6-9) in terms of known quantities so that the integration can be carried out. The area element dA_2 is known in terms of the local radius on the disk and the angle φ as $dA_2 = \rho \, d\rho \, d\varphi$. Because the integral in (6-9) must be carried out over all ρ and φ, the other quantities in the integral must be put in terms of these two variables; this is done by using auxiliary constructions. Figure 6-6b is drawn to evaluate $\cos \theta_1$ and $\cos \theta_2$, which are seen to be $\cos \theta_1 = (l + \rho \cos \varphi)/S$ and $\cos \theta_2 = h/S$. Figure 6-6c allows evaluation of the remaining unknown S as $S^2 = h^2 + B^2$, where B^2 can be evaluated by using the law of cosines on triangle $a0b$. This gives $B^2 = l^2 + \rho^2 -$

$2l\rho \cos(\pi - \varphi) = l^2 + \rho^2 + 2l\rho \cos \varphi$. Substituting these relations into (6-9) results in

$$F_{d1-2} = \int_{A_2} \frac{\cos \theta_1 \cos \theta_2}{\pi S^2} dA_2 = \int_{A_2} \frac{h(l + \rho \cos \varphi)}{\pi S^4} \rho \, d\rho \, d\varphi$$

$$= \frac{h}{\pi} \int_{\rho=0}^{r} \int_{\varphi=0}^{2\pi} \frac{\rho(l + \rho \cos \varphi)}{(h^2 + l^2 + \rho^2 + 2\rho l \cos \varphi)^2} \, d\varphi \, d\rho$$

This integration is carried out using the symmetry of the configuration and is nondimensionalized to give, after considerable manipulation,

$$F_{d1-2} = \frac{2h}{\pi} \int_{\rho=0}^{r} \int_{\varphi=0}^{\pi} \frac{\rho(l + \rho \cos \varphi)}{(h^2 + \rho^2 + l^2 + 2\rho l \cos \varphi)^2} \, d\varphi \, d\rho$$

$$= \frac{2H}{\pi} \int_{\xi=0}^{R} \int_{\varphi=0}^{\pi} \frac{\xi(1 + \xi \cos \varphi)}{(H^2 + \xi^2 + 1 + 2\xi \cos \varphi)^2} \, d\varphi \, d\xi$$

$$= \frac{H}{2} \left\{ \frac{H^2 + R^2 + 1}{[(H^2 + R^2 + 1)^2 - 4R^2]^{1/2}} - 1 \right\}$$

The nondimensionalization has been done by dividing numerator and denominator by l^4 and letting $H = h/l$, $R = r/l$, and $\xi = \rho/l$. To avoid the complicated double integration for F_{d1-2}, the analysis can be carried out more conveniently by the contour integration method of Sec. 6-4.3.

EXAMPLE 6-5 An infinitely long two-dimensional wedge cavity has an opening angle α. Derive an expression for the configuration factor from one wall of the wedge to a strip element of width dx on the other wall at a distance x from the wedge vertex as shown in Fig. 6-7a. (Such configurations approximate the geometries of long fins and ribs used in space radiators.)

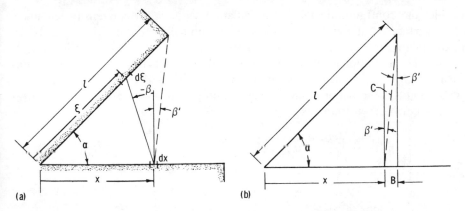

(a)   (b)

Figure 6-7 Configuration factor between one wall and strip on other wall of infinitely long wedge cavity. (a) Wedge-cavity geometry; (b) auxiliary construction to determine $\sin \beta'$.

From Example 6-2, the configuration factor between two infinitely long strip elements having parallel generating lines is $dF_{dx-d\xi} = \frac{1}{2}d(\sin \beta)$, where β is in a plane containing the normals of both strips. Note that β is measured clockwise from the normal of dx; Eq. (6-10) then gives

$$F_{dx-l} = \int_{\xi=0}^{l} dF_{dx-d\xi} = \int_{\beta=-\pi/2}^{0} \frac{1}{2}d(\sin \beta) + \int_{0}^{\beta'} \frac{1}{2}d(\sin \beta)$$

$$= \frac{\sin \beta}{2}\Big|_{\beta=-\pi/2}^{0} + \frac{\sin \beta}{2}\Big|_{\beta=0}^{\beta'} = \frac{1}{2} + \frac{\sin \beta'}{2}$$

The function $\sin \beta'$ is found by the auxiliary construction of Fig. 6-7b to be $\sin \beta' = B/C = (l \cos \alpha - x)/(x^2 + l^2 - 2xl \cos \alpha)^{1/2}$. Then

$$F_{dx-l} = \frac{1}{2} + \frac{l \cos \alpha - x}{2(x^2 + l^2 - 2xl \cos \alpha)^{1/2}}$$

The problem requires dF_{l-dx}. Using the reciprocal relation (6-15) gives

$$dF_{l-dx} = \frac{dx}{l} F_{dx-l} = dx\left[\frac{1}{2l} + \frac{\cos \alpha - x/l}{2(x^2 + l^2 - 2xl \cos \alpha)^{1/2}}\right]$$

By letting $X = x/l$, this is placed in the dimensionless form

$$dF_{l-dx} = dX\left[\frac{1}{2} + \frac{\cos \alpha - X}{2(X^2 + 1 - 2X \cos \alpha)^{1/2}}\right]$$

The only parameters are the opening angle of the wedge and the dimensionless position of dx from the vertex.

6-3.3 Configuration Factor and Reciprocity for Two Finite Areas

Consider the configuration factor for radiation with uniform intensity leaving a diffuse surface A_1 shown in Fig. 6-8 and reaching A_2. By definition, F_{1-2} is the fraction of the energy leaving A_1 that arrives at A_2. The total energy leaving A_1 is $\pi i_1' A_1$. The radiation leaving an element dA_1 that reaches dA_2 was given previously as $d^2 Q'_{d1-d2} = \pi i_1'(\cos \theta_1 \cos \theta_2 / \pi S^2) \, dA_1 \, dA_2$. If this is integrated over both A_1 and A_2, the result is the energy leaving A_1 that reaches A_2. The configuration factor is then found as

$$F_{1-2} = \frac{\displaystyle\int_{A_1}\int_{A_2} \pi i_1' \frac{\cos \theta_1 \cos \theta_2}{\pi S^2} \, dA_2 \, dA_1}{\pi i_1' A_1}$$

$$F_{1-2} = \frac{1}{A_1}\int_{A_1}\int_{A_2} \frac{\cos \theta_1 \cos \theta_2}{\pi S^2} \, dA_2 \, dA_1 \tag{6-16}$$

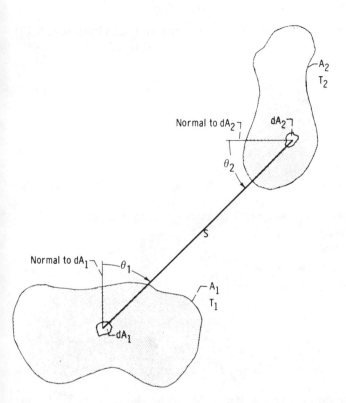

Figure 6-8 Geometry for energy exchange between finite areas.

This can be written in terms of configuration factors involving differential areas as

$$F_{1-2} = \frac{1}{A_1} \int_{A_1} \int_{A_2} dF_{d1-d2} \, dA_1 = \frac{1}{A_1} \int_{A_1} F_{d1-2} \, dA_1 \qquad (6\text{-}17)$$

In a manner similar to the derivation of (6-16), the configuration factor from A_2 to A_1 is found to be

$$F_{2-1} = \frac{1}{A_2} \int_{A_1} \int_{A_2} \frac{\cos \theta_1 \cos \theta_2}{\pi S^2} \, dA_2 \, dA_1 \qquad (6\text{-}18)$$

The reciprocity relation for configuration factors between finite areas is found by noting that the double integrals in (6-16) and (6-18) are identical. Hence the *reciprocity relation* is

$$A_1 F_{1-2} = A_2 F_{2-1} \qquad (6\text{-}19)$$

Further interrelations between configuration factors can be found by using (6-17) in conjunction with the reciprocity relations of (6-19) and (6-15),

$$F_{2-1} = \frac{A_1}{A_2} F_{1-2} = \frac{A_1}{A_2} \frac{1}{A_1} \int_{A_1} F_{d1-2} \, dA_1 = \frac{1}{A_2} \int_{A_1} dF_{2-d1} A_2 = \int_{A_1} dF_{2-d1} \qquad (6\text{-}20)$$

EXAMPLE 6-6 Two plates of the same finite width and of infinite length are joined along one edge at angle α as shown in Fig. 6-7. Using the same nondimensional parameters as in Example 6-5, derive the configuration factor between the plates.

Example 6-5 gives the configuration factor between one plate and an infinite strip on the other plate as

$$dF_{l-dx} = \left[\frac{1}{2l} + \frac{\cos \alpha - x/l}{2(x^2 + l^2 - 2xl \cos \alpha)^{1/2}} \right] dx$$

Substituting into Eq. (6-20) gives

$$F_{l-l*} = \int_{x=0}^{l*} dF_{l-dx} = \int_0^{l*} \left[\frac{1}{2l} + \frac{\cos \alpha - x/l}{2(x^2 + l^2 - 2xl \cos \alpha)^{1/2}} \right] dx$$

where, for convenience in labeling, the width of the side in Fig. 6-7 having element dx is specified as $l*$. Using the dimensionless variable $X = x/l$ and the fact that $l* = l$, this becomes

$$F_{l-l*} = \int_0^1 \left[\frac{1}{2} + \frac{\cos \alpha - X}{2(X^2 + 1 - 2X \cos \alpha)^{1/2}} \right] dX$$

Carrying out the integration yields

$$F_{l-l*} = 1 - \left(\frac{1 - \cos \alpha}{2} \right)^{1/2} = 1 - \sin \frac{\alpha}{2}$$

For the present case, in which the two plate widths are equal, the only parameter is the angle α. Also, because the areas of the two sides are equal, the reciprocity relation [Eq. (6-19)] gives, as expected from symmetry, $F_{l-l*} = F_{l*-l}$.

6-3.4 Summary of Configuration-Factor Relations

Table 6-1 summarizes the integral definitions of the configuration factors and the configuration-factor reciprocity relations.

Table 6-1 Summary of configuration-factor and reciprocity relations

Geometry	Configuration factor	Reciprocity
Elemental area to elemental area	$dF_{d1-d2} = \dfrac{\cos\theta_1 \cos\theta_2}{\pi S^2} dA_2$	$dA_1\, dF_{d1-d2} = dA_2\, dF_{d2-d1}$
Elemental area to finite area	$F_{d1-2} = \displaystyle\int_{A_2} \dfrac{\cos\theta_1 \cos\theta_2}{\pi S^2} dA_2$	$dA_1\, F_{d1-2} = A_2\, dF_{2-d1}$
Finite area to finite area	$F_{1-2} = \dfrac{1}{A_1} \displaystyle\int_{A_1}\int_{A_2} \dfrac{\cos\theta_1 \cos\theta_2}{\pi S^2} dA_2\, dA_1$	$A_1 F_{1-2} = A_2 F_{2-1}$

6-4 METHODS FOR EVALUATING CONFIGURATION FACTORS

6-4.1 Configuration-Factor Algebra

Because of the effort required for many geometries in directly computing configuration factors from their integral definitions, it is desirable to utilize shortcut methods when possible. Shortcuts can be obtained by using two concepts that have been developed in preceding sections: (1) the definition of configuration factor in terms of fractional intercepted energy and (2) the reciprocal relations. This section will show how these concepts can be used to derive configuration factors for certain geometries from known configuration factors of other geometries. The interrelation between configuration factors is termed *configuration-factor algebra*.

Consider an arbitrary isothermal area A_1 in Fig. 6-9 exchanging energy with a second area A_2. The configuration factor F_{1-2} is the fraction of all the diffuse energy leaving A_1 that is incident on A_2. If A_2 is divided into two parts A_3 and A_4, the fractions of the entire energy leaving A_1 that are incident on A_3 and A_4 must add up to F_{1-2},

$$F_{1-2} = F_{1-(3+4)} = F_{1-3} + F_{1-4} \tag{6-21}$$

If F_{1-2} and F_{1-4} are known, and configuration factor F_{3-1} is desired, then

$$F_{1-3} = F_{1-2} - F_{1-4} \tag{6-22}$$

The reciprocity relation, Eq. (6-19), gives

$$F_{3-1} = \frac{A_1}{A_3} F_{1-3} = \frac{A_1}{A_3} (F_{1-2} - F_{1-4}) \tag{6-23}$$

This is a powerful tool for obtaining new configuration factors from those previously computed. This method will be examined by use of some examples.

Please note the following with regard to notation: In an equation such as Eq. (6-21), the dashes in the subscripts mean "to" and are not to be confused with negative signs. Thus F_{1-2} means from area A_1 to area A_2. Combined areas are grouped with parentheses so that $(3 + 4)$ means the combination of areas A_3 and A_4. Thus the notation $F_{1-(3+4)}$ means from A_1 to the combination of A_3 and A_4.

EXAMPLE 6-7 An elemental area dA_1 is oriented perpendicular to a ring of outer radius r_o and inner radius r_i as shown in Fig. 6-10. Derive an expression for the configuration factor $F_{d1-\text{ring}}$.

In Example 6-4, the configuration factor between element dA_1 and the entire disk of area A_2 and outer radius r_o was found to be

$$F_{d1-2} = \frac{H}{2} \left\{ \frac{H^2 + R_o^2 + 1}{[(H^2 + R_o^2 + 1)^2 - 4R_o^2]^{1/2}} - 1 \right\}$$

where $H = h/l$ and $R_o = r_o/l$. The configuration factor to the inner disk of area A_3 and radius r_i is similarly

$$F_{d1-3} = \frac{H}{2} \left\{ \frac{H^2 + R_i^2 + 1}{[(H^2 + R_i^2 + 1)^2 - 4R_i^2]^{1/2}} - 1 \right\}$$

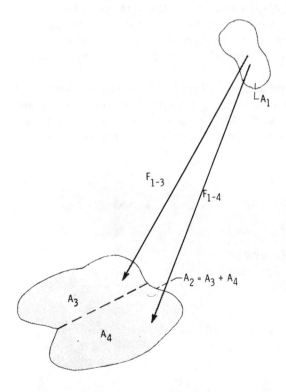

Figure 6-9 Energy exchange between finite areas with one area subdivided: $F_{1-3} + F_{1-4} = F_{1-2}$.

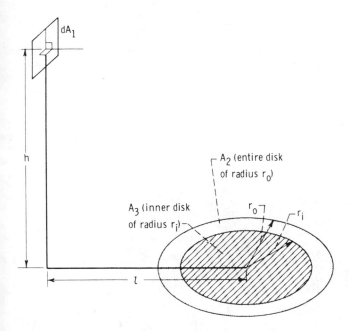

Figure 6-10 Interchange between elemental area and finite ring.

where $R_i = r_i/l$. Using configuration-factor algebra, the desired configuration factor from dA_1 to the ring A_2–A_3 is

$$F_{d1-\text{ring}} = F_{d1-2} - F_{d1-3}$$

$$= \frac{H}{2} \left\{ \frac{H^2 + R_o^2 + 1}{[(H^2 + R_o^2 + 1)^2 - 4R_o^2]^{1/2}} - \frac{H^2 + R_i^2 + 1}{[(H^2 + R_i^2 + 1)^2 - 4R_i^2]^{1/2}} \right\}$$

EXAMPLE 6-8 Suppose the configuration factor is known between two parallel disks of arbitrary size whose centers lie on the same axis. From this, derive the configuration factor between the two rings A_2 and A_3 of Fig. 6-11. Give the answer in terms of known disk-to-disk factors from disk areas on the lower surface to disk areas on the upper surface.

The factor desired is F_{2-3}. From configuration-factor algebra, F_{2-3} is equal to $F_{2-3} = F_{2-(3+4)} - F_{2-4}$. The factor $F_{2-(3+4)}$ can be found from the reciprocal relation $A_2 F_{2-(3+4)} = (A_3 + A_4)F_{(3+4)-2}$. Applying configuration-factor algebra to the right-hand side results in

$$A_2 F_{2-(3+4)} = (A_3 + A_4)(F_{(3+4)-(1+2)} - F_{(3+4)-1})$$

$$= (A_3 + A_4)F_{(3+4)-(1+2)} - (A_3 + A_4)F_{(3+4)-1}$$

Applying reciprocity to the right side gives

$$A_2 F_{2-(3+4)} = (A_1 + A_2)F_{(1+2)-(3+4)} - A_1 F_{1-(3+4)}$$

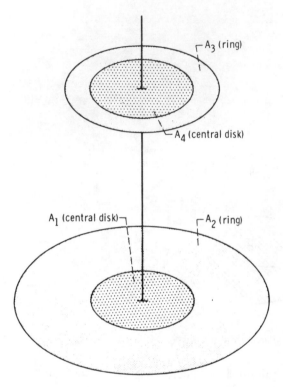

A_3 (ring)

A_4 (central disk)

A_1 (central disk)

A_2 (ring)

Figure 6-11 Interchange between parallel ring areas having a common axis.

where the F factors on the right are both disk-to-disk factors from the lower surface to the upper.

Now the factor F_{2-4} remains to be determined. Again, apply reciprocal relations and configuration-factor algebra to find

$$F_{2-4} = \frac{A_4}{A_2} F_{4-2} = \frac{A_4}{A_2} (F_{4-(1+2)} - F_{4-1}) = \frac{1}{A_2} [(A_1 + A_2) F_{(1+2)-4} - A_1 F_{1-4}]$$

Substituting the relations for F_{2-4} and $F_{2-(3+4)}$ into the first equation gives

$$F_{2-3} = \frac{A_1 + A_2}{A_2} (F_{(1+2)-(3+4)} - F_{(1+2)-4}) - \frac{A_1}{A_2} (F_{1-(3+4)} - F_{1-4})$$

and all configuration factors on the right-hand side of this equation are for exchange between two disks in the direction from disks on the lower surface to disks on the upper surface. The problem is now solved.

Because of the small differences in large numbers that can occur in obtaining an F factor by use of configuration-factor algebra (as might occur on the right side of the last equation of the preceding example), care must be taken that sufficient significant figures are retained to ensure acceptable accuracy. Feingold [5] gives one example in which an error of 0.05% in a known factor causes an error of 57% in another factor computed from it by means of angle-factor algebra.

EXAMPLE 6-9 The internal surface of a hollow circular cylinder of radius R is radiating to a disk A_1 of radius r as shown in Fig. 6-12. Express the configuration factor from the cylindrical side A_3 to the disk in terms of disk-to-disk factors for the case of r less than R.

From any position on A_1 the solid angle subtended when viewing A_3 is the difference between the solid angle when viewing A_2, or $d\omega_2$, and that viewing A_4, or $d\omega_4$. This gives the F factor from an area element dA_1 on A_1 to area A_3 as $F_{d1-3} = F_{d1-2} - F_{d1-4}$. By integrating over A_1 and using Eq. (6-17), this can be written for the entire area A_1 as $F_{1-3} = F_{1-2} - F_{1-4}$. The factors on the right are between parallel disks. The final result for F from the cylindrical side A_3 to the disk A_1 is

$$F_{3-1} = \frac{A_1}{A_3}(F_{1-2} - F_{1-4})$$

From symmetry, the factor to *any sector A_s* of A_1 is $(A_s/A_1)F_{3-1}$.

In formulating relations between configuration factors, it is sometimes useful to think in terms of *energy quantities* rather than fractions of energy leaving a surface that reach another surface. For example, in Fig. 6-9 the energy leaving

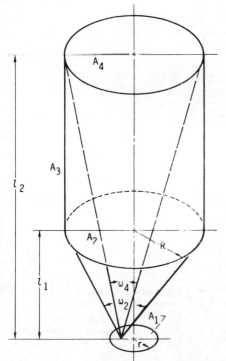

Figure 6-12 Internal surface of cylindrical cavity radiating to circular disk A_1 for $r < R$.

A_2 that arrives at A_1 is proportional to $A_2 F_{2-1}$ and is equivalent to the sums of the energies from A_3 and A_4 that arrive at A_1. Thus,

$$(A_3 + A_4) F_{(3+4)-1} = A_3 F_{3-1} + A_4 F_{4-1} \tag{6-24}$$

This can also be proved by using reciprocity relations as follows:

$$(A_3 + A_4) F_{(3+4)-1} = A_1 F_{1-(3+4)} = A_1 F_{1-3} + A_1 F_{1-4} = A_3 F_{3-1} + A_4 F_{4-1}$$

Relations found by use of symmetry There is a reciprocity relation that can be derived from the symmetry of a geometry. Consider the opposing areas in Fig. 6-13a. From symmetry it is evident that $A_2 = A_4$ and $F_{2-3} = F_{4-1}$, so that $A_2 F_{2-3} = A_4 F_{4-1}$. From reciprocity, $A_4 F_{4-1} = A_1 F_{1-4}$. Hence, there is the relation

$$A_2 F_{2-3} = A_1 F_{1-4} \tag{6-25a}$$

which relates the diagonal directions shown by the arrows on the figure. Similarly, the symmetry of Fig. 6-13b yields

$$A_2 F_{2-7} = A_3 F_{3-6} \tag{6-25b}$$

Figure 6-14a shows four areas on two perpendicular rectangles having a common edge. Since these areas are of unequal sizes, there is no apparent symmetry relation. However, it will be shown that a valid relation is

$$A_1 F_{1-2} = A_3 F_{3-4} \tag{6-26}$$

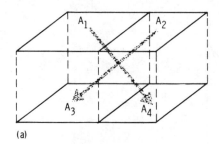

(a)

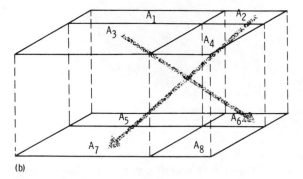

(b)

Figure 6-13 Geometry for reciprocity between opposing rectangles. (*a*) Two pairs of opposing rectangles: $A_1 F_{1-4} = A_2 F_{2-3}$; (*b*) four pairs of opposing rectangles: $A_2 F_{2-7} = A_3 F_{3-6}$.

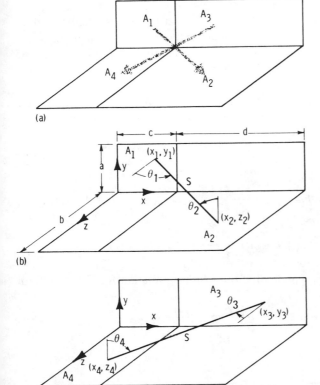

Figure 6-14 Reciprocity for diagonally opposite pairs of rectangles on two perpendicular planes having a common edge. (*a*) Representation of reciprocity: $A_1 F_{1-2} = A_3 F_{3-4}$; (*b*) construction for F_{1-2}; (*c*) construction for F_{3-4}.

To prove this, begin with the basic definition, Eq. (6-16),

$$A_1 F_{1-2} = \frac{1}{\pi} \int_{A_1} \int_{A_2} \frac{\cos \theta_1 \cos \theta_2}{S^2} \, dA_2 \, dA_1$$

From Fig. 6-14*b*, $S^2 = (x_2 - x_1)^2 + y_1^2 + z_2^2$, $\cos \theta_1 = z_2/S$, and $\cos \theta_2 = y_1/S$. Then

$$A_1 F_{1-2} = \frac{1}{\pi} \int_{x_1=0}^{c} \int_{y_1=0}^{a} \int_{x_2=c}^{c+d} \int_{z_2=0}^{b} \frac{y_1 z_2}{[(x_2 - x_1)^2 + y_1^2 + z_2^2]^2} \, dz_2 \, dx_2 \, dy_1 \, dx_1$$

$$(6\text{-}27)$$

Similarly, reference to Fig. 6-14*c* reveals that

$$A_3 F_{3-4} = \frac{1}{\pi} \int_{A_3} \int_{A_4} \frac{\cos \theta_3 \cos \theta_4}{S^2} \, dA_4 \, dA_3$$

$$= \frac{1}{\pi} \int_{x_3=c}^{c+d} \int_{y_3=0}^{a} \int_{x_4=0}^{c} \int_{z_4=0}^{b} \frac{y_3 z_4}{[(x_3 - x_4)^2 + y_3^2 + z_4^2]^2} \, dz_4 \, dx_4 \, dy_3 \, dx_3$$

$$(6\text{-}28)$$

By interchanging the dummy integration variables x_1, y_1, x_2, and z_2 for x_4, y_3, x_3, and z_4, it is found that the integrals in Eqs. (6-27) and (6-28) are identical, thus proving (6-26). In [6] these diagonal relations are extended to nonplanar surfaces. The geometries considered are two infinitely long parallel cylindrical areas of finite width, and two coaxial surfaces of rotation.

EXAMPLE 6-10 If the configuration factor is known for two perpendicular rectangles with a common edge as shown in Fig. 6-15a, derive the configuration factor F_{1-6} for Fig. 6-15b.

First consider the geometry in Fig. 6-15c and derive the factor F_{7-6} as follows:

$$F_{(5+6)-(7+8)} = F_{(5+6)-7} + F_{(5+6)-8} = \frac{A_7}{A_5 + A_6} F_{7-(5+6)} + \frac{A_8}{A_5 + A_6} F_{8-(5+6)}$$

$$F_{(5+6)-(7+8)} = \frac{A_7}{A_5 + A_6} (F_{7-5} + F_{7-6}) + \frac{A_8}{A_5 + A_6} (F_{8-5} + F_{8-6})$$

Substitute $A_7 F_{7-6}$ for $A_8 F_{8-5}$ and solve the resulting relation for F_{7-6} to obtain

$$F_{7-6} = \frac{1}{2A_7} [(A_5 + A_6) F_{(5+6)-(7+8)} - A_7 F_{7-5} - A_8 F_{8-6}]$$

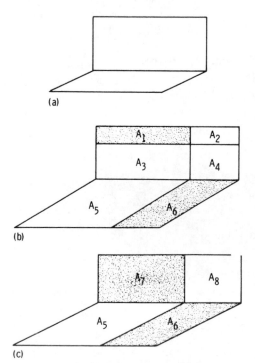

(a)

(b)

(c)

Figure 6-15 Orientation of areas for Example 6-10. (a) Perpendicular rectangles with one common edge; (b) geometry for F_{1-6}; (c) auxiliary geometry.

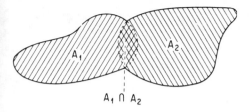

Figure 6-16 Union and intersection of finite areas.

$$A_1 \cap A_2$$

Now in Fig. 6-15*b*

$$F_{1\text{-}6} = \frac{A_6}{A_1} F_{6\text{-}1} = \frac{A_6}{A_1} F_{6\text{-}(1+3)} - \frac{A_6}{A_1} F_{6\text{-}3}$$

The factors $F_{6\text{-}(1+3)}$ and $F_{6\text{-}3}$ are of the same type as $F_{7\text{-}6}$ so that $F_{1\text{-}6}$ can finally be written as

$$F_{1\text{-}6} = \frac{A_6}{A_1} \left\{ \frac{1}{2A_6} [(A_1 + A_2 + A_3 + A_4)F_{(1+2+3+4)\text{-}(5+6)} \right.$$

$$- A_6 F_{6\text{-}(2+4)} - A_5 F_{5\text{-}(1+3)}] - \frac{1}{2A_6} [(A_3 + A_4)F_{(3+4)\text{-}(5+6)}$$

$$\left. - A_6 F_{6\text{-}4} - A_5 F_{5\text{-}3}] \right\}$$

All the F factors on the right side are for two rectangles having one common edge as in Fig. 6-15*a*.

Set-theory notation In some instances, set-theory notation can provide a convenient way of working with configuration factors. Consider two overlapping areas A_1 and A_2 as shown in Fig. 6-16. The double crosshatched area is defined in set theory as the *intersection* of A_1 and A_2 and is denoted as $A_1 \cap A_2$. The net area enclosed by the continuous solid boundary around A_1 and A_2 is the *union* of A_1 and A_2 and is denoted as $A_1 \cup A_2$.

An area A_E will have a configuration factor to $A_1 \cup A_2$ given by

$$F_{E\text{-}1\cup 2} = F_{E\text{-}1} + F_{E\text{-}2} - F_{E\text{-}1\cap 2} \tag{6-29}$$

This relation is evident by noting that the fraction of energy leaving A_E and incident on $A_1 \cup A_2$ can be derived from the fraction leaving A_E and incident on A_1 and the fraction leaving A_E and incident on A_2. However, these two fractions cover the portion $A_1 \cap A_2$ twice, so we must subtract a fraction $F_{E\text{-}1\cap 2}$.

The configuration factor between an element dA_E and a rectangle in a parallel plane when the normal to the element passes through a corner of the rectangle (Fig. 6-17) is given in Appendix C. In Figs. 6-18*a* and *b* is shown a configuration made up of two such overlapping rectangles. The configuration factor between dA_E and the resulting L-shaped area is then, from Eq. (6-29),

$$F_{dE\text{-}1\cup 2} = F_{dE\text{-}1} + F_{dE\text{-}2} - F_{dE\text{-}1\cap 2} \tag{6-30}$$

and all terms on the right are known from the factor in Fig. 6-17.

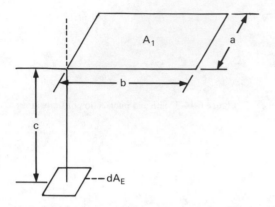

Figure 6-17 Geometry of known configuration factor.

A somewhat more complex (and thus less obvious) geometry is shown in Fig. 6-19. Again, note that

$$F_{dE-1\cup2} = F_{dE-1} + F_{dE-2} - F_{dE-1\cap2} \tag{6-31}$$

The factor F_{dE-1} between an element and a disk in parallel planes with the center of the disk on the normal to the element is given in Appendix C. F_{dE-2}, the factor between an element and a right triangle in a parallel plane with a vertex of the

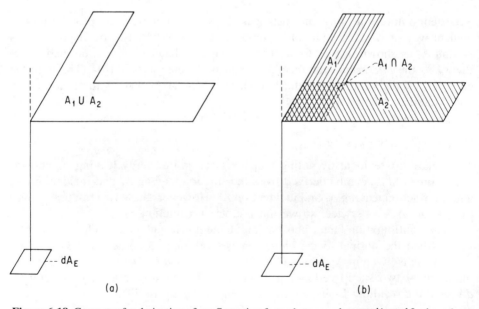

(a)   (b)

Figure 6-18 Geometry for derivation of configuration factor between element dA_E and L-shaped area $A_1 \cup A_2$. (a) Desired configuration; (b) configuration in terms of overlapping areas.

triangle on the normal to the element, is given in Example 6-17. Finally, $F_{dE-1\cap2}$ is observed to be, by symmetry,

$$F_{dE-1\cap2} = \frac{\varphi}{2\pi} F_{dE-1} \tag{6-32}$$

All terms on the right of (6-31) are known, and substituting (6-32) into (6-31) gives

$$F_{dE-1\cup2} = \left(1 - \frac{\varphi}{2\pi}\right) F_{dE-1} + F_{dE-2} \tag{6-33}$$

Note that the usual reciprocity relations apply in this notation:

$$(A_1 \cup A_2) F_{1\cup2-E} = A_E F_{E-1\cup2} \tag{6-34}$$

and

$$(A_1 \cap A_2) F_{1\cap2-E} = A_E F_{E-1\cap2} \tag{6-35}$$

The relations can be further generalized to include the combination of more than two areas. Consider the factor F_{E-4} in Fig. 6-20a. As shown in Fig. 6-20b, the A_4 can be considered as the union of three areas, each of which has a corner above dA_E and hence has a known configuration factor between itself and dA_E. The intersections of these areas are also in this same class of known configuration factors. By use of Eq. (6-29)

$$F_{E-4} = F_{E-1\cup2\cup3} = F_{E-1\cup2} + F_{E-3} - F_{E-3\cap(1\cup2)} \tag{6-36}$$

Equation (6-29) is applied to the first term on the right side, and it is noted that $A_3 \cap (A_1 \cup A_2) = A_3 \cap A_1 + A_3 \cap A_2 - A_3 \cap A_1 \cap A_2$. Then Eq. (6-36) becomes

$$F_{E-4} = F_{E-1} + F_{E-2} + F_{E-3} - (F_{E-1\cap2} + F_{E-1\cap3} + F_{E-2\cap3}) + F_{E-1\cap2\cap3} \tag{6-37}$$

The form of a general relation thus becomes evident. The first group on the right side is the sum of the factors from dA_E to all of the individual areas. Then the

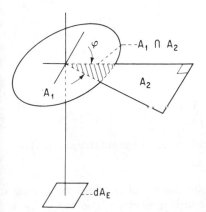

Figure 6-19 Geometry for derivation of configuration factor between element dA_E and overlapping circle A_1 and triangle A_2.

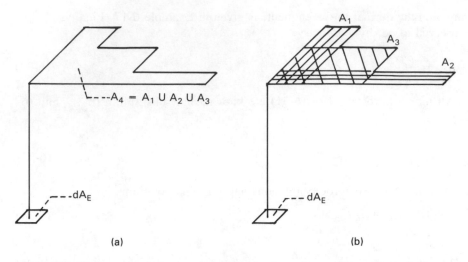

Figure 6-20 Geometry for derivation of configuration factor between element dA_E and A_4. (a) Desired configuration; (b) configuration in terms of three overlapping rectangles.

sum of factors from dA_E to all intersections of two individual areas is subtracted. Next the sum of factors to all the intersections of three areas is added. Then all factors to the intersections of four areas are subtracted. The process is continued until the last term contains the intersection of all of the component areas that combine to form the original area.

6-4.2 Configuration Factor Relations in Enclosures

To this point, only the configuration factors between two isolated surfaces have been considered, although subdivision of one or both of the surfaces into smaller portions has been examined. Consider the very useful class of problems in which the configuration factors are between surfaces that form a complete enclosure.

For an enclosure of N surfaces, such as in Fig. 6-21 (where $N = 8$ as an example), the entire energy leaving any surface inside the enclosure, for example surface A_k, must be incident on all the surfaces making up the enclosure. Thus all the fractions of energy leaving one surface and reaching the surfaces of the enclosure must total to unity; that is,

$$F_{k-1} + F_{k-2} + F_{k-3} + \cdots + F_{k-k} + \cdots + F_{k-N} = \sum_{j=1}^{N} F_{k-j} = 1 \qquad (6\text{-}38)$$

The factor F_{k-k} is included because when A_k is *concave*, it will intercept a portion of its own outgoing energy.

EXAMPLE 6-11 Two diffuse isothermal concentric spheres are exchanging energy. Find all the configuration factors for this geometry if the surface area of the inner sphere is A_1 and the surface area of the outer sphere is A_2.

All energy leaving A_1 is incident upon A_2, so immediately $F_{1-2} = 1$. Using the reciprocal relation reveals further that $F_{2-1} = A_1F_{1-2}/A_2 = A_1/A_2$. Also, from (6-38), $F_{2-1} + F_{2-2} = 1$ or $F_{2-2} = 1 - F_{2-1} = (A_2 - A_1)/A_2$.

EXAMPLE 6-12 An isothermal cavity of internal area A_1 has a plane opening of area A_2. Derive an expression for the configuration factor of the internal surface of the cavity to itself.

Assume that a black plane surface A_2 replaces the cavity opening; this will have no effect on the F factor, which is a function only of geometry. Then $F_{2-1} = 1$ and $F_{1-2} = A_2F_{2-1}/A_1 = A_2/A_1$, which is the configuration factor from the entire internal area to the opening. Since A_1 and A_2 form an enclosure, $F_{1-1} = 1 - F_{1-2} = (A_1 - A_2)/A_1$, which is the desired F factor.

EXAMPLE 6-13 An enclosure of triangular cross section is made up of three plane plates, each of finite width and infinite length (thus forming a hollow infinitely long triangular prism). Derive an expression for the configuration factor between any two of the plates in terms of the plate widths L_1, L_2, and L_3.

For plate 1, $F_{1-2} + F_{1-3} = 1$. Using similar relations for each plate and multiplying through by the respective plate areas result in

$$A_1F_{1-2} + A_1F_{1-3} = A_1$$

$$A_2F_{2-1} + A_2F_{2-3} = A_2$$

$$A_3F_{3-1} + A_3F_{3-2} = A_3$$

By applying the reciprocal relations to some of the terms, these three equations become

$$A_1F_{1-2} + A_1F_{1-3} = A_1$$

$$A_1F_{1-2} + A_2F_{2-3} = A_2$$

$$A_1F_{1-3} + A_2F_{2-3} = A_3$$

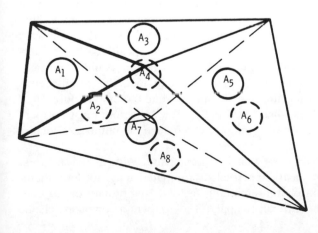

Figure 6-21 Isothermal enclosure composed of black surfaces.

thus giving three equations for the three unknown F factors. Subtracting the third from the second and adding the first give

$$F_{1\text{-}2} = \frac{A_1 + A_2 - A_3}{2A_1} = \frac{L_1 + L_2 - L_3}{2L_1}$$

For the special case of $L_1 = L_2$, this should reduce to the factor between infinitely long adjoint plates of equal width separated by an angle α as given in Example 6-6. For $L_1 = L_2$,

$$F_{1\text{-}2} = \frac{2L_1 - L_3}{2L_1} = 1 - \frac{L_3/2}{L_1} = 1 - \sin\frac{\alpha}{2}$$

which agrees with Example 6-6.

The set of three simultaneous equations from which the final result was derived in Example 6-13 will now be examined more closely. The first equation involves two unknowns, $F_{1\text{-}2}$ and $F_{1\text{-}3}$; the second equation has one additional unknown $F_{2\text{-}3}$; and the final equation has no additional unknowns. Generalizing the procedure from a three-surface enclosure to any N-sided enclosure made up of plane or convex surfaces shows that of N simultaneous equations, the first would involve $N - 1$ unknowns, the second $N - 2$ unknowns, and so forth. The total number of unknowns U is then

$$U = (N-1) + (N-2) + \cdots + 1 = N^2 - \sum_{j=1}^{N} j = N(N-1)/2 \tag{6-39}$$

Thus, $[N(N - 1)/2] - N = N(N - 3)/2$ factors must be provided. For a four-sided enclosure made up of planar or convex surfaces of known area, four equations relating $4(4 - 1)/2$ or six unknown configuration factors can be written. Specifying any two of these factors allows calculation of the rest by solving the set of four simultaneous equations.

If all the surfaces can view themselves, the factor $F_{k\text{-}k}$ must be included in each of the equations. Analyzing this situation, as previously done, shows that an N-sided enclosure allows the writing of N equations in $N(N + 1)/2$ unknowns. Thus $[N(N + 1)/2] - N = N(N - 1)/2$ factors must be specified. For a four-sided enclosure, four equations involving ten unknown F factors could be written. The specification of six factors would be required, and then the simultaneous relations could be solved to determine the remaining four factors. If only M of the surfaces can view themselves, then $[N(N - 3)/2] + M$ factors must be specified. For some geometries, the use of symmetry can reduce the number of factors that must be specified.

Sowell and O'Brien [7] point out that in an enclosure with many surfaces, the configuration factors most easily specified or computed may not lend themselves to sequential calculation of the remaining factors. Reciprocity or the conservation relation (6-38) may be difficult to apply. They present a computer scheme using matrix algebra that allows calculation of all remaining factors in an N-surface planar or convex-surface enclosure once the needed configuration factors are specified. In [8], methods for smoothing sets of "direct exchange areas" for enclosures using the constraints imposed by reciprocity and energy conservation [Eq. (6-38)] are discussed. The methods are directly applicable to configuration factors and can be used when some inaccurate or ill-defined configuration factors are present in a set. Another presentation of the compatibility conditions for enclosure configuration factors is in [9]. This will guarantee overall energy conservation for the enclosure.

6-4.3 Mathematical Techniques for the Evaluation of Configuration Factors

As shown by the summary in Table 6-1, the evaluation of the configuration factors F_{d1-2} and F_{1-2} requires integration over the finite areas involved. A number of mathematical methods are useful in evaluating configuration factors when straightforward analytical integration is not possible or becomes too cumbersome. These methods can encompass all techniques that are used in the evaluation of integrals, including numerical approaches. A few methods that are especially valuable in dealing with configuration factors will be discussed here.

Hottel's crossed-string method Consider the class of configurations such as long grooves in which all surfaces are assumed to extend infinitely far along one coordinate. Such surfaces can be generated by moving a line through space in such a way that it always remains parallel to its original position.

A typical configuration of this type is shown in cross section in Fig. 6-22. Suppose that the configuration factor is needed between A_1 and A_2 when some blockage of radiant transfer occurs because of the presence of other surfaces A_3 and A_4. To obtain F_{1-2}, first consider that A_1 may be concave. In this case, draw the dashed line agf across A_1. Then draw in the dashed lines cf and abc to complete the enclosure $abcfga$, which has three sides that are either convex or planar. The relation found in Example 6-13 for enclosures of this type can be written as

$$A_{agf}F_{agf-abc} = \frac{A_{agf} + A_{abc} - A_{cf}}{2} \tag{6-40}$$

For the three-sided enclosure $adefga$, similar reasoning gives

$$A_{agf}F_{agf-def} = \frac{A_{agf} + A_{def} - A_{ad}}{2} \tag{6-41}$$

Further, note that

$$F_{agf-abc} + F_{agf-2} + F_{agf-def} = 1 \tag{6-42}$$

Substituting (6-40) and (6-41) into (6-42) results in

$$A_{agf}F_{agf-2} = A_{agf}(1 - F_{agf-abc} - F_{agf-def}) - \frac{A_{cf} + A_{ad} - A_{abc} - A_{def}}{2} \tag{6-43}$$

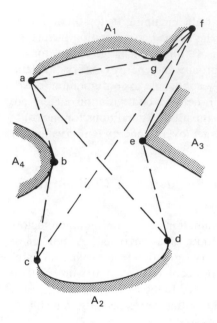

Figure 6-22 Hottel's crossed-string method for configuration-factor determination.

Now $F_{2-agf} = F_{2-1}$ since A_{agf} and A_1 subtend the same solid angle when viewed from A_2. Then, with the additional use of reciprocity, the left side of (6-43) can be written as

$$A_{agf}F_{agf-2} = A_2 F_{2-agf} = A_2 F_{2-1} = A_1 F_{1-2} \qquad (6\text{-}44)$$

Substituting (6-44) into (6-43) results in

$$A_1 F_{1-2} = \frac{A_{cf} + A_{ad} - A_{abc} - A_{def}}{2} \qquad (6\text{-}45)$$

If the dashed lines in Fig. 6-22 are imagined as being lengths of string stretched tightly between the outer edges of the surfaces, then the term on the right of (6-45) is interpreted as *one-half the total quantity formed by the sum of the lengths of the crossed strings connecting the outer edges of A_1 and A_2 minus the sum of the lengths of the uncrossed strings*. This is a very useful way for determining configuration factors *in this type of two-dimensional geometry* and was first pointed out by Hottel [10].

EXAMPLE 6-14 Two infinitely long semicylindrical surfaces of radius R are separated by a minimum distance D as shown in Fig. 6-23a. Derive the configuration factor F_{1-2} for this case.

The length of crossed string *abcde* will be denoted as L_1, and that of uncrossed string *ef* as L_2. From the symmetry of the problem, (6-45) may be written as

$$F_{1-2} = \frac{2L_1 - 2L_2}{2A_1} = \frac{L_1 - L_2}{\pi R}$$

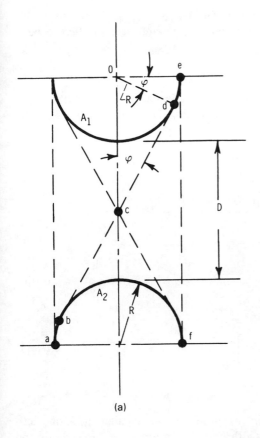

(a)

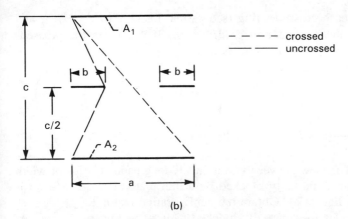

- - - - crossed
———— uncrossed

(b)

Figure 6-23 Examples of crossed-string method. (*a*) Configuration factor between infinitely long semicylindrical surfaces; (*b* partially blocked view between parallel strips.

The length L_2 is given by $L_2 = D + 2R$. The length of L_1 is twice the length cde. The segment of L_1 from c to d is found from right triangle $0cd$ to be

$$L_{1,c-d} = \left[\left(\frac{D}{2} + R\right)^2 - R^2\right]^{1/2} = \left[D\left(\frac{D}{4} + R\right)\right]^{1/2}$$

and the segment of L_1 from d to e is $L_{1,d-e} = R\varphi$. From triangle $0cd$, the angle φ is given by $\varphi = \sin^{-1}[R/(D/2 + R)]$. Combining the known relations results in

$$F_{1-2} = \frac{L_1 - L_2}{\pi R} = \frac{2(L_{1,c-d} + L_{1,d-e}) - L_2}{\pi R}$$

$$= \frac{[4D(D/4 + R)]^{1/2} + 2R \sin^{-1}[R/(D/2 + R)] - D - 2R}{\pi R}$$

Letting $X = 1 + D/2R$ gives

$$F_{1-2} = \frac{2}{\pi}\left[(X^2 - 1)^{1/2} + \sin^{-1}\frac{1}{X} - X\right] \tag{6-46}$$

This can also be put in the form

$$F_{1-2} = \frac{2}{\pi}\left[(X^2 - 1)^{1/2} + \frac{\pi}{2} - \cos^{-1}\frac{1}{X} - X\right] \tag{6-47}$$

which agrees with the result in [11].

EXAMPLE 6-15 The view between two infinitely long parallel strips of width a is partially blocked by strips of width b as shown in Fig. 6-23b. Obtain the configuration factor F_{1-2}.

The length of each crossed string (see dashed lines) is $(a^2 + c^2)^{1/2}$, and the length of each uncrossed string is $2[b^2 + (c/2)^2]^{1/2}$. From the crossed-string method the configuration factor is then

$$F_{1-2} = \frac{\sqrt{a^2 + c^2} - 2\sqrt{b^2 + (c/2)^2}}{a} = \sqrt{1 + \left(\frac{c}{a}\right)^2} - \sqrt{\left(\frac{2b}{a}\right)^2 + \left(\frac{c}{a}\right)^2}$$

As expected, $F_{1-2} \to 0$ as b is extended inward so that $b \to a/2$.

EXAMPLE 6-16 The view between two infinitely long parallel strips of width c and spaced c apart is partially blocked by a thin plate of width $c/2$ positioned as shown in Fig. 6-24. Obtain the configuration factor F_{1-2}.

Following the idea of a partially blocked view as in Fig. 6-22, area A_1 can view A_2 either to the right or to the left side of the center plate. Figure 6-24a shows the strings drawn through the opening to the right side by considering the region to the left of the center plate to be obstructed. Figure 6-24b shows the remainder of the view by drawing the strings as if the region

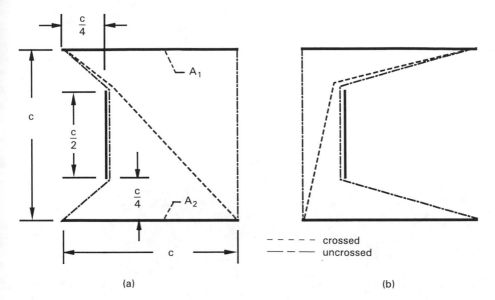

Figure 6-24 Partially blocked view between two parallel strips. (*a*) View through right side of vertical plate; (*b*) view through left side.

to the right of the center plate were obstructed. The final view factor is the sum of the results from the two parts. For each part of the figure the crossed strings are equal, so, for clarity, only one is shown. For part (*a*) the lengths of the crossed and uncrossed strings are respectively $2\sqrt{2}c$ and $c + c/2 + 2\sqrt{2}(c/4)$. Similarly, for part (*b*) the lengths are $4c[(1/4)^2 + (3/4)^2]^{1/2} - \sqrt{10}c$ and $c + c/2 + \sqrt{10}\,c/2$. Then from Eq. (6-45),

$$cF_{1-2} = \frac{1}{2}\left[2\sqrt{2}c - \left(\frac{3c}{2} + \sqrt{2}\frac{c}{2}\right)\right] + \frac{1}{2}\left[\sqrt{10}c - \left(\frac{3c}{2} + \sqrt{10}\frac{c}{2}\right)\right]$$

This simplifies to

$$F_{1-2} = \frac{1}{4}(3\sqrt{2} + \sqrt{10} - 6) = 0.35123$$

When the center plate is absent the crossed-string method yields $F_{1-2} = \sqrt{2} - 1 = 0.41421$, so the center plate has provided about 15% blockage.

Contour integration Another useful tool in the evaluation of configuration factors is the application of Stokes' theorem for reduction of the multiple integration over a surface area to a single integration around the boundary of the area. This method is treated at some length by Sparrow and Cess [11], Moon [12], who provides earlier references, de Bastos [13], and Sparrow [14]. Consider a surface area A as shown in Fig. 6-25 with its boundary designated as C (where C is piecewise-continuous). An arbitrary point on the area is located at coordinate position x, y, z. At this point the normal to A is constructed, and the angles between

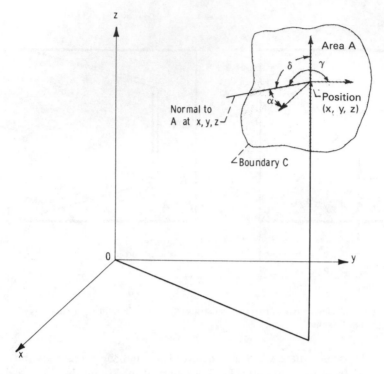

Figure 6-25 Geometry for quantities used in Stokes' theorem.

this normal and the x, y, and z axes are designated as α, γ, and δ. Let the functions P, Q, and R be any twice-differentiable functions of x, y, and z. Stokes' theorem in three dimensions provides the following relation between an integral of P, Q, and R around the boundary C of the area and an integral over the surface A of the area:

$$\oint_C (P\,dx + Q\,dy + R\,dz)$$

$$= \int_A \left[\left(\frac{\partial R}{\partial y} - \frac{\partial Q}{\partial z} \right) \cos \alpha + \left(\frac{\partial P}{\partial z} - \frac{\partial R}{\partial x} \right) \cos \gamma + \left(\frac{\partial Q}{\partial x} - \frac{\partial P}{\partial y} \right) \cos \delta \right] dA \quad (6\text{-}48)$$

Now this relation will be applied to express area integrals in configuration-factor computations in terms of integrals around the boundaries of the areas.

Configuration factor between a differential and a finite area The integrand in the configuration factor $F_{d1\text{-}2}$ is $(\cos \theta_1 \cos \theta_2 / \pi S^2)\,dA_2$ as shown in Table 6-1. In general, for the two cosines the following can be written (Fig. 6-26):

$$\cos \theta_1 = \frac{x_2 - x_1}{S} \cos \alpha_1 + \frac{y_2 - y_1}{S} \cos \gamma_1 + \frac{z_2 - z_1}{S} \cos \delta_1 \quad (6\text{-}49)$$

$$\cos\theta_2 = \frac{x_1 - x_2}{S}\cos\alpha_2 + \frac{y_1 - y_2}{S}\cos\gamma_2 + \frac{z_1 - z_2}{S}\cos\delta_2 \tag{6-50}$$

This follows from the relation that, for two vectors $\mathbf{V}_1$ and $\mathbf{V}_2$ having direction cosines (l_1, m_1, n_1) and (l_2, m_2, n_2), the cosine of the angle between the vectors is given by $l_1 l_2 + m_1 m_2 + n_1 n_2$.

Substituting Eqs. (6-49) and (6-50) into the integral relation for a configuration factor between a differential element and a finite area gives

$$F_{d1-2} = \int_{A_2} \frac{\cos\theta_1 \cos\theta_2}{\pi S^2}\, dA_2$$

$$= \frac{1}{\pi}\int_{A_2}\frac{(x_2 - x_1)\cos\alpha_1 + (y_2 - y_1)\cos\gamma_1 + (z_2 - z_1)\cos\delta_1}{S^4}$$

$$\times\, [(x_1 - x_2)\cos\alpha_2 + (y_1 - y_2)\cos\gamma_2 + (z_1 - z_2)\cos\delta_2]\, dA_2 \tag{6-51}$$

Now let,

$$l = \cos\alpha \qquad m = \cos\gamma \qquad n = \cos\delta \tag{6-52}$$

and

$$f = \frac{(x_2 - x_1)l_1 + (y_2 - y_1)m_1 + (z_2 - z_1)n_1}{\pi S^4} \tag{6-53}$$

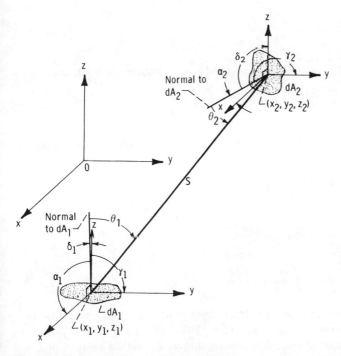

Figure 6-26 Geometry for contour integration.

Equation (6-51) can then be written in the abbreviated form

$$F_{d1-2} = \int_{A_2} [(x_1 - x_2)fl_2 + (y_1 - y_2)fm_2 + (z_1 - z_2)fn_2]\, dA_2 \tag{6-54}$$

Comparison of (6-54) with the right side of (6-48) shows that Stokes' theorem can be applied if

$$\frac{\partial R}{\partial y_2} - \frac{\partial Q}{\partial z_2} = (x_1 - x_2)f \tag{6-55a}$$

$$\frac{\partial P}{\partial z_2} - \frac{\partial R}{\partial x_2} = (y_1 - y_2)f \tag{6-55b}$$

and

$$\frac{\partial Q}{\partial x_2} - \frac{\partial P}{\partial y_2} = (z_1 - z_2)f \tag{6-55c}$$

Useful solutions to these three equations are in [14] of the form

$$P = \frac{-m_1(z_2 - z_1) + n_1(y_2 - y_1)}{2\pi S^2} \tag{6-56a}$$

$$Q = \frac{l_1(z_2 - z_1) - n_1(x_2 - x_1)}{2\pi S^2} \tag{6-56b}$$

$$R = \frac{-l_1(y_2 - y_1) + m_1(x_2 - x_1)}{2\pi S^2} \tag{6-56c}$$

Equation (6-48) is used to express F_{d1-2} in (6-54) as a contour integral; that is,

$$F_{d1-2} = \oint_{C_2} (P\, dx_2 + Q\, dy_2 + R\, dz_2) \tag{6-57}$$

Then P, Q, and R are substituted from (6-56), and the result is rearranged to obtain

$$
\begin{aligned}
F_{d1-2} = {} & \frac{l_1}{2\pi} \oint_{C_2} \frac{(z_2 - z_1)\, dy_2 - (y_2 - y_1)\, dz_2}{S^2} \\
& + \frac{m_1}{2\pi} \oint_{C_2} \frac{(x_2 - x_1)\, dz_2 - (z_2 - z_1)\, dx_2}{S^2} \\
& + \frac{n_1}{2\pi} \oint_{C_2} \frac{(y_2 - y_1)\, dx_2 - (x_2 - x_1)\, dy_2}{S^2}
\end{aligned}
\tag{6-58}
$$

The double integration over area A_2 has been replaced by a set of three line integrals for the determination of F_{d1-2}. Sparrow [14] discusses the superposition properties of Eq. (6-54) that allow additions of the configuration factors of elements aligned parallel to the x, y, and z axes to obtain the factors for arbitrary orientation. The numerical computation of the view factor for parallel directly opposed squares is discussed in [15]. The line integral method was found to be advantageous in terms of both accuracy and computing time.

EXAMPLE 6-17 Determine the configuration factor F_{d1-2} from an element dA_1 to a right triangle as shown in Fig. 6-27.

The normal to dA_1 is perpendicular to both the x and y axes and is thus parallel to z. The direction cosines for dA_1 are then $\cos \alpha_1 = l_1 = 0$, $\cos \gamma_1 = m_1 = 0$, and $\cos \delta_1 = n_1 = 1$, and (6-58) becomes

$$F_{d1-2} = \frac{1}{2\pi} \oint_{C_2} \frac{(y_2 - y_1)\, dx_2 - (x_2 - x_1)\, dy_2}{S^2}$$

Since dA_1 is situated at the origin of the coordinate system, $x_1 = y_1 = 0$ and F_{d1-2} further reduces to

$$F_{d1-2} = \frac{1}{2\pi} \oint_{C_2} \frac{y_2\, dx_2 - x_2\, dy_2}{S^2}$$

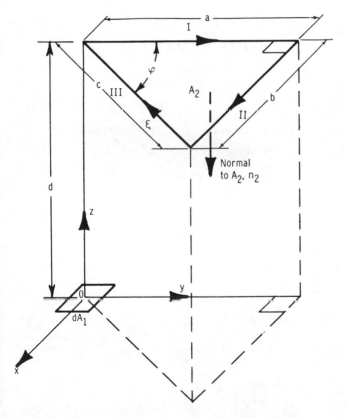

Figure 6-27 Configuration factor between plane-area element and right triangle in parallel plane.

The distance S between dA_1 and any point (x_2, y_2, z_2) on A_2 is

$$S^2 = x_2{}^2 + y_2{}^2 + z_2{}^2 = x_2{}^2 + y_2{}^2 + d^2$$

The contour integration of the configuration-factor equation must now be carried out around the three sides of the right triangle. To keep the sign of F_{d1-2} positive, the integration is performed by traveling around the boundary lines I, II, and III in a particular direction. The correct direction is that of a person walking around the boundary with their head in the direction of the normal n_2 and always keeping A_2 to their left. Along boundary line I, $x_2 = 0$, $dx_2 = 0$, and $0 \le y_2 \le a$. On boundary II, $y_2 = a$, $dy_2 = 0$, and $0 \le x_2 \le b$. On boundary III, the integration is from $\xi = 0$ to c where ξ is a co-ordinate along the hypotenuse of the triangle so that

$$x_2 = (c - \xi) \sin \varphi \qquad \qquad dx_2 = -\sin \varphi \, d\xi$$
$$\text{and}$$
$$y_2 = (c - \xi) \cos \varphi \qquad \qquad dy_2 = -\cos \varphi \, d\xi$$

Substituting these quantities into the integral for F_{d1-2} gives

$$2\pi F_{d1-2} = \oint_{C_2} \frac{y_2 \, dx_2 - x_2 \, dy_2}{S^2} = \oint_{\text{I, II, III}} \frac{y_2 \, dx_2 - x_2 \, dy_2}{x_2{}^2 + y_2{}^2 + d^2}$$

$$2\pi F_{d1-2} = 0 + \int_{x_2=0}^{b} \frac{a \, dx_2}{x_2{}^2 + a^2 + d^2}$$

$$+ \int_{\xi=0}^{c} \frac{-(c - \xi) \cos \varphi \sin \varphi \, d\xi + (c - \xi) \sin \varphi \cos \varphi \, d\xi}{(c - \xi)^2 \sin^2 \varphi + (c - \xi)^2 \cos^2 \varphi + d^2}$$

or

$$2\pi F_{d1-2} = \int_{0}^{b} \frac{a \, dx_2}{x_2{}^2 + a^2 + d^2}$$

This is integrated to yield

$$F_{d1-2} = \frac{a}{2\pi (a^2 + d^2)^{1/2}} \tan^{-1} \frac{b}{(a^2 + d^2)^{1/2}}$$

or, in dimensionless variables, $X = a/d$ and $\tan \varphi = b/a$,

$$F_{d1-2} = \frac{X}{2\pi (1 + X^2)^{1/2}} \tan^{-1} \frac{X \tan \varphi}{(1 + X^2)^{1/2}}$$

Configuration factor between finite areas For configuration factors between two finite areas, substitution of (6-58) into (6-17) gives

$$A_1 F_{1-2} = A_2 F_{2-1} = \int_{A_1} F_{d1-2}\, dA_1$$

$$= \frac{1}{2\pi} \oint_{C_2} \left[\int_{A_1} \frac{(y_2 - y_1)n_1 - (z_2 - z_1)m_1}{S^2}\, dA_1 \right] dx_2$$

$$+ \frac{1}{2\pi} \oint_{C_2} \left[\int_{A_1} \frac{(z_2 - z_1)l_1 - (x_2 - x_1)n_1}{S^2}\, dA_1 \right] dy_2$$

$$+ \frac{1}{2\pi} \oint_{C_2} \left[\int_{A_1} \frac{(x_2 - x_1)m_1 - (y_2 - y_1)l_1}{S^2}\, dA_1 \right] dz_2 \qquad (6\text{-}59)$$

where the integrals have been rearranged, and dx_2, dy_2, and dz_2 have been factored out since these are independent of the area integration over A_1.

Stokes' theorem is applied in turn to each of the three area integrals. Consider the first of the integrals,

$$\int_{A_1} \frac{(y_2 - y_1)n_1 - (z_2 - z_1)m_1}{S^2}\, dA_1$$

and compare it with the area integral in Stokes' theorem, Eq. (6-48). This gives

$$\frac{\partial R}{\partial y_1} - \frac{\partial Q}{\partial z_1} = 0$$

$$\frac{\partial P}{\partial z_1} - \frac{\partial R}{\partial x_1} = \frac{-(z_2 - z_1)}{S^2}$$

$$\frac{\partial Q}{\partial x_1} - \frac{\partial P}{\partial y_1} = \frac{y_2 - y_1}{S^2}$$

A solution to this set of partial differential equations [11] is $P = \ln S$, $Q = 0$, and $R = 0$; the area integral becomes, by use of Eq. (6-48) to convert it into a surface integral,

$$\int_{A_1} \frac{(y_2 - y_1)n_1 - (z_2 - z_1)m_1}{S^2}\, dA_1 = \oint_{C_1} \ln S\, dx_1$$

By applying Stokes' theorem in a similar fashion to the other two integrals in (6-59), that equation can be written as

$$A_1 F_{1-2} = \frac{1}{2\pi} \oint_{C_2} \left(\oint_{C_1} \ln S \, dx_1 \right) dx_2 + \frac{1}{2\pi} \oint_{C_2} \left(\oint_{C_1} \ln S \, dy_1 \right) dy_2$$

$$+ \frac{1}{2\pi} \oint_{C_2} \left(\oint_{C_1} \ln S \, dz_1 \right) dz_2$$

or more compactly as

$$F_{1-2} = \frac{1}{2\pi A_1} \oint_{C_1} \oint_{C_2} (\ln S \, dx_2 \, dx_1 + \ln S \, dy_2 \, dy_1 + \ln S \, dz_2 \, dz_1) \qquad (6\text{-}60)$$

Thus the integrations over two areas, which would involve integrating over four variables, have been replaced by integrations over the two surface boundaries. This allows considerable computational savings when numerical evaluations must be carried out, and it can sometimes facilitate analytical integration.

EXAMPLE 6-18 Using the contour-integration method, formulate the configuration factor for parallel rectangles as shown in Fig. 6-28.

Note that on both surfaces dz will be zero. First integrate (6-60) around the boundary C_2. The value of S to be used in (6-60) is measured from an

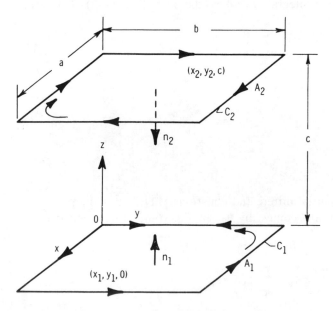

Figure 6-28 Contour integration to determine configuration between two parallel rectangles.

arbitrary point $(x_1, y_1, 0)$ on A_1 to a point on the portion of the boundary C_2 being considered. This gives

$$F_{1-2} = \frac{1}{2\pi ab} \oint_{C_1} \left\{ \int_{y_2=0}^{b} \ln [x_1^2 + (y_2 - y_1)^2 + c^2]^{1/2} \, dy_2 \right.$$

$$+ \int_{y_2=b}^{0} \ln [(a - x_1)^2 + (y_2 - y_1)^2 + c^2]^{1/2} \, dy_2 \bigg\} \, dy_1$$

$$+ \frac{1}{2\pi ab} \oint_{C_1} \left\{ \int_{x_2=0}^{a} \ln [(x_2 - x_1)^2 + (b - y_1)^2 + c^2]^{1/2} \, dx_2 \right.$$

$$+ \int_{x_2=a}^{0} \ln [(x_2 - x_1)^2 + y_1^2 + c^2]^{1/2} \, dx_2 \bigg\} \, dx_1$$

Carrying the integration out over C_1 gives, in this case, eight integrals. The first four corresponding to the first two integrals of the previous equation are written out as

$$2\pi ab F_{1-2} = \int_{y_1=0}^{b} \int_{y_2=0}^{b} \ln [a^2 + (y_2 - y_1)^2 + c^2]^{1/2} \, dy_2 \, dy_1$$

$$+ \int_{y_1=b}^{0} \int_{y_2=0}^{b} \ln [(y_2 - y_1)^2 + c^2]^{1/2} \, dy_2 \, dy_1$$

$$+ \int_{y_1=0}^{b} \int_{y_2=b}^{0} \ln [(y_2 - y_1)^2 + c^2]^{1/2} \, dy_2 \, dy_1$$

$$+ \int_{y_1=b}^{0} \int_{y_2=b}^{0} \ln [a^2 + (y_2 - y_1)^2 + c^2]^{1/2} \, dy_2 \, dy_1$$

$$+ \text{(4 integral terms in } x)$$

$$= \int_{y_1=0}^{b} \int_{y_2=0}^{b} \ln \left[\frac{a^2 + (y_2 - y_1)^2 + c^2}{(y_2 - y_1)^2 + c^2} \right] \, dy_2 \, dy_1$$

$$+ \int_{x_1=0}^{a} \int_{x_2=0}^{a} \ln \left[\frac{(x_2 - x_1)^2 + b^2 + c^2}{(x_2 - x_1)^2 + c^2} \right] \, dx_2 \, dx_1$$

and the configuration factor is now given by the sum of two integrals. These can be integrated analytically by expressing each integrand as the difference of two log functions and letting $y_2 - y_1$ and $x_2 - x_1$ be new variables to reduce the integrals to standard forms. The result is given in Appendix C.

When the contour integration method is applied to geometries in which two surfaces share a common edge, the contour integrals become indeterminate. This difficulty is overcome for two quadrilaterals as follows: For the geometry in Fig.

6-29, one of the integrals in Eq. (6-60) can be carried out analytically [16, 17], resulting in

$$F_{1-2} = \frac{1}{2\pi A_1} \sum_{p=1}^{4} \sum_{q=1}^{4} \Phi(p, q) \oint_{C_p} [(T \cos \theta_1 \ln T + S \cos \theta_2 \ln S + U\mu - V) \, dv]_{p,q}$$

$$(6\text{-}61)$$

where dv is an element on the boundary of A_2, $\Phi(p, q) = l_p l_q + m_p m_q + n_p n_q$, and the other symbols are as shown in Fig. 6-29. Equation (6-61) does not become indeterminate when A_1 and A_2 share a common edge, allowing contour integration to be applied in that case.

Differentiation of known factors An extension of configuration-factor algebra is the generation of configuration factors to differential areas by differentiating known factors to finite areas. This technique is very valuable in certain cases and is demonstrated by an example.

> **EXAMPLE 6-19** As part of the determination of radiative exchange in a square channel whose temperature varies longitudinally, it is desired to find the configuration factor dF_{d1-d2} between an element dA_1 at one corner of the channel end and a differential length of wall section dA_2 as shown in Fig. 6-30a.

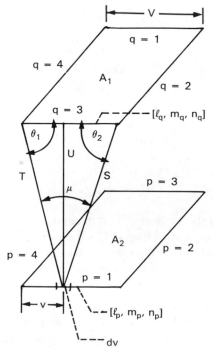

Figure 6-29 Definition of symbols used in contour integration, Eq. (6-61).

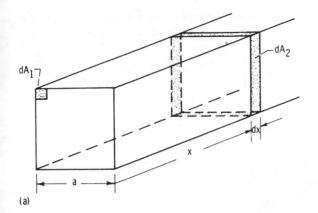

(a)

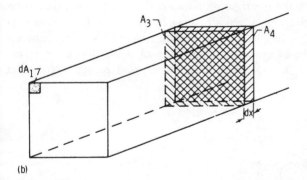

(b)

Figure 6-30 Derivation of configuration factor between differential length of square channel and element at corner of channel end. (*a*) Configuration factor between dA_1 and differential length of channel wall dA_2; (*b*) configuration factor between dA_1 and squares A_3 and A_4.

Configuration-factor algebra plus differentiation can be used to find the required factor. Refer to Fig. 6-30*b*. Since the fraction of energy leaving dA_1 that reaches dA_2 is the difference between the fractions reaching the squares A_3 and A_4, the factor dF_{d1-d2} is the difference between F_{d1-3} and F_{d1-4}. Then

$$dF_{d1-d2} = F_{d1-3} - F_{d1-4} = -\frac{\Delta F_{d1-\square}}{\Delta x} \Delta x \bigg|_{\Delta x \to 0} = -\frac{\partial F_{d1-\square}}{\partial x} dx$$

Thus, if the configuration factor $F_{d1-\square}$ between a corner element and a square in a parallel plane were known, the derivative of this factor with respect to the separation distance could be used to obtain the required factor.

From Example 6-17, the configuration factor between a corner element and a parallel isosceles right triangle is given by setting $\tan \varphi = 1$ in the expression derived for a general right triangle. This yields for the present case, where the distance $d = x$

$$F_{d1-\triangle} = \frac{a}{2\pi(a^2 + x^2)^{1/2}} \tan^{-1} \frac{a}{(a^2 + x^2)^{1/2}}$$

Inspection shows that, by symmetry, the factor between a corner element and a square is twice the factor $F_{d1-\triangle}$ The required factor dF_{d1-d2} is then

$$dF_{d1-d2} = -\frac{\partial F_{d1-\square}}{\partial x} dx = -\frac{a\,dx}{\pi}\frac{\partial}{\partial x}\left[\frac{1}{(a^2+x^2)^{1/2}}\tan^{-1}\frac{a}{(a^2+x^2)^{1/2}}\right]$$

$$= \frac{ax\,dx}{\pi(a^2+x^2)^{3/2}}\left[\tan^{-1}\frac{a}{(a^2+x^2)^{1/2}}+\frac{a(a^2+x^2)^{1/2}}{x^2+2a^2}\right]$$

$$= \frac{X\,dX}{\pi(1+X^2)^{3/2}}\left[\tan^{-1}\frac{1}{(1+X^2)^{1/2}}+\frac{(1+X^2)^{1/2}}{2+X^2}\right]$$

where $X = x/a$.

More generally, start with the configuration factor F_{1-2} for two parallel areas A_1 and A_2 that are cross sections of a cylindrical channel of arbitrary cross-sectional shape (Fig. 6-31a). This factor depends on the spacing $|x_2 - x_1|$ between the two areas and includes blockage due to the channel wall; that is, it is the factor by which A_2 is viewed from A_1 with the channel wall present. Note that

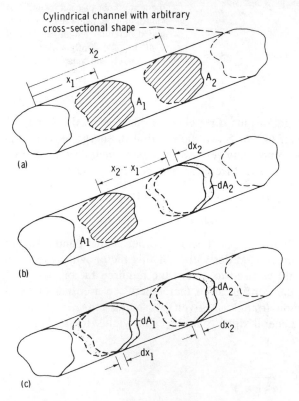

(a)

(b)

(c)

Figure 6-31 Configuration factors for differential areas as derived by differentiation of factor for finite areas. (a) Two finite areas, F_{1-2}; (b) finite to differential area, $dF_{1-d2} = -(\partial F_{1-2}/\partial x_2)\,dx_2$; (c) two differential areas, $dF_{d1-d2} = -(A_1/dA_1)(\partial^2 F_{1-2}/\partial x_1\,\partial x_2)\,dx_2\,dx_1$.

for simple geometries such as a circular tube or rectangular channel the wall block-age is zero. The factor between A_1 and dA_2 in Fig. 6-31b is then given by

$$dF_{1-d2} = -\frac{\partial F_{1-2}}{\partial x_2} dx_2 \qquad (6\text{-}62)$$

as in Example 6-19. Equation (6-62) will now be used to obtain dF_{d1-d2}, the configuration factor between the two differential area elements in Fig. 6-31c.

By reciprocity,

$$F_{d2-1} = \frac{-A_1}{dA_2} \frac{\partial F_{1-2}}{\partial x_2} dx_2$$

Then in a fashion similar to the derivation of (6-62),

$$dF_{d2-d1} = \frac{\partial F_{d2-1}}{\partial x_1} dx_1$$

Substituting F_{d2-1} results in

$$dF_{d2-d1} = -\frac{A_1}{dA_2} \frac{\partial^2 F_{1-2}}{\partial x_1 \partial x_2} dx_2\, dx_1 \qquad (6\text{-}63)$$

or after using reciprocity,

$$dF_{d1-d2} = -\frac{A_1}{dA_1} \frac{\partial^2 F_{1-2}}{\partial x_1 \partial x_2} dx_2\, dx_1 \qquad (6\text{-}64)$$

Hence, by two differentiations the factor dF_{d1-d2} can be found from F_{1-2} for the cylindrical configuration under consideration.

6-4.4 The Unit-Sphere Method

Experimental determination of configuration factors is possible by use of the *unit-sphere method* introduced by Nusselt [18]. If a hemisphere of unit radius is constructed over the area element dA_1 in Fig. 6-32, then the configuration factor from dA_1 to an area A_2 is, by Eq. (6-9),

$$F_{d1-2} = \frac{1}{\pi} \int_{A_2} \cos\theta_1 \frac{\cos\theta_2\, dA_2}{S^2} = \frac{1}{\pi} \int_{A_2} \cos\theta_1\, d\omega_1$$

Note that $d\omega_1$ is the projection of dA_2 onto the surface of the hemisphere, because

$$d\omega_1 = \frac{dA_s}{r^2} = dA_s = \frac{\cos\theta_2\, dA_2}{S^2}$$

where $r = 1$ is the radius of the unit hemisphere. The configuration factor then becomes

$$F_{d1-2} = \frac{1}{\pi} \int_{A_s} \cos\theta_1\, dA_s$$

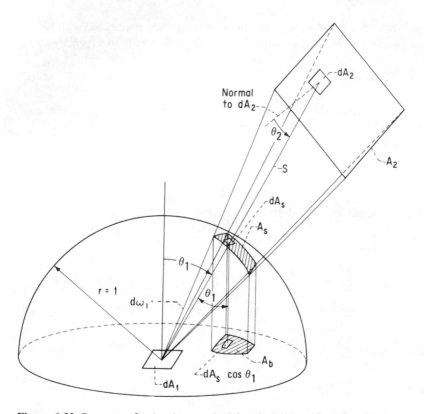

Figure 6-32 Geometry of unit-sphere method for obtaining configuration factors.

However, $dA_s \cos \theta_1$ is the projection of dA_s onto the base of the hemisphere. It follows that integrating $\cos \theta_1 \, dA_s$ gives the projection A_b of A_s onto the base of the hemisphere, or

$$F_{d1-2} = \frac{1}{\pi} \int_{A_s} \cos \theta_1 \, dA_s = \frac{A_b}{\pi} \tag{6-65}$$

This relation forms the basis of several graphic and experimental methods for configuration-factor determination. In one such method, a hemisphere mirrored on the outside is placed over the area element dA_1. A photograph taken by a camera placed above the hemisphere and precisely normal to dA_1 then shows the projection of A_2, which we have called A_b. The measurement of A_b on the photograph then leads to the determination of F_{d1-2} by

$$F_{d1-2} = \frac{A_b}{\pi r_e^2}$$

where r_e is the radius of the experimental mirrored hemisphere. A means of optical projection is given in [19]. Numerical techniques based on this method are given in [20, 21].

6-4.5 Direct Numerical Integration

It is common to determine configuration factors between finite areas by direct numerical integration of Eq. (6-16). An example for a differential-to-finite area geometry is carried out in Sec. 11-3.1. Such integrations require care to incorporate the effects of shading or blocking by intervening surfaces, and they can be quite computationally intensive if many surfaces are present in a complex system. A number of methods are available for increasing the speed or accuracy of these computations, including adaptive gridding to reduce the number of grid nodes while retaining accuracy [22, 23], and partitioning of space by various search algorithms to find and minimize the number of equal-intensity areas that must be summed in the numerical integration [24–26a].

6-4.6 Computer Programs for Evaluation of Configuration Factors

Many computer programs have been written that use one or more of the methods outlined in this chapter for numerical calculation of configuration factors. Examples are FACET [17], which uses area integration, contour integration, and Eq. (6-61) for various geometries; VIEW [27], which can be used with the NASTRAN thermal analysis code; MONTE [28], which uses a Monte Carlo algorithm for exchange factors as well as configuration factors; GLAM [29], for calculating configuration factors for axisymmetric geometries including the effect of shadowing; CNVUFAC [30]; and a program that relies on the computer-graphical analog to the unit-sphere method [31]. The latter program provides factors between a differential element and an arbitrary three-dimensional object. The program TSS (Thermal Simulation System), being developed under NASA sponsorship, incorporates an advanced graphical user interface for displaying configurations. Configuration factor values are represented by shades of color of the objects. The program TRASYS [32] is a comprehensive thermal analysis system that contains subroutines for configuration factor computation and includes effects of shadowing. Some other commercially available programs for thermal analysis also incorporate configuration factor computation. Additional references are in Appendix C.

These and other computer codes provide a means to generate configuration factors for complex geometries and are an invaluable aid to the radiation analyst. Their accuracy can be assessed by comparison of computed results with the analytical expressions developed here for simpler geometries. Several different numerical methods for calculating configuration factors are compared in [33] for computing speed, accuracy, and convenience. The geometries ranged from surfaces almost unobstructed in their view to highly obstructed intersecting surfaces. The methods compared include double integration, Monte Carlo, contour integration, and projection techniques. Monte Carlo performed well, although it can sometimes converge slowly. If the view is not too complex, methods based on contour integration were found to be successful. A good review of numerical methods applicable to complex configurations is provided in [33].

6-5 COMPILATION OF REFERENCES FOR KNOWN CONFIGURATION FACTORS

Many configuration factors for specific geometries have been given in analytical form or tabulated. These formulas and tabulations are spread throughout the literature. Appendix B provides a list of about 230 geometries for which configuration factors are available and includes a reference list to aid in finding these factors. Some factors are given in Appendix C for convenient use. A detailed compilation of factors in the form of formulas, tables, and graphs has been given by Howell [4].

6-6 HISTORICAL NOTE ON CONFIGURATION FACTORS

The first published statement that the geometric factors in the radiant interchange equations could be separated from the energy factors has proved to be elusive. Certainly Nusselt realized that this was so in his derivation of the unit-sphere technique published in 1928 [18]. He refers to the "angle-factor" in the paper and gives the modern interpretation of the factor as the fraction of radiant energy leaving one surface that is incident on another.

One of the first calculations of the radiant exchange between two surfaces was carried out by Christiansen in 1883 [34]. He analyzed radiant exchange between concentric cylinders, treating the cylinders as both diffuse or both specular. By using a ray-tracing technique, he found the relation for diffuse cylinders:

$$\frac{Q_1}{A_1} = \frac{\sigma(T_1^4 - T_2^4)}{1/\epsilon_1 + (1/\epsilon_2 - 1)A_1/A_2} \tag{6-66}$$

With such a derivation, the concept of the configuration factor is neither necessary nor obvious, and Christiansen makes no mention of it.

Sumpner, in 1894, discussed the validity of Lambert's cosine law in relation to some experiments in photometry [35]. He also came near to defining a configuration factor, but didn't make the final step. He did note a fact that remains valid today: "Terms in light are used vaguely, and it will not be deemed out of place to define those which will be here needed."

As late as 1907, Hyde discussed the geometric theory of radiation [36]. He still did not separate the geometry terms and explicitly define a configuration factor, even though he went on to evaluate the integrals appearing in the interchange equations. He examined some rather complex geometries, looking, for example, at the radiant exchange from an ellipse to an area element.

Saunders extended Christiansen's work and in so doing defined a factor he denoted as K [37]. This is the fraction of energy leaving a surface that returns to it by reflections from all other surfaces and is then reabsorbed by the surface.

Saunders applied this concept to simple geometric arrangements of two bodies and did not carry it further. It is in the period 1920–1930 that the concept of configuration factors appears in a great many references. Nusselt's paper is one example, but papers by Buckley and Yamauti make use of the idea [18, 38, 39]. On the other hand, the text by Schack does not, although a passing reference to Nusselt's paper is given [40, 41].

REFERENCES

1. Charle, M.: "Les Manuscripts de Léonard de Vinci, Manuscripts *C*, *E*, et *K* de la Bibliothèque de l'Institute Publiés en Facsimilés Phototypiques," Ravisson-Mollien, Paris, 1888. (Referenced in Middleton, W. E. Knowles: Note on the Invention of Photometry, *Am. J. Phys.*, vol. 31, no. 3, pp. 177–181, 1963.)
2. Francois d'Aguillon, S. J.: "Opticorum Libri Sex," Antwerp, 1613. (Referenced in Middleton, W. E. Knowles: Note on the Invention of Photometry, *Am. J. Phys.*, vol. 31, no. 3, pp. 177–181, 1963.)
3. Herschel, William: Investigation of the Powers of the Prismatic Colours to Heat and Illuminate Objects, *Trans. R. Soc. London*, vol. 90, pt. 2, pp. 255–283, 1800.
4. Howell, John R.: "A Catalog of Radiation Configuration Factors," McGraw-Hill, New York, 1982.
5. Feingold, A.: Radiation-Interchange Configuration Factors between Various Selected Plane Surfaces, *Proc. R. Soc. London*, ser. A, vol. 292, no. 1428, pp. 51–60, 1966.
6. Lebedev, V. A.: Equations Relating Integral Radiation Configuration Factors in Cylindrical Emitting Systems, *Sov. J. Appl. Phys.*, vol. 2, no. 6, pp. 11–17, 1988.
7. Sowell, E. F., and P. F. O'Brien: Efficient Computation of Radiant-Interchange Configuration Factors within an Enclosure, *J. Heat Transfer*, vol. 94, no. 3, pp. 326–328, 1972.
8. Larsen, M. E., and J. R. Howell: Least Squares Smoothing of Direct Exchange Areas in Zonal Analysis, *J. Heat Transfer*, vol. 108, no. 1, pp. 239–242, 1986.
9. van Leersum, J.: A Method for Determining a Consistent Set of Radiation View Factors From a Set Generated by a Nonexact Method, *Int. J. Heat Fluid Flow*, vol. 10, no. 1, pp. 83–85, 1989.
10. Hottel, H. C.: Radiant Heat Transmission, in William H. McAdams (ed.), "Heat Transmission," 3rd ed., chap. 4, McGraw-Hill, New York, 1954.
11. Sparrow, E. M., and R. D. Cess: "Radiation Heat Transfer," augmented ed., Hemisphere, Washington, D.C., 1978.
12. Moon, Parry: "The Scientific Basis of Illuminating Engineering," rev. ed., Dover, New York, 1961.
13. de Bastos, R.: "Computation of Radiation Configuration Factors by Contour Integration," M.S. thesis, Oklahoma State University, Stillwater, 1961.
14. Sparrow, E. M.: A New and Simpler Formulation for Radiative Angle Factors, *J. Heat Transfer*, vol. 85, no. 2, pp. 81–88, 1963.
15. Shapiro, A. B.: Computer Implementation, Accuracy, and Timing of Radiation View Factor Algorithms, *J. Heat Transfer*, vol. 107, no. 3, pp. 730–732, 1985.
16. Mitalas, G. P., and D. G. Stephenson: FORTRAN IV Programs to Calculate Radiant Interchange Factors, National Research Council of Canada, Division of Building Research Report DBR-25, Ottawa, 1966.
17. Shapiro, Arthur B.: FACET—A Computer View Factor Computer Code for Axisymmetric, 2D Planar, and 3D Geometries with Shadowing, UCID-19887, University of California, Lawrence Livermore National Laboratory, August 1983.

18. Nusselt, Wilhelm: Graphische Bestimmung des Winkelverhaltnisses bei der Wärmestrahlung, *VDI Z.*, vol. 72, p. 673, 1928.
19. Farrell, R.: Determination of Configuration Factors of Irregular Shape, *J. Heat Transfer*, vol. 98, no. 2, pp. 311–313, 1976.
20. Lipps, F. W.: Geometric Configuration Factors for Polygonal Zones Using Nusselt's Unit Sphere, *Solar Energy*, vol. 30, no. 5, pp. 413–419, 1983.
21. Rushmeier, H. E., D. R. Baum, and D. E. Hall: Accelerating the Hemi-Cube Algorithm for Calculating Radiation Form Factors, *J. Heat Transfer*, vol. 113, no. 4, pp. 1044–1047, 1991.
22. Campbell, A. T., III, and D. S. Fussell: Adaptive Mesh Generation for Global Diffuse Illumination, *Computer Graphics*, vol. 24, no. 4, pp. 155–164, August 1990.
23. Saltiel, C. J., and J. Kolibal: Adaptive Grid Generation for the Calculation of Radiative Configuration Factors, *AIAA J. Thermophys. Heat Transfer*, in press.
24. Thibault, W., and B. Naylor: Set Operations on Polyhedra Using Binary Space Partitioning Trees, *Computer Graphics*, vol. 21, no. 3, pp. 315–324, 1987.
25. Buckalew, C., and D. Fussell: Illumination Networks: Fast Realistic Rendering with General Reflectance Functions, *Computer Graphics*, vol. 23, no. 3, pp. 89–98, 1989.
26. Cohen, M. F., S. E. Chen, and J. R. Wallace: A Progressive Refinement Approach to Fast Radiosity Image Generation, *Computer Graphics*, vol. 22, no. 4, pp. 315–324, 1988.
26a. Hanrahan, P., D. Salzman, and L. Aupperle: A Rapid Hierarchical Radiosity Algorithm, *Comput. Graphics*, vol. 25, no. 4, pp. 197–206, 1991.
27. Emery, A. F.: VIEW—A Radiation View Factor Program with Interactive Graphics for Geometry Definition (Version 5.5.3), NASA Computer Software Management and Information Center, Atlanta, 1986.
28. Burns, P. J.: MONTE—A Two-Dimensional Radiative Exchange Factor Code, Colorado State University, Fort Collins, 1983.
29. Garelis, E., T. E. Rudy, and R. B. Hickman: GLAM—A Steady-State Numerical Solution to the Vacuum Equation of Transfer in Cylindrically Symmetric Geometries, University of California, Lawrence Livermore National Laboratory, UCID-19157, 1981.
30. Wong, R. L.: User's Manual for CNVUFAC—The General Dynamics Heat Transfer Radiation View Factor Program, University of California, Lawrence Livermore National Laboratory, UCID-17275, 1976.
31. Alciatore, D., S. Lipp, and W. S. Janna: Closed-Form Solution of the General Three Dimensional Radiation Configuration Factor Problem with Microcomputer Solution, *Proc. 26th National Heat Transfer Conf.*, Philadelphia, August 1989.
32. Jensen, C. L.: "Thermal Radiation Analysis System," TRASYS-II User's Manual, ANSI Version 1.0, Martin Marietta Aerospace Corp., Denver, Colorado, 1987. Also available from COSMIC.
33. Emery, A. F., O. Johansson, M. Lobo, and A. Abrous: A Comparative Study of Methods for Computing the Diffuse Radiation Viewfactors for Complex Structures," *J. Heat Transfer*, vol. 113, no. 2, pp. 413–422, 1991.
34. Christiansen, C.: II. Absolute Bestimmung des Emissions- und Absorptionsvermögens für Warmes, *Ann. Phys., Wied.* vol. 19, pp. 267–283, 1883.
35. Sumpner, W. E.: The Diffusion of Light, *Proc. Phys. Soc. London*, vol. 94, pp. 10–29, 1894.
36. Hyde, Edward P.: Geometrical Theory of Radiating Surfaces with Discussion of Light Tubes, *Nat. Bur. Stand. U.S. Bull.*, vol. 3, pp. 81–104, 1907.
37. Saunders, O. A.: Notes on Some Radiation Heat Transfer Formulae, *Proc. Phys. Soc. London*, vol. 41, pp. 569–575, 1928–1929.
38. Buckley, H.: Radiation from the Interior of a Reflecting Cylinder, *Philos. Mag.*, vol. 4, pp. 753–762, 1927.
39. Yamauti, Z.: Geometrical Calculation of Illumination, *Res. Electrotech. Lab. Tokyo*, vol. 148, 1924.
40. Schack, A.: "Industrielle Wärmeübergang," 1st ed., Verlag Stahleisen mbH, Düsseldorf, 1929.
41. Schack, A.: "Industrial Heat Transfer" (H. Goldschmidt and E. P. Partridge, trans.), Wiley, New York, 1933.

PROBLEMS

6-1 Derive the configuration factor F_{d1-2} between a differential area centered above a disk and a finite disk of unit radius.

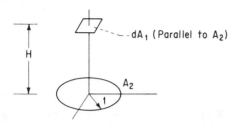

Answer: $1/(H^2 + 1)$

6-2 Find the configuration factor F_{d1-2} from a planar element to a coaxial parallel rectangle as shown below. Use any method except contour integration.

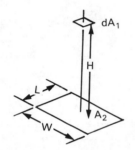

Answer: $F_{d1-2} = \left(\dfrac{2}{\pi}\right)\left\{ \left[\dfrac{X}{(1 + X^2)^{1/2}}\right] \tan^{-1}\left[\dfrac{Y}{(1 + X^2)^{1/2}}\right] + \left[\dfrac{Y}{(1 + Y^2)^{1/2}}\right] \tan^{-1}\left[\dfrac{X}{(1 + Y^2)^{1/2}}\right] \right\}$

6-3 Derive by any three methods, including use of the factors in Appendix C if you choose, the configuration factor F_{1-2} for the infinitely long geometry shown in cross section below.

Answer: $F_{1-2} = \{A + (1 + B^2)^{1/2} - [(A + B)^2 + 1]^{1/2}\}/2$

6-4 The configuration factor between two infinitely long directly opposed parallel plates of finite width L is F_{1-2}. The plates are separated by a distance D.

(a) Derive an expression for F_{1-2} by integration of the configuration factor between differential strip elements.

(b) Derive an expression for F_{1-2} by the crossed-string method.

Answer: (a), (b) $[1 + (D/L)^2]^{1/2} - (D/L)$

6-5 The configuration factor between two infinite parallel plates of finite width L is F_{1-2} in the configuration shown below in cross section.

(a) Derive an expression for F_{1-2} by the crossed-string method.

(b) Derive an expression for F_{1-2} by using the results of Problem 6-4 and configuration-factor algebra.

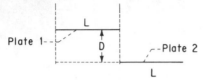

(c) Find the configuration factor F_{1-2} for the geometry of infinitely long plates shown below in cross section.

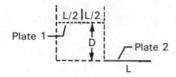

$$Answer:\ \text{(a), (b)}\ \left[1 + \left(\frac{D}{2L}\right)^2\right]^{1/2} + \frac{D}{2L} - \left[1 + \left(\frac{D}{L}\right)^2\right]^{1/2};$$

$$\text{(c)}\ F_{1-2} = 2\left[1 + \left(\frac{D}{2L}\right)^2\right]^{1/2} - \left[1 + \left(\frac{D}{L}\right)^2\right]^{1/2} - \left[\left(\frac{D}{L}\right)^2 + \left(\frac{3}{2}\right)^2\right]^{1/2} + \left[\left(\frac{D}{L}\right)^2 + \left(\frac{1}{2}\right)^2\right]^{1/2}$$

6-6 (a) For the two-dimensional geometry shown in cross section, derive a formula for F_{2-2} in terms of r_1 and r_2.

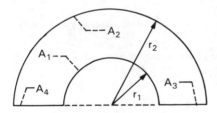

(b) Find F_{2-2} for the two-dimensional geometry. A_1 is a square (A_1 refers to the total area of the four sides) and A_2 is part of a circle.

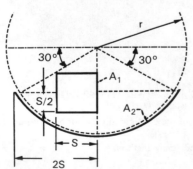

Answer: (a) $F_{2-2} = 1 - \dfrac{2}{\pi}\left[1 - \left(\dfrac{r_1}{r_2}\right)^2\right]^{1/2} - \dfrac{2}{\pi}\dfrac{r_1}{r_2}\sin^{-1}\left(\dfrac{r_1}{r_2}\right)$;

(b) $F_{2-2} = 1 - \dfrac{3\sqrt{3}}{8\pi}\left(1 + \dfrac{1}{2}\sqrt{5} + \dfrac{1}{2}\sqrt{17}\right) = 0.1359$

6-7 Derive a formula for F_{2-2} in terms of r and α. A_1 is inside the cone.

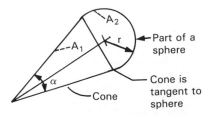

Answer: $F_{2-2} = \dfrac{1}{2}\left(1 + \sin\dfrac{\alpha}{2}\right)$

6-8 Compute the configuration factor F_{1-2} between faces A_1 and A_2 of the infinitely long parallel plates shown below in cross section when the angle β is equal to (a) $30°$ and (b) $90°$.

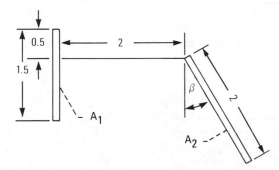

Answer: (a) 0.275; (b) 0.0377

6-9 Find the configuration factor between the two infinitely long parallel plates shown below in cross section. Use

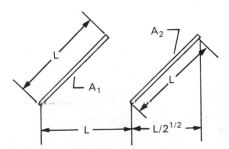

(a) the crossed-string method
(b) configuration-factor algebra with factors from Appendix C
(c) integration of differential strip–to–differential strip factors

Answer: 0.3066

6-10 A sphere of radius r is divided into two quarter-spheres and one hemisphere. Obtain the configuration factors between all areas inside the sphere, F_{1-2}, F_{2-2}, F_{3-1}, F_{1-1}, etc. From these factors find F_{4-5}, where A_4 and A_5 are equal semicircles that are at right angles to each other.

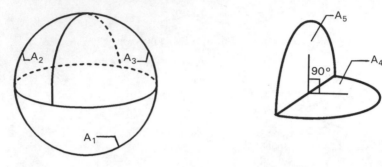

Answer: $F_{4-5} = 0.25$

6-11 For the two-dimensional geometry shown, the view between A_1 and A_2 is partially blocked by an intervening structure. Determine the configuration factor F_{1-2}.

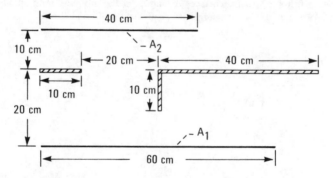

Answer: 0.2118

6-12 The cylindrical geometry shown in cross section is very long in the direction normal to the plane of the drawing. The cross section consists of two concentric three-quarter circles and two straight lines. Obtain the value of the configuration factor F_{2-2}.

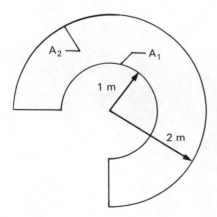

Answer: 0.35467

6-13 Derive the configuration factor F_{1-2} from a finite rectangle A_1 to an infinite plane A_2 where the rectangle is at an angle η relative to the plane.

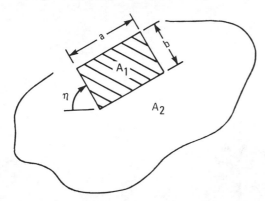

Answer: $F_{1-2} = (1/2)(1 - \cos \eta)$

6-14 The four flat plates shown in cross section are very long in the direction normal to the plane of the cross section. Obtain the value of the configuration factor F_{1-2}. What are the values of F_{2-1} and F_{1-3}?

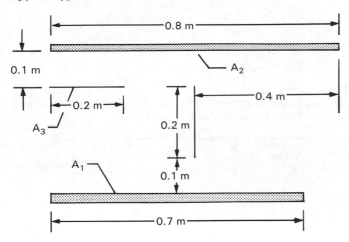

Answer: $F_{1-2} = 0.31860$, $F_{2-1} = 0.33390$, $F_{1-3} = 0.16066$

6-15 Using the crossed-string method, derive the configuration factor F_{1-2} between the infinitely long plate and cylinder shown below in cross section. Compare your result with that given for configuration C-27 in Appendix C.

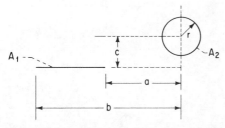

Answer: $F_{1-2} = [r/(b - a)] [\tan^{-1}(b/c) - \tan^{-1}(a/c)]$

6-16 A long tube in a tube bundle is surrounded by six other identical equally spaced tubes as shown in cross section below. What is the configuration factor from the central tube to each of the surrounding tubes?

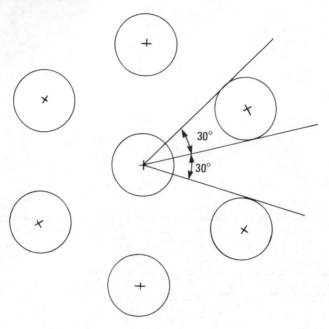

Answer: 0.0844 (>1/12)

6-17 For the two-dimensional geometry shown, the view between A_1 and A_2 is partially blocked by two identical cylinders. Determine the view factor F_{1-2}.

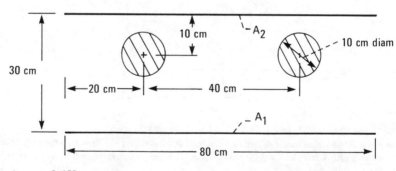

Answer: 0.459

6-18 F_{n-k} is the configuration factor between two perpendicular rectangles having a common edge. In terms of this factor, use configuration-factor relations to derive the F_{1-8} factor between the two areas A_1 and A_8.

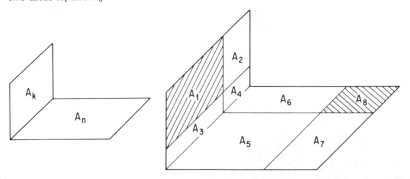

6-19 Find F_{1-2} by any two methods for the two-dimensional geometry shown in cross section.

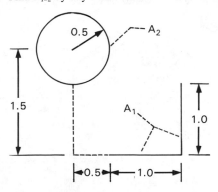

Answer: 0.232

6-20 An enclosure is formed of three mutually perpendicular isosceles triangles of short side S and hypotenuse H and an equilateral triangle of side H. Find the configuration factor F_{1-2} between two perpendicular isosceles triangles.

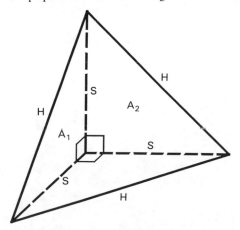

Answer: 0.2113

6-21 Find the configuration factor F_{1-2} from the quarter-disk to the planar ring shown.

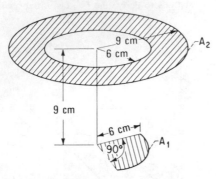

Answer: 0.195

6-22 (a) Derive the configuration factor between a sphere of radius R and a coaxial disk of radius r. (Hint: think of the disk as being a cut through a spherical envelope concentric around the sphere.)

(b) What is the F from a sphere to a sector of a disk?

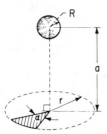

(c) What is the dF from a sphere to a portion of a ring of differential width?

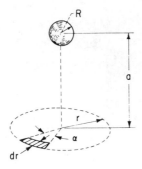

Answer: (a) $\dfrac{1}{2}\left[1 - \dfrac{a}{(a^2 + r^2)^{1/2}}\right]$; (b) $\dfrac{\alpha}{4\pi}\left[1 - \dfrac{a}{(a^2 + r^2)^{1/2}}\right]$; (c) $\dfrac{\alpha}{4\pi}\dfrac{ar}{(a^2 + r^2)^{3/2}}\,dr$

6-23 Use the crossed-string method to derive configuration factor 2 of Appendix C (infinitely long strip of differential width to infinitely long cylindrical surface).

Answer: $F_{d1-2} = (1/2)(\sin \beta_2 + \sin \beta_1)$

6-24 Use the disk-to-disk configuration factor 23 of Appendix C to obtain the factor F_{d1-d2} between the interior surfaces of two differential rings on the interior of a right circular cone.

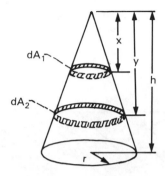

Answer: $dF_{d1-d2} = -\dfrac{\cos \alpha}{2Rx^2} ([x - 2y(R^2 + 1)] + (R^2 + 1)^{1/2}\{[x^4(R^2 + 1) + x^3y(2R^4 - R^2 - 5)$

$+ 3x^2y^2(2R^4 - R^2 + 3) + xy^3(R^2 + 1)(6R^2 - 7) + 2y^4(R^2 + 1)^2] \div [x^2(R^2 + 1) + 2xy(R^2$

$- 1) + y^2(R^2 + 1)]^{3/2}\})dy$ where $R = r/h$, $\alpha = \tan^{-1}(r/h)$, $y > x$

6-25 In terms of disk-to-disk configuration factors, derive the factor between finite ring A_1 and the finite area A_2 on the inside of the cone.

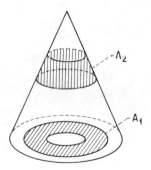

6-26 A closed right circular cylindrical shell with base diameter 1 m and height 1 m is located at the center of a spherical shell 1 m in radius.

(a) Determine the configuration factor between the sphere and itself.

(b) If the top of the cylindrical shell were removed, determine the configuration factor between the sphere and the inside of the bottom of the cylindrical shell.

Answer: (a) 5/8; (b) 0.01072

6-27 Consider a black cubic enclosure. Determine the configuration factors between (a) two adjacent walls and (b) two opposite walls. A sphere of diameter equal to one-half the length of a side of the cube is placed at the center of the cube. Determine the configuration factors between (c) the sphere and one wall of the enclosure, (d) one wall of the enclosure and the sphere, and (e) the enclosure and itself.

Answer: (a) 0.20004; (b) 0.19982; (c) 1/6; (d) 0.1309; (e) 0.8691

6-28 Obtain the configuration factor dF_{d1-2} in Problem 6-2 by using contour integration.

6-29 By use of the contour integration method of Sec. 6-4.3, obtain the final result in Example 6-4 for the configuration factor from an elemental area to a perpendicular circular disk.

$$Answer: F_{d1-2} = \frac{H}{2}\left[\frac{H^2 + R^2 + 1}{\sqrt{(H^2 + R^2 + 1)^2 - 4R^2}} - 1\right]$$

6-30 Obtain the value of the configuration factor dF_{d1-2} for the geometry shown. The areas dA_1 and A_2 are parallel. (Set theory may be helpful.)

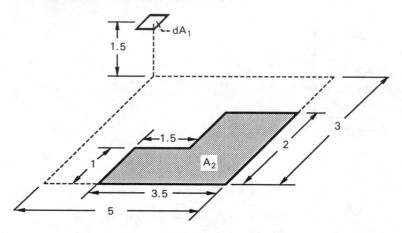

SEVEN

RADIATION EXCHANGE IN AN ENCLOSURE COMPOSED OF BLACK OR DIFFUSE-GRAY SURFACES

7-1 INTRODUCTION

7-1.1 Approximations in Black or Diffuse-Gray Enclosure Analysis

This chapter begins with a short portion on the analysis of radiation exchange in an enclosure where all surfaces are *black*. Black surfaces are perfect absorbers, and the energy exchange process is thus simplified because there are no reflections. Black surfaces emit in a diffuse fashion, so the *intensity* leaving a surface is *independent of direction*. For emission from an isothermal surface the configuration factor can be used to calculate how much radiation will reach another surface. The relation for exchange between two surfaces is then applied to multiple surfaces, each at a different uniform temperature, in an enclosure of black surfaces. The general equations for exchange within such an enclosure are developed, and illustrative examples are presented.

The major portion of the chapter contains the next step toward more complex treatments to account for the real property behavior of surfaces. The surfaces of the enclosure are taken to be *diffuse* and *gray*. When a surface is diffuse-gray, the directional spectral emissivity and absorptivity do not depend on direction or wavelength but can depend on surface temperature. At any surface temperature T_A the hemispherical total absorptivity and emissivity are equal and depend only on T_A, $\alpha(T_A) = \epsilon(T_A)$. Even though this behavior is approached by only a limited number of real materials, the diffuse-gray approximation is often made to greatly simplify enclosure theory.

Some comment is warranted as to what is meant by the individual "surfaces" or "areas" that constitute the enclosure boundary. Usually, the geometry tends to divide the enclosure into natural surface areas, such as the individual sides of a rectangular prism. In addition, it may be necessary to specify surface areas on the basis of heating conditions; for example, if one side of an enclosure is partly at one temperature and partly at a second temperature, the side would be divided into two separate areas so that this difference in boundary condition could be accounted for. An area may also be subdivided on the basis of surface charac-teristics, such as separate smooth and rough portions. The surfaces or areas in the radiation analysis are each portions of the enclosure boundary for which a heat balance is formed. These portions are selected on the basis of geometry, imposed heating or temperature conditions, or surface characteristics. A further consider-ation is the accuracy of the solution. If too few areas are designated, the accuracy may be poor because significant nonuniformity in reflected flux over an area has not been accounted for. Too many areas may require excessive computation time. Thus engineering judgment is required in selecting the size and shape of the areas and their number.

Surfaces of the enclosure can have various imposed thermal boundary con-ditions. A surface can be held at a specified temperature, have a specified heat input, or be perfectly insulated from external heat addition or removal. It is a restriction in the present analysis that, whatever conditions are imposed, each separate surface of the enclosure must be at a uniform temperature. If the imposed heating conditions are such that the temperature would vary markedly over an area, the area should be subdivided into smaller, more nearly isothermal portions; these portions can be of differential size if necessary. As a consequence of this isothermal area requirement, the analysis has a uniform emitted energy over each surface.

Because a gray surface is not a perfect absorber, part of the incident energy is reflected. With regard to reflected energy, two assumptions are made: (1) The reflected energy is diffuse, that is, the reflected *intensity* at each position *is uni-form for all directions*, and (2) the reflected energy is *uniform* over each surface. If the reflected energy is expected to vary over an area, the area should be sub-divided into smaller areas over each of which the reflected energy will not vary too much. With these restrictions reasonably met, the reflected energy for each surface has the same diffuse and uniformly distributed character as the emitted energy. Hence, the reflected and emitted energies can be combined into a single energy quantity leaving the surface.

When a surface is both diffusely emitting and reflecting, the *intensity* of all energy leaving the surface does *not* vary with direction. The geometric configu-ration factors can then be used for the enclosure analysis. The derivation of the *F* factors in Chap. 6 was based on the condition of a *diffuse uniform intensity* leaving the surface; this diffuse-uniform condition must hold for *both emitted and reflected* energies.

The radiative heat balances considered here are not limited to steady-state conditions. The radiative balances can be directly applied to situations in which transient temperature changes are occurring. The heat flux q computed in the enclosure theory can be regarded as the instantaneous net radiative loss from the location being considered on the enclosure boundary. For example, if a solid body is cooling by radiation, q provides the boundary condition for the transient heat conduction solution for the temperature distribution within the solid.

7-1.2 Summary of Restrictions in Analysis

The assumptions for the present chapter are now summarized. The enclosure boundary is divided into areas so that over each area the following restrictions are met:

1. The temperature is uniform.
2. The surface properties are uniform.
3. The ϵ'_λ, α'_λ, and ρ'_λ are independent of wavelength and direction so that $\epsilon(T_A)$ $= \alpha(T_A) = 1 - \rho(T_A)$, where ρ is the reflectivity.
4. All energy is emitted and reflected diffusely.
5. The incident and hence reflected energy flux is uniform over each individual area.

   In some instances an analysis assuming *diffuse-gray* surfaces cannot yield good results. For example, if the temperatures of the individual surfaces of the enclosure differ considerably from each other, then a surface will be emitting predominantly in the range of wavelengths characteristic of its temperature while receiving energy predominantly in a different wavelength region. If the spectral emissivity varies with wavelength, the fact that the incident radiation has a different spectral distribution than the emitted energy will make the gray assumption invalid; that is, $\epsilon(T_A) \neq \alpha(T_A)$. When polished (specular) surfaces are present, the diffuse reflection assumption will be invalid, and the directional paths of the reflected energy must be considered. The treatments of specular and other more general surfaces are the subjects of Chaps. 8 and 9. In addition, some phenomena that are rather more specialized than is the intent of this work to be can be of importance in certain situations. For example, effects of polarization and wave interference can lead to errors in energy transfer calculations if they are ignored under certain special conditions of geometry and surface finish [1, 2].

7-2 SYMBOLS

a_{kj}	matrix elements defined by Eq. (7-48)
a^{-1}	inverse matrix
A	area
$\mathscr{A}$	inverse matrix coefficients, Eq. (7-52)
dA^*	differential element on same surface area as dA
C_{kj}	matrix elements defined by Eq. (7-48)

D	diameter of tube or hole
e	emissive power
E	the quantity $(1 - \epsilon)/\epsilon$
F	configuration factor
$\mathscr{F}$	transfer factors in enclosure
G	function in integral equation, Eq. (7-80); factors in Gebhart's method, Eq. (7-38)
i	intensity
j, k	indices denoting individual surfaces
J	auxiliary variational function, Eq. (7-81)
J_1	Bessel function of the first kind
K	kernel of integral equation
l	dimensionless length
L	length of surface
m	coefficient in temperature distribution, Example 7-24
M_{kj}	minor of matrix element a_{kj}
N	number of surfaces in enclosure
q	energy flux, energy per unit area and time
Q	energy per unit time
r	radius
$\mathbf{r}$	direction vector
R	radius of sphere
S	distance between areas
t	abbreviation for T^4
T	absolute temperature
x, y, z	coordinates
α	absorptivity
β	angle away from normal in cross section
γ	polynomial coefficients, Eq. (7-82)
δ	Kronecker delta
ϵ	emissivity
θ	cone angle; angle from normal of surface
ξ, η	dimensionless coordinates
ρ	reflectivity
σ	Stefan-Boltzmann constant
φ	circumferential or azimuthal angle
Φ	dependent variable in integral equation
ω	solid angle

Subscripts

a	apparent value
A	area
b	blackbody
black	blackbody property
e	external radiation entering through opening; environment

i	incoming
j, k	property of surface j or k
o	outgoing
r	reduced temperature
s	sphere
w	wall
λ	spectrally (wavelength) dependent
1, 2	surface 1 or 2

Superscript

$'$	quantity in one direction

7-3 RADIATIVE EXCHANGE FOR BLACK SURFACES

Equations (6-3) and (6-4) can be written for black surfaces to give the total energy per unit time leaving dA_1 that is incident upon dA_2 as

$$d^2Q'_{d1-d2} = \frac{i'_{b,1}\, dA_1 \cos\theta_1\, dA_2 \cos\theta_2}{S^2} \tag{7-1}$$

and the total radiation leaving dA_2 that arrives at dA_1:

$$d^2Q'_{d2-d1} = \frac{i'_{b,2}\, dA_2 \cos\theta_2\, dA_1 \cos\theta_1}{S^2} \tag{7-2}$$

For a black receiving element all incident energy is absorbed, so Eqs. (7-1) and (7-2) give the energy from one area element that is *absorbed* by the second.

The *net* energy per unit time $d^2Q'_{d1 \rightleftharpoons d2}$ transferred from black element dA_1 to black element dA_2 *along path* S is the difference, $d^2Q'_{d1-d2} - d^2Q'_{d2-d1}$. From (2-26), the blackbody total intensity is related to the blackbody hemispherical total emissive power by $i'_b = e_b/\pi = \sigma T^4/\pi$, so the net transfer is written as

$$d^2Q'_{d1 \rightleftharpoons d2} = \sigma(T_1^4 - T_2^4) \frac{\cos\theta_1 \cos\theta_2}{\pi S^2} dA_1\, dA_2 \tag{7-3}$$

EXAMPLE 7-1 The sun emits energy at a rate that can be approximated by that of a blackbody at 5780 K. A blackbody area element in orbit around the sun at the mean radius of the earth's orbit (1.49×10^{11} m) is oriented normal to the line connecting the centers of the area element and the sun. If the sun's radius is 6.95×10^8 m, what energy flux is incident upon the element?

To the element in orbit, the sun appears as a diffuse isothermal disk element of area

$$dA_1 = \pi r_s^2 = \pi (6.95 \times 10^8)^2 = 1.52 \times 10^{18} \text{ m}^2$$

From Eq. (7-1), the incident energy flux on the element in orbit is, since $\theta_1 = \theta_2 = 0$,

$$\frac{d^2 Q'_{d1-d2}}{dA_2} = i'_{b,1} \, dA_1 \frac{\cos \theta_1 \cos \theta_2}{S^2} = \frac{\sigma T_s^4}{\pi} \frac{dA_1}{S^2}$$

$$= \frac{5.6705 \times 10^{-8} \times 5780^4}{\pi} \frac{1.52 \times 10^{18}}{(1.49 \times 10^{11})^2} = 1379 \text{ W/m}^2$$

This value is consistent with the range of measured values of the mean solar constant, 1332–1374 W/m^2 (422–436 Btu/(h · ft^2)).

An alternative procedure is to utilize the fact that the radiant energy will leave the sun in a spherically symmetric fashion. The energy radiated is $\sigma T_s^4 4\pi r_s^2$, and the area of a sphere surrounding the sun and having a radius equal to the earth's orbit is $4\pi S^2$. Hence the flux received at the earth's orbit is $\sigma T_s^4 4\pi r_s^2 / 4\pi S^2 = \sigma T_s^4 (r_s/S)^2$, giving the same result as before.

EXAMPLE 7-2 As shown in Fig. 7-1, a black square with side 0.25 cm is at temperature 1100 K and is near a tube 0.25 cm in diameter. The opening of the tube acts as a black surface, and the tube is at 700 K. What is the net

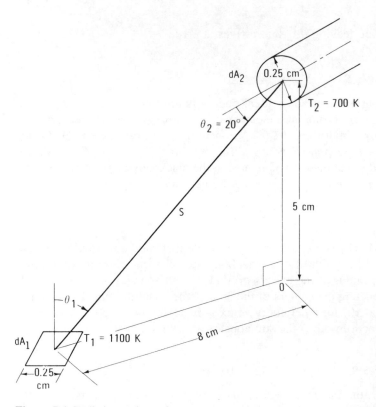

Figure 7-1 Radiative exchange between square element and circular-tube opening.

radiation exchange from dA_1 to dA_2 along the connecting path S between the square and the tube opening?

From (7-3),

$$d^2Q'_{d1 \rightleftharpoons d2} = \sigma(T_1^4 - T_2^4) \frac{\cos\theta_1 \cos\theta_2}{\pi S^2} dA_1 \, dA_2$$

The value of $\cos\theta_1$ is found from the known sides of the right triangle dA_2–0–dA_1 as $\cos\theta_1 = 5/(8^2 + 5^2)^{1/2} = 5/(89^{1/2})$. The other factors in the energy exchange equation are given, so that

$$d^2Q'_{d1 \rightleftharpoons d2} = 5.6705 \times 10^{-8}(1100^4 - 700^4) \frac{5}{89^{1/2}} \frac{\cos 20°}{\pi(89/10^4)} \frac{0.25^2}{10^4} \frac{\pi 0.25^2}{4 \times 10^4}$$

$$= 3.793 \times 10^{-5} \text{ W}$$

7-3.1 Exchange between Surfaces by Use of Configuration Factors

Using Eq. (6-5), Eq. (7-3) can be written in terms of configuration factors as

$$d^2Q'_{d1 \rightleftharpoons d2} = \sigma(T_1^4 - T_2^4) \, dF_{d1-d2} \, dA_1 = \sigma(T_1^4 - T_2^4) \, dF_{d2-d1} \, dA_2 \qquad (7\text{-}4)$$

Similarly, for radiation exchange between a differential element and a finite area,

$$dQ_{d1 \rightleftharpoons 2} = dQ_{d1-2} - dQ_{2-d1} = \sigma T_1^4 dA_1 F_{d1-2} - \sigma T_2^4 A_2 dF_{2-d1} \qquad (7\text{-}5)$$

or, using Eq. (6-15),

$$dQ_{d1 \rightleftharpoons 2} = \sigma(T_1^4 - T_2^4) \, dA_1 F_{d1-2} = \sigma(T_1^4 - T_2^4) A_2 dF_{2-d1} \qquad (7\text{-}6)$$

For two black surfaces with finite areas

$$Q_{1 \rightleftharpoons 2} = Q_{1 \rightarrow 2} - Q_{2 \rightarrow 1} = \sigma T_1^4 A_1 F_{1-2} - \sigma T_2^4 A_2 F_{2-1} \qquad (7\text{-}7)$$

or, using the reciprocity relation Eq. (6-20),

$$Q_{1 \rightleftharpoons 2} = \sigma(T_1^4 - T_2^4) A_1 F_{1-2} = \sigma(T_1^4 - T_2^4) A_2 F_{2-1} \qquad (7\text{-}8)$$

7-3.2 Radiation Exchange in a Black Enclosure

The net energy transfer between two separate surfaces or surface elements is now generalized to obtain the energy transfer *within* an enclosure of black surfaces that are individually isothermal. In practice, the interior surfaces of an enclosure, such as a furnace, may not be isothermal. In such a case, the nonisothermal surfaces are subdivided into portions that can be considered individually isothermal. The theory for a black enclosure serves as an introduction to less restrictive theory that follows.

Form a heat balance on a typical surface A_k (Fig. 7-2). The combined energy supplied to A_k by all sources, other than the radiation inside the enclosure, to maintain A_k at T_k is Q_k. The Q_k could be composed of convection to the inside of the wall and/or conduction through the wall from an energy source outside such as convection and/or radiation. For wall cooling such as by channels within the wall, the contribution to Q_k is negative. The emission from A_k is $\sigma T_k^4 A_k$. The radiant energy received by A_k from another surface A_j is $\sigma T_j^4 A_j F_{j-k}$. The heat balance is then

$$Q_k = \sigma T_k{}^4 A_k - \sum_{j=1}^{N} \sigma T_j{}^4 A_j F_{j-k} \tag{7-9}$$

where the summation includes energy arriving from all surfaces of the enclosure including A_k if A_k is concave. Equation (7-9) can be written in alternative forms. Applying reciprocity to the terms in the summation results in

$$Q_k = A_k\left(\sigma T_k^4 - \sum_{j=1}^{N} \sigma T_j^4 F_{k-j}\right) \tag{7-10}$$

For a complete enclosure, $\sum_{j=1}^{N} F_{k-j} = 1$ from (6-38), so that

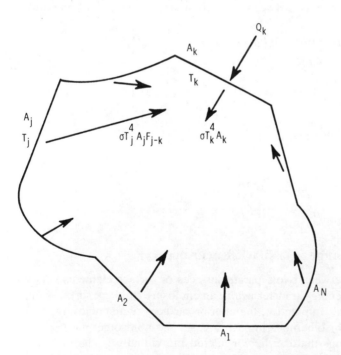

Figure 7-2 Enclosure composed of N black isothermal surfaces (shown in cross section for simplicity).

$$Q_k = A_k\left(\sigma T_k^4 \sum_{j=1}^{N} F_{k-j} - \sigma \sum_{j=1}^{N} T_j^4 F_{k-j}\right) = \sigma A_k \sum_{j=1}^{N} (T_k^4 - T_j^4) F_{k-j} \qquad (7\text{-}11)$$

This is in the form of a sum of the net energy transferred from A_k to each surface.

EXAMPLE 7-3 The three-sided black enclosure of Example 6-13 has its surfaces maintained at temperatures T_1, T_2, and T_3. Determine the amount of energy to maintain these temperatures that must be supplied to each surface per unit time by means other than radiation inside the enclosure. This is the net radiative loss from each surface resulting from radiative exchange within the enclosure.

Equation (7-11) is written for each surface as

$$Q_1 = A_1[F_{1-2}\sigma(T_1^4 - T_2^4) + F_{1-3}\sigma(T_1^4 - T_3^4)]$$

$$Q_2 = A_2[F_{2-1}\sigma(T_2^4 - T_1^4) + F_{2-3}\sigma(T_2^4 - T_3^4)]$$

$$Q_3 = A_3[F_{3-1}\sigma(T_3^4 - T_1^4) + F_{3-2}\sigma(T_3^4 - T_2^4)]$$

The configuration factors have been found in Example 6-13. Thus all factors on the right sides of the equations are known, and the Q values may be computed directly. A check on the numerical results is that, from overall energy conservation, the net Q added to the entire enclosure, $\sum_{k=1}^{N} Q_k$, must be zero to maintain steady temperatures. This is also shown by using reciprocal relations on the Q equations to obtain

$$\sum_{k=1}^{3} Q_k = A_1 F_{1-2}\sigma(T_1^4 - T_2^4) + A_1 F_{1-3}\sigma(T_1^4 - T_3^4)$$

$$+ A_1 F_{1-2}\sigma(T_2^4 - T_1^4) + A_2 F_{2-3}\sigma(T_2^4 - T_3^4)$$

$$+ A_1 F_{1-3}\sigma(T_3^4 - T_1^4) + A_2 F_{2-3}\sigma(T_3^4 - T_2^4) = 0$$

EXAMPLE 7-4 The enclosure of Example 6-13 has two sides maintained at temperatures T_1 and T_2. The third side is insulated on the outside, $Q_3 = 0$. Determine Q_1, Q_2, and T_3.

Eq. (7-11) is written for each surface as

$$Q_1 = A_1[F_{1-2}\sigma(T_1^4 - T_2^4) + F_{1-3}\sigma(T_1^4 - T_3^4)]$$

$$Q_2 = A_2[F_{2-1}\sigma(T_2^4 - T_1^4) + F_{2-3}\sigma(T_2^4 - T_3^4)]$$

$$0 = A_3[F_{3-1}\sigma(T_3^4 - T_1^4) + F_{3-2}\sigma(T_3^4 - T_2^4)]$$

The final equation is solved for T_3, the only unknown in that equation. This T_3 is then inserted into the first two equations to obtain Q_1 and Q_2.

EXAMPLE 7-5 A very long black heated tube A_1 of length L is enclosed by a concentric black split cylinder as shown in Fig. 7-3. The diameter of the split cylinder is twice that of the heated tube, and one-half as much energy *flux* is to be removed from the upper area A_3 of the split cylinder as from the lower area A_2. If $T_1 = 1700$ K and a heat flux $Q_1/A_1 = 3 \times 10^5$ W/m^2 is supplied to the heated tube, what are the values of T_2, T_3, Q_2, and Q_3? Neglect the effect of the tube ends.

Writing (7-11) for each surface gives

$$Q_1 = A_1[F_{1-2}\sigma(T_1^4 - T_2^4) + F_{1-3}\sigma(T_1^4 - T_3^4)]$$

$$Q_2 = A_2[F_{2-1}\sigma(T_2^4 - T_1^4) + F_{2-3}\sigma(T_2^4 - T_3^4)]$$

$$Q_3 = A_3[F_{3-1}\sigma(T_3^4 - T_1^4) + F_{3-2}\sigma(T_3^4 - T_2^4)]$$

From the geometry, $A_1/A_3 = A_1/A_2 = \pi D_1 L/\frac{1}{2}\pi D_2 L = 1$, since $D_2 = 2D_1$. From an energy balance, $Q_1 + Q_2 + Q_3 = 0$ and, since $A_1 = A_2 = A_3$, $(Q_1/A_1) + (Q_2/A_2) + (Q_3/A_3) = 0$. From the statement of the problem, $Q_3/A_3 = Q_2/2A_2$, and this yields

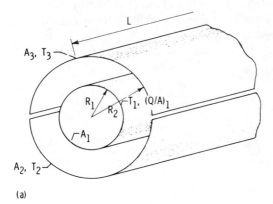

(a)

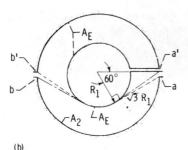

(b)

Figure 7-3 Radiant energy exchange in split-circular cylinder configuration, $L \gg R_2$. (*a*) Geometry of enclosure; (*b*) auxiliary construction to determine F_{2-2}.

$$\frac{Q_2}{A_2} = -\frac{2}{3}\frac{Q_1}{A_1} = -2.0 \times 10^5 \, \text{W/m}^2, \qquad \frac{Q_3}{A_3} = -\frac{1}{3}\frac{Q_1}{A_1} = -1.0 \times 10^5 \, \text{W/m}^2$$

From the symmetry of the geometry and configuration-factor algebra, it is known that $F_{1-2} = F_{1-3} = 1/2$, $F_{2-1} = F_{3-1} = A_1 F_{1-3}/A_3 = 1/2$, and $F_{2-3} = F_{3-2}$. To determine F_{2-3}, it is known that $F_{2-1} + F_{2-2} + F_{2-3} = 1$. Using $F_{2-1} = 1/2$ gives $F_{2-3} = (1/2) - F_{2-2}$. In the auxiliary construction of Fig. 7-3b, $F_{2-2} = 1 - F_{2-E}$. The effective area A_E has been drawn in to leave unchanged the view of surface 2 to itself and to simplify the geometry so that the crossed-string method can be used to determine F_{2-E}. The uncrossed strings extending from a to a' and b to b' have zero length. The crossed strings extend from a to b' and a' to b, and each has length $2\sqrt{3}R_1 + \pi R_1/3$. Then from Sec. 6-4.3 and the fact that $A_2 = A_1 = 2\pi R_1$, $F_{2-E} = (2\sqrt{3}R_1 + \pi R_1/3)/2\pi R_1 = (\sqrt{3}/\pi) + (1/6)$. It then follows that $F_{2-3} = (1/2) - F_{2-2} = (1/2) - (1 - F_{2-E}) = (\sqrt{3}/\pi) - (1/3) = 0.2180$.

Using this information, the energy exchange equations become

$$3 \times 10^5 = \frac{\sigma}{2}(1700^4 - T_2^4) + \frac{\sigma}{2}(1700^4 - T_3^4)$$

$$-2.0 \times 10^5 = \frac{\sigma}{2}(T_2^4 - 1700^4) + 0.2180 \, \sigma(T_2^4 - T_3^4)$$

$$-1.0 \times 10^5 = \frac{\sigma}{2}(T_3^4 - 1700^4) + 0.2180 \, \sigma(T_3^4 - T_2^4)$$

Adding the second and third equations results in the first, so only two equations are independent. Solving the first and second equations gives $T_2 = 1207$ K and $T_3 = 1415$ K.

7-4 RADIATION BETWEEN FINITE DIFFUSE-GRAY AREAS

7-4.1 Net-Radiation Method for Enclosures

Consider an enclosure composed of N discrete surface areas as shown in Fig. 7-4. The objectives are to analyze the radiation exchange between the surface areas for problems involving two types of boundary conditions: (1) The required energy supplied to a surface is to be determined when the surface temperature is specified, and (2) the temperature that a surface will achieve is to be found when a known heat input is imposed.

A complex radiative exchange occurs inside the enclosure as radiation leaves a surface, travels to other surfaces, is partially reflected, and is then rereflected many times within the enclosure with partial absorption at each contact with a surface. It would be complicated to follow the beams of radiation as they undergo this process; fortunately, this is not necessary. An analysis can be formulated in a convenient manner by using the *net-radiation method*. This method was first

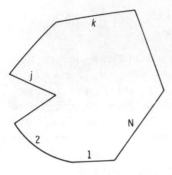

Figure 7-4 Enclosure composed of N discrete surface areas with typical surfaces j and k (shown in cross section for simplicity).

devised by Hottel [3] and later developed in a different manner by Poljak [4, 5]. An alternative approach was given by Gebhart [6, 7]. All the methods are basically equivalent (as demonstrated in [8]) and shown in an example; the Poljak approach, which the present authors usually find convenient, will now be given. The other formulations are then briefly presented.

Consider the kth *inside* surface area A_k of the enclosure shown in Figs. 7-4 and 7-5. The quantities q_i and q_o are the rates of incoming and outgoing *radiant* energy per unit *inside* area, respectively. The quantity q is the energy flux supplied *to the surface* by some means other than the radiation *inside* the enclosure to make up for the net radiative loss and thereby maintain the specified inside surface temperature. For example, if A_k is the inside surface of a wall of finite thickness, the Q_k could be the heat conducted through the wall to A_k. A heat balance for the surface area provides the relation

$$Q_k = q_k A_k = (q_{o,k} - q_{i,k})A_k \tag{7-12}$$

A second equation results from the fact that the energy flux leaving the surface is composed of emitted plus reflected energy. This gives

$$q_{o,k} = \epsilon_k \sigma T_k{}^4 + \rho_k q_{i,k} = \epsilon_k \sigma T_k{}^4 + (1 - \epsilon_k)q_{i,k} \tag{7-13}$$

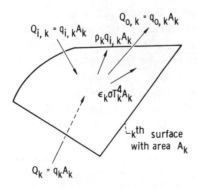

Figure 7-5 Energy quantities incident upon and leaving typical surface of enclosure.

where the relations $\rho_k = 1 - \alpha_k = 1 - \epsilon_k$ have been used for opaque gray surfaces. The term *radiosity* is often used for the quantity q_o. The incident flux $q_{i,k}$ is derived from the portions of the energy leaving the surfaces in the enclosure that arrive at the kth surface. If the kth surface can view itself (is concave), a portion of its outgoing flux will contribute directly to its incident flux. The incident energy is then equal to

$$A_k q_{i,k} = A_1 q_{o,1} F_{1-k} + A_2 q_{o,2} F_{2-k} + \cdots + A_j q_{o,j} F_{j-k} + \cdots$$
$$+ A_k q_{o,k} F_{k-k} + \cdots + A_N q_{o,N} F_{N-k} \qquad (7\text{-}14)$$

From the configuration-factor reciprocity relation (6-19),

$$A_1 F_{1-k} = A_k F_{k-1}$$

$$A_2 F_{2-k} = A_k F_{k-2}$$

$$\vdots \qquad\qquad (7\text{-}15)$$

$$A_N F_{N-k} = A_k F_{k-N}$$

Then Eq. (7-14) can be written so the only area appearing is A_k:

$$A_k q_{i,k} = A_k F_{k-1} q_{o,1} + A_k F_{k-2} q_{o,2} + \cdots + A_k F_{k-j} q_{o,j} + \cdots$$
$$+ A_k F_{k-k} q_{o,k} + \cdots + A_k F_{k-N} q_{o,N} \qquad (7\text{-}16a)$$

so that the incident flux is

$$q_{i,k} = \sum_{j=1}^{N} F_{k-j} q_{o,j} \qquad (7\text{-}16b)$$

Equations (7-12), (7-13), and (7-16) are simultaneous relations among q_k, T_k, $q_{o,k}$, and $q_{i,k}$ for each surface, and, through (7-16), to the other surfaces. One approach to a solution procedure is to note that (7-13) and (7-16) provide two different expressions for $q_{i,k}$. These are each substituted into Eq. (7-12) to eliminate $q_{i,k}$ and provide two heat balance equations for q_k in terms of T_k and $q_{o,k}$:

$$\frac{Q_k}{A_k} = q_k = \frac{\epsilon_k}{1 - \epsilon_k} (\sigma T_k^4 - q_{o,k}) \qquad (7\text{-}17)$$

$$\frac{Q_k}{A_k} = q_k = q_{o,k} - \sum_{j=1}^{N} F_{k-j} q_{o,j} = \sum_{j=1}^{N} F_{k-j}(q_{o,k} - q_{o,j}) \qquad (7\text{-}18)$$

The q_k can be regarded as either the energy flux supplied to surface k by non-radiative means (such as convection or conduction to A_k), or as the net radiative loss from surface k resulting from radiation inside the enclosure. Equation (7-17) or (7-18) is the balance between net radiative loss and energy supplied by means other than the radiation inside the enclosure.

As a first approach to becoming familiar with the analysis, consider that (7-17) and (7-18) can be written for each of the N surfaces in the enclosure. This provides $2N$ equations for $2N$ unknowns. The q_o will be N of the unknowns. The remaining unknowns will consist of q and T, depending on what boundary quantities are specified. Later, the q_o are eliminated to give N equations relating the N unknown q and T.

Other forms of (7-17) are

$$q_{o,k} = \sigma T_k^4 - \frac{1 - \epsilon_k}{\epsilon_k} q_k \qquad \text{or} \qquad \sigma T_k^4 = \frac{1 - \epsilon_k}{\epsilon_k} q_k + q_{o,k} \qquad (7\text{-}19)$$

Equations (7-17) and (7-19) were obtained by eliminating $q_{i,k}$ from (7-12) and (7-13). If, instead, $q_{o,k}$ is eliminated, the results are

$$q_k = \epsilon_k \sigma T_k^4 - \epsilon_k q_{i,k} \qquad (7\text{-}20a)$$

$$q_{i,k} = \sigma T_k^4 - \frac{q_k}{\epsilon_k} \qquad (7\text{-}20b)$$

Equation (7-20a) states that the net energy leaving the surface by radiation is the emitted energy minus the absorbed incident energy $\epsilon_k q_{i,k}$. If q_k and T_k^4 are obtained in a solution, the *absorbed incident energy* can be found for a gray surface as

$$\alpha_k q_{i,k} = \epsilon_k q_{i,k} = \epsilon_k \sigma T_k^4 - q_k \qquad (7\text{-}20c)$$

The absorbed energy added to the energy supplied by other means is equal to the emitted energy, $\alpha_k q_{i,k} + q_k = \epsilon_k \sigma T_k^4$.

Before continuing with the development, some examples will be given to illustrate the use of Eqs. (7-17) and (7-18) as simultaneous equations for each enclosure surface.

EXAMPLE 7-6 Derive the expression for the net radiative heat exchange between two infinite parallel flat plates in terms of their temperatures T_1 and T_2 ($T_1 > T_2$, Fig. 7-6).

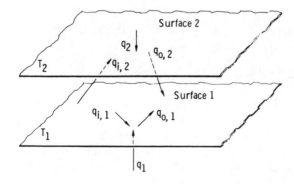

Figure 7-6 Heat fluxes for radiant interchange between infinite parallel flat plates.

Since all the radiation leaving one plate will arrive at the other plate, the configuration factors are $F_{1-2} = F_{2-1} = 1$. Equations (7-17) and (7-18) are then written for each plate:

$$q_1 = \frac{\epsilon_1}{1 - \epsilon_1} (\sigma T_1^4 - q_{o,1}), \qquad q_1 = q_{o,1} - q_{o,2} \tag{7-21a,b}$$

$$q_2 = \frac{\epsilon_2}{1 - \epsilon_2} (\sigma T_2^4 - q_{o,2}), \qquad q_2 = q_{o,2} - q_{o,1} \tag{7-21c,d}$$

By comparing (7-21b) and (7-21d), it is evident that $q_1 = -q_2$ so the heat added to surface 1 is removed from surface 2. The flux q_1 is thus the net heat transfer from 1 to 2 requested in the problem statement. Equations (7-21a) and (7-21c) yield,

$$q_{o,1} = \sigma T_1^4 - \frac{1 - \epsilon_1}{\epsilon_1} q_1, \qquad q_{o,2} = \sigma T_2^4 - \frac{1 - \epsilon_2}{\epsilon_2} q_2 = \sigma T_2^4 + \frac{1 - \epsilon_2}{\epsilon_2} q_1$$

These are substituted into (7-21b), and the result is solved for q_1:

$$q_1 = -q_2 = \frac{\sigma(T_1^4 - T_2^4)}{1/\epsilon_1(T_1) + 1/\epsilon_2(T_2) - 1} \tag{7-22a}$$

The functional notation $\epsilon(T)$ has been introduced to emphasize that ϵ_1 and ϵ_2 can be functions of temperature. Since T_1 and T_2 are specified, ϵ_1 and ϵ_2 can be evaluated at their proper temperatures and q_1 can be directly calculated.

EXAMPLE 7-7 For the parallel-plate geometry of the previous example, what temperature will surface 1 reach for a given heat input q_1 while T_2 is held at a specified value?

Equation (7-22a) still applies and when solved for T_1 gives

$$T_1 = \left\{ \frac{q_1}{\sigma} \left[\frac{1}{\epsilon_1(T_1)} + \frac{1}{\epsilon_2(T_2)} - 1 \right] + T_2^4 \right\}^{1/4} \tag{7-22b}$$

Since the emissivity $\epsilon_1(T_1)$ is a function of T_1, which is unknown, an iterative solution is necessary. A trial T_1 is selected, and ϵ_1 is chosen at this value. Equation (7-22b) is solved for T_1, and this value is used to select ϵ_1 for the next approximation. The process is continued until $\epsilon_1(T_1)$ and T_1 no longer change with further iterations.

EXAMPLE 7-8 Derive an expression for the net radiation exchange between two uniform temperature concentric diffuse-gray spheres as shown in Fig. 7-7.

This situation is more complicated than the parallel-plate geometry, as the two surfaces have unequal areas and surface 2 can partially view itself. The configuration factors were derived in Example 6-11 as $F_{1-2} = 1$, $F_{2-1} = A_1/A_2$, and $F_{2-2} = 1 - (A_1/A_2)$. Equations (7-17) and (7-18) are written for each of the two sphere surfaces as

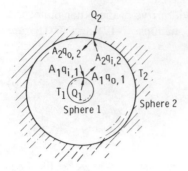

Figure 7-7 Energy quantities for radiant interchange between two concentric spheres.

$$Q_1 = A_1 \frac{\epsilon_1}{1 - \epsilon_1} (\sigma T_1^4 - q_{o,1}), \qquad Q_1 = A_1(q_{o,1} - q_{o,2}) \qquad (7\text{-}23a,b)$$

$$Q_2 = A_2 \frac{\epsilon_2}{1 - \epsilon_2} (\sigma T_2^4 - q_{o,2}) \qquad (7\text{-}24a)$$

$$Q_2 = A_2 \left[q_{o,2} - \frac{A_1}{A_2} q_{o,1} - \left(1 - \frac{A_1}{A_2}\right) q_{o,2} \right] = A_1(-q_{o,1} + q_{o,2}) \qquad (7\text{-}24b)$$

Comparing (7-23b) and (7-24b) reveals that $Q_1 = -Q_2$, as would be expected from an overall heat balance. The four equations (7-23) and (7-24) can be solved for the four unknowns $q_{o,1}$, $q_{o,2}$, Q_1, and Q_2. This yields the net heat transfer (supplied to surface 1 and removed at surface 2):

$$Q_1 = \frac{A_1\sigma(T_1^4 - T_2^4)}{1/\epsilon_1(T_1) + (A_1/A_2)[1/\epsilon_2(T_2) - 1]} \qquad (7\text{-}25)$$

When the spheres in Example 7-8 are not concentric, all the radiation leaving surface 1 is still incident on surface 2. The view factor F_{1-2} is again 1 and, with the use of the same assumptions, the analysis would follow as before, leading to (7-25). However, when sphere 1 is relatively small (say, one-half the diameter of sphere 2) and the eccentricity is large, the geometric appearance of the system is so different from the concentric case that using (7-25) would seem intuitively incorrect. The error in using (7-25) is that it was derived on the basis that q, q_i, and q_o are *uniform* over each of A_1 and A_2. These conditions are exactly met only for the concentric case. For the eccentric case the A_1 and A_2 would need to be subdivided to improve accuracy.

EXAMPLE 7-9 A gray isothermal body with area A_1 and temperature T_1 and without any concave indentations is completely enclosed by a much larger gray isothermal enclosure having area A_2. How much energy is being transferred by radiation from A_1 to A_2? The area A_1 cannot see any part of itself, and A_1 is not near A_2.

Since A_1 is completely enclosed and $F_{1-1} = 0$, the configuration factors and analysis are the same as in Example 7-8, which results in Q_1 given by Eq. (7-25). This is valid, as A_1 is specified as rather centrally located within A_2 and hence the heat fluxes tend to be uniform over A_1. For the present situation $A_1 \ll A_2$, and (7-25) reduces to (unless ϵ_2 is very small),

$$Q_1 = A_1 \epsilon_1 (T_1) \sigma (T_1^4 - T_2^4) \tag{7-26}$$

Note that this result is independent of the emissivity ϵ_2 of the enclosure (the enclosure acts like a black cavity unless ϵ_2 is very small).

EXAMPLE 7-10 Consider a long enclosure made up of three surfaces as shown in Fig. 7-8. The enclosure is long enough so that the ends can be neglected in the radiative heat balances. How much heat has to be supplied to each surface (equal to the net radiative heat loss from each surface resulting from exchange within the enclosure) to maintain the surfaces at temperatures T_1, T_2, and T_3?

To solve this problem, write (7-17) and (7-18) for each of the three surfaces:

$$\frac{Q_1}{A_1} = \frac{\epsilon_1}{1 - \epsilon_1} (\sigma T_1^4 - q_{o,1}) \tag{7-27a}$$

$$\frac{Q_1}{A_1} = q_{o,1} - F_{1-1} q_{o,1} - F_{1-2} q_{o,2} - F_{1-3} q_{o,3} \tag{7-27b}$$

$$\frac{Q_2}{A_2} = \frac{\epsilon_2}{1 - \epsilon_2} (\sigma T_2^4 - q_{o,2}) \tag{7-28a}$$

$$\frac{Q_2}{A_2} = q_{o,2} - F_{2-1} q_{o,1} - F_{2-2} q_{o,2} - F_{2-3} q_{o,3} \tag{7-28b}$$

$$\frac{Q_3}{A_3} = \frac{\epsilon_3}{1 - \epsilon_3} (\sigma T_3^4 - q_{o,3}) \tag{7-29a}$$

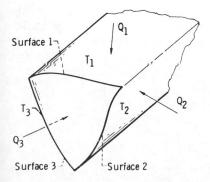

Figure 7-8 Long enclosure composed of three surfaces (ends neglected).

$$\frac{Q_3}{A_3} = q_{o,3} - F_{3-1}q_{o,1} - F_{3-2}q_{o,2} - F_{3-3}q_{o,3} \qquad (7\text{-}29b)$$

The first of each of these three pairs of equations is solved for q_o, and the q_o are substituted into the second equation of each pair to obtain

$$\frac{Q_1}{A_1}\left(\frac{1}{\epsilon_1} - F_{1-1}\frac{1-\epsilon_1}{\epsilon_1}\right) - \frac{Q_2}{A_2}F_{1-2}\frac{1-\epsilon_2}{\epsilon_2} - \frac{Q_3}{A_3}F_{1-3}\frac{1-\epsilon_3}{\epsilon_3}$$

$$= (1 - F_{1-1})\sigma T_1^4 - F_{1-2}\sigma T_2^4 - F_{1-3}\sigma T_3^4$$

$$= F_{1-2}\sigma(T_1^4 - T_2^4) + F_{1-3}\sigma(T_1^4 - T_3^4) \qquad (7\text{-}30a)$$

$$-\frac{Q_1}{A_1}F_{2-1}\frac{1-\epsilon_1}{\epsilon_1} + \frac{Q_2}{A_2}\left(\frac{1}{\epsilon_2} - F_{2-2}\frac{1-\epsilon_2}{\epsilon_2}\right) - \frac{Q_3}{A_3}F_{2-3}\frac{1-\epsilon_3}{\epsilon_3}$$

$$= -F_{2-1}\sigma T_1^4 + (1 - F_{2-2})\sigma T_2^4 - F_{2-3}\sigma T_3^4$$

$$= F_{2-1}\sigma(T_2^4 - T_1^4) + F_{2-3}\sigma(T_2^4 - T_3^4) \qquad (7\text{-}30b)$$

$$-\frac{Q_1}{A_1}F_{3-1}\frac{1-\epsilon_1}{\epsilon_1} - \frac{Q_2}{A_2}F_{3-2}\frac{1-\epsilon_2}{\epsilon_2} + \frac{Q_3}{A_3}\left(\frac{1}{\epsilon_3} - F_{3-3}\frac{1-\epsilon_3}{\epsilon_3}\right)$$

$$= -F_{3-1}\sigma T_1^4 - F_{3-2}\sigma T_2^4 + (1 - F_{3-3})\sigma T_3^4$$

$$= F_{3-1}\sigma(T_3^4 - T_1^4) + F_{3-2}\sigma(T_3^4 - T_2^4) \qquad (7\text{-}30c)$$

Since the T are known, the ϵ can be specified from surface-property data at their appropriate T values, and the three simultaneous equations solved for the Q supplied to each surface. Note that the results are only approximations, because the radiosity leaving each surface is not uniform as assumed by using (7-17) and (7-18). This is because the reflected flux is not uniform as a result of the enclosure geometry. Greater accuracy is obtained by dividing each of the three sides into more surface elements.

System of equations relating surface heating Q and surface temperature T The form of (7-30) shows that the Q and T for an enclosure of N surfaces can be related in a convenient system of N equations. Equation (7-19) for $q_{o,k}$ is substituted into (7-18). (Note that $q_{o,j}$ is found by simply changing the subscript in the relation for $q_{o,k}$.) This results in the following form for the kth surface, a result also evident from (7-30):

$$-\frac{Q_1}{A_1}F_{k-1}\frac{1-\epsilon_1}{\epsilon_1} - \frac{Q_2}{A_2}F_{k-2}\frac{1-\epsilon_2}{\epsilon_2} - \cdots + \frac{Q_k}{A_k}\left(\frac{1}{\epsilon_k} - F_{k-k}\frac{1-\epsilon_k}{\epsilon_k}\right) - \cdots$$

$$-\frac{Q_N}{A_N}F_{k-N}\frac{1-\epsilon_N}{\epsilon_N} = -F_{k-1}\sigma T_1^4 - F_{k-2}\sigma T_2^4 - \cdots$$

$$+ (1 - F_{k-k})\sigma T_k^4 - \cdots - F_{k-N}\sigma T_N^4$$

$$= F_{k-1}\sigma(T_k^4 - T_1^4) + F_{k-2}\sigma(T_k^4 - T_2^4)$$
$$+ \cdots + F_{k-N}\sigma(T_k^4 - T_N^4)$$

A summation notation is used to write this for the kth surface as

$$\sum_{j=1}^{N} \left(\frac{\delta_{kj}}{\epsilon_j} - F_{k-j}\frac{1-\epsilon_j}{\epsilon_j} \right) \frac{Q_j}{A_j} = \sum_{j=1}^{N} (\delta_{kj} - F_{k-j})\sigma T_j^4 = \sum_{j=1}^{N} F_{k-j}\sigma(T_k^4 - T_j^4) \qquad (7\text{-}31)$$

where, corresponding to each surface, k takes on one of the values 1, 2, . . . , N, and δ_{kj} is the Kronecker delta defined as

$$\delta_{kj} = \begin{cases} 1 & \text{when } k = j \\ 0 & \text{when } k \neq j \end{cases}$$

When the surface temperatures are specified, the right side of Eq. (7-31) is known and there are N simultaneous equations for the unknown Q. In general the heat inputs to some of the surfaces may be specified and the temperatures of these surfaces are to be determined. There are still a total of N unknown Q and T, and Eq. (7-31) provides the necessary number of relations. Since the values of ϵ depend on temperature, it is necessary initially to guess the unknown T. Then the ϵ values can be chosen and the system of equations solved. The T values are used to select new ϵ, and the process repeated until the T and ϵ values no longer change. Again note that the results will be approximate because the uniform radiosity assumption is not perfectly fulfilled over each finite area. Division into smaller areas is required to improve accuracy. As an example of a practical problem, the solution of an enclosure with mixed boundary conditions to simulate heat-generating electronic modules is in [9].

EXAMPLE 7-11 Consider an enclosure of three sides as shown in Fig. 7-8. Side 1 is maintained at T_1, side 2 is uniformly heated with a flux q_2, and the third side is perfectly insulated on the outside. What are the equations to determine Q_1, T_2, and T_3?

The conditions of the problem give $Q_2/A_2 = q_2$ and $Q_3 = 0$. Then (7-31) yields the following three equations, where the unknowns have been gathered on the left side:

$$\frac{Q_1}{A_1}\left(\frac{1}{\epsilon_1} - F_{1-1}\frac{1-\epsilon_1}{\epsilon_1} \right) + F_{1-2}\sigma T_2^4 + F_{1-3}\sigma T_3^4$$
$$- (1 - F_{1-1})\sigma T_1^4 + q_2 F_{1-2}\frac{1-\epsilon_2}{\epsilon_2} \qquad (7\text{-}32a)$$

$$-\frac{Q_1}{A_1}F_{2-1}\frac{1-\epsilon_1}{\epsilon_1} - (1 - F_{2-2})\sigma T_2^4 + F_{2-3}\sigma T_3^4$$
$$= -F_{2-1}\sigma T_1^4 - q_2\left(\frac{1}{\epsilon_2} - F_{2-2}\frac{1-\epsilon_2}{\epsilon_2} \right) \qquad (7\text{-}32b)$$

$$-\frac{Q_1}{A_1}F_{3-1}\frac{1-\epsilon_1}{\epsilon_1}+F_{3-2}\sigma T_2^{\,4}-(1-F_{3-3})\sigma T_3^{\,4}$$

$$=-F_{3-1}\sigma T_1^{\,4}+q_2F_{3-2}\frac{1-\epsilon_2}{\epsilon_2} \tag{7-32c}$$

Note that for this simple situation the solution could be shortened by using overall energy conservation to give $Q_1 = -Q_2$. However, it is generally a good idea to solve directly for all the unknowns and use the overall heat balance $\Sigma_{k=1}^{N} Q_k = 0$ as a check.

EXAMPLE 7-12 A hollow cylinder is heated on the outside in such a way that the cylinder is maintained at a uniform temperature. The outside of the cylinder is otherwise insulated so that the heat must be transferred away by radiation from the inside of the cylinder through the cylinder ends. The system is in a vacuum environment, so radiation is the only mode of heat transfer. As shown in Fig. 7-9, there is a disk centered on the cylinder axis and facing normal to the ends of the cylinder. The disk is exposed on both sides and hence the side facing away from the cylinder radiates to the surrounding environment at T_e. The other side is subjected to radiation coming from the inside of the cylinder. Provide equations to compute the temperature of the disk, T_d, if the inside of the cylinder is at a uniform temperature T_c.

   The heat loss from the surface of the disk facing away from the cylinder can be considered as radiation into a large environment at T_e. Hence from (7-26) the net energy leaving this surface is $q_{d,e} = \epsilon_{d,e}\sigma(T_d^4 - T_e^4)$. The exchange with the cylinder can be analyzed as a three-surface enclosure consisting of the inside of the cylinder, the side of the disk facing the cylinder, and the opening between the disk and the cylinder (a frustum of a cone) combined with the opening at the left side of the cylinder. These two open areas can be regarded as one area when computing the view factors for exchange

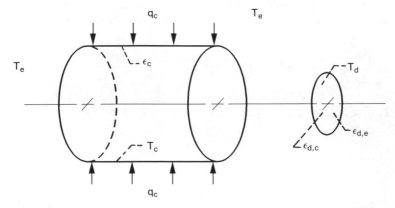

Figure 7-9 Hollow cylinder and disk configuration for Example 7-12.

to the surrounding environment. In the configuration factor designations the subscript e will designate the combination of the two open areas at T_e. Since the amount of heat lost to the environment is not asked for, only two equations are written, one for the inside of the cylinder and one for the surface of the disk facing the cylinder. The quantity $q_{d,c} = -\epsilon_{d,e}(T_d^4 - T_e^4)$ is the heat flux supplied to the left surface of the disk. From Eq. (7-31) the two equations are

$$q_c\left(\frac{1}{\epsilon_c} - F_{c-c}\frac{1 - \epsilon_c}{\epsilon_c}\right) - q_{d,c}F_{c-d}\frac{1 - \epsilon_{d,c}}{\epsilon_{d,c}}$$

$$= F_{c-d}\sigma(T_c^4 - T_d^4) + F_{c-e}\sigma(T_c^4 - T_e^4) \tag{7-33}$$

$$-q_cF_{d-c}\frac{1 - \epsilon_c}{\epsilon_c} + q_d\frac{1}{\epsilon_{d,c}} = F_{d-c}\sigma(T_d^4 - T_c^4) + F_{d-e}\sigma(T_d^4 - T_e^4) \tag{7-34}$$

The q_c is eliminated from (7-33) and (7-34); this yields an equation that can be solved for T_d.

EXAMPLE 7-13 As an example with a more complex geometry, consider the long rectangular enclosure shown in cross section in Fig. 7-10 (the end walls are neglected). There is a thin baffle along a diagonal dividing the interior into two triangular enclosures. Heat fluxes are specified along two sides, and the other two sides are each cooled to a specified uniform temperature. It is required to find the temperatures T_1 and T_2 of the heated walls and the heat fluxes q_5 and q_6 that must be removed through the cooled walls to maintain their temperatures. For simplicity the surfaces will not be subdivided into smaller areas. Subdivision will cause the results to change (see Example 7-14).

Since the temperatures T_3 and T_4 and the fluxes q_3 and q_4 are also unknown, there is a total of eight unknowns, so eight simultaneous equations are required. Two of these are the constraints at the baffle wall that yield $T_3 = T_4$ and $q_3 = -q_4$. The other six equations are found from (7-31) and are written here in abbreviated form using the notation $F_{mn} \equiv F_{m-n}$, $t \equiv T^4$, $E_n \equiv (1 - \epsilon_n)/\epsilon_n$:

$$(q_1/\epsilon_1) - q_2E_2F_{12} - q_3E_3F_{13} = \sigma(t_1 - F_{12}t_2 - F_{13}t_3)$$

$$-q_1E_1F_{21} + (q_2/\epsilon_2) - q_3E_3F_{23} = \sigma(-F_{21}t_1 + t_2 - F_{23}t_3)$$

$$-q_1E_1F_{31} - q_2E_2F_{32} + (q_3/\epsilon_3) = \sigma(-F_{31}t_1 + F_{32}t_2 + t_3)$$

$$(q_5/\epsilon_5) - q_6E_6F_{56} - q_4E_4F_{54} = \sigma(t_5 - F_{56}t_6 - F_{54}t_4)$$

$$-q_5E_5F_{65} + (q_6/\epsilon_6) - q_4E_4F_{64} = \sigma(-F_{65}t_5 + t_6 - F_{64}t_4)$$

$$-q_5E_5F_{45} - q_6E_6F_{46} + (q_4/\epsilon_4) = \sigma(-F_{45}t_5 - F_{46}t_6 + t_4)$$

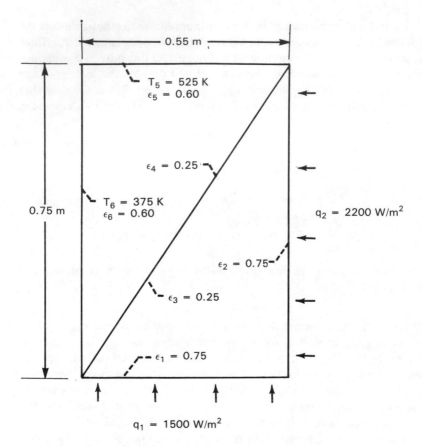

Figure 7-10 Rectangular enclosure divided into two triangular enclosures for Example 7-13.

For the three-sided enclosures, the configuration factors are found from Example 6-13. Typical values are $F_{1-2} = F_{5-6} = 0.33632$ and $F_{1-3} = F_{5-4} = 0.66369$. A matrix solver can be used to substitute the values and solve the equations to obtain $T_1 = 817.0$ K, $T_2 = 821.7$ K, $T_3 = 709.1$ K, $q_5 = -559.0$ W/m^2, and $q_6 = -2890.1$ W/m^2. Most of the energy is leaving through the more strongly cooled side A_6. The effect of various parameters can easily be examined. For example, if $\epsilon_1 = \epsilon_2$ are changed, the only results affected are T_1 and T_2. This is because the same total energy must pass through surface 4, so T_4 remains the same in order to transfer the energy to sides 5 and 6. For $\epsilon_1 = \epsilon_2 = 1.0$ the T_1 and T_2 are slightly decreased to $T_1 = 812.9$ K and $T_2 = 815.8$ K. When $\epsilon_1 = \epsilon_2 = 0.50$ these temperatures increase to $T_1 = 824.9$ K and $T_2 = 833.1$ K. (Note that the results should always be checked by determining whether $\Sigma_{k=0}^{N} q_k A_k = 0$.) In this case $(0.55 \times 1500.0) + (0.75 \times 2200.0) - (0.55 \times 559.0) - (0.75 \times 2890.1) = -0.025$ W/meter of length, which is within the roundoff accuracy of the q_k values.

EXAMPLE 7-14 A triangular enclosure has long sides normal to the cross section shown in Fig. 7-11. The triangular end walls can be neglected in the radiative exchange. Specified quantities are the temperatures of two sides and the heat flux added on the third. The q_1, q_2, and T_3 are required. The solution is obtained first with side 3 being a single area. Then this area is divided into two equal parts to refine the calculation.

Equation (7-31) is used. From Example 6-13, $F_{1-2} = (A_1 + A_2 - A_3)/2A_1 = 0.2929 = F_{2-1}$. Then $F_{1-3} = 1 - F_{1-2} = 0.7071 = F_{2-3}$. From symmetry $F_{3-1} = F_{3-2} = 0.5$. The self-view factors are all zero. The first three equations from Example 7-13 are used. If $\epsilon_1 = \epsilon_2 = 0.80$ and $\epsilon_3 = 0.90$, the results are $q_1 = -6346$ W/m², $q_2 = 1820$ W/m², and $T_3 = 649.1$ K. Thus heat must be added to maintain the high temperature of A_2. The energy added to A_2 combined with that from A_3 flows out of A_1 at a lower temperature. Now consider what happens if the emissivities are reduced by half, $\epsilon_1 = \epsilon_2 = 0.40$ and $\epsilon_3 = 0.45$. The energy supplied to A_3 has greater difficulty being transferred to the other walls, and the temperature rises to $T_3 = 733.9$ K. Since T_3 is now larger than both T_1 and T_2, heat is transferred out from both A_1 and A_2: $q_1 = -4101$ W/m² and $q_2 = -424.6$ W/m².

To refine the calculation, A_3 is divided into two equal parts A_4 and A_5. From the geometry, $F_{1-4} = 0.5 = F_{1-5} + F_{1-2} = F_{2-5}$. With F_{1-2} known, this gives $F_{1-5} = 0.2071 = F_{2-4}$. As in the previous examples, four equations are written from Eq. (7-31). With T_1, T_2, q_4, and q_5 known, these are solved for

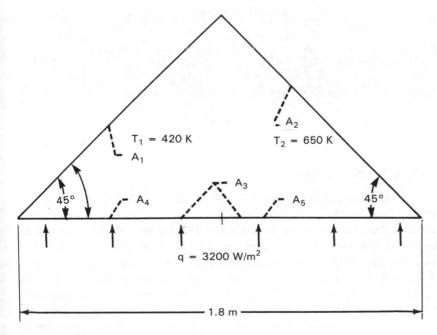

Figure 7-11 Triangular enclosure for Example 7-14. Areas A_4 and A_5 are each one-half of A_3.

q_1, q_2, T_4, and T_5. For $\epsilon_1 = \epsilon_2 = 0.80$ and $\epsilon_3 = 0.90$, this gives $q_1 = -6049$ W/m², $q_2 = 1524$ W/m², $T_4 = 626.3$ K, and $T_5 = 669.7$ K. This reveals the temperature variation along A_3 as compared with the uniform value obtained in the first part of this example. The q_2 is somewhat smaller because A_2 is adjacent to the higher-temperature portion of A_3. By further subdividing A_3, its temperature distribution would be obtained. For $\epsilon_1 = \epsilon_2 = 0.40$ and $\epsilon_3 = 0.45$, we obtain $T_4 = 726.8$ K and $T_5 = 740.8$ K, compared with $T_3 = 733.9$ K in the previous calculation. The q are changed somewhat to $q_1 = -4038$ W/m² and $q_2 = -487.2$ W/m². The q values in this example have all been verified to satisfy overall energy conservation (note, the values given have been rounded off).

Solution method in terms of outgoing radiative flux q_o An alternative approach for computing radiative exchange is to solve for q_o for each surface and then compute the q and T. When the surface is viewed with a radiation detector, it is q_o that is intercepted, that is, the *sum* of *emitted* and *reflected* radiation. Hence, it is desirable in some instances to determine the q_o values as primary quantities. Of course, in the previous formulation the q_o can be found from the q and T by using (7-19).

When the surface temperatures are all specified, the set of simultaneous equations for q_o is obtained by eliminating Q_k from (7-17) and (7-18). This yields either of the following equations for the kth surface:

$$q_{o,k} - (1 - \epsilon_k) \sum_{j=1}^{N} F_{k-j} q_{o,j} = \sum_{j=1}^{N} [\delta_{kj} - (1 - \epsilon_k) F_{k-j}] q_{o,j} = \epsilon_k \sigma T_k^4 \tag{7-35a}$$

$$q_{o,k} + \frac{1 - \epsilon_k}{\epsilon_k} \sum_{j=1}^{N} F_{k-j}(q_{o,k} - q_{o,j}) = \sigma T_k^4 \tag{7-35b}$$

To illustrate, for a system of three surfaces (7-35a) becomes

$$[1 - (1 - \epsilon_1)F_{1-1}]q_{o,1} - (1 - \epsilon_1)F_{1-2}q_{o,2} - (1 - \epsilon_1)F_{1-3}q_{o,3} = \epsilon_1 \sigma T_1^4 \tag{7-36a}$$

$$-(1 - \epsilon_2)F_{2-1}q_{o,1} + [1 - (1 - \epsilon_2)F_{2-2}]q_{o,2} - (1 - \epsilon_2)F_{2-3}q_{o,3} = \epsilon_2 \sigma T_2^4 \tag{7-36b}$$

$$-(1 - \epsilon_3)F_{3-1}q_{o,1} - (1 - \epsilon_3)F_{3-2}q_{o,2} + [1 - (1 - \epsilon_3)F_{3-3}]q_{o,3} = \epsilon_3 \sigma T_3^4 \tag{7-36c}$$

With the T given, the q_o can be found. Then, if desired, (7-17) can be used to compute q for each surface.

When q is specified for some surfaces and T for others, (7-35a) is used for the surfaces with known T in conjunction with (7-18) for the surfaces with known q to obtain the set of simultaneous equations for the unknown q_o. Once q_o is obtained for a surface, it can be combined with the given q (or T) and (7-17) used to determine the unknown T (or q). In a general form, if an enclosure has surfaces $1, 2, \ldots, m$ with specified temperature and the remaining surfaces $m + 1, m + 2, \ldots, N$ with specified heat input, then the system of equations for the q_o is, from (7-35a) and (7-18),

$$\sum_{j=1}^{N} [\delta_{kj} - (1 - \epsilon_k)F_{k-j}]q_{o,j} = \epsilon_k \sigma T_k^4 \qquad 1 \leqslant k \leqslant m \qquad (7\text{-}37a)$$

$$\sum_{j=1}^{N} (\delta_{kj} - F_{k-j})q_{o,j} = \frac{Q_k}{A_k} \qquad m + 1 \leqslant k \leqslant N \qquad (7\text{-}37b)$$

For a black surface with T_k specified, (7-37a) gives $q_{o,k} = \sigma T_k^4$ so the $q_{o,k}$ is known and the number of simultaneous equations is reduced by one. The following example will demonstrate obtaining a solution first by using Eq. (7-31) and then by using (7-37).

EXAMPLE 7-15 A frustum of a cone has its base heated as shown in Fig. 7-12. The top is held at 550 K while the side is perfectly insulated on the outside. Surfaces 1 and 2 are diffuse-gray, while surface 3 is black. What is the temperature of side 1? How important is the value of ϵ_2?

By using the configuration factor for two parallel disks (23 in Appendix C), $F_{3-1} = 0.3249$. Then $F_{3-2} = 1 - F_{3-1} = 0.6751$. From reciprocity, $A_1F_{1-3} = A_3F_{3-1}$ and $A_2F_{2-3} = A_3F_{3-2}$, so that $F_{1-3} = 0.1444$ and $F_{2-3} = 0.1310$. Then $F_{1-2} = 1 - F_{1-3} = 0.8556$. From $A_1F_{1-2} = A_2F_{2-1}$, $F_{2-1} = 0.3735$. Finally, $F_{2-2} = 1 - F_{2-1} - F_{2-3} = 0.4955$. From Eq. (7-31) and by noting that $Q_2 = 0$ and $1 - \epsilon_3 = 0$, the three equations can be written as

$$\frac{3000}{0.6} = \sigma[T_1^4 - 0.8556T_2^4 - 0.1444(550)^4]$$

$$-3000(0.3735)\frac{1 - 0.6}{0.6} = \sigma[-0.3735T_1^4 + (1 - 0.4955)T_2^4 - 0.1310(550)^4]$$

$$-3000(0.3249)\frac{1 - 0.6}{0.6} + \frac{Q_3}{A_3} = \sigma[-0.3249T_1^4 - 0.6751T_2^4 + (550)^4]$$

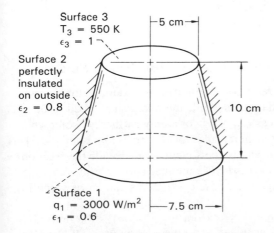

Surface 3
$T_3 = 550$ K
$\epsilon_3 = 1$

Surface 2
perfectly
insulated
on outside
$\epsilon_2 = 0.8$

5 cm

10 cm

Surface 1
$q_1 = 3000$ W/m^2
$\epsilon_1 = 0.6$

7.5 cm

Figure 7-12 Enclosure used in Example 7-15.

These three equations can be solved for T_1, T_2, and Q_3. (Note that for this particular example Q_3 can also be obtained from overall energy conservation, $Q_3 = -Q_1$.) The result requested is $T_1 = 721.6$ K ($T_2 = 667.4$ K). Since $Q_2 = 0$, all the terms involving ϵ_2 are zero so ϵ_2 does not appear in the simultaneous equations, and the emissivity of the insulated surface is of no importance. Physically, this results from the fact that for no convection or conduction, all absorbed energy must be reemitted and hence $q_{o,2} = q_{i,2}$ independently of the ϵ_2 value.

This example will now be solved using Eqs. (7-37). Because T_3 is specified and A_3 is black, Eq. (7-37a) gives $q_{o,3} = \sigma T_3^4$. Equation (7-37b) is used at A_1 and A_2 since q_1 and q_2 are specified,

$$q_{o,1} - F_{1-2}q_{o,2} = F_{1-3}\sigma T_3^4 + q_1$$

$$-F_{2-1}q_{o,1} + (1 - F_{2-2})q_{o,2} = F_{2-3}\sigma T_3^4$$

This yields $q_{o,1} = 13370$ W/m^2 and $q_{o,2} = 11250$ W/m^2. From (7-19) $\sigma T_1^4 = q_{o,1} + [(1 - \epsilon_1)/\epsilon_1]q_1$ and similarly for σT_2^4. This results in the same temperatures as in the first part of this example.

EXAMPLE 7-16 The second half of Example 7-14 is now calculated by using Eq. (7-37). A_1 and A_2 have specified T, while A_4 and A_5 have specified q. Then (7-37) yields

$$q_{o,1} - (1 - \epsilon_1)[F_{1-2}q_{o,2} + F_{1-4}q_{o,4} + F_{1-5}q_{o,5}] = \epsilon_1\sigma T_1^4$$

$$q_{o,2} - (1 - \epsilon_2)[F_{2-1}q_{o,1} + F_{2-4}q_{o,4} + F_{2-5}q_{o,5}] = \epsilon_2\sigma T_2^4$$

$$-F_{4-1}q_{o,1} - F_{4-2}q_{o,2} + q_{o,4} - F_{4-5}q_{o,5} = q_4$$

$$-F_{5-1}q_{o,1} - F_{5-2}q_{o,2} - F_{5-4}q_{o,4} + q_{o,5} = q_5$$

The solution for the q_o yields, in W/m^2, $q_{o,1} = 3277$, $q_{o,2} = 9741$, $q_{o,4} = 8370$, and $q_{o,5} = 11048$. Then from (7-17), $q_1 = [\epsilon_1/(1 - \epsilon_1)](\sigma T_1^4 - q_{o,1})$ and similarly for q_2. From (7-19), $\sigma T_4^4 = [(1 - \epsilon_4)/\epsilon_4]q_4 + q_{o,4}$ and similarly for T_5. This yields the same results as Example 7-14.

7-4.2 Enclosure Analysis in Terms of Energy Absorbed at Surface

An alternative procedure to the net-radiation method is to examine the energy *absorbed* at a surface. In the net-radiation method, the energy absorbed at each surface can be found from Eq. (7-20c) as $\alpha q_i = \epsilon q_i$ for a gray surface. A somewhat different viewpoint is briefly presented here; additional discussion is in [6, 7]. The feature of this formulation is that it yields coefficients that provide the fraction of energy emitted by a surface that is *absorbed* at another surface after reaching the absorbing surface by all possible paths. These coefficients can be of value in formulating some types of problems.

For a typical surface A_k the net energy loss from Eq. (7-20a) is the emission from the surface minus the energy that is absorbed by the surface from all incident radiation sources. Let G_{jk} be the fraction of the emission from surface A_j that reaches A_k and is *absorbed*. The G_{jk} includes all the paths for reaching A_k, that is, the direct path and paths by means of one or multiple reflections. Thus $A_j\epsilon_j\sigma T_j^4 G_{jk}$ is the amount of energy emitted by A_j that is absorbed by A_k. A heat balance on A_k then gives

$$Q_k = A_k\epsilon_k\sigma T_k^4 - (A_1\epsilon_1\sigma T_1^4 G_{1k} + A_2\epsilon_2\sigma T_2^4 G_{2k} + \cdots + A_j\epsilon_j\sigma T_j^4 G_{jk}$$
$$+ \cdots + A_k\epsilon_k\sigma T_k^4 G_{kk} + \cdots + A_N\epsilon_N\sigma T_N^4 G_{Nk})$$
$$= A_k\epsilon_k\sigma T_k^4 - \sum_{j=1}^{N} A_j\epsilon_j\sigma T_j^4 G_{jk} \tag{7-38}$$

The G_{kk} would generally not be zero, since even for a plane or convex surface some of the emission from a surface will be returned to itself by reflection from other surfaces. Equation (7-38) is written for each surface; this will relate each of the Q to the surface temperatures in the enclosure. The G factors must now be found.

The G_{jk} is the fraction of energy emitted by A_j that reaches A_k and is absorbed. The total emitted energy from A_j is $A_j\epsilon_j\sigma T_j^4$. The portion traveling directly to A_k and then absorbed is $A_j\epsilon_j\sigma T_j^4 F_{j-k}\epsilon_k$, where for a gray surface $\alpha_k = \epsilon_k$. All other radiation from A_j arriving at A_k will first undergo one reflection. The emission from A_j that arrives at a typical surface A_n and is then reflected is $A_j\epsilon_j\sigma T_j^4 F_{j-n}\rho_n$. The fraction G_{nk} reaches A_k and is absorbed. Then all the energy absorbed at A_k that originated by emission from A_j is

$$A_j\epsilon_j\sigma T_j^4 F_{j-k}\epsilon_k + (A_j\epsilon_j\sigma T_j^4 F_{j-1}\rho_1 G_{1k} + A_j\epsilon_j\sigma T_j^4 F_{j-2}\rho_2 G_{2k} + \cdots$$
$$+ A_j\epsilon_j\sigma T_j^4 F_{j-k}\rho_k G_{kk} + \cdots + A_j\epsilon_j\sigma T_j^4 F_{j-N}\rho_N G_{Nk})$$

Dividing this energy by the emission from A_j gives the fraction

$$G_{jk} = F_{j-k}\epsilon_k + F_{j-1}\rho_1 G_{1k} + F_{j-2}\rho_2 G_{2k} + \cdots + F_{j-k}\rho_k G_{kk} + \cdots + F_{j-N}\rho_N G_{Nk}$$

This is rearranged into the form

$$-F_{j-1}\rho_1 G_{1k} - F_{j-2}\rho_2 G_{2k} - \cdots + G_{jk} - \cdots - F_{j-N}\rho_N G_{Nk} = F_{j-k}\epsilon_k$$

By letting j take on all values from 1 to N, the following set of equations is obtained:

$$(1 - F_{1-1}\rho_1)G_{1k} - F_{1-2}\rho_2 G_{2k} - F_{1-3}\rho_3 G_{3k} - \cdots - F_{1-N}\rho_N G_{Nk} = F_{1-k}\epsilon_k$$

$$-F_{2-1}\rho_1 G_{1k} + (1 - F_{2-2}\rho_2)G_{2k} - F_{2-3}\rho_3 G_{3k} - \cdots - F_{2-N}\rho_N G_{Nk} = F_{2-k}\epsilon_k$$

$$-F_{3-1}\rho_1 G_{1k} - F_{3-2}\rho_2 G_{2k} + (1 - F_{3-3}\rho_3)G_{3k} - \cdots - F_{3-N}\rho_N G_{Nk} = F_{3-k}\epsilon_k \tag{7-39}$$

$$\vdots \qquad \vdots \qquad \qquad \vdots \qquad \qquad \vdots$$

$$-F_{N-1}\rho_1 G_{1k} - F_{N-2}\rho_2 G_{2k} - F_{N-3}\rho_3 G_{3k} - \cdots + (1 - F_{N-N}\rho_N)G_{Nk} = F_{N-k}\epsilon_k$$

Equations (7-39) are solved simultaneously for $G_{1k}, G_{2k}, \ldots, G_{Nk}$. This is done for each k where $1 \leq k \leq N$, and the solution is simplified by the fact that the coefficients of the G do not depend on k.

The amount of calculation can be reduced by using some relations between the G values. Since all emission by a surface is absorbed by all of the surfaces, the sum of the fractions absorbed must be unity,

$$\sum_{j=1}^{N} G_{kj} = 1 \qquad (7\text{-}40a)$$

Gebhart [6, 7] also showed that there is the reciprocity relation

$$A_k \epsilon_k G_{kj} = \epsilon_j A_j G_{jk} \qquad (7\text{-}40b)$$

Equations (7-40a) and (7-40b) can be substituted into Eq. (7-38) to yield

$$Q_k = A_k \epsilon_k \sigma T_k^4 \sum_{j=1}^{N} G_{kj} - \sum_{j=1}^{N} A_k \epsilon_k G_{kj} \sigma T_j^4$$

This is arranged into the following convenient form to obtain the Q after the G have been calculated:

$$Q_k = A_k \epsilon_k \sum_{j=1}^{N} G_{kj} \sigma (T_k^4 - T_j^4) \qquad (7\text{-}41)$$

Example 7-17 demonstrates the use of these relations.

7-4.3 Enclosure Analysis by Use of Transfer Factors

The transfer factors $\mathscr{F}$ are defined by stating that Q has the form

$$Q_k = A_k \sum_{j=1}^{N} \mathscr{F}_{kj} \sigma (T_k^4 - T_j^4) \qquad (1 \leq k \leq N) \qquad (7\text{-}42)$$

It is evident by comparing with (7-41) that

$$\mathscr{F}_{kj} = \epsilon_k G_{kj} \qquad (7\text{-}43)$$

Hence from (7-40a) and (7-40b)

$$\sum_{j=1}^{N} \mathscr{F}_{kj} = \epsilon_k \qquad (7\text{-}44)$$

$$A_k \mathscr{F}_{kj} = A_j \mathscr{F}_{jk} \qquad (7\text{-}45)$$

The $\mathscr{F}$ can obviously be calculated by first obtaining the G. An equivalent approach is to let $\sigma T_n^4 = 1$ for the nth surface and $T = 0$ for all other surfaces. Then (7-42) yields

$$\mathscr{F}_{kn} = -q_k \qquad (k \neq n)$$
$$\qquad\qquad\qquad\qquad\qquad\qquad (7\text{-}46)$$
$$\mathscr{F}_{nn} = -q_n + \epsilon_n$$

For the same conditions, Eq. (7-31) yields

$$\sum_{j=1}^{N} \left(\frac{\delta_{kj}}{\epsilon_j} - F_{k-j} \frac{1-\epsilon_j}{\epsilon_j} \right) q_j = -F_{kn} \qquad (k \neq n)$$

$$\sum_{j=1}^{N} \left(\frac{\delta_{nj}}{\epsilon_j} - F_{n-j} \frac{1-\epsilon_j}{\epsilon_j} \right) q_j = 1 - F_{nn}$$

(7-47)

If Eqs. (7-47) are solved for the q, these are a function of the F and ϵ. The $\mathscr{F}$ are found from the q by using Eq. (7-46); thus the $\mathscr{F}$ are found from the F and ϵ and do not depend on the T. Example 7-17 in the next section will show how the $\mathscr{F}$ are obtained by using matrix inversion to solve Eqs. (7-47).

7-4.4 Matrix Inversion for Enclosure Equations

When many surfaces are present in an enclosure, a large set of simultaneous equations such as (7-31), (7-39), or (7-47) will result. The equations can be solved using available computer subroutines that can accommodate hundreds of simultaneous equations.

A set of equations such as (7-31) can be written in a shorter form. Let the right side be C_k and the quantities in parentheses on the left be a_{kj}. Then the k equations can be written as

$$\sum_{j=1}^{N} a_{kj} q_j = C_k$$

(7-48a)

where

$$a_{kj} = \frac{\delta_{kj}}{\epsilon_j} - F_{k-j} \frac{1-\epsilon_j}{\epsilon_j} \qquad C_k = \sum_{j=1}^{N} F_{k-j} \sigma(T_k^4 - T_j^4)$$

(7-48b)

For an enclosure of N surfaces, the set of equations then has the form

$$a_{11}q_1 + a_{12}q_2 + \cdots + a_{1j}q_j + \cdots + a_{1N}q_N = C_1$$
$$a_{21}q_1 + a_{22}q_2 + \cdots + a_{2j}q_j + \cdots + a_{2N}q_N = C_2$$
$$\vdots \qquad \vdots \qquad \vdots \qquad \vdots \qquad \vdots$$
$$a_{k1}q_1 + a_{k2}q_2 + \cdots + a_{kj}q_j + \cdots + a_{kN}q_N = C_k$$
$$\vdots \qquad \vdots \qquad \vdots \qquad \vdots \qquad \vdots$$
$$a_{N1}q_1 + a_{N2}q_2 + \cdots + a_{Nj}q_j + \cdots + a_{NN}q_N = C_N$$

(7-49)

The array of a_{kj} coefficients is the *matrix of coefficients* and is designated by a bracket notation:

$$\text{matrix } a \equiv [a_{kj}] \equiv \begin{bmatrix} a_{11} & a_{12} & \cdots & a_{1j} & \cdots & a_{1N} \\ a_{21} & a_{22} & \cdots & a_{2j} & \cdots & a_{2N} \\ \cdot & \cdot & \cdot & \cdot & \cdot & \cdot \\ \cdot & \cdot & \cdot & \cdot & \cdot & \cdot \\ a_{k1} & a_{k2} & \cdots & a_{kj} & \cdots & a_{kN} \\ \cdot & \cdot & \cdot & \cdot & \cdot & \cdot \\ \cdot & \cdot & \cdot & \cdot & \cdot & \cdot \\ a_{N1} & a_{N2} & \cdots & a_{Nj} & \cdots & a_{NN} \end{bmatrix} \tag{7-50}$$

A method of solving a set of equations such as (7-49) is to obtain a second matrix a^{-1}, called the *inverse* of matrix a,

$$a^{-1} \equiv [\mathcal{A}_{kj}] \equiv \begin{bmatrix} \mathcal{A}_{11} & \mathcal{A}_{12} & \cdots & \mathcal{A}_{1j} & \cdots & \mathcal{A}_{1N} \\ \mathcal{A}_{21} & \mathcal{A}_{22} & \cdots & \mathcal{A}_{2j} & \cdots & \mathcal{A}_{2N} \\ \cdot & \cdot & \cdot & \cdot & \cdot & \cdot \\ \cdot & \cdot & \cdot & \cdot & \cdot & \cdot \\ \mathcal{A}_{k1} & \mathcal{A}_{k2} & \cdots & \mathcal{A}_{kj} & \cdots & \mathcal{A}_{kN} \\ \cdot & \cdot & \cdot & \cdot & \cdot & \cdot \\ \cdot & \cdot & \cdot & \cdot & \cdot & \cdot \\ \mathcal{A}_{N1} & \mathcal{A}_{N2} & \cdots & \mathcal{A}_{Nj} & \cdots & \mathcal{A}_{NN} \end{bmatrix} \tag{7-51}$$

The inverse matrix has a term $\mathcal{A}_{kj}$ corresponding to each a_{kj} in the original matrix. The $\mathcal{A}$ are found by operating on the a in a way briefly described as follows: If the kth row and jth column that contain element a_{kj} in a square matrix a are deleted, the determinant of the remaining square array is called the *minor* of element a_{kj} and is denoted by M_{kj}. The cofactor of a_{kj} is defined as $(-1)^{k+j}M_{kj}$. To obtain the inverse of a square matrix $[a_{kj}]$, each element a_{kj} is first replaced by its cofactor. The rows and columns of the resulting matrix are then interchanged. The elements of the matrix thus obtained are then each divided by the determinant $|a_{kj}|$ of the original matrix $[a_{kj}]$. The elements obtained in this fashion are the $\mathcal{A}_{kj}$. For more detailed information on matrix inversion, the reader should refer to a mathematics text such as [10]. There are computer programs that will obtain the inverse coefficients $\mathcal{A}_{kj}$ from the matrix of a_{kj} values.

After the inverse matrix has been obtained, the q_k values in (7-31) are found as the sum of products of $\mathcal{A}$ and C

$$\begin{aligned} q_1 &= \mathcal{A}_{11}C_1 + \mathcal{A}_{12}C_2 + \cdots + \mathcal{A}_{1j}C_j + \cdots + \mathcal{A}_{1N}C_N \\ q_2 &= \mathcal{A}_{21}C_1 + \mathcal{A}_{22}C_2 + \cdots + \mathcal{A}_{2j}C_j + \cdots + \mathcal{A}_{2N}C_N \\ & \cdot \qquad \cdot \qquad \cdot \qquad \cdot \qquad \cdot \qquad \cdot \\ q_k &= \mathcal{A}_{k1}C_1 + \mathcal{A}_{k2}C_2 + \cdots + \mathcal{A}_{kj}C_j + \cdots + \mathcal{A}_{kN}C_N \end{aligned} \tag{7-52a}$$

or

$$q_k = \sum_{j=1}^{N} \mathscr{A}_{kj} C_j \qquad (7\text{-}52b)$$

Thus the solution for each q_k is in the form of a weighted sum of $(T_k^4 - T_j^4)$; this is the same form as (7-42).

For a given enclosure the configuration factors F_{k-j} remain fixed. If in addition the ϵ_k are constant, then the elements a_{kj}, and hence the inverse elements $\mathscr{A}_{kj}$, remain fixed for the enclosure. The fact that the $\mathscr{A}_{kj}$ remain fixed has utility when it is desired to compute the radiation quantities within an enclosure for many different values of the T. The matrix need be inverted only once, and then (7-52b) can be applied for different values of the C.

EXAMPLE 7-17 The previous two sections have presented alternative ways of formulating the radiation network equations. These are equivalent approaches for solving the set of equations (7-31). For any appreciable number of simultaneous equations, that solution would be found with a computer subroutine or a solver in a computer package. This example gives the solution to the same problem using three different methods. This provides an example of the matrix operations and illustrates how the various transfer factors are related.

Figure 7-13 is a two-dimensional rectangular enclosure that is long in the direction normal to the cross section shown. All the surfaces are diffuse-gray and their temperatures and emissivities are given. For simplicity, the four sides will not be subdivided. The configuration factors are obtained by the

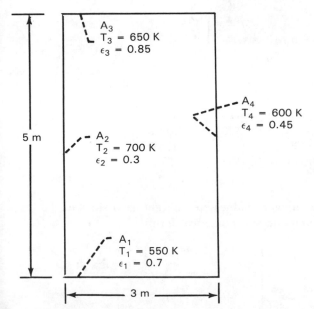

Figure 7-13 Rectangular geometry for Example 7-17.

crossed-string method to give $F_{1-3} = F_{3-1} = 0.2770$, $F_{2-4} = F_{4-2} = 0.5662$, $F_{1-2} = F_{1-4} = F_{3-2} = F_{3-4} = 0.3615$, and $F_{4-1} = F_{4-3} = F_{2-1} = F_{2-3} = 0.2169$. In what follows an abbreviated notation will be used: $E = (1 - \epsilon)/\epsilon$, $t = T^4$, and $F_{kj} = F_{k-j}$. Then the four equations from (7-31) are

$$\frac{q_1}{\epsilon_1} - q_2 F_{12} E_2 - q_3 F_{13} E_3 - q_4 F_{14} E_4 = \sigma(t_1 - F_{12} t_2 - F_{13} t_3 - F_{14} t_4)$$

$$-q_1 F_{21} E_1 + \frac{q_2}{\epsilon_2} - q_3 F_{23} E_3 - q_4 F_{24} E_4 = \sigma(-F_{21} t_1 + t_2 - F_{23} t_3 - F_{24} t_4)$$

$$-q_1 F_{31} E_1 - q_2 F_{32} E_2 + \frac{q_3}{\epsilon_3} - q_4 F_{34} E_4 = \sigma(-F_{31} t_1 - F_{32} t_2 + t_3 - F_{34} t_4)$$

$$-q_1 F_{41} E_1 - q_2 F_{42} E_2 - q_3 F_{43} E_3 + \frac{q_4}{\epsilon_4} = \sigma(-F_{41} t_1 - F_{42} t_2 - F_{43} t_3 + t_4)$$

All quantities except the q are known. Substitution and solution by a computer software package gives $q_1 = -2876.56$, $q_2 = 1612.57$, $q_3 = 1508.92$, $q_4 = -791.98$ W/m^2.

Now the method in Sec. 7-4.2 will be used. In Eq. (7-39) it is noted that the G_{jk} have the same matrix of coefficients for all k. There are four sets of four equations, one set for each k, and within each set j has values 1, 2, 3, and 4. The matrix of the G coefficients is (note that all the self-view factors $F_{kk} = 0$ for this example)

$$m = \begin{bmatrix} 1 & -\rho_2 F_{12} & -\rho_3 F_{13} & -\rho_4 F_{14} \\ -\rho_1 F_{21} & 1 & -\rho_3 F_{23} & -\rho_4 F_{24} \\ -\rho_1 F_{31} & -\rho_2 F_{32} & 1 & -\rho_4 F_{34} \\ -\rho_1 F_{41} & -\rho_2 F_{42} & -\rho_3 F_{43} & 1 \end{bmatrix}$$

When the values are substituted into this matrix it becomes

$$m = \begin{bmatrix} 1 & -0.25306 & -0.04155 & -0.19883 \\ -0.06507 & 1 & -0.03254 & -0.31140 \\ -0.08310 & -0.25306 & 1 & -0.19883 \\ -0.06507 & -0.39633 & -0.03254 & 1 \end{bmatrix}$$

The matrix of the four columns of values on the right-hand sides of the four sets of equations is, in symbolic and numerical form

$$f = \begin{bmatrix} F_{11}\epsilon_1 & F_{12}\epsilon_2 & F_{13}\epsilon_3 & F_{14}\epsilon_4 \\ F_{21}\epsilon_1 & F_{22}\epsilon_2 & F_{23}\epsilon_3 & F_{24}\epsilon_4 \\ F_{31}\epsilon_1 & F_{32}\epsilon_2 & F_{33}\epsilon_3 & F_{34}\epsilon_4 \\ F_{41}\epsilon_1 & F_{42}\epsilon_2 & F_{43}\epsilon_3 & F_{44}\epsilon_4 \end{bmatrix}$$

$$f = \begin{bmatrix} 0 & 0.10845 & 0.23544 & 0.16268 \\ 0.15183 & 0 & 0.18437 & 0.25479 \\ 0.19389 & 0.10845 & 0 & 0.16268 \\ 0.15183 & 0.16986 & 0.18437 & 0 \end{bmatrix}$$

The matrix of G_{kj} factors is obtained by matrix inversion as $G_{kj} = m^{-1}f$. This yields

$$G = \begin{bmatrix} 0.13233 & 0.18271 & 0.39315 & 0.29181 \\ 0.25579 & 0.08738 & 0.32299 & 0.33384 \\ 0.32377 & 0.19000 & 0.18278 & 0.30345 \\ 0.27236 & 0.22256 & 0.34391 & 0.16117 \end{bmatrix}$$

By summing values in each row it is evident that $\Sigma_{j=1}^{4} G_{kj} = 1$. The q are then found from Eq. (7-41), and they agree with the values given previously.

The solution will now be obtained by the method in Sec. 7-4.3. The coefficient matrix for the set of equations (7-47) is

$$D = \begin{bmatrix} \dfrac{1}{\epsilon_1} & -\rho_2 \dfrac{F_{12}}{\epsilon_2} & -\rho_3 \dfrac{F_{13}}{\epsilon_3} & -\rho_4 \dfrac{F_{14}}{\epsilon_4} \\ -\rho_1 \dfrac{F_{21}}{\epsilon_1} & \dfrac{1}{\epsilon_2} & -\rho_3 \dfrac{F_{23}}{\epsilon_3} & -\rho_4 \dfrac{F_{24}}{\epsilon_4} \\ -\rho_1 \dfrac{F_{31}}{\epsilon_1} & \rho_2 \dfrac{F_{32}}{\epsilon_2} & \dfrac{1}{\epsilon_3} & -\rho_4 \dfrac{F_{34}}{\epsilon_4} \\ -\rho_1 \dfrac{F_{41}}{\epsilon_1} & -\rho_2 \dfrac{F_{42}}{\epsilon_2} & -\rho_3 \dfrac{F_{43}}{\epsilon_3} & \dfrac{1}{\epsilon_4} \end{bmatrix}$$

and this gives in numerical form

$$D = \begin{bmatrix} 1.42857 & -0.84352 & -0.04888 & -0.44184 \\ -0.09296 & 3.33333 & -0.03828 & -0.69201 \\ -0.11871 & -0.84352 & 1.17647 & -0.44184 \\ -0.09296 & -1.32111 & -0.03828 & 2.22222 \end{bmatrix}$$

The matrix for the right side coefficients of Eq. (7-47) to determine $-q$ is (note that $F_{nn} = 0$ for this example)

$$M = \begin{bmatrix} -1 & F_{12} & F_{13} & F_{14} \\ F_{21} & -1 & F_{23} & F_{24} \\ F_{31} & F_{32} & -1 & F_{34} \\ F_{41} & F_{42} & F_{43} & -1 \end{bmatrix}$$

To relate the $-q$ and $\mathscr{F}$ from Eq. (7-46), the emissivity matrix is

$$
\epsilon = \begin{bmatrix}
\epsilon_1 & 0 & 0 & 0 \\
0 & \epsilon_2 & 0 & 0 \\
0 & 0 & \epsilon_3 & 0 \\
0 & 0 & 0 & \epsilon_4
\end{bmatrix}
$$

Then according to Eq. (7-46), the $\mathscr{F}$ factors are obtained from the matrix operations

$$\mathscr{F} = D^{-1}M + \epsilon$$

This yields the matrix of $\mathscr{F}$ values:

$$
\mathscr{F} = \begin{bmatrix}
0.09263 & 0.12790 & 0.27520 & 0.20427 \\
0.07674 & 0.02621 & 0.09690 & 0.10015 \\
0.27520 & 0.16150 & 0.15537 & 0.25793 \\
0.12256 & 0.10015 & 0.15476 & 0.07253
\end{bmatrix}
$$

From these values it is evident that there is the relation $\mathscr{F}_{kj} = \epsilon_k G_{kj}$ as in Eq. (7-43). The q are computed from Eq. (7-42), and the values are the same as before.

7-5 RADIATION ANALYSIS USING INFINITESIMAL AREAS

7-5.1 Generalized Net-Radiation Method Using Infinitesimal Areas

In the previous section the enclosure was divided into finite areas. The accuracy of the results is limited by the assumptions that the temperature and the energy incident on and leaving each surface are uniform over that surface. If these quantities are nonuniform over part of the enclosure boundary, that part must be subdivided until the variation over each area used in the analysis is not too large. Several calculations can be made in which successively smaller areas (and hence more simultaneous equations) are used until the solution results no longer change significantly when the area sizes are further diminished. In the limit, the enclosure boundary, or a portion of it, is divided into infinitesimal parts; this allows large variations in T, q, q_i, and q_o to be accounted for.

The formulation in terms of infinitesimal areas leads to heat balances in the form of integral equations. By using exact or approximate mathematical techniques for integral equations, it is sometimes possible to obtain a closed-form analytical solution. Usually an analytical solution is not obtained, and the integral equations are solved numerically.

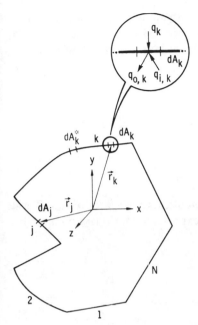

Figure 7-14 Enclosure composed of N discrete surface areas with areas subdivided into infinitesimal elements.

Consider an enclosure of N finite areas. The areas would generally be the major geometric divisions of the enclosure or the areas on which a specified boundary condition is held constant. Some or all of these areas are further subdivided into differential area elements as shown for two typical areas in Fig. 7-14. As before, the surfaces will be considered *diffuse-gray*. The additional restriction is now made that *the radiative properties are independent of temperature*.

A heat balance on element dA_k located at position $\mathbf{r}_k$ gives

$$q_k(\mathbf{r}_k) = q_{o,k}(\mathbf{r}_k) - q_{i,k}(\mathbf{r}_k) \tag{7-53}$$

The outgoing flux is composed of emitted and reflected energy:

$$q_{o,k}(\mathbf{r}_k) = \epsilon_k \sigma T_k^4(\mathbf{r}_k) + (1 - \epsilon_k) q_{i,k}(\mathbf{r}_k) \tag{7-54}$$

The incoming flux in (7-54) is composed of portions of the outgoing fluxes from the other area elements of the enclosure. This is a generalization of (7-14) in that over each finite surface an integration is performed to determine the contribution that the local flux leaving that surface makes to $q_{i,k}$:

$$dA_k q_{i,k}(\mathbf{r}_k) - \int_{A_1} q_{o,1}(\mathbf{r}_1)\, dF_{d1-dk}(\mathbf{r}_1, \mathbf{r}_k)\, dA_1 +$$

$$+ \int_{A_k} q_{o,k}(\mathbf{r}_k^*)\, dF_{dk^*-dk}(\mathbf{r}_k^*, \mathbf{r}_k)\, dA_k^* + \cdots$$

$$+ \int_{A_N} q_{o,N}(\mathbf{r}_N)\, dF_{dN-dk}(\mathbf{r}_N, \mathbf{r}_k)\, dA_N \tag{7-55}$$

The second integral on the right is the contribution that other differential elements dA_k^* on surface A_k make to the incident energy at dA_k. By using reciprocity, $dA_j \, dF_{dj-dk} = dA_k \, dF_{dk-dj}$, a typical integral in (7-55) can be transformed to obtain

$$\int_{A_j} q_{o,j}(\mathbf{r}_j) \, dF_{dj-dk}(\mathbf{r}_j,\mathbf{r}_k) \, dA_j = dA_k \int_{A_j} q_{o,j}(\mathbf{r}_j) \, dF_{dk-dj}(\mathbf{r}_j,\mathbf{r}_k)$$

By operating on all the integrals in (7-55) in this manner, the result becomes

$$q_{i,k}(\mathbf{r}_k) = \sum_{j=1}^{N} \int_{A_j} q_{o,j}(\mathbf{r}_j) \, dF_{dk-dj}(\mathbf{r}_j,\mathbf{r}_k) \tag{7-56}$$

Equations (7-54) and (7-56) provide two different expressions for $q_{i,k}(\mathbf{r}_k)$. These are each substituted into (7-53) to provide two expressions for $q_k(\mathbf{r}_k)$ comparable to (7-17) and (7-18):

$$q_k(\mathbf{r}_k) = \frac{\epsilon_k}{1-\epsilon_k} [\sigma T_k^{\,4}(\mathbf{r}_k) - q_{o,k}(\mathbf{r}_k)] \tag{7-57}$$

$$q_k(\mathbf{r}_k) = q_{o,k}(\mathbf{r}_k) - \sum_{j=1}^{N} \int_{A_j} q_{o,j}(\mathbf{r}_j) \, dF_{dk-dj}(\mathbf{r}_j,\mathbf{r}_k) \tag{7-58}$$

As shown by Eq. (6-7), the differential configuration factor dF_{dk-dj} contains the differential area dA_j. To place (7-58) in a more standard form in which the variable of integration is explicitly shown, it is convenient to define a quantity $K(\mathbf{r}_j,\mathbf{r}_k)$ by

$$K(\mathbf{r}_j,\mathbf{r}_k) \equiv \frac{dF_{dk-dj}(\mathbf{r}_j,\mathbf{r}_k)}{dA_j} \tag{7-59}$$

Then (7-58) becomes the integral equation

$$q_k(\mathbf{r}_k) = q_{o,k}(\mathbf{r}_k) - \sum_{j=1}^{N} \int_{A_j} q_{o,j}(\mathbf{r}_j) K(\mathbf{r}_j,\mathbf{r}_k) \, dA_j \tag{7-60}$$

The quantity $K(\mathbf{r}_j,\mathbf{r}_k)$ under the integral sign in (7-60) is the *kernel* of the integral equation.

As in the previous discussion for finite areas, there are two paths that can now be followed: (1) When the temperatures and imposed heat fluxes are important, (7-57) and (7-58) can be combined to eliminate the q_o. This gives a set of simultaneous relations relating the surface temperatures T and the heat fluxes q. Along each surface area, either T or q is specified. The unknown T and q are found by solving the simultaneous relations. (2) Alternatively, when q_o is an important quantity, the unknown q can be eliminated by combining (7-57) and (7-58) for each surface that does not have its q specified. For a surface where q is known, (7-58) can be used to directly relate the q_o to each other. This yields a set of simultaneous relations for the q_o in terms of the specified q and T. After solving for the q_o, Eqs. (7-57) can be used, if desired, to relate the q and T, where either the q or T will be known at each surface from the boundary conditions. Each of these procedures will now be examined.

Relations between surface temperature T and surface heating q To eliminate the q_o in the first method of solution, (7-57) is solved for $q_{o,k}(\mathbf{r}_k)$, giving

$$q_{o,k}(\mathbf{r}_k) = \sigma T_k{}^4(\mathbf{r}_k) - \frac{1 - \epsilon_k}{\epsilon_k} q_k(\mathbf{r}_k) \tag{7-61}$$

Equation (7-61) in the form shown and with k changed to j is then substituted into (7-58) to eliminate $q_{o,k}$ and $q_{o,j}$, which yields

$$\frac{q_k(\mathbf{r}_k)}{\epsilon_k} - \sum_{j=1}^{N} \frac{1 - \epsilon_j}{\epsilon_j} \int_{A_j} q_j(\mathbf{r}_j) \, dF_{dk-dj}(\mathbf{r}_j,\mathbf{r}_k) = \sigma T_k^4(\mathbf{r}_k) - \sum_{j=1}^{N} \int_{A_j} \sigma T_j^4(\mathbf{r}_j) \, dF_{dk-dj}(\mathbf{r}_j,\mathbf{r}_k)$$

$$= \sum_{j=1}^{N} \int_{A_j} \sigma [T_k^4(\mathbf{r}_k) - T_j^4(\mathbf{r}_j)] \, dF_{dk-dj}(\mathbf{r}_j,\mathbf{r}_k) \tag{7-62}$$

Equation (7-62) relates the surface temperatures to the heat fluxes supplied to the surfaces. It corresponds to Eq. (7-31) in the earlier formulation.

> **EXAMPLE 7-18** An enclosure of the general type in Fig. 7-8 is composed of three plane surfaces. Surface 1 is heated uniformly, and surface 2 is uniform in temperature. Surface 3 is black and at zero temperature. What are the governing equations needed to determine the temperature distribution along surface 1?
>
> With $T_3 = 0$, $\epsilon_3 = 1$, and the self-view factors $dF_{dj-dj*} = 0$, Eq. (7-62) can be written for the two plane surfaces 1 and 2 having uniform q_1 and T_2 as
>
> $$\frac{q_1}{\epsilon_1} - \frac{1 - \epsilon_2}{\epsilon_2} \int_{A_2} q_2(\mathbf{r}_2) \, dF_{d1-d2}(\mathbf{r}_2,\mathbf{r}_1) = \sigma T_1{}^4(\mathbf{r}_1) - \sigma T_2{}^4 \int_{A_2} dF_{d1-d2}(\mathbf{r}_2,\mathbf{r}_1) \tag{7-63a}$$
>
> $$\frac{q_2(\mathbf{r}_2)}{\epsilon_2} - q_1 \frac{1 - \epsilon_1}{\epsilon_1} \int_{A_1} dF_{d2-d1}(\mathbf{r}_1,\mathbf{r}_2) = \sigma T_2{}^4 - \int_{A_1} \sigma T_1{}^4(\mathbf{r}_1) \, dF_{d2-d1}(\mathbf{r}_1,\mathbf{r}_2) \tag{7-63b}$$
>
> A similar equation for surface 3 is not needed since Eqs. (7-63) do not involve the unknown $q_3(\mathbf{r}_3)$ as a consequence of $\epsilon_3 = 1$ and $T_3 = 0$. From the definitions of F factors, $\int_{A_2} dF_{d1-d2} = F_{d1-2}$ and $\int_{A_1} dF_{d2-d1} = F_{d2-1}$. Equations (7-63) simplify to the following relations where the unknowns have been placed on the left:
>
> $$\sigma T_1{}^4(\mathbf{r}_1) + \frac{1 - \epsilon_2}{\epsilon_2} \int_{A_2} q_2(\mathbf{r}_2) \, dF_{d1-d2}(\mathbf{r}_2,\mathbf{r}_1) = \sigma T_2{}^4 F_{d1-2}(\mathbf{r}_1) + \frac{q_1}{\epsilon_1} \tag{7-64a}$$

$$\int_{A_1} \sigma T_1^4(\mathbf{r}_1)\, dF_{d2\text{-}d1}(\mathbf{r}_1,\mathbf{r}_2) + \frac{q_2(\mathbf{r}_2)}{\epsilon_2} = \sigma T_2^4 + q_1 \frac{1-\epsilon_1}{\epsilon_1} F_{d2\text{-}1}(\mathbf{r}_2) \qquad (7\text{-}64b)$$

Equations (7-64) can be solved simultaneously for the distributions $T_1(\mathbf{r}_1)$ and $q_2(\mathbf{r}_2)$. Some solution methods will be discussed in Sec. 7-5.2.

Solution method in terms of outgoing radiative flux q_o Another method results from eliminating the $q_k(\mathbf{r}_k)$ from (7-57) and (7-58) for the surfaces where $q_k(\mathbf{r}_k)$ is unknown. This provides a relation between q_o and the T specified along a surface:

$$q_{o,k}(\mathbf{r}_k) = \epsilon_k \sigma T_k^4(\mathbf{r}_k) + (1-\epsilon_k) \sum_{j=1}^{N} \int_{A_j} q_{o,j}(\mathbf{r}_j)\, dF_{dk\text{-}dj}(\mathbf{r}_j,\mathbf{r}_k) \qquad (7\text{-}65)$$

When $q_k(\mathbf{r}_k)$, the heat supplied to surface k, is known, (7-58) can be used directly to relate q_k and q_o. The combination of (7-65) and (7-58) thus provides a complete set of relations for the unknown q_o in terms of known T and q.

This set of equations for the q_o are now formulated more explicitly. In general, an enclosure can have surfaces $1, 2, \ldots, m$ with specified temperature distributions. For these surfaces, (7-65) is utilized. The remaining $N - m$ surfaces $m + 1, m + 2, \ldots, N$ have an imposed heat flux distribution specified. For these surfaces (7-58) is applied. This results in a set of N equations for the unknown q_o distributions:

$$q_{o,k}(\mathbf{r}_k) - (1-\epsilon_k) \sum_{j=1}^{N} \int_{A_j} q_{o,j}(\mathbf{r}_j)\, dF_{dk\text{-}dj}(\mathbf{r}_j,\mathbf{r}_k) = \epsilon_k \sigma T_k^4(\mathbf{r}_k) \qquad 1 \leqslant k \leqslant m$$

$$(7\text{-}66a)$$

$$q_{o,k}(\mathbf{r}_k) - \sum_{j=1}^{N} \int_{A_j} q_{o,j}(\mathbf{r}_j)\, dF_{dk\text{-}dj}(\mathbf{r}_j,\mathbf{r}_k) = q_k(\mathbf{r}_k) \qquad m+1 \leqslant k \leqslant N \qquad (7\text{-}66b)$$

After the q_o are found from these simultaneous integral equations, (7-57) is applied to determine the unknown q or T distributions

$$q_k(\mathbf{r}_k) = \frac{\epsilon_k}{1-\epsilon_k} [\sigma T_k^4(\mathbf{r}_k) - q_{o,k}(\mathbf{r}_k)] \qquad 1 \leqslant k \leqslant m \qquad (7\text{-}67a)$$

$$\sigma T_k^4(\mathbf{r}_k) = \frac{1-\epsilon_k}{\epsilon_k} q_k(\mathbf{r}_k) + q_{o,k}(\mathbf{r}_k) \qquad m+1 \leqslant k \leqslant N \qquad (7\text{-}67b)$$

Special case when imposed heating q is specified for all surfaces An interesting special case is when the imposed energy flux q is specified for all surfaces of the enclosure and it is desired to determine the surface temperature distributions. For this case the use of the method of the previous section where the q_o are first determined has an advantage over the method given by (7-62) where the

T are directly determined from the specified q. This advantage arises from Eq. (7-66b) being independent of the radiative properties of the surfaces. This means that, for a given set of q, the q_o need be determined only once from (7-66b). Then the temperature distributions are found from (7-67b), which introduces the emissivity dependence. This would have an advantage when it is desired to examine the temperature variations for various emissivity values when there is a fixed set of q. A useful relation is obtained by considering the case in which the surfaces are all black, $\epsilon_k = 1$. Equation (7-67b) shows that $q_{o,k}(\mathbf{r}_k) = \sigma T_k^4(\mathbf{r}_k)_{\text{black}}$. Since the $q_{o,k}$ are independent of the emissivities, the solution in (7-67b) can then be written for $\epsilon_k \neq 1$ as

$$\sigma T_k{}^4(\mathbf{r}_k) = \frac{1-\epsilon_k}{\epsilon_k} q_k(\mathbf{r}_k) + \sigma T_k{}^4(\mathbf{r}_k)_{\text{black}} \tag{7-68}$$

This relates the temperature distributions in an enclosure for $\epsilon_k \neq 1$ to the temperature distributions in a black enclosure having the *same imposed heat fluxes*. Thus, once the temperature distributions have been found for the black case, the $\sigma T_k^4(\mathbf{r}_k)$ for gray surfaces are found by simply adding $[(1 - \epsilon_k)/\epsilon_k]q_k(\mathbf{r}_k)$.

To this point, a number of formulations of the governing equations of radiation interchange within an enclosure have been given. In Table 7-1, the relations are summarized for finding quantities of interest, such as Q, T, and q_o in terms of given quantities.

Table 7-1 Relations between energy flux and temperature in diffuse-gray enclosures

Type of areas	Boundary conditions	Desired quantities	Equations
Finite areas	T_k on all surfaces $1 \leq k \leq N$	Q_k $q_{o,k}$	(7-31) or (7-41) or (7-42) (7-35)
	Q_k on all surfaces $1 \leq k \leq N$	T_k $q_{o,k}$	(7-31) or (7-41) or (7-42) (7-18)
	T_k for $1 \leq k \leq m$ Q_k for $m + 1 \leq k \leq N$	Q_k for $1 \leq k \leq m$ T_k for $m + 1 \leq k \leq N$ $q_{o,k}$	(7-37) and (7-17) or (7-31) or (7-41) or (7-42) (7-37)
Infinitesimal areas	T_k on all surfaces $1 \leq k \leq N$	q_k $q_{o,k}$	(7-62) (7-65)
	q_k on all surfaces $1 \leq k \leq N$	T_k $q_{o,k}$	(7-62) (7-58)
	T_k for $1 \leq k \leq m$ q_k for $m + 1 \leq k \leq N$	q_k for $1 \leq k \leq m$ T_k for $m + 1 \leq k \leq N$ $q_{o,k}$	(7-62), or (7-66) and (7-67) (7-66a) and (7-66b)
	$q_{o,k}$ for $1 \leq k \leq N$ T_k for $1 \leq k \leq m$ q_k for $m + 1 \leq k \leq N$	q_k for $1 \leq k \leq m$ T_k for $m + 1 \leq k \leq N$	(7-67a) (7-67b)

EXAMPLE 7-19 A relatively simple example of a heated enclosure is the circular tube in Fig. 7-15, open at both ends and insulated on the outside surface [11]. For a uniform heat addition to the inside surface of the tube wall and a surrounding environment at 0 K, what is the temperature distribution along the tube? If the surroundings are at temperature T_e, how does this influence the temperature distribution?

Since the open ends of the tube are nonreflecting, they can be assumed to act as black disks at the surrounding temperature 0 K. Then with $\epsilon_1 = \epsilon_3 = 1$, (7-67b) gives $q_{o,1} = q_{o,3} = \sigma T_1^4 = \sigma T_3^4 = 0$. Consequently, the summation in (7-66b) provides only radiation from surface 2 to itself. Since the tube is axisymmetric, the two differential areas dA_k and dA_k^* can be rings located at x and y. For convenience, all lengths are nondimensionalized with respect to the tube diameter. Then Eq. (7-66b) yields

$$q_{o,2}(\xi) - \int_{\eta=0}^{\eta=l} q_{o,2}(\eta)\, dF_{d\xi-d\eta}(|\eta - \xi|) = q_2 \tag{7-69a}$$

where $\xi = x/D$, $\eta = y/D$, $l = L/D$, and $dF_{d\xi-d\eta}(|\eta - \xi|)$ is the configuration factor for two rings a distance $|\eta - \xi|$ apart and is given by 33 in Appendix C as

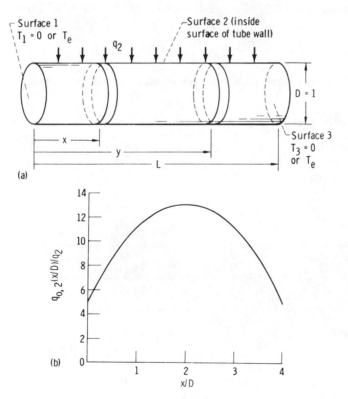

(a)

(b)

Figure 7-15 Uniformly heated tube insulated on outside and open to environment at both ends. (a) Geometry and coordinate system; (b) distribution of q_o on inside of tube for $L/D = 4$.

$$dF_{d\xi-d\eta}(|\eta - \xi|) = \left\{ 1 - \frac{|\eta - \xi|^3 + \frac{3}{2}|\eta - \xi|}{[(\eta - \xi)^2 + 1]^{3/2}} \right\} d\eta \qquad (7\text{-}69b)$$

Absolute-value signs are used on $\eta - \xi$ because the configuration factor depends only on the magnitude of the distance between the rings. When $|\eta - \xi| = 0$, $dF = d\eta$, and this is the view factor from a differential ring to itself. Equation (7-69a) can be divided by the constant q_2 and the dimensionless quantity $q_{o,2}(\xi)/q_2$ found by numerical or approximate solution methods for linear integral equations. A discussion of methods will be given in Sec. 7-5.2. The resulting $q_{o,2}(\xi)/q_2$ distribution is shown in Fig. 7-15b for a tube four diameters in length. From Eq. (7-67b), the distribution of $T_2^4(\xi)$ along the tube is given by

$$\sigma T_2^4(\xi) = \frac{1 - \epsilon_2}{\epsilon_2} q_2 + q_{o,2}(\xi)$$

Since q_2 is a constant, the $T_2^4(\xi)$ distribution has the same shape as $q_{o,2}(\xi)$. The wall temperature is high in the central region of the tube and low near the end openings where heat can be radiated more readily to the low-temperature environment.

Now consider the case where the environment is at $T_e \neq 0$. The open ends of the cylindrical enclosure can be regarded as perfectly absorbing disks at T_e, and the integral equation (7-66b) yields

$$q_{o,2}(\xi) - \int_{\eta=0}^{l} q_{o,2}(\eta) \, dF_{d\xi-d\eta}(|\eta - \xi|) - \sigma T_e^4 F_{d\xi-1}(\xi)$$

$$- \sigma T_e^4 F_{d\xi-3}(l - \xi) = q_2$$

where $F_{d\xi-1}(\xi)$ is the configuration factor from a ring element at ξ to disk 1 at $\xi = 0$,

$$F_{d\xi-1}(\xi) = \frac{\xi^2 + \frac{1}{2}}{(\xi^2 + 1)^{1/2}} - \xi$$

Since the integral equation is linear in the variable $q_{o,2}(\xi)$, let a trial solution be the sum of two parts where, for each part, either $T_e = 0$ or $q_2 = 0$:

$$q_{o,2}(\xi) = q_{o,2}(\xi)|_{T_e=0} + q_{o,2}(\xi)|_{q_2=0}$$

Substitute the trial solution into the integral equation to get

$$q_{o,2}(\xi)|_{T_e=0} + q_{o,2}(\xi)|_{q_2=0} - \int_{\eta=0}^{l} q_{o,2}(\eta)|_{T_e=0} \, dF_{d\xi-d\eta}(|\eta - \xi|)$$

$$- \int_{\eta=0}^{l} q_{o,2}(\eta)|_{q_2=0} \, dF_{d\xi-d\eta}(|\eta - \xi|) - \sigma T_e^4 F_{d\xi-1}(\xi)$$

$$- \sigma T_e^4 F_{d\xi-3}(l - \xi) = q_2$$

For $T_e = 0$, (7-69a) applies; subtract this equation to give

$$q_{o,2}(\xi)|_{q_2=0} - \int_{\eta=0}^{l} q_{o,2}(\eta)|_{q_2=0} \, dF_{d\xi-d\eta}(|\eta - \xi|)$$

$$- \sigma T_e^4 F_{d\xi-1}(\xi) - \sigma T_e^4 F_{d\xi-3}(l - \xi) = 0$$

The solution is $q_{o,2}|_{q_2=0} = \sigma T_e^4$, as can be verified by direct substitution and then integration. This would be expected physically for an unheated surface in a uniform temperature environment. The temperature distribution along the tube is found from Eq. (7-67b) as

$$\sigma T_2^4(\xi) = \frac{1 - \epsilon_2}{\epsilon_2} q_2 + q_{o,2}(\xi)|_{T_e=0} + q_{o,2}(\xi)|_{q_2=0}$$

$$= \frac{1 - \epsilon_2}{\epsilon_2} q_2 + q_{o,2}(\xi)|_{T_e=0} + \sigma T_e^4$$

where $q_{o,2}(\xi)|_{T_e=0}$ was found in the first part of this example. The superposition of an environment temperature has thus added a σT_e^4 term to the solution for $\sigma T_2^4(\xi)$, found previously for $T_e = 0$.

The final result of Example 7-19 can be written as

$$\sigma[T_2^4(\xi) - T_e^4] = \frac{1 - \epsilon_2}{\epsilon_2} q_2 + q_{o,2}(\xi)|_{T_e=0}$$

Thus for nonzero T_e the quantity $T_2^4(\xi) - T_e^4$ is equal to $T_2^4(\xi)$ for the case when $T_e = 0$. This illustrates a general way of accounting for a finite environment temperature. For heat transfer *only by radiation*, the governing equations are linear in T^4. As a result, in a cavity type of enclosure the wall temperature $T_w|_{T_e=0}$ can be calculated for a zero environment temperature. Then by superposition the wall temperature for any finite T_e is $T_w^4|_{T_e\neq 0} = T_w^4|_{T_e=0} + T_e^4$. Hence the thermal characteristics of a cavity having a wall temperature variation T_w and an external environment at T_e are the same as a cavity with wall temperature variation $(T_w^4 - T_e^4)^{1/4}$ and a zero environment temperature.

EXAMPLE 7-20 This example considers the emission from a long cylindrical hole drilled into a material that is at uniform temperature T_w (Fig. 7-16a). The hole is sufficiently long so that the surface at its bottom end can be neglected in the radiative heat balances. The environment outside the hole is at T_e. In accordance with the previous discussion, the solution is obtained by using the reduced temperature $T_r = (T_w^4 - T_e^4)^{1/4}$. If a position is viewed at x on the cylindrical side wall of the hole, the energy leaving the wall is $q_o(x)$. An apparent emissivity is defined as $\epsilon_a(x) = q_o(x)/\sigma T_r^4$. The analysis will determine how $\epsilon_a(x)$ is related to the surface emissivity ϵ, where ϵ is constant over the side of the hole. The integral equation governing the radiation exchange within a hole was first derived by Buckley [12, 13] and later by Eckert [14]; both investigators obtained approximate analytical solutions. Results were later obtained numerically by Sparrow and Albers [15].

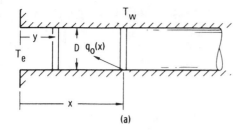

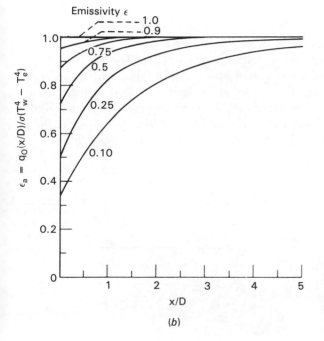

Figure 7-16 Radiant emission from cylindrical hole at uniform temperature. (*a*) Geometry and coordinate system; (*b*) apparent emissivity of cylinder wall.

By using the reduced temperature, the opening of the hole is approximated by a perfectly absorbing disk at zero reduced temperature. Then from (7-67*b*) (because $\epsilon = 1$ and $T_r = 0$ for the opening area), the q_o for the opening is zero. Hence, the governing equation for the enclosure is Eq. (7-66*a*) written for the cylindrical side wall and including only radiation from the cylindrical wall to itself. As in Example 7-19, the configuration factor is that for a ring of differential length on the cylindrical enclosure exchanging radiation with a second ring at a different axial location. Equation (7-66*a*) then yields

$$q_o(\xi) - (1 - \epsilon) \int_{\eta=0}^{\infty} q_o(\eta) \, dF_{d\xi-d\eta}(|\eta - \xi|) = \epsilon \sigma T^4 \tag{7-70}$$

where $\xi = x/D$, $\eta = y/D$, and $dF_{d\xi-d\eta}(|\eta - \xi|)$ is given by (7-69*b*). After division by σT_r^4, which is constant, the apparent emissivity $\epsilon_a(\xi)$ is found to be governed by the integral equation:

$$\epsilon_a(\xi) - (1 - \epsilon) \int_{\eta=0}^{\infty} \epsilon_a(\eta) \, dF_{d\xi - d\eta}(|\eta - \xi|) = \epsilon \qquad (7\text{-}71)$$

The solution of (7-71) was carried out for various surface emissivities ϵ, and the results for ϵ_a along the hole are in Fig. 7-16b. The radiation leaving the surface approaches that of a blackbody as the wall position is at greater depths into the hole. At the mouth of the hole, $\epsilon_a = \sqrt{\epsilon}$, as shown by Buckley [12, 13].

The radiation from a hole of finite depth was analyzed in [16], and results are given in Fig. 7-17. Approximate solutions are in [17]. The effective hemispherical emissivity ϵ_h is a different quantity than that in Fig. 7-16b; it gives the total amount of energy leaving the mouth of the cavity ratioed to that emitted from a black-walled cavity. The latter is the same as the energy emitted from a black area across the mouth of the cavity. For each surface emissivity ϵ, the ϵ_h increases to a limiting value as the depth of the cavity is increased; the limiting

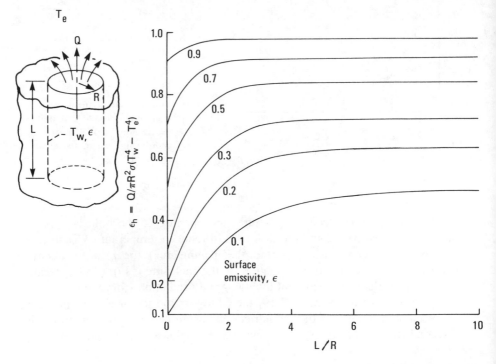

Figure 7-17 Apparent emissivity of cavity opening for a cylindrical cavity of finite length with diffuse reflecting walls at constant temperature [16].

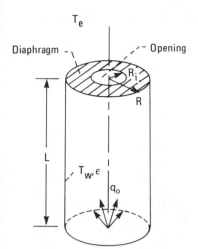

Figure 7-18 Cylindrical cavity with annular diaphragm partially covering opening [33, 34].

values are less than unity. Unless ϵ is small, a cavity more than only a few diameters deep emits the same amount of radiation as an infinitely deep cavity.

To construct a cavity that will provide very close to blackbody emission for use in calibrating measuring equipment, the cavity opening can be partially closed as shown in Fig. 7-18. The apparent emissivity ϵ_a at the center of the bottom of the cavity is given in Table 7-2 as a function of the diaphragm opening-to-outer-radius ratio and the cavity depth-to-radius ratio. The ϵ_a increases toward unity as the cavity is made deeper and the diaphragm opening is made smaller. Similar trends are shown in [18] for partially baffled conical cavities. In [19], cavities between radial fins are analyzed. A very detailed presentation of results for radiating cavities is in [20].

Some information on the directional nature of radiant emission from a cylindrical cavity is given in [21], where the emission normal to the cavity opening is calculated as would be received by a small detector along the cavity centerline and facing the opening. The normal emissivity ϵ_n' of the cavity is defined as the energy received by the detector divided by the energy that would be received if the cavity walls were entirely black. The ϵ_n' depends on the detector distance from the cavity opening; the values given in Table 7-3 are for large distances where ϵ_n' no longer changes with distance. The table compares ϵ_n' with ϵ_h from Fig. 7-17 (ϵ is the emissivity of the cavity walls). The ϵ_n' is larger than ϵ_h except for small values of cavity L/R.

EXAMPLE 7-21 What are the integral equations governing the radiation exchange between two parallel opposed plates finite in one dimension and infinite in the other, as shown in Fig. 7-19? Each plate has a specified temperature variation that depends only on the x or y coordinate shown, and the environment is at T_e.

Table 7-2 Values of apparent emissivity at center of cavity bottom, $\epsilon_a = q_o/\sigma(T_w^4 - T_e^4)$

Wall emissivity ϵ	R_i/R	ϵ_a		
		$L/R = 2$	$L/R = 4$	$L/R = 8$
0.25	0.4	0.916	0.968	0.990
	0.6	0.829	0.931	0.981
	0.8	0.732	0.888	0.969
	1.0	0.640	0.844	0.965
0.50	0.4	0.968	0.990	0.998
	0.6	0.932	0.979	0.995
	0.8	0.887	0.964	0.992
	1.0	0.839	0.946	0.989
0.75	0.4	0.988	0.997	0.999
	0.6	0.975	0.993	0.998
	0.8	0.958	0.988	0.997
	1.0	0.939	0.982	0.996

From the discussion in Example 6-2, the configuration factors between the infinitely long parallel strips dA_1 and dA_2 are

$$dF_{d1-d2} = \frac{1}{2} d(\sin \beta) = \frac{1}{2} \frac{a^2}{[(y-x)^2 + a^2]^{3/2}} dy$$

$$dF_{d2-d1} = \frac{1}{2} \frac{a^2}{[(y-x)^2 + a^2]^{3/2}} dx$$

The distribution of heat flux added to each plate can be found by applying Eq. (7-62) to each plate. As discussed before Example 7-20, reduced tem-

Table 7-3 Comparison of normal and hemispherical emissivity of cavity [21]

L/R	$\epsilon = 0.9$		$\epsilon = 0.7$		$\epsilon = 0.5$		$\epsilon = 0.3$	
	ϵ_h	ϵ_n'	ϵ_h	ϵ_n'	ϵ_h	ϵ_n'	ϵ_h	ϵ_n'
0.5	0.943	0.937	0.815	0.800	0.657	0.638	0.455	0.437
1	0.962	0.960	0.869	0.865	0.742	0.738	0.556	0.554
2	0.972	0.982	0.904	0.934	0.809	0.858	0.657	0.720
4	0.974	0.994	0.914	0.978	0.833	0.950	0.710	0.884
8	0.975	0.999	0.915	0.995	0.836	0.989	0.719	0.973

peratures are used to account for T_e. With $T_1 = (T_{w1}^4 - T_e^4)^{1/4}$ and $T_2 = (T_{w2}^4 - T_e^4)^{1/4}$, the governing equations are

$$\frac{q_1(x)}{\epsilon_1} - \frac{1-\epsilon_2}{\epsilon_2} \int_{-L/2}^{L/2} q_2(y) \frac{1}{2} \frac{a^2}{[(y-x)^2 + a^2]^{3/2}} \, dy$$

$$= \sigma T_1^4(x) - \int_{-L/2}^{L/2} \sigma T_2^4(y) \frac{1}{2} \frac{a^2}{[(y-x)^2 + a^2]^{3/2}} dy \qquad (7\text{-}72a)$$

$$\frac{q_2(y)}{\epsilon_2} - \frac{1-\epsilon_1}{\epsilon_1} \int_{-L/2}^{L/2} q_1(x) \frac{1}{2} \frac{a^2}{[(y-x)^2 + a^2]^{3/2}} \, dx$$

$$= \sigma T_2^4(y) - \int_{-L/2}^{L/2} \sigma T_1^4(x) \frac{1}{2} \frac{a^2}{[(y-x)^2 + a^2]^{3/2}} dx \qquad (7\text{-}72b)$$

An alternative formulation using (7-65) yields two equations for $q_{o,1}(x)$ and $q_{o,2}(y)$,

$$q_{o,1}(x) - (1-\epsilon_1) \int_{-L/2}^{L/2} q_{o,2}(y) \frac{1}{2} \frac{a^2}{[(y-x)^2 + a^2]^{3/2}} \, dy = \epsilon_1 \sigma T_1^4(x) \qquad (7\text{-}73a)$$

$$q_{o,2}(y) - (1-\epsilon_2) \int_{-L/2}^{L/2} q_{o,1}(x) \frac{1}{2} \frac{a^2}{[(y-x)^2 + a^2]^{3/2}} \, dx = \epsilon_2 \sigma T_2^4(y) \qquad (7\text{-}73b)$$

After the q_o are found, the desired $q_1(x)$ and $q_2(y)$ are obtained from (7-67a),

$$q_1(x) = \frac{\epsilon_1}{1-\epsilon_1} [\sigma T_1^4(x) - q_{o,1}(x)] \qquad (7\text{-}74a)$$

$$q_2(y) = \frac{\epsilon_2}{1-\epsilon_2} [\sigma T_2^4(y) - q_{o,2}(y)] \qquad (7\text{-}74b)$$

7-5.2 Methods for Solving Integral Equations

The previous examples have shown that the unknown wall heat fluxes or temperatures along the surfaces of an enclosure are found from solutions of single or simultaneous integral equations. The integral equations obtained in the formulations up to now are linear; that is, the unknown q, q_o, or T^4 variables always appear to the first power (note that T^4 is the variable rather than T). For linear integral equations there are various numerical and analytical solution methods; these are discussed in mathematics texts, for example, Chap. 4 of [10]. The use of some of these methods for radiation problems will now be discussed.

Numerical integration In most instances the functions inside the integrals of the integral equations are complicated algebraic quantities. This is because they in-

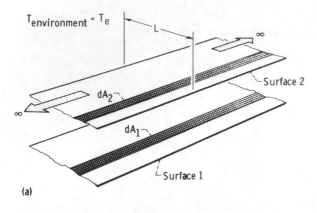

(a)

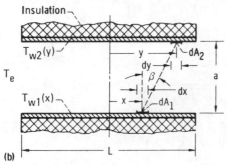

(b)

Figure 7-19 Geometry for radiation between two parallel plates infinitely long in one direction and of finite width. (*a*) Parallel plates of width L and infinite length; (*b*) coordinates in cross section of gap between parallel plates.

volve a configuration factor that, for most geometries, is not of a simple form. There is usually little chance that an exact analytical solution can be found, so a numerical solution is used. Consider, for example, the simultaneous integral equations in Eq. (7-72). Numerical integration can be used. With $T_1(x)$ and $T_2(y)$ specified, the right sides of the equations are known functions of x and y. Starting with (7-72a), a distribution for $q_2(y)$ is assumed as a first trial. Then the integration is carried out numerically for various x values to yield $q_1(x)$ at these x locations. This $q_1(x)$ distribution is inserted into Eq. (7-72b) and a $q_2(y)$ distribution is determined. This $q_2(y)$ is used to compute a new $q_1(x)$ from (7-72) and the process is continued until $q_1(x)$ and $q_2(y)$ are no longer changing as the iterations proceed.

To perform the integrations in a computer solution, an accurate integration subroutine is required. Many subroutines are available; a good method is Gaussian integration. This requires values of functions such as $q_1(x)$ and $q_2(y)$ at unevenly spaced values of the x and y coordinates. The values can be obtained by curve fitting the $q_1(x)$ and $q_2(y)$ after each iteration; standard subroutines such as cubic splines can be used. By using the spline coefficients, the $q_1(x)$ and $q_2(y)$ are calculated at the x and y values called for by the Gaussian integration routine. A precaution should be noted. A quantity such as $q_{o,j}\,dF_{dk-dj}$ may go through rapid changes in magnitude because of the geometry involved in the configuration factor; for example, dF_{dk-dj} may decrease very rapidly as the distance between dA_k and dA_j is increased (Fig. 7-20, for example). For small values of the separation distance there can be a strong peak in the integration kernel. Care should be taken to select an integration scheme that can deal properly with the behavior of the functions involved. The integration should be done in pieces on each side of a peak and not passed through the peak.

Direct solvers for a set of simultaneous equations can also be used to solve integral equations. The integrals are expressed in finite-difference form by dividing each surface into small increments. The result is a set of simultaneous equations for the unknown quantities at each increment position. This is illustrated by a simple example.

EXAMPLE 7-22 Referring to integral equation (7-69a), derive a set of simultaneous algebraic equations to determine the $q_{o,2}$ distribution for a length $l = 4$. For simplicity, divide the length into four equal increments ($\Delta\eta = 1$), and use the trapezoidal rule for integration.

When (7-69a) is applied at the end of the tube where $\xi = 0$, one obtains

$$q_{o,2}(0) - [\tfrac{1}{2}q_{o,2}(0)K(|0-0|) + q_{o,2}(1)K(|1-0|) + q_{o,2}(2)K(|2-0|)$$

$$+ q_{o,2}(3)K(|3-0|) + \tfrac{1}{2}q_{o,2}(4)K(|4-0|)](1) = q_2 \qquad (7\text{-}75)$$

The quantity in brackets is the trapezoidal-rule approximation for the integral. The $K(|\eta - \xi|) = dF(|\eta - \xi|)/d\eta$ is the algebraic expression within the braces of (7-69b). The $q_{o,2}(0)$ terms in (7-75) are grouped together to provide the first of (7-76). The other four equations are obtained by writing the finite-

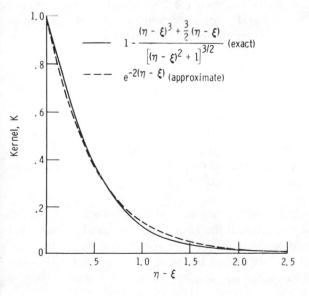

Figure 7-20 Exponential approximation to configuration-factor kernel for cylindrical enclosure.

difference equation at the other incremental positions along the cylindrical enclosure:

$$q_{o,2}(0)[1 - \tfrac{1}{2}K(0)] - q_{o,2}(1)K(1) - q_{o,2}(2)K(2) - q_{o,2}(3)K(3)$$

$$- \tfrac{1}{2}q_{o,2}(4)K(4) = q_2$$

$$- \tfrac{1}{2}q_{o,2}(0)K(1) + q_{o,2}(1)[1 - K(0)] - q_{o,2}(2)K(1) - q_{o,2}(3)K(2)$$

$$- \tfrac{1}{2}q_{o,2}(4)K(3) = q_2$$

$$- \tfrac{1}{2}q_{o,2}(0)K(2) - q_{o,2}(1)K(1) + q_{o,2}(2)[1 - K(0)] - q_{o,2}(3)K(1)$$

$$- \tfrac{1}{2}q_{o,2}(4)K(2) = q_2$$

$$- \tfrac{1}{2}q_{o,2}(0)K(3) - q_{o,2}(1)K(2) - q_{o,2}(2)K(1) + q_{o,2}(3)[1 - K(0)]$$

$$- \tfrac{1}{2}q_{o,2}(4)K(1) = q_2$$

$$- \tfrac{1}{2}q_{o,2}(0)K(4) - q_{o,2}(1)K(3) - q_{o,2}(2)K(2) - q_{o,2}(3)K(1)$$

$$+ q_{o,2}(4)[1 - \tfrac{1}{2}K(0)] = q_2 \tag{7-76}$$

This set of simultaneous equations is solved for the $q_{o,2}$ values at the five surface locations. From the symmetry of the geometry and the fact that q_2 is uniform along the enclosure, it is possible to simplify the solution for this example by using $q_{o,2}(0) = q_{o,2}(4)$ and $q_{o,2}(1) = q_{o,2}(3)$.

A set of equations such as (7-76) is first solved for a moderate number of increments along the surfaces. Then the increment size is reduced, and the solution is repeated. This process is continued until sufficiently accurate q_o values are obtained. The computer calculations would generally be programmed for an arbitrary increment size. Equations (7-76) used the trapezoidal rule as a simple numerical approximation to the integrals. More accurate numerical integration schemes can be used, which may reduce the number of increments required for sufficient accuracy.

Example 7-22 contained only one integral equation. For the situation with two integral equations described by (7-72), surfaces 1 and 2 can both be divided into increments and the equations written at each incremental location. This will yield a set of n simultaneous equations with n equal to the total number of chosen positions on both plates, and the equations can be solved simultaneously for the $q_1(x)$ and $q_2(y)$ distributions. This solution procedure is an alternative to the iterative solution described previously. The solver for the system of simultaneous equations may actually work by iteration (see Sec. 11-5).

Use of approximate separable kernel In an integral equation such as (7-69a) the solution can sometimes be simplified if the kernel is of a separable form, that is, equal to a product, or sum of products, of a function of $\mathbf{r}_j$ alone and a function of $\mathbf{r}_k$ alone. Recall from (7-59) that the kernel is $K(\mathbf{r}_j,\mathbf{r}_k) = dF_{dk-dj}(\mathbf{r}_j,\mathbf{r}_k)/dA_j$. For a separable kernel, the function of $\mathbf{r}_k$ can be taken out of the integral, thereby simplifying the integration. The general theory of integral equations with separable kernels is given in [10]. For radiation problems, K will usually not be in a separable form. However, it may be possible to find a separable function that closely approximates K and can be used to provide a simplified approximate solution.

Buckley [12, 13] demonstrated that an especially useful form for a separable kernel is an exponential function or series of exponential functions. With this type of kernel it is possible to change the integral equation into a differential equation, and sometimes an analytical solution can be obtained. A mathematical caution should be noted. Changing the integral equation into a differential equation requires taking derivatives of the approximate separable kernel. Even though the separable function may approximate the exact kernel fairly well, the approximation of the derivatives may become poor, especially for higher derivatives. The use of the separable kernel will now be demonstrated.

EXAMPLE 7-23 Determine $q_{o,2}/q_2$ from Eq. (7-69a) by use of an exponential approximate separable kernel [11].

The governing integral equation is

$$\frac{q_{o,2}(\xi)}{q_2} - \int_{\eta=0}^{l} \frac{q_{o,2}(\eta)}{q_2} K(|\eta - \xi|)\, d\eta = 1 \qquad (7\text{-}77a)$$

where

$$K(|\eta - \xi|) = 1 - \frac{|\eta - \xi|^3 + \frac{3}{2}|\eta - \xi|}{[(\eta - \xi)^2 + 1]^{3/2}} \qquad (7\text{-}77b)$$

The $K(|\eta - \xi|)$ is plotted in Fig. 7-20 and is reasonably well approximated by the function $e^{-2|\eta - \xi|}$. When the approximate kernel is substituted into (7-77a), the part of the function depending on ξ can be taken out of the integral to give

$$\frac{q_{o,2}(\xi)}{q_2} - e^{-2\xi}\int_0^\xi \frac{q_{o,2}(\eta)}{q_2} e^{2\eta}\, d\eta - e^{2\xi}\int_\xi^l \frac{q_{o,2}(\eta)}{q_2} e^{-2\eta}\, d\eta = 1 \qquad (7\text{-}78)$$

By differentiating (7-78) two times, the integrals are removed and the differential equation obtained:

$$\frac{d^2[q_{o,2}(\xi)/q_2]}{d\xi^2} = -4$$

This has the general solution, obtained by integrating twice,

$$\frac{q_{o,2}(\xi)}{q_2} = -2\xi^2 + C_1\xi + C_2 \qquad (7\text{-}79a)$$

To determine C_1 and C_2, two boundary conditions are needed. From symmetry, one boundary condition is $d(q_{o,2}/q_2)/d\xi = 0$ at $\xi = l/2$, which yields

$C_1 = 2l$. To determine C_2, a boundary condition can be obtained by evaluating (7-78) at $\xi = 0$ and $\xi = l$ and then utilizing the fact that $q_{o,2}(0) = q_{o,2}(l)$ to obtain

$$\int_0^l \frac{q_{o,2}(\eta)}{q_2} e^{-2\eta}\, d\eta = e^{-2l} \int_0^l \frac{q_{o,2}(\eta)}{q_2} e^{2\eta}\, d\eta$$

Inserting $q_{o,2}/q_2 = -2\xi^2 + 2l\xi + C_2$ and integrating yield $C_2 = l + 1$. With C_1 and C_2 thus evaluated, the final result for $q_{o,2}/q_2$ is the parabola

$$\frac{q_{o,2}(\xi)}{q_2} = l + 1 + 2(\xi l - \xi^2) \tag{7-79b}$$

More generally, the boundary conditions to evaluate C_1 and C_2 could be obtained in an asymmetric case by evaluating the integral equation at both boundaries $\xi = 0$ and $\xi = l$. This yields, from (7-78),

$$\frac{q_{o,2}(0)}{q_2} - \int_0^l \frac{q_{o,2}(\eta)}{q_2} e^{-2\eta}\, d\eta = 1, \qquad \frac{q_{o,2}(l)}{q_2} - e^{-2l} \int_0^l \frac{q_{o,2}(\eta)}{q_2} e^{2\eta}\, d\eta = 1$$

Then $q_{o,2}/q_2$ from (7-79a) is substituted into these two boundary conditions. After integrating, two simultaneous equations result for C_1 and C_2, leading to the same solution as before. The advantage of using the symmetry condition was only algebraic simplicity.

Approximate solution by variational method   An integral equation of the form

$$\Phi(\xi) = \int_a^b K(\xi, \eta)\Phi(\eta)\, d\eta + G(\xi) \tag{7-80}$$

can be solved by variational methods [10]. A restriction is that $K(\xi, \eta)$ be symmetric; that is, K is not changed when the ξ and η are interchanged. The kernel of (7-77b) is an example of a symmetric kernel since, because of the absolute-value signs, $K(|\eta - \xi|) = K(|\xi - \eta|)$.

The variational method depends on the use of an auxiliary function that is related to Eq. (7-80). This function is given by

$$J = \int\int_a^b K(\xi, \eta)\Phi(\xi)\Phi(\eta)\, d\xi\, d\eta - \int_a^b [\Phi(\xi)]^2\, d\xi + 2 \int_a^b \Phi(\xi)G(\xi)\, d\xi \tag{7-81}$$

The significance of the J function is that it will have a minimum value when the correct solution for $\Phi(\xi)$ is found.

To obtain an approximate solution, let $\Phi(\xi)$ be represented by a polynomial with unknown coefficients,

$$\Phi(\xi) = \gamma_0 + \gamma_1\xi + \gamma_2\xi^2 + \cdots + \gamma_n\xi^n \tag{7-82}$$

This is substituted into (7-81) and the integration carried out. If K is so complicated algebraically that the integration cannot be done analytically, the method is not practical. After the integration, the result is an analytical expression for J as a function of $\gamma_0, \gamma_1, \gamma_2, \ldots, \gamma_n$. These unknown coefficients are determined by differentiating J with respect to each of the individual coefficients and setting each result equal to zero, $\partial J/\partial\gamma_0 = 0$, $\partial J/\partial\gamma_1 = 0$, $\ldots$, $\partial J/\partial\gamma_n = 0$. This yields a set of $n + 1$ simultaneous equations for the $n + 1$ unknown coefficients. By differentiating J in this manner and setting the differentials equal to zero, the coefficients are found that make J a minimum value; thus the most accurate solution to the integral equation of the assumed form $\Phi(\xi) = \Sigma_{j=0}^n \gamma_j\xi^j$ is found. This method has been applied to radiation in a cylindrical tube in [11] and radiation between parallel plates of finite width and infinite length in [22].

Approximate solution by Taylor-series expansion The use of a Taylor-series expansion method for solving a radiation integral equation is demonstrated in [23, 24]. The physical idea that motivates this method is that the geometric configuration factor can often decrease quite rapidly as the distance is increased between the two elements exchanging radiation. This means that the radiative heat balance at a given location may be significantly influenced only by the radiative fluxes leaving surface elements in the immediate vicinity.

As an example, consider the type of integral equation (7-77a). The function $K(|\eta - \xi|)$ decreases rapidly as $\eta - \xi$ is increased (Fig. 7-20). If it is assumed that the important values of η are those close to the location ξ, the function $q_{o,2}(\eta)/q_2$ is expanded in a Taylor series about ξ:

$$\frac{q_{o,2}(\eta)}{q_2} = \frac{q_{o,2}(\xi)}{q_2} + (\eta - \xi)\left[\frac{d(q_{o,2}/q_2)}{d\xi}\right]_\xi + \frac{(\eta - \xi)^2}{2!}\left[\frac{d^2(q_{o,2}/q_2)}{d\xi^2}\right]_\xi + \cdots \tag{7-83}$$

The derivatives in the Taylor expansion are evaluated at ξ and hence do not contain the variable η. This means that, when (7-83) is substituted into (7-77a), the derivatives can be taken out of the integrals to yield

$$\frac{q_{o,2}(\xi)}{q_2} - \frac{q_{o,2}(\xi)}{q_2}\int_{\eta=0}^l K(|\eta - \xi|)\,d\eta - \frac{d}{d\xi}\left[\frac{q_{o,2}(\xi)}{q_2}\right]\int_{\eta=0}^l (\eta - \xi)K(|\eta - \xi|)\,d\eta$$

$$- \frac{1}{2!}\frac{d^2}{d\xi^2}\left[\frac{q_{o,2}(\xi)}{q_2}\right]\int_{\eta=0}^l (\eta - \xi)^2 K(|\eta - \xi|)\,d\eta - \cdots = 1 \tag{7-84}$$

The integrations are then carried out; if this cannot be done analytically, the method is not of practical utility because it is as easy to carry out a numerical solution of the exact integral equation. If the integrals can be carried out analytically, (7-84) becomes a differential equation for $q_{o,2}(\xi)/q_2$ that can be solved analytically or numerically if the boundary conditions can be specified. The boundary conditions are derived, as illustrated in [24], from the physical constraints in the system, such as symmetry or an overall heat balance. This method is probably of little value for enclosures involving more than one or two surfaces. The first term of the Taylor-series expansion was used in a collocation method in [25] to build a higher-order approximation. The method was applied to two parallel plates and to the inside of a cylinder.

Solution by method of Ambarzumian Crosbie and Sawheny [26, 27] applied the method of Ambarzumian [28, 29] to semi-infinite geometries. They studied the radiative exchange within two-dimensional cavities in a semi-infinite solid (Fig. 7-21) when the temperature or heat flux distribution along the cavity walls can be described by an exponential variation or a sum of exponential terms. The governing Fredholm integral equation of the second kind that describes the outgoing radiative flux $q_o(x)$ or $q_o(y)$ is transformed into an integrodifferential equation of the initial-value type.

EXAMPLE 7-24 Consider the semi-infinite rectangular cavity of Fig. 7-21 as analyzed in [26]. The walls along x and y extend to infinity, and both have the same temperature distribution, $T(x) = T_0 \exp(-mx/4a)$ and $T(y) = T_0 \exp(-my/4a)$. Both walls are diffuse-gray with emissivity $\epsilon = 1 - \rho$. Equation (7-66a) for $q_o(x)$ gives

$$q_o(x,m) = \epsilon \sigma T^4(x,m) + \rho \int_0^\infty q_o(y,m)\, dF_{dx-dy}(x,y) \tag{7-85}$$

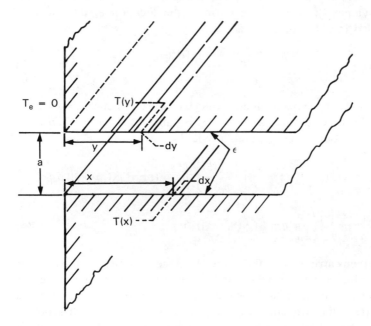

Figure 7-21 Semi-infinite rectangular cavity with exponentially varying wall temperature.

where q_o depends on the parameter m in the temperature variation. Since both walls have the same temperature distribution and properties, only one equation is required. Letting $\xi = x/a$ and $\eta = y/a$, the configuration factor between two differential strips is

$$dF_{d\xi-d\eta}(\xi,\eta) = \frac{d\eta}{2[(\xi-\eta)^2+1]^{3/2}} = \frac{1}{2}K(|\xi-\eta|)\,d\eta$$

The kernel $K(|\xi-\eta|)$ is symmetric. Equation (7-85) becomes

$$\Phi(\xi,m) = e^{-m\xi} + \frac{\rho}{2}\int_0^\infty \Phi(\eta,m)K(|\xi-\eta|)\,d\eta \qquad (7\text{-}86)$$

where $\Phi(\xi,m) = q_o(\xi,m)/\epsilon\sigma T_0^4$.

In Ambarzumian's method, (7-86) is transformed into an integrodifferential form. Equation (7-86) is rewritten as

$$\Phi(\xi,m) = e^{-m\xi} + \frac{\rho}{2}\int_0^\xi \Phi(\eta,m)K(\xi-\eta)\,d\eta + \frac{\rho}{2}\int_\xi^\infty \Phi(\eta,m)K(\eta-\xi)\,d\eta \qquad (7\text{-}87)$$

Variables are transformed in (7-87) by substituting $z = \xi - \eta$ into the first integral and $z = \eta - \xi$ into the second integral to yield

$$\Phi(\xi,m) = e^{-m\xi} + \frac{\rho}{2}\int_0^\xi \Phi(\xi-z,m)K(z)\,dz + \frac{\rho}{2}\int_0^\infty \Phi(\xi+z,m)K(z)\,dz \qquad (7\text{-}88)$$

Differentiating (7-88) with respect to ξ, again substituting to eliminate z, and gathering terms give the following integral equation for the derivative of $\Phi(\xi,m)$:

$$\frac{\partial\Phi(\xi,m)}{\partial\xi} = -me^{-m\xi} + \frac{\rho}{2}\Phi(0,m)K(\xi) + \frac{\rho}{2}\int_0^\infty \frac{\partial\Phi(\eta,m)}{\partial\eta}K(|\xi-\eta|)\,d\eta \qquad (7\text{-}89)$$

If Eq. (7-86) is multiplied by $mJ_1(m)\,dm/2$, where $J_1(m)$ is the Bessel function of the first kind, and then integrated from 0 to ∞, we find

$$\frac{1}{2}\int_0^\infty m\Phi(\xi,m)J_1(m)\,dm = \frac{1}{2}\int_0^\infty me^{-m\xi}J_1(m)\,dm$$

$$+ \frac{\rho}{4}\int_0^\infty mJ_1(m)\int_0^\infty \Phi(\eta,m)K(|\xi-\eta|)\,d\eta\,dm \qquad (7\text{-}90)$$

Multiplying (7-86) by $-m$ and (7-90) by $\rho\Phi(0,m)$ and adding give

$$-m\Phi(\xi,m) + \frac{\rho}{2}\Phi(0,m)\int_0^\infty t\Phi(\xi,t)J_1(t)\,dt$$

$$= -me^{-m\xi} + \frac{\rho}{2}\Phi(0,m)K(\xi) + \frac{\rho}{2}\int_0^\infty \frac{\partial\Phi(\eta,m)}{\partial\eta}K(|\xi-\eta|)\,d\eta \qquad (7\text{-}91)$$

after using the relation $\int_0^\infty m e^{-xm} J_1(m) \, dm = 1/(x^2 + 1)^{3/2} = K(x)$ and noting that the result of the addition gives the same integral equation as (7-89).

The right-hand sides of (7-91) and (7-89) are identical; thus, (7-89) can be written as

$$\frac{\partial \Phi(\xi, m)}{\partial \xi} = -m\Phi(\xi, m) + \frac{\rho}{2} \Phi(0, m) \int_0^\infty t\Phi(\xi, t) J_1(t) \, dt \tag{7-92}$$

The integral equation (7-87) has thus been transformed into an initial-value problem for $\Phi(\xi, m)$, but the initial condition $\Phi(0, m)$, the outgoing dimensionless wall flux at the cavity entrance $\xi = 0$, remains to be determined. References [26] and [27] derive the required relations, yielding the following equation, which was solved numerically and tabulated for the case of a prescribed exponential temperature distribution:

$$\frac{1}{\Phi(0, m)} = (1 - \rho)^{1/2} + \frac{\rho}{2} \int_0^\infty \frac{m J_1(t)}{m + t} \Phi(0, t) \, dt$$

For $m = 0$ (isothermal cavity), this gives the well-known result (see Example 7-20) $\Phi(0, 0) = 1/\epsilon^{1/2}$. After $\Phi(\xi, m)$ and thus $q_o(x, m)$ is obtained, the wall heat flux distribution is directly obtained from (7-67a).

In the previous sections, methods have been discussed for solving integral equations by numerical and some approximate analytical techniques. The analytical methods are only of value when the integral equations are relatively simple. In almost all practical cases a numerical method would be used. There are a few instances in which approximate or numerical solutions are not required since the radiation exchange integral equation has an exact analytical solution. One of these cases is now discussed.

Exact solution of integral equation for radiation from a spherical cavity Radiation from a spherical cavity as shown in Fig. 7-22a was analyzed by Jensen [30], discussed by Jakob [5], and further treated by Sparrow and Jonsson [31]. The spherical shape leads to a relatively simple integral-equation solution because there is an especially simple configuration factor between elements on the inside of a spherical cavity. The factor between the two differential elements dA_j and dA_k in Fig. 7-22b is

$$dF_{dj-dk} = \frac{\cos \theta_j \cos \theta_k}{\pi S^2} dA_k \tag{7-93}$$

Since the sphere radius is normal to both dA_j and dA_k, the distance between these elements is $S = 2R \cos \theta_j = 2R \cos \theta_k$. Then (7-93) becomes

$$dF_{dj-dk} = \frac{dA_k}{4\pi R^2} \tag{7-94}$$

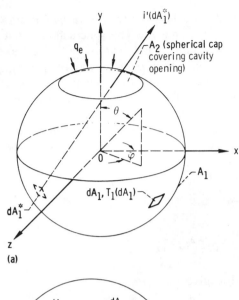

(a)

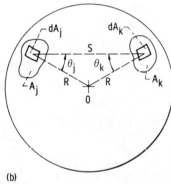

(b)

Figure 7-22 Geometry involved in radiation within spherical cavity. (*a*) Spherical cavity with diffuse entering radiation q_e and with surface at variable temperature T_1; (*b*) area elements on spherical surface.

If, instead of an infinitesimal area dA_k, the element dA_j exchanges with the finite area A_k, then (7-94) becomes

$$F_{dj-k} = \frac{1}{4\pi R^2} \int_{A_k} dA_k = \frac{A_k}{4\pi R^2} \tag{7-95}$$

Equation (7-95) is independent of the area element dA_j; hence, dA_j could be replaced by a finite area A_j so that

$$F_{j-k} = \frac{A_k}{4\pi R^2} = \frac{A_k}{A_s} \tag{7-96}$$

where A_s is the surface area of the entire sphere. The configuration factor to any area is simply the fraction of the total sphere area that the receiving area occupies.

Consider the spherical cavity in Fig. 7-22a. The cavity surface has a temperature distribution $T_1(dA_1)$ and a total surface area A_1. The spherical cap that would cover the cavity opening has area A_2. Assume there is *diffuse* radiative flux q_e (per unit area of A_2) entering from the environment through the cavity opening. The q_e can be variable over A_2. It is desired to compute the radiation intensity $i'(dA_1^*)$ leaving the cavity opening at a specified location and in a specified direction, as shown by the arrow in Fig. 7-22a. The figure shows that the desired intensity results from the diffuse flux leaving the element dA_1^* and equals $q_{o,1}(dA_1^*)/\pi$, where the factor π relates hemispherical flux q_o and intensity i'. The flux $q_{o,1}(dA_1^*)$ is found by applying (7-66a):

$$q_{o,1}(dA_1^*) - (1 - \epsilon_1) \int_{A_1} q_{o,1}(dA_1) \, dF_{d1^*-d1}$$

$$- (1 - \epsilon_1) \int_{A_2} q_e(dA_2) \, dF_{d1^*-d2} = \epsilon_1 \sigma T_1^4(dA_1^*) \tag{7-97}$$

The F factors from (7-94) are substituted to give

$$q_{o,1}(dA_1^*) - \frac{1 - \epsilon_1}{4\pi R^2} \int_{A_1} q_{o,1}(dA_1) \, dA_1 = \frac{1 - \epsilon_1}{4\pi R^2} \int_{A_2} q_e(dA_2) \, dA_2 + \epsilon_1 \sigma T_1^4(dA_1^*) \tag{7-98}$$

where the known quantities are on the right side.

To solve (7-98), a trial solution of the form $q_{o,1}(dA_1^*) = f(dA_1^*) + C$ is assumed, where f is an unknown function of the location of dA_1^*, and C is a constant. Substituting into (7-98) gives

$$f(dA_1^*) + C - \frac{1 - \epsilon_1}{4\pi R^2} \int_{A_1} f(dA_1) \, dA_1 - \frac{1 - \epsilon_1}{4\pi R^2} CA_1$$

$$= \frac{1 - \epsilon_1}{4\pi R^2} \int_{A_2} q_e(dA_2) \, dA_2 + \epsilon_1 \sigma T_1^4(dA_1^*)$$

The only two terms that are functions of local position within the cavity are the first and last, which gives $f(dA_1^*) = \epsilon_1 \sigma T_1^4(dA_1^*)$. The remaining terms are then equated to determine C. This gives the results for $q_{o,1}(dA_1^*)$ and the desired $i'(dA_1^*)$ as

$$q_{o,1}(dA_1^*) = \pi i'(dA_1^*) = \epsilon_1 \sigma T_1^4(dA_1^*)$$

$$+ \frac{\left[\dfrac{(1 - \epsilon_1)}{4\pi R^2} \right] \left[\displaystyle\int_{A_1} \epsilon_1 \sigma T_1^4(dA_1) \, dA_1 + \int_{A_2} q_e(dA_2) \, dA_2 \right]}{1 - (1 - \epsilon_1)A_1/4\pi R^2} \tag{7-99}$$

7-6 CONCLUDING REMARKS

In this chapter the basic equations were developed and solution methods presented for analyzing radiative energy exchange within enclosures having black or diffuse-gray surfaces. The surfaces can be of finite or infinitesimal size. Each enclosure surface can have a specified net energy flux added to it by some external means such as conduction or convection, or it can have a specified surface temperature. Various methods were presented for solving the array of simultaneous linear equations or the linear integral equations that result from the formulation of these interchange problems. It was pointed out that most practical problems become so complex that numerical techniques are required for solution. Computer programs such as TRASYS [32] are available for large problems. Computer subroutines such as matrix solvers and integration subroutines can be used.

In succeeding chapters, extensions are made for nonidealized surfaces to include spectral and directional property variations. Methods for incorporating *coupled* conduction, convection, and radiation transfer of energy are provided and result in nonlinear relations to be solved.

REFERENCES

1. Edwards, D. K., and R. D. Tobin: Effect of Polarization on Radiant Heat Transfer through Long Passages, *J. Heat Transfer,* vol. 89, no. 2, pp. 132–138, 1967.
2. Cravalho, E. G., C. L. Tien, and R. P. Caren: Effect of Small Spacings on Radiative Transfer between Two Dielectrics, *J. Heat Transfer,* vol. 89, no. 4, pp. 351–358, 1967.
3. Hottel, Hoyt C.: Radiant-Heat Transmission, in William H. McAdams (ed.), "Heat Transmission," 3d ed., chap. 4, McGraw-Hill, New York, 1954.
4. Poljak, G.: Analysis of Heat Interchange by Radiation between Diffuse Surfaces, *Tech. Phys. USSR,* vol. 1, nos. 5, 6, pp. 555–590, 1935.
5. Jakob, Max: "Heat Transfer," vol. II, Wiley, New York, 1957.
6. Gebhart, B.: "Heat Transfer," 2d ed., pp. 150–163, McGraw-Hill, New York, 1971.
7. Gebhart, B.: Surface Temperature Calculations in Radiant Surroundings of Arbitrary Complexity—for Gray, Diffuse Radiation, *Int. J. Heat Mass Transfer,* vol. 3, no. 4, pp. 341–346, 1961.
8. Sparrow, E. M.: On the Calculation of Radiant Interchange between Surfaces, in Warren Ibele (ed.), "Modern Developments in Heat Transfer," pp. 181–212, Academic Press, New York, 1963.
9. Arimilli, R. V., and S. P. Ketkar: Radiation Heat Transfer in Enclosures with Discrete Heat Sources, *Int. Commun. Heat Mass Transfer,* vol. 15, pp. 31–40, 1988.
10. Hildebrand, Francis B.: "Methods of Applied Mathematics," 2d ed., Prentice-Hall, Englewood Cliffs, N.J., 1965.
11. Usiskin, C. M., and R. Siegel: Thermal Radiation from a Cylindrical Enclosure with Specified Wall Heat Flux, *J. Heat Transfer,* vol. 82, no. 4, pp. 369–374, 1960.
12. Buckley, H.: Radiation from the Interior of a Reflecting Cylinder, *Philos. Mag.,* vol. 4, pp. 753–762, 1927.
13. Buckley, H.: Radiation from Inside a Circular Cylinder, *Philos. Mag.,* vol. 6, pp. 447–457, 1928.
14. Eckert, E.: Das Strahlungsverhaltnis von Flachen mit Einbuchtungen und von zylindrischen Bohrungen, *Arch. Waermewirtsch.,* vol. 16, no. 5, pp. 135–138, 1935.

15. Sparrow, E. M., and L. U. Albers: Apparent Emissivity and Heat Transfer in a Long Cylindrical Hole, *J. Heat Transfer,* vol. 82, no. 3, pp. 253–255, 1960.

16. Lin, S. H., and E. M. Sparrow: Radiant Interchange among Curved Specularly Reflecting Surfaces—Application to Cylindrical and Conical Cavities, *J. Heat Transfer,* vol. 87, no. 2, pp. 299–307, 1965.

17. Kholopov, G. K.: Radiation of Diffuse Isothermal Cavities, *Inzh. Fiz. Zh.,* vol. 25, no. 6, pp. 1112–1120, 1973.

18. Heinisch, R. P., E. M. Sparrow, and N. Shamsundar: Radiant Emission from Baffled Conical Cavities, *J. Opt. Soc. Am.,* vol. 63, no. 2, pp. 152–158, 1973.

19. Masuda, H.: Radiant Heat Transfer on Circular-Finned Cylinders, *Rep. Inst. High Speed Mech. Tohoku Univ.,* vol. 27, no. 255, pp. 67–89, 1973. (See also *Trans. Jpn. Soc. Mech. Eng.,* vol. 38, pp. 3229–3234, 1972.)

20. Bedford, R. E.: Calculation of Effective Emissivities of Cavity Sources of Thermal Radiation, in D. P. DeWitt and G. D. Nutter (eds.), "Theory and Practice of Radiation Thermometry," chap. 12, Wiley, New York, 1988.

21. Sparrow, E. M., and R. P. Heinisch: The Normal Emittance of Circular Cylindrical Cavities, *Appl. Opt.,* vol. 9, no. 11, pp. 2569–2572, 1970.

22. Sparrow, E. M.: Application of Variational Methods to Radiation Heat-Transfer Calculations, *J. Heat Transfer,* vol. 82, no. 4, pp. 375–380, 1960.

23. Krishnan, K. S., and R. Sundaram: The Distribution of Temperature along Electrically Heated Tubes and Coils, I. Theoretical, *Proc. R. Soc. London,* ser. A, vol. 257, no. 1290, pp. 302–315, 1960.

24. Perlmutter, M., and R. Siegel: Effect of Specularly Reflecting Gray Surface on Thermal Radiation through a Tube and from Its Heated Wall, *J. Heat Transfer,* vol. 85, no. 1, pp. 55–62, 1963.

25. Choi, B. C., and S. W. Churchill: A Technique for Obtaining Approximate Solutions for a Class of Integral Equations Arising in Radiative Heat Transfer, *Int. J. Heat Fluid Flow,* vol. 6, no. 1, pp. 42–48, 1985.

26. Crosbie, A. L., and T. R. Sawheny: Radiant Interchange in a Nonisothermal Rectangular Cavity, *AIAA J.,* vol. 13, no. 4, pp. 425–431, 1975.

27. Crosbie, A. L., and T. R. Sawheny: Application of Ambarzumian's Method to Radiant Interchange in a Rectangular Cavity, *J. Heat Transfer,* vol. 96, pp. 191–196, 1974.

28. Ambarzumian, V. A.: Diffusion of Light by Planetary Atmospheres, *Astron. Zh.,* vol. 19, pp. 30–41, 1942.

29. Kourganoff, V.: "Basic Methods in Transfer Problems," Dover, New York, 1963.

30. Jensen, H. H.: Some Notes on Heat Transfer by Radiation, *Kgl. Danske Videnskab. Selskab. Mat.-Fys. Medd.,* vol. 24, no. 8, pp. 1–26, 1948.

31. Sparrow, E. M., and V. K. Jonsson: Absorption and Emission Characteristics of Diffuse Spherical Enclosures, NASA TN D-1289, 1962.

32. Jensen, C. L.: "Thermal Radiation Analysis System" TRASYS-II User's Manual, ANSI Version 1.0, Martin Marietta Aerospace Corp., Denver, Colorado, February 1987. Also available from COSMIC.

33. Alfano, Gaetano: Apparent Thermal Emittance of Cylindrical Enclosures with and without Diaphragms, *Int. J. Heat Mass Transfer,* vol. 15, no. 12, pp. 2671–2674, 1972.

34. Alfano, G., and A. Sarno: Normal and Hemispherical Thermal Emittances of Cylindrical Cavities, *J. Heat Transfer,* vol. 97, no. 3, pp. 387–390, 1975.

PROBLEMS*

7-1 What is the net energy transfer from black surface dA_1 to black surface A_2?

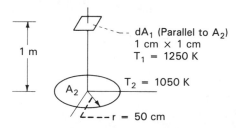

1 m

dA$_1$ (Parallel to A$_2$)
1 cm × 1 cm
T_1 = 1250 K

T_2 = 1050 K

A$_2$

r = 50 cm

Answer: $dQ_{1 \leftrightarrow 2} = 1.390$ W

7-2 A person holds her open hand (approximated by a circular disk 12 cm in diameter) 10 cm directly above and parallel to a black heater element in the form of a circular disk 20 cm in diameter (such as an electric range element). The heater element is at 700 K. How much radiant energy from the heater element is incident on the hand?

Answer: 70.12 W

7-3 A carbon steel billet $1 \times 0.5 \times 0.5$ m is initially at 1100 K and is supported in such a manner that it transfers heat by radiation from all of its surfaces to surroundings at 27°C (assume the surroundings act black). Neglect convective heat transfer and assume the billet radiates like a blackbody. Also, assume for simplicity that the thermal conductivity of the steel is infinite. How long will it take for the billet to cool to 500 K? [Take $\rho_{cs} = 7800$ kg/m^3 and $c_{cs} = 470$ J/(kg·K).]

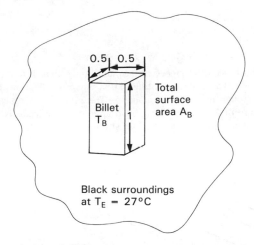

0.5   0.5

Billet
T_B

1

Total
surface
area A$_B$

Black surroundings
at T_E = 27°C

Answer: 4.62 h

7-4 A black circular disk 0.12 m in diameter and well insulated on one side is electrically heated to a uniform temperature. The electrical energy input is 850 W. The surroundings are black and are at 730 K. What fraction of the emitted energy is in the wavenumber region from 0.20 to 0.60 μm^{-1}?

Answer: 0.655

*Note: Some of the problems in this chapter use configuration factors that were derived in the problems section in Chap. 6.

7-5 A black electrically heated rod is in a black vacuum jacket. The rod must dissipate 50 W without exceeding 800 K. Calculate the maximum allowable jacket temperature (neglect end effects).

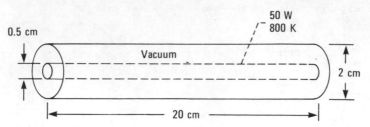

Answer: 599 K

7-6 A hollow cylindrical heating element is insulated on its outside surface. The element has a 15-cm inside diameter and is 15 cm long. The black internal surface is to be held at 1100 K. The surroundings are in vacuum and are at 800 K. Both ends of the cylinder are open to the surroundings. Estimate the energy that must be supplied to the element (in W).

Answer: 1750 W

7-7 A regular tetrahedron of side 0.25 m has black internal surfaces with the following characteristics:

Surface	Q(W)	T(K)
1	$Q_1 = 0$	T_1
2	Q_2	550
3	Q_3	1100
4	300	T_4

What are the values of T_1, Q_2, Q_3, and T_4?

Answer: $T_1 = 960.5$ K, $Q_2 = -1554.3$ W, $Q_3 = 1254.3$ W, $T_4 = 999.4$ K

7-8 A solar collector resting on the ground has a black upper surface and dimensions 1 × 2 m. The lower surface is very well insulated. The collector is tilted at an angle of 30° from the horizontal. Fluid at 80°C is passed through the collector at night. The night sky acts as a blackbody with an effective temperature of 240 K and the ground around the collector is at 30°C. Neglecting conduction and convection to the surroundings, at night:

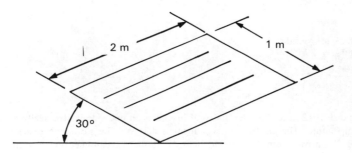

(a) at what rate is energy (W) lost by the collector?
(b) if flow through the collector is stopped, what temperature (K) will the collector attain?

Answer: (a) 1348.7 W; (b) 246.0 K

7-9 The black-surfaced three-sided enclosure shown below has infinitely long parallel sides, with the specified temperatures and energy rate additions. Find Q_1, Q_2, and T_3. (Side 3 is a circular arc.)

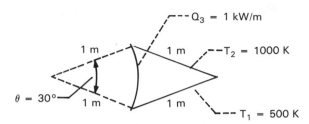

Answer: $Q_1 = -46{,}702$ W/m, $Q_2 = 45{,}702$ W/m, $T_3 = 868.0$ K

7-10 Two enclosures are identical in shape and size and have black surfaces. For one enclosure, the temperatures of the surfaces are T_1, T_2, T_3,..., T_N. For the second, the surface temperatures are $(T_1^4 + k)^{1/4}$, $(T_2^4 + k)^{1/4}$, $(T_3^4 + k)^{1/4}$,...,$(T_N^4 + k)^{1/4}$, where k is a constant. Show how the heat transfer rates Q_j at any surface A_j are related for the two enclosures.

Answer: They are the same.

7-11 An enclosure with black interior surfaces has one side open to an environment at temperature T_e. The sides of the enclosure are maintained at temperatures T_1, T_2, T_3,...,T_N. How are the rates of energy input to the sides Q_1, Q_2, Q_3,...,Q_N influenced by the value of T_e? How can the results for $T_e = 0$ be used to obtain solutions for other T_e?

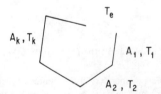

7-12 (a) A circular cylindrical enclosure has black interior surfaces, each maintained at uniform interior temperature as shown. The outside of the entire cylinder is insulated so that the outside does not radiate to the surroundings. How much Q (W) is supplied to each area as a result of the interior radiative exchange?

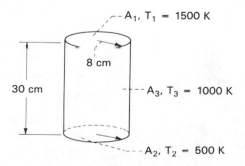

(b) For the same enclosure and the same surface temperatures, divide A_2 into two equal areas A_4 and A_5. What is the Q to each of these two areas, and how do they and their sum compare with Q_3 from part (a)?

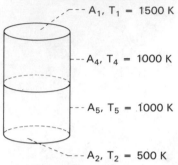

A$_1$, T$_1$ = 1500 K

A$_4$, T$_4$ = 1000 K

A$_5$, T$_5$ = 1000 K

A$_2$, T$_2$ = 500 K

(c) What are Q_6/A_6 and Q_7/A_7 for the same enclosure and surface temperatures? How do they compare with Q_3/A_3 from part (a) and Q_4/A_4 and Q_5/A_5 from part (b)?

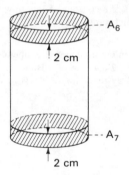

--- A$_6$

2 cm

--- A$_7$

2 cm

Answer: (a) Q_1 = 4698.6 W, Q_2 = −1358.4 W, Q_3 = −3340.2 W; (b) Q_1 = 4698.6 W, Q_2 = −1358.4 W, Q_4 = −3628.6 W, Q_5 = 288.4 W

7-13 A black 4-cm-diameter sphere at a temperature of 900 K is suspended in the center of a thin 8-cm-diameter partial sphere having a black interior surface and an exterior surface with a total hemispherical emissivity of 0.4. The surroundings are at 500 K. A 6-cm-diameter hole is cut in the outer sphere. What is the temperature of the outer sphere? What is the Q being supplied to the inner sphere? (For simplicity, do not subdivide the surface areas into smaller zones.)

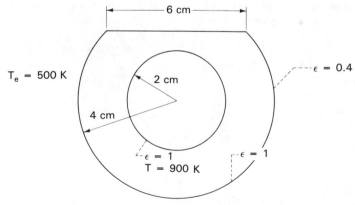

6 cm

T_e = 500 K

2 cm

4 cm

ϵ = 0.4

ϵ = 1

ϵ = 1
T = 900 K

Answer: 711 K, 123.5 W

7-14 Two infinite gray parallel plates are separated by a thin gray radiation shield. What is T_s, the temperature of the shield? What energy flux is transferred from plate 2 to plate 1? What is the ratio of the heat transferred from plate 2 to plate 1 with the shield to that transferred without the shield?

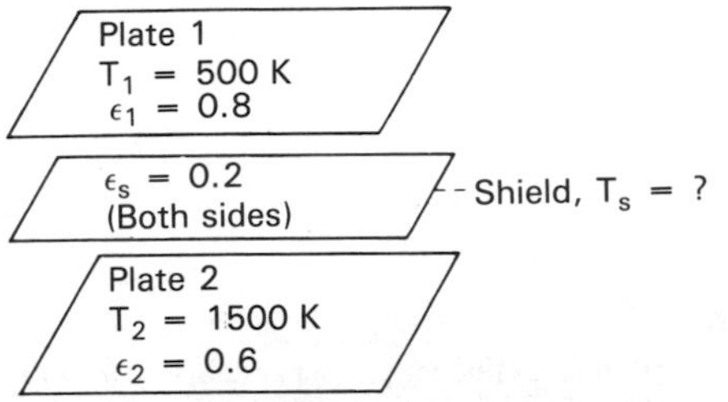

Plate 1
$T_1 = 500$ K
$\epsilon_1 = 0.8$

$\epsilon_s = 0.2$
(Both sides) — Shield, $T_s = ?$

Plate 2
$T_2 = 1500$ K
$\epsilon_2 = 0.6$

Answer: 1253 K, 26,000 W/m², 0.176

7-15 A heat flux q_0 is transferred across the gap between two gray parallel plates having the same emissivity ϵ that are at temperatures T_1 and T_2 (ϵ is independent of temperature). A single thin radiation shield also having emissivity ϵ on both sides is placed between the plates. Show that the resulting heat flux is $q_0/2$. Show that adding a second identical shield reduces the heat flux to $q_0/3$. Show that for n shields the heat flux is $q_0/(n + 1)$ when all the surface emissivities are the same.

7-16 (a) What is the effect of a single thin radiation shield on the transfer of energy between two concentric cylinders? Assume the cylinder and shield surfaces are diffuse-gray with emissivities independent of temperature. Both sides of the shield have emissivity ϵ_s, and the inner and outer cylinders have respective emissivities ϵ_1 and ϵ_2.

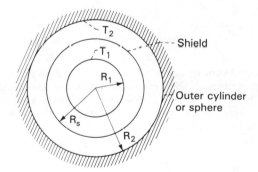

(b) What is the effect of a single thin radiation shield on the transfer of energy between two concentric spheres? Assume the sphere and shield surfaces are diffuse-gray with emissivities independent of temperature. Both sides of the shield have emissivity ϵ_s, and the inner and outer spheres have respective emissivities ϵ_1 and ϵ_2.

Answer: (a), (b)

$$\frac{Q_{1,\text{shield}}}{Q_{1,\text{no shield}}} = \frac{G_{12}}{(A_1/A_s)G_{s2} + G_{1s}} \quad \text{where } G_{ab} = \frac{1}{\epsilon_a} + \frac{A_a}{A_b}\left(\frac{1}{\epsilon_b} - 1\right)$$

$(A_a/A_b) = (R_a/R_b)$ (cylinder), $\qquad (A_a/A_b) = (R_a/R_b)^2$ (sphere)

7-17 An infinitely long enclosure is shaped as shown below. Assuming that the uniform flux restrictions are met along each surface, find Q_1 and T_2.

$\epsilon_1 = 0.5$

$T_1 = 1000$ K

A_2 ----

A_1

$R = 0.5$ m

$\epsilon_2 = 0.1$

$Q_2 = 10$ kW/m

Answer: $Q_1 = -10$ kW/m, $T_2 = 1240$ K

7-18 Two infinitely long diffuse-gray concentric circular cylinders are separated by two concentric thin diffuse-gray radiation shields. The shields have identical emissivities on both sides.

(a) Derive an expression for the energy transferred between the inner and outer cylinders in terms of their temperatures and the necessary radiative and geometric quantities. (Number the surfaces from the inside out; i.e., the inner surface is number 1, the outer surface is number 4.)

(b) Check this result by showing that in the proper limit it reduces to the correct result for four parallel plates with identical emissivities.

(c) Find the percentage reduction in heat transfer when the shields are added if the radii for the surfaces are in the ratio 1:3:5:7 and if $\epsilon_1 = \epsilon_4 = 0.5$ and $\epsilon_2 = \epsilon_3 = 0.1$

Answer:

(a) $Q = \dfrac{A_1 \sigma (T_1^4 - T_4^4)}{G_{12} + A_1 G_{23}/A_2 + A_1 G_{34}/A_3}$ where $G_{ab} = \dfrac{1}{\epsilon_a} + \dfrac{A_a}{A_b}\left(\dfrac{1}{\epsilon_b} - 1\right)$

(b) $Q = \dfrac{A\sigma (T_1^4 - T_4^4)}{3(2/\epsilon - 1)}$

(c) $Q_{\text{with}}/Q_{\text{without}} = 0.175$

7-19 Consider a diffuse-gray right circular cylindrical enclosure. The diameter is the same as the height of the cylinder, 1 m. The top is removed. If the remaining surfaces are maintained at 1000 K and have an emissivity of 0.5, determine the radiative energy escaping through the open end. Assume uniform irradiation on each surface so that subdivision of the base and curved wall is not necessary. The outside environment is at $T_e = 0$ K. How do the results compare with Fig. 7-17? Explain any difference. If $T_e = 600$ K, what percentage reduction occurs in radiative energy loss?

Answer: -3.71×10^4 W, 13% reduction in energy rate

7-20 Consider the gray cylindrical enclosure described in Problem 7-19 with the top in place. A hole 15 cm in diameter is drilled in the top. The outside environment is at $T_e = 0$ K. Determine the configuration factors between
(a) the base and the hole and
(b) the curved wall and the hole.
(c) Estimate the radiant energy escaping through the hole.

Answer: (a) 0.00448; (b) 0.00451; (c) 1002 W

7-21 A two-dimensional diffuse-gray enclosure (infinitely long into the page) has each surface at a uniform temperature. Compute the energy added per meter of enclosure length at each surface to account for the radiative exchange within the enclosure, Q_1, Q_2, Q_3. (Assume for simplicity that it is not necessary to subdivide the three areas.) The conditions are $T_1 = 1200$ K, $\epsilon_1 = 0.6$, $T_2 = 500$ K, $\epsilon_2 = 0.9$, $T_3 = 700$ K, $\epsilon_3 = 0.5$.

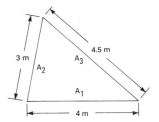

Answer: 221,190 W/m, −133,410 W/m, −87,780 W/m

7-22 A thin gray disk with emissivity 0.8 on both sides is in earth orbit. It is exposed to normally incident solar radiation (neglect radiation emitted or reflected from the earth). What is the equilibrium temperature of the disk? A single thin radiation shield having emissivity 0.1 on both sides is placed as shown. What is the disk temperature? What is the effect on both of these results of reducing the disk emissivity to 0.5? (Assume the surroundings are at zero absolute temperature and that for simplicity it is not necessary to subdivide the areas.)

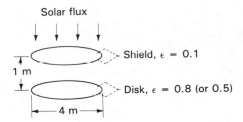

Answer: 146.6 K, 149.1 K

7-23 For the enclosure with four infinitely long parallel walls shown below, calculate the average heat flux on surface 2 (W/m²). All surfaces are diffuse-gray and may be assumed to have uniform outgoing flux distributions.

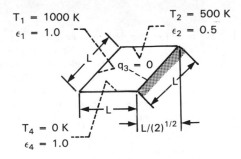

Answer: −7939 W/m²

7-24 Two infinite parallel black plates are at temperatures T_1 and T_2. A perforated thin sheet of gray material of emissivity ϵ_s is inserted between the plates. Let SF, the shading factor, be defined as the area of the solid portion of the perforated sheet over the total area of the sheet, A_g/A. Determine the temperature of the perforated plate and find the ratio of the energy transfer Q_1 between the black plates when SF = 1 (i.e., with a single gray radiation shield) to the energy transfer with a perforated shield, Q_{SF}.

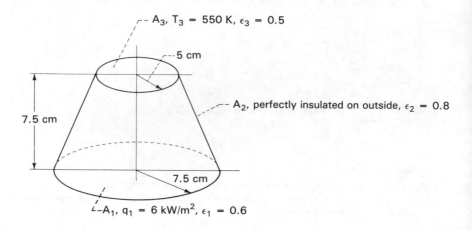

Answer: $T_g^4 = \dfrac{T_1^4 + T_2^4}{2}, \qquad \dfrac{(Q_1)_{\text{shield}}}{(Q_1)_{\text{perforated}}} = \dfrac{\epsilon_s/2}{1 + (\epsilon_s SF/2) - SF}$

7-25 A frustum of a cone has its base heated as shown. The top is held at 550 K while the side is perfectly insulated. All surfaces are diffuse-gray. What is the temperature attained by surface 1 as a result of radiative exchange within the enclosure? (For simplicity, do not subdivide the areas.)

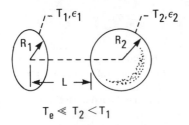

Answer: 906 K

7-26 In a metal-processing operation, a metal sphere at uniform temperature is heated in vacuum to high temperature by radiative exchange with a circular heating element. The surroundings are cool enough that they do not affect the radiative exchange and may be neglected. The surfaces are diffuse-gray. Derive an expression for the net rate of energy absorption by the sphere. The expression should be given in terms of the quantities shown. For simplicity do not subdivide the surface areas. Discuss whether this is a reasonable approximation for this geometry with regard to the distribution of reflected energy from the sphere.

7-27 An enclosure has four sides that are all equilateral triangles of the same size (i.e., it is an equilateral tetrahedron). The sides are of length $L = 2$ m and have conditions imposed as follows:

Side 1 is black and is at uniform temperature $T_1 = 500$ K.

Side 2 is diffuse-gray and is perfectly insulated on the outside.

Side 3 is black and has a uniform heat flux of 5 kW/m^2 supplied to it.

Side 4 is black and is at $T_4 = 0$ K.

Find q_1, T_2, and T_3. For simplicity, do not subdivide the areas.

Answer: -137.7 W/m^2, 503.6 K, 601.0 K

7-28 A 5-cm-diameter hole extends through the wall of a furnace having an interior temperature of 1400 K. The wall is 15-cm-thick refractory brick. Divide the wall thickness into two zones of equal length, and compute the net radiation out of the hole into a room at 295 K. (Neglect heat conduction in the wall.)

Answer: $T_1 = 1261$ K, $T_2 = 1071$ K, $Q_3 = -162$ W

7-29 A cube of side 3 m has a very small sphere placed at its center ($A_{sphere} \ll A_{side}$). The sphere has emissivity $\epsilon = 0.5$ and is maintained electrically at $T_s = 1500$ K. The interior sides of the cube have the following properties:

Side	Temperature, K	Emissivity
1	1500	1.0
2	1000	0.5
3	900	0.5
4	500	1.0
5	200	0.5
6	0	1.0

Determine the net q added or removed from each side of the cube and the q added to the sphere. Find the results in W/m^2. Tabulate all the required configuration factors. (Assume the incident radiation on each surface is uniform.)

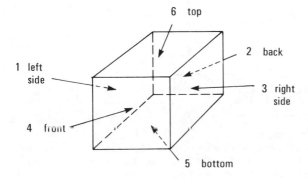

Answer: $q_1 = 252{,}815$ W/m^2, $q_2 = -10{,}738$ W/m^2, $q_3 = -21{,}374$ W/m^2, $q_4 = -87{,}416$ W/m^2, $q_5 = -41{,}618$ W/m^2, $q_6 = -91{,}669$ W/m^2, $q_s = -105{,}339$ W/m^2

7-30 A rod 1 cm in diameter and 10 cm long is at temperature $T_1 = 1800$ K and has a hemispherical total emissivity of $\epsilon_1 = 0.22$. It is within a thin-walled concentric cylinder of the same length having a diameter of 4 cm. The emissivity on the inside of the cylinder is $\epsilon_2 = 0.50$ and on the outside is $\epsilon_o = 0.17$. All surfaces are diffuse-gray. The entire assembly is suspended in a large vacuum chamber at $T_e = 300$ K. What is the temperature T_2 of the cylindrical shell? For simplicity, do not subdivide the surface areas. (*Hint:* $F_{2-1} = 0.225$, $F_{2-2} = 0.617$.)

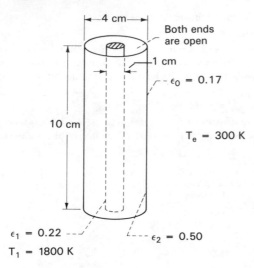

Answer: 1048 K

7-31 A thin diffuse-gray circular disk with emissivity ϵ_1 on one side and ϵ_2 on the other side is being heated in vacuum by a cylindrical electrical heater with a diameter D. The heater has a diffuse-gray interior surface and is open at both ends.

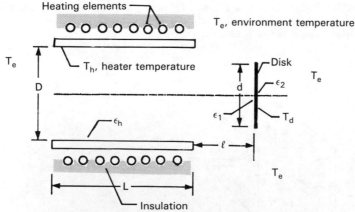

(a) Derive a formula (which can be in terms of configuration factors) for the net radiative energy rate being gained by the disk while it is heating up. The formula should be in terms of the instantaneous disk temperature and the quantities shown.

For the specific case $T_h = 1500$ K; $T_e = 850$ K; $D = 0.25$ m; $L = 0.30$ m; $d = 0.15$ m; $l = 0.05$ m; $\epsilon_1 = 0.55$; $\epsilon_2 = 0.75$; $\epsilon_h = 0.80$:

(b) What is the net gain (W) when $T_d = 1025$ K?

(c) What is the equilibrium disk temperature long after the heater is turned on?

Answer:

(a) Net gain $= \dfrac{\pi d^2}{4}\epsilon_2\sigma(T_e^4 - T_d^4) - \dfrac{\epsilon_1\pi d^2}{4}\{\sigma T_d^4[1 -(1 - \epsilon_3)(F_{3-3} + F_{3-1}F_{1-3})] - F_{1-3}\epsilon_3\sigma T_3^4$

$+ \sigma T_4^4[-F_{1-4} + (1 - \epsilon_3)(F_{3-3}F_{1-4} - F_{3-4}F_{1-3})]\}/\{1 - (1 - \epsilon_3)[F_{3-3} + F_{3-1}F_{1-3}(1 - \epsilon_1)]\}$

(b) 870.3 W; (c) 1154 K

7-32 A hollow satellite in earth orbit consists of a circular disk and a hemisphere. The disk is facing normal to the direction to the sun. The surroundings are at $T_e = 200$ K. The satellite walls are thin. All surfaces are diffuse. The properties are $\alpha_{1,solar} = 0.85$, $\epsilon_{1,infrared} = 0.20$, $\epsilon_2(gray) = 0.88$, $\epsilon_3(gray) = 0.45$, $\epsilon_4(gray) = 0.55$. What are the values of T_1 and T_4? (Do not subdivide surfaces. Neglect any radiation emitted or reflected from the earth.)

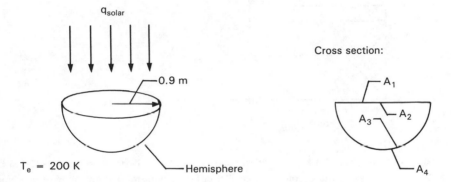

Answer: $T_2 = 437.9$ K, $T_3 = 341.8$ K

7-33 Three parallel plates of finite width are shown in cross section. The plates are maintained at $T_1 = 695$ K and $T_2 = 430$ K. The surroundings are at $T_e = 325$ K. The plates are very long in the direction normal to the cross section shown. What are the values of q_1 and q_2? All plate surfaces are diffuse-gray. (Do not subdivide the plate areas.)

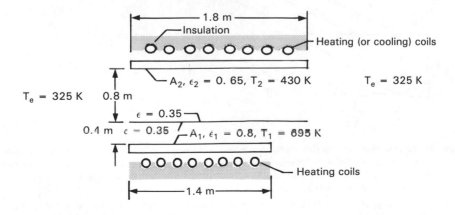

Answer: $q_1 = 6022.8$ W/m², $q_2 = 24.5$ W/m²

7-34 A space vehicle at earth orbit is in the form of a hollow cube with thin walls. It is oriented with one side always facing directly toward the sun and the other five sides in the shade. The interior is painted with a coating with $\epsilon = 0.86$. On the outside, the top is coated with a material with $\alpha_{solar} = 0.95$ and $\epsilon = 0.86$, the front and back sides are faced with aluminum foil with $\epsilon = 0.03$, and the two sides and the bottom have white paint with $\epsilon = 0.86$. The surroundings are at 0 K. Using as simple a radiation model as is reasonable, obtain the temperatures of the six faces of the cube.

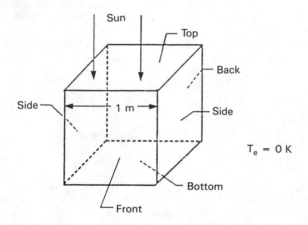

Answer: T_1 (top) = 353.4 K, T_2 (2 sides, bottom) = 242.2 K, T_3 (front and back) = 280.6 K

7-35 Two infinitely long, directly opposed, parallel diffuse-gray plates of width W have the same uniform heat flux q supplied to them. The environment is at temperature T_e. Set up the governing equation for determining the temperature distribution $T(x)$ on the bottom plate. There is no energy loss from the top side of the upper plate or from the bottom side of the lower plate.

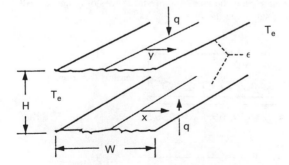

7-36 For the configuration specified in the preceding problem, solve for the temperature distribution on the bottom plate if $q = 2500$ W/m², $\epsilon = 0.4$, $H = 0.25$ m, and $W = 1$ m. A numerical solution is required. Present your results graphically in terms of appropriate dimensionless parameters. The environment temperature is $T_e = 480$ K.

7-37 Consider two parallel plates of finite extent in one direction. Both plates are perfectly insulated on the outside. Plate 1 is uniformly heated electrically with heat flux q_e. Plate 2 has no external heat input. The environment is at zero absolute temperature.

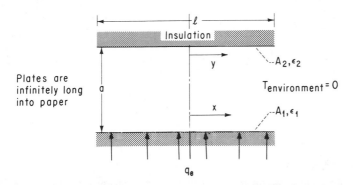

(a) For both plates *black,* show that the integral equations for the surface temperatures are

$$\theta_{b,1}(X) = 1 + \frac{1}{2} \int_{-L/2}^{L/2} \theta_{b,2}(Y) \frac{dY}{[(Y - X)^2 + 1]^{3/2}}$$

$$\theta_{b,2}(Y) = \frac{1}{2} \int_{-L/2}^{L/2} \theta_{b,1}(X) \frac{dX}{[(X - Y)^2 + 1]^{3/2}}$$

where $X = x/a$, $Y = y/a$, $\theta = \sigma T^4/q_e$, and $L = l/a$.

(b) If both plates are *gray,* show that

$$\theta_1(X) = \theta_{b,1}(X) + \frac{1 - \epsilon_1}{\epsilon_1}, \qquad \theta_2(Y) = \theta_{b,2}(Y)$$

7-38 A long groove is cut into a metal surface as shown in cross section below. The groove surface is diffuse-gray and has emissivity ϵ. The temperature profile along the groove sides, as measured from the apex, is found to be $T(x)$. The environment is at temperature T_e.

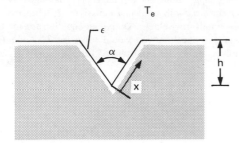

(a) Derive the equations for the heat flux distribution $q(x)$ along the groove surface.

(b) Examine the kernel of the integral equation found in part (a), and show whether it is symmetrical and/or separable.

7-39 A hemispherical cavity is in a block of lightly oxidized copper maintained at 550 K in vacuum (see Fig. 5-25a). The surroundings are at 295 K. Use the integral equation method to compute the outgoing heat flux from the cavity surface.

Answer: 3885 W/m²

7-40 A cavity having a gray interior surface S is uniformly heated electrically and achieves a surface temperature distribution $T_{w,0}(S)$ while being exposed to a zero absolute temperature environment, $T_e = 0$. If the environment is raised to T_e and the heating kept the same, what will the surface temperature distribution be?

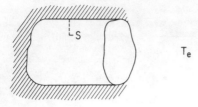

Answer: $T_w(S) = [T_{w,0}^4(S) + T_e^4]^{1/4}$

7-41 A gray circular tube insulated on the outside is exposed to an environment at $T_e = 0$ at both ends. The $q(x, T_e = 0)$ has been calculated to maintain the wall temperature at any constant value. Now, let $T_e \neq 0$, and let the wall temperature be uniform at T_w. Show that the $q(x, T_e \neq 0)$ can be obtained as the $q(x, T_e = 0)$ corresponding to the wall temperature $(T_w^4 - T_e^4)^{1/4}$.

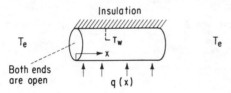

7-42 Two infinitely long plates are joined at right angles along one infinite edge. The plates are each of width W, and both have diffuse-gray surfaces with the same emissivity ϵ. The heat flux supplied to both surfaces is maintained at a uniform value q. Derive an equation for the temperature distribution on each surface, and describe how you would solve for $T(x)$ and $T(y)$. Neglect conduction in the plates and convection to the atmosphere. The surroundings are at temperature T_e.

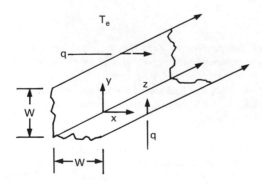

7-43 For the geometry and conditions shown below, set up the required integral equations for finding $q_1(y)$, $T_2(z)$, and $T_3(x)$. Put the equations in dimensionless form, and discuss how you would go about solving the equations. Which method of Chap. 7 appears most useful?

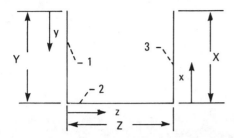

$T_1(y) = T_1 = $ constant, $q_2(z) = 0$ (insulated on the outside), $q_3(x) = q_3 = $ constant, $\epsilon_1(y) = 1.0$, and $\epsilon_2(z) = \epsilon_3(x) = 0.5$.

EIGHT

THE EXCHANGE OF THERMAL RADIATION BETWEEN NONDIFFUSE NONGRAY SURFACES

8-1 INTRODUCTION

In Chap. 7 the analysis of radiation exchange within enclosures was restricted to either black or gray surfaces. If gray, the surfaces were assumed to both emit and reflect diffusely. The additional restriction was sometimes made that the radiative properties were independent of temperature. As shown by the graphs of real properties in Chap. 5, most engineering materials deviate, in some instances radically, from the idealizations of being black, gray, or diffuse or having temperature-independent radiative properties. In many practical engineering situations the assumption of idealized surfaces is made to simplify the computations. This is often a reasonable approach for two reasons. First, the radiative properties may not be known to high accuracy, especially with regard to their detailed dependence on wavelength and direction; hence, performing a refined computation would be fruitless when only crude property data are available. Second, in an enclosure, the many reflections and rereflections tend to average out radiative nonuniformities; for example, the radiation leaving (emitted plus reflected) a directionally emitting surface may be fairly diffuse if it consists mostly of reflected energy arising from radiation incident from many directions.

In some instances more precision is required and the gray and/or diffuse assumption cannot be trusted to yield sufficiently accurate results. It is desired to carry out exchange computations using as exact a solution procedure as is reasonably possible. The results of such computations can be compared with results from simplified methods as in Chap. 7. This will yield some insight into where simplifying assumptions can be applied and still yield reasonable results. To provide techniques for refined computations, some methods of treating radiative interchange between nonideal surfaces are examined in this chapter. Analysis of such problems is inherently more difficult than for ideal surfaces, and a complete treatment of real surfaces including all types of variations, while possible in principle when all radiative properties are known, is not usually attempted or justified. As stated earlier, the directional spectral properties are often not available. Property variations with wavelength for the normal direction are available for a number of materials; the data are usually sparse at the short ($\lambda < \sim0.3$ μm) and long ($\lambda > \sim15$ μm) wavelength ranges of the spectrum. Directional variations for some materials with optically smooth surfaces can be computed using electromagnetic theory as in Chap. 4. Certain problems obviously demand inclusion of the effects of spectral and directional property variations. One example is the use of spectrally selective coatings for temperature control or solar energy collection as discussed at the end of Chap. 5.

8-2 SYMBOLS

a_0	autocorrelation distance of surface roughness
A	area
C_1, C_2	first and second constants in Planck's spectral energy distribution
D	perpendicular distance between parallel areas
e	emissive power
F	configuration factor
$F_{0-\lambda}$	fraction of blackbody intensity in spectral range $0-\lambda$
G	function of emissivities in Example 8-3
h	height of cavity
i	intensity
l	L/D, parameter in Example 8-6
L	width of infinitely long parallel plates
q	energy flux, energy per unit area and per unit time
Q	energy rate, energy per unit time
r	R/D, parameter in Example 8-7
$\mathbf{r}$	position vector
R	radius of disk in Example 8-7
S	distance between area elements
T	absolute temperature
w	width of cavity
x, y, z	cartesian coordinates
α	absorptivity
ϵ	emissivity
ζ	the variable $C_2/\lambda T$
η	angle in plane perpendicular to surface
θ	cone angle

λ wavelength

ξ distance along width of plane surface having finite width and infinite length

Ξ absorption efficiency defined in Example 8-6

ρ reflectivity

σ Stefan-Boltzmann constant

σ_0 rms amplitude of surface roughness

φ circumferential or azimuthal angle

ω solid angle

$\displaystyle\int_{\cap}$ integration over solid angle of entire enclosing hemisphere

Subscripts

a absorbed

b blackbody

e emitted

i incident, incoming

k quantity for kth surface

max   maximum

min   minimum

o outgoing

r reflected

s specular

λ spectrally dependent

$\Delta\lambda$ for a wavelength band $\Delta\lambda$

1, 2, 3   property of surface 1, 2, or 3

Superscripts

$'$ directional quantity

$''$ bidirectional quantity

8-3 ENCLOSURE THEORY FOR DIFFUSE SURFACES WITH SPECTRALLY DEPENDENT PROPERTIES

By considering diffusely emitting and reflecting surfaces we eliminate directional effects, and it is possible to see more clearly how spectral variations of properties are accounted for. The surface emissivity, absorptivity, and reflectivity are independent of direction but may depend on wavelength and temperature. These properties must be available as functions of λ and T to evaluate the radiative interchange.

For diffuse spectral surfaces, the configuration factors are valid because these factors involve only geometric effects and were computed for diffuse radiation leaving a surface. The energy-balance equations and methods in Chap. 7 remain

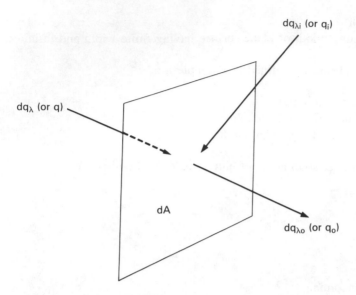

Figure 8-1 Spectral (or total) energy fluxes at a surface element.

valid as long as they are written for the energy in each *wavelength interval dλ*. Usually, however, the boundary conditions involve *total* (including all wavelengths) energy, and care must be taken to apply the boundary conditions correctly. Total energy boundary conditions cannot be applied to the spectral energies. As an example, consider the surface of Fig. 8-1 having a locally incident total radiation heat flux q_i and a radiation flux leaving by combined emission and reflection, q_o. If the surface is otherwise perfectly insulated and there is no local heat flux being added externally (an *adiabatic* surface), then q_o and q_i must be equal to each other:

$$q_o - q_i = q = 0 \tag{8-1}$$

However, in each $d\lambda$ interval, the incident and outgoing dq_λ are *not* generally equal, so

$$dq_{\lambda o} - dq_{\lambda i} = dq_\lambda \neq 0 \tag{8-2}$$

Rather, an *adiabatic* surface only has a *total* radiation gain or loss of zero, or, with Eq. (8-1) restated in terms of the quantities in (8-2),

$$q = \int_{\lambda=0}^{\infty} dq_\lambda = \int_{\lambda=0}^{\infty} (dq_{\lambda o} - dq_{\lambda i}) = 0 \tag{8-3}$$

The dq_λ is net energy flux supplied in $d\lambda$ at λ as a result of incident energy at other wavelengths. For an adiabatic surface the dq_λ can vary substantially with λ depending on the property variations with wavelength and the spectral distribution of incident energy.

Now consider a surface where a total energy flux q is being supplied by some means other than incoming and outgoing radiation. Then from Fig. 8-1

$$q = \int_{\lambda=0}^{\infty} dq_\lambda = \int_{\lambda=0}^{\infty} (dq_{\lambda o} - dq_{\lambda i}) \tag{8-4}$$

The q may be specified as an imposed condition or may be a quantity to be determined so that the surface can be locally maintained at a specified temperature. In any small wavelength interval, the net energy $dq_{\lambda o} - dq_{\lambda i}$ may be positive or negative. The boundary conditions state only that the integral of all such spectral energy values must be locally equal to the total q.

8-3.1 Parallel Plate Geometry

To build familiarity with these concepts, they are first applied to some situations for a parallel plate geometry. Then relations for a general geometry will be given.

EXAMPLE 8-1 Two infinite parallel plates of tungsten at specified temperatures T_1 and T_2 ($T_1 > T_2$) are exchanging radiant energy. Branstetter [1] determined the hemispherical spectral temperature-dependent emissivity of tungsten by using the relations of electromagnetic theory to extrapolate limited experimental data, and some of his results are in Fig. 8-2. Using these data, compare the net energy exchange between the tungsten plates with that for gray parallel plates.

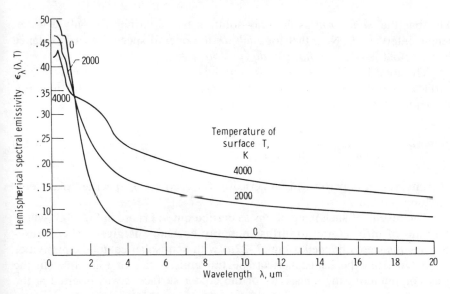

Figure 8-2 Hemispherical spectral emissivity of tungsten.

The solution for gray plates is in Example 7-6. The present case follows in the same fashion except that the equations are written spectrally. From Eqs. (7-12) and (7-13) the energy quantities at surface 1 per unit area and time in a wavelength interval $d\lambda$ are related by

$$dq_{\lambda,1} = dq_{\lambda o,1} - dq_{\lambda i,1} \tag{8-5}$$

$$dq_{\lambda o,1} = \epsilon_{\lambda,1}(\lambda,T_1)e_{\lambda b,1}(\lambda,T_1)\,d\lambda + \rho_{\lambda,1}(\lambda,T_1)\,dq_{\lambda i,1} \tag{8-6}$$

For diffuse-opaque surfaces the hemispherical properties are related by $\rho_\lambda = 1 - \alpha_\lambda = 1 - \epsilon_\lambda$, and Eq. (8-6) becomes

$$dq_{\lambda o,1} = \epsilon_{\lambda,1}(\lambda,T_1)e_{\lambda b,1}(\lambda,T_1)\,d\lambda + [1 - \epsilon_{\lambda,1}(\lambda,T_1)]\,dq_{\lambda i,1} \tag{8-7}$$

Eliminating $dq_{\lambda i,1}$ from (8-5) and (8-7) gives

$$dq_{\lambda,1} = \frac{\epsilon_{\lambda,1}(\lambda,T_1)}{1 - \epsilon_{\lambda,1}(\lambda,T_1)}\,[e_{\lambda b,1}(\lambda,T_1)\,d\lambda - dq_{\lambda o,1}] \tag{8-8}$$

For infinite parallel plates the configuration factor $F_{2-1} = 1$; then $q_{\lambda i,1} = q_{\lambda o,2}$ and (8-5) becomes

$$dq_{\lambda,1} = dq_{\lambda o,1} - dq_{\lambda o,2} \tag{8-9}$$

Equations (8-8) and (8-9) are analogous to (7-21a,b) for the gray case. After writing the equations for surface 2 in a similar fashion, the $dq_{\lambda o}$ are eliminated, and the solution for a wavelength interval $d\lambda$ follows as in Eq. (7-22a):

$$dq_{\lambda,1} = -dq_{\lambda,2} = \frac{e_{\lambda b,1}(\lambda,T_1) - e_{\lambda b,2}(\lambda,T_2)}{1/\epsilon_{\lambda,1}(\lambda,T_1) + 1/\epsilon_{\lambda,2}(\lambda,T_2) - 1}\,d\lambda \tag{8-10}$$

This has the *same form* as the gray solution but is written for only a wavelength interval $d\lambda$. Note that for *each* $d\lambda$ the overall spectral energy balance is zero, that is, $dq_{\lambda,1} + dq_{\lambda,2} = dq_{\lambda,1} - dq_{\lambda,1} = 0$.

The total heat flux exchanged (supplied to surface 1 and removed from surface 2) is found by substituting the property data of Fig. 8-2 into Eq. (8-10) and integrating over all wavelengths:

$$q_1 = -q_2 = \int_{\lambda=0}^{\infty} dq_{\lambda,1} = \int_0^{\infty} \frac{e_{\lambda b,1}(\lambda,T_1) - e_{\lambda b,2}(\lambda,T_2)}{1/\epsilon_{\lambda,1}(\lambda,T_1) + 1/\epsilon_{\lambda,2}(\lambda,T_2) - 1}\,d\lambda \tag{8-11}$$

The integration is performed numerically for each set of specified plate temperatures T_1 and T_2.

The results of such integrations as carried out in [1] are in Fig. 8-3, where the ratio of diffuse-gray to diffuse-nongray exchange is given. The diffuse-gray results were obtained using (7–22a) with hemispherical total emissivities computed from the hemispherical spectral emissivities of Fig. 8-2. (In the gray computation, the emissivity of the colder surface 2 was inserted at the

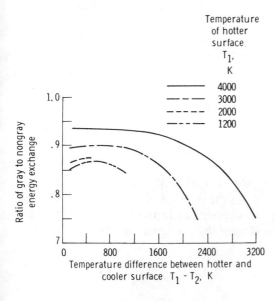

Figure 8-3 Comparison of effect of gray and nongray surfaces on computed energy exchange between infinite tungsten plates [1].

mean temperature $\sqrt{T_1 T_2}$ rather than at T_2, which is a modification based on electromagnetic theory that is sometimes recommended for metals [2].) Over the range of surface temperatures shown, the gray results deviate 25% below the nongray exchange.

The following example illustrates how the integrations in Eq. (8-11) can be carried out in a simple way, suitable for estimates by hand calculations, if the properties can be approximated by constant values in a stepwise fashion over a few spectral regions. More generally, the calculations are carried out numerically with computer integration subroutines.

EXAMPLE 8-2 Two infinite parallel plates and their approximate spectral emissivities at their respective temperatures are shown in Fig. 8-4. What is the heat flux q transferred across the gap?

From Eq. (8-11),

$$q = \int_0^3 \frac{e_{\lambda b,1}(\lambda,T_1) - e_{\lambda b,2}(\lambda,T_2)}{1/0.4 + 1/0.7 - 1} \, d\lambda + \int_3^5 \frac{e_{\lambda b,1}(\lambda,T_1) - e_{\lambda b,2}(\lambda,T_2)}{1/0.8 + 1/0.7 - 1} \, d\lambda$$

$$+ \int_5^\infty \frac{e_{\lambda b,1}(\lambda,T_1) - e_{\lambda b,2}(\lambda,T_2)}{1/0.8 + 1/0.3 - 1} \, d\lambda$$

which can be written as

$$q = \sigma T_1^4\left[\frac{0.341}{\sigma T_1^4}\int_0^3 e_{\lambda b,1}(\lambda,T_1)\,d\lambda + \frac{0.596}{\sigma T_1^4}\int_3^5 e_{\lambda b,1}(\lambda,T_1)\,d\lambda\right.$$

$$\left. + \frac{0.279}{\sigma T_1^4}\int_5^\infty e_{\lambda b,1}(\lambda,T_1)\,d\lambda\right] - \sigma T_2^4\left[\frac{0.341}{\sigma T_2^4}\int_0^3 e_{\lambda b,2}(\lambda,T_2)\,d\lambda\right.$$

$$\left. + \frac{0.596}{\sigma T_2^4}\int_3^5 e_{\lambda b,2}(\lambda,T_2)\,d\lambda + \frac{0.279}{\sigma T_2^4}\int_5^\infty e_{\lambda b,2}(\lambda,T_2)\,d\lambda\right]$$

An integral such as $(1/\sigma T_1^4)\int_3^5 e_{\lambda b,1}(\lambda,T_1)\,d\lambda$ is the fraction of blackbody radiation at T_1 between $\lambda = 3$ and 5 μm, which is $F_{3T_1-5T_1} = F_{5040-8400}$ and can be computed from the table of blackbody functions (Table A-5, Appendix A). The $F_{\lambda T}$ should not be confused with the geometric configuration factor. Then

$$q = \sigma T_1^4(0.341F_{0-3T_1} + 0.596F_{3T_1-5T_1} + 0.279F_{5T_1-\infty}) - \sigma T_2^4(0.341F_{0-3T_2}$$

$$+ 0.596F_{3T_2-5T_2} + 0.279F_{5T_2-\infty}) = 140{,}500 \text{ W/m}^2$$

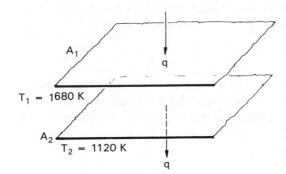

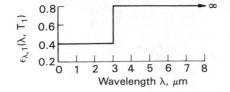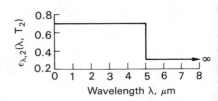

Figure 8-4 Example of heat transfer across space between infinite parallel plates having spectrally dependent emissivities.

As compared with a gray case, the previous examples illustrate the additional complication of having spectrally dependent surface properties. The enclosure theory must be carried out in wavelength intervals, and then the total quantities can be obtained by integration of the spectral energy values. The integration can be done rather easily for the parallel plate case as only one equation, (8–11), is involved. However, when many surfaces are present the length of the computations can build up rapidly, since there will be a set of exchange equations for each wavelength band. Some reduction in accuracy of the integration may be reasonable in practical applications because of the uncertainty already present in many of the spectral property values that are used. The *band-energy approximation* is the term used to designate the conceptually simple method of replacing each single integral extending over all wavelengths by a summation of smaller integrals that each extend over a portion of the spectrum in which averaged values are used. An example will further illustrate this method, which is a generalization of the calculations in Example 8-2.

EXAMPLE 8-3 Two infinite parallel plates of tungsten are at temperatures of 4000 and 2000 K. Using the data of Fig. 8-2, compute the net energy exchange between the surfaces by using the band-energy approximation.

The governing equation is (8-11). By using the substitution $G_\lambda = \{[1/\epsilon_{\lambda,1}(\lambda, T_1)] + [1/\epsilon_{\lambda,2}(\lambda, T_2)] - 1\}^{-1}$ to shorten the notation, this is written as

$$q_1 = \int_0^\infty G_\lambda e_{\lambda b,1} \, d\lambda - \int_0^\infty G_\lambda e_{\lambda b,2} \, d\lambda$$

The integrals are written as approximate sums

$$q_1 \approx \sum_l (G_{\Delta\lambda} e_{\Delta\lambda, b, 1} \Delta\lambda)_l - \sum_m (G_{\Delta\lambda} e_{\Delta\lambda, b, 2} \Delta\lambda)_m \tag{8-12}$$

where $G_{\Delta\lambda}$ and $e_{\Delta\lambda,b}$ are average values applicable to each wavelength interval $\Delta\lambda$. Depending on the way in which $G_{\Delta\lambda}$ and $e_{\Delta\lambda,b}$ are evaluated, (8-12) can have various degrees of accuracy. As a simple approximation, the terms $e_{\Delta\lambda,b}$ are evaluated using an arithmetic mean of the blackbody emissive power over $\Delta\lambda$. For better accuracy, especially for large $\Delta\lambda$ intervals, $e_{\Delta\lambda,b}$ is evaluated using the blackbody functions,

$$e_{\Delta\lambda,b}\Delta\lambda = \int_\lambda^{\lambda+\Delta\lambda} e_{\lambda b} \, d\lambda = [F_{0-(\lambda+\Delta\lambda)} - F_{0-\lambda}]\sigma T^4 \tag{8-13}$$

Equation (8-13) is used for the computations here. The $G_{\Delta\lambda}$ terms in (8-12) are approximated most simply by $(1/\epsilon_{\Delta\lambda,1} + 1/\epsilon_{\Delta\lambda,2} - 1)^{-1}$, where the $\epsilon_{\Delta\lambda}$ are appropriate mean emissivities in the wavelength interval $\Delta\lambda$.

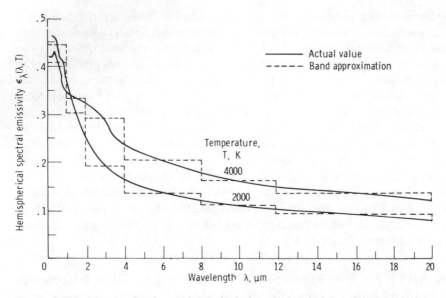

Figure 8-5 Band approximations to hemispherical spectral emissivity of tungsten.

In Fig. 8-5, the required emissivities of tungsten are plotted and *arithmetic* mean values are shown for seven $\Delta\lambda$ intervals (the seventh interval being $\lambda > 20$ μm). For temperatures of 2000 and 4000 K, the peak in the $e_{\lambda b}$ function occurs at about 1.5 and 0.75 μm, respectively. For values of $\lambda > 4$ μm in this example, $e_{\lambda b}$ is small and $G_\lambda e_{\lambda b}$ contributes little to the integrals in this wavelength region. Thus the accuracy of the averages at large values of λ is not important in this example. The computations for q_1 using these seven intervals are in the following tabulation:

$\Delta\lambda$, μm	$\epsilon_{\Delta\lambda,1}$	$\epsilon_{\Delta\lambda,2}$	$G_{\Delta\lambda}$	$e_{\Delta\lambda,b,1}\Delta\lambda$, W/m^2	$e_{\Delta\lambda,b,2}\Delta\lambda$, W/m^2	$G_{\Delta\lambda}e_{\Delta\lambda,b,1}\Delta\lambda$ W/m^2	$G_{\Delta\lambda}e_{\Delta\lambda,b,2}\Delta\lambda$ W/m^2
0–1	0.410	0.445	0.271	0.698×10^7	0.0061×10^7	1.89×10^6	0.017×10^6
1–2	0.335	0.300	0.188	0.545×10^7	0.0376×10^7	1.03×10^6	0.071×10^6
2–4	0.290	0.195	0.132	0.171×10^7	0.034×10^7	0.23×10^6	0.045×10^6
4–8	0.205	0.140	0.0907	0.032×10^7	0.011×10^7	0.03×10^6	0.010×10^6
8–12	0.160	0.115	0.0717	0.004×10^7	0.002×10^7	~0	0.002×10^6
12–20	0.140	0.095	0.0600	0.001×10^7	~0	~0	~0
>20	~0	~0	~0	~0	~0	~0	~0
Totals						3.18×10^6	0.145×10^6

$$q_1 = (3.18 - 0.15) \times 10^6 = 3030 \text{ kW/m}^2$$

Using numerical integration [1], the result is $q_1 = 3000$ kW/m^2. The approximate band solution using seven intervals differs by only a very small amount.

An examination of the tabulation shows that most of the significant energy transfer in the previous example occurs in the wavelength range 0–2 μm. If necessary, the accuracy of the band-energy approximation could be improved by dividing this range into a larger number of increments and repeating the calculation. The errors in the band-energy approximation will arise in the regions where both $e_{\lambda b}$ and ϵ_λ are large; thus the wavelength range should be divided such that most of the bands lie within these regions. If the number of intervals is increased, the exact results for energy transfer are approached. Dunkle and Bevans [3] give a calculation similar to Example 8-3 and show errors from an exact numerical result of less than 2% for the band-energy solution, as compared to about 30% error for the gray-surface approximation. They give other examples of applications in enclosures with specified temperatures or net energy fluxes.

Some additional references providing analyses of energy exchange between spectrally dependent surfaces are Love and Gilbert [4], Goodman [5], and Rolling and Tien [6]. In [4] the analytical results compare well with experimental results for a geometry closely approximating infinite parallel plates.

8-3.2 General Spectral and Band Relations for an Enclosure

For an enclosure consisting of N surfaces as in Fig. 7-4, the relation in Eq. (7-31) is written for a spectral interval $d\lambda$ for each of the k surfaces,

$$
\sum_{j=1}^{N} \left[\frac{\delta_{kj}}{\epsilon_{\lambda,j}(\lambda, T_j)} - F_{k-j} \frac{1 - \epsilon_{\lambda,j}(\lambda, T_j)}{\epsilon_{\lambda,j}(\lambda, T_j)} \right] \frac{dq_{\lambda,i}}{d\lambda}
$$

$$
= \sum_{j=1}^{N} (\delta_{kj} - F_{k-j}) e_{\lambda b,j}(\lambda, T_j) \qquad (k = 1, 2, \ldots, N) \qquad (8\text{-}14)
$$

If the N surface temperatures are all specified, the right sides of this set of simultaneous equations are known, as are the quantities in square brackets on the left. The solution yields the $dq_{\lambda,k}/d\lambda$ at λ for each of the surfaces. As a check on the calculations, the overall *spectral* energy balance for *each* $d\lambda$ yields the summation $\Sigma_{k=1}^{N} A_k \, dq_{\lambda,k} = 0$. For parallel plates this is evident from Eq. (8-10). The total energy supplied by other means to each surface is then $Q_k = A_k \int_{\lambda=0}^{\infty} (dq_{\lambda,k}/d\lambda) \, d\lambda$. If some of the q_k are specified, and hence some of the surface temperatures are unknown, a tedious iterative procedure may be required as discussed following Eq. (7-31). This can be especially difficult for the spectrally dependent case because the specified q_k is a *total* energy quantity and represents an integral of the values found from (8-14). The boundary condition does not provide the spectral values of $dq_{\lambda,k}$.

Equation (8-14) is usually used in a band form. The bands are selected as in Example 8-3, so that the properties for all surfaces are constant within each band.

Then by use of (8-13), Eq. (8-14) is integrated over a band $\Delta\lambda$ to yield

$$\sum_{j=1}^{N} \left[\frac{\delta_{kj}}{\epsilon_{\Delta\lambda,j}(T_j)} - F_{k-j} \frac{1 - \epsilon_{\Delta\lambda,j}(T_j)}{\epsilon_{\Delta\lambda,j}(T_j)} \right] q_{\Delta\lambda,j} = \sum_{j=1}^{N} (\delta_{kj} - F_{k-j}) F_{\lambda T_j - (\lambda + \Delta\lambda)T_j} \sigma T_j^4 \qquad (8\text{-}15a)$$

If the T are specified, Eq. (8-15a) provides N equations for the N unknown $q_{\Delta\lambda,k}$ for *each* $\Delta\lambda$ band. The solution is carried out for *each* $\Delta\lambda$. Then for each of the k surfaces, the total q_k is

$$q_k = \sum_{\Delta\lambda} q_{\Delta\lambda,k} \qquad (8\text{-}15b)$$

EXAMPLE 8-4 An enclosure is made up of three plates of finite width and infinite length normal to the cross section shown in Fig. 8-6. One plate is concave. The radiative properties of each surface are dependent on wavelength and temperature, and the specified temperatures of the plates are T_1, T_2, and T_3. Provide a set of band equations for the radiative energy exchange among the surfaces.

From Eq. (8-15a) for $k = 1$, 2, and 3, and for wavelength band $\Delta\lambda$, the following equations are written. As a more compact notation let $E_{\Delta\lambda,j}(T_j) = [1 - \epsilon_{\Delta\lambda,j}(T_j)]/\epsilon_{\Delta\lambda,j}(T_j)$ and $F_{\Delta\lambda T_j} = F_{\lambda T_j - (\lambda + \Delta\lambda)T_j}$ to yield

$$\left[\frac{1}{\epsilon_{\Delta\lambda,1}(T_1)} - F_{1-1}E_{\Delta\lambda,1}(T_1) \right] q_{\Delta\lambda,1} - F_{1-2}E_{\Delta\lambda,2}(T_2)q_{\Delta\lambda,2} - F_{1-3}E_{\Delta\lambda,3}(T_3)q_{\Delta\lambda,3}$$

$$= (1 - F_{1-1})F_{\Delta\lambda T_1}\sigma T_1^4 - F_{1-2}F_{\Delta\lambda T_2}\sigma T_2^4 - F_{1-3}F_{\Delta\lambda T_3}\sigma T_3^4 \qquad (8\text{-}16a)$$

$$-F_{2-1}E_{\Delta\lambda,1}(T_1)q_{\Delta\lambda,1} + \frac{1}{\epsilon_{\Delta\lambda,2}(T_2)} q_{\Delta\lambda,2} - F_{2-3}E_{\Delta\lambda,3}(T_3)q_{\Delta\lambda,3}$$

$$= -F_{2-1}F_{\Delta\lambda T_1}\sigma T_1^4 + F_{\Delta\lambda T_2}\sigma T_2^4 - F_{2-3}F_{\Delta\lambda T_3}\sigma T_3^4 \qquad (8\text{-}16b)$$

$$-F_{3-1}E_{\Delta\lambda,1}(T_1)q_{\Delta\lambda,1} - F_{3-2}E_{\Delta\lambda,2}(T_2)q_{\Delta\lambda,2} + \frac{1}{\epsilon_{\Delta\lambda,3}(T_3)} q_{\Delta\lambda,3}$$

$$= -F_{3-1}F_{\Delta\lambda T_1}\sigma T_1^4 - F_{3-2}F_{\Delta\lambda T_2}\sigma T_2^4 + F_{\Delta\lambda T_3}\sigma T_3^4 \qquad (8\text{-}16c)$$

This is a set of three simultaneous equations for the unknowns $q_{\Delta\lambda,1}$, $q_{\Delta\lambda,2}$, and $q_{\Delta\lambda,3}$. The solution is carried out for the $q_{\Delta\lambda}$ values in *each wavelength band* $\Delta\lambda$. It should be emphasized that $q_{\Delta\lambda,j}$ is the energy supplied to surface j in wavelength interval $\Delta\lambda$ as a result of external heat addition to the surface (e.g., conduction and/or convection) and energy transferred in from other wavelength bands by the radiation exchange within the enclosure. Such boundary conditions are discussed in more detail in Chap. 10 for cases including conduction, convection, or imposed heating. Finally, q at each surface is found by summing $q_{\Delta\lambda}$ for that surface over all wavelength bands, $q_k =$

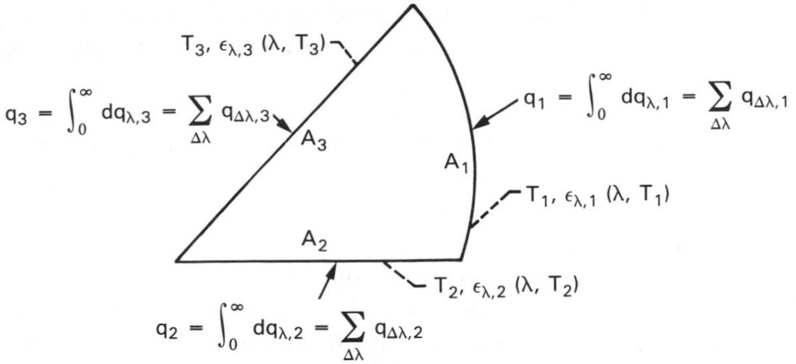

Figure 8-6 Radiant interchange in enclosure with surfaces having spectrally varying radiation properties.

$\Sigma_{\Delta\lambda} \, q_{\Delta\lambda,k}$. This is the heat flux that must be externally supplied to surface k to maintain its specified surface temperature.

EXAMPLE 8-5 Consider the geometry of Fig. 8-6. Total energy flux is supplied to the three surfaces at rates q_1, q_2, and q_3. Determine the temperatures of the plates.

The equations are the same as in Example 8-4. Now, however, the prescribed boundary conditions have made the solution much more difficult. Because the surface temperatures are unknown, the $F_{\Delta\lambda T}$ and $\epsilon_{\Delta\lambda}$ are unknown because of their temperature dependence. The solution is carried out as follows: A temperature is assumed for each surface, and $q_{\Delta\lambda}(T)$ for each surface is computed. This is done for each of the wavelength bands. The $q_{\Delta\lambda,k}(T)$ values are then summed over the $\Delta\lambda$ to find q_1, q_2, and q_3, which are compared to the specified boundary values. New temperatures are chosen, and the process repeated until the computed q values agree with the specified values. The new temperatures for successive iterations are guessed on the basis of the $F_{\Delta\lambda T}$ and $\epsilon_{\Delta\lambda}$ variations and the trends of how changes in T are reflected in changes in q throughout the system.

8-3.3 The Semigray Approximations

In some practical situations there is a natural division of the radiant energy in an enclosure into two well-defined spectral regions. An important case is an enclosure with an opening through which solar energy is entering. The solar energy has a spectral distribution concentrated in a short-wavelength region, while the energy originating by emission from the lower-temperature surfaces within the enclosure is in a longer-wavelength region. One way that has been used to treat this situation is to define a hemispherical total absorptivity for incident solar radiation and a second hemispherical total absorptivity for incident energy originating by emission within the enclosure. This approach can be carried to the point of defining N different absorptivities for surface k, one absorptivity for incident energy from each of the N enclosure surfaces.

The assumption entering these analyses is that each absorptivity $\alpha_k(T_k, T_j)$ is based on an incident *blackbody* spectral distribution at the temperature of the originating surface T_j. Of course, the incident spectrum may be quite different from the blackbody form, and this is the weakness of the method. Often the dependence of α_k on T_k is small, and its principal dependence is on T_j or, in other words, on the distribution of the incident spectrum. Because the absorptivity $\alpha_k(T_k, T_j)$ and emissivity $\epsilon_k(T_k)$ of surface k are not in general equal, this approach is often called the *semigray enclosure theory*. Reference [7] contains the formulation of a semigray analysis for a general enclosure.

Plamondon and Landram [8] compared the semigray and exact solutions for the temperature profiles along the surface of a nongray wedge cavity exposed to incident solar radiation as shown in Fig. 8-7. The wedge cavity is in vacuum with an environment at zero degrees except for the solar radiation source. The cavity has surface properties independent of temperature and the surfaces are diffuse. Three solution techniques are given. The first is an *exact* solution of the complete integral equations. The first approximation, *method I*, is the semigray analysis, which assigns an absorptivity α_{solar} for radiation (direct and reflected) that originated from the incident solar energy and a second absorptivity α_{infrared} (equal to the surface emissivity) for radiation originating by emission from the wedge surfaces. Finally, *method II* is a poorer approximation that retains these same two absorptivities but applies α_{solar} only for the incident solar energy and uses α_{infrared} for *all* energy after reflection, regardless of its source. The results are in Fig. 8-7b for a polished aluminum surface in a 30° wedge. Method I, the full semigray

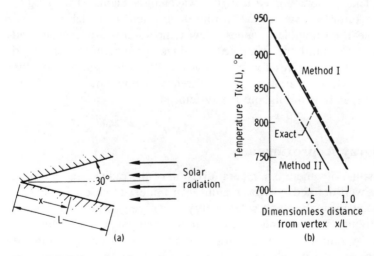

Figure 8-7 Effect of semigray approximations on computed temperature distribution in wedge cavity. (*a*) Geometry of wedge cavity; (*b*) temperature distribution along wedge, $\alpha_{\text{solar}} = 0.220$; $\alpha_{\text{infrared}} = 0.099$.

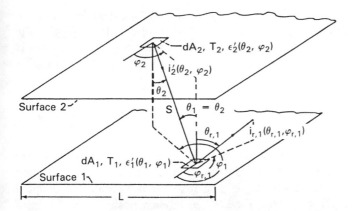

Figure 8-8 Radiant interchange between parallel directional surfaces of finite width L that are infinitely long in the direction normal to the plane of the drawing.

analysis, is in excellent agreement with the exact solution, while method II underestimates the temperatures by about 10%.

8-4 DIRECTIONAL-GRAY SURFACES

Some attention has been paid to the treatment of radiation interchange between surfaces or in enclosures, where directionally dependent properties must be considered. The bulk of radiation analyses invoke the assumption of diffuse emitting and reflecting surfaces, and some treatments include the effect of specular reflections with diffuse emission (Chap. 9). The diffuse or specular surface conditions are convenient to treat analytically, and in most instances the detailed consideration of directional emission and reflection effects is unwarranted. Nevertheless, certain materials and special situations require the examination of directional effects. In this section, some methods for considering radiant interchange between surfaces with directional properties will be presented.

The difficulty in treating the general case of directionally dependent properties can be illustrated by performing an energy balance in a simple geometry. A balance is examined for the radiative interchange between two infinitely long parallel non-diffuse-gray surfaces of finite width L (Fig. 8-8). The radiation intensity leaving element dA_1 in direction $(\theta_{r,1}, \varphi_{r,1})$ is composed of an emitted intensity $i'_{e,1}(\theta_{r,1}, \varphi_{r,1})$ and a reflected intensity $i'_{r,1}(\theta_{r,1}, \varphi_{r,1})$,

$$i'_1(\theta_{r,1}, \psi_{r,1}) = i'_{e,1}(\theta_{r,1}, \psi_{r,1}) + i'_{r,1}(\theta_{r,1}, \varphi_{r,1}) \qquad (8\text{-}17)$$

These two components are given by modifications of (3-3a) and (3-25) to yield (8-17) in the form

$$i'_1(\theta_{r,1}, \varphi_{r,1}) = \epsilon'_1(\theta_{r,1}, \varphi_{r,1})i'_{b,1}(T_1)$$

$$+ \int_{A_2} \rho''_1(\theta_{r,1}, \varphi_{r,1}, \theta_1, \varphi_1)i'_2(\theta_2, \varphi_2)\frac{\cos^2\theta_2}{S^2}\,dA_2 \qquad (8\text{-}18)$$

In the second term on the right of Eq. (8-18) the energy incident on dA_1 from each element dA_2 is multiplied by the bidirectional total reflectivity ρ_1'' to give the contribution to the reflected intensity from dA_1 into direction $(\theta_{r,1}, \varphi_{r,1})$. This is then integrated over all energy incident on dA_1 from A_2. The definition of ρ'' is given by (3-20) as

$$\rho''(\theta_r,\varphi_r,\theta,\varphi) = \frac{i_r''(\theta_r,\varphi_r,\theta,\varphi)}{i_i'(\theta,\varphi) \cos\theta \, d\omega} \tag{8-19}$$

The $\rho''(\theta_r, \varphi_r, \theta, \varphi)$ is the ratio of reflected *intensity* in the (θ_r, φ_r) direction to the *energy flux* incident from the (θ, φ) direction. A similar equation is written for an arbitrary element dA_2 on surface 2. The result is a complicated coupled pair of integral equations to be solved for $i'(\theta, \varphi)$ at each position and for each direction on the two surfaces. The integral equations are analogous to Eqs. (7-73) for diffuse-gray surfaces. Detailed property data of $\epsilon'(\theta, \varphi)$ and $\rho''(\theta_r, \varphi_r, \theta, \varphi)$ are seldom available. For the case when T_1 and T_2 are not known and the temperature dependence of the properties is considerable, the solution for the entire energy-exchange distribution becomes prohibitively tedious. To reduce the amount of computation, a number of approximations can be invoked in the situations where they are justified. Usually such approximations involve analytically simulating the real properties with simple functions, omitting certain portions of energy that are deemed negligible, or ignoring all directional effects except those expected to provide significant changes from diffuse or specular analyses. Some of these methods are outlined in [9–15]. An example will be given, and the analyst must use ingenuity in approximating the conditions for more realistic problems.

EXAMPLE 8-6 Two parallel isothermal plates of infinite length and finite width L are arranged as shown in Fig. 8-9a. The upper (plate 2) is black, while the lower is composed of a highly reflective material with parallel deep grooves of open angle 1° cut into the surface and running in the infinite direction. Such a surface might be made by stacking polished razor blades. The surroundings are at zero temperature. Compute the net energy gain by the directional surface if $T_2 > T_1$, and compare the result to the net energy gain by a diffuse surface with emissivity equivalent to the hemispherical emissivity of the directional surface.

The directional emissivity for the grooved surface is obtained from [16], where the directional emissivity at the opening of an infinitely long groove with specularly reflecting walls of surface emissivity 0.01 is calculated. The directional emissivity for the grooved surface is given by the dot-dashed line in Fig. 8-9b. The angle η is measured from the normal of the opening plane of the grooved surface and is in a plane perpendicular to the length of the groove as shown in Fig. 8-9a. The $\epsilon_1'(\eta_1)$ as given in [16] has already been averaged over all circumferential angles for a fixed η_1. Thus, it is an effective emissivity from a strip on the grooved surface to a parallel infinitely long strip element on an imaginary semicylinder over the groove and with its axis

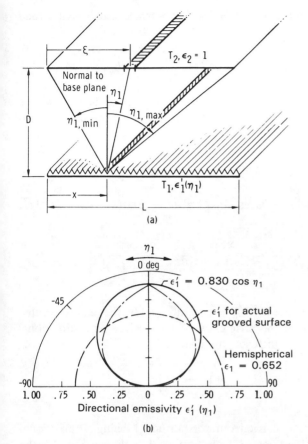

Figure 8-9 Interchange between grooved directional surface and black surface. (a) Geometry of problem (environment at zero temperature); (b) emissivity of directional surface.

parallel to the grooves. The angle η_1 is different from the usual cone angle θ_1. The actual emissivity $\epsilon_1'(\eta_1)$ of Fig. 8-9b is approximated for convenience by the analytical expression $\epsilon_1'(\eta_1) \approx 0.830 \cos \eta_1$. By using cylindrical coordinates to perform the integration over all η_1, the corresponding hemispherical emissivity of this surface is

$$\epsilon_1 = \frac{Q_1}{\sigma T_1^4} = \frac{\displaystyle\int_{-\pi/2}^{\pi/2} \epsilon_1'(\eta_1) \cos \eta_1 \, d\eta_1}{\displaystyle\int_{-\pi/2}^{\pi/2} \cos \eta_1 \, d\eta_1} = 0.830 \int_0^{\pi/2} \cos^2 \eta_1 \, d\eta_1 = 0.652$$

and this is shown in Fig. 8-9b as a dashed line.

The energy gained by surface 1 will first be determined when surface 2 is black and surface 1 is diffuse with $\epsilon_1 = 0.652$. The energy emitted by the diffuse surface 1 per unit of the infinite length and per unit time is $Q_{e,1} = 0.652\sigma T_1^4 L$. Since surface 2 is black, none of this energy is reflected to surface

1. The energy per unit length and time emitted by surface 2 that is absorbed by surface 1 is

$$Q_{a,1} = 0.652\sigma T_2^4 \int_{A_2} \int_{A_1} dF_{d2-d1} \, dA_2 = 0.652\sigma T_2^4 \int_{A_1} \int_{A_2} dF_{d1-d2} \, dA_1$$

The configuration factor between infinite parallel strips was found in Example 6-2 as $dF_{d1-d2} = d(\sin \eta_1)/2$. The double integral becomes, through integration over A_2,

$$\int_{A_1} \left(\int_{A_2} dF_{d1-d2} \right) dA_1 = \frac{1}{2} \int_{x=0}^{L} (\sin \eta_{1,\max} - \sin \eta_{1,\min}) \, dx$$

The $\sin \eta_1$ is found from Fig. 8-9a as $\sin \eta_1 = (\xi - x)/[(\xi - x)^2 + D^2]^{1/2}$, and solving for $Q_{a,1}$ gives

$$Q_{a,1} = 0.652\sigma T_2^4 \frac{1}{2} \int_{x=0}^{L} \left[\frac{L-x}{(x^2 - 2xL + L^2 + D^2)^{1/2}} + \frac{x}{(x^2 + D^2)^{1/2}} \right] dx$$

$$= 0.652\sigma T_2^4 [(L^2 + D^2)^{1/2} - D]$$

The net energy gained by surface 1, $Q_{a,1} - Q_{e,1}$, divided by the energy emitted by surface 2, is a measure of the efficiency of the surface as a directional absorber. For surface 1, being diffuse, this ratio is

$$\Xi_{\text{diffuse}} = \frac{Q_{a,1} - Q_{e,1}}{\sigma T_2^4 L} = \frac{0.652}{l} \left[(1 + l^2)^{1/2} - 1 - \frac{T_1^4}{T_2^4} l \right]$$

where $l = L/D$.

The analysis will now be carried out for surface 1 being a directional (grooved) surface. The emitted energy from surface 1 is the same as for the diffuse surface since both have the same hemispherical emissivity. The energy absorbed by the grooved surface is

$$Q_{a,1} = \sigma T_2^4 \int_{A_2} \int_{A_1} \alpha_1'(\eta_1) \, dF_{d2-d1} \, dA_2 = \frac{0.830\sigma T_2^4}{2} \int_{x=0}^{L} \int_{\eta_{1,\min}}^{\eta_{1,\max}} \cos^2 \eta_1 \, d\eta_1 \, dx$$

$$= \frac{0.830\sigma T_2^4}{2} \int_{x=0}^{L} \frac{1}{2} (\sin \eta_1 \cos \eta_1 + \eta_1) \Big|_{\eta_{1,\min}}^{\eta_{1,\max}} dx$$

$$= \frac{0.830\sigma T_2^4}{4} \int_{x=0}^{L} \left[\frac{D(L-x)}{x^2 - 2xL + L^2 + D^2} + \tan^{-1}\left(\frac{L-x}{D}\right) \right.$$

$$\left. + \frac{xD}{x^2 + D^2} + \tan^{-1}\frac{x}{D} \right] dx$$

$$Q_{a,1} = \frac{0.830\sigma T_2^4 L}{2} \tan^{-1}\frac{L}{D}$$

The absorption efficiency Ξ of the directional surface is then

$$\Xi_{\text{directional}} = \frac{0.830}{2} \tan^{-1} l - 0.652 \left(\frac{T_1}{T_2}\right)^4$$

The absorption efficiencies Ξ of the directional and diffuse surfaces are in Fig. 8-10 as a function of l with $(T_1/T_2)^4$ as a parameter. The Ξ for the directional surface is higher than that for the diffuse surface for all values of l. As l approaches zero, the configuration approaches that of infinite elemental strips, and emission from surface 1 becomes much larger than absorption from surface 2. Thus, Ξ_{diffuse} and $\Xi_{\text{directional}}$ are nearly equal since the surfaces always emit the same amount. As l approaches infinity, the directional effects are lost. At intermediate values of l, a 10% difference in absorption efficiency is attainable.

The effects of directional properties on the local heat loss can be a considerable factor in some geometries. In Fig. 8-11, a number of assumed directional distributions of reflectivity are examined for their influence on local heat loss from the walls of an infinitely long grooved cavity. The results are from [11], where for comparison the curves were gathered from original work and various sources [12, 13, 17, 18]. The walls of the groove are at 90° to each other, and the surface emissivity distributions are all normalized to give a hemispherical emissivity of 0.1. Curves are presented for diffuse reflectivity ρ, specular reflectivity assumed independent of incident angle ρ_s', specular reflectivity dependent on incident angle $\rho_s'(\theta)$ based on electromagnetic theory, and three distributions of bidirectional reflectivity $\rho''(\theta_r, \theta)$. The bidirectional distributions are based on the work of Beckmann and Spizzichino [19] for rough surfaces having various combinations of the ratio of rms optical-surface roughness amplitude to radiation wavelength, σ_0/λ,

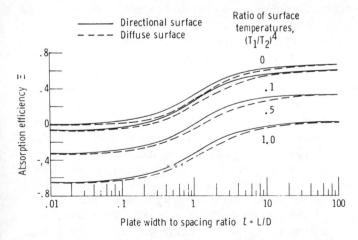

Figure 8-10 Effect of directional emissivity on absorption efficiency of surface.

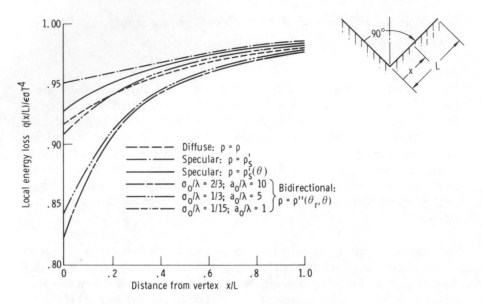

Figure 8-11 Local radiative energy loss from surface of isothermal groove cavity. Hemispherical emissivity of surface, 0.1.

and the ratio of roughness autocorrelation distance to radiation wavelength, a_0/λ. Note that the results in Fig. 8-11 for the simple specular and diffuse models do *not* provide upper and lower limits to all the solutions as is sometimes claimed. Additional work on surface roughness as it affects the directional properties of surfaces is reported in [20, 21]. Heat transfer was studied in a groove with two plane sides, each at uniform temperature, and having rough surfaces. The roughness was found to have a greater influence on the exchange of heat between the two sides than it has on the net energy radiated from the groove.

Howell and Durkee [22] compared analysis and experiment for the case of a beam of collimated radiation entering a very long low-temperature three-sided cavity. Because of the low temperature, surface emission was not important. The cavity had two surfaces with reflectivities that were diffuse with a specular component, while the third surface was a honeycomb material with a strong bidirectional reflectivity. It was found necessary to include all surface characteristics in the analysis to obtain agreement with experiment. This is characteristic of geometries interacting with collimated or point sources of incident radiation.

Black [23] analyzed the directional emission characteristics of two types of grooves. The grooves have some black and some specular sides, as shown in Fig. 8-12a and b. The V groove tends to emit in the normal direction, while the rectangular groove emits more in the grazing direction $\eta \rightarrow 90°$. Some emission results for the rectangular shape are in Fig. 8-12c, and they exhibit a strong directional characteristic for small h/w. Grooves and enclosures having *specular* surfaces are analyzed in Chap. 9.

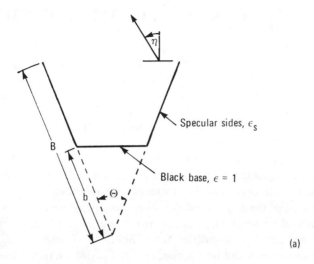

Specular sides, ϵ_s

B

b

Θ

Black base, $\epsilon = 1$

(a)

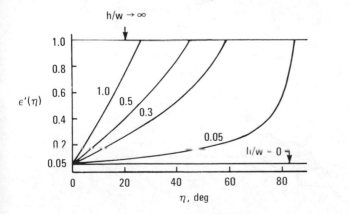

η

h

w

Black sides, $\epsilon = 1$

Specular base, ϵ_{base}

(b)

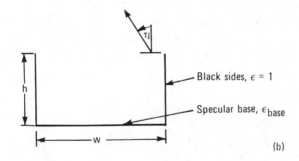

$h/w \to \infty$

$\epsilon'(\eta)$

1.0

0.8

0.6

0.4

0.2

0.05

1.0

0.5

0.3

0.05

h/w = 0

0   20   40   60   80

η, deg

(c)

Figure 8-12 Directional emission from grooves. (*a*) V-groove with specular sides; (*b*) rectangular groove with specular base; (*c*) directional emissivity for rectangular groove with specular base (ϵ_{base} = 0.05) and black sides.

8-5 SURFACES WITH DIRECTIONALLY AND SPECTRALLY DEPENDENT PROPERTIES

The general case of radiative transfer in enclosures with surfaces having radiative properties that depend on both wavelength and direction, and can be temperature dependent, is complex to treat fully. When such problems must be treated, numerical techniques are necessary. The Monte Carlo method is a likely candidate, and example applications of this method to simple directional-spectral surfaces are in Chap. 11. Toor [12] used the Monte Carlo method to study radiation interchange for various simply arranged surfaces with directional properties.

In this section, the general integral equations are formulated for radiation in such systems and a considerably simplified example problem is solved. The procedure is a combination of the previous diffuse-spectral and directional-gray analyses. The equations will be formulated at one wavelength as in Sec. 8-3 and will also be formulated in terms of intensities as in Sec. 8-4. In this manner, both spectral and directional effects can be accounted for. The interaction between two plane surfaces will be treated first. This can then be generalized to a multisurface enclosure as has been done for gray surfaces in Chap. 7.

Consider an area element dA at location $\mathbf{r}$ in an x, y, z coordinate system as shown in Fig. 8-13. The $i'_{\lambda o}$ is the outgoing spectral intensity from dA in the direction θ_r, φ_r as the result of emission and reflection. The spectral intensity emitted by dA in this direction is

$$i'_{\lambda e}(\lambda, \theta_r, \varphi_r, \mathbf{r}) = \epsilon'_\lambda(\lambda, \theta_r, \varphi_r, \mathbf{r})i'_{\lambda b}(\lambda, \mathbf{r}) \tag{8-20}$$

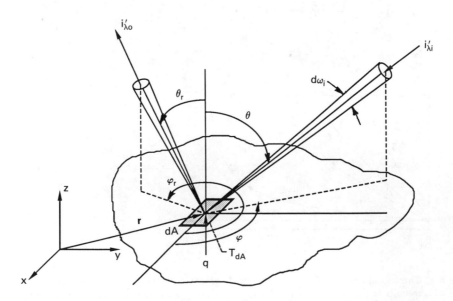

Figure 8-13 Geometry for incoming and outgoing intensities at a differential surface area.

These quantities are also functions of T_{dA}, but this designation is omitted to simplify the notation. The intensity reflected from dA into direction θ_r, φ_r results from the incident intensity from all directions of the hemisphere above dA. If the spectral intensity incident on dA within $d\omega_i$ is $i'_{\lambda i}(\lambda, \theta, \varphi, \mathbf{r})$, then the intensity reflected from dA into direction θ_r, φ_r is

$$i'_{\lambda r}(\lambda, \theta_r, \varphi_r, \mathbf{r}) = \int_{\omega_i=0}^{2\pi} \rho''_\lambda(\lambda, \theta_r, \varphi_r, \theta, \varphi, \mathbf{r}) i'_{\lambda i}(\lambda, \theta, \varphi, \mathbf{r}) \cos\theta \, d\omega_i \qquad (8\text{-}21)$$

The net energy flux supplied to dA is the difference between the outgoing and incoming fluxes,

$$q(\mathbf{r}) = \int_{\lambda=0}^{\infty} dq_{\lambda o}(\mathbf{r}) - \int_{\lambda=0}^{\infty} dq_{\lambda i}(\mathbf{r}) \qquad (8\text{-}22)$$

The $dq_{\lambda o}$ is the angular integration of the emitted and reflected spectral fluxes for all outgoing directions,

$$dq_{\lambda o} = i'_{\lambda b}(\mathbf{r}) d\lambda \int_{\varphi_r=0}^{2\pi} \int_{\theta_r=0}^{\pi/2} \epsilon'_\lambda(\lambda, \theta_r, \varphi_r, \mathbf{r}) \sin\theta_r \cos\theta_r \, d\theta_r \, d\varphi_r$$

$$+ \int_{\varphi_r=0}^{2\pi} \int_{\theta_r=0}^{\pi/2} i'_{\lambda r}(\lambda, \theta_r, \varphi_r, \mathbf{r}) \, d\lambda \sin\theta_r \cos\theta_r \, d\theta_r \, d\varphi_r \qquad (8\text{-}23)$$

The $dq_{\lambda i}$ is the result of incident spectral fluxes from all $d\omega_i$ directions,

$$dq_{\lambda i} = \int_{\varphi=0}^{2\pi} \int_{\theta=0}^{\pi/2} i'_{\lambda i}(\lambda, \theta, \varphi, \mathbf{r}) \, d\lambda \sin\theta \cos\theta \, d\theta \, d\varphi \qquad (8\text{-}24)$$

Equations (8-20) to (8-24) represent an exact formulation to obtain the heat flux $q(\mathbf{r})$ that must be supplied by other means to area dA to maintain it at temperature T_{dA} in the presence of incident radiation.

Various degrees of approximation can be made to carry these concepts over into an enclosure theory. If the enclosure is very simple, consisting for example of two plane surfaces, it might be feasible to include in the computations the variations across each surface of properties and surface temperature.

Consider the two surfaces in Fig. 8-14 and let the surrounding environment be at very low temperature so it does not contribute incident radiation. The spectral energy leaving dA_2 at $\mathbf{r}_2$ that reaches dA_1 at $\mathbf{r}_1$ is $i'_{\lambda o,2}(\lambda, \theta_2, \varphi_2, \mathbf{r}_2) \, d\lambda \, dA_2 \cos\theta_2 \, (dA_1 \cos\theta_1)/S^2$. In terms of the incident intensity, the incident spectral energy in $d\omega_1$ is $i'_{\lambda i,1}(\lambda, \theta_1, \varphi_1, \mathbf{r}_1) \, d\lambda \, dA_1 \cos\theta_1 \, d\omega_1$ where $d\omega_1 = (dA_2 \cos\theta_2)/S^2$. Thus $i'_{\lambda i,1}(\lambda, \theta_1, \varphi_1, \mathbf{r}_1) = i'_{\lambda o,2}(\lambda, \theta_2, \varphi_2, \mathbf{r}_2)$ and by using (8-20) and (8-21),

$$i'_{\lambda o,1}(\lambda, \theta_{r,1}, \varphi_{r,1}, \mathbf{r}_1) = \epsilon'_{\lambda,1}(\lambda, \theta_{r,1}, \varphi_{r,1}, \mathbf{r}_1) i'_{\lambda b,1}(\lambda, \mathbf{r}_1)$$

$$+ \int_{A_2} \rho''_{\lambda,1}(\lambda, \theta_{r,1}, \varphi_{r,1}, \theta_1, \varphi_1) i'_{\lambda o,2}(\lambda, \theta_2, \varphi_2, \mathbf{r}_2) \frac{\cos\theta_1 \cos\theta_2}{|\mathbf{r}_2 - \mathbf{r}_1|^2} dA_2 \qquad (8\text{-}25a)$$

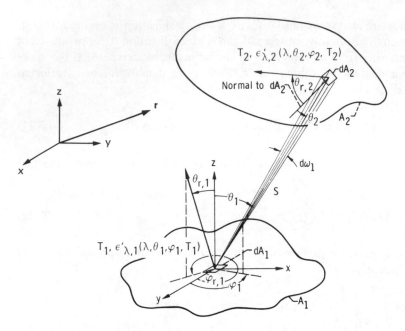

Figure 8-14 Interchange between surfaces having directional spectral properties (environment at zero temperature).

Similarly, for surface 2,

$$i'_{\lambda o,2}(\lambda,\ \theta_{r,2},\ \varphi_{r,2},\ \mathbf{r}_2) = \epsilon'_{\lambda,2}(\lambda,\ \theta_{r,2},\ \varphi_{r,2},\ \mathbf{r}_2)i'_{\lambda b,2}(\lambda,\ \mathbf{r}_2)$$

$$+ \int_{A_1} \rho''_{\lambda,2}(\lambda,\ \theta_{r,2},\ \varphi_{r,2},\ \theta_2,\ \varphi_2)i'_{\lambda o,1}(\lambda,\ \theta_1,\ \varphi_1,\ \mathbf{r}_1) \frac{\cos\theta_2\cos\theta_1}{|\mathbf{r}_1 - \mathbf{r}_2|^2}\,dA_1 \qquad (8\text{-}25b)$$

Equations (8-25) are both in terms of outgoing intensities. Thus, they form a set of simultaneous integral equations for $i'_{\lambda o,1}$ and $i'_{\lambda o,2}$. An iterative numerical solution would generally be required.

After $i'_{\lambda o,1}$ and $i'_{\lambda o,2}$ are obtained, the total energy can be determined that must be supplied to each surface element to maintain the specified local surface temperature. The total energy supplied is the difference between the total emitted and absorbed energies $dQ_{e,1} - dQ_{a,1}$. The heat flux for element dA_1 is then

$$\frac{dQ_1}{dA_1} = \int_{\lambda=0}^{\infty} \int_{\varphi_1=0}^{2\pi} \int_{\theta_1=0}^{\pi/2} \epsilon'_{\lambda,1}(\lambda,\ \theta_1,\ \varphi_1,\ \mathbf{r}_1)i'_{\lambda b,1}(\lambda,\ \mathbf{r}_1)\cos\theta_1\sin\theta_1\,d\theta_1\,d\varphi_1\,d\lambda$$

$$- \int_{\lambda=0}^{\infty} \int_{A_2} \alpha'_{\lambda,1}(\lambda,\ \theta_1,\ \varphi_1,\ \mathbf{r}_1)i'_{\lambda o,2}(\lambda,\ \theta_2,\ \varphi_2,\ \mathbf{r}_2)\cos\theta_1\frac{\cos\theta_2}{|\mathbf{r}_2 - \mathbf{r}_1|^2}\,dA_2\,d\lambda \qquad (8\text{-}26)$$

Usually in an enclosure theory the $i'_{\lambda o}$, temperature, and surface properties are assumed uniform over each surface. If we let A_k and A_j be the kth and jth surfaces of an enclosure with N surfaces, then by integrating Eq. (8-25) over A_k and summing the contributions from all of the A_j surfaces,

$$i'_{\lambda o,k}(\lambda, \theta_{r,k}, \varphi_{r,k}) = \epsilon'_{\lambda,k}(\lambda, \theta_{r,k}, \varphi_{r,k})i'_{\lambda b,k}(\lambda)$$

$$+ \frac{1}{A_k} \sum_{j=1}^{N} \int_{A_k} \int_{A_j} \rho''_{\lambda,k}(\lambda, \theta_{r,k}, \varphi_{r,k}, \theta_k, \varphi_k)i'_{\lambda o,j}(\lambda, \theta_j, \varphi_j)$$

$$\cdot \frac{\cos \theta_k \cos \theta_j}{|\mathbf{r}_j - \mathbf{r}_k|^2} \, dA_j \, dA_k \qquad (8\text{-}27)$$

When written out for each k, this yields a set of N simultaneous equations for the $i'_{\lambda o,k}$ for $k = 1, \ldots, N$.

With this degree of approximation, which is characteristic of an enclosure analysis, consider again the two plane surfaces in Fig. 8-14. The surroundings are at a very low temperature relative to the surface temperature so that radiation from the surroundings is neglected. Then, writing Eq. (8-27) for $k = 1$ and 2,

$$i'_{\lambda o,1}(\lambda, \theta_{r,1}, \varphi_{r,1}) = \epsilon'_{\lambda,1}(\lambda, \theta_{r,1}, \varphi_{r,1})i'_{\lambda b,1}(\lambda)$$

$$+ \frac{1}{A_1} \int_{A_1} \int_{A_2} \rho''_{\lambda,1}(\lambda, \theta_{r,1}, \varphi_{r,1}, \theta_1, \varphi_1)i'_{\lambda o,2}(\lambda, \theta_2, \varphi_2)$$

$$\cdot \frac{\cos \theta_1 \cos \theta_2}{S^2} \, dA_2 \, dA_1 \qquad (8\text{-}28a)$$

$$i'_{\lambda o,2}(\lambda, \theta_{r,2}, \varphi_{r,2}) = \epsilon'_{\lambda,2}(\lambda, \theta_{r,2}, \varphi_{r,2})i'_{\lambda b,2}(\lambda)$$

$$+ \frac{1}{A_2} \int_{A_2} \int_{A_1} \rho''_{\lambda,2}(\lambda, \theta_{r,2}, \varphi_{r,2}, \theta_2, \varphi_2)i'_{\lambda o,1}(\lambda, \theta_1, \varphi_1)$$

$$\cdot \frac{\cos \theta_2 \cos \theta_1}{S^2} \, dA_1 \, dA_2 \qquad (8\text{-}28b)$$

Equations (8-28) are both in terms of outgoing intensities. They form a set of simultaneous integral equations for $i'_{\lambda o,1}$ and $i'_{\lambda o,2}$. An iterative numerical solution is required.

After $i'_{\lambda o,1}(\lambda, \theta_{r,1}, \varphi_{r,1})$ and $i'_{\lambda o,2}(\lambda, \theta_{r,2}, \varphi_{r,2})$ are obtained, the total energy can be determined that must be supplied to each surface to maintain its specified temperature. This is found from the $i'_{\lambda o}$ as the difference between the energies carried away from and to the surface. For A_1 this gives

$$\frac{Q_1}{A_1} = \int_{\lambda=0}^{\infty} \int_{\theta_{r,1}=0}^{\pi/2} \int_{\varphi_{r,1}=0}^{2\pi} i'_{\lambda o,1}(\lambda, \theta_{r,1}, \varphi_{r,1}) \cos \theta_{r,1} \sin \theta_{r,1} \, d\theta_{r,1} \, d\varphi_{r,1} \, d\lambda$$

$$- \frac{1}{A_1} \int_{\lambda=0}^{\infty} \int_{A_1} \int_{A_2} i'_{\lambda o,2}(\lambda, \theta_2, \varphi_2) \frac{\cos \theta_1 \cos \theta_2}{S^2} \, dA_1 \, dA_2 \, d\lambda \qquad (8\text{-}29)$$

and similarly for A_2. For diffuse-gray surfaces, so that $i'_{\lambda o,1}$ and $i'_{\lambda o,2}$ are independent of λ, θ, and φ, this simplifies to $Q_1/A_1 = \pi i'_{o,1} - \pi i'_{o,2}F_{1-2} = q_{o,1} - q_{o,2}F_{1-2}$ as given by Eq. (7-18).

If Q_1 rather than T_1 is specified, T_1 must be determined and the solution becomes more difficult. A temperature is assumed for A_1, and the enclosure equations of the form (8-28) are solved to find the $i'_{\lambda o}$. The outgoing intensities are substituted into (8-29), and the computed Q_1 is compared to the given value. The T_1 is then adjusted and the procedure repeated until agreement between given and computed Q_1 is attained. If Q is specified for more than one surface, the solution is even more tedious.

EXAMPLE 8-7 As a simple example that can be carried out in analytical form, a small area element dA_1 is placed on the axis of, and parallel to, a black circular disk as shown in Fig. 8-15. The element is at temperature T_1, and the disk at T_2. The environment is at $T_e \approx 0$. The element has a directional spectral emissivity that is independent of φ and is approximated by

$$\epsilon'_{\lambda,1}(\lambda,\theta_1,T_1) = 0.8 \cos\theta_1(1 - e^{-C_2/\lambda T_1})$$

where C_2 is one of the constants in Planck's spectral energy distribution. As will be evident, this distribution was chosen to simplify this illustrative example. Find the energy dQ_1 added to dA_1 to maintain T_1.

The energy balance in (8-26) is written as emitted energy minus absorbed incident energy. The energy emitted by dA_1 is given by

$$dQ_{e,1} = dA_1 \int_{\lambda=0}^{\infty} \int_{\cap} \epsilon'_{\lambda,1}(\lambda,\theta_1)i'_{\lambda b,1}(\lambda)\cos\theta_1\, d\omega_1\, d\lambda$$

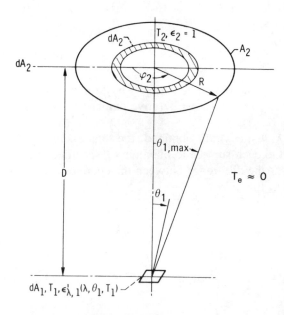

Figure 8-15 Energy exchange involving directional spectral surface element (environment at zero temperature).

Insert the expressions for $\epsilon'_{\lambda,1}$, $i'_{\lambda b,1}$ [Eq. (2-11)], and $d\omega_1 = \sin\theta_1\,d\theta_1\,d\varphi_1$ to obtain

$$dQ_{e,1} = 0.8\,dA_1 \int_{\lambda=0}^{\infty} \int_{\varphi_1=0}^{2\pi} \int_{\theta_1=0}^{\pi/2} \cos^2\theta_1(1 - e^{-C_2/\lambda T_1})$$

$$\times \frac{2C_1}{\lambda^5(e^{C_2/\lambda T_1} - 1)} \sin\theta_1\,d\theta_1\,d\varphi_1\,d\lambda$$

Carrying out the integrations for φ_1 and θ_1 gives

$$dQ_{e,1} = 0.8\,dA_1 \frac{2\pi}{3} \int_0^{\infty} \frac{2C_1}{\lambda^5 e^{C_2/\lambda T_1}}\,d\lambda$$

From the transformation $\zeta = C_2/\lambda T_1$,

$$dQ_{e,1} = 0.8\,dA_1 \frac{4C_1\pi}{3} \int_0^{\infty} \frac{T_1^4}{C_2^4} \frac{\zeta^3}{e^\zeta}\,d\zeta$$

Using the relation $\int_0^{\infty} \zeta^3 e^{-\zeta}\,d\zeta = 3!$ from [24] and the Stefan-Boltzmann constant as $\sigma = 2C_1\pi^5/15C_2^4$ gives

$$dQ_{e,1} = \frac{48}{\pi^4} \sigma T_1^4\,dA_1 = 0.493\,\sigma T_1^4\,dA_1$$

Thus the total hemispherical emission is about half that of a blackbody.
The energy absorbed by dA_1 is

$$dQ_{a,1} = dA_1 \int_{\lambda=0}^{\infty} \int_{A_2} \alpha'_{\lambda,1}(\lambda,\theta_1,\varphi_1) i'_{\lambda o,2}(\lambda,\theta_2,\varphi_2) \frac{\cos\theta_1 \cos\theta_2}{S^2}\,dA_2\,d\lambda$$

From Kirchhoff's law, the directional spectral absorptivity and emissivity can be equated without restriction. For dA_2 taken as a ring element, the solid angle $\cos\theta_2\,dA_2/S^2$ is written as $2\pi \sin\theta_1\,d\theta_1$. This is used to write the absorbed energy as (where $i'_{\lambda o,2} = i'_{\lambda b,2}$ since A_2 is specified as a black surface),

$$dQ_{a,1} = 2\pi(0.8)\,dA_1 \int_{\lambda=0}^{\infty} \int_{\theta_1=0}^{\theta_1,\max} (\cos^2\theta_1 \sin\theta_1\,d\theta_1) i'_{\lambda b,2}(1 - e^{-C_2/\lambda T_1})\,d\lambda$$

$$= -1.6\pi\,dA_1 \frac{\cos^3\theta_1}{3}\Big|_0^{\theta_1,\max} \int_0^{\infty} \frac{2C_1(1 - e^{-C_2/\lambda T_1})}{\lambda^5(e^{C_2/\lambda T_2} - 1)}\,d\lambda$$

$$dQ_{a,1} = \frac{3.2\pi C_1\,dA_1}{3}\left[1 - \frac{D^3}{(D^2 + R^2)^{3/2}}\right] \int_0^{\infty} \frac{1 - e^{-C_2/\lambda T_1}}{\lambda^5(e^{C_2/\lambda T_2} - 1)}\,d\lambda$$

Using the transformation $\zeta = C_2/\lambda T_1$, this is placed in the form

$$dQ_{a,1} = \frac{48}{\pi^4}\left[1 - \frac{1}{(1 + r^2)^{3/2}}\right] \sigma T_2^4\,dA_1\,G\!\left(\frac{T_2}{T_1}\right)$$

where $r = R/D$ and $G(T_2/T_1) = (1/6) \int_0^\infty \zeta^3 e^{-\zeta}(1 - e^{-\zeta T_2/T_1})/(1 - e^{-\zeta}) \, d\zeta$. This integral was evaluated numerically, giving $G(1.0) = 1.000$, $G(1.5) = 1.045$, and $G(2.0) = 1.063$; hence the effect of temperature ratio is small. Finally, the heat added to dA_1 to maintain it at T_1 is given by

$$dQ_1 = dQ_{e,1} - dQ_{a,1} = \frac{48\sigma}{\pi^4}\left\{ T_1^4 - T_2^4\left[1 - \frac{1}{(1 + r^2)^{3/2}} \right] G\left(\frac{T_2}{T_1}\right) \right\} dA_1$$

As shown by this example, it was difficult to construct a reasonable analytical function for ϵ_λ' that could be integrated into a convenient form over both angle and wavelength. Almost invariably numerical methods are used to solve problems of this type.

The development in this section has shown that although the formulation of radiation-exchange problems involving directional and/or spectral property effects is not conceptually difficult, it is often tedious to obtain solutions to the resulting integral equations. To simplify the equations, it is usually necessary to invoke assumptions and approximations that can vary from case to case. Numerical techniques such as iteration can be used for directional spectral problems, since closed-form analytical solutions can rarely be obtained. One very useful numerical technique is the Monte Carlo method, which is discussed in Chap. 11. For complicated directional and spectral effects, this is often a better approach than the integral equation method.

A surface with part specular and part diffuse reflectivity and a semigray analysis were used by Shimoji [25] to find local temperatures in conical and V-groove cavities exposed to incident solar radiation parallel to the cone axis or V-groove bisector plane. Toor and Viskanta [26] compared with experiment various analytical models using diffuse, specular, semigray, nongray, and combinations of these characteristics. They found for the particular geometries and materials studied that spectral effects were less important than directional effects and that the presence of one or more diffuse surfaces in an enclosure made the presence of specularly reflecting surfaces unimportant. Hering and Smith [20, 21] and Edwards and Bertak [27] applied various models of surface roughness in calculating radiant exchange between surfaces. If grooves on a surface have a size that is comparable to the wavelength of the incident or emitted radiation, there can be complex interactions of the electromagnetic waves within the grooves. This can produce unusual spectral and directional effects. The radiation behavior of materials with a grooved microstructure was studied in [28–30].

Innovative treatments of directional properties are required in modeling illumination of scenes in computer graphic representations. Representative approaches are in [31–36]. In [37] a method using two-parameter Markov chains is given for treating bidirectional surface properties.

REFERENCES

1. Branstetter, J. Robert: Radiant Heat Transfer between Nongray Parallel Plates of Tungsten, NASA TN D-1088, 1961.

2. Eckert, E. R. G., and Robert M. Drake, Jr.: "Heat and Mass Transfer," 2d ed., p. 375, McGraw-Hill, New York, 1959.
3. Dunkle, R. V., and J. T. Bevans: Part 3, A Method for Solving Multinode Networks and a Comparison of the Band Energy and Gray Radiation Approximations, *J. Heat Transfer*, vol. 82, no. 1, pp. 14–19, 1960.
4. Love, Tom J., and Joel S. Gilbert: Experimental Study of Radiative Heat Transfer between Parallel Plates, ARL-66-0103, DDC no. AD-643307, Oklahoma University, June 1966.
5. Goodman, Stanley: Radiant-Heat Transfer between Nongray Parallel Plates, *J. Res. Nat. Bur. Stand.*, vol. 58, no. 1, pp. 37–40, 1957.
6. Rolling, R. E., and C. L. Tien: Radiant Heat Transfer for Nongray Metallic Surfaces at Low Temperatures, paper no. 67-335, AIAA, April 1967.
7. Bobco, R. P., G. E. Allen, and P. W. Othmer: Local Radiation Equilibrium Temperatures in Semigray Enclosures, *J. Spacecr. Rockets*, vol. 4, no. 8, pp. 1076–1082, 1967.
8. Plamondon, J. A., and C. S. Landram: Radiant Heat Transfer from Nongray Surfaces with External Radiation. Thermophysics and Temperature Control of Spacecraft and Entry Vehicles, *Prog. Astronaut. Aeronaut.*, vol. 18, pp. 173–197, 1966.
9. Bevans, J. T., and D. K. Edwards: Radiation Exchange in an Enclosure with Directional Wall Properties, *J. Heat Transfer*, vol. 87, no. 3, pp. 388–396, 1965.
10. Hering, R. G.: Theoretical Study of Radiant Heat Exchange for Non-Gray Non-Diffuse Surfaces in a Space Environment, Rept. no. ME-TN-036-1, NASA CR-81653, Illinois University, September 1966.
11. Viskanta, Raymond, James R. Schornhorst, and Jaswant S. Toor: Analysis and Experiment of Radiant Heat Exchange between Simply Arranged Surfaces, AFFDL-TR-67-94, DDC No. AD-655335, Purdue University, June 1967.
12. Toor, J. S.: "Radiant Heat Transfer Analysis among Surfaces Having Direction Dependent Properties by the Monte Carlo Method," M.S. thesis, Purdue University, 1967.
13. Hering, R. G.: Radiative Heat Exchange between Specularly Reflecting Surfaces with Direction-dependent Properties, *Proc. 3rd Int. Heat Transfer Conf.*, Chicago, Aug. 7–12, 1966. (*AIChE J.*, vol. 5, pp. 200–206, 1966.)
14. Naraghi, M. H. N., and B. T. F. Chung: A Stochastic Approach for Radiative Exchange in Enclosures with Nonparticipating Medium, *J. Heat Transfer*, vol. 106, pp. 690–698, November 1984.
15. Naraghi, M. H. N., and B. T. F. Chung: A Stochastic Approach for Radiative Exchange in Enclosures with Directional-Bidirectional Properties, *J. Heat Transfer*, vol. 108, no. 2, pp. 264–270, 1986.
16. Howell, J. R., and M. Perlmutter: Directional Behavior of Emitted and Reflected Radiant Energy from a Specular, Gray, Asymmetric Groove, NASA TN D-1874, 1963.
17. Sparrow, E. M., J. L. Gregg, J. V. Szel, and P. Manos: Analysis, Results, and Interpretation for Radiation between Some Simply Arranged Gray Surfaces, *J. Heat Transfer*, vol. 83, no. 2, pp. 207–214, 1961.
18. Eckert, E. R. G., and E. M. Sparrow: Radiative Heat Exchange between Surfaces with Specular Reflection, *Int. J. Heat Mass Transfer*, vol. 3, no. 1, pp. 42–54, 1961.
19. Beckmann, Peter, and André Spizzichino: "The Scattering of Electromagnetic Waves from Rough Surfaces," Macmillan, New York, 1963.
20. Hering, R. G., and T. F. Smith: Surface Roughness Effects on Radiant Transfer between Surfaces, *Int. J. Heat Mass Transfer*, vol. 13, no. 4, pp. 725–739, 1970.
21. Hering, R. G., and T. F. Smith: Surface Roughness Effects on Radiant Energy Interchange, *J. Heat Transfer*, vol. 93, no. 1, pp. 88–96, 1971.
22. Howell, J. R., and R. E. Durkee: Radiative Transfer between Surfaces in a Cavity with Collimated Incident Radiation: A Comparison of Analysis and Experiment, *J. Heat Transfer*, vol. 93, no. 2, pp. 129–135, 1971.
23. Black, W. Z., Optimization of the Directional Emission from V-Groove and Rectangular Cavities, *J. Heat Transfer*, vol. 95, no. 1, pp. 31–36, 1973.
24. Dwight, Herbert B.: "Tables of Integrals and Other Mathematical Data," 4th ed., p. 230, Macmillan, New York, 1961.

25. Shimoji, S.: Local Temperatures in Semigray Nondiffuse Cones and V-Grooves, *AIAA J.*, vol. 15, no. 3, pp. 289–290, 1977.
26. Toor, J. S., and Viskanta, R.: A Critical Examination of the Validity of Simplified Models for Radiant Heat Transfer Analysis, *Int. J. Heat Mass Transfer*, vol. 15, pp. 1553–1567, 1972.
27. Edwards, D. K., and I. V. Bertak: Imperfect Reflections in Thermal Radiation Heat Transfer, in John W. Lucas (ed.), "Heat Transfer and Spacecraft Thermal Control," pp. 143–165, vol. 24 in *AIAA Progress in Astronautics and Aeronautics Series*, MIT Press, Cambridge, Mass., 1971.
28. Hesheth, P. J., B. Gebhart, and J. N. Zemel: Measurements of the Spectral and Directional Emission from Microgrooved Silicon Surfaces, *J. Heat Transfer*, vol. 110, no. 3, pp. 680–686, 1988.
29. Glass, N. E., A. A. Maradudin, and V. Celli: Surface Plasmons on a Large-Amplitude Doubly Periodically Corrugated Surface, *Phys. Rev. B*, vol. 26, no. 10, pp. 5357–5365, 1982.
30. Wirgin, A., and A. A. Maradudin: Resonant Enhancement of the Electric Field in the Grooves of Bare Metallic Gratings Exposed to S-Polarized Light, *Phys. Rev. B*, vol. 31, no. 8, pp. 5573–5576, 1985.
31. Buckalew, C., and D. Fussell: Illumination Networks: Fast Realistic Rendering with General Reflectance Functions, *Computer Graphics, SIGGRAPH '89 Proc.*, vol. 23, no. 3, pp. 89–98, 1989.
32. Immel, D. S., M. F. Cohen, and D. P. Greenberg: A Radiosity Method for Nondiffuse Environments, *Computer Graphics, SIGGRAPH '86 Proc.*, vol. 20, no. 4, pp. 133–142, 1986.
33. Kajiya, J. T.: Anisotropic Reflection Models, *Computer Graphics, SIGGRAPH '85 Proc.*, vol. 19, no. 4, pp. 15–21, 1985.
34. Wolff, L. B., and D. J. Kurlander: Ray Tracing with Polarization Parameters, *IEEE Computer Graphics and Applications*, vol. 10, no. 6, pp. 44–55, 1990.
35. He, X. D., K. E. Torrance, F. X. Sillion, and D. P. Greenberg: A Comprehensive Physical Model for Light Reflection, *Computer Graphics*, vol. 25, no. 4, pp. 175–186, 1991.
36. Sillion, F. X., J. R. Arvo, S. H. Westin, and D. P. Greenberg: A Global Illumination Solution for General Reflectance Distributions, *Computer Graphics*, vol. 25, no. 4, pp. 187–196, 1991.
37. Billings, R. L., J. W. Barnes, and J. R. Howell: Markov Analysis of Radiative Transfer in Enclosures with Bidirectional Reflections, *Numerical Heat Transfer, Part A*, vol. 19, pp. 313–326, 1991.

PROBLEMS

8-1 An antisatellite missile has a total radiation sensor built into its nose. The sensor detects total radiation emitted by a hot satellite and normally tracks to the satellite, where it impacts. The target satellite is gray with $\epsilon = 0.4$, has a surface temperature of $T_s = 400$ K, and is spherical with diameter $D_s = 2$ m. You are to design a dummy countermeasure satellite that is to be ejected from the real satellite as a decoy. The decoy material has the hemispherical spectral emissivity shown. The diameter of the decoy will be $D_d = 0.3$ m.

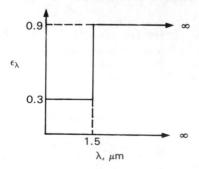

(a) What surface temperature T_d should be used for the decoy?

(b) How large a heat source (kW) is necessary in the decoy?

Answer: (a) 2342 K; (b) 253 kW

8-2 Radiative energy transfer is occurring across the space between two parallel plates with hemispherical spectral emissivities as shown. What is the net heat flux q transferred from 1 to 2?

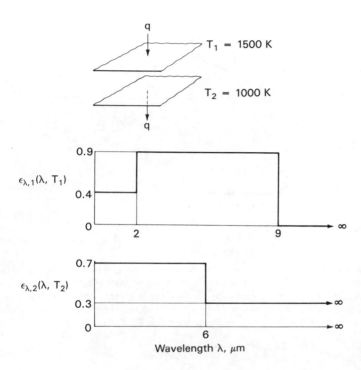

Answer: 119,100 W/m²

8-3 The two plates in Prob. 8-2 have a flat plate radiation shield placed between them so the geometry is now three parallel plates. The shield is gray and has an emissivity of 0.15 on both sides. What is the shield temperature, and what is the heat flux being transferred from plate 1 to plate 2? The spectral emissivities of plates 1 and 2 are in Prob. 8-2.

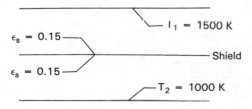

Answer: $T_s = 1317$ K, $q_1 = 15,737$ W/m²

8-4 Two large parallel plates are maintained at temperatures $T_1 = 1250$ K and $T_2 = 1050$ K. The plates are made from the same metal, and their emissivities as a function of wavelength, $\epsilon_\lambda(\lambda)$, are approximated as shown by two constant values joined by a linear decrease with wavelength. Compute the net radiant energy flux being transferred from plate 1 to plate 2. What is the energy flux if both plates are assumed gray with an average emissivity of 0.5?

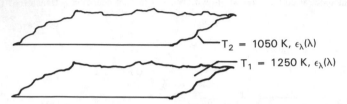

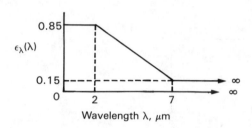

Answer: $q_1 = 36{,}933$ W/m^2, $q_{1,\text{gray}} = 23{,}172$ W/m^2

8-5 Radiation is being transferred across the gap between two large parallel plates. Both plates are made of the same metal, and the spectral emissivity of the metal does not depend on temperature. The wavelength variation of the emissivity for both plates is approximated in three steps as shown. The upper plate is maintained at $T_2 = 750$ K, and the lower plate has a uniform heat addition of $q_1 = 55{,}000$ W/m^2. What is the temperature T_1 of the lower plate?

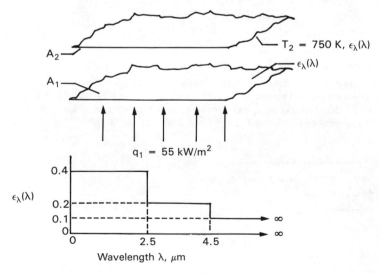

Answer: 1568 K

8-6 A polished aluminum tank in vacuum is surrounded by one thin aluminum radiation shield. The shield is painted on the outside with white paint (see Fig. 5-31). Treating the geometry as infinite parallel plates, what is the heat flux into the tank for normally incident solar radiation? Assume the tank is maintained at 170 K.

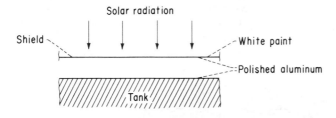

Answer: 267.2 K, 12.05 W/m²

8-7 Consider the enclosure of Prob. 7-21. Compute the heat transfer for the following values of hemispherical spectral emissivity. (For simplicity, do not subdivide the three areas into smaller regions.)

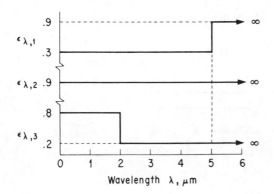

Answer: $q_1 = 37,815$ W/m², $q_2 = -38,903$ W/m², $q_3 = -7678$ W/m²

8-8 Estimate the heat flux leaking into a liquid hydrogen container from an adjacent container of liquid nitrogen. The glass walls are coated with highly polished aluminum. Use electromagnetic theory to estimate the radiative properties.

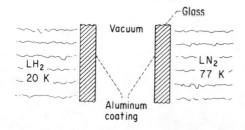

8-9 Two parallel plates are of finite width and of infinite length in the direction normal to that shown. The plates are diffuse, but their spectral emissivity varies with wavelength and is approximated by a step function as shown. The plate edge openings are exposed to the environment at $T_e = 300$ K, and there is energy exchange only from the internal surfaces. Both plates are made of the same material and hence have the same emissivity, which is independent of temperature. Energy is being added to the lower plate, and the upper plate is maintained at a fixed temperature. Obtain values for the temperature of the lower plate, T_1, and for the heat flux q_2 that must be added to or removed from the upper plate to maintain the specified temperature. For simplicity, do not subdivide the surface areas.

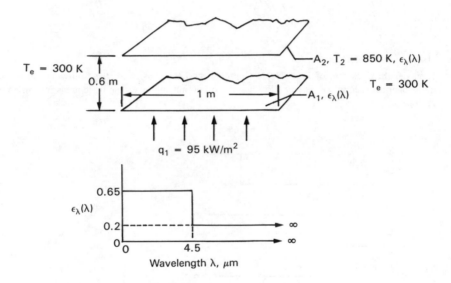

Answer: 1368 K, $-27{,}590$ W/m^2

8-10 Two circular disks are parallel and directly facing each other. The disks are diffuse, but their emissivities vary with wavelength. The properties are approximated with step functions as shown. The disks are maintained at temperatures $T_1 = 1225$ K and $T_2 = 950$ K. The surroundings are at $T_e = 600$ K. Compute the rate of energy that must be supplied to or removed from the disks to maintain the specified temperatures. The outer surfaces of the disks are insulated so there is radiation interchange only from the inner surfaces that are facing each other.

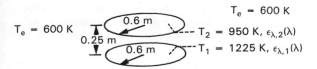

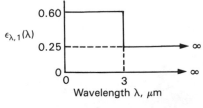

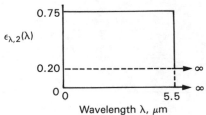

Answer: $Q_1 = 44,520$ W, $Q_2 = -5933$ W

8-11 An area element dA at temperature T is radiating out through a circular opening of radius r in a plate parallel to dA. The directional total emissivity of dA is $\epsilon'(\theta) = 0.85 \cos \theta$ as in Fig. 3-3. Obtain a relation for the radiation $Q_{\text{nondiffuse}}$ from dA through the opening. From Example 3-3 the hemispherical total emissivity of dA is 0.57. Using this ϵ and the diffuse configuration factor, compute Q_{diffuse} through the opening. Plot $Q_{\text{nondiffuse}}/Q_{\text{diffuse}}$ as a function of h/r.

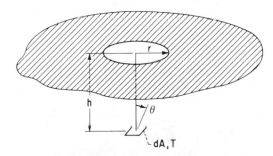

Answer: $\dfrac{Q_{\text{nondiffuse}}}{Q_{\text{diffuse}}} = \dfrac{(H^2 + 1)^{3/2} - H^3}{(H^2 + 1)^{1/2}}, \qquad H = \dfrac{h}{r}$

8-12 A sphere and area element are positioned as shown. The sphere is gray but has a nondiffuse emissivity $\epsilon_s = E \cos \theta$, where θ is the angle measured from the normal to the sphere surface and E is a constant. Set up the integral for the direct radiation from the sphere to the area element in terms of the quantities given.

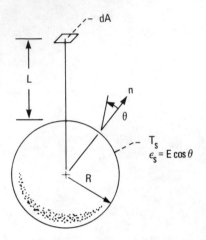

NINE

RADIATION EXCHANGE IN ENCLOSURES WITH SOME SURFACES THAT ARE SPECULARLY REFLECTING

9-1 INTRODUCTION

In Chap. 7 the surfaces considered are *diffuse emitters* and *diffuse reflectors*. In Chap. 8 surfaces with general directional emission and reflection characteristics are analyzed. In this chapter a particular type of reflection is considered that was not specifically treated in Chap. 8. This is a mirrorlike reflection and it is accounted for in a special way. All the surfaces are still assumed to emit in a diffuse fashion. Some of the surfaces in an enclosure are assumed to reflect diffusely; the remaining surfaces are assumed to be *specular,* that is, to reflect in a *mirrorlike* manner. Recall from the discussion of roughness in Chap. 5 that an important parameter is the ratio of the root-mean-square roughness height to the wavelength of the radiation (optical roughness). For long-wavelength radiation a smooth surface tends toward being optically smooth, and the reflections tend to become more specular. Thus, although a surface may not appear mirrorlike to the eye (i.e., for the short wavelengths of the visible spectrum), it may be specular for longer wavelengths in the infrared.

When reflection is diffuse, the directional history of the incident radiation is lost upon reflection: the reflected energy has the same directional distribution as if it had been absorbed and diffusely re-emitted. With specular reflection, the reflection angle relative to the surface normal is equal in magnitude to the angle of incidence. Hence, the directional history of the incident radiation is *not* lost upon reflection. Consequently, when dealing with specular surfaces, it is necessary to account for the specific *directional paths* that the reflected radiation follows between surfaces.

The specular reflectivities used in this chapter are assumed independent of incident angle of radiation; that is, the same fraction of the incident energy is reflected, regardless of the angle of incidence of the energy. In addition, all the surfaces are assumed *gray*; that is, the properties do not depend on wavelength. A nongray analysis can be done as in Chap. 8. The calculations are carried out in each significant wavelength band and the results are then summed over all bands to obtain total energy quantities.

9-2 SYMBOLS

A	area
c_p	specific heat
d	number of diffuse surfaces
D	tube diameter
f	focal length
F	configuration factor
L	length of enclosure side; tube length
N	total number of surfaces
q	energy flux, energy per unit area and per unit time
Q	energy rate, energy per unit time
T	absolute temperature
V	volume
x, X	position coordinates
α	absorptivity; angle subtended by sun
γ	distance from mirror to image
ϵ	emissivity
θ	cone angle
ρ	reflectivity
ρ_m	reflectivity of mirror
ρ_M	density of material
σ	Stefan-Boltzmann constant
τ	time

Subscripts

d	diffuse
e	emitted
F	final
i	incoming
I	initial
j, k	jth or kth surface
max	at maximum blackbody energy
nd	nondiffuse

o	outgoing
s	specular; solar
0	in vacuum
1, 2	surface 1 or 2

Superscripts

s	total specular exchange factor including all paths for specular inter-reflections plus direct exchange
$''$	bidirectional value
$*$	denotes a second portion of area on same surface

9-3 RADIATION BETWEEN PAIRS OF SURFACES WITH SPECULAR REFLECTIONS

9-3.1 Some Cases Having Simple Geometries

As an introduction, consider radiation exchange for infinite parallel plates, concentric cylinders, and concentric spheres, as shown in Fig. 9-1. Specular exchange for these cases is well understood, having been discussed by Christiansen [1] and Saunders [2] more than 60 years ago. The radiative exchange process is easy to visualize for these cases and is of practical importance in predicting the heat transfer performance of radiation shields, Dewar vessels, and cryogenic insulation.

Consider radiation between two infinite, gray, parallel *specularly reflecting* plates, Fig. 9-1a. Although the reflections are specular, the emission is *diffuse*. All emitted and reflected radiation leaving surface 1 directly reaches surface 2; similarly, all emitted and reflected radiation leaving surface 2 directly reaches surface 1. This is true whether the surfaces are specular or diffuse reflectors. For diffuse emission the absorptivity of the surface is independent of direction, so there are no directional factors to be concerned with (see Example 9-3 for a non-diffuse case). Hence, for specular reflections and diffuse emission Eq. (7-22a) continues to apply and the net radiative heat transfer from surface 1 to surface 2 is

$$Q_1 = -Q_2 = \frac{A_1 \sigma (T_1^4 - T_2^4)}{1/\epsilon_1(T_1) + 1/\epsilon_2(T_2) - 1} \tag{9-1}$$

Now consider radiation between the concentric cylinders or spheres in Fig. 9-1b and c. The emission is diffuse. Typical radiation paths for specular reflections are shown in Fig. 9-1d. As shown by path a, all the radiation emitted by surface 1 will directly reach surface 2. A portion will be reflected from surface 2 back to surface 1, and a portion of this will be rereflected from 1. This sequence of reflections continues until insignificant energy remains because radiation is partially absorbed on each contact with a surface. From the symmetry of the concentric geometry and the equal magnitudes of incidence and reflection angles for specular reflections, none of the radiation following path a can ever be reflected directly from a position on surface 2 to another location on surface 2. Thus the exchange process for radiation emitted from surface 1 is the same as though the two concentric surfaces were infinite parallel plates. However, the radiation emitted from the outer surface 2 can travel along either of two types of paths, b or c as shown in Fig. 9-1d. Since emission is assumed diffuse, the fraction F_{2-2} will follow paths of type c (F is a *diffuse* configuration factor). From the geometry of specular reflections these rays will always be reflected along surface 2 with none ever reaching surface 1. The fraction $F_{2-1} = A_1/A_2$ will be reflected back and forth between the surfaces along path b in the same fashion as radiation emitted from surface 1. The amount of radiation following this type of path is $A_2\epsilon_2 F_{2-1}\sigma T_2^4 = A_2\epsilon_2(A_1/A_2)\,\sigma T_2^4 = A_1\epsilon_2\sigma T_2^4$. The fraction of the radiation leaving surface 2 that impinges on surface 1 thus depends on A_1 and not on A_2. Hence, for specular surfaces the exchange behaves as if both surfaces were equal portions of infinite parallel plates with size equal to the area of the inner body. The net radiative heat transfer from surface 1 to surface 2 is thus given by Eq. (9-1).

EXAMPLE 9-1 A spherical vacuum bottle consists of two silvered, concentric glass spheres, the inner being 15 cm in diameter and the evacuated gap between the spheres being 0.65 cm. The emissivity of the silver coating is 0.02. If hot coffee at 368 K is in the bottle and the outside temperature is 294 K, what is the radiative heat leakage out of the bottle?

Equation (9-1) will apply for concentric *specular* spheres. For the small rate of heat leakage expected, it is assumed that the surfaces will be close to 368 K and 294 K. This gives

$$Q_1 = \frac{\pi(0.15)^2 5.671 \times 10^{-8}(368^4 - 294^4)}{(1/0.02) + (1/0.02) - 1} = 0.440 \text{ W}$$

If, instead of using the specular formulation, both surfaces are assumed diffuse, then (7-25) applies. The denominator of the Q_1 equation becomes

$$\frac{1}{\epsilon_1} + \frac{A_1}{A_2}\left(\frac{1}{\epsilon_2} - 1\right) = \frac{1}{0.02} + \left(\frac{15}{16.3}\right)^2\left(\frac{1}{0.02} - 1\right) = 91.50$$

instead of 99 as in the specular case. For diffuse surfaces the heat loss is 0.476 W.

EXAMPLE 9-2 For the previous example, how long will it take for the coffee to cool from 368 K to 322 K if the heat loss is only by radiation?

The heat capacity of the coffee is $\rho_M V c_p T_1$. Assuming the coffee is always mixed well enough that it is at uniform temperature, the cooling rate is equal to the *instantaneous* loss by radiation. The energy loss by radiation at any time τ, given by Eq. (9-1), is related to the loss of internal energy of the coffee by

$$-\rho_M V c_p \frac{dT_1}{d\tau} = \frac{A_1 \sigma [T_1^{\,4}(\tau) - T_2^{\,4}]}{1/\epsilon_1 + 1/\epsilon_2 - 1}$$

where it is assumed that surface 1 is at the coffee temperature and surface 2 is at the outside environment temperature. Then

$$-\int_{T_1=T_I}^{T_1=T_F} \frac{dT_1}{T_1^{\,4} - T_2^{\,4}} = \frac{A_1 \sigma}{\rho_M V c_p (1/\epsilon_1 + 1/\epsilon_2 - 1)} \int_0^{\tau} d\tau$$

where T_I and T_F are the initial and final temperatures of the coffee, and ϵ_1 and ϵ_2 are assumed independent of temperature. Carrying out the integration gives

$$\left(\frac{1}{4T_2^{\,3}} \ln \left| \frac{T_1 + T_2}{T_1 - T_2} \right| + \frac{1}{2T_2^{\,3}} \tan^{-1} \frac{T_1}{T_2} \right) \bigg|_{T_I}^{T_F} = \frac{A_1 \sigma \tau}{\rho_M V c_p (1/\epsilon_1 + 1/\epsilon_2 - 1)}$$

Then the cooling time from T_I to T_F is

$$\tau = \frac{\rho_M V c_p (1/c_1 + 1/c_2 - 1)}{A_1 \sigma} \left[\frac{1}{4T_2^{\,3}} \ln \left| \frac{(T_F + T_2)/(T_F - T_2)}{(T_I + T_2)/(T_I - T_2)} \right| \right.$$

$$\left. + \frac{1}{2T_2^{\,3}} \left(\tan^{-1} \frac{T_F}{T_2} - \tan^{-1} \frac{T_I}{T_2} \right) \right]$$

Substituting the values $\rho_M = 975$ kg/m^3, $V = \frac{1}{6} \pi (0.15)^3$ m^3, $c_p = 4195$ J/ (kg · K), $\epsilon_1 = \epsilon_2 = 0.02$, $A_1 = \pi (0.15)^2$ m^2, $\sigma = 5.671 \times 10^{-8}$ W/(m^2 · K^4), $T_2 = 294$ K, $T_I = 368$ K, and $T_F = 322$ K gives the cooling time as $\tau = 374.7$ h.

The coffee will be cooling for about 16 days if heat losses occur only by radiation. Conduction losses through the glass wall of the bottle neck usually cause the cooling to be considerably faster.

For surfaces with diffuse emission Eq. (9-1) applies for infinite parallel plates, infinitely long concentric cylinders, and concentric spheres when both surfaces are specular reflectors. For infinite parallel plates it also applies when both surfaces are diffuse or when one surface is diffuse and the other specular. For cylinders and spheres, (9-1) still applies if the surface of the inner body (surface 1) is diffuse as long as the outer body (surface 2) remains specular. This is because all radiation leaving surface 1 will go directly to surface 2 regardless of whether surface 1 is specular or diffuse. When surface 2 is diffuse, (7-25) applies and may be used when surface 1 is either specular or diffuse. The relations are summarized in Table 9-1.

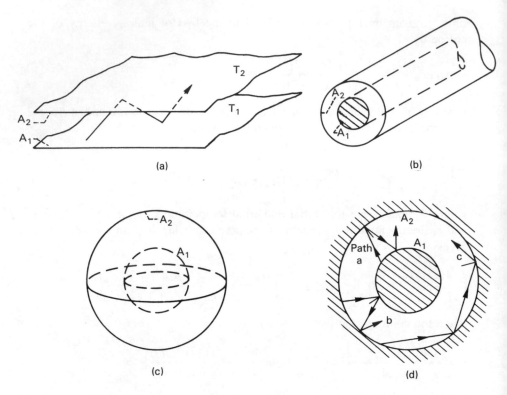

Figure 9-1 Radiation exchange for specular surfaces having simple geometries. (*a*) Infinite parallel plates; (*b*) gap between infinitely long concentric cylinders; (*c*) gap between concentric spheres; (*d*) paths for specular radiation in gap between concentric cylinders or spheres.

The previous development was for diffuse emission. Now consider an example for surfaces where the emission is directional.

EXAMPLE 9-3 Two infinite parallel plates are *specular* reflectors, but have emissivities (and hence absorptivities) that depend on angle θ (nondiffuse emitting surfaces). What is the equation of radiative exchange between the surfaces?

The radiation per unit area emitted in solid angle $d\omega$ from surface 1 in the direction θ is $\epsilon_1'(\theta)(\sigma T_1^4/\pi) \cos \theta \, d\omega = 2\epsilon_1'(\theta) \, \sigma T_1^4 \cos \theta \sin \theta \, d\theta$ and similarly for surface 2. For each direction the radiation will behave as in Eq. (9-1). Then by integrating over the exchanges within all solid angles

$$q_1 = \frac{Q_1}{A_1} = \sigma(T_1^4 - T_2^4) \, 2 \int_0^{\pi/2} \frac{\sin \theta \cos \theta}{[1/\epsilon_1'(\theta)] + [1/\epsilon_2'(\theta)] - 1} \, d\theta$$

For diffuse emission the ϵ_1 and ϵ_2 are each constant so the ratio of nondiffuse to diffuse transfer is (all reflections are specular)

$$\frac{q_{1,nd}}{q_{1,d}} = 2 \int_0^{\pi/2} \frac{\sin\theta\cos\theta}{[1/\epsilon_1'(\theta)] + [1/\epsilon_2'(\theta)] - 1}\, d\theta \bigg/ \frac{1}{(1/\epsilon_1) + (1/\epsilon_2) - 1}$$

As a simple illustration let $\epsilon_1'(\theta) = \epsilon_2'(\theta) = C\cos\theta$ as in Example 3-3. The $C \le 1$ is a constant, and from Example 3-3 the hemispherical $\epsilon = 2C/3$ for this $\epsilon'(\theta)$ variation. The integration yields

$$\frac{q_{1,nd}}{q_{1,d}} = -\left[\frac{8}{C^2}\ln\left(1 - \frac{C}{2}\right) + \frac{4}{C} + 1\right] \bigg/ \left(\frac{C}{3 - C}\right)$$

For $C = 0.5, 0.8, 1.0$ this gives respectively $q_{1,nd}/q_{1,d} = 1.029, 1.060, 1.090$. The particular $\epsilon'(\theta)$ variation chosen can increase the heat exchange up to 9%; the effect becomes small when C, the maximum emissivity, is small.

The results in Table 9-1 can be used to obtain the performance of multiple radiation shields, as shown in Figs. 9-2 and 9-3. The shields are thin, parallel, highly reflecting sheets placed between surfaces. A highly effective insulation can be formed by using many layers separated by vacuum to provide a series of alternate radiation and conduction barriers. One construction is to deposit highly reflecting metallic films on both sides of thin sheets of plastic spaced apart by placing between them a cloth net having a large open area between the fibers. A stacking of 20 radiation shields per centimeter of thickness can be obtained. An

Table 9-1 Radiant interchange between some simply arranged surfaces (emission is diffuse)

Geometry	Configuration	Type of surface reflection	Energy transfer rate Q_1
Infinite parallel plates	A_2 A_1	A_1 or A_2, either specular or diffuse	$\dfrac{A_1\sigma(T_1^4 - T_2^4)}{1/\epsilon_1 + 1/\epsilon_2 - 1}$
Infinitely long concentric cylinders	A_1 A_2	A_1, specular *or* diffuse; A_2, diffuse	$\dfrac{A_1\sigma(T_1^4 - T_2^4)}{1/\epsilon_1 + (A_1/A_2)(1/\epsilon_2 - 1)}$
		A_1, specular *or* diffuse; A_2, specular	$\dfrac{A_1\sigma(T_1^4 - T_2^4)}{1/\epsilon_1 + 1/\epsilon_2 - 1}$
Concentric spheres	A_1 A_2	A_1, specular *or* diffuse; A_2, diffuse	$\dfrac{A_1\sigma(T_1^4 - T_2^4)}{1/\epsilon_1 + (A_1/A_2)(1/\epsilon_2 - 1)}$
		A_1, specular *or* diffuse; A_2, specular	$\dfrac{A_1\sigma(T_1^4 - T_2^4)}{1/\epsilon_1 + 1/\epsilon_2 - 1}$

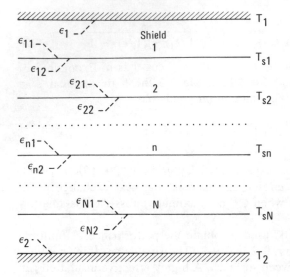

Figure 9-2 Parallel walls separated by N radiation shields.

important use of multilayer insulation is in low-temperature applications such as insulation of cryogenic storage tanks.

For the results here, the spaces between the shields are evacuated so that heat transfer is only by radiation (all emission is assumed diffuse). To analyze shield performance, consider the situation of N different radiation shields between two surfaces at temperatures T_1 and T_2 with emissivities ϵ_1 and ϵ_2. As a general case, let a typical shield n have emissivity ϵ_{n1} on one side and ϵ_{n2} on the other, as shown in Fig. 9-2. As a result of the heat flow, the nth shield will be at temperature T_{sn}. Since the same q passes through the entire series of shields, Eq. (9-1) can be written for each pair of adjacent surfaces as

$$q\left(\frac{1}{\epsilon_1} + \frac{1}{\epsilon_{11}} - 1\right) = \sigma(T_1^4 - T_{s1}^4)$$

$$q\left(\frac{1}{\epsilon_{12}} + \frac{1}{\epsilon_{21}} - 1\right) = \sigma(T_{s1}^4 - T_{s2}^4)$$

$$q\left(\frac{1}{\epsilon_{22}} + \frac{1}{\epsilon_{31}} - 1\right) = \sigma(T_{s2}^4 - T_{s3}^4)$$

$$\vdots$$

$$q\left(\frac{1}{\epsilon_{(N-1)2}} + \frac{1}{\epsilon_{N1}} - 1\right) = \sigma(T_{s(N-1)}^4 - T_{sN}^4)$$

$$q\left(\frac{1}{\epsilon_{N2}} + \frac{1}{\epsilon_2} - 1\right) = \sigma(T_{sN}^4 - T_2^4)$$

Adding these equations and dividing by the resulting factor multiplying q on the left-hand side give

$$q = \frac{\sigma(T_1^4 - T_2^4)}{1/\epsilon_1 + 1/\epsilon_{11} - 1 + \sum_{n=1}^{N-1}(1/\epsilon_{n2} + 1/\epsilon_{(n+1)1} - 1) + 1/\epsilon_{N2} + 1/\epsilon_2 - 1}$$

(9-2a)

where the sum is zero if $N = 1$. This can also be written,

$$q = \frac{\sigma(T_1^4 - T_2^4)}{1/\epsilon_1 + 1/\epsilon_2 - 1 + \sum_{n=1}^{N}(1/\epsilon_{n1} + 1/\epsilon_{n2} - 1)}$$

(9-2b)

In most instances ϵ is the same on both sides of each shield, and all the shields have the same ϵ. Let all the shield emissivities be ϵ_s; then q becomes

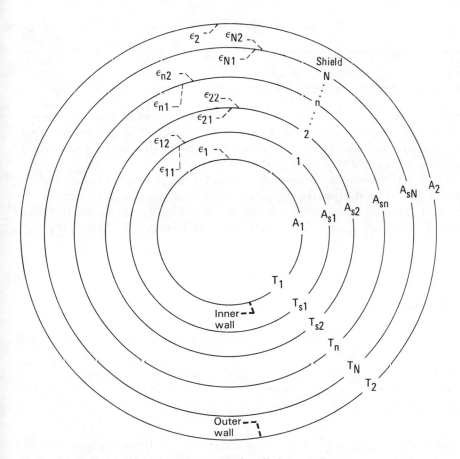

Figure 9-3 Radiation shields between concentric cylinders or spheres.

$$q = \frac{\sigma(T_1{}^4 - T_2{}^4)}{1/\epsilon_1 + 1/\epsilon_2 - 1 + N(2/\epsilon_s - 1)} \qquad (9\text{-}3)$$

If the wall emissivities are the same as the shield emissivities, $\epsilon_1 = \epsilon_2 = \epsilon_s$, (9-3) further reduces to

$$q = \frac{\sigma(T_1{}^4 - T_2{}^4)}{(N + 1)(2/\epsilon_s - 1)} \qquad (9\text{-}4)$$

In this instance q decreases as $1/(N + 1)$ as the number of shields N increases. When there are no shields, $N = 0$, and (9-3) reduces to $q = \sigma(T_1^4 - T_2^4)/[(1/\epsilon_1) + (1/\epsilon_2) - 1]$ as given in Table 9-1. As an example to illustrate the performance of the shields, if in Eq. (9-3) $\epsilon_1 = \epsilon_2 = 0.8$ and $\epsilon_s = 0.05$, then $q = \sigma(T_1^4 - T_2^4)/(1.5 + 39 N)$. The ratio $q(N \text{ shields})/q(\text{no shields})$ is $1.5/(1.5 + 39 N)$. For various numbers of shields this yields

	$N = 0$	$N = 1$	$N = 10$	$N = 100$
$q(N)/q(N = 0)$	1	0.0370	0.00383	0.00038
$1/(N + 1)$	1	0.5	0.0909	0.0099

In the table, the factor $1/(N + 1)$ is the result when the walls have the same ϵ as the shields. This illustrates that the *fractional reduction* in q is much larger when the wall ϵ are large compared to the ϵ for the shields, since for small wall ϵ values the unshielded heat flow is already low.

In a similar way to the previous derivation for flat plates, the expressions in Table 9-1 can be used to derive the heat flow through a series of concentric cylindrical or spherical radiation shields, as shown in Fig. 9-3. If the walls A_1 and A_2 and all the shields A_{sn} are diffuse reflectors, the heat flow is (emission is diffuse)

$$Q = \frac{A_1 \sigma(T_1{}^4 - T_2{}^4)}{\{1/\epsilon_1 + (A_1/A_{s1})(1/\epsilon_{11} - 1) + \sum_{n=1}^{N-1} (A_1/A_{sn})[1/\epsilon_{n2} + (A_{sn}/A_{s(n+1)})(1/\epsilon_{(n+1)1} - 1)]}$$

$$+ (A_1/A_{sN})[1/\epsilon_{N2} + (A_{sN}/A_2)(1/\epsilon_2 - 1)]\}$$
$$(9\text{-}5a)$$

$$= \frac{A_1 \sigma(T_1{}^4 - T_2{}^4)}{1/\epsilon_1 + (A_1/A_2)(1/\epsilon_2 - 1) + \sum_{n=1}^{N} (A_1/A_{sn})(1/\epsilon_{n1} + 1/\epsilon_{n2} - 1)} \qquad (9\text{-}5b)$$

The sum in (9-5a) [and in (9-6), (9-7)] is zero if $N = 1$. If the walls are diffuse and all the shields are specular, then

$$Q = \frac{A_1 \sigma (T_1{}^4 - T_2{}^4)}{\{1/\epsilon_1 + 1/\epsilon_{11} - 1 + \sum\limits_{n=1}^{N-1} (A_1/A_{sn})(1/\epsilon_{n2} + 1/\epsilon_{(n+1)1} - 1)}$$

$$+ (A_1/A_{sN})[1/\epsilon_{N2} + (A_{sN}/A_2)(1/\epsilon_2 - 1)]\}$$

$$(9\text{-}6)$$

If all the surfaces are specular, (9-6) applies if we replace A_{sN}/A_2 by unity in the last term in the denominator. In this instance, if all the shields emissivities are the same and equal to ϵ_s, then

$$Q = \frac{A_1 \sigma (T_1{}^4 - T_2{}^4)}{1/\epsilon_1 + 1/\epsilon_s - 1 + \sum\limits_{n=1}^{N-1} (A_1/A_{sn})(2/\epsilon_s - 1) + (A_1/A_{sN})(1/\epsilon_s + 1/\epsilon_2 - 1)} \qquad (9\text{-}7)$$

When using radiation shields to insulate a cryogenic tank, one or more of the shields can be cooled by using vapor boil-off from the tank. The resulting decrease of temperature within the insulation helps reduce the heat loss from the tank. The optimum utilization of vapor cooling can be examined thermodynamically by minimizing the entropy production resulting from the heat losses and the controlled cryogen boil-off used for vapor cooling of one or more shields. This optimization has been analyzed in [3] using the thermodynamic ideas in [4].

With regard to the radiative transfer between parallel surfaces, there are additional effects when reflecting layers are spaced very close to each other. These were examined by Cravalho et al. [5], who considered the geometry in Fig. 9-4, consisting of two semi-infinite dielectric media having refractive indices n_1 and n_3, separated by a vacuum gap. In the usual analysis for radiative transfer between two spectrally dependent surfaces such as surface 1 and surface 3, the heat flux transferred across the gap is given by Eq. (8-11), $q_1 = \int_0^\infty \{[e_{\lambda b,1}(\lambda, T_1) - e_{\lambda b,3}(\lambda, T_3)]/[1/\epsilon_{\lambda,1}(\lambda, T_1) + 1/\epsilon_{\lambda,3}(\lambda, T_3) - 1]\} \, d\lambda$ and the spacing between the plates does not appear. When the spacing between the surfaces is very small, however, two effects enter that are functions of spacing. The first effect is wave interference, in

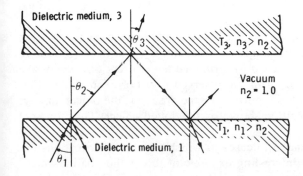

Figure 9-4 Reflection and transmission of electromagnetic wave in gap between two dielectrics; $T_3 < T_1$.

which a wave reflecting back and forth in a gap between two dielectrics may undergo cancellation or reinforcement as described in Chap. 18 for thin film layers.

The second effect is radiation tunneling. The electromagnetic fields at the outer surface of each body are able to reach to the opposite body and transfer energy if the distance is sufficiently small. For ordinary behavior at an interface as in Fig. 9-4, some of the radiation in medium 1 traveling toward region 2 will undergo total internal reflection at the interface when $n_1 > n_2$, as discussed in Sec. 18-6.2. For ordinary radiative behavior this occurs when the incidence angle θ_1 is equal to or larger than the angle for total reflection given by (18-50), $\theta_1 \geq \sin^{-1}(n_2/n_1)$. When region 2 in Fig. 9-4 is sufficiently thin, however, electromagnetic theory predicts that, even for an intensity incident at θ_1 greater than $\sin^{-1}(n_2/n_1)$, total internal reflection will not occur. Rather, part of the incident intensity will propagate across the thin region 2 and enter medium 3. This effect is radiation tunneling as viewed classically.

As shown in [5], both tunneling and interference can become important only when the spacing between radiating bodies separated by vacuum is less than about $\lambda_{0,\text{max}}(T_3)$, which is the wavelength in vacuum at the maximum blackbody spectral emissive power at the sink temperature T_3. From Wien's displacement law $\lambda_{0,\text{max}}(T_3) = C_3/T_3$ [Eq. (2-22)]. The tunneling and interference effects also depend on temperature. Even for very small spacings on the order of $\lambda_{0,\text{max}}(T_3)$, the effects become very small for dielectrics at normal temperatures, and are only important in certain cryogenic applications where temperatures of a few Kelvins are encountered. Figure 9-5 shows some representative results for conditions giving maximum effects and illustrates the influence of T_1. Note that $\lambda_{\text{max}}(T_3)$ in Fig. 9-5 is the wavelength *in medium 3* and hence is given by C_3/n_3T_3 from (2-40). The conventional solution in the figure is when wave interference and radiation tunneling are neglected in the analysis. In [6] an analysis is carried out for closely spaced *metallic* surfaces. Good agreement of augmented transfer was obtained with published data at $T \approx 315$ K and spacings in the range 1 to 10 μm.

9-3.2 Ray Tracing and the Construction of Images

When mirrorlike reflections occur in enclosures, the procedures of geometric optics can be applied to the radiative exchange process. The basic ideas are outlined in this section. More advanced ideas are in [7, 8].

An incident ray striking a specular surface is reflected in a symmetric fashion about the surface normal so that the angle of reflection is equal in magnitude to the angle of incidence. This fact is used to formulate the concept of *images*. An image is an apparent point of origin for an observed ray. For example, in Fig. 9-6a, an observer views an object in a mirror. To the observer, the object appears to be *behind* the mirror in the position shown by the dotted object. This apparent object is the *image*.

This concept is readily extended to cases in which a series of reflections occur, as shown in Fig. 9-6b. An interesting example of this is the "barber-chair"

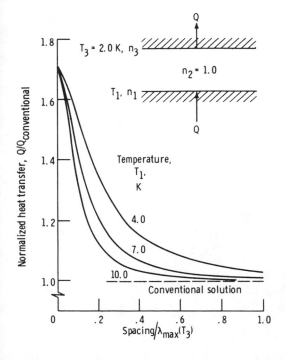

Figure 9-5 Effects of wave interference and radiation tunneling on radiative transfer between two dielectric surfaces; $n_1 = n_3 = 1.25$; $\lambda_{max}(T_3) = 2897.8/(n_3 T_3)$ μm. From [5].

geometry, where there are mirrors on opposite walls of a barber shop. If the mirrors are parallel, a large number of reflections occur and a person receiving a haircut can view many images of himself or herself.

To this point, mirrors have been discussed only with regard to their ability to change the direction of the rays. In the formulation of thermal-radiation problems, the specular surfaces will have a finite absorptivity. They will thus attenuate the energy of the rays.

In addition to reflecting and absorbing energy, mirrors can emit energy. The emission can be conveniently analyzed using an image system rather than the real mirror system. In the image system all radiation acts along straight lines without the complexity of directional changes at each reflecting surface. The attenuation at each surface is accounted for by multiplying the intensity of the ray by the specular reflectivity at each reflection. The emission from three surfaces is illustrated in Fig. 9-6c. For example, emitted energy reaching the viewer from surface 3 is considered to be coming directly from the *image of 3*, with attenuation due to reflections at surfaces 2 and 1 because of passage through these surfaces or their images.

9-3.3 Radiation Transfer by Means of Simple Specular Surfaces for Diffuse Energy Leaving a Surface

As an introduction to the analysis of radiation exchange in an enclosure having some specularly reflecting surfaces, a few examples will be considered for plane

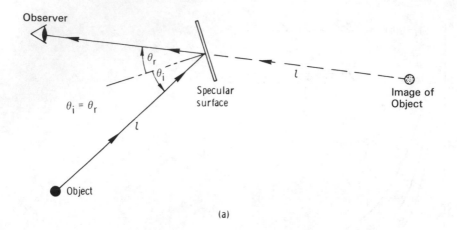

(a)

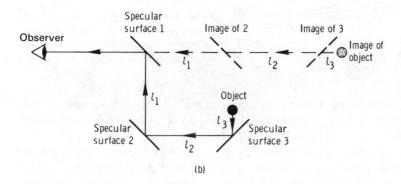

(b)

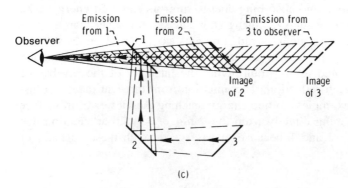

(c)

Figure 9-6 Ray tracing and images formed by specular reflections. (*a*) Image formed by single reflection; (*b*) image formed by multiple reflections; (*c*) contributions due to emission from specular surfaces.

surfaces. The examples will further demonstrate the new features that enter when mirrorlike surfaces are present.

In the enclosure analysis the *emission* from all surfaces is assumed *diffuse*. This is a fairly good assumption in most cases, as can be shown by the electromagnetic-theory predictions of the emissivity of specular surfaces (Fig. 4-6). The reflected energy is assumed either diffuse or specular. In the enclosure theory developed later, the *diffuse reflected energy is combined with the diffuse emitted energy*, and it is then necessary to know how the transfer of this diffuse energy is influenced by the presence of specularly reflecting surfaces. The exchange factors F^s that will be obtained by considering the presence of the specular surfaces will already include the specularly reflected energy, which has a directional behavior in contrast to the diffuse energy. Hence it is necessary to consider here *only* the transfer of *diffuse energy* leaving a surface.

Figure 9-7*a* shows a diffusely emitting and reflecting plane surface A_1 facing a specularly reflecting plane surface A_2. Surface 1 cannot directly view itself; the ordinary configuration factor from any part of A_1 to any other part of A_1 is thus zero. However, if A_2 is specular, then A_1 can view its image, and a path exists, by means of a reflection from the specular surface A_2, for diffuse radiation to travel from the differential area dA_1 to dA_1^*. From Fig. 9-7*a*, it is evident by ray tracing that the radiation arriving at dA_1^* from dA_1 appears to come from the image $dA_{1(2)}$. Thus the geometric configuration factor between dA_1 and dA_1^* resulting from one reflection can be obtained as $dF_{d1(2)-d1*}$ for diffuse radiation leaving dA_1. The subscript notation refers to a factor from the image of dA_1 (as seen in A_2) to dA_1^*.

There are points of similarity that should be noted when comparing the specular and diffuse reflecting cases. When A_1 and A_2 in Fig. 9-7*a* are both diffuse reflectors, the diffuse radiation leaving dA_1 is received at dA_1^* by means of diffuse reflection from A_2 and is governed by dF_{d1-d2} and then dF_{d2-d1} for each element dA_2 on A_2. The portion of the diffuse energy from dA_1 that reaches dA_1^* after one reflection from A_2 is the following, for the two cases of diffuse and specular A_2, respectively:

$$d^2 Q_{d1-d1*(2)} = dA_1 \, q_{o,1} \, \rho_2 \int_{A_2} dF_{d1-d2} \, dF_{d2-d1*}$$

$$d^2 Q_{d1-d1*(2)} = dA_1 \, q_{o,1} \, \rho_{s,2} \, dF_{d1(2)-d1*}$$

This reveals that, for $\rho_2 = \rho_{s,2}$, the difference in the two exchanges is incorporated in the configuration factors resulting from the nature of the reflection being considered; this is a purely geometric effect.

Figure 9-7*b* describes the *diffuse* radiation from dA_1 that reaches the entire area A_1 by means of one specular reflection. The reflected radiation appears to originate from the image $dA_{1(2)}$. Thus the geometric configuration factor involved from dA_1 to A_1 is $F_{d1(2)-1}$. From the symmetry revealed by the dot-dash lines in

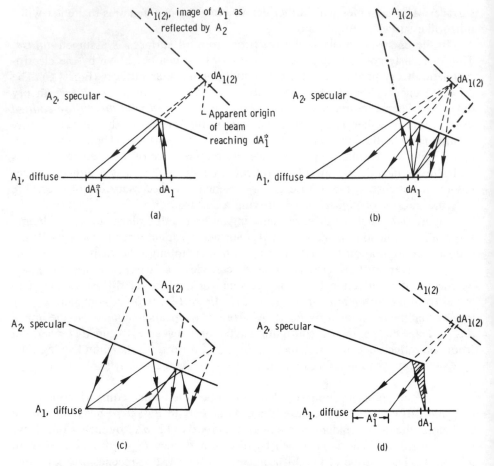

Figure 9-7 Radiation between a diffuse surface and itself by means of a specular surface. (*a*) Radiation between two differential areas with one intermediate specular reflection; (*b*) radiation from differential area to finite area by means of one intermediate specular reflection; (*c*) radiation from finite area that is reflected back to that area by means of one specular reflection; (*d*) radiation from dA_1 can reach only a portion of A_1 by means of specular reflection from A_2.

Fig. 9-7*b*, $F_{d1(2)-1} = F_{d1-1(2)}$. Thus the radiative transfer can also be expressed as a configuration factor from the first surface to the image of the second surface.

Figure 9-7*c* shows several typical rays leaving A_1 that are reflected back to A_1. These rays appear to originate from the image $A_{1(2)}$. The configuration factor from A_1 back to itself by means of one specular reflection is then $F_{1(2)-1}$. In this instance all of the image $A_{1(2)}$ is visible in A_2 from any position on A_1. In some instances this will not be true. An example is shown in Fig. 9-7*d*. The radiation from dA_1 has to be within the limited range of solid angle shown shaded in order for the radiation to be reflected back to A_1. The geometric configuration factor between dA_1 and A_1 is still $F_{d1(2)-1}$, but this factor is evaluated only over the portion of A_1 that receives reflected rays. $F_{d1(2)-1}$ is the factor by which $dA_{1(2)}$ views A_1, and it must be kept in mind that *the view may be a partial one*.

The factor from dA_1 to A_1 will have a different value as the location of dA_1 along A_1 is changed. The fact that the view between dA_1 and A_1 varies with the position of dA_1 along A_1 means that the energy from A_1 that is reflected back to A_1 will have a nonuniform distribution along A_1. The reflection of some of this energy from A_1 will provide a nonuniform q_o from A_1; this violates the assumption in the enclosure theory of uniform q_o from each surface. When partial images are present, extra caution should be exercised to subdivide the enclosure area into sufficiently small portions that the accuracy of the solution is adequate.

Now consider the geometry when there are more specular surfaces involved in the radiation exchange. This provides multiple specular reflections. At each reflection, the radiation is modified by the ρ_s of the reflecting surface. Figure 9-8 shows two specular surfaces. Energy is emitted diffusely from A_2 and travels to A_1. The fraction arriving at dA_1 is given by the diffuse factor dF_{2-d1}. This direct path is illustrated in Fig. 9-8a. A portion of the energy intercepted by A_1 will be reflected back to A_2 and then reflected back again to A_1. Hence, A_2 views dA_1 not only directly but also by means of an image formed by two reflections. This image is constructed in Fig. 9-8b. First the image $A_{1(2)}$ of A_1 reflected in A_2 is drawn. Then A_2 is reflected into this image to form $A_{2(1-2)}$. The notation $A_{2(1-2)}$ is read as the image of area 2 formed by reflections in area 1 and area 2 (in that order). The radiation paths and the shaded area shown in Fig. 9-8b reveal that the solid angle within which radiation leaving A_2 reaches dA_1 by means of two reflections is the same as the solid angle by which dA_1 views the image $A_{2(1-2)}$. Thus the configuration factor involved for two reflections is $dF_{2(1-2)-d1}$. This is interpreted as the factor from the image of surface 2 formed by reflections in surfaces 1 and 2 (in that order) to area element $d1$.

Consider the possibility of additional images. The geometric factor involved is always found by viewing dA_1 from the appropriate reflected image of A_2 as seen through the surface A_2 and all intermediate images. In the case of Fig. 9-8c, the image of A_2 after four reflections $A_{2(1-2-1-2)}$ cannot view dA_1 by looking through A_2. Hence, there is no radiation leaving A_2 that reaches dA_1 by means of four reflections, and no additional images need be considered.

EXAMPLE 9-4 The infinitely long (normal to the cross section) groove shown in Fig. 9-9 has specularly reflecting sides that emit diffusely. What fraction of the energy emitted from A_2 reaches the black receiver surface element dA_3 (an infinitely long differential strip)? Express the result in terms of diffuse geometric configuration factors.

Consider first the energy that reaches dA_3 directly from A_2 and by means of an even number of reflections. The fraction of emitted radiation that reaches dA_3 directly from A_2 is dF_{2-d3} as illustrated in Fig. 9-9a. A second portion will be emitted from A_2 to A_1, reflected back to A_2, and then reflected to dA_3. From the diagram of images in Fig. 9-9b, only part of the reflected image $A_{2(1-2)}$ can be viewed by dA_3 through A_2. The fraction of emitted energy reach-

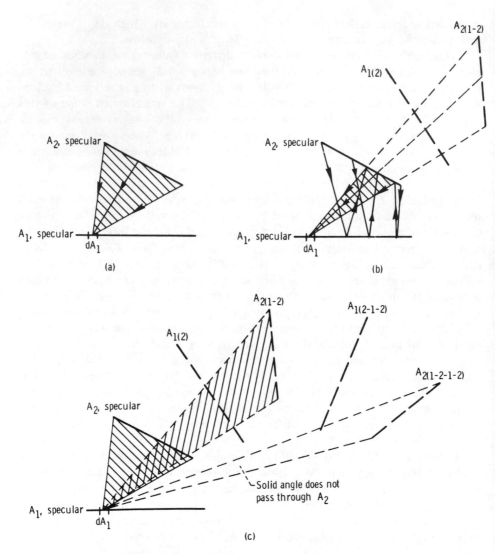

Figure 9-8 Radiant interchange between two specular reflecting surfaces. (*a*) Energy emitted from A_2 that directly reaches dA_1; (*b*) energy emitted from A_2 that reaches dA_1 after two reflections; (*c*) none of the energy emitted by A_2 reaches dA_1 by means of four reflections.

ing dA_3 by this path is the configuration factor *evaluated only over the part of $A_{2(1-2)}$ visible to dA_3* multiplied by the two specular reflectivities, $\rho_{s,1}\rho_{s,2}$ $dF_{2(1-2)-d3}$. This is not an ordinary configuration factor but *takes into account the view through the image system.* In a similar fashion there will be a contribution after two reflections from each of A_1 and A_2. This is illustrated by the shaded solid angle in Fig. 9-9*c*. The third image of A_2, $A_{2(1-2-1-2-1-2)}$, cannot be viewed by dA_3 through A_2 and hence will not make a contribution.

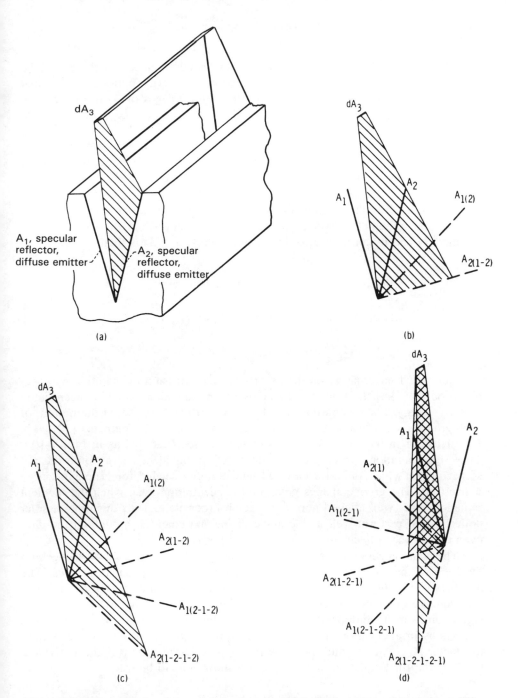

dA_3

A_1, specular
reflector,
diffuse emitter

A_2, specular
reflector,
diffuse emitter

(a)

dA_3

A_1

A_2

$A_{1(2)}$

$A_{2(1\text{-}2)}$

(b)

dA_3

A_1

A_2

$A_{1(2)}$

$A_{2(1\text{-}2)}$

$A_{1(2\text{-}1\text{-}2)}$

$A_{2(1\text{-}2\text{-}1\text{-}2)}$

(c)

dA_3

A_1

A_2

$A_{2(1)}$

$A_{1(2\text{-}1)}$

$A_{2(1\text{-}2\text{-}1)}$

$A_{1(2\text{-}1\text{-}2\text{-}1)}$

$A_{2(1\text{-}2\text{-}1\text{-}2\text{-}1)}$

(d)

Figure 9-9 Diffuse emission from one side of a specularly reflecting groove that reaches a differential-strip receiving area outside groove opening. (*a*) Geometry of direct exchange from A_2 to dA_3; (*b*) geometry of exchange for radiation from A_2 that reaches dA_3 by means of one intermediate reflection from each of A_1 and A_2; (*c*) geometry of exchange for radiation from A_2 that reaches dA_3 by means of two intermediate reflections from each of A_1 and A_2; (*d*) geometry of exchange for radiation from A_2 to A_3 by means of an odd number of reflections.

383

Also, the third image of A_2 cannot view A_1 through A_2 so there will be no additional images of A_2. The fraction of energy emitted by A_2 that reaches dA_3 both directly and by means of the images of A_2 resulting from an even number of reflections is then $dF_{2-d3} + \rho_{s,1}\rho_{s,2}\,dF_{2(1-2)-d3} + \rho_{s,1}^2\rho_{s,2}^2\,dF_{2(1-2-1-2)-d3}$.

Now consider the energy fraction that will reach dA_3 from A_2 by means of an odd number of reflections. Using Fig. 9-9d and arguments similar to those for an even number of reflections results in $\rho_{s,1}\,dF_{2(1)-d3} + \rho_{s,1}^2\rho_{s,2}\,dF_{2(1-2-1)-d3} + \rho_{s,1}^3\rho_{s,2}^2\,dF_{2(1-2-1-2-1)-d3}$. The first two of the F factors are evaluated over only the portions of the images of A_2 that can be viewed by dA_3.

The fraction of diffuse energy emitted by surface A_2 that reaches dA_3 directly and by all interreflections from both A_1 and A_2 is then

$$\frac{dQ_{2-d3}}{\epsilon_2 \sigma T_2^4 A_2} = dF_{2-d3} + \rho_{s,1}\,dF_{2(1)-d3} + \rho_{s,1}\rho_{s,2}\,dF_{2(1-2)-d3}$$

$$+ \rho_{s,1}^2\rho_{s,2}\,dF_{2(1-2-1)-d3} + \rho_{s,1}^2\rho_{s,2}^2\,dF_{2(1-2-1-2)-d3}$$

$$+ \rho_{s,1}^3\rho_{s,2}^2\,dF_{2(1-2-1-2-1)-d3}$$

Additional information on the absorption and emission of radiation by specular grooves is in [9]. Grooves may be useful in the collection and concentration of solar energy. For this purpose an absorbing surface is placed at the bottom of each groove, which can have sides in the form of parabolic segments [10] as in Fig. 9-10a or in the form of a series of straight sections [11] as in Fig. 9-10b. These configurations provide concentration of the incident energy onto the absorber plane, which provides elevated temperatures needed for efficient energy conversion. The groove shapes give good concentration even when the incident radiation is not well aligned normal to the absorber plate. Thus this type of solar collector will perform well as the angle of the sun changes throughout the day, even though the collector remains in a fixed position.

The notation adopted for the specular configuration factors allows a check on the form of the equations for radiant interchange among specular surfaces. The numbers within the subscripted parentheses of the configuration factors designate the sequence in which reflections have occurred from specular surfaces. The configuration factor is multiplied by a reflectivity for each of these specular surfaces to account for attenuation of energy by absorption at these surfaces. For example, the factor $F_{A(B-C-D)-E}$ is multiplied by $\rho_{s,B}\rho_{s,C}\rho_{s,D}$. Examination of the individual terms in the final equation of Example 9-4 shows this to be true.

9-3.4 Configuration-Factor Reciprocity for Specular Surfaces

Reciprocity relations analogous to those for configuration factors between diffuse surfaces apply for the factors involving specular surfaces under certain conditions. Consider a three-sided isothermal enclosure at temperature T, made up of two

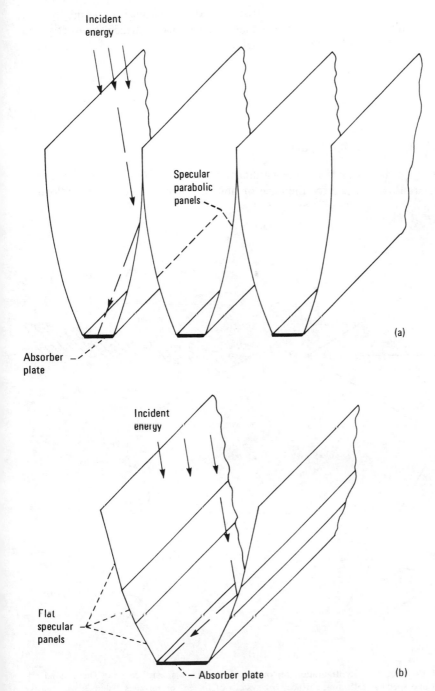

Figure 9-10 Concentrating solar collectors. (*a*) Parabolic trough concentrator; (*b*) concentrator made with flat segments.

black surfaces 1 and 2 and a specular surface 3 of reflectivity $\rho_{s,3}$ (Fig. 9-11a). The energy emitted by black surface 1 that reaches black surface 2 directly and by reflection from specular surface 3 is given by

$$Q_{1-2} = \sigma T^4(A_1F_{1-2} + A_1\rho_{s,3}F_{1(3)-2}) \tag{9-8}$$

The energy leaving surface 2 and reaching surface 1 directly and by specular reflection from surface 3 is

$$Q_{2-1} = \sigma T^4(A_2F_{2-1} + A_2\rho_{s,3}F_{2(3)-1}) \tag{9-9}$$

Since $A_1F_{1-2} = A_2F_{2-1}$ and, for the isothermal enclosure, $Q_{2-1} = Q_{1-2}$, the reciprocity relation is obtained for the case of one specular surface in the enclosure:

$$A_1F_{1(3)-2} = A_2F_{2(3)-1} \tag{9-10}$$

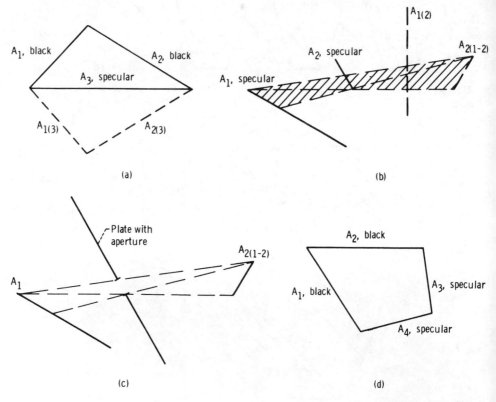

Figure 9-11 Reciprocity of configuration factors involving specular surfaces. (a) Three-sided enclosure with one specular reflecting surface; (b) system of images of surfaces 1 and 2; (c) energy-exchange analog of image system in Fig. 9-11b; (d) enclosure with two specular and two black surfaces.

This can also be deduced from the symmetry about A_3 in Fig. 9-11a and the reciprocity relations for diffuse configuration factors.

A second type of reciprocity relation exists for configuration factors involving specular surfaces. To derive this relation, we examine the energy exchange between two surfaces A_1 and A_2 contained within an isothermal enclosure. If both surfaces are specular, the image system shown in Fig. 9-11b can be constructed for the case of radiation from surface 2 to 1 by means of a reflection at surface 1 and at surface 2. For any such system, an analogous system can be constructed in which a plate with an aperture is substituted for the restraints on the ray paths that are present, as done in Fig. 9-11c. The aperture is placed to allow passage of only those rays that pass through the image system by which $A_{2(1-2)}$ can view at least a portion of A_1 through A_2 and $A_{1(2)}$.

The emitted energy leaving specular surface A_2 in the analog system and absorbed by A_1 is

$$Q_{2(1-2)-1} = Q_{e,2}\rho_{s,1}\rho_{s,2}F_{2(1-2)-1}\alpha_1 = A_{2(1-2)}\epsilon_2\sigma T^4\rho_{s,1}\rho_{s,2}F_{2(1-2)-1}\epsilon_1 \tag{9-11}$$

for the assumed gray surfaces where $\alpha_1 = \epsilon_1$. The reflectivities account for the reduction in energy by the two intermediate specular reflections. $F_{2(1-2)-1}$ is the diffuse-surface configuration factor computed for the constrained paths passing through the aperture (see Example 9-5). But these paths are exactly those through the image system, so this is also the specular configuration factor. Similarly, the energy absorbed for the reverse path is

$$Q_{1-2(1-2)} = A_1\epsilon_1\sigma T^4\rho_{s,2}\rho_{s,1}F_{1-2(1-2)}\epsilon_2 \tag{9-12}$$

Equating the energy exchanges in either direction between A_1 and $A_{2(1-2)}$ for the isothermal enclosure results in the following reciprocity relation:

$$A_1F_{1-2(1-2)} = A_{2(1-2)}F_{2(1-2)-1} = A_2F_{2(1-2)-1} \tag{9-13}$$

For nongray surfaces this will still apply as can be shown by considering the energy in each spectral region $d\lambda$. By generalizing for many intermediate reflections from surfaces A, B, C, D, and so forth, Eq. (9-13) can be written as

$$A_1F_{1-2(A-B-C-D\,\cdots)} = A_2F_{2(A-B-C-D\,\cdots)-1} \tag{9-14}$$

For two-dimensional areas, the crossed-string method (Sec. 6-4.3) can be used to obtain the configuration factors. For example, in Fig. 9-11b the A_2 and $A_{1(2)}$ are regarded as apertures in the view between A_1 and $A_{2(1-2)}$. The $F_{1-2(1-2)}$ is found by having the crossed and uncrossed strings pass through these apertures.

EXAMPLE 9-5 A black surface A_1 faces a smaller parallel mirror A_2 as in Fig. 9-12. Compute the configuration factor $F_{1-1(2)}$ between A_1 and the image of A_1 formed by one specular reflection in A_2. The surfaces are infinitely long in the direction normal to the plane of the drawing.

The factor is computed from the integral $F_{1-1(2)} = (1/A_1)\int_{A_1} F_{d1-1(2)}\,dA_1$. Consider the element dA_1 at location x on A_1. The configuration factor for

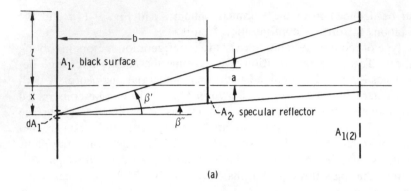

(a)

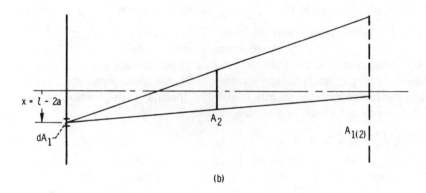

(b)

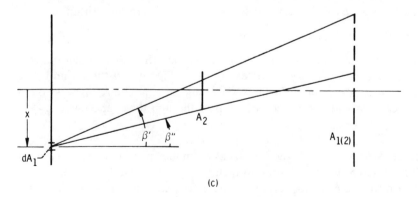

(c)

Figure 9-12 Configuration-factor computation involving partial views of surface and image (Example 9-5). (*a*) Portion of $A_{1(2)}$ in view from dA_1 through entire A_2; (*b*) limiting x for portion of $A_{1(2)}$ to be in view through entire A_2; (*c*) portion of $A_{1(2)}$ in view through part of A_2.

radiation from dA_1 to the portion of $A_{1(2)}$ in view through A_2 is (see Example 6-2)

$$F_{d1-1(2)} = \frac{1}{2} (\sin \beta' - \sin \beta'') = \frac{1}{2} \left[\frac{x+a}{\sqrt{(x+a)^2 + b^2}} - \frac{x-a}{\sqrt{(x-a)^2 + b^2}} \right]$$

This is valid until position $x = l - 2a$ is reached (Fig. 9-12b). For larger x values the geometry is as shown in Fig. 9-12c. Then

$$F_{d1-1(2)} = \frac{1}{2} (\sin \beta' - \sin \beta'') = \frac{1}{2} \left[\frac{x+l}{\sqrt{(x+l)^2 + 4b^2}} - \frac{x-a}{\sqrt{(x-a)^2 + b^2}} \right]$$

The desired configuration factor is

$$F_{1-1(2)} = \frac{1}{2l} 2 \int_0^l F_{d1-1(2)} \, dx$$

$$= \frac{1}{l} \left\{ \frac{1}{2} \int_0^{l-2a} \left[\frac{x+a}{\sqrt{(x+a)^2 + b^2}} - \frac{x-a}{\sqrt{(x-a)^2 + b^2}} \right] dx \right.$$

$$\left. + \frac{1}{2} \int_{l-2a}^l \left[\frac{x+l}{\sqrt{(x+l)^2 + 4b^2}} - \frac{x-a}{\sqrt{(x-a)^2 + b^2}} \right] dx \right\}$$

The integrations are carried out, with the result

$$F_{1-1(2)} = \sqrt{1 + \left(\frac{b}{l}\right)^2} - \sqrt{\left(1 - \frac{a}{l}\right)^2 + \left(\frac{b}{l}\right)^2}$$

This result can be found more easily by the crossed-string method. The surface views its image through the aperture of A_2. Using the crossed-string method (Sec. 6-4.3), the crossed and uncrossed strings are passed through the aperture (Fig. 9-13). Then

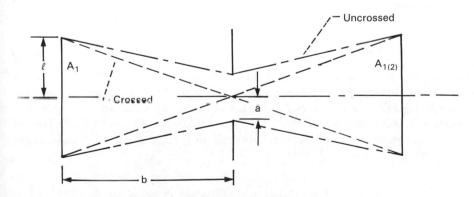

Figure 9-13 Geometry for the crossed-string method.

$$2lF_{1-1(2)} = \frac{1}{2}\left(\sum \text{crossed} - \sum \text{uncrossed}\right)$$

$$= \frac{1}{2}\left[4\sqrt{l^2 + b^2} - 4\sqrt{(l-a)^2 + b^2}\right]$$

The $F_{1-1(2)}$ is thus the same as from the previous integration method.

Reciprocity will now be considered for a situation where there is more than one specular surface in an isothermal enclosure at temperature T. For simplicity an enclosure such as Fig. 9-11d is used, where there are two specular and two black surfaces. If the heat exchange is considered between the two black surfaces by direct exchange and by all specular reflection paths, the following relations result:

$$\frac{Q_{1-2}}{\sigma T^4} = A_1(F_{1-2} + \rho_{s,3}F_{1(3)-2} + \rho_{s,4}F_{1(4)-2}$$

$$+ \rho_{s,3}\rho_{s,4}F_{1(3-4)-2} + \cdots + \rho_{s,3}^m\rho_{s,4}^n F_{1(3^m-4^n)-2} + \cdots) \tag{9-15a}$$

$$\frac{Q_{2-1}}{\sigma T^4} = A_2(F_{2-1} + \rho_{s,3}F_{2(3)-1} + \rho_{s,4}F_{2(4)-1}$$

$$+ \rho_{s,3}\rho_{s,4}F_{2(4-3)-1} + \cdots + \rho_{s,3}^m\rho_{s,4}^n F_{2(4^n-3^m)-1} + \cdots) \tag{9-15b}$$

The shorthand notation (3^m-4^n) means m reflections in 3 and n in 4; hence $F_{1(3^m-4^n)-2}$ is the configuration factor to area 2 from the image of 1 formed by these m and n reflections. For an isothermal enclosure $Q_{1-2} = Q_{2-1}$, so Eq. (9-15) can be written as

$$\frac{Q_{1-2}}{\sigma T^4} = \frac{Q_{2-1}}{\sigma T^4} = A_1 F_{1-2}^s = A_2 F_{2-1}^s \tag{9-16}$$

where the F^s are *exchange factors* equal to the quantities in parentheses in (9-15a,b), that is,

$$F_{1-2}^s = F_{1-2} + \rho_{s,3}F_{1(3)-2} + \rho_{s,4}F_{1(4)-2} + \rho_{s,3}\rho_{s,4}F_{1(3-4)-2}$$

$$+ \cdots + \rho_{s,3}^m\rho_{s,4}^n F_{1(3m-4n)-2} + \cdots \tag{9-17}$$

The *exchange factor* is the fraction of diffuse energy leaving a surface that arrives at a second surface both directly and by all possible intermediate *specular* reflections. Equation (9-17) gives F_{1-2}^s in terms of energy going from images of 1 to surface 2. An alternative form of F_{1-2}^s (or similarly of F_{2-1}^s) can be derived in terms of the radiation from surface 1 directly to surface 2 and by means of reflections to the images of 2. From (9-16),

$$F_{1-2}^s = \frac{A_2}{A_1}F_{2-1}^s = \frac{A_2}{A_1}(F_{2-1} + \rho_{s,3}F_{2(3)-1} + \rho_{s,4}F_{2(4)-1} + \rho_{s,3}\rho_{s,4}F_{2(4-3)-1} + \cdots)$$

Equation (9-14) is applied to each of the F factors in the series. The area ratio then cancels and the desired result is

$$F^s_{1-2} = F_{1-2} + \rho_{s,3}F_{1-2(3)} + \rho_{s,4}F_{1-2(4)} + \rho_{s,3}\rho_{s,4}F_{1-2(4-3)} + \cdots \qquad (9\text{-}18)$$

Now looking at (9-15) in more detail, since $A_1F_{1-2} = A_2F_{2-1}$, and from (9-10) for one reflection $A_1F_{1(3)-2} = A_2F_{2(3)-1}$ and $A_1F_{1(4)-2} = A_2F_{2(4)-1}$, the equality in (9-16) reduces to (after dividing by $\rho_{s,3}\rho_{s,4}$)

$$A_1\left(F_{1(3-4)-2} + \cdots + \rho^{m-1}_{s,3}\rho^{n-1}_{s,4}F_{1(3^m-4^n)-2} + \cdots\right)$$

$$= A_2\left[F_{2(4-3)-1} + \cdots + \rho^{m-1}_{s,3}\rho^{n-1}_{s,4}F_{2(4^n-3^m)-1} + \cdots\right) \qquad (9\text{-}19)$$

This equality must hold in the limit as $\rho_{s,3}$ and $\rho_{s,4}$ approach zero so that

$$A_1F_{1(3-4)-2} = A_2F_{2(4-3)-1} \qquad (9\text{-}20)$$

which is a geometric property of the system. A continuation of this reasoning leads to the *general reciprocity relation*

$$A_1F_{1(A-B-C-D\ldots)-2} = A_2F_{2(\ldots D-C-B-A)-1} \qquad (9\text{-}21)$$

An additional relation is found by combining (9-14) and (9-21) to give

$$A_1F_{1(A-B-C-D\ldots)-2} = A_2F_{2(\ldots D-C-B-A)-1} = A_1F_{1-2(\ldots D-C-B-A)} \qquad (9\text{-}22)$$

which shows that

$$F_{1(A-B-C-D\ldots)-2} = F_{1-2(\ldots D-C-B-A)} \qquad (9\text{-}23)$$

This can also be deduced directly from the fact that an image system can be constructed either starting with the real surface 1 and working toward image $2(\cdots D-C-B-A)$, or starting with image $1(A-B-C\ D\cdots)$ and working toward real surface 2; the geometry of the construction will be identical in both systems. Thus, the configuration factors between the initial and final surfaces must be the same.

9-4 NET-RADIATION METHOD IN ENCLOSURES HAVING SPECULAR AND DIFFUSE REFLECTING SURFACES

9-4.1 Enclosures with Plane Surfaces

In this section the radiation exchange theory will be developed in an enclosure composed of specularly and diffusely reflecting surfaces. The enclosure is composed of N surfaces, where d of the surfaces are diffuse reflectors and $N - d$ are specular. All the surfaces emit diffusely and are gray. Let the diffusely reflecting surfaces be numbered 1 through d and the specularly reflecting surfaces $d + 1$ through N. If there are no external sources of unidirectional energy entering through a window and striking a specular surface, all the energy within the enclosure originates by surface emission that is diffuse. At each diffuse surface the reflected diffuse energy is combined with the emitted energy to form q_o, which is entirely diffuse. At each specular surface the only diffuse energy leaving is $\epsilon \sigma T^4$, since the reflected energy has a directional character. The transport between surfaces of the diffuse energies q_o and $\epsilon \sigma T^4$ is determined by the exchange factors F^s. The F^s_{A-B} gives the fraction of diffuse energy leaving surface A and arriving at surface B by the direct path and by all possible paths involving intermediate specular reflections. The reflected energy from the specular surfaces is *already included* in the F^s and hence does not have to be considered after the F^s have been obtained for the enclosure surfaces. Hence, by accounting for the fractions of q_o from the diffusely reflecting surfaces and fractions of $\epsilon \sigma T^4$ from the specularly reflecting surfaces that reach a particular surface, all of the incident energy has been accounted for, including the effects of both diffuse and specular reflections. Then at any surface, the incident energy is

$$q_{i,k} A_k = \sum_{j=1}^{d} q_{o,j} A_j F^s_{j-k} + \sigma \sum_{j=d+1}^{N} \epsilon_j T_j^4 A_j F^s_{j-k} \qquad 1 \leqslant k \leqslant N$$

After applying reciprocity (9-16), the areas cancel and $q_{i,k}$ becomes

$$q_{i,k} = \sum_{j=1}^{d} q_{o,j} F^s_{k-j} + \sigma \sum_{j=d+1}^{N} \epsilon_j T_j^4 F^s_{k-j} \qquad 1 \leqslant k \leqslant N \tag{9-24}$$

A set of enclosure relations will now be derived in terms of the q_i and q_o for the surfaces. Then the q_i and q_o will be eliminated to obtain a very convenient set of equations relating the Q directly to the T. This is given by Eq. (9-35). For the diffuse surfaces it is convenient to use the heat-balance equations in the form of (7-12) and (7-13):

$$Q_k = q_k A_k = (q_{o,k} - q_{i,k}) A_k \qquad 1 \leqslant k \leqslant d \tag{9-25}$$

$$q_{o,k} = \epsilon_k \sigma T_k^4 + (1 - \epsilon_k) q_{i,k} \qquad 1 \leqslant k \leqslant d \tag{9-26}$$

For the specular reflecting surfaces, although these same heat balances apply, the $q_{o,k}$ is eliminated as this is composed of a combination of diffuse emission $\epsilon_k \sigma T_k^4$ and specular reflection $(1 - \epsilon_k) q_{i,k}$ and hence has a directional character that is not conveniently dealt with. Eliminating the $q_{o,k}$ from (9-25) and (9-26) gives

$$Q_k = A_k \epsilon_k (\sigma T_k^4 - q_{i,k}) \qquad d+1 \leqslant k \leqslant N \tag{9-27}$$

Equations (9-24) to (9-27) can be combined in various ways to obtain convenient equations to calculate the desired unknowns, depending on what quantities are specified. Consider the case where the temperatures are specified for all the surfaces and it is desired to obtain the net external energy Q_k added to each surface. Equation (9-24) is substituted into (9-26) to eliminate $q_{i,k}$ and obtain the following equation for each diffuse surface:

$$q_{o,k} - (1 - \epsilon_k) \sum_{j=1}^{d} q_{o,j} F^s_{k-j} = \epsilon_k \sigma T^4_k + (1 - \epsilon_k) \sigma \sum_{j=d+1}^{N} \epsilon_j T^4_j F^s_{k-j} \qquad 1 \leqslant k \leqslant d$$

$$(9\text{-}28)$$

This set of equations is solved for the q_o for the diffuse surfaces. This is somewhat simpler than for an enclosure having all diffuse surfaces, as there are now only d equations to solve simultaneously, rather than N equations. For each specular surface, the q_o for the diffuse surfaces are used to obtain $q_{i,k}$ directly from (9-24):

$$q_{i,k} = \sum_{j=1}^{d} q_{o,j} F^s_{k-j} + \sigma \sum_{j=d+1}^{N} \epsilon_j T^4_j F^s_{k-j} \qquad d+1 \leqslant k \leqslant N \qquad (9\text{-}29)$$

The net external energy added to each diffuse surface is obtained by eliminating $q_{i,k}$ from (9-25) and (9-26),

$$Q_k = A_k \frac{\epsilon_k}{1 - \epsilon_k} (\sigma T_k^4 - q_{o,k}) \qquad 1 \leqslant k \leqslant d \qquad (9\text{-}30)$$

and the Q_k to each specular surface is found from (9-27). Equations (9-27) to (9-30) are general energy-interchange relations for enclosures composed of diffuse reflecting surfaces and specular reflecting surfaces.

If the kth diffuse surface is black, then $q_{o,k} = \sigma T^4_k$ and $1 - \epsilon_k = 0$, so that (9-30) is indeterminate. In this case, from (9-25),

$$Q_k = A_k(\sigma T_k^4 - q_{i,k}) \qquad (9\text{-}31)$$

where $q_{i,k}$ is found from (9-29) with $1 \leqslant k \leqslant d$.

If the heat input Q_k rather than T_k is specified for a diffuse surface $1 \leqslant k \leqslant d$, then T_k is unknown in (9-28). Equation (9-30) can be used to eliminate this unknown in terms of $q_{o,k}$ and the known Q_k.

If the heat input Q_k is specified for a specular surface, $d + 1 \leqslant k \leqslant N$, then one of the T^4_j in the last term of (9-28) will be unknown. Equation (9-27) is combined with (9-29) to eliminate $q_{i,k}$, which gives

$$\sigma T_k^4 - \frac{Q_k}{A_k \epsilon_k} = \sum_{j=1}^{d} q_{o,j} F^s_{k-j} + \sigma \sum_{j=d+1}^{N} \epsilon_j T^4_j F^s_{k-j} \qquad d+1 \leqslant k \leqslant N \qquad (9\text{-}32)$$

Since Q_k is known, (9-32) can be combined with (9-28) to yield a simultaneous set of equations to determine the q_o of the diffusely reflecting surfaces and the T for the specularly reflecting surfaces having specified Q.

If *all the surfaces are specular*, a simultaneous solution is not required. The $q_{i,k}$ are given by (9-29) as

$$q_{i,k} = \sigma \sum_{j=1}^{N} \epsilon_j T^4_j F^s_{k-j} \qquad 1 \leqslant k \leqslant N \qquad (9\text{-}33)$$

and the Q_k are then found from (9-27). By substituting (9-33) into (9-27) the Q_k can be found directly from the specified surface temperatures:

$$Q_k = \sigma A_k \epsilon_k (T_k^4 - \sum_{j=1}^{N} \epsilon_j T_j^4 F_{k-j}^s) \tag{9-34}$$

A *very useful form* for the enclosure equations is found by using (9-30) and (9-27) to eliminate q_i and q_o from (9-28) and (9-29). This gives a set of equations, *all of the same form,* that *directly relate* the Q and T:

$$\frac{1}{\epsilon_k}\frac{Q_k}{A_k} - \sum_{j=1}^{d} \frac{Q_j}{A_j}\frac{1-\epsilon_j}{\epsilon_j} F_{k-j}^s = \sigma T_k^4 - \sigma \sum_{j=1}^{d} T_j^4 F_{k-j}^s - \sigma \sum_{j=d+1}^{N} \epsilon_j T_j^4 F_{k-j}^s \qquad 1 \leqslant k \leqslant N \tag{9-35}$$

Equation (9-34) is the special case of (9-35) for $d = 0$.

Equation (9-35) can be used to obtain some relations between the F^s exchange factors analogous to the relation for diffuse surfaces $\sum_{j=1}^{N} F_{k-j} = 1$. Consider the situation when the entire enclosure is at uniform temperature. Then there is no net energy exchange and all the Q are zero, so that (9-35) reduces to

$$\sum_{j=1}^{d} F_{k-j}^s + \sum_{j=d+1}^{N} \epsilon_j F_{k-j}^s = 1 \tag{9-36}$$

If all the surfaces in the enclosure are specular ($d = 0$), this further reduces to

$$\sum_{j=1}^{N} \epsilon_j F_{k-j}^s = \sum_{j=1}^{N} (1 - \rho_{s,j}) F_{k-j}^s = 1 \tag{9-37}$$

The set of enclosure equations (9-35) is not difficult to solve after the exchange factors have been found; determining these factors, however, may not be easy, depending on the complexity of the enclosure geometry. To illustrate the calculations, some specific enclosures are now considered.

Figure 9-14 shows an enclosure composed of three plane surfaces at different uniform specified temperatures; two sides are diffuse reflectors, and the third is a specular reflector. In Fig. 9-14a, the energy arriving at surface 1 comes directly from the diffuse surfaces 2 and 3 without any intermediate specular reflections. Hence, $F_{2-1}^s = F_{2-1}$ and $F_{3-1}^s = F_{3-1}$. By applying reciprocity to each side of these equations, $F_{1-2}^s = F_{1-2}$ and $F_{1-3}^s = F_{1-3}$. For surface 2 the incoming radiation is composed of four parts that originate as shown in Fig. 9-14b. The first is the diffuse energy originating from A_3 and going directly to A_2, which is $q_{o,3}A_3 F_{3-2}$. The remaining three parts arrive by means of A_1 and consist of an emitted portion $\epsilon_1 \sigma T_1^4 A_1 F_{1-2}$ and two specularly reflected portions. The latter arise from the energy leaving A_2 and A_3 that is specularly reflected to A_2 and will appear to come from the images $A_{2(1)}$ and $A_{3(1)}$ in Fig. 9-14a. The specularly reflected portions are $q_{o,2}\rho_{s,1}A_2 F_{2(1)-2} + q_{o,3}\rho_{s,1}A_3 F_{3(1)-2}$. Note that multiple specular reflections cannot occur when only one planar specular surface is present. The specular exchange factors are then

$$F_{1-2}^s = F_{1-2} \qquad F_{2-2}^s = \rho_{s,1}F_{2(1)-2} \qquad F_{3-2}^s = F_{3-2} + \rho_{s,1}F_{3(1)-2}$$

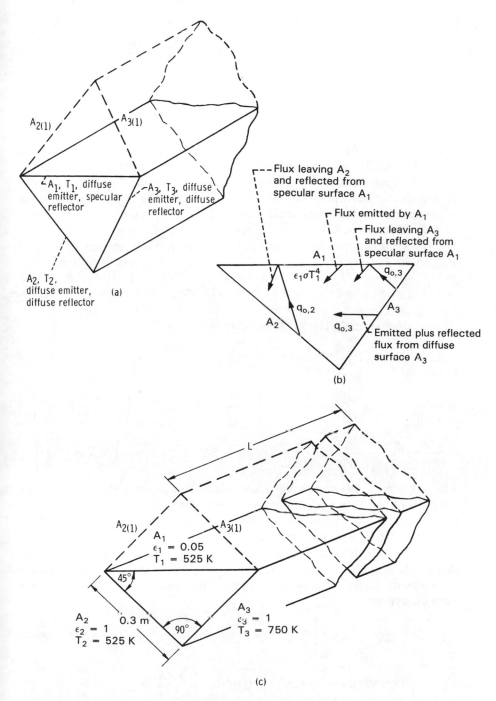

Figure 9-14 Enclosure having one specular reflecting surface and two surfaces that are diffuse reflectors. (*a*) General geometry; (*b*) energy fluxes that contribute to flux incident upon A_2; (*c*) enclosure for Example 9-6.

By using reciprocity, one obtains

$$F^s_{2-1} = F_{2-1} \qquad F^s_{2-2} = \rho_{s,1} F_{2-2(1)} \qquad F^s_{2-3} = F_{2-3} + \rho_{s,1} F_{2-3(1)}$$

Similarly, for surface 3,

$$F^s_{3-1} = F_{3-1} \qquad F^s_{3-2} = F_{3-2} + \rho_{s,1} F_{3-2(1)} \qquad F^s_{3-3} = \rho_{s,1} F_{3-3(1)}$$

A numerical example using these factors will now be given.

EXAMPLE 9-6 An enclosure having three sides is shown in Fig. 9-14c. The length L is sufficiently long that the triangular ends can be neglected in the radiative heat balances. Two of the surfaces are black, and the third is a gray diffuse emitter with emissivity $\epsilon_1 = 0.05$. Determine the heat added per meter of enclosure length to each surface for each of the two cases: (1) area 1 is a diffuse reflector and (2) area 1 is a specular reflector.

The configuration factors are computed first. From symmetry $F_{1-2} = F_{1-3}$. Also $F_{1-2} + F_{1-3} = 1$, so $F_{1-2} = F_{1-3} = \frac{1}{2}$. From reciprocity $F_{2-1} = A_1 F_{1-2}/A_2 = \sqrt{2}/2 = F_{3-1}$. Now $F_{2-1} + F_{2-3} = 1$. Hence $F_{2-3} = 1 - \sqrt{2}/2 = F_{3-2} = F_{2-2(1)} = F_{3-3(1)}$. Finally, $F_{3-2(1)} = F_{2-3(1)} = 1 - F_{3-2} - F_{3-3(1)} = \sqrt{2} - 1$.

For case 1, apply Eq. (7-30) to obtain

$$\frac{Q_1}{0.3\sqrt{2}} \frac{1}{0.05} = \sigma\left(525^4 - \frac{1}{2}525^4 - \frac{1}{2}750^4\right)$$

$$-\frac{Q_1}{0.3\sqrt{2}} \frac{\sqrt{2}}{2} \frac{1-0.05}{0.05} + \frac{Q_2}{0.3} = \sigma\left[-\frac{\sqrt{2}}{2}525^4 + 525^4 - \left(1 - \frac{\sqrt{2}}{2}\right)750^4\right]$$

$$-\frac{Q_1}{0.3\sqrt{2}} \frac{\sqrt{2}}{2} \frac{1-0.05}{0.05} + \frac{Q_3}{0.3} = \sigma\left[-\frac{\sqrt{2}}{2}525^4 - \left(1 - \frac{\sqrt{2}}{2}\right)525^4 + 750^4\right]$$

The solution of these three equations yields the Q per meter of enclosure length as $Q_1 = -144.6$ W, $Q_2 = -2571.8$ W, and $Q_3 = 2716.4$ W. The heat supplied to A_3 is removed from A_1 and A_2. The amount removed from A_1 is small because A_1 is a poor absorber.

For case 2 apply Eq. (9-28) to compute $q_{o,2}$ and $q_{o,3}$. Since $\epsilon_2 = \epsilon_3 = 1$, this yields simply $q_{o,2} = \sigma T_2^4$ and $q_{o,3} = \sigma T_3^4$, which would be expected for the outgoing fluxes from black surfaces. Then Eq. (9-24) yields the q_i for each surface as

$$q_{i,1} = \sigma\left[\frac{1}{2}525^4 + \frac{1}{2}750^4\right] = 11{,}125 \text{ W/m}^2$$

$$q_{i,2} = \sigma\left\{0.05(524)^4 \frac{\sqrt{2}}{2} + 525^4(1 - 0.05)\left(1 - \frac{\sqrt{2}}{2}\right)\right.$$

$$\left. + 750^4\left[1 - \frac{\sqrt{2}}{2} + (1 - 0.05)(\sqrt{2} - 1)\right]\right\} = 13{,}666 \text{ W/m}^2$$

$$q_{i,3} = \sigma \left\{ 0.05(525)^4 \frac{\sqrt{2}}{2} + 525^4 \left[1 - \frac{\sqrt{2}}{2} + (1 - 0.05)(\sqrt{2} - 1) \right] \right.$$

$$\left. + 750^4(1 - 0.05)\left(1 - \frac{\sqrt{2}}{2} \right) \right\} = 8101 \text{ W/m}^2$$

With q_i known for each of the surfaces, Eqs. (9-27) and (9-31) are applied to find Q. This yields, per meter of enclosure length, $Q_1 = -144.6$ W, $Q_2 = -2807.5$ W, and $Q_3 = 2952.1$ W.

Comparing cases 1 and 2 reveals that, by making A_1 specular, the heat transferred from A_3 to A_2 is increased from 2572 to 2808 W, an increase of 9%.

An alternative approach for case 2 is to use Eq. (9-35). This yields the following three equations for the Q's:

$$Q_1 = \epsilon_1 A_1 \sigma (T_1^4 - T_2^4 F_{1-2} - T_3^4 F_{1-3})$$

$$Q_2 = A_2 \sigma [\epsilon_1 T_1^4 F_{2-1} + T_2^4(1 - \rho_{s,1} F_{2(1)-2}) - T_3^4(F_{2-3} + \rho_{s,1} F_{2-3(1)})]$$

$$Q_3 = A_3 \sigma [-\epsilon_1 T_1^4 F_{3-1} - T_2^4(F_{3-2} + \rho_{s,1} F_{3-2(1)}) + T_3^4(1 - \rho_{s,1} F_{3-3(1)})]$$

Substituting values yields the same results as before.

To further demonstrate radiative analysis in an enclosure having some specularly reflecting surfaces, a rectangular geometry is shown in Fig. 9-15. All surfaces are diffuse emitters; two surfaces are diffuse reflectors, while the remaining two are specular. The reflected images are shown dashed. The reflection process continues until all the outer perimeter enclosing the composite of the original enclosure plus the reflected images is made up of either diffuse (or nonreflecting, such as an opening) surfaces or images of diffuse surfaces.

For the enclosure in Fig. 9-15, the first step is to use (9-28) to obtain $q_{o,1}$ and $q_{o,2}$ for the two diffuse areas. To obtain the exchange factors, consider first F_{1-1}^s. Part of the energy leaving A_1 returns to A_1 by three paths: direct reflection from A_3, reflection from A_3 to A_4 and then to A_1, and reflection from A_4 to A_3 and then to A_1. Thus the fraction of the energy leaving A_1 that returns to A_1 is

$$F_{1-1}^s = \rho_{s,3} F_{1(3)-1} + \rho_{s,3}\rho_{s,4} F_{1(3-4)-1} + \rho_{s,4}\rho_{s,3} F_{1(4-3)-1}$$

The $F_{1(3-4)-1}$ is the configuration factor by which $A_{1(3-4)}$ is viewed from A_1 through A_4 and then $A_{3(4)}$, which are the reflection areas by means of which the $A_{1(3-4)}$ image was formed. Similarly, $F_{1(4-3)-1}$ is the factor by which the same area $A_{1(3-4)}$ is viewed from A_1 through A_3 and then $A_{4(3)}$.

Radiation leaving A_2 reaches A_1 along four paths: direct exchange, reflection from A_3, reflection from A_4, and reflection from A_3 to A_4. No energy from A_2 reaches A_1 by means of reflections from A_4 and then A_3. This is because A_1 cannot view the image $A_{2(4-3)}$ through area A_3. This yields

$$F_{2-1}^s = F_{2-1} + \rho_{s,3} F_{2(3)-1} + \rho_{s,4} F_{2(4)-1} + \rho_{s,3}\rho_{s,4} F_{2(3-4)-1}$$

The diffuse energy leaving the specular surface A_3 (and similarly for A_4) consists only of emitted energy $\epsilon_3 A_3 \sigma T_3^4$. There are two paths by which some of this will reach A_1: direct exchange, and by means of specular reflection from A_4. This yields

$$F_{3-1}^s = F_{3-1} + \rho_{s,4} F_{3(4)-1} \qquad F_{4-1}^s = F_{4-1} + \rho_{s,3} F_{4(3)-1}$$

After applying F factor reciprocity, the factors are substituted into (9-28) to yield

$$q_{o,1} - (1 - \epsilon_1)\{q_{o,1}[\rho_{s,3} F_{1-1(3)} + \rho_{s,3}\rho_{s,4}(F_{1-1(3-4)} + F_{1-1(4-3)})]$$

$$+ q_{o,2}(F_{1-2} + \rho_{s,3} F_{1-2(3)} + \rho_{s,4} F_{1-2(4)} + \rho_{s,3}\rho_{s,4} F_{1-2(3-4)})\} \qquad (9\text{-}38)$$

$$= \epsilon_1 \sigma T_1^4 + (1 - \epsilon_1)\sigma[\epsilon_3 T_3^4(F_{1-3} + \rho_{s,4} F_{1-3(4)}) + \epsilon_4 T_4^4(F_{1-4} + \rho_{s,3} F_{1-4(3)})]$$

In a similar fashion, considering $q_{i,2}$ for surface 2 yields

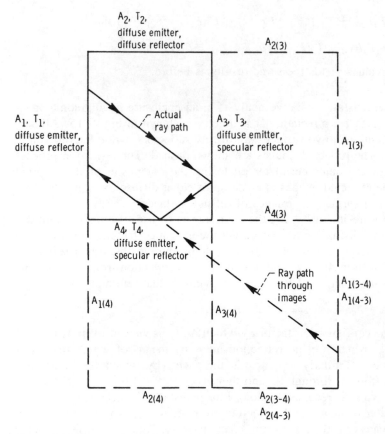

Figure 9-15 Rectangular enclosure and reflected images when two adjacent surfaces are specular reflectors and the other two are diffuse reflectors.

$$q_{o,2} - (1 - \epsilon_2)\{q_{o,1}(F_{2-1} + \rho_{s,3}F_{2-1(3)} + \rho_{s,4}F_{2-1(4)} + \rho_{s,3}\rho_{s,4}F_{2-1(4-3)})$$

$$+ q_{o,2}[\rho_{s,4}F_{2-2(4)} + \rho_{s,3}\rho_{s,4}(F_{2-2(4-3)} + F_{2-2(3-4)})]\} \qquad (9\text{-}39)$$

$$= \epsilon_2 \sigma T_2{}^4 + (1 - \epsilon_2)\sigma[\epsilon_3 T_3{}^4(F_{2-3} + \rho_{s,4}F_{2-3(4)}) + \epsilon_4 T_4{}^4(F_{2-4} + \rho_{s,3}F_{2-4(3)})]$$

Equations (9-38) and (9-39) are solved simultaneously for $q_{o,1}$ and $q_{o,2}$.

For the two specular surfaces, the $q_{i,3}$ and $q_{i,4}$ can be found as soon the q_o for the diffuse surfaces are known. From (9-29),

$$q_{i,3} = q_{o,1}(F_{3-1} + \rho_{s,4}F_{3-1(4)}) + q_{o,2}(F_{3-2} + \rho_{s,4}F_{3-2(4)}) + \epsilon_4 \sigma T_4{}^4 F_{3-4} \qquad (9\text{-}40)$$

$$q_{i,4} = q_{o,1}(F_{4-1} + \rho_{s,3}F_{4-1(3)}) + q_{o,2}(F_{4-2} + \rho_{s,3}F_{4-2(3)}) + \epsilon_3 \sigma T_3{}^4 F_{4-3} \qquad (9\text{-}41)$$

With the q_o for the diffuse surfaces and the q_i for the specular surfaces now known, the Q to maintain the specified temperatures of the surfaces can be found from (9-30) and (9-27). The solution can also be obtained by using Eq. (9-35), which directly relates the Q and T.

Determining the specular exchange factors can become tedious in enclosures with many surfaces. A more direct approach is the Monte Carlo method, wherein the direction of reflected energy can be specified for each incident small "bundle" of energy since it approaches a specular surface from a definite direction (Chap. 11). A similar statistical method involving Markov chain theory has been developed in [12] and extended in [13].

9-4.2 Curved Specular Reflecting Surfaces

In the previous discussion the specular surfaces were planar. Curved specular reflecting surfaces are now considered, and in this instance the geometry of the reflected images can become quite complex. To demonstrate some of the basic ideas, consider the relatively simple case of radiation exchange within a specular tube [14], as in Fig. 9-16.

It is assumed that the imposed temperature or heating conditions depend only on axial position and are independent of location around the tube circumference. To compute the radiative exchange within the tube for axisymmetric heating conditions, the configuration factor between two ring elements on the tube wall is needed. The direct exchange (Fig. 9-16a) is governed by the factor (see Example 7-19), and note that $|\eta - \xi|$ in that example is equal to X/D here):

$$dF_{dX_1 - dX} = \left\{ 1 - \frac{(X/D)^3 + 3X/2D}{[(X/D)^2 + 1]^{3/2}} \right\} dX$$

Figure 9-16b illustrates the configuration factor for one reflection. Because of the symmetry of the tube, all the radiation from dX_1 that reaches dX by one reflection has been reflected from a ring element halfway between dX_1 and dX. The ring at $X/2$ is only $dX/2$ wide, so the beam subtending it will spread to a

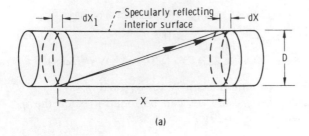

(a)

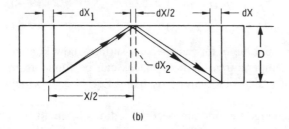

(b)

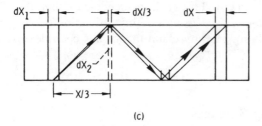

(c)

Figure 9-16 Radiation exchange within specularly reflecting cylindrical tube. (*a*) Direct exchange between two ring elements; (*b*) exchange by one reflection; (*c*) exchange by two reflections.

width dX at the location X. The configuration factor for one reflection is then the factor between dX_1 and the dashed element $dX/2$:

$$dF_{dX_1-dX/2} = \left\{1 - \frac{(X/2D)^3 + 3X/4D}{[(X/2D)^2 + 1]^{3/2}}\right\}\frac{dX}{2}$$

In a similar fashion, the geometric factor for exchange between dX_1 and dX by two reflections is

$$dF_{dX_1-dX/3} = \left\{1 - \frac{(X/3D)^3 + 3X/6D}{[(X/3D)^2 + 1]^{3/2}}\right\}\frac{dX}{3}$$

and for n reflections

$$dF_{dX_1-dX/(n+1)} = \left(1 - \frac{[X/(n+1)D]^3 + 3X/2(n+1)D}{\{[X/(n+1)D]^2 + 1\}^{3/2}}\right)\frac{dX}{n+1}$$

In general, the geometric factor for *any* number of reflections is found by considering the exchange between the originating element (dX_1 in this case) and the element (call it dX_2) from which the *first* reflection is made (the dashed element in Fig. 9-16b and c). This is because the fraction of energy leaving dX_1 in the solid angle subtended by dX_2 remains the same through the succeeding reflections along the path to dX.

At each reflection the energy must be multiplied by the specular reflectivity ρ_s. If all the contributions are summed, the fraction of energy leaving dX_1 that reaches dX by direct exchange and all reflection paths is the specular exchange factor for the inside of a tube that is open at both ends,

$$dF_{dX_1-dX}^s = \sum_{n=0}^{\infty} \rho_s^n \left(1 - \frac{[X/(n+1)D]^3 + 3X/2(n+1)D}{\{[X/(n+1)D]^2 + 1\}^{3/2}}\right) \frac{dX}{n+1} \qquad (9\text{-}42)$$

For a complete heat transfer formulation, the energy entering through the tube ends must be included, which requires the exchange factor from the end openings. This factor is obtained by a derivation similar to (9-42). The energy is considered that leaves an element on the tube wall and travels to the circular disk opening at the end of the tube by a direct path or by one or more reflections along the portion of the tube wall between the element and the end opening. This factor was obtained in [14], where it was shown that the energy exchange in the tube can be divided into two separate parts with the complete solution given by superposition. One part is for the tube being heated along its length, but with the end environments at zero temperature (Fig. 9-17). The outside of the tube is insulated, so the only heat loss is by radiation leaving through the tube ends. The heat balance on an element at x states that the heat addition $q(x)$ is equal to the emitted energy minus that absorbed as a result of incident energy arriving by specular reflections. Since there is no diffuse reflection in this analysis, all of the arriving specular energy is obtained by using the specular exchange factor, Eq. (9-42). The energy equation governing the wall temperature at x is then given by (a gray tube wall is assumed so that $\alpha = \epsilon$)

$$q(x) = \epsilon\sigma T_w^4(x) - \epsilon^2\left[\int_0^x \sigma T_w^4(y)\, dF^s(x-y) + \int_x^L \sigma T_w^4(y)\, dF^s(y-x)\right] \qquad (9\text{-}43)$$

where $dF^s(x-y)$ is given by (9-42) with $X = x - y$. Equation (9-43) is an integral equation that can be solved numerically for $T_w(x)$ if $q(x)$ is specified or can be integrated to give $q(x)$ if $T_w(x)$ is specified.

When the geometry is even slightly more involved than the cylindrical geometry, the reflection patterns can become quite complex. Some examples for radiation within a specular conical cavity and a specular cylindrical cavity with a specular end plane are in [15]; a more generalized treatment of nonplanar reflections is in [16]. An approximate analysis of the transfer through specular passages is in [17]. It is based on estimating the average number of reflections that occur during transmission.

The analyses in [18, 19] for radiation within a cylindrical and a spherical enclosure consider a more general reflection behavior obtained by dividing the reflectivity into specular and diffuse components. These components can be treated

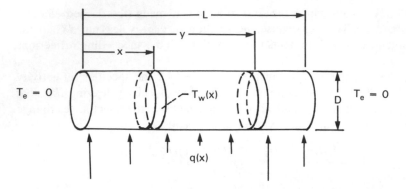

Figure 9-17 Heated tube with internal specular reflections; the outside surface of the tube is insulated.

using the methods in the enclosure theory given previously in this chapter. The governing integral equation was linearized by assuming a small axial temperature variation relative to the average cylinder temperature. This model was for application to radiation within high-temperature porous materials. The results in [19] show that for a spherical cavity the many specular reflections tend to eliminate directional effects and cause the radiation to behave in a diffuse manner.

EXAMPLE 9-7 A cylindrical cavity open at one end has a specularly reflecting cylindrical wall and base (Fig. 9-18a). Determine the fraction of radiation from ring element dX_1 that reaches dX by means of one reflection from the base with reflectivity $\rho_{s,1}$ and one reflection from the cylindrical wall with reflectivity $\rho_{s,2}$.

As shown in Fig. 9-18b, for this geometry the reflected radiation from the base can be regarded as originating from an image of dX_1. The second reflection will occur from an element of width $dX/2$ located midway between the image dX_1 and dX. The desired radiation fraction is given by the configuration factor from the image dX_1 to the dashed ring area $dX/2$ multiplied by the two reflectivities:

$$\rho_{s,1}\rho_{s,2}dF_{dX_1-dX} = \rho_{s,1}\rho_{s,2}\left(1 - \frac{[(X + X_1)/2D]^3 + 3(X + X_1)/4D}{\{[(X + X_1)/2D]^2 + 1\}^{3/2}}\right)\frac{dX}{2}$$

Another type of curved specular surface of practical importance is a paraboloidal mirror such as that in a solar furnace. The mirror axis is aligned in the direction of the sun as in Fig. 9-19a, and a receiver is placed at the focal plane. It is desired to estimate the receiver temperature.

When *parallel* rays reflect from a perfect concentrator in the form of a paraboloid, all the rays pass through the focal point. The sun's rays, however, are not quite parallel. At the earth-sun distance the solar diameter subtends about 32' of angle. The effective blackbody temperature varies somewhat across the surface of the solar disk. The radiant emission decreases somewhat at the outer edges of the disk (limb darkening), but this will not be accounted for here. Including this temperature variation can increase the calculated receiver temperature in a solar furnace by a small percentage because of the higher temperatures in the central region of the solar disk.

To obtain a sharp image of the sun, the concentrator mirror should have a long focal length relative to its diameter. Only the portion of the radiation arriving at the mirror within a diameter that is small compared with the mirror focal length reflects in a direction sufficiently close to being normal to the focal plane to create a sharp image. A long focal length relative to the diameter may not, however, be a practical design. When the focal length is short, elliptical images are produced by reflections from all but a small region near the center of the mirror. For undistorted reflection from the center of the mirror, the diameter of the sun image at the focal plane is given by $2f \tan 16' = f/107.3$, where f is the focal length. It is desired to compute the flux received within this sun image, as this is where the highest receiver temperature can be obtained. Surrounding the sun image, less flux will be received as a result of the elliptical images that are formed at the focal plane.

For the elliptical image shown in Section A–A, Fig. 9-19a, the approximate fraction of incident energy q_s in angle α that lies within the sun image (assuming uniform illumination over the ellipse) is $(\pi f^2 \alpha^2/4)/(\pi \gamma^2 \alpha^2/4 \cos \theta) = (f/\gamma)^2 \cos \theta$. The energy received on a ring element dA of the reflector is $q_s dA$

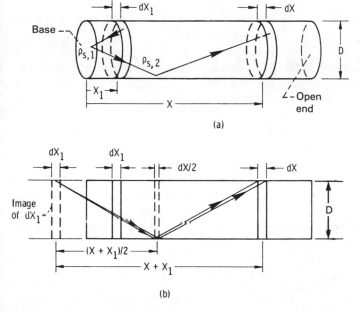

(a)

(b)

Figure 9-18 Reflection in cylindrical cavity with specular curved wall and base. (a) Cavity geometry; (b) image of dX_1 formed by reflection in cavity base.

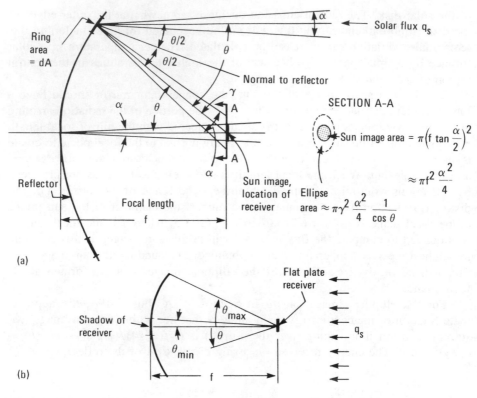

Figure 9-19 Paraboloidal-mirror solar concentrator. (*a*) Images formed at focal plane; (*b*) receiver at focal plane.

$\cos(\theta/2)$. If ρ_m is the mirror reflectivity, the resulting energy rate incident at the focal plane within the sun image is $dQ = q_s\,dA\,\cos(\theta/2)\rho_m(f/\gamma)^2\cos\theta$. The area of a ring on the mirror is $dA = 2\pi\gamma\sin\theta[(d\gamma)^2 + (\gamma d\theta)^2]^{1/2} = 2\pi\gamma\sin\theta\,\gamma\,d\theta[(d\gamma/\gamma d\theta)^2 + 1]^{1/2}$. For a paraboloid γ, f, and θ are related by $\gamma = 2f/(1 + \cos\theta)$. This yields $(1/\gamma)(d\gamma/d\theta) = \tan(\theta/2)$, so dA becomes $dA = 2\pi\gamma^2\sin\theta\,d\theta[\tan^2(\theta/2) + 1]^{1/2} = 2\pi\gamma^2\sin\theta\,d\theta\sec(\theta/2)$. Substitute into dQ to obtain $dQ = 2\pi f^2\rho_m q_s\sin\theta\cos\theta\,d\theta$.

Now consider the heat balance on a flat-plate receiver covering the sun's image at the concentrator focal plane, as shown in Fig. 9-19*b*. Assume that the absorptivity in the range of θ from $\theta_{\min}$ to $\theta_{\max}$ can be approximated by a cosine function (in many instances a constant value would be adequate). Then the energy absorbed within the sun image is

$$\int dQ_{\text{absorbed}} = 2\pi f^2\rho_m q_s\alpha_n \int_{\theta_{\min}}^{\theta_{\max}} \cos^2\theta\sin\theta\,d\theta$$

$$= \frac{2}{3}\pi f^2\rho_m q_s\alpha_n(\cos^3\theta_{\min} - \cos^3\theta_{\max}) \tag{9-44}$$

Let the receiver surface be well insulated on the back side and have no heat loss except by radiation (no convection or conduction). Since the receiver will be at high temperature and hence will emit a spectrum with considerable energy at short wavelengths, wavelength-selective effects are neglected here. Then $\epsilon(\theta) = \alpha(\theta) = \alpha_n \cos \theta$ and the emission loss is

$$2\alpha_n \sigma T_{eq}{}^4 \frac{\pi f^2 \alpha^2}{4} \int_0^{\pi/2} \cos^2 \theta \sin \theta \, d\theta = \frac{2}{3} \alpha_n \sigma T_{eq}{}^4 \frac{\pi f^2 \alpha^2}{4} \tag{9-45}$$

A heat balance is formed by equating the absorbed and emitted energies (9-44) and (9-45):

$$\frac{2}{3} \pi f^2 q_s \rho_m \alpha_n (\cos^3 \theta_{min} - \cos^3 \theta_{max}) = \frac{2}{3} \alpha_n \sigma T_{eq}{}^4 \frac{\pi f^2 \alpha^2}{4}$$

This is solved for the desired equilibrium temperature of the receiver,

$$T_{eq} = \rho_m{}^{1/4} \left(\frac{q_s}{\sigma}\right)^{1/4} \left(\frac{4}{\alpha^2}\right)^{1/4} (\cos^3 \theta_{min} - \cos^3 \theta_{max})^{1/4} \tag{9-46}$$

Additional information is in [20, 21]. An inverse analysis in mirror design to achieve a specified intensity distribution at the receiver is in [22].

In [23] the angular distribution of intensity emitted from systems of cylindrical black emitters with various reflector shapes (circular arc, parabolic, and involute) are found by ray tracing. The reflector surfaces are assumed to be either perfect or to be metallic reflecting surfaces with reflectivity predicted by electromagnetic theory.

EXAMPLE 9-8 Estimate the receiver temperature if there is 30% attenuation of the solar flux by the atmosphere, the mirror reflectivity is 0.9, and $\theta_{min} \approx 0$, $\theta_{max} = 60°$.

If T_s, r_s, and d_s are the effective temperature, radius, and distance to the sun, the q_s with 30% atmospheric attenuation is given by

$$q_s = 0.7 \sigma T_s{}^4 \left(\frac{r_s}{d_s}\right)^2 = 0.7 \sigma T_s{}^4 \tan^2 \frac{\alpha}{2} \approx 0.7 \sigma T_s{}^4 \frac{\alpha^2}{4}$$

Inserting this into Eq. (9-46) gives

$$T_{eq} = (0.7 \rho_m)^{1/4} T_s (1 - \cos^3 \theta_{max})^{1/4}$$

$$\frac{T_{eq}}{T_s} = (0.7 \times 0.9)^{1/4} (1 - \cos^3 60)^{1/4} = 0.862$$

If $T_s = 5780$ K, then $T_{eq} = 4980$ K. This is somewhat higher than achieved in practice because the collector has been assumed to be of perfect shape. There are usually losses as a result of reflector distortions.

9-5 CONCLUDING REMARKS

This chapter presents treatments of radiative interchange between specularly reflecting surfaces and in enclosures containing both specularly and diffusely reflecting surfaces. In many instances, as in Example 9-6, the interchange of energy in enclosures is modified only a small amount by the consideration of specular in place of diffuse reflecting surfaces. In certain configurations, for example those found in the design of solar concentrators and furnaces, large effects of specular reflection are present. Bobco [24], Sparrow and Lin [25], Sarofim and Hottel [26], Mahan et al. [27], and Tsai and Strieder [18] have examined radiative exchange in enclosures involving surfaces with reflectivity having both diffuse *and* specular components. Schornhorst and Viskanta [28] compared experimental and analytical results for radiant exchange among various types of surfaces and found that, *regardless* of the presence of specular surfaces, the diffuse-surface analysis agreed best with experimental results. Bobco and Drolen [29] provide a model to represent the reflectivity of a surface by diffuse and specular components. Jamaluddin

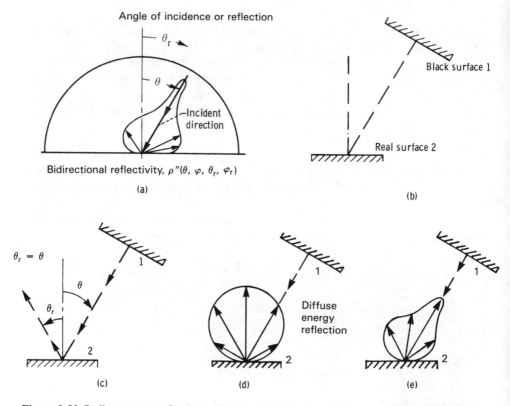

Figure 9-20 Radiant energy reflection with various idealizations of surface properties. (*a*) Bidirectional reflectivity of real surface; (*b*) geometry of surfaces; (*c*) surface 2 reflects specularly; (*d*) surface 2 reflects energy diffusely; (*e*) surface 2 reflects energy with real properties.

and Fiveland [30] analyzed the effect in a furnace of having walls with diffuse and specular components of reflectivity. The enclosures were 2- and 3-dimensional and were filled with a radiating medium, as will be discussed in Chap. 12. It was found that heat transfer performance in furnaces could be improved for some applications by using highly reflecting specular walls.

It is sometimes implied that the energy transfer between two real surfaces can be bracketed by calculating two limiting magnitudes: (1) interchange between diffuse surfaces of the same total hemispherical emissivities as the real surfaces and (2) interchange between specularly reflecting surfaces of the same total hemispherical emissivities as the real surfaces. This implication is not always true, however. Consider a surface that has a reflectivity as given by Fig. 9-20a (this is the type of reflectivity expected for the surface of the moon; see Sec. 5-3.3). Now consider the radiant interaction between the real surface 2 and a black surface 1 as shown in Fig. 9-20b. If surface 2 is given specular properties, it will return no energy to the black surface by reflection (Fig. 9-20c). If given diffuse properties, it will return a portion of the incident energy by reflection (Fig. 9-20d). If allowed to take on its real directional properties, however, it will reflect more energy to the black surface than *either* of the so-called limiting ideal surfaces (Fig. 9-20e). Thus, the ideal directional surfaces do not constitute limiting cases for energy transfer in general. Figure 8-11 demonstrates another case in which diffuse and specular properties do not provide limiting solutions. At best, calculations based on specular and diffuse assumptions for the surface characteristics give some indication of the possible magnitude of directional effects. Within enclosures, these directional effects may be small because of the many reflections taking place between the surfaces.

REFERENCES

1. Christiansen, C.: Absolute Determination of the Heat Emission and Absorption Capacity, *Ann. Phys. Wied.*, vol. 19, pp. 267–283, 1883.
2. Saunders, O. A.: Notes on Some Radiation Heat Transfer Formulae, *Proc. Phys. Soc. London*, vol. 41, pp. 569–575, 1929.
3. Chato, J. C., and J. M. Khodadadi: Optimization of Cooled Shields in Insulations, *J. Heat Transfer*, vol. 106, no. 4, pp. 871–875, 1984.
4. Schultz, W., and A. Bejan: Exergy Conservation in Parallel Thermal Insulation Systems, *Int. J. Heat Mass Transfer*, vol. 26, no. 3, pp. 335–340, 1983.
5. Cravalho, E. G., C. L. Tien, and R. P. Caren: Effect of Small Spacings on Radiative Transfer between Two Dielectrics, *J. Heat Transfer*, vol. 89, no. 4, pp. 351–358, 1967.
6. Polder, D., and M. Van Hove: Theory of Radiative Heat Transfer between Closely Spaced Bodies, *Phys. Rev. B*, vol. 4, no. 10, pp. 3303–3314, 1971.
7. Born, Max, and Emil Wolf: "Principles of Optics," 2d ed., Macmillan, New York, 1964.
8. Stone, John M.: "Radiation and Optics," McGraw-Hill, New York, 1963.
9. Howell, J. R., and M. Perlmutter: Directional Behavior of Emitted and Reflected Radiant Energy from a Specular, Gray, Asymmetric Groove, NASA TN D-1874, 1963.
10. Winston, Roland: Principles of Solar Concentrators of a Novel Design, *Sol. Energy*, vol. 16, no. 2, pp. 89–95, 1974.

11. Mannan, K. D., and L. S. Cheema: Compound-Wedge Cylindrical Stationary Concentrator, *Sol. Energy,* vol. 19, no. 6, pp. 751–754, 1977.

12. Naraghi, M. H. N., and B. T. F. Chung: A Stochastic Approach for Radiative Exchange in Enclosures with Nonparticipating Medium, *J. Heat Transfer,* vol. 106, no. 4, pp. 690–698, 1984.

13. Billings, R. L., J. W. Barnes, J. R. Howell, and O. E. Slotboom: Markov Analysis of Radiative Transfer in Specular Enclosures, *J. Heat Transfer,* vol. 113, no. 2, pp. 429–436, 1991.

14. Perlmutter, M., and R. Siegel: Effect of Specularly Reflecting Gray Surface on Thermal Radiation through a Tube and from Its Heated Wall, *J. Heat Transfer,* vol. 85, no. 1, pp. 55–62, 1963.

15. Lin, S. H., and E. M. Sparrow: Radiant Interchange among Curved Specularly Reflecting Surfaces—Application to Cylindrical and Conical Cavities, *J. Heat Transfer,* vol. 87, no. 2, pp. 299–307, 1965.

16. Plamondon, J. A., and T. E. Horton: On the Determination of the View Function to the Images of a Surface in a Nonplanar Specular Reflector, *Int. J. Heat Mass Transfer,* vol. 10, no. 5, pp. 665–679, 1967.

17. Rabl, Ari: Radiation Transfer through Specular Passages—a Simple Approximation, *Int. J. Heat Mass Transfer,* vol. 20, no. 4, pp. 323–330, 1977.

18. Tsai, D. S., and W. Strieder: Radiation across and down a Cylindrical Pore Having Both Specular and Diffuse Reflectance Components, *Ind. Eng. Chem. Fundam.,* vol. 25, pp. 244–249, 1986.

19. Tsai, D. S., and W. Strieder: Radiation across a Spherical Cavity Having Both Specular and Diffuse Reflectance Components, *Chem. Eng. Sci.,* vol. 40, no. 1, pp. 170–173, 1985.

20. Cobble, M. H.: Theoretical Concentrations for Solar Furnaces, *Sol. Energy,* vol. 5, no. 2, pp. 61–72, 1961.

21. Kamada, O.: Theoretical Concentration and Attainable Temperature in Solar Furnaces, *Sol. Energy,* vol. 9, no. 1, pp. 39–47, 1965.

22. Zakhidov, R. A.: Mirror System Synthesis for Radiant Energy Concentration—an Inverse Problem, *Sol. Energy,* vol. 42, no. 6, pp. 509–513, 1989.

23. Maruyama, S.: Uniform Isotropic Emission from an Aperture of a Collector, Proc. Fourth ASME/JSMEA Thermal Engineering Joint Conf., Reno, Nevada, vol. 4, pp. 47–53, 1991.

24. Bobco, R. P.: Radiation Heat Transfer in Semigray Enclosures with Specularly and Diffusely Reflecting Surfaces, *J. Heat Transfer,* vol. 86, no. 1, pp. 123–130, 1964.

25. Sparrow, E. M., and S. H. Lin: Radiation Heat Transfer at a Surface Having Both Specular and Diffuse Reflectance Components, *Int. J. Heat Mass Transfer,* vol. 8, no. 5, pp. 769–779, 1965.

26. Sarofim, A. F., and H. C. Hottel: Radiative Exchange among Non-Lambert Surfaces, *J. Heat Transfer,* vol. 88, no. 1, pp. 37–44, 1966.

27. Mahan, J. R., J. B. Kingsolver, and D. T. Mears: Analysis of Diffuse-Specular Axisymmetric Surfaces with Application to Parabolic Reflectors, *J. Heat Transfer,* vol. 101, no. 4, pp. 689–694, 1979.

28. Schornhorst, J. R., and R. Viskanta: An Experimental Examination of the Validity of the Commonly Used Methods of Radiant Heat Transfer Analysis, *J. Heat Transfer,* vol. 90, no. 4, pp. 429–436, 1968.

29. Bobco, R. P., and B. L. Drolen: Engineering Model of Surface Specularity: Spacecraft Design Implications, *J. Thermophys. Heat Transfer,* vol. 3, no. 3, pp. 289–296, 1989.

30. Jamaluddin, A. S., and W. A. Fiveland: Radiative Transfer in Multidimensional Enclosures with Specularly Reflecting Walls, AIAA/ASME Thermophysics and Heat Transfer Conf., Seattle, Washington, June 1990. (ASME HTD vol. 137, pp. 95–100, 1990.)

PROBLEMS

9-1 Derive Eqs. (9-5) and (9-6) for concentric radiation shields. Derive an equation for the temperature of the *n*th shield.

9-2 Two plane surfaces A_1 and A_2 are separated by 60 radiation shields having $\epsilon_s = 0.045$ on both sides. However, by accident, 17 of the shields were coated on only the upper side. The $\epsilon_s = 0.75$ for the uncoated sides. What is the rate of heat transfer for the defective system, and how does it compare with the heat flow rate when none of the shields are defective? All surfaces are gray. Give heat flow rates in W/m².

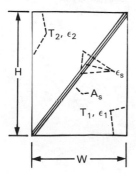

60 shields

A_1, $\epsilon_1 = 0.95$, $T_1 = 760$ K

ϵ_s

A_2, $\epsilon_2 = 0.55$, $T_2 = 525$ K

Answer: 6.484 W/m², 5.601 W/m² (no defects), 16% increase

9-3 A long rectangular enclosure is shown in cross section (two-dimensional geometry). The interior surfaces 1 and 2 are diffuse-gray with wall temperatures as shown. A group of N thin diffuse-gray radiation shields is placed on the diagonal. Derive an algebraic expression for the heat flow Q from the surfaces at T_1 to those at T_2 in terms of the quantities given. Evaluate the result for $\epsilon_1 = 0.11$, $\epsilon_2 = 0.06$, $\epsilon_s = 0.15$, $H = 0.60$ m, $W = 0.45$ m, $N = 3$, $T_1 = 525$ K, $T_2 = 465$ K, and for 1 m of length in the third dimension.

Answer: $Q = \dfrac{A_s \sigma (T_1^4 - T_2^4)}{(A_s/A_1)(1/\epsilon_1 + 1/\epsilon_2 - 2) + 1 + N(2/\epsilon_s - 1)}$, 22.60 W

9-4 Two dielectric media, both having $n = 1.25$, are spaced 0.02 cm apart in vacuum and have temperatures as shown. To what extent is the heat transfer across the gap influenced by wave interference and radiation tunneling? How large must the spacing be to reduce these effects to 5% of the ordinary heat transfer?

0.02 cm

$T = 2$ K

$T = 4$ K

Answer: 38% increase, 0.093 cm

9-5 Obtain the result for $F_{1-1(2)}$ in Example 9-5 by use of the crossed-string method. What is $F_{1-1(2)}$ if A_1 in Fig. 9-13 is rotated 45° about its center point? (The geometry remains two-dimensional.)

9-6 In Figure 9-15 the image $A_{2(3-4)}$ is shown as a dotted horizontal line on the lower right-hand side of the image diagram. By a suitable construction of rays (actual and through image surfaces) similar to the one shown in the text, show whether $A_{2(3-4)}$ is where Fig. 9-15 indicates it to be.

9-7 Obtain an analytical expression for the view factor $F_{1-1(2)}$ between the cylinder A_1 and its image $A_{1(2)}$ for each of the two situations shown (the geometries are two-dimensional).

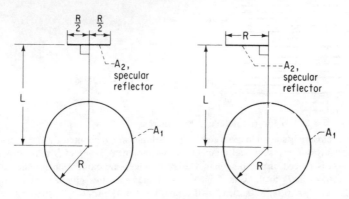

9-8 A long black cylinder is partially surrounded by two plane specular surfaces as shown. What is the rate of heat loss from the cylinder in terms of the quantities shown?

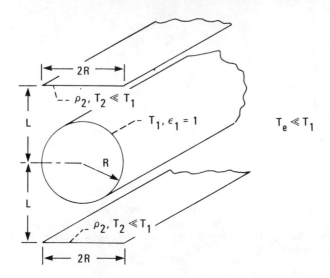

Answer: $\dfrac{Q}{\text{length}} = 2\pi R \sigma T_1^4 \left\{ 1 - \dfrac{2}{\pi}\left[\left(\dfrac{L^2}{R^2} - 1\right)^{1/2} + \sin^{-1}\dfrac{R}{L} - \dfrac{L}{R}\right]\right\}$

9-9 An enclosure is made up of two specular and two diffuse surfaces as shown. Draw a diagram of the images that are needed to determine the energy exchange process. Then write the equations for F_{1-2}^s and F_{1-3}^s in terms of the required specular configuration factors and reflectivities (that is, $F_{1-2}^s = F_{1-2} + \rho_{s,3} F_{1-2(3)} + \cdots$). Now write the set of energy exchange equations for finding Q_1, Q_2, Q_3, Q_4.

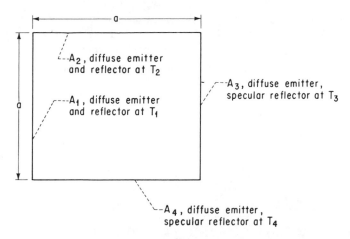

- A_2, diffuse emitter and reflector at T_2
- A_3, diffuse emitter, specular reflector at T_3
- A_1, diffuse emitter and reflector at T_1
- A_4, diffuse emitter, specular reflector at T_4

9-10 An equilateral triangular enclosure of infinite length has black surfaces 1 and 2 and a specularly reflecting surface 3 with reflectivity $\rho_{s,3} = 0.8$. Find the values of F^s_{1-1}, F^s_{1-2}, and F^s_{1-3}.

Answer: 0.1072, 0.7928, 0.5000

9-11 (a) What is the value of the following summation for Prob. 9-10?

$$\sum_{j=1}^{N} F^s_{1-j}$$

(b) What is the value of the following summation for Prob. 9-10?

$$\sum_{j=1}^{N} (1 - \rho_{s,j}) F^s_{1-j}$$

(c) Explain the results of parts (a) and (b) in terms of the definition of F^s_{1-j}. Is the result of part (b) a general relation for all specular enclosures?

Answer: (a) 1.40; (b) 1.00

9-12 An equilateral triangular enclosure has sides that extend in the normal direction infinitely far into and out of the plane of the cross section shown below.

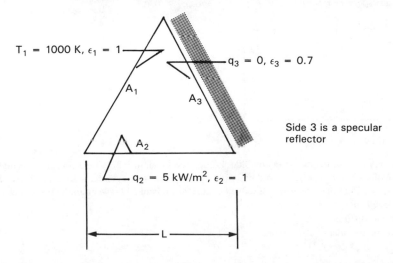

$T_1 = 1000$ K, $\epsilon_1 = 1$

$q_3 = 0$, $\epsilon_3 = 0.7$

A_1

A_3

A_2

Side 3 is a specular reflector

$q_2 = 5$ kW/m^2, $\epsilon_2 = 1$

L

(a) Find all necessary exchange factors F^s needed to solve the problem.

(b) Find the temperatures of surfaces 2 and 3.

Answer: (b) $T_2 = 1027$ K, $T_3 = 1014$ K

9-13 An enclosure is made up of three sides as shown. The length L is sufficiently long that the triangular ends can be neglected in the energy balances. Two of the surfaces are black, and the other is a diffuse-gray emitter of emissivity $\epsilon_1 = 0.05$. What is the energy added, per meter of length L, to each surface because of radiative exchange within the enclosure for each of the following two cases?

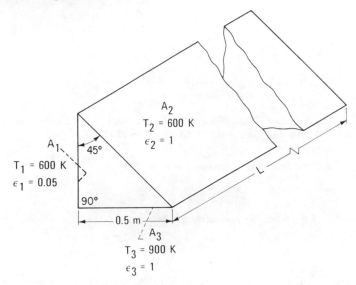

(a) Area 1 is a diffuse reflector.

(b) Area 1 is a specular reflector.

9-14 Compute the specular exchange factor F^s_{2-1} for the two-dimensional rectangular enclosure shown. All surfaces are gray and diffuse emitters. Surfaces A_1 and A_2 are diffuse reflectors, while A_3 and A_4 are specular reflectors with $\rho_{s,3} = 0.8$ and $\rho_{s,4} = 0.9$.

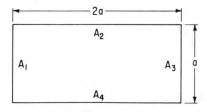

Answer: 0.349

9-15 An infinitely long square bar, 1 m on a side, is enclosed by an infinitely long concentric cylinder of 1-m radius. The temperatures and emissivities of the bar and cylinder are, respectively, T_b, ϵ_b, and T_c, ϵ_c. Find the rate at which radiant energy is exchanged between the two surfaces per unit of length if

(a) both are diffuse.

(b) both are specular.

(c) the bar is diffuse, and the cylinder is specular.

Answer: (a) $Q = \dfrac{4\sigma(T_b^4 - T_c^4)}{\dfrac{1}{\epsilon_b} + \dfrac{2}{\pi}\left(\dfrac{1}{\epsilon_c} - 1\right)}$ (b), (c) $Q = \dfrac{4\sigma(T_b^4 - T_c^4)}{\dfrac{1}{\epsilon_b} + \dfrac{1}{\epsilon_c} - 1}$

9-16 A two-dimensional rectangular enclosure has interior gray surfaces that are all diffuse emitters. Two opposing surfaces are specular reflectors, while the other two reflect diffusely. Write the set of energy equations to determine Q_1, Q_2, Q_3, and Q_4, including the expressions for the F^s factors in terms of the configuration factors F. (Note: each F^s will consist of an infinite sum.)

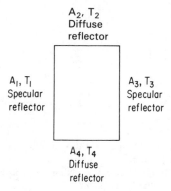

A₂, T₂
Diffuse
reflector

A₁, T₁
Specular
reflector

A₃, T₃
Specular
reflector

A₄, T₄
Diffuse
reflector

9-17 An enclosure of equilateral triangular cross section and of infinite length has two diffuse reflecting interior surfaces and one specularly reflecting interior surface that is perfectly insulated on the outside (that is, $q_3 = 0$). All surfaces are gray and are diffuse emitters. Compute T_3 and the Q added to A_1 and A_2 as a result of radiative exchange within the enclosure for the conditions shown. (For simplicity, do not subdivide the surface areas.)

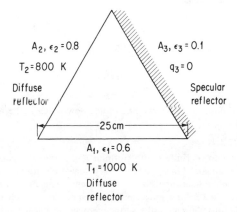

A₂, ε₂ = 0.8
T₂ = 800 K
Diffuse
reflector

A₃, ε₃ = 0.1
q₃ = 0
Specular
reflector

—25 cm—

A₁, ε₁ = 0.6
T₁ = 1000 K
Diffuse
reflector

9-18 Two parallel plates of unequal finite width are facing each other. The lower plate is diffuse-gray, and the upper plate is a specular reflector and is gray. The geometry is long in the direction normal to the cross section shown, so the configuration is two-dimensional. The lower plate is uniformly heated from below, so the energy supplied is radiated away from only its upper surface. The upper plate has its upper surface cooled so its lower surface (the specular reflector) is maintained at 525 K. The surroundings are at 425 K.

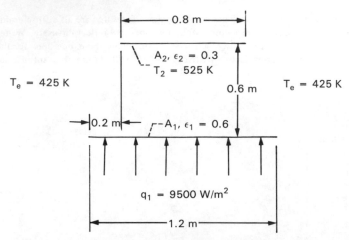

(a) Obtain the temperature of the lower plate, and determine what heat flux must be extracted from the upper plate to maintain its specified temperature.

(b) The upper plate is now roughened to make it a diffuse reflector, without changing its emissivity. How much will this alter the temperature of the lower plate?

Answer: (a) $T_1 = 783.1$ K, $q_2 = -1892.7$ W/m²; (b) $T_1 = 776.2$ K (6.9 K decrease)

9-19 A diffuse plate that is not heated or cooled by external means is facing a larger plate as shown. The spacing between the plates is small. The plates are long in the direction normal to the cross section shown, so the geometry is two-dimensional. The upper plate is gray and is a specular reflector. It is cooled to 350 K. The environment is at a higher temperature, 525 K, so radiant energy is reflected into the space between the plates. For simplicity, do not subdivide the plate areas.

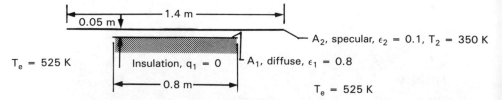

(a) What is the temperature of the lower plate?

(b) If the upper plate is made diffuse without changing its emissivity, how much is the temperature of the lower plate changed?

Answer: (a) $T_1 = 510.9$ K; (b) $T_1 = 502.1$ K

9-20 Two parallel plates shown in Prob. 7-37 are of finite width (infinite length into the paper). Both plates are perfectly insulated on the outside. Plate 1 is uniformly heated electrically with heat flux q_e. Plate 2 has no externally supplied heat input ($q_2 = 0$). The environment is at zero temperature. Plate 1 is black, while plate 2 is a diffuse gray emitter and specular reflector with emissivity ϵ_2. Derive an integral equation formulation for the surface temperature distributions. Compare with the results in Prob. 7-37a.

RADIATION COMBINED WITH CONDUCTION AND CONVECTION AT BOUNDARIES

10-1 INTRODUCTION

In the preceding chapters the enclosure theory was formulated with an emphasis on only the radiative exchange between surfaces. The local net radiation loss at the boundary surface was balanced by energy supplied by "some other means" that was not explicitly described. This chapter is concerned with interactions with this energy, supplied to the boundary by conduction from within the wall or by convection or conduction from a surrounding medium. At each location along the surface the radiation, convection, and conduction combine to form a heat flux boundary condition. The solution to the energy relations subject to this condition provides the surface temperature and heat flux distributions.

In the present analysis the same restrictions are made as in the previous theory: the surfaces are *opaque*, and the *medium* in the region between the radiating surfaces is *perfectly transparent*. Although the medium between the radiating surfaces may be conducting or convecting energy, it does not interact with radiation passing through it. The boundaries are assumed to be opaque in a special way; they are internally so strongly absorbing that absorption of incident radiation occurs within a very thin layer adjacent to the surface of the boundary material. The penetration of radiation into a partially transmitting material will be considered later; this provides a volume heat source within a boundary material that is not opaque.

The following are some examples with combined heat transfer effects. For a vapor-cycle power plant operating in outer space, the waste heat must be rejected by radiation. In the space radiator in Fig. 10-1a, the vapor of the working fluid in a thermodynamic cycle is condensed, thereby releasing its latent heat. This

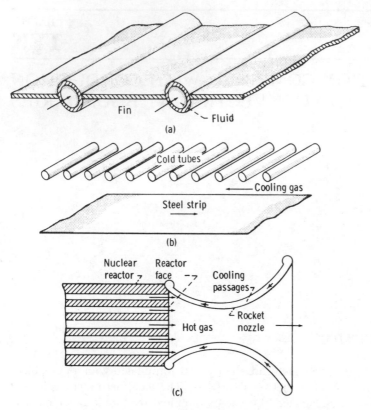

Figure 10-1 Heat transfer devices involving combined radiation, conduction, and convection. (*a*) Space radiator, or absorber plate of flat-plate solar collector; (*b*) steel-strip cooler; (*c*) nuclear rocket.

energy is conducted through the condenser wall and into fins that radiate it into space. The temperature distribution in the fins and their radiating efficiency depend on combined radiation and conduction. A fin-tube geometry is also commonly used for the absorber plate in a flat-plate solar collector. Solar energy is incident on the absorber plate through one or two transparent cover plates. Water or other fluid is heated as it flows through the tubes. The collector design involves combined radiation, conduction, and convection.

In one type of steel-strip cooler in a steel mill (Fig. 10-1*b*), a sheet of hot metal moves past a bank of cold tubes and loses energy to them by radiation. At the same time cooling gas is blown over the sheet. A combined radiation and convection analysis is required to determine the temperature distribution along the steel strip. Controlled radiative and convective cooling is also used in the tempering of sheets of high-strength plate glass used for automobile windows. Glass partially absorbs radiation passing through it; this type of partially transparent material will be considered in Chap. 18.

In a possible design for a nuclear rocket engine, illustrated by Fig. 10-1*c*, hydrogen gas is heated by flowing through a high-temperature nuclear-reactor core.

The hot gas then passes out through the rocket nozzle. The interior surface of the rocket nozzle receives energy by radiation from the exit face of the reactor core and by convection from the flowing propellant. These energy quantities are conducted through the nozzle wall and removed by flowing coolant. In later chapters other examples will be considered, such as combustion chambers, where the gaseous medium interacts with the radiation traveling through it. The medium can absorb, emit, and scatter radiation in addition to conducting and convecting energy.

The examples cited involve energy transfer by two or more modes. Energy may flow first by one mode and then by a second, as in the case of conduction through a wall followed by radiation from the surface, and the modes are in series. Heat transfer may also occur by parallel modes, such as by simultaneous conduction and radiation through a transparent medium or by simultaneous radiation and free convection from a hot surface. The modes can thus act in series, parallel, or both. Their interaction can be simple in some cases. For example, if the amounts of energy transferred by radiation and convection are independent, they can be computed separately and added. In other instances the interaction can be complex, as is the case for coupled radiation and free convection.

The various heat transfer modes depend on temperature and/or temperature differences to different powers. When radiation exchange between black surfaces is considered, the energy fluxes depend on surface temperatures to the fourth power. For nonblack surfaces the temperature exponent may differ somewhat because of emissivity variations with temperature. Heat conduction depends on the local temperature gradient, which introduces temperature derivatives. Convection depends approximately on the first power of the temperature difference. The exact power depends on the type of flow; for example, free convection depends on temperature difference to a power from 1.25 to 1.4. Physical properties that vary with temperature introduce additional temperature dependences. The various powers of temperature involved in combined energy transfer provide relations that are highly nonlinear. Except in the simplest cases, it is usually necessary to use numerical solutions. Each problem requires its own most efficient method of attack. In this chapter we concentrate on the methods of setting up the energy-balance relations and on developing insight into the physical behavior. Some common solution methods and references for available computer programs are in Chap. 11. Additional techniques are in numerical and mathematical textbooks.

10-2 SYMBOLS

a	thickness of conducting medium; thickness of fin or plate
A	area
b	spacing between fins; width of base; tube wall thickness; spacing between plates
B	parameter in Example 10-5
c	specific heat of a solid

c_p	specific heat
D	tube diameter
F	configuration factor
G	parameter in Eq. (10-22)
H	parameter defined in Example 10-8
h	heat transfer coefficient
J	parameter defined in connection with Eq. (10-18)
k	thermal conductivity
l	length of tube
L	dimensionless tube length, l/D; half length of fin; width of plate
M	parameter in Eq. (10-22)
n	normal direction
N	parameter defined in connection with Eqs. (10-18) and (10-34)
Nu	Nusselt number, hD/k
P	perimeter
Pr	Prandtl number, $c_p\mu_f/k$
q	energy flux, energy per unit area per unit time
q'''	internal heat generation per unit volume
Q	energy rate, energy per unit time
r	radius
R	dimensionless radius in Example 10-3
Re	Reynolds number, $Du_m\rho_f/\mu_f$
St	Stanton number defined in Example 10-8
t	dimensionless temperature
T	absolute temperature
u	fluid velocity
u_m	mean fluid velocity
W	width of fin in Example 10-5
x	distance from tube entrance to ring element
x, y, z	cartesian coordinate positions
X	dimensionless length, x/D
z	distance from tube entrance
Z	dimensionless length, z/D
α	absorptivity
γ, δ	dimensionless parameters of Example 10-3
ϵ	emissivity
η	fin efficiency, defined in Example 10-3
μ	dimensionless parameter defined in Example 10-5
μ_f	fluid viscosity
ξ	distance from fin base
ρ	density
ρ_f	density of fluid
σ	Stefan-Boltzmann constant
τ	time

Subscripts

a	at location $x = a$
b	evaluated at base of fin
c	conduction
e	environment; specified heating such as electrical
f	fin or fluid
fc	free convection
g	gas
i	in or inner; incoming
m	fluid medium; mean value
o	out or outer; outgoing
r	reservoir
R	radiation
w	wall
x	at position x
ξ	at position ξ
$1, 2$	evaluated at surfaces 1, 2 or at inlet and exit ends of tube

10-3 ENERGY RELATIONS AND BOUNDARY CONDITIONS

10-3.1 General Relations

In the previous enclosure theory the net radiative heat flux at any position on the boundary was balanced by the heat flux q supplied by some other means. The means considered here are *conduction* and *convection*. Since the walls are assumed *opaque* in the sense that the absorption of radiant energy is *at the surface*, the energy balance becomes a *boundary condition*. Although there can be conduction or convection in a medium situated between the radiating surfaces, the medium is assumed perfectly transparent so radiation passes through with undiminished intensity. The radiation exchange relations developed previously for an enclosure are unchanged. If, for example, convection is expressed in terms of a heat transfer coefficient, Eq. (8-1) for q can be written

$$q - h(T_g - T_w) - k \left.\frac{\partial T}{\partial n}\right|_{wall} = q_o - q_i \tag{10-1}$$

where all quantities are at **r** *on the surface* of the enclosure wall in Fig. 10-2. Equations (7-13) and (7-14) are unchanged. All of the previous enclosure relations are valid as they are written in terms of q. For example, Eq. (7-62) gives the detailed relation between T and q along the enclosure boundaries. If the T are given, (7-62) is solved for the q. Then (10-1) yields $\partial T/\partial n|_{wall}$. The T and $\partial T/\partial n$ at the wall surface are the boundary conditions for the heat conduction equation within the solid wall:

$$\rho c \frac{\partial T}{\partial \tau} = \nabla \cdot (k \, \nabla T) + q''' \tag{10-2}$$

The form of the energy equation in the medium situated between the radiating surfaces depends on the type of convection, such as a boundary layer flow, forced convection in a channel, or free convection. If the convection depends significantly on the boundary temperature or heat flux distributions, the solution may require simultaneous solution of the radiation, wall heat conduction, and convection relations.

In some problems, the net energy added to the surface by external means is specified more directly than by the normal derivative in (10-1). For example, if electrical heating is producing a specified energy flux q_e, with insulation on the exterior boundary, all of the flux q_e appears at the radiating boundary (Fig. 10-3). Then Eq. (10-1) becomes

$$q = h(T_g - T_w) + q_e = q_o - q_i \tag{10-3}$$

The term for conduction to the surface has been replaced by the specified flux produced by the electric heater. The conduction need not be expressed in terms of the thermal conductivity and temperature gradient. The q_e is often assumed uniform over the surface area, but it could have a specified variation with location. The heating might also be by direct passage of electric current through the wall such as for an electrically heated wire.

Some simple examples of uncoupled and coupled effects are now discussed.

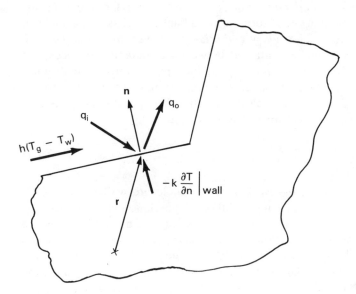

Figure 10-2 Boundary condition at surface of an opaque wall.

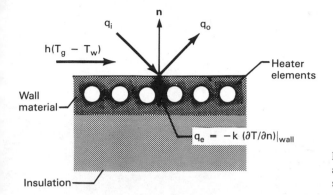

Figure 10-3 Boundary condition at surface of opaque wall with specified heat flux.

10-3.2 Uncoupled and Coupled Heat Transfer Modes

In the simplest situations, the radiation, conduction, and convection contributions to an unknown quantity, such as heat flux, are independent. The contributions are computed separately and the results combined. The heat transfer modes are *uncoupled* with regard to the desired quantity.

EXAMPLE 10-1 To illustrate an uncoupled situation, consider an enclosed region between two large gray parallel plates with a gas between them (Fig. 10-4). The internal surface temperatures T_1 and T_2 are specified. There is free convection in the gas, and the free convection heat transfer coefficient h_{fc} depends on T_1 and T_2. What is the steady-state energy transfer from plate 1 to plate 2?

The energy transferred consists of the net radiative exchange and the transfer by free convection. The net energy flux is equal to the flux q_1 that must be added to plate 1 to maintain it at its specified temperature. Since T_1 and T_2 are given, the h_{fc} can be computed from free-convection correlations and the net energy transfer is, by use of (7-22a),

$$q_1 = \frac{\sigma(T_1^4 - T_2^4)}{1/\epsilon_1(T_1) + 1/\epsilon_2(T_2) - 1} + h_{fc}(T_1, T_2)(T_1 - T_2)$$

The conductive and radiative components are *uncoupled*. The presence of one mode does not affect the other with regard to heat flow. The q for each mode is computed independently, and the two contributions added. In such problems the methods of radiative computation developed earlier can be applied *without modification*.

Unhappily, uncoupled problems are not as common as *coupled* problems. In coupled problems the *desired unknown quantity cannot be found by adding separate solutions*; the governing energy relations must be solved with the modes simultaneously included. In some situations it may be possible to assume that the modes are uncoupled because only weak coupling occurs.

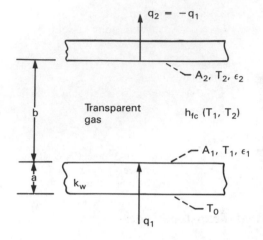

Figure 10-4 Geometry for Examples 10-1 and 10-2 $(T_1 > T_2)$.

EXAMPLE 10-2 In this instance the T_o and T_2 in Fig. 10-4 are specified. Since energy must be conserved in crossing surface 1 of the lower plate, the conduction through the lower plate must equal the transfer from surface 1 to surface 2 by combined radiation and free convection. Then for constant thermal conductivity k_w

$$q_1 = \frac{k_w}{a}(T_o - T_1) = \frac{\sigma(T_1^4 - T_2^4)}{1/\epsilon_1(T_1) + 1/\epsilon_2(T_2) - 1} + h_{fc}(T_1,T_2)(T_1 - T_2)$$

The problem is *coupled* since the unknown T_1 must be found from an equation that simultaneously incorporates all heat transfer processes. The equation for T_1 is nonlinear and can be solved iteratively.

These examples demonstrate that the type of boundary condition governs the possibility of uncoupling the calculations. When all temperatures are specified, the heat fluxes can often be uncoupled. If energy fluxes are specified, the entire problem must be treated simultaneously because of nonlinear coupling of the unknown temperatures. The treatment can become more difficult if physical property variations as functions of temperature must be included.

10-3.3 Control Volume Approach for One- or Two-Dimensional Conduction along Thin Walls

In some situations the radiating wall is thin and the temperature variation is principally along the length of the wall rather than across its thickness. An example is energy dissipation by radiating fins in devices that operate in outer space. Energy is conducted along the fin and radiated away from the fin surface. The determination of the fin temperature distribution requires a coupled conduction-radiation solution. The analysis is usually simplified by assuming a uniform temperature across the fin thickness. A control volume extending across the thickness can be used to derive the heat balance relation.

A volume element of area $dx\,dy$ and thickness a is shown in Fig. 10-5. The thickness is small, so the temperature $T(x, y)$ is uniform throughout the element. Transparent fluids at $T_{m,1}$ and $T_{m,2}$ are flowing across the upper and lower surfaces; the upper and lower convective heat transfer coefficients are h_1 and h_2. The temperature can change with time, and there can be internal heat generation within the element. A heat balance expresses that the change with time of internal energy of the element equals the energy gains by radiation exchange, conduction, convection, and internal heat sources:

$$\rho c a \frac{\partial T}{\partial \tau} = q_{i,1} - q_{o,1} + q_{i,2} - q_{o,2} + \frac{\partial}{\partial x}\left(ka\frac{\partial T}{\partial x}\right) + \frac{\partial}{\partial y}\left(ka\frac{\partial T}{\partial y}\right)$$

$$+ h_1(T_{m,1} - T) + h_2(T_{m,2} - T) + q'''a \qquad (10\text{-}4)$$

This will be used in the analyses of thin fins that follow.

10-4 RADIATION WITH CONDUCTION

Situations involving combined conduction and radiation are fairly common. Some examples are heat losses through the walls of a vacuum Dewar, heat transfer

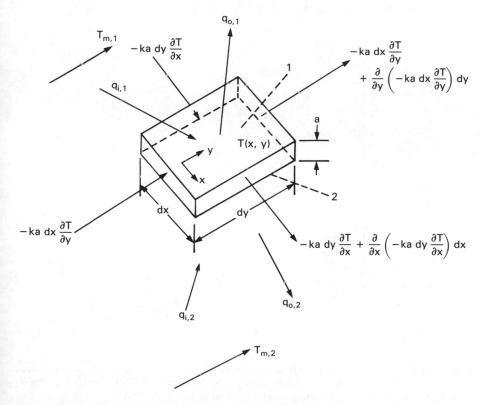

Figure 10-5 Element of thin plate for control volume derivation.

through insulation made of many separated layers of highly reflective material, heat losses from radiators in outer-space applications, and temperature distributions in satellite and spacecraft structures. The sophistication of the radiative portion of the analysis can vary considerably. The choice of radiative formulation depends on the accuracy required and the relative importance of the radiative mode in relation to heat conduction. If conduction dominates, fairly rough approximations can be invoked in the radiative portion of the analysis, and vice versa.

10-4.1 Thin Fins with One- or Two-Dimensional Conduction

One-dimensional heat flow The following analysis is for the performance of a single circular fin. From symmetry, the heat flow is one-dimensional, being only in the radial direction.

EXAMPLE 10-3 A thin annular fin in vacuum is embedded in insulation so that it is insulated on one face and around its outside edge (Fig. 10-6a,b). The disk is of thickness a, inner radius r_i, outer radius r_o, and thermal conductivity k. Energy is supplied to the inner edge from a solid rod of radius r_i that fits the central hole, and this maintains the inner edge at T_i. The ex-

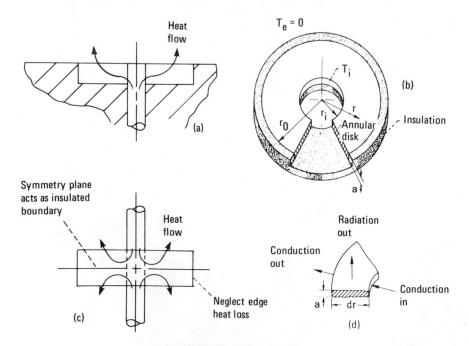

Figure 10-6 Geometry for finding temperature distribution in thin radiating annular plate insulated on one side and around outside edge. (a) Heat flow path through fin; (b) disk geometry; (c) application to annular fin; (d) portion of ring element on annular disk.

posed annular surface, which is diffuse-gray with emissivity ϵ, radiates to the environment at $T_e = 0$. Find the temperature distribution as a function of radial position along the annular disk. More generally, the results also apply to the annular fin in Fig. 10-6c if the heat loss through the end edge of the fin is neglected. There is no heat flow across the symmetry plane of the fin, and hence this plane acts as an insulated boundary.

Assume the disk is thin enough so that the local temperature can be considered constant across the thickness a. The surroundings are at zero temperature, so there is no incoming radiation. If a and k are constant, the control volume equation (10-4) for a ring element of width dr, as shown in Fig. 10-6d, gives

$$ka\frac{1}{r}\frac{d}{dr}\left(r\frac{dT}{dr}\right) - \epsilon\sigma T^4 = 0 \tag{10-5}$$

This is to be solved for $T(r)$ subject to two boundary conditions: at the inner edge $T = T_i$ at $r = r_i$, and at the insulated outer edge where there is no heat flow $dT/dr = 0$ at $r = r_o$. Using the dimensionless variables $t = T/T_i$ and $R = (r - r_i)/(r_o - r_i)$ and the two parameters $\delta = r_o/r_i$ and $\gamma = (r_o - r_i)^2\epsilon\sigma T_i^3/ka$ results in

$$\frac{d^2t}{dR^2} + \frac{1}{R + 1/(\delta - 1)}\frac{dt}{dR} - \gamma t^4 = 0 \tag{10-6}$$

with the boundary conditions $t = 1$ at $R = 0$ and $dt/dR = 0$ at $R = 1$. Equation (10-6) is a second-order differential equation that is nonlinear because it contains t raised to two different powers. The temperature distribution $t(R)$ depends on the two parameters δ and γ. Solutions can be obtained by numerical methods.

Of interest in the design of cooling fins is the *fin efficiency* η. This is the energy radiated away by the fin divided by the energy that would be radiated if the entire fin were at the temperature T_i. The fin efficiency for the circular fin is then

$$\eta = \frac{2\pi\epsilon\sigma\displaystyle\int_{r_i}^{r_o} rT^4\,dr}{\pi(r_o^2 - r_i^2)\epsilon\sigma T_i^4} = \frac{2\displaystyle\int_0^1 [R(\delta - 1) + 1]t^4\,dR}{\delta + 1}$$

This integral is carried out after t has been determined by solution of the differential equation. The fin efficiency for this annular fin has been obtained by Chambers and Somers [1] and is shown in Fig. 10-7. Keller and Holdredge [2] extended the results to fins of radially varying thickness.

In a more general situation, if the environment is at T_e and the fin is nongray with a total absorptivity α for incoming radiation, the energy balance (10-5) becomes

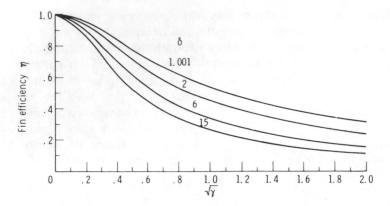

Figure 10-7 Radiation fin efficiency for fin of Example 10-3 [1].

$$ka \frac{1}{r} \frac{d}{dr}\left(r \frac{dT}{dr}\right) - \sigma(\epsilon T^4 - \alpha T_e^4) = 0$$

which can be written as

$$ka \frac{1}{r} \frac{d}{dr}\left(r \frac{dT}{dr}\right) - \epsilon\sigma\left\{T^4 - \left[\left(\frac{\alpha}{\epsilon}\right)^{1/4} T_e\right]^4\right\} = 0 \tag{10-7a}$$

The $(\alpha/\epsilon)^{1/4}T_e$ is an additional parameter, and results such as in Fig. 10-7 can be determined for various parameter values. For a gray fin, $\alpha = \epsilon$; hence a nongray fin acts like a gray fin in an effective environment temperature $(\alpha/\epsilon)^{1/4}T_e$. By using this effective temperature, results for the gray case can be used for a nongray fin. Reference [3] contains design results for rectangular fins including incident radiation from the environment.

For a transient situation where the temperature of the radiating fin changes with time, the heat storage term in (10-4) must be included. The resulting partial differential equation for $T(r, \tau)$ is

$$\rho c a \frac{\partial T}{\partial \tau} = ka \frac{1}{r} \frac{\partial}{\partial r}\left(r \frac{\partial T}{\partial r}\right) - \epsilon\sigma\left\{T^4 - \left[\left(\frac{\alpha}{\epsilon}\right)^{1/4} T_e\right]^4\right\} \tag{10-7b}$$

Results for the transient behavior of a radiating fin are in [4]. Fins of various shapes are treated in [5].

EXAMPLE 10-4 A thin plate of thickness a and length $2L$ is between two tubes in an outer-space radiator as shown in Fig. 10-8. The dimension is large in the direction normal to the cross section shown. Both sides of the plate are radiating to the vacuum of outer space, which is at very low temperature. The plate is diffuse-gray with emissivity ϵ, and has constant thermal conductivity. Find an expression for the temperature of the plate in the x direction. Neglect radiative interaction with the tubes.

From the control volume relation (10-4), the energy equation for a plate element of width dx is

$$-ka\frac{d^2T}{dx^2} + 2\epsilon\sigma T^4 = 0 \qquad (10\text{-}8)$$

The boundary conditions are $T = T_b$ specified at $x = 0$ and, from symmetry, $dT/dx = 0$ at $x = L$. To integrate the energy equation, multiply by dT/dx to obtain

$$-ka\frac{d^2T}{dx^2}\frac{dT}{dx} + 2\epsilon\sigma T^4\frac{dT}{dx} = 0$$

Integrate to yield

$$-\frac{ka}{2}\left(\frac{dT}{dx}\right)^2 + \frac{2}{5}\epsilon\sigma[T^5 - T^5(L)] = 0$$

The $T^5(L)$ (which is unknown) has been inserted to satisfy the boundary condition at $x = L$. Solve for dT/dx to yield

$$\frac{dT}{dx} = -\left(\frac{4\epsilon\sigma}{5ka}\right)^{1/2}[T^5 - T^5(L)]^{1/2}$$

The minus sign was chosen from the square root because T must be decreasing with x. By integrating again,

$$x = \left(\frac{5ka}{4\epsilon\sigma}\right)^{1/2}\int_T^{T_b}\frac{dT}{[T^5 - T^5(L)]^{1/2}} \qquad (10\text{-}9a)$$

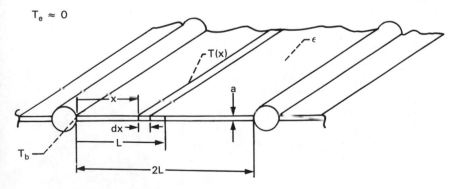

Figure 10-8 Plate geometry for Example 10-4.

which satisfies the boundary condition at $x = 0$. To obtain $T(L)$, the known length L must equal

$$L = \left(\frac{5ka}{4\epsilon\sigma}\right)^{1/2} \int_{T(L)}^{T_b} \frac{dT}{[T^5 - T^5(L)]^{1/2}} \tag{10-9b}$$

A numerical root-finding method can be used to obtain $T(L)$. The temperature distribution is then found by evaluating the integral in (10-9a) to find x for various T values between T_b and $T(L)$.

Examples 10-3 and 10-4 considered a single radiating fin. A complication of interest is the radiative interaction among fins on a multifinned surface. This introduces integral terms into the equations, as is evident from the next example.

EXAMPLE 10-5 An infinite array of thin fins of thickness a, width W, and infinite length in the z direction are attached to a black base that is held at a constant temperature T_b as pictured in Fig. 10-9. The fin surfaces are diffuse-gray and are in vacuum. Set up the equation describing the local fin temperature, assuming the environment is at $T_e \approx 0$.

Because the fins are thin, it is assumed that their local temperature is constant across the thickness a, and the control volume equation (10-4) is used for the circled differential element shown in the inset of Fig. 10-9. Since there is an infinite array of fins, the surrounding environment is identical for each fin and is the same on both sides of each fin. From symmetry, only half the fin thickness need be considered. The problem is simplified because the

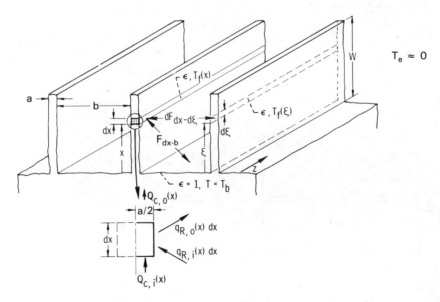

Figure 10-9 Geometry for determination of local temperatures on parallel fins.

temperature distribution $T_f(\xi)$ of the adjacent fin is the same as $T_f(x)$. Thus the energy balance need be considered for only one fin.

The net conduction into the element dx per unit time and *per unit length of fin* in the z direction is, for constant thermal conductivity, $(ka/2)$ $(d^2T_f/dx^2)\, dx$. The radiation terms are formulated from the net-radiation method. The incoming radiation to the element originates from the adjacent fin and from the base surface (since the environment is at $T_e = 0$):

$$q_{R,i}(x)\, dx = \int_{\xi=0}^{W} q_{R,o}(\xi)\, dF_{d\xi-dx}\, d\xi + b\sigma T_b^4\, dF_{b-dx}$$

$$= dx \int_{\xi=0}^{W} q_{R,o}(\xi)\, dF_{dx-d\xi} + dx\, \sigma T_b^4 F_{dx-b} \tag{10-10}$$

The outgoing radiation is composed of emission plus reflected incident radiation:

$$q_{R,o}(x)\, dx = \epsilon\sigma T_f^4(x)\, dx + (1-\epsilon)q_{R,i}(x)\, dx \tag{10-11}$$

The energy balance from Eq. (10-4) is

$$q_{R,i}(x)\, dx = q_{R,o}(x)\, dx - k\frac{a}{2}\frac{d^2T_f(x)}{dx^2}\, dx \tag{10-12}$$

Equations (10-12), (10-10), and (10-11) are three equations in the unknowns $q_{R,i}(x)$, $q_{R,o}(x)$, and $T_f(x)$. [Note that $q_{R,o}(\xi) = q_{R,o}(x)$.] Eliminating $q_{R,i}$ and $q_{R,o}$ results in

$$-\mu\frac{d^2t(X)}{dX^2} + t^4(X) = F_{dX-B} + \int_{Z=0}^{1}\left[-\mu(1-\epsilon)\frac{d^2t(Z)}{dZ^2} + t^4(Z)\right]dF_{dX-dZ} \tag{10-13}$$

where $t(X) = T_f(x)/T_b$, $B = b/W$, $\mu = ka/2\epsilon\sigma T_b^3 W^2$, $X = x/W$, and $Z = \xi/W$.

A shorter derivation of this equation follows from the direct use of the enclosure equation (7-66). Writing this for an element dx along the fin gives

$$\frac{q(x)}{\epsilon} - \frac{1-\epsilon}{\epsilon}\int_{\xi=0}^{W} q(\xi)dF_{dx-d\xi}$$

$$= \sigma T_f^4(x) - \sigma T_b^4 \int_0^b dF_{dx-db} - \int_{\xi=0}^{W}\sigma T_f^4(\xi)\, dF_{dx-d\xi}$$

Using the relations

$$\int_0^b dF_{dx-db} = F_{dx-b}, \qquad q(x) = \frac{ka}{2}\frac{d^2T_f(x)}{dx^2}, \qquad q(\xi) = \frac{ka}{2}\frac{d^2T_f(\xi)}{d\xi^2}$$

gives after rearrangement

$$-\frac{ka}{2\epsilon}\frac{d^2T_f(x)}{dx^2} + \sigma T_f^4(x) = \sigma T_b^4 F_{dx-b}$$

$$+ \int_{\xi=0}^{W} \left[-\frac{1-\epsilon}{\epsilon}\frac{ka}{2}\frac{d^2T_f(\xi)}{d\xi^2} + \sigma T_f^4(\xi) \right] dF_{dx-d\xi}$$

In dimensionless form this is the same as (10-13).

Equation (10-13) is a nonlinear integrodifferential equation and can be solved numerically. Since it is a second-order equation, two boundary conditions are needed. At the base of the fin $T_f(x = 0) = T_b$, so

$$t = 1 \qquad \text{at } X = 0 \tag{10-14a}$$

A second condition is obtained at the outer edge of the fin $x = W$. The conduction to this boundary must equal the heat radiated: $-k\,\partial T_f/\partial x|_{x=W} = \epsilon\sigma T_f^4(W)$. In terms of t,

$$-\frac{dt}{dX} = \frac{\epsilon\sigma T_b^3 W}{k}t^4 = \frac{1}{2\mu}\frac{a}{W}t^4 \qquad \text{at } X = 1 \tag{10-14b}$$

and it is evident that the fin thickness-to-width ratio a/W enters the problem as another parameter. If $(a/W)/2\mu$ is very small, dt/dX can be taken as zero. The configuration factors in (10-13) are found by the methods of Examples 6-3 and 6-5 (for $\alpha = 90°$ in those equations).

Note that for an infinite array of fins, *no benefit* is gained by adding fins to a plane black base surface because no surface can emit more than a black surface does. However, the directional characteristics of a finned surface make it attractive in some applications, as discussed in Chap. 5. Because of the interest in radiator design for application in space power systems, many conducting-radiating systems have been analyzed. Typical are [1–17]; many other references are in the literature.

Two-dimensional heat flow The configuration of many types of fins is such that they can be analyzed as having one-dimensional heat flow. Some applications use two-dimensional fins, such as in space devices to carry away excess heat from electronic equipment. The fin is a thin plate, such as aluminum, with the heat-generating equipment in good thermal contact with a portion of the plate area. The heat supplied to the plate is conducted away two-dimensionally and dissipated by radiation to cooler surroundings. Analyses for this type of cooling device are in [18] and [19]. In [18] the heat transfer modes are radiation and conduction; in [19] convection is also included.

The fin analyzed in [18] is shown in Fig. 10-10. It has uniform thickness and constant thermal properties, and its radiative properties are diffuse-gray. Heat to be dissipated is supplied to the shaded area on one side and is radiated away from

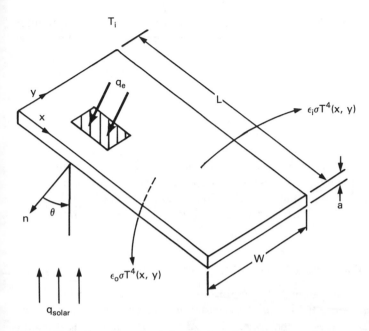

Figure 10-10 Two-dimensional fin with heat flux addition q_e to area on one side.

both sides. The plate is exposed to the sun on the outside and to surroundings at T_i on the inside. The energy equation within the zone (shown shaded) over which heat is being supplied to the plate is given by (10-4) as

$$ka\left(\frac{\partial^2 T}{\partial x^2} + \frac{\partial^2 T}{\partial y^2}\right) + \alpha_s q_{\text{solar}} \cos \theta + q_e - \epsilon_o \sigma T^4(x,y) = 0 \tag{10-15}$$

In this region an element of the plate is receiving energy by two-dimensional conduction within the plate, by absorption of solar radiation, and by heat addition from electronic equipment and/or other heat sources. The plate is losing energy by radiation from its outside surface. For the other portions of the plate where $q_e = 0$, the energy equation becomes

$$ka\left(\frac{\partial^2 T}{\partial x^2} + \frac{\partial^2 T}{\partial y^2}\right) + \alpha_s q_{\text{solar}} \cos \theta + \sigma \epsilon_i [T_i^4 - T^4(x, y)] - \epsilon_o \sigma T^4(x, y) = 0 \tag{10-16}$$

where there is a term for the net radiation loss from the surface that is inside the enclosure. Although this was the term used for the analysis in [18], it is written more generally in (10-4) as $q_i - q_o$ to account for a more complex heat exchange with a surrounding enclosure. Heat losses from the end edges of the fin were assumed small so that insulated edge boundary conditions are used,

$$\frac{\partial T}{\partial x} = 0 \quad \text{at } x = 0, L, \qquad \frac{\partial T}{\partial y} = 0 \quad \text{at } y = 0, W$$

Along the boundary between the heat addition (shaded) area and the remaining area of the plate, there is continuity within the fin of temperature and heat flow in the x and y directions.

Solutions to other fin problems involving mutual interactions are in [10–17, 20, 21]. The optimization of a fin array with respect to minimum weight is in [16, 21]. Masuda [17] analyzed the radiant interchange between diffuse gray external circular fins on cylinders and obtained the local heat flux distribution on the fins and cylinders as well as the fin effectiveness.

The examples given in this section are simplified in that no property variations have been included. When properties are variable, the basic concepts are the same as in the examples, although the inclusion of property variations adds complexity to the functional form of the equations. The usual cautions concerning the inadequacy in some cases of the diffuse-gray assumptions carry over to multimode problems.

10-4.2 Multidimensional and Transient Conduction

For a *thin* radiating fin, the temperature within the fin is assumed uniform across the fin thickness, and hence the temperature variations are only in directions parallel to the radiating surfaces. If the conducting solid is thick, however, the temperature will also vary with distance normal to the radiating surface. The conduction can be steady or transient. Since, in the current discussion, the surfaces are assumed to be opaque, radiation is absorbed at the surface and the absorbed energy acts as a boundary condition for conduction within the solid. If n is the outward normal from the surface, the conduction heat flux at the surface flowing outward from within the solid is $-k\, \partial T/\partial n$. This is the local energy flux q being supplied to the surface to balance the net loss by radiative transfer. Hence, in the absence of convection, there is the boundary condition, Eq. (10-1), $-k(\partial T/\partial n)|_{wall} = q_o - q_i = q$. The governing partial differential equation in the solid is Eq. (10-2). The distribution of $q_o - q_i$ along the conducting surfaces is found from the radiative enclosure methods previously described, for example, Eq. (7-62) for an enclosure of diffuse-gray surfaces. For a nongray surface the radiative q is the integral of the spectral fluxes as described by the enclosure theory in Chap. 8. Subject to these boundary conditions, the multidimensional heat conduction equation can be solved by finite-difference or finite-element numerical methods as discussed in Chap. 11.

When the temperature distributions within the solid are time as well as spatially dependent, there is little chance that analytical solutions can be obtained, in view of the complexity and nonlinearity of the radiative boundary conditions. Some analytical investigations were made in [9]. A transient solution is more feasible if the geometry is one-dimensional. An example is the electrical heating of a thin wire where the transient temperature varies only along the length of the wire [22]. Transient cooling of a wire was used in [23] to obtain the hemispherical-total emittance of a metal by measuring the cooling rate. The one-dimensional energy equation for a wire of radius r is provided by the control volume approach, Eq. (10-4):

$$r^2 \rho c \frac{\partial T}{\partial \tau} = kr^2 \frac{\partial^2 T}{\partial x^2} - 2r[q_o(x, \tau) - q_i(x, \tau)] \qquad (10\text{-}17)$$

The properties have been assumed constant. The radiative terms depend on the exchange with the surroundings. The q_o and q_i are functions of axial position x and τ. The q_o varies considerably as the wire transient temperature changes with time and position. The solution for $T(x, \tau)$ requires an initial temperature distribution $T(x, 0)$, and two boundary conditions in x. For example, the boundary conditions could be fixed temperatures at the ends of the wire.

A two-dimensional transient solution was carried out numerically in [24]. A hollow cylinder, insulated on its internal surface, is heated on its exterior by a time-varying radiation flux from one direction. Absorbed energy is then conducted within the cylinder in radial and circumferential directions. During the transient heating, the outer surface is losing energy by radiation and convection. When the cylinder material has low thermal conductivity and low thermal diffusivity, the temperature distributions are quite nonuniform. The surface temperatures are high, so radiative cooling is important. For high-conductivity materials the temperature levels are lower and the temperature distributions more uniform.

EXAMPLE 10-6 To illustrate a transient heating solution, consider a stainless steel plate 1 cm thick and initially at 600 K (Fig. 10-11). A wall behind the plate is maintained at 600 K, while the front surface of the plate is suddenly exposed to a radiative heat flux $q_i = 500$ kW/m^2. It is desired to find the time for the front surface to reach 1200 K and to obtain the temperature variation across the plate thickness at that time. The surfaces are diffuse-gray. The properties are $\epsilon_0 = \epsilon_a = 0.3$, $\epsilon_w = 0.8$, $k = 26$ W/(m·K), $c = 460$ J/(kg·K), and $\rho = 8500$ kg/m^3.

For illustration purposes, a simple explicit method is used. Using a forward difference in time, an energy balance on the region of thickness $\Delta x/2$ at the left surface of the plate (Fig. 10-11b) means that the change of internal energy with time results from absorbed incident radiation, radiative emission from the surface, and heat conduction into the plate,

$$\rho c \frac{\Delta x}{2} \frac{(T_{0,\tau+1} - T_{0,\tau})}{\Delta \tau} = q_i \epsilon_0 - \epsilon_0 \sigma T_{0,\tau}^4 - \frac{k}{\Delta x} (T_{0,\tau} - T_{1,\tau})$$

Similarly, at the right surface node N

$$\rho c \frac{\Delta x}{2} \frac{(T_{N,\tau+1} - T_{N,\tau})}{\Delta \tau} = \frac{k}{\Delta x} (T_{N-1,\tau} - T_{N,\tau}) - \frac{\sigma(T_{N,\tau}^4 - T_w^4)}{1/\epsilon_a + 1/\epsilon_w - 1}$$

For an interior node n

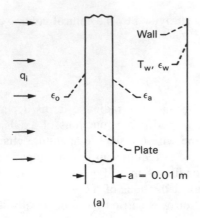

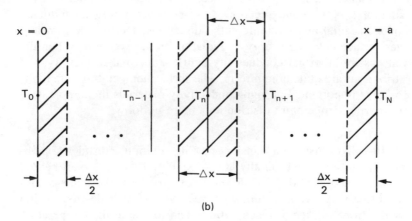

Figure 10-11 Transient heating of a plate by incident radiation with radiative losses from both sides of the plate. (a) Geometry of system; (b) finite differences in plate.

$$\rho c\, \Delta x \frac{(T_{n,\tau+1} - T_{n,\tau})}{\Delta \tau} = \frac{k}{\Delta x}(T_{n-1,\tau} - T_{n,\tau}) - \frac{k}{\Delta x}(T_{n,\tau} - T_{n+1,\tau})$$

In dimensionless form the equations at 0, n, and N become

$$t_{0,\tau+1} = t_{0,\tau} + \frac{2}{J}\left(\epsilon_0 Q - \frac{\epsilon_0}{N}t_{0,\tau}^4 - t_{0,\tau} + t_{1,\tau}\right) \tag{10-18a}$$

$$t_{n,\tau+1} = t_{n,\tau} + \frac{1}{J}(t_{n-1,\tau} - 2t_{n,\tau} + t_{n+1,\tau}) \tag{10-18b}$$

$$t_{N,\tau+1} = t_{N,\tau} + \frac{2}{J}\left[t_{N-1,\tau} - t_{N,\tau} - \frac{E}{N}(t_{N,\tau}^4 - 1)\right] \tag{10-18c}$$

where $t = T/T_w$, $Q = q_i \Delta x/kT_w$, $N = k/\sigma T_w^3 \Delta x$, $E = (1/\epsilon_a + 1/\epsilon_w - 1)^{-1}$, and $J = \Delta x^2/\alpha \Delta \tau$ where $\alpha = k/\rho c$.

For stability in a transient linear heat conduction solution, $J \geq 2$. Since there are nonlinear terms for the boundary nodes, $J = 3$ was used for the present calculations and the solution was stable. The value of J fixes $\Delta \tau$ after Δx is specified; $\Delta x = a/10$ was found sufficiently small for good accuracy in the present calculations. For the conditions given, approximately 200 s was required for the surface temperature to rise from 600 to 1200 K. In the first 100 s, the temperature rose from 600 to 972 K; the heating rate decreased with time as radiative losses increased with temperature. Throughout the heating process the temperature difference across the plate, $T_0 - T_N$, was approximately constant at 28 K.

10-5 RADIATION WITH CONVECTION AND CONDUCTION

Radiation-convection-conduction interaction problems are found in many situations, such as cooling of high-temperature components, furnace and combustion chamber design where heat transfer from surfaces occurs by simultaneous radiation and convection, cooling of hypersonic and reentry vehicles, interactions of incident solar radiation with the earth's surface to produce complex free-convection patterns and complicate the art of weather forecasting, convection cells and their effect on radiation from stars, and marine environment studies for predicting free-convection patterns in oceans and lakes.

The energy equations and boundary conditions contain temperature differences arising from convection and temperature derivatives from conduction. Additional independent parameters arise from the convective heat transfer coefficients. Results must usually be obtained by using numerical techniques. The basic ideas are developed here by discussing a few specific problems. Additional information and results are in [25–39].

10-5.1 Thin Fins with Convection

EXAMPLE 10-7 The performance of the fin in Fig. 10-12 depends on combined conduction, convection, and radiation. A gas at T_e is flowing over the fin and removing heat by convection. The environment to which the fin radiates is also assumed to be at T_e. The cross section of the fin has area A and perimeter P. The fin is nongray with absorptivity α for radiation incident from the environment.

From the control volume approach, an energy balance on a fin element of length dx yields

$$kA \frac{d^2T}{dx^2} dx = \sigma[\epsilon T^4(x) - \alpha T_e^4]P\, dx + hP\, dx[T(x) - T_e] \tag{10-19}$$

The term on the left is the net conduction into the element, and the terms on the right are the radiative and convective losses. The radiative exchange be-

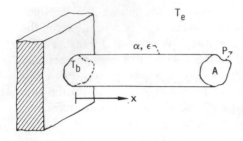

Figure 10-12 Fin of constant cross-sectional area transferring energy by radiation and convection. (Flowing gas and environment are both at T_e.)

tween the fin and its base is neglected. This equation is to be solved for T as a function of x. Multiply (10-19) by $[1/(kA\,dx)]\,dT/dx$ to obtain

$$\frac{d^2T}{dx^2}\frac{dT}{dx} = \frac{\epsilon\sigma P}{kA}\left(T^4 - \frac{\alpha}{\epsilon}T_e^4\right)\frac{dT}{dx} + \frac{hP}{kA}(T - T_e)\frac{dT}{dx}$$

This is integrated once to yield

$$\frac{1}{2}\left(\frac{dT}{dx}\right)^2 = \frac{\epsilon\sigma P}{kA}\left(\frac{T^5}{5} - \frac{\alpha}{\epsilon}TT_e^4\right) + \frac{hP}{kA}\left(\frac{T^2}{2} - TT_e\right) + C \tag{10-20}$$

where C is a constant of integration.

For convection and radiation from the end surface of the fin, or for the end of the fin assumed insulated, the dT/dx in (10-20) is expressed in terms of the appropriate end conditions. The solution can then be carried somewhat further; this is considered in Problems 9-2 and 9-3 at the end of the chapter. To examine a somewhat shorter case that leads to an analytical solution let $T_e \approx 0$ and let the fin be very long. Then for large x, $T(x) \to 0$ and $dT/dx \to 0$, and from (10-20) the constant $C = 0$. Solving for dT/dx results in

$$\frac{dT}{dx} = -\left(\frac{2}{5}\frac{P\epsilon\sigma}{kA}T^5 + \frac{hP}{kA}T^2\right)^{1/2} \tag{10-21}$$

The minus sign is used when taking the square root since T decreases as x increases. The variables in (10-21) are separated and the equation integrated with the condition that $T(x) = T_b$ at $x = 0$:

$$\int_0^x dx = -\int_{T_b}^T \frac{dT}{T[\frac{2}{5}(P\epsilon\sigma/kA)T^3 + hP/kA]^{1/2}}$$

Integration yields

$$x = \frac{1}{3}M^{-1/2}\left[\ln\frac{(GT_b^3 + M)^{1/2} - M^{1/2}}{(GT_b^3 + M)^{1/2} + M^{1/2}} - \ln\frac{(GT^3 + M)^{1/2} - M^{1/2}}{(GT^3 + M)^{1/2} + M^{1/2}}\right] \tag{10-22}$$

where $G = \frac{2}{5}P\epsilon\sigma/kA$ and $M = hP/kA$. For this simplified case a closed-form analytical solution for the temperature distribution is obtained. A detailed treatment of this type of fin problem is in [37, 38].

10-5.2 Channel Flows

An example is now considered that involves gas flow through a heated tube. Representative solutions of this type are found in [25–29, 39]. The presentation is limited to convective media that are completely transparent to radiation. The enclosure energy-balance equations, such as (7-17) and (7-18), can be used as before; the q_k will now specifically contain convective heat addition to the wall. In Example 10-8 the q_k consists of convection and electrical heat generation within the tube wall.

EXAMPLE 10-8 A transparent gas flows through a black circular tube (Fig. 10-13). The tube wall is thin, and its outer surface is perfectly insulated. The wall is heated electrically to provide a uniform input of energy q_e per unit area and time. The variation of wall temperature along the tube length is to be determined. The convective heat transfer coefficient h between the gas and the inside of the tube is assumed to be constant. The gas has a mean velocity u_m, heat capacity c_p, and density ρ_f. Axial conduction in the tube wall is neglected.

If radiation were not considered, the local heat addition to the gas would equal the local electrical heating (since the outside of the tube is insulated) and hence would be invariant with axial position x along the tube. The gas temperature and wall temperature would both rise linearly with x. On the other hand, if convection were not considered, the only means for heat removal would be by radiation out of the tube ends as in Example 7-19. In this instance, for equal environment temperatures at the ends of the tube, the wall temperature is a maximum at the center of the tube and decreases continuously toward both ends. The solution for combined radiation and convection is expected to exhibit partially the trends of both limiting solutions.

Consider a ring element of length dx on the interior of the tube wall at position x as in Fig. 10-13. The energy supplied per unit time is composed of electrical heating, energy radiated to dA_x by other wall elements of the tube interior (see Example 7-19), and energy radiated to dA_x from the inlet and exit reservoirs,

$$q_e \pi D \, dx + \int_{z=0}^{l} \sigma T_w^4(z) \, dF_{dz-dx}(|z - x|) \, \pi D \, dz$$

$$+ \sigma T_{r,1}^4 \frac{\pi D^2}{4} \, dF_{1-dx}(x) + \sigma T_{r,2}^4 \frac{\pi D^2}{4} \, dF_{2-dx}(l - x)$$

The tube ends are assumed to act as black disks at the inlet and outlet reservoir temperatures, which are assumed to be at the inlet and outlet gas temperatures. The energy transferred away from the ring element at x by convection and radiation is $\{h[T_w(x) - T_g(x)] + \sigma T_w^4(x)\}\pi D \, dx$. Neglecting axial heat conduction in the wall, the energy quantities are equated to yield (reciprocity was used on the F factors so that dx could be divided out):

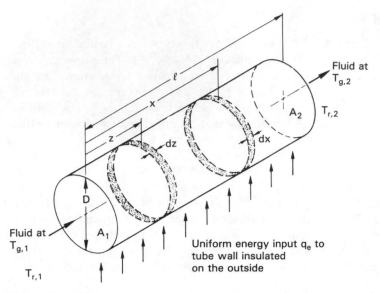

Figure 10-13 Flow through tube with uniform internal energy input to wall, and outer surface insulated.

$$h[T_w(x) - T_g(x)] + \sigma T_w^4(x) = q_e + \int_{z=0}^{l} \sigma T_w^4(z)\, dF_{dx-dz}(|x - z|)$$

$$+ \sigma T_{r,1}^4 F_{dx-1}(x) + \sigma T_{r,2}^4 F_{dx-2}(l - x) \tag{10-23}$$

This has the form of (7-10), where for (10-23), $Q_k/A_k = q_e + h[T_g(x) - T_w(x)]$. Equation (10-23) has two unknowns, $T_w(x)$ and $T_g(x)$; a second equation must be found before a solution can be obtained. This is obtained from an energy balance on a volume in the tube of length dx. The energy carried into this volume by the flowing gas is $u_m \rho_f c_p T_g(x)(\pi D^2/4)$, and that added by convection from the wall is $\pi D h[T_w(x) - T_g(x)]\, dx$. The energy carried out by the flowing gas is $u_m \rho_f c_p(\pi D^2/4)\{T_g(x) + [dT_g(x)/dx]\, dx\}$. Equating outgoing and incoming energies gives the energy balance

$$u_m \rho_f c_p \frac{D}{4} \frac{dT_g(x)}{dx} = h[T_w(x) - T_g(x)] \tag{10-24}$$

By defining the dimensionless quantities

$$\text{St} = \frac{4h}{u_m \rho_f c_p} = \frac{4\,\text{Nu}}{\text{Re Pr}} \qquad H = \frac{h}{q_e}\left(\frac{q_e}{\sigma}\right)^{1/4} \qquad t = T\left(\frac{\sigma}{q_e}\right)^{1/4}$$

and $X = x/D$, $Z = z/D$, and $L = l/D$, the energy balances on the wall and fluid are

$$t_w^4(X) + H[t_w(X) - t_g(X)] = 1 + \int_0^X t_w^4(Z) \, dF_{dX-dZ}(X - Z)$$

$$+ \int_X^L t_w^4(Z) \, dF_{dX-dZ}(Z - X) + t_{r,1}^4 F_{dX-1}(X) + t_{r,2}^4 F_{dX-2}(L - X) \qquad (10\text{-}25)$$

$$\frac{dt_g(X)}{dX} = \text{St} \, [t_w(X) - t_g(X)] \qquad (10\text{-}26)$$

The two equations have the unknowns $t_w(X)$ and $t_g(X)$, and the five parameters St, H, L, $t_{r,1}$, and $t_{r,2}$.

To solve Eqs. (10-25) and (10-26), it is noted that (10-26) is a first-order linear differential equation that can be solved in general form by use of an integrating factor. The boundary condition is that at $X = 0$ the gas temperature has a specified value $t_{g,1}$. The general solution is

$$t_g(X) = \text{St} \, e^{-\text{St}X} \int_0^X e^{\text{St}Z} t_w(Z) \, dZ + t_{g,1} e^{-\text{St}X} \qquad (10\text{-}27)$$

This is substituted into (10-25) to eliminate $t_g(X)$ and yield an integral equation for the desired variation in tube-wall temperature:

$$t_w^4 + Ht_w - H \, \text{St} \, e^{-\text{St}X} \int_{Z=0}^X e^{\text{St}Z} t_w(Z) \, dZ - Ht_{g,1} e^{-\text{St}X}$$

$$= 1 + \int_{Z-0}^X t_w^4(Z) \, dF_{dX-dZ}(X - Z) + \int_{Z=X}^L t_w^4(Z) \, dF_{dX-dZ}(Z - X)$$

$$+ t_{r,1}^4 F_{dX-1}(X) + t_{r,2}^4 F_{dX-2}(L - X) \qquad (10\text{-}28)$$

An alternative derivation of (10-25) is to use the enclosure equation (7-62). The energy added to the inner wall surface by means other than internal radiative exchange is $q_w(x) = h[T_g(x) - T_w(x)] + q_e$. The enclosure equation yields

$$q_w(x) = \sigma T_w^4(x) - \int_{z=0}^l \sigma T_w^4(z) \, dF_{dx-dz} - \sigma T_{r,1}^4 \, F_{dx-1} - \sigma T_{r,2}^4 F_{dx-2} \qquad (10\text{-}29)$$

With the expression for $q_w(x)$ inserted, this is the same as (10-25).

Solutions to (10-28) were obtained by Perlmutter and Siegel [25], and some representative results calculated by numerical integration are in Fig. 10-14. Note that the predicted temperatures for combined radiation-convection fall below the temperatures predicted for either convection or radiation acting independently. For a short tube the radiation effects are significant over the entire tube length, and for the parameters shown the combined-mode temperature distribution is similar

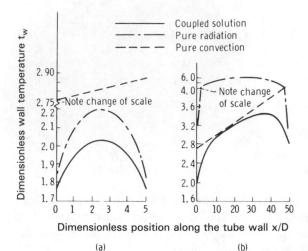

Figure 10-14 Tube wall temperatures resulting from combined radiation and convection for transparent gas flowing in uniformly heated black tube for St = 0.02, $H = 0.8$, $t_{r,1} = t_{g,1} = 1.5$, and $t_{r,2} = t_{g,2}$. (a) Tube length, $l/D = 5$; (b) tube length, $l/D = 50$.

to that for radiation alone. For a long tube, however, the combined-mode distribution is very close to that for convection alone over the central portion of the tube. The heat transfer resulting from combined convection-radiation is more efficient than by either mode alone. This means that the wall temperature distribution in the combined problem will lie below distributions predicted by using either mode alone.

EXAMPLE 10-9 What are the governing energy equations if the tube in Example 10-8 has a diffuse-gray interior surface with emissivity ϵ rather than being black?

The energy supplied to an area element at x by means other than radiation inside the tube is $q_e + h[T_g(x) - T_w(x)]$. Then from Eq. (7-19), at the interior surface of the tube wall,

$$q_o(x) = \sigma T_w^4(x) + \frac{1 - \epsilon}{\epsilon} \{h[T_w(x) - T_g(x)] - q_e\} \tag{10-30}$$

The analysis leading to Eq. (10-23) applies for the gray case if the radiation σT_w^4 leaving the surface is replaced by q_o. This gives

$$h[T_w(x) - T_g(x)] + q_o(x) = q_e + \int_{z=0}^{l} q_o(z)\, dF_{dx-dz}(|x - z|) + \sigma T_{r,1}^4 F_{dx-1}(x)$$

$$+ \sigma T_{r,2}^4 F_{dx-2}(l - x) \tag{10-31}$$

Equation (10-27) is unchanged by having the wall gray. Thus (10-30), (10-31), and (10-27) constitute a set of three equations for the unknowns $T_w(x)$, $q_o(x)$, and $T_g(x)$. Numerical solutions are in [26].

A shorter derivation is by use of the enclosure equation (7-62). The energy added to the interior surface of the wall by means other than internal radiative exchange is $q_w(x) = h[T_g(x) - T_w(x)] + q_e$. The enclosure equation yields

$$\frac{q_w(x)}{\epsilon} - \frac{1-\epsilon}{\epsilon} \int_{z=0}^{l} q_w(z) \, dF_{dx-dz}(z, x) = \sigma T_w^4(x) - \int_{z=0}^{l} \sigma T_w^4(z) \, dF_{dx-dz}(z, x)$$

$$- \sigma T_{r,1}^4 F_{dx-1} - \sigma T_{r,2}^4 F_{dx-2} \tag{10-32}$$

By use of (10-30) and the relation for q_w, a little algebra shows that (10-32) is the same as (10-31).

EXAMPLE 10-10 Consider again the tube in Example 10-8. The tube is uniformly heated, perfectly insulated on the outside, and has a black interior surface. Gas flows through the tube, and the convective heat transfer coefficient h is assumed constant. Axial heat conduction within the tube wall is now included. The tube wall has thermal conductivity k_w and thickness b, and the tube inside and outside diameters are D_i and D_o. The desired result is the wall temperature distribution along the tube length. The wall is assumed sufficiently thin that its temperature at each axial position is constant across the wall thickness.

The energy balance given by Eq. (10-23) must be modified to include axial wall conduction. From the previous control volume approach, the net gain of energy by the element from conduction is

$$k_w \pi \frac{D_o^2 - D_i^2}{4} \frac{d^2 T_w(x)}{dx^2} \, dx$$

This term is divided by the internal area of the ring $\pi D_i \, dx$ and is added to the right side of (10-23) to obtain the energy balance:

$$h[T_w(x) - T_g(x)] + \sigma T_w^4(x) = q_e + k_w \frac{D_o^2 - D_i^2}{4D_i} \frac{d^2 T_w(x)}{dx^2}$$

$$+ \int_{z=0}^{l} \sigma T_w^4(z) \, dF_{dx-dz}(|x - z|) + \sigma T_{r,1}^4 F_{dx-1}(x) + \sigma T_{r,2}^4 F_{dx\ 2}(l - x)$$

$$\tag{10-33}$$

This is of the same form as (7-10) with the nonradiative heat flux addition being

$$q_k = q_e + h[T_g(x) - T_w(x)] + k_w \frac{D_o^2 - D_i^2}{4D_i} \frac{d^2 T_w(x)}{dx^2}$$

As in connection with (10-25), all lengths are nondimensionalized by dividing by the internal tube diameter, and dimensionless parameters are introduced. The conduction term yields a new parameter

$$N = \frac{k_w}{4q_e D_i} \left[\left(\frac{D_o}{D_i} \right)^2 - 1 \right] \left(\frac{q_e}{\sigma} \right)^{1/4}$$

For thin walls where $(D_o - D_i)/2 = b \ll D_i$ this reduces to $N = (k_w b/q_e D_i^2)(q_e/\sigma)^{1/4}$, which is the parameter used in some references.

The dimensionless form of the energy equation is

$$t_w^4(X) + H[t_w(X) - t_g(X)] = 1 + N\frac{d^2 t_w(X)}{dX^2} + \int_{Z=0}^X t_w^4(Z)\,dF_{dX-dZ}(X - Z)$$

$$+ \int_{Z=X}^L t_w^4(Z)\,dF_{dX-dZ}(Z - X) + t_{r,1}^4 F_{dX-1}(X) + t_{r,2}^4 F_{dX-2}(L - X) \qquad (10\text{-}34)$$

The energy equation for the fluid is still (10-27); these equations can be combined as in (10-28). Hottel discussed this problem in terms of slightly different parameters. He obtained a numerical solution for five ring-area intervals on the tube wall, before the common use of computers. The solution required 10 h of hand computation; results are shown in Fig. 10-15 in terms of the parameters derived here.

Two additional factors that enter this problem are the conduction boundary conditions. The solution of (10-34) requires two boundary conditions because of the arbitrary constants introduced by integrating the $d^2 t_w/dX^2$ term. These boundary conditions depend on the physical construction at the ends of the tube that determine the amount of conduction present. In [30] some detailed results were obtained. It was assumed for simplicity that the tube end edges were insulated, $(dT_w/dx)|_{x=0} = (dT_w/dx)|_{x=l} = 0$. The extension was also made in [30] to have the convective heat transfer coefficient vary with position along the tube. This accounts for the variation of h in the thermal entrance region. A finite-element numerical method was used in [39] to extend the results. Results for uniform

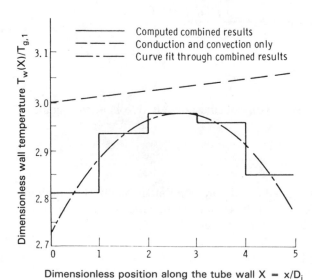

Figure 10-15 Wall temperature distribution for flow of transparent fluid through black tube with combined radiation, convection, and conduction for $L = 5$, St $= 0.005$, $N = 0.316$, $H = 1.58$, $t_{r,1} = t_{g,1} = 0.316$, and $t_{r,2} = t_{g,2}$.

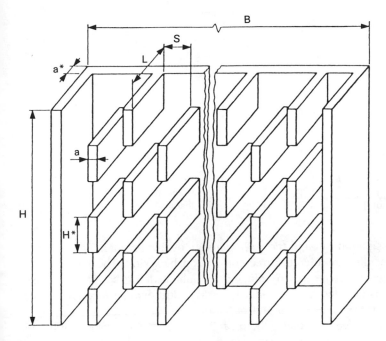

Figure 10-16 Finned heat sink with staggered array of plates cooled by combined radiation and free convection.

heating were in good agreement with those in earlier references. The calculations were extended to include a sinusoidal heat flux addition along the tube length.

10-5.3 Free Convection with Radiation

At moderate temperatures radiative fluxes are small, but in conjunction with free-convection heat transfer, the radiative terms may be comparable to the convective terms. A proposed heat sink for cooling electronic components attached to a base plate is discussed in [40]. A series of staggered fins is attached to a base plate as shown in Fig. 10-16, and cooling is by radiation and free convection. Enclosure theory was used to evaluate the radiative interaction between the fins and the base surface. The results were in good agreement with an experimental study of this type of configuration.

The free convection induced by heated walls in a two-dimensional channel was studied in [41] and [42]. Two parallel walls of length L are spaced b apart as shown in Fig. 10-17. The internal surfaces A_1 and A_2 can have specified temperatures or specified heat fluxes. Because the walls are heated unequally, an asymmetric free-convection velocity distribution develops. The radiative exchange tends to equalize the convective heat transfer from the walls, which leads to an improvement in the overall convective cooling ability. In addition, there is ra-

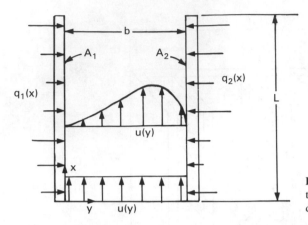

Figure 10-17 Free convection between parallel plates exchanging radiation.

diative energy dissipation through the end openings of the channel. The radiative exchange can considerably alter the free-convection behavior. The heat transfer from the plates can be *augmented* by placing an unheated vertical plate between the walls. Energy is radiated from A_1 and A_2 to this additional surface and is then transferred away by free convection and radiation.

For a single heated vertical plate, there can be an interaction of radiation with free convection depending on the heating condition of the plate. If the plate is uniformly heated, the temperature distribution along the plate depends on the local ability of the plate to dissipate the energy being supplied. The heat transfer behavior was investigated analytically and experimentally in [43]. It was found that if radiative dissipation is included along with free convection, the plate tends to have a more uniform temperature distribution than in the absence of radiation effects.

Radiation exchange can also be important in modifying free-convection instabilities. This was analyzed in [44] for a transparent fluid in a plane layer between two horizontal walls that radiate and conduct energy. The walls exchange radiation, and this modifies their temperature distributions compared with a nonradiating case. The radiation exchange tends to stabilize the fluid against the development of free-convection cells by partially equalizing temperature nonuniformities. The stability of confined plane layers has application to the design of flat-plate solar collectors. The radiation exchange between surfaces can have significant effects on crystals being grown from the vapor phase. An analysis in [45] showed that the radiative exchange can induce free convection in a normally stable configuration, thus altering the vapor transport to the crystal growth interface.

REFERENCES

1. Chambers, R. L., and E. V. Somers: Radiation Fin Efficiency for One-Dimensional Heat Flow in a Circular Fin, *J. Heat Transfer*, vol. 81, no. 4, pp. 327–329, 1959.

2. Keller, H. H., and E. S. Holdredge: Radiation Heat Transfer for Annular Fins of Trapezoid Profile, *J. Heat Transfer*, vol. 92, no. 6, pp. 113–116, 1970.
3. Mackay, Donald B.: "Design of Space Powerplants," Prentice-Hall, Englewood Cliffs, N.J., 1963.
4. Eslinger, R., and B. Chung: Periodic Heat Transfer in Radiating and Convecting Fins or Fin Arrays, *AIAA J.*, vol. 17, no. 10, pp. 1134–1140, 1979.
5. Kern, D. Q. and A. J. Kraus: "Extended Surface Heat Transfer," McGraw-Hill, New York, 1972.
6. Wilkins, J. Ernest, Jr.: Minimum-Mass Thin Fins and Constant Temperature Gradients, *J. Soc. Ind. Appl. Math.*, vol. 10, no. 1, pp. 62–73, 1962.
7. Jaeger, J. C.: Conduction of Heat in a Solid with a Power Law of Heat Transfer at Its Surface, *Cambridge Philos. Soc. Proc.*, vol. 46, pt. 4, pp. 634–641, 1950.
8. Chambré, Paul L.: Nonlinear Heat Transfer Problem, *J. Appl. Phys.*, vol. 30, no. 11, pp. 1683–1688, 1959.
9. Abarbanel, Saul S.: Time Dependent Temperature Distribution in Radiating Solids, *J. Math. Phys.*, vol. 39, no. 4, pp. 246–257, 1960.
10. Stockman, Norbert O., and John L. Kramer: Effect of Variable Thermal Properties on One-Dimensional Heat Transfer in Radiating Fins, NASA TN D-1878, 1963.
11. Sparrow, E. M., and E. R. G. Eckert: Radiant Interaction between Fin and Base Surfaces, *J. Heat Transfer*, vol. 84, no. 1, pp. 12–18, 1962.
12. Sparrow, E. M., G. B. Miller, and V. K. Jonsson: Radiating Effectiveness of Annular-Finned Space Radiators, Including Mutual Irradiation between Radiator Elements, *J. Aerosp. Sci.*, vol. 29, no. 11, pp. 1291–1299, 1962.
13. Heaslet, Max A., and Harvard Lomax: Numerical Predictions of Radiative Interchange between Conducting Fins with Mutual Irradiations, NASA TR R-116, 1961.
14. Nichols, Lester D.: Surface-Temperature Distribution on Thin-Walled Bodies Subjected to Solar Radiation in Interplanetary Space, NASA TN D-584, 1961.
15. Russell, Lynn D., and Alan J. Chapman: Analytical Solution of the "Known-Heat-Load" Space Radiator Problem, *J. Spacecr. Rockets*, vol. 4, no. 3, pp. 311–315, 1967.
16. Schnurr, N. M., A. B. Shapiro, and M. A. Townsend: Optimization of Radiating Fin Arrays with Respect to Weight, *J. Heat Transfer*, vol. 98, no. 4, pp. 643–648, 1976.
17. Masuda, H.: Radiant Heat Transfer on Circular-Finned Cylinders, *Rep. Inst. High Speed Mech. Tohōku Univ.*, vol. 27, no. 255, pp. 67–89, 1973. (See also *Trans. Jpn. Soc. Mech. Eng.*, vol. 38, pp. 3229–3234, 1972.)
18. Badari Narayana, K., and S. U. Kumari: Two-Dimensional Heat Transfer Analysis of Radiating Plates, *Int. J. Heat Mass Transfer*, vol. 31, no. 9, pp. 1767–1774, 1988.
19. Bobco, R. P., and R. P. Starkovs: Rectangular Thermal Doublers of Uniform Thickness, *AIAA J.*, vol. 23, no. 12, pp. 1970–1977, 1985.
20. Frankel, J. I., and T. P. Wang: Radiative Exchange between Gray Fins Using a Coupled Integral Equation Formulation, *J. Thermophys. Heat Transfer*, vol. 2, no. 4, pp. 296–302, 1988.
21. Chung, B. T. F., and B. X. Zhang: Optimization of Radiating Fin Array Including Mutual Irradiations between Radiator Elements, *J. Heat Transfer*, vol. 113, no. 4, pp. 814–822, 1991.
22. Carslaw, H. S., and J. C. Jaeger: "Conduction of Heat in Solids," 2d ed., pp. 154–160, Clarendon, Oxford, U.K., 1959.
23. Masuda, H., and M. Higano: Measurement of Total Hemispherical Emissivities of Metal Wires by Using Transient Calorimetric Technique, *J. Heat Transfer*, vol. 110, no. 1, pp. 166–172, 1988.
24. Sunden, B.: Transient Conduction in a Cylindrical Shell with a Time-Varying Incident Surface Heat Flux and Convective and Radiative Surface Cooling, *Int. J. Heat Mass Transfer*, vol. 32, no. 3, pp. 575–584, 1989.
25. Perlmutter, M., and R. Siegel: Heat Transfer by Combined Forced Convection and Thermal Radiation in a Heated Tube, *J. Heat Transfer*, vol. 84, no. 4, pp. 301–311, 1962.
26. Siegel, R., and M. Perlmutter: Convective and Radiant Heat Transfer for Flow of a Transparent Gas in a Tube with a Gray Wall, *Int. J. Heat Mass Transfer*, vol. 5, pp. 639–660, 1962.

27. Cess, R. D.: The Effect of Radiation Upon Forced-Convection Heat Transfer, *Appl. Sci. Res.*, vol. 10, sec. A., pp. 430–438, 1961.

28. Keshock, E. G., and R. Siegel: Combined Radiation and Convection in an Asymmetrically Heated Parallel Plate Flow Channel, *J. Heat Transfer*, vol. 86, no. 3, pp. 341–350, 1964.

29. Aziz, A., and J. Y. Benzies: Application of Perturbation Techniques to Heat Transfer Problems with Variable Thermal Properties, *Int. J. Heat Mass Transfer*, vol. 19, pp. 271–276, 1976.

30. Siegel, R., and E. G. Keshock: Wall Temperatures in a Tube with Forced Convection, Internal Radiation Exchange, and Axial Wall Heat Conduction, NASA TN D-2116, 1964.

31. Okamoto, Yoshizo: Thermal Performance of Radiative and Convective Plate-Fins with Mutual Irradiation, *Bull. JSME*, vol. 9, no. 33, pp. 150–165, 1966.

32. Okamoto, Yoshizo: Temperature Distribution and Efficiency of a Single Sheet of Radiative and Convective Fin Accompanied by Internal Heat Source, *Bull. JSME*, vol. 7, no. 28, pp. 751–758, 1964.

33. Okamoto, Yoshizo: Temperature Distribution and Efficiency of a Plate and Annular Fin with Constant Thickness, *Bull. JSME*, vol. 9, no. 33, pp. 143–150, 1966.

34. Sohal, M., and J. R. Howell: Thermal Modeling of a Plate with Coupled Heat Transfer Modes, in "Thermophysics and Spacecraft Thermal Control," vol. 35 of *Progress in Astronautics and Aeronautics*, MIT Press, Cambridge, Mass., 1974.

35. Sohal, M., and J. R. Howell: Determination of Plate Temperature in Case of Combined Conduction, Convection, and Radiation Heat Exchange, *Int. J. Heat Mass Transfer*, vol. 16, pp. 2055–2066, 1973.

36. Campo, Antonio: Variational Techniques Applied to Radiative-Convective Fins with Steady and Unsteady Conditions, *Wärme Stoffübertrag.*, vol. 9, pp. 139–144, 1976.

37. Shouman, A. R.: An Exact Solution for the Temperature Distribution and Radiant Heat Transfer along a Constant Cross Sectional Area Fin with Finite Equivalent Surrounding Sink Temperature, *Proc. Ninth Midwestern Mech. Conf.*, Madison, Wis., Aug. 16–18, 1965, pp. 175–186.

38. Shouman, A. R.: Nonlinear Heat Transfer and Temperature Distribution through Fins and Electric Filaments of Arbitrary Geometry with Temperature-Dependent Properties and Heat Generation, NASA TN D-4257, 1968.

39. Razzaque, M. M., J. R. Howell, and D. E. Klein: Finite Element Solution of Heat Transfer for Gas Flow through a Tube, *AIAA J.*, vol. 20, no. 7, pp. 1015–1019, 1982.

40. Guglielmini, G., E. Nannei, and G. Tanda: Natural Convection and Radiation Heat Transfer from Staggered Vertical Fins, *Int. J. Heat Mass Transfer*, vol. 30, no. 9, pp. 1941–1948, 1987.

41. Carpenter, J. R., D. G. Briggs, and V. Sernas: Combined Radiation and Developing Laminar Free Convection between Vertical Flat Plates with Asymmetric Heating, *J. Heat Transfer*, vol. 98, no. 1, pp. 95–100, 1976.

42. Moutsoglou, A., and Y. H. Wong: Convection-Radiation Interaction in Buoyancy-Induced Channel Flow, *J. Thermophys. Heat Transfer*, vol. 3, no. 2, pp. 175–181, 1989.

43. Webb, B. W.: Interaction of Radiation and Free Convection on a Heated Vertical Plate: Experiment and Analysis, *J. Thermophys. Heat Transfer*, vol. 4, no. 1, pp. 117–121, 1990.

44. Lienhard V, J. H.: Thermal Radiation in Rayleigh-Benard Instability, *J. Heat Transfer*, vol. 112, no. 1, pp. 100–109, 1990.

45. Kassemi, M., and W. M. B. Duval: Interaction of Surface Radiation with Convection in Crystal Growth by Vapor Transport, *J. Thermophys. Heat Transfer*, vol. 4, no. 4, pp. 454–461, 1990.

PROBLEMS

10-1 An uncovered styrofoam pan filled with water is placed outdoors on a cloudy night. The air temperature is 2° C (note: K = °C + 273.15). There is almost no wind, so the heat transfer coefficient between the air and water is $h = 11$ W/(m² · K). The cloudy night sky acts as blackbody surroundings at $T_e = 265$ K. Water is opaque in the long-wavelength region. The index of refraction of water is 1.33.

(a) Will the water start to freeze? Show calculations to prove your answer.

(b) What is the minimum air temperature required to prevent freezing?

Answer: (a) freezing does begin; (b) 3.06°C

10-2 A thin two-dimensional fin in vacuum is radiating to outer space, which is assumed at $T_e = 0$ K. The base of the fin is at T_b, and the heat loss from the end edge of the fin is negligible. The fin surface is gray with emissivity ϵ. Write the differential equation and boundary conditions in dimensionless form for determining the temperature distribution $T(x)$ of the fin. (Neglect any radiant interaction with the fin base.) Can you separate variables and indicate the integration necessary to obtain the temperature distribution?

$$\left[Hint: \int \left(\frac{d^2\theta}{dx^2}\right)\left(\frac{d\theta}{dx}\right) = \left(\frac{1}{2}\right)\left(\frac{d\theta}{dx}\right)^2 + constant. \right]$$

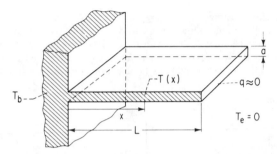

10-3 Consider the fin in Fig. 10-12 as analyzed in Example 10-7. The heat transfer coefficient at the tip of the fin is h_L, and the emissivity of the end area is ϵ as for the rest of the fin surface. Formulate the boundary condition for the end face of the fin, and apply this condition to the general solution of the fin energy equation. Formulate all of the analytical relations, and describe how you would obtain the fin efficiency.

10-4 Consider the fin in Example 10-3 and in the further development in Eq. (10-7a) that follows the example. The environment temperature is nonzero at T_e. The diffuse fin now has more general spectral properties such that the spectral emissivity is $\epsilon_\lambda(\lambda)$. However, it is assumed that $\epsilon_\lambda(\lambda)$ can be approximated as independent of temperature.

(a) Write the energy balance that now applies, of the type in Eq. (10-7a).

(b) Put the equation in dimensionless form similar to Eq. (10-6), using the same parameters where possible.

(c) Discuss how best to obtain the solution for $T(x)$, and how to present the results. Discuss whether the results with $T_e \neq 0$ can be related to those for $T_e = 0$.

10-5 Assume the fin in Example 10-7 extends from a black plane surface at T_b that is very large compared with the fin length. How would the fin formulation be modified to account for the interaction between the fin and the base surface?

10-6 Consider Prob. 7-14. The radiation shield now consists of a layer of opaque plastic 0.15 cm thick coated on each side with a thin layer of metal having the same emissivity, $\epsilon_s = 0.2$, as in Prob. 7-14. The thermal conductivity of the plastic is 0.200 W/(m·K). What is the heat transfer from plate 2 to plate 1, and how does it compare with that for the very thin shield analyzed in Prob. 7-14?

Answer: 19,900 W/m²

10-7 A space radiator is composed of a series of plane fins of thickness $2t$ between tubes. The tubes are at uniform temperature T_b. The tubes are black and the fins are gray with emissivity ϵ. The radiator operates in vacuum with an environment temperature of $T_e = 0$ K. Formulate the differential equation (including the analytical expressions for the configuration factors) and boundary conditions to obtain the temperature distribution $T(x)$ along the fin. Include the interaction between the fin and the tubes.

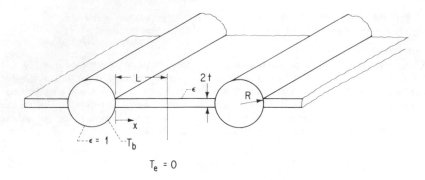

$T_e = 0$

10-8 A thin electrically heated wire is dissipating a total energy rate of Q_e W. The electrodes at the ends of the wire are water cooled, and both are maintained at T_0. The surroundings are large in size and are at a uniform temperature T_e. The wire surface is diffuse-gray with emissivity ϵ. Air at T_e is blowing across the wire, providing a uniform heat transfer coefficient h. Set up the differential equation and boundary conditions to obtain the wire temperature distribution.

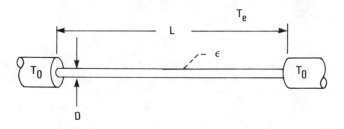

10-9 Two thin vertical posts stand immediately adjacent to a pool of molten material at temperature T_m and are diametrically across the pool from one another. The posts have a square cross section of area A_x and are of length L, and they have thermal conductivity k. The entire surface of the posts has emissivity ϵ. The pool is of radius R. A breeze blows across the posts, and the air has temperature T_a. The air motion produces a heat transfer coefficient h between the post surface and the air.

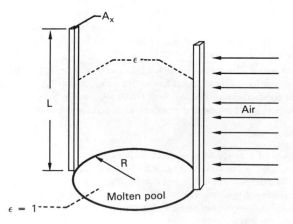

Derive an equation for the temperature distribution along the posts, including the effect of mutual radiative exchange by the posts. Assume that the temperature at the bottom of the posts is equal to the temperature of the molten pool. Also, assume that the effect of the temperature of the surroundings on radiative transfer can be neglected. The pool surface is assumed black. Show the necessary boundary conditions for the problem and relations for all of the required configuration factors. (You do not need to substitute the F into the equation, however.)

10-10 Steam is condensing inside a thin tube of radius r_i. The tube has a coating of emissivity $\epsilon_t = 1$ on the outer surface. The saturation temperature of the steam is T_b. Identical annular fins of outer radius r_o and emissivity ϵ_f are evenly spaced a distance L (between fin faces) along the tube. The fins are of thickness δ and thermal conductivity k. The environment surrounding the fin-tube assembly is at $T_e = 0$ K. Convection may be ignored. The configuration factor from a ring element on the tube to a ring element on fin 1 is $dF_{t-d1}(x, \rho_1)$ and from a ring element on fin 2 to a ring element on fin 1 is $dF_{d2-d1}(\rho_1, \rho_2)$.

Set up the governing equation for the temperature distribution of the fin, $T(\rho)$.

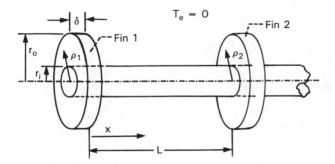

10-11 Two directly opposed parallel diffuse-gray plates of finite width W have a uniform heat flux q_e supplied to each of them. The plates are infinitely long in the direction normal to the cross section shown. They are separated by a distance H. The plates are each of thickness t ($t \ll W$) and have thermal conductivity k. The plates are in vacuum, and the surroundings are at T_e. Set up the governing integral equation for finding $T(x)$. The outer surface of each plate is insulated, so all radiative heat exchange is from the inner surface.

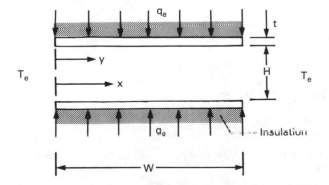

10-12 A copper-constantan thermocouple ($\epsilon = 0.15$) is in an inert gas stream at 350 K adjacent to a large blackbody surface at 900 K. The heat transfer coefficient from the gas to the thermocouple is 25 W/(m$^2 \cdot$ K). Estimate the thermocouple temperature if it is

(a) bare

(b) surrounded by a single polished aluminum radiation shield in the form of a cylinder open at both ends. The heat transfer coefficient from the gas to both sides of the shield is 15 W/(m² · K).

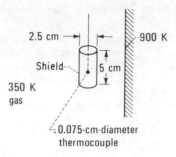

2.5 cm →

900 K

Shield--

5 cm

350 K
gas

0.075-cm-diameter
thermocouple

Answer: (a) 450 K

10-13 Thin wire is extruded at fixed velocity through a die at temperature T_0. The wire then passes through air at T_a until its temperature is reduced to T_L. The heat transfer coefficient to the air is h, and the wire emissivity is ϵ. It is desired to obtain the relation between T_L and T_0 as a function of wire velocity V and distance L. Derive a differential equation for wire temperature as a function of distance from the die, and state the boundary conditions. (*Hint:* Compute the energy balance for flow in and out of a control volume fixed in space.)

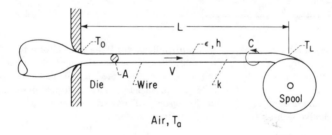

L

T_0 ϵ, h C T_L

A   V

Die   Wire   k

Spool

Air, T_a

10-14 A single circular fin is to dissipate energy from both sides in vacuum to surroundings at low temperature. The fin is on a tube with 1-cm outer diameter. The tube wall is maintained at 1000 K by vapor condensing on the inside of the tube. The fin has 10-cm outer diameter and is 0.15 cm thick. Estimate the rate of energy loss by radiation if the fin is made from

(a) copper with a polished surface (Fig. 5-25a);

(b) copper with a lightly oxidized surface;

(c) stainless steel [$k = 35$ W/(m · K)] with a clean surface ($\epsilon = 0.3$). What is the effect on energy dissipation of increasing the fin thickness to 0.30 cm?

Answer: (c) 69 W, 101 W

10-15 How would the analysis in Prob. 10-10 be modified to include the effect of a nonzero environment temperature?

10-16 The billet in Prob. 7-3 has air at 27° C blowing across it that provides an average convective heat transfer coefficient of 15 W/(m² · K). Estimate the cooling time with both radiation and convection included.

Answer: 2.74 h

10-17 Opaque liquid at temperature $T(0)$ and mean velocity $\bar{u}$ enters a long tube that is surrounded by a vacuum jacket and a concentric electric heater that is kept at uniform axial temperature T_e. The heater is black, and the tube exterior is diffuse-gray with emissivity ϵ. The convective heat transfer coefficient between the liquid and the tube wall is h, and the tube wall thermal conductivity is k_w. Derive the relations to determine the mean fluid temperature and the tube wall outer surface temperature as a function of distance x along the tube (assume that the liquid properties are constant).

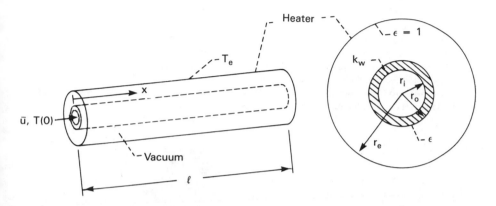

10-18 A solar collector is designed to fit onto the horizontal section of a roof as diagrammed at the left below. Flow is from the right to the left in the tubes of the collector. A white diffuse roof section helps to reflect additional solar flux onto the collector. Set up the equations for determining the local temperature of the tubes for two cases:

(a) no flow in the tubes
(b) flow of water in each tube of 2.2 kg/min. Indicate a possible solution method. Assume that the roof and collector are very long normal to the cross section shown.

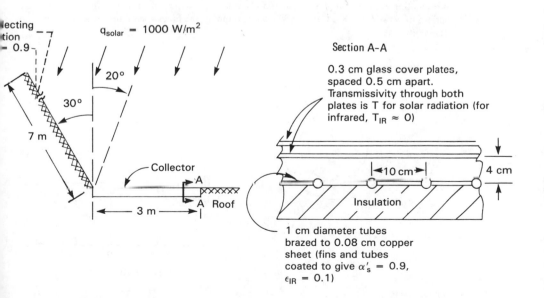

10-19 A highly polished copper tube carries condensing steam at temperature T_s. The outer tube surface is very close to the steam temperature because of the high conductivity of copper and the large condensation heat transfer coefficient, and the surrounding air and environment are at T_e. Energy losses are Q W and are deemed excessive. Insulation of thermal conductivity k and thickness t is added to the outside of the copper tube. The infrared emissivity of the insulation surface is ϵ. When the system is returned to operation, the energy losses are found to be greater than before! Over what ranges of parameters (ϵ, k, t, T_s) is this possible? Put your results in terms of dimensionless groups, and plot the results. (*Hint:* the problem can be linearized by assuming $T_s/T_e = 1 + \Delta$, where $\Delta \ll 1$.)

10-20 A rod of circular cross section extends out of a space vehicle in earth orbit into surroundings at T_e. The rod axis is normal to the direction from the sun. The rod is coated so that its infrared emissivity is ϵ_{IR} and its solar absorptivity is α_s. The base of the rod is at $T_b > T_e$. Derive a differential equation to predict the rod temperature distribution, $T(x)$. State the boundary conditions, including radiation at the circular end face. Neglect temperature variations within the rod cross section at each x. Neglect radiation to the rod from the vehicle surface, and neglect any radiation emitted or reflected from the earth.

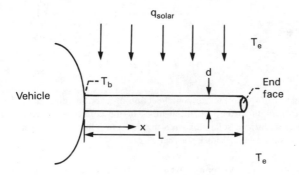

10-21 A_1 and A_2 are diffuse concentric spheres. The inner surface of the inner sphere is heated with nonradiating combustion products at 1050 K with a convective heat transfer coefficient to the surface of 60 W/(m² · K). Calculate the temperature of the inner sphere for the surface spectral emissivities shown.

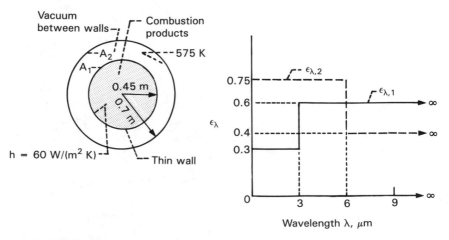

Answer: 859 K

10-22 A wire between two electrodes is heated electrically with a total of Q_e W. The wire resistivity r_e ohm-cm is constant, and the current is I. One end of the wire is at T_1 and the other is at T_2. The immediate surroundings are a vacuum, and the surroundings have a radiating temperature of T_o. The wire is gray with emissivity ϵ, diameter D, thermal conductivity k, and length L.

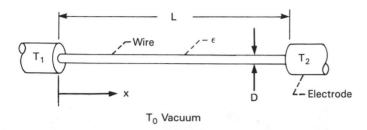

(a) Set up the differential equation for the steady-state temperature distribution along the wire, neglecting radial temperature variations within the wire. Integrate the equation, and find an expression (in the form of an integral) for the wire temperature as a function of x. The final results should contain only the quantities given and should not contain dT/dx. Explain how you would evaluate the expression to find $T(x)$.

(b) Let both the emissivity and the electrical resistivity be proportional to T [see Eqs. (4-75) and (4-76)]. Derive the solution for this case, and describe how it can be evaluated for a fixed value of Q_e.

10-23 A spherical temperature sensor, 0.4 cm in diameter, is on the axis of a short pipe, halfway between the open ends. Air at 475 K is flowing through the pipe, and the convective heat transfer coefficient on the sensor is 20 W/(m² · K). Calculate the sensor temperature. All surfaces are gray. Neglect blockage (shadowing) by the sensor when computing configuration factors between boundaries of the pipe. Do not subdivide surfaces.

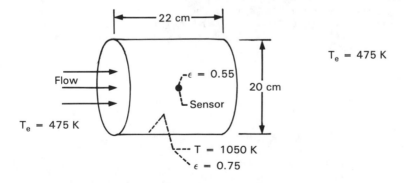

Answer: 866.4 K

ELEVEN

NUMERICAL SOLUTION TECHNIQUES FOR RADIATION PROBLEMS WITH OTHER ENERGY MODES

11-1 INTRODUCTION

In this chapter numerical methods are presented and applied to problems where radiation is combined with conduction and/or convection. The methods also apply to pure radiation problems, but these are sometimes not as difficult because, for gray surfaces, they are linear in the variable of temperature to the fourth power. In the previous derivations for radiation in systems including combined energy transfer modes, the basic energy equations were obtained from a heat balance on each element of the system used to model the real configuration. For pure radiation problems, detailed temperature distributions may not be needed and an enclosure may be divided into relatively few elements. With conduction and convection present, there is usually a need for finer detail. Because these modes depend on local temperature derivatives, accurate temperature distributions must be available so derivatives can be properly evaluated. Sometimes it is the difference between large incoming and outgoing radiation terms that is needed to determine a relatively small amount of conduction and/or convection, and this leads to convergence difficulties. If the temperature derivatives are not sufficiently accurate, erroneous solutions are obtained.

The local energy balance on each element of the system involves the net radiation that is absorbed at the surface, and this depends on the summation of all the radiative contributions from the surroundings. In a detailed formulation the radiative portion may be set up as an integration involving the temperature distributions and configuration factors from the surrounding surfaces. This type of formulation was carried out in Sec. 7-5 and provided the enclosure equations in the form of Eq. (7-62).

To set the integrals in this enclosure theory into finite-difference form for numerical calculation, a numerical integration method must be chosen, such as the trapezoidal rule, Simpson's rule, or Gaussian integration. Some of these methods are presented here and illustrated with a few examples. Integration subroutines are available in computer packages, and these can be used. Some existing thermal analysis codes are pointed out that can be applied to multidimensional problems.

Sec. 11-4 presents a discussion of the benefits of nondimensionalization in treating multimode problems. The relative sizes of the dimensionless parameters may provide insight into the best numerical approach. Some example problems are set up using finite-difference and finite-element methods. The result of the numerical formulation is a set of nonlinear algebraic equations, and methods are discussed for their solution. A common method is by successive substitutions or iteration, and because of the nonlinearity of the combined-mode equations it is usually necessary to use damping or underrelaxation factors to obtain convergence. Some current information is presented on the amount of underrelaxation to be used. Problems involving only radiation will sometimes permit overrelaxation to speed convergence. The Newton-Raphson method is presented as another very useful technique for solving a set of nonlinear algebraic equations.

The last section of the chapter presents the Monte Carlo method for multimode problems. This is a statistical method in which small quantities of energy are followed along their individual paths during radiative transfer. This method is relatively easy to set up for complex problems that involve spectral effects and/or directional surfaces. The solutions may require long computer running times but may be the only way to attack some complex problems, especially those involving directional surfaces.

11-2 SYMBOLS

A	surface area
$A, B, C, D, \ldots$	constants
A_{ij}, B_{ij}, C_i	matrix coefficients
b	thickness of wall
B	dimensionless width of base surface
c	specific heat of solid
E	exchange factor including direct exchange and all reflection paths
f	function in numerical integration
$f(\xi)$	frequency distribution of events occurring at ξ
F	configuration factor
$F_{0-\lambda}$	fraction of total blackbody energy emitted in the range 0 to λ

h	convective heat transfer coefficient
H	convection-radiation parameter
i	radiant intensity
I	total number of subsets used to compute mean
K	integration kernel
l, m	lattice indices in square mesh corresponding to x, y positions
n	individual sample index
N	conduction-radiation parameter; total number of sample bundles per unit time
P	probability density function
$\bar{P}$	mean of calculated values of P
q	heat flux
Q	energy per unit time
r	radius
$\mathbf{r}$	position vector
R	dimensionless radial coordinate; number chosen at random from evenly distributed set of numbers in range 0 to 1
$\mathbf{R}$	dimensionless position vector
S	dimensionless heat generation parameter; number of events occurring at a position
St	Stanton number
t	dimensionless temperature
T	absolute temperature
w	energy carried by sample Monte Carlo bundle
W	weighting function
x, y, z	position coordinates
X, Y, Z	dimensionless coordinates
α	surface absorptivity
β_i	coefficients in shape function in finite-element method
γ	standard deviation defined by Eq. (11-62)
$\delta, \delta', \delta''$	indices in computer program, Fig. 11-14
δ_{ij}	Kronecker delta
ϵ	surface emissivity
η	function defined by Eq. (11-61)
θ	cone angle
λ	wavelength
μ	probable error; dimensionless fin radiation-conduction parameter
ξ	radial position variable; variable in Eq. (11-54)
ρ	radial coordinate
σ	Stefan-Boltzmann constant
τ	time
ϕ	dimensionless heat flux, Eq. (11-15)
φ	circumferential angle
Φ	shape function in finite-element method
ω	the variable t^4

Subscripts

b	blackbody
e	emitted
g	gas
i	incoming; inside
o	outgoing; outside
λ	spectrally dependent
1, 2	at surface 1 or 2

Superscripts

$'$	quantity in one direction
$''$	bidirectional quantity
$*$	denotes dummy variable

11-3 NUMERICAL INTEGRATION METHODS FOR USE WITH ENCLOSURE EQUATIONS

This section provides some integration methods for use in the numerical solution of pure radiation or combined-mode problems. The integrals to be placed into numerical form are often functions of two position variables, and integration is over one or both of them. For example, the configuration factor dF_{di-dj} from position $\mathbf{r}_i$ on surface i to position $\mathbf{r}_j$ on surface j appears in the integral over surface j to obtain F_{di-j} in the form

$$F_{di-j}(\mathbf{r}_i) = \int_{A_j} dF_{di-dj}(\mathbf{r}_i, \mathbf{r}_j) = \int_{A_j} K(\mathbf{r}_i, \mathbf{r}_j)\, dA_j \qquad (11\text{-}1)$$

There are many ways to numerically approximate an integral. Because the integrands in radiative enclosure problems are usually well behaved at the end points, it is often useful to employ *closed* numerical integration forms that include values at the end points in the evaluation of the integral. Closed methods allow inclusion of known boundary values in a numerical formulation. *Open* methods do not include the end points and can be utilized when the end point values are indeterminate, e.g., for improper integrals that yield finite values when integrated. In problems including convection and/or conduction, the numerical integration will in some way usually use the grid spacing imposed by the differential terms. In some situations it is sufficient to use numerical integration methods that have a regular grid spacing. In recent work, uneven spacings have been used, for example, in adaptive gridding. *Gaussian quadrature* can be used for variable grid spacing. Simpler schemes such as the *trapezoidal rule* or *Simpson's rule* may be adequate for some problems. These usually employ uniform grid spacing and are closed, whereas Gaussian quadrature is an open method. A variable grid size is very useful for situations in which there are large variations, as in a region with large local temperature gradients. In this way more grid points can be placed in regions as required for adequate definition of the functions. The trapezoidal rule can readily be used with a nonuniform grid size. To supplement the information here, textbooks on numerical methods such as [1–3] provide detailed presentations of the many available integration methods and their relative accuracies, advantages, and disadvantages. Libraries of computer codes have available many subroutines for single or multidimensional numerical integrations; these can be directly used for many problems.

11-3.1 Trapezoidal Rule

The *trapezoidal rule* is a closed numerical integration method that usually employs equal increment spacing, but the method can easily employ a variable increment size. Consider the function in Fig. 11-1, in which an equal grid spacing of Δz is shown, and the range of integration is from z_0 to z_N. The $f_j(x)$ is the value of $f(x, z)$ evaluated at fixed x and at $z = z_j = j\,\Delta z$, where $0 \le j \le N$. In the trapezoidal rule each pair of adjacent points, such as $f(x, z_j)$ and $f(x, z_{j+1})$, is connected by a straight line. Then the integral from z_j to z_{j+1} is approximated by

$$\int_{z_j}^{z_{j+1}} f(x, z)\, dz \approx (z_{j+1} - z_j) \frac{f_j(x) + f_{j+1}(x)}{2}$$

This approximation can be made for each interval between grid points, so an irregular grid spacing can be used. For the case of *equally sized increments,* the sum over all intervals gives the approximation for the integral as (since this is for multiple intervals it is sometimes called the *extended* trapezoidal rule)

$$\int_{z_0}^{z_N} f(x, z)\, dz \approx \Delta z \left[\frac{1}{2} f_0(x) + \sum_{j=1}^{N-1} f_j(x) + \frac{1}{2} f_N(x) \right] \tag{11-2}$$

EXAMPLE 11-1 Using the ring-to-ring factor, use the extended trapezoidal rule to evaluate the configuration factor from a ring element on the interior of a right circular cylinder to the cylinder base for the geometry in Fig. 11-2 and with $x = r$. Compare the results with the known analytical solution.

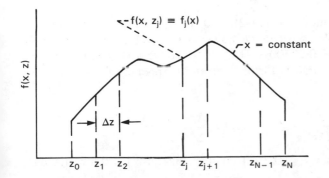

Figure 11-1 Numerical integration of the function $f(x, z)$ with respect to z for a fixed x.

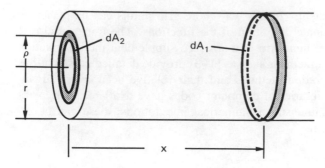

Figure 11-2 Geometry for configuration factor from ring element on interior of cylinder to ring element on base.

The factor from dA_1 to a ring dA_2 on the base surface is given by

$$dF_{d1-d2} = \frac{2XR(1 + X^2 - R^2)\,dR}{[(1 + X^2 + R^2)^2 - 4R^2]^{3/2}} \tag{11-3}$$

where $X = x/r$ and $R = \rho/r$. For this example, $X = 1$, so

$$dF_{d1-d2}(X = 1) = \frac{2R(2 - R^2)\,dR}{(4 + R^4)^{3/2}} \tag{11-4}$$

To apply (11-2), the function $f_j(X = 1)$ is given by

$$f_j(X = 1) = \frac{2R_j(2 - R_j^2)}{(4 + R_j^4)^{3/2}}$$

where $R_j = j\,\Delta R$ and $\Delta R = 1/N$. In particular,

$$f_0(1) = 0, \qquad f_j(1) = \frac{2j\,\Delta R[2 - (j\,\Delta R)^2]}{[4 + (j\,\Delta R)^4]^{3/2}},$$

$$f_N(1) = \frac{2N(1/N)\{2 - [N(1/N)]^2\}}{\{4 + [N(1/N)]^4\}^{3/2}} = \frac{2}{5^{3/2}}$$

These terms are substituted into (11-2), and the approximate integration is carried out. Choosing $N = 5$ yields $dF_{d1-d2}(1) = (1/5)[(1/2) \times 0 + 0.09794 + 0.18225 + 0.23451 + 0.23500 + (1/2) \times 0.17889] = 0.16783$. For comparison, the exact configuration factor is given in Appendix C as

$$F_{d1-2}(X = 1) = \frac{X^{*2} + \dfrac{1}{2}}{(X^{*2} + 1)^{1/2}} - X^*, \qquad X^* = \frac{x}{2r} = \frac{X}{2} = \frac{1}{2}$$

which gives $F_{d1-2}(X = 1) = F_{d1-2}(X^* = 0.5) = (0.75/1.25^{1/2}) - 0.5 = 0.17082$.

The numerical integration is -1.75% in error with respect to the exact result. Larger numbers of increments improve the accuracy as follows:

N	$F_{d1-2}(X = 1)$	% Error
5	0.16783	-1.75
10	0.17007	-0.44
50	0.17079	-0.02
100	0.17081	-0.006
200	0.17082	0

11-3.2 Simpson's Rule

The trapezoidal rule is derived by connecting each pair of adjacent points on the curve in Fig. 11-1 with a straight line. The usual Simpson's rule is obtained by passing a second-order (parabolic) curve through three adjacent points. The result for equally spaced increments is the integral from z_j to z_{j+2} approximated by

$$\int_{z_j}^{z_{j+2}} f(z)\, dz \approx \frac{\Delta z}{3}\, (f_j + 4f_{j+1} + f_{j+2})$$

Since this uses two intervals (three points), the repeated application to cover a range having many grid points requires an *even* number of intervals (odd number of points). For N *equally spaced* increments in Fig. 11-1, the result is

$$\int_{z_0}^{z_N} f(z)\, dz \approx \frac{\Delta z}{3}\, (f_0 + 4f_1 + 2f_2 + 4f_3 + \cdots + 4f_{N-1} + f_N) \tag{11-5}$$

If an odd number of increments must be used, Simpson's rule can be applied over an even number of $N - 1$ increments, and the trapezoidal rule used for the one remaining increment.

If a curve goes through a sharp, cusplike peak, it may not be accurate to apply Simpson's rule if the peak is at the central point of the three adjacent points; the cusplike behavior is not accurately approximated by a second-order curve. Simpson's rule could be used on each side of the peak, or the trapezoidal rule may be more convenient to apply. Care should be used in selecting a suitable integration scheme for each application.

Higher-order approximations have been developed by passing a cubic curve through four adjacent points, a fourth-order curve through five adjacent points, etc. This yields the set of *Newton-Cotes closed integration formulas* of which the trapezoidal and Simpson's rules are the first two. The approximation using a cubic curve through four adjacent points is called *Simpson's second rule*,

$$\int_{z_j}^{z_{j+3}} f(z)\, dz \approx \frac{3\,\Delta z}{8}\, (f_j + 3f_{j+1} + 3f_{j+2} + f_{j+3}) \tag{11-6}$$

11-3.3 Other Integration Methods and Computer Subroutines

Formulas for only a few of the most common numerical integration methods have been presented here. An excellent description of additional techniques is in [1] and in similar books on numerical analysis [2, 3]. The techniques include Romberg integration, in which the trapezoidal rule is utilized. The integration is performed with a small number of increments and is then repeated for twice the number of increments (by adding the contribution from the additional interior points), four times the number, etc. The sequence of integration results is extrapolated to an improved result using the technique of Richardson extrapolation (see [1, 2]). The process is continued until a desired accuracy of convergence is achieved in the extrapolated result.

A very useful technique is *Gaussian integration,* which is an open integration method using an array of unevenly spaced points. The uneven points can be positioned between a fixed grid of evenly or variably spaced points. Curve fitting, such as by cubic splines, is performed for the individual portions of the curve between the fixed grid points. Values of the integrand at positions between the grid points are interpolated for use in the Gaussian method.

Many numerical integration subroutines have been written for computer use (see [2, 3] for example), and the software can readily be applied. Curve-fitting software routines are also available that can be used in conjunction with Gaussian or other techniques, requiring interpolation to obtain unevenly spaced values of the function being integrated. Some subroutines perform multidimensional integrations. An example is the set of FORTRAN subroutines in [4]; others are available for both mainframe and personal computers. Most computational software packages provide numerical integration using, e.g., Romberg integration and Simpson's rule.

11-4 NUMERICAL EQUATIONS FOR COMBINED-MODE PROBLEMS

Figures 10-2 and 10-3 illustrated the boundary condition for an opaque surface where the net radiation $q_i - q_o$ provides the local addition of radiative heat flux. The $q_i - q_o$ is found by analyzing the radiative enclosure surrounding the surface. This boundary condition can be applied to obtain a three-dimensional conduction solution in the wall. As a simplified case, Fig. 10-5 gave a control volume derivation for a thin wall in which the temperature distribution is two-dimensional, as it is assumed not to vary significantly across the wall thickness. The control volume approach is further developed here to illustrate combined-mode solutions by using enclosures with thin walls as an example.

Consider an enclosure as in Fig. 7-14, and let one or more of the walls be thin and heat conducting. There is uniform heat generation within each wall and convection at the inside surfaces as in Fig. 11-3. The outside of the enclosure is assumed well insulated for simplicity, but external heat flows can be added in a similar fashion to those included here.

A local rectangular coordinate system is positioned along a typical wall A_k. For a wall element at $\mathbf{r}_k$, an energy balance is written where the net radiative loss is balanced by the decrease in internal energy and by the energy added by conduction, convection, and internal heat generation,

$$q_{o,k}(\mathbf{r}_k) - q_{i,k}(\mathbf{r}_k) = q_k(\mathbf{r}_k) = \left[-\rho c a \frac{\partial T_k}{\partial \tau} + k a \left(\frac{\partial^2 T_k}{\partial x^2} + \frac{\partial^2 T_k}{\partial y^2} \right) \right.$$

$$\left. + h(T_m - T_k) + q'''a \right]_{\mathbf{r}_k} \qquad (11\text{-}7a)$$

The local radiative heat loss $q_k(\mathbf{r}_k)$ is found from the enclosure equations (7-62),

$$\frac{q_k(\mathbf{r}_k)}{\epsilon_k} - \sum_{j=1}^{N} \frac{1 - \epsilon_j}{\epsilon_j} \int_{A_j} q_j(\mathbf{r}_j) \, dF_{dk-dj}(\mathbf{r}_j, \mathbf{r}_k)$$

$$= \sigma T_k^4(\mathbf{r}_k) - \sum_{j=1}^{N} \int_{A_j} \sigma T_j^4(\mathbf{r}_j) \, dF_{dk-dj}(\mathbf{r}_j, \mathbf{r}_k) \qquad (11\text{-}7b)$$

The local temperature of the convecting medium, $T_m(\mathbf{r}_k)$, is obtained from additional convective heat transfer relations as will be illustrated. The equations are placed in dimensionless form to yield

$$\bar{q}_k(\mathbf{R}_k) = \left[-\frac{\partial t_k}{\partial \bar{\tau}} + N \left(\frac{\partial^2 t_k}{\partial X^2} + \frac{\partial^2 t_k}{\partial Y^2} \right) + H(t_m - t_k) + S \right]_{\mathbf{R}_k} \qquad (11\text{-}8a)$$

$$\frac{\bar{q}_k(\mathbf{R}_k)}{\epsilon_k} - \sum_{j=1}^{N} \frac{1 - \epsilon_j}{\epsilon_j} \int_{A_j} \bar{q}_j(\mathbf{R}_j) \, dF_{dk-dj}(\mathbf{R}_j, \mathbf{R}_k)$$

$$= t_k^4(\mathbf{R}_k) - \sum_{j=1}^{N} \int_{A_j} t_j^4(\mathbf{R}_j) \, dF_{dk-dj}(\mathbf{R}_j, \mathbf{R}_k) \qquad (11\text{-}8b)$$

where $\bar{q} = q/\sigma T_{\text{ref}}^4$, $N = k/a\sigma T_{\text{ref}}^3$, $H = h/\sigma T_{\text{ref}}^3$, $S = q'''a/\sigma T_{\text{ref}}^4$, $\bar{\tau} = \tau\sigma T_{\text{ref}}^3/\rho c a$, $\mathbf{R} = \mathbf{r}/a$, $X = x/a$, $Y = y/a$ and $t = T/T_{\text{ref}}$.

In Eqs. (11-8a,b) the dimensionless temperature t and the dimensionless net radiative heat flux $\bar{q}$ are the dependent variables, and X, Y, and $\bar{\tau}$ are the independent variables. The dimensionless parameters N, S, and H (or slight modifications of them) appear in combined-mode problems involving radiative transfer. They provide a measure of the importance, relative to radiation, of conduction, internal energy generation, and convection, and their relative magnitudes can be a guide to the best solution method. If H is large, the problem could be solved as convective heat transfer with a small effect of radiation; thus, the solution will probably converge best if t is chosen as the dependent variable. The radiation provides a nonlinear perturbation on the linear convection problem. For small H, the problem might best be solved as a radiative transfer problem using t^4 as the dependent variable. The convection is a nonlinear perturbation on the radiation

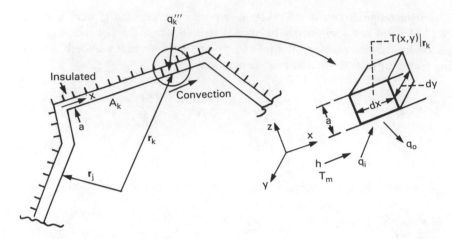

Figure 11-3 Radiative enclosure with walls in which there is two-dimensional heat conduction.

problem, which is linear in the variable t^4. In a transient problem, the temperature changes may shift the relative importance of the modes. Once the best arrangement of the equations is determined, they must be placed in a form for numerical solution. The application of finite-difference and finite-element methods is described in some examples that follow.

Equation (11-8b) is the enclosure equation for gray surfaces. If the enclosure has surface properties *that vary with wavelength*, Eq. (8-15) should be used. This is solved for $q_{\Delta\lambda,k}$ in each wavelength band for each surface. The $\bar{q}_k(\mathbf{R}_k)$ in (11-8a) is then found as the summation $\sum_{\Delta\lambda} \bar{q}_{\Delta\lambda,k}$. Equation (11-8a) is otherwise unchanged.

11-4.1 Finite-Difference Formulation

To describe the solution technique by finite differences, two examples are set up.

EXAMPLE 11-2 The fin temperature distribution in the array of fins in Fig. 10-9 is considered in Example 10-5 and is governed by the dimensionless energy equation

$$-\mu \frac{d^2t(X)}{dX^2} + t^4(X) = F_{dX-B}(X) + \int_{Z=0}^{1} \left[-\mu(1-\epsilon) \frac{d^2t(Z)}{dZ^2} + t^4(Z) \right]$$
$$\cdot dF_{dX-dZ}(X, Z) \tag{11-9}$$

where

$$F_{dX-B}(X) = \frac{1}{2} \left[1 - \frac{X}{(B^2 + X^2)^{1/2}} \right]$$

$$dF_{dX-dZ}(X, Z) = \frac{1}{2} \frac{B^2}{[B^2 + (Z - X)^2]^{3/2}} \, dZ$$

$$\mu = ka/2\epsilon\sigma T_b^3 W^2 \quad \text{and} \quad B = b/W.$$

The boundary conditions are $t = 1$ at the base $X = 0$ and $dt/dX = 0$ at $X = 1$ because it is assumed for simplicity that the end of the fin has negligible energy loss. Only the single unknown $t(X)$ appears in this equation because all fins in the infinite array have the same temperature distribution. The fourth power of the temperature at X, and its second derivative, are on the left-hand side of the energy equation; however, to evaluate the integral on the right-hand side, *the entire distribution of temperature and its second derivative must be known along the fin.* Thus, if this equation is written at various X values, each equation will involve the unknown $t(X)$ at every other X location.

To solve this equation, the fin is divided into N small elements, so that $X_i = i\,\Delta X$ and $Z_j = j\,\Delta Z$, where $0 \le i \le N$ and $0 \le j \le N$. In the energy equation, the second derivative is approximated by

$$\frac{d^2 t}{dX^2} = \frac{t_{i+1} - 2t_i + t_{i-1}}{(\Delta X)^2}$$

The integral on the right-hand side of (11-9) can be approximated using the trapezoidal rule (11-2) as

$$\int_0^1 f(X, Z)\, dZ \approx \Delta Z \left[\frac{1}{2} f_0(X) + \sum_{j=1}^{N-1} f_j(X) + \frac{1}{2} f_N(X) \right]$$

where

$$f(X, Z) = \frac{B^2}{2} \left[-\mu(1 - \epsilon)\frac{d^2 t}{dZ^2} + t^4(Z) \right] \frac{1}{[B^2 + (Z - X)^2]^{3/2}}$$

Using $\Delta Z = \Delta X$ and substituting the finite-difference form of the second derivative results in

$$f_j(X_i) = \frac{B^2}{2} \left[-\mu(1 - \epsilon)\frac{t_{j+1} - 2t_j + t_{j-1}}{(\Delta X)^2} + t_j^4 \right] \frac{1}{\{B^2 + [(j - i)\,\Delta X]^2\}^{3/2}}$$

The two limits where $j = 0$ and $j = N$ are evaluated by applying the boundary conditions. For $j = i = 0$, $t_0 = 1$ and $d^2 t/dZ^2 = 0$ (this condition is needed because the t_{i-1} term would otherwise be undefined). The latter condition arises because the net energy entering the fin at the base is all by conduction, so the temperature gradient must be linear at that location. Then, for $j = 0$,

$$f_0(X_i) = \frac{B^2}{2[B^2 + (i\,\Delta X)^2]^{3/2}}$$

For $j = N$, $t_{N+1} = t_{N-1}$ from $(dt/dX)_{X=1} = (t_{N+1} - t_{N-1})/(2\,\Delta X) = 0$ ($N + 1$ is a symmetric image point of $N - 1$). Then

$$f_N(X_i) = \frac{B^2}{2} \left[-\mu(1 - \epsilon)\frac{2(t_{N-1} - t_N)}{(\Delta X)^2} + t_N^4 \right] \frac{1}{\{B^2 + [(N - i)\,\Delta X]^2\}^{3/2}}$$

Now, the energy equation (11-9) for element i at $X_i = i\,\Delta X$ can be written in finite-difference form as

$$-\mu\frac{t_{i+1} - 2t_i + t_{i-1}}{(\Delta X)^2} + t_i^4 = \frac{1}{2}\left\{1 - \frac{1}{\{[B/(i\,\Delta X)]^2 + 1\}^{1/2}}\right\}$$

$$+ \Delta X\left[\frac{1}{2}f_0(X_i) + \sum_{j=1}^{N-1} f_j(X_i) + \frac{1}{2}f_N(X_i)\right] \qquad (11\text{-}10)$$

This equation is written for each element i for the range $1 \le i \le N$, giving N equations for the N unknown values of temperature t_i. Each equation contains every unknown t_i, which appear in the f_j terms. If the resulting set of equations is written in matrix form, the matrix of coefficients is full. This is in contrast to one-dimensional pure conduction problems, the coefficient matrix is usually tridiagonal. This reflects the fact that in radiation problems the temperature of every element can be influenced by the temperature of every other element.

To continue this illustration of a finite difference formulation, Eq. (11-10) is written for element $i = 1$:

$$-\mu\frac{t_2 - 2t_1 + 1}{(\Delta X)^2} + t_1^4$$

$$= \frac{1}{2}\left\{1 - \frac{1}{[(B/\Delta X)^2 + 1]^{1/2}}\right\} + \frac{\Delta X\,B^2}{2}\left(\frac{1}{2}\frac{1}{[B^2 + (\Delta X)^2]^{3/2}}\right.$$

$$- \sum_{j=1}^{N-1}\left[\mu(1 - \epsilon)\frac{t_{j+1} - 2t_j + t_{j-1}}{(\Delta X)^2} - t_j^4\right]\frac{1}{\{B^2 + [(j-1)\,\Delta X]^2\}^{3/2}}$$

$$+ \frac{1}{2}\left[-\mu(1 - \epsilon)\frac{2(t_{N-1} - t_N)}{(\Delta X)^2} + t_N^4\right]\frac{1}{\{B^2 + [(N-1)\,\Delta X]^2\}^{3/2}}\right)$$

$$(11\text{-}11)$$

Gathering terms that have t_j and t_j^4 and using $t_0 = 1$, (11-11) is written as (a repeated subscript denotes a summation over the values of that subscript, e.g., $A_{1j}t_j = \sum_{j=1}^{N}A_{1j}t_j$)

$$A_{1j}t_j + B_{1j}t_j^4 = C_1 \qquad (11\text{-}12)$$

where

$$A_{11} = \frac{B^2\mu(1 - \epsilon)}{2\,\Delta X}\left\{\frac{4}{(1 - \epsilon)\,\Delta X\,B^2} - \frac{2}{B^3} + \frac{1}{[B^2 + (\Delta X)^2]^{3/2}}\right\}$$

$$A_{12} = \frac{B^2\mu(1-\epsilon)}{2\,\Delta X}\left\{\frac{-2}{(1-\epsilon)\,\Delta X\,B^2} + \frac{1}{B^3} - \frac{2}{[B^2+(\Delta X)^2]^{3/2}}\right.$$

$$\left. + \frac{1}{[B^2+(2\,\Delta X)^2]^{3/2}}\right\}$$

$$\vdots$$

$$A_{1j} = \frac{B^2\mu(1-\epsilon)}{2\,\Delta X}\left(\frac{1}{\{B^2+[(j-2)\,\Delta X]^2\}^{3/2}} - \frac{2}{\{B^2+[(j-1)\,\Delta X]^2\}^{3/2}}\right.$$

$$\left. + \frac{1}{[B^2+(j\,\Delta X)^2]^{3/2}}\right) \qquad 2<j<N-1$$

$$\vdots$$

$$A_{1(N-1)} = \frac{B^2\mu(1-\epsilon)}{2\,\Delta X}\left(\frac{1}{\{B^2+[(N-3)\,\Delta X]^2\}^{3/2}} - \frac{2}{\{B^2+[(N-2)\,\Delta X]^2\}^{3/2}}\right.$$

$$\left. + \frac{1}{\{B^2+[(N-1)\,\Delta X]^2\}^{3/2}}\right)$$

$$A_{1N} = \frac{B^2\mu(1-\epsilon)}{2\,\Delta X}\left(\frac{1}{\{B^2+[(N-2)\,\Delta X]^2\}^{3/2}} - \frac{1}{\{B^2+[(N-1)\,\Delta X]^2\}^{3/2}}\right)$$

$$B_{1j} = \delta_{1j} - \frac{B^2\,\Delta X}{\beta_j\{B^2+[(j-1)\,\Delta X]^2\}^{3/2}} \qquad \begin{matrix} \beta_j = 2 & 1\le j\le N \\ \beta_j = 4 & j=N \end{matrix}$$

$$C_1 = \frac{1}{2}\left\{1 - \frac{1}{[(B/\Delta X)^2+1]^{1/2}}\right\} + \frac{\mu}{(\Delta X)^2} + \frac{B^2\,\Delta X}{4[B^2+(\Delta X)^2]^{3/2}} - \frac{\mu(1-\epsilon)}{2B\,\Delta X}$$

This is done for each element i, $1 \le i \le N$, and a matrix equation is generated of the form

$$[A_{ij}][t_j] + [B_{ij}][t_j^4] = [C_i] \tag{11-13}$$

This is a set of nonlinear algebraic equations for the unknown temperatures t_i. Methods for solving Eq. (11-13) are in Sec. 11-5.

EXAMPLE 11-3 A transparent gas with mean velocity u_m and constant physical properties flows through a circular tube of inner diameter D_i and length l (Fig. 11-4). A specified heat flux $q_e(x) = q_{max}\sin(\pi x/l)$ is applied along the length of the tube. The heat transfer coefficient h between the flowing gas and the tube interior surface is assumed independent of x. The tube wall is thin and has thermal conductivity k_w. The gas enters the tube from a large plenum at $T_{g,1}$. The gas leaves the tube at $T_{g,2}$ and enters a mixing plenum

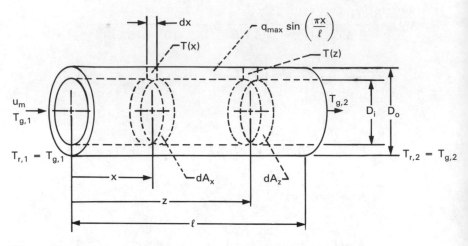

Figure 11-4 Cylindrical tube geometry with gas flow and internal radiation exchange.

that is also at $T_{g,2}$. The interior surface of the tube is diffuse-gray with emissivity ϵ. Set up the energy equations and boundary conditions to determine the wall temperature variation $T(x)$, and put the equations in dimensionless form. Then put them into a finite-difference form suitable for numerical solution.

Following Examples 10-9 and 10-10, the governing energy equations are

$$\epsilon \left\{ t^4(X) + \int_{Z=0}^{X} \left[\frac{1-\epsilon}{\epsilon} \phi(Z) - t^4(Z) \right] dF_{dX-dZ}(X-Z) \right.$$

$$+ \int_{Z=X}^{L} \left[\frac{1-\epsilon}{\epsilon} \phi(Z) - t^4(Z) \right] dF_{dX-dZ}(Z-X)$$

$$\left. - t_{r,1}^4 F_{dX-1}(X) - t_{r,2}^4 F_{dX-2}(L-X) \right\} = \phi(X) \tag{11-14}$$

$$\phi(X) = \sin\left(\frac{\pi X}{L}\right) + H[t_g(X) - t(X)] + N \frac{d^2 t}{dX^2} \tag{11-15}$$

$$t_g(X) = \text{St } e^{-\text{St}X} \int_0^X e^{\text{St}Z} t(Z) \, dZ + t_{g,1} e^{-\text{St}X} \tag{11-16}$$

In this form $t = T(\sigma/q_{max})^{1/4}$ and the other parameters are as defined in Examples 10-8 and 10-10 (the reference value of q_e is q_{max} and the reference length is D_i). Substituting the last two equations to eliminate $\phi(X)$ and $t_g(X)$ in (11-14) results in

$$\epsilon t^4(X) + \int_0^L \left[(1 - \epsilon)\left\{ \sin\left(\frac{\pi Z}{L}\right) + H\left[\text{St } e^{-\text{St}Z} \int_0^Z e^{\text{St}\xi} t(\xi) \, d\xi + t_{g,1} e^{-\text{St}Z} - t(Z) \right] \right. \right.$$

$$\left. \left. + N\frac{d^2 t}{dZ^2} \right\} - \epsilon t^4(Z) \right] dF_{dX-dZ}(|X - Z|)$$

$$= N\frac{d^2 t}{dX^2} + \sin\left(\frac{\pi X}{L}\right) + H\left[\text{St } e^{-\text{St}X} \int_0^X e^{\text{St}Z} t(Z) \, dZ + t_{g,1} e^{-\text{St}X} - t(X) \right]$$

$$+ t_{r,1}^4 F_{dX-1}(X) + t_{r,2}^4 F_{dX-2}(L - X) \tag{11-17}$$

For the boundary conditions required by the $d^2 t/dX^2$ term, the end edges of the tube are assumed to have negligible heat losses, so that $(dt/dX)_{X=0} = (dt/dX)_{X=L} = 0$. The condition $t_g(X = 0) = t_{g,1}$ was used in deriving (11-16). To proceed with the numerical solution, define

$$f(X, Z) = \left(\frac{1 - \epsilon}{\epsilon} \left\{ \sin\left(\frac{\pi Z}{L}\right) + H\left[\text{St } e^{-\text{St}Z} \int_0^Z g(\xi) \, d\xi + t_{g,1} e^{-\text{St}Z} - t(Z) \right] \right. \right.$$

$$\left. \left. + N\frac{d^2 t}{dZ^2} \right\} - t^4(Z) \right) \frac{dF_{dX-dZ}(|X - Z|)}{dZ} \tag{11-18}$$

where $g(\xi) = e^{\text{St}\xi} t(\xi)$. Equation (11-17) becomes

$$\epsilon\left[t^4(X) + \int_0^L f(X, Z) \, dZ \right] = N\frac{d^2 t}{dX^2} + \sin\left(\frac{\pi X}{L}\right) + H\left[\text{St } e^{-\text{St}X} \right.$$

$$\left. \cdot \int_0^X g(Z) \, dZ + t_{g,1} e^{-\text{St}X} - t(X) \right] + t_{r,1}^4 F_{dX-1}(X) + t_{r,2}^4 F_{dX-2}(L - X) \tag{11-19}$$

Numerical integration is applied to each integral, and a set of nonlinear algebraic equations is obtained in a similar fashion to Example 11-2. At a particular position $X_i = i \, \Delta X$, (11-19) becomes by use of the trapezoidal rule and having $\Delta X = \Delta Z$ where $\Delta X = L/N$,

$$\epsilon\left\{ t_i^4 + \Delta X\left[\frac{1}{2}f_0(X_i) + \sum_{j=1}^{N-1} f_j(X_i) + \frac{1}{2}f_N(X_i) \right] \right\} = N\frac{t_{i+1} - 2t_i + t_{i-1}}{(\Delta X)^2}$$

$$+ \sin\left(\frac{X_i}{L}\right) + H\left\{ \text{St } e^{-\text{St}X_i} \, \Delta X\left[\frac{1}{2}g_0 + \sum_{j=1}^{i-1} g_j + \frac{1}{2}g_i \right] + t_{g,1} e^{-\text{St}X_i} - t_i \right\}$$

$$+ t_{r,1}^4 F_{dX_i-1}(X_i) + t_{r,2}^4 F_{dX_i-2}(L - X_i) \tag{11-20}$$

Note that for the reservoir temperatures, $t_{r,1} = t_{g,1}$ and $t_{r,2} = t_{g,2}$ for the specified conditions of this example. By gathering terms after expansion as in Example 11-2, the full set of equations can be written as

$$
\begin{aligned}
(A_{00}t_0 + B_{00}t_0^4) &+ (A_{01}t_1 + B_{01}t_1^4) + & \cdots & + \cdots + (A_{0N}t_N + B_{0N}t_N^4) = C_0 \\
(A_{10}t_0 + B_{10}t_0^4) &+ (A_{11}t_1 + B_{11}t_1^4) + & \cdots & + \cdots + (A_{1N}t_N + B_{1N}t_N^4) = C_1 \\
\cdots \quad &+ \quad \cdots \quad + \quad \cdots \quad + \cdots + \quad \cdots \quad = \cdots \\
(A_{i0}t_0 + B_{i0}t_0^4) &+ \quad \cdots \quad + (A_{ij}t_j + B_{ij}t_j^4) + \cdots + (A_{iN}t_N + B_{iN}t_N^4) = C_i \\
&\vdots \\
(A_{N0}t_0 + B_{N0}t_0^4) &+ \quad \cdots \quad + (A_{Nj}t_j + B_{Nj}t_j^4) + \cdots + (A_{NN}t_N + B_{NN}t_N^4) = C_N
\end{aligned}
$$

$$(11\text{-}21a)$$

This can be placed in the matrix form,

$$[A_{ij}][t_j] + [B_{ij}][t_j^4] = [C_i] \tag{11-21b}$$

Once t_j is determined from the solution of Eqs. (11-21), the heat flux ϕ_j can be found from (11-15) if desired.

11-4.2 Finite-Element Method Formulation

The finite-element method (FEM) has the advantage that the approximation for the temperature within a given volume or surface element can vary across the element. Finite-difference formulations assign a single temperature to each element. The temperature variation in the FEM can be specified to various degrees of approximation (constant, linear, parabolic, etc.) at the cost of increasing the computation time. Temperatures at the boundaries of adjacent elements can be matched, temperature gradients can be forced to match, and, again at the cost of increased complexity and computer time, the second or higher derivatives can be forced to match.

Various approaches to the FEM formulation are used, but the most prevalent is the Galerkin method. It is only feasible to provide a brief description here, but there are detailed presentations of the basic FEM in a number of references [5–8]. After this brief description, which will be oriented toward the solution of the energy equation, the use of the FEM will be illustrated by formulating Example 11-3 to illustrate its application to radiation heat transfer. Further developments using the FEM for other types of radiative heat transfer problems are given in Chaps. 15 and 17.

Consider the energy equation, Eq. (14-26), for constant properties in the following form, where the radiative source term $-\nabla \cdot \mathbf{q}_r$ has been combined with the heat source by internal heat generation q''' into the term q_s,

$$\rho c_p \frac{\partial T}{\partial \tau} - k\nabla^2 T - q_s(\mathbf{r}, T, \tau) = 0 \tag{11-22}$$

The volume in which the energy equation is to be solved is divided into finite subregions; these are the *finite elements*. For one-dimensional geometries the elements are plane, cylindrical, or spherical layers that can have unequal thicknesses. In two dimensions, triangles are usually used as illustrated in Fig. 11-5, or quadrilaterals of irregular shape. In three dimensions, tetrahedrons or rectangular prisms are often used. These are examples of the types of finite elements that are used to divide geometries of irregular shape; the treatment of irregular volumes is one advantage of using the FEM.

Nodes are assigned to locations in the elements at which the unknown function, in this case the temperature, is to be determined. Nodes are often placed only at the corners of the elements, but there can be additional nodes placed internally or along the element boundaries.

The shape function The next step in the FEM is to choose *shape* or *interpolation* functions to provide an approximate variation of the dependent variable within each element between the values at the nodes. The simplest shape function is linear in the coordinates, but quadratic and higher order variations can be used. The shape function in an element is equal to unity at one node and decreases to zero at the neighboring nodes. Each shape function is zero in the elements that do not contain its particular node. Hence, each shape function is a *local interpolation function* that is finite only within elements containing a particular node.

To describe the shape function in more detail, we consider a planar one-dimensional problem with coordinate X. The shape function can be derived by using a series expansion of the unknown function, $T(X) = a_0 + a_1X + a_2X^2 + a_3X^3 + \cdots$. The number of terms used in the series is determined by two considerations: retaining more terms allows a more accurate representation of the temperature distribution within each element, and, conversely, using fewer terms will reduce computation time. The choice of the form of the shape function is a

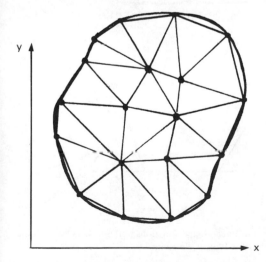

Figure 11-5 Two-dimensional region represented by triangular finite elements having nodes at vertexes.

tradeoff between the accuracy within an element, and thus the number of elements required, and the average computation time per element. Most solutions have used linear or quadratic forms.

Usually the unknown values of the function are defined only at nodes located at the corners or end points of each element, but interior nodes or intermediate nodes along the element boundaries can also be used. For a *linear* shape function in one dimension, there is a node at each end of the element; these nodes are designated here by subscripts 1 and 2 on T and X (the a_0 and a_1 are coefficients, and their subscripts do not refer to the nodes). The values of the dependent variable at the nodes are $T_1 = T(X_1) = a_0 + a_1 X_1$ and $T_2 = T(X_2) = a_0 + a_1 X_2$. Solving for a_0 and a_1 gives $a_0 = (T_1 X_2 - T_2 X_1)/(X_2 - X_1)$ and $a_1 = (T_2 - T_1)/(X_2 - X_1)$. Then in the element $T(X) = \Phi_1(X) T_1(X_1) + \Phi_2(X)T_2(X_2)$, where $\Phi_1(X) = (X_2 - X)/(X_2 - X_1)$ and $\Phi_2(X) = (X - X_1)/(X_2 - X_1)$. The Φ are thus the desired shape or interpolation functions. These functions are each unity at the node designated by their subscript, and they decrease linearly to zero at the neighboring node. Outside of the elements that do not contain its particular node, each shape function is equal to zero. This is illustrated in Fig. 11-6.

For a quadratic shape function in a one-dimensional geometry, three nodes are used per element, and, for this and higher order shape functions, the first and higher order derivatives of $T(X)$ are matched at the nodes, in addition to the T

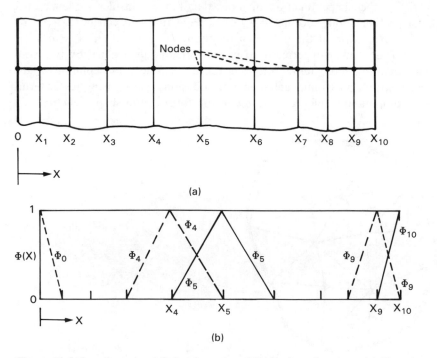

Figure 11-6 One-dimensional finite elements and linear shape functions. (*a*) Plane layer with elements of unequal size; (*b*) linear shape functions for typical elements.

values, to evaluate the a_m coefficients. For two-dimensional problems, the form of the linear shape function for triangular elements is found by using the expansion $T(X, Y) = a_0 + a_1X + a_2Y$. For square elements, the biquadratic form can be used: $T(X, Y) = a_0 + a_1X + a_{1,2}X^2 + a_{2,1}Y + a_{2,2}Y^2$. The shape functions are then derived by substituting the T values at the nodes and solving simultaneously for the unknown a_m or $a_{m,n}$ coefficients as illustrated for the linear case. For the five coefficients in the biquadratic form, there can be a node at each corner of the square and one in the center.

Galerkin form for the energy equation By using the shape functions, an approximate solution $\hat{T}(\mathbf{r}, \tau)$ for $T(\mathbf{r}, \tau)$ is now assumed in the form

$$T(\mathbf{r}, \tau) \approx \hat{T}(\mathbf{r}, \tau) = \sum_{j=1}^{N} T_j(\tau)\Phi_j(\mathbf{r}) \tag{11-23}$$

where the T_j are the values of temperature at the nodes desired from the solution and $\Phi_j(r)$ are the shape functions that equal unity at each node, as in Fig. 11-6. When this approximate solution is substituted into the energy equation, Eq. (11-22), there is a residual that depends on $\mathbf{r}$ and τ,

$$\rho c_p \frac{\partial \hat{T}}{\partial \tau} - k\nabla^2\hat{T} - q_s(\mathbf{r}, \hat{T}, \tau) = \text{Res}(\mathbf{r}, \tau) \tag{11-24}$$

It is desired to obtain a solution that, in an average sense over the entire volume, is as close as possible to the exact solution at each time. Variational principles are applied to minimize the residual. A set $W_i(\mathbf{r})$ of independent weighting functions is applied, and the residual is made orthogonal with respect to *each* of the weighting functions. This provides the following integral, which is evaluated for each of the set of independent weighting functions,

$$\int_V \text{Res}(\mathbf{r}, \tau)W_i(\mathbf{r}) \, dV = 0 \tag{11-25}$$

The integration is over the whole volume in which the solution is being obtained. Carrying out the integration for each of the set of weighting functions yields a set of simultaneous equations that can be solved for the $\hat{T}_i$ values at the nodes. In the Galerkin method the weighting functions are chosen to be the same function set as the shape functions. Since each shape function Φ_j is zero except within an element containing $\hat{T}_j$, the resulting matrix for solving the simultaneous equations for $\hat{T}_i$ is banded and sparse. By using the residual from Eq. (11-24), Eq. (11-23) for $\hat{T}$, and the $\Phi_j(r)$ as the set of weighting functions, Eq. (11-25) provides the Galerkin form of the energy equation for each of the weighting functions,

$$\int_V \left[\rho c_p \sum_{j=1}^{N} \Phi_j(\mathbf{r}) \frac{\partial T_j(\tau)}{\partial \tau} - k \sum_{j=1}^{N} T_j(\tau)\nabla^2\Phi_j(\mathbf{r}) - q_s(\mathbf{r}, \hat{T}, \tau) \right]\Phi_i(\mathbf{r}) \, dV = 0 \tag{11-26}$$

Evaluating Eq. (11-26) for each i provides N simultaneous equations for the $\hat{T}_j$. This method is now illustrated by application to an example.

EXAMPLE 11-4 Set up Example 11-3 for numerical solution using the finite-element method. For simplicity, let the tube interior surface be black.

Following the analysis in [9], it is advantageous to use the independent variable $\omega = t^4$, so (11-17) becomes, with $\epsilon = 1$,

$$\omega(X) - \int_0^L \omega(Z) \, dF_{dX-dZ}(|X - Z|) = \frac{N}{4} \frac{d}{dX} \left(\frac{1}{\omega^{3/4}} \frac{d\omega}{dX} \right)$$

$$+ \sin\left(\frac{\pi X}{L}\right) + H\left[\text{St } e^{-\text{St}X} \int_0^X e^{\text{St}Z}\omega^{1/4}(Z) \, dZ + t_{g,1}e^{-\text{St}X} - \omega^{1/4}(X) \right]$$

$$+ t_{r,1}^4 F_{dX-1}(X) + t_{r,2}^4 F_{dX-2}(L - X) \tag{11-27}$$

with boundary conditions $d\omega/dX = 0$ at $X = 0, L$. If we now define $A(\omega) \equiv N/4\omega^{3/4}$ and

$$\psi(X, \omega) \equiv \sin\left(\frac{\pi X}{L}\right) + H\left[\text{St } e^{-\text{St}X} \int_0^X e^{-\text{St}Z}\omega^{1/4}(Z) \, dZ + t_{g,1}e^{-\text{St}X} - \omega^{1/4}(X) \right]$$

$$+ t_{r,1}^4 F_{dX-1}(X) + t_{r,2}^4 F_{dX-2}(L - X) + \int_0^L \omega(Z) \, dF_{dX-dZ}(|X - Z|)$$

$$\tag{11-28}$$

Eq. (11-27) reduces to

$$-\frac{d}{dX}\left[A(\omega) \frac{d\omega}{dX} \right] + \omega = \psi(X, \omega) \tag{11-29}$$

Note that the function $\psi(X, \omega)$ contains highly nonlinear terms in the variable ω.

To solve (11-29) by the Galerkin FEM approach, the $\omega(X)$ is required to satisfy a *variational* form of (11-29) and its boundary conditions, which has the form [Eq. (11-25)]

$$\int_0^L \left\{ -\frac{d}{dX}\left[A(\omega) \frac{d\omega}{dX} \right] + \omega \right\} W(X) \, dX - \int_0^L \psi(X, \omega)W(X) \, dX = 0 \tag{11-30}$$

The $W(X)$ is a weighting function defined by [see (11-23)]

$$W(X) = \sum_{i=1}^N W_i\Phi_i(X) \tag{11-31}$$

The $\Phi_i(X)$ is the *shape function*, and the W_i are coefficients at the nodes. The first term in the first integral in Eq. (11-30) is integrated by parts and the boundary conditions of the insulated end edges of the tube wall are used: $dT/dX = d\omega/dX = 0$ at $X = 0$ and L. This gives

$$\int_0^L \left[A(\omega) \frac{d\omega}{dX} \frac{dW}{dX} + \omega W \right] dX - \int_0^L \psi(X, \omega) W(X)\, dX = 0 \tag{11-32}$$

Now, as in Eq. (11-23), an approximate solution is sought of the form

$$\omega(X) \approx \Omega(X) = \sum_{j=1}^N \Omega_j \Phi_j(X) \tag{11-33}$$

where, in the Galerkin method, the shape functions are the same as in the weighting function, Eq. (11-31), and Ω_j are the values of Ω at the nodes. Substituting (11-31) and (11-33) into (11-32) results in

$$\sum_{i=1}^N W_i \left(\sum_{j=1}^N \left\{ \int_0^L \left[A(\Omega) \frac{d\Phi_i}{dX} \frac{d\Phi_j}{dX} + \Phi_i \Phi_j \right] dX \right\} \Omega_j - \int_0^L \psi(X, \Omega) \Phi_i\, dX \right) = 0 \tag{11-34}$$

If we now define

$$K_{ij} \equiv \int_0^L \left[A(\Omega) \frac{d\Phi_i}{dX} \frac{d\Phi_j}{dX} + \Phi_i \Phi_j \right] dX \qquad \text{and} \qquad \psi_i \equiv \int_0^L \psi(X, \Omega) \Phi_i\, dX$$

Eq. (11-34) becomes

$$\sum_{i=1}^N W_i \left[\sum_{j=1}^N K_{ij} \Omega_j - \psi_i \right] = 0 \tag{11-35}$$

To satisfy this equation, the quantity inside the square brackets must equal zero, since the W_i are constant coefficients. Then the required Ω_j values can be obtained by solving the equivalent set of equations

$$K_{11}\Omega_1 + K_{12}\Omega_2 + \cdots + \cdots + K_{1N}\Omega_N = \psi_1$$
$$K_{21}\Omega_1 + K_{22}\Omega_2 + \cdots + \cdots + K_{2N}\Omega_N = \psi_2$$
$$\vdots$$
$$K_{i1}\Omega_1 + \cdots + K_{ij}\Omega_j + \cdots + K_{iN}\Omega_N = \psi_i \tag{11-36a}$$
$$\vdots$$
$$K_{N1}\Omega_1 + \cdots + K_{Nj}\Omega_j + \cdots + K_{NN}\Omega_N = \psi_N$$

In matrix form, this is

$$[K_{ij}][\Omega_j] = [\psi_i] \tag{11-36b}$$

Because each shape function (and hence each weighting function in the Galerkin method) is usually zero except within elements containing a particular node, the $[K_{ij}]$ is a sparse matrix that is banded along the diagonal. After the Ω_j are obtained, Eq. (11-33) is used to find the required temperature values from $\Omega(X) \approx \omega(X) = t^4(X)$. Note, however, that both K_{ij} and ψ_i are functions of Ω, and thus of Ω_j, so the solution is iterative.

As noted in the introduction to this section, the shape function Φ can have many forms. For example, a linear form could be used for $\Phi_i(X)$ within each element, $X_i \leq X \leq X_{i+1}$ (where $X_{i+1} = X_i + \Delta X$), of this problem to yield $\Phi_i(X) = (X_{i+1} - X)/(X_{i+1} - X_i)$. The FEM solution can now be carried out by solving (11-36) using one of the numerical techniques in Sec. 11-5.

Note that the equation for numerical solution in the FEM is (11-36), which has the same form as (11-21). Numerical methods that work well with finite-difference formulations will usually apply for solving FEM formulations.

Razzaque et al. [9] applied the FEM to the situation in Examples 11-3 and 11-4. He used a quadratic shape function and was able to extend previous results to a dimensionless tube length of $L = 20$ (limited in other methods to less than 10, and in some cases to less than 5, by numerical instabilities). Some results are in Fig. 11-7.

The finite-element method was applied to determine the three-dimensional temperature distribution inside a conducting solid exposed to conduction–radiation boundary conditions in [10]. Complex geometries representing circuit board chips and a magneto-plasma-dynamic propulsion system were modeled using this approach. Radiation interaction among surface elements was included in the solutions, which were obtained from a modified version of a commercially available code.

11-5 NUMERICAL SOLUTION TECHNIQUES

Examples 11-2 through 11-4 result in a matrix of nonlinear algebraic equations of the form (11-21). As noted in the introduction to Sec. 11-4, it is important to examine the relative values of the elements in the two coefficient matrices, $[A_{ij}]$ and $[B_{ij}]$. If the elements A_{ij} are comparatively large, the problem can be treated as linear in t_j; conversely, the problem can be treated as linear in t_j^4. When the coefficients A and B are approximately equal, other treatments are in order.

If we define $A_{ij}^* = A_{ij} + B_{ij}t_j^3$, Eq. (11-21) becomes

$$[A_{ij}][t_j] + [B_{ij}][t_j^4] = [A_{ij} + B_{ij}t_j^3][t_j] = [A_{ij}^*][t_j] = [C_i] \tag{11-37}$$

This is a set of linear algebraic equations with coefficients A_{ij}^* that are variable and highly nonlinear. The equations cannot be solved by elimination or direct matrix inversion, because the coefficients A_{ij}^* are temperature dependent and thus are not known. Some methods of numerical solution are now discussed.

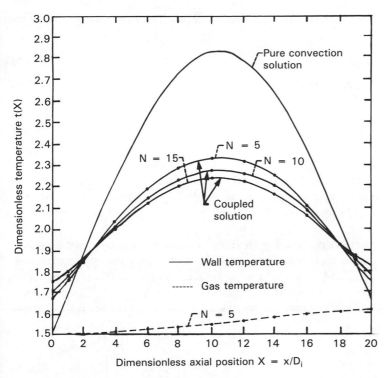

Figure 11-7 Comparison of solutions with sinusoidal wall heat flux to illustrate effects of radiation and axial wall conduction, $L = 20$, $St = 0.01$, $t_{r,1} = t_{g,1} = 1.5$, $t_{r,2} = t_{g,2}$, $H = 0.8$, N is defined in Example 10-10.

11-5.1 Successive Substitution Methods

Simple successive substitution (SSS) A simple solution concept is to assume an initial set of temperatures $[t_j^{(0)}]$ and use them to compute $[A_{ij}^*(t_j^{(0)})]$. This fixes the values of the elements in the matrix of coefficients, leaving the temperature vector $[t_j]$ as the unknown. Equation (11-37) or (11-21) is then solved for a new set of temperatures $[t_j^{(n+1)}]$ from

$$[A_{ij}^*(t_j^{(n)})][t_j^{(n+1)}] = [C_i] \tag{11-38}$$

This process is continued until the difference between successive temperature sets is less than an acceptable error, indicating convergence. Unfortunately, this method depends on an accurate initial guess for $[t_j]$. An inaccurate guess can lead to unstable iterations that may diverge rapidly.

Successive underrelaxation (SUR) The SSS method can be modified to obtain convergence in many cases if (11-21) or (11-37) is written as

$$[A_{ij}^*(t_j^{*(n)})][t_j^{(n+1)}] = [C_i] \tag{11-39a}$$

where the coefficients $A_{ij}^*(t_j^{*(n)})$ are computed at each iteration by using a modified temperature

$$t_j^{*(n)} = \alpha t_j^{(n)} + (1 - \alpha)t_j^{(n-1)} \tag{11-39b}$$

in which α is a weighting coefficient (or *relaxation parameter*) in the range $0 \leq \alpha \leq 1$. When $\alpha = 1$, the SUR method reduces to SSS; when $\alpha < 1$, the new guess is weighted toward the previous guess (i.e., underrelaxed), and oscillations between iterations are damped. If possible, the α should be chosen or found that will provide optimized convergence. Values in the range $0.1 < \alpha < 0.4$ give convergence in many cases. Decreasing α usually provides slower convergence, but greater assurance that convergence will occur. Sometimes decreasing α somewhat will increase convergence by reducing oscillatory behavior. Values of $\alpha \approx 0.3$ are reported in [11] to provide rapid convergence in many cases.

Regulated successive underrelaxation (RSUR) Cort et al. [11] proposed a method of regulated successive underrelaxation (RSUR) that allows the underrelaxation factor α to be chosen and modified for successive iterations. They recommend the following steps: (1) initialize $\alpha = 1$; (2) solve (11-39b) for $t_j^{*(n)}$ [for the first iteration, an initial guess $t_j^{(0)}$ must be provided]; (3) solve (11-39a) for $t_j^{(n+1)}$; (4) calculate

$$\nu^{(n+1)} = \left[\sum_{j=1}^{N} (t_j^{(n+1)} - t_j^{(n)})^2 \right]^{1/2} \quad \text{and} \quad R^{(n+1)} = \left[\sum_{j=1}^{N} (t_j^{(n+1)})^2 \right]^{1/2} \tag{11-40a,b}$$

and if $\nu^{(n+1)} > \nu^{(n)}$ or if $\nu^{(n+1)} > (1/3)R^{(n+1)}$, reduce α by 0.1; and (5) repeat steps 2 through 4 until convergence.

Equation (11-40a) checks for divergence of the solution between iterations, and (11-40b) is used to see whether the residual error after each iteration is smaller than a measure of the root-mean-square temperature over the region of the solution. The latter check is made to eliminate slowly oscillating but converging solutions that pass the test of (11-40a) but take a very long time to converge.

Another approach is to rewrite (11-21) in the form

$$A_{ii}t_i^{(n+1)} + B_{ii}(t_i^{(n+1)})^4 = C_i - \sum_{j=1}^{N} (1 - \delta_{ij})[A_{ij}t_j^{(n+1)} + B_{ij}(t_j^{(n+1)})^4] \equiv D_i \tag{11-41}$$

where δ_{ij} is the Kronecker delta. An initial set of temperatures $t_j^{(0)}$ is guessed, and D_i is evaluated based on this set. Then the $t_i^{(1)}$ are found by iterative solution of (11-41) and are used to evaluate the next set of D_i. This process is repeated to solve for $t_i^{(n)}$ until convergence. Tan [12] points out that for a given value of i, (11-41) is a quartic equation with a single real positive root $t_i^{(n+1)}$ given by

$$t_i^{(n+1)} = \frac{y^{1/2}}{2} \frac{p - 2}{(p - 1)^{1/2} + 1} \tag{11-42}$$

where

$$p = 2 \left(1 + \frac{4D_i}{B_{ii} y^2} \right)^{1/2}, \qquad y = \frac{2r}{(s + r)^{2/3} + (s - r)^{1/3}[(s + r)^{1/3} + (s - r)^{1/3}]}$$

and

$$r = \frac{1}{2} \left(\frac{A_{ii}}{B_{ii}} \right)^2, \qquad s = \left[r^2 + \left(\frac{4D_i}{3B_{ii}} \right)^3 \right]^{1/2}$$

Thus, for each set of D_i the $t_i^{(n+1)}$ can be found directly from the nonlinear equation (11-41) rather than by an inner iteration and then can be used to evaluate new D_i and continue to the next main iteration. This method is quite fast and can be combined with the SUR technique to determine succeeding approximations to provide a method that is *both* stable and fast.

11-5.2 Newton-Raphson-Based Methods for Nonlinear Problems

Modified Newton-Raphson (MNR) Ness [13] proposed a modified Newton-Raphson method for the class of nonlinear problems encountered here. Starting from (11-21)

$$[A_{ij}][t_j] + [B_{ij}][t_j^4] - [C_i] = 0 \tag{11-43}$$

an initial approximate temperature $t_j^{(0)}$ is guessed at each node. A correction factor δ_j is then computed so that $t_j = t_j^{(0)} + \delta_j$. This corrected temperature is used to compute a new δ_j, and this process is continued until δ_j becomes smaller than a specified value. The δ_j are found by solving the following set of linear equations:

$$[f_{ij}][\delta_j] + [f_i] = 0 \tag{11-44}$$

where

$$f_i = \sum_{j=1}^{N} [A_{ij} t_j^{(0)} + B_{ij}(t_j^{(0)})^4] - C_i \tag{11-45}$$

and

$$f_{ij} = A_{ij} + 4B_{ij}(t_j^{(0)})^3 \tag{11-46}$$

The MNR method has limits on its ability to converge if a poor initial temperature set is chosen, and a priori determination of whether a given initial set will give convergence is not possible.

Accelerated Newton-Raphson (ANR) Cort et al. [11] proposed a method in which the amount of change in t_i at each iteration is adjusted to accelerate convergence. They recommended that the f_{ij} in the MNR method be replaced by

$$f_{ij} = A_{ij} + \frac{4B_{ij}}{[(1 - (\beta/3)]} (t_j^{(0)})^{(3-\beta)} \qquad \beta \geq 0 \tag{11-47}$$

This effectively modifies the slope of the changes in t_j with respect to iteration number compared with that used in the MNR method. For $\beta = 0$, the ANR method reduces to MNR. If β is too large, oscillations and divergence between iterations may occur. For $\beta = 0.175$, the number of iterations to provide a given accuracy for a particular problem was reduced from 28 using MNR to 12 using ANR, and reductions in computer time of up to 80% were obtained. A starting value of $\beta = 0.15$ is recommended [11].

In [11] results using the above methods were compared for some typical radiation-conduction problems with temperature-dependent properties and internal energy generation. Consideration was limited to surfaces with radiative exchanges to black surroundings at a single temperature and the problems were cast in finite-element form. Because we have seen by the example problems in this section that even complicated radiation-conduction-convection problems with multiple surfaces reduce to the same general form of (11-21), the conclusions probably apply to a broader class of problems than were studied. In [14], a linearized solution of Eq. (11-36) is proposed that speeds convergence over the MNR method. For problems that are either conduction or radiation dominated, or where both modes are important, the method performed well, providing improvement in solution speed of a factor of ten.

It was found that the SUR method gave convergence with the fewest iterations and the least computer time; RSUR was useful to find the optimum value of α for use in the SUR method. For the Newton-Raphson method, ANR was always faster than MNR, but neither method was as fast as SUR. Some computation times to convergence are given for comparison in Table 11-1 for the case of a conducting plate with internal generation, insulated at both ends, and radiating to a black environment at a given temperature.

In [14a] the convergence ranges and behavior of equations of the form of Eq. (11-8) are discussed, and the various solution methods of this chapter are examined.

11-6 THE MONTE CARLO METHOD

In Chap. 8 it was found that the enclosure-theory analysis became very complex when directional- and spectral-surface property variations were accounted for. An alternative approach that can deal with these complexities is the Monte Carlo method. Since Monte Carlo is a statistical numerical method, it is first necessary to discuss some concepts of statistical theory. Then the basic procedure is outlined with regard to radiative exchange, and two example problems are formulated to demonstrate the method. The straightforward Monte Carlo approach is presented. Some refinements that can shorten computation time by increasing accuracy are discussed briefly.

Table 11-1 Comparison of CPU time to convergence for various numerical methods [11]

Numerical method	CPU time, s	Iterations
Modified Newton-Raphson	9.53	24
Accelerated Newton-Raphson	6.00	15
Regulated SUR	2.72	29
SUR	1.61	15

11-6.1 Definition of Monte Carlo Method

Herman Kahn [15] gave a definition of the Monte Carlo method that incorporates the salient ideas: "The expected score of a player in any reasonable game of chance, however complicated, can in principle be estimated by averaging the results of a large number of plays of the game. Such estimation can be rendered more efficient by various devices which replace the game with another known to have the same expected score. The new game may lead to a more efficient estimate by being less erratic, that is, having a score of lower variance, or by being cheaper to play with the equipment on hand. There are obviously many problems about probability that can be viewed as problems of calculating the expected score of a game. Still more, there are problems that do not concern probability but are nonetheless equivalent for some purposes to the calculation of an expected score. The Monte Carlo method refers simply to the exploitation of these remarks."

This definition provides a good outline for the method. What must be done for a specific problem is to set up a game or model that has the same behavior, and hence is expected to produce the same outcome as the physical problem that the model simulates; make the game as simple and fast to play as possible; use any available methods to reduce the variance of the average outcome of the game; then play the game many times and find the average outcome.

Referring to *the* Monte Carlo method is probably meaningless. Any specific problem more likely entails *a* Monte Carlo method, as the label has been placed on a large class of loosely related techniques. A number of general books and monographs are available that detail methods and/or review the literature. A valuable early outline is in [16], which is the first work to use the term Monte Carlo for the approach considered here. For clarity and usefulness, both [15] and [17] are valuable, as are the general texts [18–21] and the many papers gathered in the symposium volume [22].

No definitive method of predicting computer running time exists for most Monte Carlo problems. The time will depend on the machine used and on the ability of the programmer to pick appropriate methods and shortcuts. An example is the use of special subroutines for computation of such functions as sine and cosine. These routines sacrifice some accuracy to gain speed. If problem answers accurate to a small percentage are desired, then the use of eight-place functions from a relatively slow subroutine is not needed, especially if the subroutine is to be used tens of thousands of times.

11-6.2 Details of the Method

The random walk The background material contains the term *Markov chain*. This is a series of events in sequence with the condition that the probability of each succeeding event is uninfluenced by prior events. The usual example of this is an inebriated gentleman who begins to walk through a city. At each street corner, he becomes confused and chooses completely at random one of the streets leading from the intersection. He may walk up and down the same block several times before he chances to move off down a new street. His walk is a Markov chain, as his decision at any corner is not influenced by where he has been. Because of the randomness of choice at each intersection, it is possible to simulate a sample walk by having a roulette wheel with only four positions, each corresponding to a possible direction. The probability of the gentleman starting from his favorite bar and reaching any point in the city limits could be found by simulating a large number of histories, using the wheel to determine the direction of the walk at each decision point in each history.

The probability of the man reaching intersection (l, m) on a square grid representing the city street map is simply

$$P(l, m) = \tfrac{1}{4}[P(l + 1, m) + P(l - 1, m) + P(l, m + 1) + P(l, m - 1)] \qquad (11\text{-}48)$$

where the factors in brackets are the probabilities of his being at each of the adjacent four intersections. This is because the probability of reaching $P(l, m)$ from a given adjacent intersection is one-fourth. This type of random walk is a model for processes described by Laplace's equation; Eq. (11-48) is recognized as the finite-difference analog of the Laplace equation.

The probability of a certain occurrence for other processes is usually not as immediately obvious as Eq. (11-48). More often, the probability of an event must be determined from physical constraints, and then the decision as to what event will occur is made on the basis of this probability. Some of the basic methods of choosing an event from a known probability distribution of events are now examined. Also, means of constructing these distributions are discussed.

Choosing from probability distributions Consider small packets of radiation leaving a differential area and arriving at a disk with an outer radius 10 units in length. For many packets, the number $F(\xi)$ that have arrived at the disk within each small radial increment $\Delta\xi$ about some radius ξ can be represented by a histogram of the frequency function $f(\xi) = F(\xi)/\Delta\xi$, where $0 \le \xi \le 10$. A smooth curve can then be passed through the histogram to give a continuous frequency distribution, as in Fig. 11-8. What is now needed is a method for simulating additional packets. This method should assign an expected radius ξ to each succeeding packet. The distribution of ξ values should correspond to the distribution that the packets follow according to a known physical process. A method of assigning ξ values to individual packets is desired so the distribution for all the samples will agree with the required distribution. In addition, for a Markov pro-

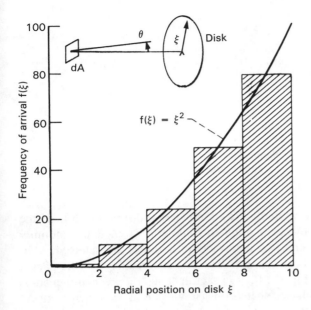

Figure 11-8 Distribution of radiation packets arriving at various disk radii.

cess the values must be assigned in a random manner so that each decision at each step in the transfer is independent.

To show how this is done, the frequency curve in Fig. 11-8 can, for this example, be approximated by

$$f(\xi) = \xi^2 \tag{11-49}$$

in the interval $0 \le \xi \le 10$, and $f(\xi) = 0$ elsewhere, because the only packets being considered for the moment are those directed toward the disk. Equation (11-49) is normalized by dividing by the area under the frequency curve (that is, the total number of packets) to obtain

$$P(\xi) = \frac{f(\xi)}{\displaystyle\int_0^{10} f(\xi)\, d\xi} = \frac{3\xi^2}{1000} \tag{11-50}$$

If the distribution with which packets have struck the target radii is taken as the basis for estimating the locations the next packets will strike, then the *probability density function* defined by (11-50) is the average distribution that must be satisfied by the ξ values determined by the simulation scheme. The $P(\xi)$ is plotted in Fig. 11-9 and is interpreted physically as the proportion of values that lie in the region $\Delta\xi$ around ξ.

To determine ξ values, the simulation scheme can proceed as follows: Choose two *random numbers* R_A and R_B from a large set of numbers evenly distributed in the range 0 to 1. The two random numbers are then used to select a point $(P(\xi), \xi)$ in Fig. 11-9 by setting $P(\xi) = R_A$ and $\xi = (\xi_{max} - \xi_{min})R_B = 10R_B$. This value

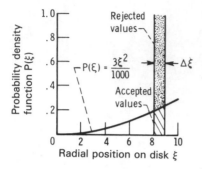

Figure 11-9 Probability density function of energy packets hitting disk.

of $P(\xi)$ is then compared to the value of $P(\xi)$ computed at ξ from (11-50). If the randomly selected value lies above the computed value of $P(\xi)$, then the randomly selected value of ξ is rejected and two new random numbers are selected. Otherwise, the value of ξ that has been found is listed as the location that the packet will strike. Referring again to Fig. 11-9, it is seen that such a procedure ensures that the correct fraction of ξ values selected for use will lie in each increment $\Delta\xi$ after a large number of completely random selections of $(P(\xi), \xi)$ is made.

The disadvantage of such an event-choosing procedure is that in some cases a large portion of the ξ values may be rejected because they lie above the $P(\xi)$ curve. A more efficient method for choosing ξ is therefore desirable. One such method is to integrate the probability density function $P(\xi)$ using the general relation

$$R(\xi) = \int_{-\infty}^{\xi} P(\xi^*) \, d\xi^* \tag{11-51}$$

where $R(\xi)$ can only have values in the range 0 to 1 because the integral under the entire $P(\xi)$ curve is unity according to Eq. (11-50). Equation (11-51) is the general definition of the *cumulative distribution function*. A plot of R against ξ from (11-51) shows the probability of an event occurring in the range $-\infty$ to ξ. For the method given here, the function R is taken to be a random number; each value of ξ is then obtained by choosing an R value at random and using the functional relation $R(\xi)$ to determine the corresponding value of ξ. To show that the probability density of ξ formed in this way corresponds to the required $P(\xi)$, the probability density function of Fig. 11-9 is used as an illustrative example. Inserting the example $P(\xi)$ of (11-50) into (11-51) and noting that $P(\xi) = 0$ for $-\infty < \xi < 0$ give

$$R = \int_{0}^{\xi} P(\xi^*) \, d\xi^* = \frac{\xi^3}{1000} \qquad 0 \leq R \leq 1 \tag{11-52}$$

Equation (11-52) is shown plotted in Fig. 11-10.

Now it will be shown that choosing R at random and determining a corresponding value of ξ from (11-52) is equivalent to taking the derivative of the cumulative distribution function and that this derivative is, by examination of (11-52) and (11-50), simply $P(\xi)$. Divide the range of ξ into a number of equal increments $\Delta\xi$. Suppose that M values of R are chosen in the range 0 to 1 at equal intervals along R. There will be M values of ξ corresponding to these M values of R. The fraction of the M values of ξ per given increment $\Delta\xi$ is then $M_{\Delta\xi}/M = \Delta R$, which gives

$$\frac{M_{\Delta\xi}/M}{\Delta\xi} = \frac{\Delta R}{\Delta\xi} \tag{11-53}$$

The quantity $\Delta R/\Delta\xi$ approaches $dR/d\xi$ if a large enough M is used and small increments $\Delta\xi$ are examined. But $dR/d\xi$ can be seen from (11-52) and (11-50) to be simply $P(\xi)$; therefore, by obtaining values of ξ as described preceding (11-52), the required probability distribution is indeed generated.

Often in physical problems the frequency distribution depends on more than one variable. If the interdependence of the variables is such that the frequency distribution can be factored into a product form, the following can be written:

$$f(\xi, \varphi) = g(\xi)h(\varphi) \tag{11-54}$$

Values of $P(\xi)$ and $P(\varphi)$ can be found by integrating out each variable to obtain

$$P(\xi) = \frac{\displaystyle\int_{\varphi_{min}}^{\varphi_{max}} f(\xi, \varphi)\, d\varphi}{\displaystyle\int_{\xi_{min}}^{\xi_{max}}\int_{\varphi_{min}}^{\varphi_{max}} f(\xi, \varphi)\, d\varphi\, d\xi} = \frac{g(\xi)\displaystyle\int_{\varphi_{min}}^{\varphi_{max}} h(\varphi)\, d\varphi}{\displaystyle\int_{\xi_{min}}^{\xi_{max}} g(\xi)\, d\xi \int_{\varphi_{min}}^{\varphi_{max}} h(\varphi)\, d\varphi} \tag{11-55}$$

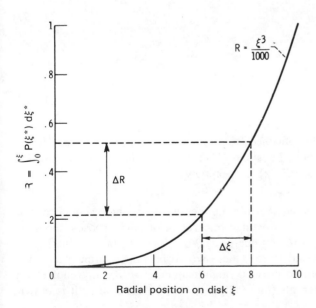

Figure 11-10 Cumulative distribution of energy packets on disk.

$$= \frac{g(\xi)}{\displaystyle\int_{\xi_{min}}^{\xi_{max}} g(\xi)\, d\xi} \qquad\qquad (11\text{-}55) \\ (Cont.)$$

and, similarly

$$P(\varphi) = \frac{\displaystyle\int_{\xi_{min}}^{\xi_{max}} f(\xi, \varphi)\, d\xi}{\displaystyle\int_{\varphi_{min}}^{\varphi_{max}} \int_{\xi_{min}}^{\xi_{max}} f(\xi, \varphi)\, d\xi\, d\varphi} = \frac{h(\varphi)}{\displaystyle\int_{\varphi_{min}}^{\varphi_{max}} h(\varphi)\, d\varphi} \qquad (11\text{-}56)$$

The methods given previously in this section are used to evaluate ξ and φ independently of one another after two random numbers are chosen.

If $f(\xi, \varphi)$ cannot be placed in the form of Eq. (11-54), it can be shown [15, 19] that ξ and φ values can be determined by choosing two random numbers R_ξ and R_φ. Note that

$$P(\xi, \varphi) = \frac{f(\xi, \varphi)}{\displaystyle\int_{\xi_{min}}^{\xi_{max}} \int_{\varphi_{min}}^{\varphi_{max}} f(\xi, \varphi)\, d\varphi\, d\xi}$$

Then ξ and φ are found from the equations

$$R_\xi = \int_{-\infty}^{\xi} \int_{\varphi_{min}}^{\varphi_{max}} P(\xi^*, \varphi)\, d\varphi\, d\xi^* \qquad (11\text{-}57)$$

and

$$R_\varphi = \int_{-\infty}^{\varphi} P(\varphi^*, \xi = \text{fixed})\, d\varphi^* \qquad (11\text{-}58)$$

where ξ in (11-58) is the value obtained from (11-57). This procedure may be extended to any number of variables. Equations (11-57) and (11-58) define the *marginal* and *conditional* distributions of $P(\xi, \varphi)$, respectively.

Random numbers A *random number* is a number chosen without sequence from a large set of numbers spaced at equivalued intervals. For our purposes, the numbers are in the range 0 to 1. If the numbers, 0, 0.01, 0.02, 0.03, . . . , 0.99, 1.00 are placed on slips of paper and mixed, there would be fair assurance that, if a few slips are picked, they would provide random numbers. If many choices are to be made, then smaller intervals (more slips) should be used; after it is drawn, each slip should be replaced and randomly mixed with the others. For a computer problem, random numbers might be needed for 10^5 or more decisions. It is desirable to have a rapid way for obtaining them and to have the numbers be truly random.

One way to obtain truly random numbers is to sample a truly random process. Such phenomena as noise in an electronic circuit or radioactive-decay particle counts per unit time have been tried, but they are too slow for direct computer use. A second means is to generate tables of random numbers [23, 24], perhaps by one of the processes mentioned, and enter these tables in the computer memory. This allows rapid access to random numbers, but for complex problems requiring a large quantity of random numbers the required storage space becomes excessive. This method has been used when a modest problem is to be solved.

The most useful method for obtaining random numbers is a pseudorandom-number generator. This is a subroutine that exploits the apparent randomness of groups of digits in large numbers. One simple example of such a routine is to take an 8-digit number, square it, and then choose the middle 8 digits of the resulting 16-digit number as the required random number. When a new random number is needed, square the previous random number and take the new random number as the middle 8 digits of the result. This process is said by Schreider [19] to degenerate after a few thousand cycles by propagating to an all-zero number. A more satisfactory routine is based on suggestions in [25]. A random number is generated by taking the low-order 36 bits of the product $R_{n-1}K$, where $K = 5^{15}$ and R_{n-1} is the previously computed random number. By always starting a given program with the same R_0, it is possible to check solutions through step-by-step tracing of a few histories. Many subroutines for generating random numbers that are available in software systems are based on this approach.

The fact that such subroutines generate *pseudo*random numbers raises the question of whether such pseudorandomness is sufficiently random for the problem being treated. Does the sequence repeat and, if so, after how many numbers? Certain standard tests that give partial answers to these questions are discussed in [17, 25, 26]. No finite set of tests is sufficient to establish randomness, although passing the tests is necessary. Perhaps the safest course is to obtain a standard subroutine whose properties have been established by such tests and use it within its proven limits.

Evaluation of error Because the solutions obtained by Monte Carlo methods are averages over the results of a number of individual samples, they will contain fluctuations about a mean value. The mean can be more accurately determined by increasing the number of values used in determining the mean as long as the budget for computer time can stand the strain. More generally, some ad hoc rules of economy and an estimate of desired accuracy in a given problem can be applied and solutions are obtained by trading off within these limits.

To establish the accuracy of the solutions, one of several tests can be applied. For example, suppose we want to know the probability of the drunken gentleman reaching a location at the city limits. To determine his success exactly, an infinite number of hypothetical paths have to be followed, and the probability $P(l, m)$ of reaching the boundary point (l, m) can be determined as

$$P(l, m) = \left[\frac{S(l, m)}{N}\right]_{N \to \infty} \tag{11-59}$$

where $S(l, m)$ is the number of samples reaching the boundary point and N is the total number of samples. In practice, a probability would be computed based on some finite number, perhaps 10^2–10^6. Then an estimate is needed of the error μ involved in using this relatively small sample size.

For a sample size greater than about $N = 20$, it is found [17, 19], from application of the central limit theorem and the relations governing normal probability distributions, that the following relation holds whenever the samples S in question can be considered to leave a source and either reach a scoring position with probability P or not reach it with probability $1 - P$. The probability that the average $S(l, m)/N$ for finite N differs by less than some value μ from $[S(l, m)/N]_{N \to \infty}$ is given by

$$P\left[\left|\frac{S}{N} - \left(\frac{S}{N}\right)_{N \to \infty}\right| \le \mu\right] = \frac{2}{\sqrt{\pi}} \int_0^{\eta/\sqrt{2}} e^{-\eta^{*2}} \, d\eta^* = \text{erf}\,\frac{\eta}{\sqrt{2}} \tag{11-60}$$

where

$$\eta \approx \mu\left[\frac{N}{(S/N)(1 - S/N)}\right]^{1/2} \tag{11-61}$$

The error function (erf) is given in standard reference tables [27, 28].

In many problems, such an error estimation cannot be applied because the samples do not originate from a single source. The radiative energy flux at a location on the boundary of an enclosure will usually depend on the energy arriving from many sources. For such situations, the most straightforward way to estimate the error (such as in the local radiative heat flux) is to subdivide the calculation of the desired statistical mean result into a group of I submeans. The *central limit theorem* then applies. This theorem states that the statistical fluctuations in the submeans are distributed in a normal or Gaussian distribution about the overall mean. For such a distribution, a measure of the fluctuations in the means can be calculated. This is called the *variance*. For example, if 2000 samples are examined, a mean result $\bar{P}$ is calculated on the basis of the samples, and 20 submeans $P_1, P_2, \ldots, P_I$ of 100 samples each are calculated. Then the *variance* γ^2 of the mean solution $\bar{P}$ is given by

$$\gamma^2 = \frac{1}{I - 1}\left[\sum_{i=1}^{I}(P_i - \bar{P})^2\right] = \frac{1}{I - 1}\left[\sum_{i=1}^{I}P_i^2 - \frac{\left(\sum_{i=1}^{I}P_i\right)^2}{I}\right] \tag{11-62}$$

This variance is an estimate of the mean-square deviation of the sample mean $\bar{P}$ from the true mean, where the true mean would be obtained by using an infinite number of samples. From the properties of the normal frequency distribution, which the fluctuations in the results computed by Monte Carlo will generally follow, it is shown in texts on statistics that the probability of the sample mean $\bar{P}$ lying within $\pm\gamma$ of the true mean is about 68%, that of its lying within $\pm 2\gamma$ is about 95%, and that of its lying within $\pm 3\gamma$ is 99.7%.

Another measure of the statistical fluctuations in the mean is γ, the *standard deviation*. Because γ is given by the square root of Eq. (11-62), it is evident that to reduce γ by half, the number of samples used in computing the results must be quadrupled (thereby quadrupling I for constant submean size). This probably means quadrupling the computer time involved unless the term in brackets can somehow be reduced by decreasing the variance (scatter) of the individual submeans. Much time and ingenuity have been expended on the latter, under labels such as stratified sampling, splitting, importance sampling, and energy partitioning. These and other variance-reducing techniques are discussed in [17, 19, 29, 30]. The savings in computer time is substantial and the user of a Monte Carlo method for any problem of significant complexity is urged to apply them.

11-6.3 Application to Thermal Radiative Transfer

As discussed in Chaps. 7–10, the formulation of radiation exchange heat balances in enclosures leads to integral equations for the unknown surface temperature or heat flux distributions. Integral equations also result when radiation exchange is considered within a radiating medium such as a gas. These equations can be difficult to solve and are a consequence of using a deterministic viewpoint when deriving the heat flow quantities. By invoking a probabilistic model of the radiative exchange process and applying Monte Carlo sampling techniques, it is possible to avoid many of the difficulties inherent in the integral equation formulations. In this way, actions of small parts of the total energy can be examined on an individual basis, in place of solving simultaneously for the entire behavior of the energy involved. A microscopic type of model for the radiative exchange process is examined; then the solutions of two examples are outlined.

Model of the radiative exchange process In radiation calculations, the usual quantities of interest are local temperatures and energy fluxes. We can model the radiative exchange process by following the progress of discrete amounts ("bundles") of energy, since local energy flux is then easily computed as the number of these bundles arriving per unit area and time at a location. The obvious bundle is the photon, but this has a disadvantage because its energy depends on its wavelength, which introduces a needless complication. A more convenient quantity is a *photon bundle* that is carrying a given amount of energy w; it can be thought of as a group of photons bound together. For spectral problems where the wavelength of the bundle is specified, enough photons of that wavelength are grouped together to make the energy of the bundle equal w. By assigning equal energies to all bundles, local energy flux is computed by counting the number of bundles arriving at a position of interest per unit time and per unit area and multiplying by the energy of the bundle. The bundle paths and histories are computed by the Monte Carlo method which will now be discussed.

To aid in showing the development of a Monte Carlo radiative model, it is helpful to consider a sample calculation involving directional spectral surface properties. It is desired to obtain the amount of energy radiated from element dA_1 at temperature T_1 that is absorbed by an infinite plane A_2 at temperature $T_2 = 0$ as shown in Fig. 11-11. Let dA_1 have directional spectral emissivity

$$\epsilon'_{\lambda,1} = \epsilon'_{\lambda,1}(\lambda, \theta_1, T_1) \tag{11-63}$$

and area 2 have directional spectral emissivity

$$\epsilon'_{\lambda,2} = \epsilon'_{\lambda,2}(\lambda, \theta_2, T_2) \tag{11-64}$$

and assume that the directional emissivities of both surfaces are independent of circumferential angle φ.

For surface element dA_1, the total emitted energy per unit time is

$$dQ_{e,1} = \epsilon_1(T_1)\sigma T_1^4 \, dA_1 \tag{11-65}$$

where $\epsilon_1(T_1)$ is the hemispherical total emissivity, given in this case by Eq. (3-6a):

$$\epsilon_1(T_1) = \frac{\int_0^\infty \int_\cap \epsilon'_{\lambda,1}(\lambda, \theta_1, T_1) i'_{\lambda b,1}(\lambda, T_1) \cos \theta \, d\omega \, d\lambda}{\sigma T_1^4} \tag{11-66}$$

where $i'_{\lambda b,1}(\lambda, T_1)$ is the blackbody spectral intensity at T_1.

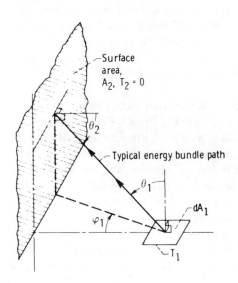

Surface area, A_2, $T_2 = 0$

θ_2

Typical energy bundle path

θ_1

dA_1

φ_1

T_1

Figure 11-11 Radiant interchange between two surfaces.

If $dQ_{e,1}$, the total energy emitted per unit time by dA_1, is composed of N energy bundles emitted per unit time, the energy assigned to each bundle is

$$w = \frac{dQ_{e,1}}{N} \tag{11-67}$$

To determine the energy radiated from element dA_1 that is absorbed by surface A_2, follow N bundles of energy after their emission from dA_1 and determine the number S_2 absorbed at A_2. If the energy reflected from A_2 back to dA_1 and then rereflected to A_2 is neglected, the energy transferred per unit time from dA_1 to A_2 is

$$dQ_{1 \rightarrow \text{absorbed by } 2} = wS_2 = \frac{\epsilon_1(T_1)\sigma T_1^4 \, dA_1}{N} S_2 \tag{11-68}$$

The next question is how to determine the path direction and wavelength assigned to each bundle. This must be done in such a way that the directions and wavelengths of the N bundles conform to the constraints given by the emissivity of the surface and the laws governing radiative processes. For example, if wavelengths are assigned to N bundles, the spectral distribution of emitted energy generated by the Monte Carlo method (consisting of the energy $wN_\lambda\Delta\lambda$ for discrete intervals $\Delta\lambda$) must closely approximate the spectrum of the actual emitted energy (plotted as $\pi\epsilon_{\lambda,1}i'_{\lambda b,1} \, d\lambda$ against λ).

The energy emitted by dA_1 per unit time in the wavelength interval $d\lambda$ about the wavelength λ and in the angular interval $d\theta_1$ about θ_1 is $d^3Q'_{\lambda e,1}(\lambda, \theta_1) = 2\pi\epsilon'_{\lambda,1}(\lambda, \theta_1, T_1)i'_{\lambda b,1}(\lambda, T_1) \cos \theta_1 \, dA_1 \sin \theta_1 \, d\theta_1 \, d\lambda$. The probability $P(\lambda, \theta_1) \, d\theta_1 \, d\lambda$ of emission in a wavelength interval about λ and in an angular interval around θ_1 is then the energy in $d\theta_1 \, d\lambda$ divided by the total emitted energy [Eq. (11-65)]:

$$P(\lambda, \theta_1) \, d\theta_1 \, d\lambda = \frac{2\pi\epsilon'_{\lambda,1}(\lambda, \theta_1)i'_{\lambda b,1}(\lambda) \cos \theta_1 \sin \theta_1 \, d\theta_1 \, d\lambda}{\epsilon_1 \sigma T_1^4} \tag{11-69}$$

The T_1 in the functional notation has been dropped for simplicity.

It is assumed here for simplicity that the directional spectral emissivity is a product function of the variables wavelength and angle; that is,

$$\epsilon'_{\lambda,1}(\lambda, \theta_1) = \Phi_1(\lambda)\Phi_2(\theta_1) \tag{11-70}$$

This assumption is probably not valid for many real surfaces because, in general, the angular distribution of emissivity depends on wavelength as shown, for example, by Fig. 5-14. For the assumed form in (11-70), it follows that the emissivity dependence on either variable may be found by integrating out the other variable [see Eq. (11-55)]. Then the normalized probability of emission occuring in the interval $d\lambda$ is

$$P(\lambda) \, d\lambda = d\lambda \int_0^{\pi/2} P(\lambda, \, \theta_1) \, d\theta_1$$

$$= \frac{2\pi \, d\lambda \int_0^{\pi/2} \epsilon'_{\lambda,1}(\lambda, \, \theta_1) i'_{\lambda b,1}(\lambda) \sin \theta_1 \cos \theta_1 \, d\theta_1}{\epsilon_1 \sigma T_1^4} \qquad (11\text{-}71)$$

Substituting into (11-51) and noting that $P(\lambda) \, d\lambda$ is zero in the range $-\infty < \lambda < 0$ give

$$R_\lambda = \frac{2\pi \int_0^\lambda \int_0^{\pi/2} \epsilon'_{\lambda,1}(\lambda^*, \, \theta_1) i'_{\lambda b,1}(\lambda^*) \sin \theta_1 \cos \theta_1 \, d\theta_1 \, d\lambda^*}{\epsilon_1 \sigma T_1^4} \qquad (11\text{-}72a)$$

where the asterisk denotes a dummy variable of integration. If the number of bundles is very large and this equation is solved for λ each time a random R_λ value is chosen, the computing time becomes too large. To circumvent this, equations like Eq. (11-72a) can be numerically integrated once over the range of λ values and a curve fitted to the result. A polynomial approximation

$$\lambda = A + BR_\lambda + CR_\lambda^2 + \cdots \qquad (11\text{-}72b)$$

is often adequate. Equation (11-72b) rather than (11-72a) is used in the computer program. Alternatively, a table of λ versus R_λ can be put into memory and interpolated to obtain λ for chosen R_λ values.

Following a similar procedure for the variable cone angle of emission θ_1 gives

$$R_{\theta_1} = \int_0^{\theta_1} \int_0^\infty P(\theta_1^*, \, \lambda) \, d\lambda \, d\theta_1^*$$

$$= \frac{2\pi \int_0^{\theta_1} \int_0^\infty \epsilon'_{\lambda,1}(\lambda, \, \theta_1^*) i'_{\lambda b,1}(\lambda) \sin \theta_1^* \cos \theta_1^* \, d\lambda \, d\theta_1^*}{\epsilon_1 \sigma T_1^4} \qquad (11\text{-}73a)$$

which is curve-fit to give

$$\theta_1 = D + ER_{\theta_1} + FR_{\theta_1}^2 + \cdots \qquad (11\text{-}73b)$$

If dA_1 is a *diffuse-gray* surface, (11-72a) reduces to

$$R_{\lambda,\text{diffuse gray}} = \frac{\pi \int_0^\lambda i'_{\lambda b,1}(\lambda^*) \, d\lambda^*}{\sigma T_1^4} = F_{0-\lambda} \qquad (11\text{-}74)$$

Table 11-2 Inverse probability function for choosing wavelength of emission from a gray or black surface (λT in $\mu m \cdot K$)

$\lambda T = D_1 + D_2 R_\lambda^{1/8} + D_3 R_\lambda^{1/4} + D_4 R_\lambda^{3/8} + D_5 R_\lambda^{1/2}$	$0.0 < R_\lambda < 0.1$
$\lambda T = D_1 + D_2 R_\lambda + D_3 R_\lambda^2 + D_4 R_\lambda^3 + D_5 R_\lambda^4$	$0.1 < R_\lambda < 0.9$
$\lambda T = \left[\dfrac{0.152886 \times 10^{12}}{D_1(1 - R_\lambda) + D_2(1 - R_\lambda)^2 + D_3(1 - R_\lambda)^3 + D_4(1 - R_\lambda)^4} \right]^{1/3}$	$0.9 < R_\lambda < 1$

Range	Coefficients				
of R_λ	D_1	D_2	D_3	D_4	D_5
0.0–0.1	503.247	230.243	5,863.85	−10,759.6	8,723.14
0.1–0.4	1,560.84	7,603.61	−15,540.1	31,257.7	−20,844.8
0.4–0.7	2,846.63	−4,430.38	27,936.0	−41,041.9	25,960.9
0.7–0.9	345,197	−1,828,567	3,674,856	−3,284,391	1,108,939
0.9–0.99	1.200	9.476	−44.84	156.9	—
0.99–1.0	1.10064	16.8148	−183.445	890.699	—

where $F_{0-\lambda}$ is the fraction of blackbody emission in the wavelength interval $0-\lambda$. Haji-Sheikh [30] presents the inverse function $\lambda T = F(R_\lambda)$ for a diffuse-gray surface at temperature T. The relations are in Table 11-2.[†]

For the diffuse-gray case Equation (11-73a) reduces to

$$R_{\theta_1, \text{diffuse gray}} - 2 \int_0^{\theta_1} \sin \theta_1^* \cos \theta_1^* \, d\theta_1^* = \sin^2 \theta_1 \tag{11-75a}$$

or

$$\sin \theta_1 = \sqrt{R_{\theta_1, \text{diffuse gray}}} \tag{11-75b}$$

The point to be made here is that computational difficulty is not greatly different in obtaining λ from either (11-72b) or (11-74), nor is it much different for obtaining θ_1 from either (11-73b) or (11-75b). The difference between directional-spectral and diffuse-gray cases is mainly in the auxiliary numerical integrations of (11-72a) and (11-73a). These integrations are performed once to obtain the curve fits; then as far as the main problem-solving program is concerned, the seemingly more difficult case might just as well be handled. Thus, increasing problem complexity leads to only gradual increases in the complexity of the Monte Carlo computer program.

For emission of an individual energy bundle from surface dA_1, a wavelength λ is obtained from (11-72b) and a cone angle of emission θ_1 from (11-73b) by choosing two random numbers R_λ and R_{θ_1}. To define the bundle path, there re-

[†]Slightly modified for $R_\lambda > 0.9$ as a result of personal communication with A. Haji-Sheikh.

mains specification of the circumferential angle φ_1. Because of the earlier assumption that emission does not depend on φ_1, it is shown by the formalism outlined and is also fairly obvious that φ_1 can be determined by

$$\varphi_1 = 2\pi R_{\varphi_1} \tag{11-76}$$

where R_{φ_1} is a random number chosen from the range 0 to 1.

Because the position of plane A_2 with respect to dA_1 is known, it is not difficult to determine whether a given energy bundle will strike A_2 after leaving dA_1 in direction (θ_1, φ_1). As shown in Fig. 11-11, it will hit A_2 whenever $\cos \varphi_1 \geq 0$. If it misses A_2 another bundle must be emitted from dA_1. If the bundle strikes A_2, it must be determined whether it is absorbed or reflected. To do this, the geometry is used to find the angle of incidence θ_2 of the bundle onto A_2,

$$\cos \theta_2 = \sin \theta_1 \cos \varphi_1 \tag{11-77}$$

Knowing the absorptivity of A_2 from Kirchhoff's law,

$$\alpha'_{\lambda,2}(\lambda, \theta_2) = \epsilon'_{\lambda,2}(\lambda, \theta_2) \tag{11-78}$$

and having determined the wavelength λ of the incident bundle from (11-72b) and the incident angle θ_2 from (11-77), the probability of absorption of the bundle at A_2 can be determined. This is simply $\alpha'_{\lambda,2}(\lambda, \theta_2)$, since this is the fraction of energy incident on A_2 in a given wavelength interval and within a given solid angle that is absorbed by the surface. The absorptivity is the probability density function for the absorption of incident energy. To determine whether a given incident energy bundle is absorbed, the surface absorptivity $\alpha'_{\lambda,2}(\lambda, \theta_2)$ is compared with a random number R_{α_2}. If

$$R_{\alpha_2} \leq \alpha'_{\lambda,2}(\lambda, \theta_2) \tag{11-79}$$

the bundle of energy is absorbed and a counter S_2 in the computer memory is increased by 1 to account for the absorbed bundles. Otherwise, the bundle is assumed to be reflected and is not further accounted for. If the bundle path were followed further, rereflections from dA_1 would be considered. Angles of reflection are chosen from known directional reflectivities, and the bundle is followed further along its path until it is absorbed by A_2 or lost from the system. The derivation of the necessary relations is similar to that presented. A new bundle is now chosen at dA_1, and its history is followed. This procedure is continued until N bundles have been emitted from dA_1. The energy absorbed at A_2 is then calculated from (11-68).

The derivation of the equations needed for the present solution is now complete. In putting together a flowchart to aid in writing a computer program (Fig. 11-12), some methods for shortening computing time can be invoked. For example, the angle φ_1 is computed first. If the bundle is not going to strike A_2 on the basis of the calculated φ_1, there is no point in computing λ and θ_1 for that bundle. Alternatively, because φ_1 values are isotropically distributed, it is noted for this geometry that exactly half the bundles must strike A_2. Therefore, the calculated φ_1 values can be constrained to the range $-\pi/2 < \varphi_1 < \pi/2$.

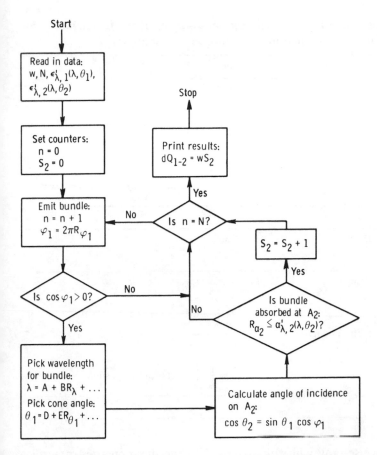

Figure 11-12 Computer flow diagram for example radiant-interchange problem.

The formulation of this problem for a Monte Carlo solution is now complete. The reader will note that the desired result could be obtained without much trouble by standard integral methods. However, extension to only slightly more difficult problems could cause difficulties for standard treatments, but would not be very difficult for Monte Carlo. For example, introduce a third surface with directional properties and account for all interactions.

Useful functions A number of useful relations for choosing angles of emission and assigning a wavelength to bundles are given in the previous section. These and other functions are summarized in Table 11-3.

EXAMPLE 11-5 A wedge is made up of two very long parallel sides of equal width joined at an angle of 90°, Fig. 11-13. The surface temperatures are $T_1 = 1000$ K and $T_2 = 2000$ K. The effects of the ends may be neglected.

Table 11-3 Convenient functions relating random numbers to variables for emission (assume no dependence on circumferential angle φ)

Variable	Type of emission	Relation
Cone angle θ	Diffuse	$\sin \theta = R_\theta^{1/2}$
	Directional-gray	$R_\theta = \dfrac{2 \displaystyle\int_0^\theta \epsilon'(\theta^*) \sin \theta^* \cos \theta^* \, d\theta^*}{\epsilon}$
	Directional-nongray	$R_\theta = \dfrac{2\pi \displaystyle\int_0^\theta \int_0^\infty \epsilon_\lambda'(\lambda,\theta^*) i_{\lambda b}'(\lambda) \sin \theta^* \cos \theta^* \, d\lambda \, d\theta^*}{\epsilon \sigma T^4}$
Circumferential angle φ	Diffuse	$\varphi = 2\pi R_\varphi$
Wavelength λ	Black or gray	$F_{0-\lambda} = R_\lambda$
	Diffuse-nongray	$R_\lambda = \dfrac{\pi \displaystyle\int_0^\lambda \epsilon_\lambda(\lambda^*) i_{\lambda b}'(\lambda^*) \, d\lambda^*}{\epsilon \sigma T^4}$
	Directional-nongray	$R_\lambda = \dfrac{2\pi \displaystyle\int_0^\lambda \int_0^{\pi/2} \epsilon_\lambda'(\lambda^*,\theta) i_{\lambda b}'(\lambda^*) \sin \theta \cos \theta \, d\theta \, d\lambda^*}{\epsilon \sigma T^4}$

Surface 1 is diffuse-gray with emissivity 0.5, while the properties of surface 2 are directional gray with directional total emissivity and absorptivity given by

$$\epsilon_2'(\theta_2) = \alpha_2'(\theta_2) = 0.5 \cos \theta_2 \tag{11-80}$$

Assume for simplicity that surface 2 reflects diffusely. Set up a Monte Carlo flowchart for determining the energy to be added to each surface to maintain its temperature. Assume that the environment is at $T_e = 0$ K.

The energy flux emitted by surface 1 is $q_{e,1} = \epsilon_1 \sigma T_1^4$. If N_1 emitted energy bundles are followed per unit time and area from surface 1, the energy per bundle is

$$w = \frac{q_{e,1}}{N_1} = \frac{\epsilon_1 \sigma T_1^4}{N_1} \tag{11-81}$$

The energy flux emitted from surface 2 is

$$q_{e,2} = 2\sigma T_2^4 \int_0^{\pi/2} \epsilon_2'(\theta) \cos \theta \sin \theta \, d\theta = \sigma T_2^4 \int_0^{\pi/2} \cos^2 \theta \sin \theta \, d\theta = \frac{\sigma T_2^4}{3}$$

If the same amount of energy w is assigned to each bundle emitted by wall 2 as was used for wall 1, then $wN_2 = \sigma T_2^4/3$. Substituting (11-81), the value of ϵ_1, and the known surface temperatures gives

$$N_2 = \frac{\sigma T_2^4}{3} \frac{N_1}{\epsilon_1 \sigma T_1^4} = \frac{32}{3} N_1 \tag{11-82}$$

Because all bundles have equal energy and 32/3 as many bundles are emitted from surface 2 as from surface 1, it is evident that surface 2 will make the major contribution to the energy transfer.

Now the distributions of directions for emitted bundles from the two surfaces will be derived. Surface 1 emits diffusely, so (11-75b) applies. For surface 2, however, (11-73a) must be used. Substituting (11-80) into (11-73a) gives, for the directional-gray case,

$$R_{\theta_2} = \frac{2\pi i'_{b,2} \displaystyle\int_0^{\theta_2} (0.5 \cos \theta_2^*) \sin \theta_2^* \cos \theta_2^* \, d\theta_2^*}{\epsilon_2 \sigma T_2^4}$$

The hemispherical total emissivity is substituted from Eq. (11-66) to give

$$R_{\theta_2} = \frac{\displaystyle\int_0^{\theta_2} \cos^2 \theta_2^* \sin \theta_2^* \, d\theta_2^*}{\displaystyle\int_0^{\pi/2} \cos^2 \theta_2 \sin \theta_2 \, d\theta_2} = 1 - \cos^2 \theta_2$$

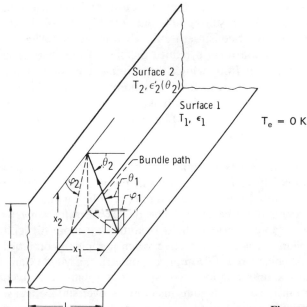

Figure 11-13 Geometry of Example 11-5.

The fact that R and $1 - R$ are both uniform random distributions in the range $0 \le R \le 1$ can be used to write this as $\cos \theta_2 = R_{\theta_2}^{1/3}$. Note that by similar reasoning (11-75b) can be written $\cos \theta_1 = R_{\theta_1}^{1/2}$. Since there is no dependence on angle φ for either surface, (11-76) applies for both surfaces.

Next the position must be determined on each surface from which each bundle will be emitted. Because the wedge sides are isothermal, the emission from a given side is uniform. In such a case, random positions x (Fig. 11-13) on a given side could be picked as points of emission. Such a procedure requires generation of a random number. The computer time required to generate a random number can be saved by noting that the bundle emission is the initial process in each Monte Carlo history; hence, there is no prior history to be eliminated by using a random number. In this case, x positions along L can be sequentially chosen as $x = (n/N)L$, where n is the sample-history index for the history being begun, $1 \le n \le N$.

The remaining calculations involve determination of whether each emitted bundle will strike the adjacent wall or will leave the cavity. Examination of Fig. 11-13 shows that, for either surface, when $\pi \le \varphi \le 2\pi$ the bundles will leave the cavity for any θ, and when $0 < \varphi < \pi$ they will leave if

$$\sin \theta < \frac{x/\sin \varphi}{[(x/\sin \varphi)^2 + L^2]^{1/2}} = \frac{1}{[1 + (L \sin \varphi/x)^2]^{1/2}}$$

The angle of incidence θ_i on a surface is given in terms of the angles θ_δ and φ_δ, at which the bundle leaves the other surface, by $\cos \theta_i = \sin \theta_\delta \sin \varphi_\delta$. All the necessary relations are now available. A flow diagram is constructed to combine these relations in the correct sequence. Diffuse reflection is assumed from both surfaces. The resulting diagram is in Fig. 11-14. Study of this figure will show one way of constructing the flow of events. The use of the indices δ, δ', and δ'' is an artifice to reduce the size of the chart. The index δ always refers to the wall from which the original emission of the bundle occurred, and δ' refers to the wall from which emission or reflection is presently occurring. The index δ'' is used to make the emitted distribution of θ angles correspond to either $R_{\theta_1}^{1/2}$ or $R_{\theta_2}^{1/3}$ and have all the reflected bundles correspond to a diffuse distribution.

Literature on radiation exchange between surfaces The value of Monte Carlo methods is that program complexity increases roughly in proportion to problem complexity, while the difficulty of carrying out conventional solutions increases roughly as the square of the complexity because of the matrix form for conventional formulations. Because a Monte Carlo method is somewhat more difficult to apply to the simplest problems, it is most effective when complex geometries and variable properties must be considered. In complex geometries, a Monte Carlo method has the advantage that simple relations will specify the path of a given energy bundle, whereas most other methods require integrations over surface areas. The integrations become difficult when various curved or skewed surfaces are present.

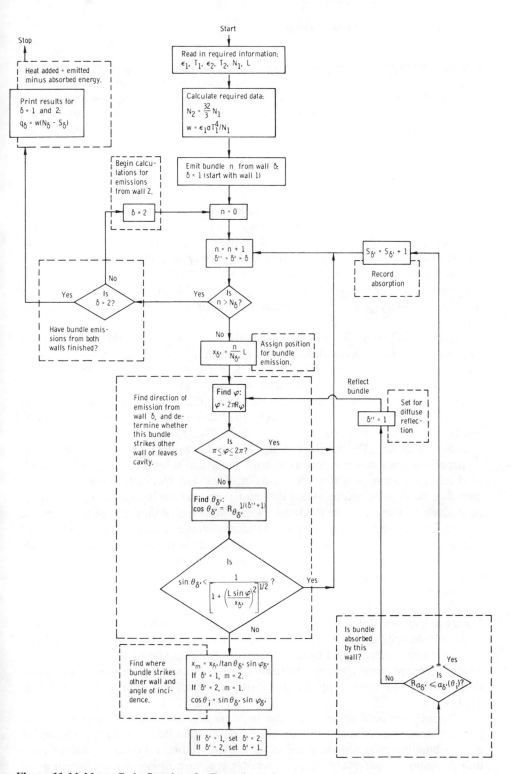

Figure 11-14 Monte Carlo flowchart for Example 11-5.

Configuration-factor computation The calculation of radiative configuration factors by standard means usually involves assumptions that restrict the application of these factors in exchange computations. The assumptions when using ordinary configuration factors as in Chap. 6 are that the surfaces are diffuse emitters and reflectors, that each surface is isothermal, and that the flux arriving at and leaving each surface is evenly distributed across the surface. Any of these assumptions may be poor; most surfaces are not diffuse, and the distribution of reflected flux usually deviates from uniformity to some extent. Where deviations from the assumptions must be considered, calculation of the configuration factors becomes difficult; and if geometries with nonplanar surfaces are involved, Monte Carlo techniques may become invaluable. It should be noted, however, that unless a parametric study of the interchange of radiant energy within an enclosure with specified characteristics is being carried out, it may be easier to compute directly the entire radiative flux distribution by a Monte Carlo method. This would be simpler than computing configuration factors by a Monte Carlo method and then using an auxiliary program to calculate energy exchange by means of these factors.

Corlett [31] computed exchange factors (as distinguished from configuration factors) for various geometries, including louvers, and circular and square ducts with various combinations of diffusely and specularly reflecting interior surfaces and ends. The factors give the fraction of energy emitted by a given surface that reaches another surface by all paths, including intermediate reflections. One set of results, the exchange factors between the black ends of a cylinder with a diffusely reflecting internal surface, is shown in Fig. 11-15.

Weiner et al. [32] carried out the Monte Carlo evaluation of some simple configuration factors for comparison with analytical solutions. They then considered energy exchange within an enclosure with five specularly reflecting sides, each side being assumed to have a directional emissivity dependent on cone angle of emission. They also worked out the interchange within a simulated optical system constructed of spherical and conical surfaces that enclose a cylindrical specular reflector with two surfaces. This interchange problem would cause many hours of analyzing integral limits in the usual formulations. Additional configuration factor calculations are in [33]. The exchange in a bed of spheres was analyzed in [34].

Cavity properties Polgar and Howell [35] analyzed the bidirectional reflectivity of a conical cavity exposed to a beam of parallel incident radiation and determined the directional emissivity of the cavity. One set of representative results is in Fig. 11-16. Hemispherical absorptivity results were obtained by integrating the directional values and were compared in [36] to analytical results from [37] (see Fig. 11-17).

The bidirectional reflectivity results computed by Monte Carlo methods in [35] illustrate the scatter of the computed points that depends on the number of energy bundles reflected from the cone interior through any area element on a unit hemisphere over the cavity. The scatter is shown in Fig. 11-18, which gives

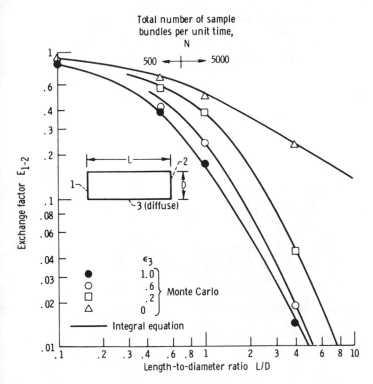

Figure 11-15 Radiation exchange factors between black ends of diffuse walled cylinder (from [31]). $\epsilon_1 = \epsilon_2 = 1$.

the standard deviation of the computed reflectivity at various angles of reflection. The solid angle subtended by area elements of equal angular increment $\Delta\theta \, \Delta\varphi$ on the hemisphere varies with the sine of the angle of reflection, so the number of sample energy bundles per unit solid angle $d\omega = \sin\theta \, d\theta \, d\varphi$ near the cone axis becomes very small. This leads to larger scatter at angles near the cone axis, where $\sin\theta \to 0$.

Monte Carlo techniques were applied in [29, 38, 39] to determine the emittance from conical and cylindrical cavities with baffles. The speed of computation was improved by using "energy partitioning." Each bundle emitted from a given position on the interior surface of a cavity is divided into two parts. The first part is the fraction that leaves the cavity directly through the aperture without reflection from any surface. This fraction is the configuration factor from the surface element to the aperture. The remaining part of the bundle may undergo one or more reflections within the cavity. This part is traced to its next point of impingement. Absorption is determined following the procedures in this chapter. If the bundle is not absorbed, it is again partitioned into a part leaving the aperture and a second part that strikes the cavity interior. The sequence is continued until the bundle is

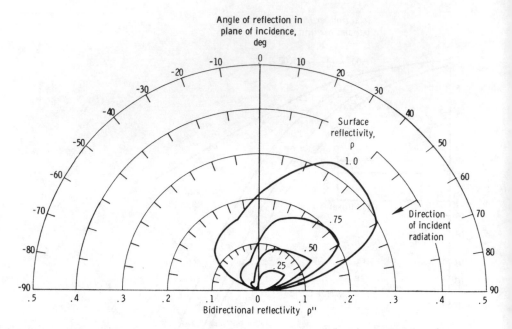

Figure 11-16 Bidirectional reflectivity of diffusely reflecting conical cavity. Cone angle of cavity, 30°; angle of incidence of radiation, 60°. From [35].

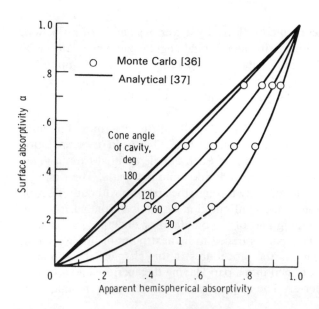

Figure 11-17 Comparison of Monte Carlo results for absorptivity of conical cavities [36] with analytical results [37].

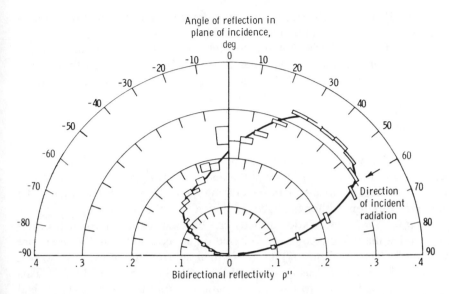

Figure 11-18 Expected standard deviation of results for bidirectional reflectivity of diffuse conical cavity. Cone angle of cavity, 30°; angle of incidence of radiation, 60°. From [35].

absorbed or its remaining energy is negligible. This procedure gives at least one statistical sample to the emitted energy from the cavity for each original bundle emission. The method improved convergence to a stable solution of low variance.

Extension to directional and spectral surfaces Few references exist that treat problems involving both directionally and spectrally dependent properties. The reasons are twofold. First, accurate and complete directional spectral properties are not often found in the literature. An analyst desiring to include such effects is usually unable to find the requisite data. Second, when solutions are obtained for such problems, they are often so specialized that little interest exists to warrant their dissemination in the open literature. As pointed out by Dunn et al. [40], when the radiative properties become available, the methods for handling such surface radiative-energy-exchange problems now exist, and the Monte Carlo technique appears to be one of the better-suited techniques. Toor and Viskanta [41, 42] applied Monte Carlo techniques to some interchange problems involving surfaces with directional and spectral property variations. Some of these results were discussed in Chap. 8. Howell and Durkee [43] compared experimental data and a Monte Carlo analysis of radiative exchange in an enclosure with directional surfaces, with good results. Naraghi and Chung [44] analyzed enclosures with directional properties. A general purpose Monte Carlo computer code for calculating radiative exchange factors in three-dimensional enclosures is described in [45]. The code includes mixed specular and diffuse reflection models for surfaces, wavelength-banded spectral surface properties, transmission of radiation through surfaces, and incident beam radiation.

Extension of Monte Carlo methods to problems with radiation and conduction, including property uncertainties, has been done in [46, 47]. In these treatments, mean values and standard deviations of both material properties and dimensional tolerances were used to choose sample systems. A large number of such systems were analyzed to predict the average expected performance of the thermal system and the variance in expected performance.

Application of Monte Carlo methods to combined-mode problems Monte Carlo methods can be utilized in multimode problems by referring to Eq. (10-4), which gives the general heat balance equation including radiation for a thin element. The net radiative flux on either surface is given by the term $q_{R,\text{net}}(x, y) = q_i(x, y) - q_o(x, y)$. Monte Carlo methods can be used as described in Sec. 11-6.3 to determine the value of $q_{R,\text{net}}(x, y)$ based on an assumed initial temperature distribution $T_j^{(0)}$. The resulting $q_{R,\text{net}}^{(0)}(x, y)$ is then substituted into (10-4), which may be solved for a new $T_j^{(n)}$ using the methods in Secs. 11-5.1 and 11-5.2. This new $T_j^{(n)}$ is then used in the Monte Carlo routine to find a new $q_{R,\text{net}}^{(n)}(x, y)$, and the process is repeated until convergence. This is an iterative approach, and an underrelaxation factor may be used between iterations to aid stability and convergence.

The advantages of using Monte Carlo methods are the ability to treat complex geometries and to include spectral and directional surface properties. A disadvantage is that the statistical fluctuations in the $q_{R,\text{net}}^{(n)}(x, y)$ values found from the Monte Carlo code may cause the $T_j^{(n)}$ values computed from (10-4) to fluctuate or diverge between iterations, and convergence to an acceptable accuracy may be difficult. Problems in which radiation dominates will converge fairly quickly to the accuracy imposed by the Monte Carlo solution, while those in which convection dominates will be insensitive to the fluctuating values of $q_{R,\text{net}}^{(n)}(x, y)$. For problems where both radiation and at least one other heat transfer mode are present and have approximately equal influence (i.e., a parameter N, S, or H is of order unity), it may be necessary to use care in the Monte Carlo solution so that smooth and accurate $q_{R,\text{net}}^{(n)}(x, y)$ values are obtained for the two-dimensional situation discussed here. The same considerations apply to more general transient or steady three-dimensional problems.

REFERENCES

1. Carnahan, B., H. A. Luther, and J. O. Wilkes: "Applied Numerical Methods," Wiley, New York, 1969.
2. Press, W. H., B. P. Flannery, S. A. Teukolsky, and W. T. Vetterling: "Numerical Recipes," Cambridge University Press, London, 1986.
3. Chapra, S. C., and R. P. Canale: "Numerical Methods for Engineers," 2d ed., McGraw-Hill, New York, 1988.
4. "FORTRAN Subroutines for Mathematical Applications," Version 1.0, IMSL, Houston, April 1987.

5. Becker, E. B., G. F. Carey, and J. T. Oden: "Finite Elements, an Introduction," vol. 1, Prentice-Hall, Englewood Cliffs, N.J., 1981.
6. Zienkiewicz, O. C.: "The Finite Element Method," 3d ed., McGraw-Hill, New York, 1977.
7. Heubner, K. H.: "The Finite Element Method for Engineers," Wiley, New York, 1975.
8. Minkowycz, W. J., E. M. Sparrow, G. E. Schneider, and R. H. Pletcher (eds.): "Handbook of Numerical Heat Transfer," Wiley, New York, 1988.
9. Razzaque, M. M., J. R. Howell, and D. E. Klein: Finite Element Solution of Heat Transfer for a Gas Flow through a Tube, *AIAA J.*, vol. 20, no. 7, pp. 1015–1019, 1982.
10. Altes, D., G. Breitbach, and J. Szimmat: Application of the Finite Element Method to the Calculation of Heat Transfer by Radiation, Proc. Int. Finite Element Method Congress, Baden-Baden, Germany, November 1986.
11. Cort, G. E., H. L. Graham, and N. L. Johnson: Comparison of Methods for Solving Non-Linear Finite Element Equations in Heat Transfer, ASME Paper 82-HT-40, August 1982.
12. Tan, Z.: Combined Radiative and Conductive Heat Transfer in Two-Dimensional Emitting, Absorbing, and Anisotropically Scattering Square Media, *Int. Comm. Heat Mass Transfer,* vol. 16, pp. 391–401, 1989.
13. Ness, A. J.: Solution of Equations of a Thermal Network on a Digital Computer, *Sol. Energy,* vol. 3, no. 2, p. 37, 1959.
14. Costello, F. A., and G. L. Schrenk: Numerical Solution to the Heat-Transfer Equations with Combined Conduction and Radiation, *J. Comput. Phys.,* vol. 1, pp. 541–543, 1966.
14a. Howell, J. R.: Modern Numerical Methods in Radiative Heat Transfer, in K. T. Yang and W. Nakayama (eds.), "Proc. U.S.–Japan Heat Transfer Seminar on Computers in Heat Transfer Science," Hemisphere, Washington, D.C., 1992.
15. Kahn, Herman: Applications of Monte Carlo, Rept. No. RM-1237-AEC (AEC No. AECU-3259), Rand Corp., April 27, 1956.
16. Metropolis, Nicholas, and S. Ulam: The Monte Carlo Method, *J. Am. Stat. Assoc.,* vol. 44, no. 247, pp. 335–341, 1949.
17. Hammersley, J. M., and D. C. Handscomb: "Monte Carlo Methods," Wiley, New York, 1964.
18. Cashwell, E. D., and C. J. Everett: "A Practical Manual on the Monte Carlo Method for Random Walk Problems," Pergamon, New York, 1959.
19. Schreider, Yu. A. (ed.): "Method of Statistical Testing–Monte Carlo Method," American El-sevier, New York, 1964.
20. Brown, G. W.: Monte Carlo Methods, in E. F. Bechenbach (ed.), "Modern Mathematics for the Engineer," pp. 279–307, McGraw-Hill, New York, 1956.
21. Halton, John H.: A Retrospective and Prospective Survey of the Monte Carlo Method, *SIAM Rev.,* vol. 12, no. 1, pp. 1–63, 1970.
22. Meyer, Herbert A. (ed.): "Symposium on Monte Carlo Methods," Wiley, New York, 1956.
23. Rand Corp.: "A Million Random Digits with 100,000 Normal Deviates," Free Press, Glencoe, Ill., 1955.
24. Kendall, M. G., and B. Babington Smith: "Tables of Random Sampling Numbers," 2d ser., Cambridge University Press, New York, 1954.
25. Taussky, O., and J. Todd: Generating and Testing of Pseudo-Random Numbers, in Herbert A. Meyer (ed.), "Symposium on Monte Carlo Methods," pp. 15–28, Wiley, New York, 1956.
26. Kendall, M. G., and B. B. Smith: Randomness and Random Sampling Numbers, *R. Stat. Soc. J.,* pt. 1, pp. 147–166, 1938.
27. Dwight, H. B.: "Mathematical Tables of Elementary and Some Higher Mathematical Functions," 2d ed., Dover, New York, 1958.
28. Jahnke, Eugene, and Fritz Emde: "Tables of Functions with Formulae and Curves," 4th ed., Dover, New York, 1945.
29. Shamsundar, N., E. M. Sparrow, and R. P. Heinisch: Monte Carlo Radiation Solutions—Effect of Energy Partitioning and Number of Rays, *Int. J. Heat Mass Transfer,* vol. 16, no. 3, pp. 690–694, 1973.
30. Haji-Sheikh, A.: Monte Carlo Methods, in W. J. Minkowycz et al. (eds.), "Handbook of Numerical Heat Transfer," pp. 673–717, Wiley, New York, 1988.

31. Corlett, R. C.: Direct Monte Carlo Calculation of Radiative Heat Transfer in Vacuum, *J. Heat Transfer*, vol. 88, no. 4, pp. 376–382, 1966.

32. Weiner, M. M., J. W. Tindall, and L. M. Candell: Radiative Interchange Factors by Monte Carlo, ASME Paper No. 65-WA/HT-51, November 1965.

33. Yarbrough, D. W., and Chon-Lin Lee: Monte Carlo Calculation of Radiation View Factors, in F. R. Payne et al. (eds.), "Integral Methods in Science and Engineering 85," pp. 563–574, Hemisphere, Washington, D.C., 1985.

34. Yang, Y. S., J. R. Howell, and D. E. Klein: Radiative Heat Transfer through Randomly Packed Bed of Spheres by the Monte Carlo Method, *J. Heat Transfer,* vol. 105, no. 2, p. 325–332, 1983.

35. Polgar, L. G., and J. R. Howell: Directional Thermal-radiative Properties of Conical Cavities, NASA TN D-2904, 1965.

36. Polgar, L. G., and J. R. Howell: Directional Radiative Characteristics of Conical Cavities and Their Relation to Lunar Phenomena, in G. B. Heller (ed.), "Thermophysics and Temperature Control of Spacecraft and Entry Vehicles," pp. 311–323, Academic, New York, 1966.

37. Sparrow, E. M., and V. K. Jonsson: Radiant Emission Characteristics of Diffuse Conical Cavities, *J. Opt. Soc. Am.,* vol. 53, no. 7, pp. 816–821, 1963.

38. Heinisch, R. P., E. M. Sparrow, and N. Shamsundar: Radiant Emission from Baffled Conical Cavities, *J. Opt. Soc. Am.,* vol. 63, no. 2, pp. 152–158, 1973.

39. Sparrow, E. M., R. P. Heinisch, and N. Shamsundar: Apparent Hemispherical Emittance of Baffled Cylindrical Cavities, *J. Heat Transfer,* vol. 96, no. 1, pp. 112–114, 1974.

40. Dunn, S. Thomas, Joseph C. Richmond, and Jerome F. Parmer: Survey of Infrared Measurement Techniques and Computational Methods in Radiant Heat Transfer, *J. Spacecr. Rockets,* vol. 3, no. 7, pp. 961–975, 1966.

41. Toor, J. S., and R. Viskanta: A Numerical Experiment of Radiant Heat Exchange by the Monte Carlo Method, *Int. J. Heat Mass Transfer,* vol. 11, no. 5, pp. 883–897, 1968.

42. Toor, J. S., and R. Viskanta: Effect of Direction Dependent Properties on Radiant Interchange, *J. Spacecr. Rockets,* vol. 5, no. 6, pp. 742–743, 1968.

43. Howell, J. R., and R. Durkee: Radiative Transfer between Surfaces with Collimated Incident Radiation: A Comparison of Analysis and Experiment, *J. Heat Transfer,* vol. 93, no. 2, pp. 129–135, 1971.

44. Naraghi, M. H. N., and B. T. F. Chung: A Stochastic Approach for Radiative Exchange in Enclosures with Directional-Bidirectional Properties, *J. Heat Transfer,* vol. 108, no. 2, pp. 264–270, 1986.

45. Maltby, J. D., and P. J. Burns: Performance, Accuracy, and Convergence in a 3-D Monte Carlo Radiative Heat Transfer Simulation, *Numerical Heat Transfer, Part B,* vol. 19, no. 3, pp. 191–209, 1991.

46. Howell, J. R.: Monte Carlo Treatment of Data Uncertainties in Thermal Analysis, *J. Spacecr. Rockets,* vol. 10, no. 6, pp. 411–414, 1973.

47. Zigrang, Denis J.: Statistical Treatment of Data Uncertainties in Heat Transfer, AIAA Paper 75-710, May 1975.

PROBLEMS

11-1 Evaluate the configuration factor between two infinite parallel plates of equal width joined along one edge and making an angle of 45°. Use a trapezoidal rule numerical integration of the integral form found in Example 6-5 to obtain the result, and compare with the exact solution.

Answer: 0.61732

11-2 Evaluate the configuration factor between two infinite parallel plates of equal width joined along one edge and making an angle of 45°. Use a Simpson's rule numerical integration of the integral form found in Example 6-5 to obtain the result, and compare with the exact solution.

Answer: 0.61732

11-3 An electrically heated nickel wire is suspended in a vacuum between two water-cooled electrodes maintained at 765 K. The wire diameter is 0.2 cm and the wire length is 3 cm. The surroundings are at a uniform temperature of 850 K and act as a black environment. Set up the combined radiation and conduction relations to determine the wire temperature $T(x)$ assuming that the radial temperature distribution within the wire is uniform at each x. Determine how many watts must be generated within the wire for its center temperature to be 975 K. (The wire thermal conductivity is constant with a value of 85 W/(m · K) and the wire emissivity is constant with value $\epsilon_w = 0.10$.) What is the required wattage if the nickel becomes oxidized so that $\epsilon_w = 0.42$?

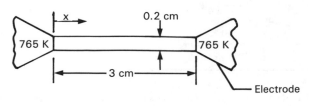

$T_e = 850$ K

Answer: 15.2 W for $\epsilon_w = 0.10$, 16.2 W for $\epsilon_w = 0.42$

11-4 A long solid rectangular region in vacuum has the cross section shown and thermal conductivity k_w. One-half of one of the long sides is heated by contact with a source of uniform flux q_e. The surroundings, which act as a black environment, are at uniform temperature T_e. The exposed surfaces of the region are gray and have emissivity ϵ_w. Using the grid shown (for simplicity), set up the finite-difference relations to be solved for the steady temperature distribution in the region.

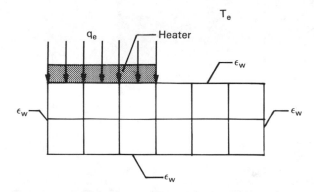

11-5 A long stainless steel tube filled with a highly insulating material is near a black hot wall. Divide the circumference of the tube symmetrically about $\theta = 0$ into eight increments and obtain an expression for the radiant energy from the hot wall to each tube increment. Then, using a finite-difference approximation, obtain an approximate temperature distribution around the tube. Neglect radial temperature distributions in the tube wall. The tube wall material has thermal conductivity $k_w = 20$ W/(m · K) and is gray with emissivity $\epsilon_w = 0.25$. The surroundings are at a low temperature that can be neglected, $T_e \approx 0$.

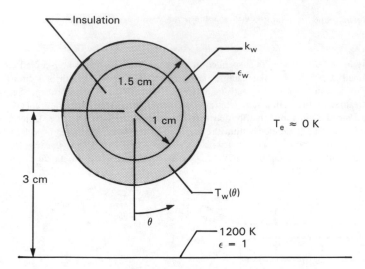

11-6 A long gray empty circular tube is in the vacuum of outer space so that the only heat exchange is by radiation. The metal tube is coated with a material that has a solar absorptivity α_s and an emissivity in the infrared region of ϵ_{IR}. The solar flux q_s is incident from a direction normal to the tube axis, and the surrounding environment is at a very low temperature T_e that can be neglected in the radiative energy balances. The geometry is as shown in cross section. The tube is empty so there is internal radiative exchange. Energy is conducted circumferentially within the tube wall. The wall thermal conductivity is k_w.

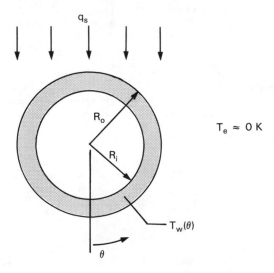

(a) Set up the energy relations required to obtain the temperature distribution around the tube circumference assuming that radial temperature variations within the tube wall can be neglected.

(b) Place the energy relations in finite-difference form and describe how a numerical solution can be obtained.

11-7 The tube in Prob. 11-6 is shielded from solar radiation by being in the shadow of a space vehicle, so that it cools to a very low temperature. It is then suddenly exposed to the solar flux. Set up the transient energy relations required to calculate the tube circumferential temperature distribution as a function of time using the same conditions and assumptions as in Prob. 11-6. Place the equations in finite-difference form and describe how a numerical solution can be obtained.

11-8 A tube has length $L = 2.00$ m and has an inside diameter of $D_i = 0.5$ m. The tube has a wall thickness of $b = 1$ cm, and the tube material has a thermal conductivity of $k = 300$ W/(m·K). An electrical heating tape is wrapped around the tube and is uniformly and carefully insulated on its outer surface. The tape imposes a uniform heat flux of $q_e = 6000$ W/m^2 at the *inner* tube surface. A transparent liquid at $T_{l,in} = 300$ K enters the tube from a large plenum at the same temperature. The liquid flows through the tube at a mass flow rate of 0.2 kg/s. The liquid has $c_p = 4100$ J/(kg·K) and density $\rho = 1000$ kg/m^3. At the given flow rate, the heat transfer coefficient between the liquid and the tube surface is given by $h_x = 1000/(x + 0.01)^{1/2}$ W/(m^2·K) where x is the distance from the tube entrance in meters. Three diffuse-gray coatings are available to cover the inside of the tube. These have emissivities of $\epsilon = 0.05, 0.25$, and 1.0.

   (a) Derive the energy equations that govern the heat transfer behavior for this problem, and place them in dimensionless form.
   (b) Find the maximum tube surface temperature and the position of that temperature, $x(T_{max})$, and the mean liquid temperature at the tube exit, $T_l(x = L)$, for each of the three possible emissivities. Show a plot of the tube inner surface and liquid temperature distributions vs x for each emissivity. These may be in dimensionless form.
   (c) Discuss your results, giving some physical discussion of the relative effects of the various heat transfer mechanisms on the shapes of both the tube wall and liquid temperature profiles.
   (d) Discuss the numerical accuracy of your results. Are you satisfied that they are converged and that they are within acceptable accuracy? What is your estimate of the possible error in your solutions?

11-9 A thin sheet of copper moves through a radiative–convective oven at a velocity of 0.1 m/s. The sheet and oven are very wide. Air flows at a mass flow rate of 0.2 kg/s per meter of oven width over the sheet in counterflow, and the heat transfer coefficient between the air and sheet surface is constant along the sheet at a value of 100 W/(m^2·K). The back of the sheet is insulated. A black radiant heater at $T_R = 1200$ K covers the top of the oven as shown. The radiant heater does not interact convectively with the air stream. Louvered curtains at each end of the oven are opaque to radiation but allow air flow. The emissivities of all surfaces are shown. Find the temperature distribution $T_s(x)$ along the strip as a function of position x within the oven, and present the result graphically. Discuss all assumptions made in the solution, and justify them by numerical argument where possible.

   Data for copper: $k = 400$ W/(m·K); $c_{p,s} = 385$ J/(kg·K); $\rho = 9000$ kg/m^3.

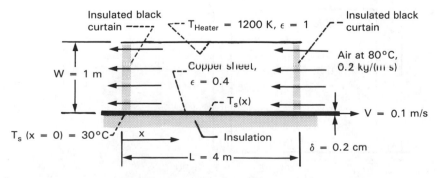

11-10 A high-temperature nuclear reactor is cooled by a transparent gas. The gas flows through cylindrical fuel elements. The elements are of length L m, inside diameter D m, and the gas has specific heat c_p kJ/(kg·K), density ρ kg/m³ (both temperature independent), and has a mass flow rate of $\dot{m}$ kg/s through each fuel element. Energy is generated in a sinusoidal distribution along the length of the tube at a rate $q(x) = q_{max} \sin(\pi x/L)$ W/m² based on the inside tube area. The heat transfer coefficient h between the gas and the tube surface is assumed constant with x and has units of W/(m²·K). The gas enters the tube at temperature $T_{g,i}$ from a large chamber at temperature $T_{r,i}$. The gas leaves the tube at $T_{g,e}$ and enters a mixing plenum that is at temperature $T_{r,e}$. The tube interior surface is diffuse-gray with emissivity ϵ. The tube exterior surface is perfectly insulated. The tube has thermal conductivity k W/(m·K).

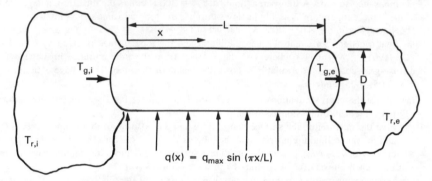

$$q(x) = q_{max} \sin(\pi x/L)$$

(a) Set up the governing equations for determining the wall and gas temperature distributions (see Example 11-3 for some help.)
(b) Using the dimensionless groups given in Example 10-8 or modifications appropriate for this problem, put the equations in dimensionless form.
(c) For values of the dimensionless parameters St = 0.02, H based on $q_{max} = 0.08$, $t_{r,i} = t_{g,i} = 1.5$, and $t_{r,e}^r = t_{g,e} = 5$, solve for the wall temperature distribution along the tube for the cases $\epsilon = 1$ and $\epsilon = 0.5$. You may wish to obtain the pure radiation and pure convection results as limiting cases.

11-11 A square enclosure of length 1 m on a side has the conditions shown below. Water is flowing along the right-hand vertical surface and enters the channel at the bottom of that wall at $T_w = 90°$ C. The flow rate of the water is large, so that its temperature changes by a negligible amount in passing along surface A_1. There is a heat transfer coefficient between the water and the thin enclosure surface of 50 W/m². Surface 4 is specular and perfectly reflecting; the remaining surfaces are diffuse-gray. Find the temperature distribution along surface 1.

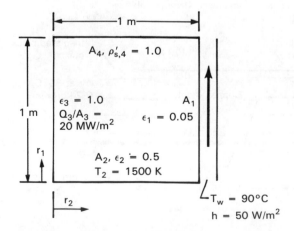

11-12 A two-dimensional problem is to be solved by the finite-element method in a rectangle that has a width significantly larger than its height. A rectangular finite element is to be used with unequal sides so the elements will conveniently scale into the problem geometry. Linear shape functions are used, so the temperature distribution is represented as $T(x, y) = a + bx + cy + dxy$. Obtain the shape functions at each of the four nodes for the element and geometry shown.

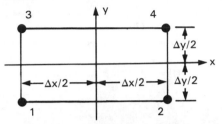

11-13 To provide a more accurate representation of the functional variation, the previous problem for a rectangular element can be extended to use quadratic shape functions. An intermediate node is inserted at the center of each side as shown in the figure. The interpolating polynomial is chosen to have the quadratic form $T(x, y) = a + bx + cy + dxy + ex^2 + fy^2 + gx^2y + hxy^2$. Obtain the shape functions at the eight nodes. Make a three-dimensional plot of the shape function for a corner node.

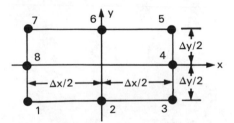

11-14 The hemispherical-spectral emissivity of a surface is approximated by $\epsilon_\lambda = 0.1\lambda^3$ for $\lambda \leq 2$ μm. Derive a closed-form relation between λ and a random number R for a surface at temperature T for energy emitted from the surface in the range $0 \leq \lambda \leq 2$ μm. [*Hint:* $\int [dx/(e^x - 1)] = -x + \ln(e^x - 1) = \ln(1 - e^{-x})$].

$$Answer:\ \lambda = -\frac{C_2}{T}\left(\ln\left\{1 - \left[1 - \exp\left(-\frac{C_2}{2T}\right)\right]^R\right\}\right)^{-1}$$

11-15 An area element dA_1 has directional emissivity given by $\epsilon_1'(\theta) = 0.8 \cos \theta$. Find, for the geometry pictured below, the fraction of energy leaving dA_1 that is absorbed by the black disk A_2. Use the Monte Carlo method. Compare your result with the analytical solution.

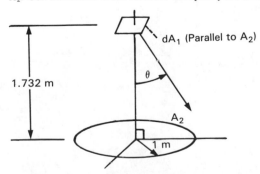

Answer: 0.35

11-16 Construct a computer flowchart for the Monte Carlo computation of the configuration factor F_{d1-2} from an area element to a perpendicular disk as shown in Example 6-4.

11-17 Construct a complete computer flow diagram for the Monte Carlo solution of the problem outlined in Prob. 7-37 for gray plates.

11-18 Construct a Monte Carlo computer flow diagram to obtain the specular exchange factor F^s_{2-1} in Prob. 9-14.

11-19 Program and solve Prob. 11-17 for $L = 1$, $\epsilon_1 = \epsilon_2 = 1$, and for $L = 1$, $\epsilon_1 = \epsilon_2 = 0.5$. By comparing the two results, verify the result of Prob. 7-37b.

11-20 Obtain a Monte Carlo solution for the nongray heat transfer between infinite parallel plates computed in Example 8-2.

11-21 Obtain a Monte Carlo solution for the second part of Example 9-6, i.e., when surface A_1 is a specular reflector.

FUNDAMENTALS AND PROPERTIES FOR RADIATION IN ABSORBING, EMITTING, AND SCATTERING MEDIA

12-1 INTRODUCTION

The study of energy transfer through media that can absorb, emit, and scatter radiation has received increased attention in the past three decades. This interest stems from phenomena associated with combustion chambers at high pressure and temperature, rocket propulsion, nuclear explosions, hypersonic shock layers, plasma generators for nuclear fusion, ablating systems, insulating layers, and heat transfer in porous regions. Although some of these applications are fairly recent, the study of radiation in gases has been of continuing interest for over 100 years. One of the early considerations was absorption and scattering of radiation in the earth's atmosphere. This has plagued astronomers when observing on earth the light from the sun and more distant stars. The arriving form of the solar spectrum was recorded by Samuel Langley over a period of years beginning in 1880 [1]. Figure 12-1 shows some more recent results [2]. The uppermost solid curve is the incident solar spectrum outside the earth's atmosphere, and a 6000-K blackbody spectrum is shown for comparison. The lowest solid curve, which has a number of sharp dips, shows the spectrum received at ground level after the solar radiation has passed through the atmosphere in a direction normal to the earth. The shaded regions show where radiation has been absorbed by various atmospheric constituents, mainly water vapor and carbon dioxide. The absorption occurs in specific wavelength regions, illustrating that gas-radiation properties vary considerably with wavelength. Extensive discussions of absorption in the atmosphere are given by Goody [3] and Kondratyev [4].

513

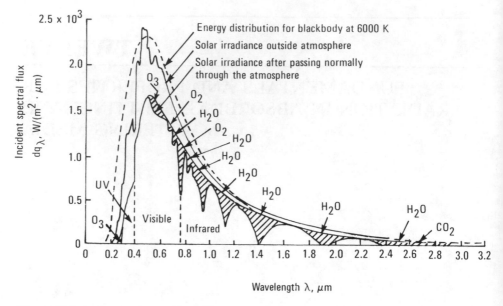

Figure 12-1 Attenuation by earth's atmosphere of incident solar spectral energy flux [2].

Gas radiation has been of interest to astrophysicists with regard to studying stellar structure. Because the spectrum observed during the emission or absorption of radiation by a gas is characteristic of only that gas, it can be used as a diagnostic tool to determine the gas temperature and concentration. Models of stellar atmospheres, such as for the sun, and the energy transfer processes within them have been constructed; then emitted energy spectra calculated from the models are compared with observed stellar spectra.

In industry the importance of gas radiation was recognized in the 1920s in connection with heat transfer inside furnaces. Carbon dioxide and water vapor formed as combustion products were found to be significant emitters and absorbers of radiant energy. Radiation can also be appreciable in engine combustion chambers, where peak temperatures can reach a few thousand K. The energy emitted from flames arises not only from the gaseous emission but also from hot carbon (soot) particles within the flame and from suspended particulate material as in pulverized-coal combustion.

Another interesting example of radiation within an absorbing-emitting medium is in a glass-melting furnace. As described in [5], the temperature distribution measured within a deep tank of molten glass was more uniform than expected from heat conduction alone. It was thought that convection might account for the discrepancy, but experimental investigations did not indicate that this was the contributing heat transfer mode. In the late 1940s it became evident that radiative transfer by absorption and reemission within the glass is a significant means of energy transport.

When radiation interacts with a substance, part of the energy may be redirected by scattering. Scattering may occur by interaction with particles or objects of any size from electrons to planets. Common examples of scattering are radiant transfer through fog, dust suspensions, and some pigmented materials.

Two difficulties in studying radiation within absorbing, emitting, and scattering media make these studies challenging. First, with a participating medium present, absorption, emission, and scattering of energy occur not only at system boundaries but at all locations within the medium. A complete solution for energy exchange requires knowledge of the temperature, radiation intensity, and physical properties throughout the medium. The mathematics describing the radiative field is inherently complex. A second difficulty is that spectral effects are often much more pronounced in gases than for solid surfaces, and a detailed spectrally dependent analysis may be required. Most of the simplifications introduced in gas-radiation problems are aimed at dealing with one or both of these complexities.

In this chapter the concepts are introduced for radiant intensity within a medium and for the effects of absorption, emission, and scattering on radiant propagation. These are the fundamentals needed in Chaps. 13–16 for the equation of radiant energy transfer and the equations of energy flux used for energy conservation relations.

The next portion of the chapter deals with spectral band absorption models and correlations for gases. These are needed for the integration of spectral energies to obtain total quantities. The band behavior is developed by first describing absorption by a single spectral line. Then models are presented for an absorption band consisting of many broadened lines. There are shown to be two effects of gas pressure: (1) broadening the spectral lines and (2) increasing the number of gas molecules that interact with the radiation along a path. Limiting cases are considered for overlapping lines and for nonoverlapping lines that are either weak or strong absorbers. The trends indicated by the band models lead to correlations for band behavior.

The final section of the chapter deals with scattering fundamentals. Scattering is the redirection of radiation by interaction with particles or molecules within the medium. The interaction can be a combination of reflection, refraction, and diffraction. The amount of energy scattered and its directional distribution are given for various types of particles.

12-2 SYMBOLS

a	absorption coefficient
A	area
$\bar{A}$	effective line- or bandwidth
A_0	width of band
c	speed of light in medium other than vacuum
c_0	speed of light in vacuum
C_i	concentration of gas in mixture

C_1, C_2, C_3	constants in Table 12-4
C_2	second constant in Planck's spectral energy distribution, hc_0/k
D	particle diameter
e	emissive power
E	energy
E_a	absorption efficiency factor
E_I	ionization potential
E_s	scattering efficiency factor
f_v	particle volume fraction
$G(\bar{n})$	function of $\bar{n}$ in Rayleigh scattering relation
h	Planck's constant
i	radiation intensity
J_n	Bessel function of the first kind of order n
k	Boltzmann constant
K	extinction coefficient, $a + \sigma_s$
l_m	mean penetration distance
M	mass of molecule
n	simple refractive index; relative refractive index, n_1/n_2
$\bar{n}$	complex refractive index, $n - i\kappa$
N	number density, number of particles per unit volume
p	partial pressure of gas in mixture
P	total pressure
P_e	effective broadening pressure
q	energy flux, energy per unit time and area
Q	energy per unit time
r	radial coordinate
R	radius of sphere
s	scattering cross section
S	coordinate along path of radiation
S_c	integrated absorption coefficient
S_{ij}	integrated line absorption coefficient
t	time
T	absolute temperature
T_0	reference temperature of 100 K in Tables 12-4 and 12-5
u	the quantity SS_c/δ
v	velocity
V	volume
x, y, z	coordinates in cartesian system
X	mass path length, ρS
α	absorptance
α_p	polarizability
β	scattering angle from forward direction; pressure broadening parameter; the parameter $4\Delta_c/\delta$
δ	average spacing between lines of absorption band
Δ	half-width of spectral line

ϵ	emittance
η	wavenumber
θ	cone angle, angle from normal of area; scattering angle measured relative to forward direction
κ	optical thickness; extinction coefficient in $\bar{n}$
λ	wavelength in medium
ν	frequency
ξ	size parameter, $\pi/D\lambda$
ρ	density; reflectivity
σ	Stefan-Boltzmann constant
σ_s	scattering coefficient
τ	transmittance
φ	circumferential angle
Φ	phase function for scattering
ω	solid angle

Subscripts

a	absorbed
b	blackbody
c	for collisional broadening
D	for Doppler broadening
g	gas
i	incoming; incident; component i
i, j	energy state i or j
l	lth band
m	mass coefficient; mean value; in particle material
N_2	nitrogen
p	particle; projected
s	source or scatter; scattering or scattered
S	at distance S
η	wavenumber dependent
λ	wavelength dependent
ν	frequency dependent

Superscripts

$'$	directional value
$+$	true value, not modified by addition of induced emission
$*$	dummy variable of integration

12-3 SOME FUNDAMENTAL PROPERTIES OF RADIATION INTENSITY IN A MEDIUM

Intensity is a convenient quantity for dealing with radiative transfer through absorbing, emitting, and scattering media. This is because of certain invariance properties. In Sec. 2-5.1 the radiation intensity leaving a surface in the direction (θ, φ) was defined as the energy leaving per unit time, per unit of projected surface area normal to the (θ, φ) direction, and per unit elemental solid angle centered around direction (θ, φ). As a result of this definition, the emitted intensity from a blackbody does not vary with direction. This invariance was useful in comparing the directional intensity emitted from nonblack surfaces with that from black surfaces. The ratio of the two directional emissions is the surface directional emissivity.

For a transmitting medium, the intensity has to be considered in terms of a local area *within* the medium. The intensity is then defined in a manner consistent with the solid surface. It is as if radiation traveling through an area within the medium originated at that area. *The intensity is defined* (see Fig. 12-2a) *as the radiation energy passing through the area per unit time, per unit of the projected area, and per unit solid angle.* The projected area is formed by taking the area that the energy is passing through and projecting it *normal to the direction of travel.* The elemental solid angle is centered about the direction of travel and has its origin at dA. The spectral intensity is the intensity per unit small wavelength interval around a wavelength λ.

The emitted intensity from a blackbody is invariant with emission angle; now a second invariant property of intensity will be examined. Consider radiation from a source dA_s, traveling in an ideal medium that is nonabsorbing, nonemitting, and nonscattering and has constant properties. Suppose that an imaginary area element dA_1 is considered at distance S_1 from dA_s and that dA_s and dA_1 are normal to S_1 as shown in Fig. 12-2b. From the definition of spectral intensity $i'_{\lambda,1}$ as the rate of energy passing through dA_1 per unit projected area of dA_1 per unit solid angle and per unit wavelength interval, the energy from dA_s passing through dA_1 in the direction of S_1 is

$$d^3Q'_{\lambda,1} = i'_{\lambda,1} dA_1 \, d\omega_1 \, d\lambda \tag{12-1a}$$

where the third differential notation d^3 emphasizes that there are three differential quantities on the right-hand side of the equation. The solid angle $d\omega_1$ equals dA_s/S_1^2 so

$$d^3Q'_{\lambda,1} = i'_{\lambda,1} dA_1 \frac{dA_s}{S_1^2} d\lambda \tag{12-1b}$$

Suppose the dA_1 is now placed a distance S_2 from the source along the same direction as the original position. The rate of energy passing through dA_1 in the new position is

$$d^3Q'_{\lambda,2} = i'_{\lambda,2} dA_1 \, d\omega_2 \, d\lambda = i'_{\lambda,2} dA_1 \frac{dA_s}{S_2^2} d\lambda \tag{12-2}$$

Dividing (12-1b) by (12-2) gives

$$\frac{d^3Q'_{\lambda,1}}{d^3Q'_{\lambda,2}} = \frac{i'_{\lambda,1}S_2^2}{i'_{\lambda,2}S_1^2} \tag{12-3}$$

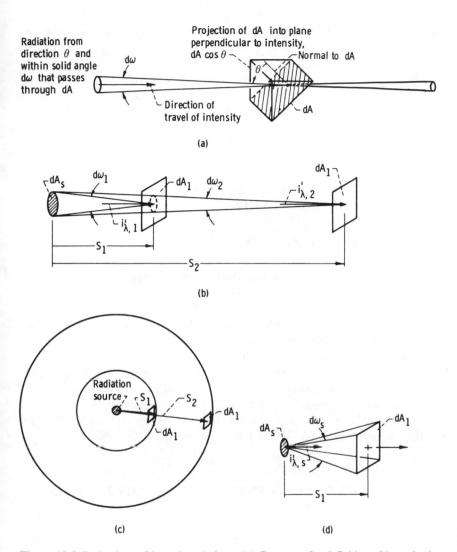

Figure 12-2 Derivations of intensity relations. (*a*) Geometry for definition of intensity in medium; (*b*) intensity from source to area element; (*c*) variation of energy flux with distance from source; (*d*) intensity of emitted radiation.

Now consider a differential source emitting energy equally in all directions, and draw two concentric spheres around it as in Fig. 12-2*c*. If $d^2Q_{\lambda,s}$ is the entire spectral energy leaving the source, the energy *flux* crossing the inner sphere is $d^2Q_{\lambda,s}/4\pi S_1^2$, and that crossing the outer sphere is $d^2Q_{\lambda,s}/4\pi S_2^2$. The ratio of the energies passing through the two elements dA_1 is

$$\frac{d^3Q'_{\lambda,1}}{d^3Q'_{\lambda,2}} = \frac{(d^2Q_{\lambda,s}/4\pi S_1^2)\,dA_1}{(d^2Q_{\lambda,s}/4\pi S_2^2)\,dA_1} = \frac{S_2^2}{S_1^2} \tag{12-4}$$

Substituting (12-4) for the left side of (12-3) gives the following important result:

$$i'_{\lambda,1} = i'_{\lambda,2} \tag{12-5}$$

Thus, *the intensity in a given direction in a nonattenuating and nonemitting medium with constant properties is independent of position along that direction.* Note that these intensities are based on the solid angles subtended by the source as viewed from dA_1 as in Fig. 12-2b. As S is increased, the decrease in solid angle by which dA_1 views the source dA_s is accompanied by a comparable decrease in energy flux arriving at dA_1. Thus the flux per unit solid angle, used in forming the intensity, remains constant.

The radiant energy passing through dA_1 can also be written in terms of the intensity leaving the source. Using Fig. 12-2d results in

$$d^3 Q'_{\lambda,1} = i'_{\lambda,s}\, dA_s\, d\omega_s\, d\lambda = i'_{\lambda,s}\, dA_s\, \frac{dA_1}{S_1^{\,2}}\, d\lambda \tag{12-6}$$

Equating this with the energy rate passing through dA_1 as given by (12-1b) results in

$$i'_{\lambda,1} = i'_{\lambda,s} \tag{12-7}$$

This relation again shows the invariance of intensity with position along a path in a nonattenuating and nonemitting medium.

The invariance of intensity when no attenuation or emission is present provides a convenient way of specifying the magnitude of any attenuation or emission, as these effects are given directly by the change of intensity with distance. By use of the foregoing intensity properties, the attenuation and emission of radiation within a medium are now considered.

12-4 THE ATTENUATION OF INTENSITY BY ABSORPTION AND SCATTERING

Consider spectral radiation of intensity i'_λ incident normally on a volume element of thickness dS as in Fig. 12-3. The medium in dV absorbs and scatters radiation. Only these two effects will be considered for the present. Later sections deal with emission and with the scattering into one direction of radiation from other directions. As the radiation passes through dS, its intensity is reduced by absorption and scattering. The change in intensity has been found experimentally to depend on the magnitude of the local intensity. If a coefficient of proportionality K_λ that depends on the local properties of the medium is introduced, then the decrease is

$$di'_\lambda = -K_\lambda(S)\, i'_\lambda\, dS \tag{12-8}$$

The K_λ is the *extinction coefficient* of the material; it is a physical property and has units of reciprocal length. It is a function of temperature T, pressure P,

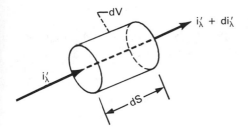

Figure 12-3 Intensity incident normally on absorbing and scattering volume element of thickness dS.

composition of the material (specified here in terms of the concentrations C_i of the i components), and wavelength of the incident radiation:

$$K_\lambda = K_\lambda(\lambda,T,P,C_i) \tag{12-9}$$

As will be shown [see Eq. (12-16)], the K_λ is the inverse of the mean penetration distance of radiation in a medium.

Integrating (12-8) over a finite path of length S gives

$$\int_{i'_\lambda(0)}^{i'_\lambda(S)} \frac{di'_\lambda}{i'_\lambda} = -\int_0^S K_\lambda(S^*)\, dS^* \tag{12-10}$$

where $i'_\lambda(0)$ is the intensity at the origin of the path and S^* is a dummy variable of integration. Carrying out the integration on the left side of (12-10) yields

$$\ln\frac{i'_\lambda(S)}{i'_\lambda(0)} = -\int_0^S K_\lambda(S^*)\, dS^* \tag{12-11}$$

or

$$i'_\lambda(S) = i'_\lambda(0) \exp\left[-\int_0^S K_\lambda(S^*)\, dS^*\right] \tag{12-12}$$

Equation (12-12) is *Bouguer's law*[†]; it shows that, as a consequence of the proportionality in (12-8), the intensity of spectral radiation along a path is attenuated exponentially while it passes through an absorbing-scattering medium (not including emission or scattering in direction S). The exponent is equal to the integral of the local extinction coefficient over the path length traversed by the radiation.

[†]Named after Pierre Bouguer (boō'gâr') (1698–1758), who first showed on a quantitative basis how light intensities could be compared. Equation (12-12) is sometimes called Lambert's law, the Bouguer-Lambert law, or Beer's law. Beer's law is more properly a restricted form of Eq. (12-9) stating that the absorption of radiation depends only on the concentration of the absorbing species along the path. To avoid confusion with Lambert's cosine law, (12-12) will be referred to herein as Bouguer's law.

12-4.1 The Extinction Coefficient

The extinction coefficient is composed of two parts, the *absorption coefficient* $a_\lambda(\lambda, T, P)$ and the *scattering* coefficient $\sigma_{s\lambda}(\lambda, T, P)$:

$$K_\lambda(\lambda,T,P) = a_\lambda(\lambda,T,P) + \sigma_{s\lambda}(\lambda,T,P) \tag{12-13}$$

For simplicity, the notation has been dropped showing dependence on the relative concentration of the constituents of the medium. These coefficients have units of reciprocal length and are therefore called *linear coefficients* (they are also called *volumetric* coefficients). Some people prefer to work with *mass coefficients*,

$$K_{\lambda,m} = a_{\lambda,m} + \sigma_{s\lambda,m} = \frac{K_\lambda}{\rho} = \frac{a_\lambda}{\rho} + \frac{\sigma_{s\lambda}}{\rho} \tag{12-14}$$

where ρ is the local density of the absorbing-scattering species. The mass coefficients have units of area per unit mass and are directly related to the concept of cross sections in molecular physics. Since the extinction coefficient K_λ increases as the density of the absorbing or scattering species is increased, use of $K_{\lambda,m} = K_\lambda/\rho$ has the advantage that it tends to remain more constant than K_λ. However, the K_λ used here also has an advantage that, when K_λ is constant, it can be interpreted as the reciprocal of the radiation mean penetration distance, as will now be shown.

12-4.2 Radiation Mean Penetration Distance

From (12-12) the fraction of the original radiation that penetrates through the path length S is

$$\frac{i_\lambda'(S)}{i_\lambda'(0)} = \exp\left[-\int_0^S K_\lambda(S^*)\, dS^*\right]$$

The fraction absorbed in the layer from S to $S + dS$ is

$$-\frac{di_\lambda'(S)}{i_\lambda'(0)} = K_\lambda(S)\frac{i_\lambda'(S)}{i_\lambda'(0)}\, dS = K_\lambda(S)\exp\left[-\int_0^S K_\lambda(S^*)\, dS^*\right] dS$$

The mean penetration distance of the radiation is obtained by multiplying the fraction absorbed at S by the distance S and integrating over all path lengths from $S = 0$ to ∞:

$$l_m = \int_{S=0}^\infty SK_\lambda(S)\exp\left[-\int_0^S K_\lambda(S^*)\, dS^*\right] dS \tag{12-15}$$

When K_λ is constant, carrying out the integration gives

$$l_m = K_\lambda \int_0^\infty S\exp\left(-K_\lambda S\right) dS = \frac{1}{K_\lambda} \tag{12-16a}$$

demonstrating that the average penetration distance before absorption or scattering is the reciprocal of K_λ when K_λ does not vary along the path. Equation (12-16) provides some insight into whether an absorbing-scattering medium is very opaque to radiation traveling through it. Mean penetration distances can also be defined before *each* process of absorption or scattering occurs,

$$l_{m,a} = \frac{1}{a_\lambda} \qquad l_{m,s} = \frac{1}{\sigma_{s\lambda}} \tag{12-16b}$$

The mean penetration distance is now discussed further in connection with the definition of optical thickness.

12-4.3 Optical Thickness

The exponential factor in (12-12) is often written in an alternative form by defining the useful dimensionless quantity

$$\kappa_\lambda(S) \equiv \int_0^S K_\lambda(S^*)\, dS^* \tag{12-17}$$

so that (12-12) becomes

$$i_\lambda'(S) = i_\lambda'(0) \exp\left[-\kappa_\lambda(S)\right] \tag{12-18}$$

The quantity $\kappa_\lambda(S)$ is the *optical thickness* or *opacity* of the layer of thickness S and is a function of all the values of K_λ between 0 and S. Because K_λ is a function of the local parameters P, T, and C_i, the optical thickness is a function of all these conditions along the path between 0 and S.[†] The optical thickness is a measure of the ability of a path length to attenuate radiation of a given wavelength. A large optical thickness means large attenuation.

For a medium with uniform composition and at uniform temperature and pressure (a *uniform* medium) or for a medium with K_λ independent of T, P, and C_i, Eq. (12-17) becomes

$$\kappa_\lambda(S) = K_\lambda S \tag{12-19}$$

The optical thickness then depends directly on the extinction coefficient and the thickness of the region. By using (12-16), $\kappa_\lambda = S/l_m$, so that the optical thickness is the number of mean penetration distances. If $\kappa_\lambda \gg 1$, the medium is *optically thick;* that is, the mean penetration distance is quite small compared to the characteristic dimension of the medium. For this condition the radiation reaching a volume element within the material is influenced only by the neighboring elements. If $\kappa_\lambda \ll 1$, the medium is *optically thin* and the mean penetration distance

---

[†]The notation for the optical thickness κ_λ should not be confused with the extinction coefficient for electromagnetic radiation κ used in (12-22) and (12-23) that follow. It is regrettable but true that the notation possibilities of the English and Greek alphabets reach saturation when interdisciplinary fields such as gas radiation are discussed.

is much larger than the medium dimension. Radiation can pass entirely through the material without significant extinction, and each element within the medium interacts directly with the medium boundary. Radiation emitted within the material is not significantly reabsorbed or scattered by the material; this is *negligible self-extinction*.

12-4.4 The Absorption Coefficient

If scattering can be neglected ($\sigma_{s\lambda} \approx 0$), then $K_\lambda = a_\lambda$ and (12-12) becomes

$$i_\lambda'(S) = i_\lambda'(0) \exp \left[-\int_0^S a_\lambda(S^*)\, dS^* \right] \tag{12-20}$$

If, in addition, a_λ is not a function of position as is the case in a uniform medium, then

$$i_\lambda'(S) = i_\lambda'(0) \exp\left(-a_\lambda S\right) \tag{12-21}$$

In the electromagnetic theory of radiant energy propagation [see discussion following Eq. (4-24a)], it is shown that intensity is attenuated in conducting media according to the relation

$$\frac{i_\lambda'(S)}{i_\lambda'(0)} = \exp\frac{-4\pi\kappa S}{\lambda} \tag{12-22}$$

where κ is the extinction coefficient from electromagnetic theory and is related to the magnetic permeability, electrical resistivity, and electrical permittivity of the medium [Eq. (4-21b)]. Thus, a_λ is related to κ by

$$a_\lambda = \frac{4\pi\kappa}{\lambda} \tag{12-23}$$

where κ is a function of λ and λ is the value in vacuum. This provides a theoretical basis for Bouguer's law, which was originally based on experimental observations. It also provides a means for obtaining a_λ from optical data available for $\kappa(\lambda)$.

The absorption coefficient $a_\lambda(\lambda, T, P)$ usually varies strongly with wavelength and often varies substantially with temperature and pressure (and hence density, for a gas). Considerable analytical and experimental effort has been expended to determine a_λ for various gases, liquids, and solids. Analytical determinations of a_λ require detailed quantum-mechanical calculations. Except for the simplest gases, such as atomic hydrogen, the calculations are very tedious and require many simplifying assumptions. For the methods used in the calculation of a_λ, [6–8] give detailed discussions.

Figure 12-4 shows a_λ for pure diamond. Strong absorption peaks exist due to crystal-lattice vibrations at certain wavelengths. Figure 12-5 shows the calculated emission spectrum of hydrogen gas at 40 atm and 11,300 K for a path length through the gas of 50 cm. The presence of "spikes" or strong emission lines is

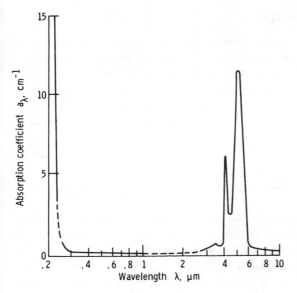

Figure 12-4 Spectral absorption coefficient of diamond. From [91].

the result of transitions between bound energy states. The continuous part of the emission spectrum is due to various photodissociations, photo-ionization, and free electron–atom–photon interactions of other types. The lines and the continuous regions are common features of both emission and absorption spectra. Figure 12-6 shows the absorption coefficient of air at 1 atm and 12,000 K. In this case, there is a merging of the contributions from the many closely spaced lines produced by vibrational and rotational transitions between energy states, and the absorption coefficient has the appearance of being continuous. Even when this merging is not complete, the resolution of experimental measurements may cause the measured spectrum to appear continuous over these closely spaced lines.

Note that Figs. 12-4 to 12-6 each have a different abscissa: wavelength, wavenumber, and frequency. For opaque surfaces (Chap. 5) the wavelength is generally used. For radiation within media, however, frequency is more common. It has the advantage that frequency does not change when radiation passes from one medium into another with a different refractive index. The wavelength does change because of the change in propagation velocity.

Dealing with spectral line emission and absorption is one of the computational difficulties in analyses of radiant energy transfer through gases and other media. Radiation at wavelengths near an individual spectral line center will be strongly absorbed, while radiation of only a slightly different wavelength may experience almost no attenuation. Integrating line absorption coefficients with respect to wavelength to obtain band or total absorption coefficients is generally tedious. Averaged coefficients are used in certain calculation methods as will be developed later.

12-4.5 True Absorption Coefficient

Equation (12-20) gives the attenuation of intensity passing through a nonemitting, nonscattering medium as would be observed by detectors of incident and emerging radiation. Such information could be used in determining a_λ. Actually, as radiative energy passes through a medium, it is not only absorbed, but there is an additional phenomenon in that its presence stimulates some of the atoms or molecules to emit energy. This is not the *ordinary* or *spontaneous* emission caused by the temperature of the gas that will be discussed in Sec. 12-5. Spontaneous emission is the result of the excited energy state of the medium being unstable and decaying spontaneously to a lower energy state. Emission resulting from the presence of the radiation field is termed *stimulated* or *induced* emission and is in a sense a negative absorption.

Physically, induced emission can be pictured as follows: A photon of a certain frequency from the radiation field encounters a particle, such as an atom or molecule, that is presently in an excited energy state above the ground state. There exists a certain probability that the incident photon will trigger a return of the particle to a lower energy state. If this occurs, the particle will emit a photon at the *same frequency* and *in the same direction* as the incident photon. Thus, the incident photon is not absorbed but is joined by a second identical photon. This process is often viewed as negative absorption when deriving the energy balance equations in Chaps. 13–15.

Induced emission constitutes a portion of the intensity that is observed emerging from a gas volume. Consequently, the amount of energy that is actually ab-

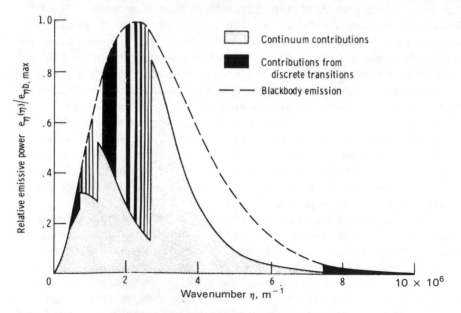

Figure 12-5 Normalized emission spectrum of hydrogen at 11,300 K and 40 atm, for a path length of 0.5 m. From [92].

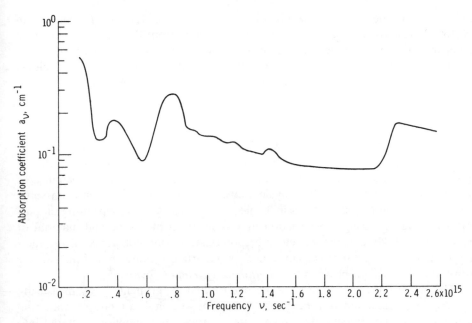

Figure 12-6 Absorption coefficient of air at 12,000 K and 1 atm. From [93].

sorbed by the gas is greater than that found by taking the difference between the entering and leaving intensities. This is because the observed emerging intensity is the result of the actual absorption modified by the addition of induced emission along the path. The actual absorbed energy should be calculated using a *true absorption coefficient* $a_\lambda^+(\lambda, T, P)$, which is larger than the absorption coefficient $a_\lambda(\lambda, T, P)$ calculated by using observed attenuation data. The law for "true" absorption along path S is then

$$i_\lambda'(S) = i_\lambda'(0) \exp\left[-\int_0^S a_\lambda^+(S^*)\, dS^*\right] \tag{12-24}$$

Statistical-mechanical considerations give the relation between $a_\lambda(\lambda, T, P)$ and $a_\lambda^+(\lambda, T, P)$ for a gas with refractive index $n = 1$ as

$$a_\lambda(\lambda, T, P) = \left[1 - \exp\left(-\frac{hc_0}{k\lambda T}\right)\right] a_\lambda^+(\lambda, T, P) = \left[1 - \exp\left(-\frac{C_2}{\lambda T}\right)\right] a_\lambda^+(\lambda, T, P) \tag{12-25}$$

Because of the negative exponential term, a_λ^+ is always larger than a_λ (hence the use of the superscript +). Because induced emission depends on the incident radiation field, it is usually grouped (as will be done here) with the true absorption, thereby yielding the absorption coefficient a_λ. The emission term in the equation of radiative transfer then includes *only* the spontaneous emission and consequently depends only on the local conditions of the gas.

The exponential term in (12-25) is small unless λT is large. Thus a_λ and a_λ^+ are nearly equal except at large values of λT (long wavelengths and/or high temperatures). The values are within 1% for λT less than 3120 μm·K, and within 5% for λT less than 4800 μm·K. When properties from the literature are used in calculations of radiative transfer in absorbing-emitting media, care must occasionally be exercised to determine whether the reported absorption coefficients include the effects of induced emission. Usually it is a_λ that is given.

12-4.6 The Scattering Coefficient

The extent of scattering can be viewed in terms of the *scattering cross section s*. This is the apparent area that an object presents to an incident beam in relation to the ability of the object to deflect radiation from the beam. This apparent area may be quite different from the physical cross-sectional area, as seen from some of the approximate cross sections in Table 12-1. In addition to depending on particle size, the scattering cross section may depend on the shape and material of the scattering body and the wavelength, polarization, and coherence of the incident radiation. The ratio of s_λ to the actual geometric projected area of the particle normal to the incident beam is the *scattering efficiency factor, $E_{s\lambda}$*.

The scattering cross section can be determined experimentally by measuring the amount of radiation in a beam that penetrates through a cloud of scattering particles. One experimental difficulty is in separating the radiation scattered into the forward direction from the radiation transmitted without any particle interaction. This difficulty can be diminished by using an incident beam with a very small divergence angle. Then the forward direction of the transmitted radiation will encompass only a small solid angle that includes only a small portion of the radiation scattered in from other directions. The scattered portion $di'_{\lambda,s}$ of the incident intensity, divided by the intensity i'_λ of the incident beam, is equal to the apparent projected scattering area $d^2 A_{s\lambda}$ of all the scattering particles that are encountered, divided by the cross-sectional area of the incident beam dA. Then for a differential distance within a medium,

$$\frac{di'_{\lambda,s}}{i'_\lambda} = \frac{d^2 A_{s\lambda}}{dA} \tag{12-26}$$

The apparent projected scattering area of the particles will usually depend on wavelength.

The effective scattering area presented by a *group* of independently scattering particles is related to the scattering areas of the individual particles as follows: Let $dN(R)$ be the number of particles per unit volume in the radius range from R to $R + dR$, and let $s_\lambda(R)$ be the scattering cross section for a particle of radius R. By integrating over all the particles, the effective scattering area in a differential volume dV (see Fig. 12-8) of the particle cloud is

$$d^2 A_{s\lambda} = dV \int_{N(R)=0}^{\infty} s_\lambda(R) \, dN(R) \tag{12-27}$$

The *effective scattering area per unit volume* is defined as the *scattering coefficient $\sigma_{s\lambda}$*,

$$\sigma_{s\lambda} \equiv \int_{N(R)=0}^{\infty} s_\lambda(R) \, dN(R) \tag{12-28}$$

If the particles are spherical and of uniform diameter D, and there are N particles per unit volume, then

$$\sigma_{s\lambda} \equiv s_\lambda N = \frac{\pi D^2}{4} E_{s\lambda} N \tag{12-29}$$

Using $d^2 A_{s\lambda} = dV \sigma_{s\lambda} = dA \, dS \, \sigma_{s\lambda}$ in (12-26) gives the change of intensity by scattering from the incident beam,

$$-\frac{di'_\lambda}{i'_\lambda} = \frac{di'_{\lambda,s}}{i'_\lambda} = \frac{\sigma_{s\lambda} \, dA \, dS}{dA} = \sigma_{s\lambda} \, dS \tag{12-30}$$

There is also intensity scattered from all incident directions *into* the S direction, which contributes to di'_λ, but this will be incorporated later.

By integrating (12-30) over a path from 0 to S, the intensity is found at S after attenuation by scattering of a beam with original intensity $i'_\lambda(0)$:

$$i'_\lambda(S) = i'_\lambda(0) \exp\left(-\int_0^S \sigma_{s\lambda} \, dS^*\right) \tag{12-31}$$

The portion of the incident intensity scattered away along the path is thus

$$i'_\lambda(0) - i'_\lambda(S) = i'_\lambda(0) \left[1 - \exp\left(-\int_0^S \sigma_{s\lambda} \, dS^*\right)\right] \tag{12-32}$$

Equation (12-31) is the pure-scattering form of Bouguer's law.

As in the interpretation of the extinction coefficient, the scattering coefficient $\sigma_{s\lambda}$ can be regarded as the reciprocal of the mean free path that radiation will traverse before being scattered, Eq. (12-16b). At particle densities near or below the molecular density of air at 1 atm ($N_s \approx 2.7 \times 10^{19}$ particles/cm^3), it can be seen that for most of the processes listed in Table 12-1 the scattering coefficient will be very small (and thus the scattering mean free path very long). This is especially true for photon-photon, Thomson, and Raman scattering, which may generally be ignored in engineering radiative transfer calculations.

All of the results in this section have been for conditions where energy scattered by one particle is assumed *not* to interact with another particle. When particle concentrations become large, multiple scattering will occur, and the simple measurement of $\sigma_{s\lambda}$ implied by (12-30) will not be correct.

Note that the *absorption* in a cloud of particles can be described in an analogous way to Eqs. (12-27), (12-28), and (12-29). If $E_{a\lambda}$ is the absorption efficiency factor for a particle, then for a cloud of independently absorbing spheres, all of diameter D, the absorption coefficient is

$$a_\lambda = \frac{\pi D^2}{4} E_{a\lambda} N \tag{12-33}$$

Table 12-1 Approximate scattering cross sections for various bodies exposed to incident photons

Body	Physical cross section, cm^2	Conditions	Type of scattering	Scattering cross section,[a] cm^2
Photon		Energy of incident photon small		$\sim 2 \times 10^{-56}$
Free electron		Energy of photon $\ll$ electron kinetic energy	Thomson	$\frac{8}{3}\pi r_0^2 = 6.65 \times 10^{-25} \equiv \sigma_T$
		Energy of photon $\gg$ electron kinetic energy	Compton	$\frac{3}{8}\sigma_T \dfrac{1}{\epsilon_p}\left(\frac{1}{2} + \ln 2\epsilon_p\right)$
Atom or molecule	0.88×10^{-16} (first Bohr electron orbit)	Elastic, $\lambda \gg$ size of molecule or atom	Rayleigh	Proportional to $\sim 6.65 \times 10^{-25}/\lambda^4$
		Energy of incident photon $\gg$ electronic binding energy	Rayleigh, approaches Thomson	$\sim 6.65 \times 10^{-25}$
		Inelastic, Energy of incident photon $\ll$ electronic binding energy	Raman	$\sim 6.65 \times 10^{-25}$
		Energy of incident photon $\gg$ electronic binding energy	Approaches Compton	$\sim \frac{3}{8}\sigma_T \dfrac{1}{\epsilon_p}\left(\frac{1}{2} + \ln 2\epsilon_p\right)$
Particles of diameter D	$\dfrac{\pi D^2}{4}$	$\lambda \gg D$, single scattering	Rayleigh	Proportional to V^2/λ^4
		$\lambda \approx D$	Mie	Varies widely
		$\lambda \ll D$	Fraunhofer and Fresnel diffraction plus reflection	$\sim 2\left(\dfrac{\pi D^2}{4}\right)$

[a] r_0 = classical electron radius 2.81794×10^{-15} m; $\epsilon_p = h\nu/m_e c_0^2$, where h is Planck's constant; σ_T = cross section for Thomson scattering; m_e = electron mass, 9.10939×10^{-31} kg.

12-5 THE INCREASE OF INTENSITY BY EMISSION

Having considered the various definitions connected with attenuation, we now turn to the emission of energy within the medium. Consider an elemental volume dV of medium as in Fig. 12-7, with absorption coefficient $a_\lambda(\lambda, T, P)$ within dV. Let dV be placed at the center of a large black hollow sphere of radius R at uniform temperature T. The space between dV and the sphere wall is filled with nonparticipating material. The spectral intensity incident at the location of dA_s on dV, from element dA on the surface of the enclosure is, by use of (12-7),

$$i_\lambda'(0) = i_{\lambda b}'(\lambda, T) \tag{12-34}$$

The change of this intensity in dV as a result of absorption is, from (12-8)

$$di_\lambda' = -i_\lambda'(0)a_\lambda \, dS = -i_{\lambda b}'(\lambda, T)a_\lambda \, dS \tag{12-35}$$

Since the absorption coefficient has been used, this includes the effect of induced emission. The energy absorbed by the differential subvolume $dS \, dA_s$ from this incident radiation is

$$d^5Q_{\lambda, a}' = -di_\lambda' \, dA_s \, d\lambda \, d\omega = i_{\lambda b}'(\lambda, T)a_\lambda \, dS \, dA_s \, d\lambda \, d\omega \tag{12-36}$$

where $d\omega = dA/R^2$ and dA_s is a projected area normal to $i_\lambda'(0)$. The energy emitted by dA and absorbed by all of dV is found by integrating over dV,

$$d^4Q_{\lambda, a}' = i_{\lambda b}'(\lambda, T)a_\lambda \, d\lambda \, d\omega \int_{dV} dA_s \, dS = a_\lambda i_{\lambda b}'(\lambda, T) \, dV \, d\lambda \, d\omega \tag{12-37}$$

where $d\omega$ is the solid angle subtended by dA when viewed from dV. To account for all energy incident upon dV from the entire spherical enclosure, integration over all such solid angles gives

$$d^3Q_{\lambda, a} = a_\lambda i_{\lambda b}'(\lambda, T) \, dV \, d\lambda \int_{\omega=0}^{4\pi} d\omega = 4\pi a_\lambda i_{\lambda b}'(\lambda, T) \, dV \, d\lambda \tag{12-38}^\dagger$$

where $i_{\lambda b}'$ is the blackbody spectral intensity, Eq. (2-11).

To maintain equilibrium in the enclosure, dV must spontaneously emit an amount of energy equal to that absorbed (true absorption minus induced emission). Hence, *the energy spontaneously emitted by an isothermal volume element is*

$$d^3Q_{\lambda, e} = 4\pi a_\lambda(\lambda, T, P)i_{\lambda b}'(\lambda, T) \, dV \, d\lambda = 4a_\lambda(\lambda, T, P)e_{\lambda b}(\lambda, T) \, dV \, d\lambda \tag{12-39}$$

This is a very important result for emission within a volume. The shape of the element dV is arbitrary, and small enough so that energy emitted within dV escapes before reabsorption within dV. Further, the gas must be in thermodynamic equilibrium with respect to its internal energy, as discussed below. An emission

---

†Note that dV is regarded as a second-order differential, that is, $dA \, dS$.

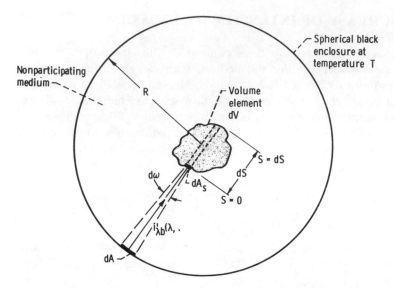

Figure 12-7 Geometry for derivation of emission from volume of medium.

coefficient could be defined in a manner similar to the absorption coefficient. However, the radiation literature has other definitions of the emission coefficient,[‡] and there is no need to add confusion by defining a different coefficient. Equation (12-39) will be used directly as the relation for emission by an infinitesimal volume element of gas.

When the spontaneous emission is the same for all directions (isotropic), which is the condition for all cases here, the radiation intensity emitted by a volume element into any direction is

$$di'_{\lambda,e}(\lambda, T, P) = \frac{d^3 Q_{\lambda,e}}{4\pi dA_p \, d\lambda} = a_\lambda(\lambda, T, P)i'_{\lambda b}(\lambda, T) \, dS \qquad (12\text{-}40a)$$

where dA_p is the projected area of dV normal to the direction of emission, and $dS = dV/dA_p$, the mean thickness of dV parallel to the direction of emission.

With $i'_{\lambda b}(\lambda, T)$ defined as in Eq. (2-11), Eq. (12-40a) is valid for a medium with index of refraction $n \approx 1$. If the index of refraction of the medium is not equal to unity, $n \neq 1$, the spectral volumetric emission is modified by an n_λ^2 factor [see Eq. (2-37)] so that (12-40a) becomes

$$di'_{\lambda,e}(\lambda, T, P) = n_\lambda^2(\lambda, T, P)a_\lambda(\lambda, T, P)i'_{\lambda b}(\lambda, T) \, dS \qquad (12\text{-}40b)$$

The n_λ^2 factor is discussed in Chap. 18.

[‡] In the astrophysical literature [9, 10], the emission coefficient is usually given the symbol j_λ defined by $j_\lambda = a_\lambda e_{\lambda b}$ and having units, therefore, of energy rate per unit volume per unit wavelength interval.

12-6 THE INCREASE OF INTENSITY BY INCOMING SCATTERING

The scattering relations in Sec. 12-4.6 are for the portion of intensity that is lost by being scattered away along a path. Also needed is the intensity increase by scattering into a path direction from other directions. To calculate this, the directional distribution of the scattered radiation is needed; this is given by an *angularly dependent phase function*.

12-6.1 The Phase Function

Consider the radiation within solid angle $d\omega_i$ that is incident on dA in Fig. 12-8. The portion of the incident intensity scattered away in dS is given by (12-30)

$$di'_{\lambda,s} = \sigma_{s\lambda} i'_\lambda \, dS \tag{12-41}$$

The $di'_{\lambda,s}$ is the spectral energy scattered within path length dS per unit incident solid angle and area normal to the incident beam:

$$di'_{\lambda,s} = \frac{d^4 Q'_{\lambda,s}}{d\omega_i \, dA \, d\lambda} \tag{12-42}$$

As shown by Fig. 12-8, the scattered energy produces an intensity distribution as a function of the circumferential angle φ and the angle θ measured relative to the forward direction. The scattered intensity in direction (θ, φ) is defined as the

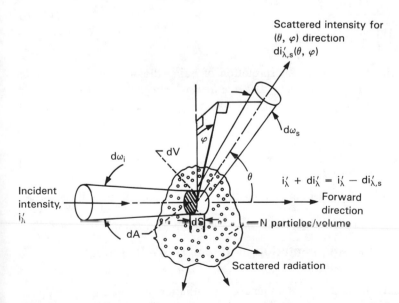

Figure 12-8 Scattering of intensity into direction (θ, φ) from incident radiation within solid angle $d\omega_i$.

energy scattered in that direction per unit solid angle of the scattered direction and per unit normal area and solid angle of the *incident* radiation [note that the functional notation (θ, φ) is used to differentiate between the two different quantities $i'_{\lambda,s}(\theta, \varphi)$ and $i'_{\lambda,s}$]

$$di'_{\lambda,s}(\theta,\varphi) \equiv \frac{\text{spectral energy scattered in direction } (\theta,\varphi)}{d\omega_s \, dA \, d\omega_i \, d\lambda} = \frac{d^5Q'_{\lambda,s}(\theta,\varphi)}{d\omega_s \, dA \, d\omega_i \, d\lambda} \qquad (12\text{-}43)$$

The phase function $\Phi(\theta, \varphi)$ is defined to describe the *angular distribution* of the scattered intensity. The phase function relates the directional magnitude of $di'_{\lambda,s}$ (θ, φ) to the entire intensity $di'_{\lambda,s}$ scattered away from the incident radiation, such that

$$di'_{\lambda,s}(\theta,\varphi) = di'_{\lambda,s} \frac{\Phi(\theta,\varphi)}{4\pi} = \sigma_{\lambda s} i'_\lambda \, dS \frac{\Phi(\theta,\varphi)}{4\pi} \qquad (12\text{-}44)$$

To better understand the phase function, note that the spectral energy per unit $d\lambda$, $d\omega_i$, and dA scattered into $d\omega_s$ is $di'_{\lambda,s}(\theta, \varphi) \, d\omega_s$, and hence, that scattered into all $d\omega_s$ is $\int_{\omega_s=0}^{4\pi} di'_{\lambda,s}(\theta, \varphi) \, d\omega_s$. However, the scattered energy per unit $d\lambda$, $d\omega_i$, and dA is $di'_{\lambda,s}$ from (12-42) so that

$$di'_{\lambda,s} = \int_{\omega_s=0}^{4\pi} di'_{\lambda,s}(\theta, \varphi) \, d\omega_s \qquad (12\text{-}45)$$

Using (12-44) to eliminate $di'_{\lambda,s}$ gives the phase function as

$$\Phi(\theta, \varphi) = \frac{di'_{\lambda,s}(\theta, \varphi)}{\left(\dfrac{1}{4\pi}\right) \displaystyle\int_{\omega_s=0}^{4\pi} di'_{\lambda,s}(\theta, \varphi) \, d\omega_s} \qquad (12\text{-}46)$$

Thus, $\Phi(\theta, \varphi)$ has the physical interpretation of being the *scattered intensity in a direction, divided by the intensity that would be scattered in that direction if the scattering were isotropic.* For isotropic scattering, $\Phi = 1$. Integrating (12-46) over all $d\omega_s$ shows that $\Phi(\theta, \varphi)$ is a function normalized such that

$$\frac{1}{4\pi} \int_{\omega_s=0}^{4\pi} \Phi(\theta, \varphi) \, d\omega_s = \frac{1}{4\pi} \int_0^{2\pi} \int_0^{\pi} \Phi(\theta, \varphi) \sin \theta \, d\theta \, d\varphi = 1 \qquad (12\text{-}47)$$

The phase function can be a complicated function of θ and φ, as will be shown in Sec. 12-9.

12-6.2 Relation for Augmentation of Intensity by Incoming Scattering

The local intensity along a path is enhanced by radiation scattered into the direction being considered. To compute the scattering from all directions into the direction of $i'_\lambda(S)$, consider the radiation incident at angle (θ, φ) as shown in Fig.

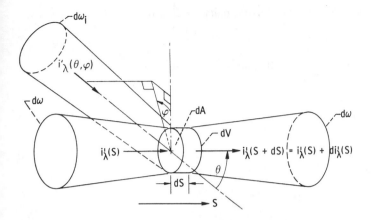

Figure 12-9 Scattering of energy into S direction.

12-9. This radiation has intensity $i'_\lambda(\theta, \varphi)$, and in the process of going through the volume element dV it will pass through a path length $dS/\cos \theta$. From Eq. (12-44) for radiation in the S direction, the intensity scattered from $i'_\lambda(\theta, \varphi)$ into the direction of i'_λ is

$$di'_{\lambda,s} = \sigma_{s\lambda} i'_\lambda(\theta,\varphi) \frac{dS}{\cos \theta} \frac{\Phi(\theta,\varphi)}{4\pi} \tag{12-48}$$

However, from (12-43), $i'_{\lambda,s}$ is an intensity defined as energy in the scattered direction per unit $d\lambda$, per unit scattered solid angle, per unit incident solid angle $d\omega_i$, and per unit area normal to the incident intensity. This is the area normal to $i'_\lambda(\theta, \varphi)$, which is $dA \cos \theta$. Then the spectral energy scattered into the S direction as a result of $i'_\lambda(\theta, \varphi)$ is, by use of (12-48),

$$d^5Q'_{\lambda,s} = di'_{\lambda,s}\, d\omega\, d\omega_i\, d\lambda\, dA \cos \theta = \sigma_{s\lambda} i'_\lambda(\theta, \varphi) \frac{dS}{\cos \theta} \frac{\Phi(\theta, \varphi)}{4\pi} d\omega\, d\omega_i\, d\lambda\, dA \cos \theta$$

$$= \sigma_{s\lambda} i'_\lambda(\theta, \varphi)\, dS\, \frac{\Phi(\theta, \varphi)}{4\pi}\, d\omega\, d\omega_i\, d\lambda\, dA$$

The contribution of this scattered energy to the spectral intensity in the S direction is then

$$\frac{d^5Q'_{\lambda,s}}{dA\, d\omega\, d\lambda} = \sigma_{s\lambda} i'_\lambda(\theta,\varphi) \frac{\Phi(\theta,\varphi)}{4\pi}\, d\omega_i\, dS \tag{12-49}$$

To account for the scattering contributions by the intensities incident from all directions this is integrated over all $d\omega_i$. The scattering particles are assumed randomly oriented so that the scattering cross section $\sigma_{s\lambda}$ is independent of the incidence direction. The *augmentation* of intensity in direction S *by incoming scattering* is then

$$\frac{di_\lambda'}{dS} = \frac{1}{dS} \int_{\omega_i=0}^{4\pi} \frac{d^5 Q_{\lambda,s}'}{dA \, d\omega \, d\lambda} = \frac{\sigma_{s\lambda}}{4\pi} \int_{\omega_i=0}^{4\pi} i_\lambda'(\theta, \varphi)\Phi(\theta, \varphi) \, d\omega_i$$

$$= \frac{\sigma_{s\lambda}}{4\pi} \int_0^{2\pi} \int_{-\pi/2}^{\pi/2} i_\lambda'(\theta, \varphi)\Phi(\theta, \varphi) \sin \theta \, d\theta \, d\varphi \tag{12-50}$$

The general situation where $\sigma_{s\lambda}$ depends on the *incidence* direction is discussed in connection with Eq. (14-38). The phase function is then a function of incidence angle.

12-7 THE EQUATION OF TRANSFER IN AN ABSORBING AND EMITTING MEDIUM

For use in the following section it is helpful to consider a *limited form* of the *equation of transfer* and some associated definitions. The equation of transfer describes how the intensity changes along a path through a medium. For the present, scattering effects are not included. Then $K_\lambda = a_\lambda$ and from Eqs. (12-8) and (12-40a) the spectral intensity along a path is attenuated by absorption and augmented by emission to yield

$$\frac{di_\lambda'}{dS} = -a_\lambda(\lambda, S)i_\lambda'(\lambda, S) + a_\lambda i_{\lambda b}'(\lambda, S) \tag{12-51}$$

This is a limited form of the general equation of transfer that is in Chap. 14.

A gas with *uniform temperature and uniform composition* is now considered. The a_λ and $i_{\lambda b}'$ are then constant throughout the volume. Equation (12-51) is integrated from $S = 0$ to S starting from an initial intensity $i_\lambda'(0)$ at $S = 0$. This yields

$$i_\lambda'(\lambda, S) = i_\lambda'(\lambda, 0) \exp[-a_\lambda(\lambda)S] + i_{\lambda b}'(\lambda)\{1 - \exp[-a_\lambda(\lambda)S]\} \tag{12-52}$$

The $\exp(-a_\lambda S)$ is the spectral *transmittance* (fraction transmitted), $\tau_\lambda'(S)$, of the initial intensity. Then $1 - \exp(-a_\lambda S)$ is the fraction of $i_\lambda'(0)$ that was absorbed; this is the spectral *absorptance* $\alpha_\lambda'(S)$ along the path. By virtue of Kirchhoff's law, this quantity appears in the spectral emission along the path as the last term of (12-52). Equation (12-52) becomes

$$i_\lambda'(\lambda, S) = i_\lambda'(\lambda, 0)\tau_\lambda'(\lambda, S) + i_{\lambda b}'(\lambda)\alpha_\lambda'(\lambda, S) \tag{12-53a}$$

where $\alpha_\lambda' = 1 - \tau_\lambda'$. Another form of (12-53a) is

$$i_\lambda'(\lambda, S) - i_\lambda'(\lambda, 0) = [i_{\lambda b}'(\lambda) - i_\lambda'(\lambda, 0)]\alpha_\lambda'(\lambda, S) \tag{12-53b}$$

By integration over the entire spectrum, a *total absorptance* along the path in a *uniform gas* (uniform composition and temperature) is defined as

$$\alpha'(S) = \frac{\displaystyle\int_0^\infty i_\lambda'(\lambda, 0)\,\alpha_\lambda'(\lambda, S)\,d\lambda}{\displaystyle\int_0^\infty i_\lambda'(\lambda, 0)\,d\lambda} = \frac{\displaystyle\int_0^\infty i_\lambda'(\lambda, 0)[1 - e^{-a_\lambda(\lambda)S}]\,d\lambda}{\displaystyle\int_0^\infty i_\lambda'(\lambda, 0)\,d\lambda} \qquad (12\text{-}54a)$$

Similarly, the *total emittance* along a path in a *uniform gas* is

$$\epsilon'(S) = \frac{\displaystyle\int_0^\infty i_{\lambda b}'(\lambda)\,\alpha_\lambda'(\lambda, S)\,d\lambda}{\displaystyle\int_0^\infty i_{\lambda b}'(\lambda)\,d\lambda} = \frac{\pi\displaystyle\int_0^\infty i_{\lambda b}'(\lambda)\alpha_\lambda'(\lambda, S)\,d\lambda}{\sigma T^4}$$

$$= \frac{\pi\displaystyle\int_0^\infty i_{\lambda b}'(\lambda)[1 - e^{-a_\lambda(\lambda)S}]\,d\lambda}{\sigma T^4} \qquad (12\text{-}54b)$$

where T is the gas temperature. A *total transmittance* along the gas path is similarly defined as

$$\tau'(S) = \frac{\displaystyle\int_0^\infty i_\lambda'(\lambda, 0)\tau_\lambda(\lambda, S)\,d\lambda}{\displaystyle\int_0^\infty i_\lambda'(\lambda, 0)\,d\lambda} = \frac{\displaystyle\int_0^\infty i_\lambda'(\lambda, 0)e^{-a_\lambda(\lambda)S}\,d\lambda}{\displaystyle\int_0^\infty i_\lambda'(\lambda, 0)\,d\lambda} \qquad (12\text{-}54c)$$

It follows that

$$1 - \alpha'(S) = \tau'(S) \qquad (12\text{-}54d)$$

12-8 ABSORPTION AND EMISSION PROPERTIES OF GASES

12-8.1 Physical Mechanisms of Absorption and Emission

Although the remaining chapters are devoted to radiation in absorbing, emitting, and scattering media in general, it should be noted that gases are a very important group of these media. If the radiation properties of gases and opaque solids are compared, a difference in spectral behavior is usually quite evident. As shown in Chap. 5, the property variations with wavelength for opaque solids range from fairly smooth to somewhat irregular. Gas properties, however, exhibit very irregular wavelength dependences. The absorption or emission by gases is significant only in certain wavelength regions, especially when the temperature is below a few thousand K. The absorbing ability of a gas layer as a function of wavelength typically looks like that for carbon dioxide (CO_2), water vapor (H_2O), and methane (CH_4) in Fig. 12-10a,b. The radiation emitted from a solid originates from within the material (that is, not at a solid surface, which has no volume), so a solid can be considered an absorbing and emitting medium like a gas. The differences in emission spectra are caused by the types of energy transitions that occur with a particular medium. The transitions for a gas usually lead to a less continuous spectrum than for a solid. Since the absorption coefficient contains the density, this is a large factor in comparing solids with gases. The energy transitions that account for radiant emission and absorption are now discussed.

A radiating gas can be composed of molecules, atoms, ions, and free electrons. These particles can have various energy levels associated with them. In a molecule, the atoms form a dynamical system with vibrational and rotational modes with specific energy levels associated with them. A schematic diagram of the energy levels for an atom or ion is in Fig. 12-11. (The levels for a molecule are diagrammed in Fig. 12-16.) Zero energy level is assigned to the ground state (lowest-energy bound state), with the higher bound states being at positive energy levels. The energy E_I is the ionization potential, which is the energy required to produce ionization from the ground state. Energies above E_I denote that ionization has taken place and free electrons have been produced.

It is convenient to discuss the radiation process by utilizing a photon or quantum point of view. The photon is the basic unit of radiative energy. Radiative emission consists of the release of photons, and absorption is the capture of photons. When a photon is emitted or absorbed, the energy of the emitting or absorbing particle is correspondingly decreased or increased. Figure 12-11 is a diagram of the three types of transitions that can occur. These are bound-bound, bound-free, and free-free. In addition to emission and absorption processes, it is possible for a photon to transfer part of its energy in certain inelastic scattering processes that are of minor engineering importance.

The magnitude of a radiative energy transition is related to the frequency of the emitted or absorbed radiation. The energy of a photon is $h\nu$, where h is Planck's constant and ν is the frequency of the photon energy. For an energy transition, for example, from bound state E_3 down to bound state E_2 in Fig. 12-11, a photon is emitted with energy $E_3 - E_2 = h\nu$. The frequency of the emitted energy is then $\nu = (E_3 - E_2)/h$, so a *fixed frequency* is associated with the transition from a specific energy level to another. Thus, in the absence of other effects, the spectrum of the emitted radiation is in the form of a spectral line. Conversely, in a transition between two bound states when a particle absorbs energy, the quantum nature of the process dictates that the absorption is such that the particle can only go to one of the discrete higher energy levels. Consequently, for a photon to be absorbed, the frequency of the photon energy must have one of certain discrete values. For example, a particle in the ground state in Fig. 12-11 may absorb photons with frequencies $(E_2 - E_1)/h$, $(E_3 - E_1)/h$, or $(E_4 - E_1)/h$ and undergo

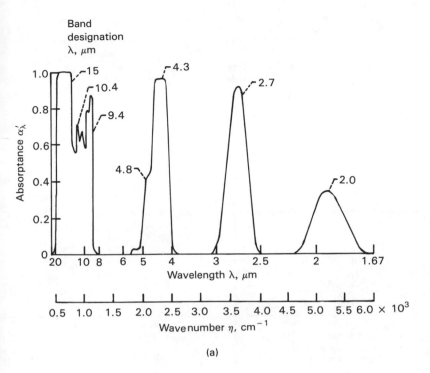

(a)

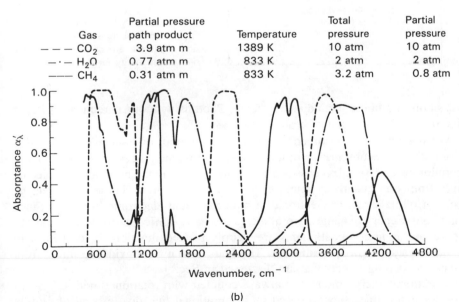

(b)

Figure 12-10 Low-resolution spectrum of absorption bands for various gases. (*a*) Carbon dioxide gas at 830 K, 10 atm, and for path length through gas of 0.388 m. (*b*) Carbon dioxide, water vapor, and methane. From Edwards [25].

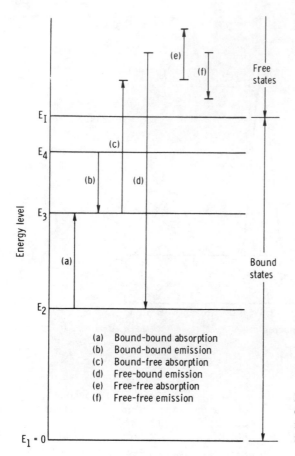

(a) Bound-bound absorption
(b) Bound-bound emission
(c) Bound-free absorption
(d) Free-bound emission
(e) Free-free absorption
(f) Free-free emission

Figure 12-11 Schematic diagram of energy states and transitions for atom or ion. E_1 is the ground state and E_I the ionization energy for the gas.

a transition to a higher bound energy level. Photons with other frequencies in the range $0 < \nu < E_I/h$ cannot be absorbed.

When a photon is absorbed or emitted by an atom or molecule and there is no ionization or recombination of ions and electrons, the process is a *bound-bound* absorption or emission (processes *a* and *b* in Fig. 12-11). The atom or molecule moves from one quantized bound energy state to another. These states can be rotational, vibrational, or electronic in molecules and electronic in atoms. Since bound-bound energy changes are associated with specific energy levels, the absorption and emission coefficients are sharply peaked functions of frequency in the form of a series of spectral lines. The lines have a finite width resulting from various broadening effects that are discussed in Sec. 12-8.3.

Vibrational energy modes are always coupled with rotational modes. The rotational spectral lines superimposed on a vibrational line give a band of closely spaced spectral lines. If these lines overlap into one continuous region, it becomes a *vibration-rotation band* (see Sec. 12-8.5). Rotational transitions within a given vibrational state are associated with energies at long wavelengths, $\sim 8–1000$ μm (see Fig. 1-2). Vibration-rotation transitions are at infrared energies of about 1.5–20 μm. Electronic transitions are at short wavelengths in the visible region 0.4–0.7 μm and at portions of the ultraviolet and infrared near the visible region. At industrial temperatures radiation is principally from vibrational and rotational transitions; at high temperatures (above several thousand degrees K), electronic transitions are important.

Process c in Fig. 12-11 is a *bound-free* absorption (photo-ionization). An atom absorbs a photon with sufficient energy to cause ionization. The resulting ion and electron are free to take on any kinetic energy; hence, the bound-free absorption coefficient is a continuous function of photon energy frequency ν as long as the photon energy $h\nu$ is sufficiently large to cause ionization. The reverse (process d in Fig. 12-11) is free-bound emission (photorecombination). Here an ion and free electron combine, a photon of energy is released, and the energy of the resulting atom drops to that of a discrete bound state. Free-bound emission produces a continuous spectrum, as the combining particles can have any initial kinetic energy.

In an ionized gas, a free electron can pass near an ion and interact with its electric field. This can produce a *free-free* transition (often called *bremsstrahlung*, meaning brake radiation). The electron can absorb a photon (process e in Fig. 12-11), thereby going to a higher kinetic energy, or it can emit a photon (process f) and drop to a lower free energy. Since the initial and final free energies can have any values, a continuous absorption or emission spectrum is produced. Bremsstrahlung can also be produced if an electron passes very close to a neutral atom, since there can be an electric field very close to an atom. This process is much less probable than electron–positive ion interactions.

12-8.2 The Condition of Local Thermodynamic Equilibrium

It has been tacitly assumed in earlier chapters that opaque solids emit energy based solely on their temperature and physical properties. The spectrum of emitted energy was assumed unaffected by the characteristics of any incident radiation. This is generally true because all the absorbed incident energy is quickly redistributed into an equilibrium distribution of internal energy states at the temperature of the solid.

In a gas, the redistribution of absorbed energy occurs by various types of collisions between the atoms, molecules, electrons, and ions that constitute the gas. Under most engineering conditions, this redistribution occurs quite rapidly, and the energy states of the gas *will* be populated in *equilibrium distributions* at any given locality. When this is true, the Planck spectral distribution and (12-39) correctly describe the emission from a gas volume element. The specification that a gas will emit according to (12-39) regardless of the spectral distribution of intensity passing through and being absorbed by dV is a consequence of the assumption of "local thermodynamic equilibrium" (LTE). When LTE is not present, the calculation of radiant transfer becomes much more complex [11, 12, 12a].

Cases in which the LTE assumption breaks down are occasionally encountered. Examples are in very rarefied gases, where the rate and/or effectiveness of interparticle collisions in redistributing absorbed radiant energy is low; when rapid transients exist so that the populations of energy states of the particles cannot fully adjust to new conditions during the transient (see [12a] for an analysis of the heating response of a metal to a very short energy pulse from a laser); where very sharp gradients occur so that local conditions depend on particles that arrive from adjacent localities at widely different conditions and which may emit before reaching local equilibrium; and where extremely large radiative fluxes exist, so that absorption of energy and therefore population of higher energy states occur so strongly that collisional processes cannot repopulate the lower states to an equilibrium density. Under any of these conditions, the spectral distribution of emitted radiation is not given by (12-39). Then the populations must be determined by detailed examination of the relation between the collision and radiation processes and their effect on the distribution of energy among the various possible states— a most formidable undertaking. It is, however, necessary in the examination of shock phenomena (sharp gradients), stellar atmospheres (extreme energy flux and low density), nuclear explosions (transients, sharp gradients, and extreme fluxes), and high-altitude and interplanetary gas dynamics (very low densities). A gas with small optical thickness can have transmitted within it radiation from regions at widely different conditions. For this reason, a nearly transparent gas is more likely to depart from LTE than an optically thick gas of the same density.

A prominent non-LTE effect is in the laser, in which a material with a metastable energy state is excited by some external means. Because the excited state is metastable and is chosen so that no competing process is trying to depopulate it, its population can reach a value well above the equilibrium value. This condition is called a *population inversion*. The material is then exposed to radiation containing photons with the same frequency as the transition frequency from the excited metastable state to a lower state. This induces or stimulates the transition to the lower state. Consequently, many photons with the transition frequency are emitted, thus amplifying the intensity of the incident radiation. This leads to the acronym *l*ight *a*mplification by *s*timulated *e*mission of *r*adiation, or *laser*.

The term *luminescence* covers additional non-LTE mechanisms of radiant emission by transitions from an excited state to a lower energy state, where the original excitation is by means *other than thermal agitation*. Common modes of excitation are by visible light, ultraviolent radiation, and electron bombardment. The absorption and reemission result in a change of wavelength of the radiation. Because the transitions are between discrete energy states, the span of wavelengths over which the emission occurs is quite small. Luminescence, therefore, does not add significant energy to the spectrum of emission in engineering situations and is neglected in heat transfer calculations. There are some situations in which effects other than total energy transport are of interest; hence a brief mention of luminescence is included here.

A common classification of luminescence is by its persistence. Luminescence that persists only during the influence of an external exciting agent, such as an ultraviolet lamp, is *fluorescence*, a name arising from the strong luminescence shown by the mineral fluorspar when so irradiated. Luminescence that continues with slowly diminishing intensity after the excitation is removed is *phosphorescence*, a word derived from the luminescence of white phosphorus.[†] Another categorization is by excitation agent. Luminescence from a chemical reaction such as the oxidation of white phosphorus is *chemiluminescence*; that caused by a beam of incident electrons as on a color TV screen is *cathodoluminescence*; a biochemical reaction producing luminosity, as in fireflies and some marine animals, is *bioluminescence*; luminous emission by the presence of an electric field, as in certain commercial panel lamps, is *electroluminescence*; and luminescence due to photon bombardment is often called *photoluminescence* (the same mechanism that causes the laser to function). Other mechanisms are proton bombardment, believed to be responsible for the luminous red patches observed on the moon, and nuclear reactions causing luminous emission.

Because luminescence is common to materials at room temperature, it obviously is not predicted by the laws for thermal radiation, as these would predict no visible radiation at such temperatures. This is the origin of the term *cold light* for fluorescent-lamp emission. The quantum-mechanical properties of such luminescent materials must be examined to explain their behavior. Detailed material on luminescence is in [13, 14].

Non-LTE problems are not within the scope of this work. It is assumed here that LTE exists and the spontaneous emission from dV is governed by (12-39).

12-8.3 Spectral Line Broadening

As discussed in Sec. 12-8.1, if a gas is not dissociated or ionized, its internal energy is in discrete vibrational, rotational, and electronic energy states of its atoms or molecules. If the energies of the upper and lower discrete states are E_j and E_i, only photons of energy E_p can cause a transition, where

$$E_p = E_j - E_i = h\nu_{ij} = hc\eta_{ij} \tag{12-55}$$

The discrete transitions result in the absorption of photons of only very definite frequencies, causing the appearance of dark lines in the transmission spectrum; this is *line absorption*. The rates at which these bound-bound transitions occur are available in tabular form for some molecules and atoms [15, 16]. The relations for transition rates are often given by semiclassical results describing radiating atoms, multiplied by a modifying factor, called the *Gaunt factor*, that provides the correction for quantum-mechanical effects.

Equation (12-55) would predict that very little energy could be absorbed from the entire incident spectrum by an absorption line, because only those photons having a single wavenumber could be absorbed. Other effects, however, cause the line to be broadened and consequently have a finite wavenumber span around the transition wavenumber η_{ij}. The wavenumber span of the broadened line, and

[†]Phosphorus itself is named from the Greek word *phosphoros,* meaning "light bearing."

the variation within it of its absorption ability, depends on the physical mechanism of the broadening. Some of the important mechanisms are natural broadening, Doppler broadening, collision broadening, and Stark broadening. Collision broadening is the most important for most engineering conditions involving infrared radiation.

The variation of the absorption coefficient with wavenumber within a broadened line is the *shape* of the line. The shape is important as it is related to the basic trends of gas absorption with temperature, pressure, and path length through the gas. The shape of a typical spectral line is illustrated by Fig. 12-12. The *integrated absorption coefficient* S_{ij} for a line is the integral under the $a_{\eta,ij}(\eta)$ curve,

$$S_{ij} = \int_0^\infty a_{\eta,ij}(\eta)\, d\eta = \int_{-\infty}^\infty a_{\eta,ij}(\eta)\, d(\eta - \eta_{ij}) \qquad (12\text{-}56)$$

The $a_{\eta,ij}(\eta)$ is essentially zero except for η close to η_{ij}. The regions away from η_{ij}, where $a_{\eta,ij}$ is small, are the "wings" of the line. The magnitudes of $a_{\eta,ij}$ and S_{ij} depend on the number of molecules in energy level i and hence depend on gas density. Taking the ratio $a_{\eta,ij}(\eta)/S_{ij}$ tends to cancel the density effect, and the ratio shows the effect of density in changing the shape of the line.

The line shape depends on the line-broadening phenomenon. One characteristic of the line shape is the *half-width of the line*, Δ. This is the half-width (in units of wavenumber for the present discussion) evaluated at half the maximum line height (Fig. 12-12). It provides a definite width to help describe the line. Since $a_{\eta,ij}$ goes to zero asymptotically as $|\eta - \eta_{ij}|$ increases, it is not possible to define a line width in terms of wavenumbers at which $a_{\eta,ij}$ becomes zero.

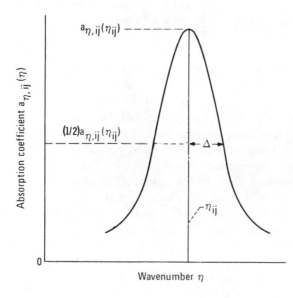

Figure 12-12 Absorption coefficient for symmetric broadened spectral line for transition between energy levels i and j.

Four mechanisms for line broadening are now discussed, along with the resulting line shapes.

Natural broadening A perfectly stationary emitter unperturbed by all external effects emits energy over a finite spectral interval about each transition wavenumber. This *natural line broadening* arises from the uncertainty in the exact levels E_i and E_j of the transition energy states, which is related to the Heisenberg uncertainty principle. Natural line broadening produces the line shape

$$\frac{a_{\eta,ij}(\eta)}{S_{ij}} = \frac{\Delta_n/\pi}{\Delta_n{}^2 + (\eta - \eta_{ij})^2} \tag{12-57}$$

where Δ_n is the half-width for natural broadening. This form is a *resonance* or *Lorentz profile*. In units of wavenumber, it provides a profile that is symmetric about η_{ij}. For engineering applications the half-width produced by natural broadening is usually quite small compared with that by other line-broadening mechanisms. Natural line broadening is therefore usually neglected.

Doppler broadening The atoms or molecules of an absorbing or emitting gas are not stationary, but have a distribution of velocities associated with their thermal energy. If an atom or molecule is emitting at wavenumber η_{ij} and at the same time is moving at velocity v toward an observer, the waves arrive at the observer at an increased η given by

$$\eta = \eta_{ij}\left(1 + \frac{v}{c}\right) \tag{12-58}$$

If the emitter is moving away from the observer, v is negative and the observed wavenumber is less than η_{ij}. In thermal equilibrium the gas molecules have a Maxwell-Boltzmann distribution of velocities. If an observer is detecting radiation along one coordinate direction, the velocity components of interest are those along the single direction either toward or away from the observer. The fraction of molecules moving in that direction within a velocity range between v and $v + dv$ is

$$\frac{dN}{N} = \sqrt{\frac{M}{2\pi kT}}\exp\left(-\frac{Mv^2}{2kT}\right)dv \tag{12-59}$$

where M is the mass of a molecule of the radiating gas and k is the Boltzmann constant. Using (12-58) in (12-59) to eliminate v gives the fractional number of molecules providing radiation in each differential wavenumber interval as a result of Doppler broadening. The result is a spectral line shape with a Gaussian distribution,

$$\frac{a_{\eta,ij}(\eta)}{S_{ij}} = \frac{1}{\Delta_D}\sqrt{\frac{\ln 2}{\pi}}\exp\left[-(\eta - \eta_{ij})^2\frac{\ln 2}{\Delta_D{}^2}\right] \tag{12-60}$$

The Δ_D is the line half-width for Doppler broadening,

$$\Delta_D = \frac{\eta_{ij}}{c}\left(\frac{2kT}{M}\ln 2\right)^{1/2}$$

(12-61)

which depends on η_{ij}, T, and M. The dependence of Δ_D on $T^{1/2}$ shows that Doppler broadening is important at high temperatures.

Collision broadening As the pressure of a gas is increased, the collision rate by an atom or molecule is increased. Collisions can perturb the energy states of the atoms or molecules, resulting in collision broadening of the spectral lines. For noncharged particles, the line has a Lorentz profile [4]

$$\frac{a_{\eta, ij}(\eta)}{S_{ij}} = \frac{\Delta_c/\pi}{\Delta_c^2 + (\eta - \eta_{ij})^2}$$

(12-62)

which is the same shape as for natural broadening. The collision half-width Δ_c is determined by the collision rate. An approximate value from kinetic theory is

$$\Delta_c = \frac{1}{2\pi c}\frac{4\sqrt{\pi}D^2 P}{(MkT)^{1/2}}$$

(12-63)

where D is the diameter of the atoms or molecules and P is the gas pressure for the single-component gas. Equation (12-63) shows that collision broadening becomes important at high pressures and low temperatures or, from the perfect gas law, high pressures and densities.

Collision broadening is often the chief contributor to line broadening for engineering infrared conditions, and the other line-broadening mechanisms can usually be neglected. The shapes of Doppler and Lorentz broadened lines are compared in Fig. 12-13 for the same half-width and area under the curves. The Lorentz profile is lower at the line center but remains of appreciable size farther out in the wings of the line. Even when Doppler broadening is dominant near the line center, collision broadening is often the important mechanism far from the center.

When both collision and Doppler broadening are important, a convolution of the collision and Doppler profiles results, called the Voigt profile [17]. This depends on two parameters, the normalized frequency separation from the line center, $[(\nu - \nu_{ij})(\ln 2)^{1/2}]/(\pi^{1/2}\Delta_D)$, and the broadening parameter, $\Delta_c(\ln 2)^{1/2}/\Delta_D$. Here, the half-widths are in terms of frequency rather than wavenumber.

Collision narrowing Collision narrowing refers to the reduction of Doppler broadening that occurs because of collisions that limit the scale of molecular motion. Such narrowing is normally important when both Doppler and collision broadening are present and results in a more complex line shape. The shape is given by the Galatry profile [17–19], which is a function of the two parameters of the Voigt profile and an additional narrowing parameter $kT(\ln 2)^{1/2}/(2M\pi c \Delta_D D)$. The definitions are as for the Voigt profile, and D is the optical diffusion coefficient, usually approximated as equal to the mass diffusion coefficient.

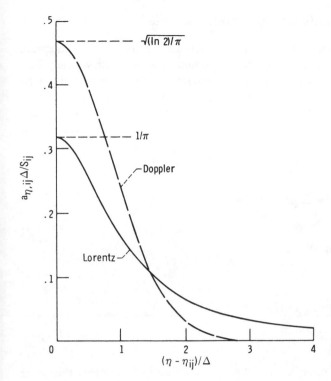

Figure 12-13 Line-shape parameter for Doppler and Lorentz broadened spectral lines (areas under two curves are equal).

Stark broadening When strong electric fields are present, the energy levels of the radiating gas particles can be greatly perturbed. This is the *Stark effect*, which can result in very large line broadening. It is often observed in partially ionized gases where radiating particle interactions with electrons and protons give large Stark effects. Calculation of the line shapes must be approached through quantum mechanics, and the resulting shapes are quite unsymmetrical and complicated.

Stark and collision broadening are often lumped under the general heading of pressure broadening. Both depend on the pressure of the broadening component of the gas. When two or more broadening effects contribute simultaneously, calculation of the resulting line shape becomes more difficult. References [6, 15, 20–22] can be consulted for additional information.

Broadening has been discussed here under the assumption that only one atomic or molecular species is present in the gas. If the gas has more than one component, collision broadening in the radiation-absorbing gas is caused by collisions with like molecules (self-broadening) and by collisions with other species. Both collision processes must be included in calculating line shapes.

Although expressions for line *shapes* as given by $a_{\eta,ij}(\eta)S_{ij}$ have now been provided, obtaining the absorption coefficient requires S_{ij}. This is briefly discussed at the end of the next section.

12-8.4 Absorption or Emission by a Spectral Line

As a step toward the evaluation of gas absorptance and emittance, the integrals in (12-54) are now considered over the spectral region of a single broadened line. For a line centered about η_{ij}, the absorption coefficient $a_{\eta,ij}(\eta)$ is essentially zero except in a narrow wavenumber range surrounding η_{ij}. Unless S is large, the integrands in the numerators of (12-54a,b) are appreciable only within this range. The $i'_\eta(0)$ or $i'_{\eta b}$ remain essentially constant within this range, and since the largest absorption is at η_{ij}, the $i'_\eta(0)$ and $i'_{\eta b}$ are ordinarily used at η_{ij}. Then Eqs. (12-54a,b) become, for a spectral line (in terms of wavenumber),

$$\alpha'_{ij} = \frac{i'_\eta(\eta_{ij}, 0) \int_{-\infty}^{\infty} \{1 - \exp[-a_{\eta,ij}(\eta)S]\}\, d(\eta - \eta_{ij})}{\int_0^\infty i'_\eta(\eta, 0)\, d\eta} \tag{12-64}$$

$$\epsilon'_{ij} = \frac{\pi i'_{\eta b}(\eta_{ij}) \int_{-\infty}^{\infty} \{1 - \exp[-a_{\eta,ij}(\eta)S]\}\, d(\eta - \eta_{ij})}{\sigma T^4} \tag{12-65}$$

The absorbed and emitted energies for the spectral line both contain the same integral. This is called the *effective line width* $\bar{A}_{ij}$

$$\bar{A}_{ij}(S) \equiv \int_{-\infty}^{\infty} \left\{1 - \exp\left[-a_{\eta,ij}(\eta)S\right]\right\} d(\eta - \eta_{ij}) \tag{12-66}$$

which has the units of the spectral variable (η in this instance). By considering a spectral line within which the gas is perfectly absorbing ($a_{\eta,ij} \to \infty$), and having no absorption outside this line, it is found from (12-66) that $\bar{A}_{ij}$ *can be interpreted as the width of a black line centered about η_{ij} that produces the same absorption or emission as the actual line*.

The evaluation of $\bar{A}_{ij}$ is now considered for some important limiting cases. First let the optical path length be small, $a_{\eta,ij}(\eta)S \ll 1$. The exponential term in Eq. (12-66) can be approximated by $1 - \exp[-a_{\eta,ij}(\eta)S] \approx a_{\eta,ij}(\eta)S$ so that by using S_{ij} from (12-56)

$$\bar{A}_{ij}(S) = S \int_{-\infty}^{\infty} a_{\eta,ij}(\eta)\, d(\eta - \eta_{ij}) = SS_{ij} \tag{12-67}$$

(The integrated absorption coefficient S_{ij} should not be confused with the path length S.) The effective line width is thus *linear* with path length S *regardless of the line shape*. A line with linear behavior is called a *weak* line.

Next consider asymptotic forms for the Lorentz line shape for collision broadening (12-62), as this is the most important broadening for engineering applications in the infrared. Substituting (12-62) into (12-66) gives

$$\bar{A}_{ij}(S) = \int_{-\infty}^{\infty} \left\{ 1 - \exp\left[-\frac{S_{ij}}{\pi} \frac{\Delta_c S}{\Delta_c^2 + (\eta - \eta_{ij})^2} \right] \right\} d(\eta - \eta_{ij}) \tag{12-68}$$

This is integrated to yield

$$\bar{A}_{ij}(S) = 2\pi\Delta_c \xi e^{-\xi} [J_0(i\xi) - iJ_1(i\xi)] \tag{12-69}$$

where $\xi = SS_{ij}/2\pi\Delta_c$ and J_n is the Bessel function of order n. For a strong line, $\xi \gg 1$, the asymptotic form of (12-69) is

$$\bar{A}_{ij}(S) = 2\sqrt{S_{ij}\Delta_c S} \tag{12-70}$$

For a weak line, $\xi \ll 1$, (12-69) reduces to (12-67). Equation (12-70) shows that, for a *strong Lorentz line*, the absorptance varies as the *square root* of the path length. This is in contrast to the results for *any weak line*, Eq. (12-67), where the absorptance varies *linearly* with path length. Experimental results bear out these dependences. Some relations that approximate (12-69) quite well are given in [23]. One useful form is

$$\bar{A}_{ij}(S) = 2\pi\Delta_c \left(\frac{2\xi}{\pi} \right)^{1/2} \{1 - \exp[-(\pi\xi/2)^{1/2}]\} \tag{12-71}$$

Some expressions for $\bar{A}_{ij}$ have now been obtained; numerical values can be calculated if Δ_c and S_{ij} are known. An expression for Δ_c was given in (12-63), and it depends on gas pressure and temperature. The S_{ij} is also a function of these variables, as it depends on the number of gas molecules occupying an energy level and on the probability of a transition occurring. This gives S_{ij} in the form

$$S_{ij} \propto \rho \exp\left(\frac{-K_{ij}}{T} \right) \propto \frac{P}{T} \exp\left(\frac{-K_{ij}}{T} \right) \tag{12-72}$$

where K_{ij} is a coefficient depending on the particular quantum state involved in the transition. As a result, in (12-70), for a fixed gas temperature the $\bar{A}_{ij}$ will depend on ρS or PS (from $S_{ij}S$) and on ρ or P alone (from Δ_c). There are thus two effects of increasing gas density or pressure. One is the increase in absorption because there are more molecules along the radiation path, and the second is the increased width of the spectral line. These trends help guide the correlation of absorption and emission behavior as many lines are superposed into an absorption band.

12-8.5 Band Absorption

Band structure The gases commonly encountered in engineering calculations are diatomic or polyatomic and possess vibrational and rotational energy states that are absent in monatomic gases. The transitions between vibrational and rotational states usually provide the main contribution to the absorption coefficient in the significant thermal radiation spectral regions at moderate temperatures. As the temperature is raised, dissociation, electron transitions, and ionization become more probable, and the contributions of these additional processes to the absorption coefficient must be included.

When the absorption coefficient of a gas is determined experimentally, contributions of all the line and continuum processes are superimposed. In computing such coefficients, each absorption process must be analyzed and the complete coefficient obtained by combining contributions from the various processes. In Fig. 12-14, the contributions to the spectral absorption coefficient as given by [24] are shown for air at 1 atm pressure and a range of temperatures. The ordinate is the maximum value of the spectral absorption coefficient (regardless of wavenumber) of the contribution from a given process divided by the sum of all such maxima at the same temperature. At low temperatures the entire absorption arises from transitions of oxygen between molecular states. As the temperature is in-

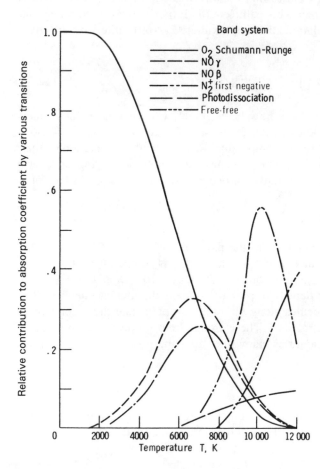

Band system

———— O_2 Schumann-Runge
— — — NO γ
— · — NO β
— ··· — N_2^+ first negative
— — — Photodissociation
— ···· — Free-free

Figure 12-14 Relative contributions of energy transitions in bands of various species to absorption coefficient of air at 1 atm [7].

creased, there is some formation of NO, which provides additional bound-bound transitions. At high temperatures continuous absorption processes are dominant; these are bound-free and free-free transitions.

The vibration-rotation bands are usually the most important absorbing and emitting spectral regions in engineering radiation calculations. The structure of such bands is now examined, and this will indicate the complexity of computing band absorption coefficients from basic principles. Some simplified band models are discussed for analysis of some band-absorption features. The correlation of experimental band absorptance data is considered to illustrate how gas properties can be presented in a manner useful for engineering applications. Often in engineering heat transfer problems a reasonable approximation to the total radiation will suffice. It is then not necessary to go into the details of the radiation associated with individual bands. For total radiation calculations, charts and formulas for gas total emittance and absorptance have been developed from total radiation measurements (see Chap. 13). The information in the following sections will aid in understanding how the physical variables influence gas radiation.

Consider the vibration-rotation transitions governing the absorption coefficient of most polyatomic gases up to a temperature of about 3000 K. The transitions are strongly dependent on wavenumber, and consequently the absorption coefficient is strongly spectrally dependent. The absorption in a vibration-rotation band consists of groups of very closely spaced spectral lines resulting from transitions between vibrational and rotational energy states. An example is in Fig. 12-15 for a portion of the carbon dioxide spectrum.[†] The absorption lines are so closely spaced in certain spectral regions that individual lines are not fully resolved by experimental measurements. The lines may overlap, as a consequence of broadening, and merge to form absorption bands. An example of absorption bands observed with low resolution are those for carbon dioxide in Fig. 12-10.

The large number of possible energy transitions that can produce spectral lines is illustrated by the many energy levels and transition arrows in Fig. 12-16. This shows the potential energy for a diatomic molecule as a function of the separation distance between its two atoms. The two curves are each for a different electronic energy state where the electron may be shared by the two atoms. The distance R_e is the mean interatomic distance corresponding to each of the electronic states. The long-dashed horizontal lines denote vibrational energy levels, while the short-dashed lines are rotational states superimposed on the vibrational states. Transitions between rotational levels of the *same* vibrational state involve small $E_j - E_i$. Hence from (12-55) these transitions give lines at low wavenumbers in the far infrared. Transitions between rotational levels in *different* vibrational states of the same electronic state give vibration-rotation bands at wavenumbers in the near infrared. If transitions occur from a rotational level of an electronic and vibrational

---

[†]The notation (01^10), etc., in Fig. 12-15 is a designation used to show the quantum state of a harmonic oscillator. In the general case $(v_1 v_2^l v_3)$, the v_i are the vibrational quantum numbers, and l is the quantum number for angular momentum. Transitions between two energy states, such as those denoted by $(00^00) \rightarrow (01^10)$, give rise to absorption lines. Certain selection rules govern the allowable transitions. A good introductory treatment is given in Chap. 3 of Goody [3].

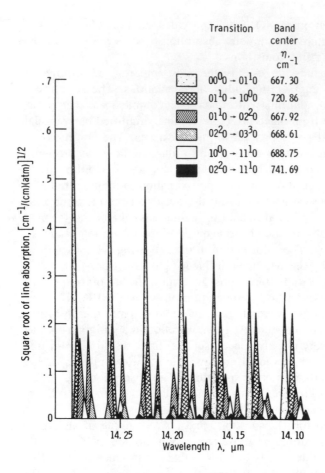

Figure 12-15 A portion of high-resolution spectrum of carbon dioxide [3].

state to a rotational level in a different electronic and vibrational state, then large $E_j - E_i$ are involved and a band system can be formed in the high-wavenumber visible and ultraviolet regions.

Band models Two types of models are employed for the absorption and emission behavior of bands of closely spaced lines. *Narrow-band* models use characterizations of individual line shapes, widths, and spacings to derive band characteristics within a defined wavenumber interval. *Wide-band* models provide correlations of band characteristics over the entire wavenumber region of the band. These models account for the increasing importance of weakly absorbing lines in the wings of the band as the radiation path length becomes large. General discussions are in [20, 25–28]. In a detailed radiation-exchange calculation the absorbed and emitted energy will be needed in each band region, for example, in the four main CO_2 bands of Fig. 12-10. These bands are separated by spectral regions that are nearly transparent. For the total absorbed or emitted energy in a *uniform* gas Eqs. (12-54a,b) are used; both involve the same type of integral.

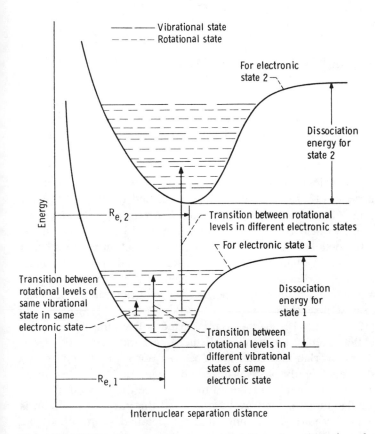

Figure 12-16 Potential energy diagram and transitions for a diatomic molecule.

Since the absorption bands usually occupy a rather narrow spectral region, an average value of $i'_\eta(0)$ or $i'_{\eta b}$ can be taken out of the integral for each band. For example, for the total emittance, (12-54b) becomes

$$\epsilon'(S) = \frac{\pi \sum_l i'_{\eta b,l} \int_l [1 - \exp(-a_\eta S)] \, d\eta}{\sigma T^4} \tag{12-73}$$

where the subscript l denotes a band, the integral is over each band, and the summation is over all the bands.

In a fashion similar to the effective line width in (12-66), the integral in (12-73) is defined as the *effective bandwidth* $\bar{A}_l$

$$\bar{A}_l(S) \equiv \int_{\substack{\text{absorption} \\ \text{bandwidth}}} \{1 - \exp\left[-a_\eta(\eta) S\right]\} \, d\eta \tag{12-74}$$

and (12-73) becomes

$$\epsilon'(T, P, S) = \frac{\pi}{\sigma T^4} \sum_l i'_{\eta b,l} \bar{A}_l \tag{12-75}$$

The $\bar{A}_l$ has units of the spectral variable, which is η for Eq. (12-74). The span of the absorption band that provides the upper and lower limits of the integral in Eq. (12-74) does not have a specific value that applies for all conditions. It can be defined as the spectral interval beyond which there is only a specified small fractional contribution to $\bar{A}_l$. The width of this interval increases slowly with path length as a result of proportionately more absorption in the wings of the band.

The total emittance in (12-75) can be used as described in Sec. 13-3.4 for engineering calculations of radiation from an isothermal uniform gas to an enclosure boundary. The $\bar{A}_l$ can also be used as shown in Sec. 13-3.4 to obtain the band absorptance α_l for detailed spectral-exchange calculations in enclosures. It is evident that for calculating quantities such as ϵ' and α_l [and, from (12-54d), τ_l] for various conditions, correlations for $\bar{A}_l$ are needed as a function of path length, pressure, temperature, etc. for the important bands of radiating gases. The correlations of $\bar{A}_l$ are now discussed.

By comparing (12-74) and (12-66) the $\bar{A}_l$ for a band is the sum of the $\bar{A}_{ij}$ for all the spectral lines that occupy the band if all the $\bar{A}_{ij}$ act independently. Generally, the spectral lines overlap and each line does not absorb as much energy as it would if it acted independently. As was observed in Fig. 12-15, an absorption band is typically composed of many broadened lines. Hence the $a_\eta(\eta)$ in (12-74) is a complicated irregular function of wavenumber, and the integration for $\bar{A}_l$ is difficult. The integration would also require that the detailed shape of all the broadened lines be known. It is evident that a simplified model for the form of $a_\eta(\eta)$ must be devised if integration over the lines to obtain band radiation properties is to be a fruitful approach. Two common narrow-band models represent extremes in specifying the individual line spacings and magnitudes (Fig. 12-17).

Elsasser [29] modeled the lines as all having the same Lorentz shape and being of equal heights and equal spacings (and hence each having the same integrated absorption coefficient S_c). As shown in Fig. 12-17a this gives a_η as a periodic function that depends on the parameters governing the shape of the Lorentz lines and on the spacing δ between them. The absorption coefficient at a particular wavenumber is found by summing the contributions from all the adjacent overlapping lines. The distance of the line centers from a position η are $|\eta - 0|$, $|\eta - \delta|$, $|\eta - 2\delta|$, and so forth. Then, summing all the contributions by use of the Lorentz shape in (12-62) gives

$$a_\eta(\eta) = \frac{S_c}{\pi} \sum_{n=-\infty}^{\infty} \frac{\Delta_c}{\Delta_c^2 + (\eta - n\delta)^2}$$

Elsasser obtained a closed form for this series,

$$a_\eta(\eta) = \frac{S_c}{\delta} \frac{\sinh{(\pi\beta/2)}}{\cosh{(\pi\beta/2)} - \cos{(\pi z/2)}} \qquad \beta = 4\Delta_c/\delta, \; z = 4(\eta - \eta_{ij})/\delta$$

where $\eta - \eta_{ij}$ varies between $-\delta/2$ and $\delta/2$ over each periodic line interval. The β is important as it specifies the effect of line structure. For large β the lines are broad compared to their spacing and the line structure is lost as the lines strongly overlap. This corresponds to large pressure [Eq. (12-63)]. Using (12-74), the effective width for a band of m lines is

$$\bar{A}_l = m \int_{-\delta/2}^{\delta/2} \left\{ 1 - \exp\left[-\frac{SS_c}{\delta} \frac{\sinh{(\pi\beta/2)}}{\cosh{(\pi\beta/2)} - \cos{(\pi z/2)}} \right] \right\} d(\eta - \eta_{ij})$$

which can be written as

$$\frac{\bar{A}_l}{m\delta} = \frac{\bar{A}_l}{A_0} = 1 - \frac{1}{2} \int_0^2 \exp\left[-\frac{u \sinh{(\pi\beta/2)}}{\cosh{(\pi\beta/2)} - \cos{(\pi z/2)}} \right] dz \tag{12-76}$$

where $u = SS_c/\delta$. The $\bar{A}_l/A_0$ is the effective width relative to the actual width A_0 of the band.

Although the integral in (12-76) cannot be evaluated in closed form, some informative limiting results are found. For *weak nonoverlapping lines*, $u \ll 1$, each line acts independently, and $\bar{A}_l/A_0$ is the same as for a single line. Using (12-67) with $S_{ij} = S_c$ in this band model gives

$$\frac{\bar{A}_l}{A_0} = \frac{\bar{A}_{ij}}{\delta} = \frac{SS_c}{\delta} = u \qquad u \ll 1 \tag{12-77}$$

For a *long radiation path*, $u \gg 1$, (12-76) yields

$$\frac{\bar{A}_l}{A_0} = 1 \qquad u \gg 1 \tag{12-78}$$

This represents total absorption within the *entire* width of the band. Three other limits, and the conditions giving them, are summarized in [26]. For *strong nonoverlapping lines*, (12-76) yields

$$\frac{\bar{A}_l}{A_0} = \text{erf}\left(\frac{1}{2} \sqrt{\pi\beta u} \right) \qquad \text{when } \beta \ll 1, \frac{u}{\beta} \gg 1 \tag{12-79}$$

If in addition βu is small so that *lines are thin and spaced well apart*,

$$\frac{\bar{A}_l}{A_0} = \sqrt{\beta u} \qquad \text{when } \beta \ll 1, \frac{u}{\beta} \gg 1, \beta u \ll 1 \tag{12-80}$$

This is called the *square-root limit*. If β is very large, $\Delta_c \gg \delta$, the lines are very broad compared with the spacing between them, and the line structure becomes spread out over the band. This yields

$$\frac{\bar{A}_l}{A_0} = 1 - \exp{(-u)} \tag{12-81}$$

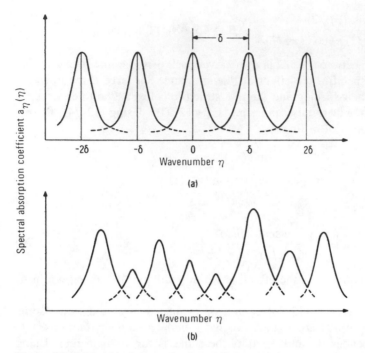

Figure 12-17 Models of absorption lines forming an absorption band. (*a*) Elsasser model, equally spaced Lorentz lines; (*b*) statistical model.

where, with the line structure spread out, the S_c/δ in u is the average absorption coefficient.

Another band model is a statistical array of lines as shown in Fig. 12-17*b* and presented by Goody [3]. There can be a random spacing of identical lines or, more generally, the lines can differ from each other. A spacing that is essentially random is typical of bands of polyatomic molecules such as CO_2 and water vapor. To apply the model, probability distributions of line strengths and positions must be assumed. These statistical assumptions remove the necessity of calculating the exact properties of the individual lines in the band. The parameters u and β are based on average values of the line widths and strengths.

Many other band models have been proposed, some of them of more utility in certain cases than the Elsasser or Goody models. Several of these band models are discussed by Goody [3], Edwards and Menard [30], and Ludwig et al. [31]. Modifications to the Elsasser model have been made by Kyle [32] and Golden [33, 34], who treated evenly spaced lines with a Doppler profile, and Golden [35], who treated the same case with a Voigt profile.

Band correlations for wide-band models The band model of Edwards and Menard [30] consists of rotation lines in a band that are reordered in wavenumber so that they form an array with exponentially decreasing line intensities moving away from the band center. This is the *exponential wide-band* model. The u and β are based on average values over the band of the line-integrated absorption coefficient and the line width. For large absorption in the band, e.g., for large path length S, the center of the band becomes completely absorbing. As S is further increased, $\bar{A}$ increases as the wings of the band gradually become more absorbing. For this reason the limit in (12-78) is not reached for this model. The limit for large u is

$$\frac{\bar{A}_l}{A_0} = \ln u \qquad u \gg 1 \tag{12-82}$$

For this model the limit for the conditions in (12-80) becomes

$$\frac{\bar{A}_l}{A_0} = 2\sqrt{\beta u} \tag{12-83}$$

Relations (12-77) to (12-83) are limiting forms that show behavior for particular conditions of u and β. Using the band correlations, however, requires that the $\bar{A}_l/A_0$ be available for all values of the parameters u and β. Then these parameters must be related to physical variables such as path length, pressure, and temperature. Using the three limiting conditions in (12-78), (12-82), and (12-83), Edwards and Menard [30] constructed the following functions that smoothly join $\bar{A}_l/A_0$ for all regions of u and β and go to the proper limiting values:

$$\beta \leqslant 1: \quad \frac{\bar{A}_l}{A_0} = u \qquad\qquad \frac{\bar{A}_l}{A_0} \leqslant \beta$$

$$\frac{\bar{A}_l}{A_0} = 2\sqrt{\beta u} - \beta \qquad \beta \leqslant \frac{\bar{A}_l}{A_0} \leqslant (2 - \beta)$$

$$\frac{\bar{A}_l}{A_0} = \ln (\beta u) + 2 - \beta \qquad \frac{\bar{A}_l}{A_0} \geqslant (2 - \beta)$$

$$\beta > 1: \quad \frac{\bar{A}_l}{A_0} = u \qquad\qquad \frac{\bar{A}_l}{A_0} \leqslant 1$$

$$\frac{\bar{A}_l}{A_0} = \ln u + 1 \qquad\qquad \frac{\bar{A}_l}{A_0} \geqslant 1$$

Tien and Lowder [36] devised the continuous correlation

$$\frac{\bar{A}_l}{A_0} - \ln \left[u f(\beta) \frac{u + 2}{u + 2 f(\beta)} + 1 \right] \tag{12-84}$$

where

$$f(\beta) = 2.94 \, [1 - \exp (- 2.60\beta)]$$

This does not satisfy the square-root limit (12-80) [37]. Other correlations have been constructed, such as those of Cess and Tiwari [37]

$$\frac{\bar{A}_l}{A_0} = 2 \ln \left[1 + \frac{u}{2 + u^{1/2}(1 + 1/\beta)^{1/2}} \right] \qquad (12\text{-}85)$$

and Goody and Belton [24]

$$\frac{\bar{A}_l}{A_0} = 2 \ln \left[1 + \frac{\sqrt{\beta}\, u}{(u + 4\beta)^{1/2}} \right] \qquad (12\text{-}86)$$

A more complicated expression that covers all ranges is given by Morizumi [27], who also summarizes some of the other correlations in the literature. Tiwari [26] compares the results of using various narrow-band models in developing the exponential wide-band model. In addition, the correlations presented by Edwards and Balakrishnan [38, 39], Edwards and Menard [30], Tien and Lowder [36], Tien and Ling [40], Cess and Tiwari [37], Felske and Tien [41], Goody and Belton [42], and Hsieh and Greif [43] are compared and analyzed for regions of accurate use.

Since the effective bandwidth $\bar{A}_l$ is calculated from models utilizing the line structure, it is evident that $\bar{A}_l$ depends on the line spacing, the line half-width, and the line-integrated absorption, as well as other quantities, when a statistical model is used. To utilize these analytical results for radiative calculations involving a real gas mixture, it is necessary to know how all these factors are influenced by conditions such as gas temperature, partial pressure of the absorbing gas, and total pressure of the gas mixture. If relations between these quantities are specified, the correlation of experimental data based on the theoretically indicated dependences of the band integration can be attempted. The background for these correlations is given by Goody [3]. Edwards and co-workers [44–52] assembled a large body of data on common important radiating gases. By comparing their band-correlation relations with data over a large range of pressure and temperature, they empirically determined how the required correlation quantities are related to physical variables. This includes the effective bandwidth $A_0(T)$ and the pressure broadening parameter $\beta(T, P_e)$, where P_e is the effective broadening pressure. The u was expressed in terms of the mass path length $X = \rho S$ of the absorbing gas component. The data used are summarized in Table 12-2, which also contains some more recent data. Some of the correlation results are now given.

The effective bandwidth $\bar{A}_l$ can be calculated from the relations in Table 12-3 as derived from the exponential wide-band model. These results are in units of wavenumber, cm^{-1}. The correlations are in terms of mass path length $X = \rho S$ and the pressure broadening parameter β. These quantities account for the two effects of pressure; the number of molecules interacting with the radiation along a path and the broadening of the spectral lines. The quantities b, n, C_1, C_2, and $C_3{}^\dagger$ needed to evaluate these relations are in Table 12-4 for CO_2, CH_4, H_2O, and

---

†These C values should not be confused with the fundamental radiation constants given in Table A-4.

Table 12-2 Available band absorptance correlations for isothermal media

Gas	Bands	Reference	Comments	Type of correlation
CO_2	All important	45, 57	$300 \leqslant T \leqslant 1400$ K	Equivalent bandwidth
	2.7, 4.3, and 15 μm	46, 94[a]	$300 \leqslant T \leqslant 1400$ K $0.1 \leqslant X \leqslant 23{,}000$ g/m^2	Exponential wide band
	9.4 and 10.4 μm	47, 94[a]	$300 \leqslant T \leqslant 1400$ K $0.1 \leqslant X \leqslant 23{,}000$ g/m^2	Exponential wide band
	All important	95–98	$T \sim 300$ K	Equivalent bandwidth
	2.7 μm	53, 56	$T \sim 300$ K $0.1 \leqslant X \leqslant 23{,}000$ g/m^2	Basic spectroscopic theory
H_2O	All important	48, 94[a]	$300 \leqslant T \leqslant 1100$ K $1 \leqslant X \leqslant 38{,}000$ g/m^2	Exponential wide band
	2.7 and 6.3 μm	51	$300 \leqslant T \leqslant 1100$ K $1 \leqslant X \leqslant 21{,}000$ g/m^2	Equivalent line
	All important	95–98	$T \sim 300$ K	Equivalent bandwidth
	2.7 μm	53	$T \sim 300$ K $0.1 \leqslant X \leqslant 23{,}000$ g/m^2	Basic spectroscopic theory
	Pure rotational ($\lambda > 10$ μm)	99	$500 \leqslant T \leqslant 2400$ K $1.5 \leqslant pS \leqslant 30$ atm-cm, $P = 1$ atm	Exponential wide band
CH_4	7.6 and 3.3 μm	46, 88[a]	$300 \leqslant T \leqslant 830$ K $0.1 \leqslant X \leqslant 1200$ g/m^2	Exponential wide band
CO	2.35 and 4.67 μm	49, 94[a]	$300 \leqslant T \leqslant 1800$ K $19 \leqslant X \leqslant 650$ g/m^2	Exponential wide band
	4.7 μm	43, 55, 56	$300 \leqslant T \leqslant 1800$ K $19 \leqslant X \leqslant 650$ g/m^2	Basic spectroscopic theory
HCl		100	Not correlated—presented in terms of spectral emittance	

(*table continues on next page*)

Table 12-2 Available band absorptance correlations for isothermal media (*Continued*)

Gas	Bands	Reference	Comments	Type of correlation
H_2		92	Not correlated— presented in terms of spectral and total emittance	
Atmospheric gases—N_2, O_2, CO_2, O_3, H_2O, CH_3, and nitrogen oxides		3	Discussion of literature up to 1960	
Air	All important contributing bands	15 (Table 11-2)	References to literature up to 1965 for needed data to calculate band absorptance	
NH_3	3.0, 10.5, 2.9, and 6.15 μm	101	$T \sim 300$ K $2 \leqslant X \leqslant 312$ g/m^2	Exponential wide band
NO	5.35 μm	55, 56	$T \sim 300$ K $0.845 \leqslant PS \leqslant 12.56$ atm-cm	Basic spectroscopic theory
N_2O	4.5 μm	56	$T = 303$ K $0.0186 \leqslant PS \leqslant 76.4$ atm-cm	Basic spectroscopic theory
CCl_4 (liquid)	All important	61		Two-parameter narrow band
H_2, N_2, O_2, CH_4, CO, A (liquids only)	All important far-infrared bands in 40 to 500-μm region; for H_2, 16.7 to 500 μm	102	T near normal boiling point at 1 atm, $S = 1.27$ and 2.54 cm, plus 3.25 cm for H_2	Data only of absorption coefficient versus wavenumber
C_2H_2	wavenumber cm^{-1} 3287, 1328, 729	103	Discussion of literature Measurements from $T = 290$ to 600K	Exponential wide band
C_2H_4	4 in infrared	104	Measurements and correlation	Statistical narrow band, wide band

[a]Correlations are given in Tables 12-3 and 12-4.

Table 12-3 Effective bandwidth correlation equations for isothermal gas[a]

Pressure broadening parameter $\beta = \dfrac{C_2^2 P_e}{4C_1 C_3}$	Effective bandwidth $\bar{A}$, η, cm^{-1}	Limits of $\bar{A}$, η, cm^{-1}
$\beta \leqslant 1$	$\bar{A} = C_1 X$	$0 \leqslant \bar{A} \leqslant \beta C_3$
	$\bar{A} = C_2 (X P_e)^{1/2} - \beta C_3$	$\beta C_3 \leqslant \bar{A} \leqslant C_3 (2 - \beta)$
	$\bar{A} = C_3 \left(\ln \dfrac{C_2^2 X P_e}{4 C_3^2} + 2 - \beta \right)$	$C_3 (2 - \beta) \leqslant \bar{A} \leqslant \infty$
$\beta > 1$	$\bar{A} = C_1 X$	$0 \leqslant \bar{A} \leqslant C_3$
	$\bar{A} = C_3 \left(\ln \dfrac{C_1 X}{C_3} + 1 \right)$	$C_3 \leqslant \bar{A} \leqslant \infty$

[a] C_1, C_2, C_3, b, and n are in Table 12-4. X is mass path length ρS, g/m^2. $P_e = [(bp + p_{N_2})/P_0]^n$, where $P_o = 1$ atm, p is partial pressure of absorbing gas, and p_{N_2} is partial pressure of N_2 broadening gas in atmospheres. The l subscript on $\bar{A}$ has been dropped for convenience.

CO, each in a mixture with nitrogen. The C_1 is the integrated band intensity (integral of S_{ij}/δ over the band), and C_3 is a bandwidth parameter. The method of using these band correlations is shown by two examples.

EXAMPLE 12-1 Find the effective bandwidth A of the 9.4-μm band of pure CO_2 at 1 atm and 500 K for a path length S of 0.364 m.

To obtain $\bar{A}$ from the relations in Table 12-3, the constant C_1 must be evaluated. From Table 12-3 at the 9.4-μm CO_2 band, $C_1 = 0.76\varphi_1(T)$, where

$$\varphi_1(T) = \left[1 - \exp \frac{-hc(\eta_3 - \eta_1)}{kT} \right] \left[\exp \left(-\frac{hc\eta_1}{kT} \right) - \tfrac{1}{2} \exp \left(-\frac{2hc\eta_1}{kT} \right) \right]$$

$$\times \left[1 - \exp \left(-\frac{hc\eta_1}{kT} \right) \right]^{-1} \left[1 - \exp \left(-\frac{hc\eta_3}{kT} \right) \right]^{-1}$$

Substitute the values $\eta_1 = 1351$ cm^{-1}, $\eta_2 = 667$ cm^{-1}, $\eta_3 = 2396$ cm^{-1}, $h = 6.626 \times 10^{-34}$ J·s, $k = 1.381 \times 10^{-23}$ J/K, $c = 2.998 \times 10^{10}$ cm/s, and $T = 500$ K. This gives $\varphi_1 = 0.0196$ so that $C_1 = 0.0149$ m^2/(cm·g). Table 12-3 gives the quantity β as $\beta = C_2^2 P_e/4 C_1 C_3$. For the 9.4-$\mu$m CO_2 band, Table 12-4 gives C_2 and C_3 as

$$C_2 = 1.6 \left(\frac{T}{T_0} \right)^{0.5} C_1^{0.5} \qquad \text{and} \qquad C_3 = 12.4 \left(\frac{T}{T_0} \right)^{0.5}$$

Table 12-4 Exponential wide-band model correlation quantities[a]

Gas	Band, μm	Band center η, cm^{-1}	Pressure parameters b	Pressure parameters n	C_1, cm^{-1}/(g$\cdot$m^{-2})	C_2,[b] cm^{-1}/ [(g$\cdot$m^{-2})]$^{1/2}$	C_3,[b] cm^{-1}
CO$_2$[c]	15	667	1.3	0.7	19	$6.9(T/T_0)^{0.5}$	$12.9(T/T_0)^{0.5}$
	10.4	960	1.3	0.8	$0.76\varphi_1(T)$	$1.6(T/T_0)^{0.5}C_1^{0.5}$	$12.4(T/T_0)^{0.5}$
	9.4	1060	1.3	0.8	$0.76\varphi_1(T)$	$1.6(T/T_0)^{0.5}C_1^{0.5}$	$12.4(T/T_0)^{0.5}$
	4.3	2350	1.3	0.8	110	$31(T/T_0)^{0.5}$	$11.5(T/T_0)^{0.5}$
	2.7	3715	1.3	0.65	$4.0\varphi_2(T)$	$8.6\varphi_3(T)$	$24(T/T_0)^{0.5}$
CH$_4$	7.6	1310	1.3	0.8	28	$10(T/T_0)^{0.5}$	$23(T/T_0)^{0.5}$
	3.3	3020	1.3	0.8	46	$14.5(T/T_0)^{0.5}$	$55(T/T_0)^{0.5}$
H$_2$O[d]	6.3	1600	5.0	1.0	41.2	44	$52(T/T_0)^{0.5}$
	2.7	3750	5.0	1.0	23.3	39	$65(T/T_0)^{0.5}$
	1.87	5350	5.0	1.0	$3.0\varphi_{011}(T)$	$6.0C_1^{0.5}$	$46(T/T_0)^{0.5}$
	1.38	7250	5.0	1.0	$2.5\varphi_{101}(T)$	$8.0C_1^{0.5}$	$46(T/T_0)^{0.5}$
CO[e]	4.67	2143	1.1	0.8	20.9	$\varphi_5(T)$	$22(T/T_0)^{0.5}$
	2.35	4260	1.0	0.8	0.14	$0.08\varphi_5(T)$	$22(T/T_0)^{0.5}$

[a] For limits on T and X, see Table 12-2.
[b] T_0 is taken as 100 K for all cases.
[c] For CO$_2$,

$$\varphi_1 = \left\{1 - \exp\left[-\frac{hc}{kT}(\eta_3 - \eta_1)\right]\right\}\left[\exp\left(-\frac{hc\eta_1}{kT}\right) - \frac{1}{2}\exp\left(-\frac{2hc\eta_1}{kT}\right)\right]$$

$$\times \left[1 - \exp\left(-\frac{hc\eta_1}{kT}\right)\right]^{-1}\left[1 - \exp\left(-\frac{hc\eta_3}{kT}\right)\right]^{-1}$$

$$\varphi_2 = \left\{1 - \exp\left[-\frac{hc}{kT}(\eta_1 + \eta_3)\right]\right\}\left[1 - \exp\left(-\frac{hc\eta_1}{kT}\right)\right]^{-1}\left[1 - \exp\left(-\frac{hc\eta_3}{kT}\right)\right]^{-1}$$

$$\varphi_3 = \left[1 + 0.053\left(\frac{T}{T_0}\right)^{3/2}\right]$$

where $\eta_1 = 1351$ cm^{-1}, $\eta_2 = 667$ cm^{-1}, and $\eta_3 = 2396$ cm^{-1}.
[d] For H$_2$O,

$$\varphi_{\nu_1\nu_2\nu_3} = \left[1 - \exp\left(-\frac{hc}{kT}\sum_{i=1}^{3}\nu_i\eta_i\right)\right]\prod_{i=1}^{3}\left[1 - \exp\left(-\frac{hc\eta_i}{kT}\right)\right]^{-1}$$

where $\eta_1 = 3652$ cm^{-1}, $\eta_2 = 1595$ cm^{-1}, and $\eta_3 = 3756$ cm^{-1}
[e] For CO,

$$\varphi_5 = \left[15.15 + 0.22\left(\frac{T}{T_0}\right)^{3/2}\right]\left[1 - \exp\left(-\frac{hc\eta}{kT}\right)\right]$$

where $\eta = 2143$ cm^{-1}.

so that

$$\beta = \frac{(1.6)^2 P_e (T/T_0)^{0.5}}{4 \times 12.4} = 0.0516 P_e \left(\frac{T}{T_0}\right)^{0.5}$$

From Table 12-3, P_e for pure CO_2 at 1 atm is

$$P_e = \left(\frac{1.3 + 0}{1}\right)^{0.8} = 1.234$$

Then, since $T_0 = 100$ K, $\beta = 0.516(1.234)(500/100)^{0.5} = 0.142$. Also $C_3 = 12.4(T/T_0)^{0.5} = 12.4(500/100)^{0.5} = 27.7$ cm^{-1}. Because $\beta \leq 1$, the correlation equations for the specified conditions are the first set in Table 12-3. The mass path length is given by $X = \rho S = 0.364\rho$ g/m^2. The gas density is

$$\rho = \frac{1}{22.42l/\text{g}\cdot\text{mol}} \frac{44 \text{ g}}{\text{g}\cdot\text{mol}} \frac{1000l}{\text{m}^3} \frac{273}{500} = 1.07 \times 10^3 \text{ g/m}^3$$

so that the mass path length is $X = 390$ g/m^2. The choice of correlation equation depends on the limits into which X causes $\bar{A}$ to fall. The first equation in Table 12-3 gives $\bar{A} = C_1 X = 0.0149 \times 390 = 5.8$ cm^{-1}, but this falls well outside the prescribed upper limit of the bandwidth given by $\beta C_3 = 0.142 \times 27.7 = 3.93$ cm^{-1} for the $\beta \leq 1$ part of the correlation. For intermediate X, the second line of Table 12-3 gives $\bar{A} = C_2 (X P_e)^{1/2} - \beta C_3$ or $\bar{A} = 1.6(500/100)^{1/2} (0.0149)^{1/2} \times (481)^{1/2} - 3.93 = 5.6$ cm^{-1} and this lies within the range $\beta C_3 \leq \bar{A} \leq C_3(2 - \beta)$ or $3.93 \leq \bar{A} \leq 51.5$ cm^{-1} for this part of the correlation. The result for $\bar{A}$ compares reasonably well with an experimental value of 5.9 cm^{-1} from [45] for similar conditions.

EXAMPLE 12-2 Determine the emittance in the axial direction for the 9.4-μm band for a thin column of CO_2 gas at 1 atm pressure and 500 K if the column is 0.364 m long.

Using Eq. (12-54b) and integrating only over the 9.4-μm band gives

$$\epsilon' = \frac{\pi \displaystyle\int_{\Delta\eta} i'_{\eta b}(\eta)(1 - e^{-a_\eta S}) \, d\eta}{\sigma T^4} \approx \frac{\pi \bar{A} i'_{\eta b} \, (\eta_{\text{band center}})}{\sigma T^4}$$

where (12-74) has been substituted and $\eta_{\text{band center}}$ is the wavenumber of the band center. For this band Table 12-4 gives $\eta_{\text{band center}} - 1060$ cm^{-1}. Using Eq. (2-14) for $i'_{\eta b}$ and $\bar{A}$ from Example 12-1 give the result

$$\bar{A} i'_{\eta b}(\eta_{\text{band center}}) = 5.6 \left(\frac{2 C_1 \eta^3}{e^{C_2 \eta/T} - 1}\right)_{\text{band center}}$$

$$= 5.6 \frac{10^2}{\text{m}} \frac{2 \times 0.59552 \times 10^{-16} \text{ W} \cdot \text{m}^2 (1060)^3}{e^{1.4388 \times 1060/500} - 1} \frac{10^6}{\text{m}^3} = 3.95 \text{ W/m}^2$$

$$\epsilon' = \frac{3.95\pi \text{ W/m}^2}{(5.6705 \times 10^{-8} \text{ W/m}^2 \cdot \text{K}^4)500^4 \text{ K}^4} = 0.00350$$

The band correlations in Tables 12-3 and 12-4 are convenient to use. They are, however, based somewhat on empirical results. To improve the theoretical foundation of the correlations and to extrapolate beyond the range of experimental results, a correlation was given in [39] based on the quantum-mechanical behavior of a vibration-rotation band. A brief description of the correlation is given here for the four gases in Table 12-4. Information on NO and SO_2 is in [39] (also see [25]). (To be consistent with the notation used here, the η, β, and ν in [39, 25] are changed to β, γ, and η).

The correlation is in terms of three quantities: α, the integrated band intensity (called C_1 in Table 12-4); γ, the line-width parameter;[†] and ω, the bandwidth parameter (called C_3 in Table 12-4). The desired total band absorption $\bar{A}$ is found from the following correlations (the l subscript on $\bar{A}$ has been dropped for convenience):

For $\beta < 1$:
$$\bar{A} = \omega u \qquad\qquad 0 \leqslant u \leqslant \beta$$

$$\bar{A} = \omega(2\sqrt{\beta u} - \beta) \qquad\qquad \beta \leqslant u \leqslant \frac{1}{\beta}$$

$$\bar{A} = \omega(\ln \beta u + 2 - \beta) \qquad\qquad \frac{1}{\beta} \leqslant u \leqslant \infty$$

For $\beta \geqslant 1$:
$$\bar{A} = \omega u \qquad\qquad 0 \leqslant u \leqslant 1$$

$$\bar{A} = \omega(\ln u + 1) \qquad\qquad 1 \leqslant u \leqslant \infty$$

The β is π times the ratio of mean line width to spacing, including pressure broadening; $\beta = \gamma P_e$, where $P_e = [P/P_0 + (p/P_0)(b - 1)]^n$, P is the total pressure of radiating and nonradiating gas, $P_0 = 1$ atm, and p is the partial pressure of the radiating gas ($P_e \rightarrow 1$ as $p \rightarrow 0$ and $P \rightarrow P_0$). The b and n are in Table 12-5 for each gas band. The $u = X\alpha/\omega$, where X is the mass path length of the radiating gas. The ω is found from $\omega = \omega_0(T/T_0)^{1/2}$, where ω_0 is in Table 12-5 and $T_0 = 100$ K. The table also gives the quantities necessary to obtain α and γ from the following:

$$\alpha(T) = \alpha_0 \frac{1 - \exp\left(-\sum_{k=1}^{m} u_k \delta_k\right)}{1 - \exp\left(-\sum_{k=1}^{m} u_{0,k}\delta_k\right)} \frac{\Psi(T)}{\Psi(T_0)}$$

$$\gamma(T) = \gamma_0 \left(\frac{T_0}{T}\right)^{1/2} \frac{\Phi(T)}{\Phi(T_0)}$$

[†]The γ is π times the ratio of mean line width to spacing for a dilute mixture of the radiating component at a total pressure of 1 atm so that there is negligible pressure broadening.

Table 12-5 Exponential wide-band parameters [39, 25]

Gas m, η (cm^{-1}), g	Band, μm	Band center η, cm^{-1}	$\delta_1, \ldots \delta_m$	Pressure parameters ($T_0 = 100$ K) b	n	α_0, cm^{-1}/g·m^{-2}	γ_0	ω_0, cm^{-1}
CO_2 $m = 3$, $\eta_1 = 1351$, $g_1 = 1$	15	667	0, 1, 0	1.3	0.7	19.0	0.06157	12.7
$\eta_2 = 667$, $g_2 = 2$	10.4	960	-1, 0, 1	1.3	0.8	2.47×10^{-9}	0.04017	13.4
$\eta_3 = 2396$, $g_3 = 1$	9.4	1060	0, -2, 1^b	1.3	0.8	2.48×10^{-9b}	0.11888^b	10.1
	4.3	2410^a	0, 0, 1	1.3	0.8	110.0	0.24723	11.2
	2.7	3660	1, 0, 1	1.3	0.65	4.0	0.13341	23.5
	2.0	5200	2, 0, 1	1.3	0.65	0.066	0.39305	34.5
CH_4 $m = 4$, $\eta_1 = 2914$, $g_1 = 1$	7.66	1310	0, 0, 0, 1	1.3	0.8	28.0	0.08698	21.0
$\eta_2 = 1526$, $g_2 = 2$	3.31	3020	0, 0, 1, 0	1.3	0.8	46.0	0.06973	56.0
$\eta_3 = 3020$, $g_3 = 3$	2.37	4220	1, 0, 0, 1	1.3	0.8	2.9	0.35429	60.0
$\eta_4 = 1306$, $g_4 = 3$	1.71	5861	1, 1, 0, 1	1.3	0.8	0.42	0.68598	45.0
H_2O $m = 3$, $\eta_1 = 3652$, $g_1 = 1$	6.3	1600	0, 1, 0	$8.6(T_0/T)^{1/2} + 0.5$	1	41.2	0.09427	56.4
$\eta_2 = 1595$, $g_2 = 1$	2.7	3760^b	0, 2, 0	$8.6(T_0/T)^{1/2} + 0.5$	1	0.19	0.13219	60.0
$\eta_3 = 3756$, $g_3 = 1$			1, 0, 0			2.30		
			0, 0, 1			22.40		
	1.87	5350	0, 1, 1	$8.6(T_0/T)^{1/2} + 0.5$	1	3.0	0.08169	43.1
	1.38	7250	1, 0, 1	$8.6(T_0/T)^{1/2} + 0.5$	1	2.5	0.11628	32.0
CO $m = 1$, $\eta_1 = 2143$, $g = 1$	4.7	2143	1	1.1	0.8	20.9	0.07506	25.5
	2.35	4260	2	1.0	0.8	0.14	0.16758	20.0

^{a}Upper band limit.
^{b}See notes in [39].

where

$$\Psi(T) = \frac{\prod\limits_{k=1}^{m} \sum\limits_{v_k=v_{0,k}}^{\infty} [(v_k + g_k + |\delta_k| - 1)!/(g_k - 1)!v_k!]\,e^{-u_k v_k}}{\prod\limits_{k=1}^{m} \sum\limits_{v_k=0}^{\infty} [(v_k + g_k - 1)!/(g_k - 1)!v_k!]\,e^{-u_k v_k}}$$

$$\Phi(T) = \frac{\left(\prod\limits_{k=1}^{m} \sum\limits_{v_k=v_{0,k}}^{\infty} \left\{[(v_k + g_k + |\delta_k| - 1)!/(g_k - 1)!v_k!]\,e^{-u_k v_k}\right\}^{1/2}\right)^2}{\prod\limits_{k=1}^{m} \sum\limits_{v_k=v_{0,k}}^{\infty} [(v_k + g_k + |\delta_k| - 1)!/(g_k - 1)!v_k!]\,e^{-u_k v_k}}$$

in which

$$u_k = \frac{hc\eta_k}{kT} \qquad u_{0,k} = \frac{hc\eta_k}{kT_0}$$

and $v_{0,k} = 0$ if δ_k is positive and is $|\delta_k|$ if δ_k is negative. Some illustrative numerical examples are in [25].

Greif and co-workers [43, 53–56] developed wide-band correlations from basic spectroscopic theory, finding good agreement with experimental measurements in many cases. With this approach, no arbitrary constants are introduced, and recourse to experimental data is not needed to evaluate the constants. Hottel and Sarofim [57] discuss in detail total absorptance curves of the type discussed in Chap. 13. Such curves are available for a number of gases, and their accuracy has been confirmed by many measurements. The use of total absorptances and effective bandwidths for various engineering problems are discussed in Chap. 13. To determine the effect of the band models on the final radiative transfer results, several band models were applied to two problems [58] involving radiative transfer in gases with internal heat sources and heat transfer. In most instances good agreement was obtained by using the various models, but it is necessary to look at [58] to appreciate the detailed comparisons. A model to apply the exponential wide-band properties in a multidimensional geometry is in [59].

The preceding discussion is concerned with gases with a single radiating and absorbing component. If two gases are present that both absorb energy, their band absorptances may overlap in some spectral regions. In this case, Hottel and Sarofim [57] show that, for two gases a and b in an overlapping band of width $\Delta\eta$,

$$\bar{A}_{a+b} = \Delta\eta \left[1 - \left(1 - \frac{\bar{A}_a}{\Delta\eta}\right)\left(1 - \frac{\bar{A}_b}{\Delta\eta}\right)\right] = \bar{A}_a + \bar{A}_b - \frac{\bar{A}_a\bar{A}_b}{\Delta\eta} \tag{12-87}$$

Thus the simple sum of the two $\bar{A}$ is reduced by the quantity $\bar{A}_a\bar{A}_b/\Delta\eta$ [see also the equations following Eq. (13-81)]. Restriction is to wavenumber intervals over which both $\bar{A}_a$ and $\bar{A}_b$ are applicable average values and in which there is no correlation between the positions of the individual lines of gases a and b.

Many additional complexities are introduced when a gas mixture is considered. For example, the partial pressure p of an absorbing gas in a multicomponent system varies with T and P, the populations of the energy states vary with T, and the overlapping of spectral lines changes with P. It is thus very complex to formulate analytically the dependence of A_l on T, p, and P for a real gas mixture. Useful results must depend heavily on experiment while theory is used as a guide. Some calculations for mixtures are given in [39]. Negrelli et al. [60] applied the exponential wide-band model to a radiating flame layer of nonuniform composition and temperature. The predicted temperature distributions were in excellent agreement with those measured.

A small amount of band-correlation information has been developed for liquids; reference [61] is an example that gives band absorption models for carbon tetrachloride. This information is important in evaluating the radiative contribution during the experimental determination of the thermal conductivity of an absorbing liquid.

12-9 SCATTERING OF ENERGY BY PARTICLES

Scattering is an encounter between a photon and one or more other particles during which the photon does not lose its *entire* energy. It may undergo a change in direction and a loss or a gain of energy. The *scattering coefficient* $\sigma_{s\lambda}$ is the inverse of the mean free path that a photon of wavelength λ will travel before undergoing scattering (strictly true only when $\sigma_{s\lambda}$ does not vary along the path). Scattering can be characterized by four types of events: *elastic* scattering in which the energy (and, therefore, frequency and wavelength) of the photon is unchanged by the scattering, *inelastic* scattering in which the energy is changed, *isotropic* scattering in which scattering into any direction is equally likely, and *anisotropic* scattering in which there is a distribution of scattering directions. Elastic-isotropic scattering is most amenable to analysis. Most scattering of importance in engineering is essentially elastic, and the analysis here will consider only the elastic case.

When incident radiation strikes a particle, some of the incident radiation may be reflected from the particle surface. The remaining radiation will penetrate into the particle, where part can be absorbed. If the particle is not a strong internal absorber, some of this radiation will leave. This may occur after travel along a single path through the particle, or radiation may undergo multiple internal reflections before escaping. When interacting with the particle boundary, the radiation will be refracted and will also have its direction changed by subsequent internal reflections. The redirection by these processes of the energy penetrating into the particle and then escaping is *scattering by refraction*. Additional scattering is by *diffraction*, which produces the interference patterns observed when light passes through an aperture in a screen. Diffraction is the result of slight bending of the paths for radiation passing near the edges of an obstruction.

For a process with elastic scattering and without emission or absorption, there is no exchange of energy between the radiation field and the medium. Therefore, the thermodynamic conditions of the medium are *not affected* by the radiation field. Scattering calculations for this special case are more tractable than when the internal energy of the medium and the radiation field interact by absorption and emission.

In principle, the scattering behavior can be obtained from solutions of the Maxwell electromagnetic equations that govern the radiation field for the medium-particle system. These solutions are documented in detail in books such as [62–64]. The solutions often provide complicated relations; hence in many instances simplifications or limiting cases have been used, as will be described.

A common simplification is to let the scattering particles have the properties of spheres. This is not as restrictive as it might appear, since results for spheres have wider geometric applicability. Consider an array of irregularly shaped particles, the surfaces of which are assumed composed of convex portions (no concave indentations). Because the particles are in a random orientation, an equal portion of surface elements will face each angular direction, which is the same angular distribution of elements as for a spherical particle. The net result is that the angular distribution of scattered radiation viewed at a distance from the particles is the same as that scattered from spherical particles.

The interaction of electromagnetic waves with a particle produces scattering by reflection, refraction, and diffraction, and these effects depend on two parameters containing the particle properties and size. The important optical properties are the components of the complex index of refraction of the particle material relative to those of the surrounding medium. A common situation is when the medium surrounding the particles has a unity refractive index, $n = 1$, and a zero extinction coefficient, $\kappa = 0$. In this instance the surrounding medium does not enter into the optical behavior of the medium-particle system. If dielectric particles are suspended in a dielectric liquid (both with $\kappa \approx 0$), then the relative index $n_{particle}/n_{liquid}$ is used as in Fig. 4-6.

The second important parameter is the particle size relative to the wavelength λ_m of the radiation within the particle. This is usually expressed as a *size parameter* $\xi = \pi D/\lambda_m$ where D is the spherical particle diameter. The limiting behavior for large and small spheres is of interest. For large spheres, ξ greater than about 5, the scattering is chiefly a reflection process and can be calculated from relatively simple geometric reflection relations. There is also diffraction of the radiation passing near the sphere, which will be accounted for separately. For small spheres, ξ less than about 0.3, the approximation of Rayleigh scattering can be used as will be given.

Another type of limiting case is with regard to the optical constants of the particle. For metals n and κ are often large, and simplifications can be introduced as in Sec. 4-6.3. For a dielectric particle ($\kappa = 0$) a limiting case is where $n \approx 1$. In this instance the reflectivity of the particle surface is small. Limiting types of surface conditions are also considered (that is, specular and diffuse surfaces). The particle surface can act diffuse only if its dimensions are large compared with the wavelength of the incident radiation. Starting in the next section, the theory and results are discussed for several useful scattering relations.

If the scattering behavior is known for a single particle, there are some additional considerations in analyzing the radiation characteristics for a particle cloud. As indicated by Eqs. (12-28) and (12-29), it is usually considered that each particle scatters independently, and the cross sections of the individual particles are added. *Independent* scattering applies for conditions when the clearance c between two particles is sufficiently large relative to both the wavelength of the radiation and the particle diameter D. Since the index of refraction of the particle material differs from unity, the wavelength λ_m in the particle differs from that in vacuum, λ. The significant wavelength is that within the particle material. The conditions for independent scatter were tentatively suggested [65, 66] as clearances between particles exceeding $0.4\lambda_m$ and clearance-diameter ratios exceeding 0.4. These limits have been further investigated, and current results are summarized in [67] by Fig. 12-18, where a boundary separates the regions of independent and dependent scattering. The position of the boundary shows that the assumption of independent scattering can be made in many practical situations. The boundary is given as a function of $\pi D/\lambda_m$ and particle volume fraction f_v. There is a lower limit such that there are no dependent scattering effects below $f_v = 0.006$ ($c/D = 4$). For $f_v > 0.006$ the critical c/λ_m value is 0.5, which is in good agreement with experimental data. It is interesting at the upper right corner of the figure that independent scattering can exist in packed beds up to large volume fractions, $f_v \approx 0.7$. The boundary between independent and dependent scattering was examined in [68] for particles with various complex indices of refraction, and somewhat different values were obtained for f_v.

An effect of dependent scattering is that the scattering cross section for particles in a dispersion can decrease as the volume fraction of particles increases. Thus increasing the number of particles does not result in a proportionate increase in the extinction by scattering. It has been observed that an excessive amount of pigment in paint can decrease the hiding ability of the paint.

Another consideration in analyzing scattering effects is *single* versus *multiple scattering*. In multiple scattering, energy scattered from one particle can hit other particles and be scattered additional times. To account for this, the incoming scattering term (12-50) is included in the equation of radiative transfer. Although this complicates the analysis, it must be included in many engineering problems including particle scattering. From Eq. (12-16) the mean penetration distance for scattering alone is $l_{m,s} = 1/\sigma_s$ (if σ_s is not a function of position). A suggested limit for single scattering is $l_{m,s}/S > 10$, where S is a characteristic path length of the region. This corresponds to requiring an optical thickness for scattering $\sigma_s S < 0.1$ in order to neglect multiple scattering interactions [69]. If single scattering prevails, the scattered intensity from an illuminated sample will increase in direct proportion to the particle concentration.

Another effect of increasing particle density is *dependent absorption*. This is analyzed in [70]. For a case with highly absorbing particles, the particle absorption efficiency was found to increase as the spacing between the particles decreased, while the scattering efficiency declined. The total extinction by combined absorption and scattering increased. About 5% increase in absorption efficiency

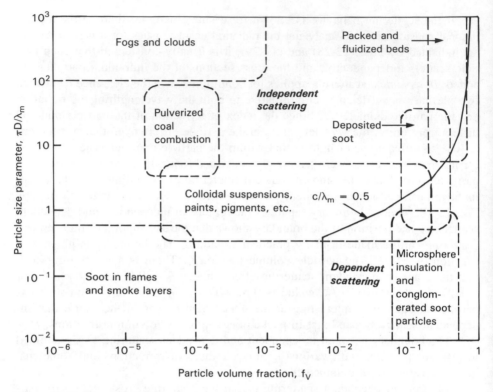

Figure 12-18 Map of independent and dependent scattering regimes as a function of particle size parameter and volume fraction [67].

was found for a particle volume fraction of 0.06. Dependent effects tend to increase the absorption over that predicted by an analysis using independent absorption, while dependent scattering tends to decrease scattering from that predicted by using independent scattering. Enhanced absorption resulting from dependent scattering is analyzed in [71].

12-9.1 Scattering by a Large Specularly Reflecting Sphere

One of the simplest scattering geometries is a large spherical particle ($\xi >\approx 5$) with a *specularly* reflecting surface. The details of the scattering by reflection are shown in Fig. 12-19. The energy intercepted by the projected area of a band of width $R\,d\theta$ and circumference $2\pi R \sin\theta$ on the surface of the sphere is $i'_\lambda\,d\omega_i\,d\lambda\,2\pi R^2 \sin\theta\cos\theta\,d\theta$. The fraction $\rho'_\lambda(\theta)$ is reflected, where $\rho'_\lambda(\theta)$ is the directional specular reflectivity for incidence at angle θ. The energy reflected from the entire sphere is found by integrating over the sphere area:

$$\text{Reflected energy} = i'_\lambda\,d\omega_i\,d\lambda\,\pi R^2 \int_0^{\pi/2} 2\rho'_\lambda(\theta)\sin\theta\,d(\sin\theta)$$

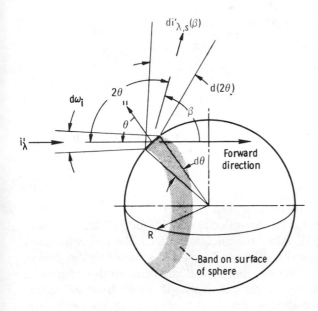

Figure 12-19 Reflection of incident radiation by surface of specular sphere.

From Eq. (3-37) the integral is the hemispherical reflectivity ρ_λ. Hence the energy scattered by reflection from the entire sphere is $i'_\lambda \, d\omega_i \, d\lambda \, \pi R^2 \rho_\lambda$. By using the scattering cross section s_λ for the particle, the scattered energy is $i'_\lambda \, d\omega_i \, d\lambda \, s_\lambda$. Hence, the particle scattering cross section is

$$s_\lambda(R) = \pi R^2 \rho_\lambda \tag{12-88}$$

Thus s_λ is equal to the projected area of the particle times the hemispherical reflectivity. For a cloud of particles that scatter independently, the $s_\lambda(R)$ is used in Eq. (12-28) or (12-29). The latter yields for spheres all of the same size with radius R

$$\sigma_{s\lambda} = \rho_\lambda \pi R^2 N_s \tag{12-89a}$$

In a similar fashion, the absorption coefficient for the cloud is

$$a_\lambda = (1 - \rho_\lambda)\pi R^2 N_s \tag{12-89b}$$

To obtain the phase function, Fig. 12-19 shows that the energy specularly reflected from the band of the sphere at angle θ is reflected into direction 2θ and into a solid angle $d\omega_s = 2\pi \sin 2\theta \, d(2\theta) = 8\pi \sin \theta \cos \theta \, d\theta$. The scattered intensity is the reflected energy per unit incident solid angle, projected area, and wavelength,

$$i'_{\lambda,s} = \frac{i'_\lambda \, d\omega_i \, d\lambda \, \pi R^2 \rho_\lambda}{d\omega_i \, \pi R^2 \, d\lambda} = i'_\lambda \rho_\lambda$$

The energy scattered away by a particle into $d\omega_s$ is $i'_\lambda \, d\omega_i \, d\lambda \, 2\pi R^2 \sin \theta \cos \theta \, d\theta \, \rho'_\lambda(\theta)$. The intensity scattered into direction 2θ is then

$$i'_{\lambda,s}(2\theta) = \frac{i'_\lambda \, d\omega_i \, d\lambda \, 2\pi R^2 \sin\theta \cos\theta \, d\theta \, \rho'_\lambda(\theta)}{d\omega_i \, \pi R^2 \, d\omega_s \, d\lambda} = \frac{i'_\lambda \rho'_\lambda(\theta)}{4\pi} = \frac{i'_{\lambda,s}}{4\pi} \frac{\rho'_\lambda(\theta)}{\rho_\lambda}$$

Inserting this into (12-44) gives

$$\Phi(2\theta) = \frac{\rho'_\lambda(\theta)}{\rho_\lambda} \tag{12-90}$$

The angle 2θ is related to the angle β in Fig. 12-19 by $\beta = \pi - 2\theta$ so that, relative to the forward scattering direction,

$$\Phi(\beta) = \frac{\rho'_\lambda((\pi-\beta)/2)}{\rho_\lambda} \tag{12-91}$$

For unpolarized incident radiation the reflectivity $\rho'_\lambda(\theta)$ for a dielectric sphere can be obtained from Eq. (4-58). Also, the directional-hemispherical reflectivity is equal to unity minus the emissivity values in Fig. 4-6. As shown by this figure, the $\rho'_\lambda(\theta)$ for normal incidence is usually quite small compared to that at grazing angles ($\rho'_\lambda \to 1$ at $\theta = 90°$). Consequently, the forward scatter from the sphere (at $\beta = 0$) is unity, and the backward scatter (at $\beta = \pi$) is small. By use of Fig. 4-6 the quantity $\rho_\lambda\Phi(\beta)$ can be plotted as shown in Fig. 12-20 for various refractive index ratios n_2/n_1 of particle to surroundings (as explained in connection with Fig. 4-6).

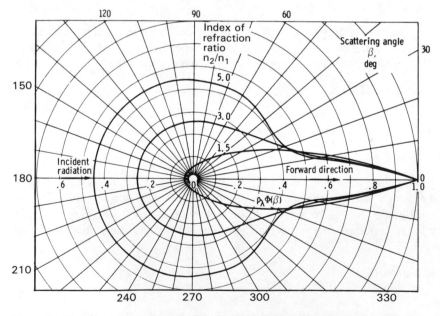

Figure 12-20 Scattering diagram for *specular* reflecting dielectric sphere that is large compared with radiation wavelength within the sphere (n_2 for particle, n_1 for surrounding medium).

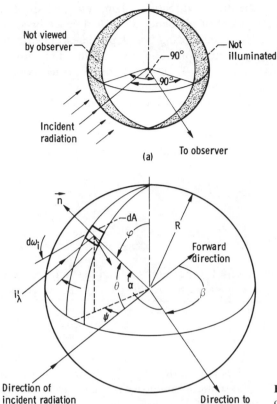

Figure 12-21 Scattering by reflection from diffuse sphere. (a) Illuminated region visible to observer; (b) geometry on sphere.

12-9.2 Reflection from a Large Diffuse Sphere

For a large specularly reflecting sphere, the energy scattered into each direction results from reflection at a specific single location on the sphere. If the sphere is *diffuse*, each surface element that intercepts incident radiation will reflect into the entire 2π solid angle above that element. Thus the radiation scattered into a specified direction will arise from the entire region of the sphere that receives radiation and is also visible from the specified direction. This is illustrated by Fig. 12-21a. The shaded portion of the sphere will not contribute radiation in the direction of the observer because it either does not receive radiation or is hidden from the direction of observation.

Consider the sphere of radius R in Fig. 12-21b. A typical surface-area element dA is located at angles ψ and φ. The observer is at angle β from the forward direction. The normal to dA is at θ and α relative to the directions of incidence and observation. The incident spectral energy flux within the incident solid angle $d\omega_i$ is $i'_\lambda\, d\omega_i\, d\lambda$. The projected area of dA normal to the incident direction is $dA \cos \theta$, so the energy received by dA is $i'_\lambda\, d\omega_i\, d\lambda \cos \theta$. The amount that is reflected is $\rho_\lambda' i_\lambda' \, d\omega_i \, d\lambda \, dA \cos \theta$, where ρ_λ' is the diffuse directional-hemispherical spectral reflectivity. The ρ_λ' is assumed independent of incidence angle and hence is equal to the hemispherical reflectivity ρ_λ. Using the cosine-law dependence for diffuse reflection gives the reflected energy per unit solid angle $d\omega_s$ in the direction of the observer as $\rho_\lambda i_\lambda' \, d\omega_i \, d\lambda \, dA \cos \theta \cos \alpha / \pi$. To integrate the reflected contributions that are received by the observer from all elements on the sphere surface, the dA, $\cos \theta$, and $\cos \alpha$ are expressed in terms of the spherical coordinates R, ψ, and φ, which gives $dA = R^2 \sin \varphi \, d\varphi \, d\psi$, $\cos \theta = \sin \varphi \cos \psi$, and $\cos \alpha = \sin \varphi \cos(\psi + \pi - \beta)$. Then the energy scattered by reflection into the β direction per unit solid angle $d\omega_s$ about that direction is found by integrating,

$$\frac{\rho_\lambda i_\lambda' \, d\omega_i \, d\lambda \, R^2}{\pi} \int_{\varphi=0}^{\pi} \int_{\psi=-\pi/2}^{\beta-(\pi/2)} \sin^3 \varphi \cos \psi \cos(\psi + \pi - \beta) \, d\psi \, d\varphi$$

$$= \frac{\rho_\lambda i_\lambda' \, d\omega_i \, d\lambda \, R^2}{\pi} \frac{2}{3}(\sin \beta - \beta \cos \beta)$$

The energy per unit $d\lambda$ scattered in direction β per unit $d\omega_s$ and per unit area and solid angle of the incident radiation is obtained by dividing the scattered energy by $\pi R^2 \, d\omega_i \, d\lambda$, giving

$$i_{\lambda,s}'(\beta) = \frac{\rho_\lambda i_\lambda'}{\pi^2} \tfrac{2}{3}(\sin \beta - \beta \cos \beta)$$

The entire amount of incident intensity that is scattered is $i_{\lambda,s}' = \rho_\lambda i_\lambda'$. Then from (12-44), the directional magnitude of the scattered intensity is related to the entire scattered intensity times the phase function by

$$\frac{\rho_\lambda i_\lambda'}{\pi^2} \tfrac{2}{3}(\sin \beta - \beta \cos \beta) = \rho_\lambda i_\lambda' \frac{\Phi(\beta)}{4\pi}$$

so the phase function for a *diffuse* sphere is

$$\Phi(\beta) = \frac{8}{3\pi}(\sin \beta - \beta \cos \beta) \tag{12-92}$$

The $\Phi(\beta)$ from (12-92) is plotted in Fig. 12-22. The largest scattering is for $\beta = 180°$, that is, toward an observer back in the same direction as the origin of the incident radiation. In this instance, the entire illuminated surface of the sphere is observed.

12-9.3 Large Dielectric Sphere with Refractive Index Near Unity

For a large nonattenuating dielectric ($\kappa = 0$) sphere with refractive index $n \approx 1$, the reflectivity of the particle surface approaches zero. The incident radiation can thus pass with unchanged amplitude into the sphere, and there is no scattering by reflection. With the extinction coefficient zero, the radiation will pass out of the

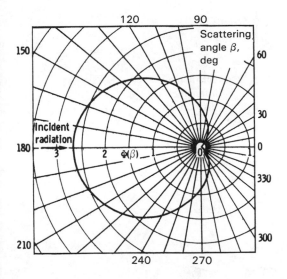

Figure 12-22 Scattering phase function for diffuse reflecting sphere, large compared with wavelength of incident radiation and with constant reflectivity.

sphere with unchanged amplitude. However, the velocity $c = c_0/n$ within the sphere is slightly less than that outside, so radiation passing through different portions of the sphere having different thicknesses will have different phase lags. The resulting interference of the waves passing out of the sphere yields a scattering cross section

$$s_\lambda = \frac{\pi D^2}{4} \left[2 - \frac{4}{W} \sin W + \frac{4}{W^2} (1 - \cos W) \right] \tag{12-93}$$

where $W - 2(\pi D/\lambda)(n - 1)$. Additional information for this situation is in [62].

12-9.4 Diffraction from a Large Sphere

For large spheres there is diffraction of the radiation passing in the vicinity of the particle. The effects of diffraction and reflection must be added to obtain the total scattering behavior. Fortunately diffraction is predominantly in the forward scattering direction. This means that diffraction can be included in the radiative transfer as if it were transmitted past the particle without interacting with the particle; hence, diffraction *can often be neglected* in considering energy exchange within a scattering medium.

The most familiar form of diffraction is when light passes through a small hole or slit. As shown in Fig. 12-23a, the result is a diffraction pattern of alternate illuminated and dark rings or strips. If a spherical particle is in the path of incident radiation, Babinet's principle states that the diffracted intensities are the same as for a hole. This is a consequence of the fact that a hole and a particle produce complementary disturbances in the amplitude of an incident electromagnetic wave. The energy diffracted by a spherical particle is thus the same as that diffracted by a hole of the same diameter. As a consequence, the entire projected area of a

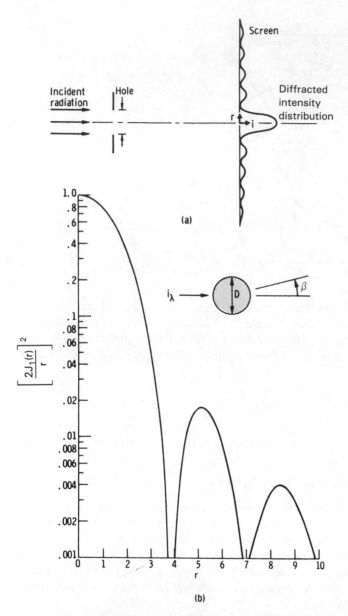

Figure 12-23 Diffraction by hole or large spherical particle. (*a*) Diffraction of radiation by hole; (*b*) phase function for diffraction from large sphere.

sphere takes part in the diffraction process, and the scattering cross section for diffraction is equal to the projected area $\pi D^2/4$. Since diffraction and reflection occur simultaneously, the total scattering cross section can approach $2\pi D^2/4$ when a sphere is highly reflecting.

The distribution of energy diffracted by a large sphere onto a vertical screen, as shown in Fig. 12-23*a*, is given by the relation $[2J_1(r)/r]^2$, where J_1 is a Bessel function of the first kind of order one (see pages 107–108 of [62]). This is plotted in Fig. 12-23b. The function is normalized such that $2 \int_{r=0}^{\infty} \{[J_1(r)]^2/r\} \, dr = 1$. The integration showing this is clarified on page 398 of [72]. The amount of energy diffracted depends on the particle size. From diffraction theory the distance r is related to the angle β by $r = \xi \tan \beta$. Using this transformation, and in view of the above normalization, the phase function in terms of β is

$$\Phi(\beta) = \xi^2 \left[\frac{2J_1 (\xi \tan \beta)}{\xi \tan \beta} \right]^2 \frac{1}{\cos^3 \beta} \qquad (12\text{-}94)$$

For particles with large ξ the diffracted radiation lies within a narrow angular region in the forward scattering direction. For smaller particles where ξ is of order unity, the theory leading to (12-94) is invalid and general Mie scattering theory must be applied.

12-9.5 Rayleigh Scattering by Small Spheres

A limiting case is when the diameter of the scattering particles is considerably smaller than the wavelength of the radiation within the particle. Scattering from such particles is termed *Rayleigh* scattering after Lord Rayleigh, who investigated this situation. Rayleigh scattering is important in the atmosphere, where scattering is by gas molecules. Originally, Rayleigh derived the functional dependence by dimensional analysis and found that the scattered energy is proportional to $G^2(\bar{n})V^2/\lambda_m^4$, where V is the volume of a particle and $G(\bar{n})$ is a function of the complex refractive indices of the scattering material and surrounding medium. The important result is that for Rayleigh scattering, *the scattered energy in any direction is proportional to the inverse fourth power of the wavelength of the radiation*. This inverse dependence shows that when the incident radiation covers a wavelength spectrum, the shorter-wavelength radiation will be Rayleigh scattered with a strong preference.

Rayleigh scattering by the molecules of the atmosphere accounts for the background of the sky being blue and for the sun becoming red in appearance at sunset. The blue portion of the incident sunlight is at the short-wavelength end of the visible spectrum. Hence it undergoes strong Rayleigh scattering into all directions, giving the sky its overall blue background. Without molecular scattering, the sky would appear black except for the direct view of the sun. As the sun is setting, the path length for direct radiation through the atmosphere becomes much longer than during the middle of the day and the sunlight becomes less intense. In traversing this longer path, proportionately more of the short-wavelength part of the visible spectrum is scattered away from the direct path of the sun's rays. As a result, at sunset the sun takes on a red color as the longer-wavelength red rays are able to penetrate the atmosphere with less attenuation than the rest of the visible spectrum. If many dust particles are present, a deep red sunset may be seen.

If particles with a very limited range of sizes are present in the atmosphere, unusual scattering effects may be observed. Following the eruption of Krakatoa in 1883, the occurrence of blue and green suns and moons was noted over a period of many years. This effect was attributed to particles in the atmosphere of such a size range as to scatter only the red portion of the visible spectrum. On September 26, 1950, a blue sun and moon were observed in Europe, a phenomenon believed due to finely dispersed smoke particles of uniform size carried from forest fires burning in Canada. A green moon was observed following the 1982 eruption of El Chichon in Mexico.

Scattering cross sections for Rayleigh scattering From electromagnetic theory quantitative information has been obtained for the particle scattering cross section and the angular distribution of the scattered intensity. An approximate size limit for Rayleigh scattering [64] is that the ratio of particle radius to the wavelength λ_m within the particle be less than 0.05; hence the size parameter $\xi_m = \pi D/\lambda_m <$ ~0.3. This is approximate, and more precise limits are in [64, 73]. The limits depend on both ξ_m and the refractive indices of the particle and surrounding medium. Consider first the scattering from small nonabsorbing ($\kappa = 0$) spherical particles in a nonabsorbing ($\kappa = 0$) medium. The particle material is designated by subscript 1 and the medium by subscript 2. Then $\kappa_1 = 0$ and $\kappa_2 = 0$, so that $\bar{n}_1 = n_1$ and $\bar{n}_2 = n_2$. Let n be the relative refractive index n_1/n_2. The Rayleigh scattering cross section for unpolarized incident radiation is

$$s_\lambda = \frac{24\pi^3 V^2}{\lambda^4}\left(\frac{n^2 - 1}{n^2 + 2}\right)^2 = \frac{8}{3}\frac{\pi D^2}{4}\xi^4\left(\frac{n^2 - 1}{n^2 + 2}\right)^2 \qquad n = \frac{n_1}{n_2} \qquad (12\text{-}95)$$

where $\xi = \pi D/\lambda$ and λ is the wavelength in the medium surrounding the particle. The scattering efficiency factor is the cross section divided by the actual geometric cross section so that

$$E_{s\lambda} = \frac{s_\lambda}{\pi D^2/4} = \frac{8}{3}\xi^4\left(\frac{n^2 - 1}{n^2 + 2}\right)^2$$

Often Rayleigh-scattering cross sections are given in terms of the polarizability α_p of the particles. This relates the dipole moment per unit volume (defined as the polarization) produced in the particle material to the external electromagnetic field. For the case under discussion here, the polarizability is

$$\alpha_p = \frac{3}{4\pi} V \frac{n^2 - 1}{n^2 + 2} \qquad (12\text{-}96)$$

so that (12-95) can be written

$$s_\lambda = \frac{2^7\pi^5\alpha_p^2}{3\lambda^4} \qquad (12\text{-}97)$$

In this more general form, cross sections for various particles that follow the Rayleigh scattering relations can be introduced by substituting the requisite form

Table 12-6 Polarizability for various scattering conditions

Scattering particles	Restrictions	Polarizability α_p, length[3]
Electrons (Thomson scattering)	Energy of incident photon is small, $h\nu \ll m_e c_0^2$	$\dfrac{e^2}{m_e c_0^2}\left(\dfrac{\lambda}{2\pi}\right)^{2\dagger}$
Small dielectric particle (Rayleigh)	Particle diameter is small compared with wavelength in medium and in particle	$\dfrac{n^2-1}{n^2+2}\left(\dfrac{D}{2}\right)^3$
Medium containing small particles (Lorentz-Lorenz)	Spacing between particles is small compared with wavelength ($<\lambda$). Particle diameter is very small ($D \ll \lambda$) compared with λ in both medium and particle. Spacing between particles $>D$.	$\dfrac{3}{4\pi N}\left\|\dfrac{\bar{n}^2-1}{\bar{n}^2+2}\right\|^{\ddagger}$
Medium containing small particles	Spacing between particles is large ($\gg\lambda$). Particle diameter is very small ($D \ll \lambda$).	$\dfrac{\|\bar{n}^2-1\|^{\ddagger}}{4\pi N}$
Medium containing small particles	Spacing between particles is large ($\gg\lambda$). Particle diameter is very small ($D \ll \lambda$). The $\bar{n}$ is close to 1.	$\dfrac{\|\bar{n}-1\|^{\ddagger}}{2\pi N}$

$\dagger e^2/m_e c_0^2 = $ classical electron radius, 2.818×10^{-13} cm.

$\ddagger$In this instance the $\bar{n}$ is the refractive index of the particulate medium as a whole and depends on the particle number density and volume; see Ref. [62].

tor α_p into (12-97). Table 12-6 gives some quantities for individual particles and for particles in a nonparticipating medium.

The actual scattering cross section for particles in a medium may vary with λ in a manner somewhat different from a $1/\lambda^4$ dependence. In air at standard temperature and pressure, the restrictions are satisfied and Rayleigh scattering from the gas molecules governs. However, the variation of refractive index with wavelength causes the variation of the scattering cross section to depart somewhat from the $1/\lambda^4$ dependence. This is shown in Fig. 12-24, where the actual scattering dependence on wavelength is compared with $1/\lambda^4$.

When the particles are of a conducting material with complex index of refraction $\bar{n}_1 = n_1 - i\kappa_1$, and the surrounding medium is nonabsorbing so that $\bar{n}_2 = n_2$, the scattering cross-section and efficiency factor have the following more general forms:

$$s_\lambda = \frac{24\pi^3 V^2}{\lambda^4}\left|\frac{\bar{n}^2-1}{\bar{n}^2+2}\right|^2 = \frac{8}{3}\frac{\pi D^2}{4}\xi^4\left|\frac{\bar{n}^2-1}{\bar{n}^2+2}\right|^2 \qquad \bar{n} = \frac{n_1-i\kappa_1}{n_2} \qquad (12\text{-}98a)$$

$$E_{s\lambda} = \frac{8}{3}\xi^4\left|\frac{\bar{n}^2-1}{\bar{n}^2+2}\right|^2 \qquad (12\text{-}98b)$$

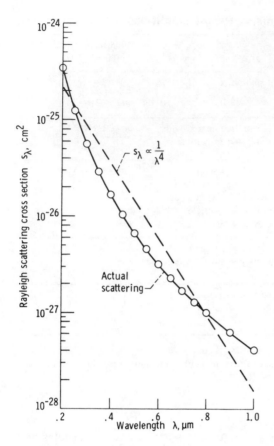

Figure 12-24 Comparison of actual Rayleigh scattering cross section for air at standard temperature and pressure with $1/\lambda^4$ variation. From [3].

Inserting $\bar{n} = n - i\kappa$ and taking the square of the absolute value as indicated give

$$s_\lambda = \frac{24\pi^3 V^2}{\lambda^4} \frac{[(n^2 - \kappa^2 - 1)(n^2 - \kappa^2 + 2) + 4n^2\kappa^2]^2 + 36n^2\kappa^2}{[(n^2 - \kappa^2 + 2)^2 + 4n^2\kappa^2]^2}$$

$$n = (n_1/n_2) \qquad \kappa = (\kappa_1/\kappa_2)$$

(12-99)

For $\kappa = 0$ this reduces to (12-95). The quantity $s_\lambda/(24\pi^3 V^2/\lambda^4)$ from (12-99) is given in Fig. 12-25 for various n and κ.

For use later in connection with soot radiation, the *absorption* efficiency factor is included here for a small absorbing sphere ($\kappa_1 \neq 0$) in a nonabsorbing medium ($\kappa_2 = 0$),

$$E_{a\lambda} = -4\xi \, \mathrm{Im}\!\left(\frac{\bar{n}^2 - 1}{\bar{n}^2 + 2}\right) \qquad \bar{n} = \frac{n_1}{n_2} - i\frac{\kappa_1}{n_2} = n - i\kappa$$

(12-100)

$$E_{a\lambda} = 24\xi \frac{n\kappa}{(n^2 - \kappa^2 + 2)^2 + 4n^2\kappa^2}$$

where $\xi = \pi D/\lambda$ and λ is the wavelength in the medium surrounding the sphere.

Phase function for Rayleigh scattering For incident unpolarized radiation, electromagnetic theory gives, for Rayleigh scattering,

$$\Phi(\beta,\varphi) = \tfrac{3}{4}(1 + \cos^2 \beta) \tag{12-101}$$

This is independent of the circumferential angle φ.

The phase functions for Rayleigh scattering and for *isotropic* scattering (a circle of unit radius) are plotted in Fig. 12-26. For Rayleigh scattering, the scattered energy is directed preferentially along the forward direction of the incident radiation and strongly back toward the radiation source.

12-9.6 Mie Scattering Theory

When the scattering particles are not large, as in Secs. 12-9.1 to 12-9.4, and are not small enough to fall into the range adequately described by Rayleigh scattering, recourse must be taken to more complicated treatments. This is required for the approximate range $0.3 < \xi < 5$. Gustav Mie [74] originally applied electromagnetic theory to derive the properties of the electromagnetic field when a plane spectral wave is incident on a spherical surface across which the optical properties n and κ change abruptly. Energy absorption by the medium, absorption by the scattering particles, or both can be accounted for, and the results apply over the entire range of particle diameters. Strong polarization effects can be present. In certain cases, the phase function becomes very complicated, as illustrated by Fig. 12-27.

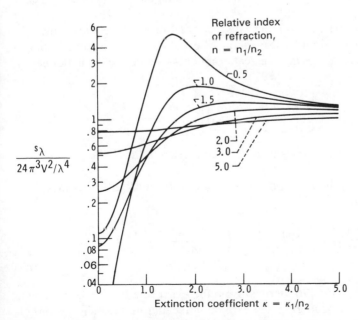

Figure 12-25 Rayleigh-scattering cross section as a function of relative index of refraction and extinction coefficient.

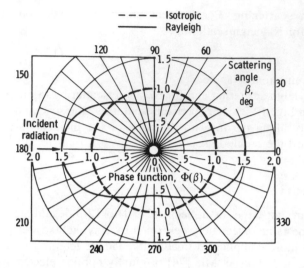

Figure 12-26 Phase functions for Rayleigh and isotropic scattering.

Van de Hulst [62] gives a detailed treatment of the Mie theory. Limiting cases for small and large particles are examined, and working formulas are presented for all size ranges. Cross sections and phase functions are discussed for dielectric and metallic particles of various shapes, including spheres and cylinders. Additional references on absorbing particles and using detailed Mie scattering theory are [63, 64, 75].

One of the simpler results from the Mie theory is for small spheres somewhat larger than the Rayleigh limit. The general Mie solutions are expanded into a power series in terms of ξ, giving the scattering cross section as [$\bar{n}$ is defined in (12-98a)],

$$s_\lambda = \frac{8}{3}\frac{\pi D^2}{4}\xi^4 \left|\frac{\bar{n}^2-1}{\bar{n}^2+2}\left(1+\frac{3}{5}\frac{\bar{n}^2-2}{\bar{n}^2+2}\xi^2+\cdots\right)\right|^2 \tag{12-102}$$

The second term in the parentheses is the first correction to the Rayleigh scattering relation in (12-98).

In a similar fashion the *absorption* efficiency in Eq. (12-100) becomes, with the second-order term

$$E_{a\lambda} = -4\xi\,\mathrm{Im}\left\{\left(\frac{\bar{n}^2-1}{\bar{n}^2+2}\right)\left[1+\frac{\xi^2}{15}\left(\frac{\bar{n}^2-1}{\bar{n}^2+2}\right)\left(\frac{\bar{n}^4+27\bar{n}^2+38}{2\bar{n}^2+3}\right)+\cdots\right]\right\} \tag{12-103}$$

For small spheres, Penndorf [76] derived from the exact Mie theory the following series expansions for the scattering and extinction (scattering plus absorption) efficiencies:

$$E_{s\lambda} = \frac{8\xi^4}{3z_1^2}\{[(n^2 + \kappa^2)^2 + n^2 - \kappa^2 - 2]^2 + 36n^2\kappa^2\}$$

$$\times \left\{1 + \frac{6}{5z_1}[(n^2 + \kappa^2)^2 - 4]\xi^2 - \frac{8n\kappa}{z_1}\xi^3\right\}, \tag{12-104}$$

$$E_{s\lambda} + E_{a\lambda} = \frac{24n\kappa}{z_1}\xi + \left\{\frac{4}{15} + \frac{20}{3z_2} + \frac{4.8}{z_1^2}[7(n^2 + \kappa^2)^2 + 4(n^2 - \kappa^2 - 5)]\right\}n\kappa\xi^3$$

$$+ \frac{8}{3z_1^2}\{[(n^2 + \kappa^2)^2 + n^2 - \kappa^2 - 2]^2 - 36n^2\kappa^2\}\xi^4, \tag{12-105}$$

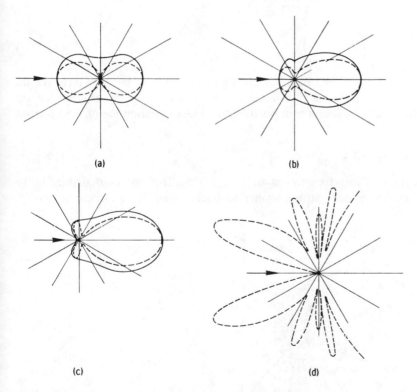

Figure 12-27 Phase functions for Mie scattering from metallic and dielectric spheres (on arbitrary scales). (From [62] and [105]). (a) $\pi D/\lambda \to 0$, metallic sphere, $\bar{n} = 0.57 - 4.29i$; (b) $\pi D/\lambda = 9.15$, metallic sphere, $\bar{n} = 0.57 - 4.29i$; (c) $\pi D/\lambda = 10.3$, metallic sphere, $\bar{n} = 0.57 - 4.29i$; (d) $\pi D/\lambda = 8$, dielectric sphere, $n = 1.25$.

where

$$z_1 = (n^2 + \kappa^2)^2 + 4(n^2 - \kappa^2) + 4, \qquad z_2 = 4(n^2 + \kappa^2)^2 + 12(n^2 - \kappa^2) + 9$$

The lowest-order terms in (12-104) and (12-105) are the same as (12-98b) and (12-100). Equations (12-104) and (12-105) can be used up to $\xi = 0.8$ in the range $n = 1.25$ to 1.75 and $\kappa < 1$. For larger particles the exact Mie formulas are required. Some information on computational codes for Mie scattering is in [77]. The use of the Penndorf correction to extend the Rayleigh limit is further developed in [78].

For small spheres the limit where n becomes very large can also be considered. The scattering particles then form an idealized cloud of very highly reflecting dielectric particles. The result for this case cannot be obtained by letting $\bar{n}$ in (12-102) approach $\infty + i0$. As n becomes large, the small part of the incident radiation that does penetrate the particle becomes almost totally internally reflected. This creates standing waves within the particle, which provide resonance peaks in the scattering. The expansion used to obtain Eq. (12-102) does not account for this behavior. In the limit for $n \to \infty$ the scattering cross section for small spheres is

$$s_\lambda = \frac{\pi D^2}{4} \left(\frac{10}{3} \xi^4 + \frac{4}{5} \xi^6 + \cdots \right) \tag{12-106}$$

If, in addition to $n \to \infty$, the particles are so small that only the first term within the parentheses of (12-106) is significant (the use of only the first term is accurate within 2% for $\xi < 0.2$), then for unpolarized incident radiation the phase function is given by

$$\Phi(\beta) = \tfrac{3}{5}[(1 - \tfrac{1}{2}\cos\beta)^2 + (\cos\beta - \tfrac{1}{2})^2] \tag{12-107}$$

A polar diagram of this function is in Fig. 12-28. This shows that the highly reflecting particles produce strong scattering back toward the source.

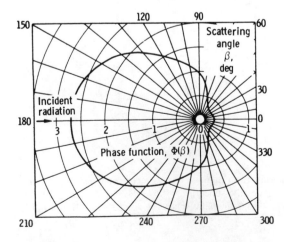

Figure 12-28 Phase function for scattering of unpolarized incident radiation from small nonabsorbing sphere with $n \to \infty$.

12-9.7 Dependent Anisotropic Scattering from Fibers

In the production of structures by the winding of closely spaced fibers in a resin matrix, it is possible to initiate the resin polymerization by imposing a radiation field from an external source. In such a case, the fibers are very closely spaced, and the effects on the scattering phase function of both dependent scattering and dependence of scattering on incident angle must be considered. In [79], Maxwell's equations are solved for two cases: (1) normal incidence on a layer of evenly spaced coplanar fibers and (2) randomly oriented fibers with a prescribed distribution of fiber spacing. For a single cylinder at normal incidence with $\xi \leq 1$, the phase function in a plane normal to the cylinder is given by

$$\Phi(\theta) = \frac{|\bar{n}^2 + 1|^2 + 4\cos^2\theta}{|\bar{n}^2 + 1|^2 + 2} \tag{12-108a}$$

which for $\bar{n} \approx 1$ reduces to

$$\Phi(\theta) = \frac{2}{3}(1 + \cos^2\theta) \tag{12-108b}$$

For *evenly spaced fibers of diameter D with spacing a between fiber edges*, the scattering efficiency E_N relative to that for a single fiber E_1 is found for small size parameter $\xi = \pi D/\lambda$ to be

$$\frac{E_{\lambda,N}}{E_{\lambda,1}} = 1 + \frac{1}{N}\sum_{k=1}^{N}\sum_{\substack{j=1 \\ j \neq k}}^{N}\left\{J_0[2\xi A(k-j)] + \frac{1}{3}J_2[2\xi A(k-j)]\right\} \tag{12-109}$$

where $A = 1 + (a/D)$. For *randomly oriented fibers* with a solid volume fraction of f_v, two cases are considered. For *small f_v*,

$$\frac{E_{\lambda,N}}{E_{\lambda,1}} = 1 - 4f_v\left\{J_0^2(2\xi) + J_1^2(2\xi) + \frac{1}{3}[J_2^2(2\xi) - J_1(2\xi)J_3(2\xi)]\right\} \tag{12-110}$$

and for densely packed systems (*large f_v*)

$$\frac{E_{\lambda,N}}{E_{\lambda,1}} = 1 - 4f_v\left\{J_0^2(2\xi) + J_1^2(2\xi) + \frac{1}{3}[J_2^2(2\xi) - J_1(2\xi)J_3(2\xi)]\right\}$$
$$+ 8\left\{0.4204 f_v\left[J_0^2(2\xi) + \frac{1}{3}J_2^2(2\xi)\right]\right\} \tag{12-111}$$

Comparison of the analysis with experimental measurement for transmission (forward scattering less absorption) for silica fibers showed excellent agreement. The rms error was ± 0.03 at wavelengths greater than 8 μm for a fiber diameter of 9 μm ($\xi \approx 1$), and within ± 0.22 at shorter wavelengths ($\xi > 1$). Local maxima and minima in the spectral transmittance curves were well predicted by the analysis.

12-9.8 Approximate Anisotropic Scattering Phase Functions

Because the exact Mie phase function is complicated in form and difficult to use in radiative transfer analysis, a number of approximate phase functions have been proposed. These simpler forms retain the anisotropic characteristics of the exact phase function, while being easier to use. The cosine of the scattering angle $\mu_0 = \cos \theta_0$ can be expressed as

$$\mu_0 = \mu\mu' + (1 - \mu^2)^{1/2}(1 - \mu'^2)^{1/2} \cos (\varphi - \varphi') \tag{12-112}$$

where $\mu = \cos \theta$ (θ, φ is the angle between the direction of the incident intensity and a coordinate) and $\mu' = \cos \theta'$ (θ', φ' is the angle of scatter with respect to the same coordinate.) The exact Mie scattering phase for single scattering into the direction θ_0 relative to the direction of the incident intensity can be expressed by an expansion in terms of Legendre polynomials of the first kind, $P_m(\mu_0)$, by

$$\Phi(\mu_0) = 1 + \sum_{m=1}^{\infty} a_m P_m(\mu_0) \tag{12-113}$$

where [80]

$$P_1(\mu_0) = \mu_0, \qquad P_2(\mu_0) = \frac{1}{2}(3\mu_0^2 - 1),$$

$$P_3(\mu_0) = \frac{1}{2}(5\mu_0^3 - 3\mu_0), \ldots \tag{12-114}$$

The a_m are constants to be determined by fitting to the Mie results.

Forward-scattering phase function For a highly forward-scattering medium, the phase function can be approximated [81, 82] by

$$\Phi_\delta(\theta, \varphi, \theta', \varphi') = 4\pi\delta(\cos \theta - \cos \theta')\delta(\varphi - \varphi') \tag{12-115}$$

$$= 4\pi\delta(\mu - \mu')\delta(\varphi - \varphi')$$

where δ is the Dirac delta function, equal to unity when the argument is zero ($\mu = \mu'$ and $\varphi = \varphi'$ for forward scattering) and equal to zero otherwise.

The linear-anisotropic phase function Some researchers [83] have used a phase function of the form

$$\Phi_{la}(\theta, \theta') = 1 + g\mu_0 \tag{12-116}$$

where g is a dimensionless asymmetry factor that can vary between $\pm\infty$, and $g = 0$ produces the isotropic phase function. This phase function is rotationally symmetric around the direction of travel of the incident intensity and contains the first two terms of the general phase function in terms of Legendre polynomials (12-113). Integration over all solid angles shows that this phase function is normalized for any value of g.

The delta-Eddington phase function The delta-Eddington approximation [85] uses a two-term Legendre polynomial expansion [from Eqs. (12-113) and (12-114)] of the actual phase function plus a Dirac delta term to account for forward scattering. For the cosine of the angle of scatter between the incident and scattered directions μ_0, the phase function becomes

$$\Phi_{\delta E}(\mu_0) = 2f\delta(1 - \mu_0) + (1 - f)(1 + 3g'\mu_0) \tag{12-117}$$

where $f = (1/2) \int_{-1}^{1} \Phi(\mu_0)P_2(\mu_0) \, d\mu_0$. P_2 is the second term of the Legendre polynomial, and $g' = (g - f)/(1 - f)$ where $g = (1/2) \int_{-1}^{1} \mu_0\Phi(\mu_0) \, d\mu_0$. When f and g are calculated by this procedure, $\Phi_{\delta E}(\mu_0)$ can have negative values for certain ranges of μ_0. This has no physical meaning, and in such a case f and g can be chosen by methods discussed in [82a]. For isotropic scattering, f and g are zero, and hence g' is zero.

The Henyey-Greenstein phase function Henyey and Greenstein [84] proposed the phase function

$$\Phi_{HG}(\mu_0) = \frac{(1 - g^2)}{(1 + g^2 - 2g\mu_0)^{3/2}} \tag{12-118}$$

where g is a dimensionless asymmetry factor that can vary from 0 (isotropic scattering) to 1 (for forward scattering). If g is negative, a backscattering phase function is produced. This approximate phase function has been shown to produce good agreement with complete Mie scattering calculations when used in the discrete ordinates [85] and P-N methods.

The effect of phase function on radiative flux was investigated by Love et al. [86, 87], who studied radiation at plane and cylindrical boundaries with given reflectivities adjacent to volumes of absorbing, emitting, and scattering media. Some experimentally determined values of the scattering phase functions for glass beads and aluminum, carbon, iron, and silica particles were used for comparison of their effect in various energy exchange cases. In some cases there was not much difference in the energy transfer results using these experimental phase functions and the results using either Rayleigh or isotropic phase functions. Other examples did show significant anisotropic effects. In Fig. 12-29, results from [86] are shown for the energy flux, as a result of emitted energy from a black disk, that is scattered back to the base plane from a cylinder of scattering medium adjacent to the disk. The results using various scattering phase functions in the medium are in reasonably good agreement for this situation. Some of the other cases resulted in larger effects. In parallel-plane geometries the various phase functions gave energy transfer results that had less variation than in the cylindrical geometry. The phase function is important for beam transmission or other situations in which strong sources transmit directionally into a scattering atmosphere. It may also have a larger effect near the boundary of a medium, where there is less equalization by reflections in all directions, than in the interior of a medium.

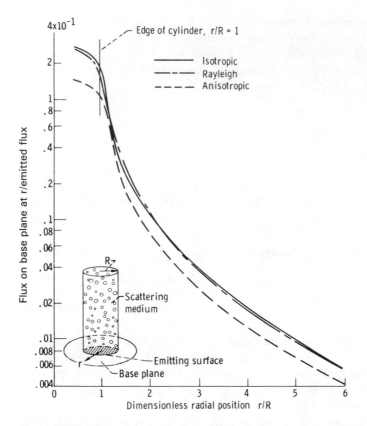

Figure 12-29 Effect of scattering phase function on energy scattered back to base plane by cylinder of scattering medium [86]. Optical diameter of cylinder, 2; height to diameter ratio, 5.

In [88] the effect of anisotropic scattering was analyzed for a medium within a long cylinder. Detailed results are given for the medium at uniform temperature. As might be expected, particles with strong forward scattering tended to enhance the heat flow from the central regions toward the wall, which was at a temperature one-fifth that of the medium. Strong backscattering by the particles decreased the heat flux to the wall. For an optical radius of 2 and a scattering albedo of 0.5 the anisotropy influenced the wall heat flux by about 10%. Larger effects were found when the albedo was increased to 0.9 and 0.95. A similar analysis was carried out in [89] for a medium between two concentric spheres at uniform temperature. Backscattering was found to increase the heat flux at the inner wall and decrease the heat flux at the outer wall. The fibrous media studied in [90] are examples of materials with scattering that is highly anisotropic with a strong forward peak in the direction of the incident radiation.

REFERENCES

1. Langley, S. P.: Experimental Determination of Wave-Lengths in the Invisible Prismatic Spectrum, *Mem. Natl. Acad. Sci.*, vol. 2, pp. 147–162, 1883.
2. Thekaekara, M. P.: Survey of the Literature on the Solar Constant and the Spectral Distribution of Solar Radiant Flux, NASA SP-74, 1965.
3. Goody, R. M.: "Atmospheric Radiation, Theoretical Basis," vol. 1, Clarendon Press, Oxford, 1964.
4. Kondratyev, K. Ya.: "Radiation in the Atmosphere," Academic Press, New York, 1969.
5. Gardon, Robert: A Review of Radiant Heat Transfer in Glass, *J. Am. Ceram. Soc.*, vol. 44, no. 7, pp. 305–312, 1961.
6. Penner, S. S.: "Quantitative Molecular Spectroscopy and Gas Emissivities," Addison-Wesley, Reading, Mass., 1959.
7. Bond, John W., Jr., Kenneth M. Watson, and Jasper A. Welch, Jr.: "Atomic Theory of Gas Dynamics," Addison-Wesley, Reading, Mass., 1965.
8. Bates, David R. (ed.): "Atomic and Molecular Processes," Academic Press, New York, 1962.
9. Chandrasekhar, S.: "Radiative Transfer," Dover, New York, 1960.
10. Kourganoff, Vladimir: "Basic Methods in Transfer Problems; Radiative Equilibrium and Neutron Diffusion," Dover, New York, 1963.
11. Kulander, John L.: Non-Equilibrium Radiation, Rept. R64SD41, General Electric Company, June 1965. (Available from DDC as AD-617383.)
12. Thomas, Richard N.: "Some Aspects of Nonequilibrium Thermodynamics in the Presence of a Radiation Field," University of Colorado Press, Boulder, 1965.
12a. Qiu, T. Q., and C. L. Tien: Short-Pulse Laser Heating on Metals, *Int. J. Heat Mass Transfer*, vol. 35, no. 3, pp. 719–726, 1992.
13. Curie, Daniel (G. F. K. Garlick, trans.): "Luminescence in Crystals," Wiley, New York, 1963.
14. Pringsheim, Peter: "Fluorescence and Phosphorescence," Interscience, New York, 1949.
15. Bond, John W., Jr., Kenneth M. Watson, and Jasper A. Welch, Jr.: "Atomic Theory of Gas Dynamics," Addison-Wesley, Reading, Mass., 1965.
16. Wiese, W. L., M. W. Smith, and B. M. Glennon: Atomic Transition Probabilities, vol. I. Hydrogen through Neon—a Critical Data Compilation. Rept. NSRDS-NBS-4, *Natl. Bur. Std.*, vol. 1, May 20, 1966.
17. Varghese, P. L., and R. K. Hanson: Measured Collisional Narrowing Effect on Spectral Line Shapes at High Resolution, *Appl. Opt.*, vol. 23, p. 2376, 1984.
18. Galatry, L.: Simultaneous Effect of Doppler and Foreign Gas Broadening on Spectral Lines, *Phys. Rev.*, vol. 122, p. 1218, 1961.
19. Ouyang, X., and P. L. Varghese: Reliable and Efficient Program for Fitting Galatry and Voight Profiles to Spectral Data on Multiple Lines, *Appl. Opt.*, vol. 28, no. 8, pp. 1538–1545, 1989.
20. Herzberg, G.: "Molecular Spectra and Molecular Structure," Van Nostrand, New York, 1951.
21. Griem, Hans R.: "Plasma Spectroscopy," McGraw-Hill, New York, 1964.
22. Breene, Robert G.: "The Shift and Shape of Spectral Lines," Pergamon Press, New York, 1961.
23. Tien, C. L.: Thermal Radiation Properties of Gases, in T. F. Irvine, Jr., and J. P. Hartnett (eds.), "Advances in Heat Transfer," vol. 5, pp. 253–324, Academic Press, New York, 1968.
24. Armstrong, B. H., J. Sokoloff, R. W. Nicholls, D. H. Holland, and R. E. Meyerott: Radiative Properties of High Temperature Air, *J. Quant. Spectrosc. Radiat. Transfer*, vol 1, no 2, pp 143–162, 1961.
25. Edwards, D. K.: Molecular Gas Band Radiation, in T. F. Irvine, Jr., and J. P. Hartnett (eds.), "Advances in Heat Transfer," vol. 12, pp. 115–193, Academic Press, New York, 1976.
26. Tiwari, S. N.: Band Models and Correlations for Infrared Radiation, in "Radiative Transfer and Thermal Control," vol. 49 of *Progress in Astronautics and Aeronautics Series*, pp. 155–182, AIAA, 1976.
27. Morizumi, S. J.: Comparison of Analytical Model with Approximate Models for Total Band

Absorption and Its Derivative, *J. Quant. Spectrosc. Radiat. Transfer*, vol. 22, no. 5, pp. 467–474, 1979.

28. Sarofim, A. F., I. H. Farag, and H. C. Hottel: Radiative Heat Transmission from Non-Luminous Gases. Computational Study of the Emissivities of Carbon Dioxide, ASME Paper 78-HT-16, May 1978.

29. Elsasser, Walter M.: "Heat Transfer by Infrared Radiation in the Atmosphere," Harvard Meteorological Studies no. 6, Harvard University Press, Cambridge, Mass., 1942.

30. Edwards, D. K., and W. A. Menard: Comparison of Models for Correlation of Total Band Absorption, *Appl. Opt.*, vol. 3, no. 5, pp. 621–625, 1964.

31. Ludwig, C. B., W. Malkmus, J. E. Reardon, and J. A. L. Thomson: Handbook of Infrared Radiation from Combustion Gases, NASA SP-3080, 1973.

32. Kyle, T. G.: Absorption of Radiation by Uniformly Spaced Doppler Lines, *Astrophys. J.*, vol. 148, no. 3, pp. 845–848, 1967.

33. Golden, S. A.: The Doppler Analog of the Elsasser Band Model, *J. Quant. Spectrosc. Radiat. Transfer*, vol. 7, no. 3, pp. 483–494, 1967.

34. Golden, S. A.: The Doppler Analog of the Elsasser Band Model, II, *J. Quant. Spectrosc. Radiat. Transfer*, vol. 8, pp. 877–897, 1968.

35. Golden, S. A.: The Voigt Analog of an Elsasser Band, *J. Quant. Spectrosc. Radiat. Transfer*, vol. 9, no. 8, pp. 1067–1081, 1969.

36. Tien, C. L., and J. E. Lowder: A Correlation for Total Band Absorptance of Radiating Gases, *Int. J. Heat Mass Transfer*, vol. 9, no. 7, pp. 698–701, 1966.

37. Cess, R. D., and S. N. Tiwari: Infrared Radiative Energy Transfer in Gases, in T. F. Irvine, Jr., and J. P. Hartnett, (eds.), "Advances in Heat Transfer," vol. 8, pp. 229–283, Academic Press, New York, 1972.

38. Edwards, D. K., and A. Balakrishnan: Slab Band Absorptance for Molecular Gas Radiation, *J. Quant. Spectrosc. Radiat. Transfer*, vol. 12, pp. 1379–1387, 1972.

39. Edwards, D. K., and A. Balakrishnan: Thermal Radiation by Combustion Gases, *Int. J. Heat Mass Transfer*, vol. 16, no. 1, pp. 25–40, 1973.

40. Tien, C. L., and G. R. Ling: On a Simple Correlation for Total Band Absorptance of Radiative Gases, *Int. J. Heat Mass Transfer*, vol. 12, no. 9, pp. 1179–1181, 1969.

41. Felske, J. D., and C. L. Tien: A Theoretical Closed-Form Expression for the Total Band Absorptance of Infrared-Radiating Gases, *Int. J. Heat Mass Transfer*, vol. 17, pp. 155–158, 1974.

42. Goody, R. M., and M. J. S. Belton: Radiative Relaxation Times for Mars (Discussion of Martian Atmospheric Dynamics), *Planet. Space Sci.*, vol. 15, no. 2, pp. 247–256, 1967.

43. Hsieh, T. C., and R. Greif: Theoretical Determination of the Absorption Coefficient and the Total Band Absorptance Including a Specific Application to Carbon Monoxide, *Int. J. Heat Mass Transfer*, vol. 15, pp. 1477–1487, 1972.

44. Edwards, D. K.: Radiant Interchange in a Nongray Enclosure Containing an Isothermal Carbon-Dioxide–Nitrogen Gas Mixture, *J. Heat Transfer*, vol. 84, no. 1, pp. 1–11, 1962.

45. Edwards, D. K.: Absorption of Infrared Bands of Carbon Dioxide Gas at Elevated Pressures and Temperatures, *J. Opt. Soc. Am.*, vol. 50, no. 6, pp. 617–626, 1960.

46. Edwards, D. K., and W. A. Menard: Correlations for Absorption by Methane and Carbon Dioxide Gases, *Appl. Opt.*, vol. 3, no. 7, pp. 847–852, 1964.

47. Edwards, D. K., and W. Sun: Correlations for Absorption by the 9.4-μ and 10.4-μ CO_2 Bands, *Appl. Opt.*, vol. 3, no. 12, pp. 1501–1502, 1964.

48. Edwards, D. K., B. J. Flornes, L. K. Glassen, and W. Sun: Correlation of Absorption by Water Vapor at Temperatures from 300 K to 1100 K, *Appl. Opt.*, vol. 4, no. 6, pp. 715–721, 1965.

49. Edwards, D. K.: Absorption of Radiation by Carbon Monoxide Gas According to the Exponential Wide-Band Model, *Appl. Opt.*, vol. 4, no. 10, pp. 1352–1353, 1965.

50. Edwards, D. K., and K. E. Nelson: Rapid Calculation of Radiant Energy Transfer between Nongray Walls and Isothermal H_2O or CO_2 Gas, *J. Heat Transfer*, vol. 84, no. 4, pp. 273–278, 1962.

51. Weiner, Michael M.: "Radiant Heat Transfer in Non-Isothermal Gases," Ph.D. thesis, University of California at Los Angeles, 1966.

52. Hines, W. S., and D. K. Edwards: Infrared Absorptivities of Mixtures of Carbon Dioxide and Water Vapor, *Chem. Eng. Prog. Symp. Ser.*, vol. 64, no. 82, pp. 173–180, 1968.

53. Lin, J. C., and R. Greif: Total Band Absorptance of Carbon Dioxide and Water Vapor Including Effects of Overlapping, *Int. J. Heat Mass Transfer*, vol. 17, pp. 793–795, 1974.

54. Lin, J. C., and R. Greif: Approximate Method for Absorption at High Temperatures, *Int. J. Heat Mass Transfer*, pp. 1805–1807, 1973.

55. Hashemi, A., T. C. Hsieh, and R. Greif: Theoretical Determination of Band Absorption with Specific Application to Carbon Monoxide and Nitric Oxide, *J. Heat Transfer*, vol. 98, pp. 432–437, 1976.

56. Chu, K. H., and R. Greif: Theoretical Determination of Band Absorption for Nonrigid Rotation with Applications to CO, NO, N_2O and CO_2, *J. Heat Transfer*, vol. 100, pp. 230–234, 1978.

57. Hottel, Hoyt C., and Adel F. Sarofim: "Radiative Transfer," McGraw-Hill, New York, 1967.

58. Tiwari, S. N.: Applications of Infrared Band Model Correlations to Nongray Radiation, *Int. J. Heat Mass Transfer*, vol. 20, no. 7, pp. 741–751, 1977.

59. Modest, M. F.: Evaluation of Spectrally-Integrated Radiative Fluxes of Molecular Gases in Multi-Dimensional Media, *Int. J. Heat Mass Transfer*, vol. 26, no. 10, pp. 1533–1546, 1983.

60. Negrelli, D. E., J. R. Lloyd, and J. L. Novotny: A Theoretical and Experimental Study of Radiation-Convection Interaction in a Diffusion Flame, *J. Heat Transfer*, vol. 99, pp. 212–220, 1977.

61. Novotny, J. L., D. E. Negrelli, and T. Van der Driessche: Total Band Absorption Models for Absorbing-Emitting Liquids: CCl_4, *J. Heat Transfer*, vol. 96, pp. 27–31, 1974.

62. Van de Hulst, Hendrick C.: "Light Scattering by Small Particles," Wiley, New York, 1957.

63. Bohren, C. F., and D. R. Huffman: "Absorption and Scattering of Light by Small Particles," Wiley, New York, 1983.

64. Kerker, M.: "The Scattering of Light and Other Electromagnetic Radiation," Academic Press, New York, 1961.

65. Hottel, H. C., A. F. Sarofim, I. A. Vasalos, and W. H. Dalzell: Multiple Scatter: Comparison of Theory with Experiment, *J. Heat Transfer*, vol. 92, no. 2, pp. 285–291, 1970.

66. Hottel, H. C., A. F. Sarofim, W. H. Dalzell, and I. A. Vasalos: Optical Properties of Coatings, Effect of Pigment Concentration, *AIAA J.*, vol. 9, pp. 1895–1898, 1971.

67. Tien, C. L., and B. L. Drolen: Thermal Radiation in Particulate Media with Dependent and Independent Scattering, "Annual Review of Numerical Fluid Mechanics and Heat Transfer," vol. 1, pp. 1–32, Hemisphere, Washington, D.C., 1987.

68. Kamiuto, K.: Study of the Scattering Regime Diagrams, *J. Thermophys. Heat Transfer*, vol. 4, no. 4, pp. 432–435, 1990.

69. Bayvel, L. P., and A. R. Jones: "Electromagnetic Scattering and Its Applications," pp. 5–6, Applied Science, London, 1981.

70. Kumar, S., and C. L. Tien: Dependent Absorption and Extinction of Radiation by Small Particles, *J. Heat Transfer*, vol. 112, no. 1, pp. 178–185, 1990.

71. Ma, Y., V. K. Varadan, and V. V. Varadan: Enhanced Absorption Due to Dependent Scattering, *J. Heat Transfer*, vol. 112, no. 2, pp. 402–407, 1990.

72. Born, Max, and Emil Wolf: Principles of Optics, 2d ed., Pergamon Press, New York, 1964.

73. Ku, J. C., and J. D. Felske: The Range of Validity of the Rayleigh Limit for Computing Mie Scattering and Extinction Efficiencies, *J. Quant. Spectrosc. Radiat. Transfer*, vol. 31, no. 6, pp. 569–574, 1984.

74. Mie, Gustav: Optics of Turbid Media, *Ann. Phys.*, vol. 25, no. 3, pp. 377–445, 1908.

75. Plass, Gilbert N.: Mie Scattering and Absorption Cross Sections for Absorbing Particles, *Appl. Opt.*, vol. 5, no. 2, pp. 279–285, 1966.

76. Penndorf, R. B.: Scattering and Extinction Coefficients for Small Absorbing and Nonabsorbing Aerosols, *J. Opt. Soc. Am.*, vol. 52, no. 8, pp. 896–904, 1962.

77. Felske, J. D., Z. Z. Chu, and J. C. Ku: Mie Scattering Subroutines (DBMIE and MIEVO):

A Comparison of Computational Times, *Appl. Opt.*, vol. 22, no. 15, pp. 2240–2241, 1983.

78. Selamet, A., and V. S. Arpaci: Rayleigh Limit—Penndorf Extension, *Int. J. Heat Mass Transfer*, vol. 32, no. 110, pp. 1809–1820, 1989.

79. White, S. M., and Kumar, S.: Interference Effects on Scattering by Parallel Fibers at Normal Incidence, *J. Thermophys. Heat Transfer*, vol. 4, no. 3, pp. 305–310, 1990.

80. Abramowitz, M., and I. A. Stegun: "Handbook of Mathematical Functions," Dover, New York, 1965.

81. Viskanta, R., and J. S. Toor: Effect of Multiple Scattering on Radiant Energy Transfer in Water, *J. Geophys. Res.*, vol. 78, pp. 3538–3551, 1973.

82. Houf, W. G., and F. P. Incropera: An Assessment of Techniques for Predicting Radiation Heat Transfer in Aqueous Media, *J. Quant. Spectrosc. Radiat. Transfer*, vol. 23, pp. 101–115, 1980.

82a. Crosbie, A. L., and G. W. Davidson: Dirac-Delta Function Approximations to the Scattering Phase Function, *J. Quant. Spectrosc. Radiat. Transfer*, vol. 33, no. 4, pp. 391–409, 1985.

83. Mengüç, M. P., and R. K. Iyer: Modeling of Radiative Transfer Using Multiple Spherical Harmonics Approximations, *J. Quant. Spectrosc. Radiat. Transfer*, vol. 39, no. 6, pp. 445–461, 1988.

84. Henyey, L. G., and J. L. Greenstein: Diffuse Radiation in the Galaxy, *Astrophys. J.*, vol. 88, pp. 70–83, 1940.

85. Joseph, J. H., W. J. Wiscombe, and J. A. Weinman: The Delta-Eddington Approximation for Radiative Flux Transfer, *J. Atmos. Sci.*, vol. 33, no. 12, pp. 2452–2459, 1976.

86. Love, Tom, Leo W. Stockham, Fu C. Lee, William A. Munter, and Yih W. Tsai: Radiative Heat Transfer in Absorbing, Emitting and Scattering Media, ARL-67-0210 (DDC no. AD-666427), Oklahoma University, December, 1967.

87. Stockham, L. W., and T. J. Love: Radiative Heat Transfer From a Cylindrical Cloud of Particles, *AIAA J.*, vol. 6, no. 10, pp. 1935–1940, 1968.

88. Azad, F. H., and M. F. Modest: Evaluation of the Radiative Heat Flux in Absorbing, Emitting and Linear-Anisotropically Scattering Cylindrical Media, *J. Heat Transfer*, vol. 103, pp. 350–356, 1981.

89. Tong, T. W., and P. S. Swathi: Radiative Heat Transfer in Emitting-Absorbing-Scattering Spherical Media, *J. Thermophys. Heat Transfer*, vol. 1, no. 2, pp. 162–170, 1987.

90. Lee, S. C.: Scattering Phase Function for Fibrous Media, *Int. J. Heat Mass Transfer*, vol. 33, no. 10, pp. 2183–2190, 1990.

91. Garbuny, Max: "Optical Physics," Academic Press, New York, 1965.

92. Aroeste, Henry, and William C. Benton: Emissivity of Hydrogen Atoms at High Temperatures, *J. Appl. Phys.*, vol 27, pp. 117–121, 1956.

93. Meyerott, R. E., J. Sokoloff, and R. A. Nicholls: Absorption Coefficients of Air, Rept. LMSD-288052. Lockheed Aircraft Corporation (AFCRC-TR-59-296), September 1959.

94. Edwards, D. K., L. K. Glassen, W. C. Hauser, and J. S. Tuchscher: Radiation Heat Transfer in Nonisothermal Nongray Gases, *J. Heat Transfer*, vol. 89, no. 3, pp. 219–229, 1967.

95. Howard, John N., Darrell E. Burch, and Dudley Williams: Near-Infrared Transmission through Synthetic Atmospheres, *Geophys. Res.*, no. 40, AFCRL-TR-55-213, Air Force Cambridge Research Center, 1955.

96. Howard, J. N., D. E. Burch, and Dudley Williams: Infrared Transmission of Synthetic Atmospheres, I. Instrumentation, *J. Opt. Soc. Am.*, vol. 46, no. 3, pp. 186–190, 1956.

97. Howard, J. N., D. E. Burch, and Dudley Williams: Infrared Transmission of Synthetic Atmospheres. II. Absorption by Carbon Dioxide, *J. Opt. Soc. Am.*, vol. 46, no. 4, pp. 237–241, 1956.

98. Howard, J. N., D. E. Burch, and Dudley Williams: Infrared Transmission of Synthetic Atmospheres. IV. Application of Theoretical Band Models, *J. Opt. Soc. Am.*, vol. 46, no. 5, pp. 334–338, 1956.

99. Charalampopoulos, T. T., and J. D. Felske: Total Band Absorptance, Emissivity, and Absorptivity of the Pure Rotational Band of Water Vapor, *J. Quant. Spectrosc. Radiat. Transfer*, vol. 30, no. 1, pp. 89–96, 1983.

100. Stull, V. Robert, and Gilbert N. Plass: Spectral Emissivity of Hydrogen Chloride from 1000–3400 cm⁻¹, *J. Opt. Soc. Am.*, vol. 50, no. 12, pp. 1279–1285, 1960.
101. Tien, C. L.: Band and Total Emissivity of Ammonia, *Int. J. Heat Mass Transfer*, vol. 16, pp. 856–857, 1973.
102. Jones, M. C.: Far Infrared Absorption in Liquefied Gases, NBS Technical Note 390, National Bureau of Standards, Boulder, 1970.
103. Brosmer, M. A., and C. L. Tien: Thermal Radiation Properties of Acetylene, *J. Heat Transfer*, vol. 107, pp. 943–948, 1985.
104. Tuntomo, A., S. H. Park, and C. L. Tien: Infrared Radiation Properties of Ethylene, *Exp. Heat Transfer*, vol. 2, pp. 91–103, 1989.
105. Kerker, M. (ed.): "Proceedings of the Interdisciplinary Conference on Electromagnetic Scattering," August 1962, Pergamon Press, New York, 1963.

PROBLEMS

12-1 A spectral beam of radiation at $\lambda = 2.5$ μm and with intensity 5 kW/(m²·μm·sr) enters a gas layer 20 cm thick. The gas is at 1000 K and has an absorption coefficient $a_{2.5\mu m} = 0.5$ m⁻¹. What is the intensity of the beam emerging from the gas layer? Neglect scattering, but include emission from the gas.

Answer: 4.893 kW/(m²·μm·sr)

12-2 Radiation from a blackbody source at 3000 K is passing through a layer of air at 12,000 K and 1 atm. Considering only the transmitted radiation (that is, not accounting for emission by the air), what path length is required to attenuate by 25% the energy at the wavelength corresponding to the peak of the blackbody radiation?

Answer: 2.21 cm

12-3 Radiation with a wavelength of 1.5 μm is passing through a gas at a temperature of 10,000 K. What is the ratio of the true absorption coefficient to the absorption coefficient?

Answer: 1.621

12-4 A gas layer at constant pressure P has a linearly decreasing temperature across the layer and a constant mass absorption coefficient a_m (no scattering). For radiation passing in a normal direction through the layer, what is the ratio i_2'/i_1' as a function of T_1, T_2, and L? The temperature range T_2 to T_1 is low enough that emission from the gas can be neglected, and the gas constant is R.

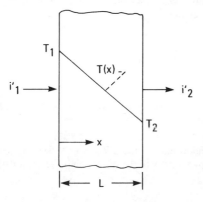

Answer: $\dfrac{i_2'}{i_1'} = \exp\left[-\dfrac{PLa_m}{R(T_1 - T_2)} \ln \dfrac{T_1}{T_2}\right]$

12-5 A thin cylinder of gas contains a uniform suspension of scattering particles giving a scattering coefficient of $\sigma_{s\lambda}$. The phase function is independent of circumferential angle φ and depends only on the angle θ away from the path of the incident radiation, $\Phi(\theta) = (3/4) (1 + \cos^2 \theta)$. A beam of radiation with spectral intensity i'_λ is incident in a normal direction on the end of the cylinder. As a function of L, what fraction of the energy exits in the forward direction ($0 \leq \theta < \pi/2$) and what fraction is backscattered ($\pi/2 < \theta \leq \pi$)? There is no absorption or emission.

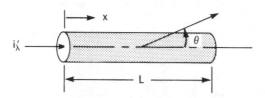

Answer: $(1/2) [1 + \exp(-\sigma_{s\lambda}L)]$, $(1/2) [1 - \exp(-\sigma_{s\lambda}L)]$

12-6 Compute the half-width for Doppler broadening of neon at a wavelength of 0.6 μm and for $T = 300$ K.

Answer: 0.0231 cm^{-1}

12-7 Two absorption lines have the same transition (centerline) wavenumber, $\eta_{ij} = 550$ cm^{-1}. Both have the same half-width, 0.15 cm^{-1}. One line has the Doppler profile; one has the Lorentz profile. Draw the two line shapes, $a_{\eta,ij}(\eta)/S_{ij}$, as a function of η, on the same plot.

12-8 A gas is composed of pure atomic hydrogen at a temperature of 500 K. Calculate the half-width of the hydrogen Lyman alpha line (transition centerline frequency = 2.4675×10^{15} Hz) for the case of Doppler broadening. Then plot the line shape $a_{\eta,ij}(\eta)/S_{ij}$ for this line as a function of wavenumber. The mass of the hydrogen atom is 1.66×10^{-24} g.

Answer: $\Delta_D = 0.659$ cm^{-1}

12-9 For the same gas and temperature as Prob. 12-8, compute the half-width of the line for collision broadening at a pressure of 1 atm. Assume the diameter of the hydrogen atom is about 1.06×10^{-8} cm. Plot $a_{\eta,ij}(\eta)/S_{ij}$ for collision broadening on the same wavenumber plot as Prob. 12-8.

Answer: $\Delta_c = 0.0127$ cm^{-1}

12-10 For $\beta = 0.2$, prepare a plot comparing various band correlation functions of $\bar{A}_l/A_0$ for $0.01 \leq u \leq 100$. Compare the correlation functions of Edwards and Menard, Tien and Lowder, Cess and Tawari, and Goody and Belton.

12-11 Find the effective bandwidth $\bar{A}$ of the 9.4-μm band of CO_2 at a partial pressure of 0.4 atm in a mixture with nitrogen at a total pressure of 1 atm. The temperature is 500 K, and the path length S is 0.364 m. Compare with the result in Example 12-1.

Answer: 2.32 cm^{-1}

12-12 For pure CO gas at 1 atm pressure, determine the effective bandwidth for the 4.67-μm band at $T = 600$ K for a path length of $S = 0.5$ m.

Answer: 221 cm^{-1}

12-13 From Fig. 12-10a, estimate the effective bandwidth for the 2.7-μm CO_2 band at 830 K, 10 atm, and a path length of 38.8 cm. Compare this with the result computed from the correlations in Tables 12-3 and 12-4.

Answer: 421 cm^{-1}

12-14 Prove that for weak, nonoverlapping Lorentz lines, the Elsasser band model predicts the weak absorption band relation $\bar{A}_l = C_l S$, where C_l is a constant for the particular band.

12-15 A collimated beam of red light ($\lambda \approx 0.65\ \mu$m) is to be attenuated by scattering. A proposed scheme is to use very small spherical particles of copper having a characteristic diameter of 200 Å (optical data for copper are in Table 4-2). Absorption by the particles is neglected. The particles are to be suspended in a nonscattering, nonabsorbing medium. Assuming Rayleigh scattering is applicable, what is the particle scattering cross section s_λ? For the beam to be 10% attenuated by scattering in a path length of 1 m, approximately what number density of particles would be required? What are the volume fraction of the particles and the mass of the particles per cubic centimeter of the scattering medium?

Answer: $13.2 \times 10^{-8}\ \mu$m^2, 79.5×10^{10} cm^{-3}, 3.33×10^{-6}, 29.8×10^{-6} g

12-16 The beam in Prob. 12-15 is changed to green light ($\lambda \approx 0.55\ \mu$m) while the scattering particle size and number density are kept the same. Assuming the same optical constants apply at this wavelength, what is the percentage attenuation for this beam for a 1-m path length?

Answer: 18.5%

12-17 Consider Rayleigh scattering for very small gold particles at the two wavelengths shown for gold in Table 4-2. How do the scattering cross sections s_λ differ at these two λ (note Fig. 12-25)?

Answer: Ratio of $s_\lambda = 266$

12-18 Verify that the phase function in Eq. (12-107) satisfies the normalization specified by Eq. (12-47).

12-19 The complex index of refraction for a particular sample of carbon is given by $\bar{n} = 2.2 - 1.2i$ at a wavelength of 2.5 μm. The carbon is ground into fine particles (assumed spherical) that are 0.1 μm in diameter. Compute the efficiency factors for absorption and scattering for these particles. What is the ratio of the absorption cross section to the scattering cross section? The surrounding medium is air.

Answer: $E_a = 0.1396$, $E_s = 3.9217 \times 10^{-4}$, $E_a/E_s = 355.9$

12-20 In Prob. 12-19, the scattering for very small carbon particles is found to be small compared with absorption. A dispersion of these particles in air is formed with a particle concentration of 10^{10} particles/cm^3. The dispersion is at low temperature.

(a) Radiation at $\lambda = 2.5\ \mu$m enters the dispersion with an intensity of 0.4×10^4 W/m$^2 \cdot \mu$m $\cdot$ sr and travels through a path length of 0.15 m. What is the intensity at the end of this path length?

(b) The dispersion is now heated to 800 K. What is the intensity leaving for this condition?

(c) The dispersion is now heated to a different uniform temperature, and the leaving intensity is measured as 0.4×10^4 W/m$^2 \cdot \mu$m $\cdot$ sr. What is the temperature of the dispersion?

Answer: (a) 772 W/m$^2 \cdot \mu$m $\cdot$ sr; (b) 2015 W/m$^2 \cdot \mu$m $\cdot$ sr; (c) 922.2 K

THIRTEEN

THE ENGINEERING TREATMENT OF GAS RADIATION IN ENCLOSURES

13-1 INTRODUCTION

An extensive body of engineering literature deals with radiation exchange between solid surfaces when no absorbing and/or scattering medium is between them. Methods for treating such problems are highly developed and have been presented in Chaps. 6–11. The additional complication of having an intervening absorbing-emitting and/or scattering gas between surfaces can be included by building on this foundation. Some of the relations in Chap. 12 will be used in the derivation of engineering methods for solving gas-radiation problems. In many engineering situations of heat transfer in furnaces and combustion chambers, the effect of scattering is small; hence only absorption and emission are considered here. Scattering is included in the chapters that follow.

Most of this chapter is concerned with gas that is *isothermal* and has *uniform* composition and density. This is useful, as in many instances combustion products are well mixed. This provides the simplification that the gas temperature *distribution* need not be computed to obtain the radiative behavior. The *mean radiative beam length* is introduced as an engineering approximation to account for the geometry of a gas radiating to its boundary. The beam length is found to be relatively insensitive to the wavelength dependence of the absorption coefficient. This enables the effects of the geometry to be separated from the effects of spectral property variations.

The chapter concludes with a section on flames, both nonluminous flames and those containing luminous particles, mainly soot. A nonluminous hydrocarbon flame contains carbon dioxide and water vapor as its chief radiating constituents.

Radiation by these gases is fairly well understood. When soot is present and the flame thereby becomes luminous, the radiation is dependent on the radiative properties of the soot, the soot concentration, and its distribution. Soot radiative properties are discussed; additional information is needed on this subject. Determining the soot concentration is a serious difficulty in flame radiation computations. The concentration depends on the particular fuel, the flame chemistry and geometry, and the mixing within the flame. At present the soot concentration and distribution cannot be predicted accurately from basic parameters such as the burner geometry, the fuel-air ratio, and the particular fuel.

13-2 SYMBOLS

a	absorption coefficient
a_i	coefficients in gray gas emittance model
a, b, c	dimensions in system of two rectangles
A	area
$AF\bar{\alpha}$	geometric absorption factor
$AF\bar{\tau}$	geometric transmission factor
$\bar{A}$	effective bandwidth
$b_{j,i}$	coefficients in emittance model for mixture
C	ratio $L_e/L_{e,0}$; volume fraction of particles in medium
C_{CO_2}, C_{H_2O}	pressure-correction coefficients
D	spacing between parallel plates; diameter
e	emissive power
E_n	exponential integral
E_λ	spectral absorption efficiency factor
F	geometric configuration factor
$\overline{gg}$	gas-gas direct-exchange area
$\overline{gs}$	gas-surface direct-exchange area
h	height of cylinder; height of rectangle
i	radiation intensity
k, k_1, k_2	constants in equations for soot absorption
L_e	mean beam length of gas volume
$L_{e,0}$	mean beam length for limiting case of small absorption
N	total number of surfaces in enclosure
p	partial pressure
P	total pressure of gas or gas mixture
q	energy flux, energy per unit area and time
Q	energy per unit time
r	inner radius of annulus
R	radius of hemisphere, semicylinder, cylinder, or sphere
$\overline{sg}$	surface-gas direct-exchange area
$\overline{ss}$	surface-surface direct-exchange area
S	coordinate along path of radiation
$\bar{S}$	geometric-mean beam length

S_c	line absorption coefficient
T	absolute temperature
V	volume; voltage signal
w	width of rectangle
W	width of plate
X	mass path length, ρS; shortest dimension of rectangular parallelepiped
$\alpha(S)$	absorptance
$\bar{\alpha}(S)$	geometric-mean absorptance
α_l	band absorptance
δ	spacing of absorption lines
δ_{kj}	Kronecker delta
$\Delta\alpha, \Delta\epsilon$	correction for spectral overlap
ϵ	emissivity of surface
$\epsilon(S)$	emittance of medium
η	wavenumber
θ	cone angle, angle from normal of area
κ	optical depth
λ	wavelength
ρ	reflectivity; density
σ	Stefan-Boltzmann constant
$\tau(S)$	transmittance
$\bar{\tau}(S)$	geometric-mean transmittance
ω	solid angle

Subscripts

b	blackbody
CO_2	carbon dioxide
e	environment
g	gas
H_2O	water vapor
i	incident, incoming
j, k	surfaces j or k
$j-k$	from surface j to surface k
l	absorption band l
m	maximum value
o	outgoing
u	uniform
w	wall
λ	spectrally (wavelength) dependent
η	wavenumber-dependent

Superscripts

*, **	dummy variable of integration
+	quantities defined after Eq. (13-88)
'	directional quantity

13-3 NET-RADIATION METHOD FOR ENCLOSURE FILLED WITH ISOTHERMAL UNIFORM GAS

The radiation-exchange equations were developed in Sec. 8-3 for an enclosure that does not contain an absorbing-emitting medium and has surfaces with spectrally dependent properties. Since the absorption properties of gases and other absorbing media are almost always strongly wavelength dependent, the present development is carried out at a single wavelength. Integrations over all wavelengths will then yield the total radiative behavior. It is assumed that surface directional property effects are sufficiently unimportant that surfaces can be treated as diffuse emitters and reflectors.

Often in a gas-filled enclosure, such as in an engine combustion chamber or industrial furnace, there is sufficient mixing that the entire gas is essentially isothermal and of uniform composition. In this instance the analysis is simplified, as it is unnecessary to compute the gas temperature distribution. Sometimes the gas temperature is specified, but if not, it is necessary only to compute a single gas temperature from the governing heat balances. Even with this simplification, a detailed radiation exchange computation between the gas and bounding surfaces can become quite involved.

Consider an enclosure of N surfaces, each at a uniform temperature, as in Fig. 13-1. Typical surfaces are designated by j and k. Some of the surfaces can be open boundaries such as a window or hole. This is usually modeled as a perfectly absorbing (black) surface at the temperature of the surroundings outside the

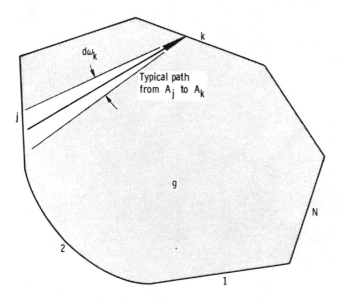

Figure 13-1 Enclosure composed of N discrete surface areas and filled with uniform gas g (enclosure shown in cross section for simplicity).

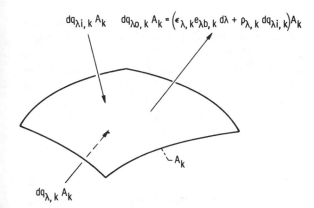

$$dq_{\lambda i,\,k}\,A_k \qquad dq_{\lambda o,\,k}\,A_k = \left(\epsilon_{\lambda,\,k}e_{\lambda b,\,k}\,d\lambda + \rho_{\lambda,\,k}\,dq_{\lambda i,\,k}\right)A_k$$

$$dq_{\lambda,\,k}\,A_k$$

$$A_k$$

Figure 13-2 Spectral energy quantities incident on and leaving typical surface area of enclosure.

opening. The enclosure is filled with an absorbing-emitting medium at uniform temperature T_g. The quantity Q_g is the amount of heat that it is necessary to supply by means other than radiation to the entire absorbing medium to maintain this temperature. A common source of Q_g is combustion. If in the solution of a problem Q_g is negative, the medium is gaining a net amount of radiative energy from the enclosure walls, and energy must be removed from the gas to maintain it at its steady temperature T_g. The Q_g is analogous to the Q_k at a surface, which is the energy supplied by nonradiative means to area A_k.

The enclosure theory yields equations relating Q_k and T_k for each surface to Q_g and T_g for the gas or other absorbing isothermal medium filling the enclosure. Considering all the surfaces and the gas, if half the Q and T are specified, then the radiative heat balance equations can be solved for the remaining unknown Q or T values. If the heat input to the gas from external sources Q_g is given, the analysis will yield the steady gas temperature T_g. Conversely, if T_g is given, the analysis will yield the energy that must be supplied to the gas to maintain this gas temperature.

The net-radiation method in Chaps. 7 and 8 is now extended to include gas-radiation terms. At the kth surface of an enclosure, as shown in Fig. 13-2, a heat balance gives

$$dQ_{\lambda,k} = dq_{\lambda,k}A_k = (dq_{\lambda o,k} - dq_{\lambda i,k})A_k \tag{13-1}$$

The $dq_{\lambda o,k}$ and $dq_{\lambda i,k}$ are respectively the outgoing and incoming radiative fluxes in wavelength interval $d\lambda$. The $dQ_{\lambda,k}$ is the energy supplied to the surface in wavelength region $d\lambda$. Note that, as discussed in connection with (8-4), the external energy supplied to A_k by some means such as conduction and/or convection equals $\int_{\lambda=0}^{\infty} dQ_{\lambda,k}$.

The outgoing spectral flux is composed of emitted and reflected energy:

$$dq_{\lambda o,k} = \epsilon_{\lambda,k}(\lambda,T_k)e_{\lambda b,k}(\lambda,T_k)\,d\lambda + \rho_{\lambda,k}(\lambda,T_k)\,dq_{\lambda i,k} \tag{13-2}$$

The functional notation will usually be omitted to shorten the form of the equations that follow. The $e_{\lambda b,k}(\lambda, T_k)\,d\lambda$ is the blackbody spectral emission at T_k in the wavelength region $d\lambda$ about wavelength λ.

The $dq_{\lambda i,k}$ in (13-1) is the incoming spectral flux to A_k. It is equal to the sum of the contributions from all the surfaces that reach the kth surface after allowance for absorption in passing through the intervening gas, plus the contribution due to emission from the gas. The equation of transfer allows for both attenuation and emission as radiation passes along a path through a gas. A typical path from A_j to A_k within an incident solid angle $d\omega_k$ is shown in Fig. 13-1. If all such paths and solid angles by which radiation can pass from all the surfaces (including A_k if it is concave) to A_k are accounted for, the solid angles $d\omega_k$ will encompass all of the gas region that can radiate to A_k. Thus by using the equation of transfer, which includes the gas emission term, to compute the energy transported along all paths between surfaces, the gas emission is automatically included. The radiation passing from one surface to another, including emission and absorption by the intervening gas, is now considered.

A typical pair of surfaces is shown in Fig. 13-3. In enclosure theory $dq_{\lambda o}$ is assumed uniform over each surface. Since the surfaces are assumed diffuse, the spectral intensity leaving dA_j is $i'_{\lambda o,j} = (1/\pi) \, dq_{\lambda o,j}/d\lambda$. By use of the equation of transfer [Eq. (12-52)], the intensity arriving at dA_k after traversing path S is, for a gas with uniform temperature and composition (constant spectral absorption coefficient),

$$i'_{\lambda i,j-k} = i'_{\lambda o,j} \exp(-a_\lambda S) + i'_{\lambda b,g}[1 - \exp(-a_\lambda S)] \tag{13-3}$$

Using the definitions introduced in Chap. 12, $\tau_\lambda(S) \equiv \exp(-a_\lambda S)$ is the spectral transmittance of the gas along path length S and $\alpha_\lambda(S) \equiv 1 - \exp(-a_\lambda S)$ is the spectral absorptance along the path, (13-3) is written as

$$i'_{\lambda i,j-k} = i'_{\lambda o,j}\tau_\lambda(S) + i'_{\lambda b,g}(T_g)\alpha_\lambda(S) \tag{13-4}$$

This intensity arriving at dA_k in solid angle $d\omega_k$ provides the energy $i'_{\lambda i,j-k} \, dA_k \cos\theta_k \, d\omega_k \, d\lambda$. Using $d\omega_k = dA_j \cos\theta_j/S^2$, the arriving spectral energy is

$$d^3 Q_{\lambda i,j-k} = [i'_{\lambda o,j}\tau_\lambda(S) + i'_{\lambda b,g}(T_g)\alpha_\lambda(S)] \frac{dA_k \, dA_j \cos\theta_k \cos\theta_j}{S^2} \, d\lambda \tag{13-5}$$

For a diffuse surface $dq_{\lambda o,j} = \pi i'_{\lambda o,j} \, d\lambda$ and $e_{\lambda b,g} = \pi i'_{\lambda b,g}$, so (13-5) can be written as

$$d^3 Q_{\lambda i,j-k} = [dq_{\lambda o,j}\tau_\lambda(S) + e_{\lambda b,g}(T_g) \, d\lambda \, \alpha_\lambda(S)] \frac{dA_k \, dA_j \cos\theta_k \cos\theta_j}{\pi S^2} \tag{13-6}$$

Equation (13-6) is integrated over all of A_k and A_j to give the spectral energy along all paths from A_j that is incident upon A_k:

$$dQ_{\lambda i,j-k} \equiv \int_{A_k} \int_{A_j} [dq_{\lambda o,j}\tau_\lambda(S) + e_{\lambda b,g}(T_g) \, d\lambda \, \alpha_\lambda(S)] \frac{\cos\theta_k \cos\theta_j}{\pi S^2} \, dA_j \, dA_k \tag{13-7}$$

The first term of the double integral is the spectral energy leaving A_j that is transmitted to A_k. The second term is the spectral energy received at A_k as a result of

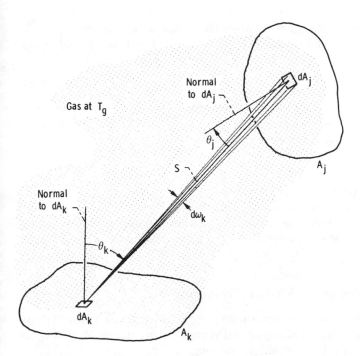

Figure 13-3 Radiation between two surfaces with isothermal uniform gas between them.

emission by the constant-temperature gas in the envelope between A_j and A_k. This envelope is the volume occupied by all straight paths between any part of A_j and A_k.

13-3.1 Definitions of Spectral Geometric-Mean Transmission and Absorption Factors

The double integral in (13-7) has some similarity to the double integral in (6-16) for the configuration factor between two surfaces. By analogy, define the factor $\bar{\tau}_{\lambda,j-k}$ such that

$$F_{j-k}\bar{\tau}_{\lambda,j-k} \equiv \frac{1}{A_j} \int_{A_k} \int_{A_j} \frac{\tau_\lambda(S)\cos\theta_k \cos\theta_j}{\pi S^2}\, dA_j\, dA_k \qquad (13\text{-}8)$$

where F_{j-k} is the geometric configuration factor. With no absorbing medium, $\tau_\lambda(S)$ = 1, and the right side of (13-8) becomes F_{j-k} so that in this instance $\bar{\tau}_{j-k}$ − 1. For complete absorption between A_j and A_k, $\bar{\tau}_{\lambda,j-k}$ = 0. The $\bar{\tau}_{\lambda,j-k}$ is the *geometric-mean transmittance* from A_j to A_k. Similarly, from the second quantity in brackets in Eq. (13-7), a *geometric-mean absorptance* $\bar{\alpha}_{\lambda,j-k}$ is defined so that

$$F_{j-k}\bar{\alpha}_{\lambda,j-k} \equiv \frac{1}{A_j} \int_{A_k} \int_{A_j} \frac{\alpha_\lambda(S)\cos\theta_k \cos\theta_j}{\pi S^2}\, dA_j\, dA_k \qquad (13\text{-}9)$$

For a nonabsorbing medium $\bar{\alpha}_{\lambda,j-k} = 0$, while for complete absorption $\bar{\alpha}_{\lambda,j-k} = 1$. From the definitions of τ_λ and α_λ and Eqs. (13-8) and (13-9), the $\bar{\tau}_\lambda$ and $\bar{\alpha}_\lambda$ are related by

$$\bar{\alpha}_{\lambda,j-k} = 1 - \bar{\tau}_{\lambda,j-k} \tag{13-10}$$

An alternative terminology is also used wherein the entire quantity $A_j F_{j-k}\bar{\tau}_{\lambda,j-k}$ is called the *geometric transmission factor* and $A_j F_{j-k}\bar{\alpha}_{\lambda,j-k}$ the *geometric absorption factor*. Equation (13-7) can now be written as

$$dQ_{\lambda i,j-k} = dq_{\lambda o,j} A_j F_{j-k}\bar{\tau}_{\lambda,j-k} + e_{\lambda b,g}(T_g) \, d\lambda \, A_j F_{j-k}\bar{\alpha}_{\lambda,j-k} \tag{13-11}$$

In computing the heat exchange in an enclosure, it is necessary to determine $\bar{\tau}_\lambda$ and $\bar{\alpha}_\lambda$. It is necessary to perform only one double integration to obtain both $\bar{\tau}_\lambda$ and $\bar{\alpha}_\lambda$ because of relation (13-10). In the present discussion the enclosure-theory formulation will be completed first. Then the evaluation of $\bar{\tau}_\lambda$ and $\bar{\alpha}_\lambda$ is considered.

13-3.2 Matrix of Enclosure-Theory Equations

For an enclosure with N surfaces bounding an isothermal gas at T_g of uniform composition and density, the incident spectral energy on any surface A_k will equal that arriving from the directions of all surrounding surfaces:

$$dQ_{\lambda i,k} = A_k \, dq_{\lambda i,k} = \sum_{j=1}^{N} (dq_{\lambda o,j} A_j F_{j-k}\bar{\tau}_{\lambda,j-k} + e_{\lambda b,g} \, d\lambda \, A_j F_{j-k}\bar{\alpha}_{\lambda,j-k}) \tag{13-12}$$

From reciprocity (6-19), $A_j F_{j-k} = A_k F_{k-j}$, so A_k can be eliminated to give

$$dq_{\lambda i,k} = \sum_{j=1}^{N} (dq_{\lambda o,j} F_{k-j}\bar{\tau}_{\lambda,j-k} + e_{\lambda b,g} \, d\lambda \, F_{k-j}\bar{\alpha}_{\lambda,j-k}) \tag{13-13}$$

Equations (13-1), (13-2), and (13-13) form a set of three equations relating $dq_{\lambda o}$, $dq_{\lambda i}$, and dq_λ for each of the surfaces in the enclosure to the $e_{\lambda b}$ for that surface and to $e_{\lambda b,g}$. The $dq_{\lambda i}$ is eliminated by combining (13-1) and (13-2) and by substituting (13-13) into (13-1). This yields a set of two equations for $dq_{\lambda o}$ and dq_λ for each surface in terms of $e_{\lambda b,k}$ and $e_{\lambda b,g}$,

$$dq_{\lambda,k} = \frac{\epsilon_{\lambda,k}}{1 - \epsilon_{\lambda,k}} (e_{\lambda b,k} \, d\lambda - dq_{\lambda o,k}) \tag{13-14}$$

$$dq_{\lambda,k} = dq_{\lambda o,k} - \sum_{j=1}^{N} (dq_{\lambda o,j} F_{k-j}\bar{\tau}_{\lambda,j-k} + e_{\lambda b,g} \, d\lambda \, F_{k-j}\bar{\alpha}_{\lambda,j-k}) \tag{13-15}$$

If $\epsilon_{\lambda,k} = 1$ Eq. (13-14) becomes $dq_{\lambda o,k} = e_{\lambda b,k} \, d\lambda$. Equation (13-14) is the same as for an enclosure without an absorbing gas [see (8-8)]. Equations (13-14) and (13-15) are analogous to (7-17) and (7-18) for a gray enclosure without an absorbing gas. From the symmetry of the integrals in (13-8) and (13-9) and the reciprocity relation $A_j F_{j-k} = A_k F_{k-j}$, it is found that

$$\bar{\tau}_{\lambda, j-k} = \bar{\tau}_{\lambda, k-j} \tag{13-16}$$

and

$$\bar{\alpha}_{\lambda, j-k} = \bar{\alpha}_{\lambda, k-j} \tag{13-17}$$

Then (13-15) can also be written as

$$dq_{\lambda, k} = dq_{\lambda o, k} - \sum_{j=1}^{N} (dq_{\lambda o, j} F_{k-j} \bar{\tau}_{\lambda, k-j} + e_{\lambda b, g} \, d\lambda \, F_{k-j} \bar{\alpha}_{\lambda, k-j}) \tag{13-18}$$

As in Sec. 7-4.1, the set of equations (13-14) and (13-18) can be further reduced by solving (13-14) for $dq_{\lambda o}$ and inserting it into (13-18). This results in the relation

$$\sum_{j=1}^{N} \left(\frac{\delta_{kj}}{\epsilon_{\lambda, j}} - F_{k-j} \frac{1 - \epsilon_{\lambda, j}}{\epsilon_{\lambda, j}} \bar{\tau}_{\lambda, k-j} \right) dq_{\lambda, j}$$

$$= \sum_{j=1}^{N} [(\delta_{kj} - F_{k-j} \bar{\tau}_{\lambda, k-j}) e_{\lambda b, j} \, d\lambda - F_{k-j} \bar{\alpha}_{\lambda, k-j} e_{\lambda b, g} \, d\lambda] \tag{13-19}$$

The Kronecker delta δ_{kj} has values $\delta_{kj} = 1$ when $k = j$, and $\delta_{kj} = 0$ when $k \neq j$. This equation is analogous to (7-31). If (13-19) is written for each k from 1 to N, a set of N equations is obtained relating the $2N$ quantities dq_λ and $e_{\lambda b}$ for the surfaces. If the gas temperature (and hence $e_{\lambda b, g}$) is assumed known, then one-half of the dq_λ and $e_{\lambda b}$ values need to be specified, and the equations can be solved for the remaining unknowns. To determine total energy quantities, this set of equations must be solved in a number of wavelength intervals and integration of each energy quantity then be performed over all wavelengths. This is the same as for the band equations (Sec. 8-3.2). If the gas temperature is unknown, an additional equation is needed as given in the next section.

13-3.3 Heat Balance on Gas

A heat balance on the gas provides a relation between the gas temperature and the energy supplied to the gas by means other than radiative exchange within the enclosure. From the energy balance on the entire enclosure, the energy that must be supplied to the gas by combustion, for example, is equal to the net energy escaping from the boundaries. The total energy escaping at surface k is $-A_k \int_{\lambda=0}^{\infty} dq_{\lambda, k}$. Then the energy added to the gas is found by summing over all enclosure surfaces:

$$Q_g = - \sum_{k=1}^{N} A_k \int_{\lambda=0}^{\infty} dq_{\lambda, k} \tag{13-20}$$

This can be evaluated after the dq_λ are found for each surface in a sufficient number of wavelength intervals from the matrix of Eqs. (13-19). This matrix can be solved if the gas temperature and hence $e_{\lambda b, g}$ is known. If the gas temperature is unknown and instead Q_g is specified, a less direct solution is required. A gas temperature is guessed and Q_g is found. The solution is repeated for several gas temperatures and the resulting relation between Q_g and T_g yields the T_g corresponding to the specified Q_g.

EXAMPLE 13-1 As a simple but useful situation, consider a gas at T_g completely surrounded by a single surface at T_1. The heat transfer from the gas to the wall is to be found in terms of the wall and gas temperatures and properties.

For an enclosure consisting of a single wall, Eq. (13-19) yields

$$\left(\frac{1}{\epsilon_{\lambda,1}} - F_{1-1} \frac{1 - \epsilon_{\lambda,1}}{\epsilon_{\lambda,1}} \bar{\tau}_{\lambda,1-1} \right) dq_{\lambda,1}$$

$$= (1 - F_{1-1}\bar{\tau}_{\lambda,1-1})e_{\lambda b,1}\, d\lambda - F_{1-1}\bar{\alpha}_{\lambda,1-1}e_{\lambda b,g}\, d\lambda$$

Using the relations $F_{1-1} = 1$ and $1 - \bar{\tau}_{\lambda,1-1} = \bar{\alpha}_{\lambda,1-1}$, this equation is simplified and then integrated over all λ to obtain

$$Q_g = -Q_1 = -A_1 \int_{\lambda=0}^{\infty} dq_{\lambda,1} = A_1 \int_0^{\infty} \frac{e_{\lambda b,g} - e_{\lambda b,1}}{1/\epsilon_{\lambda,1} + 1/\bar{\alpha}_{\lambda,1-1} - 1}\, d\lambda \qquad (13\text{-}21)$$

EXAMPLE 13-2 As a further example of the net-radiation method, consider the heat transfer for an enclosure of two infinite parallel plates at temperatures T_1 and T_2 ($T_1 > T_2$) bounding a gas at uniform temperature T_g.

Equation (13-19) applied to a two-surface enclosure gives for $k = 1$ and 2 (note that $F_{1-1} = F_{2-2} = 0$):

$$\frac{1}{\epsilon_{\lambda,1}} dq_{\lambda,1} - F_{1-2}\frac{1 - \epsilon_{\lambda,2}}{\epsilon_{\lambda,2}}\bar{\tau}_{\lambda,1-2}\, dq_{\lambda,2}$$

$$= e_{\lambda b,1}\, d\lambda - F_{1-2}\bar{\tau}_{\lambda,1-2}e_{\lambda b,2}\, d\lambda - F_{1-2}\bar{\alpha}_{\lambda,1-2}e_{\lambda b,g}\, d\lambda \qquad (13\text{-}22a)$$

$$-F_{2-1}\frac{1 - \epsilon_{\lambda,1}}{\epsilon_{\lambda,1}}\bar{\tau}_{\lambda,2-1}\, dq_{\lambda,1} + \frac{1}{\epsilon_{\lambda,2}} dq_{\lambda,2}$$

$$= -F_{2-1}\bar{\tau}_{\lambda,2-1}e_{\lambda b,1}\, d\lambda - F_{2-1}\bar{\alpha}_{\lambda,2-1}e_{\lambda b,g}\, d\lambda + e_{\lambda b,2}\, d\lambda \qquad (13\text{-}22b)$$

For the infinite parallel-plate geometry, $F_{1-2} = F_{2-1} = 1$, and from (13-16) and (13-17) $\bar{\tau}_{\lambda,2-1} = \bar{\tau}_{\lambda,1-2}$ and $\bar{\alpha}_{\lambda,2-1} = \bar{\alpha}_{\lambda,1-2}$. For simplicity, the numerical subscripts on $\bar{\tau}$ and $\bar{\alpha}$ are omitted. Then (13-22a) and (13-22b) become

$$\frac{1}{\epsilon_{\lambda,1}} dq_{\lambda,1} - \frac{1 - \epsilon_{\lambda,2}}{\epsilon_{\lambda,2}}\bar{\tau}_\lambda\, dq_{\lambda,2} = (e_{\lambda b,1} - \bar{\tau}_\lambda e_{\lambda b,2} - \bar{\alpha}_\lambda e_{\lambda b,g})\, d\lambda \qquad (13\text{-}23a)$$

$$-\frac{1 - \epsilon_{\lambda,1}}{\epsilon_{\lambda,1}}\bar{\tau}_\lambda\, dq_{\lambda,1} + \frac{1}{\epsilon_{\lambda,2}} dq_{\lambda,2} = (-\bar{\tau}_\lambda e_{\lambda b,1} + e_{\lambda b,2} - \bar{\alpha}_\lambda e_{\lambda b,g})\, d\lambda \qquad (13\text{-}23b)$$

Equations (13-23a) and (13-23b) are solved simultaneously for $dq_{\lambda,1}$ and $dq_{\lambda,2}$. After using the relation $\bar{\alpha}_\lambda = 1 - \bar{\tau}_\lambda$, this yields

$$
dq_{\lambda,1} = \frac{d\lambda}{1 - (1 - \epsilon_{\lambda,1})(1 - \epsilon_{\lambda,2})\bar{\tau}_\lambda^2}
$$
$$
\times \{\epsilon_{\lambda,1}\epsilon_{\lambda,2}\bar{\tau}_\lambda(e_{\lambda b,1} - e_{\lambda b,2}) + \epsilon_{\lambda,1}(1 - \bar{\tau}_\lambda)
$$
$$
\times [1 + (1 - \epsilon_{\lambda,2})\bar{\tau}_\lambda](e_{\lambda b,1} - e_{\lambda b,g})\} \tag{13-24a}
$$

$$
dq_{\lambda,2} = \frac{d\lambda}{1 - (1 - \epsilon_{\lambda,1})(1 - \epsilon_{\lambda,2})\bar{\tau}_\lambda^2}
$$
$$
\times \{\epsilon_{\lambda,1}\epsilon_{\lambda,2}\bar{\tau}_\lambda(e_{\lambda b,2} - e_{\lambda b,1}) + \epsilon_{\lambda,2}(1 - \bar{\tau}_\lambda)
$$
$$
\times [1 + (1 - \epsilon_{\lambda,1})\bar{\tau}_\lambda](e_{\lambda b,2} - e_{\lambda b,g})\} \tag{13-24b}
$$

The total energy fluxes added to surfaces 1 and 2 are, respectively,

$$
q_1 = \int_{\lambda=0}^{\infty} dq_{\lambda,1} \quad \text{and} \quad q_2 = \int_{\lambda=0}^{\infty} dq_{\lambda,2} \tag{13-25}
$$

The total energy added to the gas to maintain its temperature T_g is equal to the net energy leaving the parallel plates. Hence, per unit area of the plates,

$$
q_g = -(q_1 + q_2) \tag{13-26}
$$

When the medium between the plates does not absorb or emit radiation, then $\bar{\tau}_\lambda = 1$ and (13-24a) and (13-24b) reduce to (8-10). With an absorbing-radiating gas present, the numerical integration of (13-24a) and (13-24b) over all wavelengths to obtain the q_1 and q_2 is difficult because of the very irregular variations of the gas absorption coefficient with wavelength.

13-3.4 Band Equations for Solution and Integration of Spectral Enclosure Equations

An approach for integrating over wavelength is developed by dividing the spectrum into bands in which the gas is either absorbing or nonabsorbing. For situations where both the geometry and thermal boundary conditions are simple, the enclosure equations can be solved in closed form such as in Examples 13-1 and 13-2. Considering Eq. (13-21), for example, if $\bar{\alpha}_\lambda$ is zero within a wavelength region this nonabsorbing band does not contribute to Q_g. Let l designate an absorbing band; then by summation

$$
\frac{Q_g}{A_1} = \sum_l \left(\frac{e_{\lambda b,g} - e_{\lambda b,1}}{1/\epsilon_{\lambda,1} + 1/\bar{\alpha}_{1-1} - 1} \right)_l \Delta\lambda_l \tag{13-27}
$$

The $\bar{\alpha}_{1-1}$ for each band is found from (13-43) as will be discussed. Usually the bandwidth is small so that $e_{\lambda b,g}$, $e_{\lambda b,1}$, and $\epsilon_{\lambda,1}$ can be considered constant over each band. If the $e_{\lambda b}(\lambda)$ variation is significant over a band, the integrated black-body functions can be used as in Eq. (8-13).

For more complex enclosures, the solution of (13-19) in matrix form can be obtained. For a band of width $\Delta\lambda$ the integration of Eq. (13-19) gives, as in (8-15a),

$$\int_{\Delta\lambda} \sum_{j=1}^{N} \left(\frac{\delta_{kj}}{\epsilon_{\lambda,j}} - F_{k-j} \frac{1-\epsilon_{\lambda,j}}{\epsilon_{\lambda,j}} \bar{\tau}_{\lambda,k-j} \right) dq_{\lambda,j}$$

$$= \int_{\Delta\lambda} \sum_{j=1}^{N} [(\delta_{kj} - F_{k-j}\bar{\tau}_{\lambda,k-j})e_{\lambda b,j} - F_{k-j}\bar{\alpha}_{\lambda,k-j}e_{\lambda b,g}] d\lambda \qquad (13\text{-}28)$$

It is assumed that the bands are sufficiently narrow so that $dq_{\lambda,j}$, $\epsilon_{\lambda,j}$, $\bar{\tau}_{\lambda,k-j}$, $\bar{\alpha}_{\lambda,k-j}$, $e_{\lambda b,j}$, and $e_{\lambda b,g}$ can be regarded as constants over the bandwidth, being characteristic of some mean wavelength within the band, or, in the case of $\bar{\tau}$ and $\bar{\alpha}$, being averaged over the band. Then (13-28) is written for band l as

$$\sum_{j=1}^{N} \left(\frac{\delta_{kj}}{\epsilon_{l,j}} - F_{k-j} \frac{1-\epsilon_{l,j}}{\epsilon_{l,j}} \bar{\tau}_{l,k-j} \right) \Delta q_{l,j}$$

$$= \sum_{j=1}^{N} [(\delta_{kj} - F_{k-j}\bar{\tau}_{l,k-j})e_{lb,j} - F_{k-j}\bar{\alpha}_{l,k-j}e_{lb,g}] \Delta\lambda_l \qquad (13\text{-}29)$$

In a spectral region where the gas is essentially nonabsorbing, $\bar{\tau}_l = 1$ and $\bar{\alpha}_l = 0$ so that (13-29) reduces to

$$\sum_{j=1}^{N} \left(\frac{\delta_{kj}}{\epsilon_{l,j}} - F_{k-j} \frac{1-\epsilon_{l,j}}{\epsilon_{l,j}} \right) \Delta q_{l,j} = \sum_{j=1}^{N} (\delta_{kj} - F_{k-j})e_{lb,j} \Delta\lambda_l \qquad (13\text{-}30)$$

which is the same form as (8-15a). Equations (13-29) and (13-30) are the *enclosure equations* that provide a set of simultaneous equations to determine Δq for each band at each boundary. The $\bar{\tau}_{l,k-j}$ in (13-29) is found from (13-8) by taking an integrated average over the band:

$$\bar{\tau}_{l,k-j} = \frac{1}{A_k F_{k-j}} \int_{A_j} \int_{A_k} \frac{\left[\frac{1}{\Delta\lambda_l} \int_{\Delta\lambda_l} \tau_\lambda(S) \, d\lambda \right] \cos\theta_j \cos\theta_k}{\pi S^2} \, dA_k \, dA_j \qquad (13\text{-}31)$$

Similarly, the $\bar{\alpha}_{l,k-j}$ is

$$\bar{\alpha}_{l,k-j} = \frac{1}{A_k F_{k-j}} \int_{A_j} \int_{A_k} \frac{\left[\frac{1}{\Delta\lambda_l} \int_{\Delta\lambda_l} \alpha_\lambda(S) \, d\lambda \right] \cos\theta_j \cos\theta_k}{\pi S^2} \, dA_k \, dA_j \qquad (13\text{-}32)$$

For each small bandwidth, $\bar{\alpha}_{l,k-j} = 1 - \bar{\tau}_{l,k-j}$, and to evaluate $\bar{\alpha}_l$ and $\bar{\tau}_l$ only the single integral is needed:

$$A_k F_{k-j} \bar{\alpha}_{l, k-j} = \int_{A_l} \int_{A_k} \frac{\alpha_l(S) \cos \theta_j \cos \theta_k}{\pi S^2} dA_k \, dA_j \qquad (13\text{-}33)$$

where $\alpha_l(S)$ is the integrated band absorption.

$$\alpha_l(S) = \frac{1}{\Delta\lambda_l} \int_{\Delta\lambda_l} \alpha_\lambda(S) \, d\lambda = \frac{1}{\Delta\lambda_l} \int_{\Delta\lambda_l} [1 - \exp(-a_\lambda S)] \, d\lambda \qquad (13\text{-}34)$$

If desired, the α_l can be expressed from Eq. (12-66) in terms of the effective bandwidth as

$$\alpha_l(S) = \frac{\bar{A}_l(S)}{\Delta\lambda_l} \qquad (13\text{-}35)$$

A detailed development is given by Nelson [1] of the band relations for a nongray isothermal gas in an enclosure with diffuse walls.

13-3.5 Gray Medium in a Gray Enclosure

If a gas contains many suspended particles or droplets it may be reasonable to neglect spectral variations of the properties of the suspension. In addition, the walls in a furnace or combustion chamber may be partially soot covered, so an examination of the radiation transfer with gray boundaries may yield worthwhile results.

From Example 13-1 if a gray gas at T_g is bounded by a chamber consisting of one gray wall with area A_1 at T_1, the energy supplied to the gas by combustion or other nonradiative means and transferred to the wall is

$$Q_g = \frac{A_1 \sigma(T_g^4 - T_1^4)}{1/\epsilon_1 + 1/\bar{\alpha}_{1-1} - 1} = \frac{\epsilon_1 \bar{\alpha}_{1-1}}{1 - (1 - \epsilon_1)(1 - \bar{\alpha}_{1-1})} A_1 \sigma(T_g^4 - T_1^4) \qquad (13\text{-}36)$$

From Example 13-2, the q_1 from Eq. (13-24a) becomes for a gray medium between infinite gray parallel plates,

$$q_1 = \frac{\epsilon_1 \epsilon_2 \bar{\tau} \sigma(T_1^4 - T_2^4) + \epsilon_1(1 - \bar{\tau})[1 + (1 - \epsilon_2)\bar{\tau}]\sigma(T_1^4 - T_g^4)}{1 - (1 - \epsilon_1)(1 - \epsilon_2)\bar{\tau}^2} \qquad (13\text{-}37)$$

and similarly for q_2 from Eq. (13-24b).

Now consider a chamber made up of two finite surfaces that enclose a gas. From Eq. (13-19),

$$\left(\frac{1}{\epsilon_1} - F_{1-1} \frac{1 - \epsilon_1}{\epsilon_1} \bar{\tau}_{1-1}\right) q_1 - F_{1-2} \frac{1 - \epsilon_2}{\epsilon_2} \bar{\tau}_{1-2} q_2$$

$$= (1 - F_{1-1}\bar{\tau}_{1-1})\sigma T_1^4 - F_{1-2}\bar{\tau}_{1-2}\sigma T_2^4 - (F_{1-1}\bar{\alpha}_{1-1} + F_{1-2}\bar{\alpha}_{1-2})\sigma T_g^4 \qquad (13\text{-}38a)$$

$$-F_{2-1} \frac{1 - \epsilon_1}{\epsilon_1} \bar{\tau}_{2-1} q_1 + \left(\frac{1}{\epsilon_2} - F_{2-2} \frac{1 - \epsilon_2}{\epsilon_2} \bar{\tau}_{2-2} \right) q_2$$

$$= -F_{2-1} \bar{\tau}_{2-1} \sigma T_1^4 + (1 - F_{2-2} \bar{\tau}_{2-2}) \sigma T_2^4 - (F_{2-1} \bar{\alpha}_{2-1} + F_{2-2} \bar{\alpha}_{2-2}) \sigma T_g^4 \quad (13\text{-}38b)$$

If T_1, T_2, and T_g are specified, these equations can be solved for q_1 and q_2. Then the energy supplied to the gas by some means other than radiation within the enclosure is $Q_g = -(q_1 A_1 + q_2 A_2)$.

If one of the walls (wall 2) is adiabatic (for example, an uncooled refractory wall), then $q_2 = 0$. The T_2 can be eliminated from (13-38) and the result solved for q_1 in terms of T_1 and T_g to yield,

$$\frac{q_1 A_1}{\sigma(T_1^4 - T_g^4)} = \left(\frac{1 - \epsilon_1}{A_1 \epsilon_1} + \frac{1}{1/f_{1g} + 1/(f_{12} + f_{2g})} \right)^{-1} \quad (13\text{-}39)$$

where

$$f_{1g} = [A_1(F_{1-1} \bar{\alpha}_{1-1} + F_{1-2} \bar{\alpha}_{1-2})]^{-1}$$

$$f_{12} = (A_1 F_{1-2} \bar{\tau}_{1-2})^{-1}$$

$$f_{2g} = [A_2(F_{2-1} \bar{\alpha}_{2-1} + F_{2-2} \bar{\alpha}_{2-2})]^{-1}$$

Returning to Eqs. (13-38), if the gas is adiabatic ($Q_g = 0$) then the two walls are exchanging energy through the medium that will come to an equilibrium temperature. The $Q_2 = -Q_1$ so that $q_2 = -q_1 A_1/A_2$. The q_2 is then eliminated from (13-38) and the two equations are combined to eliminate T_g. The resulting $Q_1 = q_1 A_1$ is

$$\frac{q_1 A_1}{\sigma(T_1^4 - T_2^4)} = \left(\frac{1 - \epsilon_1}{A_1 \epsilon_1} + \frac{1}{1/f_{12} + 1/(f_{1g} + f_{2g})} + \frac{1 - \epsilon_2}{A_2 \epsilon_2} \right)^{-1} \quad (13\text{-}40)$$

where the f quantities are defined in (13-39). With Q_1 known, the equations can be solved for the gas temperature if desired.

For the general case of a gray gas in a gray enclosure where there are more than a few surfaces, a closed-form algebraic equation would usually not be obtained. Rather the enclosure equations would be solved numerically as a set of simultaneous relations. For the general situation, Eq. (13-29) is written for the gray case (one band) as

$$\sum_{j=1}^{N} \left(\frac{\delta_{kj}}{\epsilon_j} - F_{k-j} \frac{1 - \epsilon_j}{\epsilon_j} \bar{\tau}_{k-j} \right) q_j = \sum_{j=1}^{N} (\delta_{kj} - F_{k-j} \bar{\tau}_{k-j}) \sigma T_j^4 - F_{k-j} \bar{\alpha}_{k-j} \sigma T_g^4 \quad (13\text{-}41)$$

13-4 EVALUATION OF SPECTRAL GEOMETRIC-MEAN TRANSMITTANCE AND ABSORPTANCE FACTORS

To compute values from the enclosure equations, the $\bar{\tau}$ and $\bar{\alpha}$ or $AF\bar{\tau}$ and $AF\bar{\alpha}$ must be evaluated. These quantities depend on both geometry and wavelength. By use of the definitions in (13-8) and (13-9),

$$A_j F_{j-k} \bar{\tau}_{\lambda, j-k} = \int_{A_k} \int_{A_j} \frac{\exp{(-a_\lambda S)} \cos \theta_k \cos \theta_j}{\pi S^2} \, dA_j \, dA_k \qquad (13\text{-}42)$$

$$A_j F_{j-k} \bar{\alpha}_{\lambda, j-k} = \int_{A_k} \int_{A_j} \frac{[1 - \exp{(-a_\lambda S)}] \cos \theta_k \cos \theta_j}{\pi S^2} \, dA_j \, dA_k$$

$$= A_j F_{j-k} (1 - \bar{\tau}_{\lambda, j-k}) \qquad (13\text{-}43)$$

The double integral in (13-42) must be carried out for various orientations of the surfaces A_j and A_k. The evaluation of some specific geometries are now considered.

13-4.1 Hemisphere to Differential Area at Center of Its Base

As shown in Fig. 13-4, let A_j be the surface of a hemisphere of radius R, and dA_k be a differential area at the center of the hemisphere base. Then (13-42) becomes, since $S = R = \theta_j = 0$ (path R is normal to hemisphere surface),

$S = R$ and $\theta_j = 0$ (path R is normal to hemisphere surface),

$$A_j \, dF_{j-dk} \, \bar{\tau}_{\lambda, j-dk} = dA_k \int_{A_j} \frac{\exp{(-a_\lambda R)} \cos \theta_k \cos (0)}{\pi R^2} \, dA_j$$

The convenient dA_j is a ring element $dA_j = 2\pi R^2 \sin \theta_k \, d\theta_k$, and the factors involving R can be taken out of the integral. This gives

$$A_j \, dF_{j-dk} \, \bar{\tau}_{\lambda, j-dk} = dA_k \frac{\exp{(-a_\lambda R)} \, 2\pi R^2}{\pi R^2} \int_{\theta_k = 0}^{\pi/2} \cos \theta_k \sin \theta_k \, d\theta_k$$

$$= dA_k \exp{(-a_\lambda R)}$$

With $A_j \, dF_{j-dk} = dA_k \, F_{dk-j}$ and $F_{dk-j} = 1$, this reduces to

$$\bar{\tau}_{\lambda, j-dk} = \exp{(-a_\lambda R)} \qquad (13\text{-}44)$$

This especially simple relation will be used later in the concept of mean beam length. This is an approximation in which radiation from an actual gas volume is replaced by that from an effective hemispherical volume.

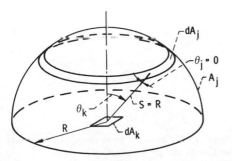

Figure 13-4 Hemisphere filled with isothermal gas.

13-4.2 Top of Right Circular Cylinder to Center of Its Base

This geometry is shown in Fig. 13-5. Since $\theta_j = \theta_k = \theta$, the integral in Eq. (13-42) becomes, for the top of the cylinder A_j radiating to the element dA_k at the center of its base,

$$A_j \, dF_{j-dk} \, \bar{\tau}_{\lambda,j-dk} = dA_k \int_{A_j} \frac{\exp(-a_\lambda S) \cos^2 \theta}{\pi S^2} \, dA_j \tag{13-45}$$

Since $\xi^2 = S^2 - h^2$, $dA_j = 2\pi\xi \, d\xi = 2\pi S \, dS$. Then using $\cos\theta = h/S$, Eq. (13-45) becomes

$$A_j \, dF_{j-dk} \, \bar{\tau}_{\lambda,j-dk} = dA_k \, 2h^2 \int_h^{\sqrt{R^2+h^2}} \frac{\exp(-a_\lambda S)}{S^3} \, dS \tag{13-46}$$

Now let $a_\lambda S = \kappa_\lambda$ to obtain

$$A_j \, dF_{j-dk} \, \bar{\tau}_{\lambda,j-dk} = dA_k \, 2h^2 a_\lambda^2 \int_{a_\lambda h}^{a_\lambda \sqrt{R^2+h^2}} \frac{\exp(-\kappa_\lambda)}{\kappa_\lambda^3} \, d\kappa_\lambda \tag{13-47}$$

This integral can be expressed in terms of a tabulated function, the *exponential integral function* defined in Appendix E, by writing

$$\int_{a_\lambda h}^{a_\lambda \sqrt{R^2+h^2}} \frac{\exp(-\kappa_\lambda)}{\kappa_\lambda^3} \, d\kappa_\lambda = \int_\infty^{a_\lambda \sqrt{R^2+h^2}} \frac{\exp(-\kappa_\lambda)}{\kappa_\lambda^3} \, d\kappa_\lambda - \int_\infty^{a_\lambda h} \frac{\exp(-\kappa_\lambda)}{\kappa_\lambda^3} \, d\kappa_\lambda$$

$$\tag{13-48}$$

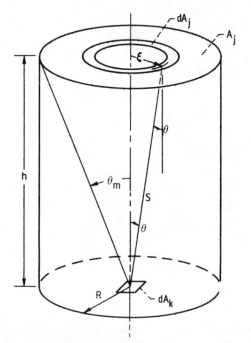

Figure 13-5 Geometry for exchange from top of gas-filled cylinder to center of its base.

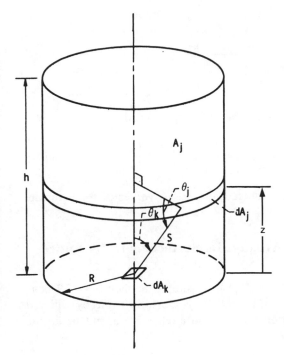

Figure 13-6 Geometry for exchange from side of gas-filled cylinder to center of its base.

Letting $\kappa_\lambda = (a_\lambda \sqrt{R^2 + h^2})/\mu$ and $a_\lambda h/\mu$, respectively, in the two integrals gives

$$-\frac{1}{(a_\lambda\sqrt{R^2 + h^2})^2} \int_0^1 \mu \exp\left(-\frac{a_\lambda\sqrt{R^2 + h^2}}{\mu}\right) d\mu + \frac{1}{(a_\lambda h)^2} \int_0^1 \mu \exp\left(-\frac{a_\lambda h}{\mu}\right) d\mu$$

The integral in (13-47) can then be written in terms of the exponential integral function as

$$\int_{a_\lambda h}^{a_\lambda\sqrt{R^2+h^2}} \frac{\exp(-\kappa_\lambda)}{\kappa_\lambda^3}\, d\kappa_\lambda = \frac{1}{(a_\lambda h)^2} E_3(a_\lambda h)$$

$$-\frac{1}{[a_\lambda h\sqrt{(R/h)^2 + 1}]^2} E_3\left[a_\lambda h\sqrt{\left(\frac{R}{h}\right)^2 + 1}\right] \quad (13\text{-}49)$$

so it can be readily evaluated for various values of the parameters R/h and $a_\lambda h$.

13-4.3 Side of Cylinder to Center of Its Base

Let dA_j be a ring around the wall of a cylinder as shown in Fig. 13-6, and note that $dA_j = 2\pi R\, dz$, $\cos\theta_k = z/S$, $\cos\theta_j = R/S$, and $z\, dz = S\, dS$. Then (13-42) can be written for the side of the cylinder to dA_k as

$$A_j\, d\Gamma_{j-dk}\, \bar{\tau}_{\lambda, j-dk} = 2\, dA_k R^2 \int_R^{\sqrt{R^2+h^2}} \frac{\exp(-a_\lambda S)}{S^3}\, dS \quad (13\text{-}50)$$

This is of the same form as (13-46). Let $a_\lambda S = \kappa_\lambda$ to obtain

$$A_j\, dF_{j-dk}\, \bar{\tau}_{\lambda, j-dk} = 2dA_k\, R^2 a_\lambda{}^2 \int_{a_\lambda R}^{a_\lambda \sqrt{R^2 + h^2}} \frac{\exp{(-\kappa_\lambda)}}{\kappa_\lambda{}^3}\, d\kappa_\lambda$$

$$= 2dA_k \left(\frac{R}{h}\right)^2 (a_\lambda h)^2 \left\{ \frac{1}{[a_\lambda h(R/h)]^2} E_3\left[a_\lambda h \left(\frac{R}{h}\right)\right] \right.$$

$$\left. - \frac{1}{[a_\lambda h \sqrt{(R/h)^2 + 1}]^2} E_3\left[a_\lambda h \sqrt{\left(\frac{R}{h}\right)^2 + 1}\right] \right\} \qquad (13\text{-}51)$$

As for (13-49), this can be readily evaluated for various values of the parameters R/h and $a_\lambda h$.

13-4.4 Entire Sphere to Any Area on Its Surface or to Its Entire Surface

From Fig. 13-7, since $\theta_k = \theta_j$ let them both be simply θ; then $S = 2R \cos \theta$. Starting with (13-45) it is noted that $dA_j \cos \theta / S^2$ is the solid angle by which dA_j is viewed from dA_k. The intersection of the solid angle with a unit hemisphere shows that this equals $2\pi \sin \theta\, d\theta$. Then

$$A_j\, dF_{j-dk}\, \bar{\tau}_{\lambda, j-dk} = dA_k \int_{\theta=0}^{\pi/2} \exp(-a_\lambda S) 2 \cos \theta \sin \theta\, d\theta = \frac{2dA_k}{4R^2} \int_{S=0}^{2R} \exp(-a_\lambda S) S\, dS$$

Integrating gives

$$A_j\, dF_{j-dk}\, \bar{\tau}_{\lambda, j-dk} = \frac{2dA_k}{(2a_\lambda R)^2} \left[1 - (2a_\lambda R + 1) \exp{(-2a_\lambda R)}\right] \qquad (13\text{-}52)$$

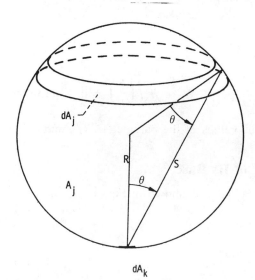

dA_k

Figure 13-7 Geometry for exchange from surface of gas-filled sphere to itself.

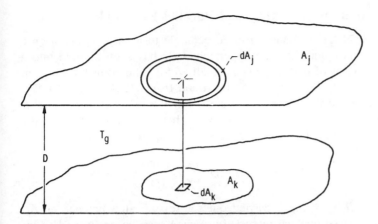

Figure 13-8 Isothermal gas layer between infinite parallel plates.

which is in terms of the single parameter $2a_\lambda R$, the optical diameter of the sphere.

Equation (13-52) is integrated over any finite area A_k to give $\bar{\tau}_\lambda$ from the entire sphere to A_k as

$$A_j F_{j-k} \bar{\tau}_{\lambda,j-k} = \frac{2A_k}{(2a_\lambda R)^2} [1 - (2a_\lambda R + 1) \exp(-2a_\lambda R)]$$

Since $F_{j-k} = A_k/A_j$ [from (7-96)],

$$\tau_{\lambda,j-k} = \frac{2}{(2a_\lambda R)^2} [1 - (2a_\lambda R + 1) \exp(-2a_\lambda R)] \qquad (13\text{-}53)$$

which also holds for the entire sphere to its entire surface.

13-4.5 Infinite Plate to Any Area on Parallel Plate

Consider on one plate an element dA_k (Fig. 13-8), and on the other plate a concentric ring element dA_j centered about the normal to dA_k. The geometry is like that in Fig. 13-5 for a ring on the top of a cylinder to the center of its base. Then from (13-47)

$$A_j \, dF_{j-dk} \, \bar{\tau}_{\lambda,j-dk} = dA_k \, 2(a_\lambda D)^2 \int_{a_\lambda D}^{\infty} \frac{\exp(-\kappa_\lambda)}{\kappa_\lambda^3} \, d\kappa_\lambda$$

where $a_\lambda D$ is the optical spacing between the plates. By use of the procedure leading to (13-49), the integral is transformed to $E_3(a_\lambda D)/(a_\lambda D)^2$. Then integrating over any finite area A_k as shown in Fig. 13-8 gives $A_j F_{j-k} \bar{\tau}_{\lambda,j-k} = A_k 2E_3(a_\lambda D)$. With $A_j F_{j-k} = A_k F_{k-j}$ and $F_{k-j} = 1$, this reduces to

$$\bar{\tau}_{\lambda,j-k} = 2E_3(a_\lambda D) \qquad (13\text{-}54)$$

13-4.6 Rectangle to a Directly Opposed Parallel Rectangle

Consider as in Fig. 13-9 the exchange from a rectangle to an area element on a directly opposed parallel rectangle. The upper rectangle has been divided into a circular region and a series of partial rings of small width. The contribution from the circle of radius R to $A_j \, dF_{j-dk} \, \bar{\tau}_{\lambda,j-dk}$ can be found from (13-47) and (13-49), for the top of a cylinder to the center of its base. For the nth partial ring, let f_n be the fraction it occupies of a full circular ring. Then by use of (13-45), the contribution of all the partial rings to $A_j \, dF_{j-dk} \, \bar{\tau}_{\lambda,j-dk}$ is approximated by

$$dA_k \sum_n f_n \frac{\exp(-a_\lambda S_n)}{\pi S_n^2} \left(\frac{D}{S_n}\right)^2 2\pi R_n \, \Delta R_n = dA_k \, 2D^2 \sum_n f_n \frac{\exp(-a_\lambda S_n)}{(D^2 + R_n^2)^2} R_n \, \Delta R_n$$

This evaluation of $A_j \, dF_{j-dk} \, \bar{\tau}_{\lambda,j-dk}$ is carried out for several area patches on A_k. This is usually sufficient so that the integration over A_k can be performed as indicated by (13-9) to yield

$$A_j F_{j-k} \bar{\tau}_{\lambda,j-k} = \int_{A_k} A_j \, dF_{j-dk} \, \bar{\tau}_{\lambda,j-dk}$$

EXAMPLE 13-3 A nongray, absorbing-emitting, and partially transmitting plane layer with a_λ as shown in Fig. 13-10 is 1.2 cm thick and is on top of an opaque diffuse-gray infinite plate. The plate temperature has been raised suddenly so that the layer remains at its initial uniform temperature below the temperature of the plate. What is the net heat flux being lost from the plate-layer system? Neglect heat conduction effects in the layer, and neglect any reflections at the upper boundary of the layer.

The environment above the layer acts as a black enclosure at T_e. Equation (13-24b) can be applied directly by integrating in two bands and using $\epsilon_{\lambda,2} = 1$. The geometric-mean transmittance factors are obtained from

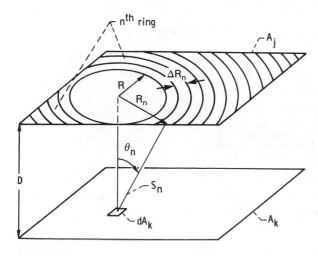

Figure 13-9 Geometry for exchange between two directly opposed parallel rectangles with intervening gas.

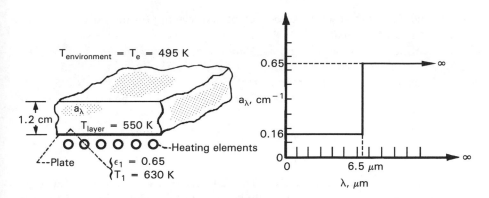

Figure 13-10 Plane layer of nongray absorbing-emitting material.

Eq. (13-54) as $\bar{\tau}_{\lambda,1} = 2E_3(0.16 \times 1.2) = 0.7142$, $0 \le \lambda \le 6.5$ μm; $\bar{\tau}_{\lambda,2} = 2E_3(0.65 \times 1.2) = 0.2974$, 6.5 μm $\le \lambda \le \infty$. This yields

$$q = -q_2 = \epsilon_1 \bar{\tau}_{\lambda,1}(\sigma T_1^4 F_{0-6.5T_1} - \sigma T_e^4 F_{0-6.5T_e})$$

$$+ (1 - \bar{\tau}_{\lambda,1})[1 + (1 - \epsilon_1)\bar{\tau}_{\lambda,1}](\sigma T_g^4 F_{0-6.5T_g} - \sigma T_e^4 F_{0-6.5T_e})$$

$$+ \epsilon_1 \bar{\tau}_{\lambda,2}[\sigma T_1^4(1 - F_{0-6.5T_1}) - \sigma T_e^4(1 - F_{0-6.5T_e})]$$

$$+ (1 - \bar{\tau}_{\lambda,2})[1 + (1 - \epsilon_1)\bar{\tau}_{\lambda,2}][\sigma T_g^4(1 - F_{0-6.5T_g}) - \sigma T_e^4(1 - F_{0-6.5T_e})]$$

Inserting the values $F_{0-6.5\times630} = 0.4979$, $F_{0-6.5\times550} = 0.3984$ and $F_{0-6.5\times495} = 0.3220$ yields $q = 2954$ W/m^2.

13-4.7 Geometric-Mean Beam Length for Band Equations

For use in the band equations (13-29), the integral in (13-33) must be evaluated between pairs of enclosure surfaces for the various wavelength bands involved. When more than a few bands absorb appreciably, the solution requires considerable computational effort. A simplification was developed by Dunkle [2] that reduces labor and yields good accuracy. Dunkle assumes that the integrated band absorption $\alpha_l(S)$ is a *linear function* of path length. This has some physical basis, as it holds exactly for a band of weak nonoverlapping lines [Eq. (12-77)]. Also, it is the form of some of the effective bandwidths in the correlations of Table 12-3 and those following Example 12-2. As shown in [2] by means of a few examples, reasonable values of the energy exchange are obtained by use of this approximation. Hence, let α_l in (13-33) have the linear form from (12-77) [note that $\Delta\lambda_l = A_0$ in (12-77)]

$$\alpha_l(S) = \frac{\bar{A}_l}{\Delta\lambda_l} = \frac{\bar{A}_l}{A_0} = \frac{S_c}{\delta} S \tag{13-55}$$

where S_c and δ are the absorption coefficient and spacing of the individual weak lines as defined in Chap. 12.

Now define a mean path length $\bar{S}_{k-j}$ called the *geometric-mean beam length*. This mean length is such that α_l evaluated from (13-55) by using $S = \bar{S}_{k-j}$ will yield $\bar{\alpha}_{l,k-j}$ as found from the integral in (13-34). After substitution of $\bar{\alpha}_{l,k-j} = (S_c/\delta)\bar{S}_{k-j}$ and $\alpha_l = (S_c/\delta)S$ into (13-33), the S_c/δ drops out and the relation to obtain $\bar{S}_{k-j}$ is

$$\bar{S}_{k-j} = \bar{S}_{j-k} = \frac{1}{A_k F_{k-j}} \int_{A_j} \int_{A_k} \frac{\cos\theta_j \cos\theta_k}{\pi S} \, dA_k \, dA_j \tag{13-56}$$

which is dependent only on geometry. This integral is also obtained in Eq. (13-43) when $a_\lambda S$ is small (optically thin limit). In [2] S_{k-j} values are tabulated for parallel equal rectangles, for rectangles at right angles, and for a differential sphere and a rectangle. Analytical relations for rectangles are in Eqs. (13-57a,b). For directly opposed parallel equal rectangles with sides of length a and b and spaced a distance c apart,

$$\frac{\bar{S}_{k-j} A_k F_{k-j}}{abc} = \frac{4}{\pi} \left\{ \tan^{-1}\frac{\eta\beta}{\sqrt{1+\eta^2+\beta^2}} + \frac{1}{\eta}\ln\left[\frac{\beta+\sqrt{1+\eta^2+\beta^2}}{\sqrt{1+\eta^2}(\beta+\sqrt{1+\beta^2})}\right] \right.$$

$$+ \frac{1}{\beta}\ln\left[\frac{\eta+\sqrt{1+\eta^2+\beta^2}}{\sqrt{1+\beta^2}(\eta+\sqrt{1+\eta^2})}\right]$$

$$\left. + \frac{1}{\eta\beta}[\sqrt{1+\eta^2}+\sqrt{1+\beta^2}-1-\sqrt{1+\eta^2+\beta^2}] \right\} \tag{13-57a}$$

where $\eta = a/c$ and $\beta = b/c$. The F_{k-j} can be obtained from factor 11 in Appendix C. For rectangles ab and bc at right angles with a common edge b,

$$\frac{\bar{S}_{k-j} A_k F_{k-j}}{abc} = \frac{1}{\pi}\left\{ \frac{\gamma}{\alpha}\ln\frac{(1+\sqrt{1+\gamma^2})\sqrt{\alpha^2+\gamma^2}}{\gamma(1+\sqrt{1+\alpha^2+\gamma^2})} \right.$$

$$+ \frac{\alpha}{\gamma}\ln\frac{(1+\sqrt{1+\alpha^2})\sqrt{\alpha^2+\gamma^2}}{\alpha(1+\sqrt{1+\alpha^2+\gamma^2})}$$

$$+ \frac{1}{3\gamma\alpha}[(1+\gamma^2)^{3/2}+(1+\alpha^2)^{3/2}+(\alpha^2+\gamma^2)^{3/2}-(1+\alpha^2+\gamma^2)^{3/2}]$$

$$+ \frac{\alpha}{\gamma}[\sqrt{1+\alpha^2+\gamma^2}-\sqrt{\alpha^2+\gamma^2}-\sqrt{1+\alpha^2}]$$

$$+ \frac{\gamma}{\alpha}[\sqrt{1+\alpha^2+\gamma^2}-\sqrt{\alpha^2+\gamma^2}-\sqrt{1+\gamma^2}]$$

$$\left. + \frac{2}{3}\left[\frac{\gamma^2}{\alpha}+\frac{\alpha^2}{\gamma}-\frac{1}{2\gamma\alpha}\right] \right\} \tag{13-57b}$$

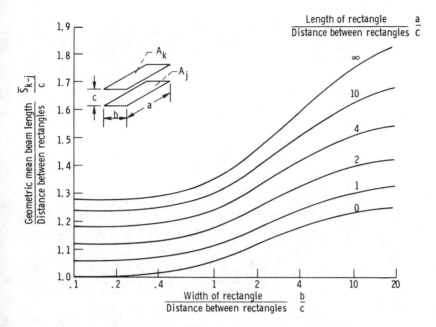

Figure 13-11 Geometric mean beam lengths for equal parallel rectangles [2].

where $\alpha = a/b$ and $\gamma = c/b$. The F_{k-j} can be obtained from factor 15 in Appendix C. Results for equal opposed parallel rectangles are in Fig. 13-11. Values for equal parallel rectangles and rectangles at right angles are given in Tables 13-1 and 13-2. Other $\bar{S}_{k-j}$ values are referenced by Hottel and Sarofim [3]. In [4] values are obtained for gas in the space between two infinitely long coaxial cylinders.

For a given gas at uniform conditions, the geometric-mean beam length can be used in the effective-bandwidth correlations discussed in Chap. 12 to obtain $\bar{A}_l$. Using the $\Delta\lambda_l$ as obtained in the next paragraph yields $\bar{\alpha}_l = \bar{A}_l/\Delta\tau_l$ from Eq. (13-55) and $\bar{\tau}_l$ from $1 - \bar{\alpha}_l$. Then (13-29) and (13-30) can be solved for Δq_k for each wavelength band l. The total energies at each surface k are found from a summation over all wavelength bands,

$$q_k = \sum_{\substack{\text{absorbing} \\ \text{bands}}} \Delta q_{l,k} + \sum_{\substack{\text{nonabsorbing} \\ \text{bands}}} \Delta q_{l,k} \qquad (13\text{-}58)$$

The wavelength span $\Delta\lambda_l$ of each band is needed to carry out the solution. As discussed after (12-74), this span can increase with path length. Edwards and Nelson [5] and Edwards [6] give recommended spans for CO_2 and H_2O vapor; these values are reproduced in Table 13-3 for the parallel-plate geometry. Note that these values are in wavenumber units. For other geometries, Edwards and Nelson give methods for choosing approximate spans for CO_2 and H_2O bands. Briefly, the method is to use approximate band spans based on the longest important mass path length in the geometry being studied. With this in mind, the limits of Table 13-3 are probably adequate for problems involving CO_2 and H_2O vapor.

Table 13-1 Geometric mean-beam-length ratios and configuration factors for parallel equal rectangles [2]

a/c		0	0.1	0.2	0.4	0.6	1.0	2.0	4.0	6.0	10.0	20.0
0	$\bar{S}_{k-j}/c$	1.000	1.001	1.003	1.012	1.025	1.055	1.116	1.178	1.205	1.230	1.251
	F_{k-j}											
0.1	$\bar{S}_{k-j}/c$	1.001	1.002	1.004	1.013	1.026	1.056	1.117	1.179	1.207	1.233	1.254
	F_{k-j}		0.00316	0.00626	0.01207	0.01715	0.02492	0.03514	0.04210	0.04463	0.04671	0.04829
0.2	$\bar{S}_{k-j}/c$	1.003	1.004	1.006	1.015	1.028	1.058	1.120	1.182	1.210	1.235	1.256
	F_{k-j}		0.00626	0.01240	0.02391	0.03398	0.04941	0.06971	0.08353	0.08859	0.09271	0.09586
0.4	$\bar{S}_{k-j}/c$	1.012	1.013	1.015	1.024	1.037	1.067	1.129	1.192	1.220	1.245	1.267
	F_{k-j}		0.01207	0.02391	0.04614	0.06560	0.09554	0.13513	0.16219	0.17209	0.18021	0.18638
0.6	$\bar{S}_{k-j}/c$	1.025	1.026	1.028	1.037	1.050	1.080	1.143	1.206	1.235	1.261	1.282
	F_{k-j}		0.01715	0.03398	0.06560	0.09336	0.13627	0.19341	0.23271	0.24712	0.25896	0.26795
1.0	$\bar{S}_{k-j}/c$	1.055	1.056	1.058	1.067	1.080	1.110	1.175	1.242	1.272	1.300	1.324
	F_{k-j}		0.02492	0.04941	0.09554	0.13627	0.19982	0.28588	0.34596	0.36813	0.38638	0.40026
2.0	$\bar{S}_{k-j}/c$	1.116	1.117	1.120	1.129	1.143	1.175	1.246	1.323	1.359	1.393	1.421
	F_{k-j}		0.03514	0.06971	0.13513	0.19341	0.28588	0.41525	0.50899	0.54421	0.57338	0.59563
4.0	$\bar{S}_{k-j}/c$	1.178	1.179	1.182	1.192	1.206	1.242	1.323	1.416	1.461	1.505	1.543
	F_{k-j}		0.04210	0.08353	0.16219	0.23271	0.34596	0.50899	0.63204	0.67954	0.71933	0.74990
6.0	$\bar{S}_{k-j}/c$	1.205	1.207	1.210	1.220	1.235	1.272	1.359	1.461	1.513	1.564	1.609
	F_{k-j}		0.04463	0.08859	0.17209	0.24712	0.36813	0.54421	0.67954	0.73258	0.77741	0.81204
10.0	$\bar{S}_{k-j}/c$	1.230	1.233	1.235	1.245	1.261	1.300	1.393	1.505	1.564	1.624	1.680
	F_{k-j}		0.04671	0.09271	0.18021	0.25896	0.38638	0.57338	0.71933	0.77741	0.82699	0.86563
20.0	$\bar{S}_{k-j}/c$	1.251	1.254	1.256	1.267	1.282	1.324	1.421	1.543	1.609	1.680	1.748
	F_{k-j}		0.04829	0.09586	0.18638	0.26795	0.40026	0.59563	0.74990	0.81204	0.86563	0.90785
∞	$\bar{S}_{k-j}/c$	1.272	1.274	1.277	1.289	1.306	1.349	1.452	1.584	1.660	1.745	1.832
	F_{k-j}		0.04988	0.09902	0.19258	0.27698	0.41421	0.61803	0.78078	0.84713	0.90499	0.95125

If all surface temperatures are given in a problem, the results found from Eq.
(13-58) complete the solution. If q_k is given for n surfaces and T_k for the remaining
$N - n$ surfaces, then the n unknown surface temperatures are guessed, the equations are solved for all q, and then the calculated q_k are compared to the given
values. If they do not agree, new values of T_k for the n surfaces are assumed and
the calculation is repeated. This procedure is continued until there is agreement
between given and calculated q_k for all k. Equation (13-20) expressed as a sum

**Table 13-2 Configuration factors and mean beam-length functions for rect-
angles at right angles [2]**

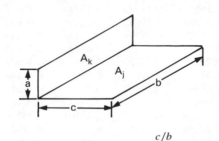

a/b		0.05	0.10	0.20	0.4	0.6	1.0
0.02	$A_k F_{k-j}/b^2$	0.007982	0.008875	0.009323	0.009545	0.009589	0.009628
	$A_k F_{k-j}\tilde{S}_{k-j}/abc$	0.17840	0.12903	0.08298	0.04995	0.03587	0.02291
0.05	$A_k F_{k-j}/b^2$	0.014269	0.018601	0.02117	0.02243	0.02279	0.02304
	$A_k F_{k-j}\tilde{S}_{k-j}/abc$	0.21146	0.18756	0.13834	0.08953	0.06627	0.04372
0.10	$A_k F_{k-j}/b^2$		0.02819	0.03622	0.04086	0.04229	0.04325
	$A_k F_{k-j}\tilde{S}_{k-j}/abc$		0.20379	0.17742	0.12737	0.09795	0.06659
0.20	$A_k F_{k-j}/b^2$			0.05421	0.06859	0.07377	0.07744
	$A_k F_{k-j}\tilde{S}_{k-j}/abc$			0.18854	0.15900	0.13028	0.09337
0.40	$A_k F_{k-j}/b^2$				0.10013	0.11524	0.12770
	$A_k F_{k-j}\tilde{S}_{k-j}/abc$				0.16255	0.14686	0.11517
0.60	$A_k F_{k-j}/b^2$					0.13888	0.16138
	$A_k F_{k-j}\tilde{S}_{k-j}/abc$					0.14164	0.11940
1.0	$A_k F_{k-j}/b^2$						0.20004
	$A_k F_{k-j}\tilde{S}_{k-j}/abc$						0.11121
2.0	$A_k F_{k-j}/b^2$						
	$A_k F_{k-j}\tilde{S}_{k-j}/abc$						
4.0	$A_k F_{k-j}/b^2$						
	$A_k F_{k-j}\tilde{S}_{k-j}/abc$						
6.0	$A_k F_{k-j}/b^2$						
	$A_k F_{k-j}\tilde{S}_{k-j}/abc$						
10.0	$A_k F_{k-j}/b^2$						
	$A_k F_{k-j}\tilde{S}_{k-j}/abc$						
20.0	$A_k F_{k-j}/b^2$						
	$A_k F_{k-j}\tilde{S}_{k-j}/abc$						

(*Table continues on next page*)

Table 13-2 Configuration factors and mean beam-length functions for rectangles at right angles [2] (*Continued*)

| | | \multicolumn{6}{c}{c/b} | | | | | |
a/b		2.0	4.0	6.0	10.0	20.0	∞
0.02	$A_k F_{k-j}/b^2$	0.009648	0.009653	0.009655	0.009655	0.009655	0.009655
	$A_k F_{k-j}\bar{S}_{k-j}/abc$	0.01263	0.006364	0.004288	0.002594	0.001305	
0.05	$A_k F_{k-j}/b^2$	0.02316	0.02320	0.02321	0.02321	0.02321	0.02321
	$A_k F_{k-j}\bar{S}_{k-j}/abc$	0.02364	0.01234	0.008342	0.005059	0.002549	
0.10	$A_k F_{k-j}/b^2$	0.04376	0.04390	0.04393	0.04394	0.04394	0.04395
	$A_k F_{k-j}\bar{S}_{k-j}/abc$	0.03676	0.01944	0.013184	0.008018	0.004049	
0.20	$A_k F_{k-j}/b^2$	0.07942	0.07999	0.08010	0.08015	0.08018	0.08018
	$A_k F_{k-j}\bar{S}_{k-j}/abc$	0.05356	0.02890	0.01972	0.012047	0.006103	
0.40	$A_k F_{k-j}/b^2$	0.13514	0.13736	0.13779	0.13801	0.13811	0.13814
	$A_k F_{k-j}\bar{S}_{k-j}/abc$	0.07088	0.03903	0.02666	0.01697	0.008642	
0.60	$A_k F_{k-j}/b^2$	0.17657	0.18143	0.18239	0.18289	0.18311	0.18318
	$A_k F_{k-j}\bar{S}_{k-j}/abc$	0.07830	0.04467	0.03109	0.02025	0.010366	
1.0	$A_k F_{k-j}/b^2$	0.23285	0.24522	0.24783	0.24921	0.24980	0.25000
	$A_k F_{k-j}\bar{S}_{k-j}/abc$	0.08137	0.04935	0.03502	0.02196	0.01175	
2.0	$A_k F_{k-j}/b^2$	0.29860	0.33462	0.34386	0.34916	0.35142	0.35222
	$A_k F_{k-j}\bar{S}_{k-j}/abc$	0.07086	0.04924	0.03670	0.02401	0.01325	
4.0	$A_k F_{k-j}/b^2$		0.40544	0.43104	0.44840	0.45708	0.46020
	$A_k F_{k-j}\bar{S}_{k-j}/abc$		0.04051	0.03284	0.02320	0.01300	
6.0	$A_k F_{k-j}/b^2$			0.46932	0.49986	0.51744	0.52368
	$A_k F_{k-j}\bar{S}_{k-j}/abc$			0.02832	0.02132	0.01272	
10.0	$A_k F_{k-j}/b^2$				0.5502	0.5876	0.6053
	$A_k F_{k-j}\bar{S}_{k-j}/abc$				0.01759	0.01146	
20.0	$A_k F_{k-j}/b^2$					0.6608	0.7156
	$A_k F_{k-j}\bar{S}_{k-j}/abc$					0.008975	

over the wavelength bands gives the required energy input to the gas for the given T_g.

13-5 MEAN BEAM-LENGTH APPROXIMATION FOR SPECTRAL RADIATION FROM AN ENTIRE GAS VOLUME TO ALL OR PART OF ITS BOUNDARY

In some practical situations it is desired to determine the energy radiated from a volume of isothermal gas with uniform composition to all or part of its boundaries, without considering emission and reflection from the boundaries. An example is radiation from hot furnace gases to walls that are cool enough so their emission is small and that are rough and soot covered so they are essentially nonreflecting. In this section the energy will be considered in a range $d\lambda$. In the next section an integration will incorporate all λ to obtain the total energy. For the specified conditions the $dq_{\lambda o,j}$ in Eq. (13-12), which is the spectral outgoing flux from a typical surface A_j, is zero. The spectral incoming energy at A_k is then

Table 13-3 Approximate band limits for parallel-plate geometry [5, 6, 109]

Gas	Band λ, μm	Band center η, cm^{-1}	Band limits η, cm^{-1a} Lower	Band limits η, cm^{-1a} Upper
CO_2	15	667	$667 - (\bar{A}_{15}/1.78)$	$667 + (\bar{A}_{15}/1.78)$
	10.4	960	849	1013
	9.4	1060	1013	1141
	4.3	2350	$2350 - (\bar{A}_{4.3}/1.78)$	2430
	2.7	3715	$3715 - (\bar{A}_{2.7}/1.78)$	3750
H_2O	6.3	1600	$1600 - (\bar{A}_{6.3}/1.6)$	$1600 + (\bar{A}_{6.3}/1.6)$
	2.7	3750	$3750 - (\bar{A}_{2.7}/1.4)$	$3750 + (\bar{A}_{2.7}/1.4)$
	1.87	5350	4620	6200
	1.38	7250	6200	8100

$^a\bar{A}$ are found for various bands from Tables 12-3 and 12-4. Terms such as $\bar{A}_{15}/1.78$ are $\bar{A}/2(1 - \tau_g)$ from Eq. (17) and Tables 1 and 2 of [5].

$$A_k \, dq_{\lambda i,k} = \sum_{j=1}^{N} e_{\lambda b,g} \, d\lambda \, A_j F_{j-k} \bar{\alpha}_{\lambda,j-k} \tag{13-59}$$

If the geometry is a *hemisphere of gas* radiating to an *area element dA_k at the center of its base* as shown in Fig. 13-4, Eq. (13-59) has an especially simple form. Since the hemispherical boundary is the only surface in view of dA_k, and dA_k is a differential element, (13-59) reduces to

$$dA_k \, dq_{\lambda i,k} = e_{\lambda b,g} \, d\lambda \, A_j \, dF_{j-dk} \, \bar{\alpha}_{\lambda,j-dk} = e_{\lambda b,g} \, d\lambda \, dA_k \, F_{dk-j} \bar{\alpha}_{\lambda,j-dk} \tag{13-60}$$

where, from Eq. (13-44),

$$\bar{\alpha}_{\lambda,j-dk} = 1 - \bar{\tau}_{\lambda,j-dk} = 1 - \exp\left(-a_\lambda R\right)$$

For radiation between dA_k and the surface of a hemisphere, $F_{dk-j} = 1$, so (13-60) reduces to the following simple expression giving the *incident* heat flux from a hemisphere of gas to the center of the hemisphere base:

$$dq_{\lambda i,k} = [1 - \exp\left(-a_\lambda R\right)]e_{\lambda b,g} \, d\lambda \tag{13-61}$$

From the form of this equation the $1 - \exp(-a_\lambda R)$ is the *spectral emittance* of the gas $\epsilon_\lambda(\lambda, T, P, R)$ for path length R.[†] This spectral emittance definition was obtained previously [see Eq. (12-54b)] by use of Kirchhoff's law in connection with Eq. (12-52). Then (13-61) becomes

$$dq_{\lambda i,k} = \epsilon_\lambda(a_\lambda R)e_{\lambda b,g} \, d\lambda \qquad [\epsilon_\lambda(a_\lambda R) = 1 - \exp(-a_\lambda R)] \tag{13-62}$$

The incident energy depends on the optical radius of the hemisphere $a_\lambda R$.

It would be very convenient if a relation having the simple form of (13-62) could be used to determine the value of $dq_{\lambda i,k}$ on A_k for *any* geometry of radiating

[†]For simplicity, the prime notation used for a directional quantity will be omitted; in this instance ϵ_λ is independent of direction.

gas volume and for A_k being all or part of its boundary. Because the geometry of the gas enters (13-62) only through $\epsilon_\lambda(a_\lambda R)$, it is possible to define a fictitious value of R, say L_e, that would yield a value of $\epsilon_\lambda(a_\lambda L_e)$ such that (13-62) would give the correct $dq_{\lambda i}$ for a given geometry. This fictitious length L_e is called the *mean beam length*. Then for an arbitrary geometry of gas let

$$dq_{\lambda i, k} = \epsilon_\lambda(a_\lambda L_e)e_{\lambda b, g}\, d\lambda = [1 - \exp(-a_\lambda L_e)]e_{\lambda b, g}\, d\lambda \tag{13-63}$$

The mean beam length is the required radius of a gas hemisphere such that it radiates a flux to the center of its base equal to the average flux radiated to the area of interest by the actual volume of gas.

13-5.1 Mean Beam Length for Gas between Parallel Plates Radiating to Area on Plate

Consider two black infinite parallel plates at zero absolute temperature separated by a distance D. The plates enclose a uniform gas at temperature T_g with absorption coefficient a_λ. The rate at which spectral energy is incident upon A_k on one plate (Fig. 13-8) is, from (13-59) and (13-54),

$$dQ_{\lambda i, k} = A_k\, dq_{\lambda i, k} = e_{\lambda b, g}\, d\lambda\, A_j F_{j-k}\bar{\alpha}_{\lambda, j-k} = e_{\lambda b, g}\, d\lambda\, A_j F_{j-k}[1 - 2E_3(a_\lambda D)] \tag{13-64a}$$

For infinite plates $F_{k-j} = 1$ and, by reciprocity, $F_{j-k} = A_k/A_j$ so (13-64a) reduces to

$$dq_{\lambda i, k} = [1 - 2E_3(a_\lambda D)]e_{\lambda b, g}\, d\lambda \tag{13-64b}$$

Comparing (13-64b) and (13-63) provides the mean beam length as

$$L_e = -\frac{1}{a_\lambda}\ln 2E_3(a_\lambda D)$$

or in terms of the optical thickness $a_\lambda D$,

$$\frac{L_e}{D} = -\frac{1}{a_\lambda D}\ln 2E_3(a_\lambda D) \tag{13-65}$$

13-5.2 Mean Beam Length for Sphere of Gas Radiating to Any Area on Boundary

Consider gas in a nonreflecting sphere of radius R where the sphere boundary A_j is at $T_j = 0$. From (13-59) the radiation flux incident on an element dA_k is

$$dq_{\lambda i, dk} = e_{\lambda b, g}\, d\lambda \frac{A_j}{dA_k}\, dF_{j-dk}\, \bar{\alpha}_{\lambda, j-dk}$$

For a sphere, $dF_{j-dk} = dA_k/A_j$ [by use of (7-95)]. Then substituting (13-53)

$$dq_{\lambda i, dk} = e_{\lambda b, g}\, d\lambda \left\{ 1 - \frac{2}{(2a_\lambda R)^2}[1 - (2a_\lambda R + 1)\exp(-2a_\lambda R)] \right\}$$

Equate this to dq_λ from (13-63) and solve for L_e to obtain

$$\frac{L_e}{2R} = -\frac{1}{2a_\lambda R} \ln\left\{ \frac{2}{(2a_\lambda R)^2} [1 - (2a_\lambda R + 1)\exp(-2a_\lambda R)] \right\} \tag{13-66}$$

In view of the general applicability of Eq. (13-53), Eq. (13-66) gives the correct mean beam length for the entire sphere radiating to any portion of its boundary. Some additional results for spheres are in [7].

13-5.3 Radiation from Entire Gas Volume to Its Entire Boundary in Limit When Gas is Optically Thin

Because of the integrations involved, the mean beam length for an entire gas volume radiating to all or part of its boundary will usually be difficult to evaluate for any but the simplest of shapes. It is fortunate that some practical approximations can be found by looking first at the optically thin limit. By expanding the exponential term in a series for small $a_\lambda S$, the transmittance becomes

$$\lim_{a_\lambda S \to 0} \tau_\lambda = \lim_{a_\lambda S \to 0} \exp(-a_\lambda S) = \lim_{a_\lambda S \to 0} \left[1 - a_\lambda S + \frac{(a_\lambda S)^2}{2!} - \cdots \right] = 1$$

Any differential volume of the uniform-temperature gas emits spectral energy $4a_\lambda e_{\lambda b,g}$ $d\lambda\, dV$. Since $\tau_\lambda = 1$, there is no attenuation of the emitted radiation, and all of it reaches the enclosure boundary. For the entire radiating volume the energy reaching the boundary is $4a_\lambda e_{\lambda b,g}\, d\lambda\, V$ so the *average* spectral flux received at the boundary having entire area A is in the optically thin limit

$$dq_{\lambda i} = 4a_\lambda e_{\lambda b,g}\, d\lambda \frac{V}{A} \tag{13-67}$$

By use of the mean beam length the average flux reaching the boundary is given by (13-63). For the special case of small absorption let L_e be designated by $L_{e,0}$. Then expand the exponential term in (13-63) in a series to obtain, for small $a_\lambda L_{e,0}$,

$$dq_{\lambda i} = \left\{ 1 - \left[1 - a_\lambda L_{e,0} + \frac{(a_\lambda L_{e,0})^2}{2!} - \cdots \right] \right\} e_{\lambda b,g}\, d\lambda = a_\lambda L_{e,0} e_{\lambda b,g}\, d\lambda \tag{13-68}$$

Equating this to the $dq_{\lambda i}$ in (13-67) gives the desired result for the mean beam length of an *optically thin gas* radiating to its entire boundary:

$$L_{e,0} = \frac{4V}{A} \tag{13-69}$$

To give a few examples, for a sphere of diameter D,

$$L_{e,0} = \frac{4\pi D^3/6}{\pi D^2} = \tfrac{2}{3}D \tag{13-70}$$

For an infinitely long circular cylinder of diameter D,

$$L_{e,0} = \frac{4\pi D^2/4}{\pi D} = D \tag{13-71}$$

For gas between infinite parallel plates spaced D apart,

$$L_{e,0} = \frac{4D}{2} = 2D \tag{13-72}$$

For an infinitely long rectangular parallelepiped with cross section dimensions h and w,

$$L_{e,0} = \frac{4hw}{2(h+w)} = \frac{2hw}{h+w} \tag{13-73}$$

13-5.4 Correction for Mean Beam Length When Gas Is Not Optically Thin

For an optically thick gas it would be very convenient if L_e could be obtained by applying a simple correction factor to the $L_{e,0}$ computed from (13-69). It has been found that a useful technique is to introduce a correction coefficient C so that

$$L_e = CL_{e,0} \tag{13-74}$$

Then the incoming heat flux in (13-63) can be obtained as

$$dq_{\lambda i} = [1 - \exp(-a_\lambda CL_{e,0})]e_{\lambda b,g}\, d\lambda \tag{13-75}$$

The coefficient C is now examined by considering the example of a radiating gas between infinite parallel plates spaced D apart. Using (13-72) in (13-75) gives $dq_{\lambda i} = [1 - \exp(-a_\lambda C2D)]e_{\lambda b,g}\, d\lambda$. From (13-64b) the actual flux received is $dq_{\lambda i} = [1 - 2E_3(a_\lambda D)]e_{\lambda b,g}\, d\lambda$. To demonstrate how well these fluxes compare, the ratio $[1 - 2E_3(a_\lambda D)]/[1 - \exp(-2Ca_\lambda D)]$ is plotted in Fig. 13-12 for a range of

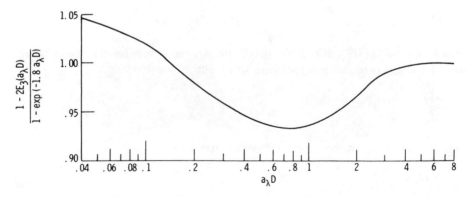

Figure 13-12 Ratio of emission by gas layer to that calculated using a mean beam length $L_e = 1.8D$.

$a_\lambda D$ using $C = 0.9$. This value of C was found to yield a ratio close to unity for all $a_\lambda D$, and hence is a valid correction coefficient for this geometry.

Table 13-4 gives the mean beam length $L_{e,0}$ for a number of geometries, along with values of L_e that provide reasonably good radiative fluxes for nonzero optical thicknesses. The values of C are found to be in a range near 0.9 [3, 8, 9]. Hence, it is recommended that for a geometry for which exact L_e values have not been calculated, the approximation

$$L_e = 0.9L_{e,0} = 0.9 \frac{4V}{A} \tag{13-76}$$

be used *for an entire uniform isothermal gas volume radiating to its entire boundary.*

The different optical thicknesses of absorption bands may make it desirable to use a somewhat different mean beam length for each band. Mean beam lengths based on various band absorption models have been studied for slab geometries [1] and in spheres and cylinders [1, 10]. Variations in mean beam length from those predicted in Table 13-4 were greatest for the slab geometry, approaching $0.82L_{e,0}$ at optical thicknesses near 100.

It is shown in [11] that the concept of mean beam length applies to *scattering* media, and in the limit of pure scattering and small optical thickness, (13-76) applies.

EXAMPLE 13-4 Parallel nongray plates are 2.5 cm apart and are at temperatures of $T_1 = 1100$ K and $T_2 = 550$ K. Pure CO_2 gas at 10 atm pressure and $T_g = 550$ K is between the plates. The plate hemispherical spectral emissivity as a function of wavenumber is approximated by the following table:

η, cm^{-1}	ϵ_η	η, cm^{-1}	ϵ_η
0–500	0.37	1150–2200	0.45
500–750	0.26	2200–2500	0.65
750–850	0.32	2500–3600	0.61
850–1000	0.37	3600–3750	0.69
1000–1150	0.46	3750–∞	0.73

Assume that only the 15-, 10.4-, 9.4-, 4.3-, and 2.7-μm CO_2 bands cause significant attenuation in the gas. Compute the total heat flux being added to plate 2.

In Example 13-2 the spectral exchange was found for radiation between infinite parallel plates with a gas between them. The total energy added to plate 2 is found by integrating (13-24b) over all wavenumbers:

$$q_2 = \int_{\eta=0}^{\infty} \frac{\{\epsilon_{\eta,1}\epsilon_{\eta,2}\bar{\tau}_\eta(e_{\eta b,2} - e_{\eta b,1}) + \epsilon_{\eta,2}(1 - \bar{\tau}_\eta)[1 + (1 - \epsilon_{\eta,1})\bar{\tau}_\eta](e_{\eta b,2} - e_{\eta b,g})\} \, d\eta}{1 - (1 - \epsilon_{\eta,1})(1 - \epsilon_{\eta,2})\bar{\tau}_\eta^2}$$

Table 13-4 Mean beam lengths for radiation from entire gas volume

Geometry of radiating system	Characterizing dimension	Mean beam length for optical thickness $a_\lambda L_e \to 0$, $L_{e,0}$	Mean beam length corrected for finite optical thickness,[a] L_e	$C = L_e/L_{e,0}$
Hemisphere radiating to element at center of base	Radius R	R	R	1
Sphere radiating to its surface	Diameter D	$\frac{2}{3}D$	$0.65D$	0.97
Circular cylinder of infinite height radiating to concave bounding surface	Diameter D	D	$0.95D$	0.95
Circular cylinder of semiinfinite height radiating to:				
Element at center of base	Diameter D	D	$0.90D$	0.90
Entire base	Diameter D	$0.81D$	$0.65D$	0.80
Circular cylinder of height equal to diameter radiating to:				
Element at center of base	Diameter D	$0.77D$	$0.71D$	0.92
Entire surface	Diameter D	$\frac{2}{3}D$	$0.60D$	0.90
Circular cylinder of height equal to two diameters radiating to:				
Plane end	Diameter D	$0.73D$	$0.60D$	0.82
Concave surface	Diameter D	$0.82D$	$0.76D$	0.93
Entire surface	Diameter D	$0.80D$	$0.73D$	0.91
Circular cylinder of height equal to one-half the diameter radiating to:				
Plane end	Diameter D	$0.48D$	$0.43D$	0.90
Concave surface	Diameter D	$0.52D$	$0.46D$	0.88
Entire surface	Diameter D	$0.50D$	$0.45D$	0.90
Cylinder of infinite height and semicircular cross section radiating to element at center of plane rectangular face	Radius R		$1.26R$	
Infinite slab of gas radiating to:				
Element on one face	Slab thickness D	$2D$	$1.8D$	0.90
Both bounding planes	Slab thickness D	$2D$	$1.8D$	0.90
Cube radiating to a face	Edge X	$\frac{2}{3}X$	$0.6X$	0.90
Rectangular parallelepipeds 1 × 1 × 4 radiating to:				
1 × 4 face	Shortest edge X	$0.90X$	$0.82X$	0.91

Table 13-4 Mean beam lengths for radiation from entire gas volume (*Continued*)

1 × 1 face		0.86X	0.71X	0.83
all faces		0.89X	0.81X	0.91
1 × 2 × 6 radiating to:				
2 × 6 face		1.18X		
1 × 6 face		1.24X		
1 × 2 face		1.18X		
all faces		1.20X		
Gas between infinitely long parallel concentric cylinders	Radius of outer cylinder R and of inner cylinder r	$2(R - r)$	See [4]	
Gas volume in the space between the outside of the tubes in an infinite tube bundle and radiating to a single tube:				
Equilateral triangular array:	Tube diameter			
$S = 2D$	D, and spacing	$3.4(S - D)$	$3.0(S - D)$	0.88
$S = 3D$	between tube	$4.45(S - D)$	$3.8(S - D)$	0.85
Square array:	centers, S			
$S = 2D$		$4.1(S - D)$	$3.5(S - D)$	0.85

aCorrections are those suggested by Hottel et al. [3, 8] or Eckert [9]. Corrections were chosen to provide maximum L_e where these references disagree.

In this example $\epsilon_{\eta,1} = \epsilon_{\eta,2}$ and $T_g = T_2$ so q_2 simplifies to

$$q_2 = - \int_0^\infty \frac{\epsilon_{\eta,1}{}^2 \bar{\tau}_\eta (e_{\eta b,1} - e_{\eta b,2})}{1 - (1 - \epsilon_{\eta,1})^2 \bar{\tau}_\eta{}^2} \, d\eta$$

The integration is expressed in finite-difference form as a sum over wavenumber bands. For the lth band let $\epsilon_{\eta,1} = \epsilon_l$ and $\bar{\tau}_\eta = \bar{\tau}_l$. Then

$$q_2 = - \sum_l \frac{\epsilon_l{}^2 \bar{\tau}_l [e_b(T_1) - e_b(T_2)]_l \, \Delta\eta_l}{1 - (1 - \epsilon_l)^2 \bar{\tau}_l{}^2}$$

where $(e_b)_l \, \Delta\eta_l$ is the blackbody radiation in the lth band. From (13-35), $\bar{\tau}_l$ can be written as

$$\bar{\tau}_l = 1 - \bar{\alpha}_l = 1 - \frac{\bar{A}_l}{\Delta\eta_l}$$

where $\bar{A}_l$ is the integrated bandwidth that includes the integrated path-length variation for a parallel-plate geometry. The q_2 now becomes

$$q_2 = - \sum_l \frac{\epsilon_l{}^2 (1 - \bar{A}_l/\Delta\eta_l)[e_b(T_1) - e_b(T_2)]_l \, \Delta\eta_l}{1 - (1 - \epsilon_l)^2 (1 - \bar{A}_l/\Delta\eta_l)^2}$$

The needed quantities and results are shown in the tables that follow. Values of $\bar{A}_l$ were computed from the exponential wide-band correlations of Tables 12-3 and 12-4 using the mean beam length from Table 13-4 as the effective path length. The wavenumber spans $\Delta\eta_l$ were computed from Table 13-3. For the nonabsorbing regions the $[e_b(T_1) - e_b(T_2)]_l \, \Delta\eta_l$ were computed using $F_{0-\lambda T}$, that is, $[e_b(T_1)]_l \, \Delta\eta_l = F_{\lambda_1 T_1 - \lambda_2 T_1} e_b(T_1)$ where λ_1 and λ_2 correspond to the wavenumber limits of the band. For the absorbing regions, $[e_b(T_1)]_l \, \Delta\eta_l = [e_b(T_1)]_{\text{bandcenter}} \, \Delta\eta_l$. When the band correlations gave $\bar{A}_l > \Delta\eta_l$ then $\bar{A}_l/\Delta\eta_l = 1.0$ was used, as physically $\bar{A}_l$ cannot exceed $\Delta\eta_l$.

Band λ, μm	Band center η, cm^{-1}	C_1, cm^{-1}/(g$\cdot$m^{-2})	C_2, cm^{-1}/[(g$\cdot$m^{-2})]$^{1/2}$	C_3, cm^{-1}	β
15	667	19	16.3	30.4	0.691
10.4	960	0.0218	1.76	29.2	9.50
9.4	1060	0.0218	1.76	29.2	9.50
4.3	2350	110	73.0	27.1	3.49
2.7	3715	4.14	14.6	56.5	1.20

Band η, cm^{-1}	ϵ_l	$\bar{A}_l$, cm^{-1}	$\Delta\eta_l$, cm^{-1}	$[e_b(T_1) - e_b(T_2)]_l \, \Delta\eta_l$, W/m^2	$-q_{l,2}$, W/m^2
0–555	0.37	0	555	774	176
555–779 (15 μm)	0.26	199	224	1,259	10
779–849	0.32	0	70	553	105
849–1013 (10.4 μm)	0.37	9.6	164	1,683	334
1013–1141 (9.4 μm)	0.46	9.6	128	1,521	397
1141–2221	0.45	0	1080	20,777	6,032
2221–2430 (4.3 μm)	0.65	230	209	4,704	0
2430–3573	0.61	0	1143	22,523	9,884
3573–3750 (2.7 μm)	0.69	253	177	2,634	0
3750–∞	0.73	0	∞	21,373	12,285
					29,223

Edwards and Nelson [5] use the network method of Oppenheim [12] in deriving the energy transfer equation, which gives the same result as the method used here. Partial emittances were used in place of the band correlations for computing gas properties, and these led to slightly different wavenumber spans for the bands used in solving a problem similar to this example in [5]. Note that for this example most of the radiative transfer is in the transparent regions between the CO_2 absorption bands. Results for the exact and geometric-mean beam-length solutions are compared in [13] for hydrogen plasma between parallel plates.

13-6 EXCHANGE OF TOTAL RADIATION IN AN ENCLOSURE BY APPLICATION OF MEAN BEAM LENGTH

The mean beam length was obtained in the previous section at one wavelength. This concept will now be applied to obtain the exchange of *total* energy within an enclosure. The use of the mean beam length simplifies the geometric considerations, but it remains to integrate the spectral relations to obtain total energy transfer.

13-6.1 Total Radiation from Entire Gas Volume to All or Part of Boundary

The mean beam length was found to be *approximately independent* of a_λ as evidenced by (13-76). This means that L_e can be used as a characteristic dimension of the gas volume and regarded as constant during an integration over wavelength. The total heat flux from the gas that is incident on a surface is found by integrating (13-63) over all λ:

$$q_i = \int_0^\infty [1 - \exp(-a_\lambda L_e)] e_{\lambda b,g} \, d\lambda \tag{13-77}$$

where L_e is independent of λ. Now define a gas *total emittance* ϵ_g such that

$$q_i = \epsilon_g \sigma T_g^4 \tag{13-78}$$

Equating the last two relations gives

$$\epsilon_g = \frac{\displaystyle\int_{\lambda=0}^\infty [1 - \exp(-a_\lambda L_e)] e_{\lambda b,g} \, d\lambda}{\sigma T_g^4} \tag{13-79}$$

The ϵ_g in (13-79) is a convenient quantity that can be presented in graphical form for each gas in terms of the variables L_e and T_g. Then for a particular geometry and gas state, the ϵ_g is taken from the graphs and applied by use of (13-78). Analytical forms for ϵ_g will also be given that are convenient for computer use.

Charts for gas total emittance Graphical presentations of the total emittance of the important radiating gases have been developed from total radiation measurements, spectral measurements of absorption lines and bands, and the theories of line and band absorption as in Chap. 12. As a very simple illustration of Eq. (13-79) relating spectral and total properties, consider the following example for a very thick absorbing layer.

EXAMPLE 13-5 As a rough approximation, idealize the absorptance of CO_2 at $T_g = 830$ K and 10 atm as in Fig. 12-10a so that it consists of four bands having vertical boundaries at the values 1.8 and 2.2, 2.6 and 2.8, 4.0 and 4.6, and 9 and 19 μm. What is the total emittance of a thick layer of gas at this temperature?

For a thick layer, Eq. (13-79) indicates that $\epsilon'_\lambda = 1 - \exp(-a_\lambda L_e)$ goes to unity in the absorbing regions. Hence, the gas will emit like a blackbody in the four absorption bands. In the nonabsorbing regions between the bands, ϵ'_λ is very small and is neglected in this simplified model. The total emittance becomes

$$\epsilon'(T_g,P,S) = \frac{\int_0^\infty \epsilon'_\lambda(\lambda,T_g,P,S)e_{\lambda b,g}\,d\lambda}{\sigma T_g^4} = \frac{\int_{\text{absorbing bands}} e_{\lambda b,g}\,d\lambda}{\sigma T_g^4}$$

The emittance is thus the fractional emission of a blackbody over the wavelength intervals of the absorbing bands, which can be obtained from the $F_{0-\lambda T_g}$ factors in Table A-5 in the appendix. The required values are as follows:

λ, μm	λT_g, μm·K	$F_{0-\lambda T_g}$	λ, μm	λT_g, μm·K	$F_{0-\lambda T_g}$
1.8	1,500	0.01285	4.0	3,320	0.34734
2.2	1,830	0.04338	4.6	3,820	0.44977
2.6	2,160	0.09478	9	7,470	0.83435
2.8	2,320	0.12665	19	15,800	0.97302

Then the emittance is

$$\epsilon'(T_g, P, S) = \sum_{\substack{\text{absorbing} \\ \text{bands}}} \left(F_{0-(\lambda T_g)_{\text{upper}}} - F_{0-(\lambda T_g)_{\text{lower}}}\right)_{\text{band}}$$

$$\epsilon' = (0.04338 - 0.01285) + (0.12665 - 0.09478) + (0.44977 - 0.34734)$$
$$+ (0.97302 - 0.83435) = 0.304$$

At industrial furnace and combustion chamber temperatures, it is only heteropolar gases that absorb and emit significantly, such as CO_2, H_2O, CO, SO_2, NO, and CH_4. Gases with symmetric diatomic molecules, such as N_2, O_2, and H_2 are transparent to infrared radiation and do not emit significantly. Charts for ϵ_g were developed by Hottel [8] from experimental measurements. The thickness of the gas enters through the parameter L_e. The gas pressure enters as a parameter because of the dependence of a_λ on gas density. If the gas is in a mixture, both the pressure of the mixture and the partial pressure of the radiating constituent under consideration will be parameters. Charts for CO_2 and water vapor are in Figs. 13-13 to 13-17. Additional charts for sulfur dioxide, ammonia, carbon monoxide, methane, and a few other gases are in [3]. The discussion here is limited to CO_2 and water vapor.

To compute the radiation to area A from the gas volume by using (13-78),

$$Q_i = q_i A = A\epsilon_g \sigma T_g^4 \tag{13-80}$$

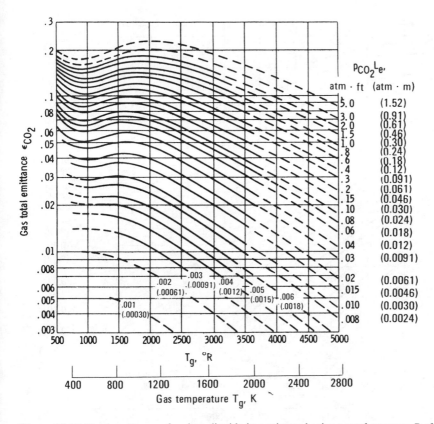

Figure 13-13 Total emittance of carbon dioxide in a mixture having a total pressure P of 1 atm [8].

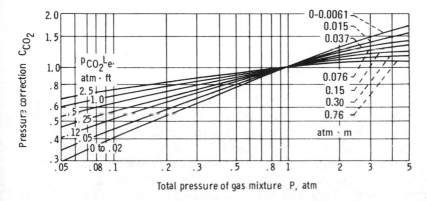

Figure 13-14 Pressure correction for CO_2 total emittance for values of P other than 1 atm [8].

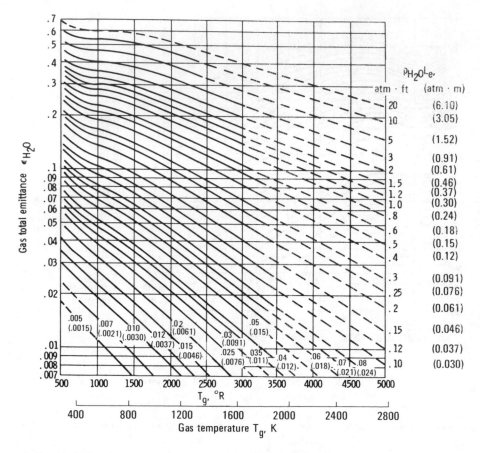

Figure 13-15 Total emittance of water vapor in limit of zero partial pressure in a mixture having a total pressure P of 1 atm [8].

the mean beam length for the gas geometry is first obtained from Table 13-4 or Eq. (13-76). Then, with the partial pressure of the gas and its temperature known, the gas emittance is found by using Figs. 13-13 to 13-17 (also see 13-18 to 13-21). Figure 13-13 gives the total emittance of CO_2 obtained experimentally using a mixture with nonabsorbing gases so the total pressure of the mixture was 1 atm while the partial pressure of the CO_2 was varied. The dashed lines indicate regions unverified by experimental data. For a mixture at total pressure other than 1 atm, a pressure-broadening correction is applied [8]. This is a multiplying coefficient C_{CO_2} in Fig. 13-14. In the case of water vapor, the emittance depends in a more complex manner on both the partial pressure of the water vapor and the total pressure of the gas mixture. For correlation purposes the values in Fig. 13-15 are emittances that are "reduced," by using a factor depending on p_{H_2O} and $p_{H_2O}L_e$, to limiting values as the partial pressure p_{H_2O} approaches zero in a mixture having a total pressure $P = 1$ atm. A multiplying correction coefficient C_{H_2O} is given in Fig. 13-16 to account for the actual partial and total pressures.

If CO_2 and water vapor are both present in the gas mixture, an additional quantity $\Delta\epsilon$ must be included to account for an emittance reduction resulting from spectral overlap of the CO_2 and H_2O absorption bands (for discussion of individual band behavior, see [14, 15]). This correction is found from Fig. 13-17. For a mixture of CO_2 and water vapor in a nonabsorbing carrier gas, the emittance is then

$$\epsilon_g = C_{CO_2}\epsilon_{CO_2} + C_{H_2O}\epsilon_{H_2O} - \Delta\epsilon \tag{13-81}$$

To better understand (13-81), consider (13-79) when there are two radiating constituents having spectral absorption coefficients $a_{\lambda1}(\lambda)$ and $a_{\lambda2}(\lambda)$. Then

$$\epsilon_g = \frac{1}{\sigma T_g^4}\int_0^\infty [1 - e^{-(a_{\lambda1}+a_{\lambda2})L_e}]e_{\lambda b,g}\,d\lambda$$

$$= \frac{1}{\sigma T_g^4}\int_0^\infty [1 - e^{-a_{\lambda1}L_e} + 1 - e^{-a_{\lambda2}L_e} - (1 - e^{-a_{\lambda1}L_e})(1 - e^{-a_{\lambda2}L_e})]e_{\lambda b,g}\,d\lambda$$

The first four terms under the integral integrate to yield the total emittances of the individual radiating components so that

$$\epsilon_g = \epsilon_1 + \epsilon_2 - \frac{1}{\sigma T_g^4}\int_0^\infty (1 - e^{-a_{\lambda1}L_e})(1 - e^{-a_{\lambda2}L_e})e_{\lambda b,g}\,d\lambda$$

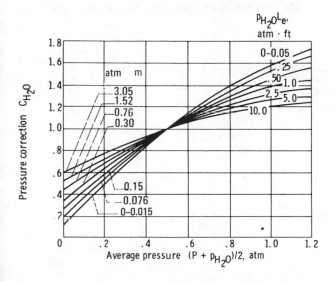

Figure 13-16 Pressure correction for water vapor total emittance for values of partial pressure p_{H_2O} and total pressure P other than 0 and 1 atm, respectively [8].

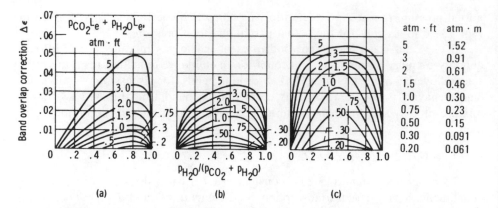

Figure 13-17 Correction on total emittance for band overlap when both CO_2 and water vapor are present [8]. (*a*) Gas temperature $T_g = 400$ K (720°R); (*b*) gas temperature $T_g = 810$ K (1460°R); (*c*) gas temperature, $T_g \geq 1200$ K (2160°R).

The last term on the right is the correction $\Delta\epsilon$ for *spectral overlap*. The integrand is nonzero only in the wavelength regions where both $a_{\lambda 1}$ and $a_{\lambda 2}$ are nonzero; these are the spectral overlap regions.

The charts in Figs. 13-13 to 13-17 have been subject to reevaluations to incorporate information from more recent detailed spectral data and to extend to larger pL_e and T_g [16–22]. Results for higher pressures are needed for gas turbine combustors as compared with industrial furnaces. Figure 13-18, from [19], shows the reevaluated CO_2 emittance calculated for line broadening at 1 bar total pressure

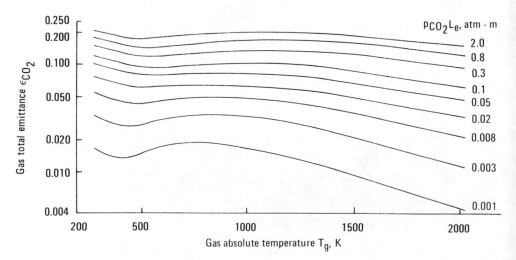

Figure 13-18 Total emittance of carbon dioxide in limit of zero partial pressure in a mixture at total pressure of 1 bar [19].

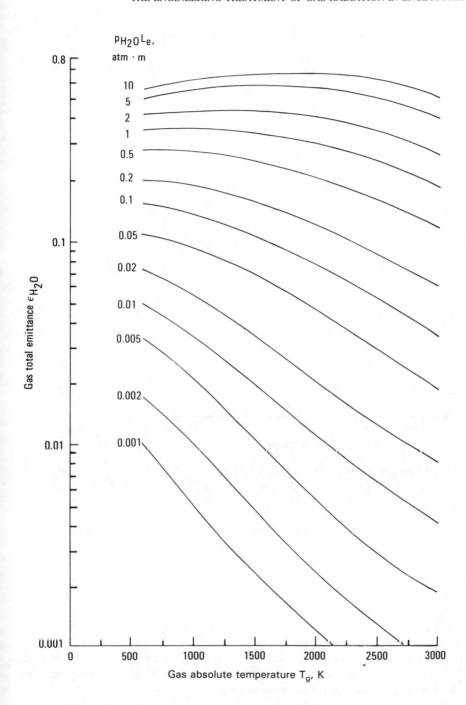

Figure 13-19 Total emittance of water vapor in limit of zero partial pressure in a mixture at total pressure of 1 atm [17, 18].

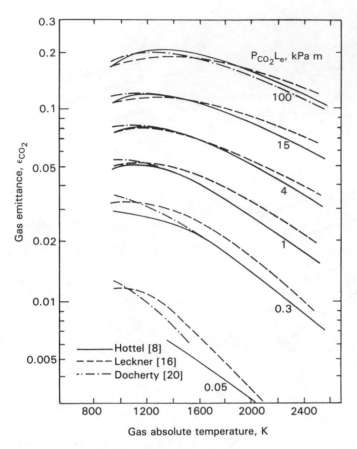

Figure 13-20 Comparison of models for emittance of carbon dioxide at a total pressure of 1 atm [110].

and zero partial pressure (the partial-pressure effect is in reality not significant in the range of atmospheric pressure). Results for water vapor are given in [17] and [18], and the curves comparable to Fig. 13-15 are shown in Fig. 13-19. There is a significant difference in the shape of the curves for large $p_{H_2O}L_e$. The references also give recomputed results on pressure corrections and the effect of H_2O and CO_2 band overlap. The results from [20] and [22] are compared with the older results in Figs. 13-20 and 13-21. At the higher temperature levels some of the deviations are quite significant.

In [23] the effect of particulate scattering on the total hemispherical emittance of CO_2 and H_2O was analyzed. Emittance charts were calculated using the exponential wide-band model for gaseous absorption.

EXAMPLE 13-6 A cooled right cylindrical tank 4 m in diameter and 4 m long has a black interior surface and is filled with hot gas at a total pressure of 1 atm. The gas is composed of a transparent gas at partial pressure 0.75 atm, and carbon dioxide. The gas is uniformly mixed at a temperature of 1100 K. Compute how much energy must be removed from the surface of the tank to keep it cool if the tank walls are all at sufficiently low temperature so that only radiation from the gas is significant.

The geometry is a finite circular cylinder of gas, and the radiation to its walls will be computed. Using Table 13-4, the corrected mean beam length for this geometry is $L_e = 0.60D = 2.4$ m. The partial pressure of the CO_2 is 0.25 atm, so that $p_{CO_2}L_e = 0.25 \times 2.4 = 0.6$ atm·m. From Fig. 13-13, $\epsilon_{CO_2}(p_{CO_2}L_e, T_g) = 0.185$, and C_{CO_2} from Fig. 13-14 is 1.0, since the mixture total pressure is unity. The energy to be removed is, from (13-80),

$$Q_i = \epsilon_{CO_2}\sigma T_g^4 A = 0.185 \times 5.6705 \times 10^{-8}(1100)^4 24\pi = 1158 \text{ kW}$$

Representation of total emittance in analytical form For computer use in solving the radiative exchange equations, it is much more convenient to have the gas

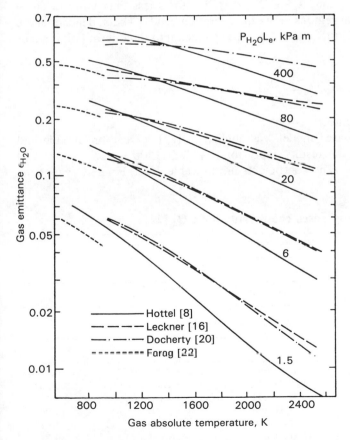

Figure 13-21 Comparison of models for emittance of water vapor at a total pressure of 1 atm [110].

total emittance in the form of analytical expressions rather than in chart form. This can be accomplished by various curve-fitting procedures accounting for the variation in mass path length and temperature. Much of the data has been obtained at a total pressure of 1 atm, which has wide application in furnaces. For applications such as gas turbine combustors, the pressures are elevated, and scaling rules such as in [24] have been used to extrapolate known information. The curve fitting has been done with polynomials or by use of a model called the *weighted sum of gray gases*.

For the sum of gray gases model, the gas is assumed to behave like a mixture of gray gases and a transparent (nonemitting) medium to account for the windows between the absorption bands. Then

$$\epsilon_g = a_1(1 - e^{-k_1 pL_e}) + a_2(1 - e^{-k_2 pL_e}) + \cdots \tag{13-82}$$

$$\epsilon_g = \sum_{i=1}^{n} a_i - \sum_{i=1}^{n} a_i e^{-k_i pL_e}$$

When the path length is long $\epsilon_g \rightarrow \sum_{i=1}^{n} a_i < 1$; this is less than unity to allow for the clear gas spectral regions between absorption bands. The a_i depend on the gas temperature. In [22] the a_i and $\sum a_i$ are expanded into three-term polynomials of T,

$$a_i = b_{1,i} + b_{2,i}\tau + b_{3,i}\tau^2, \qquad A_s \equiv \sum_{i=1}^{n} a_i = c_1 + c_2\tau + c_3\tau^2$$

where $\tau = T/1000$ (T in K).

The sum of gray gases model was applied in [22] for carbon dioxide and water vapor. For CO_2 the results apply for $T = 300\text{--}1800$ K, $pL_e = 0.01\text{--}10$ atm-m, and $P = 1$ atm. The coefficients are in Table 13-5. For H_2O vapor the

Table 13-5 Sum of gray gases coefficients for CO_2 [22]

	Coefficients of A_s		
	c_1	c_2	c_3
	2.7769×10^{-1}	3.869×10^{-2}	1.4249×10^{-5}

	Coefficients of a_i and k_i (atm-m)$^{-1}$			
i	$b_{1,i}$	$b_{2,i}$	$b_{3,i}$	k_i
1	0.1074	−0.10705	0.072727	0.03647
2	0.027237	0.10127	−0.043773	0.3633
3	0.058438	−0.001208	0.0006558	3.10
4	0.019078	0.037609	−0.015424	14.96
5	0.056993	−0.025412	0.0026167	103.61
6	0.0028014	0.038826	−0.020198	780.7

Table 13-6 Sum of gray gas coefficients for H_2O [22]

	Coefficients of A_s		
	c_1	c_2	c_3
	0.70892	-0.39834	0.21919

	Coefficients of a_i and k_i (atm-m)$^{-1}$			
i	$b_{1,i}$	$b_{2,i}$	$b_{3,i}$	k_i
1	0.44004	-0.303388	0.24443	0.82352
2	0.085585	0.23875	-0.23723	7.1972
3	0.15451	-0.30955	0.22091	50.574
4	0.027313	-0.016281	-0.015629	417.88

ranges are $T = 300$–700 K, $pL_e = 0.01$–2 atm-m, $P = 1$ atm, and $p \to 0$ as in Figs. 13-15 and 13-21. The coefficients are in Table 13-6.

Another set of coefficients was obtained in [25]. This is for the emittance of CO_2, H_2O, and a mixture of these two gases. In this instance the $b_{j,i}$ coefficients are defined by $a_i = \Sigma_{j=1}^{N} b_{j,i}T^{j-1}$ and the $\Sigma_{i=1}^{n} a_i$ is found by summing the a_i as obtained using these $b_{i,j}$ coefficients. The results are in Table 13-7.

The application of the weighted sum of gray gases model was examined in detail in [26] and compared with results obtained with exact integral equations and the P-1 approximation. Good results were obtained with a substantial reduction in computer time. The applications were limited to nonscattering media and enclosures with black walls.

An analytical representation for the ϵ_{H_2O} in Fig. 13-15 is given in [27] by

$$\epsilon_{H_2O} = a_0[1 - \exp(-a_1\sqrt{X})] \tag{13-83}$$

where $X = p_{H_2O} L_e P(300/T)$ and a_0 and a_1 are functions of temperature given by:

T, K	a_0	a_1, m$^{-1/2} \cdot$ atm^{-1}
300	0.683	1.17
600	0.674	1.32
900	0.700	1.27
1200	0.673	1.21
1500	0.624	1.15

For H_2O-air mixtures the parameter $X = p_{H_2O}L_e(300/T)(p_{air} + bp_{H_2O})$, where T is in K, p in atm, and L_e in m. The self-broadening coefficient b for water vapor is given by $b = 5.0(300/T)^{1/2} + 0.5$. A comparison with the Hottel chart is in Fig. 13-22.

Additional curve-fit representations for emittance properties have been made. Some references are in the review article [28].

Table 13-7 Coefficients for emittance

i	k_i	$b_{1,i} \times 10^1$	$b_{2,i} \times 10^4$	$b_{3,i} \times 10^7$	$b_{4,i} \times 10^{11}$
			Carbon dioxide, $p_c \to 0$ atm		
1	0.3966	0.4334	2.620	−1.560	2.565
2	15.64	−0.4814	2.822	−1.794	3.274
3	394.3	0.5492	0.1087	−0.3500	0.9123
			Water vapor, $p_w \to 0$ atm		
1	0.4098	5.977	−5.119	3.042	−5.564
2	6.325	0.5677	3.333	−1.967	2.718
3	120.5	1.800	−2.334	1.008	−1.454
			Water vapor, $p_w = 1.0$ atm		
1	0.4496	6.324	−8.358	6.135	−13.03
2	7.113	−0.2016	7.145	−5.212	9.868
3	119.7	3.500	−5.040	2.425	−3.888
			Mixture, $p_w/p_c = 1$		
1	0.4303	5.150	−2.303	0.9779	−1.494
2	7.055	0.7749	3.399	−2.297	3.770
3	178.1	1.907	−1.824	0.5608	−0.5122
			Mixture, $p_w/p_c = 2$		
1	0.4201	6.508	−5.551	3.029	−5.353
2	6.516	−0.2504	6.112	−3.882	6.528
3	131.9	2.718	−3.118	1.221	−1.612

$P = 1$ atm, $0.001 \leq pS \leq 10.0$ atm-m, $600 \leq T \leq 2400$ K

13-6.2 Exchange between Entire Gas Volume and Emitting Boundary

In the previous section the temperature of the black enclosure wall was small enough that emission from the wall could be neglected. If wall emission is significant, the *average* heat flux *removed* at the wall is the gas emission to the wall (which is all absorbed because the *wall is black*) minus the emitted flux from the wall that is absorbed by the gas. The total energy removed from the entire boundary must equal the energy supplied to the gas by some other means, such as by combustion. A heat balance gives

$$\frac{Q_g}{A} = -\frac{Q_w}{A} = \sigma[\epsilon_g T_g^4 - \alpha_g(T_w)T_w^4] \tag{13-84}$$

This also follows from Eq. (13-21) when the wall is black. The $\alpha_g(T_w)$ is the total absorptance of the gas for radiation emitted from the wall at temperature T_w. The

$\alpha_g(T_w)$ depends on the spectral properties of the gas and on T_w, as this determines the spectral distribution of the radiation received by the gas. To illustrate how α_g can differ from ϵ_g, consider the absorptance of the gas in Example 13-5 for incident radiant energy from the sun. The spectral distribution of solar radiation obviously differs considerably from that emitted by a blackbody at T_g. The resulting α_g is much different from the ϵ_g obtained in Example 13-5.

EXAMPLE 13-7 What fraction of incident solar radiation is absorbed by a thick layer of CO_2 at 10 atm and 830 K? Use the approximate absorption bands of Example 13-5.

The effective radiating temperature of the sun is $T_s = 5780$ K. The desired result is the fraction of the solar spectrum that lies within the four CO_2 bands, as this is the only portion of the incident radiation that will be absorbed. Using the $F_{0-\lambda T_s}$ factors as in Example 13-5, but obtained using the solar temperature, gives the values

λ, μm	λT_s, $\mu m \cdot K$	$F_{0-\lambda T_s}$	λ, μm	λT_s, $\mu m \cdot K$	$F_{0-\lambda T_s}$
1.8	10,400	0.92166	4.0	23,120	0.99024
2.2	12,720	0.95255	4.6	26,590	0.99336
2.6	15,030	0.96892	9	52,020	0.99901
2.8	16,180	0.97446	19	109,800	~1.00000

The fraction absorbed is then

$$\alpha' = \sum_{\substack{absorbing \\ bands}} (F_{0-(\lambda T_s)_{upper}} - F_{0-(\lambda T_s)_{lower}})_{band}$$

$$= (0.95255 - 0.92166) + (0.97446 - 0.96892) + (0.99336 - 0.99024)$$

$$+ (1.00000 - 0.99901) = 0.041$$

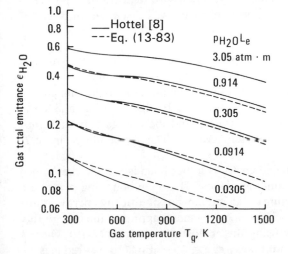

Figure 13-22 Comparison of Eq. (13-83) with Hottel chart [8] for $P = 1$ atm and $p_{H_2O} \rightarrow 0$.

Even though the gas layer is thick, only 4.1% of the incident energy is absorbed since the gas transmits well in the region between the absorption bands.

For radiative exchange calculations in furnaces, a convenient *approximate procedure* for determining α_g is given in Hottel and Sarofim [3]. The α_g is obtained from the gas total emittance charts by use of the following:

$$\alpha_g = \alpha_{CO_2} + \alpha_{H_2O} - \Delta\alpha \tag{13-85}$$

where

$$\alpha_{CO_2} = C_{CO_2}\epsilon_{CO_2}^+ \left(\frac{T_g}{T_w}\right)^{0.5} \tag{13-86}$$

$$\alpha_{H_2O} = C_{H_2O}\epsilon_{H_2O}^+ \left(\frac{T_g}{T_w}\right)^{0.5} \tag{13-87}$$

$$\Delta\alpha = (\Delta\epsilon)_{at\ T_w} \tag{13-88}$$

The $\epsilon_{CO_2}^+$ and $\epsilon_{H_2O}^+$ are, respectively, ϵ_{CO_2} and ϵ_{H_2O} obtained from Figs. 13-13 and 13-15 evaluated at the abscissa T_w and at the respective parameters $p_{CO_2}L_e'$ and $p_{H_2O}L_e'$ where $L_e' = L_e T_w/T_g$. It is pointed out in [24] that the exponent 0.5 is now becoming more accepted to replace the values 0.65 and 0.45 originally used in (13-86) and (13-87). At high temperatures and pressures when there is overlapping of absorption lines in the infrared spectrum, the $L_e' = L_e(T_w/T_g)^{3/2}$, which is discussed in [24].

Section 13-6.1 considered a uniform isothermal gas bounded by a black enclosure. If the bounding is not black and hence is reflecting, radiation can pass through the gas by means of multiple reflections from the boundary. For an enclosure with a single wall, this can be included by integrating Eq. (13-21) over all wavelengths. In [24] a procedure is proposed to use total emittance values for a multiple reflection situation. The solution of enclosure heat transfer problems by integration over the wavelength absorption bands was discussed in Sec. 13-3.4.

EXAMPLE 13-8 Two black parallel plates are separated by $D = 1$ m. The plates are of width $W = 1$ m and have infinite length normal to the cross section shown (Fig. 13-23). The space between the plates is filled with carbon dioxide gas at $p_{CO_2} = 1$ atm and $T_g = 1000$ K. If plate 1 is maintained at 2000 K and plate 2 is maintained at 500 K, find the energy flux that must be supplied to plate 2 to maintain its temperature. The surroundings are at $T_e \ll 500$ K.

As shown by Fig. 13-23, the geometry is a four-boundary enclosure formed by two plates and two open bounding planes. The open bounding planes are perfectly absorbing (nonreflecting) and radiate no significant energy as the temperature of the surroundings is low. As a general approach the energy flux added to surface 2 is found by using the enclosure equation (13-19) where $k = 2$ and $N = 4$. All surfaces are black, $\epsilon_{\lambda,j} = 1$, so Eq. (13-19) reduces to

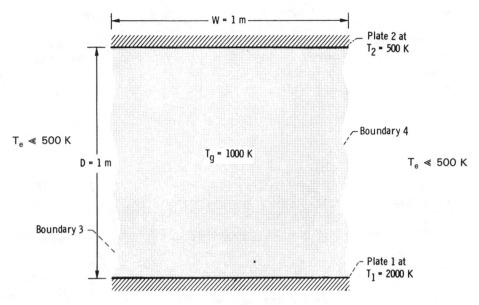

Figure 13-23 Isothermal carbon dioxide contained between black plates (Example 13-8).

$$\sum_{j=1}^{4} \delta_{2j}\, dq_{\lambda,j} = \sum_{j=1}^{4} [(\delta_{2j} - F_{2-j}\bar{\tau}_{\lambda,2-j})e_{\lambda b,j}\, d\lambda - F_{2-j}\bar{\alpha}_{\lambda,2-j}e_{\lambda b,g}\, d\lambda] \qquad (13\text{-}89)$$

The self-view factor $F_{2-2} = 0$ and $e_{\lambda b,3} = e_{\lambda b,4} \approx 0$, so this becomes

$$dq_{\lambda,2} = [\quad F_{2-1}\bar{\tau}_{\lambda,2-1}e_{\lambda b,1} \mid e_{\lambda b,2} \quad (F_{2-1}\bar{\alpha}_{\lambda,2-1} \mid F_{2-3}\bar{\alpha}_{\lambda,2-3}$$
$$+ F_{2-4}\bar{\alpha}_{\lambda,2-4})e_{\lambda b,g}]\, d\lambda \qquad (13\text{-}90)$$

To simplify the example, it is carried out by considering the entire wavelength region as a single band. To obtain the total energy supplied to plate 2, integrate over all wavelengths to obtain

$$q_2 = -F_{2-1}\int_0^\infty \bar{\tau}_{\lambda,2-1}e_{\lambda b,1}\, d\lambda + \sigma T_2^4$$

$$- \int_0^\infty (F_{2-1}\bar{\alpha}_{\lambda,2-1} + F_{2-3}\bar{\alpha}_{\lambda,2-3} + F_{2-4}\bar{\alpha}_{\lambda,2-4})e_{\lambda b,g}\, d\lambda$$

By use of the definitions of total transmission and absorption factors, which are

$$\bar{\tau}_{2-1}\sigma T_1^4 = \int_0^\infty \bar{\tau}_{\lambda,2-1}e_{\lambda b,1}\, d\lambda \qquad \bar{\alpha}_{2-1}\sigma T_g^4 = \int_0^\infty \bar{\alpha}_{\lambda,2-1}e_{\lambda b,g}\, d\lambda$$

and so forth, q_2 becomes

$$q_2 = \sigma T_2^4 - F_{2-1}\bar{\tau}_{2-1}\sigma T_1^4 - (F_{2-1}\bar{\alpha}_{2-1} + F_{2-3}\bar{\alpha}_{2-3} + F_{2-4}\bar{\alpha}_{2-4})\sigma T_g^4 \qquad (13\text{-}91)$$

To determine $\bar{\tau}$ and $\bar{\alpha}$, the geometric-mean beam length will be used. For opposing rectangles (Fig. 13-11) at an abscissa of 1.0 and on the curve for a length-to-spacing ratio of ∞, the $\bar{S}_{2-1}/D = 1.34$ or $\bar{S}_{2-1} = 1.34$ m. To determine $\bar{\alpha}_{2-1}$, which determines the emission of the gas, use the emittance chart in Fig. 13-13 at a pressure of 1 atm, a beam length of 1.34 m, and $T_g = 1000$ K. This gives $\bar{\alpha}_{2-1} = 0.22$. When obtaining $\bar{\tau}_{2-1}$, note from (13-90) that the radiation in the $\bar{\tau}_{2-1}$ term is $e_{\lambda b,1}$ and is coming from wall 1. Therefore it has a spectral distribution different from that of the gas radiation. To account for this nongray effect, (13-86) is used with ϵ^+ evaluated at $p_{CO_2}\bar{S}_{2-1}(T_1/T_g) = 1.34(2000/1000) = 2.68$ atm · m and $T_1 = 2000$ K. Then, using Fig. 13-13 (extrapolated) and Eq. (13-86) result in $\bar{\tau}_{2-1} \approx 1 - 0.2(\frac{1}{2})^{0.5} = 0.86$.

From factor 11 in Appendix C the configuration factor F_{2-1} is given by

$$F_{2-1} = \frac{(D^2 + W^2)^{1/2} - D}{W} = \sqrt{2} - 1 = 0.414$$

Then $F_{2-3} = F_{2-4} = \frac{1}{2}(1 - 0.414) = 0.293$.

The $\bar{\alpha}_{2-3} = \bar{\alpha}_{2-4}$, and they remain to be found. For adjoint planes as in the geometry for Table 13-2, the following expression from Eq. (12) of [2] can be used, obtained for the present case where $b \to \infty$, $a = 1$, and $c = 1$:

$$\bar{S}_{2-3} = \frac{1}{\pi F_{2-3}} (2 \ln \sqrt{2}) = \frac{2 \times 0.347}{\pi \times 0.293} = 0.753 \text{ m}$$

Using Fig. 13-13 at $p\bar{S} = 0.753$ atm · m and $T_g = 1000$ K gives $\bar{\alpha}_{2-3} = \bar{\alpha}_{2-4} = 0.19$. Then

$$q_2 = \sigma T_2^4 - 0.414(0.86)\sigma T_1^4 - (0.414 \times 0.22 + 2 \times 0.293 \times 0.19)\sigma T_g^4$$

$$= 5.6705 \times 10^{-12}(500^4 - 0.36 \times 2000^4 - 0.20 \times 1000^4)$$

$$= -33.4 \text{ W/cm}^2$$

The solution is now complete. Note that the largest contribution to q_2 is by energy leaving surface 1 and being absorbed by surface 2. Emission from the gas to surface 2 and emission from surface 2 are negligible.

An alternative approach that is simpler for this particular example is to note that the term involving T_g in (13-91) is the flux received by surface 2 as a result of emission by the entire gas. This can be calculated from (13-80) using the mean beam length. Then $q_2 = \sigma T_2^4 - F_{2-1}\bar{\tau}_{2-1}\sigma T_1^4 - \epsilon_g\sigma T_g^4$. For this symmetric geometry the average flux from the gas to one side of the enclosure is the same as that to the entire enclosure boundary. Consequently, the mean beam length can be obtained from (13-76), which gives $L_e = 0.9(4)V/A = 0.9(4)(1 \text{ m})^2/4$ m $= 0.9$ m. Then from Fig. 13-13 at $T_g = 1000$ K and $p_{CO_2}L_e = 0.9$ atm · m, the $\epsilon_g = 0.20$. This gives the same q_2 as previously calculated.

13-7 THE ZONE METHOD FOR RADIATION BY NONISOTHERMAL GASES

The previous material in this chapter considered an isothermal gas in an enclosure. Chapters 14–17 deal with theory and solution methods, both analytical and numerical, for emitting and scattering media where the temperature distribution in the medium is to be found. One method that has been used extensively in engineering problems is the *zoning method,* and it is discussed here.

In the zoning method a nonisothermal enclosure filled with nonisothermal gas is subdivided into areas and volumes (zones) that can each be approximated as isothermal. An energy balance is written for each zone. This provides a set of simultaneous equations for the unknown heat fluxes or temperatures in the same manner as the procedure in Sec. 13-3 for an isothermal gas. The method is practical and powerful; Hottel and Sarofim [3] discuss it at some length. Multidimensional applications have been carried out by Hottel and Cohen [29], Einstein [30, 31], and Hottel and Sarofim [32]. References [29] and [32] provide a useful introduction to the method. Scattering can be included as shown by Noble [33]. The development in this section is limited to radiation exchange only; extensions to include conduction and convection are in Chaps. 16 and 17 and in [3].

The basic concepts of the zoning method are now developed for a uniform nonisothermal gas with constant absorption coefficient (gray gas). Consider volume V_γ in Fig. 13-24 and surface A_k. From (12-38) the emissive power from a volume element dV_γ is $4\pi a_\lambda i'_{\lambda b}\, dV_\gamma\, d\lambda$, or per unit solid angle around dV_γ it is

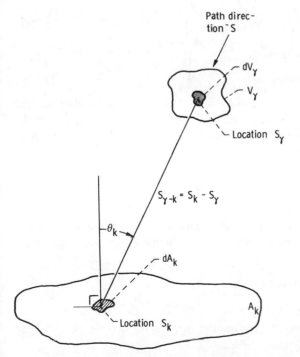

Figure 13-24 Radiation from gas volume V_γ to area A_k.

$a_\lambda i'_{\lambda b} \, dV_\gamma \, d\lambda$. The surface element dA_k subtends the solid angle $dA_k \cos \theta_k / S^2_{\gamma-k}$ when viewed from dV_γ. The fraction of radiation transmitted through the path length $S_{\gamma-k}$ is $\exp[-\int_{S_\gamma}^{S_k} a_\lambda(S^*) \, dS^*]$. Multiplying these factors and integrating over V_γ and A_k gives the spectral energy arriving at A_k from gas volume V_γ as

$$dq_{\lambda i,\gamma-k} A_k = d\lambda \int_{V_\gamma} \int_{A_k} \frac{a_\lambda(\gamma) i'_{\lambda b}(\gamma) \cos \theta_k}{S^2_{\gamma-k}} \exp\left[-\int_{S_\gamma}^{S_k} a_\lambda(S^*) \, dS^*\right] dA_k \, dV_\gamma$$

(13-92)

For uniform $a_\lambda(\gamma)$, the exponential factor becomes $\exp[-a_\lambda(S_k - S_\gamma)] = \tau_\lambda(S_{\gamma-k})$. The entire gas volume is divided into finite subvolumes V_γ and the assumption is made that conditions are uniform over each V_γ. Then (13-92) simplifies to

$$dq_{\lambda i,\gamma-k} A_k = d\lambda \, a_\lambda i'_{\lambda b}(\gamma) \int_{V_\gamma} \int_{A_k} \frac{\cos \theta_k}{S^2_{\gamma-k}} \tau_\lambda(S_{\gamma-k}) \, dA_k \, dV_\gamma$$

(13-93)

For a gray gas, (13-93) integrated over all wavelengths gives the total energy incident on A_k as

$$q_{i,\gamma-k} A_k = a \frac{\sigma T_\gamma^4}{\pi} \int_{V_\gamma} \int_{A_k} \frac{\cos \theta_k}{S^2_{\gamma-k}} \tau(S_{\gamma-k}) \, dA_k \, dV_\gamma$$

(13-94)

Now define the *gas-surface direct-exchange area* $\overline{g_\gamma s_k}$ as

$$\overline{g_\gamma s_k} \equiv \frac{a}{\pi} \int_{V_\gamma} \int_{A_k} \frac{\cos \theta_k}{S^2_{\gamma-k}} \tau(S_{\gamma-k}) \, dA_k \, dV_\gamma$$

(13-95)

Equation (13-94) can then be written as

$$q_{i,\gamma-k} A_k = \overline{g_\gamma s_k} \sigma T_\gamma^4$$

(13-96)

Thus the energy $q_{i,\gamma-k} A_k$ arriving at A_k can be regarded as the blackbody emissive power σT_γ^4 of the gas in V_γ that is radiated from an *effective* area $\overline{g_\gamma s_k}$. Let the entire gas volume be divided into Γ finite regions. The energy flux incident upon A_k from all the gas-volume regions is then

$$(q_{i,k})_{\text{from gas}} = \frac{1}{A_k} \sum_{\gamma=1}^{\Gamma} \overline{g_\gamma s_k} \sigma T_\gamma^4$$

(13-97)

Now consider the interchange between the bounding surface zones of the enclosure. The energy leaving surface area A_j that reaches A_k is, for a gas with uniform gray properties,

$$q_{i,j-k} A_k = \frac{q_{o,j}}{\pi} \int_{A_k} \int_{A_j} \tau(S_{j-k}) \frac{\cos \theta_j \cos \theta_k \, dA_j \, dA_k}{S^2_{j-k}}$$

(13-98)

where $q_{o,j}$ is assumed uniform over A_j. Define the *surface-surface direct-exchange area* as

$$\overline{s_j s_k} \equiv \int_{A_k} \int_{A_j} \tau(S_{j-k}) \frac{\cos \theta_j \cos \theta_k \, dA_j \, dA_k}{\pi S_{j-k}^2} \tag{13-99}$$

Equation (13-98) is then written as

$$q_{i,j-k} A_k = \overline{s_j s_k} q_{o,j} \tag{13-100}$$

Thus the energy $q_{i,j-k} A_k$ from A_j arriving at A_k is the flux $q_{o,j}$ leaving A_j times an *effective* area $\overline{s_j s_k}$. The flux incident upon A_k as a result of fluxes leaving all N surfaces of the enclosure is then

$$(q_{i,k})_{\text{from surfaces}} = \frac{1}{A_k} \sum_{j=1}^{N} \overline{s_j s_k} q_{o,j} \tag{13-101}$$

Now the total energy flux incident upon surface A_k can be obtained as

$$q_{i,k} = (q_{i,k})_{\text{from surfaces}} + (q_{i,k})_{\text{from gas}} = \frac{1}{A_k} \left(\sum_{j=1}^{N} \overline{s_j s_k} q_{o,j} + \sum_{\gamma=1}^{\Gamma} \overline{g_\gamma s_k} \sigma T_\gamma^4 \right) \tag{13-102}$$

The usual net-radiation equations [(7-12) and (7-13)] also apply at surface A_k:

$$q_k = q_{o,k} - q_{i,k} \tag{13-103}$$

$$q_{o,k} = \epsilon_k \sigma T_k^4 + (1 - \epsilon_k) q_{i,k} \tag{13-104}$$

For problems in which the temperatures T_γ are given for all gas volume elements V_γ, (13-102) to (13-104) are sufficient to solve for N unknown values of T_k and q_k; the other N values of T_k and q_k must be provided as known boundary conditions. The methods of Sec. 7-4 can be directly applied.

When the T_γ of the Γ gas elements are unknowns, then Γ additional equations are required. These are obtained by taking an energy balance on each gas zone. For each gas element V_γ the emission must equal the absorption of energy plus the local heat source in the gas. For a gray gas with constant properties a heat balance on the volume region V_γ gives

$$4 a \sigma T_\gamma^4 V_\gamma = \sum_{\text{all } V_{\gamma*}} \int_{V_\gamma} \int_{V_{\gamma*}} \frac{4 a \sigma T_{\gamma*}^4 \, dV_{\gamma*}}{4\pi} \tau(S_{\gamma*-\gamma}) \frac{a \, dV_\gamma}{S_{\gamma*-\gamma}^2}$$

$$+ \sum_{\text{all } A_k} \int_{V_\gamma} \int_{A_k} \frac{q_{o,k} \cos \theta_k}{\pi} \, dA_k \, \tau(S_{k-\gamma}) \frac{a \, dV_\gamma}{S_{k-\gamma}^2} + q_\gamma''' V_\gamma$$

$$4 a \sigma T_\gamma^4 V_\gamma = a^2 \sum_{\gamma*=1}^{\Gamma} \sigma T_{\gamma*}^4 \int_{V_\gamma} \int_{V_{\gamma*}} \frac{\tau(S_{\gamma*-\gamma}) \, dV_{\gamma*} \, dV_\gamma}{\pi S_{\gamma*-\gamma}^2}$$

$$+ a \sum_{k=1}^{N} q_{o,k} \int_{V_\gamma} \int_{A_k} \frac{\cos \theta_k}{\pi S_{k-\gamma}^2} \tau(S_{k-\gamma}) \, dA_k \, dV_\gamma + q_\gamma''' V_\gamma \tag{13-105}$$

It is assumed that each V_γ is isothermal. Define the *surface-gas direct-exchange area* as

$$\overline{s_k g_\gamma} \equiv \frac{a}{\pi} \int_{V_\gamma} \int_{A_k} \frac{\cos \theta_k}{S_{k-\gamma}^2} \tau(S_{k-\gamma}) \, dA_k \, dV_\gamma \tag{13-106}$$

Comparing (13-106) with (13-95) shows that there is reciprocity between the surface-gas and gas-surface direct-exchange areas:

$$\overline{s_k g_\gamma} = \overline{g_\gamma s_k} \tag{13-107}$$

Define the *gas-gas direct-exchange area* as

$$\overline{g_{\gamma^*} g_\gamma} = \overline{g_\gamma g_{\gamma^*}} \equiv \frac{a^2}{\pi} \int_{V_\gamma} \int_{V_{\gamma^*}} \frac{\tau(S_{\gamma^*-\gamma}) \, dV_{\gamma^*} \, dV_\gamma}{S_{\gamma^*-\gamma}^2} \tag{13-108}$$

Substituting (13-106) to (13-108) into (13-105) gives

$$4a\sigma T_\gamma^4 V_\gamma = \sum_{\gamma^*=1}^{\Gamma} \sigma T_{\gamma^*}^4 \, \overline{g_{\gamma^*} g_\gamma} + \sum_{k=1}^{N} q_{o,k} \overline{g_\gamma s_k} + q_\gamma''' V_\gamma \tag{13-109}$$

The $\overline{g_{\gamma^*} g_\gamma}$ have been tabulated [29] so that (13-109), written for each V_γ, provides the additional set of Γ equations required to obtain the gas temperature distribution.

The notation developed by Hottel and co-workers has been used in this section with only slight modification. A comparison with Sec. 13-3 shows that, in terms of the notation used there [Eq. (13-8)], the following identity exists:

$$F_{j-k} \bar{\tau}_{j-k} A_j = \overline{s_j s_k} \tag{13-110}$$

The gas absorptance factor in (13-9), $F_{j-k} \bar{\alpha}_{j-k} A_j$, is generally not related in a useful way to $\overline{g_\gamma s_k}$. The latter quantity is derived for an element of gas volume, while $F_{j-k} \bar{\alpha}_{j-k} A_j$ is concerned with the entire gas volume.

Hottel and co-workers [3, 8, 29] have further developed the approach outlined in this section. It is possible to make approximate allowances for spectral variations in gas properties. Variations in properties with position in the enclosure are handled by defining a suitable mean absorption coefficient between each set of zones. Einstein [30, 31] modified the $\overline{gs}$ and $\overline{gg}$ factors to give better accuracy when strong gradients are present. All these approximations become difficult to carry through if the absorption coefficient is a strong function of temperature. Noble [33] shows how matrix theory can be used as a computational aid in the zone method using exchange areas.

Values of $\overline{s_j s_k}$ and $\overline{g_\gamma s_k}$ have been tabulated for cubical isothermal volumes and square isothermal boundary elements by Hottel and Cohen [29]. Exchange areas are available for elements in rectangular [34, 35], cylindrical [31, 34], and conical [36] enclosures. Hottel and Sarofim [3] present an extensive tabulation of factors for the cylindrical geometry and a table of references for several other geometries. In [37] numerical integration was used to evaluate direct-exchange areas for examples involving squares, cubes, circular cylinders, and elliptic cylinders. Some relations for a cylindrical enclosure are in [37a]. Useful results are given by Tucker [38] for application in rectangular furnaces. Since many furnaces can be conveniently divided into cubic gas zones and square surface zones, the $\overline{ss}$, $\overline{gs}$, and $\overline{gg}$ factors were evaluated for these geometries and the results extended to a value of 18 for the optical dimension of the side of the square or cube. The factors are given on charts and are curve fitted with exponential functions that can be readily used for computer calculations. An example of the correlations is in Table 13-8. Results are given for squares and cubes that are close to each other. For larger separation distances it is a good approximation to assume that the view and path length for absorption are the same for all points within each zone and thus treat the zones as differential elements. Then differential forms can be used as follows, with θ and S based on the center-to-center orientation and separation distance:

$$\overline{s_j s_k} = \overline{s_k s_j} = \frac{\tau(S_{j-k}) \cos \theta_j \cos \theta_k \, dA_j \, dA_k}{\pi S_{j-k}^2} \tag{13-111}$$

Table 13-8 Correlation coefficients for direct-exchange areas between perpendicularly oriented square surfaces (a is absorption coefficient); $\overline{ss}/B^2 = C \exp(-AaB)$; $A = a_0 + a_1 aB + \cdots + a_4(aB)^4$ [38]

X/B	Y/B	Z/B	C	a_0	a_1	a_2	a_3	a_4
1	1	1	.2000	.5390	−.0615	.00429	$-.151 \times 10^{-3}$	$.206 \times 10^{-5}$
2	1	1	.0406	.9965	−.0878	.00419	$-.773 \times 10^{-4}$	
3	1	1	.0043	1.906				
1	2	1	.0328	1.571	−.0391	.00208		
2	2	1	.0189	1.751				
3	2	1	.0059	2.384				
1	3	1	.0089	2.502				
2	3	1	.0069	2.665				
3	3	1	.0036	3.129				
1	2	2	.0329	2.055				
2	2	2	.0230	2.245				
3	2	2	.0101	2.780				
1	3	2	.0159	2.860				
2	3	2	.0129	3.010				
3	3	2	.0076	3.435				
1	3	3	.0124	3.481				
2	3	3	.0107	3.609				
3	3	3	.0073	3.976				

$$\overline{s_k g_\gamma} = \overline{g_\gamma s_k} = \frac{a\tau(S_{k-\gamma})\cos\theta_k \, dA_k \, dV_\gamma}{\pi S_{k-\gamma}^2} \tag{13-112}$$

$$\overline{g_{\gamma*} g_\gamma} = \overline{g_\gamma g_{\gamma*}} = \frac{a^2\tau(S_{\gamma*-\gamma}) \, dV_{\gamma*} \, dV_\gamma}{\pi S_{\gamma*-\gamma}^2} \tag{13-113}$$

If the gas in the enclosure is nonhomogeneous in temperature and/or composition, the exchange areas must account for variations in absorption coefficient in the gas between any two elements. In [39] exchange areas are derived for cubes in nonhomogeneous media. Values are presented for narrow bands with overlapped lines and with nonoverlapped lines for the cases of adjacent and remote pairs of cubes, assuming a linear temperature profile between the cubes.

Numerical techniques for application of the zoning method to multimode problems are in Chap. 17.

13-8 FLAMES, LUMINOUS FLAMES, AND PARTICLE RADIATION

Under certain conditions, gases emit much more radiation in the visible region than would be expected from the absorption coefficients discussed to this point. For example, the typical almost transparent blue flame of a bunsen burner can be made into a smoky yellow-orange flame by changing only the fuel-air ratio. Such luminous emission is usually ascribed to soot (hot carbon) particles formed because of incomplete combustion in hydrocarbon flames. There is room for argument even here. Echigo, Nishiwaki, and Hirata [40] and others have advanced the hypothesis, supported by some experimental facts, that luminous emission from some flames is due to emission from vibration-rotation bands of chemical species that appear during the combustion process *prior* to the formation of soot particles. However, since soot formation is the most widely accepted view, soot radiation will be emphasized in this discussion of luminous flames.

Combustion is very complicated, often consisting of chemical reactions in series and in parallel and involving various intermediate species. The composition and concentration of these species cannot be predicted well unless knowledge is available of the flame reaction kinetics; this knowledge will not usually be at hand. Because the radiation properties of the flame depend on the distributions of temperature and species within the flame, a detailed prediction of radiation from flames is not often possible from knowledge only of the original combustible constituents and the flame geometry. Because of these difficulties, it is usually necessary to resort to empirical methods for predicting radiation from systems involving combustion. The following discussion will examine the calculation of a theoretical flame temperature from the chemical energy release (without accounting for radiative heat loss), the radiation from the nonluminous gas products, and the more complex problem of radiation from a gas containing particles.

13-8.1 Radiation from Nonluminous Flames

Radiation from the nonluminous portion of the combustion products is fairly well understood. The complexities of the chemical reaction are not as important here, since it is the gaseous end products above the active burning region that are considered. In most instances hydrocarbon combustion is being considered, and the radiation is from the CO_2 and H_2O bands in the infrared. For flames a meter or more thick, as in commercial furnaces, the emission leaving the flame within the CO_2 and H_2O vibration-rotation bands can be close to blackbody emission. The gaseous radiation properties and methods in this chapter can be used to compute the radiative heat transfer. The analysis is greatly simplified if the gas is well mixed and can be assumed isothermal. For a nonisothermal condition the gas can be divided into approximately isothermal zones, and convection can be included if the circulation pattern in the combustion chamber is known. A nonisothermal analysis with convection was carried out in [32] for cylindrical flames. In [41–45] radiation is treated from various types of nonluminous flames (laminar or turbulent, mixed or diffusion). The flame shape for an open diffusion flame is considered in [46]. The local absorption coefficient in nonluminous flames is calculated in [46a] as a function of mixture fraction and fuel composition.

When considering the radiation from flames, a characteristic parameter is the average temperature of a well-mixed flame as a result of the chemical energy addition. Well-developed methods exist [47–50] for computing the theoretical flame temperature from available thermodynamic data. The effect of preheating either the fuel or oxidizer, or both, can be included. Complete combustion and no heat losses are assumed. The theoretical flame temperature T is computed using energy conservation. The energy in the combustion constituents plus the energy of combustion is equated to the energy of the combustion products to give

$$T - T_{ref}$$

$$= \frac{(\text{energy in feed air and fuel above } T_{ref}) + (\text{energy released by combustion})}{(\text{total mass of products}) \times (\text{mean heat capacity of products})}$$

$$(13\text{-}114)$$

The computation is illustrated by an example.

EXAMPLE 13-9 Using the mean specific heat data of Fig. 13-25 (adapted from [47]) and the heat of combustion from Table 13-9, calculate the theoretical temperature of an ethane flame burning with 100% excess air (by volume). The ethane is supplied at room temperature (25° C), and the air feed is preheated to 500° C. The flame is burning in an environment at a pressure of 1 atm.

For ethane, assuming complete combustion, the reaction is $2C_2H_6 + 7O_2 \rightarrow 4CO_2 + 6H_2O$. Assume that 2 kg-mol of ethane is burned; then 7 mol of oxygen is consumed. Since there is 100% excess air, only one-half the feed air contributes oxygen to the combustion process, so 14 mol of oxygen is introduced in the feed air. Oxygen makes up 21% by volume of the feed air,

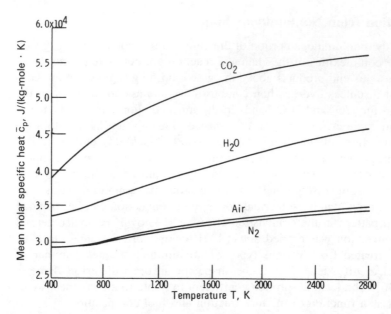

Figure 13-25 Mean molar specific heat of various gases averaged between T and 298 K.

and since mole fraction is equal to part by volume, the moles of air are m_{air} $= 14/0.21 = 66.67$ mol. The sensible heat in the feed components above a reference temperature is $H_i = \Sigma_k[m\bar{c}_p(T_i - T_{ref})]_k$ where $\bar{c}_p$ is the mean molar specific heat between the reference temperature and the input temperature T_i. Using Fig. 13-25, which has a reference temperature of 298 K (25° C), gives

$$H_i = [m\bar{c}_p(T_i - T_{ref})]_{ethane} + [m\bar{c}_p(T_i - T_{ref})]_{air}$$
$$= 0 + 66.67 \times 2.93 \times 10^4(773 - 298) = 9.28 \times 10^8 \, J$$

where the ethane contributes nothing, as it is supplied at T_{ref}. The heat ΔH released by combustion is found by using the heat of combustion from Table 13-9 and the fact that the molecular weight of ethane is 30:

$$\Delta H = 2 \text{ mol} \times 30 \text{ kg/mol} \times 4.74 \times 10^7 J/kg = 28.4 \times 10^8 \, J$$

The numerator of (13-114) is then $H_i + \Delta H = 37.7 \times 10^8$ J.

Nitrogen, which makes up about 79% by volume of air, remains from that portion of the feed air that supplied the oxygen for combustion (in addition to that carried in the product air). This amount of nitrogen in the combustion products is then $33.3 \times 0.79 = 26.3$ mol. The total quantity of products after combustion is (in moles):

Product	Quantity, mol
Carbon dioxide (CO_2)	4
Water vapor (H_2O)	6
Air	33.3 (half of feed air)
Nitrogen (N_2)	26.3

To find the denominator of (13-114), the individual quantities are summed: $H_{prod} = \Sigma_j \, m_j \bar{c}_{p,j}$. However, $\bar{c}_p$ depends on the temperature T, which is the flame temperature and is not yet known. We must estimate T to determine the $\bar{c}_{p,j}$ values and substitute the quantities into (13-114). If the calculated flame temperature agrees with the assumed value, the solution is finished. Otherwise, a new temperature is estimated, and H_{prod} is recalculated and substituted into (13-114); this procedure is continued until assumed and calculated temperatures agree. A table of calculations is as follows:

Assumed flame temperature, K	Mean molar specific heat $\bar{c}_p$, J/(kg-mol · K)				Product enthalpy H_{prod}, J	Calculated temperature, K
	H_2O	CO_2	Air	N_2		
2400	4.44×10^4	5.51×10^4	3.35×10^4	3.30×10^4	2.47×10^6	1825
2000	4.25×10^4	5.36×10^4	3.33×10^4	3.28×10^4	2.44×10^6	1845

Because a 400-K change in the assumed flame temperature produced only about a 20-K change in the calculated flame temperature, a value of 1853 K is estimated as within a few degrees of the converged result.

In this example, it was assumed there was complete combustion and no dissociation of combustion products. No consideration was given to energy loss by radiation, which would lower the flame temperature. Methods for including these

Table 13-9 Heat of combustion and flame temperature for hydrocarbon fuels [49, 50]

Fuel	Heat of Combustion J/kg	Maximum flame temperature, K (combustion with dry air at 298 K)		
		Theoretical (complete combustion)	Theoretical (with dissociation and ionization)	Experimental
Carbon monoxide (CO)	4.83×10^7	2615		
Hydrogen (H_2)	12.0×10^7	2490		
Methane (CH_4)	5.0×10^7	2285	2191	2158
Ethane (C_2H_6)	4.74×10^7	2338	2222	2173
Propane (C_3H_8)	4.64×10^7	2629	2240	2203
n-Butane (C_4H_{10})	4.56×10^7	2357	2246	2178
n Pentane (C_5H_{12})	4.53×10^7	2360		
Ethylene (C_2H_4)	4.72×10^7	2523	2345	2253
Propylene (C_3H_6)	4.57×10^7	2453	2323	2213
Butylene (C_4H_8)	4.53×10^7	2431	2306	2208
Amylene (C_5H_{10})	4.50×10^7	2477		
Acetylene (C_2H_2)	4.82×10^7	2859		
Benzene (C_6H_6)	4.06×10^7	2484		
Toluene ($C_6H_5CH_3$)	4.09×10^7	2460		

effects are discussed in [49]. A list of theoretical flame temperatures (no radiation included) is in Table 13-9 for various hydrocarbon flames. Results for complete combustion with dry air are shown, followed by calculated results modified to allow for product dissociation and ionization. The latter are compared with experimental results. In addition, the heats of combustion of the substances are shown. All data are from [49, 50]. Extensive tabulations of similar data for more than 200 hydrocarbons are in [47, 50]. Now that its average temperature has been estimated, consider the radiation emitted by a nonluminous flame.

EXAMPLE 13-10 In Example 13-9 the combustion products were 4 mol of CO_2, 6 mol of H_2O vapor, 33.3 mol of air, and 26.3 mol of N_2. Assume these products are in a cylindrical region 4 m high and 2 m in diameter and are uniformly mixed at the theoretical flame temperature 1853 K. The pressure is 1 atm. Compute the radiation from the gaseous region.

The partial pressure of each constituent is equal to its mole fraction:

$$p_{CO_2} = \left(\frac{4}{69.6}\right)(1 \text{ atm}) = 0.0575 \text{ atm},$$

$$p_{H_2O} = \left(\frac{6}{69.6}\right)(1 \text{ atm}) = 0.0862 \text{ atm}$$

The gas mean beam length for negligible self-absorption is, from (13-69),

$$L_{e,0} = \frac{4V}{A} = \frac{4[\pi(2^2/4)]4}{(2\pi \times 4) + 2\pi(2^2/4)} = 1.6 \text{ m}$$

To include self-absorption, a correction factor of 0.9 is applied to give $L_e = 0.9(1.6) = 1.44$ m. Then $p_{CO_2}L_e = 0.0575 \times 1.44 = 0.0828$ atm m, and $p_{H_2O}L_e = 0.0862 \times 1.44 = 0.124$ atm m. Using Figs. 13-13 to 13-17 at the flame temperature (1853 K) results in $\epsilon_{CO_2} = 0.060$ and $\epsilon_{H_2O} = 0.070 \times 1.08 = 0.076$. The 1.08 factor in ϵ_{H_2O} is a correction for the partial pressure of the water vapor being nonzero. There is also a negative correction from spectral overlap of the CO_2 and H_2O radiation bands. This is obtained from Fig. 13-17c at the values of the parameters:

$$\frac{p_{H_2O}}{p_{CO_2} + p_{H_2O}} = \frac{0.0862}{0.0575 + 0.0862} = 0.60$$

$$p_{CO_2}L_e + p_{H_2O}L_e = 0.0828 + 0.124 = 0.207 \text{ atm m}$$

The correction is $\Delta\epsilon = 0.026$. Then the gas emittance is $\epsilon_g = \epsilon_{CO_2} + \epsilon_{H_2O} - \Delta\epsilon = 0.060 + 0.076 - 0.026 = 0.110$. The radiation from the gas region at the theoretical flame temperature is then

$$Q = \epsilon_g A\sigma T_g^4 = 0.110(10\pi)5.6705 \times 10^{-8}(1853)^4 = 2310 \text{ kW}$$

13-8.2 Radiation from and through Luminous Flames and Gases with Particles

Several factors complicate the radiative transfer in the region of the flame that is actively burning. The simultaneous production and loss of energy produces a temperature variation and thus a variation of properties and emission within the flame. The intermediate combustion products resulting from the complex reaction chemistry can significantly alter the radiation characteristics from those of the final products. Soot is the most important radiating product formed in burning hydrocarbons. Soot emits in a continuous spectrum in the visible and infrared regions and can often double or triple the heat radiated by the gaseous products alone. A method for increasing the flame emission, if desired, is to promote slow initial mixing of the oxygen with the fuel so that large amounts of soot will form at the base of the flame. Ash particles in the combustion gases can also contribute to the radiation [28, 51, 52].

Determining the effect of soot on flame radiation resolves into two requirements. One is to somehow obtain the soot distribution in the flame; this is the biggest obstacle to the calculations. The distribution depends on the type of fuel, the mixing of fuel and oxidant, and the flame temperature (see, for example, experimental results in [53]). The soot distribution would be complicated to calculate from basic principles, so some experimental knowledge of a given combustion system is needed. The second requirement is to know the radiative properties of the soot. Then if the soot concentration and distribution are known, a radiation computation can be attempted. The radiant properties of soot are only approximately known.

If flames found in both the laboratory and industry are included, the individual soot particles produced in hydrocarbon flames generally range in diameter from 0.005 μm to more than 0.3 μm. Typical diameters measured in [54] were 0.02 0.7 μm, and most probable diameters in [55] were 0.08–0.1 μm. The soot can be in the form of spherical particles, agglomerated masses, or long filaments. The experimental determination of the physical form of the soot is difficult, as a probe used to gather soot for photomicrographic analysis may cause agglomeration of particles or otherwise alter the soot characteristics. The nucleation and growth of the soot particles are not well understood. Some soot can be nucleated in less than a millisecond after the fuel enters the flame, and the rate at which soot continues to form does not seem influenced much by the residence time of the fuel in the flame. An unknown precipitation mechanism governs soot production. In typical gaseous diffusion flames, the volume of soot per total volume of combustion products has been found experimentally to be in the range from 10^{-8} to 10^{-5} [51, 53, 56–60].

For a beam of radiation passing through a transparent carrier gas containing suspended soot, it has been found experimentally that the attenuation obeys Bouguer's law,

$$i'_\lambda(S) = i'_\lambda(0) \exp\left[-\int_0^S a_\lambda(S^*)\, dS^* \right] \tag{13-115}$$

From the Mie theory the radiative properties of soot depend on the size parameter $\pi D / \lambda$ (where D is the particle diameter) and the optical constants n and κ of the particles, which depend on the chemical composition of the soot. The n and κ depend somewhat on λ, as will be shown later, but do not depend strongly on temperature [55, 61, 62]. At the temperatures in combustion systems, the radiation is mostly in a wavelength range of 1 μm and larger; hence, for the small soot particles $\pi D / \lambda$ is generally much less than 0.3. In this range of size parameter the Mie theory implies in Eq. (12-102) that the scattering cross section depends on $(\pi D / \lambda)^4$ and that the absorption cross section (12-100) depends on $\pi D / \lambda$ to the first power. Thus scattering is very small compared with absorption, and a_λ in (13-115) is the absorption coefficient of the soot rather than the extinction coefficient of (12-13). Then, as a consequence of (13-63), the spectral emittance of an isothermal luminous gas volume, composed of soot *uniformly* distributed in a *nonradiating* carrier gas, is

$$\epsilon_\lambda = 1 - \exp\left(-a_\lambda L_e\right) \tag{13-116}$$

where L_e is the mean beam length for the volume. Radiation by the carrier gas will be included later.

Experimental correlation of soot spectral absorption coefficient A relatively simple empirical relation for the absorption coefficient a_λ, found experimentally in some instances, is

$$a_\lambda = Ck\lambda^{-\alpha} \tag{13-117}$$

where C is the soot volume concentration (volume of particles per unit volume of cloud), and k is a constant. The λ is in μm for the numerical quantities given here. Hottel [8] recommends the use of

$$a_\lambda = \frac{Ck_1}{\lambda^{0.95}} \tag{13-118a}$$

in the infrared region for $\lambda \geq 0.8$ μm. In some experiments, Siddall and McGrath [63] also found the functional relation of (13-118a) to hold approximately. They give, in the range $\lambda = 1$–7 μm, the following mean values of α:

Source of soot	Mean α for $\lambda = 1$–7 μm
Amyl acetate	0.89, 1.04
Avtur kerosene	0.77
Benzene	0.94, 0.95
Candle	0.93
Furnace samples	0.96, 1.14, 1.25
Petrotherm	1.06
Propane	1.00

Thus the 0.95 exponent recommended by Hottel appears reasonable.

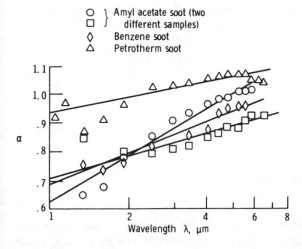

Figure 13-26 Experimental values of the exponent α plotted against λ for cases where α varies approximately linearly with $\ln \lambda$ [63].

In [63] the data were inspected in more detail to see if α had a functional variation with λ that would provide a more accurate correlation than using a constant α. In some instances α took the form

$$\alpha = c_1 + c_2 \ln \lambda$$

where c_1 and c_2 are positive constants. Examples are shown in Fig. 13-26. In other cases, as in Fig. 13-27, a more general polynomial was required to express α as a function of λ. Thus, as a generalization of (13-118a) for the infrared region,

$$a_\lambda = \frac{Ck_1}{\lambda^{\alpha(\lambda)}} \tag{13-118b}$$

and letting α be a constant is only an approximation.

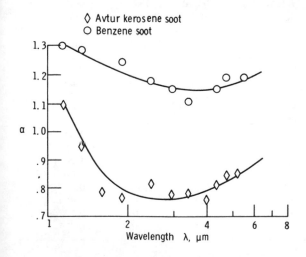

Figure 13-27 Experimental values of the exponent α as a function of λ where α does not vary linearly with $\ln \lambda$ [63].

In the visible range an inspection of experimental data led to the recommended form [8]

$$a_\lambda = \frac{Ck_2}{\lambda^{1.39}}$$ (13-119)

for the wavelength region around $\lambda = 0.6$ μm (say $\lambda \approx 0.3$–0.8 μm).

Electromagnetic-theory prediction of soot spectral absorption To examine from a more fundamental basis the absorption coefficient of a soot cloud in a nonabsorbing gas, electromagnetic theory can be employed [54, 55, 62–68]. From Eq. (12-33) for particles all of the same size, the absorption coefficient is

$$a_\lambda = E_\lambda AN$$ (13-120)

where N is the number of particles per unit volume and A is the projected area of a particle ($= \pi D^2/4$, as particles are assumed spherical). The E_λ is the spectral absorption efficiency factor, which is the ratio of the spectral absorption cross section to the physical cross section of the particle (ratio of energy absorbed to that incident on the particle). From Eq. (12-100), for the limit of small $\pi D/\lambda$, the Mie equations give E_λ for a small absorbing sphere as

$$E_\lambda = 24 \frac{\pi D}{\lambda} \frac{n\kappa}{(n^2 - \kappa^2 + 2)^2 + 4n^2\kappa^2} \equiv 24 \frac{\pi D}{\lambda} F(\lambda)$$ (13-121)

Table 13-10 Optical constants of acetylene and propane soots [62]

	Acetylene soot			Propane soot		
Wavelength λ, μm	Index of refraction n	Extinction coefficient κ	Absorption coefficient per particle volume fraction a_λ/C μm^{-1}	Index of refraction n	Extinction coefficient κ	Absorption coefficient per particle volume fraction a_λ/C μm^{-1}
0.4358	1.56	0.46	9.37	1.57	0.46	9.29
0.4500	1.56	0.48	9.45	1.56	0.50	9.83
0.5500	1.56	0.46	7.42	1.57	0.53	8.44
0.6500	1.57	0.44	5.96	1.56	0.52	7.07
0.8065	1.57	0.46	5.02	1.57	0.49	5.34
2.5	2.31	1.26	1.97	2.04	1.15	2.34
3.0	2.62	1.62	1.44	2.21	1.23	1.75
4.0	2.74	1.64	0.998	2.38	1.44	1.24
5.0	2.88	1.82	0.747	2.07	1.72	1.30
6.0	3.22	1.84	0.505	2.62	1.67	0.727
7.0	3.49	2.17	0.383	3.05	1.91	0.484
8.5	4.22	3.46	0.213	3.26	2.10	0.357
10.0	4.80	3.82	0.143	3.48	2.46	0.271

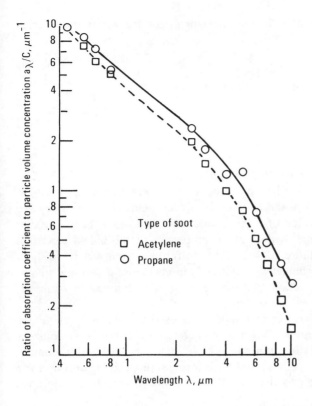

Figure 13-28 Ratio of spectral absorption coefficient to particle volume concentration for soot particles.

where n and κ are functions of λ. Then from (13-120)

$$\frac{a_\lambda}{C} = \frac{36\pi}{\lambda} F(\lambda) = \frac{36\pi}{\lambda} \frac{n\kappa}{(n^2 - \kappa^2 + 2)^2 + 4n^2\kappa^2} \tag{13-122}$$

where $C = N\pi D^3/6$ is the volume of particles per unit volume of cloud. The $a_\lambda(\lambda)/C$ can be evaluated if the n and κ of soot are known as functions of λ.

In [62] the n and κ of acetylene and propane soot were measured by collecting soot and then compressing it on a brass plate. The n and κ are obtained by using the equations for reflection from an interface in conjunction with measurements of the reflected intensity of polarized light. The values are in Table 13-10. Using these values a_λ/C was evaluated from (13-122), yielding the points in Fig. 13-28. Although the a_λ/C decreases with λ as expected from the form of (13-117) it is evident that an approximate curve fit by straight lines on the logarithmic plot would yield exponents on λ somewhat different than those of (13-118a) and (13-119).

The form of the exponent $\alpha(\lambda)$ predicted by Mie theory is examined in the infrared region by equating (13-118b) and (13-122),

$$\frac{k_1}{\lambda^{\alpha(\lambda)}} = \frac{36\pi}{\lambda} F(\lambda) \tag{13-123}$$

By evaluating this relation at $\lambda = 1$, the constant k_1 for the infrared region is found as

$$k_1 = 36\pi F(1) \tag{13-124}$$

Then $\lambda^{\alpha(\lambda)} = \lambda F(1)/F(\lambda)$ and solving for $\alpha(\lambda)$ gives

$$\alpha(\lambda) = 1 + \frac{\ln\left[F(1)/F(\lambda)\right]}{\ln \lambda} \tag{13-125}$$

The optical properties of soot are used in $F(1)$ and $F(\lambda)$ to evaluate $\alpha(\lambda)$. This was done in [63] using the properties of a baked electrode carbon at 2250 K. The results are in Fig. 13-29. The trend is the same as in the experimental curves of Fig. 13-26, but the α values are larger than the experimental values. They are also larger than the average value of 0.95 recommended in (13-118a). The discrepancy is probably partly due to the optical properties of the baked electrode carbon being different from those of soot. This is further discussed in [69], in which optical properties of carbonaceous materials more like real soot are given, and good comparisons of predictions with experiment are obtained in some instances.

In [70] the compressed powder technique was used to obtain n and κ for propane soot in the infrared region. When soot is compressed, the layer adjacent to the surface has a finite void fraction that may influence the reflection characteristics. In [70] the soot was compressed with a pressure of 2000 atm and the resulting reflections conformed to the Fresnel equations. Electron micrographs

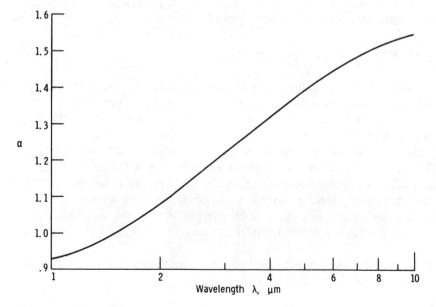

Figure 13-29 Calculated variation of the exponent α with wavelength using properties of baked electrode carbon at 2250 K [63].

Table 13-11 Complex refractive indices ($\bar{n} = n - i\kappa$) of propane soot corrected for surface layer void fraction of 0.18 [70].

$\lambda(\mu m)$	n	κ
2.0	2.33	0.78
3.0	2.31	0.71
4.0	2.36	0.66
4.5	2.37	0.71
5.0	2.39	0.66
6.0	2.40	0.80
7.0	2.49	0.85
8.0	2.66	0.93
9.0	2.63	0.92
10.0	2.62	1.00

were used to examine the surface void fraction and a correction for it was made in the results as given in Table 13-11. The values of n do not vary as much with λ as those in Table 13-10, and the κ values are smaller. The compressed soot technique yields a specular reflection for infrared radiation, but difficulty is encountered with the method in the visible region where the surface may not be optically smooth. For the visible region an in situ laser light-scattering technique is described in [54]. By use of this method for a methane-oxygen flame, the $\bar{n}$ at $\lambda = 0.488$ μm (blue line) was found as $\bar{n} = n - i\kappa = 1.60 - i0.59$. This was compared with $n = 1.57 - i0.56$ from [60] and $\bar{n} = 1.90 - i0.55$ from [55]. Additional information is provided by the in situ measurements in [71] using premixed flat flames of methane, propane, and ethylene.

Total emittance of soot cloud   A total emittance can be found for a path S through an *isothermal* cloud of suspended soot having a *uniform* concentration. The emittance given now accounts for soot absorptance alone; the suspending gas is non-emitting. The effect of an emitting gas will be included later. The total emittance is found from (12-54b) as

$$\epsilon(T, S) = \frac{\int_0^\infty e_{\lambda b}[1 - \exp(-a_\lambda S)]\, d\lambda}{\sigma T^4}$$

$$= \frac{\int_0^\infty e_{\lambda b}\{1 - \exp[-(a_\lambda/C)CS]\}\, d\lambda}{\sigma T^4} \tag{13-126}$$

By using a_λ/C from (13-122) the ϵ can be evaluated numerically and is a function of the cloud temperature and CS. This is shown in Fig. 13-30 for propane soot using properties in Table 13-10. By integrating over a distribution of particle sizes,

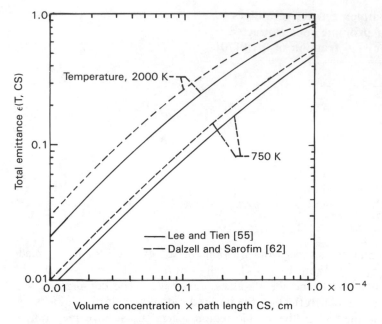

Figure 13-30 Total emittance of soot particles in the Rayleigh limit [55].

it was found in [63] that the individual particle sizes were unimportant, and thus at a fixed T and S the ϵ depends only on the soot volume concentration in the cloud. An electron oscillator model for soot optical constants was developed in [55]. The constants were used in the Rayleigh equation for the absorption coefficient, and the integration in (13-126) then carried out. The results are shown for two temperatures in Fig. 13-30, and they tend to be somewhat below those in Fig. 13-31.

With certain assumptions a convenient expression can be derived for the total emittance of a soot cloud in a nonemitting gas. As indicated by (13-122), if n and κ are weak functions of λ, then $a_\lambda/C = k/\lambda$, where k is a constant depending on the type of soot. As shown by the previous discussion, this is only an approximation, as the exponent α can deviate appreciably from unity. However, in some instances this was found to be a good approximation, for example, for the soot data in [70, 72], and in Fig. 13-28 for λ up to about 5. The coefficient k was found as the value of a_λ/C at $\lambda = 1$ μm, and Fig. 13-28 gave k (propane) ≈ 4.9 and k (acetylene) ≈ 4. For soot from coal flames, k has been found to range from 3.7 to 7.5; for oil flames $k \approx 6.3$ [73]. The value $k = 5$ was used for the calculations in [74].

The above approximation for a_λ is inserted into (13-126). Also, for the wavelength regions of interest here, the $e_{\lambda b}$ can be approximated quite well by Wien's formula, $e_{\lambda b} \approx 2\pi C_1/\lambda^5 \exp(C_2/\lambda T)$. Equation (13-126) then becomes

$$\epsilon(T, CS) = \frac{\int_0^\infty (2\pi C_1/\lambda^5)\, e^{-C_2/\lambda T} (1 - e^{-kCS/\lambda})\, d\lambda}{\int_0^\infty (2\pi C_1/\lambda^5)\, e^{-C_2/\lambda T}\, d\lambda}$$

where, to be consistent in the degree of approximation, Wien's formula has also been used in the denominator. For an integral of the form $\int_0^\infty \lambda^{-5} e^{-K/\lambda} \, d\lambda$, where K is independent of λ, the substitution of $\eta = K/\lambda$ yields $K^{-4} \int_0^\infty \eta^3 e^{-\eta} \, d\eta$. Then ϵ can be written as

$$\epsilon(T, CS) = \frac{[(C_2/T)^{-4} - (C_2/T + kCS)^{-4}] \int_0^\infty \eta^3 e^{-\eta} \, d\eta}{(C_2/T)^{-4} \int_0^\infty \eta^3 e^{-\eta} \, d\eta}$$

The integrals cancel, giving ϵ as

$$\epsilon(T, CS) = 1 - \frac{1}{[1 + (kCS/C_2) T]^4} \tag{13-127}$$

With $k = 4.9$ for propane soot, a comparison with values from the numerical integration in [62] is shown in Table 13-12 and further results are in Fig. 13-31. A comparison with the results of [55] is in Fig. 13-30. From Table 13-12, Eq. (13-127) is shown to be a useful approximation. The $k/C_2 = 4.9/(0.01439 \text{ m} \cdot \text{K}) = 341 \text{ m}^{-1} \cdot \text{K}^{-1}$ for this soot. Reference [51] recommends $k/C_2 = 350 \text{ m}^{-1} \cdot \text{K}^{-1}$ as a mean value for all types of soot, and this coefficient does not vary significantly with temperature over the temperature range of interest in combustion chambers.

A somewhat more complicated expression results from not using the Wien formula in (13-126). Then, letting $a_\lambda = Ck/\lambda$ and using (2-11) give

$$\epsilon(T, CS) = 1 - \frac{\int_0^\infty e_{\lambda b} \, e^{-kCS/\lambda} \, d\lambda}{\sigma T^4} = 1 - \frac{2\pi C_1}{\sigma T^4} \int_0^\infty \frac{e^{-kCS/\lambda} \, d\lambda}{\lambda^5 (e^{C_2/\lambda T} - 1)}$$

If we let $z = 1 + kCST/C_2$ and $t = C_2/\lambda T$, this becomes

$$\epsilon(T, CS) = 1 - \frac{2\pi C_1}{\sigma C_2^4} \int_0^\infty \frac{t^3 e^{-tz}}{1 - e^{-t}} \, dt$$

From pages 260 and 271 of [75], the integral is the pentagamma function $\psi^{(3)}(z)$, for which tabulated values are available. Then, with (2-27), ϵ becomes

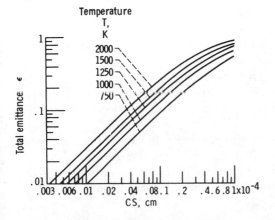

Figure 13-31 Total emittance of soot suspensions as a function of temperature and volume fraction–path length product for propane soot [62].

Table 13-12 Comparison of approximate suspension emittance for propane soot with values from reference [62]

Concentration–path length product CS, μm	Suspension temperature T, K	Suspension emittance ϵ		
		Equation (13-127)	Reference [62]	Equation (13-128)
0.01	1000	0.0135	0.013	0.0127
0.10	1000	0.125	0.125	0.120
1.00	1000	0.690	0.64	0.671
0.01	2000	0.0268	0.030	0.0254
0.10	2000	0.232	0.250	0.223
1.00	2000	0.875	0.90	0.857

$$\epsilon(T, CS) = 1 - \frac{15}{\pi^4} \, \psi^{(3)}\left(1 + \frac{kCST}{C_2}\right) \tag{13-128}$$

Results from this expression are included in Table 13-12 (using $k = 4.9$); they are not better than those from the simpler expression (13-127). Yuen and Tien [76] show that (13-128) can be approximated by the convenient form $\epsilon(T, CS) = 1 - \exp(-3.6kCST/C_2)$.

Stull and Plass [64] give results for the scattering coefficient of soot, which show (in agreement with the Mie theory) that in most instances scattering has little effect on emittance in the wavelength range that contains significant energy at hydrocarbon combustion temperatures. Erickson et al. [77] have experimentally studied scattering from a luminous benzene-air flame. Their results agreed with the predictions of Stull and Plass if the soot particles are taken to be of two predominant diameters. This indicates that small particles on the order of 250 Å in diameter are formed along with agglomerated particles with an equivalent diameter of 1850 Å. These sizes were observed by gathering soot with a probe and using electron microscopy. A comparison of some of the experimental results of [77] with the analysis of [64] is shown in Fig. 13-32.

Thring, Beer, and Foster [78] have put some of the results of Stull and Plass [64], along with their own extensive experimental results, into useful graphs of emittance, extinction coefficient, and soot concentration for flames applicable in industrial practice. They note, however, that soot concentration can only be predicted for flames geometrically similar and with the same control variables as those that have already been studied. Their paper contains a useful review of the worldwide effort to gather information on and give methods for the prediction of radiation from luminous industrial flames. Other such information is in [28, 51, 79–83].

In addition to the uncertainties in optical properties and hence in a_λ and ϵ for soot, it is noted that a_λ and ϵ are given in terms of the soot concentration. To use Eq. (13-117), the Ck is needed. To use Fig. 13-31 to determine ϵ for a given flame size, the C in the abscissa must be known. At present the C cannot be accurately computed from first principles, knowing the fuel and burner geometry. Hence some indication of C or Ck must be obtained by examining flames experimentally. It may be possible to extrapolate performance for a particular application by examining a similar flame. Examples of mathematical models to predict soot formation are in [84] and [85]. For gas turbine combustors, because of the complexity in estimating the luminous emissivity, a luminosity factor has been introduced into the expression for the emittance of a nonluminous flame [86].

Experimental examination of soot concentration   Various techniques are used to examine the soot concentration in flames. Radiometric instruments can be used to obtain the flame emittance, and laser scattering from the soot particles can be used. Some of the methods will be briefly described.

One technique to obtain information on the soot-concentration quantity Ck_2 [as in Eq. (13-119) for the visible region] is to sight through the flame onto a cold black background with a pyrometer and match the brightness of the pyrometer filament to that of the flame while using first a red filter and then a green filter. With each of the filters, the pyrometer is also sighted on a blackbody source, and the source temperatures are obtained that produce the same brightness as when the flame was viewed. As a convenient simplification at the small λT for red and green wavelengths when considering typical flame temperatures, the blackbody intensity can be approximated very well by Wien's formula, Eq. (2-18):

$$i'_{\lambda b} = \frac{2C_1}{\lambda^5 \exp{(C_2/\lambda T)}}$$

(13-129)

Then if T_r is the blackbody temperature producing the same brightness as the flame did when the red filter was used, this intensity is

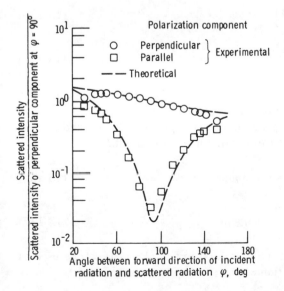

Figure 13-32 Comparison of experiment with Mie scattering theory for radiation scattered from benzene-air flame at wavelength $\lambda = 5461$ Å. Theoretical curves based on spheres of diameter 250 Å with 0.002% spheres of diameter 1850 Å, all with complex refractive index $n - i\kappa = 1.79 - 0.79i$. From [77].

$$i'_{\lambda b,r} = \frac{2C_1}{\lambda_r^5 \exp\left(C_2/\lambda_r T_r\right)} \tag{13-130}$$

where λ_r is the red wavelength, 0.665 μm. This intensity can also be written as a spectral emittance of the flame times the blackbody intensity at the flame temperature, which gives

$$\frac{2C_1}{\lambda_r^5 \exp\left(C_2/\lambda_r T_r\right)} = \epsilon_{\lambda_r} \frac{2C_1}{\lambda_r^5 \exp\left(C_2/\lambda_r T_f\right)}$$

By grouping the exponential terms and taking the logarithm, the result can be rearranged into

$$\frac{1}{T_f} - \frac{1}{T_r} = \frac{\lambda_r}{C_2} \ln \epsilon_{\lambda_r} \tag{13-131}$$

Similarly, with a green filter ($\lambda_g = 0.555$ μm),

$$\frac{1}{T_f} - \frac{1}{T_g} = \frac{\lambda_g}{C_2} \ln \epsilon_{\lambda_g} \tag{13-132}$$

Now as a simple approximation, (13-119) is used for a_λ in the visible region. Then ϵ_λ from (13-116) is

$$\epsilon_\lambda = 1 - \exp\left(-\frac{Ck_2 S}{\lambda^{1.39}}\right) \tag{13-133}$$

where S is the path length sighted through the flame. Substitute (13-133) into (13-131) and (13-132) to obtain

$$\frac{1}{T_f} - \frac{1}{T_r} = \frac{\lambda_r}{C_2} \ln \left[1 - \exp\left(-\frac{Ck_2 S}{\lambda_r^{1.39}}\right)\right] \tag{13-134}$$

$$\frac{1}{T_f} - \frac{1}{T_g} = \frac{\lambda_g}{C_2} \ln \left[1 - \exp\left(-\frac{Ck_2 S}{\lambda_g^{1.39}}\right)\right] \tag{13-135}$$

These two equations are solved for T_f and Ck_2 to yield the needed measure of the soot concentration as well as the flame temperature.

As an approximation, Ck_2 is assumed independent of wavelength and is used in (13-116), (13-118a), and (13-119) to yield, for a path length S,

$$\epsilon_\lambda = 1 - \exp\left(-\frac{Ck_2 S}{\lambda^{1.39}}\right) \qquad \text{visible, } \lambda < 0.8 \ \mu\text{m} \tag{13-136}$$

$$\epsilon_\lambda = 1 - \exp\left(-\frac{Ck_2 S}{\lambda^{0.95}}\right) \qquad \text{infrared, } \lambda > 0.8 \ \mu\text{m} \tag{13-137}$$

Then with these spectral emittances the definition in Eq. (13-126) can be used to evaluate the total emittance of the flame as

$$\epsilon(T_f,S) = \frac{\int_0^\infty e_{\lambda b}(T_f)\epsilon_\lambda(\lambda,T_f,S)\,d\lambda}{\sigma T_f^4} \tag{13-138}$$

Some convenient graphs for use in this procedure are in [8]. The hope is that the Ck_2 obtained in this way can be applied to "similar" flames. This is a very rough approximation; so many variables affect the flow and mixing in the flame that it is difficult to know when the flames will have a similar character. The detailed nature of flames and their radiating characteristics is a continuing area of active research [45, 84, 87–90].

Another way to measure flame emittance is to use a radiometer and a black-body source with a blackened chopper in front of it that periodically covers and uncovers the blackbody aperture (Fig. 13-33). With the flame turned off, the voltage signal at the dectector from the uncovered blackbody is V_b, and with the blackbody covered by the chopper it is V_0. With the flame turned on these signals become $V_{b,f}$ and $V_{0,f}$. The measurements are made through a filter so that they are for a small spectral region $\Delta\lambda$ centered about λ. The transmittance of a path through the flame is $\tau_\lambda = 1 - \epsilon_\lambda$. Then $V_{b,f}$ is the result of transmitted radiation from the blackbody and emission by the flame so that $V_{b,f} = (1 - \epsilon_\lambda)V_b + \epsilon_\lambda e_{\lambda b}(T_f)$. Similarly, $V_{0,f} = (1 - \epsilon_\lambda)V_0 + \epsilon_\lambda e_{\lambda b}(T_f)$. The two signals are subtracted to yield an expression for the spectral emittance along the path viewed through the flame,

$$\epsilon_\lambda = 1 - \frac{V_{b,f} - V_{0,f}}{V_b - V_0} = \frac{(V_b - V_{b,f}) - (V_0 - V_{0,f})}{V_b - V_0} \tag{13-139}$$

Scattering of laser light can be used to measure local soot concentrations as discussed in [53, 56]. Although scattering by soot is small, it can be accurately detected with sensitive instrumentation. The Mie scattering theory provides the scattering intensity in a given direction as a function of the soot concentration and optical properties. If values are assumed for the optical properties, the measured scattering intensities can be used to obtain local soot concentrations. Usually spherical soot particles are assumed, and the agglomeration of some of the soot provides a source of error. Agglomeration was found in [56a] to have little effect on the absorption coefficient, but it did have an influence on scattering. Another difficulty is knowing what complex refractive index to use for the soot. In [56] the complex refractive index was chosen as $\bar{n} = 1.57 - i0.56$ for soot in an ethene flame at the argon ion laser wavelength of 0.5145 μm.

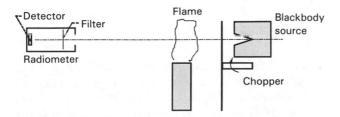

Figure 13-33 Radiometric measurement of flame emittance.

Uniform soot cloud in an isothermal radiating gas Generally in a flame or in combustion products containing soot, there will be gaseous constituents that radiate energy (the previous treatment was for soot in a nonradiating carrier gas). To simplify the present discussion, it is assumed that only three radiating constituents are present—carbon dioxide, water vapor, and soot—but the method can be extended to more constituents. As a spectral beam of radiation passes through the gas-soot suspension, the local attenuation depends on the sum of the absorption coefficients so that from (13-54) the spectral emittance is

$$\epsilon_\lambda = \alpha_\lambda = 1 - e^{-(a_{\lambda C} + a_{\lambda H} + a_{\lambda S})S} \tag{13-140}$$

where the subscripts C, H, and S refer to CO_2, H_2O, and soot. Then from (13-126) the total emittance is

$$\epsilon = \frac{1}{\sigma T^4} \int_0^\infty (1 - e^{-(a_{\lambda C} + a_{\lambda H} + a_{\lambda S})S}) e_{\lambda b}\, d\lambda \tag{13-141}$$

This can be rewritten in the equivalent forms

$$\epsilon = \frac{1}{\sigma T^4} \int_0^\infty [1 - (1 - \epsilon_{\lambda C})(1 - \epsilon_{\lambda H})(1 - \epsilon_{\lambda S})] e_{\lambda b}\, d\lambda \tag{13-142}$$

$$= \frac{1}{\sigma T^4} \int_0^\infty (\epsilon_{\lambda C} + \epsilon_{\lambda H} + \epsilon_{\lambda S} - \epsilon_{\lambda C}\epsilon_{\lambda H} - \epsilon_{\lambda C}\epsilon_{\lambda S} - \epsilon_{\lambda S}\epsilon_{\lambda H} + \epsilon_{\lambda C}\epsilon_{\lambda H}\epsilon_{\lambda S}) e_{\lambda b}\, d\lambda$$

where $\epsilon_{\lambda J} = 1 - e^{-a_{\lambda J}S}$ with $J = C, H, S$. The first three terms in the last integral yield the total emittances of the three constituents, so that

$$\epsilon = \epsilon_C + \epsilon_H + \epsilon_S - \frac{1}{\sigma T^4} \int_0^\infty (\epsilon_{\lambda C}\epsilon_{\lambda H} + \epsilon_{\lambda C}\epsilon_{\lambda S} + \epsilon_{\lambda S}\epsilon_{\lambda H} - \epsilon_{\lambda C}\epsilon_{\lambda H}\epsilon_{\lambda S}) e_{\lambda b}\, d\lambda \tag{13-143}$$

A term such as $\epsilon_{\lambda C}\epsilon_{\lambda H}$ is nonzero only in spectral regions where both $a_{\lambda C}$ and $a_{\lambda H}$ are nonzero, that is, where both constituents are radiating. Thus the terms in the integral represent overlap regions in the spectrum in which two or three components are radiating. The total emittance ϵ is then the sum of the three individual emittances, computed as if the other constituents were absent (see Figs. 13-13 and 13-15) minus a correction term for spectral overlap. This is the same type of correction that was discussed in connection with Eq. (13-81). For the simplified case of gray constituents,

$$\epsilon = 1 - (1 - \epsilon_C)(1 - \epsilon_H)(1 - \epsilon_S) \tag{13-144}$$

Results for the spectral-overlap terms were calculated in [66] by using existing information on the form of the CO_2 and H_2O vapor absorption bands and the soot absorption coefficient $a_\lambda = Ck/\lambda$ discussed earlier. Typical emittances are shown in Fig. 13-34. For low values of CS in the figure, the soot concentration is low especially when S is high. Hence the left sides of the curves are dominated by the gas emittance, and the vertical displacement of the curves shows the increase

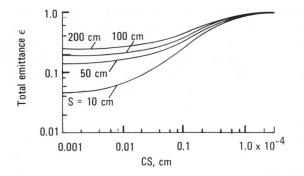

Figure 13-34 Total emittance of gas-soot suspension as a function of volume fraction of soot times path length. Gas temperature, 1600 K; total pressure, 1 atm; partial pressures: $p_{H_2O} = 0.19$ atm, $p_{CO_2} = 0.09$ atm [66].

in gas emittance with path length. As CS is increased from 0.001, the curves are somewhat horizontal (especially at high S, which corresponds to low C) as the soot concentration is not sufficient to increase ϵ for the mixture significantly. For larger CS the soot begins to dominate, and for all the path lengths shown, ϵ approaches unity when CS is about 3×10^{-4} cm. This is consistent with the results in Fig. 13-31.

Since soot has a rather continuous emission spectrum, it is a reasonable assumption (as shown in [74]) to consider gray emission of soot suspended in a nongray gas. This leads to a particularly simple form relating the emittance of the suspension to the individual emittances of the soot and gas. If there is no soot, Eq. (13-142) becomes

$$\epsilon_g = 1 - \frac{1}{\sigma T^4} \int_0^\infty (1 - \epsilon_{\lambda C})(1 - \epsilon_{\lambda H}) e_{\lambda b}\, d\lambda$$

Then from (13-142) if the soot is gray

$$\epsilon = 1 - (1 - \epsilon_s)\frac{1}{\sigma T^4} \int_0^\infty (1 - \epsilon_{\lambda C})(1 - \epsilon_{\lambda H}) e_{\lambda b}\, d\lambda = 1 - (1 - \epsilon_s)(1 - \epsilon_g)$$

$$\epsilon = \epsilon_s + \epsilon_g(1 - \epsilon_s) \tag{13-145}$$

The sum of gray gases model [see Eq. (13-82)] was used in [74] to represent the emittance for a mixture of soot, CO_2, and H_2O vapor. The model is

$$\epsilon = \sum_{i=1}^n a_i(T)(1 - e^{-K_i L_e})$$

where K_i is a function of the soot volume fraction and the partial pressures of the CO_2 and H_2O. This model was used in [91] to compute radiative transfer for a gas-soot mixture between parallel plates. Kunitomo [89] gives results for the ratio, in a flame, of the soot-cloud emittance to the emittance of the nonluminous suspending gas; these results are for a liquid fuel. The ratio increases as the fuel carbon-hydrogen ratio is increased and as the excess air is decreased. Babikian et al. [92] made mass absorption coefficient measurements of soot in spray combustor flames. The values were found to be between those in [62] and [55].

Gases containing particles other than soot In addition to hydrocarbon flames containing soot, there are applications involving radiation from gases containing other types of particles. A common example is ash particles in pulverized coal flames [28, 51]. In [28] it is discussed that the complex refractive index for fly ash ranges from about $\bar{n} = 1.43 - i0.307$ to $1.5 - i0.005$. When the imaginary part is small the particles are acting as a nonabsorbing dielectric.

Another example is the luminosity in the exhaust plume of solid-fueled and some liquid-fueled rockets. For a solid fuel the luminosity may be caused by metal particles added to promote combustion stability. Williams and Dudley [93] present calculations for a rocket exhaust with entrained liquid aluminium oxide particles. Edwards et al. [94, 95] modeled the radiation from a solid rocket motor plume with two groups of particles. The very small particles cool rapidly and are assumed to be cold scattering particles. The large particles that remain hot are emitting and absorbing. The calculations in [96] demonstrate the combination of radiating particles and gases where all the constituents have spectrally dependent properties. Radiation by a cylindrical region at uniform temperature was analyzed in [97] for either CO_2 or H_2O gas containing soot and gray particles. The absorption of the nongray gas was obtained from the exponential wide-band model. The absorption coefficient of the soot was inversely proportional to wavelength, and the scattering from the gray particles had an anisotropic component. Experiments were performed in [98] with a heated mixture of carbon dioxide, nitrogen, and particles of BNi-2, a boron nickel alloy. The particulate scattering increased the radiation in the wings of the 4.3-μm CO_2 radiation band. Experiments in [99] show the effect of highly scattering Al_2O_3 particles in CO_2 gas. The radiative behavior of lycopodium particles during combustion in air is considered in [100].

The presence of particles in an otherwise weakly absorbing medium can cause the mixture to be strongly absorbing. *Seeding* of a gas with particles, such as finely divided carbon, can increase gas absorption and heating by incident radiation [101], or the particles can shield a surface from incident radiation [57, 102, 103]. These techniques have possible application in connection with advanced energy systems and for the collection of solar energy [104].

Another use for seeding is in the direct determination of flame temperatures for a nonluminous flame by the *line-reversal technique*. In this method, a seeding material such as a sodium or cadmium salt is introduced into an otherwise transparent flame. These materials produce a strong line in the visible spectrum because of an electronic transition; cadmium gives a red line and sodium a bright yellow line. A spectrally continuous radiation source is placed so that it may be viewed *through* the seeded flame with a spectroscope. The intensity seen in the spectroscope at the line wavelength is, from (14-13),

$$i'_{\lambda,\text{scope}} = i'_{\lambda,\text{cont source}} \exp(-\kappa_\lambda) + \int_0^{\kappa_\lambda} i'_{\lambda b}(\kappa_\lambda^*) \exp\left[-(\kappa_\lambda - \kappa_\lambda^*)\right] d\kappa_\lambda^* \qquad (13\text{-}146)$$

If the flame is assumed isothermal and of diameter D and no attenuation occurs along the remainder of the path between the continuous source and the spectroscope, (13-146) becomes

$$i'_{\lambda,\text{scope}} = i'_{\lambda,\text{cont source}} \exp\left(-\kappa_{\lambda,D}\right) + i'_{\lambda b,\text{flame}} \left[1 - \exp\left(-\kappa_{\lambda,D}\right)\right] \qquad (13\text{-}147)$$

where

$$\kappa_{\lambda,D} = \int_0^D a_\lambda(S^*)\,dS^*$$

In the wavelength region adjacent to the absorbing and emitting spectral line, the flame is essentially transparent so the background radiation observed adjacent to the line is $i'_{\lambda,\text{cont source}}$. Hence, by subtracting $i'_{\lambda,\text{cont source}}$ from (13-147) the line intensity relative to the adjacent background is found as

$$i'_{\lambda,\text{scope}} - i'_{\lambda,\text{cont source}} = \left(i'_{\lambda b,\text{flame}} - i'_{\lambda,\text{cont source}}\right)\left[1 - \exp\left(-\kappa_{\lambda,D}\right)\right] \qquad (13\text{-}148)$$

If the flame is at a higher temperature than the continuous source, (13-148) shows that the line intensity in the spectroscope will be greater than the continuous background intensity. The line will appear as a bright line imposed upon a less bright continuous spectrum. Increasing the temperature of the continuous source causes the source term to override. The line then appears as a dark line on a brighter continuous spectrum. If the continuous source is a blackbody and its temperature is made *equal* to the flame temperature, then $i'_{\lambda,\text{cont source}} = i'_{\lambda b,\text{flame}}$, and (13-148) reduces to $i'_{\lambda,\text{scope}} = i'_{\lambda,\text{cont source}}$. The line will then disappear into the continuum in the spectroscope because the absorption by the flame and the flame emission exactly compensate. If the continuous source is a tungsten lamp, the source temperature measurement is usually made with an optical pyrometer.

In the derivation of (13-148) it was assumed that the flame is transparent except within the spectral line produced by the cadmium or sodium seeding. If soot is in the flame, the soot particles emit and scatter radiation in a continuous spectrum along the path of the incident beam. The line-reversal technique is of less utility in this instance, as it then depends on the soot behavior. The effect of soot is analyzed in [105].

Another instance of radiation attenuation by means of particles is found in the effect of dust or "grains" that are believed to exist in interstellar space and cause reductions in the observed intensity of radiation from stars [106–108].

REFERENCES

1. Nelson, D. A.: Band Radiation of Isothermal Gases within Diffuse-Walled Enclosures, *Int. J. Heat Mass Transfer*, vol. 27, no. 10, pp. 1759–1769, 1984.
2. Dunkle, R. V.: Geometric Mean Beam Lengths for Radiant Heat-Transfer Calculations, *J. Heat Transfer*, vol. 86, no. 1, pp. 75–80, 1964.
3. Hottel, Hoyt C., and Adel F. Sarofim: "Radiative Transfer," McGraw-Hill, New York, 1967.
4. Anderson, K. M., and S. Hadvig: Geometric Mean Beam Lengths for a Space between Two Coaxial Cylinders, *J. Heat Transfer*, vol. 111, no. 3, pp. 811–813, 1989.
5. Edwards, D. K., and K. E. Nelson: Rapid Calculation of Radiant Energy Transfer between Nongray Walls and Isothermal H_2O and CO_2 Gas, *J. Heat Transfer*, vol. 84, no. 4, pp. 273–278, 1962.
6. Edwards, D. K.: Radiation Interchange in a Nongray Enclosure Containing an Isothermal Carbon-Dioxide-Nitrogen Gas Mixture, *J. Heat Transfer*, vol. 84, no. 1, pp. 1–11, 1962.

7. Koh, J. C. Y.: Radiation of Spherically Shaped Gases, *Int. J. Heat Mass Transfer*, vol. 8, no. 2, pp. 373–374, 1965.

8. Hottel, H. C.: Radiant-Heat Transmission, in William H. McAdams (ed.), "Heat Transmission," 3d ed., chap. 4, McGraw-Hill, New York, 1954.

9. Eckert, E. R. G., and Robert M. Drake, Jr.: "Heat and Mass Transfer," 2d ed., McGraw-Hill, New York, 1959.

10. Wassel, A. T., and D. K. Edwards: Mean Beam Lengths for Spheres and Cylinders, *J. Heat Transfer*, vol. 98, pp. 308–309, 1976.

11. Cartigny, J. D.: Mean Beam Length for a Scattering Medium, *Proc. 8th Int. Heat Transfer Conf.*, vol. 2, pp. 769–772, Hemisphere, Washington, D.C., 1986.

12. Oppenheim, A. K.: Radiation Analysis by the Network Method, *Trans. ASME*, vol. 78, no. 4, pp. 725–735, 1956.

13. Mandell, David A., and Forest A. Miller: Comparison of Exact and Mean Beam Length Results for a Radiating Hydrogen Plasma, *J. Quant. Spectrosc. Radiat. Transfer*, vol. 13, pp. 49–56, 1973.

14. Felske, J. D., and C. L. Tien: Wide Band Characterization of the Total Band Absorptance of Overlapping Infrared Gas Bands, *Combust. Sci. Technol.*, vol. 11, p. 111, 1975.

15. Saido, K., and W. H. Giedt: Spectral Absorption of Water Vapor and Carbon Dioxide Mixtures in the 2.7 Micron Band, *J. Heat Transfer*, vol. 99, no. 1, pp. 53–59, 1977.

16. Leckner, Bo: Spectral and Total Emissivity of Water Vapor and Carbon Dioxide, *Combust. Flame*, vol. 19, pp. 33–48, 1972.

17. Boynton, Frederick P., and Claus B. Ludwig: Total Emissivity of Hot Water Vapor—II, Semi-Empirical Charts Deduced from Long-Path Spectral Data, *Int. J. Heat Mass Transfer*, vol. 14, pp. 963–973, 1971.

18. Ludwig, C. B., W. Malkmus, J. E. Reardon, and J. A. L. Thomson: Handbook of Infrared Radiation from Combustion Gases, NASA SP-3080, 1973.

19. Sarofim, A. F., I. H. Farag, and H. C. Hottel: Radiative Heat Transmission from Non-Luminous Gases. Computational Study of the Emissivities of Carbon Dioxide, ASME paper 78-HT-16, May, 1978.

20. Docherty, P.: Prediction of Gas Emissivity for a Wide Range of Conditions, *Proc. 7th Int. Heat Transfer Conf.*, vol. R5, pp. 481–485, 1982.

21. Cheng, S. C., and C. Nguyen: Emissivity of Water Vapour at Elevated Pressures, *Int. Comm. Heat Mass Transfer*, vol. 16, no. 5, pp. 723–729, 1989.

22. I. H. Farag, Non-Luminous Gas Radiation: Approximate Emissivity Models, *Proc. 7th Int. Heat Transfer Conf.*, Munchen, vol. R6, pp. 487–492, 1982.

23. Skocypec, R. D., and R. O. Buckius: Total Hemispherical Emittances for CO_2 or H_2O Including Particulate Scattering, *Int. J. Heat Mass Transfer*, vol. 27, no. 1, pp. 1–13, 1984.

24. Edwards, D. K., and R. Matavosian: Scaling Rules for Total Absorptivity and Emissivity of Gases, *J. Heat Transfer*, vol. 106, pp. 684–689, 1984.

25. Smith, T. F., Z. F. Shen, and J. N. Friedman: Evaluation of Coefficients for the Weighted Sum of Gray Gases Model, *J. Heat Transfer*, vol. 104, no. 4, pp. 602–608, 1982.

26. Modest, M. F.: The Weighted-Sum-of-Gray-Gases Model for Arbitrary Solution Methods in Radiative Transfer, *J. Heat Transfer*, vol. 113, no. 3, pp. 650–656, 1991.

27. Cess, R. D., and M. S. Lian: A Simple Parameterization for the Water Vapor Emissivity, *J. Heat Transfer*, vol. 98, no. 4, pp. 676–678, 1976.

28. Viskanta R., and M. P. Menguc: Radiation Heat Transfer in Combustion Systems, *Prog. Energy Combust. Sci.*, vol. 13, pp. 97–160, 1987.

29. Hottel, H. C., and E. S. Cohen: Radiant Heat Exchange in a Gas-Filled Enclosure: Allowance for Nonuniformity of Gas Temperature, *AIChE J.*, vol. 4, no. 1, pp. 3–14, 1958.

30. Einstein, Thomas H.: Radiant Heat Transfer to Absorbing Gases Enclosed between Parallel Flat Plates with Flow and Conduction, NASA TR R-154, 1963.

31. Einstein, Thomas H.: Radiant Heat Transfer to Absorbing Gases Enclosed in a Circular Pipe with Conduction, Gas Flow, and Internal Heat Generation, NASA TR R-156, 1963.

32. Hottel, H. C., and A. F. Sarofim: The Effect of Gas Flow Patterns on Radiative Transfer in Cylindrical Furnaces, *Int. J. Heat Mass Transfer*, vol. 8, no. 8, pp. 1153–1169, 1965.

33. Noble, James J.: The Zone Method: Explicit Matrix Relations for Total Exchange Areas, *Int. J. Heat Mass Transfer,* vol. 18, no. 2, pp. 261–269, 1975.

34. Nelson, D. A.: A Study of Band Absorption Equations for Infrared Radiative Transfer in Gases. I, Transmission and Absorption Functions for Planar Media, *J. Quant. Spectrosc. Radiat. Transfer,* vol. 14, pp. 69–80, 1974.

35. Scholand, E. and P. Schenkel: A Solution of Mean Beam Lengths of Radiating Gases in Rectangular Parallelepiped Enclosures, *Proc. 8th Int. Heat Transfer Conf.,* vol. 2, pp. 763–768, Hemisphere, Washington, D.C., 1986.

36. Bannerot, R. B., and F. Wierum: Approximate Configuration Factors for a Gray Nonisothermal Gas-Filled Conical Enclosure, pp. 41–63, in "Thermophysics and Spacecraft Thermal Control," vol. 35 of *Progress in Astronautics and Aeronautics,* MIT Press, Cambridge, Mass., 1974.

37. Mihail, R., and Gh. Maria: The Emitter-Receptor Geometrical Configuration Influence on Radiative Heat Transfer, *Int. J. Heat Mass Transfer,* vol. 26, no. 12, pp. 1783–1789, 1983.

37a. Sika, J.: Evaluation of Direct-Exchange Areas for a Cylindrical Enclosure, *J. Heat Transfer,* vol. 113, no. 4, pp. 1040–1044, 1991.

38. Tucker, R. J.: Direct Exchange Areas for Calculating Radiation Transfer in Rectangular Furnaces, *J. Heat Transfer,* vol. 108, pp. 707–710, 1986.

39. Edwards, D. K., and A. Balakrishnan: Volume Interchange Factors for Nonhomogeneous Gases, *J. Heat Transfer,* vol. 94, no. 2, pp. 181–188, 1972.

40. Echigo, R., N. Nishiwaki, and M. Hirata: A Study on the Radiation of Luminous Flames, *Eleventh Symp. (Int.) Combust.,* pp. 381–389. The Combustion Institute, 1967.

41. Dayan, A., and C. L. Tien: Radiant Heating from a Cylindrical Fire Column, *Combust. Sci. Technol.,* vol. 9, pp. 41–47, 1974.

42. Edwards, D. K., and A. Balakrishnan: Thermal Radiation by Combustion Gases, *Int. J. Heat Mass Transfer,* vol. 16, pp. 25–40, 1973.

43. Modak, Ashok: Thermal Radiation from Pool Fires, *Combust. Flame,* vol. 29, no. 2, pp. 177–192, 1977.

44. Modak, Ashok: Nonluminous Radiation from Hydrocarbon-Air Diffusion Flames, *Combust. Sci. Technol.,* vol. 10, pp. 245–259, 1975.

45. Taylor, P. B., and P. J. Foster: The Total Emissivities of Luminous and Non-Luminous Flames, *Int. J. Heat Mass Transfer,* vol. 17, pp. 1591–1605, 1974

46. Annamali, K., and P. Durbetaki: Characteristics of an Open Diffusion Flame, *Combust. Flame,* vol. 25, pp. 137–139, 1975.

46a. Grosshandler, W. L., and E. M. Thurlow: Generalized State-Property Relations for Nonluminous Flame Absorption Coefficients, *J. Heat Transfer,* vol. 114, no. 1, pp. 243–249, 1992.

47. Perry, Robert H., Cecil H. Chilton, and Sidney D. Kirkpatrick (eds.): "Chemical Engineers' Handbook," 4th ed., McGraw-Hill, New York, 1963.

48. Hougen, Olaf A., Kenneth M. Watson, and Roland A. Ragatz: Material and Energy Balance, in "Chemical Process Principles," 2d ed., vol. 1, Wiley, New York, 1954.

49. Gaydon, A. G., and H. G. Wolfhard: "Flames, Their Structure, Radiation, and Temperature," 2d ed., Macmillan, New York, 1960.

50. Barnett, Henry C., and Robert R. Hibbard (eds.): Basic Considerations in the Combustion of Hydrocarbon Fuels with Air, NACA Rept. 1300, 1957.

51. Sarofim, A. F., and H. C. Hottel: Radiative Transfer in Combustion Chambers: Influence of Alternative Fuels, *Sixth Int. Heat Transfer Conf.,* Toronto, vol. 6, pp. 199–217, August, 1978.

52. Boothroyd, S. A., and A. R. Jones: A Comparison of Radiative Characteristics for Fly Ash and Coal, *Int. J. Heat Mass Transfer,* vol. 29, no. 11, pp. 1649–1654, 1986.

53. Santoro, R. J., T. T. Yeh, J. J. Horvath, and H. G. Semerjian: The Transport and Growth of Soot Particles in Laminar Diffusion Flames, *Combust. Sci. Technol.,* vol. 53, pp. 89–115, 1987.

54. Charalampopoulos, T. T., and J. D. Felske: Refractive Indices of Soot Particles Deduced from In-Situ Laser Light Scattering Measurements, *Combust. Flame,* vol. 68, pp. 283–294, 1987.

55. Lee, S. C., and C. L. Tien: Optical Constants of Soot in Hydrocarbon Flames, *Eighteenth Symp. (Int.) Combust.,* pp. 1159–1166, The Combustion Institute, 1981.

56. Santoro, R. J., H. G. Semerjian, and R. A. Dobbins: Soot Particle Measurements in Diffusion Flames, *Combust. Flame,* vol. 51, pp. 203–218, 1983.

56a. Ku, J. C., and K.-H. Shim: Optical Diagnostics and Radiative Properties of Simulated Soot Agglomerates, *J. Heat Transfer,* vol. 113, no. 4, pp. 953–958, 1991.

57. Lee, K. Y., Z. Y. Zhong, and C. L. Tien: Blockage of Thermal Radiation by the Soot Layer in Combustion of Condensed Fuels, *Twentieth Symp. (Int.) Combust.,* 1629–1636, The Combustion Institute, 1984.

58. Ang, J. A., P. J. Pagni, T. G. Mataga, J. M. Margle, and V. L. Lyons: Temperature and Velocity Profiles in Sooting Free Convection Diffusion Flames, *AIAA J.* vol. 26, no. 3, pp. 323–329, 1988.

59. Sato, T., T. Kunitomo, S. Yoshi, and T. Hashimoto: On the Monochromatic Distribution of the Radiation from the Luminous Flame, *Bull. Jpn. Soc. Mech. Eng.,* vol. 12, pp. 1135–1143, 1969.

60. Kunugi, M., and H. Jinno: Determination of Size and Concentration of Soot Particles in Diffusion Flames by a Light-Scattering Technique, *Eleventh Symp. (Int.) Combust.,* pp. 257–266, 1966.

61. Howarth, C. R., P. J. Foster, and M. W. Thring: The Effect of Temperature on the Extinction of Radiation by Soot Particles, *Third Int. Heat Transfer Conf. AIChE,* vol. 5, pp. 122–128, 1966.

62. Dalzell, W. H., and A. F. Sarofim: Optical Constants of Soot and Their Application to Heat-Flux Calculations, *J. Heat Transfer,* vol. 91, no. 1, pp. 100–104, 1969.

63. Siddall, R. G., and I. A. McGrath: The Emissivity of Luminous Flames, in W. G. Berl, ed., *Ninth Symp. (Int.) Combust.,* pp. 102–110, 1963.

64. Stull, V. Robert, and Gilbert N. Plass: Emissivity of Dispersed Carbon Particles, *J. Opt. Soc. Am.,* vol. 50, no. 2, pp. 121–129, 1960.

65. Hawksley, P. G. W.: The Methods of Particle Size Measurements, pt. 2, Optical Methods and Light Scattering, *Brit. Coal Utilization Res. Assoc. Monthly Bull.,* vol. 16, nos. 4 and 5, pp. 134–209, 1952.

66. Felske, J. D., and C. L. Tien: Calculation of the Emissivity of Luminous Flames, *Combust. Sci. Technol.,* vol. 7, no. 1, pp. 25–31, 1973.

67. Felske, J. D., and C. L. Tien: The Use of the Milne-Eddington Absorption Coefficient for Radiative Heat Transfer in Combustion Systems, *J. Heat Transfer,* vol. 99, no. 3, pp. 458–465, 1977.

68. Buckius, R. O., and C. L. Tien: Infrared Flame Radiation, *Int. J. Heat Mass Transfer,* vol. 20, no. 2, pp. 93–106, 1977.

69. Kunitomo, Takeshi, and Takashi Sato: Experimental and Theoretical Study on the Infrared Emission of Soot Particles in Luminous Flame, *4th Int. Heat Transfer Conf.,* Paris-Versailles, September, 1970.

70. Felske, J. D., T. T. Charalampopoulos, and H. S. Hura: Determination of the Refractive Indices of Soot Particles from the Reflectivities of Compressed Soot Pellets, *Combust. Sci. Technol.,* vol. 37, pp. 263–284, 1984.

71. Habib, Z. G., and P. Vervisch: On the Refractive Index of Soot at Flame Temperature, *Combust. Sci. Technol.,* vol. 59, pp. 261–274, 1988.

72. Liebert, Curt H., and Robert R. Hibbard: Spectral Emittance of Soot, NASA TN D-5647, 1970.

73. Gray, William A., and R. Muller: "Engineering Calculations in Radiative Heat Transfer," Pergamon Press, New York, 1974.

74. Felske, J. D., and T. T. Charalampopoulos: Gray Gas Weighting Coefficients for Arbitrary Gas-Soot Mixtures, *Int. J. Heat Mass Transfer,* vol. 25, no. 12, pp. 1849–1855, 1982.

75. Abramowitz, Milton, and Irene A. Stegun: "Handbook of Mathematical Functions," NBS Applied Mathematics Series, no. 55, 1967.

76. Yuen, W. W., and C. L. Tien: A Simple Calculation Scheme for the Luminous-Flame Emissivity, *Sixteenth Symp. (Int.) Combust.,* The Combustion Institute, pp. 1481–1487, 1977.

77. Erickson, W. D., G. C. Williams, and H. C. Hottel: Light Scattering Measurements on Soot in a Benzene-Air Flame, *Combust. Flame,* vol. 8, no. 2, pp. 127–132, 1964.

78. Thring, M. W., J. M. Beer, and P. J. Foster: The Radiative Properties of Luminous Flames, *Third Int. Heat Transfer Conf. AIChE*, vol. 5, pp. 101-111, 1966.
79. Thring, M. W., P. J. Foster, I. A. Mc Grath, and J. S. Ashton: Prediction of the Emissivity of Hydrocarbon Flames, *Int. Develop. Heat Transfer, ASME*, pp. 796–803, 1963.
80. Sato, Takashi, and Ryuichi Matsumoto: Radiant Heat Transfer from Luminous Flame, *Int. Develop. Heat Transfer, ASME*, pp. 804–811, 1963.
81. Yagi, S., and H. Inoue: Radiation from Soot Particles in Luminous Flames, *Eighth Symp. (Int.) Combust.*, pp. 288–293, 1962.
82. Bone, William A., and Donald T. A. Townsend: "Flame and Combustion in Gases," Longmans, Green, London, 1927.
83. Leckner, B.: Radiation from Flames and Gases in a Cold Wall Combustion Chamber, *Int. J. Heat Mass Transfer*, vol. 13, no. 1, pp. 185–197, 1970.
84. Magnussen, B. F., and B. H. Hjertager: On Mathematical Modeling of Turbulent Combustion with Special Emphasis on Soot Formation and Combustion, *Sixteenth Symp. (Int.) Combust.*, The Combustion Institute, pp. 719–729, 1977.
85. Frenklach, M., and J. Warnatz: Detailed Modeling of PAH Profiles in a Sooting Low-Pressure Acetylene Flame, *Combust. Sci. Technol.*, vol. 51, pp. 265–283, 1987.
86. Rizk, N. K. and H. C. Mongia: Three-Dimensional Analysis of Gas Turbine Combustors, *J. Prop. Power*, vol. 7, no. 3, pp. 445–451, 1991.
87. Gibson, M. M., and J. A. Monahan: A Simple Model of Radiation Heat Transfer from a Cloud of Burning Particles in a Confined Gas Stream, *Int. J. Heat Mass Transfer*, vol. 14, pp. 141–147, 1971.
88. Selcuk, Nevin, and R. G. Siddall: Two-Flux Spherical Harmonic Modelling of Two-Dimensional Radiative Transfer in Furnaces, *Int. J. Heat Mass Transfer*, vol. 19, pp. 313–321, 1976.
89. Kunitomo, Takeshi: Luminous Flame Emission under Pressure up to 20 Atm, in N. H. Afgan and J. M. Beer (eds.), "Heat Transfer in Flames," pp. 271–281, Scripta, Washington, D.C., 1974.
90. Chin, J. S. and A. H. Lefebvre: Influence of Fuel Composition on Flame Radiation in Gas Turbine Combustors, *J. Prop.*, vol. 6, no. 4, pp. 497–503, 1990.
91. Smith, T. F., A. M. Al-Turki, K.-H. Byun, and T. K. Kim: Radiative and Conductive Transfer for a Gas/Soot Mixture between Diffuse Parallel Plates, *J. Thermophys. Heat Transfer*, vol. 1, no. 1, pp. 50–55, 1987.
92. Babikian, D. S., D. K. Edwards, S. E. Karam, C. P. Wood, and G. S. Samuelsen: Experimental Mass Absorption Coefficients of Soot in Spray Combustor Flames, *J. Thermophys. Heat Transfer*, vol. 4, no. 1, pp. 8–15, 1990.
93. Williams, J. J., and D. P. Dudley: The Radiative Contribution to Heat Transfer in Metalized Propellant Exhausts, *Fifth Symp. Thermophys. Properties, ASME*, Boston, Sept. 30–Oct. 2, 1970.
94. Edwards, D. K., Y. Sakurai, and D. S. Babikian: A Two-Particle Model for Rocket Plume Radiation, *J. Thermophys. Heat Transfer*, vol. 1, no. 1, pp. 13–20, 1987.
95. Edwards, D. K. and D. S. Babikian: Radiation from a Nongray Scattering, Emitting, and Absorbing Solid Rocket Motor Plume, *J. Thermophysics Heat Transfer*, vol. 4, no. 4, pp. 446–453, 1990.
96. Tabanfar, S., and M. F. Modest: Radiative Heat Transfer in a Cylindrical Mixture of Non-Gray Particulates and Molecular Gases, *J. Quant. Spectrosc. Radiat. Transfer*, vol. 30, no. 6, pp. 555–570, 1983.
97. Thynell, S. T.: Radiation Due to CO_2 or H_2O and Particulates in Cylindrical Media, *J. Thermophys. Heat Transfer*, vol. 4, no. 4, pp. 436–445, 1990.
98. Skocypec, R. D., D. V. Walters, and R. O. Buckius: Spectral Emission Measurements from Planar Mixtures of Gas and Particulates, *J. Heat Transfer*, vol. 109, pp. 151–158, 1987.
99. Walters, D. V., and R. O. Buckius: Normal Spectral Emission From Nonhomogeneous Mixtures of CO_2 Gas and Al_2O_3 Particulate, *J. Heat Transfer*, vol. 113, no. 1, pp. 174–184, 1991.
100. Berlad, A. L., V. Tangirala, H. Ross, and L. Facca: Radiative Structures of Lycopodium-Air Flames in Low Gravity, *J. Prop. Power*, vol. 7, no. 1, pp. 5–8, 1991.

101. Lanzo, Chester, D., and Robert G. Ragsdale: Heat Transfer to a Seeded Flowing Gas from an Arc Enclosed by a Quartz Tube, *Proc. 1964 Heat Transfer Fluid Mech. Inst.* Warren H. Giedt and Salomon Levy (eds.), pp. 226–244, 1964.

102. Howell, J. R., and H. E. Renkel: Analysis of the Effects of a Seeded Propellant Layer on Thermal Radiation in the Nozzle of a Gaseous-Core Nuclear Propulsion System, NASA TN D-3119, 1965.

103. Siegel, R.: Radiative Behavior of a Gas Layer Seeded with Soot, NASA TN D-8278, July, 1976.

104. Abdelrahman, M., P. Fumeaux, and P. Suter: Study of Solid-Gas Suspensions Used for Direct Absorption of Concentrated Solar Radiation, *Sol. Energy*, vol. 22, no. 1, pp. 45–48, 1979.

105. Thomas, D. Lyddon: Problems in Applying the Line Reversal Method of Temperature Measurement to Flames, *Combust. Flame*, vol. 12, no. 6, pp. 541–549, 1968.

106. Greenberg, J. M., and T. P. Roark (eds.): Interstellar Grains, NASA SP-140, 1967.

107. Donn, B., and K. S. Krishna Swamy: Extinction by Interstellar Grains, Mie Particles and Polycyclic Aromatic Molecules, *Physica*, vol. 41, no. 1, pp. 144–150, 1969.

108. Huffman, Donald R.: Optical Properties of Particulates, *Astrophys. Space Sci.*, vol. 34, pp. 175–184, 1975.

109. Edwards, D. K., L. K. Glassen, W. C. Hauser, and J. S. Tuchscher: Radiation Heat Transfer in Nonisothermal Nongray Gases, *J. Heat Transfer*, vol. 89, no. 3, pp. 219–229, 1967.

110. Lefebvre, A. H.: Flame Radiation in Gas Turbine Combustion Chambers, *Int. J. Heat Mass Transfer*, vol. 27, no. 9, pp. 1493–1510, 1984.

PROBLEMS

13-1 Evaluate the geometric mean transmittance $\bar{\tau}_{d1-2}$ from the element dA_1 to the area A_2 in Prob. 7-1. The region between the areas is filled with a gray medium at uniform temperature having an absorption coefficient $a = 0.01$ cm^{-1}.

Answer: 0.341

13-2 From the spectral absorptance α_λ' in Fig. 12-10a, estimate the total emittance of CO_2 for the temperature, pressure, and path length given in the figure caption. Do not assume that the bands are black. Compare the result with the value obtained from extrapolation of the values in Figure 13-13.

Answer: 0.259

13-3 A thin black plate 1×1 cm is at the center of a sphere of CO_2-air mixture at a uniform temperature of 2400 K and 1 atm total pressure. The partial pressure of the CO_2 is 0.6 atm and the sphere diameter is 1 m. How much energy is absorbed by the plate? What will the plate temperature be? (Assume the boundary of the sphere is black and kept cool so that it does not enter into the radiative exchange.)

Answer: 24.5 W, 1212 K

13-4 A spherical cavity is filled with an isothermal gray medium having an absorption coefficient a. Set up the relations needed to determine the geometric mean transmittance $\bar{\tau}_{1-d2}$ from the cavity surface A_1 to the area element dA_2 at the center of the opening A_2. (*Note:* A similar situation is considered by Koh, J. C. Y.: *Int. J. Heat Mass Transfer*, vol. 8, no. 2, pp. 373–374, 1965).

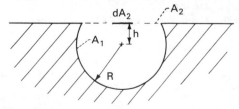

13-5 A sphere of gray gas at uniform temperature is situated above a surface with the region between the sphere and surface being nonabsorbing. Derive a relation for the radiative energy incident on the circular area as shown. (*Hint:* Consider the circular area as being a cut through a concentric sphere surrounding the gas sphere, and make use of the symmetry of the geometry.)

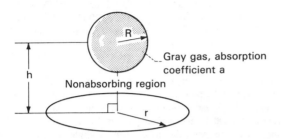

Gray gas, absorption coefficient a

Nonabsorbing region

13-6 For the radiating sphere of gas in Prob. 13-5, derive an expression for the local energy flux incident along the plane surface as a function of distance r.

13-7 Pure carbon dioxide at 1 atm and 2220 K is contained between parallel plates 0.15 m apart. What is the radiative flux received at the plates as a result of radiation by the gas? (Use the CO_2 total emittance chart.)

Answer: 99.2 kW/m²

13-8 A rectangular furnace of dimensions 0.3 × 0.3 × 1.3 m has soot-covered interior walls that can be considered black. The furnace is filled with well-mixed combustion products at a temperature of 2000 K composed of 40% by volume CO_2, 30% by volume water vapor, and the remainder N_2. The total pressure is 2 atm. Compute the radiative flux to the walls using the CO_2 and H_2O emittance charts.

Answer: 265 kW/m²

13-9 A pipe 10 cm in diameter is carrying superheated steam at 1.2 atm pressure and at a uniform temperature of 1110 K. What is the radiative flux from the steam received at the pipe wall?

Answer: 17.9 kW/m²

13-10 A furnace at atmospheric pressure with interior in the shape of a cube having a 0.7-m edge dimension is filled with a 50:50 mixture by volume of CO_2 and N_2. The gas temperature is uniform at 1750 K and the walls are cooled to 1200 K. The interior surfaces are black. At what rate is energy being supplied to the gas (and removed from the walls) to maintain these conditions? Use the method in Sec. 13-6.2.

Answer: 98.4 kW

13-11 Consider the same conditions and furnace volume as in Prob. 13-10. The furnace is now a cylinder with height equal to two times its diameter. What is the supplied energy rate?

Answer: 96.8 kW

13-12 A furnace being designed by a chemical company will be used to burn toxic waste composed of hydrocarbons. Complete elimination of the hydrocarbons with oxygen requires that the temperature of the combustion products in the furnace (60% CO_2, 40% H_2O by volume) be maintained at 1500 K. To prevent leakage to the surroundings, the furnace interior is kept at 0.5 atm. The furnace is in the shape of a right circular cylinder of height equal to its diameter of 3 m. What average radiative flux is incident on the interior surface of the furnace? Note any assumptions used in obtaining the result.

Answer: 71,800 W/m²

13-13 Estimate the maximum radiative flux incident on any position on the surface of the furnace described in Prob. 13-12.

Answer: 77,200 W/m²

13-14 Carbon dioxide at 5 atm pressure and 1600 K is contained between parallel nongray plates 5 cm apart. The plates are both at 1000 K and have hemispherical emissivity as a function of wavelength as given in Example 13-4. Considering the five CO_2 absorption bands as in Example 13-4, compute the rate of energy loss by the gas to the plates.

13-15 A cubical black-walled enclosure with 1-m edges contains a mixture of gases at $T = 1800$ K and pressure of 2 atm. The gas has volume fractions of $0.4/0.4/0.2$ for $CO_2/H_2O/N_2$. Cooling water is passed over one face of the cube. If the water is initially at 30°C and has a maximum allowable temperature rise of 10°C, what mass flow rate of water (kg/s) is required? (Neglect reradiation from the wall, and assume all other walls are cool.)

Answer: 4.81 kg/s

13-16 A large furnace has a large number of tubes arranged in an equilateral triangular array. The tubes are of 2.5-cm outside diameter, and the tube centers are spaced 5 cm apart. The furnace gas is composed of 75% CO_2 and 25% H_2O by volume, and the combustion process in the gas maintains the gas temperature at 1200 K and a total pressure of 0.9 atm. What is the radiative flux incident on a tube in the interior of the bundle (i.e., completely surrounded by other tubes) per meter of tube length?

Answer: 1.10 kW/m

13-17 Two opposed parallel rectangles are separated by a distance of 0.3 m. The rectangles are of size 0.6×0.9 m. The space between the rectangles is filled with H_2O vapor at $P = 1$ atm and $T = 1100$ K. (Assume that for this problem only the 2.7-μm band of H_2O participates in radiative absorption and emission by the gas, and use the data of Tables 12-4 and 12-5 to compute α_g.) Rectangle 1 has $T_1 = 1360$ K, $\epsilon_1 = 1.0$. Rectangle 2 has $T_2 = 530$ K, $\epsilon_1 = 0.5$. Assume that the surroundings are at low temperature. Compute the total energy being added to each plate, using the method of Sec. 13-3.4.

13-18 A furnace is in the form of a cube with an edge length of 50 cm and with black walls that are cooled so that they emit negligible energy. The furnace is filled with nitrogen maintained at 1500 K. It is desired that the average energy flux received by the walls be 20% of the blackbody flux emitted at the gas temperature. To accomplish this, the gas is seeded with propane soot (Fig. 13-31). What volume fraction of soot is required? If the edge dimension of the furnace is doubled, what soot volume fraction is required? For the edge length of 50 cm, what soot volume fraction is required to double the energy flux reaching the walls?

Answer: 3.3×10^{-7}, 1.7×10^{-7}, 7.7×10^{-7}

13-19 A wall is to be shielded from normally incident radiation by a flow along it of a layer of cool nitrogen seeded with soot particles. The seeded layer is 0.5 cm thick. For design purposes, it is desired to examine the attenuation of the incident radiation as a function of soot volume concentration in the nitrogen. For incident green light and for infrared radiation at $\lambda = 5$ μm, prepare a plot of percentage transmission as a function of soot volume concentration. [Use the Mie theory, Eq. (13-122), and the optical properties of propane soot. Neglect scattering.]

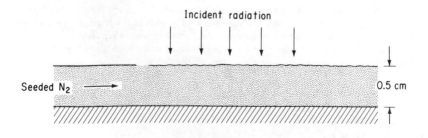

Incident radiation

Seeded N_2 ⟶   0.5 cm

13-20 An infinitely long rectangular enclosure has diffuse-gray walls at conditions shown below and encloses a gray gas at $T_g = 1500$ K. The gas has an absorption coefficient of 0.5 m^{-1}. Find the average net radiative flux at each surface and the energy necessary to maintain the gas at 1500 K.

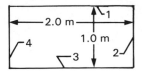

Surface	ϵ	T_w (K)
1	1.0	2000
2	0.5	1500
3	0.1	1000
4	0	500

Answer: $q_1 = 607$ kW/m^2, $q_2 = -75.1$ kW/m^2, $q_3 = -39.3$ kW/m^2, energy added $= 1060$ kW/m

FOURTEEN

THE EQUATIONS OF ENERGY TRANSFER FOR ABSORBING, EMITTING, AND SCATTERING MEDIA

14-1 INTRODUCTION

In Chap. 12 basic concepts and definitions were presented for intensity, emission, absorption, and scattering within a medium. Radiation traveling along a path is attenuated by absorption and scattering and is enhanced by emission and by scattering in from other directions. These concepts will be employed to develop an integrodifferential equation governing the radiation intensity along a path through a medium. This is the *equation of transfer*. In obtaining a solution to the equation of transfer, the constant of integration accounts for the intensity at the origin of the radiation path being considered. Because the origin is usually at the boundary of the radiating medium, the radiation at the boundaries is thereby coupled into the radiation distribution within the medium.

The intensity given by the equation of transfer is the local radiation traveling in a single direction per unit solid angle and wavelength and crossing a unit area normal to the direction of travel. To obtain the net *energy* crossing an area requires integrations to include the contributions by the intensities crossing in all directions and at all wavelengths. This results in an equation for local radiative flux that is used in forming an energy balance within the medium. The energy balance is needed to obtain the temperature distribution in the medium.

Since the local intensities depend on scattering, if it is present, the energy balance depends on the net energy scattered into a volume element. The scattering into a direction along a path is often combined with the local emission in that direction to form the *source function*. When scattering is present, a source-function equation must be solved simultaneously with the energy equation to determine two unknowns: the temperature distribution and the source-function distribution. This is a consequence of the two coupled transfer processes absorption (and hence emission) and scattering. For a nonabsorbing medium, the scattering becomes decoupled from the energy equation, considerably simplifying the analysis.

In this chapter the equation of transfer for local intensities is derived first. Then the radiative flux and its divergence are considered, and the energy equation is presented in three dimensions. To keep the geometry simple, the formulation is then developed in detail for a plane layer between infinite parallel boundaries. The most general equations include anisotropic scattering. Simplified cases are presented for isotropic scattering, a gray medium, and a medium bounded by diffuse walls.

Having gained experience with the plane layer, the development is generalized into three dimensions. The radiative flux relations are carried out in detail for the nonplanar geometries of long rectangular and cylindrical regions. These illustrate the geometric complexities that arise and how they are treated analytically to obtain the governing energy and transfer relations.

14-2 SYMBOLS

a	absorption coefficient
A	area
b, d	dimensions of rectangle
c	speed of light in a medium; specific heat of solid
c_p	specific heat
D	spacing between parallel plates; thickness of layer
e	emissive power
E_n	exponential integral function, Eq. (14-66)
f	photon distribution function
F	function in Eq. (14-89a)
$F(r, \beta)$	function in Eq. (14-150)
G	function in Eq. (14-89b); quantity defined in Eq. (14-105)
h	Planck's constant; enthalpy
i	radiation intensity
$\mathbf{i}, \mathbf{j}, \mathbf{k}$	unit vectors in x, y, z coordinate directions
I	source function, Eq. (14-9)
k	thermal conductivity
K	extinction coefficient, $a + \sigma_s$
$\mathbf{n}$	unit normal vector
P	pressure
$\mathbf{q}_r$	radiative flux vector
q	energy flux, energy per unit area and time
q'''	volume heat source

Q	energy per unit time
$\tilde{Q}$	dimensionless variable, Eq. (14-84)
r	position vector
R	radius of cylinder
s	unit vector in S direction
S	coordinate along path of radiation
S_n	two-dimensional radiation function, Eq. (14-134)
T	absolute temperature
U	radiant energy density
V	volume
x, y, z	coordinates in cartesian system
α, δ, γ	direction cosines
β	coefficient of volume expansion; angle in Fig. 14-17
Γ	integral in Eq. (14-151)
ϵ	emissivity
θ	cone angle, angle from normal of area
κ	optical thickness
κ_D	optical thickness for path of length D
λ	wavelength
μ	the quantity $\cos\theta$
ν	frequency
ρ	density
ρ^*	two-dimensional distance, Eq. (14-133)
σ	Stefan-Boltzmann constant
σ_s	scattering coefficient
τ	time
φ	circumferential angle
ϕ	dimensionless temperature ratio
Φ	phase function for scattering
Φ_d	viscous dissipation function
ψ	dimensionless energy flux
ω	solid angle
Ω	albedo for scattering

Subscripts

a	absorbed
b	blackbody
D	at the boundary at position D
e	emitted; environment
fd	fully developed
g	gas; gray medium
i	incident; incoming; initial value
n	normal direction
o	outgoing
P	Planck mean value, Eq. (14-95)

r	radiative
s	scattering
w	wall
x, y, z	in x, y, z directions
λ, ν	spectrally dependent
1, 2	surface 1 or 2

Superscripts

$+$	along directions having positive cosine θ
$-$	along directions having negative cosine θ
$*$	dummy variable of integration
$^-$	(overbar) averaged over all incident or outgoing solid angles
$'$	directional quantity

14-3 THE TRANSFER EQUATION AND SOURCE FUNCTION

The equation of transfer in a medium is now derived. This describes the radiation intensity at any position along a path through an absorbing, emitting, and scattering medium. Bouguer's law, Eq. (12-12), accounts for attenuation by absorption and scattering. The equation of transfer also includes contributions to the radiation intensity by emission and incoming scattering along a path.

14-3.1 Equation of Transfer

Consider radiation of intensity $i'_\lambda(S)$ within an absorbing, emitting, and scattering medium (Fig. 14-1). Attention is directed to the change of intensity as the radia-

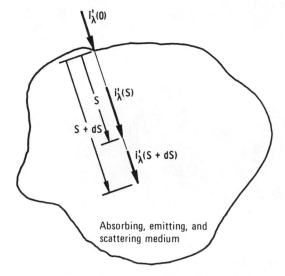

Figure 14-1 Geometry for derivation of equation of transfer.

tion passes through distance dS. Using (12-8) gives the decrease due to absorption as

$$di'_{\lambda,a} = -a_\lambda(S)i'_\lambda(S)\,dS \tag{14-1}$$

Note that a_λ is used in (14-1) rather than the "true" absorption coefficient a_λ^+. Thus the change in intensity along the path is not only the result of true absorption but also includes the contribution of induced emission, which is along the path direction as discussed in Sec. 12-4.5.

If the radiation along the path is in local thermodynamic equilibrium, the contribution by spontaneous emission in the medium along dS to the intensity in the S direction is given by Eq. (12-40a) as

$$di'_{\lambda,e} = a_\lambda(S)i'_{\lambda b}(S)\,dS \tag{14-2}$$

From (12-30), (12-50), and Fig. 12-9, the sum of the attenuation by scattering and the gain by incoming scattering in direction S can be written as

$$di'_{\lambda,s} = -\sigma_{s\lambda}(S)i'_\lambda(S)\,dS + \frac{dS\,\sigma_{s\lambda}}{4\pi}\int_{\omega_i=0}^{4\pi} i'_\lambda(S,\omega_i)\Phi(\lambda,\omega,\omega_i)\,d\omega_i \tag{14-3}$$

Adding (14-1) to (14-3) gives, for the *change in intensity with S* in the solid angle $d\omega$ about the S direction,

$$\frac{di'_\lambda}{dS} = -a_\lambda i'_\lambda(S) + a_\lambda i'_{\lambda b}(S) - \sigma_{s\lambda}i'_\lambda(S) + \frac{\sigma_{s\lambda}}{4\pi}\int_{\omega_i=0}^{4\pi} i'_\lambda(S,\omega_i)\Phi(\lambda,\omega,\omega_i)\,d\omega_i \tag{14-4}$$

Loss by absorption (including the contribution by induced emission)	Gain by emission (not including induced emission)	Loss by scattering	Gain by scattering into S direction

The two terms for decreases by absorption and scattering are combined, giving the *equation of transfer* for absorbing, emitting, and scattering media (for *elastic anisotropic scattering*):

$$\frac{di'_\lambda}{dS} = -K_\lambda i'_\lambda(S) + a_\lambda i'_{\lambda b}(S) + \frac{\sigma_{s\lambda}}{4\pi}\int_{\omega_i=0}^{4\pi} i'_\lambda(S,\omega_i)\Phi(\lambda,\omega,\omega_i)\,d\omega_i \tag{14-5}$$

where $K_\lambda = a_\lambda + \sigma_{s\lambda}$ is the extinction coefficient (Sec. 12-4.1) and in general is a function of position S.

The albedo for single scattering, Ω, is the ratio of the scattering coefficient to the extinction coefficient,

$$\Omega_\lambda \equiv \frac{\sigma_{s\lambda}}{K_\lambda} = \frac{\sigma_{s\lambda}}{a_\lambda + \sigma_{s\lambda}} \qquad \left(1 - \Omega_\lambda = \frac{a_\lambda}{a_\lambda + \sigma_{s\lambda}}\right) \tag{14-6}$$

For scattering alone $\Omega_\lambda \to 1$, while for absorption alone $\Omega_\lambda \to 0$. The *optical depth* or opacity when both scattering and absorption are present was defined in (12-17):

$$\kappa_\lambda(S) = \int_0^S K_\lambda(S^*)\, dS^* = \int_0^S [a_\lambda(S^*) + \sigma_{s\lambda}(S^*)]\, dS^* \tag{14-7}$$

where S^* is a dummy variable of integration. Equation (14-5) now becomes

$$\frac{di_\lambda'}{d\kappa_\lambda} = -i_\lambda'(\kappa_\lambda) + (1 - \Omega_\lambda)i_{\lambda b}'(\kappa_\lambda) + \frac{\Omega_\lambda}{4\pi}\int_{\omega_i=0}^{4\pi} i_\lambda'(\kappa_\lambda, \omega_i)\Phi(\lambda, \omega, \omega_i)\, d\omega_i \tag{14-8}$$

where $d\kappa_\lambda = K_\lambda(S)\, dS$ is the *optical differential thickness*. The final two terms in (14-8) are combined into the source function $I_\lambda'(\kappa_\lambda, \omega)$ defined as

$$I_\lambda'(\kappa_\lambda, \omega) \equiv (1 - \Omega_\lambda)i_{\lambda b}'(\kappa_\lambda) + \frac{\Omega_\lambda}{4\pi}\int_{\omega_i=0}^{4\pi} i_\lambda'(\kappa_\lambda, \omega_i)\Phi(\lambda, \omega, \omega_i)\, d\omega_i \tag{14-9}$$

This is the source of intensity along the optical path from both emission and incoming scattering. For anisotropic scattering I_λ' is a function of ω (that is, the direction of S). The *equation of transfer* (14-8) then becomes

$$\frac{di_\lambda'}{d\kappa_\lambda} + i_\lambda'(\kappa_\lambda) = I_\lambda'(\kappa_\lambda, \omega) \tag{14-10}$$

This is an integrodifferential equation since i_λ' is within the integral of the source function.

An integrated form of (14-10) is obtained by use of an integrating factor. Multiplying through by $\exp \kappa_\lambda$ gives

$$\exp \kappa_\lambda \frac{di_\lambda'}{d\kappa_\lambda} + i_\lambda'(\kappa_\lambda) \exp \kappa_\lambda = \frac{d}{d\kappa_\lambda}[i_\lambda'(\kappa_\lambda) \exp \kappa_\lambda] = I_\lambda'(\kappa_\lambda, \omega) \exp \kappa_\lambda \tag{14-11}$$

Integrating over an optical thickness from $\kappa_\lambda = 0$ to $\kappa_\lambda(S)$ gives

$$i_\lambda'(\kappa_\lambda, \omega) \exp \kappa_\lambda - i_\lambda'(0, \omega) = \int_0^{\kappa_\lambda} I_\lambda'(\kappa_\lambda^*, \omega) \exp \kappa_\lambda^*\, d\kappa_\lambda^* \tag{14-12}$$

or

$$i_\lambda'(\kappa_\lambda, \omega) = i_\lambda'(0, \omega) \exp(-\kappa_\lambda) + \int_0^{\kappa_\lambda} I_\lambda'(\kappa_\lambda^*, \omega) \exp[-(\kappa_\lambda - \kappa_\lambda^*)]\, d\kappa_\lambda^* \tag{14-13}$$

where κ_λ^* is a dummy variable of integration. Equation (14-13) is the *integrated form of the equation of transfer*.

Equation (14-13) is interpreted physically as the intensity at optical depth κ_λ being composed of two terms. The first is the attenuated initial intensity arriving at κ_λ. The second is the intensity at κ_λ resulting from emission and incoming scattering in the S direction by all thickness elements along the path, reduced by exponential attentuation between each point of emission and incoming scattering κ_λ^* and the location κ_λ.

14-3.2 Source Function Equation

Equations (14-13) and (14-9) each contain the intensity and source function. It is useful to eliminate i_λ' to obtain an equation for I_λ'. This is the *source-function equation* and is obtained by using (14-13) in (14-9) to eliminate i_λ' to yield,

$$I_\lambda'(\kappa_\lambda, \omega) = (1 - \Omega_\lambda)i_{\lambda b}'(\kappa_\lambda)$$

$$+ \frac{\Omega_\lambda}{4\pi} \int_{\omega_i=0}^{4\pi} \left[i'(0)e^{-\kappa_\lambda} + \int_0^{\kappa_\lambda} I_\lambda'(\kappa_\lambda^*, \omega_i)e^{-(\kappa_\lambda - \kappa_\lambda^*)} \, d\kappa_\lambda^* \right] \Phi(\lambda, \omega, \omega_i) \, d\omega_i$$

$$(14\text{-}14)$$

Equations (14-14) and (14-13) cannot generally be used by themselves to solve for the source function and intensity distributions because they contain the unknown temperature distribution of the medium in the blackbody intensity $i_{\lambda b}'(\lambda, T)$. The temperature distribution is also needed to determine the absorption and scattering coefficients so that the local optical depth $\kappa_\lambda(S)$ can be computed from (14-7), and the physical coordinate S thereby related to the optical coordinate κ_λ. The temperature distribution depends on energy conservation within the medium, which in turn depends on the total absorbed radiation in each volume element along the path. This total energy will be obtained by utilizing the spectral intensity passing through a location and integrating over all incident solid angles and all wavelengths. As will be shown, the energy equation is solved in conjunction with the intensity and source-function equations to yield temperature and scattering distributions. The resulting equations are sufficiently complex that numerical solutions are almost always required. In some instances advanced analytical techniques have been used to obtain closed-form solutions for a limited range of geometries and conditions. The energy equation will be considered in the next section, which, along with the equations for the intensity and source function, completes the set of basic equations required for analysis.

EXAMPLE 14-1 A black surface element dA is 10 cm from an element of gas dV (Fig. 14-2). The gas element is a part of a gas volume V that is iso-

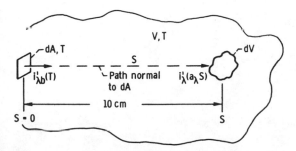

Figure 14-2 Geometry for Example 14-1.

thermal and at the same temperature T as dA. If the gas has an absorption coefficient $a_\lambda = 0.1$ cm^{-1} at $\lambda = 1$ μm and there is no scattering, what is the spectral intensity at $\lambda = 1$ μm that arrives at dV along the path S from dA to dV?

Because dA is black and at temperature T, the intensity at $S = 0$ is $i'_\lambda(0) = i'_{\lambda b}(T)$. Since the gas is isothermal, the emitted blackbody intensity in the gas is $i'_{\lambda b}(\kappa_\lambda) = i'_{\lambda b}(T)$. Substituting into the integrated equation of transfer [Eq. (14-13)] with $\sigma_{s\lambda} = 0$ gives

$$i'_\lambda(\kappa_\lambda) = i'_{\lambda b}(T) \exp(-\kappa_\lambda) + i'_{\lambda b}(T) \exp(-\kappa_\lambda) \int_0^{\kappa_\lambda} \exp(\kappa_\lambda^*)\, d\kappa_\lambda^*$$

After integration, this reduces to

$$i'_\lambda(\kappa_\lambda) = i'_{\lambda b}(T)$$

The $i'_{\lambda b}(T)$ is given by (2-11) for a gas with refractive index $n = 1$. The intensity arriving at dV along an isothermal path from a black surface element at the same temperature as the gas is thus equal to the blackbody intensity emitted by the wall and does not depend on a_λ or S. The attenuation by gas absorption of the intensity emitted by the wall is exactly compensated by emission from the gas along the path from dA to dV.

EXAMPLE 14-2 An absorbing-emitting gray gas layer is adjacent to a black wall. As a result of internal heat generation, the medium has a parabolic temperature distribution decreasing from the wall temperature, 650 K, to 425 K at the boundary $x = D$ (Fig. 14-3). What is the intensity $i'(D)$?

The parabolic temperature distribution is given by $T(x) = T_w - (T_w - 425)(x/D)^2$. From Eq. (14-13) the intensity is given by

$$i'(D) = \frac{\sigma T_w^4}{\pi} \exp(-aD) + \int_0^{aD} \frac{\sigma}{\pi} T^4(x) \exp[-a(D - x)] a\, dx$$

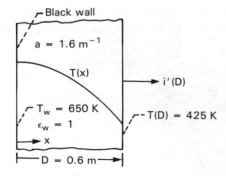

Figure 14-3 Conditions for Example 14-2.

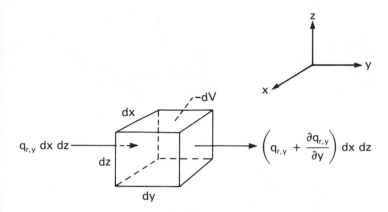

Figure 14-4 Radiative fluxes for volume element.

Numerical integration using the values in Fig. 14-3 yields $i'(D) = 2510$ W/$(m^2 \cdot sr)$.

14-4 ENERGY CONSERVATION WITHIN A MEDIUM

14-4.1 The Radiative Flux Vector

For an energy balance on a volume element, the net radiative energy supplied to dV is desired. If a volume element $dx\, dy\, dz$ is considered, the radiative energies in and out of the $dx\, dz$ faces are shown in Fig. 14-4. The energies are similarly written for the other faces, the outgoing energies are subtracted from the incoming, and the result divided by $dx\, dy\, dz$. The result is the net radiative energy supplied per unit volume:

$$-\left[\frac{\partial q_{r,x}}{\partial x} + \frac{\partial q_{r,y}}{\partial y} + \frac{\partial q_{r,z}}{\partial z}\right] = -\nabla \cdot \mathbf{q}_r \tag{14-15}$$

which is the negative of the divergence of the radiative heat flux vector

$$\mathbf{q}_r = \mathbf{i}q_{r,x} + \mathbf{j}q_{r,y} + \mathbf{k}q_{r,z} \tag{14-16}$$

The vector form $-\nabla \cdot \mathbf{q}_r$ can be used for coordinate systems other than the rectangular coordinates considered here.

The flux vector is now related to the intensity. Consider the area dA in Fig. 14-5. This could be one of the faces of the volume element in Fig. 14-4, and $\mathbf{n}$ is then along a coordinate direction. In general, $\mathbf{n}$ is in any direction and the flux $q_{r,n}$ is through an area normal to the $\mathbf{n}$ direction. Let $\mathbf{s}$ be a unit vector in the S direction. Then $\cos\theta = \mathbf{s} \cdot \mathbf{n}$, and in Fig. 14-5 the intensity is the energy per unit solid angle crossing dA per unit area normal to i'. Hence the energy crossing dA

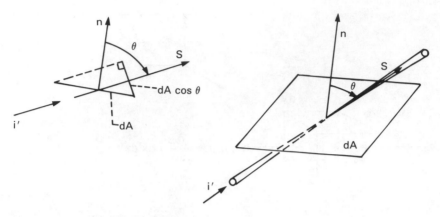

Figure 14-5 Quantities in derivation of radiative flux vector.

as a result of i' is i' dA cos θ $d\omega$. The radiative flux crossing dA as a result of intensities incident from all directions is thus

$$q_{r,n} = \int_{\omega=0}^{4\pi} i' \cos \theta \, d\omega = \int_{\omega=0}^{4\pi} i' \mathbf{s} \cdot \mathbf{n} \, d\omega \tag{14-17}$$

where θ is the angle from the normal of dA to the direction of i', and i' is a function of direction. The $q_{r,n}$ depends on the direction of $\mathbf{n}$ relative to $\mathbf{s}$ (cos θ becomes negative for $\theta > \pi/2$ so the sign of the portion of the energy flux traveling in the direction opposite to the positive $\mathbf{n}$ direction is automatically included). The $q_{r,n}$ is the component in the $\mathbf{n}$ direction of a flux vector given by

$$\mathbf{q}_r = \int_{\omega=0}^{4\pi} i' \mathbf{s} \, d\omega \tag{14-18}$$

that is, $q_{r,n} = \mathbf{n} \cdot \mathbf{q}_r$. If the direction cosines of $\mathbf{n}$ are α', δ', and γ' and those for $\mathbf{s}$ (the direction of i') are α, δ, and γ, then $\mathbf{n} = \mathbf{i}\alpha' + \mathbf{j}\delta' + \mathbf{k}\gamma'$ and $\mathbf{s} = \mathbf{i}\alpha + \mathbf{j}\delta + \mathbf{k}\gamma$ so that $\mathbf{s} \cdot \mathbf{n} = \cos \theta = \alpha\alpha' + \delta\delta' + \gamma\gamma'$ and

$$q_{r,n} = \int_{\omega=0}^{4\pi} i'(\alpha, \delta, \gamma)(\alpha\alpha' + \delta\delta' + \gamma\gamma') \, d\omega$$

Thus

$$q_{r,n} = \alpha' q_{r,x} + \beta' q_{r,y} + \gamma' q_{r,z} \tag{14-19}$$

The $q_{r,x}$, $q_{r,y}$, and $q_{r,z}$ are fluxes across areas normal to the x, y, and z directions and are the components of the *radiative flux vector*. Each component is obtained from the integral in (14-17) with $\mathbf{n}$ oriented in that coordinate direction. For example, $q_{r,x} = \int_{\omega=0}^{4\pi} i'(\alpha, \delta, \gamma)\alpha \, d\omega$, where α is the cosine of the angle between the x axis and the direction of i' (the direction of $\mathbf{s}$).

The radiative flux vector can be written more explicitly by using Eq. (14-18) and a spherical coordinate system as shown in Fig. 14-6. The unit vector **s** is then written as

$$\mathbf{s} = \mathbf{i} \cos \varphi \sin \theta + \mathbf{j} \sin \varphi \sin \theta + \mathbf{k} \cos \theta \qquad (14\text{-}20)$$

Substituting **s** and $d\omega = \sin \theta \, d\theta \, d\varphi$ into (14-18) gives the vector $\mathbf{q}_r$ in terms of its three components:

$$\mathbf{q}_r = \mathbf{i} q_{r,x} + \mathbf{j} q_{r,y} + \mathbf{k} q_{r,z} = \mathbf{i} \int_{\varphi=0}^{2\pi} \int_{\theta=0}^{\pi} i'(\theta, \varphi) \cos \varphi \sin^2 \theta \, d\theta \, d\varphi$$

$$+ \mathbf{j} \int_{\varphi=0}^{2\pi} \int_{\theta=0}^{\pi} i'(\theta, \varphi) \sin \varphi \sin^2 \theta \, d\theta \, d\varphi$$

$$+ \mathbf{k} \int_{\varphi=0}^{2\pi} \int_{\theta=0}^{\pi} i'(\theta, \varphi) \cos \theta \sin \theta \, d\theta \, d\varphi \qquad (14\text{-}21)$$

The $q_{r,n}$ in Eq. (14-19) is $q_{r,n} = \mathbf{n} \cdot \mathbf{q}_r$. The divergence of the radiant heat-flux vector, $\nabla \cdot \mathbf{q}_r$, is used in the energy equation as discussed in Sec. 14-4.2.

In some situations, such as extremely rapid transients, it is necessary to account for the density of radiative energy within a local volume element. This is related to the average local intensity. For this derivation it is useful to consider radiation as a collection of photons, and the conditions at any location in a medium are given by the photon distribution function f. Since photon energy is related more directly to frequency, frequency is used here rather than wavelength. In volume dV at position $\mathbf{r}$, let

$$d^3 N_p = f(\nu, \mathbf{r}, S) \, d\nu \, dV \, d\omega \qquad (14\text{-}22)$$

be the number of photons traveling in direction S in frequency interval $d\nu$ centered about ν and within solid angle $d\omega$ about the direction S (see Fig. 14-7). Each

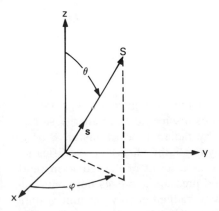

Figure 14-6 Spherical coordinate system for radiative heat flux vector.

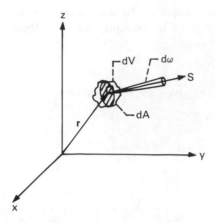

Figure 14-7 Quantities in derivation of radiative energy density.

photon has energy $h\nu$. The energy per unit volume per unit frequency interval is then $h\nu f\,d\omega$ integrated over all solid angles. This is the *spectral radiant energy density*:

$$U_\nu(\nu,\mathbf{r}) = h\nu \int_{\omega=0}^{4\pi} f(\nu,\mathbf{r},S)\,d\omega \tag{14-23}$$

To obtain the intensity, the energy flux in the S direction is needed across area dA normal to the S direction, Fig. 14-7. The photons have velocity c, and the number density traveling in the normal direction across dA is $f\,d\nu\,d\omega$. The number of photons crossing dA in the S direction per unit time is then $cf\,d\nu\,d\omega\,dA$. The energy carried by these photons is $h\nu cf\,d\nu\,d\omega\,dA$. The spectral intensity is the energy in a single direction per unit time, unit frequency interval, and unit solid angle crossing a unit area normal to that direction. This gives the intensity at location $\mathbf{r}$ and in direction S as

$$i'_\nu = h\nu cf(\nu,\mathbf{r},S) \tag{14-24}$$

The relation between energy density and intensity is then obtained by using (14-24) to eliminate f in Eq. (14-23),

$$U_\nu(\nu,\mathbf{r}) = \frac{1}{c} \int_{\omega=0}^{4\pi} i'_\nu\,d\omega = \frac{4\pi}{c}\bar{\imath}_\nu \tag{14-25}$$

14-4.2 General Energy Conservation

A general energy balance on a volume element includes conduction, convection, internal heat sources, compression work, viscous dissipation, and energy storage due to transients, as well as the contribution by radiative heat transfer. Storage of radiant energy within an element is generally negligible; hence no modification of the usual transient terms is considered here as a result of the radiation field. Radiation pressure is negligible relative to fluid pressure and does not contribute to the compression work term. The net inflow of radiant energy per unit volume can be written as the negative of the divergence of a radiant flux vector $\mathbf{q}_r$ [Eq.

(14-15)]. The net heat conduction into a volume element can be written as the divergence of a vector $\nabla \cdot (k \, \nabla T)$. Thus the energy equation for a single-component fluid can be modified for the effect of radiative transfer by adding $-\mathbf{q}_r$ to $k\nabla T$ to yield

$$\rho c_p \frac{DT}{D\tau} = \beta T \frac{DP}{D\tau} + \nabla \cdot (k \, \nabla T - q_r) + q''' + \Phi_d \tag{14-26}$$

where $D/D\tau$ is the substantial derivative. The β is the thermal coefficient of volume expansion of the fluid, q''' is the local heat source per unit volume and time, and Φ_d is the heat production by viscous dissipation. An alternative form in terms of enthalpy is

$$\rho \frac{Dh}{D\tau} = \frac{DP}{D\tau} + \nabla \cdot (k \, \nabla T - q_r) + q''' + \Phi_d \tag{14-27}$$

To obtain the temperature distribution in the medium by solving (14-26), an expression for $\nabla \cdot \mathbf{q}_r$ is needed to relate $\nabla \cdot \mathbf{q}_r$ to the temperature distribution and the scattered radiation within the medium. One approach is to derive $\mathbf{q}_r$ and then differentiate to obtain $\nabla \cdot \mathbf{q}_r$. This will be given later for a plane layer and a long cylinder. Another approach is to obtain $\nabla \cdot \mathbf{q}_r$ directly by considering the local radiative interaction with a differential volume in the medium. This is now carried out. A derivation will be given based on physical reasoning with scattering absent. Then a more mathematical derivation is given with scattering included.

14-4.3 Divergence of Radiative Flux for Absorption Alone (No Scattering)

In many instances scattering can be neglected, so the relations obtained now are quite useful. For no scattering $\Omega_\lambda \to 0$, and the source function reduces to $I'_\lambda = i'_{\lambda b}$. Equation (14-13) then becomes

$$i'_\lambda(\kappa_\lambda) = i'_\lambda(0) \exp{(-\kappa_\lambda)} + \int_0^{\kappa_\lambda} i'_{\lambda b}(\kappa_\lambda^*) \exp{[-(\kappa_\lambda - \kappa_\lambda^*)]} \, d\kappa_\lambda^* \tag{14-28}$$

Equation (14-28) describes the spectral intensity along a single path within a medium. To obtain the net radiative energy supplied to a volume element, consider the energy absorbed by volume element dV in Fig. 14-8. The energy absorbed from the incident intensity $i'_\lambda(\lambda, \omega, \kappa_\lambda)$ arriving within solid angle $d\omega$ is, by analogy with (12-37),

$$d^4 Q'_{\lambda,a} = a_\lambda (dV) i'_\lambda(\lambda, \omega, \kappa_\lambda) dV \, d\lambda \, d\omega \tag{14-29}$$

The incident intensity $i'_\lambda(\lambda, \omega, \kappa_\lambda)$ is given by (14-28) as

$$i'_\lambda(\lambda, \omega, \kappa_\lambda) = i'_\lambda(\lambda, \omega, 0) \exp{(-\kappa_\lambda)} + \int_0^{\kappa_\lambda} i'_{\lambda b}(\lambda, \kappa_\lambda^*) \exp{[-(\kappa_\lambda - \kappa_\lambda^*)]} \, d\kappa_\lambda^* \tag{14-30}$$

where $i'_\lambda(\lambda, \omega, 0)$ is the spectral intensity directed toward dV from the boundary.

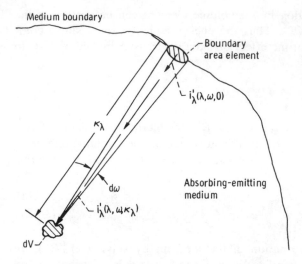

Medium boundary

Boundary area element

$i'_\lambda(\lambda,\omega,0)$

κ_λ

$d\omega$

Absorbing-emitting medium

$i'_\lambda(\lambda,\omega,\kappa_\lambda)$

dV

Figure 14-8 Geometry for derivation of energy-conservation relation.

The energy absorbed by dV from all incident directions is found by integrating (14-29) with respect to ω:

$$d^3Q_{\lambda,a} = a_\lambda(dV)\, dV\, d\lambda \int_{\omega=0}^{4\pi} i'_\lambda(\lambda,\,\omega,\,\kappa_\lambda)\, d\omega \tag{14-31}$$

For convenience in writing the equations more compactly, a *mean incident intensity* $\bar{i}_\lambda(\lambda)$ at dV is defined as

$$\bar{i}_\lambda(\lambda) \equiv \frac{1}{4\pi} \int_0^{4\pi} i'_\lambda(\lambda,\omega,\kappa_\lambda)\, d\omega \tag{14-32}$$

Then (14-31) becomes

$$d^3Q_{\lambda,a} = 4\pi a_\lambda(dV)\bar{i}_\lambda(\lambda)\, dV\, d\lambda \tag{14-33}$$

By integrating (14-33) over all wavelengths, the *total energy absorbed* by dV from the radiation field is

$$d^2Q_a = 4\pi\, dV \int_0^\infty a_\lambda(dV)\bar{i}_\lambda(\lambda)\, d\lambda \tag{14-34}^\dagger$$

The *total energy emitted* spontaneously from dV is obtained from (12-39) by integrating over all wavelengths:

$$d^2Q_e = 4\, dV \int_0^\infty a_\lambda(dV)e_{\lambda b}(\lambda,\,T)\, d\lambda \tag{14-35}$$

---

†Note that the differential on the left is of second order since, in the nomenclature used, dV is of order $dA\, dS$.

Since dV is very small, all the energy emitted by dV escapes before any reabsorption within dV can occur. The net outflow of radiant energy per unit volume, which is the desired *divergence of the radiant heat-flux vector*, is then, from (14-34) and (14-35),

$$\nabla \cdot \mathbf{q}_r = 4 \int_0^\infty a_\lambda(\lambda,T,P) [e_{\lambda b}(\lambda,T) - \pi \bar{\imath}_\lambda(\lambda)] \, d\lambda \tag{14-36}$$

Consider briefly how an analysis would be carried out. The energy equation requires the term $-\nabla \cdot \mathbf{q}_r$, which is the energy supplied locally by the radiative intensities. This term is obtained by using Eq. (14-36). However, this equation contains the average local intensity $\bar{\imath}_\lambda$ received from all directions, which is obtained by carrying out the integral (14-32). The intensity for evaluating this integral is obtained from the transfer equation (14-30) that gives the spectral intensity at each location and in each direction in terms of the initial intensity. The initial intensity accounts for the boundary conditions. A general solution requires determining the spectral intensities at each location and in each direction. Since the temperature is in the blackbody spectral intensity, a simultaneous solution with the energy equation is required. The solution procedure is further elaborated as the chapter proceeds.

14-4.4 Divergence of Radiative Flux Including Scattering

This is an extension of Sec. 14-4.3. Starting with the equation of transfer (14-4), the first term is written as

$$\frac{di'_\lambda}{dS} = \frac{\partial i'_\lambda}{\partial x}\frac{dx}{dS} + \frac{\partial i'_\lambda}{\partial y}\frac{dy}{dS} + \frac{\partial i'_\lambda}{\partial z}\frac{dz}{dS}$$

The $dx/dS = \alpha$, $dy/dS = \delta$, and $dz/dS = \gamma$ where α, δ, γ are the direction cosines of i'_λ in the S direction in an x, y, z coordinate system. Equation (14-4) is integrated at location S over all ω,

$$\int_{\omega=0}^{4\pi} \frac{di'_\lambda}{dS} \, d\omega = \int_{\omega=0}^{4\pi} \left(\frac{\partial i'_\lambda}{\partial x}\alpha + \frac{\partial i'_\lambda}{\partial y}\delta + \frac{\partial i'_\lambda}{\partial z}\gamma \right) d\omega = -(a_\lambda + \sigma_{s\lambda}) \int_{\omega=0}^{4\pi} i'_\lambda(S, \omega) \, d\omega$$

$$+ a_\lambda \int_{\omega=0}^{4\pi} i'_{\lambda b}(S) \, d\omega + \frac{\sigma_{s\lambda}}{4\pi} \int_{\omega=0}^{4\pi} \int_{\omega_i=0}^{4\pi} i'_\lambda(S, \omega_i)\Phi(\lambda, \omega, \omega_i) \, d\omega_i \, d\omega \tag{14-37}$$

Let

$$\bar{\Phi}(\lambda, \omega_i) \equiv \frac{1}{4\pi} \int_{\omega=0}^{4\pi} \Phi(\lambda, \omega, \omega_i) \, d\omega \tag{14-38}$$

The $\bar{\Phi}(\lambda, \omega_i)$ is a measure of how much scattering occurs for radiation incident from the ω_i direction. If $\sigma_{s\lambda}(\omega_i)$ is the linear scattering coefficient for this incident direction, then $\bar{\Phi}(\lambda, \omega_i) = \sigma_{s\lambda}(\omega_i)/\sigma_{s\lambda}$, where $\sigma_{s\lambda} = (1/4\pi) \int_{\omega_i=0}^{4\pi} \sigma_{s\lambda}(\omega_i) \, d\omega_i$. This is for general anisotropic scattering. There are two important simpler cases that apply in practically all instances: (1) anisotropic scattering with scattering *independent* of ω_i and (2) isotropic scattering. The first of these applies for the common situation where scattering particles are randomly oriented in a medium. In this instance $\bar{\Phi}(\lambda) = (1/4\pi) \int_{\omega=0}^{4\pi} \Phi(\lambda, \omega) \, d\omega$ and by use of (12-47), $\bar{\Phi}(\lambda) = 1$. For isotropic scattering $\Phi(\lambda, \omega, \omega_i) = 1$ and hence $\bar{\Phi}(\lambda) = 1$. Hence for both anisotropic scattering with Φ independent of incident direction ω_i and isotropic scattering

$$\bar{\Phi}(\lambda) = 1 \tag{14-39}$$

From Sec. 14-4.1, $\int_{\omega=0}^{4\pi} i'_\lambda(S, \omega)\alpha \, d\omega = q_{r\lambda,x}(S)$, and similarly in the y and z directions. Note that $i'_{\lambda b}$ is independent of angular direction ω. Then (14-37) becomes

$$\int_{\omega=0}^{4\pi} \frac{di'_\lambda}{dS} \, d\omega = \frac{\partial q_{r\lambda,x}}{\partial x} + \frac{\partial q_{r\lambda,y}}{\partial y} + \frac{\partial q_{r\lambda,z}}{\partial z} = \nabla \cdot \mathbf{q}_{r\lambda}$$

$$= -(a_\lambda + \sigma_{s\lambda})4\pi \bar{i}'_\lambda(S) + a_\lambda 4\pi i'_{\lambda b}(S) + \sigma_{s\lambda} \int_{\omega_i=0}^{4\pi} i'_\lambda(S, \omega_i)\bar{\Phi}(\lambda, \omega_i) \, d\omega_i \tag{14-40}$$

To obtain the *local divergence of the total radiative flux*, (14-40) is integrated over all λ to obtain

$$\nabla \cdot \mathbf{q}_r = 4 \int_0^\infty \left\{ a_\lambda(\lambda)e_{\lambda b}(\lambda) - \pi[a_\lambda(\lambda) + \sigma_{s\lambda}(\lambda)]\bar{i}_\lambda(\lambda) \right.$$

$$\left. + \frac{\sigma_{s\lambda}(\lambda)}{4} \int_{\omega_i=0}^{4\pi} i'_\lambda(\lambda, \omega_i)\bar{\Phi}(\lambda, \omega_i) \, d\omega_i \right\} d\lambda \tag{14-41}$$

This is the *generalized form* of (14-36) that *includes anisotropic scattering*.

For *anisotropic scattering with the scattering independent of incident direction* ω_i, and for *isotropic scattering*, it was shown in Eq. (14-39) that $\bar{\Phi}(\lambda) = 1$. Then (14-41) reduces to

$$\nabla \cdot \mathbf{q}_r = 4 \int_0^\infty \left\{ a_\lambda(\lambda)e_{\lambda b}(\lambda) - \pi[a_\lambda(\lambda) + \sigma_{s\lambda}(\lambda)]\bar{i}_\lambda(\lambda) \right.$$

$$\left. + \frac{\sigma_{s\lambda}(\lambda)}{4} \int_{\omega_i=0}^{4\pi} i'_\lambda(\lambda, \omega_i) \, d\omega_i \right\} d\lambda$$

Since $\int_{\omega_i=0}^{4\pi} i'_\lambda(\lambda, \omega_i) \, d\omega_i = 4\pi\bar{i}_\lambda(\lambda)$, the two terms involving scattering cancel and the equation for $\nabla \cdot \mathbf{q}_r$ is,

$$\nabla \cdot \mathbf{q}_r = 4 \int_0^\infty a_\lambda(\lambda)[e_{\lambda b}(\lambda) - \pi\bar{i}_\lambda(\lambda)] \, d\lambda \tag{14-42}$$

which is the same as (14-36). Although there is no scattering coefficient in (14-42), the $\bar{\iota}_\lambda(\lambda)$ is a *function of the scattering* as it is obtained by integrating (14-13) over all ω and λ.

For the two types of scattering that yield Eq. (14-42), consider the situation for radiative propagation in a purely scattering medium that does not absorb or emit. Then $a_\lambda(\lambda) = 0$, and Eq. (14-42) shows that

$$\nabla \cdot \mathbf{q}_r = 0 \tag{14-43}$$

From Eq. (14-41), this is not evident for general anisotropic scattering where scattering depends on incidence angle. However, from energy conservation for pure scattering, Eq. (14-43) must be valid.

For isotropic scattering another form of (14-42) is obtained by using (14-9), which simplifies to (since $\Phi = 1$),

$$I'_\lambda(\kappa_\lambda) = \frac{1}{a_\lambda + \sigma_{s\lambda}} [a_\lambda i'_{\lambda b}(\kappa_\lambda) + \sigma_{s\lambda}\bar{\iota}_\lambda(\kappa_\lambda)] = (1 - \Omega_\lambda)i'_{\lambda b}(\kappa_\lambda) + \Omega_\lambda\bar{\iota}_\lambda(\kappa_\lambda) \tag{14-44}$$

The $\bar{\iota}_\lambda$ in (14-42) is then eliminated to obtain $\nabla \cdot \mathbf{q}_r$ as

$$\nabla \cdot \mathbf{q}_r = 4\pi \int_0^\infty [a_\lambda(\lambda) + \sigma_{s\lambda}(\lambda)] \frac{a_\lambda(\lambda)}{\sigma_{s\lambda}(\lambda)} [i'_{\lambda b}(\lambda) - I'_\lambda(\lambda)] \, d\lambda$$

$$= 4 \int_0^\infty \frac{a_\lambda(\lambda)}{\Omega_\lambda(\lambda)} [e_{\lambda b}(\lambda) - \pi I'_\lambda(\lambda)] \, d\lambda \tag{14-45}$$

14-5 EQUATIONS OF TRANSFER, RADIATIVE FLUX, AND SOURCE FUNCTION FOR A PLANE LAYER

To evaluate the influence of some of the many variables in gas radiation problems, it is convenient to consider a simple geometry. A plane layer is often used, and this will illustrate the use of the preceding general relations. There is a considerable literature for this geometry in engineering and astrophysical publications. The astrophysical interest [1–3] stems from the fact that the earth's atmosphere or the outer radiating layers of the sun can be approximated as a plane layer. A plane layer of thickness D between two boundaries is shown in Fig. 14-9. The temperature and properties of the medium vary *only in the x direction*.

14-5.1 Equation of Transfer

The arbitrary paths S in Fig. 14-9 are at angles θ from the positive x direction. It is convenient to adopt a new notation here. The prime denoting a directional quantity is replaced by $+$ or $-$, corresponding respectively to directions with positive or negative $\cos \theta$; i^+ corresponds to $0 \le \theta \le \pi/2$, and i^- to $\pi/2 \le \theta \le \pi$.

The optical depth $\kappa_\lambda(x)$ is now defined *along the x coordinate as*

$$\kappa_\lambda(x) = \int_0^x K_\lambda(x^*) \, dx^* \tag{14-46}$$

For the + directions the relation between optical positions along the S and x directions is given by

$$\kappa_\lambda(S) = \int_0^S K_\lambda(S^*) \, dS^* = \int_0^{x/\cos\theta} K_\lambda\left(\frac{x^*}{\cos\theta}\right) d\left(\frac{x^*}{\cos\theta}\right) = \frac{1}{\cos\theta} \int_0^x K_\lambda(x^*) \, dx^* = \frac{\kappa_\lambda(x)}{\cos\theta} \tag{14-47}$$

For the minus direction $dS = -dx/\cos(\pi - \theta) = dx/\cos\theta$, so the derivation in (14-47) again applies. Then, with $d\kappa_\lambda(S) = d\kappa_\lambda(x)/\cos\theta$, the equation of transfer (14-10) becomes, for i_λ^+ and i_λ^-,

$$\cos\theta \frac{\partial i_\lambda^+}{\partial\kappa_\lambda(x)} + i_\lambda^+(\kappa_\lambda(x), \theta) = I_\lambda'(\kappa_\lambda(x), \theta) \qquad 0 \le \theta \le \pi/2 \tag{14-48a}$$

$$\cos\theta \frac{\partial i_\lambda^-}{\partial\kappa_\lambda(x)} + i_\lambda^-(\kappa_\lambda(x), \theta) = I_\lambda'(\kappa_\lambda(x), \theta) \qquad \pi/2 \le \theta \le \pi \tag{14-48b}$$

The I_λ' depends on θ because of the angular dependence of the phase function Φ for anisotropic scattering. A partial derivative is used to emphasize that i_λ^+ and i_λ^- depend on both $\kappa_\lambda(x)$ and θ. A convenient substitution is to let $\mu = \cos\theta$; then (14-48a,b) become

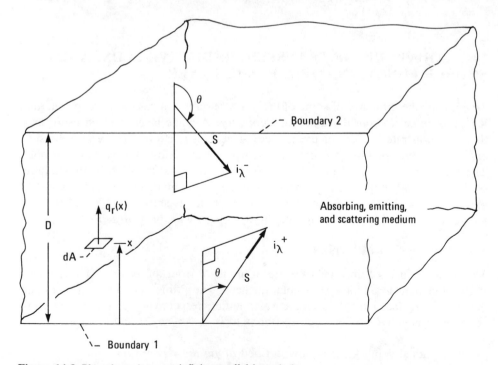

Figure 14-9 Plane layer between infinite parallel boundaries.

$$\mu \frac{\partial i_\lambda^+}{\partial \kappa_\lambda(x)} + i_\lambda^+(\kappa_\lambda(x),\mu) = I_\lambda'(\kappa_\lambda(x),\mu) \tag{14-49a}$$

$$\mu \frac{\partial i_\lambda^-}{\partial \kappa_\lambda(x)} + i_\lambda^-(\kappa_\lambda(x),\mu) = I_\lambda'(\kappa_\lambda(x),\mu) \tag{14-49b}$$

Using an integrating factor as in (14-11), Eqs. (14-49) are integrated subject to the boundary conditions

$$i_\lambda^+(\kappa_\lambda,\mu) = i_*^+(0,\mu) \qquad \text{at } \kappa_\lambda = 0 \tag{14-50a}$$

$$i_\lambda^-(\kappa_\lambda,\mu) = i_\lambda^-(\kappa_{D\lambda},\mu) \qquad \text{at } \kappa_\lambda = \kappa_{D\lambda} \tag{14-50b}$$

where $\kappa_{D\lambda} = \int_0^D K_\lambda \, dx = \int_0^D (a_\lambda + \sigma_{s\lambda}) \, dx$. This gives the integrated forms

$$i_\lambda^+(\kappa_\lambda,\mu) = i_\lambda^+(0,\mu) \exp \frac{-\kappa_\lambda}{\mu} + \int_0^{\kappa_\lambda} I_\lambda'(\kappa_\lambda^*,\mu) \exp \frac{-(\kappa_\lambda - \kappa_\lambda^*)}{\mu} \frac{d\kappa_\lambda^*}{\mu} \tag{14-51a}$$

$$i_\lambda^-(\kappa_\lambda,\mu) = i_\lambda^-(\kappa_{D\lambda},\mu) \exp \frac{\kappa_{D\lambda} - \kappa_\lambda}{\mu} - \int_{\kappa_\lambda}^{\kappa_{D\lambda}} I_\lambda'(\kappa_\lambda^*,\mu) \exp \frac{\kappa_\lambda^* - \kappa_\lambda}{\mu} \frac{d\kappa_\lambda^*}{\mu} \tag{14-51b}$$

Note that in (14-51b) θ is between $\pi/2$ and π so $\mu = \cos\theta$ is negative.

14-5.2 Radiative Flux

The total radiative energy flux in the positive x direction is equal to the integral of the spectral flux over all λ,

$$q_r(x) = \int_0^\lambda dq_{r\lambda}(x) - \int_0^\lambda dq_{r\lambda}(\kappa_\lambda) = \int_0^\lambda \frac{dq_{r\lambda}(\kappa_\lambda)}{d\lambda} \, d\lambda \tag{14-52}$$

Note that for a fixed x the κ_λ will vary with λ since K_λ is a function of λ.

The spectral flux in the positive x direction crossing the plane at x in Fig. 14-9 is found in two parts, one from i_λ^+ and one from i_λ^-. Since intensity represents energy per unit solid angle crossing an area normal to the direction of i', the projection of the area dA must be considered normal to either i_λ^+ or i_λ^-. The spectral energy flux in the positive x direction from i_λ^+ is then (using $d\omega = 2\pi \sin\theta \, d\theta$)

$$\frac{dq_{r\lambda}^+(\kappa_\lambda)}{d\lambda} = 2\pi \int_{\theta=0}^{\pi/2} i_\lambda^+(\kappa_\lambda, \theta) \cos\theta \sin\theta \, d\theta \tag{14-53a}$$

which agrees with the z component in (14-21). Similarly, the spectral flux in the negative x direction arising from i_λ^- is

$$\frac{dq_{r\lambda}^-(\kappa_\lambda)}{d\lambda} = 2\pi \int_{\pi-\theta=0}^{\pi-\theta=\pi/2} i_\lambda^-(\kappa_\lambda, \theta) \cos(\pi - \theta) \sin(\pi - \theta) \, d(\pi - \theta)$$

$$= -2\pi \int_{\theta=\pi/2}^{\pi} i_\lambda^-(\kappa_\lambda, \theta) \cos\theta \sin\theta \, d\theta \tag{14-53b}$$

The net spectral flux in the positive x direction is then

$$\frac{dq_{r\lambda}(\kappa_\lambda)}{d\lambda} = \frac{dq_{r\lambda}^+(\kappa_\lambda)}{d\lambda} - \frac{dq_{r\lambda}^-(\kappa_\lambda)}{d\lambda}$$

$$= 2\pi\left[\int_{\theta=0}^{\pi/2} i_\lambda^+(\kappa_\lambda,\theta)\cos\theta\sin\theta\,d\theta + \int_{\theta=\pi/2}^{\pi} i_\lambda^-(\kappa_\lambda,\theta)\cos\theta\sin\theta\,d\theta\right]$$

(14-54)

Using $\mu = \cos\theta$ yields

$$\frac{dq_{r\lambda}(\kappa_\lambda)}{d\lambda} = 2\pi\left[\int_0^1 i_\lambda^+(\kappa_\lambda,\mu)\mu\,d\mu + \int_{-1}^0 i_\lambda^-(\kappa_\lambda,\mu)\mu\,d\mu\right]$$

$$= 2\pi\int_0^1 [i_\lambda^+(\kappa_\lambda,\mu) - i_\lambda^-(\kappa_\lambda,-\mu)]\mu\,d\mu$$

(14-55)

The intensities are substituted from (14-51) to yield

$$\frac{dq_{r\lambda}(\kappa_\lambda)}{d\lambda} = 2\pi\int_0^1 i_\lambda^+(0,\mu)\exp\left(\frac{-\kappa_\lambda}{\mu}\right)\mu\,d\mu - 2\pi\int_0^1 i_\lambda^-(\kappa_{D\lambda},-\mu)\exp\left(\frac{\kappa_{D\lambda}-\kappa_\lambda}{-\mu}\right)\mu\,d\mu$$

$$+ 2\pi\int_0^1\int_0^{\kappa_\lambda} I_\lambda'(\kappa_\lambda^*,\mu)\exp\frac{\kappa_\lambda-\kappa_\lambda^*}{-\mu}\,d\kappa_\lambda^*\,d\mu$$

$$- 2\pi\int_0^1\int_{\kappa_\lambda}^{\kappa_{D\lambda}} I_\lambda'(\kappa_\lambda^*,-\mu)\exp\frac{\kappa_\lambda^*-\kappa_\lambda}{-\mu}\,d\kappa_\lambda^*\,d\mu$$

(14-56)

14-5.3 Divergence of Radiative Flux

For use in the energy equation (14-26), $\nabla\cdot\mathbf{q}_r$ is needed. For a plane layer $\mathbf{q}_r$ depends only on x so that

$$\nabla\cdot\mathbf{q}_r = \frac{dq_r}{dx} = \frac{d}{dx}\int_{\lambda=0}^{\infty}\frac{dq_{r\lambda}}{d\lambda}\,d\lambda = \int_{\lambda=0}^{\infty}\frac{d^2q_{r\lambda}}{dx\,d\lambda}\,d\lambda$$

With $d\kappa_\lambda = K_\lambda(x)\,dx$ this becomes

$$\nabla\cdot\mathbf{q}_r(x) = \int_0^{\infty} K_\lambda(\kappa_\lambda)\frac{d^2q_{r\lambda}}{d\kappa_\lambda\,d\lambda}\,d\lambda$$

(14-57)

where throughout the integration over λ, the κ_λ *corresponds to the specified* x. Differentiating (14-56) with respect to κ_λ yields

$$\frac{d^2q_{r\lambda}}{d\kappa_\lambda\,d\lambda} = -2\pi \int_0^1 i_\lambda{}^+(0,\mu)\exp\frac{-\kappa_\lambda}{\mu}\,d\mu - 2\pi \int_0^1 i_\lambda{}^-(\kappa_{D\lambda},-\mu)\exp\frac{\kappa_{D\lambda}-\kappa_\lambda}{-\mu}\,d\mu$$

$$-2\pi \int_0^1 \int_0^{\kappa_\lambda} I_\lambda'(\kappa_\lambda^*,\mu)\exp\frac{\kappa_\lambda-\kappa_\lambda^*}{-\mu}\,\frac{d\kappa_\lambda^*}{\mu}\,d\mu + 2\pi \int_0^1 I_\lambda'(\kappa_\lambda,\mu)\,d\mu$$

$$-2\pi \int_0^1 \int_{\kappa_\lambda}^{\kappa_{D\lambda}} I_\lambda'(\kappa_\lambda^*,-\mu)\exp\frac{\kappa_\lambda^*-\kappa_\lambda}{-\mu}\,\frac{d\kappa_\lambda^*}{\mu}\,d\mu + 2\pi \int_0^1 I_\lambda'(\kappa_\lambda,-\mu)\,d\mu$$

$$(14\text{-}58)$$

14-5.4 Equation for Source Function

The source-function equation is obtained from Eq. (14-14). The substitutions are made that $\kappa_\lambda(S) = \kappa_\lambda(x)/\cos\theta$ and $d\omega_i = 2\pi\sin\theta_i\,d\theta_i$. Then by following the procedure in the derivation of the previous plane layer relations, the source-function equation becomes

$$I_\lambda'(\kappa_\lambda,\mu) = (1-\Omega_\lambda)i_{\lambda b}'(\kappa_\lambda) + \frac{\Omega_\lambda}{2}\left[\int_0^1 i_\lambda{}^+(0,\mu_i)\exp\left(\frac{-\kappa_\lambda}{\mu_i}\right)\Phi(\lambda,\mu,\mu_i)\,d\mu_i\right.$$

$$+ \int_0^1 i_\lambda{}^-(\kappa_{D\lambda},-\mu_i)\exp\left(\frac{\kappa_{D\lambda}-\kappa_\lambda}{-\mu_i}\right)\Phi(\lambda,\mu,-\mu_i)\,d\mu_i$$

$$+ \int_0^1 \int_0^{\kappa_\lambda} I_\lambda'(\kappa_\lambda^*,\mu_i)\exp\frac{\kappa_\lambda-\kappa_\lambda^*}{-\mu_i}\,\frac{d\kappa_\lambda^*}{\mu_i}\,\Phi(\lambda,\mu,\mu_i)\,d\mu_i$$

$$\left.+ \int_0^1 \int_{\kappa_\lambda}^{\kappa_{D\lambda}} I_\lambda'(\kappa_\lambda^*,-\mu_i)\exp\frac{\kappa_\lambda^*-\kappa_\lambda}{-\mu_i}\,\frac{d\kappa_\lambda^*}{\mu_i}\,\Phi(\lambda,\mu,-\mu_i)\,d\mu_i\right]\qquad(14\text{-}59)$$

This is a complicated integral equation for the source function. The equation involves $i_{\lambda b}'$, which depends on temperature obtained by solving the energy equation. The energy equation is obtained by substituting (14-58) into (14-57) and inserting the result into Eq. (14-26). It is evident that for a general case involving anisotropic scattering, the simultaneous solution of the energy equation with the source-function equation to obtain T (or $i_{\lambda b}'$) and I_λ' will be difficult. For this reason a number of simplified cases will be considered.

14-5.5 Relations for Anisotropic Scattering Independent of Incidence Angle

For this type of scattering the phase function Φ is independent of μ_i but is still a function of μ. This would exist, for example, when scattering particles are randomly oriented in a medium. Equation (14-58) remains the same but (14-9) and (14-59) reduce to

$$I_\lambda'(\kappa_\lambda, \mu) = (1 - \Omega_\lambda)i_{\lambda b}'(\kappa_\lambda) + \Omega_\lambda \Phi(\lambda, \mu)\bar{i}_\lambda(\kappa_\lambda) \tag{14-60}$$

$$I_\lambda'(\kappa_\lambda, \mu) = (1 - \Omega_\lambda)i_{\lambda b}'(\kappa_\lambda) + \frac{\Omega_\lambda \Phi(\lambda, \mu)}{2}\left[\int_0^1 i_\lambda^+(0, \mu_i) \exp \frac{-\kappa_\lambda}{\mu_i}\, d\mu_i\right.$$

$$+ \int_0^1 i_\lambda^-(\kappa_{D\lambda}, -\mu_i) \exp \frac{\kappa_{D\lambda} - \kappa_\lambda}{-\mu_i}\, d\mu_i$$

$$+ \int_0^1 \int_0^{\kappa_\lambda} I_\lambda'(\kappa_\lambda^*, \mu_i) \exp \frac{\kappa_\lambda - \kappa_\lambda^*}{-\mu_i}\frac{d\kappa_\lambda^*}{\mu_i}\, d\mu_i$$

$$\left. + \int_0^1 \int_{\kappa_\lambda}^{\kappa_{D\lambda}} I_\lambda'(\kappa_\lambda^*, -\mu_i) \exp \frac{\kappa_\lambda^* - \kappa_\lambda}{-\mu_i}\frac{d\kappa_\lambda^*}{\mu_i}\, d\mu_i\right] \tag{14-61}$$

Another form for (14-58) is obtained by noting that the integrals in (14-61) are the same as in (14-58) and by comparison with (14-60) are equal to $2\bar{i}_\lambda(\kappa_\lambda)$. Then (14-58) becomes

$$\frac{d^2 q_{r\lambda}}{d\kappa_\lambda\, d\lambda} = 2\pi\left[\int_0^1 I_\lambda'(\kappa_\lambda, \mu)\, d\mu + \int_0^1 I_\lambda'(\kappa_\lambda, -\mu)\, d\mu - 2\bar{i}_\lambda(\kappa_\lambda)\right]$$

The $I_\lambda'(\kappa_\lambda, \mu)$ is now substituted from (14-60) to yield

$$\frac{d^2 q_{r\lambda}}{d\kappa_\lambda\, d\lambda} = 2\pi\left\{2(1 - \Omega_\lambda)i_{\lambda b}'(\kappa_\lambda) + \Omega_\lambda \bar{i}_\lambda(\kappa_\lambda)\left[\int_0^1 \Phi(\lambda, \mu)\, d\mu\right.\right.$$

$$\left.\left. + \int_0^1 \Phi(\lambda, -\mu)\, d\mu\right] - 2\bar{i}_\lambda(\kappa_\lambda)\right\} \tag{14-62}$$

From the definition of $\bar{\Phi}$ the two integrals on the right are equal to $2\bar{\Phi}$, and from Eq. (12-47) $\bar{\Phi} = 1$. Equation (14-62) then simplifies to

$$\frac{d^2 q_{r,\lambda}}{d\kappa_\lambda\, d\lambda} = 4\pi(1 - \Omega_\lambda)[i_{\lambda b}'(\kappa_\lambda) - \bar{i}_\lambda(\kappa_\lambda)] = 4\pi \frac{a_\lambda}{a_\lambda + \sigma_{s\lambda}}[i_{\lambda b}'(\kappa_\lambda) - \bar{i}_\lambda(\kappa_\lambda)] \tag{14-63}$$

Use $d\kappa_\lambda = (a_\lambda + \sigma_{s\lambda})\, dx$ and integrate over all λ to obtain

$$\frac{dq_r}{dx} = 4\pi \int_0^\infty a_\lambda(\lambda)\,[i_{\lambda b}'(\lambda, x) - \bar{i}_\lambda(\lambda, x)]\, d\lambda \tag{14-64}$$

This is the form in (14-42) and is valid for anisotropic scattering with Φ independent of incident angle and for isotropic scattering.

14-5.6 Relations for Isotropic Scattering

For isotropic scattering the phase function Φ is unity and the source function in (14-9) reduces to

$$I'_\lambda(\kappa_\lambda) = (1 - \Omega_\lambda)i'_{\lambda b}(\kappa_\lambda) + \frac{\Omega_\lambda}{4\pi} \int_{\omega=0}^{4\pi} i'_\lambda(\kappa_\lambda, \omega)\,d\omega \tag{14-65}$$

where it is no longer necessary to use an i subscript on ω. The I'_λ is *independent of angle* in this instance.

The spectral radiative flux (14-56) reduces to

$$\begin{aligned}
\frac{dq_{r\lambda}(\kappa_\lambda)}{d\lambda} = 2\pi\Bigg[&\int_0^1 i_\lambda^+(0, \mu) \exp\left(\frac{-\kappa_\lambda}{\mu}\right)\mu\,d\mu \\
&- \int_0^1 i_\lambda^-(\kappa_{D\lambda}, -\mu) \exp\left(\frac{\kappa_{D\lambda} - \kappa_\lambda}{-\mu}\right)\mu\,d\mu \\
&+ \int_0^{\kappa_\lambda} I'_\lambda(\kappa_\lambda^*) \int_0^1 \exp\frac{\kappa_\lambda - \kappa_\lambda^*}{-\mu}\,d\mu\,d\kappa_\lambda^* \\
&- \int_{\kappa_\lambda}^{\kappa_{D\lambda}} I'_\lambda(\kappa_\lambda^*) \int_0^1 \exp\frac{\kappa_\lambda^* - \kappa_\lambda}{-\mu}\,d\mu\,d\kappa_\lambda^* \Bigg]
\end{aligned}$$

The *exponential integral function* is now introduced. This is defined as

$$E_n(\xi) = \int_0^1 \mu^{n-2} \exp\frac{-\xi}{\mu}\,d\mu \tag{14-66}$$

The $dq_{r\lambda}(\kappa_\lambda)$ is then written as

$$\begin{aligned}
\frac{dq_{r\lambda}(\kappa_\lambda)}{d\lambda} = 2\pi\Bigg[&\int_0^1 i_\lambda^+(0, \mu) \exp\left(\frac{-\kappa_\lambda}{\mu}\right)\mu\,d\mu - \int_0^1 i_\lambda^-(\kappa_{D\lambda}, -\mu) \exp\left(\frac{\kappa_{D\lambda} - \kappa_\lambda}{-\mu}\right)\mu\,d\mu \\
&+ \int_0^{\kappa_\lambda} I'_\lambda(\kappa_\lambda^*)E_2(\kappa_\lambda - \kappa_\lambda^*)\,d\kappa_\lambda^* - \int_{\kappa_\lambda}^{\kappa_{D\lambda}} I'_\lambda(\kappa_\lambda^*)E_2(\kappa_\lambda^* - \kappa_\lambda)\,d\kappa_\lambda^* \Bigg] \tag{14-67}
\end{aligned}$$

The exponential integral functions are discussed in detail by Kourganoff [1] and Chandrasekhar [2]. The important relations are given in Appendix E.

Similarly, by use of the exponential integral function, Eq. (14-58) for the flux divergence reduces to

$$\begin{aligned}
\frac{d^2q_{r\lambda}}{d\kappa_\lambda\,d\lambda} = -2\pi\Bigg[&\int_0^1 i_\lambda^+(0, \mu) \exp\frac{-\kappa_\lambda}{\mu}\,d\mu + \int_0^1 i_\lambda^-(\kappa_{D\lambda}, -\mu) \exp\frac{\kappa_{D\lambda} - \kappa_\lambda}{-\mu}\,d\mu \\
&+ \int_0^{\kappa_{D\lambda}} I'_\lambda(\kappa_\lambda^*)E_1(|\kappa_\lambda - \kappa_\lambda^*|)\,d\kappa_\lambda^* \Bigg] + 4\pi I'_\lambda(\kappa_\lambda) \tag{14-68}
\end{aligned}$$

for isotropic scattering. The source-function equation (14-59) reduces to

$$I'_\lambda(\kappa_\lambda) = (1 - \Omega_\lambda)\, i'_{\lambda b}(\kappa_\lambda) + \frac{\Omega_\lambda}{2} \left[\int_0^1 i_\lambda{}^+(0,\mu) \exp \frac{-\kappa_\lambda}{\mu}\, d\mu \right.$$

$$+ \int_0^1 i_\lambda{}^-(\kappa_{D\lambda}, -\mu) \exp \frac{\kappa_{D\lambda} - \kappa_\lambda}{-\mu}\, d\mu$$

$$\left. + \int_0^{\kappa_{D\lambda}} I'_\lambda(\kappa_\lambda^*)\, E_1(|\kappa_\lambda^* - \kappa_\lambda|)\, d\kappa_\lambda^* \right] \tag{14-69}$$

The expressions in square brackets are the same in Eqs. (14-68) and (14-69). The two equations can then be combined to eliminate the integrals to obtain, as in (14-45)

$$\frac{d^2 q_{r\lambda}}{d\kappa_\lambda\, d\lambda} = 4\pi\, \frac{1 - \Omega_\lambda}{\Omega_\lambda}\, [i'_{\lambda b}(\kappa_\lambda) - I'_\lambda(\kappa_\lambda)] \tag{14-70}$$

Equations (14-63) and (14-64) are also valid.

14-5.7 Diffuse Boundary Fluxes for Plane Layer with Isotropic Scattering

For diffuse boundaries $i_\lambda^+(0, \mu)$ and $i_\lambda^-(\kappa_{D\lambda}, -\mu)$ do not depend on angle (they are independent of μ) and can be expressed in terms of outgoing diffuse fluxes:

$$i_\lambda^+(0, \mu) = i_\lambda^+(0) = \frac{1}{\pi}\frac{dq_{\lambda o,1}}{d\lambda} \qquad i_\lambda^-(\kappa_{D\lambda}, -\mu) = i_\lambda^-(\kappa_{D\lambda}) = \frac{1}{\pi}\frac{dq_{\lambda o,2}}{d\lambda} \tag{14-71}$$

where 1 and 2 correspond to boundaries $\kappa_\lambda = 0$ and $\kappa_{D\lambda}$. Then, in Eqs. (14-67) to (14-69), the $i_\lambda^+(0)$ and $i_\lambda^-(\kappa_{D\lambda})$ are taken out of the integrals over μ and the integrals are expressed in terms of exponential integral functions:

$$\int_0^1 i_\lambda{}^+(0,\mu) \exp\left(\frac{-\kappa_\lambda}{\mu}\right) \mu\, d\mu = \frac{1}{\pi}\frac{dq_{\lambda o,1}}{d\lambda}\, E_3(\kappa_\lambda) \tag{14-72a}$$

$$\int_0^1 i_\lambda{}^-(\kappa_{D\lambda}, -\mu) \exp\left(\frac{\kappa_{D\lambda} - \kappa_\lambda}{-\mu}\right) \mu\, d\mu = \frac{1}{\pi}\frac{dq_{\lambda o,2}}{d\lambda}\, E_3(\kappa_{D\lambda} - \kappa_\lambda) \tag{14-72b}$$

$$\int_0^1 i_\lambda{}^+(0,\mu) \exp\frac{-\kappa_\lambda}{\mu}\, d\mu = \frac{1}{\pi}\frac{dq_{\lambda o,1}}{d\lambda}\, E_2(\kappa_\lambda) \tag{14-73a}$$

$$\int_0^1 i_\lambda{}^-(\kappa_{D\lambda}, -\mu) \exp\frac{\kappa_{D\lambda} - \kappa_\lambda}{-\mu}\, d\mu = \frac{1}{\pi}\frac{dq_{\lambda o,2}}{d\lambda}\, E_2(\kappa_{D\lambda} - \kappa_\lambda) \tag{14-73b}$$

To obtain equations for $dq_{\lambda o,1}$ and $dq_{\lambda o,2}$ in terms of the boundary emissivities, Eq. (14-67) is evaluated at surface 1, $\kappa_\lambda = 0$, to yield

$$\frac{dq_{r\lambda,1}}{d\lambda} = \frac{dq_{\lambda o,1}}{d\lambda} - \left[2\frac{dq_{\lambda o,2}}{d\lambda} E_3(\kappa_{D\lambda}) + 2\pi \int_0^{\kappa_{D\lambda}} I'_\lambda(\kappa_\lambda^*) E_2(\kappa_\lambda^*) \, d\kappa_\lambda^* \right]$$

By comparison with (8-5), $dq_{\lambda,1} = dq_{\lambda o,1} - dq_{\lambda i,1}$, and the term in square brackets is $dq_{\lambda i,1}/d\lambda$. Then using (8-7) gives the equation for $dq_{\lambda o,1}$:

$$\frac{dq_{\lambda o,1}}{d\lambda} = \epsilon_{\lambda,1} e_{\lambda b,1} + 2(1 - \epsilon_{\lambda,1}) \left[\frac{dq_{\lambda o,2}}{d\lambda} E_3(\kappa_{D\lambda}) + \pi \int_0^{\kappa_{D\lambda}} I'_\lambda(\kappa_\lambda^*) E_2(\kappa_\lambda^*) \, d\kappa_\lambda^* \right]$$

$$(14\text{-}74a)$$

Similarly, at surface 2 [by use of (14-67) evaluated at $\kappa_{D\lambda}$]

$$\frac{dq_{\lambda o,2}}{d\lambda} = \epsilon_{\lambda,2} e_{\lambda b,2} + 2(1 - \epsilon_{\lambda,2}) \left[\frac{dq_{\lambda o,1}}{d\lambda} E_3(\kappa_{D\lambda}) + \pi \int_0^{\kappa_{D\lambda}} I'_\lambda(\kappa_\lambda^*) E_2(\kappa_{\lambda D} - \kappa_\lambda^*) \, d\kappa_\lambda^* \right]$$

$$(14\text{-}74b)$$

Thus the boundary $dq_{\lambda o}$ depend on $I'_\lambda(\kappa_\lambda)$, and they enter into the simultaneous solution of the system of equations. The preceding spectral relations are now simplified for the special case of a gray medium.

14-6 PLANE LAYER OF A GRAY ABSORBING AND EMITTING MEDIUM WITH ISOTROPIC SCATTERING

A medium having absorption and scattering coefficients independent of wavelength is a *gray medium*. From the discussion of gas-property spectral variations, Secs. 12-4.4 and 12-8.1, it is evident that gases are usually far from gray. However, in some instances gases may be considered gray over all or a portion of the spectrum. If the temperatures are such that this spectral region contains an appreciable portion of the energy being exchanged, the approximation will be a good one. When particles of soot or other material are present or are injected into a gas to enhance its absorption or emission of radiation, the gas-particle mixture may act nearly gray; for soot, scattering is usually small as discussed in Chap. 13. Examination of the radiative behavior of a gray medium provides an understanding of some features of a real medium without the additional complications that real media introduce. These reasons, along with the mathematical simplifications introduced, account for the gray medium receiving considerable attention in the literature.

The equations for local flux and the source function are now written for a *gray medium* with *isotropic scattering*. The equations are expressed in terms of the total quantities:

$$i' = \int_0^\infty i'_\lambda(\lambda) \, d\lambda, \qquad \frac{\sigma T^4}{\pi} = i'_b = \int_0^\infty i'_{\lambda b}(\lambda) \, d\lambda, \qquad I' = \int_0^\infty I'_\lambda(\lambda) \, d\lambda$$

In Eq. (14-65) for the source function, the Ω_λ and κ_λ become Ω and κ. Then integrating over all wavelengths gives

$$I'(\kappa) = (1 - \Omega)\frac{\sigma T^4(\kappa)}{\pi} + \frac{\Omega}{4\pi}\int_{\omega=0}^{4\pi} i'(\kappa, \omega)\, d\omega = (1 - \Omega)\frac{\sigma T^4(\kappa)}{\pi} + \Omega\bar{i}(\kappa)$$

(14-75)

In the flux equation (14-67), note that the boundary values $i_\lambda^+(0, \mu)$ and $i_\lambda^-(\kappa_D, -\mu)$ can still have a spectral dependency (the boundaries have not been assumed gray). Integrating over all λ gives

$$q_r(\kappa) = 2\pi \int_0^1 i^+(0,\mu) \exp\left(\frac{-\kappa}{\mu}\right)\mu\, d\mu - 2\pi \int_0^1 i^-(\kappa_D,-\mu) \exp\left(\frac{\kappa_D - \kappa}{-\mu}\right)\mu\, d\mu$$

$$+ 2\pi \int_0^\kappa I'(\kappa^*) E_2(\kappa - \kappa^*)\, d\kappa^* - 2\pi \int_\kappa^{\kappa_D} I'(\kappa^*) E_2(\kappa^* - \kappa)\, d\kappa^* \qquad (14\text{-}76)$$

Equation (14-68) for the divergence of the radiative flux becomes

$$\frac{dq_r}{d\kappa} = -2\pi\left[\int_0^1 i^+(0, \mu) \exp\frac{-\kappa}{\mu}\, d\mu + \int_0^1 i^-(\kappa_D, -\mu) \exp\frac{\kappa_D - \kappa}{-\mu}\, d\mu\right.$$

$$\left. + \int_0^{\kappa_D} I'(\kappa^*) E_1(|\kappa - \kappa^*|)\, d\kappa^*\right] + 4\pi I'(\kappa) \qquad (14\text{-}77)$$

The integral equation (14-69) for the source function is

$$I'(\kappa) = (1 - \Omega)\frac{\sigma T^4(\kappa)}{\pi} + \frac{\Omega}{2}\left[\int_0^1 i^+(0, \mu) \exp\frac{-\kappa}{\mu}\, d\mu\right.$$

$$\left. + \int_0^1 i^-(\kappa_D, -\mu) \exp\frac{\kappa_D - \kappa}{-\mu}\, d\mu + \int_0^{\kappa_D} I'(\kappa^*) E_1(|\kappa^* - \kappa|)\, d\kappa^*\right] \qquad (14\text{-}78)$$

By eliminating the integrals from (14-77) and (14-78) or from (14-70) the flux divergence has the convenient form

$$\frac{dq_r}{d\kappa} = 4\pi\frac{1 - \Omega}{\Omega}\left[\frac{\sigma T^4(\kappa)}{\pi} - I'(\kappa)\right]$$

(14-79)

By use of (14-75) the $I'(\kappa)$ can be eliminated to provide the form

$$\frac{dq_r}{d\kappa} = 4\pi(1 - \Omega)\left[\frac{\sigma T^4(\kappa)}{\pi} - \bar{i}(\kappa)\right]$$

(14-80)

An interesting result for a *purely scattering medium* is obtained by first eliminating $\sigma T^4(\kappa)$ from (14-79) and (14-80) to give

$$\frac{dq_r}{d\kappa} = 4\pi[I'(\kappa) - \bar{\iota}(\kappa)] \tag{14-81}$$

It is now observed from (14-75) and then from (14-81) that in the limit of zero absorption, so there is only isotropic scattering ($\Omega = 1$),

$$I'(\kappa) = \bar{\iota}(\kappa) \qquad \text{and} \qquad \frac{dq_r}{d\kappa} = 0 \qquad (\Omega = 1) \tag{14-82a,b}$$

The $dq_r/d\kappa = 0$ was shown earlier by Eq. (14-43).

EXAMPLE 14-3 Starting with Eq. (14-81), derive Eq. (14-77) for a gray absorbing and isotropically scattering one-dimensional layer.

From (14-32) and in a fashion similar to (14-55),

$$4\pi\bar{\iota} = \int_0^{4\pi} i' \, d\omega = 2\pi \int_{\theta=0}^{\pi} i' \sin\theta \, d\theta = 2\pi \left[\int_0^1 i^+(\mu) \, d\mu + \int_0^1 i^-(-\mu) \, d\mu \right]$$

For a gray absorbing and isotropically scattering medium, (14-51) gives i^+ and i^- as

$$i^+(\mu) = i^+(0,\mu) \exp\frac{-\kappa}{\mu} + \int_0^\kappa I'(\kappa^*) \exp\frac{-(\kappa - \kappa^*)}{\mu} \frac{d\kappa^*}{\mu}$$

$$i^-(\mu) = i^-(\kappa_D,\mu) \exp\frac{\kappa_D - \kappa}{\mu} - \int_\kappa^{\kappa_D} I'(\kappa^*) \exp\frac{\kappa^* - \kappa}{\mu} \frac{d\kappa^*}{\mu}$$

Then

$$4\pi\bar{\iota} = 2\pi \left\{ \int_0^1 i^+(0,\mu) \exp\frac{-\kappa}{\mu} \, d\mu + \int_0^1 i^-(\kappa_D,-\mu) \exp\frac{\kappa_D - \kappa}{-\mu} \, d\mu \right.$$

$$+ \int_0^1 \int_0^\kappa I'(\kappa^*) \exp\left[\frac{-(\kappa - \kappa^*)}{\mu}\right] \frac{1}{\mu} \, d\mu \, d\kappa^*$$

$$\left. + \int_0^1 \int_\kappa^{\kappa_D} I'(\kappa^*) \exp\left(\frac{\kappa^* - \kappa}{-\mu}\right) \frac{1}{\mu} \, d\mu \, d\kappa^* \right\} \tag{14-83}$$

Equation (14-83) is inserted into (14-81), and the result is (14-77).

To illustrate the use of the transfer equations for a plane layer, the dimensionless heat emission $\bar{Q}$ is calculated for a plane layer of gray medium at uniform temperature T_g with emission, absorption, and isotropic scattering, in surroundings at T_e (Fig. 14-10). The absorption and scattering coefficients are constant throughout the layer. This situation has application to the dissipation of waste heat in outer space by use of radiation from sheets of liquid droplets traveling through space [4, 5]. The $\bar{Q}$ is defined as $\bar{Q} \equiv q_r/[\sigma(T_g^4 - T_e^4)]$. This definition

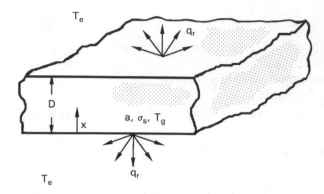

Figure 14-10 Plane layer of emitting, absorbing, and scattering medium at uniform temperature.

is used as it yields $\tilde{Q}$ independent of T_e. The q_r can be found by evaluating (14-76) at $\kappa = \kappa_D$ and having the surroundings act like a black background at T_e so that $i^+(0, \mu) = i^-(\kappa_D, -\mu) = \sigma T_e^4/\pi$. The result can be arranged into the form

$$\tilde{Q} = 2 \int_0^{\kappa_D} \frac{[\pi I'(\kappa) - \sigma T_e^4]}{\sigma(T_g^4 - T_e^4)} E_2(\kappa_D - \kappa) \, d\kappa \tag{14-84}$$

The $I'(\kappa)$ is obtained from the integral equation (14-78), which becomes

$$I'(\kappa) = (1 - \Omega) \frac{\sigma T_g^4}{\pi} + \frac{\Omega}{2} \left\{ \frac{\sigma T_e^4}{\pi} [E_2(\kappa) + E_2(\kappa_D - \kappa)] \right.$$

$$\left. + \int_0^{\kappa_D} I'(\kappa^*) E_1(|\kappa^* - \kappa|) \, d\kappa^* \right\}$$

This is arranged into the form

$$\frac{\pi I'(\kappa) - \sigma T_e^4}{\sigma(T_g^4 - T_e^4)} = 1 - \Omega + \frac{\Omega}{2} \int_0^{\kappa_D} \frac{\pi I'(\kappa^*) - \sigma T_e^4}{\sigma(T_g^4 - T_e^4)} E_1(|\kappa^* - \kappa|) \, d\kappa^* \tag{14-85}$$

It is thus convenient to use the dimensionless variable $[\pi I'(\kappa) - \sigma T_e^4]/[\sigma(T_g^4 - T_e^4)]$, and this integral equation can be solved numerically by iteration. The results are inserted into (14-84) for $\tilde{Q}$. For the special case when $\Omega = 0$, Eq. (14-85) shows that $I'(\kappa) = \sigma T_g^4/\pi$, so $\tilde{Q} = 2 \int_0^{\kappa_D} E_2(\kappa_D - \kappa) \, d\kappa = 1 - 2E_3(\kappa_D)$. The $\tilde{Q}$ is a function of κ_D and Ω where $\kappa_D = (a + \sigma_s)D$. Some results from [5] are in Table 14-1. For each Ω the $\tilde{Q}$ values increase with κ_D asymptotically to a maximum value. This maximum decreases as the scattering albedo Ω is increased.

Another situation of interest is the *transient radiative cooling* of a layer such as in Fig. 14-10. This will show the extent to which the layer emissive ability decreases as the outer portions of the layer become cool and do not radiate as well as the inner regions. If the layer consists of a dispersion of particles or drops, heat conduction in the medium is small, and if convection is also small, radiative transfer will dominate. From the energy equation (14-26),

$$\rho c_p \frac{\partial T}{\partial \tau} = -\frac{\partial q_r}{\partial x} = -(a + \sigma_s)\frac{\partial q_r}{\partial \kappa} \tag{14-86}$$

Using (14-79) for $\partial q_r/\partial \kappa$ yields

$$\frac{\rho c_p}{a + \sigma_s}\frac{\partial T(\kappa, \tau)}{\partial \tau} = 4\frac{1 - \Omega}{\Omega}[\pi I'(\kappa, \tau) - \sigma T^4(\kappa, \tau)] \tag{14-87}$$

The $I'(\kappa)$ is related to $T^4(\kappa)$ by the integral equation (14-78)

$$I'(\kappa, \tau) = (1 - \Omega)\frac{\sigma T^4(\kappa, \tau)}{\pi} + \frac{\Omega}{2}\left\{\frac{\sigma T_e^4}{\pi}[E_2(\kappa) + E_2(\kappa_D - \kappa)]\right.$$

$$\left. + \int_0^{\kappa_D} I'(\kappa^*, \tau)E_1(|\kappa^* - \kappa|)\, d\kappa^*\right\} \tag{14-88}$$

Eqs. (14-87) and (14-88) can be solved starting from a specified initial temperature distribution $T(\kappa, 0)$. This is inserted into (14-88), which is solved numerically by iteration for $I'(\kappa, 0)$. By using this along with $T^4(\kappa, 0)$, Eq. (14-87) is used to extrapolate to $T(\kappa, \tau + \Delta\tau)$. This is inserted into (14-88) to find $I'(\kappa, \tau + \Delta\tau)$ and thereby continue the transient solution. Numerical techniques are in Chap. 17.

An interesting special solution is found when the surroundings are at low temperature, $(T_e/T)^4 \ll 1$. This solution yields the shape of the $I'(\kappa)$ and $T(\kappa)$ distributions and the layer emittance values that are ultimately achieved during transient radiative cooling. A solution is tried of the form

$$[\sigma T^4(\kappa, \tau)]^{1/4} = \Theta(\tau)F(\kappa), \qquad [\pi I'(\kappa, \tau)]^{1/4} = \Theta(\tau)G(\kappa) \tag{14-89a,b}$$

Then (14-87) and (14-88) become

$$\frac{\rho c_p}{\sigma^{1/4}(a + \sigma_s)}\frac{d\Theta}{d\tau}\frac{1}{\Theta^4(\tau)} = 4\frac{1 - \Omega}{\Omega}\frac{G^4(\kappa) - F^4(\kappa)}{F(\kappa)} \tag{14-90}$$

Table 14-1 $\bar{Q}$ values for plane layer at uniform temperature [5]

Optical thickness, κ_D	Scattering albedo, Ω					
	0	0.30	0.60	0.80	0.90	0.95
0.2	0.296	0.225	0.140	0.0748	0.0386	0.0197
0.5	0.557	0.449	0.303	0.172	0.0926	0.0401
1.0	0.781	0.667	0.490	0.304	0.173	0.0926
2	0.940	0.846	0.681	0.475	0.297	0.170
3	0.982	0.900	0.757	0.566	0.382	0.233
4	0.994	0.918	0.786	0.612	0.436	0.281
5	0.998	0.924	0.798	0.637	0.470	0.317
10	1.000	0.933	0.808	0.659	0.518	0.389

$$G^4(\kappa) = (1 - \Omega)F^4(\kappa) + \frac{\Omega}{2} \int_0^{\kappa_D} G^4(\kappa^*)E_1(|\kappa^* - \kappa|) \, d\kappa^* \qquad (14\text{-}91)$$

In (14-90) the functions of κ and τ have been separated, so the functions on each side of the equation must be a constant. Then from the right side of (14-90)

$$\frac{G^4(\kappa) - F^4(\kappa)}{F(\kappa)} = \frac{G^4(0) - F^4(0)}{F(0)} \qquad (14\text{-}92)$$

This is solved simultaneously with (14-91) to obtain $F(\kappa)$ and $G(\kappa)$. The details of the numerical solution are in [6]. The emittance reached in this "fully developed" transient region is defined as $\epsilon_{fd} = q_r(\kappa = \kappa_D, \tau)/\sigma T_m^4(\tau)$ where $T_m(\tau)$ is the integrated mean temperature across the layer. The ϵ_{fd} is independent of time and is a function only of κ_D and Ω. Values computed in [6] are in Table 14-2. The ϵ_{fd} are lower than the ϵ for uniform temperature in Table 14-1 because of the relatively larger cooling of the outer portions of the layer during transient cooling. This results in poor radiative ability compared with that characterized by the mean temperature of the layer. For each Ω the ϵ_{fd} first increases with increasing κ_D and then decreases as κ_D becomes large and the effect of cooling the outer regions of the layer becomes more important.

14-7 RADIATIVE EQUILIBRIUM

The energy equation (14-26) is now considered for the special conditions of steady state without significant heat conduction, convection, viscous dissipation, or internal heat sources. For this steady state situation where all other energy exchange mechanisms are negligible compared with radiation, the total emitted energy from each volume element must equal its total absorbed energy. This condition is termed *radiative equilibrium* and is simply a statement of steady-state energy conservation in the absence of any other exchange mechanism but radiation. With only radiation present, Eq. (14-26) yields for steady state

$$\nabla \cdot \mathbf{q}_r = 0 \qquad (14\text{-}93)$$

Table 14-2 Values of layer emittance in fully developed transient region, $\epsilon_{fd}(\kappa_D, \Omega)$ [6]

Optical thickness, κ_D	Scattering albedo, Ω						
	0	0.30	0.60	0.80	0.90	0.95	0.98
1	0.772	0.662	0.489	0.304	0.173	0.093	0.039
2	0.894	0.816	0.669	0.472	0.297	0.170	0.075
3	0.882	0.830	0.722	0.555	0.379	0.232	0.107
5	0.777	0.753	0.696	0.592	0.456	0.313	0.161
7	0.672	0.659	0.627	0.563	0.468	0.351	0.200
10	0.551	0.544	0.529	0.496	0.440	0.360	0.233
14	0.440	0.437	0.430	0.414	0.385	0.338	0.248

Then from (14-42), which includes effects of scattering (isotropic or anisotropic with Φ independent of ω_i),

$$\int_0^\infty a_\lambda(\lambda,T,P)\,e_{\lambda b}(\lambda,T)\,d\lambda = \pi \int_0^\infty a_\lambda(\lambda,T,P)\,\bar{\imath}_\lambda(\lambda)\,d\lambda \qquad (14\text{-}94)$$

14-7.1 Some Mean Absorption Coefficients

For the emission integral on the left of (14-94) it is convenient to define the *Planck mean absorption coefficient* $a_P(T, P)$:

$$a_P(T,P) \equiv \frac{\int_0^\infty a_\lambda(\lambda,T,P)\,e_{\lambda b}(\lambda,T)\,d\lambda}{\int_0^\infty e_{\lambda b}(\lambda,T)\,d\lambda} = \frac{\int_0^\infty a_\lambda(\lambda,T,P)\,e_{\lambda b}(\lambda,T)\,d\lambda}{\sigma T^4} \qquad (14\text{-}95)$$

The a_P is the mean of the spectral coefficient weighted by the blackbody (Planckian) emission spectrum. It is useful in considering *emission* from a volume, and in certain special cases of radiative transfer.

Substituting (14-95) into Eq. (14-94) results in the radiative equilibrium relation,

$$a_P(T,P)\,\sigma T^4 = \pi \int_0^\infty a_\lambda(\lambda,T,P)\,\bar{\imath}_\lambda(\lambda)\,d\lambda \qquad (14\text{-}96)$$

If $\bar{\imath}_\lambda(\lambda)$ is known at a location, (14-96) can be solved for T at that location. The Planck mean a_P is convenient since it depends only on the properties at dV. It can be tabulated and is especially useful where the pressure is constant over the volume.

For the absorption integral on the right side of (14-94), an *incident mean absorption coefficient* $a_i(T, P)$ can be defined,

$$a_i(T,P) = \frac{\int_0^\infty a_\lambda(\lambda,T,P)\,\bar{\imath}_\lambda(\lambda)\,d\lambda}{\int_0^\infty \bar{\imath}_\lambda(\lambda)\,d\lambda} \qquad (14\text{-}97)$$

However, this has little value for general use. A tabulation of a_i would be needed for many combinations of incident spectral distributions and spectral variations of local absorption coefficients. Except in very limited special cases, this would not be warranted. Further discussion of the physical interpretation of various mean absorption coefficients is in Sec. 15-4.1.

For the special case of a gray gas, $a_\lambda = a_P$ and (14-96) reduces to the convenient relation *for radiative equilibrium*,

$$\sigma T^4 = \pi \int_0^\infty \bar{\imath}_\lambda\,d\lambda = \frac{1}{4}\int_{\omega=0}^{4\pi}\int_0^\infty i'_\lambda\,d\lambda\,d\omega = \frac{1}{4}\int_{\omega=0}^{4\pi} i'\,d\omega \qquad (14\text{-}98)$$

where i' is the total intensity, $\int_0^\infty i'_\lambda\,d\lambda$.

14-7.2 The Plane Layer of Gray Medium with $\nabla \cdot \mathbf{q}_r = 0$

For no convection, conduction, or internal heating, the energy equation (14-26) or (14-93) gives for a plane layer of gray medium at steady state

$$\nabla \cdot \mathbf{q}_r = \frac{dq_r(x)}{dx} = 0 \quad \text{or} \quad \frac{dq_r(\kappa)}{d\kappa} = 0 \tag{14-99}$$

In an absorbing-emitting medium with or without scattering this is *radiative equilibrium*. The q_r is the total heat flux being transferred, since radiation is the only transfer mode present.

14-7.3 Absorbing Medium in Radiative Equilibrium with Isotropic Scattering

From Eq. (14-79) with $dq_r/d\kappa = 0$,

$$I'(\kappa) = \frac{\sigma T^4(\kappa)}{\pi} \tag{14-100}$$

so that for radiative equilibrium with a nonzero absorption coefficient the source function is equal to the local blackbody intensity. With this relation used to eliminate $I'(\kappa)$, and with $dq_r/d\kappa = 0$, the same integral equation for $T^4(\kappa)$ results from either (14-77) or (14-78):

$$4\sigma T^4(\kappa) = 2\pi \int_0^1 i^+(0,\mu) \exp\frac{-\kappa}{\mu} \, d\mu + 2\pi \int_0^1 i^-(\kappa_D,-\mu) \exp\frac{\kappa_D - \kappa}{-\mu} \, d\mu$$

$$+ 2 \int_0^{\kappa_D} \sigma T^4(\kappa^*) E_1(|\kappa - \kappa^*|) \, d\kappa^* \tag{14-101}$$

The radiative flux equation (14-76) becomes, with $I' = \sigma T^4/\pi$,

$$q_r = 2\pi \int_0^1 i^+(0,\mu) \exp\left(\frac{-\kappa}{\mu}\right) \mu \, d\mu - 2\pi \int_0^1 i^-(\kappa_D,-\mu) \exp\left(\frac{\kappa_D - \kappa}{-\mu}\right) \mu \, d\mu$$

$$+ 2 \int_0^\kappa \sigma T^4(\kappa^*) E_2(\kappa - \kappa^*) \, d\kappa^* - 2 \int_\kappa^{\kappa_D} \sigma T^4(\kappa^*) E_2(\kappa^* - \kappa) \, d\kappa^* \tag{14-102}$$

Thus for a medium in radiative equilibrium with isotropic scattering or without scattering, and in which the absorption coefficient is nonzero, a *single* integral equation (14-101) governs the temperature distribution. After the temperature distribution has been obtained, it is inserted into (14-102) to obtain q_r. Since $dq_r/d\kappa = 0$, q_r is constant (does not depend on κ). Hence (14-102) can be evaluated at any convenient κ, such as $\kappa = 0$. Now the special case is considered in which there is scattering, but absorption is zero.

14-7.4 Isotropically Scattering Medium with Zero Absorption

Consider a medium with zero absorption coefficient so energy transfer is only by scattering, which is assumed isotropic. For scattering only, $\Omega \to 1$ and from Eq. (14-82) $I'(\kappa) = \bar{i}(\kappa)$ where $\bar{i}(\kappa)$ is the average scattered intensity at κ, and $\kappa = \int_0^x \sigma_s \, dx$. Equation (14-78) reduces to an integral equation for $\bar{i}(\kappa)$:

$$
\bar{i}(\kappa) = \frac{1}{2} \left[\int_0^1 i^+(0, \mu) \exp \frac{-\kappa}{\mu} \, d\mu + \int_0^1 i^-(\kappa_D, -\mu) \exp \frac{\kappa_D - \kappa}{-\mu} \, d\mu \right.
$$

$$
\left. + \int_0^{\kappa_D} \bar{i}(\kappa^*) E_1(|\kappa - \kappa^*|) \, d\kappa^* \right] \tag{14-103}
$$

The radiative flux equation (14-76) becomes

$$
q_r = 2\pi \left[\int_0^1 i^+(0, \mu) \exp\left(\frac{-\kappa}{\mu}\right) \mu \, d\mu - \int_0^1 i^-(\kappa_D, -\mu) \exp\left(\frac{\kappa_D - \kappa}{-\mu}\right) \mu \, d\mu \right.
$$

$$
\left. + \int_0^{\kappa} \bar{i}(\kappa^*) E_2(\kappa - \kappa^*) \, d\kappa^* - \int_{\kappa}^{\kappa_D} \bar{i}(\kappa^*) E_2(\kappa^* - \kappa) \, d\kappa^* \right] \tag{14-104}
$$

Equation (14-82) shows that the divergence of the radiative flux is zero, $dq_r/d\kappa = 0$, so q_r in (14-104) is constant. This result for pure scattering does not require radiative equilibrium; it does not rely on the absence of other heat transfer modes. The physical meaning is that the absence of absorption and emission removes any means to add or subtract from the one-dimensional transfer of radiant energy by isotropic scattering.

14-7.5 Gray Medium with $dq_r/dx = 0$ between Diffuse Gray Boundaries

For *diffuse gray boundaries* $i^+(0, \mu) = q_{o,1}/\pi$, $i^-(\kappa_D, -\mu) = q_{o,2}/\pi$, and (14-74) can be integrated over all λ, which eliminates the λ subscripts ($dq_{\lambda o}/d\lambda$ becomes q_o). For an absorbing medium with or without scattering, relations (14-74) are used with (14-101) and (14-102). For a scattering medium without absorption, (14-103) and (14-104) are used. The result is the following equation for the net radiative flux in the x direction:

$$
q_r = 2 \left[q_{o,1} E_3(\kappa) - q_{o,2} E_3(\kappa_D - \kappa) + \int_0^{\kappa} G(\kappa^*) E_2(\kappa - \kappa^*) \, d\kappa^* \right.
$$

$$
\left. - \int_{\kappa}^{\kappa_D} G(\kappa^*) E_2(\kappa^* - \kappa) \, d\kappa^* \right] \tag{14-105}
$$

where

$$
G(\kappa) =
\begin{cases}
\sigma T^4(\kappa) \\[2em]
\pi \bar{\imath}(\kappa)
\end{cases}
\qquad
\kappa =
\begin{cases}
\displaystyle\int_0^x (a + \sigma_s)\, dx & \text{when } a > 0,\ \sigma_s \geqslant 0 \\[2em]
\displaystyle\int_0^x \sigma_s\, dx & \text{when } a = 0,\ \sigma_s > 0
\end{cases}
$$

The heat flux is constant across the layer so q_r can be evaluated at any convenient location such as $\kappa = 0$. The $G(\kappa)$ is found from the integral equation

$$
G(\kappa) = \frac{1}{2}\left[q_{o,1}E_2(\kappa) + q_{o,2}E_2(\kappa_D - \kappa) + \int_0^{\kappa_D} G(\kappa^*)E_1(|\kappa - \kappa^*|)\, d\kappa^*\right] \tag{14-106}
$$

The $q_{o,1}$ and $q_{o,2}$ are obtained by starting with (14-74)

$$
q_{o,1} = \epsilon_1 \sigma T_1^4 + 2(1 - \epsilon_1)\left[q_{o,2}E_3(\kappa_D) + \int_0^{\kappa_D} G(\kappa^*)E_2(\kappa^*)\, d\kappa^*\right] \tag{14-107a}
$$

$$
q_{o,2} = \epsilon_2 \sigma T_2^4 + 2(1 - \epsilon_2)\left[q_{o,1}E_3(\kappa_D) + \int_0^{\kappa_D} G(\kappa^*)E_2(\kappa_D - \kappa^*)\, d\kappa^*\right] \tag{14-107b}
$$

By use of (14-105) evaluated at $\kappa = 0$ for (14-107a), and at $\kappa = \kappa_D$ for (14-107b), Eqs. (14-107) become

$$
q_{o,1} = \epsilon_1 \sigma T_1^4 - (1 - \epsilon_1)(q_r - q_{o,1}), \qquad q_{o,2} = \epsilon_2 \sigma T_2^4 + (1 - \epsilon_2)(q_r + q_{o,2})
$$

Solving for $q_{o,1}$ and $q_{o,2}$ gives the convenient relations

$$
q_{o,1} = \sigma T_1^4 - \frac{1 - \epsilon_1}{\epsilon_1} q_r, \qquad q_{o,2} = \sigma T_2^4 + \frac{1 - \epsilon_2}{\epsilon_2} q_r \tag{14-108a,b}
$$

14-7.6 Solution for Gray Medium with $dq_r/dx = 0$ between Black or Diffuse-Gray Walls at Specified Temperatures

Let the layer of gray medium with absorption coefficient $a(T)$ and scattering coefficient $\sigma_s(T)$ be between two infinite parallel plates at specified temperatures T_1 and T_2; the plates are D apart (Fig. 14-11). It is desired to obtain the temperature distribution in the medium and the energy transfer q between the plates. Since the energy transfer is only by radiation, the total heat flow q is equal to q_r, and the r subscript is omitted in what follows.

Medium between black walls For black walls $\epsilon_1 = \epsilon_2 = 1$, $q_{o,1}$ and $q_{o,2}$ in (14-108) become σT_1^4 and σT_2^4. Evaluating (14-105) at the convenient location $\kappa = 0$ then gives the net heat flux from wall 1 to wall 2 as

$$q = \sigma T_1^4 - 2\sigma T_2^4 E_3(\kappa_D) - 2 \int_0^{\kappa_D} G(\kappa^*) E_2(\kappa^*) \, d\kappa^* \tag{14-109}$$

where from (14-106), $G(\kappa)$ is found from the integral equation

$$G(\kappa) = \frac{1}{2}\left[\sigma T_1^4 E_2(\kappa) + \sigma T_2^4 E_2(\kappa_D - \kappa) + \int_0^{\kappa_D} G(\kappa^*) E_1(|\kappa - \kappa^*|) \, d\kappa^*\right] \tag{14-110}$$

Equations (14-109) and (14-110) are placed in dimensionless forms by defining

$$\phi_b(\kappa) = \frac{G(\kappa)/\sigma - T_2^4}{T_1^4 - T_2^4} \qquad \psi_b = \frac{q}{\sigma(T_1^4 - T_2^4)} \tag{14-111}$$

where the b subscript emphasizes that this is for black walls. This yields

$$\phi_b(\kappa) = \frac{1}{2}\left[E_2(\kappa) + \int_0^{\kappa_D} \phi_b(\kappa^*) E_1(|\kappa - \kappa^*|) \, d\kappa^*\right] \tag{14-112}$$

$$\psi_b = 1 - 2 \int_0^{\kappa_D} \phi_b(\kappa) E_2(\kappa) \, d\kappa \tag{14-113}$$

The solution for $\phi_b(\kappa)$ is obtained from (14-112) by numerical or other means and then $\phi_b(\kappa)$ is used in (14-113) to obtain ψ_b.

For the limiting case as both absorption and scattering in the medium become very small, $\kappa_D \to 0$ and $E_3(\kappa_D) \to 1/2$ so (14-109) reduces to

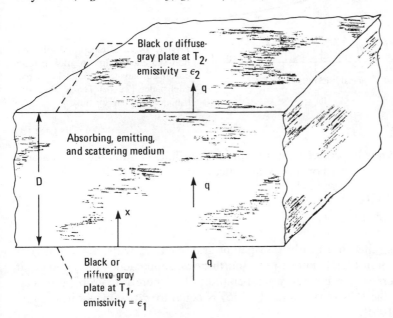

Figure 14-11 Plane layer of medium in radiative equilibrium between infinite parallel diffuse surfaces.

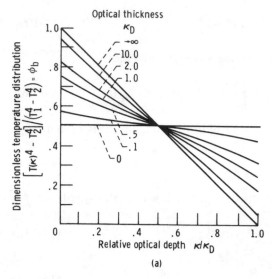

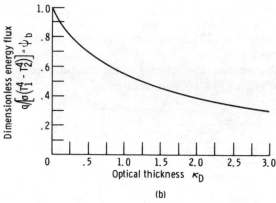

Figure 14-12 Temperature distribution and energy flux in gray medium contained between infinite black parallel plates [7]. (*a*) Temperature distribution; (*b*) energy flux.

$$q|_{\kappa_D \to 0} = \sigma(T_1^4 - T_2^4) \tag{14-114}$$

(or $\psi_b = 1$) which is the correct solution for black infinite parallel plates separated by a transparent medium. For this limit, since $E_2(0) = 1$, (14-110) yields

$$\frac{G(\kappa)}{\sigma}\bigg|_{\kappa_D \to 0} = \frac{T_1^4 + T_2^4}{2} \tag{14-115}$$

(or $\phi_b = 1/2$) so that in a nearly transparent gray medium with $a > 0$ [$G(\kappa) = \sigma T^4(\kappa)$] the medium temperature to the fourth power approaches the average of the fourth powers of the boundary temperatures. In a purely isotropically scattering medium, the $[G(\kappa)/\sigma]_{\kappa_D \to 0}$ in (14-115) is equal to $\pi \bar{i}(\kappa)/\sigma$ within the medium [Eq. (14-105)].

For convenience in the following discussion, consider the case where $a > 0$ so the results can be interpreted in terms of the medium temperature [$G(\kappa) =$

$\sigma T^4(\kappa)$]. The same results apply to $\pi \bar{i}(\kappa)$ for scattering alone. Numerical results from (14-112) and (14-113) for the temperature distribution and heat flux in a gray gas with constant properties between infinite black parallel plates have been obtained by many investigators. Some solution methods will be discussed in succeeding chapters. Heaslet and Warming [7] presented results accurate to four significant figures. These are in Fig. 14-12, and values for the dimensionless energy flux are in Table 14-3.

The temperature distributions of Fig. 14-12a show there is a discontinuity between the wall temperature and the temperature of the medium at the wall. This is called a temperature "slip" or "jump." If the slip were not present, the curves would all go to 1 at $\kappa/\kappa_D = 0$, and to 0 at $\kappa/\kappa_D = 1$. The slip disappears when heat conduction is present. To determine the magnitude of the slip, the temperature of the medium is evaluated at $\kappa = 0$. This gives, using (14-110),

$$T^4(\kappa = 0) = \frac{1}{2}\left[T_1^4 + T_2^4 E_2(\kappa_D) + \int_0^{\kappa_D} T^4(\kappa^*) E_1(\kappa^*)\,d\kappa^*\right]$$

which can be written as

$$\frac{T_1^4 - T^4(\kappa = 0)}{T_1^4 - T_2^4} = \frac{1}{2}\left[\frac{T_1^4}{T_1^4 - T_2^4} - \frac{T_2^4}{T_1^4 - T_2^4}E_2(\kappa_D) - \int_0^{\kappa_D}\frac{T^4(\kappa^*)}{T_1^4 - T_2^4}E_1(\kappa^*)\,d\kappa^*\right]$$

$$(14\text{-}116)$$

As κ_D approaches zero the integral vanishes and $E_2(\kappa_D) \to 1$ so Eq. (14-116) reduces to

$$\frac{T_1^4 - T^4(\kappa = 0)}{T_1^4 - T_2^4}\bigg|_{\kappa_D \to 0} = \frac{1}{2} \qquad (14\text{-}117)$$

The magnitude of the slip for a gray medium with constant absorption coefficient is in Fig. 14-13 as a function of the layer optical thickness. [Note that from symmetry $T_1^4 - T^4(\kappa = 0) = T^4(\kappa = \kappa_D) - T_2^4$.]

Table 14-3 Dimensionless energy flux $\dfrac{q}{\sigma(T_1^4 - T_2^4)} = \psi_b$ [7][a]

Optical thickness κ_D	ψ_b	Optical thickness κ_D	ψ_b
0.1	0.9157	0.8	0.6046
0.2	0.8491	1.0	0.5532
0.3	0.7934	1.5	0.4572
0.4	0.7458	2.0	0.3900
0.5	0.7040	2.5	0.3401
0.6	0.6672	3.0	0.3016

[a]For $\kappa_D \gg 1$, $\psi_b = \dfrac{4/3}{1.42089 + \kappa_D}$

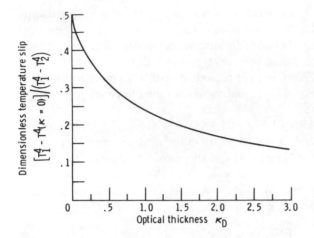

Figure 14-13 Discontinuity at wall between gray gas and black wall temperature [7].

Medium between diffuse gray walls   An extension can now be readily made to the situation where the plates are gray rather than black. From (14-105) and (14-106) the equations have the same form as (14-109) and (14-110) except that the outgoing fluxes $q_{o,1}$ and $q_{o,2}$ replace σT_1^4 and σT_2^4. Hence, as in (14-111) for the case with gray walls, let

$$\phi(\kappa) = \frac{G(\kappa) - q_{o,2}}{q_{o,1} - q_{o,2}} \qquad \psi = \frac{q}{q_{o,1} - q_{o,2}}$$

where

$$G(\kappa) = \begin{cases} \sigma T^4(\kappa) & \text{when } a > 0, \sigma_s \geqslant 0 \\ \pi \bar{i}(\kappa) & \text{when } a = 0, \sigma_s > 0 \end{cases}$$

The equations for $\phi(\kappa)$ and ψ are then the same as (14-112) and (14-113) so that $\phi = \phi_b$ and $\psi = \psi_b$, and for gray walls the $G(\kappa)$ and q are given by

$$G(\kappa) = \phi_b(\kappa)(q_{o,1} - q_{o,2}) + q_{o,2} \tag{14-118}$$

$$q = \psi_b(q_{o,1} - q_{o,2}) \tag{14-119}$$

Hence, assuming that ϕ_b and ψ_b have been obtained for the black case, only $q_{o,1}$ and $q_{o,2}$ are needed. From (14-108) with $q_r = q$,

$$q_{o,1} = \sigma T_1^4 - \frac{1 - \epsilon_1}{\epsilon_1} q \tag{14-120}$$

$$q_{o,2} = \sigma T_2^4 + \frac{1 - \epsilon_2}{\epsilon_2} q \tag{14-121}$$

Substitute these relations into (14-119) and solve for q to obtain

$$\frac{q}{\sigma(T_1^4 - T_2^4)} = \frac{\psi_b}{1 + \psi_b(1/\epsilon_1 + 1/\epsilon_2 - 2)} \tag{14-122}$$

Then substitute the $q_{o,1}$ and $q_{o,2}$ from (14-120) and (14-121) into (14-118) and eliminate the q by using (14-122) to obtain

$$\frac{G(\kappa)/\sigma - T_2^4}{T_1^4 - T_2^4} = \frac{\phi_b(\kappa) + [(1 - \epsilon_2)/\epsilon_2]\psi_b}{1 + \psi_b(1/\epsilon_1 + 1/\epsilon_2 - 2)} \qquad (14\text{-}123)$$

These relations will also be obtained in Sec. 16-4.6 by using the exchange-factor concept.

EXAMPLE 14-4 Let a medium between diffuse-gray walls have only scattering, $a = 0$, $\sigma_s > 0$, but in addition let it conduct heat with a constant thermal conductivity. Determine the heat transfer from plate 1 to plate 2 by combined heat conduction and pure scattering.

For pure scattering the temperature distribution within the layer does not enter into the radiative solution as given by (14-122) and (14-123). The $\nabla \cdot \mathbf{q}_r$ drops out of the energy equation as shown by (14-43). Hence the energy equation that determines the heat conduction is decoupled from the scattering process. The heat flux transferred is then found by adding to (14-122) the heat conduction as if scattering were not present. This gives

$$q = \frac{k(T_1 - T_2)}{D} + \frac{\sigma(T_1^4 - T_2^4)\psi_b}{1 + \psi_b(1/\epsilon_1 + 1/\epsilon_2 - 2)} \qquad (14\text{-}124)$$

The ψ_b is given in Fig. 14-12b, where the abscissa is $\kappa_D = \int_0^D \sigma_s\, dx$ (also see Table 14-3).

EXAMPLE 14-5 For the geometry in Example 14-4, determine the heat transfer from plate 1 to plate 2 by heat conduction combined with absorption and scattering in the optically thin limit.

From Fig. 14-12b, as $\kappa_D \to 0$, $\psi_b \to 1$, so Eq. (14-122) reduces to $q = \sigma(T_1^4 - T_2^4)/(1/\epsilon_1 + 1/\epsilon_2 - 1)$, which is the same as the result for no radiating medium between the plates. Equation (14-117) shows that in the optically thin limit, the radiative flux does not produce a temperature gradient in the medium and hence would not influence heat conduction. The ordinary heat conduction can thus be added to the radiation to give, in the optically thin limit,

$$q = \frac{k(T_1 - T_2)}{D} + \frac{\sigma(T_1^4 - T_2^4)}{1/\epsilon_1 + 1/\epsilon_2 - 1} \qquad (14\text{-}125)$$

Another approach is to obtain from the energy equation (14-27) $\nabla \cdot (k\, \nabla T - \mathbf{q}_r) = 0$. Then $q = \text{constant} = -k\, dT/dx + q_r$. From Eq. (14-76), for $\kappa_D = 0$ and diffuse-gray walls, $q_r = q_{o,1} - q_{o,2} = \sigma T_1^4 - q_r(1 - \epsilon_1)/\epsilon_1 - \sigma T_2^4 - q_r(1 - \epsilon_2)/\epsilon_2$ by use of (14-108). This is solved for q_r, and since q_r in the optically thin limit did not alter the temperature distribution, $-k\, dT/dx = k(T_1 - T_2)/D$. The q is then the same as (14-125).

A plane layer problem was considered in [8] to obtain an inverse solution. The observed data are the characteristics of the radiation leaving the layer. The solution is to obtain the optical thickness of the radiating layer and its scattering behavior. This has application to remote sensing and to measuring properties of materials.

14-8 MULTIDIMENSIONAL RADIATION IN AN ISOTROPICALLY SCATTERING GRAY MEDIUM

Sections 14-5 to 14-7 developed the radiative relations for plane layers. Other geometries will now be considered, such as rectangular regions. Since the relations become more complicated, the development will be restricted to a gray medium with constant properties and isotropic scattering. From Eqs. (14-18), (14-42), (14-44), (14-45), and (14-13) there are the relations

$$\mathbf{q}_r = \int_{\omega=0}^{4\pi} i's \, d\omega \tag{14-126}$$

$$\nabla \cdot \mathbf{q}_r = 4\pi a \left(\frac{\sigma T^4}{\pi} - \bar{\imath} \right) = 4\pi K (I' - \bar{\imath}) = 4\pi K \frac{a}{\sigma_s} \left(\frac{\sigma T^4}{\pi} - I' \right) \tag{14-127}$$

$$I' = \frac{1}{K} \left(a \frac{\sigma T^4}{\pi} + \sigma_s \bar{\imath} \right) = (1 - \Omega) \frac{\sigma T^4}{\pi} + \Omega \bar{\imath} \tag{14-128}$$

$$\bar{\imath} = \frac{1}{4\pi} \int_{\omega=0}^{4\pi} i' \, d\omega = \frac{1}{4\pi} \int_{\omega=0}^{4\pi} \left[i'(0)e^{-KS} + K \int_{S^*=0}^{S} I'(S^*)e^{-K(S-S^*)} dS^* \right] d\omega \tag{14-129}$$

where a and σ_s are constants ($K = a + \sigma_s$). The temperature is a function of position within the radiating medium and is found from the energy equation, which requires $\nabla \cdot \mathbf{q}_r$.

14-8.1 General Relations in Three Dimensions

The three-dimensional geometry is in Fig. 14-14b. Location vectors $\mathbf{r}_0$, $\mathbf{r}$, and $\mathbf{r}^*$ are used to give the positions of dA, dV, and dV^*. Then the path length $S = |\mathbf{r} - \mathbf{r}_0|$, and the unit vector along S is $\mathbf{s} = (\mathbf{r} - \mathbf{r}_0)/|\mathbf{r} - \mathbf{r}_0|$ or $(\mathbf{r} - \mathbf{r}^*)/|\mathbf{r} - \mathbf{r}^*|$. The cos θ is the dot product of the unit vector $\mathbf{n}$ and the unit vector $\mathbf{s}$ along S so cos $\theta = \mathbf{n} \cdot (\mathbf{r} - \mathbf{r}_0)/|\mathbf{r} - \mathbf{r}_0|$. The solid angle that dA subtends when viewed from dV is $d\omega = dA \cos \theta/S^2 = dA \, \mathbf{n} \cdot (\mathbf{r} - \mathbf{r}_0)/|\mathbf{r} - \mathbf{r}_0|^3$. In the integral of $I'(S^*)$ in Eq. (14-129), the integration over all $d\omega$ includes all volume elements of the medium ($dV^* = dS^* \, d\omega(S - S^*)^2$). By using these vector relations and $dS^* \, d\omega = dV^*/|\mathbf{r} - \mathbf{r}^*|^2$, Eq. (14-129) becomes

$$\bar{\imath}(\mathbf{r}) = \frac{1}{4\pi} \int_A i'(\mathbf{r}_0) \frac{\mathbf{n} \cdot (\mathbf{r} - \mathbf{r}_0)}{|\mathbf{r} - \mathbf{r}_0|^3} e^{-K|\mathbf{r} - \mathbf{r}_0|} \, dA + \frac{K}{4\pi} \int_V I'(\mathbf{r}^*) \frac{e^{-K|\mathbf{r} - \mathbf{r}^*|}}{|\mathbf{r} - \mathbf{r}^*|^2} \, dV^* \tag{14-130}$$

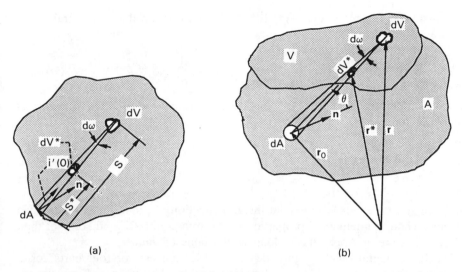

Figure 14-14 Geometry for three-dimensional medium.

The expression for $\mathbf{q}_r$ is obtained in the same manner. From Eq. (14-126) this contains the unit vector $\mathbf{s}$ and is otherwise similar to the definition for $\bar{\imath}$. By using Eq. (14-130) and inserting the additional $\mathbf{s}$, the $\mathbf{q}_r$ is

$$
\mathbf{q}_r(\mathbf{r}) = \int_A i'(\mathbf{r}_0)[\mathbf{n} \cdot (\mathbf{r} - \mathbf{r}_0)] \frac{\mathbf{r} - \mathbf{r}_0}{|\mathbf{r} - \mathbf{r}_0|^4} e^{-K|\mathbf{r} - \mathbf{r}_0|} \, dA
$$

$$
+ K \int_V I'(\mathbf{r}^*) \frac{\mathbf{r} - \mathbf{r}^*}{|\mathbf{r} - \mathbf{r}^*|^3} e^{-K|\mathbf{r} - \mathbf{r}^*|} \, dV^* \tag{14-131}
$$

Relations of this type are given in [9].

14-8.2 Two-Dimensional Transfer in a Rectangular Region

To illustrate the geometrical manipulations for applying Eqs. (14-130) and (14-131), radiative transfer is considered in a rectangular region with uniform conditions along the axial, z, direction, Fig. 14-15. The length in the z direction is large compared with dimensions b and d, and the axial location of dV is arbitrary; hence it is specified to be at $z = 0$. Then the volume integral in (14-130) can be written as

$$
\int_V I'(\mathbf{r}^*) \frac{e^{K|\mathbf{r} - \mathbf{r}^*|}}{|\mathbf{r} - \mathbf{r}^*|^2} \, dV^*
$$

$$
= 2 \int_{x^*=0}^{d} \int_{y^*=0}^{b} I'(x^*, y^*) \int_{z^*=0}^{\infty} \frac{e^{-K[(x-x^*)^2 + (y-y^*)^2 + z^{*2}]^{1/2}}}{(x - x^*)^2 + (y - y^*)^2 + z^{*2}} \, dz^* \, dx^* \, dy^*
$$

$$
\tag{14-132}
$$

Let $\xi^* = z^*/[(x - x^*)^2 - (y - y^*)^2]^{1/2} = z^*/\rho^*$ to obtain the z^* integral in the form

$$\int_{\xi=0}^{\infty} \frac{e^{-K\rho^*(1+\xi^{*2})^{1/2}}}{\rho^{*2}(1 + \xi^{*2})} \rho^* \, d\xi^* = \frac{1}{\rho^*} \int_{t=1}^{\infty} \frac{e^{-K\rho^* t}}{t(t^2 - 1)^{1/2}} \, dt = \frac{1}{\rho^*} \frac{\pi}{2} S_1(K\rho^*) \tag{14-133}$$

The transformation $t = (1 + \xi^{*2})^{1/2}$ has been used, and the function S_1 is one of the class of functions

$$S_n(x) \equiv \frac{2}{\pi} \int_1^{\infty} \frac{e^{-xt}}{t^n(t^2 - 1)^{1/2}} \, dt = \frac{2}{\pi} \int_0^{\pi/2} e^{-x/\cos \theta} \cos^{n-1} \theta \, d\theta \qquad n = 0, 1, 2, \ldots \tag{14-134}$$

The S_n are similar to the exponential integral functions E_n that are in plane layer problems. Their mathematical properties are examined in [10], and some of their characteristics are in Appendix E along with a table of values.

The first integral on the right side of (14-130) consists of four parts corresponding to the four sides of the rectangle. Consider the top side. From the triangle in Fig. 14-15, $\cos \theta = \mathbf{n} \cdot (\mathbf{r} - \mathbf{r}_0)/|\mathbf{r} - \mathbf{r}_0| = (b - y)/(\rho_0^2 + z_0^2)^{1/2}$ where $\rho_0(x_0, y_0) = [(x - x_0)^2 + (y - y_0)^2]^{1/2}$. Then for the upper surface the area integral becomes, after integration over the z variable,

$$2 \int_{x_0=0}^{d} i'(x, b) \int_{z_0=0}^{\infty} e^{-K(\rho_0^2+z_0^2)^{1/2}} \frac{b - y}{(\rho_0^2 + z_0^2)^{3/2}} \, dz_0 \, dx_0$$

$$= \pi(b - y) \int_{x_0=0}^{d} \frac{i'(x_0, b)}{\rho_0^2} S_2(K\rho_0) \, dx_0 \tag{14-135}$$

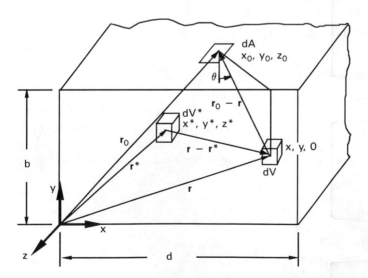

Figure 14-15 Geometry for radiation in two-dimensional rectangular region.

The $\bar{i}(\mathbf{r})$ in Eq. (14-130) is then given by

$$4\bar{i}(x, y) = (b - y) \int_{x_0=0}^{d} i'(x_0, b) \frac{S_2[K\rho_0(x_0, b)]}{\rho_0^2(x_0, b)} dx_0$$

$$+ (d - x) \int_{y_0=0}^{b} i'(d, y_0) \frac{S_2[K\rho_0(d, y_0)]}{\rho_0^2(d, y_0)} dy_0$$

$$+ y \int_{x_0=0}^{d} i'(x_0, 0) \frac{S_2[K\rho_0(x_0, 0)]}{\rho_0^2(x_0, 0)} dx_0 + x \int_{y_0=0}^{b} i'(0, y_0) \frac{S_2[K\rho_0(0, y_0)]}{\rho_0^2(0, y_0)} dy_0$$

$$+ K \int_{x^*=0}^{d} \int_{y^*=0}^{b} I'(x^*, y^*) \frac{S_1(K\rho^*)}{\rho^*} dx^* dy^* \tag{14-136}$$

When calculating results from these equations, it is sometimes advantageous to transform them into cylindrical coordinates. Then a $dx\, dy$ increment becomes $\rho\, d\rho\, d\theta$, and this cancels the ρ that is in the denominator. This eliminates any computational difficulty with the denominator becoming zero or close to zero; this is illustrated in [11] for a rectangular region of gray radiating medium with internal heat generation.

EXAMPLE 14-6 A hot rectangular bar of emitting, absorbing, and scattering material with nonreflecting boundaries such as in Fig. 14-15 is initially at uniform temperature T_i. It is suddenly placed in a very low temperature environment. The medium is gray and scatters isotropically. It has density ρ and specific heat c; heat conduction can be neglected relative to radiation. Derive the energy relations needed to obtain the transient temperature distribution.

By use of (14-127) for $\nabla \cdot \mathbf{q}_r$ the energy equation (14-26) and the initial condition for the temperature distribution $T(x, y)$ are

$$\rho c \frac{\partial T(x, y, \tau)}{\partial \tau} = -4K \frac{a}{\sigma_s} [\sigma T^4(x, y, \tau) - \pi I'(x, y, \tau)] \qquad T(x, y, 0) = T_i$$

$$\tag{14-137}$$

This is solved simultaneously with the integral equation found by substituting $\bar{i}$ from Eq. (14-136) into Eq. (14-128). The incoming intensities at the boundaries are all zero as the surroundings are at low temperature and the boundaries are nonreflecting. Then

$$I'(x, y, \tau) = \frac{a}{K} \frac{\sigma T^4(x, y, \tau)}{\pi} + \frac{\sigma_s}{4} \int_{x^*=0}^{d} \int_{y^*=0}^{b} I'(x^*, y^*, \tau) \frac{S_1(K\rho^*)}{\rho^*} dx^* dy^*$$

$$\tag{14-138}$$

where $\rho^* = [(x - x^*)^2 + (y - y^*)^2]^{1/2}$

Equations (14-137) and (14-138) can be placed in dimensionless form and solved numerically. Starting at $\tau = 0$, the $T(x, y, 0)$ in (14-138) is set equal to T_i. Then the integral equation is solved numerically for $I'(x, y, 0)$. The $T(x, y, 0)$ and $I'(x, y, 0)$ are substituted into the right side of (14-137), and the resulting temperature derivative is used to extrapolate forward in time to obtain $T(x, y, \Delta\tau)$. With this temperature distribution, the $I'(x, y, \Delta\tau)$ distribution is found by iteration of (14-138), and the process is continued to move forward in time. The numerical method is discussed in Chap. 17.

EXAMPLE 14-7 For the transient cooling problem in the previous example, derive an expression for the instantaneous emittance of the rectangular region.

The emittance can be obtained from the local heat fluxes leaving the boundaries of the region. Since for this example the externally incident intensities are zero at the boundaries, only the volume integral in Eq. (14-131) is needed. Along the upper boundary

$$q_r(x, b) = 2K$$

$$\times \int_{x^*=0}^{d} \int_{y^*=0}^{b} \int_{z^*=0}^{\infty} I'(x^*, y^*) \frac{(b - y^*)e^{-K[(x-x^*)^2+(b-y^*)^2+z^{*2}]^{1/2}}}{[(x - x^*)^2 + (b - y^*)^2 + z^{*2}]^{3/2}} \, dz^* \, dx^* \, dy^*$$

$$\tag{14-139}$$

By using the same transformation as in Eq. (14-133), the integration over z^* is expressed as an S_2 function. Then

$$q_r(x, b) = \pi K \int_{x^*=0}^{d} \int_{y^*=0}^{b} I'(x^*, y^*) \frac{b - y^*}{\rho_1^{*2}} S_2(K\rho_1^*) \, dx^* \, dy^* \tag{14-140}$$

where $\rho_1^* = [(x - x^*)^2 + (b - y^*)^2]^{1/2}$. Similarly, along the boundary $y = d$ in Fig. 14-15,

$$q_r(d, y) = \pi K \int_{x^*=0}^{d} \int_{y^*=0}^{b} I'(x^*, y^*) \frac{d - x^*}{\rho_2^{*2}} S_2(K\rho_2^*) \, dx^* \, dy^* \tag{14-141}$$

where $\rho_2^* = [(d - x^*)^2 + (y - y^*)^2]^{1/2}$. The overall emittance of the rectangle will be based on its instantaneous mean temperature,

$$T_m(\tau) = \frac{1}{bd} \int_{x=0}^{d} \int_{y=0}^{b} T(x, y, \tau) \, dx \, dy \tag{14-142}$$

The total heat loss $Q(\tau)$ is found by integrating the local q_r over their respective boundaries around the rectangle and using symmetry of the upper and lower boundaries and the two vertical sides. Then the overall emittance is

$$\epsilon(\tau) = \frac{Q(\tau)}{\sigma T_m^4(\tau)} = \frac{2}{\sigma T_m^4(\tau)} \left[\int_0^d q_r(x, b) \, dx + \int_0^d q_r(d, y) \, dy \right] \tag{14-143}$$

A useful case for testing the accuracy of numerical solutions is radiation to cold black boundaries by a two-dimensional rectangular region at a *uniform* temperature T_m. The region contains an absorbing-emitting medium that does not scatter. The fluxes at the boundary can be found from Eqs. (14-140) and (14-141) by using the source function for this case as $\sigma T_m^4/\pi$, which is a constant, and $K = a$. Then Eqs. (14-140) and (14-141) become

$$\frac{q_r(x, b)}{\sigma T_m^4} = a \int_{x^*=0}^{d} \int_{y^*=0}^{b} \frac{b - y^*}{\rho_1^{*2}} S_2(a\rho_1^*) \, dx^* \, dy^* \qquad (14\text{-}144a)$$

$$\frac{q_r(d, y)}{\sigma T_m^4} = a \int_{x^*=0}^{d} \int_{y^*=0}^{b} \frac{d - x^*}{\rho_2^{*2}} S_2(a\rho_2^*) \, dx^* \, dy^* \qquad (14\text{-}144b)$$

In [12] a series of transformations was used, and the integrations in Eq. (14-144) were carried out to yield the following forms that are readily evaluated with high precision:

$$\frac{q_r(x, b)}{\sigma T_m^4} = 1 - S_1(BX) + S_3(BX) - S_1[B(R - X)] + S_3[B(R - X)]$$

$$- B \int_{X^*=0}^{R} \{S_0(B[(X - X^*)^2 + 1]^{1/2})$$

$$- S_2(B[(X - X^*)^2 + 1]^{1/2})\} \, dX^* \qquad (14\text{-}145a)$$

$$\frac{q_r(d, y)}{\sigma T_m^4} = 1 - S_1(BY) + S_3(BY) - S_1[B(1 - Y)] + S_3[B(1 - Y)]$$

$$- B \int_{Y^*=0}^{1} \{S_0(B[R^2 + (Y - Y^*)^2]^{1/2})$$

$$- S_2(B[R^2 + (Y - Y^*)^2]^{1/2})\} \, dY^* \qquad (14\text{-}145b)$$

where $B = ab$ (optical dimension), $R = d/b$ (aspect ratio), $X = x/b$, $Y = y/b$. Tables of local heat flux along the boundary are in [12] for a wide range of aspect ratios and optical thicknesses of the rectangle. Figure 14-16 shows results for a square region. For large optical thicknesses the dimensionless heat flux is unity along the boundary away from the corners, and equals $1/2$ at the corners. An extension for a nongray medium is in [13].

Some references that provide solutions for rectangular geometries are [14–21]. Some approximate techniques and numerical solution methods for the radiative energy transfer equations are in Chaps. 15 and 17.

14-8.3 One-Dimensional Transfer in a Cylindrical Region

The cylinder is another important nonplanar geometry. It has been studied in relation to radiation in tubular furnaces, cylindrical combustion chambers, and radiating flows in pipes. For simplicity, the development here is for a gray medium,

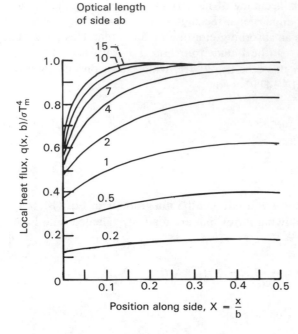

Optical length
of side ab

Figure 14-16 Local heat flux along boundary of square absorbing-emitting region as a function of optical length of side.

and isotropic scattering is included. The energy equation requires an expression for $\nabla \cdot \mathbf{q}_r$. The radiative heat flow at or across a boundary is obtained from $\mathbf{q}_r$. These expressions can be found by integrating, at any location, the contributions supplied by the radiation intensity as a result of radiation leaving boundary and volume elements. For a cylinder, these integrations become geometrically complicated. This derivation is limited to axisymmetric conditions with no variations along the cylinder axis and is for constant radiative properties. The formulation is consequently one-dimensional, and $\mathbf{q}_r$ depends only on radius.

The geometric aspects for dealing with the cylindrical shape are given by Heaslet and Warming [22] and Kesten [23], and the results have been used in various forms to obtain solutions in [24–28]. The approach in [23] is used to illustrate the derivation of the radial radiative heat flux. A typical path AOB is considered, Fig. 14-17, and the intensities along this path are integrated at location O in the required manner to obtain the radiative flux, $q_r(r)$. The $d\kappa$ along the path is related to the distance dx projected on a cross section normal to the cylinder axis by $d\kappa = dx/\cos\alpha = dx/\mu$, where $\mu = \cos\alpha$ and $d\kappa$ and dx are *optical* coordinates (physical coordinates multiplied by K). Then from (14-10) the equation of transfer is

$$\mu \frac{di'}{dx} + i'(x) = I'(x) \tag{14-146}$$

The law of cosines is used to relate x to the radial coordinate r,

$$r^2 = x^2 + R^2 - 2xR\cos\beta \tag{14-147}$$

The angle β remains constant along the path x so by differentiating (14-147) dx and dr are related along the path by

$$r\,dr = (x - R\cos\beta)\,dx \tag{14-148}$$

Equations (14-147) and (14-148) are combined to eliminate x, with the result that

$$dx = \pm\frac{r\,dr}{(r^2 - R^2\sin^2\beta)^{1/2}} \quad \begin{array}{l} - \text{ for } \quad 0 \le x < R\cos\beta \\ + \text{ for } \quad R\cos\beta < x \le 2R\cos\beta \end{array} \tag{14-149}$$

Letting $F(r, \beta) = (r^2 - R^2\sin^2\beta)^{1/2}/r$, Eq. (14-146) becomes, in terms of r,

$$\frac{di'}{dr} \pm \frac{1}{\mu F(r,\beta)}\,i' = \pm\frac{I'}{\mu F(r,\beta)} \tag{14-150}$$

with the signs used as previously defined. Equation (14-150) can be integrated as in Eq. (14-13). It is assumed that a diffuse intensity $i'(R)$, in the direction of κ, is leaving the inside of the wall at location A. To shorten the relations, define $\Gamma(a, b) \equiv \int_a^b d\xi/F(\xi, \beta)$. Then from (14-13), the i' along the path x is given by

$$i'(r) = i'(R)e^{(1/\mu)\Gamma(R,r)}$$
$$- \frac{1}{\mu}e^{(1/\mu)\Gamma(R,r)}\int_R^r \frac{e^{-(1/\mu)\Gamma(R,r^*)}\,I'(r^*)}{F(r^*, \beta)}\,dr^* \quad (0 \le x < R\cos\beta) \tag{14-151a}$$

$$i'(r) = i'(R\sin\beta)e^{-(1/\mu)\Gamma(R\sin\beta,r)} + \frac{1}{\mu}e^{-(1/\mu)\Gamma(R\sin\beta,r)}$$
$$\times \int_{R\sin\beta}^r \frac{e^{(1/\mu)\Gamma(R\sin\beta,r^*)}\,I'(r^*)}{F(r^*, \beta)}\,dr^* \quad (R\cos\beta < x \le 2R\cos\beta) \tag{14-151b}$$

where $i'(R\sin\beta)$ is Eq. (14-151a) evaluated at $r = R\sin\beta$.

The heat flux in the radially outward direction is obtained by integrating the intensities passing through a cylindrical surface with constant r, with each intensity weighted by the projected area of the surface normal to it, $q_r(r) = \int_{\omega=0}^{4\pi} i'(r)\cos\theta\,d\omega$. The angle θ is between $i'(r)$ and the outward normal to the cylindrical surface. Using $\cos\theta = -\cos\alpha\cos\gamma$ and the solid angle as obtained by projecting an element of surface area onto a unit sphere, $d\omega = \sin(\pi/2 - \alpha)\,d\gamma\,d(\pi/2 - \alpha) = -\cos\alpha\,d\gamma\,d\alpha$, the $q_r(r)$ becomes

$$q_r(r) = -4\int_{\alpha=0}^{\pi/2}\int_{\gamma=0}^{\pi} i'(r)\cos^2\alpha\cos\gamma\,d\gamma\,d\alpha \tag{14-152}$$

The $q_r(r)$ can be expressed in terms of β rather than γ by using the relation from Fig. 14-17,

$$r\sin\gamma = R\sin\beta \quad \begin{array}{l} 0 \le \gamma < \dfrac{\pi}{2},\ 0 \le x < R\cos\beta \\[2mm] \dfrac{\pi}{2} < \gamma \le \pi,\ R\cos\beta < x \le 2R\cos\beta \end{array} \tag{14-153}$$

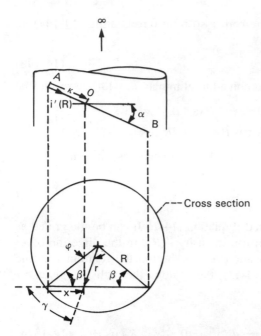

Figure 14-17 Geometry for radiative energy transfer in an infinitely long cylinder.

Let i^- and i^+ be the intensities corresponding to Eqs. (14-151a) and (14-151b). Then

$$q_r(r) = -4\frac{R}{r} \int_{\alpha=0}^{\pi/2} \left(\int_0^{\sin^{-1}(r/R)} i^- \cos \beta \, d\beta + \int_{\sin^{-1}(r/R)}^0 i^+ \cos \beta \, d\beta \right) \cos^2 \alpha \, d\alpha$$

$$= 4\frac{R}{r} \int_{\alpha=0}^{\pi/2} \left[\int_0^{\sin^{-1}(r/R)} (i^+ - i^-) \cos \beta \, d\beta \right] \cos^2 \alpha \, d\alpha \tag{14-154}$$

Eqs. (14-151a) and (14-151b) are now substituted for i^+ and i^-. The integration over α can be carried out as previously done for the rectangular cross section. This accounts for all axial locations along the cylinder and results in S_n functions as defined in Eq. (14-134). The expression for $q_r(r)$ becomes

$$q_r(r) = 2\pi \frac{R}{r} \int_{\beta=0}^{\sin^{-1}(r/R)} \left\{ i'(R)(S_3[2\Gamma(R \sin \beta, r) + \Gamma(r, R)] - S_3[\Gamma(r, R)]) \right.$$

$$+ \int_{r^*=R \sin \beta}^{r} \frac{I'(r^*)}{F(r^*, \beta)} S_2[\Gamma(r^*, r)] \, dr^* - \int_{r^*=r}^{R} \frac{I'(r^*)}{F(r^*, \beta)} S_2[\Gamma(r, r^*)] \, dr^*$$

$$\left. + \int_{r^*=R \sin \beta}^{R} \frac{I'(r^*)}{F(r^*, \beta)} S_2[\Gamma(R \sin \beta, r) + \Gamma(R \sin \beta, r^*)] \, dr^* \right\} \cos \beta \, d\beta$$

$$\tag{14-155}$$

The physical meaning of the four terms in Eq. (14-155) can be visualized by considering the effect on the heat flux at r of radiation traveling along a diameter of the cylinder. The first term is the heat flux from the cylinder wall attenuated as it passes through the medium. The positive portion is the radiation passing from the wall through the center region of the cylinder to r, while the negative portion is from the wall on the opposite side. The second term is the positive contribution by the emitting and scattering medium between the center of the cylinder and r, while the third term is the negative heat flux contribution by the medium between the wall and r. The fourth term is the radiation emitted and scattered by the medium that is between the wall and the center of the cylinder and then attenuated as it passes from the source location through the center of the cylinder to r.

The $\nabla \cdot \mathbf{q}_r$ for substitution in the energy equation can then be obtained in cylindrical coordinates for the one-dimensional axisymmetric case by carrying out the differentiation $(1/r)d(rq_r)/dr$. The $\nabla \cdot \mathbf{q}_r$ can also be derived by carrying out the integral form for $\bar{\imath}$ in Eq. (14-130) and then using (14-127). This approach was used in [22]. Various forms for $\mathbf{q}_r$ and $\nabla \cdot \mathbf{q}_r$ are also given in [24–26]. For cylindrical coordinates, the expressions are rather long and hence are not repeated here.

14-8.4 Additional Nonplanar and Multidimensional Geometries

The quantities for the one-dimensional axisymmetric cylindrical case in the previous section depend only on radius. An extension to another radially dependent case is the axisymmetric cylindrical annulus [29, 30]. In [30] the effect of radiation is studied on solidification of an annular region bounded by black surfaces. The equations are for a constant absorption coefficient, and scattering is not included.

Other radially dependent problems are a sphere [31] and the region between two uniform concentric spheres [32–34]. In [32], the bounding spheres are diffuse-gray, and the medium can emit, absorb, and scatter isotropically. Reference [34] considers the same geometry and includes anisotropic scattering. In [33], an isothermal gray medium is analyzed without scattering. A literature review is also given for spherical geometries.

Radiative transfer within a two-dimensional corner region was analyzed in [35]. The geometry is a layer of fixed thickness that extends around the corner along the exterior surface of a rectangle. An application would be the performance of a coating placed on the outside of a corner region.

For three dimensional situations, a cylinder can be considered with both circumferential and axial variations. In [36] an axisymmetric cylinder of finite length was analyzed, including anisotropic scattering. The long rectangular region discussed previously can also be generalized to a region of finite length [37, 38]. The radiative transfer relations for a three-dimensional rectangular geometry are in [39, 40].

REFERENCES

1. Kourganoff, Vladimir: "Basic Methods in Transfer Problems," Dover, New York, 1963.
2. Chandrasekhar, Subrahmanyan: "Radiative Transfer," Dover, New York, 1960.
3. Goody, R. M.: "Atmospheric Radiation. Theoretical Basis," vol. I, Clarendon Press, Oxford, 1964.
4. Taussig, R. T., and A. T. Mattick: Droplet Radiator Systems for Spacecraft Thermal Control, *J. Spacecr. Rockets*, vol. 23, no. 1, pp. 10–17, 1986.
5. Siegel, R.: Transient Radiative Cooling of a Droplet-Filled Layer, *J. Heat Transfer*, vol. 109, no. 1, pp. 159–164, 1987.
6. Siegel, R.: Separation of Variables Solution for Non-Linear Radiative Cooling, *Int. J. Heat Mass Transfer*, vol. 30, no. 5, pp. 959–965, 1987.
7. Heaslet, Max A., and Robert F. Warming: Radiative Transport and Wall Temperature Slip in an Absorbing Planar Medium, *Int. J. Heat Mass Transfer*, vol. 8, pp. 979–994, 1965.
8. Ho, C.-H., and M. N. Ozisik: An Inverse Radiation Problem, *Int. J. Heat Mass Transfer*, vol. 32, no. 2, pp. 335–341, 1989.
9. Lin, J.-D.: Radiative Transfer within an Arbitrary Isotropically Scattering Medium Enclosed by Diffuse Surfaces, *J. Thermophys. Heat Transfer*, vol. 2, no. 1, pp. 68–74, 1988.
10. Yuen, W. W., and L. W. Wong: Numerical Computation of an Important Integral Function in Two-Dimensional Radiative Transfer, *J. Quant. Spectrosc. Radiat. Transfer*, vol. 29, no. 2, pp. 145–149, 1983.
11. Yuen, W. W., and C. F. Ho: Analysis of Two-Dimensional Radiative Heat Transfer in a Gray Medium with Internal Heat Generation, *Int. J. Heat Mass Transfer*, vol. 28, no. 1, pp. 17–23, 1985.
12. Siegel, R.: Analytical Solution for Boundary Heat Fluxes from a Radiating Rectangular Medium, *J. Heat Transfer*, vol. 113, no. 1, pp. 258–261, 1991.
13. Siegel, R.: Boundary Fluxes for Spectral Radiation From a Uniform Temperature Rectangular Medium, *J. Thermophys. Heat Transfer*, vol. 6, no. 3, pp. 543–546, 1992.
14. Thynell, S. T., and M. N. Ozisik: Radiation Transfer in Isotropically Scattering, Rectangular Enclosures, *J. Thermophys. Heat Transfer*, vol. 1, no. 1, pp. 69–76, 1987.
15. Fiveland, W. A.: Discrete-Ordinates Solutions of the Radiative Transport Equation for Rectangular Enclosures, *J. Heat Transfer*, vol. 106, pp. 699–706, 1984.
16. Yuen, W. W., and Wong, L. W.: Analysis of Radiative Equilibrium in a Rectangular Enclosure with Gray Medium, *J. Heat Transfer*, vol. 106, pp. 433–440, 1984.
17. Razzaque, M. M., J. R. Howell, and D. E. Klein: Coupled Radiative and Conductive Heat Transfer in a Two-Dimensional Rectangular Enclosure with Gray Participating Media Using Finite Elements, *J. Heat Transfer*, vol. 106, pp. 613–619, 1984.
18. Siegel, R.: Some Aspects of Transient Cooling of a Radiating Rectangular Medium, *Int. J. Heat Mass Transfer*, vol. 32, no. 10, pp. 1955–1966, 1989.
19. Siegel, R.: Emittance Bounds for Transient Radiative Cooling of a Scattering Rectangular Region, *J. Thermophys. Heat Transfer*, vol. 4, no. 1, pp. 106–114, 1990.
20. Siegel, R.: Transient Cooling of a Square Region of Radiating Medium, *J. Thermophys. Heat Transfer*, vol. 5, no. 4, pp. 495–501, 1991.
21. Yuen, W. W. and E. E. Takara: Superposition Technique for Radiative Equilibrium in Rectangular Enclosures with Complex Boundary Conditions, *Int. J. Heat Mass Transfer*, vol. 33, no. 5, pp. 901–915, 1990.
22. Heaslet, M. A., and R. F. Warming: Theoretical Predictions of Radiative Transfer in a Homogeneous Cylindrical Medium, *J. Quant. Spectrosc. Radiat. Transfer*, vol. 6, no. 6, pp. 751–774, 1966.
23. Kesten, A. S.: Radiant Heat Flux Distribution in a Cylindrically Symmetric Nonisothermal Gas with Temperature-Dependent Absorption Coefficient, *J. Quant. Spectrosc. Radiat. Transfer*, vol. 8, no. 1, pp. 419–434, 1968.
24. Azad, F. H., and M. F. Modest: Evaluation of the Radiative Heat Flux in Absorbing, Emitting,

and Linear Anisotropically Scattering Cylindrical Media, *J. Heat Transfer*, vol. 103, no. 2, pp. 350–356, 1981.

25. Fernandes, R., and J. Francis: Combined Conductive and Radiative Heat Transfer in an Absorbing, Emitting, and Scattering Cylindrical Medium, *J. Heat Transfer*, vol. 104, no. 4, pp. 594–601, 1982.

26. Siegel, R.: Transient Radiative Cooling of an Absorbing and Scattering Cylinder—A Separable Solution, *J. Thermophys. Heat Transfer*, vol. 2, no. 2, pp. 110–117, 1988.

27. Siegel, R.: Transient Radiative Cooling of an Absorbing and Scattering Cylinder, *J. Heat Transfer*, vol. 111, no. 1, pp. 199–203, 1989.

28. Siegel, R.: Solidification by Radiative Cooling of a Cylindrical Region Filled with Drops, *J. Thermophys. Heat Transfer*, vol. 3, no. 3, pp. 340–344, 1989.

29. Viskanta, R., and E. E. Anderson: Heat Transfer in Semitransparent Solids, "Advances in Heat Transfer," vol. 11, pp. 348–350, Academic Press, New York, 1975.

30. Habib, I. S.: Solidification of a Semitransparent Cylindrical Medium by Conduction and Radiation, *J. Heat Transfer*, vol. 95, no. 1, pp. 37–41, 1973.

31. Thynell, S. T., and M. N. Ozisik: Radiation Transfer in an Isotropically Scattering Solid Sphere with Space Dependent Albedo $\omega(r)$, *J. Heat Transfer*, vol. 107, pp. 732–734, 1985.

32. Viskanta, R., and R. L. Merriam: Heat Transfer by Combined Conduction and Radiation between Concentric Spheres Separated by Radiating Medium, *J. Heat Transfer*, vol. 90, no. 2, pp. 248–256, 1968.

33. Crosbie, A. L., and H. K. Khalil: Radiative Transfer in a Gray Isothermal Spherical Layer, *J. Quant. Spectrosc. Radiat. Transfer*, vol. 12, pp. 1465–1486, 1972.

34. Tong, T. W., and P. S. Swathi: Radiative Heat Transfer in Emitting-Absorbing-Scattering Spherical Media, *J. Thermophys. Heat Transfer*, vol. 1, no. 2, pp. 162–170, 1987.

35. Wu, C. Y. and M. N. Fu: Radiative Transfer in a Coating on a Rectangular Corner: Allowance for Shadowing, *Int. J. Heat Mass Transfer*, vol. 33, no. 12, pp. 2735–2741, 1990.

36. Mengüc, M. P., R. Viskanta, and C. R. Ferguson: Multidimensional Modeling of Radiative Heat Transfer in Diesel Engines, SAE Technical Paper 850503, 1985.

37. Mengüc, M. P., and R. Viskanta: Radiative Transfer in Three-Dimensional Rectangular Enclosures Containing Inhomogeneous, Anisotropically Scattering Media, *J. Quant. Spectrosc. Radiat. Transfer*, vol. 33, no. 6, pp. 533–549, 1985.

38. Fiveland, W. A.: Three-Dimensional Radiative Heat Transfer Solutions by the Discrete-Ordinates Method, *J. Thermophys. Heat Transfer*, vol. 2, no. 4, pp. 309–316, 1988.

39. Crosbie, A. L., and R. G. Schrenker: Exact Expressions for Radiative Transfer in a Three-Dimensional Rectangular Geometry, *J. Quant. Spectrosc. Radiat. Transfer*, vol. 28, pp. 507–526, 1982.

40. Lin, J.-D.: Exact Expressions for Radiative Transfer in an Arbitrary Geometry Exposed to Radiation, *J. Quant. Spectrosc. Radiat. Transfer*, vol. 37, no. 6, pp. 591–601, 1987.

PROBLEMS

14-1 A slab of nonscattering solid material has a gray absorption coefficient of $a = 0.4$ cm^{-1} and refractive index $n \approx 1$. It is 2 cm thick and has an approximately linear temperature distribution within it as established by thermal conduction. What is the emitted intensity normal to the slab? What average slab temperature would give the same emitted normal intensity?

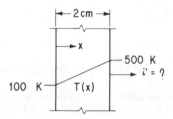

Answer: 195 W/(m$^2 \cdot$ sr), 374 K

14-2 Estimate the Planck mean absorption coefficient for air at 12,000 K and 1 atm by use of Fig. 12-6.

14-3 Using Fig. 12-6, estimate the total absorptance of air at 12,000 K and 1 atm for solar radiation passing through a path length of 5 cm.

Answer: 0.58

14-4 A nonscattering stagnant gray medium with absorption coefficient $a = 0.2$ cm^{-1} is contained between black parallel plates 10 cm apart as shown (assume constant density and $n \approx 1$). Plot the temperature distribution $T(z)$ (neglect thermal conduction). What is the net energy flux being transferred by radiation from the lower to the upper plate? If the plates are gray with $\epsilon_1 = 0.8$ and $\epsilon_2 = 0.4$, what is the energy flux being transferred?

Answer: 1484 W/m^2, 882 W/m^2

14-5 Consider the plane layer of absorbing and nonscattering gray gas between black parallel plates at temperatures T_1 and T_2 as in Sec. 14-7.6. A chemical reaction is producing a uniform energy generation per unit volume in the gas. Derive the equations for the temperature distribution in the gas and the local energy flux in the z direction. What is the equation to obtain the net energy flux supplied to each plate? What does the temperature distribution become for the limiting case when $\kappa_D = aD \rightarrow 0$?

14-6 An isothermal enclosure is filled with a nonscattering absorbing and emitting medium. The medium and enclosure are at the same temperature. Show that $\nabla \cdot \mathbf{q}_r$ must be zero for this condition.

14-7 A semi-infinite, isotropically scattering, absorbing-emitting medium maintained at uniform temperature T_g is in contact with a black wall at T_w. The medium is gray with constant scattering and absorption coefficients σ_s and a. The medium is not moving, and heat conduction is neglected. Show that the heat flux transferred to the wall can be expressed in the form $H\sigma(T_g^4 - T_w^4)$. Provide the relations needed to obtain values for H.

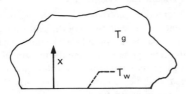

14-8 A scattering, nonabsorbing, nonconducting medium is contained between parallel plates 8 cm apart. The scattering is assumed to be isotropic and independent of wavelength, and the scattering coefficient is $\sigma_s = 0.2$ cm^{-1}. The plate temperatures are $T_1 = 700$ K and $T_2 = 600$ K. Compute the net heat flux transferred from plate 1 to plate 2 if the plates are black, or if the plates are gray with $\epsilon_1 = 0.7$ and $\epsilon_2 = 0.3$. How do these transfers compare with the values for the plates in vacuum?

Answer: 6266 W/m^2 (black, vacuum), 1666 W/m^2 (gray, vacuum), 2781 W/m^2 (black, medium), 1249 W/m^2 (gray, medium)

14-9 In Prob. 14-8, for the situation of a scattering medium between gray plates, it is desired to double the amount of energy being transferred by having the scattering particles suspended in a thermally conducting but nonabsorbing medium. What thermal conductivity of the medium would be required to accomplish this?

Answer: 1.00 W/(m·K)

14-10 A gray absorbing and scattering medium is contained between gray parallel plates 5 cm apart and with $T_1 = 800$ K, $\epsilon_1 = 0.2$, $T_2 = 600$ K, $\epsilon_2 = 0.6$. The scattering is isotropic and independent of wavelength, and heat conduction is neglected. Compute the net energy transfer from plate 1 to plate 2 if the scattering and absorption coefficients are respectively $\sigma_s = 0.1$ cm^{-1} and $a = 0.2$ cm^{-1}.

Answer: 2317 W/m^2.

14-11 In Prob. 14-7 the plane wall at $x = 0$ is modified to be diffuse-gray instead of black. Derive the integral equation relations that can be numerically evaluated to determine the heat flux transferred to the wall. Place the equations in a convenient dimensionless form.

14-12 A semi-infinite medium is absorbing, emitting, and isotropically scattering. It is gray and has absorption coefficient a and scattering coefficient σ_s. The medium is initially at uniform temperature T_i. The plane surface of the medium is suddenly subjected to radiative exchange with a large environment at a lower uniform temperature T_e. It is proposed to carry out a numerical solution to obtain the transient temperature distributions in the medium as it cools. Provide the energy and scattering equations that are to be placed in numerical form for solution in a convenient dimensionless form. Heat conduction is neglected and the medium is stationary. The density ρ and specific heat c are assumed constant.

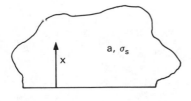

14-13 A plane layer of absorbing, emitting, isotropically scattering material is being uniformly heated throughout its volume with a volumetric heating rate q'''. The layer is of thickness D and has absorption coefficient a and scattering coefficient σ_s. The layer is confined between walls that are diffuse-gray, and both walls have emissivity ϵ_w and temperature T_w. Derive the energy and scattering equations required to perform a numerical solution for the steady temperature distribution across the layer from $x = 0$ to D. Heat conduction is neglected.

14-14 A rectangular region is filled with an absorbing, emitting, isotropically scattering medium. The medium is gray and has absorption coefficient a and scattering coefficient σ_s. The geometry is long in the z-direction, and the side dimensions are b and d. The four bounding walls are black. The vertical walls are both at uniform temperature T_v, and the horizontal walls are at T_h. The medium is heated uniformly throughout its volume with a volumetric heating rate q'''. The effect of heat conduction is neglected, and the medium is not moving. Provide the energy equation and scattering relations that are required to solve for the temperature distribution in the rectangular cross-section. Place the equations in convenient dimensionless forms.

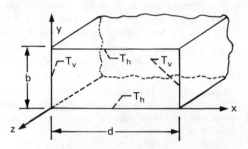

FIFTEEN

SOLUTION METHODS FOR THE EQUATIONS OF RADIATIVE TRANSFER

15-1 INTRODUCTION

Simple solutions are usually not possible for the equations of radiative transfer and energy conservation. In most practical problems in which temperatures or heat flows must be found, the solution can require moderate to considerable effort. In this chapter, some methods are presented that have been found for certain conditions to provide reasonable accuracy with a reasonable investment of time. The purpose of this chapter is to point out some simplified procedures that can be used and to set up methods with various degrees of approximation that can be solved numerically as will be discussed in Chap. 17. Some of the "exact" numerical methods, such as finite-difference procedures, have only a small section in this chapter although they are quite important; they are considered more in Chap. 17.

The first section is based on simplifications from physical reasoning. If a situation involves a medium that is weakly attenuating, or the medium boundaries are cold, or there is a cold medium exposed to a strong radiation source, then in each case certain terms can be neglected in the transfer equations. Simplified solutions can then be obtained. When the medium is optically very dense, the diffusion approximation can be used. In this instance the radiative behavior is similar to energy transport by heat conduction with a temperature-dependent thermal conductivity.

Sometimes a radiative solution of reduced accuracy is acceptable, for example, when the influence of the radiation terms is rather small in the overall energy balance. In such cases, a method that is easy to apply, but of limited accuracy, may be useful. An example is the method in Sec. 15-4, in which a mean absorption coefficient is used in solutions obtained for a gray medium.

More general methods are in Secs. 15-5 to 15-8. These sections include a discussion of the range of application of the methods and their strengths and weaknesses. The P-N methods in Sec. 15-5 utilize a set of moment equations of the equation of transfer and an expansion of the intensity in terms of spherical harmonics. The discrete ordinate method in Sec. 15-6 carries out angular integrations in the energy balance relations by dividing the 4π solid angle about each location into finite angular regions and letting the intensity be constant within each region. Another integration technique is to utilize finite elements as developed in Sec. 15-7. The Monte Carlo method in Sec. 15-8 is relatively easy to formulate as it requires following only individual bundles of energy as they travel within the geometry and are absorbed or scattered. The method is very versatile but may require long computation times to follow a sufficient number of energy bundles to yield good statistical accuracy.

Section 15-9 discusses some additional approximate and direct numerical methods. The direct numerical method by use of finite differences, which can yield very accurate solutions, will be presented and discussed in Chap. 17.

15-2 SYMBOLS

a	absorption coefficient; side length of square
A	area
$\bar{A}_l$	effective bandwidth
A_l^m	coefficients in Eq. (15-105)
C_1, C_2	constants in Planck's spectral energy distribution
$C_{1,l}, C_{2,l}$	coefficients for weak and strong bands
D	spacing between parallel plates; diameter
e	emissive power
E	ratio $(1 - \epsilon)/\epsilon$
E_1, E_2, E_3	exponential integrals
G	volumetric energy-generation rate; functions defined in Eq. (14-105)
H	length over which temperature changes significantly
i	radiation intensity
I	source function; mean absorption value defined by Eq. (15-54)
K	extinction coefficient
l_1, l_2, l_3	direction cosines
l_m	extinction mean free path, $1/(a_\lambda + \sigma_{s\lambda})$
$l_{m,s}$	scattering mean free path, $1/\sigma_{s\lambda}$
n	ordinate directions in S_n approximation
P	pressure; probability
P_l^m	Legendre polynomials Eq. (15-107)
q	energy flux, energy per unit area and time
Q	energy per unit time
r	radial coordinate

$\mathbf{r}$	position vector
R	sphere radius; random number
$\mathbf{s}$	unit vector in S direction
S	coordinate along path of radiation
T	absolute temperature
U	approximate solution
U_i	coefficients in U
V	volume; test function
w	energy per photon bundle
W	weighting factor
W_i	coefficients in W
$\left.\begin{array}{l} x, y, z \\ x_1, x_2, x_3 \end{array}\right\}$	distances measured along Cartesian coordinates
Y_l^m	normalized spherical harmonics Eq. (15-106)
β	constant in Eq. (15-175)
Γ	gamma function
δ_{kj}	Kronecker delta
ϵ	hemispherical emissivity
η	wavenumber
θ	cone angle, angle from normal of area
κ	optical depth
κ_D	optical thickness for path length D
λ	wavelength
μ	$\cos \theta$
ρ	density
σ	Stefan-Boltzmann constant
σ_s	scattering coefficient
φ	circumferential angle
ϕ	dimensionless temperature ratio, Tables 15-2 and 15-4
Φ	shape function
ψ	dimensionless heat flux, Tables 15-2 and 15-4
ω	solid angle
Ω	function defined by Eq. (15-39); scattering albedo

Subscripts

b	blackbody
D	mean absorption coefficient in Eq. (15-54)
e	emitted; effective
$g-g$	evaluated at interface between gas regions 1 and 2
i	incident; incoming
o	outgoing
P	Planck mean value
r	net value in r direction; radiative
R	Rosseland mean value in Eq. (15-44)

s	sphere
u	uniform
w	evaluated on the wall
z	net value in the z direction
$+z, -z$	propagating in positive or negative z direction, respectively
η	wavenumber dependent
$\Delta\lambda$	value integrated over wavelength interval $\Delta\lambda$
λ	wavelength dependent
0	evaluated at point of origin; initial value
1, 2	boundary 1 or 2, respectively

Superscripts

$(0), (1), (2)$	zeroth-, first-, or second-order term or moment
$(i), (ij), \ldots$	moments
$+, -$	propagating in positive or negative direction
$*$	dummy variable of integration
$**$	dummy variable of integration
$-$	(overbar) average over all incident solid angles
$'$	directional quantity

15-3 SOME LIMITING CASES FOR OPTICALLY THIN AND THICK MEDIA

In Chap. 14 the equation of transfer (14-13) gave the spectral intensity variation along a path as

$$i'_\lambda(S) = i'_\lambda(0) \exp\left[-\int_0^S K_\lambda \, dS^*\right] + \int_0^S K_\lambda I'_\lambda(S^*) \exp\left[-\int_{S^*}^S K_\lambda \, dS^{**}\right] dS^* \qquad (15\text{-}1)$$

where $K_\lambda = a_\lambda + \sigma_{s\lambda}$ may vary with S. The $I'_\lambda(S)$ is the source function that depends on the local blackbody intensity and the locally scattered radiation and is found from an integral equation such as (14-59). Since $I'_\lambda(S)$ depends on $i'_{\lambda b}(S)$, the temperature distribution must also be found, so a simultaneous solution with the energy equation is required. A number of limiting cases for the governing equations can be examined, such as the medium being optically thick or thin, the scattering being strong or weak relative to absorption, and internal emission being strong or weak relative to radiation incident at the boundaries. A few simple approximations given by the first three entries in Table 15-1 are discussed in Secs. 15-3.1 to 15-3.3. In Sec. 15-3.4 an optically thick medium is considered.

15-3.1 Nearly Transparent Optically Thin Medium

When the optical depth along a path is small, (15-1) is simplified, as the two exponential attenuation terms each approach unity, thereby giving

Table 15-1 Approximations to equation of transfer

Approximation	Form of equation of transfer	Conditions
Nearly transparent, optically thin medium	$i_\lambda'(S) = i_\lambda'(0)$	The medium has such a low extinction coefficient that intensity does not change by absorption, emission, or scattering along a path within the medium.
Optically thin medium with cold boundaries (emission approximation)	$i_\lambda'(S) = \displaystyle\int_0^S a_\lambda(S^*)i_{\lambda b}'(S^*)\, dS^*$	No energy is incident from the boundaries, and the gas is relatively transparent so that emitted energy from the gas passes within the system without significant attenuation.
Cold medium with small scattering	$i_\lambda'(S) = i_\lambda'(0) \exp\left[-\displaystyle\int_0^S K_\lambda\, dS^*\right]$	Radiation emitted and scattered by the medium is negligible compared to that incident from boundaries or external sources.
Optically thick medium (nearly opaque) (diffusion approximation)	$-\dfrac{4\pi}{3K_\lambda}\dfrac{\partial i_{\lambda b}'}{\partial S} = \dfrac{dq_\lambda(S)}{d\lambda}$	The optical depth of the gas is sufficiently large and the temperature gradients sufficiently small that the local intensity results only from local emission.

$$i_\lambda'(S) = i_\lambda'(0) + \int_0^S K_\lambda I_\lambda'(S^*)\, dS^* \tag{15-2}$$

There is no attenuation along the path of the intensity that is locally emitted or scattered. The intensity that enters at $S = 0$ is also not appreciably attenuated. In the special case when the optical depth is small and $i_\lambda'(0)$ is not small, the integral in (15-2) is small relative to $i_\lambda'(0)$ and (15-2) reduces to

$$i_\lambda'(S) = i_\lambda'(0) \tag{15-3}$$

as in Table 15-1. An incident intensity is essentially unchanged as it travels through the medium. The local energy balances based on this simple intensity relation are obviously much easier to carry out than those involving the complete transfer equation. The use of this approximation is now demonstrated.

EXAMPLE 15-1 Two infinite parallel black plates at temperatures T_1 and T_2 as in Fig. 14-11 are a distance D apart, and the space between them is filled with an absorbing, emitting, and scattering gas with absorption coefficient $a_\lambda(\lambda)$. Assuming that the nearly transparent approximation holds, derive an expression for the gas temperature as a function of position between the plates.

Equation (14-96) is the general expression for energy conservation when there is radiative equilibrium in the gas (this also includes scattering with the minor restriction that the scattering does not depend on ω_i),

$$a_P(T,P)\sigma T^4 = \pi \int_0^\infty a_\lambda(\lambda, T, P)\bar{\imath}_{\lambda,i}(\lambda)\, d\lambda \tag{15-4}$$

The $\bar{\imath}_{\lambda,i}$ is obtained from the contributions approaching a volume element from below and above:

$$4\pi\bar{\imath}_{\lambda,i}(\lambda) = \int_\cap i_\lambda^+(\lambda, \omega)\, d\omega + \int_\cup i_\lambda^-(\lambda, \omega)\, d\omega \tag{15-5}$$

Since the walls are black, the nearly transparent approximation gives, for any position between the plates, $i_\lambda^+(\lambda, \omega) = i'_{\lambda b}(\lambda, T_1)$ and $i_\lambda^-(\lambda, \omega) = i'_{\lambda b}(\lambda, T_2)$. Then, since the black intensity is independent of angle, (15-5) yields

$$4\pi\bar{\imath}_{\lambda,i}(\lambda) = 2\pi i'_{\lambda b}(\lambda, T_1)\int_0^{\pi/2} \sin\theta\, d\theta + 2\pi i'_{\lambda b}(\lambda, T_2)\int_0^{\pi/2} \sin\theta\, d\theta$$

$$= 2\pi[i'_{\lambda b}(\lambda, T_1) + i'_{\lambda b}(\lambda, T_2)] \tag{15-6}$$

Substituting (15-6) into (15-4) gives, at any x position between the plates,

$$\sigma T^4(x) = \frac{\pi}{2a_P(x)}\int_0^\infty a_\lambda(\lambda, x)[i'_{\lambda b}(\lambda, T_1) + i'_{\lambda b}(\lambda, T_2)]\, d\lambda \tag{15-7}$$

Equation (15-7) is solved iteratively for $T(x)$ because a_λ depends on local temperature.

If a_λ does not depend on gas temperature, using (14-95) results in

$$\int_0^\infty a_\lambda(\lambda)i'_{\lambda b}(\lambda, T_1)\, d\lambda = \frac{a_P(T_1)\sigma T_1^4}{\pi} \qquad \int_0^\infty a_\lambda(\lambda)i'_{\lambda b}(\lambda, T_2)\, d\lambda = \frac{a_P(T_2)\sigma T_2^4}{\pi}$$

and $a_P(T(x)) = \int_0^\infty a_\lambda(\lambda)e_{\lambda b}(T(x))\, d\lambda / [\sigma T^4(x)]$. Then (15-7) reduces to

$$T^4(x) = \frac{1}{2a_P(T(x))}[a_P(T_1)T_1^4 + a_P(T_2)T_2^4] \tag{15-8}$$

The local temperature solution, although still requiring an iterative solution on $T(x)$ and $a_P(T(x))$, is relatively easily found by using tabulated or curve fit values of $a_P(T)$. For a gray gas with temperature-independent properties, a_P is constant and (15-8) further reduces to $T^4(x) = (T_1^4 + T_2^4)/2$. Hence, the entire gray gas approaches a fourth-power temperature that is the average of the fourth powers of the boundary temperatures. This limit was also found in (14-115).

As a more formal derivation for a nearly transparent medium between infinite parallel plates, consider the heat flux equation (14-67) used in conjunction with (14-74) for diffuse boundaries that are not black. From the series expansions in Appendix E the exponential integrals can be approximated for small arguments by $E_2(x) = 1 + O(x)$ and $E_3(x) = \frac{1}{2} - x + O(x^2)$. Then (14-67) becomes [using (14-72a) and (14-72b) for diffuse boundaries in the first two integrals on the right]

$$\frac{dq_{r\lambda}(\kappa_\lambda)}{d\lambda} = \frac{dq_{\lambda o,1}}{d\lambda}(1 - 2\kappa_\lambda) - \frac{dq_{\lambda o,2}}{d\lambda}(1 - 2\kappa_{D\lambda} + 2\kappa_\lambda)$$

$$+ 2\pi \int_0^{\kappa_\lambda} I_\lambda'(\kappa_\lambda^*)\, d\kappa_\lambda^* - 2\pi \int_{\kappa_\lambda}^{\kappa_{D\lambda}} I_\lambda'(\kappa_\lambda^*)\, d\kappa_\lambda^*$$

If $\kappa_{D\lambda} \ll 1$, the terms of order κ_λ are neglected and this reduces to

$$\frac{dq_{r\lambda}(\kappa_\lambda)}{d\lambda} = \frac{dq_{\lambda o,1}}{d\lambda} - \frac{dq_{\lambda o,2}}{d\lambda} \tag{15-9}$$

Thus the local spectral flux is the difference between the fluxes leaving the boundaries; the fluxes have been unattenuated by the medium, and this corresponds to the nearly transparent approximation.

In a similar fashion the source-function equation (14-69) reduces to

$$I_\lambda'(\kappa_\lambda) = (1 - \Omega_\lambda)i_{\lambda b}'(\kappa_\lambda) + \frac{\Omega_\lambda}{2\pi}\left(\frac{dq_{\lambda o,1}}{d\lambda} + \frac{dq_{\lambda o,2}}{d\lambda}\right) \tag{15-10}$$

The flux derivative (14-68) simplifies to

$$\frac{d^2 q_{r\lambda}}{d\kappa_\lambda\, d\lambda} = -2\left(\frac{dq_{\lambda o,1}}{d\lambda} + \frac{dq_{\lambda o,2}}{d\lambda}\right) + 4\pi I_\lambda'(\kappa_\lambda)$$

Substituting $I_\lambda'(\kappa_\lambda)$ from (15-10) and noting that $1 - \Omega_\lambda = a_\lambda/K_\lambda$ and $dK_\lambda = K_\lambda d\lambda$ yield

$$\frac{d^2 q_{r\lambda}}{dx\, d\lambda} = 2a_\lambda\left[2\pi i_{\lambda b}'(\kappa_\lambda) - \left(\frac{dq_{\lambda o,1}}{d\lambda} + \frac{dq_{\lambda o,2}}{d\lambda}\right)\right] \tag{15-11}$$

The diffuse fluxes at the boundaries (14-74) reduce to

$$\frac{dq_{\lambda o,1}}{d\lambda} = \epsilon_{\lambda,1}e_{\lambda b,1} + (1 - \epsilon_{\lambda,1})\frac{dq_{\lambda o,2}}{d\lambda}$$

$$\frac{dq_{\lambda o,2}}{d\lambda} = \epsilon_{\lambda,2}e_{\lambda b,2} + (1 - \epsilon_{\lambda,2})\frac{dq_{\lambda o,1}}{d\lambda}$$

Solving simultaneously for the spectral fluxes leaving the walls yields, for the nearly transparent approximation,

$$\frac{dq_{\lambda o,1}}{d\lambda} = \frac{\epsilon_{\lambda,1}e_{\lambda b,1} + \epsilon_{\lambda,2}e_{\lambda b,2}(1 - \epsilon_{\lambda,1})}{1 - (1 - \epsilon_{\lambda,1})(1 - \epsilon_{\lambda,2})} \tag{15-12a}$$

$$\frac{dq_{\lambda o,2}}{d\lambda} = \frac{\epsilon_{\lambda,2}e_{\lambda b,2} + \epsilon_{\lambda,1}e_{\lambda b,1}(1 - \epsilon_{\lambda,2})}{1 - (1 - \epsilon_{\lambda,1})(1 - \epsilon_{\lambda,2})} \tag{15-12b}$$

EXAMPLE 15-2 A nearly transparent medium with extinction coefficient K_λ is between two diffuse parallel plates a distance D apart. The plate temperatures are T_1 and T_2. What total heat flux is being transferred across the plates in the absence of conduction and convection?

For the nearly transparent approximation, Eq. (15-9) applies. Substituting $dq_{\lambda o,1}$ and $dq_{\lambda o,2}$ from (15-12) yields

$$\frac{dq_{r\lambda}}{d\lambda} = \frac{\epsilon_{\lambda,1}e_{\lambda b,1} + \epsilon_{\lambda,2}e_{\lambda b,2}(1 - \epsilon_{\lambda,1}) - \epsilon_{\lambda,2}e_{\lambda b,2} - \epsilon_{\lambda,1}e_{\lambda b,1}(1 - \epsilon_{\lambda,2})}{1 - (1 - \epsilon_{\lambda,1})(1 - \epsilon_{\lambda,2})}$$

Simplifying and integrating with respect to λ gives the required result,

$$q_r = \int_{\lambda=0}^{\infty} \frac{e_{\lambda b,1} - e_{\lambda b,2}}{1/\epsilon_{\lambda,1} + 1/\epsilon_{\lambda,2} - 1} \, d\lambda$$

This is the same as (8-11). The transferred flux is thus uninfluenced by the presence of the nearly transparent medium.

EXAMPLE 15-3 What is the medium temperature distribution in Example 15-2?

For radiative equilibrium $\nabla \cdot \mathbf{q}_r = 0$, and from (14-57),

$$\int_0^\infty K_\lambda(\kappa_\lambda) \frac{d^2 q_{r\lambda}}{d\kappa_\lambda \, d\lambda} \, d\lambda = 0$$

Substituting (15-11) gives

$$\int_0^\infty a_\lambda \left[2\pi i'_{\lambda b}(\kappa_\lambda) - \left(\frac{dq_{\lambda o,1}}{d\lambda} + \frac{dq_{\lambda o,2}}{d\lambda} \right) \right] d\lambda = 0$$

Substituting (15-12) yields

$$2\pi \int_0^\infty a_\lambda i'_{\lambda b}(\kappa_\lambda) \, d\lambda = \int_0^\infty a_\lambda \frac{2(\epsilon_{\lambda,1}e_{\lambda b,1} + \epsilon_{\lambda,2}e_{\lambda b,2}) - \epsilon_{\lambda,1}\epsilon_{\lambda,2}(e_{\lambda b,1} + e_{\lambda b,2})}{\epsilon_{\lambda,1} + \epsilon_{\lambda,2} - \epsilon_{\lambda,1}\epsilon_{\lambda,2}} \, d\lambda$$

Using (14-95), the integral on the left is expressed in terms of the Planck mean absorption coefficient to yield

$$\sigma T^4(x) = \frac{1}{2a_P(x)} \int_0^\infty a_\lambda(\lambda,x) \frac{2(\epsilon_{\lambda,1}e_{\lambda b,1} + \epsilon_{\lambda,2}e_{\lambda b,2}) - \epsilon_{\lambda,1}\epsilon_{\lambda,2}(e_{\lambda b,1} + e_{\lambda b,2})}{\epsilon_{\lambda,1} + \epsilon_{\lambda,2} - \epsilon_{\lambda,1}\epsilon_{\lambda,2}} \, d\lambda \tag{15-13}$$

This can be solved for $T(x)$ by iteration as discussed in connection with Eq. (15-7), and it reduces to (15-7) when $\epsilon_{\lambda,1} = \epsilon_{\lambda,2} = 1$. If all properties are independent of both wavelength and temperature, (15-13) reduces to

$$T^4 = \frac{1}{2}\frac{2(\epsilon_1 T_1^4 + \epsilon_2 T_2^4) - \epsilon_1 \epsilon_2 (T_1^4 + T_2^4)}{\epsilon_1 + \epsilon_2 - \epsilon_1 \epsilon_2} \tag{15-14}$$

This is a limiting form of (14-123) when $G = \sigma T^4$ and when $\phi_b \to \frac{1}{2}$ and $\psi_b \to 1$ for the nearly transparent approximation.

15-3.2 Optically Thin Medium with Cold Boundaries (Emission Approximation)

In the nearly transparent approximation, the gas is optically thin and the local intensity is dominated by the intensities incident at the boundaries. In the emission approximation the gas is again optically thin but there is negligible incoming energy at the boundaries. For these conditions there is only energy emission within the medium and no attenuation by either absorption or scattering [note that in (15-11) scattering does not enter into the energy equation]. The equation of transfer becomes

$$i_\lambda'(S) = \int_0^S a_\lambda(S^*) i_{\lambda b}'(S^*) \, dS^* \tag{15-15}$$

so that $i_\lambda'(S)$ is the integrated contribution by all the emission along a path as the emitted intensity travels through the gas without attenuation.

Equation (15-15) is integrated over all wavelengths to give the total intensity

$$i'(S) = \int_0^\infty i_\lambda'(S) \, d\lambda = \int_0^S \left[\int_0^\infty a_\lambda(S^*) i_{\lambda b}'(S^*) \, d\lambda \right] dS^*$$

The definition of a_P [Eq. (14-95)] is now applied to give

$$i'(S) = \int_0^S a_P(S^*) \frac{\sigma T^4(S^*)}{\pi} \, dS^* \tag{15-16}$$

The fact that (15-16) contains the Planck mean absorption coefficient and was derived for optically thin conditions has sometimes led to the statement that the Planck mean absorption coefficient is applicable *only* in optically thin situations. However, the Planck mean absorption coefficient was defined in general for the *emission* from a volume element in connection with (14-35) and hence can be applied for emission in a gas of *any* optical thickness.

EXAMPLE 15-4 Use the emission approximation to find the flux emerging from an isothermal slab of gas with a Planck mean absorption coefficient of 0.010 cm^{-1} and a thickness of $D = 1.5$ cm if the slab is bounded by transparent nonradiating walls (Fig. 15-1a).

If $i'(\theta)$ is the emerging total intensity in direction θ, the emerging flux is $q = 2\pi \int_{\theta=0}^{\pi/2} i'(\theta) \cos \theta \sin \theta \, d\theta$. Since the slab is isothermal with constant

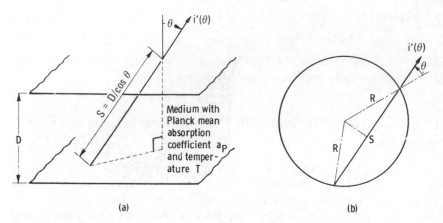

Figure 15-1 Examples for emission approximation. (*a*) Slab geometry for Example 15-4; (*b*) emission from spherical gas-filled balloon with transparent skin.

a_P, Eq. (15-16) can be integrated over any path $D/\cos\theta$ through the slab to yield $i'(\theta) = (\sigma/\pi)a_P T^4 (D/\cos\theta)$. Then

$$q = 2 \int_0^{\pi/2} a_P \sigma T^4 D \sin\theta \, d\theta = 2a_P \sigma T^4 D = 0.03\sigma T^4 \tag{15-17}$$

It should be realized that (15-17) is not a precise result even though the slab thickness is optically thin: $a_P D = 0.015 \ll 1$. This is because the radiation reaching the slab boundary along each path has passed through the thickness $D/\cos\theta$. For θ approaching $\pi/2$ the path length becomes very large so the emission approximation cannot hold. A more accurate solution including the proper path lengths gives (see Sec. 13-5.4)

$$q = 1.8a_P \sigma T^4 D \tag{15-18}$$

EXAMPLE 15-5 A spherical balloon of radius R is in orbit around the earth and enters the earth's shadow. The balloon has a perfectly transparent wall and is filled with a gray gas with constant absorption coefficient a, such that $aR \ll 1$. Neglecting radiant exchange with the earth, derive a relation for the initial rate of energy loss from the balloon if the initial temperature of the gas is T_0.

Using the emission approximation, Eq. (15-16), Fig. 15-1*b* shows that the intensity at the surface is $i'(\theta) = (a\sigma T_0^4/\pi)S = (a\sigma T_0^4/\pi)2R\cos\theta$, since a and T_0 are constants. The flux leaving the surface is

$$q = 2\pi \int_0^{\pi/2} i'(\theta)\cos\theta \sin\theta \, d\theta = 4\,a\sigma T_0^4 R \int_0^{\pi/2} \cos^2\theta \sin\theta \, d\theta = \tfrac{4}{3}a\sigma T_0^4 R$$

The initial rate of energy loss from the entire sphere is then

$$Q = \tfrac{4}{3} a o T_0^4 R (4\pi R^2) = 4 a o T_0^4 V_s \qquad (15\text{-}19)$$

where V_s is the sphere volume. This is what is expected—it was found that *any* isothermal gas volume with no internal absorption radiates according to this relation (see Sec. 12-5).

15-3.3 Cold Medium with Small Scattering

This approximation applies when both the local blackbody emission and scattering are small, such as for radiative transfer within an absorbing cryogenic fluid. The scattering is small enough so that scattering *into* the S direction can be neglected. The equation of transfer (15-1) reduces to

$$i_\lambda'(S) = i_\lambda'(0) \exp\left[-\int_0^S (a_\lambda + \sigma_{s\lambda})\, dS^*\right] \qquad (15\text{-}20)$$

so the local intensity consists only of attenuated incident intensity.

> **EXAMPLE 15-6** 100 W of radiant energy leave a spherical light bulb 10 cm in diameter enclosed in a fixture having a flat glass plate as shown in Fig. 15-2. If the glass is 2 cm thick and has a gray extinction coefficient of 0.05 cm^{-1}, find the intensity directly from the bulb leaving the fixture at an angle of 60°. Neglect interface-interaction effects resulting from the difference in the index of refraction between the glass and surrounding air (these effects are discussed in Chap. 18).
>
> Integrating (15-20) over λ and S results in the total intensity $i'(S, \theta) = i'(0, \theta) \exp[-(a + \sigma_s)S]$. To obtain $i'(0, \theta)$, consider the bulb a diffuse sphere. The emissive power at the surface of the sphere is 100 W divided by the sphere area. The intensity is this diffuse emissive power divided by π, $i'(0, \theta) = 100 \text{ W}/(\pi 10^2 \text{ cm}^2 \times \pi \text{ sr}) = 0.101 \text{ W}/(\text{cm}^2 \cdot \text{sr})$. Then $i'(S, \theta) = 0.101 \exp[-0.05(2/\cos 60°)] = 0.0827 \text{ W}/(\text{cm}^2 \cdot \text{sr})$. Note that this solution involves only a simple attenuated transmission.

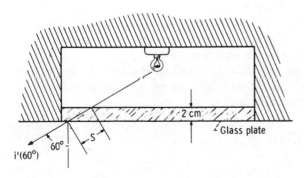

Figure 15-2 Intensity of beam from light fixture (Example 15-6).

15-3.4 Optically Thick Medium: The Diffusion Approximation

In an optically dense medium, radiation travels only a short distance before being scattered or absorbed. The local intensity is the result of radiation from only nearby locations where emission and scattering are similar to that of the location under consideration. Radiation from locations where conditions are appreciably different is greatly attenuated before reaching the location being considered. For this situation it is possible to transform the integral equation for the radiative energy balance into a diffusion equation like that for heat conduction. It is found that the energy transfer depends only on the conditions in the immediate vicinity of the position being considered and can be expressed in terms of the gradient of the conditions at that position. The use of the diffusion approximation leads to a substantial simplification. Standard techniques, including well-developed finite-difference schemes, can be used for solving the diffusion differential equation.

Real gases are usually transparent in some wavelength regions. The diffusion approach can be applied only in wavelength regions where the optical thickness of the medium is greater than about 5; this value depends on the geometry and conditions of the problem and may need to be larger in some instances. The fact that some *mean* optical thickness meets this criterion is not sufficient. Sometimes good results are found for much smaller optical thicknesses, as will be shown in Fig. 15-6. Wavelength-band applications of the diffusion method can be made in the optically thick regions.

As will be shown in the derivations, the diffusion approximation requires that the intensity within the medium be nearly isotropic [this will be revealed by Eq. (15-29)]. This can occur in the interior of an optically thick medium with small temperature gradients but cannot be valid near certain types of boundaries. For example, at a boundary adjacent to a vacuum at low temperature, radiation will leave the medium but there will be little incident radiation from the vacuum. As a result of this large anisotropy, the diffusion approximation is not valid near this type of boundary. To apply diffusion methods at interfaces between such regions, "radiation slip" or "jump" boundary conditions are employed.

Simplified derivation of the diffusion equation A simplified derivation for a one-dimensional layer will be used to illustrate the spirit of the diffusion approximation. The situation is for an absorbing-emitting medium in radiative equilibrium with isotropic scattering. In the diffusion approximation, the medium is optically dense. Let H be a path length over which the radiant energy density changes appreciably, and let the extinction mean free path be $l_m = 1/(a_\lambda + \sigma_{s\lambda})$. For the diffusion approximation to apply, $l_m/H \ll 1$. Consider the layer of gas in Fig. 15-3a. By using $dS = dx/\cos\theta$ in Eq. (14-10), the equation of transfer giving the change of i_λ' with x for a fixed θ direction is

$$-\frac{\cos\theta}{\sigma_{s\lambda} + a_\lambda} \frac{\partial i_\lambda'(x,\theta)}{\partial x} = i_\lambda'(x,\theta) - I_\lambda'(x) \tag{15-21}$$

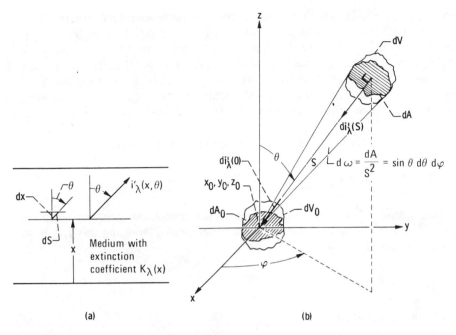

Figure 15-3 Geometry for derivation of diffusion equations. (a) One-dimensional plane layer of medium; (b) general three-dimensional region.

where the source function I'_λ does not depend on angle for isotropic scattering. Using H, nondimensionalize (15-21) to obtain

$$-\frac{l_m}{H}\cos\theta\,\frac{\partial i'_\lambda(x,\theta)}{\partial(x/H)} = i'_\lambda(x,\theta) - I'_\lambda(x) \tag{15-22}$$

We now proceed to solve (15-22) in the form of a series. Using the diffusion approximation, $l_m/H \ll 1$, the intensity is written as a series of functions $i'^{(n)}_\lambda$ multiplied by powers of l_m/H:

$$i'_\lambda = i'^{(0)}_\lambda + \frac{l_m}{H}\,i'^{(1)}_\lambda + \left(\frac{l_m}{H}\right)^2 i'^{(2)}_\lambda + \cdots \tag{15-23}$$

Insert (15-23) into the transfer equation (15-22) to obtain [I'_λ is given by (14-9) with $\Phi = 1$]

$$-\mu\frac{l_m}{H}\left[\frac{\partial i'^{(0)}_\lambda}{\partial(x/H)} + \frac{l_m}{H}\frac{\partial i'^{(1)}_\lambda}{\partial(x/H)} + \cdots\right] = i'^{(0)}_\lambda + \frac{l_m}{H}\,i'^{(1)}_\lambda + \cdots - \left(1 - \frac{l_m}{l_{m,s}}\right)i'_{\lambda b}$$

$$-\frac{l_m}{l_{m,s}}\left[\frac{1}{4\pi}\int_{\omega_i=4\pi}\left(i'^{(0)}_\lambda + \frac{l_m}{H}\,i'^{(1)}_\lambda + \cdots\right)d\omega_i\right] \tag{15-24}$$

where $l_{m,s} = 1/\sigma_{s\lambda}$. In addition to the expansion parameter l_m/H, there has appeared the quantity $l_m/l_{m,s}$, which characterizes the extinction by scattering relative to the total extinction. For the diffusion approximation including absorption *and* scattering, this parameter should be of order $1/2$. For small $l_m/l_{m,s}$ the problem degenerates to one of diffusion by absorption alone, as in a dense gas containing a few scattering particles. For $l_m/l_{m,s}$ approaching unity, there is scattering alone, as in a thin gas with many scattering particles (note that $l_m/l_{m,s}$ is the scattering albedo Ω).

In (15-24) collect the terms of zero order in l_m/H to obtain

$$i_\lambda'^{(0)} = \left(1 - \frac{l_m}{l_{m,s}}\right) i_{\lambda b}' + \frac{l_m}{l_{m,s}} \frac{1}{4\pi} \int_{\omega_i=0}^{4\pi} i_\lambda'^{(0)} \, d\omega_i \tag{15-25}$$

The terms $i_{\lambda b}'$ and $\int_{\omega_i=0}^{4\pi} i_\lambda'^{(0)} \, d\omega_i$ on the right do not depend on incidence angle $d\omega_i$. Hence $i_\lambda'^{(0)}$ on the left cannot depend on incidence angle. By using this fact in the integral, (15-25) reduces to

$$i_\lambda'^{(0)} = \left(1 - \frac{l_m}{l_{m,s}}\right) i_{\lambda b}' + \frac{l_m}{l_{m,s}} \frac{1}{4\pi} i_\lambda'^{(0)} 4\pi$$

which further reduces to

$$i_\lambda'^{(0)} = i_{\lambda b}' \tag{15-26}$$

Now collect the terms in (15-24) of first order in l_m/H to obtain

$$-\mu \frac{\partial i_\lambda'^{(0)}}{\partial(x/H)} = i_\lambda'^{(1)} - \frac{l_m}{l_{m,s}} \frac{1}{4\pi} \int_{\omega_i=0}^{4\pi} i_\lambda'^{(1)} \, d\omega_i$$

Substitute (15-26) for $i_\lambda'^{(0)}$ to obtain

$$-\mu \frac{di_{\lambda b}'}{d(x/H)} = i_\lambda'^{(1)} - \frac{l_m}{l_{m,s}} \frac{1}{4\pi} \int_{\omega_i=0}^{4\pi} i_\lambda'^{(1)} \, d\omega_i \tag{15-27}$$

To find $i_\lambda'^{(1)}$ multiply by $d\omega_i = 2\pi \sin\theta \, d\theta = -2\pi \, d\mu$ and integrate over all solid angles:

$$\frac{di_{\lambda b}'}{d(x/H)} \int_{\mu=-1}^{1} 2\pi\mu \, d\mu = \int_{\omega_i=0}^{4\pi} i_\lambda'^{(1)} \, d\omega_i - \frac{l_m}{l_{m,s}} \left(\frac{1}{4\pi} \int_{\omega_i=0}^{4\pi} i_\lambda'^{(1)} \, d\omega_i\right) \int_{\omega_i=0}^{4\pi} d\omega_i$$

The integral on the left is zero so

$$0 = \int_{\omega_i=0}^{4\pi} i_\lambda'^{(1)} \, d\omega_i - \frac{l_m}{l_{m,s}} \int_{\omega_i=0}^{4\pi} i_\lambda'^{(1)} \, d\omega_i$$

Hence, since $l_m/l_{m,s} \neq 1$, it must be true that $\int_{\omega_i=0}^{4\pi} i_\lambda'^{(1)} \, d\omega_i = 0$, and (15-27) reduces to

$$i_\lambda'^{(1)} = -\mu \frac{di_{\lambda b}'}{d(x/H)} \tag{15-28}$$

Substitute (15-26) and (15-28) into (15-23) to obtain

$$i'_\lambda = i'_{\lambda b} - \frac{\cos\theta}{a_\lambda + \sigma_{s\lambda}} \frac{di'_{\lambda b}}{dx} \tag{15-29}$$

This result reveals the important feature that *in the diffusion solution the local intensity depends only on the magnitude and gradient of the local blackbody intensity*. Since temperature gradients are small and $a_\lambda + \sigma_{s\lambda}$ is large, the last term on the right is small and i'_λ is nearly isotropic.

The local spectral energy flux at x flowing in the x direction is found by multiplying i'_λ by $\cos\theta \, d\lambda$ and integrating over all solid angles as in (14-53a):

$$\frac{dq_\lambda(x)}{d\lambda} = 2\pi \int_{\theta=0}^{\pi} i'_\lambda(x,\theta) \cos\theta \sin\theta \, d\theta = 2\pi \int_{\mu=-1}^{1} i'_\lambda(x,\mu) \, \mu \, d\mu \tag{15-30}$$

Using (15-29) in (15-30) gives, after noting that $i'_{\lambda b}$ does not depend on μ,

$$\frac{dq_\lambda(x)}{d\lambda} = 2\pi i'_{\lambda b}(x) \int_{\mu=-1}^{1} \mu \, d\mu - \frac{2\pi}{a_\lambda + \sigma_{s\lambda}} \frac{di'_{\lambda b}}{dx} \int_{\mu=-1}^{1} \mu^2 \, d\mu = -\frac{4\pi}{3K_\lambda(x)} \frac{di'_{\lambda b}}{dx}$$

$$= -\frac{4}{3K_\lambda(x)} \frac{de_{\lambda b}}{dx} \tag{15-31}$$

where $K_\lambda = a_\lambda + \sigma_{s\lambda}$. Equation (15-31) is the *Rosseland diffusion equation*. It relates the local energy flux to local conditions only; it does not involve integrals of contributions from other regions and thus, where applicable, provides a considerable simplification over the exact formulation.

In a gray medium, (15-31) becomes

$$q(x) = -\frac{4\sigma}{3(a+\sigma_s)} \frac{d(T^4)}{dx} = -\frac{16}{3K} \sigma T^3 \frac{dT}{dx} \tag{15-32}$$

For $a = 0$ (pure scattering), σT^4 in (15-32) is replaced by $\pi\bar{i}$ as in Sec. 14-7.4.

General radiation-diffusion equation in an absorbing-emitting medium In the previous section the diffusion equation was derived for a simplified situation. Only first-order terms were retained in the series (15-23), and boundaries were not considered. The general equations including second-order terms are now derived. Boundary conditions are introduced so the diffusion equations can be applied to finite regions. The derivation follows the general outline in [1]; for simplicity, scattering is not included. The intermediate equations in the derivation become somewhat complex because of their general form. The final equations [such as Eq. (15-42)] are relatively simple.

Rosseland equation for local radiative flux In Fig. 15-3b there is a volume element dV_0 at x_0, y_0, z_0 having cross-sectional area dA_0 in the xy plane. The energy flux crossing dA_0 originates from all surrounding volume elements such as dV. From (13-41), the emission from dV produces an intensity $a_\lambda(\lambda, T, P)i'_{\lambda b}(\lambda, T)\, dS$. The intensity reaching dV_0 is

$$di'_\lambda(0) = a_\lambda(\lambda)i'_{\lambda b}(\lambda, T)\, dS \exp[-a_\lambda(\lambda)S] \tag{15-33}$$

This accounts for attenuation along S but not emission along S, which will be accounted for by integrating the contributions of (15-33) from all elements of the volume. A spatially constant a_λ has been used. This is not restrictive, as in the diffusion approximation the gas temperature does not change significantly over the region contributing significant radiation to a location. The solid angle subtended by dV when viewed from dA_0 is dA/S^2, where dA is the projected area of dV normal to S. The energy per unit time incident on dA_0 as a result of the intensity in (15-33) is then

$$d^4Q_{\lambda,i}(0) = a_\lambda(\lambda)i'_{\lambda b}(\lambda,T)\, dS \exp[-a_\lambda(\lambda)S]\, \frac{dA}{S^2}\, dA_0 \cos\theta\, d\lambda \tag{15-34}$$

Since the gas is optically dense, the radiation at dV_0 originates only from locations close to dV_0. Then $i'_{\lambda b}(\lambda, T)$ in (15-34) can be obtained by expanding in a three-dimensional Taylor series about $S = 0$ in the hope that truncating the series after a few terms will give an adequate representation of the $i'_{\lambda b}$ distribution near dV_0. The general Taylor series in three dimensions is written as

$$i'_{\lambda b}(\lambda,T) = \sum_{n=0}^{\infty} \left\{ \frac{1}{n!} \left[(z-z_0)\left(\frac{\partial}{\partial z}\right)_0 + (y-y_0)\left(\frac{\partial}{\partial y}\right)_0 + (x-x_0)\left(\frac{\partial}{\partial x}\right)_0 \right]^n i'_{\lambda b}(\lambda,T) \right\} \tag{15-35}$$

This is carried for the next several steps and then truncated to a few terms. By applying the binomial theorem twice to expand the factor in brackets, (15-35) becomes

$$i'_{\lambda b}(\lambda,T) = \sum_{n=0}^{\infty} \sum_{v=0}^{n} \sum_{s=0}^{v} \frac{(z-z_0)^{n-v}(y-y_0)^{v-s}(x-x_0)^s}{(n-v)!\,(v-s)!\,s!} \left(\frac{\partial^n i'_{\lambda b}}{\partial z^{n-v}\,\partial y^{v-s}\,\partial x^s} \right)_0 \tag{15-36}$$

This is substituted into Eq. (15-34), and then the solid angle dA/S^2 is set equal to $\sin\theta\, d\theta\, d\varphi$. The result is integrated over the half space for positive z values to give all the energy traveling in the negative z direction that is incident on dA_0 as

$$\frac{d^2Q_{\lambda,-z}}{d\lambda} = a_\lambda(\lambda)\, dA_0 \sum_{n=0}^{\infty} \sum_{v=0}^{n} \sum_{s=0}^{v} \frac{1}{(n-v)!\,(v-s)!\,s!}$$

$$\times \left(\frac{\partial^n i'_{\lambda b}}{\partial z^{n-v}\,\partial y^{v-s}\,\partial x^s} \right)_0 \int_{\varphi=0}^{2\pi} \int_{\theta=0}^{\pi/2} \int_{S=0}^{\infty} (S\cos\theta)^{n-v}$$

$$\times (S\sin\varphi\sin\theta)^{v-s}(S\sin\theta\cos\varphi)^s \cos\theta \sin\theta$$

$$\times \exp[-a_\lambda(\lambda)S]\, dS\, d\theta\, d\varphi \tag{15-37}$$

The integral and summation signs have been interchanged, and spherical coordinates have been introduced: $x - x_0 = S \sin \theta \cos \varphi$; $y - y_0 = S \sin \theta \sin \varphi$; and $z - z_0 = S \cos \theta$. The following assumptions have been used in integrating Eq. (15-37) over the half space: (1) a_λ is constant within the region that contributes significantly to the energy flux at dA_0; and (2) there are no bounding surfaces that contribute significant radiation energy at dA_0. Otherwise, a_λ would have to be retained as a variable in the integration, and the integration would have to be over a finite region with a specified intensity along the boundaries.

Carrying out the integration of (15-37) gives

$$\frac{d^2 Q_{\lambda, -z}}{d\lambda} = \frac{dA_0}{4} \sum_{n=0}^{\infty} \sum_{v=0}^{n} \sum_{s=0}^{v} \Omega(n,v,s) \frac{1}{a_\lambda{}^n} \left(\frac{\partial^n i'_{\lambda b}}{\partial z^{n-v} \partial y^{v-s} \partial x^s} \right)_0 \tag{15-38}$$

where (Γ is the gamma function)

$$\Omega(n,v,s) = \frac{[1 + (-1)^{v-s}][1 + (-1)^s] n! \Gamma[(n-v+2)/2] \Gamma[(v-s+1)/2] \Gamma[(s+1)/2]}{(n-v)! (v-s)! s! \Gamma[(n+4)/2]} \tag{15-39}$$

A similar derivation for the energy incident on dA_0 from below gives, for energy traveling in the positive z direction,

$$\frac{d^2 Q_{\lambda, +z}}{d\lambda} = \frac{dA_0}{4} \sum_{n=0}^{\infty} \sum_{v=0}^{n} \sum_{s=0}^{v} (-1)^{n-v} \Omega(n,v,s) \frac{1}{a_\lambda{}^n} \left(\frac{\partial^n i'_{\lambda b}}{\partial z^{n-v} \partial y^{v-s} \partial x^s} \right)_0 \tag{15-40}$$

The *net* energy flux passing through dA_0 in the positive z direction is then

$$\frac{dq_{\lambda, z}}{d\lambda} = \frac{d^2 Q_{\lambda, +z} - d^2 Q_{\lambda, -z}}{dA_0 \, d\lambda}$$

$$= -\frac{1}{4} \sum_{n=0}^{\infty} \sum_{v=0}^{n} \sum_{s=0}^{v} [1 - (-1)^{n-v}] \Omega(n,v,s) \frac{1}{a_\lambda{}^n} \left(\frac{\partial^n i'_{\lambda b}}{\partial z^{n-v} \partial y^{v-s} \partial x^s} \right)_0 \tag{15-41}$$

Similar relations can be derived for the x and y directions.

In the diffusion approximation the optical depth is large and temperature must change slowly with respect to optical distance. Quantities such as $(1/a_\lambda^n)$ $(\partial^n i'_{\lambda b}/\partial^n z)$ become small as n is increased and the series in (15-41) can be truncated. *Retaining only terms through the second derivative* causes Eq. (15-41) to reduce to

$$\frac{dq_{\lambda, z}}{d\lambda} = -\frac{4\pi}{3a_\lambda} \left(\frac{\partial i'_{\lambda b}}{\partial z} \right)_0 = -\frac{4}{3u_\lambda} \left(\frac{\partial e_{\lambda b}}{\partial z} \right)_0 \tag{15-42}$$

This is the *general relation* for local energy flux in terms of the emissive power gradient and agrees with (15-31); it is the *Rosseland diffusion equation* for radiative energy transfer. Retaining only first-order derivatives, as in the derivation of (15-31), gives the same equation because the second-order terms cancel. Equation (15-42) has the same form as the Fourier law of heat conduction; this allows solution of some radiation problems by heat conduction methods. As indicated by

(15-31), if scattering is included in the present derivation, the only effect on (15-42) is to replace a_λ by K_λ.

To obtain the energy flux in a wavelength range, integrate (15-42) over the band $\Delta\lambda$ (the parentheses and 0 subscript are omitted for simplicity):

$$q_{\Delta\lambda,z} = \int_{\Delta\lambda} -\frac{4}{3a_\lambda} \frac{\partial e_{\lambda b}}{\partial z} \, d\lambda \equiv -\frac{4}{3a_{R,\Delta\lambda}} \int_{\Delta\lambda} \frac{\partial e_{\lambda b}}{\partial z} \, d\lambda$$

$$= -\frac{4}{3a_{R,\Delta\lambda}} \frac{\partial}{\partial z} \int_{\Delta\lambda} e_{\lambda b} \, d\lambda = -\frac{4}{3a_{R,\Delta\lambda}} \frac{\partial e_{\Delta\lambda b}}{\partial z} \tag{15-43}$$

This defines the mean absorption coefficient $a_{R,\Delta\lambda}$ as

$$\frac{1}{a_{R,\Delta\lambda}} = \frac{\int_{\Delta\lambda}(1/a_\lambda)(\partial e_{\lambda b}/\partial z) \, d\lambda}{\int_{\Delta\lambda}(\partial e_{\lambda b}/\partial z) \, d\lambda}$$

By multiplying the numerator and denominator by $\partial z/\partial e_b$, this is written as

$$\frac{1}{a_{R,\Delta\lambda}} = \frac{\int_{\Delta\lambda}(1/a_\lambda)(\partial e_{\lambda b}/\partial e_b) \, d\lambda}{\int_{\Delta\lambda}(\partial e_{\lambda b}/\partial e_b) \, d\lambda} \tag{15-44}$$

For the entire λ range

$$q_z = -\frac{4}{3a_R} \frac{\partial e_b}{\partial z} = -\frac{4}{3a_R} \frac{\partial(\sigma T^4)}{\partial z} = -\frac{16}{3a_R} \sigma T^3 \frac{\partial T}{\partial z} \tag{15-45}$$

where

$$\frac{1}{a_R} = \int_0^\infty \frac{1}{a_\lambda(\lambda)} \frac{\partial e_{\lambda b}}{\partial e_b} \, d\lambda$$

The a_R is called the *Rosseland mean absorption coefficient*, after S. Rosseland, who was the first to make use of diffusion theory in studying radiation effects in astrophysics [2]. The $\partial e_{\lambda b}/\partial e_b$ is found by differentiating Planck's law (2-11) after letting $T = (e_b/\sigma)^{1/4}$:

$$\frac{\partial e_{\lambda b}}{\partial e_b} = \frac{\partial}{\partial e_b} \left(\frac{2\pi C_1}{\lambda^5 \{\exp\left[(C_2/\lambda)(\sigma/e_b)^{1/4}\right] - 1\}} \right) = \frac{\pi}{2} \frac{C_1 C_2}{\lambda^6} \frac{\sigma^{1/4}}{e_b^{5/4}} \frac{\exp\left[(C_2/\lambda)(\sigma/e_b)^{1/4}\right]}{\{\exp\left[(C_2/\lambda)(\sigma/e_b)^{1/4}\right] - 1\}^2} \tag{15-46}$$

Emissive power jump boundary condition Up to now the position considered in the gas was far enough from any boundary so the boundary did not enter the diffusion relations. Now the interaction with a diffuse wall is considered. Let the wall bounding the gas from above, Fig. 15-4, have a hemispherical spectral emissivity $\epsilon_{\lambda w2}$. All quantities pertaining to the wall itself have the subscript w to differentiate them from quantities in the gas at the wall, which do not have a w subscript. Consider area dA_2 *in the gas* parallel and immediately adjacent to the

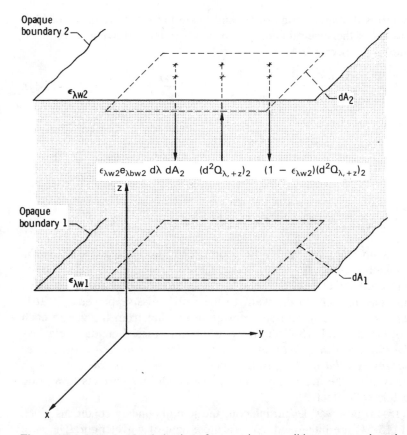

Figure 15-4 Geometry for derivation of energy-jump condition at opaque boundary.

wall. The spectral energy quantities passing through dA_2 are shown in Fig. 15-4 so that the net spectral flux across dA_2 in the positive z direction is

$$(dq_{\lambda,z})_2 = \epsilon_{\lambda w2} \left[\frac{(d^2Q_{\lambda,+z})_2}{dA_2 \, d\lambda} - e_{\lambda bw2} \right] d\lambda \tag{15-47}$$

This is rearranged into the form

$$-e_{\lambda bw2} = \frac{(dq_{\lambda,z})_2}{\epsilon_{\lambda w2} \, d\lambda} - \frac{(d^2Q_{\lambda,+z})_2}{dA_2 \, d\lambda} \tag{15-48}$$

Now (15-40) is substituted for $(d^2Q_{\lambda,+z})_2$. The first term of (15-40) for $n = 0$ and $dA_0 = dA_2$ is $(dA_2/4)4\pi i'_{\lambda b2} = dA_2 \, e_{\lambda b2}$. Then (15-48) becomes, at dA_2 in the gas adjacent to the wall,

$$e_{\lambda b2} - e_{\lambda bw2} = \frac{(dq_{\lambda,z})_2}{\epsilon_{\lambda w2} \, d\lambda} - \frac{1}{4\pi} \sum_{n=1}^{\infty} \sum_{v=0}^{n} \sum_{s=0}^{v} (-1)^{n-v} \Omega(n,v,s) \frac{1}{a_\lambda^n} \left(\frac{\partial^n e_{\lambda b}}{\partial z^{n-v} \partial y^{v-s} \partial x^s} \right)_2 \tag{15-49}$$

Retaining only terms through second order and using (15-42) to remove the first derivatives in terms of the spectral energy flux result in the following relation for the jump in emissive power at dA_2:

$$e_{\lambda b2} - e_{\lambda bw2} = \left(\frac{1}{\epsilon_{\lambda w2}} - \frac{1}{2}\right)\frac{(dq_{\lambda,z})_2}{d\lambda} - \frac{1}{2a_\lambda^2}\left(\frac{\partial^2 e_{\lambda b}}{\partial z^2} + \frac{1}{2}\frac{\partial^2 e_{\lambda b}}{\partial y^2} + \frac{1}{2}\frac{\partial^2 e_{\lambda b}}{\partial x^2}\right)_2 \qquad (15\text{-}50)$$

All quantities without a w subscript are evaluated at dA_2 *in the gas* adjacent to the wall. The quantities with a w subscript are on wall 2, and $dq_{\lambda,z}$ is the net flux in the positive z direction. Similarly, the jump in emissive power at dA_1 in Fig. 15-4 is

$$e_{\lambda bw1} - e_{\lambda b1} = \left(\frac{1}{\epsilon_{\lambda w1}} - \frac{1}{2}\right)\frac{(dq_{\lambda,z})_1}{d\lambda} + \frac{1}{2a_\lambda^2}\left(\frac{\partial^2 e_{\lambda b}}{\partial z^2} + \frac{1}{2}\frac{\partial^2 e_{\lambda b}}{\partial y^2} + \frac{1}{2}\frac{\partial^2 e_{\lambda b}}{\partial x^2}\right)_1 \qquad (15\text{-}51)$$

where quantities with a w subscript are on wall 1, and those without w are *in the gas* adjacent to wall 1.

Equations (15-50) and (15-51) are boundary conditions that relate the emissive power $e_{\lambda b}$ in the gas adjacent to the wall, to the wall emissive power $e_{\lambda bw}$. It is evident that there is a jump in emissive power in passing from the gas to each wall. The use of (15-38) and (15-40) in the derivation of these boundary relations implies that *the proportionality between local radiative flux and emissive power gradient in the gas is valid in the gas very near a bounding surface.* Although this is not strictly true, the use of jump boundary conditions corrects to a good approximation for wall effects.

To apply (15-43) in a wavelength interval, the jump boundary conditions given in (15-50) and (15-51) are integrated over an increment of wavelength. The wall emissivities are assigned averaged values in this range, and the integration is carried out in [1] to yield

$$e_{\Delta\lambda b2} - e_{\Delta\lambda bw2} = \left(\frac{1}{\epsilon_{\Delta\lambda w2}} - \frac{1}{2}\right)(q_{\Delta\lambda,z})_2 - \left\{\frac{1}{2a_{D,\Delta\lambda}^2}\left(\frac{\partial^2 e_{\Delta\lambda b}}{\partial z^2} + \frac{1}{2}\frac{\partial^2 e_{\Delta\lambda b}}{\partial y^2} + \frac{1}{2}\frac{\partial^2 e_{\Delta\lambda b}}{\partial x^2}\right)\right.$$
$$\left. + \frac{I_{\Delta\lambda}}{2}\left[\left(\frac{\partial e_{\Delta\lambda b}}{\partial z}\right)^2 + \frac{1}{2}\left(\frac{\partial e_{\Delta\lambda b}}{\partial y}\right)^2 + \frac{1}{2}\left(\frac{\partial e_{\Delta\lambda b}}{\partial x}\right)^2\right]\right\}_2 \qquad (15\text{-}52)$$

$$e_{\Delta\lambda bw1} - e_{\Delta\lambda b1} = \left(\frac{1}{\epsilon_{\Delta\lambda w1}} - \frac{1}{2}\right)(q_{\Delta\lambda,z})_1 + \left\{\frac{1}{2a_{D,\Delta\lambda}^2}\left(\frac{\partial^2 e_{\Delta\lambda b}}{\partial z^2} + \frac{1}{2}\frac{\partial^2 e_{\Delta\lambda b}}{\partial y^2} + \frac{1}{2}\frac{\partial^2 e_{\Delta\lambda b}}{\partial x^2}\right)\right.$$
$$\left. + \frac{I_{\Delta\lambda}}{2}\left[\left(\frac{\partial e_{\Delta\lambda b}}{\partial z}\right)^2 + \frac{1}{2}\left(\frac{\partial e_{\Delta\lambda b}}{\partial y}\right)^2 + \frac{1}{2}\left(\frac{\partial e_{\Delta\lambda b}}{\partial x}\right)^2\right]\right\}_1 \qquad (15\text{-}53)$$

where $q_{\Delta\lambda} = \int_{\Delta\lambda} dq_\lambda$. In these equations, the two mean coefficients are [1],

$$\frac{1}{a^2_{D,\Delta\lambda}} = \frac{\int_{\Delta\lambda}\left(\frac{1}{a^2_\lambda}\right)\left(\frac{\partial e_{\lambda b}}{\partial e_b}\right)d\lambda}{\int_{\Delta\lambda}\left(\frac{\partial e_{\lambda b}}{\partial e_b}\right)d\lambda}, \qquad I_{\Delta\lambda} = \frac{\int_{\Delta\lambda}\left(\frac{1}{a^2_\lambda}\right)\left(\frac{\partial^2 e_{\lambda b}}{\partial e^2_b}\right)d\lambda}{\left[\int_{\Delta\lambda}\left(\frac{\partial e_{\lambda b}}{\partial e_b}\right)d\lambda\right]^2} \qquad (15\text{-}54a,b)$$

The $I_{\Delta\lambda}$ has units of length squared divided by energy flux.

Emissive power jump between two absorbing-emitting regions When internal energy sources or sinks are present in absorbing-emitting media, it is possible in the absence of energy conduction to have a discontinuity in emissive power at the interface of two media. The jump is calculated by considering a volume element at the interface. The lower region has absorption coefficient $a_{\lambda 1}$, and the upper $a_{\lambda 2}$. The net flux in the z direction passing through the element is, by use of Eqs. (15-38) and (15-40) in media 2 and 1, respectively,

$$\frac{(dq_{\lambda,z})_{g\text{-}g}}{d\lambda} = \frac{(d^2 Q_{\lambda,+z})_1 - (d^2 Q_{\lambda,-z})_2}{dA\,d\lambda}$$

$$= \frac{1}{4\pi}\sum_{n=0}^{\infty}\sum_{v=0}^{n}\sum_{s=0}^{v}\Omega(n,v,s)\left[\frac{(-1)^{n-v}}{a_{\lambda 1}{}^n}\left(\frac{\partial^n e_{\lambda b}}{\partial z^{n-v}\,\partial y^{v-s}\,\partial x^s}\right)_1\right.$$

$$\left. - \frac{1}{a_{\lambda 2}{}^n}\left(\frac{\partial^n e_{\lambda b}}{\partial z^{n-v}\,\partial y^{v-s}\,\partial x^s}\right)_2\right]_{g\text{-}g} \qquad (15\text{-}55)$$

Neglecting terms of order higher than two gives the emissive power jump as

$$(e_{\lambda b2} - e_{\lambda b1})_{g\text{-}g} = -\frac{(dq_{\lambda,z})_{g\text{-}g}}{d\lambda} + \left\{-\frac{2}{3}\left[\frac{1}{a_{\lambda 1}}\left(\frac{\partial e_{\lambda b}}{\partial z}\right)_1 + \frac{1}{a_{\lambda 2}}\left(\frac{\partial e_{\lambda b}}{\partial z}\right)_2\right]\right.$$

$$+ \frac{1}{2a^2_{\lambda 1}}\left(\frac{\partial^2 e_{\lambda b}}{\partial z^2} + \frac{1}{2}\frac{\partial^2 e_{\lambda b}}{\partial y^2} + \frac{1}{2}\frac{\partial^2 e_{\lambda b}}{\partial x^2}\right)_1$$

$$\left. - \frac{1}{2a^2_{\lambda 2}}\left(\frac{\partial^2 e_{\lambda b}}{\partial z^2} + \frac{1}{2}\frac{\partial^2 e_{\lambda b}}{\partial y^2} + \frac{1}{2}\frac{\partial^2 e_{\lambda b}}{\partial x^2}\right)_2\right\}_{g\text{-}g} \qquad (15\text{-}56)$$

The integrated form of Eq. (15-56) for a wavelength interval is in [3]. The jump $e_{b2} - e_{b1}$ is nonzero under certain conditions.

Use of the diffusion solution The general radiation-diffusion equation has been derived for a single wavelength as (15-41) or (15-42), and for a wavelength band as (15-43). The boundary conditions at solid boundaries with normals into the gas in the negative and positive coordinate directions are given at a single wavelength by (15-50) and (15-51). The interface condition between two absorbing-emitting media in the absence of heat conduction is (15-56). When the diffusion equation is utilized, it is assumed to apply throughout the entire medium including the region adjacent to a boundary. The boundary effect is accounted for by using a jump boundary condition. If a real gas is considered, three coefficients given by (15-44) and (15-54a,b) must be evaluated. Each of these depends only on local conditions, so they can be tabulated.

Gray stagnant gas between parallel plates Most gases have strong variations of properties with wavelength, and it is necessary to solve the diffusion equation in a number of wavelength regions. For illustrative purposes it is not reasonable to consider an involved spectral solution. There are some situations, such as soot-filled flames and high-temperature uranium gas, for which a gray-gas approximation can be made. The equations reduce considerably in this case. Let us examine the case of a gray gas between infinite parallel gray plates at unequal temperatures (Fig. 15-5).

For a gray gas the absorption coefficient is independent of wavelength. Then the previous definitions give $a_R = a_D = a$ and

$$I = \frac{(1/a^2)(\partial^2/\partial e_b^2) \int_0^\infty e_{\lambda b} \, d\lambda}{[(\partial/\partial e_b)(\int_0^\infty e_{\lambda b} \, d\lambda)]^2} = \frac{(1/a^2)(\partial^2 e_b/\partial e_b^2)}{(\partial e_b/\partial e_b)^2} = 0 \tag{15-57}$$

Equation (15-45) gives $q_z = -(4/3a)(de_b/dz)$. This can be integrated directly because, with no sources or sinks in the gas, q_z is constant for this geometry.

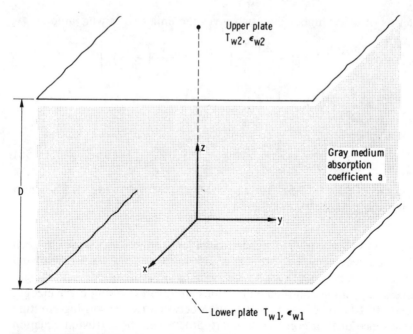

Figure 15-5 Radiant interchange between infinite parallel gray plates enclosing gray medium.

With the additional assumption that a does not depend on temperature and is therefore independent of z, the result of integrating from 0 to z is

$$e_b(z) - e_{b1} = -\frac{3a}{4} q_z z \tag{15-58}$$

Evaluating (15-58) at $z = D$ yields

$$\frac{e_{b2} - e_{b1}}{q_z} = -\frac{3aD}{4} \tag{15-59}$$

The e_{b1} and e_{b2} are *in the gas* adjacent to the walls. To connect the unknown e_{b1} and e_{b2} with the specified wall conditions, jump boundary conditions are applied. Differentiating (15-58) twice with respect to z shows that the second-derivative terms are zero in the boundary-condition equations (15-52) and (15-53). Equations (15-52) and (15-53) then become

$$\frac{e_{b2} - e_{bw2}}{q_z} = \frac{1}{\epsilon_{w2}} - \frac{1}{2} \qquad \frac{e_{bw1} - e_{b1}}{q_z} = \frac{1}{\epsilon_{w1}} - \frac{1}{2} \tag{15-60a,b}$$

To eliminate the unknown gas emissive powers e_{b1} and e_{b2}, add Eqs. (15-60a) and (15-60b),

$$\frac{e_{bw1} - e_{bw2}}{q_z} + \frac{e_{b2} - e_{b1}}{q_z} = \frac{1}{\epsilon_{w1}} + \frac{1}{\epsilon_{w2}} - 1$$

Then substitute $e_{b2} - e_{b1}$ from (15-59) and take the reciprocal to yield

$$\frac{q_z}{e_{bw1} - e_{bw2}} = \frac{1}{3aD/4 + 1/\epsilon_{w1} + 1/\epsilon_{w2} - 1} = \frac{1}{3\kappa_D/4 + 1/\epsilon_{w1} + 1/\epsilon_{w2} - 1} \tag{15-61}$$

Equation (15-61) gives the radiative energy transfer through a gray-gas layer as a function of the optical thickness aD and the plate emissivities. It is ratioed to the difference in the black emissive powers of the plates, which is the maximum possible energy transfer. A comparison of this diffusion solution with the solution of the exact integral equations [4] is in Fig. 15-6 for equal wall emissivities. Agreement is excellent for all optical thicknesses. The distribution of emissive power $e_b(z)$ is found from (15-58) by eliminating the unknown e_{b1} by use of Eq. (15-60b), or in another form by eliminating e_{b1} and e_{b2} from (15-58) to (15-60). The result is in Table 15-2.

Discontinuity in emissive power between two gas regions Consider two adjacent semi-infinite regions (Fig. 15-7). Determine the discontinuity in emissive power, if any, at the interface between the regions in the absence of heat conduction. First, consider the media in the two regions to have no internal heat sources or sinks. Both media are gray and stagnant; the lower region has a constant absorption coefficient a_1, and the upper, a_2. The emissive power jump at the interface between the two media is found by integrating (15-56) over all λ. This gives,

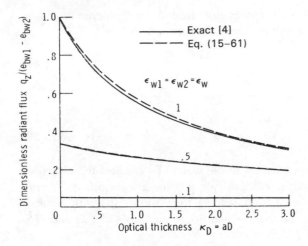

Figure 15-6 Validity of diffusion solution for energy transfer through gray gas between parallel gray plates.

after noting that $a_{\lambda 1} = a_1$, that $a_{\lambda 2} = a_2$, and that derivatives with respect to x and y are zero for a one-dimensional layer,

$$(e_{b2} - e_{b1})_{g-g} = -(q_z)_{g-g} - \frac{2}{3}\left[\frac{1}{a_1}\left(\frac{de_b}{dz}\right)_1 + \frac{1}{a_2}\left(\frac{de_b}{dz}\right)_2\right]_{g-g} \tag{15-62}$$

Second derivatives with z are zero because in *either* region (15-43) gives

$$q_z = -\frac{4}{3a}\frac{de_b}{dz} \tag{15-63}$$

The q_z must be contant, since no heat sources or sinks are present. Therefore, in either region, $d^2 e_b/dz^2 = 0$. Also, q_z must be the *same in either region* because the radiative flux is continuous across the interface. Therefore (15-63) can be substituted for the derivatives in (15-62) to give

$$(e_{b2} - e_{b1})_{g-g} = -(q_z)_{g-g} - \frac{2}{3}\left[\frac{1}{a_1}\left(-\frac{3a_1}{4}\right)q_z + \frac{1}{a_2}\left(-\frac{3a_2}{4}\right)q_z\right]_{g-g} \tag{15-64}$$

This reduces to

$$(e_{b2} - e_{b1})_{g-g} = -(q_z)_{g-g} + (q_z)_{g-g} = 0 \tag{15-65}$$

so that *no discontinuity in emissive power exists in this case.*

Now consider uniform volumetric energy sources G_1 and G_2 in regions 1 and 2. The flux gradient in either region is

$$\frac{dq_z}{dz} = G = -\frac{4}{3a}\frac{d^2 e_b}{dz^2} \tag{15-66}$$

In this case, the second derivatives of e_b with respect to z are finite, and (15-56) becomes

$$(e_{b2} - e_{b1})_{g-g} = -(q_z)_{g-g} + \left\{ -\frac{2}{3}\left[\frac{1}{a_1}\left(\frac{de_b}{dz}\right)_1 + \frac{1}{a_2}\left(\frac{de_b}{dz}\right)_2 \right] \right.$$

$$\left. + \frac{1}{2}\left[\frac{1}{a_1{}^2}\left(\frac{d^2 e_b}{dz^2}\right)_1 - \frac{1}{a_2{}^2}\left(\frac{d^2 e_b}{dz^2}\right)_2 \right] \right\}_{g-g}$$ (15-67)

Table 15-2 Diffusion-theory predictions of energy transfer and temperature distribution for a gray gas between gray surfaces

Geometry	Relations[a]
Infinite parallel plates 	$\psi = \dfrac{1}{(3aD/4) + E_1 + E_2 + 1}$ $\phi(z) = \psi\left[\dfrac{3a}{4}(D - z) + E_2 + \dfrac{1}{2} \right]$
Infinitely long concentric cylinders 	$\psi = \dfrac{1}{\dfrac{3}{8}\left[aD_1 \ln\left(\dfrac{D_2}{D_1}\right) + \dfrac{1 - (D_1/D_2)^2}{aD_1} \right] + \left(E_1 + \dfrac{1}{2}\right) + \dfrac{D_1}{D_2}\left(E_2 + \dfrac{1}{2}\right)}$ $\phi(r) = \psi\left\{ -\dfrac{3}{8}\left[aD_1 \ln\left(\dfrac{D}{D_2}\right) + \dfrac{D_1}{aD_2^2} \right] + \left(E_2 + \dfrac{1}{2}\right)\dfrac{D_1}{D_2} \right\}$
Concentric spheres 	$\psi = \dfrac{1}{\dfrac{3}{8}\left[aD_1\left(1 - \dfrac{D_1}{D_2}\right) + 2\dfrac{1 - (D_1/D_2)^3}{aD_1} \right] + \left(E_1 + \dfrac{1}{2}\right) + \dfrac{D_1^2}{D_2^2}\left(E_2 + \dfrac{1}{2}\right)}$ $\phi(r) = \psi\left\{ -\dfrac{3}{8}\left[aD_1\left(\dfrac{D_1}{D_2} - \dfrac{D_1}{D}\right) + \dfrac{2D_1^2}{aD_2^3} \right] + \left(E_2 + \dfrac{1}{2}\right)\dfrac{D_1^2}{D_2^2} \right\}$

[a]Definitions: $E_N = (1 - \epsilon_{wN})/\epsilon_{wN}$, $\psi = Q_1/A_1\sigma(T_{w1}^4 - T_{w2}^4)$, $\phi(\xi) = [T^4(\xi) - T_{w2}^4]/(T_{w1}^4 - T_{w2}^4)$, $D = 2r$.

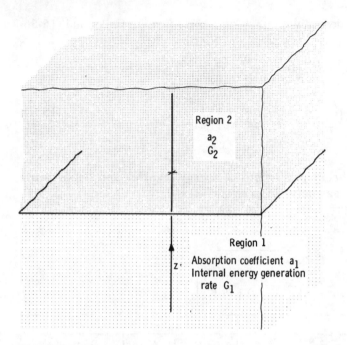

Figure 15-7 Geometry for derivation of interface emissive power discontinuity.

Again, (15-63) must hold in either region. At the interface between the two media, since the flux is continuous,

$$q_{z,g-g} = -\frac{4}{3a_1}\left(\frac{de_b}{dz}\right)_{1,g-g} = -\frac{4}{3a_2}\left(\frac{de_b}{dz}\right)_{2,g-g} \tag{15-68}$$

Substituting (15-66) and (15-68) into (15-67) to eliminate the first and second derivatives of e_b gives

$$(e_{b2} - e_{b1})_{g-g} = -(q_z)_{g-g} + \left[-\frac{2}{3}\left(\frac{1}{a_1}\frac{-3a_1}{4}q_z + \frac{1}{a_2}\frac{-3a_2}{4}q_z\right) \right.$$

$$\left. + \frac{1}{2}\left(\frac{1}{a_1{}^2}\frac{-3a_1}{4}G_1 - \frac{1}{a_2{}^2}\frac{-3a_2}{4}G_2\right) \right]_{g-g}$$

which reduces to

$$(e_{b1} - e_{b2})_{g-g} = \frac{3}{8}\left(\frac{G_1}{a_1} - \frac{G_2}{a_2}\right) \tag{15-69}$$

A discontinuity in emissive power exists when G_1/a_1 and G_2/a_2 are not equal.

If the present problem is formulated with exact integral equations [5] rather than by diffusion methods, there results $(e_{b1} - e_{b2})_{g-g} = (1/4)(G_1/a_1 - G_2/a_2)$. The diffusion solution, although giving the correct functional dependence of the emissive-power discontinuity, differs from the exact solution by a significant factor of $\frac{3}{2}$.

Other diffusion solutions for gray gases Table 15-2 contains solutions for temperature distribution and energy transfer in simple geometries involving gray gases between gray walls (see Example 15-7 for further analytical details). These results are derived from the diffusion equations; caution is advised in their application since real gases are usually not gray and are not optically thick in all wavelength regions. Agreement with exact solutions is sometimes not as good for cylindrical or spherical geometries as for the infinite parallel-plate case. Good agreement has been found in cylindrical and spherical geometries if the optical thickness is greater than about 7, with better agreement as wall emissivities become lower and diameter ratios ($D_{\text{inner}}/D_{\text{outer}}$) become larger. A comparison for cylindrical geometry is discussed later in connection with Fig. 15-10.

EXAMPLE 15-7 The space between two diffuse-gray spheres (Fig. 15-8) is filled with an optically dense stagnant medium having constant absorption coefficient a. Compute the heat flow Q_1 across the gap from sphere 1 to sphere 2 and the temperature distribution $T(r)$ in the gas, using the diffusion method with jump boundary conditions.

For a gray medium with constant a, Eq. (15-42) gives the net heat flux in the positive r direction as $q_r = -(4/3a) \, de_b/dr$. From energy conservation, q_r varies with r as $q_r = Q_1/4\pi r^2$. Combine these relations and integrate from R_1 to R_2 to obtain

$$\frac{Q_1}{4\pi} \int_{R_1}^{R_2} \frac{dr}{r^2} = -\frac{4}{3a} \int_{e_{b1}}^{e_{b2}} de_b \qquad (15\text{-}70)$$

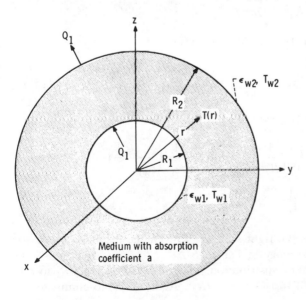

Medium with absorption coefficient a

Figure 15-8 Radiation across gap between concentric spheres with intervening medium of constant absorption coefficient.

$$\frac{Q_1}{4\pi}\left(\frac{1}{R_2}-\frac{1}{R_1}\right)=\frac{4}{3a}(e_{b2}-e_{b1}) \tag{15-71}$$

The e_{b1} and e_{b2} are *in the gas* adjacent to the boundaries, and jump boundary conditions are needed to express these quantities in terms of wall values. The jump boundary conditions are given by (15-52) and (15-53) and involve second derivatives which will now be found. By integrating (15-70) from R_1 to r, one obtains

$$e_b(r)-e_{b1}=\frac{3aQ_1}{16\pi}\left(\frac{1}{r}-\frac{1}{R_1}\right) \tag{15-72}$$

Substitute $r=(x^2+y^2+z^2)^{1/2}$ and differentiate twice with respect to x to obtain

$$\frac{\partial^2 e_b(r)}{\partial x^2}=-\frac{3aQ_1}{16\pi}\frac{(x^2+y^2+z^2)^{3/2}-3x^2(x^2+y^2+z^2)^{1/2}}{(x^2+y^2+z^2)^3} \tag{15-73}$$

and similarly for the y and z directions.

In the boundary condition (15-52) the point 2 can be conveniently taken in Fig. 15-8 at $x=y=0$ and $z=R_2$. This gives

$$\left[\frac{\partial^2 e_b(r)}{\partial x^2}\right]_2=\left[\frac{\partial^2 e_b(r)}{\partial y^2}\right]_2=-\frac{3aQ_1}{16\pi}\frac{1}{R_2^3} \qquad \left[\frac{\partial^2 e_b(r)}{\partial z^2}\right]_2=\frac{3aQ_1}{8\pi}\frac{1}{R_2^3}$$

Also $(q_z)_2=Q_1/4\pi R_2^2$. Substituting into (15-52) gives

$$e_{b2}-e_{bw2}=\left(\frac{1}{\epsilon_{w2}}-\frac{1}{2}\right)\frac{Q_1}{4\pi R_2^{\,2}}-\frac{3Q_1}{32a\pi}\frac{1}{R_2^{\,3}} \tag{15-74}$$

Similarly, at the inner sphere boundary, from (15-53),

$$e_{bw1}-e_{b1}=\left(\frac{1}{\epsilon_{w1}}-\frac{1}{2}\right)\frac{Q_1}{4\pi R_1^{\,2}}+\frac{3Q_1}{32a\pi}\frac{1}{R_1^{\,3}} \tag{15-75}$$

Adding (15-74) and (15-75) gives

$$e_{b2}-e_{b1}=e_{bw2}-e_{bw1}+\frac{Q_1}{4\pi}\left[\frac{1}{R_2^{\,2}}\left(\frac{1}{\epsilon_{w2}}-\frac{1}{2}\right)+\frac{1}{R_1^{\,2}}\left(\frac{1}{\epsilon_{w1}}-\frac{1}{2}\right)\right.$$
$$\left.+\frac{3}{8a}\left(\frac{1}{R_1^{\,3}}-\frac{1}{R_2^{\,3}}\right)\right]$$

After substituting this into the right side of (15-71), the result is solved for Q_1 to give the ψ in the last entry in Table 15-2.

To obtain the temperature distribution, integrate (15-70) from R_2 to r to obtain $e_b(r)-e_{b2}=(3aQ_1/16\pi)(1/r-1/R_2)$. Add (15-74) to eliminate e_{b2}:

$$e_b(r) - e_{bw2} = \frac{3aQ_1}{16\pi}\left(\frac{1}{r} - \frac{1}{R_2}\right) + \left(\frac{1}{\epsilon_{w2}} - \frac{1}{2}\right)\frac{Q_1}{4\pi R_2{}^2} - \frac{3Q_1}{32a\pi}\frac{1}{R_2{}^3} \tag{15-76}$$

This gives the last expression for ϕ in Table 15-2.

The diffusion method is useful in treating difficult problems by standard analytical techniques and can be used when the assumptions in its derivation are justified. The most stringent assumption is that of *optically thick conditions*. Because most gases have band spectra, the optically thick regions arise within band limits. Here, if the radiation-absorption mean free path is small, the assumption that only local conditions affect the spectral radiant flux is quite good. At other wavelengths the gas is often transparent, and diffusion methods are not justified. Care must be taken to apply the diffusion equation only in geometrical and spectral regions where the optically thick assumption is valid.

The Rosseland mean absorption coefficient for the entire range of λ should not be used as the criterion for optical thickness. It may have a large value, but the spectral absorption coefficient may be small in certain spectral regions. Use of the Rosseland mean coefficient in such cases may lead to large errors. The remedy is to use wavelength bands in which the spectral absorption coefficient is everywhere large and to evaluate a Rosseland mean for each of these regions. Howell and Perlmutter [6] applied the diffusion method to a real gas and compared the results to an exact solution by the Monte Carlo method. The agreement was not as good as when comparing results for gray gases. Bobco [7] used a modified diffusion solution to find the directional emissivity of a semi-infinite slab of isothermal gray scattering-absorbing medium with isotropic scattering. The directional emissivities were found to differ considerably from a diffuse distribution.

15-4 USE OF MEAN ABSORPTION COEFFICIENTS

A mean absorption coefficient that is spectrally averaged over all wavelengths has been used to approximate the correct procedure of carrying out a spectral analysis and then integrating the spectral energy over all wavelengths to obtain the total energy. A question is whether it is possible to decide in advance for a particular situation if using a mean absorption coefficient will yield a sufficiently accurate solution.

15-4.1 Some Mean Absorption Coefficients

To this point, three mean absorption coefficients have been defined: the Planck mean (14-96)

$$a_P(T,P) = \frac{\int_0^\infty a_\lambda(\lambda,T,P)e_{\lambda b}(\lambda,T)\,d\lambda}{\sigma T^4} \tag{15-77}$$

the incident mean (14-97)

$$a_i(T,P) = \frac{\int_0^\infty a_\lambda(\lambda, T, P) i_{\lambda,i}(\lambda) \, d\lambda}{\int_0^\infty i_{\lambda,i}(\lambda) \, d\lambda} \tag{15-78}$$

and the Rosseland mean (15-45)

$$a_R(T, P) = \frac{1}{\int_0^\infty \frac{1}{a_\lambda(\lambda, T, P)} \frac{\partial e_{\lambda b}(\lambda, T)}{\partial e_b(T)} \, d\lambda} \tag{15-79}$$

The incident mean can only be used conveniently under restrictive conditions when the incident intensity has a spectral form that remains fixed so that a_i can be evaluated and tabulated. For example, incident energy having a solar spectrum occurs sufficiently often that a_i could be tabulated for this case. The a_i is useful in the transparent approximation when the spectral intensity leaving the boundaries is known, as this spectrum will remain unchanged while radiation travels through the medium. If the mean intensity $\bar{i}_{\lambda,i}$ is proportional to a blackbody spectrum *at the temperature of the position for which $a_\lambda(\lambda, T, P)$ is evaluated*, that is, $\bar{i}_{\lambda,i} \propto i'_{\lambda b}(\lambda, T)$, then the incident mean is equal to a_P:

$$a_i(T,P) = \frac{\int_0^\infty a_\lambda(\lambda, T, P) i'_{\lambda b}(\lambda, T) \, d\lambda}{\int_0^\infty i'_{\lambda b}(\lambda, T) \, d\lambda} = a_P(T,P) \tag{15-80}$$

At first glance, the Rosseland mean appears to be entirely different in character from a_P and a_i, which are weighted by spectral distributions of energy or intensity. However, one may write Eq. (15-42) for a one-dimensional diffusion case as

$$dq_{\lambda,z} = -\frac{4}{3a_\lambda} \frac{de_{\lambda b}(\lambda, T)}{dz} \, d\lambda = -\frac{4}{3a_\lambda} \frac{\partial e_b}{\partial z} \frac{\partial e_{\lambda b}}{\partial e_b} \, d\lambda \tag{15-81}$$

Then

$$\int_0^\infty a_\lambda \, dq_{\lambda,z} = -\frac{4}{3} \frac{\partial e_b}{\partial z} \quad \text{and} \quad \int_0^\infty dq_z = -\frac{4}{3} \frac{\partial e_b}{\partial z} \int_0^\infty \frac{1}{a_\lambda} \frac{de_{\lambda b}}{de_b} \, d\lambda$$

Substituting into (15-79) gives for the *diffusion case*

$$a_R(T,P) = \frac{\int_0^\infty a_\lambda \, dq_{\lambda,z}}{\int_0^\infty dq_{\lambda,z}} \tag{15-82}$$

The Rosseland mean is thus an average value of a_λ weighted by the local spectral energy flux $dq_{\lambda,z}$ through the assumption that the local flux depends only on the local gradient of emissive power and the local a_λ.

For a gray gas the absorption coefficient is independent of wavelength, $a_\lambda(\lambda, T, P) = a(T, P)$, and (15-77) to (15-79) reduce to $a_P(T, P) = a_i(T, P) = a_R(T, P) = a(T, P)$.

Determining any of the mean coefficients from spectral absorption coefficients usually requires tedious detailed numerical integrations. Even so, if appropriate spectral mean values can be applied to yield reasonably accurate solutions, the solution time is often considerably decreased.

15-4.2 Approximate Solutions of the Radiative Transfer Equations Using Mean Absorption Coefficients

Some references are now reviewed where mean absorption coefficients have been used in radiative transfer calculations. Solving the transfer and energy equations is considerably simplified because integrations over wavelength are not needed. The most common approximation is that *gray* gas relations can be applied for a *real* gas by substituting an appropriate mean absorption coefficient in place of the *a* in the gray gas solution. In Sec. 14-7.1 it was shown that, although the Planck mean may be used in the *part* of the energy balance equation dealing with local emission, the use of this mean coefficient in the absorption and attenuation terms is invalid except in special cases. Patch [8, 9] showed, by examining 40 cases, that simple substitution of the Planck mean in gray-gas solutions leads to errors in total intensities that varied from -43 to 881% from the correct solutions obtained by using spectral properties in the transfer equations and then integrating the spectral results. Reductions in error were found by dividing the intensity into two or more spectral bands and using an individual Planck mean for each band.

In an effort to improve this situation, a number of other mean absorption coefficients have been introduced. Sampson [10] synthesized a coefficient that varies from the Planck mean to the Rosseland mean as the optical depth increases along a path. He found agreement, within a factor of two, with exact solutions for various problems. Abu-Romia and Tien [11] applied a weighted Rosseland mean over optically thick portions of the spectrum and a Planck mean over optically thin regions and obtained relations for energy transfer between bounding surfaces. Planck and Rosseland mean absorption coefficients for carbon dioxide, carbon monoxide, and water vapor are given as an aid to such computations.

Patch [8, 9] defined an *effective mean absorption* coefficient as

$$a_e(S, T, P) = \frac{\displaystyle\int_0^\infty a_\lambda(\lambda, T, P) i'_{\lambda b}(\lambda, T) \exp[-a_\lambda(\lambda, T, P)S]\, d\lambda}{\displaystyle\int_0^\infty i'_{\lambda b}(\lambda, T) \exp[-a_\lambda(\lambda, T, P)S]\, d\lambda} \tag{15-83}$$

The values of $a_e(S, T, P)$ can be tabulated as a function of temperature and pressure as for the other mean absorption coefficients. In addition, a_e depends on the path length and must be tabulated as a function of that variable. For small S, a_e

approaches a_P. For large S, the exponential term in the integrals causes a_e to approach the minimum value of a_λ in the spectrum considered. In radiative transfer calculations the approximation is made that the real gas, with T and P known variables along S, is replaced along any path by an effective uniform gas with absorption coefficient a_e. The computations are then performed using a_e in the gray-gas equation of transfer. The a_e value used is found by equating $a_e S$ at the T and P of the point to which S is measured to the optical depth of that point in the real gas. For 40 cases Patch [8, 9] shows agreement of total intensities within -25 to 28% of the integrated spectral solutions, as compared with the -43 to 881% agreement using a_P. This method has value in computer-oriented solutions, where tabulated values of $a_e(S, T, P)$ can be effectively manipulated. Other methods of using mean coefficients are in [12–16].

15-4.3 The Curtis-Godson Approximation

A useful method for thermal radiation analysis in nonuniform gases is the Curtis-Godson approximation [17–21]. In this method, the transmittance of a given path through a nonisothermal gas is related to the transmittance through an *equivalent isothermal* gas. Then the solution is obtained by using isothermal-gas methods. The relation between the nonisothermal and the isothermal gas is found by assigning an equivalent amount of isothermal absorbing material to act in place of the nonisothermal gas. The amount is based on a scaling temperature and a mean density or pressure obtained in the analysis. These mean quantities are found by having the transmittance of the uniform gas be equal to the transmittance of the nonuniform gas in the weak and strong absorption limits.

Goody [20], Krakow et al. [18], and Simmons [21] discussed the Curtis-Godson method for attenuation in a narrow vibration-rotation band. Excellent comparisons with exact numerical results were obtained. Weiner and Edwards [22] applied the method for environments with steep temperature gradients in gases with overlapping band structures. Comparison of the analysis with experimental data was again excellent. The Curtis-Godson technique is most useful when the gas *temperature distribution is specified*. If the gas temperature distribution is not known, an iterative procedure would need to be developed for its determination. This is not considered here, as the method is not too practical for that type of calculation.

In this development, spectral variations are in terms of wavenumber, as is usual for band correlations. For a nonuniform gas the absorption coefficient a_η is variable along the path. An effective bandwidth $\bar{A}_l(S)$ is defined, analogous to Eq. (12-74) but using an integrated absorption coefficient:

$$\bar{A}_l(S) = \int_{\substack{\text{absorption} \\ \text{bandwidth}}} \left\{ 1 - \exp\left[-\int_0^S a_\eta(\eta, S^*)\, dS^* \right] \right\} d\eta$$

$$= \Delta \eta_l - \int_l \left\{ \exp\left[-\int_0^S a_\eta(\eta, S^*)\, dS^* \right] \right\} d\eta \tag{15-84}$$

Similarly, for a path length extending from S^* to S, the effective bandwidth is

$$\bar{A}_l(S - S^*) = \int_{\substack{\text{absorption} \\ \text{bandwidth}}} \left\{ 1 - \exp\left[-\int_{S^*}^{S} a_\eta(\eta, S^{**}) \, dS^{**} \right] \right\} d\eta \qquad (15\text{-}85)$$

The equation of transfer is now placed in a form utilizing $\bar{A}_l(S)$ and $\bar{A}_l(S - S^*)$.

The integrated equation of transfer for intensity at S as a result of radiation traveling from 0 to S is, from Eq. (15-1),

$$i'_\eta(\eta, S) = i'_\eta(\eta, 0) \exp\left[-\int_0^S a_\eta(\eta, S^*) \, dS^* \right]$$

$$+ \int_0^S a_\eta(\eta, S^*) i'_{\eta b}(\eta, S^*) \exp\left[-\int_{S^*}^S a_\eta(\eta, S^{**}) \, dS^{**} \right] dS^*$$

An equivalent form is

$$i'_\eta(\eta, S) = i'_\eta(\eta, 0) \exp\left[-\int_0^S a_\eta(\eta, S^*) \, dS^* \right]$$

$$- \int_0^S i'_{\eta b}(\eta, S^*) \frac{\partial}{\partial S^*} \left\{ 1 - \exp\left[-\int_{S^*}^S a_\eta(\eta, S^{**}) \, dS^{**} \right] \right\} dS^* \qquad (15\text{-}86)$$

Equation (15-86) is integrated over the bandwidth $\Delta\eta_l$ of the lth band, and the order of integration is changed on the last term. The $i'_\eta(\eta, S)$, $i'_\eta(\eta, 0)$ and $i'_{\eta b}(\eta, S)$ are approximated by average values within the band to yield

$$i'_l(S) \Delta\eta_l = i'_l(0) \int_l \left\{ \exp\left[-\int_0^S a_\eta(\eta, S^*) \, dS^* \right] \right\} d\eta$$

$$- \int_0^S i'_{l,b}(S^*) \frac{\partial}{\partial S^*} \int_l \left\{ 1 - \exp\left[-\int_{S^*}^S a_\eta(\eta, S^{**}) \, dS^{**} \right] \right\} d\eta \, dS^* \qquad (15\text{-}87)$$

Equations (15-84) and (15-85) are substituted into (15-87) to obtain the equation of transfer in terms of the $\bar{A}_l$:

$$i'_l(S) \Delta\eta_l = i'_l(0)[\Delta\eta_l - \bar{A}_l(S)] - \int_0^S i'_{l,b}(S^*) \frac{\partial \bar{A}_l(S - S^*)}{\partial S^*} \, dS^* \qquad (15\text{-}88)$$

An alternative form is found by integrating Eq. (15-88) by parts to obtain

$$i'_l(S) \Delta\eta_l = i'_l(0)[\Delta\eta_l - \bar{A}_l(S)] + i'_{l,b}(0)\bar{A}_l(S) + \int_0^S \bar{A}_l(S - S^*) \frac{di'_{l,b}(S^*)}{dS^*} \, dS^* \qquad (15\text{-}89)$$

Equations (15-88) and (15-89) are nearly exact forms of the integrated equation of transfer in terms of the band properties. The only approximation is that the intensity in each term does not vary significantly across the wavenumber span of the band. For a *uniform* gas, (15-89) gives (since $di'_{l,b}/dS = 0$)

$$i'_{l,u}(S)\,\Delta\eta_l = i'_l(0)[\Delta\eta_l - \bar{A}_{l,u}(S)] + i'_{l,b,u}\bar{A}_{l,u}(S) \tag{15-90}$$

where the u subscript denotes a uniform gas.

To compute $i'_l(S)$ or $i'_{l,u}(S)$ from (15-88), (15-89), or (15-90), expressions are needed for the effective bandwidth $\bar{A}_l$ for nonuniform and uniform gases. From (12-77) and (12-80), the limiting cases of $\bar{A}_l$ for bands of independent weak or strong absorption lines in a uniform gas have the form

$$\bar{A}_{l,u}(S) = C_{1,l}\rho_u S_u \qquad \text{weak} \tag{15-91a}$$

$$\bar{A}_{l,u}(S) = C_{2,l}\rho_u S_u^{1/2} \qquad \text{strong} \tag{15-91b}$$

where $C_{1,l}$ and $C_{2,l}$ are coefficients of proportionality for the lth band, and S_c and Δ_c for the lines have been taken as proportional to gas density. For the nonuniform gas the effective bandwidth depends on the variation of properties along the path. The effective bandwidths are obtained by using (15-91a) and (15-91b) locally along the path. This gives, for a band of weak lines,

$$\bar{A}_l(S) = C_{1,l}\int_0^S \rho(S^*)\,dS^* \qquad \text{weak} \tag{15-92a}$$

Similarly, for a band of strong lines, after first squaring Eq. (15-91b), $\bar{A}_l^2(S) = C_{2,l}^2\int_0^S \rho^2(S^*)\,dS^*$, so that

$$\bar{A}_l(S) = C_{2,l}\left[\int_0^S \rho^2(S^*)\,dS^*\right]^{1/2} \qquad \text{strong} \tag{15-92b}$$

It has been assumed that the $C_{1,l}$ and $C_{2,l}$ do not vary along the path.

In the Curtis-Godson method the *nonuniform gas is replaced by an effective amount of uniform gas such that the correct intensity is obtained at the weak and strong absorption limits*. To have the uniform intensity equal the nonuniform intensity, equate (15-90) and (15-89) and simplify to obtain

$$[i'_{l,b,u}(T_u) - i'_l(0)]\bar{A}_{l,u}(S)$$

$$= [i'_{l,b}(0) - i'_l(0)]\bar{A}_l(S) + \int_0^S \bar{A}_l(S - S^*)\frac{di'_{l,b}(S^*)}{dS^*}\,dS^* \tag{15-93}$$

To have (15-93) valid at the weak absorption limit, substitute $\bar{A}_{l,u}$ from (15-91a) and $\bar{A}_l$ from (15-92a) to obtain the following after canceling the $C_{1,l}$:

$$[i'_{l,b,u}(T_u) - i'_l(0)]\rho_u S_u$$

$$= [i'_{l,b}(0) - i'_l(0)]\int_0^S \rho(S^*)\,dS^* + \int_0^S\left[\int_{S^*}^S \rho(S^{**})\,dS^{**}\right]\frac{di'_{l,b}(S^*)}{dS^*}\,dS^* \tag{15-94a}$$

Similarly, at the strong absorption limit, insert Eqs. (15-91b) and (15-92b) into

Eq. (15-93) to obtain

$$[i'_{l,b,u}(T_u) - i'_l(0)]\rho_u S_u^{1/2}$$

$$= [i'_{l,b}(0) - i'_l(0)]\left[\int_0^S \rho^2(S^*)\,dS^*\right]^{1/2} + \int_0^S\left[\int_{S^*}^S \rho^2(S^{**})\,dS^{**}\right]^{1/2}\frac{di'_{l,b}(S^*)}{dS^*}\,dS^*$$

$$(15\text{-}94b)$$

For a *known* distribution of temperature and density in a nonuniform gas, Eqs. (15-94a) and (15-94b) are solved simultaneously for ρ_u and S_u, which are the equivalent uniform gas density and path length for that particular band. The $i'_{l,b,u}(T_u)$ is not an additional unknown since the temperature T_u corresponds to ρ_u through the perfect gas law. Then (15-90) can be used for any effective bandwidth dependence on ρ_u and S_u (that is, not only at the weak and strong limits) to solve for $i'_{l,u}(S)$. This exactly equals the intensity $i'_l(S)$ in the nonuniform gas in the weak and strong limits and is usually a good approximation for intermediate absorption values. Once the intensities are found, the heat transfer is obtained by using the relations for a uniform gas. The evaluation of (15-94a) and (15-94b) usually requires numerical integration. Because the Curtis-Godson method requires evaluation of at least two integrals for each band along each path, it may in many cases be equally feasible to evaluate by computer the exact Eq. (15-88) or (15-89).

As originally formulated (see Goody [20]), the Curtis-Godson approximation was limited to application over a small wavenumber span in an absorption band. The limitation was due to line overlapping and the change in the spectral position of important lines with temperature. It has been shown (Weiner and Edwards [22] and Plass [23]), however, that the method gives good results even for situations having large temperature gradients with the use of fairly wide wavenumber spans. These references also account for overlapping absorption bands. A band absorptance formulation analogous to the Curtis-Godson approximation but involving three parameters was developed by Cess and Wang [24]. The additional parameter enables the equivalent isothermal gas to give the correct behavior in the linear and square-root limits, and also in the logarithmic limit [see Eq. (12-82)] for very strong absorption.

The Curtis-Godson technique has application to multidimensional problems, although it was originally applied to one-dimensional atmospheric problems. For a known field of temperature and density, the boundaries are subdivided into convenient, nearly isothermal zones. Between each two zones, an equivalent uniform path length and density are found for each important band by using (15-94a) and (15-94b). Based on these parameters, $\bar{A}_l$ is obtained from one of the correlations of gas properties. The uniform gas analysis of Sec. 13.6 is then used to obtain intensities and heat flows.

15-5 P-N (DIFFERENTIAL) METHODS

This section presents a method that involves multiplying the differential equation of transfer by various powers of the direction cosines of the intensity to form a set of *moment equations* and then finding a solution to this set. Unfortunately, the procedure provides one less equation than the number of unknowns generated. To overcome this difficulty, the intensity is approximated by a series expansion in terms of *spherical harmonics,* denoted by P, and the series is truncated after a selected number of terms, N. When the series is truncated after one or three terms, the method is called P-1 or P-3; in general, it is the P-N method. It is also referred to as the *moment method* and the *differential approximation.* Before proceeding to the general case, the method is introduced by examining a special case that is still used for one-dimensional problems.

15-5.1 The Milne-Eddington Approximation

With respect to the intensity, the approximation made independently by Eddington [25] and Milne [26] is that, for radiation crossing a unit area oriented normal to the *x* direction, all intensities with positive directional components in *x* have a value independent of angle, and all intensities with a negative *x* directional component have a different constant value; i.e., the local radiation in each coordinate direction is considered isotropic (see Fig. 15-13 in Sec. 15-6.1).

Start with the one-dimensional equation of transfer (14-48) without scattering. Multiply by $d\omega$ and by $\cos \theta \, d\omega$ to obtain the two moment equations

$$-\frac{\cos \theta}{a_\lambda} \frac{\partial i'_\lambda}{\partial x} d\omega = (i'_\lambda - i'_{\lambda b}) d\omega \tag{15-95a}$$

$$-\frac{\cos^2 \theta}{a_\lambda} \frac{\partial i'_\lambda}{\partial x} d\omega = \cos \theta (i'_\lambda - i'_{\lambda b}) d\omega \tag{15-95b}$$

The motivation for this is that $i'_\lambda \cos \theta$ is related to the heat flux, and (15-95) will thus yield a pair of equations involving q_λ. Equations (15-95a,b) are integrated over all solid angles to obtain

$$-\frac{1}{a_\lambda} \int_{\omega=0}^{4\pi} \cos \theta \, \frac{\partial i'_\lambda(\theta, x)}{\partial x} d\omega = -\frac{1}{a_\lambda} \frac{d^2 q_\lambda}{d\lambda \, dx} = \int_{\omega=0}^{4\pi} i'_\lambda(\theta, x) \, d\omega - 4\pi i'_{\lambda b} \tag{15-96a}$$

$$\int_{\omega=0}^{4\pi} i'_\lambda(\theta, x) \cos \theta \, d\omega = \frac{dq_\lambda}{d\lambda} = -\frac{1}{a_\lambda} \int_{\omega=0}^{4\pi} \cos^2 \theta \, \frac{\partial i'_\lambda(\theta, x)}{\partial x} d\omega \tag{15-96b}$$

The assumption is now introduced that i'_λ is isotropic in each hemisphere surrounding the positive and negative *x*-axes,

$$-\frac{1}{a_\lambda} \frac{d^2 q_\lambda}{d\lambda \, dx} = i_\lambda^+ \int_0^{\pi/2} 2\pi \sin \theta \, d\theta + i_\lambda^- \int_{\pi/2}^{\pi} 2\pi \sin \theta \, d\theta - 4\pi i'_{\lambda b} \tag{15-97a}$$

$$\frac{dq_\lambda}{d\lambda} = -\frac{1}{a_\lambda} \left(\frac{di_\lambda^+}{dx} \int_0^{\pi/2} 2\pi \cos^2 \theta \sin \theta \, d\theta + \frac{di_\lambda^-}{dx} \int_{\pi/2}^{\pi} 2\pi \cos^2 \theta \sin \theta \, d\theta \right) \tag{15-97b}$$

Integrating gives

$$-\frac{1}{a_\lambda}\frac{d^2 q_\lambda}{d\lambda\,dx} = 4\pi\left(\frac{i_\lambda^+ + i_\lambda^-}{2} - i'_{\lambda b}\right) \tag{15-98a}$$

$$\frac{dq_\lambda}{d\lambda} = -\frac{2\pi}{a_\lambda}\frac{d}{dx}\left(\frac{i_\lambda^+ + i_\lambda^-}{3}\right) \tag{15-98b}$$

Eliminating $i_\lambda^+ + i_\lambda^-$ between these two expressions gives

$$\frac{1}{a_\lambda^2}\frac{d^3 q_\lambda(x)}{d\lambda\,dx^2} = 3\frac{dq_\lambda(x)}{d\lambda} + \frac{4\pi}{a_\lambda}\frac{di'_{\lambda b}(x)}{dx} \tag{15-99}$$

For the situation of a gray gas layer with no internal heat sources, (15-99) is integrated over all wavelengths, and, noting that $d^2 q/dx^2 = 0$, this gives

$$q(x) = -\frac{4}{3a}\frac{de_b(x)}{dx} \tag{15-100}$$

This is the same relation obtained previously by the diffusion approximation. The 'boundary conditions to be used with this relation are considered in the next section, which is a generalization of the Milne-Eddington approximation.

15-5.2 The General P-N Method

The P-N approximation reduces the integral equations of radiative transfer in media to differential equations by approximating the transfer relations with a finite set of moment equations. The moments are generated by multiplying the equation of transfer by powers of the cosine between the coordinate direction and the direction of the intensity. This is a generalization of the Milne-Eddington method, as Eqs. (15-95) are the equation of transfer multiplied by $(\cos\theta)^0$ and $(\cos\theta)^1$, respectively. The first three moment equations each have a physical significance, so this technique has some physical basis. The development will be given in three dimensions so that general geometries can be treated. The treatments due to Ratzel [27] and Higenyi and Bayatzitoglu [28, 29] will be followed closely. Other pertinent references are [30–46].

A rectangular coordinate system x_1, x_2, x_3 is shown in Fig. 15-9a. The variation of intensity at position $\mathbf{r}$ along the S direction in the direction of the unit vector $\mathbf{s}$ is given by the equation of transfer (14-4) as

$$\frac{di'_\lambda}{dS} = a_\lambda i'_{\lambda b} - (a_\lambda + \sigma_{s\lambda})i'_\lambda + \frac{\sigma_{s\lambda}}{4\pi}\int_{\omega_i=0}^{4\pi} i'_\lambda(S,\,\omega_i)\Phi(\lambda,\,\omega,\,\omega_i)d\omega_i \tag{15-101}$$

where $i'_\lambda = i'_\lambda(\lambda, S, \omega)$ and $i'_{\lambda b} = i'_{\lambda b}(\lambda, S)$.

If the medium is assumed gray, with uniform scattering and absorption coefficients, and to scatter isotropically, (15-101) is integrated over all λ to give

$$\frac{di'}{dS} = ai'_b - (a + \sigma_s)\,i' + \frac{\sigma_s}{4\pi}\int_{\omega_i=0}^{4\pi} i'(S,\,\omega_i)\,d\omega_i \tag{15-102}$$

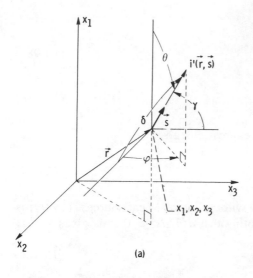

(a)

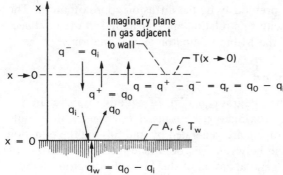

(b)

Figure 15-9 P-N approximation. (*a*) Coordinate system showing intensity as a function of position and angle for P-N approximation; (*b*) heat fluxes in boundary condition.

It is useful to express the direction S of i' in terms of the spherical angles θ and φ, or in terms of direction cosines l_i of the coordinate system:

$$\frac{d}{dS} = \cos\theta\,\frac{\partial}{\partial x_1} + \sin\theta\cos\varphi\,\frac{\partial}{\partial x_2} + \sin\theta\sin\varphi\,\frac{\partial}{\partial x_3} = l_1\frac{\partial}{\partial x_1} + l_2\frac{\partial}{\partial x_2} + l_3\frac{\partial}{\partial x_3}$$

(15-103)

where $l_1 = \cos\theta$, $l_2 = \cos\delta$, and $l_3 = \cos\gamma$ (Fig. 15-9*a*). Using the optical thickness $\kappa = (a + \sigma_s)x$ and albedo Ω, (15-102) becomes

$$\sum_{i=1}^{3} l_i\frac{\partial i'}{\partial \kappa_i} + i' = (1 - \Omega)i_b' + \frac{\Omega}{4\pi}\int_{\omega_i=0}^{4\pi} i'(S, \omega_i)\,d\omega_i$$

(15-104)

To develop the P-N method, the intensity i' is expanded in an orthogonal series of spherical harmonics,

$$i'(S, \omega) = \sum_{l=0}^{\infty} \sum_{m=-l}^{l} A_l^m(S) Y_l^m(\omega) \tag{15-105}$$

where the $A_l^m(S)$ are position-dependent coefficients to be determined by the solution, and $Y_l^m(\omega)$ are angularly dependent normalized spherical harmonics given by

$$Y_l^m(\omega) = \left[\frac{2l + 1}{4\pi} \frac{(l - m)!}{(l + m)!} \right]^{1/2} e^{jm\varphi} P_l^m(\cos \theta) \tag{15-106}$$

where $j = \sqrt{-1}$. The $P_l^m(\cos \theta)$ are associated Legendre polynomials of the first kind, defined by

$$P_l^m(\mu) = \frac{(1 - \mu^2)^{m/2}}{2^l l!} \frac{d^{l+m}}{d\mu^{l+m}} (\mu^2 - 1)^l \tag{15-107}$$

where $\mu = \cos \theta$. The values of $P_l^m(\mu)$ and $P_l^{-m}(\mu)$ are related by[†]

$$P_l^{-m}(\mu) = (-1)^m \frac{(l - m)!}{(l + m)!} P_l^m(\mu) \tag{15-108}$$

Values of $P_l^m(\cos \theta)$ for $0 \le l \le 3$ are in Table 15-3. In the limit $l \to \infty$, the spherical harmonics approximation for the intensity, (15-105), is exact.

To apply the P-N method for a practical solution, Eq. (15-105) is truncated after a finite number of terms N, so $A_l^m(S) = 0$ for $l > N$. Generally, in engineering radiative transfer problems, terms are retained for $l = 0$ and 1 (P-1 approximation) or for $l = 0, 1, 2$, and 3 (P-3 approximation). It is possible to retain higher-order terms, but with each increase in N the double series of (15-105) gives many more terms in the intensity relation. For engineering problems the P-3 approximation has been found adequate. Even-order approximations (P-2, P-4, etc.) give little increase in accuracy over the next lower order odd-numbered expansion, and they

---

[†]Two definitions of normalized spherical harmonics are in common use. One definition (not used here) includes a factor of $(-1)^m$. The factor introduces an alternating sign in spherical harmonics with positive m. If this definition is used, (15-107) and (15-108) must also correctly incorporate this factor.

Table 15-3 Associated Legendre polynomials, $P_l^m(\cos \theta)$ [27]

l	$m = 0$	$m = 1$	$m = 2$	$m = 3$
0	1.0	—	—	—
1	$\cos \theta$	$\sin \theta$	—	—
2	$(3 \cos^2 \theta - 1)/2$	$3 \cos \theta \sin \theta$	$3 \sin^2 \theta$	—
3	$(1/2) (5 \cos^2 \theta - 3) \cos \theta$	$(3/2)(5 \cos^2 \theta - 1) \sin \theta$	$15 \cos \theta \sin^2 \theta$	$15 \sin^3 \theta$

are difficult to apply when engineering boundary conditions of known temperature or energy flux are present. Thus, odd-order expansions are used.

The coefficients $A_l^m(S)$ retained in the expansion must be evaluated. This is done by generating *moment equations* by multiplying the local intensity by powers of the direction cosines l_i ($i = 1, 2, 3$), either individually or in combination, and then integrating over all solid angles. This results in the zeroth, first, second, and Nth moments as

$$i'^{(0)}(S) = \int_{\omega=0}^{4\pi} i'(S, \omega)\, d\omega \tag{15-109}$$

$$i'^{(i)}(S) = \int_{\omega=0}^{4\pi} l_i i'(S, \omega)\, d\omega \qquad (i = 1, 2, 3) \tag{15-110}$$

$$i'^{(ij)}(S) = \int_{\omega=0}^{4\pi} l_i l_j i'(S, \omega)\, d\omega \qquad (i, j = 1, 2, 3) \tag{15-111}$$

$$i'^{(ijkl\ldots)}(S) = \int_{\omega=0}^{4\pi} l_i l_j l_k l_l \ldots i'(S, \omega)\, d\omega \qquad (i, j, k, l, \ldots, = 1, 2, 3) \tag{15-112}$$

The zeroth moment divided by the speed of light is the radiation energy density, the first moment is the radiative energy flux in the i coordinate direction, and the second moment divided by the speed of light is the local stress and pressure tensor. Higher-order moments have no physical meaning and are generated by analogy with the others.

Equation (15-105) for the local intensity is substituted into the integrals of the moment equations, the series is truncated at the desired level, and the integrations are carried out. For the P-3 approximation, this results in 20 coupled algebraic equations in the 20 moments of intensity. Ratzel [27] solved this set and presents forms for $A_l^m(S)$ coefficients in terms of the moments. Substituting these expressions for $A_l^m(S)$ into (15-105) results in the relation for the P-3 local intensity:

$$4\pi i'(S, \theta, \varphi) = i'^{(0)} + 3i'^{(1)} \cos\theta + 3i'^{(2)} \sin\theta \cos\varphi + 3i'^{(3)} \sin\theta \sin\varphi$$

$$+ \frac{5}{4}(3i'^{(11)} - i'^{(0)})(3\cos^2\theta - 1)$$

$$+ 15(i'^{(12)} \cos\varphi + i'^{(13)} \sin\varphi)\cos\theta \sin\theta$$

$$+ \frac{15}{4}[(i'^{(22)} - i'^{(33)})\cos 2\varphi + 2i'^{(23)} \sin 2\varphi]\sin^2\theta$$

$$+ \frac{7}{4}(5i'^{(111)} - 3i'^{(1)})(5\cos^3\theta - 3\cos\theta) \tag{15-113}$$

$$+ \frac{21}{8} [(5i''^{(211)} - i''^{(2)}) \cos \varphi + (5i''^{(311)} - i''^{(3)}) \sin \varphi](5 \cos^2 \theta - 1) \sin \theta$$

$$+ \frac{105}{4} [(i''^{(122)} - i''^{(133)}) \cos 2\varphi + 2i''^{(123)} \sin 2\varphi] \cos \theta \sin^2 \theta$$

$$+\frac{35}{8} [(i''^{(222)} - 3i''^{(233)}) \cos 3\varphi - (i''^{(333)} - 3i''^{(322)}) \sin 3\varphi] \sin^3 \theta$$

$$(15\text{-}113)$$
$$(\text{cont.})$$

along with the identities: $i''^{(0)} = i''^{(11)} + i''^{(22)} + i''^{(33)}$; $i''^{(1)} = i''^{(111)} + i''^{(222)} + i''^{(333)}$; $i''^{(2)} = i''^{(211)} + i''^{(222)} + i''^{(233)}$; $i''^{(3)} = i''^{(311)} + i''^{(322)} + i''^{(333)}$. For the P-1 local intensity, only the first four terms in (15-113) appear,

$$i'(S, \theta, \varphi) = \frac{1}{4\pi} (i''^{(0)} + 3i''^{(1)} \cos \theta + 3i''^{(2)} \sin \theta \cos \varphi + 3i''^{(3)} \sin \theta \sin \varphi)$$

$$(15\text{-}114)$$

The remaining part of the solution is to develop expressions for the moments of intensity so that an explicit relation for intensity can be obtained from (15-113) or (15-114). This is done by generating *moment differential equations* from the differential equation of transfer, (15-104). The integral in (15-104) is the zeroth moment $i''^{(0)}$, so

$$\sum_{i=1}^{3} l_i \frac{\partial i'}{\partial \kappa_i} + i' = (1 - \Omega)i'_b + \frac{\Omega}{4\pi} i''^{(0)})$$

$$(15\text{-}115)$$

Equation (15-115) is multiplied by powers of the direction cosines individually and in combination, and the results are integrated over all solid angles. Because the derivative terms on the left of (15-115) are with respect to coordinate position while the integrals are over solid angle, the integral and derivative of each term may be interchanged. The results of carrying out these operations for the P-3 approximation are [27]

$$\sum_{i=1}^{3} \frac{\partial i''^{(i)}}{\partial \kappa_i} = (1 - \Omega)(4\pi i'_b - i''^{(0)}$$

$$(15\text{-}116)$$

$$\sum_{i=1}^{3} \frac{\partial i''^{(ij)}}{\partial \kappa_i} = - i''^{(j)} \qquad \text{(3 equations: } j = 1, 2, 3)$$

$$(15\text{-}117)$$

$$\sum_{i=1}^{3} \frac{\partial i''^{(ijk)}}{\partial \kappa_i} = - i''^{(jk)} + \frac{4\pi}{3} \delta_{jk} \left[(1 - \Omega)i'_b + \frac{\Omega}{4\pi} i''^{(0)} \right] \qquad \text{(9 equations: } j, k = 1, 2, 3)$$

$$(15\text{-}118)$$

$$\sum_{i=1}^{3} \frac{\partial i''^{(ijkl)}}{\partial \kappa_i} = - i''^{(jkl)} \qquad \text{(27 equations: } j, k, l = 1, 2, 3)$$

$$(15\text{-}119)$$

The δ_{jk} is the Kronecker delta. For the P-1 approximation, only Eqs. (15-116) and (15-117) are used.

Note that (15-119) for the P-3 approximation contains a new set of unknowns, the fourth-order moments $i'^{(ijkl)}$. For the P-1 approximation, the second-order moments $i'^{(ij)}$ are present. To close the set of equations, values for these moments must be found. This is done by substituting (15-113) [or (15-114) for the P-1 approximation] into the general moment equation (15-112) to generate a relation for the fourth moment $i'^{(ijkl)}$ (or $i'^{(ij)}$ for P-1). This relation is not exact because (15-113) and (15-114) were truncated to $N = 3$ and $N = 1$, respectively. However, an approximate closure condition is generated. For P-3, the result is

$$i'^{(ijkl)} = \frac{1}{7}(i'^{(ij)}\delta_{kl} + i'^{(ik)}\delta_{jl} + i'^{(jk)}\delta_{il} + i'^{(il)}\delta_{jk} + i'^{(kl)}\delta_{ij} + i'^{(jl)}\delta_{ik})$$

$$-\frac{1}{35}(\delta_{ij}\delta_{kl} + \delta_{il}\delta_{jk} + \delta_{ik}\delta_{jl}) \tag{15-120}$$

and for P-1,

$$i'^{(ij)} = \frac{1}{3}\delta_{ij}i'^{(0)} \tag{15-121}$$

The formulation is now complete, in that a number of equations equal to the number of unknowns is available for the moments of intensity. Once these are determined, (15-113) or (15-114) provides the local angular distribution of intensity and the problem is in principle solved. For engineering problems, there remains to define the correct boundary conditions. After these are obtained, some applications of the P-N method are presented.

15-5.3 Boundary Conditions for the P-N Method

The most useful boundary conditions for engineering application are due to Marshak [43]; these work well for odd-order expansions. Other discussions of boundary conditions are in [47, 48].

For a diffuse gray surface, the outgoing intensity i'_o from a boundary by emission and reflection is

$$i'_o = \epsilon i'_{b,w} + \frac{1 - \epsilon}{\pi} \int_{\omega=0}^{2\pi} i'(\omega)l_i\, d\omega \tag{15-122}$$

where l_i is the direction cosine of $i'(\omega)$ relative to the surface normal and the integration is over the hemisphere of incident solid angles. The general Marshak boundary condition is

$$\int_{\omega=0}^{2\pi} i'_o(\omega)Y_l^m(\omega)\, d\omega = \int_{\omega=0}^{2\pi} \left[\epsilon i'_{b,w} + \frac{1 - \epsilon}{\pi} \int_{\omega^*=0}^{2\pi} i'(\omega^*)l_i\, d\omega^*\right]Y_l^m(\omega)\, d\omega \tag{15-123}$$

For one-dimensional cases, Y_l^m may be replaced by $Y_l^m = l_i^{2n-1}$ ($n = 1, 2, 3, \ldots$). For multidimensional rectangular geometries, $Y_1^m = l_i$ ($i = 1, 2, 3$) and $Y_3^m = l_i l_j l_k$ ($i, j, k = 1, 2, 3$). These equations provide more relations than are necessary, and it is suggested in [27] that combinations be chosen with moments of the intensity normal to the boundary (see [27] for additional information).

For the one-dimensional case with $n = 1$, the left side of (15-123) reduces to

$$\int_{\omega=0}^{2\pi} i'_o(\omega) Y_l^m(\omega)\, d\omega = \int_{\omega=0}^{2\pi} i'_o l_i^{2n-1}\, d\omega = \int_{\omega=0}^{2\pi} i'_o l_i\, d\omega = q_0 \tag{15-124}$$

so this Marshak condition is directly related to the outgoing radiative flux q_o (Fig. 15-9b). The other boundary terms do not have a physical interpretation.

15-5.4 Application of the P-N Approximation

The use of the P-N relations is now demonstrated in an example. The P-1 relations are used, but the P-3 relations can be applied with an increase in complexity. Other applications are then discussed.

> **EXAMPLE 15-8** Using the P-1 approximation, derive relations for the temperature distribution and energy transfer between parallel plates at $T_{w,1}$ and $T_{w,2}$, each with the same diffuse-gray emissivity ϵ, when they are separated by an emitting, absorbing, and isotropically scattering medium with albedo Ω and optical thickness κ_D based on the plate spacing.
>
> Because the geometry requires the radiative energy flux to be only in the κ_1 direction, it follows that the first moments (equal to the fluxes) $i'^{(j)} = 0$ for $j = 2, 3$. Then (15-114) for the P-1 approximation becomes

$$i'(S, \theta, \varphi) = \frac{1}{4\pi} [i'^{(0)} + 3i'^{(1)} \cos \theta] \tag{15-125}$$

and Equations (15-116) and (15-117) become

$$\frac{di'^{(1)}}{d\kappa_1} = (1 - \Omega)(4\pi i'_b - i'^{(0)}) \tag{15-126}$$

$$\frac{di'^{(11)}}{d\kappa_1} = -i'^{(1)}, \qquad \frac{di'^{(12)}}{d\kappa_1} = 0, \qquad \frac{di'^{(13)}}{d\kappa_1} = 0 \tag{15-127}$$

The closure condition for P-1, (15-121), gives $i'^{(11)} = (1/3)i'^{(0)}$. Substituting into the second moment differential equation (15-127) gives

$$\frac{1}{3} \frac{di'^{(0)}}{d\kappa_1} = -i'^{(1)} \tag{15-128}$$

Now, because $i'^{(1)} = q_r$ and in this case q_r is constant, it follows from (15-126) that $i'^{(0)} = 4\pi i'_b = 4\sigma T^4(\kappa_1)$. Substituting this into (15-128) gives

$$q_r = -\frac{4\sigma}{3} \frac{dT^4}{d\kappa_1} \tag{15-129}$$

This is the same relation as for the diffusion solution. However, for other geometries the P-N solution does not generally provide the diffusion result. Integrating (15-129) results in a linear fourth-power temperature distribution within the medium.

$$\sigma T^4(\kappa_1) = -\frac{3q_r}{4} \kappa_1 + C \tag{15-130}$$

The boundary conditions must now be applied to relate the temperature distribution in the medium to the known boundary temperatures. Measuring κ_1 from the wall at $T_{w,1}$, Eq. (15-123) for this boundary becomes

$$q_0(\kappa_1 = 0) = 2\pi \int_{\theta=0}^{\pi/2} i'_0 \cos\theta \sin\theta \, d\theta = 2\,\epsilon\sigma T^4_{w,1} \int_{\theta=0}^{\pi/2} \cos\theta \sin\theta \, d\theta$$

$$+ 2\pi(1 - \epsilon) \int_{\theta=0}^{\pi/2} \cos\theta \sin\theta \, d\theta \; 2 \int_{\theta^*=\pi}^{\pi/2} i'(\kappa_1 = 0, \theta^*) \cos\theta^* \sin\theta^* \, d\theta^* \tag{15-131}$$

The incident intensity $i'(\kappa_1 = 0, \theta)$ in (15-131) is expressed by (15-125), so that from the moments that have been found,

$$i'(\kappa_1 = 0, \theta) = \frac{1}{4\pi} (i'^{(0)} + 3i'^{(1)} \cos\theta) = \frac{1}{4\pi} [4\sigma T^4(\kappa_1 = 0) + 3q_r \cos\theta]$$

which is independent of φ. Substituting this into the boundary condition equation (15-131) gives

$$q_0(\kappa_1 = 0) = \epsilon\sigma T^4_{w,1} + \frac{1 - \epsilon}{2} \int_{\theta^*=\pi}^{\pi/2} [4\sigma T^4(\kappa_1 = 0)$$

$$+ 3q_r \cos\theta^*] \cos\theta^* \sin\theta^* \, d\theta^* \tag{15-132}$$

From the net radiation expression (7-19) at wall 1 (*note:* $q_r = q_o - q_i = q_1$)

$$q_0 = \sigma T^4_{w,1} - \frac{1 - \epsilon}{\epsilon} q_r \tag{15-133}$$

Integrating (15-132) and using (15-133) to eliminate q_o result in

$$\sigma T^4(\kappa_1 = 0) = \sigma T^4_{w,1} - \left(\frac{1}{\epsilon} - \frac{1}{2}\right) q_r \tag{15-134}$$

A similar analysis applied at wall 2 provides the condition

$$\sigma T^4(\kappa_1 = \kappa_D) = \sigma T_{w,2}^4 + \left(\frac{1}{\epsilon} - \frac{1}{2}\right)q_r \tag{15-135}$$

Equations (15-134) and (15-135) are the same as the results found by the diffusion solution, (15-60). Equation (15-134) is used to evaluate the constant in Eq. (15-130), resulting in the fourth-power temperature distribution relation

$$\sigma T^4(\kappa_1) = \sigma T_{w,1}^4 - \left(\frac{1}{\epsilon} - \frac{1}{2}\right)q_r - \frac{3q_r}{4}\kappa_1 \tag{15-136}$$

Evaluating (15-136) at $\kappa_1 = \kappa_D$, substituting (15-135) to eliminate $T^4(\kappa_1 = \kappa_D)$, and then solving the result for q_r yields

$$\frac{q_r}{\sigma(T_{w,1}^4 - T_{w,2}^4)} = \frac{1}{3\kappa_D/4 + 2/\epsilon - 1} \tag{15-137}$$

This expression for q_r can be substituted into (15-136) to obtain $T^4(\kappa_1)$ in terms of the known boundary conditions, completing the problem (see Table 15-4).

Although the P-1 approximation provides the same result as the diffusion solution for infinite parallel gray plates, for other geometries it will generally provide a more accurate solution than the diffusion result. Table 15-4 provides the P-1 predictions for parallel plates, concentric cylinders, and concentric spheres. Comparison with Table 15-2 for the diffusion approximation using second-order slip shows that for cylinders and spheres the results differ by the presence in the diffusion results of a term in the denominator with a factor of $1/a$. This term results from the second-order slip boundary condition and causes the diffusion results for the cylindrical and spherical cases to approach $\psi = 0$ (rather than the correct value of 1) as a approaches zero.

Figures 15-10 and 15-11 compare the exact solution for energy transfer with the diffusion, P-1, and P-3 approximations for the cases of concentric cylinders and concentric spheres [29]. Physically, ψ cannot be larger than unity and should approach unity for small optical thicknesses when the bounding surfaces are black. This is the limit achieved by the Monte Carlo solution in Fig. 15-10 and the exact solution in Fig. 15-11. The diffusion approximation is based on the assumption of an optically thick medium; hence, it is in error for small optical thicknesses. It is found to give good results for optical thicknesses greater than unity for $D_{\text{inner}}/D_{\text{outer}} = 0.5$. For smaller $D_{\text{inner}}/D_{\text{outer}}$ the diffusion results are not as good, especially for spheres, and larger optical thicknesses are required for good agreement with the exact solution. The P-3 approximation is better than the P-1 approximation and provides good results for $D_{\text{inner}}/D_{\text{outer}} = 0.5$. However, the results are very poor for smaller diameter ratios, as shown in Fig. 15-11 (recall that physically $\psi \leq 1$). Additional information is in [49–58].

Table 15-4 P-1 approximations for energy transfer and temperature distribution for a gray gas between gray surfaces

Geometry	Relations[a]
Infinite parallel plates	$$\psi = \frac{1}{(3aD/4) + E_1 + E_2 + 1}$$ $$\phi(z) = \psi\left[\frac{3a}{4}(D - z) + E_2 + \frac{1}{2}\right]$$
Infinitely long concentric cylinders	$$\psi = \frac{1}{\frac{3}{8}aD_1\ln\left(\frac{D_2}{D_1}\right) + \left(E_1 + \frac{1}{2}\right) + \frac{D_1}{D_2}\left(E_2 + \frac{1}{2}\right)}$$ $$\phi(r) = \psi\left[-\frac{3}{8}aD_1\ln\left(\frac{D}{D_2}\right) + \left(E_2 + \frac{1}{2}\right)\frac{D_1}{D_2}\right]$$
Concentric spheres	$$\psi = \frac{1}{\frac{3}{8}aD_1\left(1 - \frac{D_1}{D_2}\right) + \left(E_1 + \frac{1}{2}\right) + \frac{D_1^2}{D_2^2}\left(E_2 + \frac{1}{2}\right)}$$ $$\phi(r) = \psi\left[-\frac{3}{8}aD_1\left(\frac{D_1}{D_2} - \frac{D_1}{D}\right) + \left(E_2 + \frac{1}{2}\right)\frac{D_1^2}{D_2^2}\right]$$

[a]Definitions: $E_N = (1 - \epsilon_{wN})/\epsilon_{wN}$, $\psi = Q_1/A_1\sigma(T_{w1}^4 - T_{w2}^4)$, $\phi(\xi) = [T^4(\xi) - T_{w2}^4]/(T_{w1}^4 - T_{w2}^4)$, $D = 2r$.

A more accurate solution for ψ than is provided by any of the approximate solutions alone can probably be obtained by using the optically thin solution, which is exact in the limit of small optical dimension, and either the diffusion or differential approximations at large optical dimension. A curve faired between these limiting solutions should provide acceptable accuracy at intermediate optical thicknesses.

A significant increase in accuracy is reported for the P-3 over the P-1 approximation in [27, 42, 43, 57, 58]. Although there is an increase in the complexity of the solutions, the equations remain algebraic and of closed form for parallel flat plates. For concentric cylinders and spheres, a numerical solution of the resulting fourth-order linear ordinary differential equations is necessary. Stone and Gaustad [35] give a formulation for nongray gases for the astrophysical boundary condition of zero incident flux at one boundary.

Ratzel [27] presents solutions for radiative transfer in a rectangular enclosure with diffuse-gray or specular walls at differing isothermal temperatures containing a gray isotropically scattering gas. Profiles of wall heat flux distribution and gas temperature distribution are given, based on the P-1 and P-3 approximations. Results are in Fig. 15-12 for the emissive power along the centerline of a square enclosure with one hot and three cold sides. All sides have the same emissivity, and results are shown for three values. The medium has an optical thickness of unity based on the side length. Comparisons are made of the P-1, the P-3, and a numerical solution based on the zone method. The P-3 and zone results are in good agreement.

Menguc and Viskanta [57, 58] extended the results to three-dimensional rectangular enclosures and to the P-5 approximation. They also used the P-3 approximation including delta-Eddington anisotropic scattering in three dimensions. The results indicate that the P-3 approximation is sufficiently accurate for most applications.

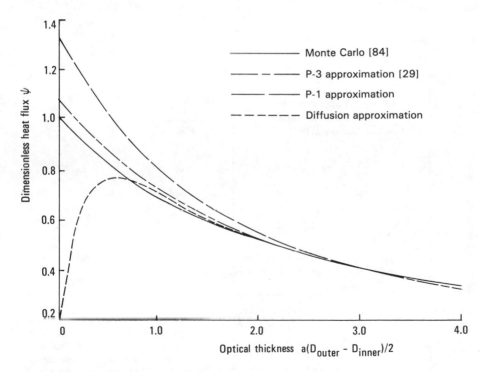

Figure 15-10 Comparison of solutions of energy transfer between infinitely long concentric black cylinders enclosing gray medium; $D_{inner}/D_{outer} = 0.5$.

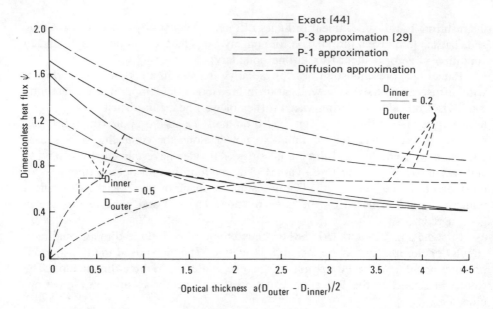

Figure 15-11 Comparison of solutions of energy transfer between black concentric spheres enclosing a gray medium.

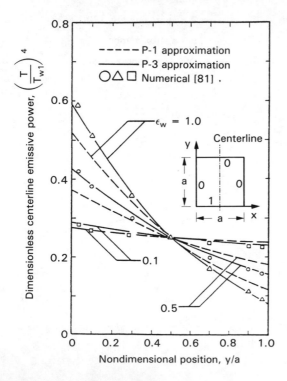

Figure 15-12 Dimensionless centerline emissive power in a gray medium within a square, gray-walled (wall emissivity = ϵ_w) enclosure of optical side dimension unity. Enclosure has surface with dimensionless temperature = 1 at $y/a = 0$; all other surfaces are at dimensionless temperature = 0. From [27].

Seiwert [59] discusses the F_N method, which is related to the P-N method. The intensity in a given direction is approximated by an expansion, but rather than using spherical harmonics, a Fourier expansion in the direction cosines is used. Kumar and Felske [60] use the F_N method to obtain a set of coupled equations from the differential equation of transfer and then express the solution as a singular eigenfunction expansion. They solved for the energy flux and intensity distributions at the boundaries of a slab of absorbing–anisotropically scattering medium exposed to collimated incident radiation.

Applications of the P-N method in the presence of additional heat transfer modes are in Chap. 17.

15-6 DISCRETE ORDINATE METHOD

The discrete ordinate method is an extension of a method, now called the *two-flux method,* proposed independently by Schuster [61] and Schwarzschild [62] for studying radiative transfer in stellar atmospheres. Chandrasekhar [63] extended the method for use with anisotropic scattering and made it applicable to multi-dimensional situations. When the solid angle about a location is divided into more than the two hemispheres, each with uniform intensities as used in the two-flux method, the method is known as the *multiflux, discrete ordinate,* or S_n method [63, 64]. Fiveland [65–67] has developed and implemented the discrete ordinate method for the analysis of heat transfer in coal-fired furnaces. Viskanta and Menguc [68] present a review of recent work using the method. Krook [30] compared the discrete ordinate method with the P-N method and showed that, in the limit of many terms, they become mathematically equivalent. The method is introduced in this section by deriving the Schuster-Schwarzschild relations, which have achieved wide use in one-dimensional engineering problems because of their simplicity. The more general discrete ordinate method is then developed.

15-6.1 Two-Flux Method: The Schuster-Schwarzschild Approximation

The simplest approximation is to assume that, for one-dimensional energy transfer, the intensity in the positive direction is isotropic and that in the negative direction is also isotropic but has a different value. This model is illustrated in Fig. 15-13. Using (14-48) without scattering, the equation of transfer for the intensity in each hemisphere is written as

$$-\frac{\cos \theta}{a_\lambda} \frac{di_\lambda{}^+(x)}{dx} = i_\lambda{}^+(x) - i'_{\lambda b}(x) \qquad 0 \leqslant \theta \leqslant \frac{\pi}{2} \qquad (15\text{-}138a)$$

$$-\frac{\cos \theta}{a_\lambda} \frac{di_\lambda{}^-(x)}{dx} = i_\lambda{}^-(x) - i'_{\lambda b}(x) \qquad \frac{\pi}{2} \leqslant \theta \leqslant \pi \qquad (15\text{-}138b)$$

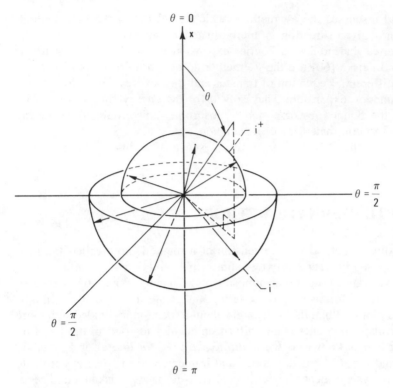

Figure 15-13 Approximation of intensities being isotropic in positive and negative directions.

From the isotropic assumption, the i_λ^+ and i_λ^- do not depend on θ. These equations are integrated over their respective hemispheres to give

$$-\frac{1}{a_\lambda}\frac{di_\lambda{}^+(x)}{dx}\int_0^{\pi/2}\cos\theta\sin\theta\,d\theta = i_\lambda{}^+(x)\int_0^{\pi/2}\sin\theta\,d\theta - i'_{\lambda b}(x)\int_0^{\pi/2}\sin\theta\,d\theta$$

$$(15\text{-}139a)$$

$$-\frac{1}{a_\lambda}\frac{di_\lambda{}^-(x)}{dx}\int_{\pi/2}^{\pi}\cos\theta\sin\theta\,d\theta = i_\lambda{}^-(x)\int_{\pi/2}^{\pi}\sin\theta\,d\theta - i'_{\lambda b}(x)\int_{\pi/2}^{\pi}\sin\theta\,d\theta$$

$$(15\text{-}139b)$$

The integrations yield

$$-\frac{1}{2a_\lambda}\frac{di_\lambda^+(x)}{dx} = i_\lambda^+(x) - i'_{\lambda b}(x), \qquad \frac{1}{2a_\lambda}\frac{di_\lambda^-(x)}{dx} = i_\lambda^-(x) - i'_{\lambda b}(x) \qquad (15\text{-}140a,b)$$

Equations (15-140) with appropriate boundary conditions can be solved using an integrating factor as in Sec. 14-3.1. For the geometry of Fig. 15-5, if $i_\lambda^+(0)$

and $i_\lambda^-(D)$ are the spectral intensities at the walls for a gas layer between parallel planes, the $i_\lambda^+(x)$ and $i_\lambda^-(x)$ are (letting $\kappa_\lambda = \int_0^x a_\lambda \, dx$)

$$i_\lambda^+(\kappa_\lambda) = i_\lambda^+(0) \exp(-2\kappa_\lambda) + 2 \int_0^{\kappa_\lambda} i'_{\lambda b}(\kappa_\lambda^*) \exp[2(\kappa_\lambda^* - \kappa_\lambda)] \, d\kappa_\lambda^* \qquad (15\text{-}141a)$$

$$i_\lambda^-(\kappa_\lambda) = i_\lambda^-(\kappa_{\lambda D}) \exp[2(\kappa_\lambda - \kappa_{\lambda D})] + 2 \int_{\kappa_\lambda}^{\kappa_{\lambda D}} i'_{\lambda b}(\kappa_\lambda^*) \exp[2(\kappa_\lambda - \kappa_\lambda^*)] \, d\kappa_\lambda^* \qquad (15\text{-}141b)$$

To illustrate the use of these relations, consider the simplified situation of a *gray* gas with no internal heat sources between parallel plates. Equations (15-141) retain the same form, but the λ subscripts are omitted. The temperature distribution and heat flux are found from (14-98) and (14-54). Since i^+ and i^- do not depend on θ in the present case, these equations become

$$\sigma T^4(\kappa) = \frac{1}{4} \left[i^+(\kappa) \int_0^{\pi/2} 2\pi \sin\theta \, d\theta + i^-(\kappa) \int_{\pi/2}^{\pi} 2\pi \sin\theta \, d\theta \right]$$

$$= \frac{\pi}{2} [i^+(\kappa) + i^-(\kappa)] \qquad (15\text{-}142a)$$

$$q_r = 2\pi \left[i^+(\kappa) \int_0^{\pi/2} \cos\theta \sin\theta \, d\theta + i^-(\kappa) \int_{\pi/2}^{\pi} \cos\theta \sin\theta \, d\theta \right]$$

$$= \pi[(i^+(\kappa) - i^-(\kappa)] \qquad (15\text{-}142b)$$

The i^+ and i^- are substituted from (15-141), and $i'_b = \sigma T^4 / \pi$ is used to obtain

$$\sigma T^4(\kappa) = \frac{1}{2} \left[\pi i^+(0) \exp(-2\kappa) + 2 \int_0^{\kappa} \sigma T^4(\kappa^*) \exp[2(\kappa^* - \kappa)] \, d\kappa^* \right.$$

$$\left. + \pi i^-(\kappa_D) \exp[2(\kappa - \kappa_D)] + 2 \int_{\kappa}^{\kappa_D} \sigma T^4(\kappa^*) \exp[2(\kappa - \kappa^*)] \, d\kappa^* \right] \qquad (15\text{-}143a)$$

$$q_r = \pi i^+(0) \exp(-2\kappa) + 2 \int_0^{\kappa} \sigma T^4(\kappa^*) \exp[2(\kappa^* - \kappa)] \, d\kappa^*$$

$$- \pi i^-(\kappa_D) \exp[2(\kappa - \kappa_D)] - 2 \int_{\kappa}^{\kappa_D} \sigma T^4(\kappa^*) \exp[2(\kappa - \kappa^*)] \, d\kappa^* \qquad (15\text{-}143b)$$

Since in the absence of heat sources the radiative flux q_r does not vary with κ, (15-143b) can be evaluated at any convenient location. Choosing $\kappa = 0$ gives

$$q_r = \pi i^+(0) - \pi i^-(\kappa_D) \exp(-2\kappa_D) - 2 \int_0^{\kappa_D} \sigma T^4(\kappa^*) \exp(-2\kappa^*) \, d\kappa^* \qquad (15\text{-}143c)$$

The integral equation (15-143a) for the gas temperature distribution and the flux equation (15-143b) are analogous to the exact formulation (14-101) and (14-102).

If Eq. (15-142a) is differentiated with respect to κ, the result is

$$\frac{d(\sigma T^4)}{d\kappa} = -\frac{\pi}{2}\left[\frac{di^+(\kappa)}{d\kappa} + \frac{di^-(\kappa)}{d\kappa}\right]$$

Now substitute (15-140) for the terms on the right to obtain

$$\frac{d(\sigma T^4)}{d\kappa} = \pi[-i^+(\kappa) + i_b'(\kappa) + i^-(\kappa) - i_b'(\kappa)] = -\pi[i^+(\kappa) - i^-(\kappa)]$$

Comparing this with Eq. (15-142b) yields a diffusion type of relation for the Schuster-Schwarzchild approximation:

$$q_r = -\frac{d(\sigma T^4)}{d\kappa} = -\frac{1}{a}\frac{de_b}{dx} \tag{15-144}$$

The two-flux method is based on a simple model with the intensities in the positive and negative coordinate directions each assumed isotropic over their respective hemisphere of solid angles. Anisotropic scattering can be considered but is used as an integrated average fraction for forward and backward scattering. The method works well in one-dimensional systems. It has been used in analysis of packed beds [69], radiative transfer through fibers and powders [70, 71], and porous layers with penetrating flow and an external radiation source [72]. Viskanta [73] reviewed the literature on the two-flux model through 1982. The size of the backward and forward scattering fractions to be used in a two-flux model is analyzed in [74] and compared with other approximations in the literature. Various approximations are compared with exact scattering fractions evaluated from Mie scattering theory. Results are also obtained for use in four- and six-flux models as required for two- and three-dimensional solutions.

EXAMPLE 15-9 Using the two-flux approximation, derive relations for the temperature distribution and energy transfer between parallel plates at $T_{w,1}$ and $T_{w,2}$, each with the same diffuse-gray emissivity ϵ, when they are separated by an emitting, absorbing, isotropically scattering medium with albedo Ω and optical thickness κ_D based on plate spacing.

The emissive power distribution is found by integrating the two-flux relation (15-144) for q_r, which is constant in this case, to obtain

$$\sigma T^4(\kappa) = -(a + \sigma_s)q_r x + C = -q_r\kappa + C$$

where scattering has been included as in Eq. (15-32). To evaluate the constant C, slip boundary conditions are employed as derived for the diffusion or P-N approximations. Using (15-60b) results in

$$\sigma T^4(\kappa) = -q_r\kappa + \sigma T_{w,1}^4 - \left(\frac{1}{\epsilon} - \frac{1}{2}\right)q_r$$

Evaluating this equation at $\kappa = \kappa_D$ and substituting the remaining boundary condition (15-60a) to eliminate $T^4(\kappa_D)$ results in

$$\frac{q_r}{\sigma(T_{w,1}^4 - T_{w,2}^4)} = \frac{1}{\kappa_D + 2/\epsilon - 1}$$

Compared with the P-1 solution (Example 15-8), a factor of $4/3$ in the κ_D term is absent in the two-flux results. Because the P-1 and diffusion solution are identical for this particular problem and agree well with the exact solution, we expect the two-flux solution to give poorer agreement for q_r at large optical thickness, and, because q_r appears in the expression for the local emissive power, poorer agreement for the temperature distribution in the medium.

15-6.2 General Discrete Ordinate Method

In the discrete ordinate method, the general transfer relations are represented by a discrete set of equations for the average intensity over a finite number of ordinate directions. Integrals over solid angles are replaced by sums over the ordinate directions. For two ordinates, the method becomes the two-flux method. Fiveland [65–67] and Kumar, Majumdar, and Tien [75] present outlines of the method. Starting with the general differential form of the equation of transfer for isotropic scattering (15-104), the integral over an incident solid angle is approximated by a summation

$$\int_{\omega_i=0}^{4\pi} i'(\omega_i)\,d\omega_i = \sum_{m'} w_{m'} i'_{m'} \tag{15-145}$$

where w is a weighting factor and m' denotes an ordinate direction. The sum is over all ordinate directions so that 4π of solid angle is included. Then (15-104) is written for the intensity in an ordinate direction m as

$$\left[\sum_{i=1}^{3} l_i \frac{\partial i'}{\partial \kappa_i}\right]_m + i'_m = (1 - \Omega)i'_b + \frac{\Omega}{4\pi}\sum_{m'} w_{m'} i'_{m'} \tag{15-146}$$

where, as before, l_i is the direction cosine of the m direction relative to coordinate direction i. Usually, Gaussian quadrature is used for the summation. Chandrasekhar [63] discusses various quadrature schemes for use with (15-146).

Because intensities along each positive and negative ordinate direction around a node must be found, an even number of simultaneous equations must be solved at each node. A solution of this type is denoted $S_2, S_4, S_6, \ldots$. The subscript indicates the order of the solution method, and the order is in turn related to the number of discrete ordinates required. For example, in a three-dimensional problem, each octant of a sphere of solid angles around a specific nodal point contains $(n/2) + [(n/2) - 1] + [(n/2) - 2] + \cdots + 1 = n(n + 2)/8$ corresponding quadrature points (ordinates). For the eight octants, there is a total of $8n(n + 2)/8 = n(n + 2)$ ordinates, requiring solution of $n(n + 2)$ simultaneous equations at each node. The S_4 discrete ordinates formulation thus requires solution of $4(4 + 2) = 24$ simultaneous equations in a three-dimensional problem. In general for a D-dimensional problem ($D = 1, 2, 3$), the number of simultaneous equations is found to be $2^D n(n + 2)/8$. Thus in a two-dimensional problem, an S_4 analysis requires solving $2^2 4(4 + 2)/8 = 12$ simultaneous equations; in one dimension only $2^1 4(4 + 2)/8 = 6$ equations are required. Each additional physical dimension doubles the number of ordinates and equations to be solved at each node for a given order of solution.

15-6.3 Boundary Conditions for the Discrete Ordinate Method

Boundary conditions are generated by expressing the intensity leaving the surface along ordinate direction m as the sum of emitted and reflected intensities:

$$i'_m(\kappa_i = 0) = \epsilon i'_{b,w} + \frac{1 - \epsilon}{\pi} \sum_{m'} l_{i,m'} w_{m'} i'_{m'} \tag{15-147}$$

where $l_{i,m'}$ is the direction cosine between m' and the coordinate direction i that is normal to the surface. Equation (15-146) is written for each volume element in the medium, and (15-147) is written for each boundary element. This provides a set of linear differential equations and boundary conditions sufficient to solve for the unknown i'_m values at every location in the medium. Once the i'_m values are known, the local energy flux in the i coordinate direction in the medium or at the boundaries is found from

$$q_{r,i} = \sum_m l_{i,m} w_m i'_m \tag{15-148}$$

Fiveland [65–67] presents an iterative method of solving the discrete ordinate equations, while Kumar et al. [75] use available linear differential equation solvers. The weights w_m used in the equations depend on the quadrature scheme for approximating the integrals. Fiveland [67] uses equal weights of value $1/n$ for all ordinate directions. He argues that Gaussian quadrature, which uses adjustable weights for best approximation of integrals over the full range of solid angles as needed in the equation of transfer, is not appropriate for the boundary condition equations, which require integration over the half range. If Gaussian quadrature is used for the boundary condition equations, unnecessarily high orders of quadrature are required, resulting in much computation time.

Kumar et al. [75] discuss various weighting options, and their relative accuracy, for the solution of a one-dimensional highly anisotropic scattering problem. They show that, for the cases studied, Gaussian quadrature provided convergence more quickly than Fiveland's scheme, but Fiveland's results using equal weights for all ordinate directions tended to be more accurate. For problems in which highly anisotropic scattering is present, the full-range integrals are more important to solution accuracy than the half-range integrals in the boundary conditions; Gaussian quadrature may then provide better overall accuracy.

Some of Fiveland's results for a medium with scattering only are compared in Fig. 15-14 with P-3 and zone method results for a two-dimensional enclosure.

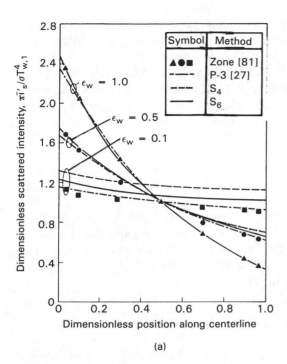

(a)

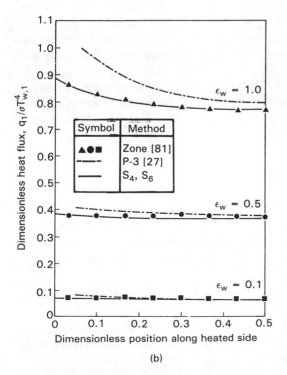

(b)

Figure 15-14 Radiative behavior in a gray square enclosure with wall emissivity ϵ_w, filled with a medium that scatters only. One side of the enclosure is at uniform temperature $T_{w,1}$; the other three sides are at zero temperature [67]. (a) Average incident scattered intensity along centerline; (b) Local heat transfer rate at the hot surface.

The S_6 discrete ordinate results are quite good. However, the S_4 solutions gave acceptable accuracy for much smaller computation times. The S_4 results for heat flux were usually more accurate than the P-3 results.

Kim and Lee [76] used an S_{14} solution with a general Mie-scattering medium in a two-dimensional enclosure exposed to a collimated incident radiation source on one face. The higher-order discrete ordinates solution is necessary for problems with scattering phase functions that are as highly anisotropic as were treated. An integral formulation is given in [77] in terms of moments of the intensity. The analysis is for a three-dimensional medium that is inhomogeneous and has aniso-tropic scattering. An example is given for a two-dimensional rectangular medium exposed to collimated radiation. A nonaxisymmetric cylindrical enclosure is ana-lyzed in [77a].

Menguc and Iyer [78] combined the P-N and discrete ordinates methods. They expressed the intensity in each ordinate direction in terms of a spherical-harmonics expansion. Using two-flux and eight-flux discrete ordinates with P-1 expansions and Marshak boundary conditions, they solved problems with anisotropically scat-tering media in one- and two-dimensional geometries.

15-7 FINITE-ELEMENT METHOD

The finite-element method (FEM) holds promise for solving radiative transfer problems because it can provide a match with the computational grid used when other energy modes are present [79, 80]. It can also provide an exact solution, in the sense that the exact transfer equations can be solved. In this section, the approach for pure radiation problems by Razzaque [79] is followed; the addition of other energy modes is discussed in Sec. 17-3. Introductory information is in Sec. 11-4.2.

The FEM was applied in [79] to the integral form of the equation of transfer, and some results were obtained for two-dimensional rectangular geometries. The formulation starts with the energy equation (14-98) for a gray medium in radiative equilibrium in the form

$$4\sigma T^4(x, y) = \int_{\omega_i=0}^{4\pi} i_i'(x, y)d\omega_i \tag{15-149}$$

The incoming intensity i_i' is evaluated from the equation of transfer so the inte-gration in (15-149) includes conditions over the entire region. The value of the integral is a function of (x, y, e_b) in a two-dimensional cartesian system, where $e_b = \sigma T^4(x, y)$. Defining $F(x, y, e_b)$ as the integral, (15-149) becomes

$$\sigma T^4(x, y) = e_b(x, y) = \frac{1}{4} F(x, y, e_b) \tag{15-150}$$

As in Chap. 11, the Galerkin FEM is applied. Equation (15-150) is multiplied by a weighting function $W(x, y)$ as in Eq. (11-30), and the resulting equation is in-tegrated over all solid angles to give

$$\int_{\omega=0}^{4\pi} e_b(x, y)W(x, y) \, d\omega = \frac{1}{4} \int_{\omega=0}^{4\pi} F(x, y, e_b)W(x, y) \, d\omega \tag{15-151}$$

As in Eq. (11-33), an approximate solution $U(x, y)$ is assumed in the form

$$e_b(x, y) = \sigma T^4(x, y) \approx U(x, y) = \sum_{j=1}^{N} U_j \Phi_j(x, y) \tag{15-152}$$

The weighting function is assumed to have the following form, where, for the Galerkin method, the same $\Phi(x, y)$ functions as in $U(x, y)$ are used:

$$W(x, y) = \sum_{i=1}^{N} W_i \Phi_i(x, y) \tag{15-153}$$

As discussed in Sec. 11-4.2, the $\Phi(x, y)$ is the *shape function*. It describes the nondimensional form of the emissive power variation that is assumed over each finite element, and it can be constant, linear, or curvilinear. Its particular form is chosen as a compromise between accuracy and solution time.

Substituting (15-152) and (15-153) into (15-151) and rearranging results in

$$\sum_{i=1}^{N} W_i \left[\sum_{j=1}^{N} \left(\int_{\omega=0}^{4\pi} \Phi_j(x, y)\Phi_i(x, y) \, d\omega \right) U_j - \frac{1}{4} \int_{\omega=0}^{4\pi} F(x, y, U)\Phi_i(x, y) \, d\omega \right] = 0 \tag{15-154}$$

Defining

$$K_{ij}(x, y) = \int_{\omega=0}^{4\pi} \Phi_j(x, y)\Phi_i(x, y) \, d\omega \tag{15-155}$$

and

$$f_i(x, y) = \frac{1}{4} \int_{\omega=0}^{4\pi} F(x, y, U)\Phi_i(x, y) \, d\omega \tag{15-156}$$

we can write (15-154) as

$$\sum_{i=1}^{N} W_i \left[\sum_{j=1}^{N} K_{ij}(x, y)U_j - f_i(x, y) \right] = 0 \tag{15-157}$$

Note that $K_{ij}(x, y)$ and $f_i(x, y)$ are integrals over the region surrounding each x, y location. The quantities under the integral sign depend on the x, y location and on an integration variable over the region [see Eq. (17-19), for example]. Because the W_i are arbitrary, (15-157) represents N equations to be satisfied by the N values of U_j. Thus, (15-157) provides the simultaneous relations

$$\sum_{j=1}^{N} K_{ij}(x, y)U_j = f_i(x, y) \qquad i = 1, 2, 3, \ldots N \tag{15-158}$$

The $K_{ij}(x, y)$ in this instance is associated with local emission; the $f_i(x, y)$ is associated with locally incident intensity. In some terminology $[K_{ij}]$ is called the stiffness matrix and $[f_i]$ the load vector, arising from the original application of FEM to problems in solid mechanics. After simultaneous solution of (15-158), the values of U_j are used in (15-152) to obtain the solution for $e_b(x, y)$.

Finite-element calculations for pure radiation problems are time consuming because they require the solution of the integral equation of transfer to determine the $i_i'(x, y)$ needed in (15-149). Numerical accuracy can in principle be improved by increasing the number of elements N. Razzaque [79] used FEM to analyze an absorbing, emitting, and isotropically scattering medium in a rectangular enclosure with gray walls. A shape function $\Phi(x, y)$ was used that was biquadratic in x and y and had nine nodes for each quadrilateral element. The node distribution and some possible shapes of $\Phi(x, y)$ for an element are in Fig. 15-15. Curvilinear shape functions are usually necessary to achieve good accuracy in combined-mode problems. Many commercially available FEM solvers use linear shape functions.

Some results are in Fig. 15-16 for the radiative flux at the hot surface of a square enclosure with black walls containing a medium of given optical thickness. The sets of curves are for various optical lengths of the enclosure side. The dimensionless temperature is unity on the hot surface and zero on the other three walls. Comparison with zone method results [81] is excellent. The P-3 results of [27] do not compare well, especially for small optical thicknesses. For these results, an array of only 2×2 elements of nine nodes each was used. The accuracy was checked by using an array of 4×4 elements, which roughly increased the computational time by a factor of four, and the results changed by less than 1%. Thus, for pure radiation problems, FEM required only a small element array to

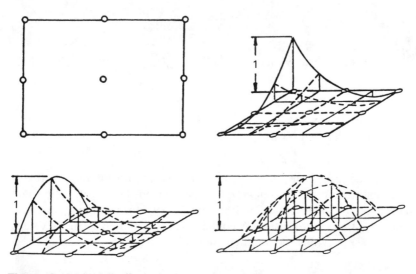

Figure 15-15 Node distribution and shape functions for a quadrilateral element [79].

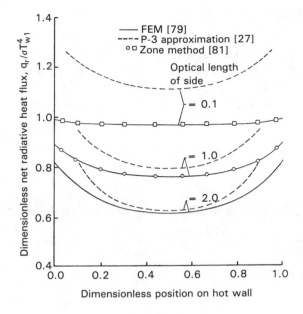

Figure 15-16 Dimensionless net local radiative heat flux at the hot wall in square enclosures with various optical thicknesses and black walls. Hot wall at dimensionless temperature $t = T/T_{w1} = 1.0$, other walls at $t = 0$.

produce excellent accuracy. When additional heat transfer modes are present, more elements are required (see Sec. 17-3).

15-8 MONTE CARLO TECHNIQUE FOR PARTICIPATING MEDIA

Monte Carlo is a method of statistical simulation to determine the average behavior of a system. In Chap. 11 it was applied to radiative transfer between surfaces without a participating medium between them. The information in that chapter is necessary to that here. Based on the transfer model in Chap. 11, extensions are now made for absorbing, emitting, and scattering media. The model consists of following a finite number of energy bundles through their transport histories. The radiative behavior of the system is determined from the average behavior of the bundles.

Monte Carlo techniques can be very useful for problems of radiative transfer through absorbing, emitting, and scattering media. The local radiation balance in the medium requires integrating the incoming radiation not only from the surrounding surfaces but from the volume elements of the surrounding medium. By extending the Monte Carlo model developed for surface-to-surface exchange in Chap. 11, it is possible to account for a large variety of effects in problems with radiating media. This can be done without resorting to the simplifying assumptions that are often necessary for numerical solutions based on analytical formulations.

15-8.1 Discussion of the Method

An additional factor required in the model in Chap. 11 is the path length traveled in the medium by an individual energy bundle before it is absorbed or leaves the system (scattering is included in Sec. 15-8.3). The required relations are in Table 15-5 as related to a random number. It is possible to allow for variations in medium properties along the path. In principle it is even possible to account for variations in refractive index of the medium by having the bundles travel curved paths.

If a problem is solved in which radiative equilibrium exists, then whenever a bundle is absorbed in the medium, a new bundle must be emitted from the same

Table 15-5 Useful relations for Monte Carlo solution of radiation problems in a medium

Phenomenon	Variable	Relation
Emission from a volume element with absorption coefficient a_λ	Cone angle θ	$\cos \theta = 1 - 2R_\theta$
	Circumferential angle φ	$\varphi = 2\pi R_\varphi$
	Wavelength λ Gray medium	$F_{0-\lambda} = R_\lambda$
	Nongray medium	$\dfrac{\displaystyle\int_0^\lambda a_\lambda(\lambda^*)i'_{\lambda b}(\lambda^*)\, d\lambda^*}{\displaystyle\int_0^\infty a_\lambda(\lambda^*)i'_{\lambda b}(\lambda^*)\, d\lambda^*} = R_\lambda$
Attenuation by medium with extinction coefficient K_λ	Path length l Uniform medium properties	$l = -\dfrac{1}{K_\lambda} \ln R_l$
	Nonuniform medium properties	$-\displaystyle\int_0^l K_\lambda(S)\, dS = \ln R_l$
Isotropic scattering from a volume element	Cone angle θ	$\cos \theta = 1 - 2R_\theta$
	Circumferential angle φ	$\varphi = 2\pi R_\varphi$
Anisotropic scattering in a gray medium with phase function Φ independent of incidence angle and circumferential scattering angle	Cone angle θ	$R_\theta = \dfrac{1}{2}\displaystyle\int_0^\theta \Phi(\theta^*) \sin \theta^* \, d\theta^*$
	Circumferential angle φ	$\varphi = 2\pi R_\varphi$

location to ensure no accumulation of energy. The functions required for determination of the angles and wavelengths of emission are in Table 15-5. The emitted bundle in the medium may be considered as the continuation of the history of the absorbed bundle, and the history continues until the energy reaches a bounding surface.

The total energy d^2Q_e emitted by a volume element dV is given by (12-39) integrated over all λ:

$$d^2Q_e = 4dV \int_0^\infty a_\lambda e_{\lambda b} \, d\lambda \tag{15-159a}$$

For radiative equilibrium the energy in the bundles emitted by a volume must equal the energy in the bundles absorbed,

$$d^2Q_e = wS_{dV} \tag{15-159b}$$

where w is the energy per bundle and S_{dV} is the number of bundles absorbed per unit time in dV. Then, if we note from Eq. (14-95) that $a_P \equiv \int_0^\infty a_\lambda e_{\lambda b} \, d\lambda / \sigma T_{dV}^4$, the Planck mean absorption coefficient a_P can be substituted into (15-159a) to eliminate the integral. Then equating (15-159a) and (15-159b) gives

$$T_{dV} = \left(\frac{wS_{dV}}{4a_P\sigma \, dV} \right)^{1/4} \tag{15-160}$$

This allows determination of the local temperature in the medium from the medium properties and the Monte Carlo quantities found in the solution. If a_P depends on local temperature T_{dV}, an iteration is required. A temperature distribution is assumed for a first iteration to obtain the bundle histories. The Monte Carlo quantities are used in (15-160) to obtain a new temperature distribution, which is then used for the second iteration. The process is repeated until the temperatures converge. Some references for situations in which the medium is not in radiative equilibrium are discussed in Sec. 17-7.

There are many variations on the Monte Carlo model to try to increase efficiency. One most frequently suggested is the fractional absorption of energy when a bundle reaches a surface of known absorptivity. The bundle energy is reduced after each reflection. The bundle history is followed until a sufficient number of reflections have occurred to reduce the bundle energy below some predetermined level where the effect of the bundle in succeeding reflections would be negligible. Such a procedure leads to better accuracy for many problems because a bundle history extends on the average through many more events, and a given number of bundles provides a larger number of events for compiling averages. Other shortcuts for reducing programming difficulties of problems involving spectral and directional properties are in [82].

EXAMPLE 15-10 A gray gas with constant absorption coefficient a is between two infinite parallel black plates spaced D apart. Plate 1 is at T_1, and plate 2 is at $T_2 = 0$. Construct a Monte Carlo flow chart for determining the energy transfer and the gas temperature distribution.

The emission per unit time and area from surface 1 is σT_1^4. If N bundles are to be emitted per unit time, then each must carry an amount of energy $w = \sigma T_1^4/N$. The bundles are emitted at cone angles θ given by the first line of Table 11-3, $\sin \theta = \sqrt{R_\theta}$, where R_θ is a random number in the range 0–1. A typical bundle will travel path length l after emission. The probability of traveling a given distance S before absorption in a medium of constant absorption coefficient a is $P(S) = e^{-aS}/\int_0^\infty e^{-aS} \, dS = ae^{-aS}$. Using Eq. (11-51), this is put in the form of a cumulative distribution

$$R_l = \frac{\int_0^l e^{-aS} \, dS}{\int_0^\infty e^{-aS} \, dS} = 1 - e^{-al} \qquad \text{or} \qquad l = -\frac{1}{a} \ln (1 - R_l)$$

Because R_l is uniformly distributed between 0 and 1, this relation may be written as $l = -(1/a) \ln R_l$ or $L = -(1/\kappa_D) \ln R_l$, where $L = l/D$ and $\kappa_D = aD$.

The dimensionless distance normal to the plate $X = x/D$ that a bundle will travel when moving through a path length L is $X = L \cos \theta = -(\cos \theta/\kappa_D) \ln R_l$. Divide the distance D between the plates into k equal increments of dimensionless width $\Delta X = \Delta x/D$, and number the increments $j = 1, 2, 3, \ldots, k$. The increment number at which absorption occurs is $j = \text{TRUNC} (X/\Delta X) + 1$, where TRUNC denotes truncating the value of $X/\Delta X$ to its integer. At each absorption, a tally is kept of j by increasing a counter S_j in the computer memory by one, $S_j = S_j + 1$.

If the bundle is absorbed in a gas element, it is immediately emitted from the same element to conserve energy for a problem in radiative equilibrium. This is done by choosing an angle of emission θ from the probability for emission into all cone angles of a unit sphere surrounding dV: $P(\theta) = \sin \theta / \int_0^\pi \sin \theta \, d\theta$. Using the cumulative distribution function $R_\theta = \int_0^\theta P(\theta^*) \, d\theta^* = (1 - \cos \theta)/2$ gives the emission angle in terms of a random number as $\theta = \cos^{-1} (1 - 2R_\theta)$. The distance from the wall to the next absorption point is then $X = X_0 - (\cos \theta/\kappa_D) \ln R_l$, where X_0 is the position of the previous absorption.

The process of absorptions and emissions is continued until the energy bundle reaches a black boundary. This occurs when $X \geq 1$ or $X \leq 0$, and a counter S_{w1} or S_{w2} is then increased by one unit to record the absorption at the surface.

A new bundle is emitted, and the process repeated until all N bundles have been emitted. The dimensionless net energy flux leaving surface 1 is then found from the total bundles emitted minus those reabsorbed at surface 1,

$$\frac{q_1}{\sigma T_1^4} = \frac{\sigma T_1^4 - wS_{w1}}{\sigma T_1^4} = 1 - \frac{S_{w1}}{N}$$

The net energy flux arriving at surface 2, $-q_2$, is given by

$$-\frac{q_2}{\sigma T_1^4} = \frac{wS_{w2}}{wN} = \frac{S_{w2}}{N}$$

The temperature at each gas increment is found from (15-160) as

$$\Theta_j = \frac{T_j}{T_1} = \left(\frac{wS_j}{4\kappa_D\sigma\,\Delta X\,T_1^4}\right)^{1/4} = \left(\frac{S_j}{4\kappa_D N\,\Delta X}\right)^{1/4}$$

and the formulation is complete.

The flowchart is in Fig. 15-17. Note that, since $S_{w1} + S_{w2} = N$, $q_1/\sigma T_1^4 = 1 - S_{w1}/N = S_{w2}/N = -q_2/\sigma T_1^4$ and, as expected, $q_1 = -q_2$: the only

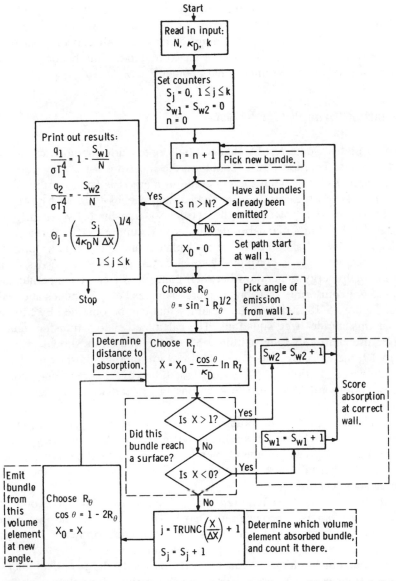

Figure 15-17 Flowchart for Monte Carlo solution of radiant transfer between infinite parallel black plates.

reason for obtaining both quantities is to check the results. By noting the linearity of this problem with T^4, it is possible to obtain solutions for any combination of surface temperatures by use of this flowchart [83]. By use of the relations between solutions for black and gray surfaces in Eq. (14-123), solutions can be obtained for any combination of gray emissivities.

This example illustrates the power of the Monte Carlo method. Fig. 15-17 gives a fairly complete diagram of the logic required for programming the problem of energy transfer through a nonisothermal gray gas between infinite parallel black plates at different temperatures. A comparison of this diagram with the analyses of, say, [4] or Chap. 14 illustrates the simplifications in both concept and formulation that may be inherent in the Monte Carlo method. This becomes even more evident in a two- or three-dimensional geometry.

15-8.2 Radiation through Gray Gases

Infinite parallel plates Some results obtained by Monte Carlo methods will now be examined. Because solutions are available in the literature for a gray gas between infinite parallel plates, almost every solution method is tried for this situation and compared with the results of analytical approaches typified by [4]. In [83] the local gas emissive power and the net energy transfer between diffuse-gray plates are calculated in a manner quite similar to Example 15-10. Parameters are the plate emissivity ϵ (taken equal for both plates) and the gas-layer optical thickness $\kappa_D = aD$, where a is constant. Two cases are examined; the first is a gas with no internal energy generation between plates at different temperatures, the second a gas with uniformly distributed energy sources between plates at equal temperatures. Figure 15-18 indicates the accuracy that can be obtained by Monte Carlo solutions in such idealized situations. The calculated energy transfer values have a 99.99% probability of lying within ±5% of the midpoints shown.

In Fig. 15-19, the emissive power distribution within the gas is shown. Comparison with the exact solutions of [4] is quite good; however, some trends common to all straightforward Monte Carlo solutions in gas radiation problems are as follows:

First, the calculated individual points in Fig. 15-19 reveal increasing error with decreasing optical thickness. This reflects the smaller fraction of energy bundles being absorbed in a given volume element as the optical thickness of the gas decreases. As the number of absorbed bundles decreases, the accuracy of the local emissive power also decreases.

Second, the computing time required for problems involving large optical thickness, say larger than 10, becomes quite large. This is because the free path of an energy bundle, $L = -(1/\kappa_D) \ln R_l$, is short for large optical thickness; therefore, many absorptions occur during a typical bundle history.

Hence, for small optical thickness, accuracy decreases; for large optical thickness, computing time becomes large. These are not serious limitations, as the

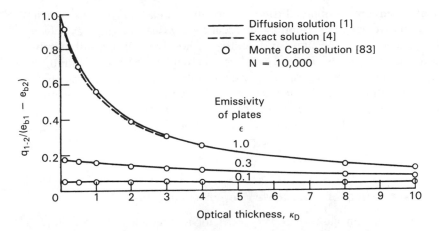

Figure 15-18 Net heat transfer between infinite gray parallel plates separated by gray gas.

transparent and diffusion approximations become valid in those regions. In addition, the range of optical thickness over which a Monte Carlo solution can be effectively utilized can be extended by various techniques such as biasing, splitting, Russian roulette, and a large number of specialized schemes for specific solutions. Many of these involve biasing the path length to increase the number of bundles absorbed in otherwise weakly absorbing regions.

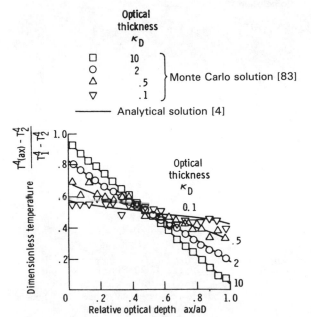

Figure 15-19 Emissive power distribution in gray gas between infinite parallel black plates.

Cylindrical geometry A more difficult problem to treat analytically is determining the emissive power distribution and local energy flux in a gray gas between concentric cylinders. The Monte Carlo approach, however, differs only slightly from that for parallel planes. The only additional complication is determining the bundle position in terms of cylindrical coordinates. Some results from [84] for an annular region are in Fig. 15-20 and are compared with a modified diffusion solution [1] from Sec. 15-3.4. Trends in accuracy are similar to those for infinite plates. In [85] finite cylinders of homogeneous nongray gas were analyzed to obtain the local source function and the profiles of the spectral lines for radiation emerging from the cylinder.

15-8.3 Consideration of Radiative Property Variations

For radiative transfer in gases it is very difficult to account accurately for the strong spectral, temperature, and pressure dependence of the radiative absorption coefficient. The coefficients can sometimes be computed with reasonable accuracy by quantum-mechanical methods, but few analyses have included the effects of many variables. Many analyses are for gray gases or use various mean absorption

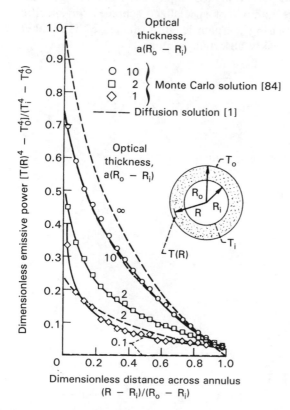

Figure 15-20 Dimensionless emissive power distribution in gray gas in annulus between black concentric cylinders of radius ratio $R_i/R_o = 0.1$.

coefficients (Sec. 15-4.2). Monte Carlo is well suited to the consideration of property variations with many variables. It requires relatively little extra effort to assign wavelengths to individual energy bundles and to allow the paths of the bundles to depend on the local spectral absorption coefficient. The relations necessary for this are in Table 15-5.

If property variations with temperature are considered, an iterative solution is usually required because the temperature distribution within the medium is not generally known a priori. Determining the path length to absorption becomes more difficult because the absorption coefficient varies with position. By applying the formalism outlined in Sec. 11-6.2, the path length l is given by

$$\ln R_l = - \int_0^l a_\lambda(S)\, dS \tag{15-161}$$

To evaluate this integral to determine l along a fixed line after choosing a random number R_l is time-consuming but at least feasible. Howell and Perlmutter [86] used this approach for including temperature and wavelength-dependent absorption coefficients of hydrogen between infinite parallel plates. They considered energy transfer through the gas between plates at different temperatures and the case of a parabolic distribution of internal energy generation in the gas. To evaluate the path length, Eq. (15-161) was approximated by dividing the gas into plane increments of thickness Δx. The path length through a given increment was then $\Delta l = \Delta x / \cos\theta$, where θ is the angle between the bundle path and the perpendicular to the plates. Equation (15-161) was then replaced by $\ln R_l + \Delta l \sum_{j=1}^{p} a_{\lambda,j} > 0$, and the summation was carried out until a value of the integer p was reached that satisfied the inequality. The value of p is related to the increment number in which absorption occurs. Values of $a_{\lambda,j}$ were assumed for the first iteration and then recalculated in successive iterations on the basis of the newly computed local temperatures. This procedure was continued until convergence.

Figure 15-21 shows the property variations used, and Fig. 15-22 shows a set of calculated emissive power distributions. The accuracy becomes poorer in the regions of low temperature because of the decrease with temperature of the absorption coefficient and, therefore, number of absorptions in the low-temperature regions.

An improved zone method using a Monte Carlo technique is in [87]. Scattering can be readily included for a given angular scattering distribution. It is analyzed by considering it as absorption and nonisotropic re-emission in a gas volume. Stockham and Love [88] and Scofield and Love [89] used Monte Carlo techniques to study problems with combined absorption and scattering including transfer in fog. House and Avery [90] give a general discussion of Monte Carlo applications in nonequilibrium radiative transport. Anisotropic scattering has been treated by a Monte Carlo method in [91], and an inverse problem is analyzed in [92]. The effect of nondiffuse boundaries is examined in [93]. Inhomogeneous media are treated in different ways in [94, 95]. In [95a] a three-dimensional enclosure containing anisotropically scattering particles and a band-absorbing gas (CO_2) was analyzed by a Monte Carlo method.

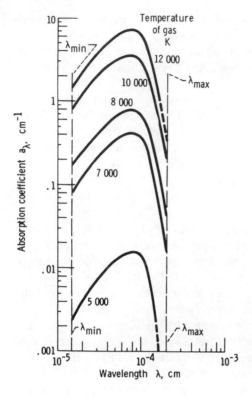

Figure 15-21 Spectral absorption coefficient of hydrogen at 1000 atm [86].

Surveys of the Monte Carlo method are in [96, 97]. Some possible methods of implementing Monte Carlo methods on parallel computers are discussed in [98].

15-9 OTHER SOLUTION METHODS

15-9.1 Exponential Kernel Approximation

In Sec. 14-7.6 the solution for a gray medium between diffuse-gray plates was derived in terms of the functions ψ_b and ϕ_b that have been obtained numerically and are in Table 14-3 and Fig. 14-12. An approximate solution for ψ_b and ϕ_b is found here using an approximation for the exponential integrals in the radiative flux equation. The ψ_b and ϕ_b are for radiative equilibrium between parallel black plates. For these conditions, the radiative flux q_r is equal to the flux q transferred between the plates, $q_{o,1} = \sigma T_1^4$, and $q_{o,2} = \sigma T_2^4$. Then with the dimensionless forms in Eq. (14-111), Eq. (14-105) becomes

$$\psi_b = 2E_3(\kappa) + 2\int_0^\kappa \phi_b(\kappa^*)E_2(\kappa - \kappa^*)\,d\kappa^* - 2\int_\kappa^{\kappa_D} \phi_b(\kappa^*)E_2(\kappa^* - \kappa)\,d\kappa^* \quad (15\text{-}162)$$

From Appendix E, the E_2 and E_3 can be approximated by the exponential functions

$$E_2(\kappa) \cong \frac{3}{4} \exp\left(-\frac{3}{2}\kappa\right) \qquad E_3(\kappa) \cong \frac{1}{2} \exp\left(-\frac{3}{2}\kappa\right) \tag{15-163}$$

These approximations are inserted into the integral equation (15-162) to yield

$$\psi_b = e^{-3\kappa/2} + \frac{3}{2}e^{-3\kappa/2} \int_0^\kappa \phi_b(\kappa^*)e^{3\kappa^*/2}\,d\kappa^* - \frac{3}{2}e^{3\kappa/2} \int_\kappa^{\kappa_D} \phi_b(\kappa^*)e^{-3\kappa^*/2}\,d\kappa^* \tag{15-164}$$

To solve for ψ_b and ϕ_b, Eq. (15-164) is differentiated twice with respect to κ, and to satisfy the heat flux being constant across the distance between the plates, $d\psi_b/d\kappa = 0$. This yields

$$0 = e^{-3\kappa/2} + \frac{4}{3}\frac{d\phi_b}{d\kappa} + \frac{3}{2}e^{-3\kappa/2} \int_0^\kappa \phi_b(\kappa^*)e^{3\kappa^*/2}\,d\kappa^* - \frac{3}{2}e^{3\kappa/2} \int_\kappa^{\kappa_D} \phi_b(\kappa^*)e^{-3\kappa^*/2}\,d\kappa^* \tag{15-165}$$

Equation (15-165) is subtracted from (15-164) to give $\psi_h = -(4/3)(d\phi_b/d\kappa)$. Since ψ_b is a constant, this is integrated to yield

$$\phi_b = -\frac{3}{4}\psi_b\kappa + C \tag{15-166}$$

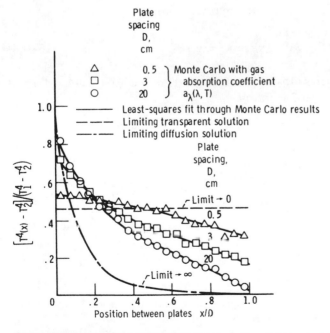

Figure 15-22 Emissive power distribution in hydrogen between infinite parallel plates at temperatures $T_1 = 9500$ K and $T_2 = 4500$ K [86].

where C is an integration constant. The ψ_b and C are found by substituting Eq. (15-166) back into the integral equation (15-164) to obtain

$$\psi_b = e^{-3\kappa/2} + \frac{3}{2} e^{-3\kappa/2} \int_0^\kappa \left(-\frac{3}{4} \psi_b \kappa^* + C \right) e^{3\kappa^*/2} \, d\kappa^*$$

$$- \frac{3}{2} e^{3\kappa/2} \int_\kappa^{\kappa_D} \left(-\frac{3}{4} \psi_b \kappa^* + C \right) e^{-3\kappa^*/2} \, d\kappa^*$$

The integrations are carried out, and after simplification this becomes

$$0 = e^{-3\kappa/2} \left(1 - \frac{1}{2} \psi_b - C \right) + e^{3(\kappa - \kappa_D)/2} \left(-\frac{3}{4} \psi_b \kappa_D - \frac{1}{2} \psi_b + C \right)$$

Thus there are two simultaneous equations for ψ_b and C:

$$1 - \frac{1}{2} \psi_b - C = 0, \qquad -\frac{3}{4} \psi_b \kappa_D - \frac{1}{2} \psi_b + C = 0$$

The solution yields

$$\psi_b = \frac{1}{\frac{3}{4}\kappa_D + 1} \tag{15-167}$$

and $C = 1 - \psi_b/2$, which is substituted into (15-166) to give

$$\phi_b(\kappa) = 1 - \frac{\psi_b}{2} \left(1 + \frac{3}{2} \kappa \right) = \psi_b \left[\frac{3}{4} (\kappa_D - \kappa) + \frac{1}{2} \right] \tag{15-168}$$

The following table compares (15-167) with the exact results from Table 14-1 and the agreement is within about 3% for all κ_D.

| κ_D | Dimensionless radiative flux, $\psi_b(\kappa_D)$ | |
	Eq. (15-167)	Table 14-1
0.2	0.8696	0.8491
0.4	0.7692	0.7458
0.6	0.6897	0.6672
1	0.5714	0.5532
1.5	0.4706	0.4572
2	0.4000	0.3900
3	0.3077	0.3016

Reference [99] discusses the use of the approximate kernel for nongray media and for cases in which radiative equilibrium is not present.

Some numerical techniques that can facilitate exact numerical solutions are now discussed. Coupled with the increasing speed and decreasing cost of numerical computation, these methods make the solution of the integral form of the radiative transfer equation feasible, eliminating the need for approximate approaches. Numerical methods such as finite differences and finite elements for combined mode problems are in Chap. 17.

15-9.2 Direct Numerical Integration

With larger and faster computers it is now more feasible to solve the radiative transfer and energy equations by direct numerical integration. This is possible for multidimensional cases, although it can become time-consuming for three-dimensional geometries, especially when properties vary with wavelength. Some two-dimensional solutions have been obtained. An example is the transient cooling of a gray gas in a square region [100]. Since this is a transient solution, it required solving the two-dimensional equations at each time increment, so that about 20 solutions were required for each case considered. The computer time was about 30 s per time increment on a Cray XMP. The procedure utilized a grid of points over the geometry. The grid should have points that are more closely spaced in regions where the temperatures are spatially varying most strongly. The integrations to obtain the radiative source term were performed with two-dimensional Gaussian integration subroutines. This required an array of nonuniformly spaced integration points. These were obtained by curve fitting the values at the grid points with two-dimensional cubic splines and interpolating intermediate values as called for by the integration subroutine. A library curve-fitting subroutine was used.

The advantage of direct numerical integration is that high accuracy can be obtained for detailed features such as temperature distributions. This is very important in a transient situation, where the solution will otherwise tend to drift, as time proceeds, into inaccuracy. Some of the approximate procedures in this chapter yield accurate boundary heat fluxes if the temperature distribution is known, but the prediction of the temperature distribution is of insufficient accuracy for some problem requirements. The details for using direct integration in connection with finite difference methods are in Sec. 17-4. Methods of increasing the speed for multiple integration by using parallel processors are discussed in [101, 102].

15-9.3 Reduction of the Integral Order

To solve efficiently the integral equation of transfer and obtain the radiative source term for the energy equation, it is useful to reduce the order of the integrals in the equation. If the integrals over a volume are reduced from triple to double integrals, computation time is greatly decreased. At least two approaches are available. Yuen and Wong [103] use the point allocation method, which reduces the order of integrals over volume by one.

Tan [104] applied the product integration method [105] to radiative transfer. In this method, the medium emissive power, the surface radiosity, and the radiative flux divergence are found at prescribed points. Their forms are assumed to be given by a series with unknown interpolation functions as coefficients. The series are substituted into the radiative transfer equation, which then takes the form of a matrix of discretized equations that is solved for the interpolating functions. The matrix elements involve multiple integrals that must be evaluated before solution, but they do not contain the unknowns and therefore need to be evaluated only once. The method is closely related to the finite-element method but reduces the dimension of the required integrations, thus reducing computation time. For a given degree of approximation of the temperature distribution within each of N finite volume or surface elements, the number of calculations in zonal or finite-element methods increases as N^2, while in the point allocation or product integration method it increases as N.

In [104] results using product integration include linear-anisotropic scattering in the analysis of a two-dimensional emitting, absorbing, scattering medium. A square enclosure divided into 8×8 volume elements was analyzed on an IBM XT computer, while in contrast the finite-element analysis [79, 106] was limited to 4×4 elements by computer time requirements on a CDC Cyber system.

15-9.4 YIX Method

The YIX method [107] is a numerical approach that reduces the order of the multiple integrals and has other important attributes. The method is named from the shape of the pattern of the integration points for three, two, and four angular directions in the two-dimensional case. The integrals over distance are constructed such that results are stored for use in subsequent integrations, allowing integrals to be computed as simple sums.

Consider the integral from the one-dimensional transfer equation,

$$I = \int_{x=0}^{L} f(x)E_1(x)\, dx \tag{15-169}$$

This is subdivided into

$$I = \sum_{i=1}^{n} \int_{x=x_{i-1}}^{x_i} f(x)E_1(x)\, dx + \int_{x=x_n}^{L} f(x)E_1(x)\, dx \tag{15-170}$$

where $0 = x_0 < x_1 < \cdots < x_n \leq L$, and the individual values of x_i are to be found. The first quantity on the right of (15-170) accounts for the contribution to I from the subregions from $x = 0$ to x_n. The second term is the contribution in the final interval from x_n to L. Each of the integrals over a subregion, $x_{i-1} \leq x \leq x_i$, is expressed as a two-point approximation so that

$$I_i = \int_{x=x_{i-1}}^{x_i} f(x)E_1(x)\, dx \approx a_i f(x_{i-1}) + b_i f(x_i) \tag{15-171}$$

For $f(x) = 1$ and for $f(x) = x$, (15-171) gives the two equations:

$$a_i + b_i = \int_{x=x_{i-1}}^{x_i} E_1(x)\, dx = E_2(x_{i-1}) - E_2(x_i) \qquad (15\text{-}172a)$$

$$a_i x_{i-1} + b_i x_i = \int_{x=x_{i-1}}^{x_i} x E_1(x)\, dx = x_{i-1} E_2(x_{i-1}) - x_i E_2(x_i) + E_3(x_{i-1}) - E_3(x_i)$$

$$(15\text{-}172b)$$

The two equations (15-172) are solved for a_i and b_i to give

$$a_i = E_2(x_{i-1}) - D(x_i) \qquad b_i = D(x_i) - E_2(x_i) \qquad \text{and}$$

$$D(x_i) = \frac{E_3(x_{i-1}) - E_3(x_i)}{x_i - x_{i-1}} \qquad (15\text{-}173)$$

This approach could be extended to higher-order approximations of the integrals, but if the intervals $(x_i - x_{i-1})$ are small enough that the integrand has only a moderate variation within the interval, the two-point approximation is adequate. Substituting (15-173) into (15-171) and the result into (15-170) give

$$I \approx [1 - D(x_1)] f(0) + \sum_{i=1}^{n-1} [D(x_i) - D(x_{i+1})] f(x_i)$$

$$+ [D(x_n) - D(L)] f(x_n) + [D(L) - E_2(L)] f(L) \qquad (15\text{-}174)$$

Now, the final simplification that makes this method useful is to let $[1 - D(x_1)] = [D(x_i) - D(x_{i+1})] \equiv \beta = \text{constant}$. Substituting into (15-174) gives

$$I \approx \beta \left[f(0) + \sum_{i=1}^{n-1} f(x_i) \right] + [D(x_n) - D(L)] f(x_n) + [D(L) - E_2(L)] f(L) \qquad (15\text{-}175)$$

Note that the term $[D(x_n) - D(L)]$ is *not* equal to β except in the special case when x_{n+1} falls exactly on the boundary at L. This term must be treated separately.

The number of evaluations of the kernel necessary in the original form (15-169) has been reduced, and, over most of the increments in the integration, only a summation over the unknown function $f(x_i)$ is necessary. The kernel evaluations can be further reduced by approximating the contribution of the final element where $x_n < x < L < x_{n+1}$ by

$$\int_{x=x_n}^{L} E_1(x) f(x)\, dx \approx \frac{L - x_n}{x_{n+1} - x_n} \int_{x=x_n}^{x_{n+1}} E_1(x) f(x)\, dx$$

$$\approx \frac{L - x_n}{x_{n+1} - x_n} [E_2(x_n) - E_2(x_{n+1})] f(x_n) \qquad x_n \leq L < x_{n+1}$$

$$(15\text{-}176)$$

This linear interpolation provides an estimate of the contribution of the element lying next to the boundary with $x_n < x \leq L$, in terms of the more easily calculated contribution of the whole fictitious element with $x_n < x < x_{n+1}$. This approximation is of the same order as that used for the terms in (15-170). Eq. (15-175) then becomes

$$I \approx \beta \left[f(0) + \sum_{i=1}^{n-1} f(x_i) \right] + \frac{L - x_n}{x_{n+1} - x_n} [E_2(x_n) - E_2(x_{n+1})] f(x_n) \tag{15-177}$$

Now, define $Q_i \equiv [E_2(x_{i-1}) - E_2(x_i)]/(x_i - x_{i-1})$ and $P_i \equiv D(x_{i-1}) - E_2(x_{i-1}) - x_{i-1}Q_i - \beta$, and note that $[E_3(x_{n+1}) - E_3(x_n)]/(x_{n+1} - x_n) \approx dE_3(x_n)/dx = - E_2(x_n)$. These relations are substituted into Eq. (15-177), which becomes [after some algebra using $\beta = D(x_n) - D(x_{n+1})$],

$$I \approx \beta \left[f(0) + \sum_{i=1}^{n-1} f(x_i) \right] + [P_{n+1} + LQ_{n+1}] f(x_n) \tag{15-178}$$

Once x_1 is specified, the constants D, P, and Q are computed and stored. Evaluation of I is then a straightforward summation, resulting in reduced computation time. The grid spacing in this method is nonuniform. The spacing is chosen so the contribution to the integral from each increment is roughly the same. The integration grid is thus uncoupled from the choice of increment spacing used, for example, to compute local temperatures, giving the numerical analyst more freedom in obtaining a fast and accurate solution.

An advantage of this method is that $f(x)$ contains the local properties of the medium; if the properties are nonhomogeneous but temperature independent, they are readily incorporated in the solution. Some cases of this type are treated in [107]. If the properties are temperature dependent, the solution is iterative.

The method is readily extended to multidimensional geometries; the exponential integral function E_n is replaced by S_n in the two-dimensional formulation as discussed in [107]. A two-dimensional enclosure with an internal partition is treated in [108], and an anisotropically scattering square medium exposed to a collimated source is analyzed in [109]. Applications including conduction and convection are discussed in Chap. 17.

15-10 CONCLUDING REMARKS

This chapter has presented techniques used for solving radiative transfer problems in absorbing, emitting, and scattering media. Completely numerical methods, such as finite differences, are given in Chap. 17. As shown by the basic equations in Chap. 14, a solution involves obtaining intensities through a medium so that the radiative energy source distribution can be found. The source is inserted into the energy equation, which is used to obtain the temperature distribution. The inclusion of scattering introduces the source-function equation, which must be carried along simultaneously. With variable properties and spectral effects, the system of equations requires a complex simultaneous solution. Additional complexities are introduced by including anisotropic scattering.

The chapter began by presenting some simplified cases for which terms can be neglected in the transfer relations. The diffusion approximation was then developed in detail. The use of mean absorption coefficients is another simplification that may be useful for including spectral effects. The main sections presented the P-N, discrete ordinate, and Monte Carlo methods. As discussed, use of finite-element and finite-difference methods is currently increasing because of the availability of larger and faster computers. These methods are given in more detail in Chap. 17. Integration of the radiative source terms by available multidimensional integration packages is also feasible for carrying out portions of the solution.

Much of the analytical work that has included scattering has used the assumption of isotropic scattering. In recent years more attention has been paid to anisotropic scattering because of its importance in atmospheric radiation transfer, furnaces with particulate matter (ash, soot), and other technologies involving particles and fibers. Almost without exception, a basic assumption is that the individual particles scatter as if they were independent point scatterers; i.e., particle interactions do not affect the distribution of scattered intensity. The papers by Tien and co-workers [69, 110, 111] show the effect of wavelength, particle number density and size, and particle scattering and absorption cross section on the regions of dependent and independent scattering and give confidence that independent scattering occurs in most (but not all) engineering situations (see Sec. 12-9). Buckius [112] reviews work that deals with the properties of scattering media and outlines the independent scattering properties of various particles of regular and irregular shape.

Methods of solution that can incorporate anisotropic effects in the radiative transfer equation, including those discussed in this chapter, are being investigated and, at the same time, realistic scattering phase functions are being tried that will allow treatment of scattering without the complexity necessary if the Mie scattering phase functions are used. The delta-Eddington function and the Henyey-Greenstein approximation offer reasonable degrees of simplification while allowing good accuracy (Sec. 12-9.8). The various Dirac-delta phase function approximations are reviewed by Crosbie and Davidson [113]. Lee and Buckius [114, 115] and Kim and Lee [116] have shown that scaling laws allow some anistropic scattering problems to be reduced to nonscattering problems.

Anisotropic scattering has been almost universally treated with the scattering independent of incidence angle. This is the case for many practical systems with particle suspensions, because the particles are randomly oriented even if individually they have nonspherical form. However in radiative transfer within fixed solid matrices such as foamed ceramics, the scattering material may have a fixed orientation. The scattering phase function then depends on the angle of incidence as well as the angle of reflection, providing another degree of difficulty to be dealt with in the analysis.

REFERENCES

1. Deissler, R. G.: Diffusion Approximation for Thermal Radiation in Gases with Jump Boundary Condition, *J. Heat Transfer*, vol. 86, no. 2, pp. 240–246, 1964.
2. Rosseland, S.: "Theoretical Astrophysics: Atomic Theory and the Analysis of Stellar Atmospheres and Envelopes," Clarendon Press, Oxford, 1936.
3. Howell, J. R.: Radiative Interactions between Absorbing-Emitting and Flowing Media with Internal Energy Generation, NASA TN D-3614, 1966.
4. Heaslet, Max A., and Robert F. Warming: Radiative Transport and Wall Temperature Slip in an Absorbing Planar Medium, *Int. J. Heat Mass Transfer*, vol. 8, no. 7, pp. 979–994, 1965.
5. Howell, J. R.: On the Radiation Slip between Absorbing-Emitting Regions with Heat Sources, *Int. J. Heat Mass Transfer*, vol. 10, no. 3, pp. 401–402, 1967.
6. Howell, J. R., and M. Perlmutter: Monte Carlo Solution of Radiant Heat Transfer in a Nongrey Nonisothermal Gas with Temperature Dependent Properties, *AIChE J.*, vol. 10, no. 4, pp. 562–567, 1964.
7. Bobco, R. P.: Directional Emissivities from a Two-Dimensional, Absorbing-Scattering Medium, the Semi-Infinite Slab, *J. Heat Transfer*, vol. 89, no. 4, pp. 313–320, 1967.
8. Patch, R. W.: Effective Absorption Coefficients for Radiant Energy Transport in Nongrey, Nonscattering Gases, *J. Quant. Spectrosc. Radiat. Transfer*, vol. 7, no. 4, pp. 611–637, 1967.
9. Patch, R. W.: Approximation for Radiant Energy Transport in Nongray, Nonscattering Gases, NASA TN D-4001, 1967.
10. Sampson, D. H.: Choice of an Appropriate Mean Absorption Coefficient for Use in the General Grey Gas Equations, *J. Quant. Spectrosc, Radiat. Transfer*, vol. 5, no. 1, pp. 211–225, 1965.
11. Abu-Romia, M. M., and C. L. Tien: Appropriate Mean Absorption Coefficients for Infrared Radiation in Gases, *J. Heat Transfer*, vol. 89, no. 4, pp. 321–327, 1967.
12. Grant, Ian P.: On the Representation of Frequency Dependence in Non-Grey Radiative Transfer, *J. Quant. Spectrosc. Radiat. Transfer*, vol. 5, no. 1, pp. 227–243, 1965.
13. Stewart, John C.: Non-Grey Radiative Transfer, *J. Quant. Spectrosc. Radiat. Transfer*, vol. 4, no. 5, pp. 723–729, 1964.
14. Thomas, M., and W. S. Rigdon: A Simplified Formulation for Radiative Transfer, *AIAA J.*, vol. 2, no. 11, pp. 2052–2054, 1964.
15. Lick, Wilbert: Energy Transfer by Radiation and Conduction, *Proc. 1963 Heat Transfer Fluid Mech. Inst.* (Anatol Roshko, Bradford Sturtevant, and D. R. Bartz, eds.), pp. 14–26, 1963.
16. Howe, John T., and Yvonne S. Sheaffer: Spectral Radiative Transfer Approximations for Multicomponent Gas Mixtures, *J. Quant. Spectrosc. Radiat. Transfer*, vol. 7, no. 4, pp. 695–701, 1967.
17. Edwards, D. K., and M. M. Weiner: Comment on Radiative Transfer in Nonisothermal Gases, *Combust. Flame*, vol. 10, no. 2, p. 202–203, 1966.
18. Krakow, Burton, Harold J. Babrov, G. Jordan Maclay, and Abraham L. Shabott: Use of the Curtis-Godson Approximation in Calculations of Radiant Heating by Inhomogeneous Hot Gases, *Appl. Opt.*, vol. 5, no. 11, pp. 1791–1800, 1966.
19. Simmons, F. S.: Band Models for Non-Isothermal Radiating Gases, *Appl. Opt.*, vol. 5, no. 11, pp. 1801–1811, 1966.
20. Goody, R. M.: "Atmospheric Radiation, Theoretical Basis," vol. I, Clarendon Press, Oxford, 1964.
21. Simmons, F. S.: Application of Band Models to Inhomogeneous Gases. Molecular Radiation and Its Application to Diagnostic Techniques (R. Goulard, ed.), NASA TM X-53711, pp. 113–133, 1968.
22. Weiner, M. M., and D. K. Edwards: Nonisothermal Gas Radiation in Superposed Vibration-Rotation Bands, *J. Quant. Spectrosc. Radiat. Transfer*, vol. 8, no. 5, pp. 1171–1183, 1968.
23. Plass, Gilbert N.: Radiation from Nonisothermal Gases, *Appl. Opt.*, vol. 6, no. 11, pp. 1995–1999, 1967.
24. Cess, R. D., and L. S. Wang: A Band Absorptance Formulation for Nonisothermal Gaseous Radiation, *Int. J. Heat Mass Transfer*, vol. 13, no. 3, pp. 547–555, 1970.
25. Eddington, A. S.: "The Internal Constitution of the Stars," Dover, New York, 1959.

26. Milne, F. A.: Thermodynamics of the Stars, "Handbuch der Astrophysik," vol. 3, pp. 65–255, Springer-Verlag, OHG, Berlin, 1930.
27. Ratzel, A. G.: P-N Differential Approximation for Solution of One- and Two-Dimensional Radiation and Conduction Energy Transfer in Gray Participating Media, Ph.D. dissertation, Department of Mechanical Engineering, University of Texas, Austin, 1981.
28. Higenyi, J.: Higher Order Differential Approximation of Radiative Energy Transfer in a Cylindrical Gray Medium, Ph.D. dissertation, Rice University, 1979.
29. Bayazitoglu, Y., and J. Higenyi: The Higher-Order Differential Equations of Radiative Transfer: P_3 Approximation, *AIAA J.*, vol. 17, no. 4, pp. 424–431, 1979.
30. Krook, Max: On the Solution of Equations of Transfer, I, *Astrophys. J.*, vol. 122, no. 3, pp. 488–497, 1955.
31. Cheng, Ping: Two-Dimensional Radiating Gas Flow by a Moment Method, *AIAA J.*, vol. 2, no. 9, pp. 1662–1664, 1964.
32. Cheng, Ping: Dynamics of a Radiating Gas with Application to Flow over a Wavy Wall, *AIAA J.*, vol. 4, no. 2, pp. 238–245, 1966.
33. Cheng, Ping: Exact Solutions and Differential Approximation for Multi-Dimensional Radiative Transfer in Cartesian Coordinate Configurations, *Prog. Astronaut. Aeronaut.*, vol. 31, p. 269, 1972.
34. Cess, Robert D.: On The Differential Approximation in Radiative Transfer, *Z. Angew. Math. Phys.*, vol. 17, pp. 776–781, 1966.
35. Stone, Peter H., and John E. Gaustad: The Application of a Moment Method to the Solution of Non-Gray Radiative-Transfer Problems, *Astrophys. J.*, vol. 134, no. 2, pp. 456–468, 1961.
36. Traugott, S. C.: A Differential Approximation for Radiative Transfer with Application to Normal Shock Structure, *Proc. 1963 Heat Transfer Fluid Mech. Inst.* (Anatol Roshko, Bradford Sturtevant, and D. R. Bartz, eds.), pp. 1–13, 1963.
37. Adrianov, V. N., and G. I. Polyak: Differential Methods for Studying Radiant Heat Transfer, *Int. J. Heat Mass Transfer*, vol. 6, no. 5, pp. 355–362, 1963.
38. Traugott, S. C., and K. C. Wang: On Differential Methods for Radiant Heat Transfer, *Int. J. Heat Mass Transfer*, vol. 7, no. 2, pp. 269–273, 1964.
39. Dennar, E. A., and M. Sibulkin: An Evaluation of the Differential Approximation for Spherically Symmetric Radiative Transfer, *J. Heat Transfer*, vol. 91, no. 1, pp. 73–76, 1969.
40. Modest, M. F.: The Improved Differential Approximation for Radiative Transfer in Multidimensional Media, *J. Heat Transfer*, vol. 112, no. 3, pp. 819–821, 1990.
41. Selcuk, Nevin, and R. G. Siddall: Two-Flux Spherical Harmonic Modelling of Two-Dimensional Radiative Transfer in Furnaces, *Int. J. Heat Mass Transfer*, vol. 19, pp. 313–321, 1976.
42. Arpaci, V. S., and D. Gözüm: Thermal Stability of Fluids: The Benard Problem, *Phys. Fluids*, vol. 16, no. 5, pp. 581–588, 1973.
43. Marshak, R. E.: Note on the Spherical Harmonics Methods as Applied to the Milne Problem for a Sphere, *Phys. Rev.*, vol. 71, pp. 443–446, 1947.
44. Rhyming, I. L.: Radiative Transfer between Two Concentric Spheres Separated by an Absorbing-Emitting Gas, *Int. J. Heat Mass Transfer*, vol. 9, pp. 315–324, 1966.
45. Chou, Y. S., and C. L. Tien: A Modified Moment Method for Radiative Transfer in Non-Planar Systems, *J. Quant. Spectrosc. Radiat. Transfer*, vol. 8, p. 919–933, 1968.
46. Shvartsburg, A. M.: Error Estimation in Differential Approximation to Equation of Transfer, *Zh. Prikl. Mekh. Tekh. Fiz. No. 5*, pp. 9–13, 1976.
47. Shokair, I. R., and G. C. Pomraning: Boundary Conditions for Differential Approximations, *J. Quant. Spectrosc. Radiat. Transfer*, vol. 25, no. 4, pp. 325–337, 1981.
48. Mark, J. C.: The Spherical Harmonic Method—Part I and II, National Research Council of Canada, Atomic Energy Reports MT 92-1944 and MT 97-1945, 1945.
49. Heaslet, Max A., and R. F. Warming: Theoretical Predictions of Radiative Transfer in a Homogenous Cylindrical Medium, *J. Quant. Spectrosc. Radiat. Transfer*, vol. 6, pp. 751–774, 1966.
50. Schmid-Burgk, Johannes: Radiant Heat Flow through Cylindrically Symmetric Media, *J. Quant. Spectrosc. Radiat. Transfer*, vol. 14, pp. 979–987, 1974.
51. Modest, M. F., and D. S. Stevens: Two-Dimensional Radiative Equilibrium of a Gray Medium

between Concentric Cylinders. *J. Quant. Spectrosc. Radiat. Transfer,* vol. 19, pp. 353–365, 1978.

52. Loyalka, S. K.: Radiative Heat Transfer between Parallel Plates and Concentric Cylinders, *Int. J. Heat Mass Transfer,* vol. 12, pp. 1513–1517, 1969.

53. Kesten, Arthur A.: Radiant Heat Flux Distribution in a Cylindrically Symmetric Nonisothermal Gas with Temperature-Dependent Absorption Coefficient, *J. Quant. Spectrosc. Radiat. Transfer,* vol. 8, p. 419–434, 1968.

54. Dua, Shyam S., and Ping Cheng: Multi-Dimensional Radiative Transfer in Non-Isothermal Cylindrical Media with Non-Isothermal Bounding Walls, *Int. J. Heat Mass Transfer,* vol. 18, pp. 254–259, 1975.

55. Amlin, D. W., and S. A. Korpela: Influence of Thermal Radiation on the Temperature Distribution in a Semi-Transparent Solid, *ASME J. Heat Transfer,* vol. 101, no. 1, pp. 76-80, 1979.

56. Modest, M. F.: A Simple Differential Approximation for Radiative Transfer in Non-Gray Gases, *J. Heat Transfer,* vol. 101, no. 4, pp. 735–736, 1979.

57. Menguc, M. P.: Modeling of Radiative Heat Transfer in Multidimensional Enclosures Using Spherical Harmonics Approximation, Ph.D. thesis, Purdue University, 1985.

58. Menguc, M. P., and R. Viskanta: Radiative Transfer in Three-Dimensional Enclosures Containing Inhomogeneous, Anisotropically Scattering Media, *J. Quant. Spectrosc. Radiat. Transfer,* vol. 33, no. 6, pp. 533–549, 1985.

59. Siewert, C. E.: The F_N Method for Solving Radiative Transfer Problems in Plane Geometry, *Astrophys. Space Sci.,* vol. 58, pp. 131–137, 1978.

60. Kumar, S., and J. D. Felske: Radiative Transport in a Planar Medium Exposed to Azimuthally Unsymmetric Incident Radiation, *J. Quant. Spectrosc. Radiat. Transfer,* vol. 35, no. 3, pp. 187–212, 1986.

61. Schuster, A.: Radiation through a Foggy Atmosphere, *Astrophys. J.,* vol. 21, pp. 1–22, 1905.

62. Schwarzschild, K.: Equilibrium of the Sun's Atmosphere, *Ges. Wiss.* Gottingen. *Nachr., Math-Phys. Klasse,* vol. 1, pp. 41–53, 1906.

63. Chandrasekhar, Subrahmanyan: "Radiative Transfer," Dover, New York, 1960.

64. Lathrop, K. D.: Use of Discrete Ordinate Methods for Solution of Photon Transport Problems, *Nucl. Sci. Eng.,* vol. 24, pp. 381–388, 1966.

65. Fiveland, W. A.: Three-Dimensional Radiative Heat Transfer Solutions by the Discrete-Ordinates Method, *J. Thermophys. Heat Transfer,* vol. 2, no. 4, pp. 309–316, 1988.

66. Fiveland, W. A.: Discrete Ordinate Methods for Radiative Heat Transfer in Isotropically and Anisotropically Scattering Media, *J. Heat Transfer,* vol. 109, no. 3, pp. 809–812, 1987.

67. Fiveland, W. A.: Discrete Ordinates Solutions of the Radiative Transport Equation for Rectangular Enclosures. *J. Heat Transfer,* vol. 106, no. 4, pp. 699–706, 1984.

68. Viskanta, R., and M. P. Menguc: Radiation Heat Transfer in Combustion Systems, *Prog. Energy Combust. Sci.,* vol. 13, pp. 97–160, 1987.

69. Brewster, M. Q., and C. L. Tien: Radiative Transfer in Packed Fluidized Beds: Dependent versus Independent Scattering, *J. Heat Transfer,* vol. 104, no. 4, pp. 573–579, 1982.

70. Wang, K. Y., and C. L. Tien: Radiative Transfer through Opacified Fibers and Powders, *J. Quant. Spectrosc. Radiat. Transfer,* vol. 30, no. 3, pp. 213–223, 1983.

71. Tong, T., and C. L. Tien: Radiative Heat Transfer in Fibrous Insulations—Part 1: Analytical Study, *J. Heat Transfer,* vol. 105, no. 1, pp. 70–75, 1983.

72. Lee, K.-B., and J. R. Howell: Effect of Radiation on the Laminar Convective Heat Transfer through a Layer of Highly Porous Medium, in T. Tong and M. Modest (eds.), "Radiation in Energy Systems", ASME Publication HTD-vol. 55, pp. 51–59, 1986.

73. Viskanta, R.: Radiation heat Transfer: Interaction with Conduction and Convection and Approximate Methods in Radiation, in "Heat Transfer 1982" *Proc. 7th Int. Heat Conf.,* vol. 2, pp. 103–121, Hemisphere, Washington, D.C., 1982.

74. Koenigsdorff, R., F. Miller, and R. Ziegler: Calculation of Scattering Fractions for Use in Radiative Flux Models, *Int. J. Heat Mass Transfer,* vol. 34, no. 10, pp. 2673–2676, 1991.

75. Kumar, S., A. Majumdar, and C. L. Tien: The Differential-Discrete-Ordinate Method for So-

lutions of the Equation of Radiative Transfer, *J. Heat Transfer,* vol. 112, no. 2, pp. 424–429, 1990.

76. Kim, T.-K., and H. S. Lee: Radiative Transfer in Two-Dimensional Anisotropic Scattering Media with Collimated Incidence, *J. Quant. Spectrosc. Radiat. Transfer.,* vol. 42, no. 3, pp. 225–238, 1989.

77. Wu, C.-Y.: Exact Integral Formulation for Radiative Transfer in an Inhomogeneous Scattering Medium, *J. Thermophys. Heat Transfer,* vol. 4, no. 4, pp. 425–431, 1990.

77a. Jamaluddin, A. S., and P. J. Smith: Discrete-Ordinates Solution of Radiative Transfer Equation in Nonaxisymmetric Cylindrical Enclosures, *J. Thermophys. Heat Transfer,* vol. 6, no. 2, pp. 242–245, 1992.

78. Menguc, M. P., and R. K. Iyer: Modeling of Radiative Transfer Using Multiple Spherical Harmonics Approximations, *J. Quant. Spectrosc. Radiat. Transfer,* vol. 39, no. 6, pp. 445–461, 1988.

79. Razzaque, M. M., D. E. Klein, and J. R. Howell: Finite Element Solution of Radiative Heat Transfer in a Two-Dimensional Rectangular Enclosure with Gray Participating Media, *J. Heat Transfer,* vol. 105, no. 4, pp. 933–934, 1983.

80. Nice, M. L.: Application of Finite Element Method to Heat Transfer in a Participating Medium, in T. M. Shih (ed.), "Numerical Properties and Methodologies in Heat Transfer", pp. 497–514, Hemisphere, Washington, D.C., 1983.

81. Larsen, M. E., and J. R. Howell: The Exchange Factor Method: An Alternative Zonal Formulation of Radiating Enclosure Analysis, *J. Heat Transfer,* vol. 107, no. 4, pp. 936–942, 1985.

82. Haji-Sheikh, A., and E. M. Sparrow: Probability Distributions and Error Estimates for Monte Carlo Solutions of Radiation Problems, *Prog. Heat Mass Transfer,* vol. 2, pp. 1–22, 1969.

83. Howell, J. R., and M. Perlmutter: Monte Carlo Solution of Thermal Transfer through Radiant Media between Gray Walls, *J. Heat Transfer,* vol. 86, no. 1, pp. 116–122, 1964.

84. Perlmutter, M., and J. R. Howell: Radiant Transfer through a Gray Gas between Concentric Cylinders Using Monte Carlo, *J. Heat Transfer,* vol. 86, no. 2, pp. 169–179, 1964.

85. Avery, L. W., L. L. House, and A. Skumanich: Radiative Transport in Finite Homogeneous Cylinders by the Monte Carlo Technique, *J. Quant. Spectrosc. Radiat. Transfer,* vol. 9, pp. 519–531, 1969.

86. Howell, J. R., and M. Perlmutter: Monte Carlo Solution of Radiant Heat Transfer in a Nongrey Nonisothermal Gas with Temperature Dependent Properties, *AIChE J.,* vol. 10, no. 4, pp. 562–567, 1964.

87. Vercammen, H. A. J., and G. F. Froment: An Improved Zone Method using Monte Carlo Techniques for the Simulation of Radiation in Industrial Furnaces, *Int. J. Heat Mass Transfer,* vol. 23, no. 3., pp. 329–337, 1980.

88. Stockham, Leo W., and Tom J. Love: Radiative Heat Transfer from a Cylindrical Cloud of Particles, *AIAA J.,* vol. 6, no. 10, pp. 1935–1940, 1968.

89. Scofield, Gordon L., and Tom J. Love: Radiative Transfer Analysis from a Heated Airport Runway to Fog, *Int. J. Heat Mass Transfer,* vol. 13, no. 2, pp. 345–358, 1970.

90. House, L. L., and L. L. Avery: The Monte Carlo Technique Applied to Radiative Transfer, *J. Quant. Spectrosc. Radiat. Transfer,* vol. 9, pp. 1579–1591, 1969.

91. Gupta, R. P., T. F. Wall, and J. S. Truelove: Radiative Scatter by Fly Ash in Pulverized-Coal-Fired Furnaces: Application of the Monte Carlo Method to Anisotropic Scatter, *Int. J. Heat Mass Transfer,* vol. 26, no. 11, p. 1649–1660, 1983.

92. Subramaniam, S., and M. P. Mengüç: Solution of the Inverse Radiation Problem for Inhomogeneous Anisotropically Scattering Media Using a Monte Carlo Technique, *Int. J. Heat Mass Transfer,* vol. 34, no. 1, pp. 253–266, 1991.

93. Zavargo, Z., M. Djuric, and M. Novakovic: Radiation Heat Exchange between Non-Diffuse Gray Surfaces Separated by Isothermal Absorbing-Emitting Gas, *Int. J. Heat Mass Transfer,* vol. 34, no. 4/5, pp. 1003–1008, 1991.

94. Dunn, W. L.: Inverse Monte Carlo Solutions for Radiative Transfer in Inhomogeneous Media, *J. Quant. Spectrosc. Radiat. Transfer,* vol. 29, no. 1, pp. 19–26, 1983.

95. Kobiyama, M.: Reduction of Computing Time and Improvement of Convergence Stability of the Monte Carlo Method Applied to Radiative Transfer with Variable Properties, *J. Heat Transfer*, vol. 111, no. 1, pp. 135–140, 1989.

95*a*. Farmer, J. T., and J. R. Howell: Monte Carlo Solution of Radiative Heat Transfer in a Three-Dimensional Enclosure with an Anisotropically Scattering, Spectrally Dependent, Inhomogeneous Medium, ASME National Heat Transfer Conf., San Diego, August 1992.

96. Haji-Sheikh, A.: Monte Carlo Methods, Chap. 16 in "Handbook of Numerical Heat Transfer," W. J. Minkowycz, E. M. Sparrow, R. H. Pletcher, and G. E. Schneider (eds.), Wiley, New York, 1988.

97. Halton, J. H.: A Retrospective and Prospective Review of the Monte Carlo Method, *SIAM Rev.*, vol. 12, no. 1, pp. 1–63, 1970.

98. Moriarity, K. M.: "Vectorizing the Monte Carlo," Dept. of Mathematics, Royal Holloway College, Egham, Surrey, UK, 1981.

99. Gilles, Scott E., Ailen C. Cogley, and Walter G. Vincenti: A Substitute-Kernel Approximation for Radiative Transfer in a Non-Grey Gas Near Equilibrium, with Application to Radiative Acoustics, *Int. J. Heat Mass Transfer*, vol. 12, pp. 445–458, 1969.

100. Siegel, R.: Transient Cooling of a Square Region of Radiating Medium, *J. Thermophys. Heat Transfer*, vol. 5, no. 4, pp. 495–501, 1991.

101. Genz, A. C.: Numerical Multiple Integration on Parallel Computers, *Comput. Phys. Comm.*, vol. 26, pp. 349–352, 1982.

102. Shih, T.-M., L. J. Hayes, W. J. Minkowycz, K.-T. Yang, and W. Aung: Parallel Computations in Heat Transfer, *Numer. Heat Transfer*, vol. 9, pp. 639–662, 1986.

103. Yuen, W. W., and L. W. Wong: Analysis of Radiative Equilibrium in a Rectangular Enclosure with a Gray Medium, *J. Heat Transfer*, vol. 106, no. 2, pp. 433–440, 1984.

104. Tan, Z.: Radiative Heat Transfer in Multidimensional Emitting, Absorbing and Anisotropic Scattering Media—Mathematical Formulation and Numerical Method, *J. Heat Transfer*, vol. 111, no. 1, pp. 141–147, 1989.

105. Baker, C. T. H.: "The Numerical Treatment of Integral Equations," Clarendon Press, Oxford, Chaps. 4 and 5, 1977.

106. Razzaque, M. M., D. E. Klein, and J. R. Howell: Coupled Radiative and Conductive Heat Transfer in a Two-Dimensional Rectangular Enclosure with Gray Participating Media Using Finite Elements, *J. Heat Transfer*, vol. 106, no. 3, pp. 613–619, 1984.

107. Tan, Z., and J. R. Howell: New Numerical Method for Radiation Heat Transfer in Nonhomogeneous Participating Media, *J. Thermophys. Heat Transfer*, vol. 4, no. 4, pp. 419–424, 1990.

108. Tan, Z., and J. R. Howell: Radiation Heat Transfer in a Partially Divided Square Enclosure with a Participating Medium, ASME HTD vol. 106, pp. 199–203, 1989 (AIChE/ASME Heat Transfer Conf., Philadelphia, August 1989).

109. Tan, Z., and J. R. Howell: Two-Dimensional Radiative Heat Transfer in an Absorbing, Emitting, and Linearly Anisotropically Scattering Medium Exposed to a Collimated Source, ASME HTD vol. 137, pp. 101–106, 1990 (ASME/AIAA Thermophysics and Heat Transfer Conf., Seattle, June 1990).

110. Tien, C. L., and B. L. Drolen: Thermal Radiation in Particulate Media with Dependent and Independent Scattering, in "Annual Review of Numerical Heat Transfer and Fluid Mechanics," vol. 1, pp. 1–32, Hemisphere, Washington, D.C., 1987.

111. Yamada, Y., J. D. Cartigny, and C. L. Tien: Radiative Transfer with Dependent Scattering by Particles: Part 2—Experimental Investigation, *J. Heat Transfer*, vol. 108, no. 3, pp. 614–618, 1986.

112. Buckius, R. O.: Radiative Heat Transfer in Scattering Media: Real Property Contributions, *Proc. 7th Int. Heat Transfer Conf.*, vol. 1, pp. 141–150, San Francisco, 1986.

113. Crosbie, A. L., and G. W. Davidson: Dirac-Delta Function Approximations to the Scattering Phase Function, *J. Quant. Spectrosc. Radiat. Transfer*, vol. 33, no. 4, pp. 391–409, 1985.

114. Lee, H., and R. O. Buckius: Scaling Anisotropic Scattering in Radiative Heat Transfer for a Planar Medium, *J. Heat Transfer*, vol. 104, no. 1, pp. 68–75, 1982.

115. Lee, H., and R. O. Buckius: Combined Mode Heat Transfer Utilizing Radiating Scaling, *J. Heat Transfer*, vol. 108, no. 3, pp. 626–631, 1986.
116. Kim, T.-K., and H. S. Lee: Scaled Isotropic Results for Two-Dimensional Anisotropic Scattering Media, *J. Heat Transfer*, vol. 112, no. 3, pp. 721–727, 1990.

PROBLEMS

15-1 Two infinite parallel plates at temperatures T_1 and T_2, having respective emissivities ϵ_1 and ϵ_2, are separated by a distance D. The space between them is filled with a gray medium having a constant absorption coefficient a. Obtain expressions for the heat flux being transferred and the temperature distribution in the medium using the strong transparent approximation.

Answer: $(T^4 - T_2^4)/(T_1^4 - T_2^4) = \left(\dfrac{1}{\epsilon_2} - \dfrac{1}{2}\right) \Big/ \left(\dfrac{1}{\epsilon_1} + \dfrac{1}{\epsilon_2} - 1\right)$

15-2 A sphere of high-temperature optically thin gray gas of fixed volume is being cooled by radiation loss to a cool black boundary (neglect emission from the boundary). At any instant the entire gas may be considered isothermal and the emission approximation used to compute the radiative loss. Heat conduction is neglected. Write the transient energy equation and solve to obtain the gas temperature as a function of time, starting from an initial temperature T_i.

Answer: $[(12\sigma a/\rho c_v)\tau + (1/T_i^3)]^{-1/3}$

15-3 A gray gas is contained between infinite parallel plates. The plates both have emissivity $\epsilon = 0.25$. Plate 1 is held at temperature $T_1 = 1000$ K, and plate 2 is at $T_2 = 500$ K. The medium between the plates is also gray but is nonscattering, and has a uniform absorption coefficient of $a = 0.5$ m^{-1}. The plate geometry is shown below.

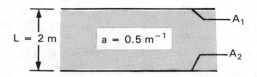

Predict the net radiative heat flux between the surfaces (W/m^2) and plot the temperature profile $[T^4(\kappa) - T_2^4]/(T_1^4 - T_2^4)$ in the gas, where $\kappa = ax$. Solve the problem using:
(a) The first-order diffusion method
(b) The exponential kernel approximation
(c) The P-1 approximation (differential approximation)
(d) The Milne-Eddington approximation
(e) The nearly transparent optically thin approximation

15-4 A gray gas is contained between infinite parallel plates. The plates both have emissivity $\epsilon = 0.25$. Plate 1 is held at temperature $T_1 = 1000$ K, and plate 2 is at $T_2 = 500$ K. The medium between the plates is also gray but is nonscattering and has a uniform absorption coefficient of $a = 0.5$ m^{-1}. The plate geometry is shown below.

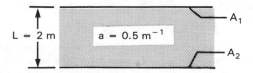

Predict the heat flux between the surfaces (W/m^2) and plot the temperature profile $[T^4(\kappa) - T_2^4]/(T_1^4 - T_2^4)$ in the gas, where $\kappa = ax$. Solve the problem using the two-flux method.

Answer: 6645 W/m^2

15-5 Energy is being transferred by radiation from a gray plate with $T_1 = 1000$ K, $\epsilon_1 = 0.8$, to a second plate parallel to it with $T_2 = 900$ K, $\epsilon_2 = 0.5$. The plates are 10 cm apart, and the space between them is filled with stagnant nitrogen gas. Heat conduction by the gas is negligible. It is desired to reduce the net heat transfer between the plates by 35%. One idea is to suspend propane soot uniformly in the gas. Estimate the required volume concentration of soot. (*Hint:* See Sec. 14-7.6.)

Answer: 1.67×10^{-5}

15-6 A spherical cavity 20 cm in diameter is filled with a gray medium having an absorption coefficient of 0.1 cm^{-1}. The cavity surface is black and is at a uniform temperature of 500 K. When the medium is first placed in the cavity, the medium is cold. For this condition, use the cold-medium approximation to estimate the heat flux radiated from the small opening as shown.

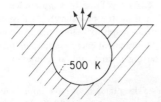

Answer: 1053 W/m^2

15-7 The space between two concentric diffuse-gray cylinders is filled with an optically dense stagnant medium having a constant absorption coefficient a. Compute the heat transfer across the gap from the inner to the outer cylinder and the radial temperature distribution in the medium by using the diffusion method with jump boundary conditions.

15-8 Two parallel gray plates are 5 cm apart. Their temperatures and emissivities are $T_1 = 700$ K, $\epsilon_1 = 0.8$; $T_2 = 500$ K, $\epsilon_2 = 0.3$. Compute the heat transferred by radiation across the gap between the plates when the gap is a vacuum and when the gap is filled with a gray medium of absorption coefficient $a = 0.6$ cm^{-1}. (Use the diffusion method as an approximation.)

Answer: 2.810 kW/m^2, 1.726 kW/m^2

15-9 Two gray parallel plates at temperatures T_1 and T_2 and with emissivities ϵ_1 and ϵ_2 are separated by an optically thick gray medium. The medium has within it a uniform volumetric energy source of G W/m^3. Compute the temperature distribution in the medium by use of the diffusion method with jump boundary conditions.

15-10 A large plate of translucent glass is laid upon a sheet of polished aluminum. The aluminum is kept at a temperature of 500 K and has an emissivity of 0.03. The glass is 10 cm thick and has a mean absorption coefficient of $a_R = 4$ cm^{-1}. A transparent liquid flows over the exposed face of the glass and maintains the face at a temperature of 250 K.

(a) What is the heat flux through the glass plate?

(b) What is the temperature distribution in the glass plate?

Neglect conduction in your calculations, and for simplicity, assume that the refractive indices of the glass and liquid are both unity.

15-11 Consider a gray absorbing medium between parallel diffuse gray plates at temperatures T_1 and T_2. In the optically thick limit, show that the exponential kernel approximation yields the same result for radiative heat transfer as obtained by the diffusion solution with jump boundary conditions.

15-12 A long cylinder 10 cm in diameter is surrounded by another cylinder 20 cm in diameter. The surfaces are gray, the inner cylinder is at $T_1 = 850$ K, $\epsilon_1 = 0.4$, and the outer cylinder is at $T_2 = 1025$ K, $\epsilon_2 = 0.7$. What is the heat transfer to the inner cylinder per unit length for vacuum between the cylinders? The space between the cylinders is filled with a gray medium having absorption coefficient $a = 0.8$ cm^{-1}. Compute the energy transfer using the P-1 method and the diffusion method.

Answer: 3.819 kW/m, 2.281 kW/m, 2.264 kW/m

15-13 A spherical fusion reactor core of radius R_1 is contained within a shell blanketed by a layer of gray gas of thickness L (that is, $R_2 = R_1 + L$). The spherical surfaces that contain the gas blanket are black. The outer spherical surface is at T_{w2}, and the gas has absorption coefficient a. If the reactor generates power Q, find expressions for the inner surface temperature T_{w1} using the diffusion and P-1 differential methods. Plot T_{w1} as a function of aL and R_1/R_2.

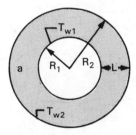

15-14 For a nongray nonscattering gas with nonuniform properties, derive the following relations for the wavelength of emission λ and the path length to absorption l:

$$R_\lambda = \int_0^\lambda a_\lambda i'_{\lambda b}\, d\lambda \Big/ \int_0^\infty a_\lambda i'_{\lambda b}\, d\lambda, \qquad \ln R_l = -\int_0^l a_\lambda(S)\, dS$$

15-15 In Example 15-10 let both plate temperatures T_1 and T_2 be nonzero. Plate 1 is black but plate 2 has a spectrally varying hemispherical emissivity $\epsilon_{\lambda,2}(\lambda, T_2)$. Modify the flowchart in Fig. 15-17 to account for this in determining the energy transfer between the plates and the gas temperature distribution.

15-16 A gray nonscattering gas with constant absorption coefficient a is between parallel black plates of finite width L and infinite length. The plate temperatures are T_1 and $T_2 = 0$ and the gas is at constant temperature T_g. The plates are a distance D apart, and the side boundaries are black and are at zero temperature as shown. Construct a Monte Carlo flowchart to obtain the energy transfer to plate 2.

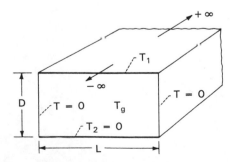

15-17 Derive the cumulative distribution function for the angle of scatter for the cases of isotropic and Rayleigh scattering.

15-18 Derive the cumulative distribution function for the angle of scatter for a gas that scatters according to the Henyey-Greenstein function.

15-19 Two infinite parallel black plates are spaced a distance D apart. The plate temperatures are T_1 and T_2, and the space between them is filled with a gray isotropically scattering medium. The medium does not absorb or emit radiation. Draw a Monte Carlo flowchart to obtain the energy transfer from plate 1 to plate 2.

15-20 Generalize Prob. 15-19 to account for an anisotropic phase function that is independent of both incidence angle and circumferential scattering angle.

SIXTEEN

ENERGY TRANSFER BY RADIATION COMBINED WITH CONDUCTION AND/OR CONVECTION

16-1 INTRODUCTION

When appreciable heat conduction and/or convection occurs simultaneously with radiation in an absorbing, emitting, and scattering medium, there are mathematical complications in addition to those for radiation alone. Unless it can be shown that both conduction and convection have a negligible effect compared with radiation or vice versa, a nonlinear integrodifferential equation results for the energy equation having combined modes of heat transfer. Numerical methods are usually required to obtain temperature distributions, distributions of scattered energy, and the heat fluxes of interest. The formulation can be simplified for some circumstances when all modes must be included. For example, when the gas is optically thick it may be possible to apply the diffusion approximation. The radiation integrals are then replaced by differential terms, and a nonlinear differential equation results. Other approximations, such as the nearly transparent gas or optically thin approximation with cold boundaries (Secs. 15-3.1 and 15-3.2), can be applied under suitable conditions to simplify the radiative terms.

Because combined-mode problems are mathematically complex, it is not usually possible to obtain analytical solutions even for seemingly simple physical cases. Consequently, for each situation discussed here the analysis will be formulated and some of the intermediate steps in the solution outlined; then the results of numerical or other solution techniques are presented and discussed. Radiation combined with conduction is considered first. The solution for a plane layer is presented with scattering included. Then the formulation is given for an absorbing-emitting rectangular region. The additive approximation is examined in which heat flows by each transfer mode are calculated independently and added.

Convection is then introduced and a boundary layer flow is examined; the optically thin and thick limits of the absorbing layer are discussed. The energy relations and some solutions are presented for flow in parallel-plate channels and tubes. Another section on convection deals with free convection as influenced by the presence of radiation. The final section of the chapter considers transient problems. The transient thermal behavior of a scattering slab and the heating of a semi-infinite region subjected to external radiation are presented.

16-2 SYMBOLS

a	absorption coefficient
A	area
b, d	height and width of rectangular region
c	propagation speed in medium; specific heat of solid
c_p, c_v	heat capacity
D	distance between parallel planes; tube diameter
e	emissive power
E_n	exponential integral function
f	function of η in Blasius boundary layer solution; gas-to-gas exchange factor
F	unidirectional flux
$\bar{F}$	exchange factor
g	surface-to-gas exchange factor; gravity
i	radiation intensity
I	radiative source function
k	thermal conductivity
K	extinction coefficient, $a + \sigma_s$
K_ν	function defined in Eq. (16-33)
l	length
N_j	conduction-radiation parameter based on temperature T_j, $kK/4\sigma T_j^3$
Nu	Nusselt number
P	pressure
Pr	Prandtl number
q	energy flux, energy per unit area and time
q'''	heat generation per unit volume and time
$\mathbf{q}_r$	radiative heat flux vector
Q	energy per unit time
r	radial coordinate
$\mathbf{r}$	position vector
S	surface area
S_1, S_2	two-dimensional radiation integral functions
t	dimensionless temperature, T/T_1
T	absolute temperature
u, v	velocities in x, y directions

$\bar{u}$	mean velocity
U	radiant energy density
V	volume
x, y, z	rectangular coordinates
α	thermal diffusivity
β	coefficient of volume expansion
δ	boundary-layer thickness
ϵ	emissivity
ϵ_h	eddy diffusivity for turbulent heat transfer
η	Blasius similarity variable
θ	cone angle, angle from normal of area
κ	optical depth
λ	wavelength
μ	$\cos \theta$; fluid viscosity
ν	kinematic viscosity
ρ	density of fluid; separation distance
σ	Stefan-Boltzmann constant
σ_s	scattering coefficient
τ	time
τ_0	surface transmissivity
ϕ	dimensionless temperature group, $(T^4 - T_2^4)/(T_1^4 - T_2^4)$
Φ	phase function for scattering
Φ_d	viscous dissipation function
ψ	slip coefficient defined by Eq. (16-32); boundary-layer streamfunction
ω	solid angle
Ω	scattering albedo

Subscripts

b	blackbody
c	conduction
D	evaluated at $x = D$
e	emitted
g	gas
i	incident; inlet
j	jth surface
m	mean value
o	outlet; outgoing; in outer region
p	particle
P	Planck mean value
r	radiative; in radial direction
R	Rosseland mean value
S	surface
V	volume
w	evaluated at wall
λ	spectrally dependent

0   free stream value; initial value

1, 2   surface 1 or 2

Superscripts

$'$ directional quantity

$*$ dummy variable of integration

$^{-}$ (overbar) average over all angles; mean value

16-3 ENERGY RELATIONS FOR ABSORBING, EMITTING, AND SCATTERING MEDIUM

For convenience, some of the energy relations previously derived are summarized here. They are applied in this chapter for various situations involving radiation combined with conduction and/or convection.

The energy equation at any location within the radiating medium is given by Eq. (14-26) as

$$\rho c_p \frac{DT}{D\tau} = \beta T \frac{DP}{D\tau} + \nabla \cdot (k \nabla T - \mathbf{q}_r) + q''' + \Phi_d \tag{16-1}$$

The not very restrictive assumption is made that the scattering phase function does not depend on the incident angle of the radiation, or alternatively the scattering can be assumed isotropic. Then from Eq. (14-42) the local divergence of the radiative flux is related to the local intensities by

$$\nabla \cdot \mathbf{q}_r = 4\pi \int_0^\infty a_\lambda(\lambda)[i'_{\lambda b}(\lambda) - \bar{i}_\lambda(\lambda)] \, d\lambda \tag{16-2}$$

The average incident intensity $\bar{i}_\lambda$ at an optical location κ_λ is obtained from Eq. (14-13) as

$$\bar{i}_\lambda(\kappa_\lambda) = \frac{1}{4\pi} \int_{\omega_i=0}^{4\pi} i'_\lambda(\kappa_\lambda) \, d\omega_i$$

$$= \frac{1}{4\pi} \int_{\omega_i=0}^{4\pi} \left[i'_\lambda(0)e^{-\kappa_\lambda} + \int_0^{\kappa_\lambda} I'_\lambda(\kappa_\lambda^*, \omega_i)e^{-(\kappa_\lambda - \kappa_\lambda^*)} \, d\kappa_\lambda^* \right] d\omega_i \tag{16-3}$$

From Eq. (14-9) the radiative source function is the sum of emitted and scattered energy

$$I'_\lambda(\kappa_\lambda, \omega) = (1 - \Omega_\lambda)i'_{\lambda b}(\kappa_\lambda) + \Omega_\lambda \Phi(\lambda, \omega)\bar{i}_\lambda(\kappa_\lambda) \tag{16-4}$$

The $\bar{i}_\lambda(\kappa_\lambda)$ is eliminated by combining (16-2) and (16-4) to obtain an alternative form of $\nabla \cdot \mathbf{q}_r$ in terms of the source function I'_λ,

$$\nabla \cdot \mathbf{q}_r = 4\pi \int_0^\infty \frac{a_\lambda}{\Omega_\lambda \Phi(\lambda, \omega)} [\{1 - \Omega_\lambda[1 - \Phi(\lambda, \omega)]\}i'_{\lambda b}(\kappa_\lambda) - I'_\lambda(\kappa_\lambda, \omega)] \, d\lambda \quad (16\text{-}5)$$

Equations (16-3) and (16-4) can be combined to yield an integral equation for I'_λ or for $\bar{\imath}_\lambda$,

$$I'_\lambda(\kappa_\lambda, \omega) = (1 - \Omega_\lambda)i'_{\lambda b}(\kappa_\lambda) + \Omega_\lambda \Phi(\lambda, \omega) \frac{1}{4\pi} \int_{\omega_i=0}^{4\pi} \left[i'_\lambda(0)e^{-\kappa_\lambda} \right.$$

$$\left. + \int_0^{\kappa_\lambda} I'_\lambda(\kappa_\lambda^*, \omega_i)e^{-(\kappa_\lambda - \kappa_\lambda^*)} \, d\kappa_\lambda^* \right] d\omega_i \quad (16\text{-}6)$$

$$\bar{\imath}_\lambda(\kappa_\lambda) = \frac{1}{4\pi} \int_{\omega=0}^{4\pi} \left\{ i'_\lambda(0)e^{-\kappa_\lambda} + \int_0^{\kappa_\lambda} [(1 - \Omega_\lambda)i'_{\lambda b}(\kappa_\lambda) \right.$$

$$\left. + \Omega_\lambda \Phi(\lambda, \omega)\bar{\imath}_\lambda(\kappa_\lambda)]e^{-(\kappa_\lambda - \kappa_\lambda^*)} \, d\kappa_\lambda^* \right\} d\omega \quad (16\text{-}7)$$

The preceding energy relations are complicated to solve for temperature distributions and heat fluxes because of their general form. Simplifications are usually made so that situations of interest can be examined more readily. Some of these cases are now considered.

16-4 RADIATION WITH CONDUCTION

There are important situations in which heat is transported within a medium by only radiation and conduction. These usually involve solid or highly viscous media, so convection in the medium is not important. In a liquid or gas, forced or free convection are usually of sufficient importance that they must be included. Radiation with conduction is considered in this section; convection is added later. The following are some practical situations in which combined radiation-conduction transport is important.

Glass can absorb significant amounts of radiation in certain wavelength regions (see Figs. 5-32 to 5-34). The absorbed energy is re-emitted within the glass, thereby providing a radiative transport traveling layer by layer through the medium. Ordinary thermal conduction is thus augmented by *radiative conduction*. Radiative effects are quite important in influencing the temperature distribution within molten glass in a furnace and during heat treatment of glass plates. These effects have been analyzed in [1–7].

Another application is in coatings on surfaces. Glassy materials are sometimes used as an ablating coating to protect the interior of a body from high external temperatures by sacrifice of the ablating coating. The radiation-conduction process is important in regulating the temperature distribution within the ablating layer

[8–10]. The temperature distribution is influential in determining how the ablating material will soften, melt, or vaporize. These processes ultimately govern how efficiently the material will protect the surface. Nonablating semitransparent ceramic coatings are also used for surface protection. Another type of coating is formed by cryodeposits of solidified gas on a very cold surface. The surface may be on a space vehicle orbiting at the upper fringe of the atmosphere or may be part of a cryopump used to produce high vacuum by condensing the gas within a chamber. The cryodeposit coating changes the radiative properties of the cold surface and can thus significantly influence the radiative exchange with this surface. Radiative transfer in cryodeposits is considered in [11–13].

Radiation can be a significant mode of energy transfer in fibrous insulation materials [14, 15], foam insulations [16], and high-temperature porous insulating materials [17]. It was usually thought that heat conduction and a little free convection were the only significant heat transfer modes in insulation such as the lightweight fibrous (such as fiberglass) materials used for building insulation. Radiation was thought to be very small because of the low temperatures involved. However, for moderate temperatures of 300–400 K, conduction in the air and radiation are the significant modes of transfer in lightweight fibrous insulation. Radiation can also be significant in polyurethane foam insulation because the cell walls in the foam are partially transparent for infrared radiation. The foam scatters radiation anisotropically, but scattering is more significant in fiberglass than in foam insulation. Low-density, high-temperature insulations such as hafnia, thoria, and zirconia are used for high-technology applications such as the space shuttle, re-entry vehicles, and solar central receivers. These materials are semitransparent to near-infrared and visible radiation, and the radiative transfer in them can equal or exceed conduction for these applications.

In this section, some methods are examined for treating combined radiation and conduction energy transfer. The conduction-radiation parameter is introduced and energy relations formulated. Methods for numerical solution are in Chap. 17. Some approximations are considered here, the simplest being the addition of separately computed radiation and conduction transfers to obtain the combined transfer. Where applicable, approximate solution methods to the transfer equations presented in Chap. 15 can be applied to simplify the radiation terms in multimode problems, and the diffusion method is applied here as an example.

16-4.1 Energy Balance

For combined conduction-radiation energy transfer in an absorbing-emitting and scattering medium, the energy equation (16-1) is used. After neglecting terms that do not apply, this is solved subject to the boundary conditions to obtain the temperature distribution in the medium; the heat flow can then be found. The terms involving convection, viscous dissipation, and volume expansion are omitted from (16-1) to yield

$$\rho c_p \frac{\partial T}{\partial \tau} = \nabla \cdot (k \nabla T - \mathbf{q}_r) + q''' \tag{16-8}$$

If the $\nabla \cdot \mathbf{q}_r$ is substituted from (16-2) the local energy balance is

$$\rho c_p \frac{\partial T}{\partial \tau} = \nabla \cdot (k \nabla T) + q''' - 4\pi \int_{\lambda=0}^{\infty} a_\lambda i'_{\lambda b}(\lambda, T) \, d\lambda$$

$$+ \int_{\lambda=0}^{\infty} \int_{\omega=0}^{4\pi} a_\lambda i'_\lambda(\lambda, \omega) \, d\omega \, d\lambda \tag{16-9}$$

Since the radiation terms in (16-9) depend not only on the local temperature but also on the entire surrounding radiation field, the energy equation is an integro-differential equation for the temperature distribution in the medium. The conduction and heat-storage terms depend on a different power of the temperature than the radiation terms, and the energy equation is thus nonlinear. The following discussion will show how the energy equation is treated for a plane layer without scattering and with scattering included.

16-4.2 Plane-Layer Geometry

Absorbing-emitting medium without scattering Consider a layer of conducting-radiating material between parallel black plates at temperatures T_1 and T_2, and spaced D apart as in Fig. 16-1. The medium between the plates is gray and has a constant thermal conductivity k and absorption coefficient a. The integrodifferential equation for steady energy transfer without scattering will be developed.

For one-dimensional heat conduction and constant k, the $\nabla \cdot (k\nabla T)$ in (16-8) reduces to $k \, d^2T/dx^2$, and $\nabla \cdot \mathbf{q}_r$ becomes $dq_{r,x}/dx$. The temperature distribution is at steady state, $\partial T/\partial \tau = 0$, and $q''' = 0$. Then with $\kappa = ax$, (16-8) reduces to

$$ka \frac{d^2T}{d\kappa^2} = \frac{dq_{r,\kappa}}{d\kappa} \tag{16-10}$$

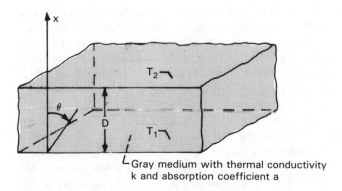

$^{\llcorner}$Gray medium with thermal conductivity
k and absorption coefficient a

Figure 16-1 Plane layer of conducting and radiating medium between walls at different uniform temperatures.

For a plane layer the $dq_{r,\kappa}/d\kappa$ is given by (14-77). For black walls the boundary intensities are $i^+(0, \mu) = \sigma T_1^4/\pi$ and $i^-(\kappa_D, -\mu) + \sigma T_2^4/\pi$. For zero scattering in a gray medium $I'(\kappa) = \sigma T^4(\kappa)/\pi$ from (16-4). Then (16-10) becomes

$$ka\frac{d^2T}{d\kappa^2} = -2\sigma T_1^4 \int_0^1 \exp\frac{-\kappa}{\mu}d\mu - 2\sigma T_2^4 \int_0^1 \exp\frac{\kappa_D - \kappa}{-\mu}d\mu$$

$$-2\int_0^{\kappa_D} \sigma T^4(\kappa^*)E_1(|\kappa - \kappa^*|)\,d\kappa^* + 4\sigma T^4(\kappa) \tag{16-11}$$

where $\kappa_D = aD$ and $\mu = \cos\theta$. Two boundary conditions are needed to solve (16-11) for $T(\kappa)$ as this equation contains a second derivative:

$$T(x) = T_1 \quad \text{at } x = 0, \qquad T(x) = T_2 \quad \text{at } x = D \tag{16-12}$$

Some further simplifications in form are possible. Define the nondimensional quantities $t = T/T_1$, $t_2 = T_2/T_1$, and $N_1 = ka/4\sigma T_1^3$. By using the exponential integral function defined in Appendix E, (16-11) is written as

$$N_1\frac{d^2t(\kappa)}{d\kappa^2} = t^4(\kappa) - \frac{1}{2}\left[E_2(\kappa) + t_2^4 E_2(\kappa_D - \kappa) \right.$$

$$\left. + \int_0^{\kappa_D} t^4(\kappa^*)E_1(|\kappa - \kappa^*|)\,d\kappa^* \right] \tag{16-13}$$

Equation (16-13) is the desired integrodifferential equation for the temperature distribution $t(\kappa)$. It is a nonlinear equation since t is raised to the first power in the conduction term while it is raised to the fourth power in the radiation terms. The boundary conditions in dimensionless form are

$$t = 1 \quad \text{at } \kappa = 0, \qquad t = t_2 \quad \text{at } \kappa = \kappa_D \tag{16-14}$$

Examining (16-13) and (16-14) shows that the solution depends on the parameters N_1, κ_D, and t_2.

The parameter $N_j \equiv ka/4\sigma T_j^3$ that arose in the nondimensionalization is called the *conduction-radiation parameter* for a nonscattering medium (or *Stark number*) based on the jth temperature. The N_j does *not* directly give the relative values of conduction to emission because the ratio of these values depends on both temperature difference and temperature level. When scattering is included, the definition of N_j includes the scattering coefficient to become $N_j \equiv k(a + \sigma_s)/4\sigma T_j^3 = kK/4\sigma T_j^3$.

The heat transfer across the layer from plate 1 to plate 2 can be obtained from the temperature distribution. From energy conservation q will not depend on κ for the conditions considered. Equation (14-109) with $G = \sigma T^4$ gives the net heat flux expression for radiation alone across a gray gas between black plates. This radiative flux equation was obtained for convenience at $\kappa = 0$. In addition, at the same location there is now a conduction flux $-k(dT/dx)|_{x=0} = -ka(dT/d\kappa)|_{\kappa=0}$, so that the heat flux relation becomes

$$q = -ka\left.\frac{dT}{d\kappa}\right|_{\kappa=0} + \sigma T_1^4 - 2\sigma T_2^4 E_3(\kappa_D) - 2\int_0^{\kappa_D} \sigma T^4(\kappa^*)E_2(\kappa^*)\,d\kappa^* \qquad (16\text{-}15a)$$

On the right, the first term is the conduction away from wall 1, the second is the radiation leaving wall 1, the third is the radiation leaving wall 2 that is then attenuated by the medium and reaches wall 1, and the last term is the radiation from the medium to wall 1. By using the previously defined dimensionless variables, the heat flux is written as

$$\frac{q}{\sigma T_1^4} = -4N_1\left.\frac{dt}{d\kappa}\right|_{\kappa=0} + 1 - 2\left[t_2^4 E_3(\kappa_D) + \int_0^{\kappa_D} t^4(\kappa^*)E_2(\kappa^*)\,d\kappa^*\right] \qquad (16\text{-}15b)$$

Since there are no heat sources in the medium, this q computed at the lower wall is the same at all κ within the medium.

Viskanta and Grosh [18, 19] obtained solutions of (16-13) by numerical integration and iteration. Some of their temperature distributions are in Fig. 16-2. For $N_1 \to \infty$, conduction dominates, and the solution reduces to the linear profile for conduction through a plane layer. When $N_1 = 0$, the conduction term drops out, and the temperature profile has a discontinuity (temperature slip) at each wall as discussed for the case of radiation alone in Sec. 14-7.6. When conduction is present, there is no temperature slip. Some heat flux results from [19] are in Table 16-1 as obtained from Eq. (16-15b). Results for cylindrical and spherical geometries are in [20–22].

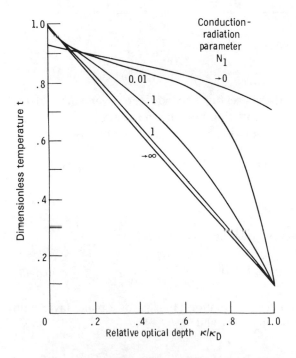

Figure 16-2 Dimensionless temperature distribution in gray gas between infinite parallel black plates with conduction and radiation. Plate temperature ratio $t_2 = 0.1$; optical spacing $\kappa_D = 1.0$. Data from [18].

Table 16-1 Heat flux between parallel black plates by combined radiation and conduction through a gray medium [19]

Optical thickness κ_D	Plate-temperature ratio t_2	Conduction-radiation parameter N_1	Dimensionless energy flux $q/\sigma T_1^4$
0.1	0.5	0	0.859
		0.01	1.074
		0.1	2.880
		1	20.88
		10	200.88
1.0	0.5	0	0.518
		0.01	0.596
		0.1	0.798
		1	2.600
		10	20.60
1.0	0.1	0	0.556
		0.01	0.658
		0.1	0.991
		1	4.218
		10	36.60
10	0.5	0	0.102
		0.01	0.114
		0.1	0.131
		1	0.315
		10	2.114

Absorbing-emitting medium with scattering To add some generality to the previous solution, let the medium be scattering in addition to being absorbing and emitting. The medium is gray and scattering is assumed isotropic. Scattering is conveniently included by utilizing the source function. For isotropic scattering the phase function $\Phi(\lambda, \omega) = 1$, so for gray properties Eq. (16-5) becomes

$$\nabla \cdot \mathbf{q}_r = 4 \frac{a}{\Omega} [\sigma T^4(\kappa) - \pi I'(\kappa)] \tag{16-16}$$

Then with $\kappa = (a + \sigma_s)x$ the energy equation (16-10) becomes

$$k(a + \sigma_s) \frac{d^2 T}{d\kappa^2} = 4 \frac{(1 - \Omega)}{\Omega} [\sigma T^4(\kappa) - \pi I'(\kappa)] \tag{16-17}$$

The $\pi I'(\kappa)$ is found by using the integral equation (14-78) for a plane layer with black boundaries at uniform temperatures T_1 and T_2,

$$\pi I'(\kappa) = (1 - \Omega)\sigma T^4(\kappa) + \frac{\Omega}{2}\left[\sigma T_1^4 E_2(\kappa) + \sigma T_2^4 E_2(\kappa_D - \kappa)\right.$$

$$\left. + \int_0^{\kappa_D} \pi I'(\kappa^*)E_1(|\kappa^* - \kappa|)\, d\kappa^* \right] \tag{16-18}$$

Equations (16-17) and (16-18) can be placed in dimensionless form by using the same quantities as in Eq. (16-13) and by using $\pi I'/\sigma T_1^4$ as a dimensionless intensity. The $N_1 = k(a + \sigma_s)/4\sigma T_1^3$ includes the scattering coefficient, and the scattering albedo Ω is an additional parameter. The boundary conditions are the same as in (16-14). The solution is obtained by numerical integration and iteration, and results are in [23].

Figure 16-3 shows the effect of scattering on some temperature profiles. For $\Omega = 0$ the results correspond to those in Fig. 16-2. Increased scattering tends to make the profiles linear as for heat conduction only. When $\Omega \rightarrow 1$ there is no absorption or emission by the medium, and hence there is no mechanism that can convert radiant energy into internal energy of the medium. Hence the temperature

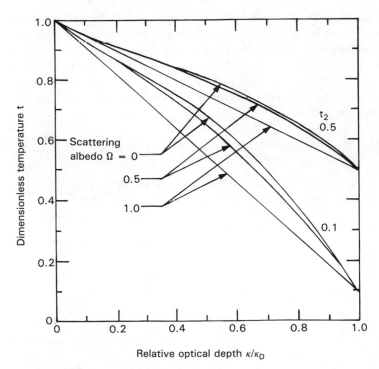

Figure 16-3 Effect of scattering albedo on temperature distribution in gray medium between infinite parallel black plates. Conduction-radiation parameter $N_1 = 0.1$, optical spacing $\kappa_D = 1.0$, plate temperature ratios $t_2 = 0.1$ and 0.5. From [23].

has the linear distribution for conduction alone. Heat is then transferred by conduction and by scattering of radiant energy through the medium from the high-temperature boundary to the lower-temperature boundary. The effect of partially reflecting walls (ϵ_1 and/or $\epsilon_2 < 1$) has not been discussed here but is included in [23]. In [24] results were obtained for a layer between diffuse gray walls with isotropic scattering in the absorbing and emitting medium. From a literature survey, some typical heat fluxes by combined radiation and conduction from the boundary at T_1 to that at T_2 are in Table 16-2, which demonstrates the influence of the parameters. In [25] the additional complexity was investigated of having strong forward or backward scattering rather than using the isotropic assumption.

Some experimental results are given by Schimmel et al. [26] for layers of CO_2, N_2O, and mixtures CO_2-CH_4 and CO_2-N_2O. The conduction heat fluxes at the walls were compared with analytical results using gray-gas, box, and wide-band gas models. The wide-band model of Edwards and Menard (see Sec. 12-8.5) was found to give the best results. An analysis is in [27] for a nongray medium consisting of CO_2, H_2O vapor, and soot between gray diffuse plates having different surface emissivities.

16-4.3 Nonplanar and Multidimensional Geometries

The previous section considered a plane layer of infinite extent. Some results for other shapes are now considered. The steady-state energy transfer in a rectangular region has been analyzed in [28] and [29] for a gray absorbing-emitting medium without scattering. The solutions are two-dimensional and depend only on the x and y shown in Fig. 14-15. Internal heat generation and gray boundaries are included in [29], while in [28] the boundaries are black. From Eq. (16-1) the energy equation for constant properties is

$$k\nabla^2 T - \nabla \cdot \mathbf{q}_r + q''' = 0 \qquad (16\text{-}19)$$

Table 16-2 Dimensionless heat fluxes across a plane layer by combined radiation and conduction with isotropic scattering, $T_1/T_2 = 2$, [24]

$\kappa_D = KD$	ϵ_1	ϵ_2	N_1	Ω	$(q_r + q_c)/\sigma T_1^4$
1.0	0.1	0.1	0.1	0	0.461
1.0	0.1	0.1	0.1	0.5	0.350
1.0	0.1	0.1	0.1	1.0	0.248
1.0	0.5	0.5	0.1	0	0.571
1.0	0.5	0.5	0.1	0.5	0.513
1.0	0.5	0.5	0.1	1.0	0.449
1.0	1	0	1.0	0.5	2.303
1.0	0	1	1.0	0.5	2.210
1.0	0.1	0	1.0	0.5	2.143
1.0	0	0.1	1.0	0.5	2.132

and for a gray medium Eq. (16-2) gives

$$\nabla \cdot \mathbf{q}_r = 4a\sigma T^4 - 4\pi a \bar{\imath} \tag{16-20}$$

For a two-dimensional rectangular region $\bar{\imath}(x, y)$ is given by Eq. (14-136). If for simplicity an enclosure with black walls is considered, the energy equation has the form, since $i_b' = \sigma T_b^4/\pi$,

$$
\begin{aligned}
-\frac{k}{a}\left(\frac{\partial^2 T}{\partial x^2} + \frac{\partial^2 T}{\partial y^2}\right) + 4\sigma T^4 &= \frac{q'''}{a} + (b - y)\int_{x_0=0}^{d} \sigma T_b^4(x_0, b)\, \frac{S_2[a\rho_0(x_0, b)]}{\rho_0^2(x_0, b)}\, dx_0 \\
&+ (d - x)\int_{y_0=0}^{b} \sigma T_b^4(d, y_0)\, \frac{S_2[a\rho_0(d, y_0)]}{\rho_0^2(d, y_0)}\, dy_0 \\
&+ y\int_{x_0=0}^{d} \sigma T_b^4(x_0, 0)\, \frac{S_2[a\rho_0(x_0, 0)]}{\rho_0^2(x_0, 0)}\, dx_0 \\
&+ x\int_{y_0=0}^{b} \sigma T_b^4(0, y_0)\, \frac{S_2[a\rho_0(0, y_0)]}{\rho_0^2(0, y_0)}\, dy_0 \\
&+ a\int_{x^*=0}^{d}\int_{y^*=0}^{b} \sigma T^4(x^*, y^*)\, \frac{S_1(a\rho^*)}{\rho^*}\, dx^*\, dy^*
\end{aligned}
\tag{16-21}
$$

where

$$\rho^* = [(x - x^*)^2 + (y - y^*)^2]^{1/2}, \qquad \rho_0(x_0, y_0) = [(x - x_0)^2 + (y - y_0)^2]^{1/2}$$

A specific situation is now considered as in [28]. The radiating medium has no internal heat generation and three walls of the enclosure are at the same temperature T_2 with the remaining wall at a different temperature T_1 as in Fig. 16-4. With $q''' = 0$, Eq. (16-21) is written for all the walls at T_2 and is then subtracted from Eq. (16-21) to yield

$$
\begin{aligned}
-\frac{k}{a}\left(\frac{\partial^2 T}{\partial x^2} + \frac{\partial^2 T}{\partial y^2}\right) + 4\sigma(T^4 - T_2^4) &= y\int_{x_0=0}^{d} \sigma(T_1^4 - T_2^4)\, \frac{S_2[a\rho_0(x_0, 0)]}{\rho_0^2(x_0, 0)}\, dx_0 \\
&+ a\int_{x^*=0}^{d}\int_{y^*=0}^{b} \sigma(T^4 - T_2^4)\, \frac{S_1(a\rho^*)}{\rho^*}\, dx^*\, dy^*
\end{aligned}
\tag{16-22}
$$

Then let $t = T/T_1$ and $t_2 = T_2/T_1$ to obtain

$$
\begin{aligned}
-\frac{k}{a\sigma T_1^3}\left[\frac{\partial^2 t}{\partial x^2} + \frac{\partial^2 t}{\partial y^2}\right] + 4(t^4 - t_2^4) &= y\int_{x_0=0}^{d} (1 - t_2^4)\, \frac{S_2[a\rho_0(x_0, 0)]}{\rho_0^2(x_0, 0)}\, dx_0 \\
&+ a\int_{x^*=0}^{d}\int_{y^*=0}^{b} (t^4 - t_2^4)\, \frac{S_1(a\rho^*)}{\rho^*}\, dx^*\, dy^*
\end{aligned}
\tag{16-23}
$$

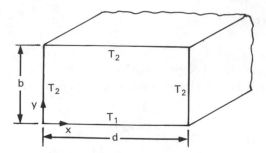

Figure 16-4 Two-dimensional rectangular region with black interior boundaries.

This was further nondimensionalized in terms of optical dimensions in [28] and solved with a numerical technique utilizing a Taylor series expansion of the temperature about each local temperature value. In [29] cases were considered for gray walls and with internal heat generation in the radiating medium. Solutions were obtained with finite-element methods. Numerical methods are described in Chap. 17.

Another nonplanar geometry that has been extensively studied is the cylindrical shape that includes infinitely long cylinders [21], concentric cylindrical regions [20, 30], and cylinders of finite length. Some examples are [31], for an infinite cylinder with isotropic scattering (some transient cases are also included), and [32], which includes a model for anisotropic scattering. Reference [33] treats anisotropic scattering in an annular region by the P-1 and P-3 methods; [34] deals with the more complex situation of a cylindrical enclosure of finite length. Results for a sphere are in [35], and concentric spheres were studied in [22].

16-4.4 Simple Addition of Radiation and Conduction Energy Transfers

A relatively simple idea for obtaining energy transfer by radiation and conduction is to assume that the interaction between the two transfer processes is so weak that each process can be considered to act independently. Then the conduction and radiation transfers are each formulated as if the other mechanism were not present. Einstein [36] and Cess [37] investigated this approximation for an absorbing-emitting gray medium between infinite parallel plates. When the plates are black, the energy transfer is within 10% of the exact solution. Exact results are approached in the optically thin and thick limits. Larger errors are possible if highly reflecting surfaces are present. Yuen and Wong [25] showed that for an absorbing-emitting medium with scattering, the presence of low-emissivity boundaries increases the error of the additive solution. In [28] the additive solution was examined for a square enclosure with black walls filled with a gray absorbing-emitting medium. Three of the enclosure boundaries are at a single temperature, and the remaining boundary is at a different temperature. The additive solution gave good results. Howell [38] showed that the additive solution is also fairly accurate for a gray gas between black concentric cylinders.

An additive solution *cannot be used to predict temperature profiles*. It is a simple method for predicting energy transfer by combined modes, although the accuracy obtained becomes doubtful in some situations. The use of the additive method is not advisable for problems where the accuracy has not been established by comparison with more exact solutions.

EXAMPLE 16-1 By using the additive approximation, obtain a relation for the energy transfer from a gray infinite plate at temperature T_1 and emissivity ϵ_1 to a parallel infinite gray plate at T_2 with emissivity ϵ_2. The spacing between the plates is D, and the region between the plates is filled with gray material having constant absorption coefficient a and thermal conductivity k. Use the diffusion approximation for the radiative transfer.

The energy flux by pure conduction from surface 1 to surface 2 is

$$q_c = \frac{k(T_1 - T_2)}{D} \tag{16-24}$$

The diffusion solution for pure radiation from 1 to 2 is given in Table 15-2 as

$$q_r = \frac{\sigma(T_1^4 - T_2^4)}{3aD/4 + 1/\epsilon_1 + 1/\epsilon_2 - 1} \tag{16-25}$$

Since the two modes are assumed independent, the additive solution gives

$$q = q_c + q_r \tag{16-26}$$

Using the dimensionless variables defined in connection with Eq. (16-13) gives q as

$$\frac{q}{\sigma T_1^4} = \frac{4N_1(1 - t_2)}{\kappa_D} + \frac{1 - t_2^4}{3\kappa_D/4 + 1/\epsilon_1 + 1/\epsilon_2 - 1} \tag{16-27}$$

Equation (16-27) must give correct results for $N_1 = 0$ (pure radiation) within the accuracy of the diffusion solution, and as $N_1 \rightarrow \infty$ (pure conduction), because the solution simply adds these two limiting cases. As shown by (15-31), the effect of scattering is included in (16-27) if $\kappa_D = (a + \sigma_s)D$. Examples 14-4 and 14-5 show that superposition provides the exact solution for the limits of a purely scattering medium or an optically thin medium.

A comparison of $q/\sigma T_1^4$ from (16-27) with exact numerical solutions for $\epsilon_1 = \epsilon_2 = 1$ and $t_2 = 0.5$ from the work of Viskanta and Grosh [18, 19] is in Fig. 16-5. For this geometry and black surfaces, the results of the additive solution are very accurate. The additive method appears even better here because of the fortuitous benefit that the diffusion solution gives a pure-radiation heat transfer that is slightly above the exact pure-radiation solution (Fig. 15-6) while the pure-conduction result is too low. This is because the conduction solution is based on

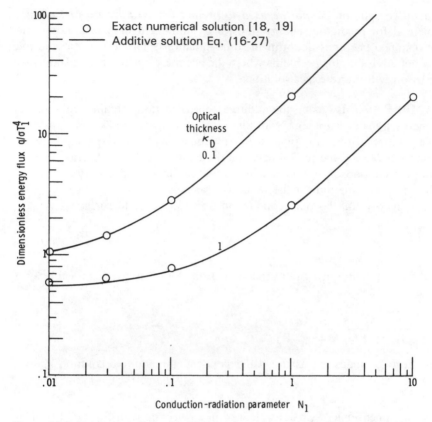

Figure 16-5 Comparison of simple additive and exact numerical solutions of combined conduction-radiation energy transfer between black parallel plates. Plate temperature ratio $t_2 = 0.5$.

the linear gradient of T while the actual gradients at the boundaries are larger when radiation is present (see Fig. 16-9a, for example). The errors in the two solutions tend to cancel, giving an accurate combined solution for this geometry. Nelson [39] shows that superposition provides good results for a nongray gas with a single absorption band.

16-4.5 The Diffusion Method

This method has an advantage over the additive method in that a solution is obtained to the coupled energy equation and this yields the temperature distribution in the medium. In Sec. 15-3.4, it was shown that the diffusion heat flux relation for radiative transfer has the same form as the Fourier conduction law. By using the Rosseland mean absorption coefficient defined in (15-44), the radiative flux vector can be written as

$$q_r = -\frac{4}{3a_R} \nabla e_b = -\frac{16\sigma T^3}{3a_R} \nabla T = -k_r \nabla T \tag{16-28}$$

where k_r is the *radiative conductivity* defined by

$$k_r \equiv \frac{16\sigma T^3}{3a_R} \tag{16-29}$$

Then the energy flux vector by combined radiation and conduction at any position in the medium can be expressed as

$$\mathbf{q} = \mathbf{q}_r + \mathbf{q}_c = -(k_r + k)\nabla T = -\left(\frac{16\sigma T^3}{3a_R} + k\right)\nabla T \tag{16-30}$$

This can be used, as in the heat conduction equation, to obtain an energy balance on a differential volume element within an absorbing-emitting medium. For example, in two-dimensional rectangular coordinates, with internal heat sources, the transient energy equation is

$$-\nabla \cdot \mathbf{q} = \frac{\partial}{\partial x}\left[\left(\frac{16\sigma T^3}{3a_R} + k\right)\frac{\partial T}{\partial x}\right] + \frac{\partial}{\partial y}\left[\left(\frac{16\sigma T^3}{3a_R} + k\right)\frac{\partial T}{\partial y}\right]$$

$$= -q'''(x,y) + \rho c_p \frac{\partial T}{\partial \tau} \tag{16-31}$$

The medium behaves like a conductor with thermal conductivity dependent on temperature.

To obtain the temperature distribution in the medium, an equation such as (16-31) must be integrated subject to the initial and boundary conditions. The boundary conditions would often be specified temperatures of the enclosure surfaces. However, near a boundary the diffusion approximation is not valid; consequently, the solution is incorrect near the wall and cannot be matched directly to the boundary conditions. To overcome this difficulty, the boundary condition at the edge of the absorbing-emitting medium is modified so the resulting solution to the diffusion equation with this effective boundary condition will be correct in the region away from the boundaries where the diffusion approximation is valid.

In the pure radiation case, a temperature slip was introduced to overcome the difficulty of matching the diffusion solution in the medium to the wall temperature. For combined conduction-radiation, a similar concept was introduced by Goldstein and Howell [40, 41]. By using asymptotic expansions to match linearized solutions for intensity, flux, and temperature near the wall with the diffusion solution for these quantities far from the wall, an effective slip condition was derived. As shown by Fig. 16-6, this slip gives the boundary condition $T(x \to 0)$ that the diffusion solution must have if the solution is to extend to the wall. The slip is given in terms of the *slip coefficient* ψ, which is a function only of the conduction-radiation parameter N. In terms of quantities at wall 1, ψ_1 is given by

$$\psi_1 = \frac{\sigma[T_1^4 - T^4(x \to 0)]}{q_{r,1}} \tag{16-32}$$

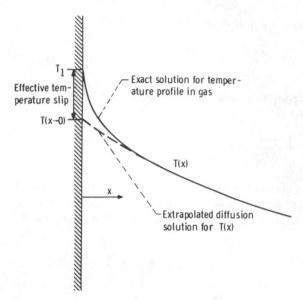

Figure 16-6 Use of effective temperature slip as boundary condition for diffusion solution in combined conduction and radiation.

where $q_{r,1}$ is the radiative flux at the boundary as evaluated by the diffusion approximation; T_1 is the wall temperature; and $T(x \rightarrow 0)$ is the extrapolated temperature *in the medium* at the wall, which is the effective slip temperature to be used in the diffusion solution. The ψ_1 is computed from the relations of [40] as

$$\psi_1 = \frac{3}{4\pi} \int_0^1 \tan^{-1} \frac{1}{K_\nu} \, dv \qquad \text{where } K_\nu = \frac{1}{\pi}\left(\frac{N_1}{2v^3} - \frac{2}{v} - \ln \frac{1-v}{1+v}\right) \tag{16-33}$$

A graph of ψ as a function of N that can be used for any geometry is in Fig. 16-7.

With the diffusion approximation, results for combined radiation and conduction can be obtained for both energy transfer and temperature profiles, as will be shown by an example. Other solutions of this general type are in [42, 43].

EXAMPLE 16-2 Using the diffusion method, find an equation for the steady temperature profile in a medium of constant absorption coefficient a and thermal conductivity k, with $q''' = 0$, and contained between infinite parallel black plates at T_1 and T_2 spaced D apart with the lower plate 1 at $x = 0$. What is the heat transfer across the layer?

For this geometry (16-30) becomes in dimensionless form (note that $a_R = a$ in this case)

$$\frac{q}{\sigma T_1^4} = -\left(\frac{4}{3a}\frac{dt^4}{dx} + \frac{k}{\sigma T_1^3}\frac{dt}{dx}\right) = -\left(\frac{4}{3}\frac{dt^4}{d\kappa} + 4N_1\frac{dt}{d\kappa}\right) \tag{16-34}$$

From energy conservation, with no internal heat sources, the q is constant across the space between plates. Equation (16-34) can then be integrated from 0 to κ_D to yield

$$\frac{q}{\sigma T_1^4} \kappa_D = -\left\{ \frac{4}{3}[t^4(\kappa_D) - t^4(0)] + 4N_1[t(\kappa_D) - t(0)] \right\} \tag{16-35}$$

where $t(0)$ and $t(\kappa_D)$ are *in the medium* at the lower and upper boundaries. These two temperatures are eliminated by using the slip boundary conditions to relate them to the specified wall temperatures T_1 and T_2.

Consider first the boundary condition at wall 1. For the particular $N = N_1$ of the problem, the ψ_1 is found from Fig. 16-7 and set equal to $\psi_1 = \sigma[T_1^4 - T^4(0)]/q_{r,1}$. From (16-30) the radiative flux $q_{r,1}$ at the wall can be written as

$$q_{r,1} = -\frac{16\sigma T_1^3}{3a}\frac{dT}{dx}\bigg|_1 = \frac{16\sigma T_1^3}{3a}\frac{q}{16\sigma T_1^3/3a + k}$$

Then ψ_1 becomes

$$\psi_1 = \frac{\sigma[T_1^4 - T^4(0)]}{(4\sigma/3a)q/(4\sigma/3a + k/4T_1^3)}$$

This is rearranged into

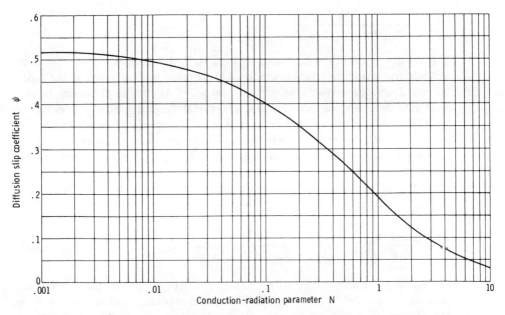

Figure 16-7 Slip coefficient for combined conduction-radiation solutions by the diffusion approximation.

$$\frac{4}{3a} q = \frac{1}{\psi_1} \left\{ \frac{4\sigma}{3a} [T_1^4 - T^4(0)] + \frac{k}{4T_1^3} [T_1^4 - T^4(0)] \right\} \tag{16-36}$$

As shown in the derivation of ψ [40], the conditions for which the diffusion solution is valid lead to the jump $T_1 - T(0)$ being small. For convenience a portion of (16-36) can then be linearized. With $T_1 - T(0) = \delta$ where δ is small,

$$\frac{T_1^4 - T^4(0)}{4T_1^3} = \frac{T_1^4 - (T_1 - \delta)^4}{4T_1^3} \approx \frac{T_1^4 - T_1^4 + 4T_1^3\delta}{4T_1^3} = \delta = T_1 - T(0)$$

Then (16-36) becomes $(4/3a)q \approx (1/\psi_1)\{(4\sigma/3a)[T_1^4 - T^4(0)] + k[T_1 - T(0)]\}$ or, in dimensionless form,

$$\frac{4}{3} \psi_1 \frac{q}{\sigma T_1^4} = \frac{4}{3} [1 - t^4(0)] + 4N_1[1 - t(0)] \tag{16-37}$$

Similarly, at wall 2 (note that ψ_2 corresponds to $N = N_2$ in Fig. 16-7)

$$\frac{4}{3} \psi_2 \frac{q}{\sigma T_1^4} = \frac{4}{3} [t^4(\kappa_D) - t_2^4] + 4N_1[t(\kappa_D) - t_2] \tag{16-38}$$

Now add (16-35), (16-37), and (16-38) to eliminate the unknown temperatures in the medium $t(0)$ and $t(\kappa_D)$. This yields the energy flux transferred across the layer as

$$\frac{q}{\sigma T_1^4} = \frac{1 - t_2^4 + 3N_1(1 - t_2)}{3\kappa_D/4 + \psi_1 + \psi_2} \tag{16-39}$$

The results of (16-39) are plotted in Fig. 16-8 and compared with the exact and additive solutions (the exchange-factor approximation shown in the figure will be discussed in the next section). At $\kappa_D = 1$, the results compare very well with the exact solution. For a small optical thickness $\kappa_D = 0.1$; however, the diffusion-slip procedure breaks down for intermediate values of N_1, and the simple additive solution provides much better energy transfer values.

An advantage of the diffusion solution is that it yields the temperature distribution in the medium. Temperature profiles are predicted by integrating (16-34) from 0 to κ (note that q is constant) and then using (16-37) and (16-39) to eliminate $t(0)$ and q. This yields

$$\frac{1 - t^4(\kappa) + 3N_1[1 - t(\kappa)]}{1 - t_2^4 + 3N_1(1 - t_2)} = \frac{3\kappa/4 + \psi_1}{3\kappa_D/4 + \psi_1 + \psi_2} \tag{16-40}$$

Profiles are shown in Fig. 16-9. For $\kappa_D = 1$ (Fig. 16-9a), the profiles are poor except for the largest N_1 shown. Better results are obtained for all N_1 in Fig. 16-9b for $\kappa_D = 10$ because the assumptions in the diffusion solution become more valid at larger κ_D. For $N_1 \rightarrow 0$ and $N_1 \rightarrow \infty$, the diffusion-slip

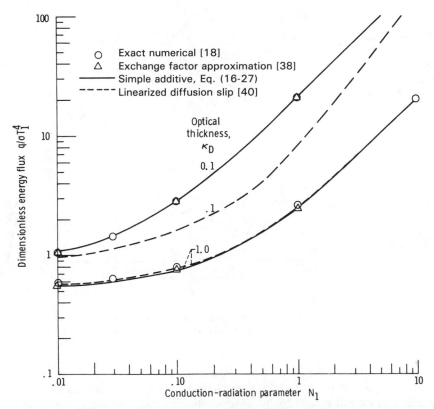

Figure 16-8 Comparison of various methods for predicting energy transfer by conduction and radiation across a layer between parallel black plates. Plate temperature ratio $T_2/T_1 = t_2 = 0.5$.

method goes to the correct limiting solutions. The diffusion method may be poor for transient solutions that require accurate time-dependent temperature distributions.

Within their limits of applicability, diffusion methods provide a useful interpretation of the conduction-radiation parameter. The ratio of molecular conductivity to radiative conductivity given by (16-29) is

$$\frac{k}{k_r} = \frac{k}{16\sigma T^3/3a_R} = \frac{3}{4}\frac{ka_R}{4\sigma T^3} = \frac{3}{4}N \tag{16-41}$$

Therefore, *in the diffusion limit*, N is a direct measure of the ratio of k to k_r and consequently in this limit is also a direct measure of the ratio of the energy transferred by the two modes.

16-4.6 The Exchange Factor Approximation

Introduction for transfer by radiation only In this section a *nonisothermal* gray gas is considered between parallel plates, concentric cylinders, and concentric

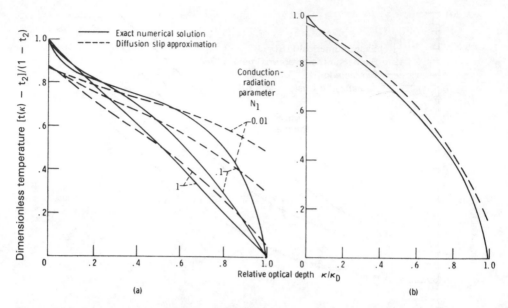

Figure 16-9 Comparison of temperature profile by exact solution [18] with diffusion-slip approximation. Plate temperature ratio $t_2 = 0.5$; plate emissivities $\epsilon_1 = \epsilon_2 = 1.0$. (a) Optical thickness $\kappa_D = 1$; (b) optical thickness $\kappa_D = 10$, conduction-radiation parameter $N_1 = 0.02916$.

spheres. The purpose is to show how engineering concepts of exchange factors can be applied to extend results for black bounding surfaces to surfaces that are diffuse-gray. Perlmutter and Howell [44] showed that once results are available for black boundaries, the diffuse-gray-wall results can be obtained from simple algebraic relations. This is also demonstrated in Sec. 14-7.6 for a gray gas layer between parallel plates. Following this section on radiation only, the method is extended for combined modes.

The theory follows the general development of the net-radiation method (Sec. 7-4.1). A heat balance at surface A_k gives

$$Q_k = q_k A_k = (q_{o,k} - q_{i,k})A_k \tag{16-42}$$

The energy flux leaving A_k is composed of emitted and reflected energy:

$$q_{o,k} = \epsilon_k \sigma T_k^4 + (1 - \epsilon_k)q_{i,k} \tag{16-43}$$

If $q_{i,k}$ is eliminated from (16-42) and (16-43), the result is

$$Q_k = A_k \frac{\epsilon_k}{1 - \epsilon_k}(\sigma T_k^4 - q_{o,k}) \tag{16-44}$$

The $q_{i,k}$ in (16-43) can be found in terms of exchange areas (13-102). However, we define a different quantity, the *exchange factor* $\bar{F}_{j-k}$, as *the fraction of the energy leaving surface j that is incident on surface k when all boundaries are black and the intervening medium is in radiative equilibrium* (transfer is only by radiation *without heat sources or sinks*). When the gas is transparent, $\bar{F}_{j-k}$ becomes the configuration factor F_{j-k}. For radiative equilibrium, energy conservation requires that all energy leaving A_j must reach other enclosure surfaces or return to A_j. The $\bar{F}_{j-k}$ includes all interactions with the gas by means of which energy leaving A_j arrives at A_k.

For a general enclosure of N surfaces containing a gas in radiative equilibrium, the incident energy on surface k can be written in terms of exchange factors as

$$Q_{i,k} = \sum_{j=1}^{N} Q_{o,j} \bar{F}_{j-k} \tag{16-45}$$

Note that using exchange areas as in Sec. 13-7 requires an additional term to account for energy emitted by the gas and reaching the wall. This energy is included by definition within the $\bar{F}_{j-k}$. Substituting (16-45) into (16-43) to eliminate $q_{i,k}$ gives

$$Q_{o,k} = q_{o,k} A_k = \epsilon_k \sigma T_k^4 A_k + (1 - \epsilon_k) \sum_{j=1}^{N} Q_{o,j} \bar{F}_{j-k} \tag{16-46}$$

Because the medium is in *radiative equilibrium*, all energy leaving a surface must finally reach a surface so that

$$\sum_{k=1}^{N} \bar{F}_{j-k} = 1 \tag{16-47}$$

Because $\bar{F}_{j-k}$ is the fraction of energy leaving A_j that arrives at A_k for *black boundaries*, it is obtained from the black-walled solution, which we assume has already been found,

$$\bar{F}_{j-k} = \left(\frac{Q_{i,k \, (\text{from } j)}}{A_j \sigma T_j^4} \right)_{\text{black surfaces}} \equiv \psi_{j-k,b} \tag{16-48}$$

where the b subscript is used to emphasize that this is obtained from the black solution.

For two-surface enclosures, convenient closed-form solutions are obtained. The procedure will be outlined for infinite parallel plates. With a constant gas absorption coefficient, the *fraction* $\bar{F}_{1-2}$ of energy leaving surface 1 that reaches surface 2 must equal the *fraction* $\bar{F}_{2-1}$ from surface 2 to 1. The $\bar{F}_{1-2}$ is found from the black solution (16-48) as

$$\bar{F}_{1-2} = \frac{Q_{i,2}}{A_1 \sigma T_1^4} = \left(\frac{\sigma T_2^4 - q_2}{\sigma T_1^4} \right)_{\text{black surfaces}} = \psi_{1-2,b}$$

where $A_1 = A_2$ has been used. Then $\bar{F}_{2-1} = \bar{F}_{1-2}$, and for simplicity call them ψ_b. From Eq. (16-47), $\bar{F}_{1-1} = \bar{F}_{2-2} = 1 - \psi_b$, and (16-46) becomes

$$q_{o,1} = \epsilon_1 \sigma T_1^4 + (1 - \epsilon_1)(q_{o,1} - q_{o,1} \psi_b + q_{o,2} \psi_b)$$

$$q_{o,2} = \epsilon_2 \sigma T_2^4 + (1 - \epsilon_2)(q_{o,1}\psi_b + q_{o,2} - q_{o,2}\psi_b)$$

Solving simultaneously for $q_{o,1}$ and $q_{o,2}$ yields the symmetric relations

$$q_{o,1} = \frac{\epsilon_1\epsilon_2\sigma T_1^4 + \epsilon_1(1 - \epsilon_2)\psi_b\sigma T_1^4 + \epsilon_2(1 - \epsilon_1)\psi_b\sigma T_2^4}{\psi_b(\epsilon_1 + \epsilon_2 - 2\epsilon_1\epsilon_2) + \epsilon_1\epsilon_2} \qquad (16\text{-}49a)$$

$$q_{o,2} = \frac{\epsilon_1\epsilon_2\sigma T_2^4 + \epsilon_2(1 - \epsilon_1)\psi_b\sigma T_2^4 + \epsilon_1(1 - \epsilon_2)\psi_b\sigma T_1^4}{\psi_b(\epsilon_1 + \epsilon_2 - 2\epsilon_1\epsilon_2) + \epsilon_1\epsilon_2} \qquad (16\text{-}49b)$$

The $q_{o,1}$ is substituted into (16-44) to yield, after rearrangement,

$$\frac{q_1}{\sigma(T_1^4 - T_2^4)} = \frac{\psi_b}{(E_2 + E_1)\psi_b + 1} \qquad (16\text{-}50a)$$

where $E_1 = (1 - \epsilon_1)/\epsilon_1$ and $E_2 = (1 - \epsilon_2)/\epsilon_2$. Evaluating (16-50a) for the black case, $E_1 = E_2 = 0$, shows that

$$\psi_b = \frac{q_{1b}}{\sigma(T_1^4 - T_2^4)} \qquad (16\text{-}50b)$$

Equation (16-50a) gives the energy supplied to surface 1 and removed from surface 2. Since the gas absorption coefficient is independent of temperature, energy leaving a location on the boundary and reaching a given point in the gas will be attenuated by the same amount independent of the gas temperature distribution. Any energy absorbed along the path is balanced by local isotropic emission. With these facts, a synthesis of black-wall enclosure solutions and gas-element exchange factors is used to find the temperature distribution in the gas for diffuse-gray walls.

The energy emitted by a volume element of gas of area A and thickness dx (Fig. 16-10) is $Q_e = 4a\sigma T^4(x)A\,dx$. For radiative equilibrium, this must equal the heat absorbed by the volume element, which is written as (per unit area)

$$4a\sigma T^4(x)\,dx = \sum_{j=1}^{N} q_{o,j}\,d\bar{F}_{j\text{-}dx} = q_{o,1}\,d\bar{F}_{1\text{-}dx} + q_{o,2}\,d\bar{F}_{2\text{-}dx} \qquad (16\text{-}51a)$$

or

$$\phi = \frac{T^4(x) - T_2^4}{T_1^4 - T_2^4} = \frac{1}{4a\sigma(T_1^4 - T_2^4)}\left(\frac{q_{o,1}\,d\bar{F}_{1\text{-}dx}}{dx} + \frac{q_{o,2}\,d\bar{F}_{2\text{-}dx}}{dx}\right) - \frac{T_2^4}{T_1^4 - T_2^4} \qquad (16\text{-}51b)$$

The $d\bar{F}_{j\text{-}dx}$ is the fraction of energy flux leaving boundary surface A_j that is *absorbed* in volume element dx when the boundaries are black. To determine $d\bar{F}_{j\text{-}dx}$, consider an isothermal system. Equation (16-51a) reduces to

$$4a\,dx = d\bar{F}_{1\text{-}dx} + d\bar{F}_{2\text{-}dx} \qquad (16\text{-}52)$$

This is used to eliminate $d\bar{F}_{2\text{-}dx}$ from (16-51b). The resulting equation is solved for $d\bar{F}_{1\text{-}dx}$,

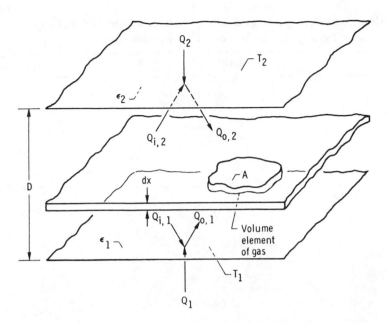

Figure 16-10 Energy quantities for gas between infinite parallel gray plates.

$$d\bar{F}_{1-dx} = 4a\, dx\, \phi_b \tag{16-53}$$

where, from (16-51b),

$$\phi_b = \frac{T_1^4\, d\bar{F}_{1-dx} + T_2^4\, d\bar{F}_{2-dx}}{4a\, dx(T_1^4 - T_2^4)} - \frac{T_2^4}{T_1^4 - T_2^4}$$

Then substituting (16-53) into (16-52) gives

$$d\bar{F}_{2-dx} = 4a\, dx(1 - \phi_b) \tag{16-54}$$

Substituting (16-53), (16-54), (16-49a), and (16-49b) to eliminate $d\bar{F}_{1-dx}$, $d\bar{F}_{2-dx}$, $q_{o,1}$, and $q_{o,2}$ from (16-51b) results in the gas temperature distribution

$$\phi = \frac{T^4(x) - T_2^4}{T_1^4 - T_2^4} = \frac{\phi_b + E_2\psi_b}{1 + \psi_b(E_1 + E_2)} \tag{16-55}$$

Equations (16-50) and (16-55) relate the energy transfer and fourth-power temperature distribution for a gray gas between gray walls to that for a gray gas between black walls for the infinite parallel plate geometry. Similar relations for infinitely long concentric cylinders are in [20, 44], and these plus the relations for concentric spheres (which can be used with the results of [22, 45, 46]) are in Table 16-3. Reference [44] also includes exchange factors when heat sources are in the gas.

Table 16-3 Relations between gray- and black-wall solutions for radiation between surfaces enclosing a gray gas in radiative equilibrium

Geometry	Relation[a]
Infinite parallel plates 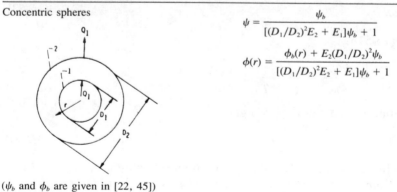 (ψ_b and ϕ_b are given in [7] of Chap. 14 and Fig. 14-12a and b)	$$\psi = \frac{\psi_b}{(E_2 + E_1)\psi_b + 1}$$ $$\phi(x) = \frac{\phi_b(x) + E_2\psi_b}{(E_2 + E_1)\psi_b + 1}$$
Infinitely long concentric cylinders (ψ_b and ϕ_b are given in [20, 44])	$$\psi = \frac{\psi_b}{[(D_1/D_2)E_2 + E_1]\psi_b + 1}$$ $$\phi(r) = \frac{\phi_b(r) + E_2(D_1/D_2)\psi_b}{[(D_1/D_2)E_2 + E_1]\psi_b + 1}$$
Concentric spheres (ψ_b and ϕ_b are given in [22, 45])	$$\psi = \frac{\psi_b}{[(D_1/D_2)^2E_2 + E_1]\psi_b + 1}$$ $$\phi(r) = \frac{\phi_b(r) + E_2(D_1/D_2)^2\psi_b}{[(D_1/D_2)^2E_2 + E_1]\psi_b + 1}$$

[a] Definitions: $E_N = (1 - \epsilon_{wN})/\epsilon_{wN}$, $\psi = Q_1/A_1\sigma(T_{w1}^4 - T_{w2}^4)$, $\psi_b = Q_{1b}/A_1\sigma(T_{w1}^4 - T_{w2}^4)$, $\phi(\xi) = [T^4(\xi) - T_{w2}^4]/(T_{w1}^4 - T_{w2}^4)$.

EXAMPLE 16-3 A gray gas with absorption coefficient 0.5 cm^{-1} is between gray parallel plates 2 cm apart. The plate temperatures and emissivities are $T_1 = 1000$ K, $T_2 = 840$ K, $\epsilon_1 = 0.1$, and $\epsilon_2 = 0.2$. What are the energy transfer between the plates and the gas temperature 0.5 cm from surface 1? For black walls, Fig. 14-12b gives for $aD = 0.5 \times 2 = 1.0$, $q_{1b}/\sigma(T_1^4 - T_2^4) = 0.56$ so that, from (16-50b), $\psi_b = 0.56$. From Table 16-3,

$$\psi = \frac{q_1}{\sigma(T_1^4 - T_2^4)} = \frac{\psi_b}{(E_2 + E_1)\psi_b + 1}$$

For this example, $E_1 = (1 - 0.1)/0.1 = 9$ and $E_2 = (1 - 0.2)/0.2 = 4$ so

$$q_1 = \sigma(T_1^4 - T_2^4) \frac{0.56}{13 \times 0.56 + 1} = 1.93 \text{ kW/m}^2$$

The temperature at $x = 0.5$ cm is calculated from the result in Table 16-3:

$$\phi(x) = \frac{T^4(x) - T_2^4}{T_1^4 - T_2^4} = \frac{\phi_b(x) + E_2\psi_b}{(E_2 + E_1)\psi_b + 1}$$

From Fig. 14-12a, at the abscissa $\kappa/\kappa_D = ax/aD = 0.25$ and on the curve for $\kappa_D = aD = 1$, $\phi_b = 0.62$, so that $\phi|_{x=0.5} = (0.62 + 4 \times 0.56)/(13 \times 0.56 + 1) = 0.345$. Then $T^4(0.5) = T_2^4 + 0.345(T_1^4 - T_2^4) = 0.671 \times 10^{12}$, which gives $T(0.5) = 905$ K. Note that for gray walls, a curve of ϕ against ax/aD will have an antisymmetrical shape about $ax/aD = 0.5$ as in Fig. 14-12a only when $\epsilon_1 = \epsilon_2$.

Radiation with conduction and/or convection Exchange factors are now applied to combined-mode problems. The total energy emitted by a volume element in the *presence* of heat conduction can be equated to three terms: (1) the energy that would be emitted in the absence of conduction, (2) the *net* energy supplied to the element by conduction (which must be radiated away), and (3) any additional energy d^2Q_{extra} added to the element by radiation over that given by the radiative-equilibrium (zero-conduction) case because of the change in temperature profile by including conduction. This is written as

$$4a_P\sigma T^4 \, dV = 4a_P\sigma T_R^4 \, dV + k \nabla^2 T \, dV + d^2 Q_{\text{extra}} \tag{16-56}$$

The temperature T_R for radiative equilibrium is written, as in (16-51a), in terms of exchange factors: $4a_P\sigma T_R^4 \, dV = \Sigma_j \, Q_{o,j} \, d^2\bar{F}_{j-dV}$. The $Q_{o,j}$ is the energy leaving the jth surface of the enclosure. The $d^2\bar{F}_{j-dV}$ is the fraction of energy leaving A_j that is *absorbed* in dV for radiative equilibrium in a black enclosure, and it includes the effects of gas absorption and emission while the energy is in transit from A_j to dV. Substituting this relation and assuming that d^2Q_{extra} is small reduce (16-56) to

$$4a_P\sigma T^4 \approx \frac{1}{dV} \sum_j Q_{o,j} \, d^2\bar{F}_{j-dV} + k \nabla^2 T \tag{16-57}$$

Note that the exchange factors include the effect of gas-to-gas volume-element radiant interchange based on the temperature profile for *no conduction*. The approximation in (16-57) is that the gas-to-gas radiant exchange is not significantly affected by the new temperature profile because of the presence of gas conduction ($d^2Q_{\text{extra}} \approx 0$). The approximation is also made that the $Q_{o,j}$ from the solution without conduction can be used. If radiation predominates, this is a good assumption; if radiation is small, it will not matter that the radiative terms are somewhat inaccurate.

Equation (16-57) is a nonlinear differential equation for the local gas temperature. Howell [38] applied this approach to gray gases in annular enclosures and between infinite parallel plates. Accuracy was comparable to the simple additive solution when the heat flux through the gas was computed, as shown in Fig. 16-8. The advantage of the method is that accurate temperature distributions can be obtained for combined-mode energy transfer problems with little effort. The procedure is demonstrated with an example.

EXAMPLE 16-4 Find an expression for the steady-state temperature profile in a gray gas without internal heat sources between infinite parallel plates D apart if the gas has absorption coefficient a, thermal conductivity k, and zero scattering, and the plates are black at T_1 and T_2. Use the exchange-factor approximation.

For a layer dx thick, (16-57) becomes, for this geometry,

$$k \frac{d^2T(x)}{dx^2} dx + d\bar{F}_{1-dx} \sigma T_1^4 + d\bar{F}_{2-dx} \sigma T_2^4 = 4a\sigma T^4(x) dx \tag{16-58}$$

The exchange factors are given by (16-53) and (16-54) as $d\bar{F}_{1-dx} = 4a\,dx\,\phi_b(x)$ and $d\bar{F}_{2-dx} = 4a\,dx\,(1 - \phi_b)$. Using the diffusion approximation, ϕ_b is obtained from Table 15-2 as $\phi_b(\kappa) = [\frac{3}{4}(\kappa_D - \kappa) + \frac{1}{2}]/(3\kappa_D/4 + 1)$. Substituting into (16-58) and using nondimensional quantities give

$$t^4(\kappa) - N_1 \frac{d^2t(\kappa)}{d\kappa^2} - \frac{1}{3\kappa_D/4 + 1}\left[\frac{3}{4}(\kappa_D - \kappa) + \frac{1}{2} + t_2^4\left(\frac{3}{4}\kappa + \frac{1}{2}\right)\right] = 0 \tag{16-59}$$

which is to be solved subject to the boundary conditions $t = 1$ at $\kappa = 0$ and $t = t_2$ at $\kappa = \kappa_D$. This relation gives the correct limiting diffusion solution for $N_1 \to 0$ and the correct conduction solution for $N_1 \to \infty$. Howell [38] solved for $[t(\kappa) - t_2]/(1 - t_2)$ numerically, using exchange factors from the numerical solutions of the pure-radiation problem, which are more accurate than the diffusion exchange factors in (16-59). Comparison with the numerical solution of the coupled conduction-radiation problem is in Fig. 16-11.

The energy transfer by conduction was found by numerically evaluating $dt/d\kappa|_{\kappa=0}$ and using this to evaluate the conduction flux at the boundary. The radiation flux was assumed unaffected by the conduction process in the spirit of the exchange-factor approximation. The approximate results are compared with the numerical solution in Fig. 16-8. Since $dt/d\kappa$ varies with κ while the radiative flux without conduction is constant with κ, evaluating $dt/d\kappa$ at a location other than $\kappa = 0$ would give different results. For the most accurate calculation of heat flux, it was found in [38] that the temperature gradient should be evaluated at the boundary with highest temperature. Using the gradient at the coldest wall yields a flux prediction that is too large.

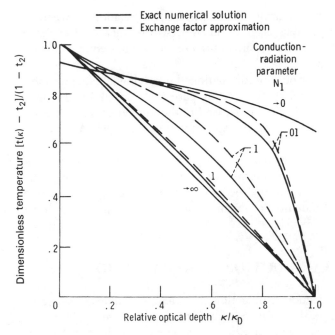

Figure 16-11 Comparison of temperature profiles by exact solution with exchange-factor approximation. Optical thickness $\kappa_D = 1.0$; plate temperature ratio $t_2 = 0.1$; plate emissivities $\epsilon_1 = \epsilon_2 = 1.0$.

The exchange-factor approximation is not limited to any range of geometry, surface emissivity, or optical thickness; it can be applied where the exchange factors have been previously obtained or where they can be obtained by a simplified pure-radiation solution. The resulting nonlinear energy relations can usually be cast in the form of a matrix of nonlinear difference equations that can be solved by the numerical technique of Ness [47] (see Sec. 11-5.2).

In situations where the boundaries are not the chief contributors to the radiant energy flux in the gas, the gas-temperature profile becomes important to the radiative flux distribution. Then the exchange-factor approximation may become inaccurate; however, [38] gives results for parallel plates and concentric cylinders that compare well with exact numerical solutions for both energy transfer and temperature profiles. Fulkerson and Bannerot [48] extended the approximation to cases involving both convection and conduction. Limiting cases for heat transfer through optically thick and thin media are presented along with an asymptotic solution uniformly valid at all optical thicknesses. In [49] exchange factors are used to develop complete relations for energy exchange in absorbing-emitting-scattering media without introducing the approximation of (16-57). The matrix of equations developed for enclosures is shown to be equivalent to Hottel's exchange areas (Sec. 13-7), although the form of the equations is somewhat different. Exchange factors may be obtained experimentally for complex geometries [50], allowing analysis of enclosures of arbitrary geometry. The relation between exchange factors and exchange areas is given in [49], so that measured exchange factors may be converted to exchange areas for use in existing computer codes based on the zoning method.

Lick [51] has presented various approximations for solving conduction-radiation problems. He develops methods for treating gases with spectral and temperature-dependent properties. Goldstein and Howell [40] outline methods of treating temperature-dependent properties using the apparent-slip technique. A gray scattering medium is treated in [52].

Although numerical methods (Chap. 17) are usually the only way to obtain exact solutions to combined-mode problems, and their use is becoming more prevalent, the approximate methods outlined here often provide acceptable accuracy for engineering conduction-radiation problems. The methods can be applied to problems in two and three dimensions.

16-5 CONVECTION, CONDUCTION, AND RADIATION

Examples of combined convection, conduction, and radiation in absorbing, emitting, and scattering media are found in combustion chambers, atmospheric phenomena, shock wave problems, rocket nozzles, high-temperature heat exchangers, radiation devices for outer-space applications, and industrial furnaces. As a consequence, a large amount of literature is available. Review articles and comprehensive books are in [37, 53–59]. These problems are often difficult to solve but considerable progress has been made. In this section, some of the basic equations are given and solution methods in use are outlined. Numerical methods are in Chap. 17.

16-5.1 Boundary-Layer Problems

Radiant emission, absorption, and scattering can affect the heat transfer in a convection boundary layer [60–66]. Boundary layer problems are approached by writing the usual continuity and momentum equations; these do not contain radiation terms and are not influenced by heat transfer if constant fluid properties are assumed, as is the case here. The energy equation includes an energy source term $-\nabla \cdot \mathbf{q}_r$ to account for the net thermal radiation gained by a volume element.

For a two-dimensional laminar boundary layer flowing over a flat plate (Fig. 16-12), assuming a gray fluid with constant properties, negligible viscous dissipation, and no internal heat sources such as those from combustion, the energy equation is

$$\rho c_p \left(u \frac{\partial T}{\partial x} + v \frac{\partial T}{\partial y} \right) = k \frac{\partial^2 T}{\partial y^2} - \frac{\partial q_{r,y}}{\partial y} \tag{16-60}$$

where $q_{r,y}$ is the radiation flux in the positive y direction. It is assumed that convection in the x direction is dominant over the radiative contribution in that di-

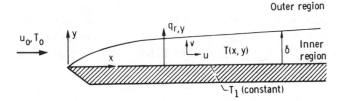

Figure 16-12 Boundary layer flow over flat plate.

rection, that is, $\rho c_p u \, \partial T/\partial x \gg -\partial q_{r,x}/\partial x$, where $q_{r,x}$ is the radiative flux in the x direction; hence the $-\partial q_{r,x}/\partial x$ term has been neglected on the right side of (16-60). In the diffusion limit this assumption states that

$$\rho c_p \, u \, \frac{\partial T}{\partial x} \gg \frac{\partial}{\partial x} \left(-\frac{16\sigma T^3}{3a_R} \frac{\partial T}{\partial x} \right) \tag{16-61}$$

must be true. The problem becomes one of introducing into (16-60) one of the formulations developed for $q_{r,y}$, and then solving the resulting energy equation together with the momentum and continuity equations. In [61, 62] the diffusion solution is used for $q_{r,y}$, and the last two terms in (16-60) can then be combined as was done in (16-30).

Various techniques can be used to solve the resulting energy equation. References [48, 60, 67] use matched asymptotic expansions of the energy equation assuming a known flow field. A solution of the linearized energy equation near the surface is matched to an asymptotic solution far from the surface. Viskanta and Grosh [61] had earlier applied the diffusion approach; they assumed that the diffusion solution was valid all the way to the boundary. Cess [62] and others assumed the boundary layer to be optically thin, so it emits but does not significantly absorb radiation. By introducing some other assumptions, Cess was able to treat nongray gas effects. Neglecting absorbed radiation from the surroundings can be a useful approximation when the boundary layer is being heated by frictional dissipation while the surface and surrounding gas are cool. A multilayer method is used in [68] and gives good agreement with the results of Cess.

Fritsch et al. [63] studied the shielding of a body by a radiant absorbing layer. The boundary layer was assumed to absorb but not emit radiation. The effects of transpiration and an external radiation field were included. Howe [65] had treated a similar problem under somewhat different conditions for the external radiation field. In [69] the influence of radiation on mixed forced and free convection was analyzed. A real gas mixture of air and water vapor was considered on a vertical plate at uniform wall temperature. An implicit finite-difference forward marching solution procedure was used.

Optically thin thermal layer   To analyze laminar forced-flow heat transfer on a flat plate, an expression is needed for the radiative source term $-\partial q_{r,y}/\partial y$ in (16-60). Within the boundary layer it is assumed that the thermal conditions are changing slowly enough in the x direction, as compared with the y direction, so that the conditions contributing to $q_{r,y}$ at a specific x, say x^+, are all at that x^+ and hence at the temperature distribution $T(x^+, y)$. Then $\partial q_{r,y}/\partial y$ can be evaluated using one-dimensional relations. From (16-10) and (16-11), which are for a region between two black walls,

$$\frac{\partial q_{r,y}}{\partial y} = 4a\sigma T^4 - 2a\sigma \left[T_1^4 E_2(\kappa) + T_2^4 E_2(\kappa_D - \kappa) + \int_0^{\kappa_D} T^4(\kappa^*) E_1(|\kappa - \kappa^*|) \, d\kappa^* \right]$$

(16-62)

For only one bounding wall, the T_2 term is not present and the upper limit of the integral is extended to infinity. Also the $T^4(\kappa^*)$ is replaced by $T^4(x, \kappa^*)$ to emphasize the approximation for the radiation term that the temperatures surrounding position x are all assumed at $T(x, y)$. Then for flow over a black wall, the laminar boundary-layer energy equation (16-60) for $T(x, y)$ becomes

$$\rho c_p \left(u \frac{\partial T}{\partial x} + v \frac{\partial T}{\partial y} \right) = k \frac{\partial^2 T}{\partial y^2} - 4a\sigma T^4$$

$$+ 2a\sigma \left[T_1^4 E_2(\kappa) + \int_0^\infty T^4(x, \kappa^*) E_1(|\kappa - \kappa^*|) \, d\kappa^* \right] \quad (16\text{-}63)$$

where $\kappa = ay$ and there is no scattering.

The temperature field can be considered as composed of two regions. Near the wall in the usual thermal boundary layer of thickness δ that would be present in the absence of radiation, there are large temperature gradients, and heat conduction is important. This layer is usually of small thickness; hence, it can be assumed optically thin so that radiation will pass through it without attenuation. For larger y than in this layer, the temperature gradients are small and heat conduction is neglected compared with radiation transfer. The approximate analysis can now proceed, for example, along the path developed in [37].

In the outer region the velocity in the x direction has the free-stream value u_0, and with the neglect of heat conduction the boundary layer energy equation reduces to

$$\rho c_p u_0 \frac{\partial T}{\partial x} = -4a\sigma T^4 + 2a\sigma \left[T_1^{\,4} E_2(\kappa) + \int_0^\infty T^4(x, \kappa^*) E_1(|\kappa - \kappa^*|) \, d\kappa^* \right] \quad (16\text{-}64)$$

To obtain an approximate solution by iteration, substitute the incoming free-stream temperature T_0 for the temperature on the right side as a first approximation, and then carry out the integral to obtain a second approximation. This yields, for the outer region to first-order terms,

$$T(x, y) = T_0 + \sigma(T_1^4 - T_0^4) E_2(ay) \frac{2ax}{\rho c_p u_0} + \cdots \quad (16\text{-}65)$$

where at $x = 0$, $T = T_0$.

At the edge of the thermal layer $ay = a\delta$, which is small, so that $E_2(a\delta) \approx E_2(0) = 1$. Hence, at $y = \delta$, (16-65) becomes

$$T(x, \delta) = T_0 + \sigma(T_1{}^4 - T_0{}^4)\frac{2ax}{\rho c_p u_0} + \cdots \tag{16-66}$$

Equation (16-66) is the edge boundary condition that the outer radiation layer imposes on the inner thermal layer. The $T(x, \delta)$ to this approximation is increasing linearly with x. This is the result of the flowing gas absorbing a net radiation from the plate in proportion to the difference $T_1^4 - T_0^4$ and the absorption coefficient a.

To solve the boundary layer equation in the inner thermal layer region, the last integral in (16-63) is divided into two parts, one from $\kappa = 0$ to $a\delta$ and the second from $a\delta$ to ∞. The first portion is neglected as the thermal layer is optically thin, and the second is evaluated by using the outer solution (16-65). By retaining only first-order terms, the boundary layer energy equation is reduced to

$$u\frac{\partial T}{\partial x} + v\frac{\partial T}{\partial y} = \alpha\frac{\partial^2 T}{\partial y^2} + \frac{2a\sigma}{\rho c_p}(T_1{}^4 + T_0{}^4 - 2T^4) \tag{16-67}$$

The boundary conditions are given by (16-66) at $y = \delta$, and the specified wall temperature $T = T_1$ at $y = 0$. The solution becomes rather complex and is not developed further here. The reader interested in this particular topic is referred to [37, 70] for additional information.

Optically thick thermal layer At the opposite extreme from the situation in the previous section, if the thermal layer has become very thick or the medium is highly absorbing, the boundary layer can be optically thick. The analysis is then simplified, as the diffusion approximation can be employed. With reference to (16-30), radiative diffusion adds a radiative conductivity to the ordinary thermal conductivity. Then the laminar boundary layer energy equation (16-60) is written

$$\rho c_p \left(u\frac{\partial T}{\partial x} + v\frac{\partial T}{\partial y}\right) = \frac{\partial}{\partial y}\left[\left(\frac{16\sigma T^3}{3a_R} + k\right)\frac{\partial T}{\partial y}\right] \tag{16-68}$$

With the assumption of constant fluid properties, the momentum and continuity equations do not depend on temperature. Consequently, the flow is unchanged by heat transfer, and the velocity distribution is given by the Blasius solution [71]. The Blasius solution is in terms of a similarity variable $\eta = y\sqrt{u_0/\nu x}$; the streamfunction and velocity components are,

$$\psi = \sqrt{\nu x u_0}\, f(\eta) \qquad u = \frac{\partial \psi}{\partial y} = u_0\frac{df}{d\eta} \qquad v = -\frac{\partial \psi}{\partial x} = \frac{1}{2}\sqrt{\frac{\nu u_0}{x}}\left(\eta\frac{df}{d\eta} - f\right) \tag{16-69}$$

where the function $f(\eta)$ is given in [68]. These quantities are substituted into the energy equation (16-68), which is then placed in the form

$$-\frac{\mathrm{Pr}}{2}f\frac{dt}{d\eta} = \frac{d}{d\eta}\left[\left(\frac{4t^3}{3N_0} + 1\right)\frac{dt}{d\eta}\right] \qquad \text{where } t = T/T_0 \qquad \text{and } N_0 = ka_R/4\sigma T_0^3 \tag{16-70}$$

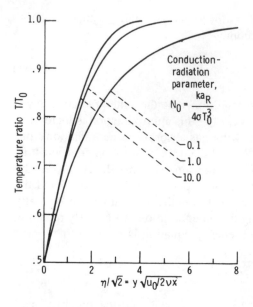

Figure 16-13 Boundary-layer temperature profiles for laminar flow on flat plate [61]. Prandtl number Pr = 1.0; temperature ratio $T_1/T_0 = 0.5$.

The boundary conditions that were used in the numerical solution are

$$t = t_1 = T_1/T_0 \quad \text{at } \eta = 0, \qquad t = 1 \quad \text{at } \eta = \infty$$

To be more precise, a diffusion-slip condition should be used at the wall, but this has not been formulated for combined radiation, convection, and conduction.

   Numerical solutions were carried out in [61], and typical temperature profiles are in Fig. 16-13. For $N_0 = 10$ the profile was found to be within 2% of that for conduction and convection alone (that is, for $N_0 \rightarrow \infty$). The effect of radiation is found to thicken the thermal boundary layer similar to the effect of decreasing the Prandtl number. This would be expected, since the Prandtl number is the ratio of viscous to thermal diffusion ν/α. Radiation has supplied an additional means for thermal diffusion, thereby effectively increasing α.

16-5.2 Forced Convection Channel Flows

Of engineering interest in some high-temperature heat exchange devices is combined radiative and convective energy transfer for flow of an absorbing-emitting gas in a channel. For laminar flow, the energy equation (16-60) (with $v = 0$ for fully developed flow) applies within the flow. References [36, 72–81] treat problems with varying degrees of approximation. Viskanta [72] gives approximate numerical solutions for the flow of a gray absorbing-emitting medium in a parallel-plate channel. Temperature-independent properties are assumed. In addition to κ_D, N, and temperature ratios, a new parameter enters; it is a Nusselt number defined to include a radiative contribution and thus differing from the usual parameter.

Einstein [36, 73] applied the gas-to-surface and gas-to-gas direct-exchange area methods of Hottel (Sec. 13-7) to a solution of the energy equation in a finite-length parallel-plate channel and circular tube. Internal heat generation in the gas was included. Comparison was made with the work of Adrianov and Shorin [74], who had used the cold-material approximation (Sec. 15-3.3) so that absorption but not emission from the gas was included. Chen [75] included scattering in an analysis for flow between parallel plates, but assumed a slug-flow (uniform) velocity profile.

The preceding analyses are for gray gases flowing in channels with gray or black walls, and for temperature-independent properties. DeSoto and Edwards [82] presented a tube-flow heat transfer analysis that accounts for nongray gases with temperature-dependent radiative properties. An exponential band model (Sec. 12-8.5) was used to account for the spectral effects. Entrance region flows were included. A similar study by Pearce and Emery [83] employed a *box model* for the nongray absorption properties. In this model an absorption band is approximated by a constant absorption coefficient within an effective bandwidth and zero absorption elsewhere. Jeng et al. [84, 85] also analyzed the laminar flow of a radiating gas in a tube at constant wall temperature for both gray and nongray media. In the latter case they used the expression of Tien and Lowder ([36] in Chap. 12) for band radiation. In [85] they looked at the optically thin and thick limits for real gases. In the thick limit the logarithmic limit was used for a vibration-rotation band as given by (12-82). In [86] a narrow-band model was used to analyze emitting and absorbing laminar gas flow between two parallel walls at uniform temperature. A water vapor–air mixture was studied for conditions with strong temperature gradients. Variable properties were included by coupling the continuity, momentum, and energy equations. An analysis of and experiment with combined radiation and convection were done in [87]. This was for laminar flow of a mixture of CO_2 and water vapor. The radiative properties were calculated using a statistical narrow-band model. In [88] a nongray analysis was made of a laminar channel flow of high-temperature CO_2 and combustion gases. The effect of installing an additional plate in the channel was studied. For some conditions this serves as a means of heat transfer enhancement by absorption and reradiation.

An example of a channel flow analysis is heat transfer by an absorbing, emitting, and scattering medium flowing between parallel walls at different uniform temperatures (Fig. 16-14). The medium and walls are gray, scattering in the medium is isotropic, and the walls are diffuse and opaque. All physical properties are constant. The velocity distribution is specified as fully developed when the

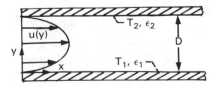

Figure 16-14 Flow of radiating and scattering medium between parallel walls at unequal temperatures.

flow enters the heat transfer section at $x = 0$; hence it is only the thermal behavior that develops as a function of x. This has been investigated in [89].

The energy equation is, from (16-1) with viscous dissipation neglected,

$$\rho c_p u(y) \frac{\partial T}{\partial x} = k\left(\frac{\partial^2 T}{\partial x^2} + \frac{\partial^2 T}{\partial y^2}\right) - \left(\frac{\partial q_r}{\partial x} + \frac{\partial q_r}{\partial y}\right) + q'''(x, y) \tag{16-71}$$

Axial heat conduction is usually small, and in addition it is assumed that temperatures change slowly enough in the axial direction that $\partial q_r/\partial x \ll \partial q_r/\partial y$. For this example there are no internal heat sources such as combustion in the flow. The $\partial q_r/\partial y$ is found from (16-5) as

$$\frac{\partial q_r}{\partial y} = 4\pi \frac{a}{\Omega}\left(\frac{\sigma T^4}{\pi} - I'\right) \tag{16-72}$$

so the energy equation becomes

$$\rho c_p u(y) \frac{\partial T}{\partial x} = k \frac{\partial^2 T}{\partial y^2} - 4 \frac{a}{\Omega}[\sigma T^4(x, y) - \pi I'(x, y)] \tag{16-73}$$

The source function $I'(x, y)$ is found as a function of y at each x from Eq. (14-78) written for a gray medium where $\kappa = (a + \sigma_s)y$,

$$\pi I'(\kappa) = (1 - \Omega)\sigma T^4(\kappa)$$

$$+ \Omega\left[q_{o,1}E_2(\kappa) + q_{o,2}E_2(\kappa_D - \kappa) + \pi \int_0^{\kappa_D} I'(\kappa^*)E_1(|\kappa^* - \kappa|)\, d\kappa^*\right] \tag{16-74}$$

The outgoing fluxes $q_{o,1}$ and $q_{o,2}$ are each equal to the sum of emitted and reflected energies and are given by (14-74) as

$$q_{o,1} = \epsilon_1 \sigma T_1^4 + 2(1 - \epsilon_1)\left[q_{o,2}E_3(\kappa_D) + \pi \int_0^{\kappa_D} I'(\kappa^*)E_2(\kappa^*)\, d\kappa^*\right] \tag{16-75a}$$

$$q_{o,2} = \epsilon_2 \sigma T_2^4 + 2(1 - \epsilon_2)\left[q_{o,1}E_3(\kappa_D) + \pi \int_0^{\kappa_D} I'(\kappa^*)E_2(\kappa_D - \kappa^*)\, d\kappa^*\right] \tag{16-75b}$$

Equations (16-73) and (16-74) are placed in dimensionless form and then solved numerically with the $q_{o,1}$ and $q_{o,2}$ used from (16-75) for Poiseuille flow the $u(y) = 6\bar{u}[y/D - (y/D)^2]$, and at $x = 0$, $T(0, y) = T_i$, a uniform initial value; other initial temperature distributions could be used. Using $T(0, y) = T_i$, Eq. (16-74) is solved by iteration for $I'(0, y)$ subject to the constraints of Eq. (16-75). The $T(0, y)$ and $I'(0, y)$ are substituted into the right side of Eq. (16-73) and the resulting values of $\partial T/\partial x$ as a function of y are used to obtain new T values for a grid of y points a small Δx down the channel. The $T(\Delta x, y)$ are then used in Eqs. (16-74) and (16-75) to obtain $I'(\Delta x, y)$, and the process is continued to move forward in the x direction.

Numerical solutions were obtained in [89] for the flow initially at uniform temperature, with both walls at the same uniform temperature T_1 and with the same emissivity ϵ_1. For this case the local Nusselt number is [$T_m(x)$ is the local mean fluid temperature] $\mathrm{Nu}(x) = q_w(x)2D/k[T_1 - T_m(x)]$. The $q_w(x)$ is the local heat flux that must be added at the walls to maintain their uniform temperature. The heat loss from the wall interior surfaces is composed of conduction and radiation components. Using Eq. (7-17) for the radiative flux,

$$q_w = -k\left(\frac{\partial T}{\partial y}\right)_{y=0} + \frac{\epsilon_1}{1 - \epsilon_1}(\sigma T_1^4 - q_{o,1}) \tag{16-76}$$

Results are in [89] for the variation of bulk mean temperature $T_m(x)$ and $\mathrm{Nu}(x)$ with axial distance as a function of scattering albedo, wall emissivity, and conduction-radiation parameter. For pure convection, $\mathrm{Nu}(x)$ approaches a constant value for large x. When radiation is present, $\mathrm{Nu}(x)$ was found to pass through a minimum and then increase farther downstream. The increase in $\mathrm{Nu}(x)$ became more significant as the optical thickness of the channel increased and as the conduction-radiation parameter decreased.

Kim and Lee [90] and Huang and Lin [90a] included anisotropic scattering and used a two-dimensional radiative analysis to determine heat transfer for Poiseuille flow in a channel. They found a significant difference from the result using a one-dimensional radiative analysis. Significant radiative preheating of the entering fluid caused the difference.

A special case of the previous problem was examined in detail in [91] with application to a liquid drop radiator for outer-space applications. This radiator consists of a layer of small hot liquid drops traveling through space and cooling by radiative loss. In this instance, because of the vacuum of space, there is no heat conduction within the layer and the layer has no solid boundaries. The surroundings are assumed to be at very low temperature. The convective velocity originates with a uniform distribution at $x = 0$ and remains that way in the absence of wall friction. Equations (16-73) and (16-74) then have the form

$$\rho c_p \bar{u} \frac{\partial T}{\partial x} = 4 \frac{a}{\Omega} [\sigma T^4(x, y) - \pi I'(x, y)] \tag{16-77}$$

$$\pi I'(x, \kappa) = (1 - \Omega)\sigma T^4(x, \kappa) + \Omega\pi \int_0^{\kappa_D} I'(x, \kappa^*)E_1(|\kappa^* - \kappa|)\,d\kappa^* \tag{16-78}$$

The numerical solution of these equations with $T(0, y) = T_i$ (a constant) revealed that, for large x, the local heat flux being radiated away, $q_r(x, 0)$, is proportional to the fourth power of the local mean temperature of the droplet layer, that is, $q_r(x, 0)/T_m^4(x) = \mathrm{const}$. In [92] this limit was examined in more detail and it was found that Eqs. (16-77) and (16-78) could be solved by a separation of variables solution, which proved analytically that the emittance $q_r/\sigma T_m^4$ of the layer becomes constant as x is increased. An extension of the analysis in [93] allowed for solidification of the medium within the radiating layer. Since radiation can escape from throughout the medium, there is the interesting feature that solidification occurs in a distribution throughout the volume rather than at a well-defined interface.

In some tube-flow analyses such as [82], the radiative terms in the energy equation (16-71) have also been simplified, as in the previous discussion, by assuming that the chief radiative contribution to the gas at an axial location is the result of temperatures only in the immediate surroundings. The axial temperature variation is thus neglected in determining the radiative terms. The radiative fluxes are calculated as those from an infinitely long cylinder of gas having a radial temperature distribution the same as that at the axial location for which the flux is being evaluated.

The neglect of the axial radiation flux component would not be accurate for a short channel or tube or when conditions are changing rapidly in the flow direction. In this instance a zone method could be used to account for radiation in both the transverse and axial directions. To examine a specific situation, consider the analysis of Einstein [73] for flow of an absorbing-emitting gas without scattering in a tube of diameter D as shown in Fig. 16-15a. Gas enters the tube at temperature T_i and leaves at T_o. The tube wall temperature is constant at T_w. The surrounding environments at the inlet and exit ends of the tube are assumed to be at the inlet and exit gas temperatures, respectively. The governing energy equation at position $\mathbf{r}$ in the tube for laminar flow can be written as

$$\rho c_p u \left.\frac{\partial T}{\partial x}\right|_{\mathbf{r}} = \frac{k}{r}\frac{\partial}{\partial r}\left(r\frac{\partial T}{\partial r}\right)\bigg|_{\mathbf{r}} - 4a\sigma T^4(\mathbf{r})$$

$$+ a\left[\iiint_V \sigma T^4(\mathbf{r}^*)f(\mathbf{r}^* - \mathbf{r})\,dV + \iint_S \sigma T_s^4(\mathbf{r}^*)g(\mathbf{r}^* - \mathbf{r})\,dS\right] \qquad (16\text{-}79)$$

The triple integral is the radiation absorbed at $\mathbf{r}$ as a result of emission from all the gas in the tube. The $f(\mathbf{r}^* - \mathbf{r})$ is a gas-to-gas exchange factor from position $\mathbf{r}^*$ to position $\mathbf{r}$. The double integral is the radiation absorbed at $\mathbf{r}$ as a result of emission from the boundaries, which include the tube wall and the end planes of the tube. The $g(\mathbf{r}^* - \mathbf{r})$ is a surface-to-gas exchange factor; the f and g are given in [73].

Some typical results of the numerical solution are given in Fig. 16-15b for a Poiseuille flow velocity profile. These results show how well the gas obtains energy from the wall, since the ordinate is a measure of how close the exit gas temperature approaches the wall temperature. The results are given in terms of gas optical thickness based on tube diameter and a conduction-radiation parameter based on wall temperature. As the optical thickness is increased from zero, the amount of radiated energy from the wall that is absorbed by the gas increases to a maximum value. Then for large κ_D the heat absorbed by the gas decreases. The decrease is caused by the self-shielding of the gas, which means that for high κ_D most of the direct radiation from the tube wall is absorbed in a thin gas layer near the wall. Since gas emission is isotropic, about one-half the energy re-emitted by this thin layer goes back toward the wall. Thus the gas in the center of the tube is shielded from direct radiation, and the heat transfer efficiency decreases.

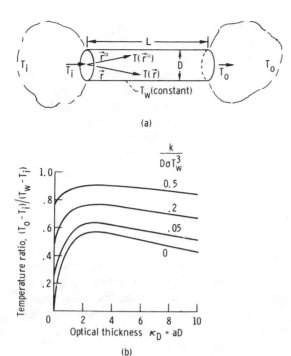

(a)

(b)

Figure 16-15 Combined radiation and convection for absorbing gas flowing in tube with constant wall temperature [73]. (*a*) Tube geometry and boundary conditions; (*b*) exit temperature for $T_i/T_w = 0.4$, $l/D = 5$, and $\bar{\rho}\bar{u}c_p/\sigma T_w^3 = 33$.

A two-dimensional (radial and axial) analysis was made in [94] for laminar flow in a circular tube. Uniform heating was imposed along a portion of the tube length, while the remainder of the tube wall was insulated. An absorbing-emitting gray medium was flowing through the tube, which had black internal surfaces. Temperature distributions were obtained by solving the energy equation by iteration using a Crank-Nicolson finite-difference method. A finite-element node approximation was used to evaluate the radiative source term in the energy equation.

The axial radiative component could be important if the channel is significantly diverging or converging. A two-dimensional analysis is in [95] for a radiating medium flowing between diverging or converging flat plates. Scattering is included, the walls are black, and they are at unequal uniform temperatures. The inlet and outlet boundary planes are treated by assuming they act like porous black surfaces. This may not always be a good approximation because of scattering back into the channel by the medium in the end reservoirs. The numerical solution showed that including radiation effects had a more significant effect on temperature profiles in a diverging channel than in a converging channel.

For fully developed turbulent flow in a tube, the energy equation using a turbulent eddy diffusivity is

$$\rho c_p u \frac{\partial T}{\partial x} = \frac{1}{r} \frac{\partial}{\partial r}\left[(k + \rho c_p \epsilon_h)r \frac{\partial T}{\partial r}\right] - \nabla \cdot \mathbf{q}_r \tag{16-80}$$

Landram et al. [96] used the optically thin limit to approximate $-\nabla \cdot \mathbf{q}_r$. It was evaluated for a volume element as equal to the energy absorbed from the wall by use of an incident mean absorption coefficient minus the energy emitted by use of a Planck mean absorption coefficient (see Sec. 14-7.1). If axial diffusion of radiation is neglected, then $\nabla \cdot \mathbf{q}_r = (1/r)\, \partial(rq_{r,r})/\partial r$. Expressions for $q_{r,r}$ (radiative flux in radial direction) are in [97, 98], and detailed temperature-distribution and Nusselt-number results are given in [97] for heat transfer to turbulent flow in a tube. Detailed calculations using an exponential band absorption model are in [99] for flow in the thermal entrance region of a flat-plate duct downstream of a step in wall temperature. Both laminar and turbulent flow are considered.

There are various applications where solid particles or liquid drops are suspended in gas streams in tubes or channels with conditions providing significant radiation. These include pulverized-coal combustion, steam-water droplet mixtures, and seeded gases in magnetohydrodynamic (MHD) generators. Radiative analyses for turbulent particulate flow in a tube are in [100] and [101]. The first of these includes anisotropic scattering by the particles, which are gray emitters and absorbers. The gas, however, is transparent and transfers energy to or from the particles by convection. The gas-phase energy equation has no radiation terms but has a term for convection to the particles. The particulate energy equation contains this convection term as well as the divergence of the radiative flux in cylindrical coordinates. In [101] the gas-particulate suspension is nonscattering, but the effect of a nongray gas is included. The gas and particle temperatures are assumed to be locally equal, so only one energy equation is used ($N_p V_p \ll 1$),

$$(N_p V_p \rho_p c_{p,p} + \rho_g c_{p,g}) u(r) \frac{\partial T}{\partial x} = \frac{1}{r} \frac{\partial}{\partial r}\left[(k_g + \rho_g c_{p,g} \epsilon_{h,g}) r \frac{\partial T}{\partial r} \right] - \nabla \cdot \mathbf{q}_r(r) \qquad (16\text{-}81)$$

with boundary conditions $\partial T/\partial r = 0$ at $r = 0$, $T = T_w$ at $r = D/2$, and $T = T_i$ at $x = 0$. The $\mathbf{q}_r$ is for a cylinder as given in Sec. 14-8.3. As briefly discussed earlier, the presence of radiation causes the Nusselt number, as a function of axial length, to pass through a minimum and then increase. In contrast, for convection only in a hot tube at uniform wall temperature the Nu(x) approaches a constant asymptotic value. This behavior is for the wall being at a higher temperature than the gas. As the particulate radiation is increased, such as by having a larger particle density, the minimum Nu moves toward the tube entrance. In [101] the case of a hot gas and cool wall was also studied. In this instance the fluid temperature decreases along the tube length and the importance of radiation is thereby reduced with axial distance. The Nu values are higher than for convection only, but Nu(x) does not pass through a minimum along the axial length.

16-5.3 Free Convection Flow, Heat Transfer, and Stability

When free convection is significant, the buoyancy force appears in the momentum equation while the continuity and energy equations are unchanged. A review of literature to 1986 is in [102]. The effect of free convection on laminar forced upward flow of carbon dioxide in a vertical tube was studied analytically and experimentally in [98]. For fully developed tube flow the momentum equation is

$$\frac{dP}{dx} + \rho g = \mu \frac{1}{r} \frac{d}{dr} \left(r \frac{du}{dr} \right) \tag{16-82}$$

This must be solved in conjunction with the energy equation (16-80), as the buoyancy term ρg is temperature dependent; the solution is in [98]. As is usual in free convection, the density was linearized by letting $\rho = \rho_w[1 + \beta(T_w - T)]$. In the same spirit the spectral blackbody function was linearized by letting $e_{\lambda b}(T) = e_{\lambda b}(T_w) - (\partial e_{\lambda b}/\partial T)_w(T_w - T)$. Good agreement was obtained between predicted and measured temperature distributions. A more complex channel flow problem was studied analytically and experimentally in [103]. The geometry is parallel vertical walls with unequal uniform temperatures and with unequal heat addition to them. The heating results in upward free convection flow between the walls. Spectral variations are included for the wall emissivities and for the radiation properties of the flowing medium. Combined free convection and radiation heat transfer was studied in [104] for a vertical surface with a staggered array of fins attached to it. The analytical results were verified by an experimental study.

The effect of radiation on the boundary layer development in free convection of an absorbing-emitting gas is examined in [66, 105–108]. Reference [105] treats the development of the boundary layer on a horizontal cylinder, while [66, 106–108] examine layer growth on a vertical plate. References [66, 109] experimentally determine the onset of free convection in a gas exposed to thermal radiation.

Numerical techniques have been applied to analyze combined radiation and free convection in enclosed rectangular regions filled with an absorbing-emitting medium. In [110] an enclosure with a square cross section was studied. The third dimension is large so the geometry is two-dimensional. The top and bottom horizontal boundaries of the square are insulated, and the two vertical side walls are at unequal uniform temperatures. Some results were also obtained when partial vertical partitions were within the square, partially obstructing the view between the two vertical sides. For an enclosure without internal partitions, radiation increased the temperature of the gas within the enclosure except in the immediate vicinity of the cold wall and in a small region at the lower corner close to the cold wall. The net effect was that natural convection was somewhat reduced by the radiation interaction with the gas. However, with regard to overall heat transfer, this effect was considerably overcompensated by the increase in heat transfer by radiation. The analysis in [111] was for a similar two-dimensional rectangular geometry. A semitransparent fluid is in the form of a vertical layer that is bounded by four sides. The two horizontal boundaries are adiabatic. One vertical boundary is opaque and is maintained at uniform temperature in order to cool the fluid. The other vertical boundary is transparent and serves to confine the fluid and to transmit incident external radiation. The fluid is cold so it absorbs incident radiation but does not reradiate. The analysis provided the free-convection flow patterns. An increase in fluid layer opacity promotes the formation of a boundary layer region near the transmitting boundary because of the high volumetric absorption of energy in that region. The effects of internal radiation and free convection on crystal growth were analyzed in [112]. A numerical investigation was made in

[113, 114] of the radiative interaction with laminar and turbulent free convection in a square enclosure filled with nongray gas. In [113] the radiation calculations were made with a P-1 model and a weighted sum of nongray gases model. A control-volume-based finite-difference method was used in [114] and comparisons were made with experimental data. In [115], the product integration method (Sec. 15-9.3) was used to find temperature and velocity profiles in an absorbing-emitting and isotropically scattering medium in a gray-walled square enclosure with free convection. A two-dimensional radiative analysis was necessary even for the case where both horizontal boundaries are adiabatic; otherwise, the variation in radiative flux on the boundaries and in the medium was not accurately included, and significant changes in streamlines and heat fluxes on the constant temperature boundaries were found.

The more complex geometry of a participating medium between concentric cylinders with free convection was analyzed in [116]. The flow streamlines and heat transfer were found using the YIX method (Sec. 15-9.4) to evaluate the radiative integrals. In [115] and [116] it was found that, for the limited range of parameters studied, the total heat flux calculated by the analysis of coupled radiation and free convection could be obtained to within a small percentage by adding the independent free-convective and radiative fluxes.

Another aspect of free convection is to know the stability conditions for which the onset of free convection can occur. Arpaci and Gozum [107] used the P-1 approximation (Sec. 15-5) to include radiation in an analysis of the thermal convective stability of a nongray fluid between horizontal parallel confining boundaries (the Benard problem with radiation.) Following [107], the radiative flux equation for a nongray fluid is

$$\int_0^\infty \left[\frac{\partial}{\partial x_k} \left(\frac{1}{a_\lambda} \sum_{j=1}^3 \frac{\partial^2 q_{r\lambda,j}}{\partial x_j \, d\lambda} \right) - 4 \frac{\partial e_{\lambda b}}{\partial x_k} - 3a_\lambda \frac{\partial q_{r\lambda,k}}{\partial \lambda} \right] d\lambda = 0 \tag{16-83}$$

In the limit $a_\lambda x \to 0$, (16-83) reduces to the thin-gas approximation with the term $-3a_\lambda \, \partial q_{r\lambda,k}/\partial \lambda \to 0$ so that

$$\int_0^\infty \left[\frac{\partial}{\partial x_k} \left(\frac{1}{a_\lambda} \sum_{j=1}^3 \frac{\partial^2 q_{r\lambda,j}}{\partial x_j \, \partial \lambda} \right) - 4 \frac{\partial e_{\lambda b}}{\partial x_k} \right] d\lambda = 0$$

By setting the integrand equal to zero, integrating with respect to x_k, and then integrating with respect to λ, this is transformed to

$$\sum_{j=1}^3 \frac{\partial q_{r,j}}{\partial x_j} = \int_0^\infty 4a_\lambda e_{\lambda b} \, d\lambda + C = 4a_P \sigma T^4 + C \tag{16-84}$$

The $a_P = \int_0^\infty a_\lambda e_{\lambda b} \, d\lambda / \sigma T^4$ from (14-95). For the optically thick limit, Eq. (16-83) reduces to [as in (15-32) for no scattering]

$$q_{r,k} = -\frac{4\sigma}{3a_R} \frac{\partial T^4}{\partial x_k} \tag{16-85}$$

where (15-82) has been used.

Because (16-83) should apply at both the optically thin and thick limits, (16-84) and (16-85) can be substituted into (16-83) to yield

$$\frac{\partial}{\partial x_k} \left(\sum_{j=1}^{3} \frac{\partial q_{r,j}}{\partial x_j} \right) - 4a_p\sigma \frac{\partial T^4}{\partial x_k} - 3a_p a_R q_{r,k} = 0 \tag{16-86}$$

where from (16-84) the Planck mean was used to approximate a_λ in the first term of (16-83). For the one-dimensional case

$$\frac{d^2 q_{r,x}}{dx^2} - 3a_p a_R q_{r,x} = 4a_p\sigma \frac{dT^4}{dx} \tag{16-87}$$

For the problem of an initial steady temperature profile in the fluid (no free convection, heat generation or dissipation), the general energy equation (14-26) becomes

$$\frac{d}{dx} \left(k\frac{dT}{dx} \right) - \frac{dq_{r,x}}{dx} = 0 \tag{16-88}$$

Equations (16-88) and (16-87) and the boundary conditions on $q_{r,x}$ specified by (15-134) and (15-135), along with the prescribed wall temperatures, constitute the formulation for the initial fluid-temperature profile. This is used in the stability problem in [107], which requires only the temperature gradient in the initial state to determine whether the fluid will become unstable. In [107], the P-3 and P-5 approximations were also used, with increasing accuracy in comparison with an exact solution.

The effect of radiation on fluid stability in enclosed two-dimensional slender vertical cavities (height/width ≈ 10) is analyzed in [117]. The fluid is emitting, absorbing, and nonscattering, and the side walls are each isothermal but at unequal temperatures. The stability of the convective region was investigated. The stability of the conduction regime before the onset of convection was studied by Hassab and Özisik [118].

With regard to other multimode problems, radiation effects in rocket exhaust plumes have been examined by DeSoto [119]. References [8–10] treat the influence of radiation in ablating bodies. A very large body of literature deals with re-entry of bodies into the atmosphere [111] and radiation within and from hypersonic shocks. A rigorous treatment of these problems is difficult because of the nonequilibrium chemical reactions that are coupled with the radiation effects. References [54, 56, 57, 120] give a good introduction and discussion of shock problems. Radiation interaction with a layer of gas including transpiration is treated in [121]

16-6 TRANSIENT PROBLEMS

In a transient situation there can be two types of internal energy storage per unit volume and time; one is the local variation with time of the radiant energy density; the second is the ordinary heat capacity of the material as encountered in conventional heat conduction and convection problems. From (14-25) the first of these is

$$\frac{\partial}{\partial \tau} \int_0^\infty U_\lambda \, d\lambda = \frac{\partial}{\partial \tau} \left(\frac{1}{c} \int_0^\infty \int_0^{4\pi} i'_\lambda \, d\omega \, d\lambda \right)$$

The second is $\rho c_v \, \partial T / \partial \tau$. Because of the large value of the propagation speed c in a medium, the storage of radiant energy is usually neglected. However, some problems dealing with the effects of nuclear weapons and some situations in astrophysics require consideration of transient variations in the radiant energy.

Consider first only the effect of transient radiation. The equation of transfer derived in Chap. 14 neglected changes in radiation intensity with time. The equation of transfer is written for radiation of intensity i'_λ traveling in the S direction. As the radiation travels through the differential length from S to $S + dS$, its intensity is increased by emission and decreased by absorption. Also during the residence of the radiation within dS, the intensity can change with time. The residence time is $d\tau = dS/c$. Hence, the change in i'_λ can be written as

$$di'_\lambda = \frac{\partial i'_\lambda}{\partial \tau} \frac{dS}{c} + \frac{\partial i'_\lambda}{\partial S} dS$$

By substituting for di'_λ, the equation of transfer (14-4) for the case of no scattering becomes

$$\frac{1}{c} \frac{\partial i'_\lambda(S, \tau)}{\partial \tau} + \frac{\partial i'_\lambda(S, \tau)}{\partial S} = a_\lambda(S, \tau) [i'_{\lambda b}(S, \tau) - i'_\lambda(S, \tau)] \tag{16-89}$$

Since the conditions such as temperature within the medium are changing with time, the absorption coefficient is a function of time as well as position.

To better understand the transient term, consider as a simple illustration what the radiative behavior would be if a thick medium at temperature T_1 instantaneously had its temperature increased to a higher uniform value T_2. The medium would then be at T_2 but the intensity within the medium would need to change from $i'_{\lambda b}(T_1)$ to $i'_{\lambda b}(T_2)$. During this process, the radiation would not be in equilibrium. The equation of transfer reduces to (assuming as an approximation that a_λ can be used in the emission term, which is an equilibrium assumption)

$$\frac{1}{c} \frac{\partial i'_\lambda(\tau)}{\partial \tau} = a_\lambda(T_2) [i'_{\lambda b}(T_2) - i'_\lambda(\tau)] \tag{16-90}$$

After integration with the condition $i'_\lambda = i'_{\lambda b}(T_1)$ at $\tau = 0$, the result is

$$\frac{i'_{\lambda b}(T_2) - i'_\lambda(\tau)}{i'_{\lambda b}(T_2) - i'_{\lambda b}(T_1)} = \exp[-ca_\lambda(T_2)\tau] \tag{16-91}$$

The radiation relaxation time (time to change by a factor of $e = 2.718$) for equilibrium to be re-established is thus $1/ca_\lambda(T_2)$, which is usually very short for reasonable values of a_λ in view of the large value of the propagation velocity c in the medium.

In the preceding illustration, it was assumed that the medium temperature could be instantaneously raised so that at the beginning of the transient the radiation intensity was not in equilibrium at the black radiation value corresponding to T_2. Generally the temperature change of a medium would be governed by the heat capacity of the medium, and consequently transient temperature changes would be much slower than the radiation relaxation time. Hence, when used with the transient energy-conservation equation containing the heat capacity term, the unsteady radiation term in the equation of transfer would be negligible. This is why the steady form of the equation of transfer, as derived in Chap. 14, can be instantaneously applied (as in the following example) during almost all transient heat transfer processes involving radiation.

EXAMPLE 16-5 A gray medium is in a slab configuration originally at a uniform temperature T_0. The absorption coefficient is a, and the slab thickness is D. The heat capacity of the medium is c_v and its density is ρ. At $\tau = 0$, the slab is placed in surroundings at zero temperature. Neglecting conduction and convection, discuss the solutions for the temperature profiles for radiative cooling when a is very large and when a is very small.

At the slab center, $x = D/2$ (boundaries are at $x = 0, D$), the condition of symmetry provides the relation for any time: $\partial T/\partial x = 0$ for $x = D/2$, τ. At time $\tau = 0$ there is the condition for any x: $T = T_0$ for x, $\tau = 0$. As discussed in Sec. 14-7.6 for radiation only being included, there will be a temperature slip at the boundaries $x = 0, D$, so that the temperature at the boundaries will be finite rather than being equal to the zero outside temperature. If heat conduction were present at the boundary, the temperature slip would not exist.

For large a the diffusion approximation can be employed, and from (15-32) the heat flux in the x direction is

$$q(x, \tau) = -\frac{4}{3a} \frac{\partial e_b(x, \tau)}{\partial x} = -\frac{4\sigma}{3a} \frac{\partial T^4(x, \tau)}{\partial x}$$

By conservation of energy $-\partial q(x, \tau)/\partial x = \rho c_v \, \partial T/\partial \tau$. Combining these two equations to eliminate q gives the transient energy diffusion equation for the temperature distribution in the slab with constant absorption coefficient:

$$\rho c_v \frac{\partial T}{\partial \tau} = \frac{4\sigma}{3a} \frac{\partial^2 T^4(x, \tau)}{\partial x^2}$$

Defining dimensionless variables as $\bar{\tau} = a \, \sigma T_0^3 \tau/\rho c_v$, $\kappa = ax$, and $t = T/T_0$ gives

$$\frac{\partial t}{\partial \bar{\tau}} = \frac{4}{3} \frac{\partial^2 t^4(\kappa, \bar{\tau})}{\partial \kappa^2}$$

The initial condition and the symmetry condition at $x = D/2$ become, respectively, $t(\kappa, 0) = 1$ and $(\partial t/\partial \kappa)(0, \bar{\tau}) = 0$. At the boundary $\kappa = aD$, a slip condition must be used. From (15-50), when the surroundings are empty space at zero temperature, $e_{bw} = 0$ and $\epsilon_w = 1$ (boundary assumed nonreflecting), so that at the exposed boundary of the medium for any time

$$\sigma T^4 \big|_{x=D} = \frac{1}{2}\left(-\frac{4\,\sigma}{3\,a}\frac{\partial T^4}{\partial x}\right)_{x=D} - \frac{\sigma}{2a^2}\frac{\partial^2 T^4}{\partial x^2}\bigg|_{x=D}$$

or

$$0 = \left(2t^4 + \frac{4}{3}\frac{\partial t^4}{\partial \kappa} + \frac{\partial^2 t^4}{\partial \kappa^2}\right)_{\kappa=aD}$$

Similar relations apply at $x = 0$. For these conditions, a numerical solution is necessary.

For a small absorption coefficient, and since there are no enclosing radiating boundaries, the emission approximation (Sec. 15-3.2) can be applied. For very small a the medium is optically so thin that it is at uniform temperature throughout its thickness at any instant. From the results of Example 15-4, the heat flux emerging from both boundaries of the layer is $q = 4a\sigma T^4 D$. The energy equation then becomes $\rho c_v\, dT/d\tau = -4a\sigma T^4$, or, in dimensionless form, $dt/d\bar{\tau} = -4t^4$. Integrating with the condition that $t = 1$ at $\bar{\tau} = 0$ gives the transient temperature throughout the slab as

$$t = \frac{1}{(1 + 12\bar{\tau})^{1/3}}$$

Viskanta and Bathla [122] obtained numerical solutions to the transient form of the transfer equations, along with the limiting solutions derived here. Some of their results for an optical thickness of unity are in Fig. 16-16.

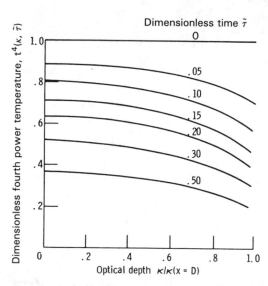

Figure 16-16 Dimensionless temperature profiles as a function of time for radiative cooling of a gray slab; optical thickness $\kappa(x = D) = 1.0$. From [122].

EXAMPLE 16-6 A plane layer with thickness extending from $x = 0$ to D is initially at temperature $T_i(x)$ and is in a vacuum. The layer is gray and is composed of a partially transparent porous solid material that emits, absorbs, and scatters radiation. Scattering is assumed isotropic, and the thermal conductivity of the layer is a constant, k. The absorption and scattering coefficients, a and σ_s, are uniform within the layer. The surfaces of the porous layer are assumed nonreflecting. For time $\tau > 0$ an internal heat generation $q'''(x, \tau)$ is applied. The surrounding environment for $x < 0$ is at $T_{e,1}$, and for $x > D$ it is at $T_{e,2}$. Provide the energy relations to solve for the temperature distribution within the layer as a function of position and time, $T(x, \tau)$.

From Eq. (16-1) the energy equation is

$$\rho c \frac{\partial T}{\partial \tau} = k \frac{\partial^2 T}{\partial x^2} - \nabla \cdot \mathbf{q}_r + q'''(x, \tau)$$

The $\nabla \cdot \mathbf{q}_r$ is found from Eq. (16-5) where the phase function $\Phi = 1$ for isotropic scattering. For a gray material

$$\nabla \cdot \mathbf{q}_r = 4\pi \frac{a}{\Omega} \left[\frac{\sigma T^4}{\pi}(x, \tau) - I'(x, \tau) \right]$$

so the energy equation becomes

$$\rho c \frac{\partial T}{\partial \tau} = k \frac{\partial^2 T}{\partial x^2} - 4 \frac{a}{\Omega} [\sigma T^4(x, \tau) - \pi I'(x, \tau)] + q'''(x, \tau)$$

Let $T_{i,0}$ be an arbitrary reference temperature such as $T_i(0, 0)$ and $t = T/T_{i,0}$, $\kappa = (a + \sigma_s)x$, and $\kappa_D = (a + \sigma_s)D$. The energy equation is then placed in the dimensionless form

$$\frac{\partial t}{\partial \bar{\tau}} = 4N \frac{\partial^2 t}{\partial \kappa^2} - \frac{4(1 - \Omega)}{\Omega} [t^4(\kappa, \bar{\tau}) - \bar{I}(\kappa, \bar{\tau})] + \bar{q}(\kappa, \bar{\tau}) \tag{16-92}$$

where

$$\bar{\tau} = \frac{(a + \sigma_s)\sigma T_{i,0}^3}{\rho c} \tau, \qquad N = \frac{k(a + \sigma_s)}{4\sigma T_{i,0}^3}, \qquad \bar{I} = \frac{\pi I'}{\sigma T_{i,0}^4}, \qquad \bar{q} = \frac{q'''}{(a + \sigma_s)\sigma T_{i,0}^4}$$

The equation for $\bar{I}$ is obtained from Eq. (14-78), where for nonreflecting boundaries $i^+(0, \mu) = \sigma T_{e,1}^4/\pi$ and $i^-(\kappa_D, -\mu) = \sigma T_{e,2}^4/\pi$, with the surroundings assumed to act as sources of black radiation,

$$I'(\kappa, \tau) = (1 - \Omega) \frac{\sigma T^4(\kappa, \tau)}{\pi}$$

$$+ \frac{\Omega}{2} \left[\frac{\sigma T_{e,1}^4(\tau)}{\pi} E_2(\kappa) + \frac{\sigma T_{e,2}^4(\tau)}{\pi} E_2(\kappa_D - \kappa) + \int_0^{\kappa_D} I'(\kappa^*, \tau) E_1(|\kappa^* - \kappa|) \, d\kappa^* \right]$$

This has the dimensionless form

$$\tilde{I}(\kappa, \bar{\tau}) = (1 - \Omega)t^4(\kappa, \bar{\tau})$$

$$+ \frac{\Omega}{2} \left[t_{e,1}^4(\bar{\tau})E_2(\kappa) + t_{e,2}^4(\bar{\tau})E_2(\kappa_D - \kappa) + \int_0^{\kappa_D} \tilde{I}(\kappa^*, \bar{\tau})E_1(|\kappa^* - \kappa|) \, d\kappa^* \right]$$

$$(16\text{-}93)$$

Equations (16-92) and (16-93) are solved numerically subject to the boundary conditions $\partial t(0, \bar{\tau})/\partial\kappa = 0$, $\partial t(\kappa_D, \bar{\tau})/\partial\kappa = 0$ (no heat conduction into surrounding vacuum), and the specified initial condition $t(\kappa, 0) = T_i(x)/T_{i,0}$. Numerical solution methods are presented in Chap. 17.

Transient heat transfer in a layer with Arrhenius heat generation is studied in [123]. Thermal behavior of a porous material is studied analytically and experimentally in [124]. Numerical solutions for spherical geometries are found in [125–127], and results for the region between coaxial cylinders are in [128]. Wendlandt [7] derives expressions for the transient temperature distribution in a semi-infinite absorbing medium exposed to a laser pulse. The transient heating of a number of semitransparent solid geometries is reviewed in [129]. Consider, for example, a unidirectional flux q_i that is incident at angle θ on a semi-infinite gray solid as shown in Fig. 16-17a. A fraction $\tau_0(\theta)$ is transmitted through the interface, and within the material the energy is refracted into direction χ (a smooth interface has been assumed). The absorption coefficient of the solid is a. The fraction of entering radiation that reaches depth x is $\exp(-ax/\cos \chi)$, and the amount $(a/\cos \chi) \exp(-ax/\cos \chi)$ is absorbed at x per unit volume. The solid is assumed to be in its initial state of heating so that its temperature is low, and hence emission within the solid and heat loss from its surface can be neglected. The energy equation, initial condition, and boundary conditions are then, where for convenience $\bar{a} = a/\cos \chi$,

$$\rho c \frac{\partial T}{\partial \tau} = k \frac{\partial^2 T}{\partial x^2} + q_i \cos \theta \, \tau_0(\theta)\bar{a}e^{-\bar{a}x} \tag{16-94}$$

$$T(x, 0) = T_0 \qquad \text{(initial condition)}$$

$$\frac{\partial T}{\partial x}(0, \tau) = 0 \qquad \lim_{x \to \infty} T(x, \tau) = T_0$$

The solution is in [130] and can be placed in the form

$$\frac{[T(x, \tau) - T_0]\sqrt{k\rho c}}{q_i \cos \theta \, \tau_0(\theta)\sqrt{\tau}} = \frac{2}{\sqrt{\pi}} e^{-X^2} - 2X \operatorname{erfc} X - \frac{1}{A} e^{-2AX}$$

$$+ \frac{1}{2A} \left[e^{A(A-2X)} \operatorname{erfc}(A - X) + e^{A(A+2X)} \operatorname{erfc}(A + X) \right]$$

$$(16\text{-}95)$$

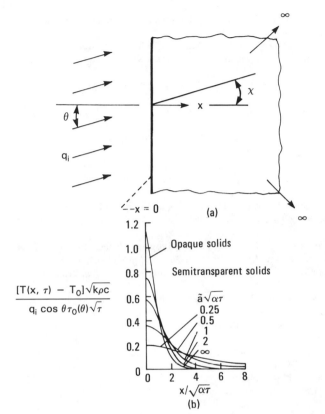

$$\frac{[T(x, \tau) - T_0]\sqrt{k\rho c}}{q_i \cos \theta \tau_0(\theta)\sqrt{\tau}}$$

Figure 16-17 Transient heating of semi-infinite solid by incident unidirectional radiative flux. (*a*) Flux incident on semi-infinite slab; (*b*) dimensionless temperature distribution.

where $X = x/2\sqrt{\alpha\tau}$, $A = \tilde{a}\sqrt{\alpha\tau}$, $\tilde{a} = a/\cos \chi$, and $\alpha = k/\rho c$. (Note that $\tau =$ time, $\tau_0 =$ surface transmissivity.)

Some results are in Fig. 16-17*b*. Since the parameter on the curves depends on the absorption coefficient a, an increase of a will increase the surface temperature that is reached at any time. A small a increases the penetration of the temperature distribution into the solid. As $\tilde{a}\sqrt{\alpha\tau} \to \infty$, the temperature distribution approaches that for an opaque solid. Thus no matter what a is, given sufficient time, the semi-infinitely thick material will act as an opaque solid.

REFERENCES

1. Kellett, B. 3.. Transmission of Radiation through Glass in Tank Furnaces, *J. Soc. Glass Tech.*, vol. 36, pp. 115–123, 1952.
2. Gardon, Robert: The Emissivity of Transparent Materials, *J. Am. Ceram. Soc.*, vol. 39, no. 8, pp. 278–287, 1956.
3. Condon, Edward U.: Radiative Transport in Hot Glass, *J. Quant. Spectrosc. Radiat. Transfer*, vol. 8, no. 1, pp. 369–385, 1968.
4. Eryou, N. D., and L. R. Glicksman: An Experimental and Analytical Study of Radiative and Conductive Heat Transfer in Molten Glass, *J. Heat Transfer*, vol. 94, no. 2, pp. 224–230, 1972.

5. Anderson, E. E., R. Viskanta, and W. H. Stevenson: Heat Transfer through Semitransparent Solids, *J. Heat Transfer*, vol. 95, no. 2, pp. 179–186, 1973.
6. Kuriyama, Masaaki, et al.: The Effect of Radiation Heat Transfer in the Measurement of Thermal Conductivity for the Semitransparent Medium, *Bull. JSME*, vol. 19, no. 134, pp. 973–979, 1976.
7. Wendlandt, B. C. H.: Temperature in an Irradiated Thermally Conducting Medium, *J. Phys. D*, vol. 6, pp. 657–660, 1973.
8. Kadanoff, Leo P.: Radiative Transport within an Ablating Body, *J. Heat Transfer*, vol. 83, no. 2, pp. 215–225, 1961.
9. Nelson, H. F.: Radiative Transfer through Carbon Ablation Layers, *J. Quant. Spectrosc. Radiat. Transfer*, vol. 13, pp. 427–445, 1973.
10. Boles, M. A., and M. N. Özişik: Simultaneous Ablation and Radiation in an Absorbing, Emitting and Isotropically Scattering Medium, *J. Quant. Spectrosc. Radiat. Transfer*, vol. 12, pp. 838–847, 1972.
11. Merriam, R. L., and R. Viskanta: Radiative Characteristics of Cryodeposits for Room Temperature Black Body Radiation, *Cryogenic Eng. Conf.*, Case-Western Reserve University, Cleveland, August 1968.
12. McConnell, Dudley G.: Radiant Energy Transport within Cryogenic Condensates, *Inst. Environmental Sci. Ann. Meeting Equipment Exposition*, San Diego, April 11–13, 1966.
13. Gilpin, R. R., R. B. Roberton, and B. Singh: Radiative Heating in Ice, *J. Heat Transfer*, vol. 99, pp. 227–232, 1977.
14. Tong, T. W., and C. L. Tien: Radiative Heat Transfer in Fibrous Insulations—Part I: Analytical Study, *J. Heat Transfer*, vol. 105, no. 1, pp. 70–75, 1983.
15. Mathes, R., J. Blumenberg, and K. Keller: Radiative Heat Transfer in Insulations with Random Fibre Orientation, *Int. J. Heat Mass Transfer*, vol. 33, no. 4, pp. 767–770, 1990.
16. Glicksman, L., M. Schuetz, and M. Sinofsky: Radiation Heat Transfer in Foam Insulation, *Int. J. Heat Mass Transfer*, vol. 30, no. 1, pp. 187–197, 1987.
17. Matthews, L. K., R. Viskanta, and F. P. Incropera: Combined Conduction and Radiation Heat Transfer in Porous Materials Heated by Intense Solar Radiation, *J. Sol. Energy Eng.*, vol. 107, pp. 29–34, 1985.
18. Viskanta, R., and R. J. Grosh: Heat Transfer by Simultaneous Conduction and Radiation in an Absorbing Medium, *J. Heat Transfer*, vol. 84, no. 1, pp. 63–72, 1962.
19. Viskanta, R., and R. J. Grosh: Effect of Surface Emissivity on Heat Transfer by Simultaneous Conduction and Radiation, *Int. J. Heat Mass Transfer*, vol. 5, pp. 729–734, 1962.
20. Greif, Ralph, and Gean P. Clapper: Radiant Heat Transfer between Concentric Cylinders, *Appl. Sci. Res.*, sec. A, vol. 15, pp. 469–474, 1966.
21. Men, A. A.: Radiative-Conductive Heat Transfer in a Medium with a Cylindrical Geometry. I, *Inz. Fiz. Zh.*, vol. 24, no. 6, pp. 984–991, 1973.
22. Viskanta, R., and R. L. Merriam: Heat Transfer by Combined Conduction and Radiation between Concentric Spheres Separated by Radiating Medium, *J. Heat Transfer*, vol. 90, no. 2, pp. 248–256, 1968.
23. Viskanta, R.: Heat Transfer by Conduction and Radiation in Absorbing and Scattering Materials, *J. Heat Transfer*, vol. 87, no. 1, pp. 143–150, 1965.
24. Enoch, I. E., E. Ozil, and R. C. Birkebak: Polynomial Approximation Solution of Heat Transfer by Conduction and Radiation in a One-Dimensional Absorbing, Emitting, and Scattering Medium, *Numer. Heat Transfer*, vol. 5, pp. 353–358, 1982.
25. Yuen, W. W., and L. W. Wong: Heat Transfer by Conduction and Radiation in a One-Dimensional Absorbing, Emitting and Anisotropically Scattering Medium, *J. Heat Transfer*, vol. 102, no. 2, pp. 303–307, 1980.
26. Schimmel, W. P., J. L. Novotny, and F. A. Olsofka: Interferometric Study of Radiation-Conduction Interaction, *Fourth Int. Heat Transfer Conf.*, Paris, September 1970.
27. Smith, T. F., A. M. Al-Turki, K.-H. Byun, and T. K. Kim: Radiative and Conductive Transfer for a Gas/Soot Mixture between Diffuse Parallel Plates, *J. Thermophys. Heat Transfer*, vol. 1, no. 1, pp. 50–55, 1987.

28. Yuen, W. W., and E. E. Takara: Analysis of Combined Conductive-Radiative Heat Transfer in a Two-Dimensional Rectangular Enclosure with a Gray Medium, *J. Heat Transfer*, vol. 110, no. 2, pp. 468–474, 1988.

29. Razzaque, M. M., J. R. Howell, and D. E. Klein: Coupled Radiative and Conductive Heat Transfer in a Two-Dimensional Rectangular Enclosure with Gray Participating Media Using Finite Elements, *J. Heat Transfer*, vol. 106, no. 3, pp. 613–619, 1984.

30. Pandey, D. K.: Combined Conduction and Radiation Heat Transfer in Concentric Cylindrical Media, *J. Thermophys. Heat Transfer*, vol. 3, no. 1, pp. 75–82, 1989.

31. Fernandes, R., and J. Francis: Combined Conductive and Radiative Heat Transfer in an Absorbing, Emitting, and Scattering Cylindrical Medium, *J. Heat Transfer*, vol. 104, no. 4, pp. 594–601, 1982.

32. Azad, F. H., and M. F. Modest: Evaluation of the Radiative Heat Flux in Absorbing, Emitting and Linear-Anisotropically Scattering Cylindrical Media, *J. Heat Transfer*, vol. 103, no. 2, pp. 350–356, 1981.

33. Harris, J. A.: Solution of the Conduction/Radiation Problem with Linear-Anisotropic Scattering in an Annular Medium by the Spherical Harmonics Method, *J. Heat Transfer*, vol. 111, no. 1, pp. 194–197, 1989.

34. Yucel, A., and M. L. Williams: Heat Transfer by Combined Conduction and Radiation in Axisymmetric Enclosures, *J. Thermophys. Heat Transfer*, vol. 1, no. 4, pp. 301–306, 1987.

35. Thynell, S. T.: Interaction of Conduction and Radiation in Anisotropically Scattering, Spherical Media, *J. Thermophys. Heat Transfer*, vol. 4, no. 3, pp. 299–304, 1990.

36. Einstein, Thomas H.: Radiant Heat Transfer to Absorbing Gases Enclosed between Parallel Flat Plates with Flow and Conduction, NASA TR R-154, 1963.

37. Cess, R. D.: The Interaction of Thermal Radiation with Conduction and Convection Heat Transfer, in Thomas F. Irvine, Jr., and James P. Hartnett (eds.), "Advances in Heat Transfer," vol. 1, pp. 1–50, Academic Press, New York, 1964.

38. Howell, J. R.: Determination of Combined Conduction and Radiation of Heat through Absorbing Media by the Exchange Factor Approximation, *Chem. Eng. Progr. Symp. Ser.*, vol. 61, no. 59, pp. 162–171, 1965.

39. Nelson, D. A.: On the Uncoupled Superposition Approximation for Combined Conduction-Radiation through Infrared Radiating Gases, *Int. J. Heat Mass Transfer*, vol. 18, no. 5, pp. 711–713, 1975.

40. Goldstein, M. E., and J. R. Howell: Boundary Conditions for the Diffusion Solution of Coupled Conduction-Radiation Problems, NASA TN D-4618, 1968.

41. Howell, J. R., and M. E. Goldstein: Effective Slip Coefficients for Coupled Conduction-Radiation Problems, *J. Heat Transfer*, vol. 91, no. 1, pp. 165–166, 1969.

42. Taitel, Yehuda, and J. P. Hartnett: Application of Rosseland Approximation and Solution Based on Series Expansion of the Emission Power to Radiation Problems, *AIAA J.*, vol. 6, no. 1, pp. 80–89, 1968.

43. Wang, L. S., and C. L. Tien: A Study of Various Limits in Radiation Heat-Transfer Problems, *Int. J. Heat Mass Transfer*, vol. 10, no. 10, pp. 1327–1338, 1967.

44. Perlmutter, M., and J. R. Howell: Radiant Transfer through a Gray Gas between Concentric Cylinders Using Monte Carlo, *J. Heat Transfer*, vol. 86, no. 2, pp. 169–179, 1964.

45. Rhyming, I. L.: Radiative Transfer between Two Concentric Spheres Separated by an Absorbing and Emitting Gas, *Int. J. Heat Mass Transfer*, vol. 9, no. 4, pp. 315–324, 1966.

46. Sparrow, E. M., C. M. Usiskin, and H. A. Hubbard: Radiation Heat Transfer in a Spherical Enclosure Containing a Participating, Heat Generating Gas, *J. Heat Transfer*, vol. 83, no. 2, pp. 199–206, 1961.

47. Ness, A. J.: Solution of Equations of a Thermal Network on a Digital Computer, *Sol. Energy*, vol. 3, no. 2, p. 37, 1959.

48. Fulkerson, G. D., and R. B. Bannerot: An Approximation for Combined Heat Transfer in a Radiatively Absorbing and Emitting Gas, paper 73-750, AIAA, July 16, 1973.

49. Larsen, M. E., and J. R. Howell: The Exchange Factor Method: An Alternative Basis for Zonal Analysis of Radiating Enclosures, *J. Heat Transfer*, vol. 107, no. 4, pp. 936–942, 1985.

50. Liu, H.-P., and J. R. Howell: Scale Modeling of Radiation in Enclosures with Absorbing/Emitting and Isotropically Scattering Media, *J. Heat Transfer*, vol. 109, no. 2, pp. 470–477, 1987.

51. Lick, Wilbert: Energy Transfer by Radiation and Conduction, in Anatol Roshko, Bradford Sturtevant, and D. R. Bartz (eds.), *Proc. 1963 Heat Transfer Fluid Mech. Inst.*, pp. 14–26, 1963.

52. Naraghi, M. H. N., B. T. F. Chung, and B. Litkouhi: A Continuous Exchange Factor Method for Radiative Exchange in Enclosures with Participating Media, *J. Heat Transfer*, vol. 110, no. 2, pp. 456–462, 1988.

53. Viskanta, R.: Radiation Transfer and Interaction of Convection with Radiation Heat Transfer, in Thomas F. Irvine, Jr., and James P. Hartnett (eds.), "Advances in Heat Transfer" vol. 3, pp. 175–251, Academic Press, New York, 1966.

54. Pai, Shih-I: "Radiation Gas Dynamics," Springer-Verlag OHG, Berlin, 1966.

55. Bond, John W., Jr., Kenneth M. Watson, and Jasper, A. Welch, Jr.: "Atomic Theory of Gas Dynamics," chaps. 10–13, Addison-Wesley, Reading, Mass., 1965.

56. Zel'dovich, Ya. B., and Yu. P. Raizek: "Physics of Shock Waves and High-Temperature Hydrodynamic Phenomena," vol. I, pt. II, Academic Press, New York, 1966.

57. Vincenti, Walter G., and Charles H. Kruger, Jr.: "Introduction to Physical Gas Dynamics," chaps. 11–12, Wiley, New York, 1965.

58. Viskanta, R.: Radiation Heat Transfer: Interaction with Conduction and Convection and Approximate Methods in Radiation, in "Heat Transfer 1982," *Proc. 7th Int. Heat Transfer Conf.*, vol. 1, pp. 103–121, Hemisphere, Washington, D.C., 1982.

59. Edwards, D. K.: Numerical Methods in Radiation Heat Transfer, in T. M. Shih (ed.), "Numerical Properties and Methodologies in Heat Transfer," Hemisphere, Washington, D.C., 1983.

60. Novotny, J. L., and Kwang-Tzu Yang: The Interaction of Thermal Radiation in Optically Thick Boundary Layers, *J. Heat Transfer*, vol. 89, no. 4, pp. 309–312, 1967.

61. Viskanta, R., and R. J. Grosh: Boundary Layer in Thermal Radiation Absorbing and Emitting Media, *Int. J. Heat Mass Transfer*, vol. 5, pp. 795–806, 1962.

62. Cess, R. D.: Radiation Effects upon Boundary-Layer Flow of an Absorbing Gas, *J. Heat Transfer*, vol. 86, no. 4, pp. 469–475, 1964.

63. Fritsch, C. A., R. J. Grosh, and M. W. Wildin: Radiative Heat Transfer through an Absorbing Boundary Layer, *J. Heat Transfer*, vol. 88, no. 3, pp. 296–304, 1966.

64. Kubo, Syozo: Stagnation Point Flow of a Radiating Gas of a Large Optical Thickness II, *Trans. Jpn. Soc. Aeronaut. Space Sci.*, vol. 18, no. 41, pp. 141–151, 1975.

65. Howe, John T.: Radiation Shielding of the Stagnation Region by Transpiration of an Opaque Gas, NASA TN D-329, 1960.

66. Hasegawa, Shu, Ryozo Echigo, and Kenji Fukuda: Analytical and Experimental Studies on Simultaneous Radiative and Free Convective Heat Transfer along a Vertical Plate, *Proc. Jpn. Soc. Mech. Eng.*, vol. 38, no. 315, pp. 2873–2882, 1972; vol. 39, no. 317, pp. 250–257, 1973.

67. Doura, S., and J. R. Howell: An Approximate Solution for the Energy Equation with Radiant Participating Media, paper 77-HT-70, ASME, August, 1977.

68. Lee, H. S., J. A. Menart, and A. Fakheri: Multilayer Radiation Solution for Boundary-Layer Flow of Gray Gases, *J. Thermophys. Heat Transfer*, vol. 4, no. 2, pp. 180–185, 1990.

69. Zhang, L., A. Soufiani, J. P. Petit, and J. Taine: Coupled Radiation and Laminar Mixed Convection in an Absorbing and Emitting Real Gas Mixture along a Vertical Plane, *Int. J. Heat Mass Transfer*, vol. 33, no. 2, pp. 319–329, 1990.

70. Cess, Robert D.: The Interaction of Thermal Radiation in Boundary Layer Heat Transfer, *Third Int. Heat Transfer Conf. AIChE*, vol. 5, pp. 154–163, 1966.

71. Schlichting, Hermann (J. Kestin, trans.): "Boundary Layer Theory," 4th ed., p. 116, McGraw-Hill, New York, 1960.

72. Viskanta, R.: Interaction of Heat Transfer by Conduction, Convection, and Radiation in a Radiating Fluid, *J. Heat Transfer*, vol. 85, no. 4, pp. 318–328, 1963.

73. Einstein, Thomas H.: Radiant Heat Transfer to Absorbing Gases Enclosed in a Circular Pipe with Conduction, Gas Flow, and Internal Heat Generation, NASA TR R-156, 1963.

74. Adrianov, V. N., and S. N. Shorin: Radiative Transfer in the Flow of a Radiating Medium, *Trans.* TT-1, Purdue University, February 1961.

75. Chen, John C.: Simultaneous Radiative and Convective Heat Transfer in an Absorbing, Emitting, and Scattering Medium in Slug Flow between Parallel Plates, Rept. BNL-6876-R, Brookhaven National Laboratory, March 18, 1963.

76. Edwards, D. K., and A. Balakrishnan: Self-Absorption of Radiation in Turbulent Molecular Gases, *Combust. Flame*, vol. 20, pp. 401–417, 1973.

77. Chiba, Z., and R. Greif: Heat Transfer to Steam Flowing Turbulently in a Pipe, *Int. J. Heat Mass Transfer*, vol. 16, pp. 1645–1648, 1973.

78. Nakra, N. K., and T. F. Smith: Combined Radiation-Convection for a Real Gas, paper 76-HT-58, ASME, August, 1976.

79. Bergquam, J. B., and N. S. Wang: Heat Transfer by Convection and Radiation in an Absorbing, Scattering Medium Flowing between Parallel Plates, paper 76-HT-50, ASME, August, 1976.

80. Greif, Ralph, and D. R. Willis: Heat Transfer between Parallel Plates including Radiation and Rarefaction Effects, *Int. J. Heat Mass Transfer*, vol. 10, pp. 1041–1048, 1967.

81. Martin, J. K., and C. C. Hwang: Combined Radiant and Convective Heat Transfer to Laminar Stream Flow between Gray Parallel Plates with Uniform Heat Flux, *J. Quant. Spectrosc. Radiat. Transfer*, vol. 15, pp. 1071–1081, 1975.

82. De Soto, Simon, and D. K. Edwards: Radiative Emission and Absorption in Nonisothermal Nongray Gases in Tubes, in A. F. Charwat (ed.), *Proc. 1965 Heat Transfer Fluid Mech. Inst.*, pp. 358–372, 1965.

83. Pearce, B. L., and A. F. Emery: Heat Transfer by Thermal Radiation and Laminar Forced Convection to an Absorbing Fluid in the Entry Region of a Pipe, *J. Heat Transfer*, vol. 92, no. 2, pp. 221–230, 1970.

84. Jeng, D. R., E. J. Lee, and K. J. DeWitt: Simultaneous Conductive and Radiative Heat Transfer for Laminar Flow in Circular Tubes with Constant Wall Temperature, *Proc. 5th Int. Heat Transfer Conf.*, Tokyo, vol. I, pp. 118–122, 1974.

85. Jeng, D. R., E. J. Lee, and K. J. DeWitt: A Study of Two Limiting Cases in Convective and Radiative Heat Transfer with Nongray Gases, *Int. J. Heat Mass Transfer*, vol. 19, no. 6, pp. 589–596, 1976.

86. Soufiani, A., and J. Taine: Application of Statistical Narrow-Band Model to Coupled Radiation and Convection at High Temperature, *Int. J. Heat Mass Transfer*, vol. 30, no. 3, pp. 437–447, 1987.

87. Soufiani, A., and J. Taine: Experimental and Theoretical Studies of Combined Radiative and Convective Transfer in CO_2 and H_2O Laminar Flows, *Int. J. Heat Mass Transfer*, vol. 32, no. 3, pp. 477–486, 1989.

88. Hirano, M.: Enhancement of Radiative Heat Transfer in the Laminar Channel Flow of Non-Gray Gases, *Int. J. Heat Mass Transfer*, vol. 31, no. 2, pp. 367–374, 1988.

89. Chawla, T. C., and S. H. Chan: Combined Radiation Convection in Thermally Developing Poiseuille Flow with Scattering, *J. Heat Transfer*, vol. 102, no. 2, pp. 297–302, 1980.

90. Kim, T.-K., and H. S. Lee: Two-Dimensional Anisotropic Scattering Radiation in a Thermally Developing Poiseuille Flow, *J. Thermophys. Heat Transfer*, vol. 4, no. 3, pp. 292–298, 1990.

90a. Huang, J. M., and J. D. Lin: Combined Radiative and Forced Convective Heat Transfer in Thermally Developing Laminar Flow through a Circular Pipe, *Chem. Eng. Comm.*, vol. 101, pp. 147–164, 1991.

91. Siegel, R.: Transient Radiative Cooling of a Droplet-Filled Layer, *J. Heat Transfer*, vol. 109, no. 1, pp. 159–164, 1987.

92. Siegel, R.: Separation of Variables Solution for Nonlinear Radiative Cooling, *Int. J. Heat Mass Transfer*, vol. 30, no. 5, pp. 959–965, 1987.

93. Siegel, R.: Transient Radiative Cooling of a Layer Filled with Solidifying Drops, *J. Heat Transfer*, vol. 109, no. 4, pp. 977–982, 1987.

94. Huang, J. M., and J. D. Lin: Radiation and Convection in Circular Pipe with Uniform Wall Heat Flux, *J. Thermophys. Heat Transfer*, vol. 5, no. 4, pp. 502–507, 1991.

95. Chung, T. J., and J. Y. Kim: Two-Dimensional Combined Mode Heat Transfer by Conduction,

Convection and Radiation in Emitting, Absorbing and Scattering Media—Solution by Finite Elements, *J. Heat Transfer*, vol. 106, no. 2, pp. 448–452, 1984.

96. Landram, C. S., R. Greif, and I. S. Habib: Heat Transfer in Turbulent Pipe Flow with Optically Thin Radiation, *J. Heat Transfer*, vol. 91, no. 3, pp. 330–336, 1969.

97. Wassel, A. T., and D. K. Edwards: Molecular Gas Radiation in a Laminar or Turbulent Pipe Flow, *J. Heat Transfer*, vol. 98, no. 1, pp. 101–107, 1976.

98. Greif, R.: Laminar Convection with Radiation: Experimental and Theoretical Results, *Int. J. Heat Mass Transfer*, vol. 21, no. 4, pp. 477–480, 1978.

99. Balakrishnan, A., and D. K. Edwards: Molecular Gas Radiation in the Thermal Entrance Region of a Duct, *J. Heat Transfer*, vol. 101, no. 3, pp. 489–495, 1979.

100. Azad, F. H., and M. F. Modest: Combined Radiation and Convection in Absorbing, Emitting and Anisotropically Scattering Gas—Particulate Tube Flow, *Int. J. Heat Mass Transfer*, vol. 24, no. 10, pp. 1681–1698, 1981.

101. Tabanfar, S., and M. F. Modest: Combined Radiation and Convection in Absorbing, Emitting, Nongray Gas–Particulate Tube Flow, *J. Heat Transfer*, vol. 109, no. 2, pp. 478–484, 1987.

102. Yang, K. T.: Numerical Modelling of Natural Convection-Radiation Interactions in Enclosures, in "Heat Transfer 1986," *Proc. 8th International Heat Transfer Conf*, vol. 1, pp. 131–140, Hemisphere, Washington, D.C., 1986.

103. Yamada, Y.: Combined Radiation and Free Convection Heat Transfer in a Vertical Channel with Arbitrary Wall Emissivities, *Int. J. Heat Mass Transfer*, vol. 31, no. 2, pp. 429–440, 1988.

104. Guglielmini, G., E. Nannei, and G. Tanda: Natural Convection and Radiation Heat Transfer from Staggered Vertical Fins, *Int. J. Heat Mass Transfer*, vol. 30, no. 9, pp. 1941–1948, 1987.

105. Novotny, J. L., and M. D. Kelleher: Free-Convection Stagnation Flow of an Absorbing-Emitting Gas, *Int. J. Heat Mass Transfer*, vol. 10, no. 9, pp. 1171–1178, 1967.

106. Cess, R. D.: The Interaction of Thermal Radiation with Free Convection Heat Transfer, *Int. J. Heat Mass Transfer*, vol. 9, no. 11, pp. 1269–1277, 1966.

107. Arpaci, V. S., and D. Gözüm: Thermal Stability of Radiating Fluids: The Benard Problem, *Phys. Fluids*, vol. 16, no. 5, pp. 581–588, 1973.

108. Arpaci, V. S., and Y. Bayazitoglu: Thermal Stability of Radiating Fluids: Asymmetric Slot Problem, *Phys. Fluids*, vol. 16, no. 5, pp. 589–593, 1973.

109. Gille, John, and Richard Goody: Convection in a Radiating Gas, *J. Fluid Mech.*, vol. 20, pt. 1, pp. 47–79, 1964.

110. Chang, L. C., K. T. Yang, and J. R. Lloyd: Radiation–Natural Convection Interactions in Two-Dimensional Complex Enclosures, *J. Heat Transfer*, vol. 105, no. 1, pp. 89–95, 1983.

111. Webb, B. W., and R. Viskanta: Analysis of Radiation-Induced Natural Convection in Rectangular Enclosures, *J. Thermophys. Heat Transfer*, vol. 1, no. 2, pp. 146–153, 1987.

112. Matsushima, H., and R. Viskanta: Effects of Internal Radiative Transfer on Natural Convection and Heat Transfer in a Vertical Crystal Growth Configuration, *Int. J. Heat Mass Transfer*, vol. 33, no. 9, pp. 1957–1968, 1990.

113. Fusegi, T., and B. Farouk: Laminar and Turbulent Natural Convection–Radiation Interactions in a Square Enclosure Filled with a Nongray Gas, *Numer. Heat Transfer*, vol. 15, no. 3, pp. 303–322, 1989.

114. Fusegi, T., and B. Farouk: A Computational and Experimental Study of Natural Convection and Surface/Gas Radiation Interactions in a Square Cavity, *J. Heat Transfer*, vol. 112, no. 3, pp. 802–804, 1990.

115. Tan, Z., and J. R. Howell: Combined Radiation and Natural Convection in a Two-Dimensional Participating Square Medium, *Int. J. Heat and Mass Trans.*, vol. 34, no. 3, pp. 785–793, 1991.

116. Tan, Z., and J. R. Howell: Combined Radiation and Natural Convection in a Participating Medium between Horizontal Concentric Cylinders, ASME HTD vol. 106, pp. 87–93, 1989 (AIChE/ASME National Heat Transfer Conf., Philadelphia, August 1989).

117. Desrayaud, G., and G. Lauriat: Radiative Influence on the Stability of Fluids Enclosed in Vertical Cavities, *Int. J. Heat Mass Transfer*, vol. 31, no. 5, pp. 1035–1048, 1988.
118. Hassab, M. A., and M. N. Özisik: Effects of Radiation and Convective Boundary Conditions on the Stability of Fluid in an Inclined Slender Slot, *Int. J. Heat Mass Transfer*, vol. 22, pp. 1095–1105, 1979.
119. DeSoto, Simon: The Radiation from an Axisymmetric, Real Gas System with a Nonisothermal Temperature Distribution, *Chem. Eng. Progr. Symp. Ser.*, vol. 61, no. 59, pp. 138–154, 1965.
120. Penner, S. S., and D. B. Olfe: "Radiation and Reentry," Academic Press, New York, 1968.
121. Viskanta, R., and R. L. Merriam: Shielding of Surfaces in Couette Flow against Radiation by Transpiration of an Absorbing-Emitting Gas, *Int. J. Heat Mass Transfer*, vol. 10, no. 5, pp. 641–653, 1967.
122. Viskanta, Raymond, and Pritam S. Bathla: Unsteady Energy Transfer in a Layer of Gray Gas by Thermal Radiation, *Z. Angew. Math. Phys.*, vol. 18, no. 3, pp. 353–367, 1967.
123. Crosbie, A. L., and M. Pattabongse: Transient Conductive and Radiative Transfer in a Planar Layer with Arrhenius Heat Generation, *J. Quant. Spectrosc. Radiat. Transfer*, vol. 37, pp. 319–329, 1987.
124. Yoshida, H., J. H. Yun, and R. Echigo: Transient Characteristics of Combined Conduction, Convection, and Radiation Heat Transfer in Porous Media, *Int. J. Heat Mass Transfer*, vol. 33, no. 5, pp. 847–857, 1990.
125. Viskanta, R., and P. S. Lall: Transient Cooling of a Spherical Mass of High-Temperature Gas by Thermal Radiation, *J. Appl. Mech.*, vol. 32, no. 4, pp. 740–746, 1965.
126. Viskanta, R., and P. S. Lall: Transient Heating and Cooling of a Spherical Mass of Gray Gas by Thermal Radiation, in M. A. Saad and J. A. Miller (eds.), *Proc. Heat Transfer Fluid Mech. Inst.*, pp. 181–197, 1966.
127. Tsai, J. R., and M. N. Ozisik: Transient, Combined Conduction and Radiation in an Absorbing, Emitting, and Isotropically Scattering Solid Sphere, *J. Quant. Spectrosc. Radiat. Transfer*, vol. 33, no. 4, pp. 243–251, 1987.
128. Chang, Yan-Po, and R. Scott Smith, Jr.: Steady and Transient Heat Transfer by Radiation and Conduction in a Medium Bounded by Two Coaxial Cylindrical Surfaces, *Int. J. Heat Mass Transfer*, vol. 13, no. 1, pp. 69–80, 1970.
129. Viskanta, R., and E. E. Anderson: Heat Transfer in Semitransparent Solids, in Thomas F. Irvine, Jr., and James P. Hartnett (eds.), "Advances in Heat Transfer," vol. 11, pp. 317–441, Academic Press, New York, 1975.
130. Carslaw, H. S., and J. C. Jaeger: "Conduction of Heat in Solids," 2d ed., p. 80, Clarendon Press, Oxford, 1959.

PROBLEMS

16-1 A gray gas with absorption coefficient 0.2 cm^{-1} is contained between concentric gray spheres. The inner sphere has a diameter of 5 cm, a temperature of 1100 K, and emissivity of 0.2. The outer sphere has a diameter of 10 cm, a temperature of 830 K, and emissivity of 0.4. Compute the heat transfer rate from the inner to the outer sphere by using the result of the exchange factor approximation. (*Hint:* ψ_b can be found from Table 1 of [45].)

Answer: 80.2 W

16-2 The region between two infinite black parallel plates is filled with a stationary gray medium. The plate temperatures are 1000 K and 500 K, the spacing between the plates is 5 cm, and the absorption coefficient of the medium is 0.2 cm^{-1}. If the medium thermal conductivity is 0.011 W/(cm · K), what is the net energy flux being transferred between the plates?

Answer: 44.8 kW/m^2

16-3 A long chamber has opposite sides at T_1 and T_2, and the other two sides are insulated. The chamber is filled with a stationary medium that is scattering and is optically thin. The medium has thermal conductivity k. The surfaces are diffuse-gray, with emissivities as shown; the insulated walls both have the same emissivity. Develop an approximate expression for the heat transfer from A_1 to A_2. (*Hint:* see Example 14-5.)

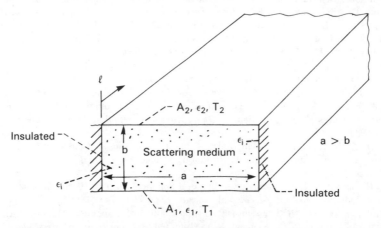

Answer:

$$\frac{Q}{al} = \frac{k}{b}(T_1 - T_2) + \frac{\sigma(T_1^4 - T_2^4)}{1/\epsilon_1 + 1/\epsilon_2 - 2F_{1-2}/(1 + F_{1-2})} \quad \text{where } F_{1-2} = \sqrt{1 + B^2} - B, B = \frac{b}{a}$$

16-4 Two infinite parallel plates are separated by an optically thick absorbing, emitting, and conducting gas in which a chemical reaction is occurring that produces a uniform energy generation rate of q''' per unit volume. The plates are gray and have temperatures and emissivities T_1, ϵ_1 and T_2, ϵ_2. Determine the heat fluxes q_1 and q_2 that must be supplied to each of the plates as a result of the combined radiation exchange and conduction between them. Use the radiative diffusion approximation, and assume that the gas is stationary and that the absorption coefficient and thermal conductivity are both constant.

16-5 A chamber wall consists of three parallel plates as shown. The surfaces are all gray, and plate 2 is sufficiently thin so that it can be considered to have uniform temperature across its thickness. What is the net heat flux from plate 1 to plate 3 if the regions between the plates are both vacuum? An absorbing, emitting, and conducting medium is now introduced between plates 2 and 3. Use the additive approximation with the diffusion method for the radiative transfer to estimate the net flux from 1 to 3 for this situation.

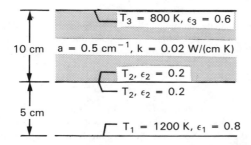

Answer: 8.66 kW/m², 9.57 kW/m²

16-6 Two infinite parallel plates are separated by a distance D. The plates are both at local temperature T_0 and have emissivity ϵ. A gas with Rosseland mean absorption coefficient a_R is flowing between the plates at a uniform velocity U parallel to the plates. The gas has density ρ and specific heat c_p. Energy is added to both plates at a rate q_w per unit area. Neglecting conduction in the gas, derive an expression for the temperature distribution in the gas using the diffusion method. A fully developed temperature profile may be assumed as an approximation.

16-7 Repeat Prob. 16-6 above, but include conduction in the gas with thermal conductivity k. For this problem, also assume (to provide a simplification) that $a_R(T) = a_{R,0}(T/T_0)^3$.

16-8 A stagnant gray gas with absorption coefficient $a = 6 \text{ m}^{-1}$ is contained between parallel plates spaced 0.15 m apart. The plates are nongray and have hemispherical spectral emissivities and temperatures as shown. The gas thermal conductivity is $k = 0.35 \text{ W}/(\text{m} \cdot \text{K})$. Compute the energy flux transferred from plate 1 to plate 2. Use the additive approximation and the diffusion solution for the radiative transfer.

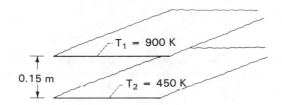

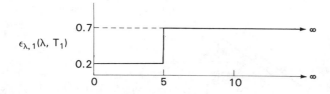

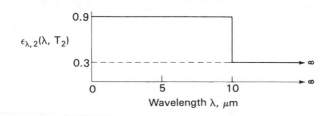

Wavelength λ, μm

Answer: 10,322 W/m²

16-9 Two gray parallel plates with emissivities $\epsilon_1 = 0.6$ and $\epsilon_2 = 0.9$ are spaced 0.15 m apart. The plate temperatures are $T_1 = 500$ K and $T_2 = 300$ K. A stagnant nongray gas with absorption coefficient as shown is in the space between the plates. The thermal conductivity of the gas is $k = 0.32$ W/(m·K). Using the additive method with the radiative diffusion solution, find the heat transfer between the plates. How significant is the heat conduction?

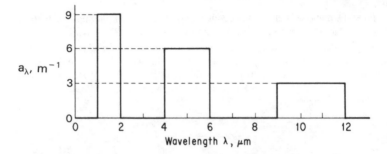

Answer: 2004 W/m^2

16-10 An absorbing liquid with absorption coefficient a and thermal conductivity k_l is flowing down a vertical flat plate. The plate is diffuse gray with absorptivity α and has insulation with thickness B and thermal conductivity k_i on the back as shown. A solar flux q_s is incident on the plate at angle θ. Formulate the relations necessary to determine the mean temperature of the liquid at the bottom of the plate. The result should be in terms of the liquid mass flow rate, plate length, plate absorptivity, and the insulation thickness and conductivity. The temperatures are low enough that emission from the plate and the liquid may be neglected. The back side of the insulation layer is subjected to a heat transfer coefficient h_a.

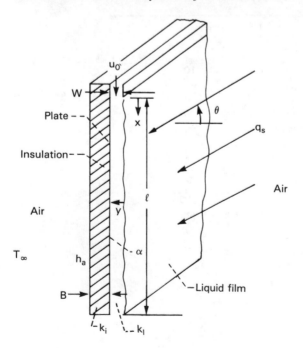

16-11 An optically thin gray gas with constant absorption coefficient a is contained in a long transparent cylinder of diameter D. The surrounding environment is at low temperature that can be considered zero. Initially, the cylinder is at the environment temperature. Then an electrical discharge is passed through the cylinder, continuously producing in the gas a uniform energy source q''' per unit volume and time. Derive a relation for the transient gas temperature variation if radiation is assumed to be the only significant mode of heat transfer. What is the maximum temperature T_m that the gas will achieve?

$$Answer: \quad \frac{8a\sigma T_{max}^3}{\rho c_v}\tau = \frac{1}{2}\ln\frac{1+\theta}{1-\theta} + \tan^{-1}\theta, \qquad T_{max} = \left(\frac{q'''}{4a\sigma}\right)^{1/4}, \qquad \theta = \frac{T}{T_{max}}$$

16-12 A partially transparent solid absorbing sphere is placed in a black cavity at T_b. The sphere temperature is initially low. The sphere has surface transmissivity τ_o, thermal conductivity k, and absorption coefficient a. Transparent gas at T_g is circulating in the cavity, providing a heat transfer coefficient h at the sphere surface. Give the equations and boundary conditions to compute the transient temperature distribution in the sphere for the initial period during which emission by the sphere is small. Neglect any refraction effects and any dependence of τ_o on angle.

16-13 A plane layer of gray isotropically scattering, absorbing, emitting, and heat conducting medium is being heated uniformly throughout its volume by chemical action that starts at time zero. The volumetric heating rate is q'''. The medium is in contact with black walls at $x = 0$ and D that are cooled to maintain them at T_w. Initially the entire medium is also at T_w. The medium has absorption coefficient a, scattering coefficient σ_s, thermal conductivity k, specific heat c, and constant density ρ. Derive the transient energy equation and the integral equation relation for the scattering source function that could be placed in finite-difference form to solve numerically for the transient temperature distributions as a function of position between the walls. Place the equations in a convenient dimensionless form that contains a parameter involving the thermal conductivity.

16-14 Modify the formulation in Prob. 16-13 to include the effect of both boundary walls being diffuse-gray with the same emissivity ϵ_w, rather than being black.

16-15 A semi-infinite medium is heat conducting, absorbing, emitting, and isotropically scattering. It is gray and has constant absorption and scattering coefficients a and σ_s and a constant thermal conductivity k. Initially the medium is at a uniform temperature T_i. The plane boundary of the medium at $x = 0$ is then suddenly exposed to a large environment at a uniform temperature $T_e < T_i$. A numerical solution is to be carried out to obtain transient temperature distributions in the medium as it cools. Provide the energy and scattering equations in convenient dimensionless forms to be placed in numerical form for solution. All thermal properties are constant, and the medium is not moving.

16-16 A plane layer of absorbing, emitting, isotropically scattering, and heat conducting material is being uniformly heated throughout its volume with volumetric heating rate q'''. The layer is of thickness D and has constant absorption and scattering coefficients a and σ_s and a constant thermal conductivity k. The layer is confined between walls that are diffuse gray. The walls at $x = 0$ and D have emissivities $\epsilon_{w,0}$ and $\epsilon_{w,D}$ and temperatures $T_{w,0}$ and $T_{w,D}$. Provide the energy and scattering equations in a convenient dimensionless form required to perform a numerical solution for the steady temperature distribution across the layer from $x = 0$ to D. (*Hint*: Integrate the energy equation two times with respect to x.)

16-17 A rectangular region is filled with an absorbing, emitting, and heat conducting medium. The medium is gray and has constant absorption coefficient a and constant thermal conductivity k. The geometry is long in the z direction, and the side dimensions in the x and y directions are d and b. The four bounding walls are black. Both vertical walls are at uniform temperatures

T_v, and the horizontal walls are at T_h. The medium is heated throughout its volume with volumetric energy generation rate $q'''(x, y)$ and is not moving. Provide the energy equation required to solve for the steady temperature distribution within the rectangular region. Place the equation in a convenient dimensionless form for use in a numerical solution.

16-18 Provide the relations needed to include isotropic scattering in the medium in Prob. 16-17. The scattering coefficient is a constant, σ_s.

SEVENTEEN

NUMERICAL SOLUTION METHODS FOR RADIATION IN PARTICIPATING MEDIA WITH OTHER ENERGY TRANSFER MODES

17-1 INTRODUCTION

In this chapter some numerical methods are described for solving the equations of radiative transfer and the energy equation in participating media when energy transfer modes or sources are present in addition to radiation. This includes heat transfer in a medium by conduction, forced convection, and free convection, and having temperature-dependent volumetric heat sources. The presentation builds on and supplements the methods in Chap. 15 and to some extent builds on the numerical methods in Chap. 11. Additional information on numerical methods is in [1, 2].

At the beginning of Chap. 16 are summarized the energy equation, equation of transfer, and the source-function equation needed to include scattering. The radiative intensities are in general spectrally dependent, and an integration of radiant energy over the spectrum is required to obtain the total energy quantities in the energy equation. The equation of transfer provides the local intensities, and these must be integrated over all directions surrounding a location to obtain the local radiative source term in the energy equation. Thus for a general situation both spectral and directional integrations are required. With scattering present, the source function must be obtained by solving the source-function equation simultaneously with the energy equation. The two unknowns are the temperature and source-function distributions.

As discussed at the beginning of Chap. 16, obtaining a solution becomes quite involved for a problem having both spectrally and spatially dependent variations.

Consequently the governing equations were developed in more detail for various simplified situations, such as a gray medium between parallel boundaries or in a two-dimensional rectangular configuration. In some instances scattering was included. Conduction and convection terms were also included. Some numerical techniques are considered here that can be used for obtaining solutions. Equations developed previously will be used as examples to show how numerical solutions can be carried out. The combination of radiation with conduction, convection, and/or temperature-dependent volumetric energy generation terms provides a non-linear energy equation, and some of the standard numerical methods may not work well. For example, the net heat conduction may be the result of a difference between large incoming and outgoing radiation terms, which can result in loss of accuracy for the conduction terms. Strongly temperature-dependent volumetric sources, as in some combustion problems, are particularly sensitive to small differences in local temperatures, and solutions are difficult to bring to convergence. Extra care is generally required to obtain a method that gives good accuracy, converges well, and can be carried out with a reasonable amount of computer time.

The energy equation contains the radiative energy source term, which depends on the energy arriving from the surrounding medium. Evaluation of the local radiative source requires an integration by some means over the medium surrounding that location. This can be done by such methods as finite elements, finite differences, discrete ordinates, Monte Carlo, or product integration methods. The integration must be carried out surrounding a sufficient number of locations within the medium to define the radiative source distribution accurately; this can be quite time consuming.

With the radiative source term evaluated, the energy equation must be solved including conduction and/or convection terms. This could be carried out, for example, by a finite-difference or finite-element method. Solution methods can involve over- or underrelaxation techniques, implicit methods, or matrix solutions of the set of simultaneous equations developed for the grid locations within the medium. It may be necessary to use two sets of grid points. A coarse grid can be used for the locations at which the radiative source term is evaluated and intermediate values of the source term within the grid can then be obtained by interpolation. The conduction and convection portions may require a locally finer grid to resolve large temperature variations in regions such as near boundaries. A helpful technique is to use variable grid size. The iterative method that can be used will depend on whether a steady-state or transient solution is required.

Each of the methods for obtaining the local radiative source distribution has advantages and disadvantages when it is applied to multimode problems. The finite-differencing schemes for the finite-element, discrete ordinate, finite-difference, and P-N methods can be made to mesh with standard finite-differencing schemes for most other heat transfer modes. The zone (using a small number of zones) and Monte Carlo methods are less easy to match with other heat transfer modes. However, the Monte Carlo scheme can be easier to formulate for a more complete treatment of the effects of nonuniform and spectral property variations in problems where these are important. The zone technique can be applied to a broad variety of practical problems because its long history of development has provided numerical values for various exchange factors.

To conclude the chapter, some comments are provided on areas of ongoing research.

17-2 SYMBOLS

a, b, c	elements of tridiagonal matrix, Eq. (17-18)
a	absorption coefficient
A	area
A_R	aspect ratio of rectangle, W/D
c_p, c_v	specific heat
D	spacing between parallel plates; height of rectangle
E_n	nth exponential integral function
f	forcing function in finite-element solution, Eq. (17-5)
F	local energy source produced by radiation absorption and internal generation
$\overline{gg}$	gas-gas element direct-exchange area
i	radiative intensity
I	dimensionless intensity
k	thermal conductivity
K	coefficient matrix in finite element solution, Eq. (17-4)
N	conduction/radiation parameter, $k(a + \sigma_s)/(4\sigma T_{\text{ref}}^3)$ [in Sec. 17-4, $N = k(a + \sigma_s)/(4\kappa_D \sigma T_{\text{ref}}^3)$] (*note:* $\sigma_s = 0$ for pure absorption)
N_x, N_y	number of x points, number of y points
q	energy flux; energy/(area · time)
q'''	volumetric energy generation rate
R	dimensionless radius in cylindrical coordinates
s	vector elements in Eq. (17-18)
$\overline{sg}$	surface-gas element direct-exchange area
$\overline{ss}$	surface-surface element direct-exchange area
S	path length; dimensionless energy source
S_r	radiative source
S_n	discrete ordinate solution method
S_1	two-dimensional radiation integral function
t	dimensionless temperature, T/T_{ref}
T	absolute temperature
U	approximation to dimensionless temperature in finite-element solution
V	volume
w	weighting factor in discrete ordinates method
W	width of rectangle; weighting function in finite-element method
x, y, z	coordinate positions

X, Y, Z	dimensionless coordinate positions, $X = x/D$, $Y = y/D$, $Z = z/D$
α, γ, δ	direction cosines
ϵ	emissivity
η	spatial weighting factor in discrete ordinates method
θ	angle in cylindrical or spherical coordinates
κ	optical depth or coordinate
κ_D	optical thickness of dimension D
σ	Stefan-Boltzmann constant
σ_s	scattering coefficient
τ	dimensionless time, $[4(a + \sigma_s)\sigma T_{\text{ref}}^3/\kappa_D \rho c_p](\text{time})$ (*note:* $\sigma_s = 0$ for pure absorption and c_v is used as appropriate)
Φ	shape function in finite-element solution; scattering phase function
ω	solid angle
Ω	albedo for single scattering; $\sigma_s/(a + \sigma_s)$

Subscripts

b	blackbody
c	conduction; corner
e	surrounding environment
i	incident; ith grid point; ith wall
m, m'	outgoing and incoming angular directions
N	total number of grid points
N, S, E, W	north, south, east, west boundaries
o	outgoing
P	at point P
r	radiation
ref	reference value
w	wall
0	initial value
$1, 2, \ldots$	various walls

Superscripts

n	nth time interval
$(0), (1), \ldots$	moments

17-3 FINITE-ELEMENT METHOD

The finite-element method discussed in Sec. 11-4.2 has been applied to various combined-mode radiative transfer problems. Introductory information is in [2, 3] and a further development is in [4]. Most of the solutions have used the Galerkin method. The finite-element method in principle gives exact solutions within the errors introduced by the numerical solution itself; that is, no approximations need to be made in formulating the physical problem. The medium is divided into convenient subvolumes (finite elements). Nodes are placed along element boundaries, and some can also be located within each element. The values of temperature (its fourth power, or other dependent variable) at the nodes are to be obtained from the solution. The distribution of temperature (or other dependent variable) is described, to the degree of accuracy desired, by using interpolation functions based on the expected form of the solution. If each element is taken to be isothermal, the method is equivalent to a zoning technique and, for small elements, the finite-difference method. In two-dimensional problems the temperature distributions in the elements are usually described by biquadratic functions. This allows a continuous temperature profile to be prescribed in the medium by matching the element boundary node temperatures to the temperatures at the boundary nodes of adjacent elements. The use of higher-order functions for the element temperatures allows additional matching of temperature slopes at the element boundary nodes.

Finite-element solutions have been obtained in the literature for problems of conduction and/or convection combined with radiation including scattering and for boundary conditions of specified temperature or heat flux. One-dimensional geometries of a plane layer or cylinder were analyzed in [5–7] for combined conduction and radiation. The latter included scattering and transient conditions. Two-dimensional rectangular geometries were analyzed in [8] for radiation only, in a gray, nonscattering medium. In [9] conduction was also included, and in [10] the solutions incorporated isotropic scattering. A boundary layer flow was analyzed in [11] using a one-dimensional model for the radiation. The medium was gray, and scattering was not included. Another analysis combining radiation with convection is for a two-dimensional flow in a diverging or converging channel with scattering included [12].

Further development of the finite-element approach appears worthwhile. The method offers the possibility of high accuracy and can facilitate using a different numerical grid for evaluating the radiation source term than is used for solving the energy equation. Most solutions to date have used a large grid size (few elements) because computer running times for this method tend to be long, especially for problems in which conduction is small compared with radiation. If a creative way can be found to use the generated temperature profiles within the individual elements to carry out analytically all or part of the radiative field integrals, the method could be greatly speeded up and perhaps used to develop radiative transfer subroutines to be incorporated into existing thermal analysis programs. Perhaps this can be done by using a partial analytical integration of the general form of the radiation integrals, followed by numerical evaluation of the remaining simplified form.

17-3.1 Finite-Element Application to the Transfer Equations

There are finite-element computer codes that can be used for the numerical solution of the sets of simultaneous equations resulting from the discretized finite-element formulation. Many codes are limited to linear problems, and an iterative solution may be necessary for nonlinear problems. This can be done by assuming an initial temperature distribution to determine the unknown temperature-dependent coefficients in a linearized formulation and then iterating until convergence. The method for carrying out such a formulation is outlined here.

For an illustrative case with heat conduction and an energy source in the medium, the energy equation for $T(x, y)$ is written for a two-dimensional gray medium with no scattering as

$$4a\sigma T^4(x, y) - k \nabla^2 T = F(x, y, T) \tag{17-1}$$

The $F(x, y, T)$ includes the local absorption of incident radiation and the specified local internal energy generation in the volume. The energy equation is written in dimensionless form as

$$(4t^3)t - 4N \nabla^2 t = F(\kappa_1, \kappa_2, t) \tag{17-2}$$

where $N = ka/4\sigma T_{\text{ref}}^3$ and ∇ is now with respect to κ. Following the method of Sec. 15-7, Eq. (17-2) is multiplied by a weighting function $W(\kappa_1, \kappa_2)$ and integrated over all solid angles to include all of the region surrounding each κ_1, κ_2 location. As in the derivation of Eq. (15-154), the approximate solution $t(\kappa_1, \kappa_2) \approx U(\kappa_1, \kappa_2)$ and the assumed form for $W(\kappa_1, \kappa_2)$ using the same shape functions as in $U(\kappa_1, \kappa_2)$ are substituted to obtain the form

$$\sum_{i=1}^{N} W_i \left(\sum_{j=1}^{N} \left\{ \int_{\omega=0}^{4\pi} [-N \nabla\Phi_j \nabla\Phi_i + t^3\Phi_j\Phi_i] \, d\omega \right\} U_j \right.$$
$$\left. - \frac{1}{4} \int_{\omega=0}^{4\pi} F(\kappa_1, \kappa_2, U)\Phi_i(\kappa_1, \kappa_2) \, d\omega \right) = 0 \tag{17-3}$$

The t^3 remains as a result of the linearized iterative method being used. Defining

$$K_{ij}(\kappa_1, \kappa_2) = \int_{\omega=0}^{4\pi} [-N \nabla\Phi_j \nabla\Phi_i + t^3\Phi_j\Phi_i] \, d\omega \tag{17-4}$$

and

$$f_i(\kappa_1, \kappa_2) = \frac{1}{4} \int_{\omega=0}^{4\pi} F(\kappa_1, \kappa_2, U)\Phi_i(\kappa_1, \kappa_2) \, d\omega \tag{17-5}$$

Eq. (17-3) is written as

$$\sum_{i=1}^{N} W_i \left(\sum_{j=1}^{N} K_{ij}(\kappa_1, \kappa_2)U_j - f_i(\kappa_1, \kappa_2) \right) = 0 \tag{17-6}$$

The quantities under the integrals in (17-4) and (17-5) depend on κ_1, κ_2 and on dummy integration coordinates over the two-dimensional region surrounding the location κ_1, κ_2 [see, for example, Eq. (17-19)]. Because the W_i are arbitrary, (17-6) provides N simultaneous equations to be satisfied by the values of U_j. Thus, (17-6) can also be written as

$$\sum_{j=1}^{N} K_{ij}(\kappa_1, \kappa_2)U_j = f_i(\kappa_1, \kappa_2) \qquad i = 1, 2, 3, \ldots, N \qquad (17\text{-}7)$$

The solution to the set of equations (17-7) can be obtained using an available finite-element routine, as the [K] matrix and the **f** vector are defined. Note, however, that unlike the pure radiation case in (15-155), the [K] matrix contains the unknown t^3 values; thus, the problem is iterative as well as nonlinear. In the iteration the t from the previous iteration is used in the evaluation of the K_{ij}. The numerical solution methods outlined in Sec. 11-5 can be used.

17-3.2 Some Results of Finite-Element Analyses

Razzaque et al. [9] obtained solutions for combined radiation and conduction in a medium in a rectangular enclosure with prescribed boundary temperatures and emissivities. For the particular conditions studied, the finite-element method required modification. The square enclosures had a fixed temperature on one wall and a lower fixed temperature on the other three walls. These conditions were chosen because solutions for the pure radiation case by the P-N and zone methods are available for comparison [13, 14]. However, the finite-element method requires boundary conditions as continuous functions and, in this instance, the boundary conditions are discontinuous at the enclosure corners. The boundary conditions were modified so that in the finite elements containing corners, the temperatures on the element boundary were forced to approach the average of the two wall temperatures. For a real situation including conduction, this is more realistic than the original problem, because with conduction in the medium it is not possible to have a temperature discontinuity on the boundary. This change in the boundary conditions should be noted in comparing these solutions with other work. The solutions are exact within the accuracy of the numerical method, because the energy equation contained the exact integral for the local energy source. Results for centerline temperatures in the medium with N as a parameter are in Fig. 17-1. Comparisons with the P-3 and zone results indicate that the change in corner conditions had little effect on the centerline profiles for the parameters shown.

Chung and Kim [12] included isotropic scattering and convection in an analysis of heat transfer in convergent and divergent channels. The finite-element scheme parallels that outlined above but includes convection. Shape functions with four nodes were used. A Newton-Raphson method was used to solve the nonlinear matrix of the form given by (17-7). Solutions are presented for a set of cases with velocity profiles specified, so the energy equation could be solved directly.

A compressible boundary layer with combined conduction, convection, and radiation was studied in [11] using an optimal control penalty method to speed convergence. This method reduces the order of the differential equations by replacing them with a set of lower-order equations and a set of constraint equations, which can be solved in finite-element form. For boundary layer flows, this method produces a well-conditioned positive-definite matrix of the K_{ij} coefficients.

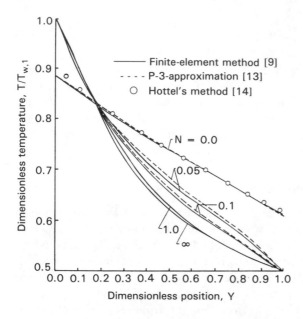

Figure 17-1 Nondimensional centerline temperature profiles in a square enclosure for various conduction-radiation parameters N and with black walls; optical length of side $\kappa_D = 1.0$; hot wall nodes at dimensionless temperature, $t_{w,1} = 1.0$, corner nodes at hot-cold intersections at $t_c = 0.75$, other wall nodes at $t_{w,2} = 0.5$. From [9].

17-4 FINITE-DIFFERENCE METHODS

The energy relations to determine the temperature distribution in an absorbing, emitting, and scattering medium were summarized in Sec. 16-3. Some simplified situations, such as a gray absorbing medium between parallel plates or in a two-dimensional rectangular shape, provided Eqs. (14-87), (16-92), and (14-137). If the local radiative heat source term is called S_r, the energy equation for transient temperature variations in a gray absorbing medium without scattering or internal heat generation has the general form (∇ is with respect to X, Y, and Z)

$$\frac{\partial t}{\partial \tau} = N \nabla^2 t - S_r(t) \tag{17-8}$$

where $N = ka/4\kappa_D \sigma T_{\text{ref}}^3$ (in Sec. 17-4 only), $\tau = 4$ (time)$(a + \sigma_s)\sigma T_{\text{ref}}^3/\kappa_D \rho c_v$, and $t = T/T_{\text{ref}}$. The most convenient set of dimensionless quantities to use depends on the problem under study. An implicit finite-difference method is now discussed for transient and steady-state solutions. This is an example of a numerical procedure that is stable for forward integration in time. A subscript and superscript notation is used, $t_{i,j,k}^n$, where n refers to the nth time interval and i, j, k are grid locations in the x, y, z directions.

In the solution, two operations are necessary, a forward integration in time or iteration for a steady solution, and a spatial integration to evaluate the local radiative source S_r. To integrate forward in time, the relation $t = \int(\partial t/\partial \tau)\,d\tau$ can be approximated in various ways. The trapezoidal rule is convenient, and this gives

$$t^{n+1} = t^n + \frac{\Delta\tau}{2}\left(\frac{\partial t^{n+1}}{\partial\tau} + \frac{\partial t^n}{\partial\tau}\right)$$

$$= t^n + \frac{\Delta\tau}{2}(N\nabla^2 t^{n+1} - S_r^{n+1} + N\nabla^2 t^n - S_r^n) \tag{17-9}$$

as described in [15]. The S_r^{n+1} is now expressed by a linearized expansion away from S_r^n,

$$S_r^{n+1} = S_r^n + \frac{dS_r^n}{dt}\Delta t, \qquad \Delta t = t^{n+1} - t^n \tag{17-10}$$

The $\nabla^2 t^{n+1}$ is rewritten in terms of t^n by using the identity

$$\nabla^2 t^{n+1} = \nabla^2(t^{n+1} - t^n) + \nabla^2 t^n = \nabla^2(\Delta t) + \nabla^2 t^n \tag{17-11}$$

Equations (17-10) and (17-11) are substituted into (17-9), and the result simplifies to

$$\left(1 + \frac{\Delta\tau}{2}\frac{\partial S_r^n}{\partial t} - \frac{\Delta\tau}{2}N\nabla^2\right)\Delta t = \Delta\tau(N\nabla^2 t^n - S_r^n) \tag{17-12}$$

Using the values at time increment n, Eq. (17-12) yields a matrix of equations to solve for Δt at each grid point. Then the new t at each grid point is obtained as $t^{n+1} = t^n + \Delta t$.

Before giving some illustrative examples, the equation corresponding to (17-12) is given that can be used to obtain a *steady-state solution* by iteration from an initial guessed temperature distribution. The steady energy equation is

$$N\nabla^2 t = S_r(t) \tag{17-13}$$

The linearized expansion for the $(n + 1)$st iteration in terms of the nth iteration gives

$$N\nabla^2 t^{n+1} = S_r(t^n) + \frac{dS_r^n}{dt}\Delta t \tag{17-14}$$

By equating the $\nabla^2 t^{n+1}$ given by Eqs. (17-11) and (17-14),

$$N\nabla^2(\Delta t) + N\nabla^2 t^n = S_r^n + \frac{dS_r^n}{dt}\Delta t$$

This gives the equation to solve for Δt as

$$\left(\frac{dS_r^n}{dt} - N\nabla^2\right)\Delta t = N\nabla^2 t^n - S_r^n \tag{17-15}$$

17-4.1 Analysis for a Plane Layer

To illustrate the use of Eq. (17-12), consider a plane layer of gray radiating and conducting medium between parallel plates as in Sec. 16-4.2 and Fig. 17-2. Initially the temperature has a uniform value T_0. Let this be the reference temperature so that at $\tau = 0$, $t_0 = 1$. The lower and upper wall temperatures are then suddenly changed to $t_w(0)$ and $t_w(\kappa_D)$. The transient temperature distribution is to be obtained.

From Eq. (16-13) (note different definition of N), the radiation source at $X = x/D$ for this situation is

$$S_r(t) = \kappa_D \left\{ t^4(X) - \frac{1}{2} \left[t_1^4 E_2(\kappa_D X) + t_2^4 E_2[\kappa_D(1 - X)] \right] \right.$$

$$\left. + \kappa_D \left\{ \int_0^X t^4(X^*) E_1[\kappa_D(X - X^*)] \, dX^* + \int_X^1 t^4(X^*) E_1[\kappa_D(X^* - X)] \, dX^* \right\} \right\}$$

(17-16a)

and by differentiation,

$$\frac{dS_r}{dt} = 4\kappa_D \left[t^3(X) - \frac{\kappa_D}{2} \left\{ \int_0^X t^3(X^*) E_1[\kappa_D(X - X^*)] \, dX^* \right. \right.$$

$$\left. \left. + \int_X^1 t^3(X^*) E_1[\kappa_D(X^* - X)] \, dX^* \right\} \right]$$

(17-16b)

The heat conduction term is placed in finite-difference form,

$$\frac{\partial^2 t_i}{\partial X^2} = \frac{t_{i-1} - 2t_i + t_{i+1}}{(\Delta X)^2}$$

The specified boundary conditions give $t_1 = t_w(0)$ and $t_N = t_w(\kappa_D)$. Since these values are fixed, $\Delta t_1 = \Delta t_N = 0$. Equation (17-12) is then applied at each of the interior points $i = 2, 3, \ldots, N - 1$. For example, at $i = 3$,

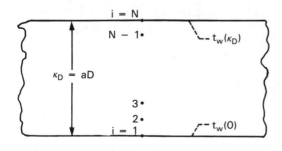

Figure 17-2 Plane layer for transient analysis using finite differences.

$$\left(1 + \frac{\Delta\tau}{2}\frac{dS^n_{r,3}}{dt}\right)\Delta t^n_3 - \frac{\Delta\tau}{2}N\frac{\Delta t^n_2 - 2\Delta t^n_3 + \Delta t^n_4}{(\Delta X)^2}$$

$$= \Delta\tau\left(N\frac{t^n_2 - 2t^n_3 + t^n_4}{(\Delta X)^2} - S^n_{r,3}\right) \tag{17-17}$$

Similar equations for all of the interior points yield the following tridiagonal matrix for the Δt_i values:

$$\begin{bmatrix} 1 + \dfrac{\Delta\tau}{2}\dfrac{dS^n_{r,2}}{dt} + \Delta\tau\dfrac{N}{(\Delta X)^2} & -\dfrac{\Delta\tau}{2}\dfrac{N}{(\Delta X)^2} & & & \\[3ex] -\dfrac{\Delta\tau}{2}\dfrac{N}{(\Delta X)^2} & 1 + \dfrac{\Delta\tau}{2}\dfrac{dS^n_{r,3}}{dt} + \Delta\tau\dfrac{N}{(\Delta X)^2} & -\dfrac{\Delta\tau}{2}\dfrac{N}{(\Delta X)^2} & & \\[3ex] & & \ddots & & \\[1ex] & & -\dfrac{\Delta\tau}{2}\dfrac{N}{(\Delta X)^2} & 1 + \dfrac{\Delta\tau}{2}\dfrac{dS^n_{r,N-1}}{dt} + \Delta\tau\dfrac{N}{(\Delta X)^2} \end{bmatrix} \begin{bmatrix} \Delta t^n_2 \\[2ex] \Delta t^n_3 \\[2ex] \vdots \\[2ex] \Delta t^n_{N-1} \end{bmatrix}$$

$$= \begin{bmatrix} \Delta\tau\left(N\dfrac{t_w(0) - 2t^n_2 + t^n_3}{(\Delta X)^2} \quad S^n_{r,2}\right) \\[3ex] \Delta\tau\left(N\dfrac{t^n_2 - 2t^n_3 + t^n_4}{(\Delta X)^2} - S^n_{r,3}\right) \\[3ex] \vdots \\[3ex] \Delta\tau\left(N\dfrac{t^n_{N-2} - 2t^n_{N-1} + t_w(\kappa_D)}{(\Delta X)^2} - S^n_{r,N-1}\right) \end{bmatrix}$$

This can be written in the more compact form

$$\begin{bmatrix} b_2 & c_2 & & & \\ a_3 & \ddots & \ddots & & \\ & \ddots & \ddots & c_{N-2} \\ & & a_{N-1} & b_{N-1} \end{bmatrix} \begin{bmatrix} \Delta t^n_2 \\ \Delta t^n_3 \\ \vdots \\ \Delta t^n_{N-1} \end{bmatrix} = \begin{bmatrix} s_2 \\ s_3 \\ \vdots \\ s_{N-1} \end{bmatrix} \tag{17-18}$$

$$a_i = -\frac{\Delta\tau}{2}\frac{N}{(\Delta X)^2}, \qquad b_i = 1 + \frac{\Delta\tau}{2}\frac{dS^n_{r,i}}{dt} + \Delta\tau\frac{N}{(\Delta X)^2}, \qquad c_i = -\frac{\Delta\tau}{2}\frac{N}{(\Delta X)^2}$$

$$s_i = \Delta\tau\left(N\frac{t^n_{i-1} - 2t^n_i + t^n_{i+1}}{(\Delta X)^2} - S^n_{r,i}\right) \qquad 2 \le i \le N - 1$$

where for $i = 2$, $t_{i-1} = t_1 = t_w(0)$ and for $i = N - 1$, $t_{i+1} = t_N = t_w(\kappa_D)$. The tridiagonal matrix (17-18) is solved by the following well-known algorithm where f and g are first computed from

$$f_2 = \frac{c_2}{b_2} \qquad\qquad g_2 = \frac{s_2}{b_2}$$

$$f_i = \frac{c_i}{b_i - a_i f_{i-1}} \qquad g_i = \frac{s_i - a_i g_{i-1}}{b_i - a_i f_{i-1}} \qquad 3 \le i \le N - 1$$

and the Δt_i values are then obtained recursively from the relations

$$\Delta t_{N-1} = g_{N-1}, \qquad \Delta t_i = g_i - f_i \, \Delta t_{i+1}, \qquad 2 \le i \le N - 2$$

The values of t at the next time step are $t^{n+1} = t_i^n + \Delta t_i$ for $2 \le i \le N - 2$.

For a steady situation, the transient solution can be continued to steady state, or Eq. (17-15) can be used. For the latter, the iteration is started by making a guess of the steady temperature distribution. The implicit iteration of Eq. (17-15) yields a tridiagonal matrix of the same form as (17-18). The coefficients are

$$a_i = - \frac{N}{(\Delta X)^2}, \qquad b_i = \frac{dS_{r,i}^n}{dt} + \frac{2N}{(\Delta X)^2}, \qquad c_i = - \frac{N}{(\Delta X)^2}$$

$$s_i = N \frac{t_{i-1}^n - 2t_i^n + t_{i+1}^n}{(\Delta X)^2} - S_{r,i}^n$$

At each step of the forward integration or iteration in these solutions, the S_r and dS_r/dt values must be obtained at each grid location. These can be evaluated by numerical integration of (17-16a) and (17-16b), or other techniques such as finite elements or discrete ordinates can be used as described in this chapter. Evaluation by numerical integration requires an accurate integration technique. Since $E_1(0) = \infty$, care must be taken as X^* approaches X. Since the integral of E_1 is $-E_2$, and $E_2(0) = 1$, the integration is carried out analytically over a very small region near the singularity, with t constant over this region. The integration in the regions away from the singularity can be done by Gaussian integration using library subroutines. The routines require values at unevenly spaced points. These are obtained by interpolation using a cubic spline curve-fitting subroutine and the values of the function at the grid points. A one-dimensional transient analysis using a Crank-Nicolson type of procedure and trapezoidal integration is in [16]. A transient cooling analysis for a plane layer in [16a] used Gaussian integration to obtain the local radiative source term and a finite-difference procedure with variable space and time increments to solve the transient energy equation.

17-4.2 Analysis for a Two-Dimensional Rectangular Region

As another example, consider a two-dimensional transient analysis of a gray absorbing-emitting rectangular region, as in Fig. 17-3a, that is initially at uniform

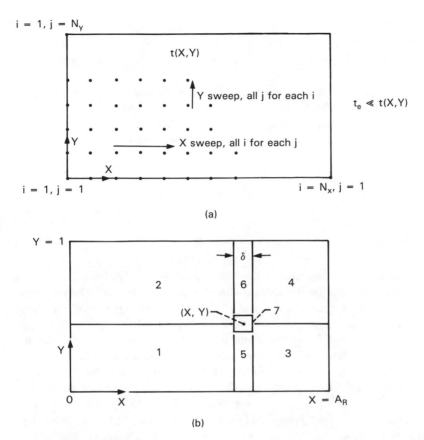

Figure 17-3 Two-dimensional rectangular region for cooling analysis by finite differences of combined radiation and conduction. (a) Grid for ADI method; (b) Regions for double integration in energy equation.

temperature. The rectangle has height D, width W (aspect ratio $A_R = W/D$), and optical height $aD = \kappa_D$. To initiate the transient, the region is placed into vacuum at a much lower temperature. For these conditions, energy can leave the rectangle only by radiation because there is no surrounding cooling medium. As the cooling proceeds, heat conduction partially equalizes the transient temperature distribution.

The transient energy equation including conduction and radiation is obtained from Eqs. (17-8) and (14-138) as

$$
\frac{\partial t}{\partial \tau} = N \left(\frac{\partial^2 t}{\partial X^2} + \frac{\partial^2 t}{\partial Y^2} \right) - \kappa_D \Bigg[t^4(X, Y, \tau)
$$

$$
- \frac{\kappa_D}{4} \int_{X^*=0}^{A_R} \int_{Y^*=0}^{1} t^4(X^*, Y^*, \tau) \frac{S_1(\kappa_D R)}{R(X, Y, X^*, Y^*)} \, dX^* \, dY^* \Bigg] \tag{17-19}
$$

This has the same form as (17-8) and hence

$$S_r(X, Y, \tau) = \kappa_D \left[t^4(X, Y, \tau) - \frac{\kappa_D}{4} \int_{X^*=0}^{A_R} \int_{Y^*=0}^{1} t^4(X^*, Y^*, \tau) \frac{S_1(\kappa_D R)}{R} dX^* dY^* \right]$$

Equation (17-12) can now be directly applied to find $\Delta t = t^{n+1} - t^n$ at each grid point within the region. The alternating direction implicit (ADI) method will be used. This is described in [17]; the method here differs somewhat as given in [15]. The ∇^2 operator is split into each of the coordinate directions, and Eq. (17-12) is approximated by

$$\left(1 + \frac{\Delta \tau}{2} \frac{dS_r^n}{dt} - \frac{\Delta \tau}{2} N \frac{\partial^2}{\partial X^2} \right) \Delta \varphi = \Delta \tau \left(N \nabla^2 t^n - S_r^n \right) \tag{17-20a}$$

$$\left(1 - \frac{\Delta \tau}{2} N \frac{\partial^2}{\partial Y^2} \right) \Delta t = \Delta \varphi \tag{17-20b}$$

To move ahead one time increment, the first equation is solved for $\Delta \varphi$; this is then used on the right side of the second equation to solve for Δt. The rectangle is covered with a square grid as shown in Fig. 17-3a. A sweep is made in the X direction for each j to obtain the $\Delta \varphi_{i,j}^n$ for all i. In this situation the boundary temperatures are unknown. Radiation from within the volume passes through the boundaries but cannot interact with conduction exactly at the boundary because the boundary itself has no volume. Heat transfer by conduction is thus zero at all boundaries, so the normal temperature derivative is zero at the boundaries. The matrix for each X sweep for $\Delta \varphi_{i,j}^n$, $1 \leq i < N_x$, has the same form as (17-18). A few coefficients are different because (17-18) was written for known boundary temperatures. The values are

$$a_i = -\frac{N}{2} \frac{\Delta \tau}{(\Delta X)^2}, \quad 2 \leq i \leq N_x - 1 \qquad a_{N_x} = -N \frac{\Delta \tau}{(\Delta X)^2}$$

$$b_i = 1 + \frac{\Delta \tau}{2} \frac{dS_{r,i}^n}{dt} + N \frac{\Delta \tau}{(\Delta X)^2} \qquad 1 \leq i \leq N_x$$

$$c_1 = -N \frac{\Delta \tau}{(\Delta X)^2} \qquad c_i = -\frac{N}{2} \frac{\Delta \tau}{(\Delta X)^2} \qquad 2 \leq i \leq N_x - 1$$

The $s_{i,j}$ on the right side of the equation become

$$s_{1,j} = \Delta \tau \left[\frac{2N}{(\Delta X)^2} (t_{2,j}^n - t_{1,j}^n) + \frac{N}{(\Delta Y)^2} (t_{1,j-1}^n - 2t_{1,j}^n + t_{1,j+1}^n) - S_{r,1,j}^n \right],$$

$$2 \leq j \leq N_y - 1$$

$$s_{i,j} = \Delta \tau \left[\frac{N}{(\Delta X)^2} (t_{i-1,j}^n - 2t_{i,j}^n + t_{i+1,j}^n) + \frac{N}{(\Delta Y)^2} (t_{i,j-1}^n - 2t_{i,j}^n + t_{i,j+1}^n) - S_{r,i,j}^n \right],$$

$$2 \leq i \leq N_x - 1 \qquad 2 \leq j \leq N_y - 1$$

$$s_{N_x,j} = \Delta\tau\left[\frac{2N}{(\Delta X)^2}(t^n_{N_x-1,j} - t^n_{N_x,j}) + \frac{N}{(\Delta Y)^2}(t^n_{N_x,j-1} - 2t^n_{N_x,j} + t^n_{N_x,j+1}) - S^n_{N_x,j}\right],$$

$$2 \le j \le N_y - 1$$

For $j = 1$, replace $t^n_{i,j-1} - 2t^n_{i,j} + t^n_{i,j+1}$ by $2(t^n_{i,2} - t^n_{i,1})$

For $j = N_y$, replace $t^n_{i,j-1} - 2t^n_{i,j} + t^n_{i,j+1}$ by $2(t^n_{i,N_y-1} - t^n_{i,N_y})$

After the $\Delta\varphi^n_{i,j}$ are obtained, they are used to sweep along each set of grid points in the Y direction by use of Eq. (17-20b). The matrix form for the Y sweep is

$$
\begin{bmatrix}
1 + N\dfrac{\Delta\tau}{(\Delta Y)^2} & -N\dfrac{\Delta\tau}{(\Delta Y)^2} & & & \\
-\dfrac{N}{2}\dfrac{\Delta\tau}{(\Delta Y)^2} & 1 + N\dfrac{\Delta\tau}{(\Delta Y)^2} & -\dfrac{N}{2}\dfrac{\Delta\tau}{(\Delta Y)^2} & & \\
& & \ddots & & \\
& & & -N\dfrac{\Delta\tau}{(\Delta Y)^2} & 1 + N\dfrac{\Delta\tau}{(\Delta Y)^2}
\end{bmatrix}
\begin{bmatrix}
\Delta t_{i,1} \\
\Delta t_{i,2} \\
\vdots \\
\Delta t_{i,N_y}
\end{bmatrix}
=
\begin{bmatrix}
\Delta\varphi_{i,1} \\
\Delta\varphi_{i,2} \\
\vdots \\
\Delta\varphi_{i,N_y}
\end{bmatrix}
$$

This yields all the $\Delta t_{i,j}$; the t at the new time $\tau + \Delta\tau$ are then $t^{n+1}_{i,j} = t^n_{i,j} + \Delta t^n_{i,j}$.

Since the function $S_1(\kappa_D R)$ in the integrands for the local radiative source S_r is well behaved as $R \to 0$ (see Appendix E-2), the integrands appear to be singular because of the $1/R$ factor when the integration variables X^*, Y^* approach grid point X, Y. The integrands are actually not singular, as is evident by using cylindrical coordinates R, θ about X, Y. The $dX\,dY$ becomes $R\,dR\,d\theta$, and the $1/R$ is thus removed. However, when using rectangular coordinates for purposes of numerical integration, the apparent singularity must be dealt with for small R. For this reason, the integration about each grid point X, Y is divided into seven regions as shown in Fig. 17-3b. Region 7 is a small square of width less than one grid spacing. The integration in this region is carried out in cylindrical coordinates over a circle of equal area. For the other integrations, the cross section can be covered with a square grid of equally spaced points. Two-dimensional cubic spline fits are made of $t^4(X, Y, \tau)$ and $t^3(X, Y, \tau)$ using IMSL routines BSNAK and BS2IN. The spline coefficients are used to interpolate values at locations between the grid points as called for by the two-dimensional integration subroutines. A routine such as QAND from IMSL can be used, which inserts additional integration points (up to 256 in each coordinate direction) to try to evaluate the integral within a specified error limit.

Near boundaries there can be larger temperature gradients than in the interior of a medium. This would occur during transient cooling by radiation and conduction or with convection boundary layers present. Additional grid points are necessary in the large-gradient regions for an accurate representation of the conduction or convection portions of the solution. The radiative source term can be evaluated using a coarser grid and intermediate values interpolated at the finer grid point locations for the local solution of the energy equation. It will usually save computer time to use a variable grid size [17a] having relatively more points in regions where the temperature is changing most rapidly. A square grid was used here for simplicity in presentation.

17-5 DISCRETE ORDINATE METHOD

The numerical methods for multidimensional discrete ordinate solutions were originally developed for neutron transport calculations in reactor physics [18, 19]. In the past 15 years, the methods have been applied to some multidimensional radiative transport problems; examples are [20–25]. A neutron transport code (DOT-IV) was used in [26] for solving an axisymmetric combined conduction and radiation problem with internal heat generation. The discrete ordinate equations were introduced in Sec. 15-6.2 for problems with radiation as the only mode of heat flow. The solution yields intensities within a radiating medium as a function of location and direction. At each location, the integration of intensities over all angular directions provides the local radiative heat source. This can be used in the energy equation for multimode problems with conduction and/or convection included. The discrete ordinate solution is carried out simultaneously with the solution of the energy equation to provide the distribution of the local radiative source, $-\nabla \cdot \mathbf{q}_r$, and the temperature distribution.

Solutions to *multimode* problems using the discrete ordinate method are rather few in number, but this is an area of current investigation. Kumar et al. [24] provide a scheme for coupling the solution of the radiative transfer equation by the discrete ordinate method to the energy equation by use of the local emissive power of the medium. Effectively, the energy differential equation is solved simultaneously with the set of first-order differential equations that result from the discrete ordinate representation of the equation of transfer in the region.

Khalil [27] applied the S_4 discrete ordinate method to predict incident radiative fluxes in a rectangular test furnace $2 \times 2 \times 6$ m in size, with flow along the long wall direction and with a known temperature profile in the flowing gas. Good agreement was obtained with experimental heat fluxes and with calculations using the zone method. A second comparison gave good agreement for an axisymmetric furnace, again calculating heat fluxes from known temperature distributions. When the S_4 model was used to predict both the temperature distribution and surface fluxes in a $4 \times 4 \times 11$ m test furnace, comparisons with experiments were less accurate. For this comparison, the furnace gas was assumed to be gray with an absorption coefficient $a = 0.3$ m^{-1}.

In a numerical solution by the S_n method, the intensity is obtained in discrete directions covering the 4π solid angle about each coordinate direction at each location in the volume. As discussed in Sec. 15-6.2, intensities along each positive and negative ordinate direction around each node must be found so an even number of simultaneous equations must be solved at each node. The solution is denoted S_2, S_4, $S_6, \ldots$, where the subscript gives the order of the solution. The order is related to the number of discrete ordinates required. For a D-dimensional problem ($D = 1, 2, 3$) the number of simultaneous equations is related to n by $2^D n(n + 2)/8$. The angular integrals of intensity are expressed as a weighted finite sum in terms of these directions by using an appropriate integration method as indicated by Eq. (15-145). The resulting discrete ordinate equations for intensity are applied locally within the medium by using a control volume technique, such as is used for convection problems [28]. By this technique, the discrete ordinate method can be incorporated into existing computer codes based on the control volume method. This provides a means to analyze processes having radiation combined with convection and/or conduction.

To illustrate the numerical method, consider a two-dimensional medium as in Fig. 17-4a. The volume has been divided into rectangular regions, with a typical control volume centered at $\mathbf{r}_v$. A typical wall element is at $\mathbf{r}_w$. For simplicity, the walls are assumed diffuse and gray. The boundary condition for the intensity leaving the wall is then given by

$$
i'_o(0) = \epsilon_w i'_{b,w} + \frac{1 - \epsilon_w}{\pi} \int_{\omega_i=0}^{2\pi} i'_i \cos\theta \, d\omega_i \tag{17-21}
$$

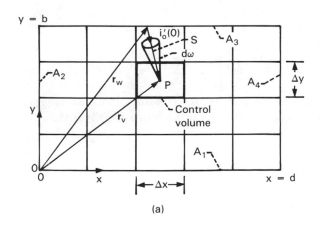

(a)

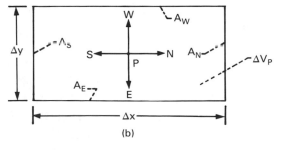

(b)

Figure 17-4 Two-dimensional radiating medium and typical control volume. (*a*) Region divided into control volumes; (*b*) typical control volume.

where $i'_o(0)$ is the intensity leaving location $\mathbf{r}_w$ in the direction $d\omega$ toward the typical control volume. The two terms on the right are the contributions to the outgoing intensity by emission from the wall and by diffuse reflection of incoming energy. The i'_i is the incoming intensity arriving at angle θ relative to the normal to the wall. The $i'_o(0)$ is at the origin of a path S going from the wall element to the control volume. The intensity arriving at the control volume depends on this outgoing initial intensity $i'_o(0)$ and on the emission and scattering along the path. The equation of transfer (Eq. 14-4)) along the path in the direction of $d\omega$ is (α and δ are direction cosines relative to the x and y directions)

$$\frac{di'}{dS} = \frac{\partial i'}{\partial x}\frac{dx}{dS} + \frac{\partial i'}{\partial y}\frac{dy}{dS} = \alpha\frac{\partial i'}{\partial x} + \delta\frac{\partial i'}{\partial y}$$

$$= ai'_b(S) - (a + \sigma_s)i'(S, \omega) + \frac{\sigma_s}{4\pi}\int_{\omega_i=0}^{4\pi} i'(S, \omega_i)\Phi(\omega, \omega_i)\,d\omega_i \tag{17-22}$$

17-5.1 Discrete Ordinate Form of Equations

In the discrete ordinates method the previous two equations are each written for a finite number of n ordinate directions. The integrals over angular directions are approximated by a weighted sum of the angular quantities as in Eqs. (15-145) and (15-146). Let m and m' correspond to outgoing and incoming angular directions. Then the equation of transfer is written as

$$\alpha_m\frac{\partial i'_m}{\partial x} + \delta_m\frac{\partial i'_m}{\partial y} = ai'_b(S) - (a + \sigma_s)i'_m(S, \omega_m) + \frac{\sigma_s}{4\pi}\sum_{m'} w_{m'}i'_{m'}\Phi_{m'm} \tag{17-23}$$

The α_m and δ_m are the direction cosines of i' for the mth direction. Similarly, the equation at the boundary is typically given as (for the horizontal boundary at $y = b$ in Fig. 17-4a)

$$i'_m = \epsilon_w i'_{b,w} + \frac{1 - \epsilon_w}{\pi}\sum_{m'} w_{m'}i'_{m'}|\delta_{m'}| \tag{17-24}$$

These equations are m coupled partial differential equations for the m intensities i'_m at each location. The coupling occurs through scattering and through the energy equation that governs the i'_b emission term. In terms of the intensity distribution over angle, the average intensity for the local radiant heat source term in the energy equation is written as

$$\bar{i} = \frac{1}{4\pi}\int_{\omega=0}^{4\pi} i'(\omega)\,d\omega = \frac{1}{4\pi}\sum_{m'} w_{m'}i'_{m'} \tag{17-25}$$

where $w_{m'}$ are quadrature weighting functions and $i'_{m'}$ are incoming intensities at the location selected.

17-5.2 Control Volume Equations

A typical control volume in the radiating medium is in Fig. 17-4b. The four sides are labeled N, E, S, W (as has been used in the literature), corresponding to the directions north, east, south, and west, where north is in the x direction. To set up the numerical solution, a control volume form for the equation of transfer is derived by multiplying Eq. (17-23) by $dx\,dy$ and integrating over the control volume to obtain

$$\alpha_m(i'_{mN}A_N - i'_{mS}A_S) + \delta_m(i'_{mW}A_W - i'_{mE}A_E)$$

$$= a\,\Delta V_P\,i'_{bP} - (a + \sigma_s)\,\Delta V_P\,i'_{mP} + \Delta V_P\frac{\sigma_s}{4\pi}\sum_{m'}w_{m'}i'_{m'P}\Phi_{m'm} \tag{17-26}$$

The scattering source term for the ordinate directions is computed from the scattering phase function. If the scattering is highly anisotropic, a high-order S_n solution may be required for accuracy. The numerical solution to the equation of transfer for each of the m directions is iterative because the volume emission, scattering source terms, and boundary conditions depend on the intensities. The calculations can be started by assuming the boundaries are black, the medium is at uniform temperature, and the scattering terms are zero. The radiant intensities are then computed as initiated by wall emission. For the iterations that follow, the full boundary equations and source terms are used. During an iteration, a solution of the control volume equation together with the boundary conditions is found for each x and y by traversing from point to point. The solution proceeds by computing intensities at all the x values for a given y and then advancing to a new y. At each y the intensities are obtained at each x for all of the m directions before going on to a new x value. Values for all the control volumes are recalculated until the intensity and wall flux values converge.

Consider the iterative process at a location and for directions for which both direction cosines are positive. Before using the control volume equation, the number of unknowns is reduced by relating the radiant fluxes at the control volume boundaries to the radiant flux at the center of the control volume. A spatially weighted approximation is written as

$$i'_{mP} = \eta i'_{mN} + (1 - \eta)i'_{mS} = \eta i'_{mW} + (1 - \eta)i'_{mE} \tag{17-27}$$

A value of $\eta = 1/2$ is usually used, which corresponds to the so called "diamond difference" relations proposed by Carlson and Lathrop [19]. If the calculation is going in the direction of positive direction cosines, the i'_{mS} and i'_{mE} are assumed known, then Eq. (17-27) is used to eliminate the intensities i'_{mN} and i'_{mW} in Eq. (17-26). Solving for the intensity i'_{mP} at the center of the control volume yields

$$i'_{mP} = \{\alpha_m[A_N(1 - \eta) + A_S\eta]i'_{mS} + \delta_m[A_W(1 - \eta) + A_E\eta]i'_{mE} + \eta a\,\Delta V_P\,i'_{bP}$$

$$+ \eta\,\Delta V_P\frac{\sigma_s}{4\pi}\sum_{m'}w_{m'}i'_{m'P}\Phi_{m'm}\} \div [\alpha_m A_N + \delta_m A_W + \eta(a + \sigma_s)\,\Delta V_P]$$

$$\tag{17-28}$$

Equation (17-28) is used when both direction cosines are positive and the integration proceeds in a direction of increasing x and y. For negative direction cosines, the solution direction is reversed. If $\delta_m < 0$ the integration proceeds from the upper to the lower boundary; the subscript S is replaced by N. Similar considerations occur when α_m is negative and when both α_m and δ_m are negative.

There are some recommendations in the literature to help obtain physically valid solutions and minimize errors. Because Eq. (17-27) is an extrapolation across a control volume, negative intensities can result when the extinction coefficient is large and the control volume is not sufficiently small. Negative intensities are replaced by zero. A recommended set of weights and ordinate directions for a one-dimensional geometry is provided in [29] for S_2, S_4, S_6, S_8, S_{10}, and S_{12} approximations; the set is in Table 17-1. In [21] a set of ordinates is recommended to improve accuracy for the S_2 and S_4 approximations. The ordinates have been arranged to satisfy a half-range flux condition which states that for a uniform intensity, the weighting factors must obey the sums $\Sigma_{\alpha_m>0}\, w_m\alpha_m = \pi$ and $\Sigma_{\alpha_m>0}\, w_m\delta_m = \pi$. The accuracy of a six-flux model is examined in [30].

Results for the S_4, S_6, and S_8 discrete ordinates method are in [20] for an anisotropically scattering gray participating medium in an idealized three-dimensional furnace enclosure. Comparisons are made in Fig. 17-5 with P-3 and zone method results for the same problem except that the scattering is isotropic. Combined radiation and conduction in a two-dimensional rectangular enclosure was analyzed in [31] using the S_4 method. A description of the discrete ordinates method was given, and linear anisotropic scattering was included. Three walls of the rectangular enclosure were at one temperature and the fourth wall was at a different uniform temperature. Comparisons were made with results from other methods given in [9, 32, 33], and good agreement was obtained. Only small amounts of computer time were required.

Table 17-1 Weight (w) and ordinate ($\mu = \cos\theta$) values for S_n method [29]

S_n	w	$\pm\mu$	S_n	w	$\pm\mu$
S_2	1	0.500000	S_{10}	1/5	0.083752
				1/5	0.312729
S_4	1/2	0.211325		1/5	0.500000
	1/2	0.788675		1/5	0.687270
				1/5	0.916248
S_6	1/3	0.146446			
	1/3	0.500000	S_{12}	1/6	0.066877
	1/3	0.853554		1/6	0.366693
				1/6	0.288732
S_8	1/4	0.102672		1/6	0.711267
	1/4	0.406205		1/6	0.633307
	1/4	0.593795		1/6	0.933123
	1/4	0.897327			

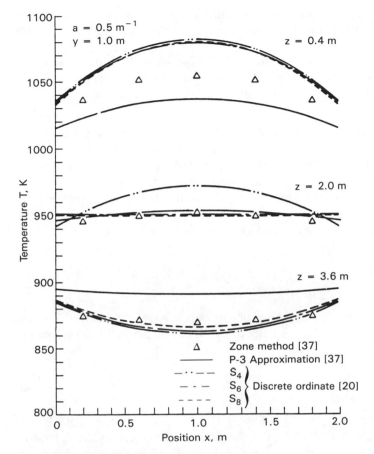

Figure 17-5 Comparison between the discrete-ordinates solutions, the P-3 approximation, and the zone method for temperature distributions at three axial locations ($z = 0.4$, 2.0, and 3.6 m) in a rectangular enclosure. From [20].

17-6 P-N (DIFFERENTIAL) METHODS

The P-N method provides an expression for the local divergence of the radiative flux (radiative source term) that is *differential* in form. The expressions can be directly incorporated into the energy equation in differential form that includes other modes of energy transfer. The method thus naturally fits whatever grid size is used for a numerical solution of the energy equation.

17-6.1 The P-N Approximation Including Other Energy Transfer Modes

Ratzel [13] applied the P-1 and P-3 methods to rectangular enclosures containing a participating medium with an internal volumetric source and finite thermal conductivity. Menguc [34] extended the application to three-dimensional rectangular geometries and two-dimensional axisymmetric geometries of finite length with anisotropic scattering.

The general relations (15-101) through (15-124) are valid in the presence of other energy transfer modes. However, the presence of more terms in the energy equation complicates the problem. Consider the following example.

EXAMPLE 17-1 A plane layer of radiating and scattering medium with constant thermal conductivity k, optical thickness κ_D, and albedo Ω is between infinite parallel gray plates of emissivity ϵ. The medium has a uniform internal heat source q'''. The bounding plates have temperatures $T_{w,1}$ and $T_{w,2}$. Find the heat flux to each bounding plate and the temperature distribution in the medium.

For this case, the energy equation (16-1) is $\nabla \cdot (k\,\nabla T - \mathbf{q}_r) + q''' = 0$ or, using a summation form,

$$\sum_{i=1}^{3} \frac{\partial q_{r,i}}{\partial x_i} + \sum_{i=1}^{3} \frac{\partial q_{c,i}}{\partial x_i} = q''' \tag{17-29}$$

For the present one-dimensional problem, since $q_r = i'^{(1)}$ where $i'^{(1)}$ is the first moment of intensity, the energy equation can be put in the dimensionless form

$$\frac{dI^{(1)}}{d\kappa_1} = 4N \frac{d^2 t}{d\kappa_1^2} + \frac{S}{\kappa_D} \tag{17-30}$$

where $N = k(a + \sigma_s)/4\sigma T_{w,1}^3$, $S = q'''D/\sigma T_{w,1}^4$, $\kappa_D = (a + \sigma_s)D$, $I^{(1)} = i'^{(1)}/\sigma T_{w,1}^4$, and $t = T/T_{w,1}$. This is the defining equation for the derivative of $I^{(1)}$. In the pure radiation problem of Example 15-8, the derivative of the moment (15-127) was set equal to zero, because for pure radiation $I^{(1)} = q_r/\sigma T_{w,1}^4 = \text{const}$ and therefore $dI^{(1)}/d\kappa_1 = 0$. This is not the case here.

The presence of the second derivative of dimensionless temperature now requires two boundary conditions for the energy equation. These are $t(\kappa_1 = 0) = 1$ and $t(\kappa_1 = \kappa_D) = T_{w,2}/T_{w,1}$.

To proceed with the solution, two coupled second-order differential equations are derived for $I^{(0)}$ and $I^{(1)}$ (the nondimensional radiative flux). The first is obtained by equating (17-30) and the first moment differential equation, (15-126). For the present one-dimensional problem, this yields:

$$\frac{4N}{1 - \Omega} \frac{d^2 t}{d\kappa_1^2} + \frac{S}{\kappa_D(1 - \Omega)} - 4t^4 = -I^{(0)} \tag{17-31}$$

The second equation is found by substituting the closure equation, (15-121), into the second moment differential equation (15-117) to obtain

$$\frac{dI^{(11)}}{d\kappa_1} = \frac{1}{3} \frac{dI^{(0)}}{d\kappa_1} = -I^{(1)} \tag{17-32}$$

Now, (17-32) is differentiated with respect to κ_1, and the result is substituted into (17-30) to yield

$$\frac{d^2 I^{(0)}}{d\kappa_1^2} + 12N \frac{d^2 t}{d\kappa_1^2} + \frac{3S}{\kappa_D} = 0 \tag{17-33}$$

Equations (17-31) and (17-33) can be combined into a single fourth-order equation in t by differentiating (17-31) twice with respect to κ_1. The result is substituted into (17-33) to eliminate the second derivative of $I^{(0)}$.

   The resulting fourth-order equation (or the two second-order equations) requires two boundary conditions in addition to the known boundary surface temperatures. These are generated from the Marshak boundary conditions (15-123) and (15-124) using (15-133) to eliminate q_0, which results in

$$\frac{I^{(0)}}{4}(\kappa_1 = 0) = 1 - \left(\frac{1}{\epsilon} - \frac{1}{2}\right) I^{(1)}(\kappa_1 = 0)$$

$$\frac{I^{(0)}}{4}(\kappa_1 = \kappa_D) = t_{w,2}^4 + \left(\frac{1}{\epsilon} - \frac{1}{2}\right) I^{(1)}(\kappa_1 = \kappa_D) \tag{17-34}$$

or

$$\frac{I_i^{(0)}}{4} = t_{w,i}^4 \pm \left(\frac{1}{\epsilon} - \frac{1}{2}\right) I_i^{(1)} \tag{17-35}$$

where i denotes walls 1 or 2 and the positive sign applies at wall $i = 2$. Inserting (17-32) to eliminate $I^{(1)}$ results in the final boundary relation

$$\pm \left(\frac{dI^{(0)}}{d\kappa_1}\right)_i = -\frac{3}{4\left(\frac{1}{\epsilon} - \frac{1}{2}\right)} [I_i^{(0)} - 4t_{w,i}^4] \tag{17-36}$$

where, as above, $i = 1, 2$ and the positive sign applies at surface 2. Now, (17-36) can be directly applied as the boundary conditions for (17-33). The problem is completely specified with two second-order nonlinear differential equations and the requisite four boundary conditions.

17-6.2 Numerical Approaches to P-N Solutions

Ratzel [13] solved Example 17-1 as a limiting case to the P-3 formulation of the problem. The two coupled equations were solved by the collocation method [35] for sets of mixed-order boundary value problems in ordinary differential equations, which automatically generates the grid until a specified tolerance is met in solution accuracy. A modified successive overrelaxation method is used for nonlinear equations. It was found that the particular commercial routine, which accommodates equations up to fourth order and thus should solve the single P-1 equation outlined above, would not converge for large N to a specified accuracy on t of 10^{-3}. When the problem was expressed as a set of two second-order equations rather than a single fourth-order equation, the package worked satisfactorily. For the P-3 solution of the same problem, the P-3 equations were used as six coupled nonlinear equations, and convergence was obtained with little difficulty.

Figure 17-6 compares the P-1 and P-3 results for the centerline temperature profile (with the conduction/radiation number N as a parameter) in a square enclosure with one hot surface and three cold surfaces. Figure 17-7 shows the P-3 results for the temperature distribution in the entire medium for the same case, with $N = 1$.

For the P-1 approximation for radiation in a medium with conduction and internal energy sources in a two-dimensional enclosure, the finite-element code TWODPEP [36] was used from the IMSL packages. Triangular elements with six nodes and quadratic shape functions were used. For the P-3 approximation in two dimensions, the finite-element approach became too time consuming, so a modified successive under-relaxation method (Sec. 11-5.1) was used instead. Run times for combined-mode problems increased by about 20% over pure radiation prob-

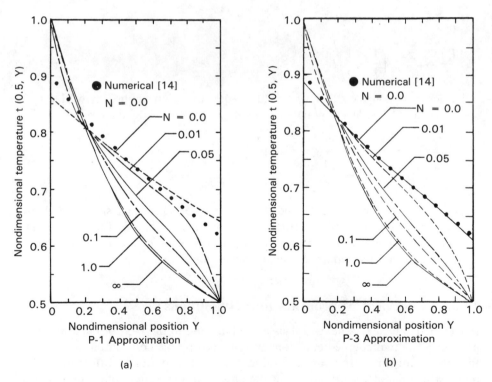

Figure 17-6 Comparison of P-1 and P-3 approximation results for the nondimensional centerline temperature distribution in a square enclosure for various values of the conduction-radiation parameter N: optical length of side $\kappa_D = 1.0$, $t_{w,1} = 1.0$, $t_{w,i} = 0.5$ ($i = 2, 3, 4$), $\epsilon_i = 1.0$ ($i = 1–4$), $S = 0$. From [13]. (*a*) P-1 approximation; (*b*) P-3 approximation.

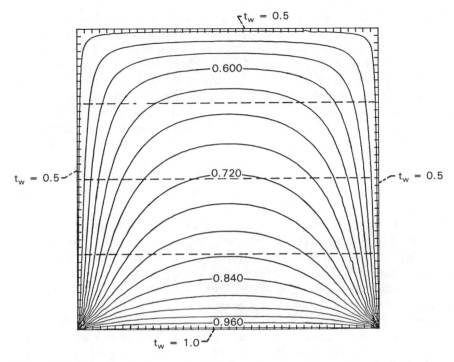

Figure 17-7 P-3 approximation results for the nondimensional temperature distribution in a square enclosure for $N = 1.0$: optical length of side $\kappa_D = 1.0$, $t_{w,1} = 1.0$, $t_{w,i} = 0.5$ ($i = 2, 3, 4$), $\epsilon_i = 1.0$ ($i = 1$–4), $S = 0$. From [13].

lems for all values of N. Rapid convergence was obtained if P-1 results were used as initial guesses for the iterative procedures.

Menguc and Viskanta [37] used a proprietary code at Purdue University to obtain P-3 results for radiative transfer in three-dimensional geometries with anisotropic scattering and a uniform internal energy source. Conduction was not included. For radiative transfer in a medium within a cylinder of finite length bounded by gray walls at known temperature, a finite-element solution took up to 10 times longer to run than a finite-difference solution for the same accuracy.

17-7 MONTE CARLO METHODS FOR COMBINED MODES

Sufficiently low computer costs will allow solution of almost any problem if even an inefficient method is available for exact modeling. The Monte Carlo method can in principle be programmed to include an exact simulation of the important radiation processes. Applications of the Monte Carlo method have appeared that exploit its flexibility and power to examine difficult problems. Reviews of the method are in [38, 39].

An iterative way to use a Monte Carlo method in a multimode problem is to use it to evaluate the local radiative energy source (the radiative flux divergence) based on an assumed temperature distribution in the medium. This source distribution is then used in the energy equation, which is solved by any convenient numerical method for a new temperature distribution. This temperature distribution is used to repeat the Monte Carlo calculation for flux divergence, and the procedure is continued until convergence. A difficulty with this method is that the statistical fluctuations inherent in Monte Carlo evaluations can provide irregular behavior in the spatial variations of the radiative flux divergence. This may produce numerical instabilities in the solution of the energy equation. This may be a difficulty when conduction is present because the fluctuations can cause the numerical approximations for second derivatives in the conduction term to be erratic.

Work on reducing computation time includes that by Mishkin and Kowalski [40], who used the transient form of the energy equation to predict a temperature field and then used the predicted temperature field in a Monte Carlo evaluation of the local radiation source. The radiative flux divergence from the Monte Carlo solution was substituted into the energy equation at the succeeding time step, and the procedure was repeated until convergence to steady-state.

The computational element size required for statistical accuracy in the Monte Carlo radiative source evaluation may be larger than the grid size necessary for numerical solution of the energy equation. Given enough computer power, this grid difference problem can be overcome by taking a sufficiently small Monte Carlo computation element size (with a resulting increase in the number of statistical simulations) to match other grid size requirements. An alternative approach that is usually accurate is to use a grid size compatible with the Monte Carlo statistical requirements and then interpolate intermediate values where the flux divergence is needed for grid matching to the energy equation. This should not introduce significant error, as the radiative flux divergence is often a slowly varying function compared with the temperature and its derivatives, and a large grid may be adequate for the local radiative source evaluation.

17-8 THE ZONE METHOD

Considerable literature exists on the zone method and its variations. The basis of this method is given in Sec. 13-7. The method has been modified in [14], which used the same assumptions inherent in the zone method of Hottel [41] but derived the transfer equations in terms of *exchange factors*. These factors are defined in terms of the fraction of radiative energy leaving one element that is absorbed by a receiving element, including all possible paths of intermediate scattering as well as intermediate absorption and re-emission in a medium in radiative equilibrium. Exchange factors are measurable, and [42] reports measured values for a rectangular enclosure. Measuring exchange factors in scale models of furnaces eliminates the restrictions that limit zone analyses to cases where the exchange areas are known or can be calculated. In [14] it is shown that the exchange areas of the Hottel method can be calculated from the exchange factors and vice versa, so the methods are interchangeable. Exchange factors defined somewhat differently were used to analyze a rectangular enclosure in [43].

17-8.1 Developments of the Zone Method

Smoothing of exchange area sets To achieve accurate numerical results with the zone method, the calculated exchange areas must satisfy the necessary conditions of reciprocity and energy conservation. Generally, the exchange areas described in Sec. 13-7 must satisfy the reciprocity constraints

$$\overline{s_i s_j} = \overline{s_j s_i} \qquad \overline{s_i g_\gamma} = \overline{g_\gamma s_i} \qquad \overline{g_\gamma g_\mu} = \overline{g_\mu g_\gamma} \tag{17-37}$$

as well as the conservation relations,

$$(4aV)_\gamma = \sum_{i=1}^{N} \overline{g_\gamma s_i} + \sum_{\mu=1}^{\Gamma} \overline{g_\gamma g_\mu} \qquad A_i = \sum_{j=1}^{N} \overline{s_i s_j} + \sum_{\gamma=1}^{\Gamma} \overline{s_i g_\gamma} \tag{17-38}$$

where N is the number of surface elements in the enclosure and Γ is the number of volume elements. However, if the various exchange areas are computed independently, there is no guarantee that (17-37) and (17-38) will be satisfied.

In general, exchange areas between every pair of surface and volume elements must be known to carry through a complete analysis by the zone method. If reciprocity is used to compute as many factors as possible, there may remain as many as $M(M + 1)/2$ independent exchange areas to be evaluated, where $M = N + \Gamma$. Symmetry may reduce this number in a given geometry. Sowell and O'Brien [44] use (17-38) to evaluate M additional exchange areas, leaving $M(M - 1)/2$ independent areas to be evaluated. However, as pointed out in [44], this approach may lump all the errors of the independently evaluated exchange areas into the M areas found by applying (17-38). Vercammen and Froment [45] obtained exchange areas by a Monte Carlo approach and found the usual statistical scatter in the results, so the constraints of (17-37) and (17-38) were not met exactly. They present a regression method for smoothing all unique and nonzero factors.

In [46] a method of least-squares smoothing is used that utilizes Lagrangian multipliers with (17-37) and (17-38) as constraints. This method ensures that constraints are met with a minimum disturbance to the original set of factors. The tendency of this method is to adjust each exchange area in proportion to its original magnitude. This method can be incorporated into general zone codes to assure the "best" set of exchange areas is provided. The method appears to work best for large M; for small M, the methods of [44] and [45] may be more appropriate. The use of smoothing, however, was found to have little effect on the zone predictions for two problems carried out in [47]. Additional work is needed on the allowable accuracy of the exchange factors.

Other formulations of the zone method Noble [48] presents a set of explicit matrix relations for the calculation of total exchange areas from the direct exchange areas, reducing the time required for solution. Naraghi and Chung [49] use a stochastic approach to recast the zone equations into a third basis (counting the Hottel [41] and Larsen [14] approaches as two others). They claim increased coding efficiency for the method, which uses exchange factors defined somewhat differently from those of [14].

The imaginary planes method is a technique directed toward decreasing the computation time required for the zone method. Considerable reductions in computer time are reported in [50], where an outline of the method is given for three-dimensional geometries. In the method, each volume zone is linked only to the adjacent zones by the net radiative heat fluxes passing through its zone boundaries (imaginary planes). The radiative transfer is therefore modeled in terms of only the interactions from the immediately adjacent zones, as opposed to using direct interactions with more distant zones as in the classical zone method. Although each volume zone has a direct view of only its own boundaries, the transfer with all other zones is linked in a chain fashion through the radiative heat fluxes crossing the imaginary planes. Hence, the interaction between all zones is included. This technique in formulation provided appreciable savings in computation time for the demonstration cases in [50].

17-8.2 Numerical Zone Method Results

Larsen [14] calculated temperature distributions in the medium, and surface heat fluxes, in two- and three-dimensional enclosures with gray and black bounding walls containing an absorbing, emitting, and isotropically scattering medium. Results were obtained for various values of the conduction-radiation parameter N. The exchange-factor method was used with exchange-factor smoothing techniques developed to ensure energy conservation [46]. The resulting set of nonlinear algebraic equations for medium temperature was solved using IMSL routine ZSCNT. Once the medium temperature distribution was found, the radiative surface flux was computed, and this was added to the conductive flux computed from the temperature gradient at the boundary to obtain the total energy flux. An 11×11 set of volume elements and corresponding surface elements was used for the two-dimensional solution and a $5 \times 5 \times 5$ set was used for the three-dimensional solution.

Comparisons of Larsen's zone results with the two-dimensional exact solution by Crosbie and Schrenker [51] for pure radiation (Table 17-2) are quite good. Errors are within 1.5%, and most values of emissive power and surface flux are within 0 to 0.5%. Some comparisons with the P-3 and finite-element solutions are in Fig. 17-8 for a square two-dimensional enclosure with black surfaces. Some results for a cubic enclosure (Fig. 17-9) with and without conduction are in Tables 17-3 and 17-4 [14].

Table 17-2 Comparison of dimensionless surface heat flux and dimensionless centerplane emissive power in a two-dimensional rectangular enclosure as computed by zone analysis [14] and exact formulation [51][a]

X or Y	$\bar{q}_{side}$		$\bar{q}_{top}$		$\bar{q}_{bottom}$		$\bar{e}_{centerplane}$	
	Zone	Exact	Zone	Exact	Zone	Exact	Zone	Exact
0.1	0.524	0.518(1.2)	0.832	0.827(0.6)	0.189	0.190(0.5)	0.521	0.519(0.4)
0.2	0.437	0.431(1.4)	0.798	0.796(0.3)	0.212	0.213(0.5)	0.434	0.433(0.2)
0.3	0.368	0.366(0.5)	0.778	0.777(0.1)	0.229	0.230(0.4)	0.361	0.361(0.0)
0.4	0.310	0.308(0.6)	0.768	0.767(0.1)	0.240	0.240(0.0)	0.300	0.299(0.3)
0.5	0.260	0.259(0.4)	0.764	0.764(0.0)	0.243	0.244(0.4)	0.250	0.250(0.0)
0.6	0.218	0.217(0.5)	0.768	0.767(0.1)	0.240	0.240(0.0)	0.208	0.208(0.0)
0.7	0.182	0.181(0.6)	0.778	0.777(0.1)	0.229	0.230(0.4)	0.173	0.173(0.0)
0.8	0.149	0.149(0.0)	0.798	0.796(0.3)	0.212	0.213(0.5)	0.142	0.142(0.0)
0.9	0.119	0.119(0.0)	0.832	0.827(0.6)	0.189	0.190(0.5)	0.114	0.115(0.9)

[a]Numbers in parentheses are percentage differences; $\bar{q}$ = dimensionless heat flux, $\bar{e}$ = dimensionless emissive power.

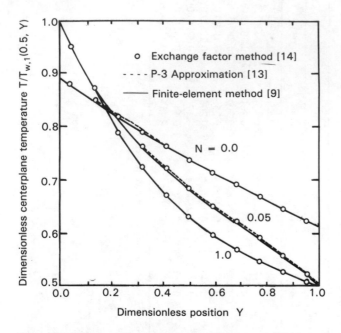

Figure 17-8 Centerplane temperature profiles for an infinitely long enclosure of square cross section with black walls, optical side length $\kappa_D = 1.0$, $t_{w,1} = 1.0$, $t_{w,i} = 0.5$ ($i = 2, 3, 4$), at various conduction/radiation parameters, N.

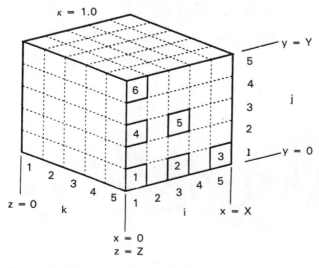

Figure 17-9 Black-walled cube with five zones in each direction.

Table 17-3 Three-dimensional zone analysis results, pure radiation [14][a]

k	j	Dimensionless emissive power, $(T_w/T_{wX})^4$			Zone	Dimensionless heat flux, $q_w/\sigma T_{wX}^4$
		$i = 1$	$i = 3$	$i = 5$		
5	5	0.105	0.177	0.368	1	0.499
5	3	0.177	0.298	0.500	2	0.695
5	1	0.368	0.500	0.632	3	0.821
					4	0.292
3	5	0.177	0.298	0.500	5	0.483
3	3	0.298	0.500	0.702	6	0.177
3	1	0.500	0.702	0.823		
1	5	0.368	0.500	0.632		
1	3	0.500	0.702	0.823		
1	1	0.632	0.823	0.895		

[a]Surfaces for which $z = 0$, $y = 0$, $x = 1$ have unit emissive power; others are cold. All walls are black, and no internal source is present.

17-9 OTHER NUMERICAL SOLUTION TECHNIQUES

Yuen and Takara [32] give solutions for combined conduction and radiation in two-dimensional enclosures, using an iterative solution to a finite-difference form of the energy equation. The method is based on an earlier method for radiation without conduction [52]. They use a numerical approximation to the integral form of the equation of transfer for determining the local radiative flux divergence. The

Table 17-4 Three-dimensional zone analysis results, combined radiation and conduction [14][a]

k	j	Dimensionless emissive power, $(T_w/T_{wX})^4$			Zone	Dimensionless heat flux $q_w/\sigma T_{wX}^4$
		$i = 1$	$i = 3$	$i = 5$		
5	5	0.069	0.181	0.479	1	0.663
5	3	0.181	0.431	0.679	2	0.959
5	1	0.479	0.679	0.779	3	1.165
					4	0.298
3	5	0.181	0.431	0.679	5	0.589
3	3	0.431	0.807	0.917	6	0.158
3	1	0.679	0.917	0.960		
1	5	0.479	0.679	0.779		
1	3	0.679	0.917	0.960		
1	1	0.779	0.960	0.980		

[a]Surface conditions are as in Table 17-3; $N = 0.01$.

approximation assumes a linear variation of the emissive power distribution within each difference element and approaches an exact solution for small elements. The only physical approximation is in the radiative boundary condition, which is generated by extrapolation of the linear function normal to the boundary. Good accuracy was found for an 11×11 grid in the square enclosure, and comparisons with the finite-element results of [9] were excellent. Comparisons with the P-3 results of [13] were good for emissive power in the medium but were poor for surface energy flux in some cases; this is likely due to inaccuracy in the P-3 method. Yuen and Takara [32] also compared results with the diffusion solution in two dimensions and found deviations up to 17% in local emissive power of the medium and very large errors in surface energy flux. Much better predictions of total surface energy flux (within 10% for all cases reported and within 1% in most cases) were obtained by simply adding the energy flux for a pure conduction solution to that for a pure radiation solution.

Tan [33] used the product integration method [53] to evaluate the integrals in the equation of transfer. The method is outlined in Sec. 15-9.3 and reduces the order of the multiple integrals by one, resulting in much faster execution than for the finite-element method. Results for a two-dimensional enclosure with a uniform energy source compared well with the zone method, Fig. 17-10. Results were also obtained including highly anisotropic scattering.

The YIX method (Sec. 15-9.4) has been applied to free convection–radiation problems [54], and solutions of the resulting set of equations were obtained without difficulty using standard linear equation solvers. The method requires precomputation of a number of coefficients used in the solution but greatly reduces the time involved in computing the integrals in the radiative source terms.

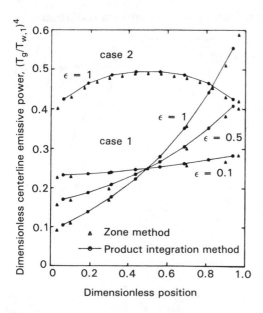

Figure 17-10 Centerline dimensionless emissive power in a square enclosure, optical side length $\kappa_D = 1$. Case 1: effect of surface emissivity for pure radiation with surfaces at $t_{w,1} = 1$, $t_{w,i} = 0$ for $i = 2, 3, 4$; case 2: uniform source in medium with $t_{w,i} = 0$ for $i = 1$–4. From [33].

DeMarco and Lockwood [55] modified the two-flux model of Sec. 15-6 into a four-flux or six-flux model. This is a hybrid of the discrete ordinate method and the differential (P-1) approximation. It has certain computational advantages over the more complete models and generally predicts energy fluxes in good agreement with the other models.

A finite-volume method is developed in [56]. The method can be applied on the same grid that is used to compute fluid flow and heat transfer. A Taylor series expansion is used to relate the local intensity to that at adjacent grid points. Good agreement was obtained with results in [25] for surface fluxes by emission of a square region of radiating medium at uniform temperature. The fluxes also agree with the exact solution by Siegel [57]. Good agreement was also obtained with a solution by Crosbie and Schrenker [51] for a rectangular cavity having three walls at a single temperature and the fourth wall at a higher temperature. The cavity contains a medium in radiative equilibrium.

Some solutions for combined radiation and conduction were obtained by Ho and Ozisik [58, 59] and Tsai and Ozisik [60]. These included effects such as transient conditions, scattering, spherical geometry, and a two-layer planar region. The conduction part of the solution was obtained with finite differences. The radiation source term in the energy equation was obtained by expanding the locally incident radiation in a power series and substituting into the integral equation of transfer. A collocation or Galerkin procedure was used to obtain the unknown coefficients in the power series expansion. The source function is then known in terms of the incident intensities.

17-10 CONCLUDING REMARKS

Considerable effort is still needed to improve numerical methods for solving the radiation transfer relations for problems with radiation only and for those in which radiation is combined with other heat transfer modes. The difficulties result from the multidimensional integrations required for the radiative source terms in multidimensional geometries and the additional integration over the spectrum needed to obtain total energy quantities. The nonlinearity of combined-mode problems leads to convergence difficulties in some instances. Local regions having rapidly changing temperatures require small grid sizes to obtain good accuracy in the radiation terms. Combustion problems present an energy equation that is extremely stiff because of the strong dependence of the combustion reaction rates on temperature. In some cases the radiation terms can lead to either slow, or a lack of, convergence. The increase in size and speed of computers may result in a gradual change in emphasis on the various methods used for solutions. There may be an increase in the use of direct numerical integration to evaluate the radiative source term. Available computer subroutines can be used for multidimensional integration, such as by Gaussian integration, which is quite accurate. Integration methods such as product integration or the YIX method are under development and may provide improvements. This may result in increased use of finite-difference and finite-element methods compared with the P-N method.

The increasing ability of computers to use vector and parallel processing will have an influence [61–63]. Adaptation of radiative transfer methods to these capabilities is a current research area and could have a major impact in the next several years. The best methodology will need to be found to take advantage of parallel processing for multimode heat transfer calculations. One method is to solve the radiative transfer equation using an initial assumed temperature field on one processor while simultaneously solving the energy equation on another to compute the temperature field from an initial assumed radiation field. Then the calculated radiation field and temperature information from each parallel path are traded and used as new guesses, and iteration proceeds until convergence. This could significantly reduce computation time.

Denning [64] reviewed the state of parallel processing and stated that we are presently in *stage I* of its use. Present computers make parallel processing available by having their operating systems access more than one processor, so successive tasks need not await completion of prior tasks. Newer machines are structured so their compilers analyze programs in a way that keeps the pipeline of instructions busy much of the time; this provides much faster execution. To use the parallel-processing capability efficiently, software must be structured to make use of the inherent features of the parallel-processing system. The software must have the capability of *microtasking*, i.e., telling the computer to access a parallel processor for a given task.

Stage II of parallel processing will make use of advanced languages that allow structuring programs so information for continuation of a given calculation is supplied from a parallel processor as soon as it is available, and the calculation proceeds. Such a language allows parallel processing even on separate machines, as long as a communication link is established. Although these systems are well along in development, their application to practical engineering computations has not been exploited. Monte Carlo methods in particular stand to improve significantly in reduced computer time. A vectorized Monte Carlo program in [65] provided a speed-up factor of approximately 16. The use of multiprocessor systems to handle the multiple paths describing the distribution of radiation in complex systems should reduce the computing time.

REFERENCES

1. Edwards, D. K.: Numerical Methods in Radiation Heat Transfer, in T. M. Shih (ed.), "Numerical Properties and Methodologies in Heat Transfer," pp. 479–496, Hemisphere, Washington, D.C., 1983.
2. Minkowycz, W. J., E. M. Sparrow, G. E. Schneider, and R. H. Pletcher: "Handbook of Numerical Heat Transfer," Wiley, New York, 1988.
3. Jaluria, Y., and K. E. Torrance: "Computational Heat Transfer," Hemisphere, Washington, D.C., 1986.
4. Chung, T. J.: Integral and Integrodifferential Systems, in W. J. Minkowycz , E. M. Sparrow, G. E. Scheider, and R. H. Pletcher (eds.), "Handbook of Numerical Heat Transfer," chap. 14, Wiley, New York, 1988.

5. Fernandes, R., J. Francis, and J. N. Reddy: A Finite Element Approach to Combined Conductive and Radiative Heat Transfer in a Planar Medium, in A. L. Crosbie (ed.), "Heat Transfer and Thermal Control," pp. 93–109, AIAA, New York, 1981.

6. Wu, S. T., R. E. Ferguson, and L. L. Altgilbers: Application of Finite Element Techniques to the Interaction of Conduction and Radiation in Participating Media, in A. L. Crosbie (ed.), "Heat Transfer and Thermal Control," pp. 61–92, AIAA, New York, 1981.

7. Fernandes, R., and J. Francis: Combined Conductive and Radiative Heat Transfer in an Absorbing, Emitting, and Scattering Cylindrical Medium, *J. Heat Transfer*, vol. 104, no. 4, pp. 594–601, 1982.

8. Razzaque, M. M., D. E. Klein, and J. R. Howell: Finite Element Solution of Radiative Heat Transfer in a Two-Dimensional Rectangular Enclosure with Gray Participating Media, *J. Heat Transfer*, vol. 105, no. 4, pp. 933–935, 1983.

9. Razzaque, M. M., J. R. Howell, and D. E. Klein: Coupled Radiative and Conductive Heat Transfer in a Two-Dimensional Rectangular Enclosure with Gray Participating Media Using Finite Elements, *J. Heat Transfer*, vol. 106, no. 3, pp. 613–619, 1984.

10. Sokman, C. N., and M. M. Razzaque: Finite Element Analysis of Conduction-Radiation Heat Transfer in an Absorbing-Emitting and Scattering Medium Contained in an Enclosure with Heat Flux Boundary Conditions, in "Radiation, Phase Change Heat Transfer, and Thermal Systems," ASME HTD-vol. 81, Y. Jaluria, V. P. Carey, W. A. Fiveland, and W. Yuen (eds.), pp. 17–23, 1987.

11. Utreja, L. R., and T. J. Chung: Combined Convection-Conduction-Radiation Boundary Layer Flows Using Optimal Control Penalty Finite Elements, *J. Heat Transfer*, vol. 111, no. 2, pp. 433–437, 1989.

12. Chung, T. J., and J. Y. Kim: Two-Dimensional Combined-Mode Heat Transfer by Conduction, Convection and Radiation in Emitting, Absorbing and Scattering Media—Solution by Finite Elements, *J. Heat Transfer*, vol. 106, no. 2, pp. 448–452, 1984.

13. Ratzel, A. C., III: P-N Differential Approximation for Solution of One- and Two-Dimensional Radiation and Conduction Energy Transfer in Gray Participating Media, Ph.D. dissertation, Department of Mechanical Engineering, University of Texas, Austin, 1981.

14. Larsen, M. E., and J. R. Howell: The Exchange Factor Method: An Alternative Basis for Zonal Analysis of Radiating Enclosures, *J. Heat Transfer*, vol. 107, no. 4, pp. 936–942, 1985.

15. Warming, R. F., and R. M. Beam: On the Construction and Application of Implicit Factored Schemes for Conservation Laws, *SIAM-AMS Proc.*, vol. 11, "Computational Fluid Dynamics," H. B. Keller (ed.), pp. 85–129, 1978.

16. Yoshida, H., J. H. Yun, R. Echigo, and T. Tomimura: Transient Characteristics of Combined Conduction, Convection and Radiation Heat Transfer in Porous Media, *Int. J. Heat Mass Transfer*, vol. 33, no. 5, pp. 847–857, 1990.

16a. Siegel, R.: Finite Difference Solution for Transient Cooling of a Radiating-Conducting Semitransparent Layer, *J. Thermophys. Heat Transfer*, vol. 6, no. 1, pp. 77–83, 1992.

17. Press, W. H., B. P. Flannery, S. A. Teukolsky, and W. T. Vetterling: "Numerical Recipes," Cambridge University Press, London, 1989.

17a. Siegel, R., and F. B. Molls: Finite Difference Solution for Transient Radiative Cooling of a Conducting Semitransparent Square Region, *Int. J. Heat Mass Transfer*, 1992 (in press).

18. Lathrop, K. D.: Use of Discrete-Ordinates Methods for Solution of Photon Transport Problems, *Nucl. Sci. Eng.*, vol. 24, pp. 381–388, 1966.

19. Carlson, B. G., and K. D. Lathrop: Transport Theory—The Method of Discrete Ordinates, chap. 3 in H. Greenspan, C. N. Kelber, and D. Okrent (eds.), "Computing Methods in Reactor Physics," Gordon & Breach, New York, 1968.

20. Fiveland, W. A.: Three-Dimensional Radiative Heat Transfer Solutions by the Discrete-Ordinates Method, *J. Thermophys. Heat Transfer*, vol. 2, no. 4, pp. 309–316, 1988.

21. Truelove, J. S.: Discrete-Ordinate Solutions of the Radiation Transport Equation, *J. Heat Transfer*, vol. 109, no. 4, pp. 1048–1051, 1987.

22. Fiveland, W. A., D. K. Cornelius, and W. J. Oberjohn: COMO: A Numerical Model for Predicting Furnace Performance in Axisymmetric Geometries, ASME Paper 84-HT-103, August 1984.

23. Fiveland, W. A., and A. S. Jamaluddin: Three-Dimensional Spectral Radiative Heat Transfer Solutions by the Discrete-Ordinates Method, *J. Thermophys. Heat Transfer*, vol. 5, no. 3, pp. 335–339, 1991.

24. Kumar, S., A. Majumdar, and C. L. Tien: The Differential-Discrete Ordinate Method for Solutions of the Equation of Radiative Transfer, *J. Heat Transfer*, vol. 112, no. 2, pp. 424–428, 1990.

25. Fiveland, W. A.: Discrete-Ordinates Solutions of the Radiative Transport Equation for Rectangular Enclosures, *J. Heat Transfer*, vol. 106, no. 4, pp. 699–706, 1984.

26. Yucel, A., and M. L. Williams: Heat Transfer by Combined Conduction and Radiation in Axisymmetric Enclosures, *J. Thermophys. Heat Transfer*, vol. 1, no. 4, pp. 301–306, 1987.

27. Khalil, E. E.: Modelling of Furnaces and Combustors, in "Energy and Engineering Science Series." Abacus Press, Tunbridge Wells, England, 1982.

28. Patankar, S. V.: "Numerical Heat Transfer and Fluid Flow," Hemisphere, Washington, D.C., 1980.

29. Fiveland, W. A.: Discrete Ordinate Methods for Radiative Heat Transfer in Isotropically and Anisotropically Scattering Media, *J. Heat Transfer*, vol. 109, no. 3, pp. 809–812, 1987.

30. Selcuk, N.: Evaluation of Flux Models for Radiative Transfer in Rectangular Furnaces, *Int. J. Heat Mass Transfer*, vol. 31, pp. 1477–1482, 1988.

31. Kim, T. Y., and S. W. Baek: Analysis of Combined Conductive and Radiative Heat Transfer in a Two-Dimensional Rectangular Enclosure Using the Discrete Ordinates Method, *Int. J. Heat Mass Transfer*, vol. 34, no. 9, pp. 2265–2273, 1991.

32. Yuen, W. W., and E. E. Takara: Analysis of Combined Conductive-Radiative Heat Transfer in a Two-Dimensional Rectangular Enclosure with a Gray Medium, *J. Heat Transfer*, vol. 110, no. 2, pp. 468–474, 1988.

33. Tan, Z.: Radiative Heat Transfer in Multidimensional Emitting, Absorbing, and Anisotropically Scattering Media—Mathematical Formulation and Numerical Method, *J. Heat Transfer*, vol. 111, no. 1, pp. 141–147, 1989.

34. Menguc, M. P.: Modeling of Radiative Heat Transfer in Multidimensional Enclosures Using Spherical Harmonics Approximation, Ph.D. Dissertation, Department of Mechanical Engineering, Purdue University, 1985.

35. Ascher, U., J. Christiansen, and R. D. Russell: Collocation Software for Boundary Value ODE's, in B. Childs et al. (eds.), "Codes for Boundary Value Problems in Ordinary Differential Equations," pp. 164–185, Springer-Verlag, Berlin, 1979.

36. "TWODEPEP Users Manual," International Mathematical and Statistical Libraries Manual, IMSL TDP-0003, Houston, 1981.

37. Menguc, M. P., and R. Viskanta: Radiative Transfer in Three-Dimensional Rectangular Enclosures Containing Inhomogeneous, Anisotropically Scattering Media, *J. Quant. Spectrosc. Radiat. Transfer*, vol. 33, no. 6, pp. 533–549, 1985.

38. Halton, J. H.: A Retrospective and Prospective Review of the Monte Carlo Method, *SIAM Rev.*, vol. 12, no. 1, pp. 1–63, 1970.

39. Haji-Sheikh, A.: Monte Carlo Methods, chap. 16 in W. J. Minkowycz, E. M. Sparrow, R. H. Pletcher, and G. E. Schneider (eds.), "Handbook of Numerical Heat Transfer," Wiley, New York, 1988.

40. Mishkin, M., and G. Kowalski: Application of Monte Carlo Techniques to the Steady State Radiative and Conductive Heat Transfer Problem through a Participating Medium, ASME Paper 83-WA/HT-27, December 1983.

41. Hottel, H. C., and A. F. Sarofim: "Radiative Transfer," McGraw-Hill, New York, 1967.

42. Liu, H.-P., and J. R. Howell: Scale Modeling of Radiation in Enclosures with Absorbing/Emitting and Isotropically Scattering Media, *J. Heat Transfer*, vol. 109, no. 2, pp. 470–477, 1987.

43. Naraghi, M. H. N., and M. Kassemi: Analysis of Radiative Transfer in Rectangular Enclosures Using a Discrete Exchange Factor Method, *J. Heat Transfer*, vol. 111, no. 4, pp. 1117–1119, 1989.

44. Sowell, E. F., and R. F. O'Brien: Efficient Computation of Radiant-Interchange Configuration Factors within an Enclosure, *J. Heat Transfer*, vol. 94, no. 3, pp. 326–328, 1972.

45. Vercammen, H. A., and G. F. Froment: An Improved Zone Method Using Monte Carlo Techniques for the Simulation of Radiation in Industrial Furnaces, *Int. J. Heat Mass Transfer*, vol. 23, no. 3, pp. 329–336, 1980.

46. Larsen, M. E., and J. R. Howell: Least-Squares Smoothing of Direct-Exchange Areas in Zonal Analysis, *J. Heat Transfer*, vol. 108, no. 1, pp. 239–242, 1986.

47. Murty, C. V. S., and B. S. N. Murty: Significance of Exchange Area Adjustment in Zone Modelling, *Int. J. Heat Mass Transfer*, vol. 34, no. 2, pp. 499–503, 1991.

48. Noble, J. J.: The Zone Method: Explicit Matrix Relations for Total Exchange Areas, *Int. J. Heat Mass Transfer*, vol. 18, no. 2, pp. 261–269, 1975.

49. Naraghi, M. H. N., and B. T. F. Chung: A Unified Matrix Formulation for the Zone Method: A Stochastic Approach, *Int. J. Heat Mass Transfer*, vol. 28, no. 1, pp. 245–251, 1985.

50. Charette, A., A. Larouche, and Y. S. Kocaefe: Application of the Imaginary Planes Method to Three-Dimensional Systems, *Int. J. Heat Mass Transfer*, vol. 33, no. 12, pp. 2671–2681, 1990.

51. Crosbie, A. L., and R. G. Schrenker: Radiative Transfer in a Two-Dimensional Rectangular Medium Exposed to Diffuse Radiation, *J. Quant. Spectrosc. Radiat. Transfer*, vol. 31, no. 4, pp. 339–372, 1984.

52. Yuen, W. W., and L. W. Wong: Analysis of Radiative Equilibrium in a Rectangular Enclosure with Gray Medium, *J. Heat Transfer*, vol. 106, no. 2, pp. 433–440, 1984.

53. Baker, C. T. H.: "The Numerical Treatment of Integral Equations," Clarendon Press, Oxford, 1977.

54. Tan, Z., and J. R. Howell: Combined Radiation and Natural Convection in a Participating Medium between Horizontal Concentric Cylinders, in "Heat Transfer Phenomena in Radiation, Convection, and Fires," ASME HTD vol. 106, pp. 87–94, 1989 (1989 National Heat Transfer Conference).

55. DeMarco, A. G., and F. C. Lockwood: A New Flux Model for the Calculation of Three-Dimensional Radiation Heat Transfer, *Riv. Combust.*, vol. 5, p. 184, 1975.

56. Raithby, G. D., and E. H. Chui, A Finite-Volume Method for Predicting a Radiant Heat Transfer in Enclosures with Participating Media, *J. Heat Transfer*, vol. 112, no. 2, pp. 415–423, 1990.

57. Siegel, R.: Analytical Solution for Boundary Heat Fluxes from a Radiating Rectangular Medium, *J. Heat Transfer*, vol. 113, no. 1, pp. 258–261, 1991.

58. Ho, C.-H., and M. N. Ozisik: Simultaneous Conduction and Radiation in a Two-Layer Planar Medium, *J. Thermophys. Heat Transfer*, vol. 1, no. 2, pp. 154–161, 1987.

59. Ho, C.-H., and M. N. Ozisik: Combined Conduction and Radiation in a Two-Dimensional Rectangular Enclosure, *Numer. Heat Transfer*, vol. 13, pp. 229–239, 1988.

60. Tsai, J. R., and M. N. Ozisik: Transient, Combined Conduction and Radiation in an Absorbing, Emitting, and Anisotropically Scattering Solid Sphere, *J. Quant. Spectrosc. Radiat. Transfer*, vol. 38, no. 4, pp. 243–251, 1987.

61. Shih, T. M., L. J. Hayes, W. J. Minkowycz, K. T. Yang, and W. Aung: Parallel Computations in Heat Transfer, *Numer. Heat Transfer*, vol. 9, pp. 639–662, 1986.

62. Howell, J. R.: Improving the Monte Carlo Method for Radiative Transfer by the Use of Parallel Processors, *Proc. Heat Transfer in Thermal Systems Seminar, Phase II* (Coordinating Council for North American Affairs), pp. 53–58, National Cheng Kung University, Taiwan, 1986.

63. Hennebutte, V. R., and E. E. Lewis: A Massively Parallel Algorithm for Radiative Transfer Calculations, ASME Paper 91-WA-HT-10, December 1991.

64. Denning, P. J.: The Science of Computing: The Evolution of Parallel Processing, *Am. Sci.*, vol. 73, no. 5, pp. 414–416, 1985.

65. Burns, P. J., and D. V. Pryor: Vector and Parallel Monte Carlo Radiative Heat Transfer Simulation, *Numer. Heat Transfer, Part B*, vol. 16, no. 1, pp. 97–124, 1989.

PROBLEMS

17-1 A rectangular enclosure infinitely long normal to the cross section shown has the conditions and properties listed in the table. The enclosure is filled with an absorbing, emitting, non-scattering gray gas in radiative equilibrium (no heat conduction or convection and no internal sources). Find the heat flux that must be supplied to each surface to maintain the specified temperatures. Use a finite-difference numerical solution of the energy equation including radiative transfer.

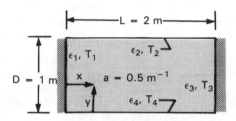

Surface	Type	Emissivity	T, K	q
1	Specular	0		0
2	Diffuse	0.6	750	
3	Specular	0		0
4	Diffuse	0.3	1500	

17-2 A gray nonscattering medium is contained between infinite parallel black plates at temperatures $T_1 = 1500$ K, $T_2 = 900$ K. The medium has absorption coefficient $a = 0.5$ m^{-1} and thermal conductivity 1.54 W/(m·K). The walls are separated by a distance of 2 m, and there is uniform volumetric energy generation in the medium of 2.27 kW/m^3. The medium is not moving. Find the fluxes q_1 and q_2 at the walls, and plot the temperature distribution in the medium. Solve the problem by numerical finite-difference solution of the complete integro-differential equation that applies.

17-3 The finite-difference formulation for a plane layer in Sec. 17-4.1 is for a grid of X points with a uniform increment size ΔX. An improvement is to use a variable grid size so that the distribution of grid points can be modified to place more of the points in regions where the temperature gradients are large. For unequal ΔX increments on either side of an X location, the finite-difference representation for the second derivative of temperature is

$$\left.\frac{\partial^2 t}{\partial X^2}\right|_i = \frac{2t_{i+1}}{\Delta X^+(\Delta X^+ + \Delta X^-)} - \frac{2t_i}{\Delta X^+\Delta X^-} + \frac{2t_{i-1}}{\Delta X^-(\Delta X^+ + \Delta X^-)}$$

where $\Delta X^+ = X_{i+1} - X_i$ and $\Delta X^- = X_i - X_{i-1}$. Using this representation, modify the matrix coefficients in Eq. (17-18) to incorporate a variable grid size.

17-4 Using the solution ideas for the two-dimensional transient analysis in Sec. 17-4.2 and those in the steady one-dimensional analysis at the end of Sec. 17-4.1, set up the finite-difference method for the ADI solution for the following situation with an emitting, absorbing, and conducting gray medium in a two-dimensional rectangular enclosure that is infinitely long in directions normal to the cross section shown. There is no scattering in the medium. The rectangular region has a uniform heating q''' W/m^3 throughout its volume. The steady two-dimensional temperature distribution is to be determined. For simplicity, use a grid having the same increment size in both the x and y directions. All of the boundary walls are black and are at the same temperature T_w.

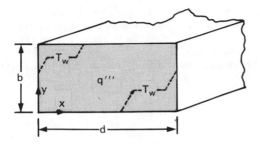

17-5 Modify Prob. 17-4 to incorporate a variable grid size into the solution for the steady temperature distribution within a square cross section. Use the ideas in Prob. 17-3 extended into two dimensions.

17-6 Obtain the governing fourth-order differential equation for t for Example 17-1. Also express the governing relation in the alternative form of two coupled second-order equations. Write out the explicit forms of the four boundary conditions required for solution of either form.

17-7 A plane layer of thickness D has its boundaries maintained at fixed temperatures T_0 and T_D. It is heated internally with a volumetric heat source distribution $q'''(x)$ that depends on the x location within the layer. As an approximate trial solution by finite elements, it is decided to use only two elements, but to use quadratic shape functions so that the temperature distribution can better adjust to the $q'''(x)$ distribution. To use quadratic shape functions, an additional node is placed at the center of each element. The two elements are of equal width Δx_e and the nodes are all equally spaced with spacing Δx. Obtain expressions for the shape functions at nodes 1, 2, and 3, and make a plot of these functions.

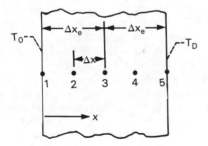

17-8 For use in a finite-element solution in a two-dimensional region, consider a three-node triangular element. The dimensionless temperature distribution $t(X, Y)$ is to be obtained. A linear interpolation function for $t(X, Y)$ is to be used in the form $t(X, Y) = aX + bY + c$. Find relations for the shape functions at each of the nodes.

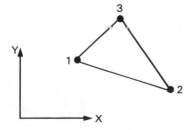

17-9 A plane layer of thickness D has a constant thermal conductivity k and has uniform internal heat generation q''' W/m³ throughout its volume. The boundaries are maintained at fixed temperatures T_0 and T_D. It is desired to set up the solution method by finite elements for a pure conduction situation. Linear shape functions are to be used. The layer is divided into four equal elements as shown. Derive the required finite-element equations to be solved for the temperatures at the interior nodes.

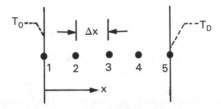

17-10 Modify Prob. 17-9 to have the Δx increments be of arbitrary unequal sizes across the width of the layer.

17-11 The layer in Prob. 17-9 is now a gray gas that is optically thin and has absorption coefficient a. Set up the finite-element equations to be solved for the temperatures at the internal nodes. The boundaries are transparent and do not radiate into the gas. The external surroundings are at a very low temperature.

17-12 The plane layer of optically thin gray gas in Prob. 17-11 now undergoes a transient heating process. The layer is initially unheated and is at a low temperature in equilibrium with its surroundings. Then the uniform internal heat generation q''' is suddenly applied. Develop the finite-element equations required to solve for the transient temperatures at the internal nodes 2, 3, and 4 as shown in the figure for Prob. 17-9. The thermal properties are assumed constant.

17-13 Reformulate Prob. 17-9 using the finite-element method with quadratic shape functions. Use the shape functions derived in Prob. 17-7, and the 2 elements and 5 nodes in that problem. For this problem, the $q'''(x)$ distribution is specified as $q'''(x) = 2 + \cos(\pi x/D)$.

17-14 A plane layer of semi-transparent absorbing, emitting, and heat conducting medium is between two thin metal (highly conducting) walls. The medium is not moving, and it is gray with absorption coefficient a and thermal conductivity k. The wall surfaces are diffuse-gray with emissivities ϵ_1 on both sides of the wall at $x = 0$ and ϵ_2 on both sides of the wall at $x = D$. The large isothermal surroundings below the lower wall are filled with gas at T_{g1}, and there is a heat transfer coefficient h_1 between the gas and the outside of the lower wall. Outside the upper wall there is a large isothermal reservoir filled with gas at T_{g2} that provides a convection coefficient h_2. Set up the necessary finite-difference relations to determine the wall temperatures T_{w1} and T_{w2}, and the heat flow q from T_{g1} to T_{g2}.

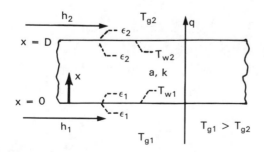

17-15 A parallel plate channel is heated with a uniform heat flux q along the outside of both of its walls. A semi-transparent absorbing, emitting, heat conducting medium is flowing between the walls with fully developed Poiseuille flow having a parabolic velocity distribution $u(y)$. The medium has absorption coefficient a and thermal conductivity k. Thermal properties are assumed constant. The channel wall interior surfaces are black. The region being considered is downstream from the channel entrance region. Axial heat conduction and axial radiative transfer are to be neglected in the formulation, although axial radiation can be considerable for some conditions. Set up a numerical procedure to determine the distribution of temperature within the medium across the channel. List the assumptions made. Note that the mean temperature of the medium will rise linearly along the channel length as a result of the uniform wall heat addition and the assumptions of negligible axial conduction and axial radiation.

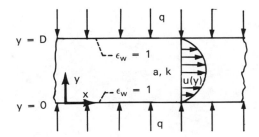

17-16 Obtain the solution to Prob. 17-1 by using a Monte Carlo numerical method.

EIGHTEEN

RADIATIVE BEHAVIOR OF WINDOWS, COATINGS, AND SEMITRANSPARENT SOLIDS

18-1 INTRODUCTION

In previous chapters most of the discussion of radiative transfer within a medium has been for materials that have a simple refractive index n of unity. This has application when the absorbing, emitting, and scattering medium is a gas, since almost all gases have a refractive index very close to unity (see Table 18-1). However, many important effects result from the nonunity refractive indices possessed by many common materials, such as those listed in the table. The most obvious effects are reflection and refraction at an interface; these phenomena determine the behavior of coatings, thin films, and multiple windows. Another important effect of nonunity n, as discussed in Sec. 2-5.12, is to increase the blackbody emission *within the medium* by a factor of n^2.

The previous discussions did not include enclosures in which radiation passed through windows. The analytical development was for enclosures composed of opaque walls or, in cavity geometries, with an opening through which the enclosure could directly interact with the environment. In some instances, single or multilayered windows form a major portion of an enclosure, such as in a flat-plate solar collector. Multiple reflections from the window surfaces can be appreciable, especially if more than one window layer is present, as in a two-cover-plate solar collector. Transmission losses through the windows can also be significant. The windows are usually thin, and hence it is often unnecessary to account for a temperature variation through the window thickness.

If one or more reflecting partially transparent layers are directly attached to an opaque surface or to other reflecting partially transparent layers, a coating system can be formed that has desirable radiative properties. The coating can be either thick or thin compared to the wavelength of the radiation. Thin films produce wave interference effects between incident and reflected waves in the film, thus influencing their reflection and transmission characteristics. Thin-film coating systems with high or low reflectivity can be produced. A composite window can be formed that is selectively transmitting, or a coated surface that is selectively absorbing can be produced. Examples are the transparent heat mirror and solar selective surfaces discussed in Sec. 5-5. Films that are thick relative to the radiation wavelength can also be used to modify surface characteristics.

The development in this chapter provides analytical tools for treating enclosures with windows and for obtaining the radiative behavior of coatings. Since the windows and coatings are usually thin, temperature variations within them are not considered in most of the development. There are applications, however, in which the temperature distribution must be obtained within a partially transmitting medium, including the effect of interface reflections. This requires solving the transfer equations within the medium as in Chap. 14 with the inclusion of reflection boundary conditions at the interfaces. Some important applications are heat treating of glass plates, temperature distributions in glass melting tanks, high-temperature solar components, laser heating of windows and lenses, and heating of spacecraft and aircraft windows. The theory for these applications is introduced in the last two sections of this chapter; additional development is in [2].

18-2 SYMBOLS

a	absorption coefficient
A	ratio of energy absorbed in plate system to energy incident; area
c	speed of light; specific heat of solid
E	amplitude of electric intensity wave; overall emittance
E_n	exponential integral function
F	configuration factor
i	intensity
I	source function
k	thermal conductivity
L	plate thickness
n	simple index of refraction
$\bar{n}$	complex index of refraction, $n - i\kappa$
N	number of surfaces in enclosure
q	energy flux, rate of energy per unit area
r	ratio of reflected to incident electric intensity
R	ratio of energy reflected from plate system to energy incident
S	path length
t	ratio of transmitted to incident electric intensity

T	ratio of energy transmitted through plate system to energy incident (overall transmittance); absolute temperature
V	volume
x	coordinate in medium
α	absorptivity
γ	the quantity $4\pi nL/\lambda_0$
θ	angle measured away from normal of surface
κ	extinction coefficient
λ	wavelength
μ	$\cos\theta$
ν	frequency
ρ	reflectivity; density
σ	Stefan-Boltzmann constant
σ_s	scattering coefficient
τ	transmissivity of layer; interface transmissivity; time
φ	circumferential angle
χ	angle of refraction
ω	solid angle; angular frequency
Ω	scattering albedo

Subscripts

c	at collector plate
e	emitted; entering
i	incoming; incident direction at boundary in Fig. 18-18
j	at surface j
k	at surface k
l	lost from outer surface
m	in a medium with $n > 1$
m, n	number of identical plates in system
max	maximum refraction angle
M	maximum value
n	groups of plates defined in Eqs. (18-12), (18-13)
o	outgoing
r	reflected
s	surroundings
sub	substrate
t	transmitted
w	window
λ, ν	spectrally dependent
0	in vacuum
1, 2, 3, 4	interface or medium 1, 2, 3, or 4
$\perp$	perpendicular component
$\parallel$	parallel component
I, II	layers I and II

Superscripts

(N)	group of N plates
$'$	for reciprocal path; directional quantity
$''$	bidirectional quantity
$*$	complex conjugate
$+$	in positive x direction
$-$	in negative x direction

18-3 TRANSMISSION, ABSORPTION, AND REFLECTION OF WINDOWS

Enclosures can have windows that are partially transparent to radiation. A window can be of a single material, or it can have one or more transmitting coatings on it. The transparency is a function of the window and coating thicknesses and can be quite wavelength dependent as illustrated by the transmittance of glass in Fig.

Table 18-1 Refractive indices of some common substances (from [1])

Material	Refractive index n
Gases (at $\lambda = 0.589$ μm)	
Air	1.00029
Argon	1.00028
Carbon dioxide	1.00045
Chlorine	1.00077
Hydrogen	1.00013
Methane	1.00044
Nitrogen	1.00030
Oxygen	1.00027
Water vapor	1.00025
Liquids (at $\lambda = 0.589$ μm)	
Chlorine	1.385
Ethyl alcohol	1.36 (25° C)
Oxygen	1.221
Water	1.33–1.32 (15–100° C)
Solids (at λ given)	
Glass:	
Crown	1.50–1.55 ⎫
Light flint	1.55–1.61 ⎬ (2–0.36 μm)
Heavy flint	1.62–1.71 ⎭
Ice	1.31 (0.589 μm)
Quartz	1.52–1.68 (2.3–0.19 μm)
Rock salt	1.5–1.9 (8.8–0.19 μm)

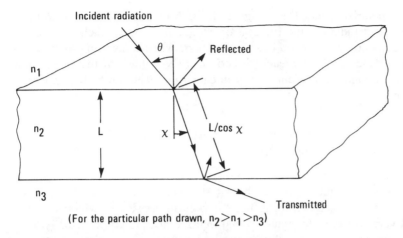

(For the particular path drawn, $n_2 > n_1 > n_3$)

Figure 18-1 Reflection and transmission of radiation by a window.

5-33. Some applications are glass or plastic cover plates for flat-plate solar collectors and thick- and thin-film coatings to provide modified reflection and transmission properties for camera lenses and solar cells. This section is concerned only with the transmission of incident radiation as influenced by surface reflections and absorption within the layer (no scattering).

As discussed in Sec. 4-5 and illustrated in Fig. 18-1, radiation incident on a surface is reflected and refracted. For a layer of thickness L, the refracted portion travels a distance $L/\cos \chi$ and is then partially reflected from the inside of the second surface. To analyze the radiative behavior of a plate such as in Fig. 18-1, the reflectivities are needed for radiation striking the outside and the inside of an interface. For a smooth interface, reflectivity relations are given by Eqs. (4-55), since windows are often dielectrics with a small extinction coefficient. Since these relations depend only on the square of the terms containing $\theta - \chi$, the θ and χ can be interchanged and hence along the incident and refracted paths the reflectivity is the same for radiation incident on an interface either from the outside or from within the material. For a constant absorption coefficient within the material, the transmittance along the path length within the window is $\tau = \exp(-aL/\cos \chi)$ where a is a function of wavelength as shown by the transmission behavior of glass in Figs. 5-32 and 5-33.[†]

The Fresnel reflection relations in (4-55) make it evident that the reflectivity at a smooth interface not only is a function of incidence angle but also is different for the two components of polarization. In what follows the path of radiation will be traced through partially transparent plates where the interface reflections are specular. If precise results are required, the resulting formulas should be applied at each incidence angle and for each component of polarization. If the incident

---

[†]For simplicity in notation, the functional dependence on λ is omitted in this section.

radiation is nonpolarized, half the energy is in each component of polarization. For diffuse incident radiation the fraction of energy incident in each θ direction within the increment $d\theta$ is $2 \sin \theta \cos \theta \, d\theta$. Then, when integrating to find the total reflectivity, the results for each direction are weighted in the integration according to the amount of incident energy in each $d\theta$ at θ and in each component of polarization.

18-3.1 Single-Layer Window or Transmitting Layer with Thickness $L > \lambda$ (No Wave Interference Effects)

When a window or transmitting layer is thin such that its thickness is comparable to the radiation wavelength, there can be interference between incident and reflected waves. This is discussed later. For the present consider a window where L is at least several wavelengths thick so that interference effects are usually lost. The actual thickness required for this to occur depends on the coherence length of the radiation in the material, which is beyond the scope of the present discussion.

Ray-tracing method Referring to Fig. 18-2, consider a unit intensity incident on the upper boundary, and apply a ray-tracing technique. At contact with the first interface, an amount ρ is reflected so that $1 - \rho$ enters the material. Of this, $(1 - \rho)\tau$ is transmitted [hence $(1 - \rho)(1 - \tau)$ is absorbed along the path] to interface 3 where $\rho(1 - \rho)\tau$ is reflected, and consequently $(1 - \rho)^2\tau$ passes out of the plate through the lower boundary. As the process continues, as shown in Fig. 18-2, the fraction of incident energy reflected by the plate is the sum of the terms leaving surface 1:

$$R = \rho\left[1 + (1 - \rho)^2\tau^2(1 + \rho^2\tau^2 + \rho^4\tau^4 + \cdots)\right] = \rho\left[1 + \frac{(1 - \rho)^2\tau^2}{1 - \rho^2\tau^2}\right] \qquad (18\text{-}1)$$

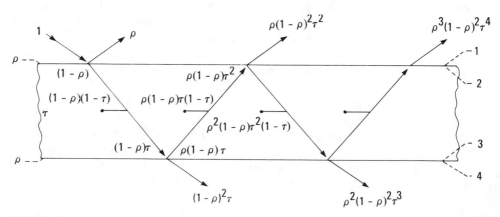

Figure 18-2 Multiple internal reflections for radiation incident on a window.

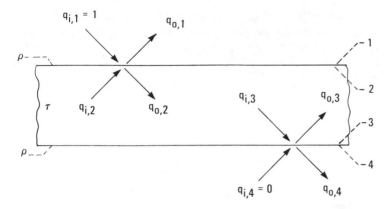

Figure 18-3 Net-radiation method applied to partially transmitting layer.

The fraction transmitted is the sum of terms leaving surface 4:

$$T = \tau(1-\rho)^2 \left[1 + \rho^2\tau^2 + \rho^4\tau^4 + \cdots\right] = \frac{\tau(1-\rho)^2}{1-\rho^2\tau^2} = \tau \frac{1-\rho}{1+\rho} \frac{1-\rho^2}{1-\rho^2\tau^2} \tag{18-2}$$

The last term on the right is often close to unity; in this instance $T \approx \tau(1-\rho)/(1+\rho)$. The fraction of energy absorbed is

$$A = (1-\rho)(1-\tau)\left[1 + \rho\tau + \rho^2\tau^2 + \rho^3\tau^3 + \cdots\right] = \frac{(1-\rho)(1-\tau)}{1-\rho\tau} \tag{18-3}$$

In the limit when absorption in the plate can be neglected, $\tau = 1$ and $R = 2\rho/(1+\rho)$, $T = (1-\rho)/(1+\rho)$, and $A = 0$.

Net-radiation method From the enclosure theory in Chap. 7, it is evident that the net radiation method is a powerful analytical tool that can, in many situations, be much less difficult to apply than the ray-tracing method. The net-radiation method [3] is now applied to derive the radiation characteristics of a partially absorbing layer. Referring to Fig. 18-3, the outgoing flux at each interface can be written in terms of the incoming fluxes to yield the following equations for the conditions of a unit incoming flux at surface 1 and a zero incoming flux at surface 4:

$$q_{o,1} = \rho q_{i,1} + (1-\rho)q_{i,2} = \rho + (1-\rho)q_{i,2} \tag{18-4a}$$

$$q_{o,2} = (1-\rho)q_{i,1} + \rho q_{i,2} = (1-\rho) + \rho q_{i,2} \tag{18-4b}$$

$$q_{o,3} = \rho q_{i,3} + (1-\rho)q_{i,4} = \rho q_{i,3} \tag{18-4c}$$

$$q_{o,4} = (1-\rho)q_{i,3} + \rho q_{i,4} = (1-\rho)q_{i,3} \tag{18-4d}$$

The transmittance of the layer is used to relate the internal q_i and q_o to give $q_{i,2} = q_{o,3}\tau$ and $q_{i,3} = q_{o,2}\tau$. These are used to eliminate the q_i from Eqs. (18-4), and the resulting equations are solved for the q_o to yield

$$q_{o,1} = \rho \left[1 + \frac{(1-\rho)^2 \tau^2}{1 - \rho^2 \tau^2} \right] \qquad q_{o,2} = \frac{1-\rho}{1 - \rho^2 \tau^2}$$

$$q_{o,3} = \frac{\rho\tau(1-\rho)}{1 - \rho^2 \tau^2} \qquad q_{o,4} = \frac{\tau(1-\rho)^2}{1 - \rho^2 \tau^2}$$

For $q_{i,1} = 1$, the fractions reflected and transmitted by the plate are $q_{o,1}$ and $q_{o,4}$ so that

$$R = q_{o,1} = \rho \left[1 + \frac{(1-\rho)^2 \tau^2}{1 - \rho^2 \tau^2} \right] = \rho(1 + \tau T) \tag{18-5a}$$

$$T = q_{o,4} = \frac{\tau(1-\rho)^2}{1 - \rho^2 \tau^2} = \tau \frac{1-\rho}{1+\rho} \frac{1-\rho^2}{1 - \rho^2 \tau^2} \tag{18-5b}$$

The fraction absorbed is

$$A = (q_{o,2} + q_{o,3})(1-\tau) = \frac{(1-\rho)(1-\tau)}{1 - \rho\tau} \tag{18-6}$$

These results agree, as they should, with those obtained by the ray-tracing method. If the ρ at the upper and lower surfaces are not equal, the results for R and T are in Prob. 18-6.

EXAMPLE 18-1 What is the fraction of externally incident unpolarized radiation that is transmitted through a glass plate in air? The plate is 0.75 cm thick, the radiation is incident at $\theta = 50°$, $n_{glass} = 1.53$, and $a_{glass} = 0.1$ cm^{-1}.

To find the path length through the glass write $\chi = \sin^{-1}(\sin \theta/n) = \sin^{-1}(\sin 50°/1.53) = 30°$. The path length is $S = 0.75/\cos \chi = 0.866$ cm. The transmittance is $\tau = \exp(-aS) = \exp(-0.1 \times 0.866) = 0.917$. The surface reflectivities for the two components of polarization are

$$\rho_\| = \frac{\tan^2(\theta - \chi)}{\tan^2(\theta + \chi)} = 0.00412 \qquad \rho_\perp = \frac{\sin^2(\theta - \chi)}{\sin^2(\theta + \chi)} = 0.1206$$

Then the overall transmittance for each component is

$$T_\| = \tau \frac{1-\rho_\|}{1+\rho_\|} \frac{1-\rho_\|^2}{1-\rho_\|^2 \tau^2} = 0.917 \frac{0.9959}{1.0041} \frac{1-0.00002}{1-0.00001} = 0.9095$$

$$T_\perp = \tau \frac{1-\rho_\perp}{1+\rho_\perp} \frac{1-\rho_\perp^2}{1-\rho_\perp^2 \tau^2} = 0.917 \frac{0.8794}{1.1206} \frac{1-0.0145}{1-0.0122} = 0.7179$$

For unpolarized incident radiation, one-half the energy is in each component. Hence $T = (T_\| + T_\perp)/2 = 0.814$.

18-3.2 Multiple Parallel Windows

Ray-tracing method As a more general case, consider a system of windows composed of a group of m identical plates and a group of n identical plates; the m plates can be different from the n plates. Since there are so many possible reflection paths, it might seem that ray tracing would be very complicated. Figure 18-4 shows, however, how the analysis by ray tracing can be organized so it can be carried out. An amount R_m of the incident unit energy is reflected from the group of m plates, and an amount T_m is transmitted. The R_m and T_m are found by building up results for a system of m plates with the formulas that are now obtained. Continuing the reflection and transmission process yields the terms shown in Fig. 18-4. Summing the reflected and the transmitted terms gives, for the system of $m + n$ plates,

$$R_{m+n} = R_m + R_n T_m^2 (1 + R_m R_n + R_m^2 R_n^2 + \cdots) = R_m + \frac{R_n T_m^2}{1 - R_m R_n} \tag{18-7a}$$

$$T_{m+n} = T_m T_n (1 + R_m R_n + R_m^2 R_n^2 + \cdots) = \frac{T_m T_n}{1 - R_m R_n} \tag{18-7b}$$

The absorbed energy can be found by summing the terms in Fig. 18-4, and the total absorbed energy will equal $1 - R_{m+n} - T_{m+n}$.

To illustrate how R_m (or R_n) and T_m (or T_n) are obtained, R_1 and T_1 are first found from (18-1) and (18-2), which are written for a single plate. Then from (18-7) the R_m and T_m for two plates are

$$R_2 = R_{1+1} = R_1 + \frac{R_1 T_1^2}{1 - R_1^2} \qquad T_2 = T_{1+1} = \frac{T_1^2}{1 - R_1^2}$$

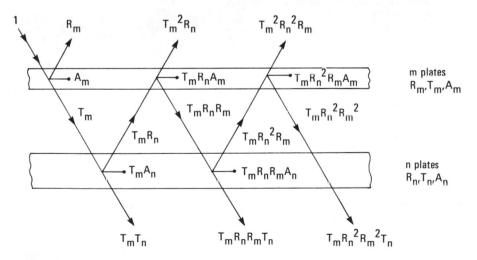

Figure 18-4 Ray-tracing method for multiple parallel transmitting layers.

where T_1 is from (18-2) for a single plate. Then, for three plates,

$$R_3 = R_{1+2} = R_1 + \frac{R_2 T_1^2}{1 - R_1 R_2} \qquad T_3 = T_{1+2} = \frac{T_1 T_2}{1 - R_1 R_2}$$

The process can be continued to any value m.

Net-radiation method The system of m and n plates is now analyzed by the net-radiation method. Using the notation in Fig. 18-5, the outgoing radiation terms are written in terms of the incoming fluxes as

$$q_{o,m1} = R_m + q_{i,m2} T_m \qquad (18\text{-}8a)$$

$$q_{o,m2} = q_{i,m2} R_m + T_m \qquad (18\text{-}8b)$$

$$q_{o,n1} = q_{i,n1} R_n \qquad (18\text{-}8c)$$

$$q_{o,n2} = q_{i,n1} T_n \qquad (18\text{-}8d)$$

The q_i are further related to the q_o by using the relations $q_{i,m2} = q_{o,n1}$ and $q_{i,n1} = q_{o,m2}$. The q_i are then eliminated, and solving for the q_o yields

$$T_{m+n} = q_{o,n2} = \frac{T_m T_n}{1 - R_m R_n} \qquad (18\text{-}9)$$

$$R_{m+n} = q_{o,m1} = R_m + \frac{R_n T_m^2}{1 - R_m R_n} \qquad (18\text{-}10)$$

The fractions of energy absorbed in the group of m plates and in the group of n plates within the system of $m + n$ plates are

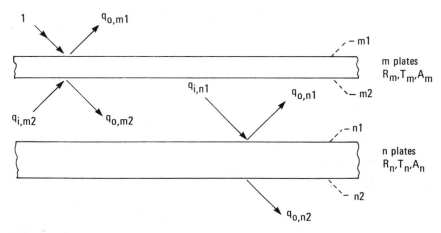

Figure 18-5 Net-radiation method for system of multiple parallel transmitting plates.

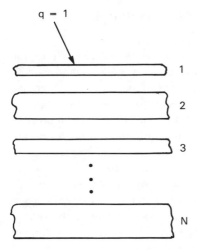

$q = 1$

1

2

3

N

Figure 18-6 A stack of N partially transmitting plates.

$$A_m{}^{(m+n)} = A_m + q_{i,m2} A_m = A_m \left(1 + \frac{T_m R_n}{1 - R_m R_n}\right) \qquad (18\text{-}11a)$$

$$A_n{}^{(m\mid n)} = q_{i,n1} A_n = \frac{T_m A_n}{1 - R_m R_n} \qquad (18\text{-}11b)$$

Note that T_{m+n} is symmetric; that is, the m and n subscripts can be exchanged and the expression remains the same, hence $T_{m+n} = T_{n+m}$. The T_{m+n} is the transmission for the system of $m + n$ plates for radiation incident first on the m plates, while T_{n+m} is for incidence first on the n plates. From (18-10), however, $R_{m+n} \neq R_{n+m}$; that is, the system reflectance depends on whether the radiation is incident first on the group of m plates or on the group of n plates.

If there is a stack of N plates that can all be different, as in Fig. 18-6, then the fraction of incident energy absorbed by the top plate (plate 1) is, from Eq. (18-11a),

$$A_1^{(N)} = A_1 \left(1 + \frac{T_1 R_n}{1 - R_1 R_n}\right) \qquad n = 2 + 3 + \cdots + N \qquad (18\text{-}12)$$

The fraction absorbed in plate 2 is

$$A_2^{(N)} = A_{1+2}^{(N)} - A_1^{(N)} = A_{1+2}\left(1 + \frac{T_{1+2} R_n}{1 - R_{1+2} R_n}\right) - A_1^{(N)} \qquad n = 3 + 4 + \cdots + N$$
$$(18\text{-}13)$$

This process can be continued to obtain the energy absorbed in each plate. The absorption in each plate is analyzed in [4] for a system of partially transmitting plates and an opaque absorber plate.

18-3.3 Results for Transmission through Multiple Parallel Glass Plates

Transmission through multiple glass plates is of interest in the design of flat-plate solar collectors. The glass surface reflectivity as given by (4-55a) and (4-55b) depends on the angle of incidence and the component of polarization. Since reflections from within the glass are assumed to be specular, the same angles of reflection and refraction are maintained throughout the multiple reflection process. Equation (18-9) was used to calculate the overall transmittance through one and three parallel glass plates with an index of refraction $n = 1.5$. Neglecting absorption within the glass, results (from [5]) are in Fig. 18-7 for the two components of polarization and as a function of the incidence angle of the radiation. As $\theta \to 90°$, the T goes to zero; this is because dielectrics have perfect reflectivity at grazing incidence (Fig. 4-6). For incidence at Brewster's angle, the overall transmittance for the parallel component becomes unity, as there is zero reflection at this angle, and losses by absorption are being neglected.

Incident solar radiation is unpolarized and hence has equal energy in the parallel and perpendicular components. The transmittance is then the average of the two T values computed by individually using $\rho_{\parallel}$ and $\rho_{\perp}$. This is shown in Fig. 18-8 for the limiting case of nonabsorbing plates and for absorbing plates having a product of absorption coefficient and thickness of 0.0524 per plate. For angles near normal incidence, the effect of absorption reduces transmission by about 5% for each plate.

18-3.4 Interaction of Transmitting Plates with Absorbing Plate

A flat-plate solar collector usually consists of one or more parallel transmitting windows covering an opaque absorber plate, as in Fig. 18-9. It is desired to obtain the fraction A_c of incident energy that is absorbed by the opaque plate. At the collector plate,

$$A_c = q_{i,c} - q_{o,c} \tag{18-14a}$$

$$q_{o,c} = (1 - \alpha_c)q_{i,c} \tag{18-14b}$$

Across the space between the transmitting plates and the opaque collector plate,

$$q_{o,c} = q_{i,n} \tag{18-14c}$$

$$q_{o,n} = q_{i,c} \tag{18-14d}$$

For the system of n plates,

$$q_{o,n} = T_n + q_{i,n}R_n \tag{18-14e}$$

The system of equations (18-14) is solved to yield the fraction of incident energy absorbed by the opaque plate:

$$A_c = \frac{\alpha_c T_n}{1 - (1 - \alpha_c)R_n} \tag{18-15}$$

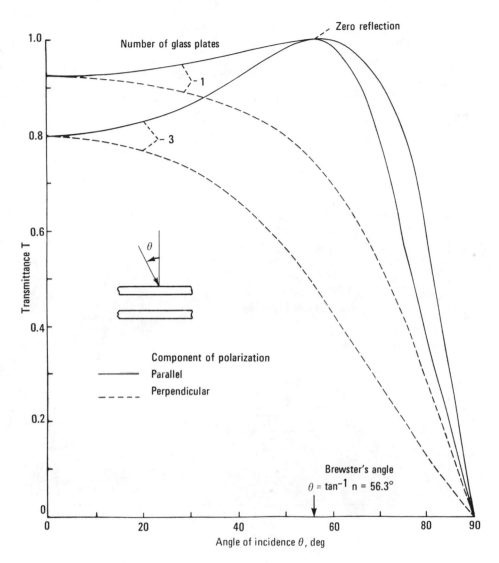

Figure 18-7 Overall transmittance of radiation in two components of polarization for nonabsorbing parallel glass plates; $n = 1.5$. From [5].

In this section, the fundamentals have been developed for analyzing the reflection, transmission, and absorption behavior of window systems for incident radiation, and some numerical results have been given. Many additional aspects are treated in the literature [3–10]. In [10] the model considers both beam and diffuse incident energy and the conversion of beam to diffuse energy that occurs in some of the semitransparent layers.

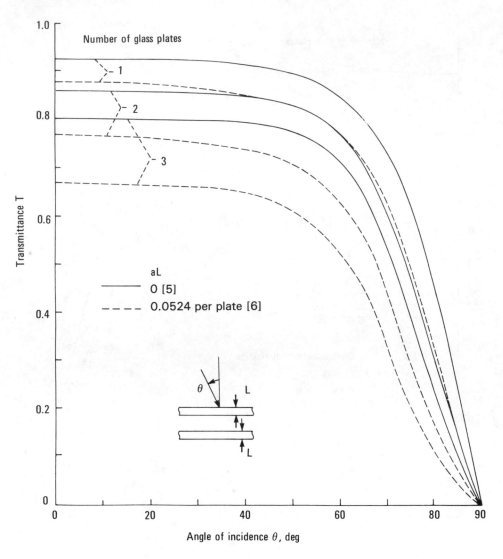

Figure 18-8 Effect of incidence angle and absorption on overall transmittance of multiple glass plates; $n = 1.5$.

18-4 ENCLOSURE ANALYSIS WITH PARTIALLY TRANSPARENT WINDOWS

The enclosures considered in Chaps. 6–11 have opaque walls or an opening such as at the end of a cylindrical cavity. A more general enclosure could contain partially transmitting windows, as in Fig. 18-10. Only a simplified case for such an enclosure is considered here; more information is in [8]. For this simplified

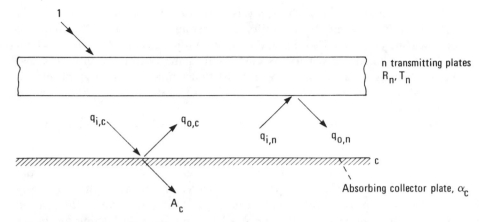

Figure 18-9 Interaction of transmitting windows and an absorbing plate.

case, the window properties are assumed independent of wavelength, and the radiation transmitted through and reflected from the window is assumed diffuse. A window such as a smooth glass plate reflects in a specular fashion so that the reflected portion of an individual beam of radiation leaving the surface will not be diffuse. However, within an enclosure there are usually multiple reflections, and the directionality of each reflection loses its importance in contributing to the heat fluxes on the boundaries. Hence the assumption of diffuse reflection is often satisfactory when the enclosure has multiple surfaces. When the window is hot

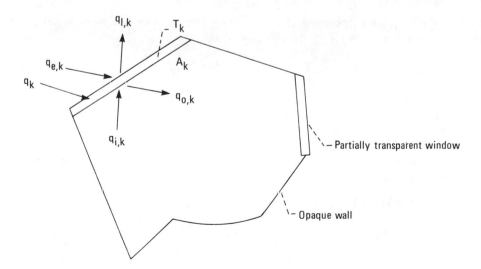

Figure 18-10 Enclosure with partially transparent windows.

enough to radiate appreciably, the analysis is restricted to a window that is thin enough so that it is essentially at a uniform temperature throughout. A symmetric window will be considered so that the radiative properties are the same on both sides; for example, if the window is coated, the same coating is on both sides. A two-band semigray analysis is given in [8] to account for the large difference in transmission properties on either side of the infrared cutoff wavelength as shown for glass in Fig. 5-33.

For the quantities in Fig. 18-10, an overall energy balance at window k yields

$$q_k = q_{o,k} - q_{i,k} + q_{l,k} - q_{e,k} \tag{18-16a}$$

The q_k is energy supplied to the window by a means other than radiation from *both* the inside and outside of the window, such as by convective heating or by heating wires within the semitransparent material. For convective *cooling* of the window the contribution to q_k is negative. The radiative flux leaving the inside surface of the window consists of emitted energy, reflected incoming energy, and transmitted externally incident flux:

$$q_{o,k} = E_{w,k}\sigma T_k^4 + R_{w,k}q_{i,k} + T_{w,k}q_{e,k} \tag{18-16b}$$

where E_w, R_w, and T_w are the overall emittance, reflectance, and transmittance of the window (T without a w subscript is the window temperature). Similarly, the radiative flux leaving the outside surface of the window is

$$q_{l,k} = E_{w,k}\sigma T_k^4 + R_{w,k}q_{e,k} + T_{w,k}q_{i,k} \tag{18-16c}$$

For a gray window $E_{w,k} = A_{w,k} = 1 - T_{w,k} - R_{w,k}$. The use of an emittance implies that the window has a temperature that can be considered uniform throughout its thickness. The $q_{l,k}$ is eliminated from Eqs. (18-16a) and (18-16c) to give

$$q_{o,k} = q_k - E_{w,k}\sigma T_k^4 + (1 - T_{w,k})q_{i,k} + (1 - R_{w,k})q_{e,k} \tag{18-16d}$$

where q_k and $q_{e,k}$ are specified. The incoming flux is obtained from the usual enclosure relation

$$q_{i,k} = \sum_{j=1}^{N} q_{o,j}F_{k-j} \tag{18-16e}$$

Equations (18-16b), (18-16d), and (18-16e) provide three equations relating q_o, q_i, and T for each partially transparent surface.

The $q_{i,k}$ can be eliminated to reduce the three relations to two equations. The first equation is obtained by solving Eq. (18-16b) for $q_{i,k}$ and inserting it into Eq. (18-16d). The second results from substituting (18-16e) into (18-16d). After rearrangement, the two relations are

$$q_k - E_{w,k}\sigma T_k^4 + (1 - R_{w,k})q_{e,k}$$
$$= \frac{E_{w,k}}{R_{w,k}}[(1 - T_{w,k})\sigma T_k^4 - q_{o,k}] + (1 - T_{w,k})\frac{T_{w,k}}{R_{w,k}}q_{e,k} \tag{18-17a}$$

$$q_k - E_{w,k}\sigma T_k^4 + (1 - R_{w,k})q_{e,k} = q_{o,k} - (1 - T_{w,k})\sum_{j=1}^{N} q_{o,j}F_{k-j} \tag{18-17b}$$

The left sides are the external heat input to the inside surface of the window. This is q_k minus the emission leaving through the outside surface and augmented by the amount of external radiation that passes into the window. For an opaque wall, q_k was defined to include all the energy quantities other than radiation at the surface inside the enclosure. Hence, for an opaque wall the left sides of Eqs. (18-17) become q_k, and, with the window transmittance $T_w = 0$ on the right side, the equations become the same as Eqs. (7-17) and (7-18). By eliminating the q_o from (18-17a) and (18-17b), a result that directly reduces to (7-31) is obtained,

$$\sum_{j=1}^{N} \frac{1}{E_{w,j}} (\delta_{kj} - R_{w,j} F_{k-j})[q_j - E_{w,j}\sigma T_j^4 + (1 - R_{w,j})q_{e,j}]$$

$$= \sum_{j=1}^{N} \left(\sigma T_j^4 + \frac{T_{w,j}}{E_{w,j}} q_{e,j} \right)[\delta_{kj} - (1 - T_{w,j})F_{k-j}] \qquad (18\text{-}18)$$

18-5 EFFECT OF COATINGS OR THIN FILMS ON SURFACES

The radiative behavior of a surface can be modified by depositing on it one or more very thin layers of other materials. The layers can be dielectric or metallic and can be thick or thin relative to the wavelength of the radiation. As will be shown, it is possible to obtain high or low absorption by the coated surface, and these characteristics will depend on the wavelength of the radiation. Thus it is possible to use thin-film coatings to tailor surfaces to have a desired wavelength-selective behavior.

18-5.1 Coating without Wave Interference Effects

The geometry is shown in Fig. 18-11, and it consists of a coating of thickness L on a thick substrate. The coating has a transmittance τ and reflectivities ρ_1 and ρ_2 at the first and second interfaces. The film is thick enough so it is not necessary

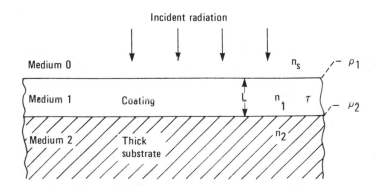

Figure 18-11 Coating of thickness L on a thick substrate.

to consider interference between waves reflected from the two interfaces. It will be determined to what extent the film alters the reflection characteristics from that of the substrate alone. By the net-radiation method, as derived for multiple layers, the fraction of incident radiation that is reflected is

$$R = \frac{\rho_1 + \rho_2(1 - 2\rho_1)\tau^2}{1 - \rho_1\rho_2\tau^2} \tag{18-19}$$

This result is now used to calculate the behavior for a few different types of coatings.

Nonabsorbing dielectric coating on nonabsorbing dielectric substrate Consider the relatively simple case of a dielectric film on a dielectric substrate. There is normally incident radiation through a surrounding dielectric medium having an index of refraction n_s. The film and substrate have refractive indices n_1 and n_2. Although the film is thick relative to the radiation wavelength, it is still physically quite thin, and since it is also a dielectric, the effect of absorption within it is generally quite small. Then $\tau \cong 1$, and R in (18-19) reduces to

$$R = \frac{\rho_1 + \rho_2(1 - 2\rho_1)}{1 - \rho_1\rho_2} \tag{18-20}$$

For normal incidence the interface reflectivities are given by ($\kappa_1 = \kappa_2 = \kappa_s \approx 0$)

$$\rho_1 = \left(\frac{n_1 - n_s}{n_1 + n_s}\right)^2, \qquad \rho_2 = \left(\frac{n_2 - n_1}{n_2 + n_1}\right)^2 \tag{18-21a,b}$$

Substituting (18-21a) and (18-21b) into (18-20) gives, after simplification,

$$R = 1 - \frac{4n_s n_1 n_2}{(n_1^2 + n_s n_2)(n_s + n_2)} \tag{18-22}$$

Dielectric coatings can be used to provide reduced reflection at the surface and hence maximize the amount of radiation passing into the substrate. The proper n_1 to minimize reflection is obtained by letting $dR/dn_1 = 0$. This yields

$$n_1 = \sqrt{n_s n_2} \tag{18-23}$$

Hence, minimum reflection is obtained if the n_1 of the coating is the geometric mean of the n values on either side of the coating. Using this n_1 in (18-22) gives the R for minimum reflection:

$$R = 1 - \frac{2\sqrt{n_s n_2}}{n_s + n_2} \tag{18-24}$$

If there is no coating, the reflectivity of the substrate by itself for normal incidence is $\rho_{sub} = (n_2 - n_s)^2/(n_2 + n_s)^2$. The ratio of the minimum R to ρ_{sub} can be simplified to

$$\frac{R}{\rho_{sub}} = \frac{n_s + n_2}{n_s + 2\sqrt{n_s n_2} + n_2} \tag{18-25}$$

For an optimum antireflection coating on glass in air, $n_s \approx 1$ and $n_2 \approx 1.53$, which yields $R/\rho_{sub} = 0.506$. Thus for these materials the dielectric coating can, at best, reduce surface reflection to about half the uncoated value. As will be seen later, better results can be obtained by using thin films. The square root of $n_2 = 1.53$ is 1.24, which from (18-23) is the optimum value of n_1, and it is difficult to find a suitable coating material with a refractive index this low. Some commonly used materials are magnesium fluoride, $n = 1.38$, or cryolite (sodium aluminum fluoride), $n = 1.36$. Also used are lithium fluoride, $n = 1.36$, and aluminum fluoride, $n = 1.39$.

Absorbing coating on metal substrate The same geometry is considered as in Fig. 18-11, but now the coating is attenuating ($\tau < 1$) and is on a metallic substrate. An example is a coated metal absorber plate for a flat-plate solar collector. The external radiation is normally incident in air, with $n_s \approx 1$. The complex index of refraction for the coating is $n_1 - i\kappa_1$, and for the substrate $n_2 - i\kappa_2$. The film transmittance is $\tau = \exp(-a_1 L)$ where $a_1 = 4\pi\kappa_1/\lambda$. This equality is useful for obtaining absorption coefficient values from handbook values of the optical coefficient κ. The a, κ, and n are wavelength dependent so the following can be regarded as a spectral calculation. For normal incidence [see Eq. (4-49)]

$$\rho_1 = \frac{(n_1 - 1)^2 + \kappa_1^2}{(n_1 + 1)^2 + \kappa_1^2} \tag{18-26a}$$

$$\rho_2 = \frac{(n_2 - n_1)^2 + (\kappa_2 - \kappa_1)^2}{(n_2 + n_1)^2 + (\kappa_2 + \kappa_1)^2} \tag{18-26b}$$

Equation (18-19) applies, and this becomes

$$R = \frac{\rho_1 + \rho_2(1 - 2\rho_1)\exp(-2a_1 L)}{1 - \rho_1\rho_2 \exp(-2a_1 L)} \tag{18-27}$$

As an example, if $a_1 = 10^3$ cm^{-1} and $\lambda = 0.7$ μm, the results in Fig. 18-12 are obtained for various coating thicknesses, for two substrate reflectivities and two indices of refraction of the coating. For high reflectance R of the coated metal, it is evident that the substrate reflectivity should be high and the absorptivity of the coating low; the index of refraction of the coating is not important. To obtain a low reflectance, $a_1 L > 1$ and $n_1 < 2$ are desired; the substrate reflectivity ρ_2 is not important when $a_1 L$ becomes greater than about 1.

18-5.2 Thin Film with Wave Interference Effects

Nonabsorbing dielectric thin film on nonabsorbing dielectric substrate When a film coating is thin, of order λ thick, there are interference effects between

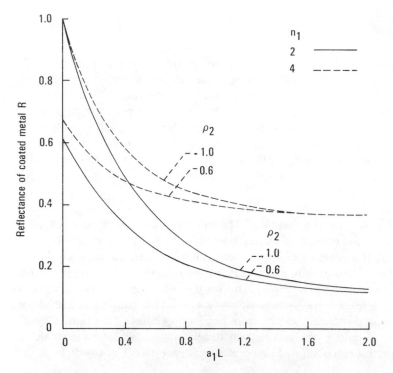

Figure 18-12 Reflectivity of thick attenuating film on metal substrate; no wave interference effects; $a_1 = 10^3$ cm^{-1}; $\lambda = 0.7$ μm.

waves reflected from the first and second surfaces of the film. As given by Eqs. (4-41), the amplitude reflection coefficients for the two components of polarization of the incident radiation are

$$\frac{E_{\|,r}}{E_{\|,i}} = r_\| = \frac{\tan(\theta - \chi)}{\tan(\theta + \chi)}, \qquad \frac{E_{\perp,r}}{E_{\perp,i}} = r_\perp = -\frac{\sin(\theta - \chi)}{\sin(\theta + \chi)} \qquad (18\text{-}28a,b)$$

When $n_2 > n_1$ for a wave incident from medium 1, then $\theta > \chi$ and $r_\perp$ is negative; that is, there is a phase change of π upon reflection. The $\tan(\theta - \chi)$ is positive, but $\tan(\theta + \chi)$ becomes negative for $\theta + \chi > \pi/2$, and $r_\|$ then yields a phase change of π. In a similar fashion, for transmitted radiation,

$$\frac{E_{\|,t}}{E_{\|,i}} = t_\| = \frac{2 \sin \chi \cos \theta}{\sin(\theta + \chi) \cos(\theta - \chi)}, \qquad \frac{E_{\perp,t}}{E_{\perp,i}} = t_\perp = \frac{2 \sin \chi \cos \theta}{\sin(\theta + \chi)} \qquad (18\text{-}29a,b)$$

In going from medium 2 to medium 1, the χ and θ values are interchanged in these relations (the χ and θ remain the angles in 2 and 1, respectively), and the reflection coefficients are equal to $-r_\|$ and $-r_\perp$. In this instance, the transmission coefficients are called t' and are equal to

$$t'_\| = \frac{2 \sin \theta \cos \chi}{\sin(\chi + \theta) \cos(\chi - \theta)}, \qquad t'_\perp = \frac{2 \sin \theta \cos \chi}{\sin(\chi + \theta)} \qquad (18\text{-}30a,b)$$

For a simplified case of normal incidence, and for radiation going from medium 1 to medium 2, these expressions reduce to

$$\frac{E_r}{E_i} = r_{\parallel} = r_{\perp} = r = \frac{n_1 - n_2}{n_1 + n_2}, \qquad \frac{E_t}{E_i} = t_{\parallel} = t_{\perp} = t = \frac{2n_1}{n_1 + n_2} \qquad \text{(18-31a,b)}$$

Formally the $r_{\parallel}$ will have a negative sign as $E_{\parallel,r}$ and $E_{\parallel,i}$ point in opposite directions as shown in Fig. 4-4. This sign is not significant in the present discussion. For normally incident radiation going from medium 2 into medium 1,

$$r = \frac{n_2 - n_1}{n_2 + n_1}, \qquad t' = \frac{2n_2}{n_2 + n_1} \qquad \text{(18-32a,b)}$$

For simplicity, the following discussion is limited to *normal incidence* on the thin film. Figure 18-13 shows the radiation reflected from the first and second interfaces. The beams a and b can interfere with each other. For normal incidence, beam b reflected from the second interface travels $2L$ farther than beam a, which is reflected from the first interface. Hence reflected beam b originated at time $2L/c_1$ earlier than reflected beam a, where c_1 is the propagation speed in the film. If beam a originated at time 0, then beam b originated at time $-2L/c_1$. If the two waves originated from the same vibrating source, the phase of b relative to a is $e^{i\omega\tau} = e^{-i\omega 2L/c_1}$. The circular frequency can be written as $\omega = 2\pi c_0/\lambda_0$ where λ_0 is the wavelength in vacuum. Then $e^{i\omega\tau} = e^{-i4\pi n_1 L/\lambda_0}$, where $n_1 = c_0/c_1$ is the refractive index of the film.

These results are now applied to analyze the performance of a thin film. Consider a thin nonabsorbing film of refractive index n_1 on a substrate with index n_2. For a normally incident wave of unit amplitude, the reflected radiation is shown in Fig. 18-14. Taking into account the phase relationships, the reflected amplitude is

$$R_M = r_1 + t_1 t_1' r_2 \exp\left(-i\frac{4\pi n_1 L}{\lambda_0}\right) - t_1 t_1' r_1 r_2{}^2 \exp\left(-i\frac{8\pi n_1 L}{\lambda_0}\right)$$

$$+ t_1 t_1' r_1{}^2 r_2{}^3 \exp\left(-i\frac{12\pi n_1 L}{\lambda_0}\right) - \cdots = r_1 + \frac{t_1 t_1' r_2 \exp\left(-i\,4\pi n_1 L/\lambda_0\right)}{1 + r_1 r_2 \exp\left(-i\,4\pi n_1 L/\lambda_0\right)} \qquad \text{(18-33)}$$

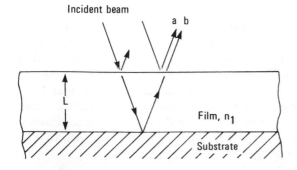

Incident beam

a b

L

Film, n_1

Substrate

Figure 18-13 Reflection from the first and second interfaces of a thin film. This is for *normal* incidence; paths are drawn at an angle for clarity.

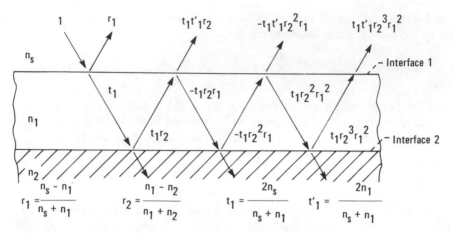

Figure 18-14 Multiple reflections within a thin nonattenuating film.

Note that $t_1 t_1' = 1 - r_1^2$ so this can be further reduced to

$$R_M = \frac{r_1 + r_2 \exp\left(-i 4\pi n_1 L/\lambda_0\right)}{1 + r_1 r_2 \exp\left(-i 4\pi n_1 L/\lambda_0\right)} \tag{18-34}$$

One application for this type of thin coating is to obtain low reflection from a surface for use in reducing reflection losses during transmission through a series of lenses in optical equipment. To have zero reflected amplitude, $R_M = 0$, requires $r_1 = -r_2 \exp(-i4\pi n_1 L/\lambda_0)$. This can be obtained if $r_1 = r_2$ and $\exp(-i4\pi n_1 L/\lambda_0) = -1$. Since $\exp(-i\pi) = -1$, this yields $L = \lambda_0/4n_1$. The quantity λ_0/n_1 is the wavelength of the radiation within the film. Hence the film thickness for zero reflection at normal incidence is one-quarter of the wavelength within the film. The required condition $r_1 = r_2$ gives $(n_s - n_1)/(n_s + n_1) = (n_1 - n_2)/(n_1 + n_2)$, which reduces to $n_1 = \sqrt{n_s n_2}$. Thus for normal incidence onto a quarter-wave film from a dielectric medium with index of refraction n_s, the index of refraction of the film should be $n_1 = \sqrt{n_s n_2}$, that is, the geometric mean of the n values on either side of the film. The thin film provides better performance than the thick film, as it is possible to achieve zero reflectivity. However, this result is only for normal incidence at one wavelength. To obtain more than one condition of zero reflectivity, it is necessary to use multilayer films. The optimization of the design of a low-reflectivity multilayer coating is discussed in [11]. For a system of two nonabsorbing quarter-wave films, zero reflection is obtained for the condition $n_1^2 n_3 = n_2^2 n_s$, where n_1 and n_2 are for the coatings (the coating with n_2 is next to the substrate) and n_3 is for the substrate.

The previous expressions have been derived for the reflected amplitude from a thin film; now the reflected energy for normal incidence is considered. From Sec. 4-4.3 the reflected energy depends on $|\mathbf{E}|^2$. Since $R_M = E_r/E_i$, the reflectivity for energy is $R = |R_M|^2 = R_M R_M^*$, where R_M^* is the complex conjugate of R_M. From (18-34), the reflectivity of the film is

$$R = \frac{r_1 + r_2 e^{-i\gamma_1}}{1 + r_1 r_2 e^{-i\gamma_1}} \frac{r_1 + r_2 e^{i\gamma_1}}{1 + r_1 r_2 e^{i\gamma_1}} \qquad \gamma_1 = \frac{4\pi n_1 L}{\lambda_0}$$

After multiplication and simplification this becomes

$$R = \frac{r_1^2 + r_2^2 + 2r_1 r_2 \cos \gamma_1}{1 + r_1^2 r_2^2 + 2r_1 r_2 \cos \gamma_1} \tag{18-35}$$

Another form of this equation is obtained by inserting $r_1 = (n_s - n_1)/(n_s + n_1)$ and $r_2 = (n_1 - n_2)/(n_1 + n_2)$ and using the identity $\cos \gamma_1 = 1 - 2 \sin^2(\gamma_1/2)$. After simplification R takes the form

$$R = \frac{n_1^2 (n_s - n_2)^2 - (n_s^2 - n_1^2)(n_1^2 - n_2^2) \sin^2(2\pi n_1 L/\lambda_0)}{n_1^2 (n_s + n_2)^2 - (n_s^2 - n_1^2)(n_1^2 - n_2^2) \sin^2(2\pi n_1 L/\lambda_0)} \tag{18-36}$$

For a quarter-wave film, $L = \lambda_0/4n_1$, and this reduces to

$$R = \left(\frac{n_s n_2 - n_1^2}{n_s n_2 + n_1^2}\right)^2 \tag{18-37}$$

As deduced before by looking at R_M, the reflectivity becomes zero when $n_1 = \sqrt{n_s n_2}$. If n_1 is high, R is increased, and this behavior can be used to obtain *dielectric mirrors*. For various film materials of refractive index n_1 on glass ($n_2 = 1.5$), the reflectivity becomes, for incidence in air ($n_s \approx 1$),

Film	n_1	n_2	R
None	1	1.5	0.04
ZnS	2.3	1.5	0.31
Ge	4.0	1.5	0.69
Te	5.0	1.5	0.79

Multilayer films can be used to obtain reflectivities very close to unity. An application is for the reflection of high-intensity laser beams where the fractional absorption of energy must be kept very small to avoid damage from heating the mirror.

Thin films of superconducting material are predicted to have absorptivity approaching zero out to very large wavelengths [12]. This high reflectivity gives some promise of producing nearly ideal (zero heat loss) radiation shields, particularly for preventing energy loss from space cryogenic storage systems. This technology will be of increasing usefulness as the critical temperature T_c of superconductors is brought to higher values with new materials.

Absorbing thin film on a metal substrate In this instance the complex indices of refraction $n_1 - i\kappa_1$ and $n_2 - i\kappa_2$ replace the n_1 and n_2 of the previous section. For *normal incidence* the $R = |R_M|^2$ becomes, by use of Eq. (18-34),

$$R = \left| \frac{r_1 + r_2 \exp(-i4\pi\bar{n}_1 L/\lambda_0)}{1 + r_1 r_2 \exp(-i4\pi\bar{n}_1 L/\lambda_0)} \right|^2 \tag{18-38}$$

where r_1 and r_2 now contain $\bar{n}_1$ and $\bar{n}_2$. An investigation is made in [13] of the behavior of Eq. (18-38) to obtain a selective surface for absorption of solar energy. The behavior of some coatings on an aluminum substrate is in Fig. 18-15.

The two complex indices of refraction of the film illustrate a trade-off between a low reflectivity for solar energy (high solar absorption) and the width of the transition region between the visible and infrared regions. Coating 1, which is thicker and has a lower n, has a significantly higher solar absorptance than coating 2, but it also has a wider transition region and would re-emit more in the infrared region. The lower n for coating 1 provides lower reflection losses at the front surface of the coating, but its higher value of κL produces a wider transition region. A method for improving the performance of the film-substrate combination as a solar collection surface is to have the index of refraction n vary within the film from the substrate value at the back surface (the interface with the substrate) to the value for the incident medium (air) at the front surface. This is explored in [13] and a list of references is given for analytical solutions of the graded index of refraction situation.

This development has shown how the net-radiation method and ray tracing can be used to analyze the radiative behavior of partially transparent coatings. The material presented is rather brief in view of the extensive information available on coatings and thin films. In addition to the references already discussed,

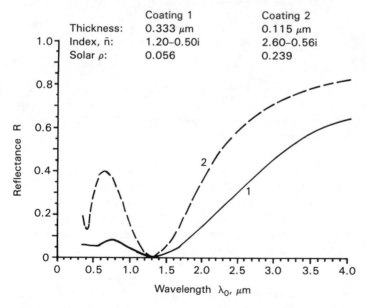

Figure 18-15 Spectral reflectance for two homogeneous films on an aluminum ($\bar{n} = 1.50 - 10i$) substrate for normal incidence [13].

the reader is referred to [14–21] for further information. Reference [21] provides a variety of spectral data for n and κ for metals, semiconductors, and dielectrics. By use of the relation $a(\lambda) = 4\pi\kappa(\lambda)/\lambda$ the spectral absorption coefficient can be found from the $\kappa(\lambda)$ values.

18-5.3 Films with Partial Coherence

When an incident wave enters a thin film it is refracted and propagates into the film. The incident wave is joined at the interface by the wave that has been reflected from the substrate and re-reflected from the film–air interface. This re-reflected wave has a time lag relative to the refracted incident wave. The two portions of the forward-propagating wave are thus partially coherent. In the limits of a thick film, the two portions become completely incoherent, and the theory presented in Sec. 18-5.1 (geometric optics) applies. If the film is quite thin (thickness $\approx \lambda$ or less), then the theory presented in Sec. 18-5.2 for a coherent wave applies. However, there is a significant portion of the wavelength range for a given film thickness where neither approach provides accurate results. This portion is called the region of *partial coherence,* and it must be treated in more depth than either of the limiting cases. Chen and Tien [22] have analyzed this region, and they present relations for determining the film transmittance and absorptance in the partially coherent range. They show that the region of partial coherence for nearly monochromatic incident radiation with frequency range $\Delta\nu$ is bounded by $1.13 \leq f \leq 2.59$, where $f = 4\pi n L \Delta\nu/c_0$ and c_0 is the speed of light in a vacuum. Below this range for f, coherent optics apply, and above it geometric optics may be used. In [22a] information is given on the effect of thickness on the radiative performance of a superconducting thin film on a substrate.

18-6 EFFECTS OF NONUNITY REFRACTIVE INDEX ON RADIATIVE BEHAVIOR WITHIN A MEDIUM

The partially transmitting layers considered in Secs. 18-3 to 18-5 have a uniform temperature throughout their thickness, or the temperatures are low enough that internal radiation is not significant. The behavior of layers with significant temperature levels and temperature variations are analyzed using the radiative equations in Chap. 14 modified for $n \neq 1$, along with appropriate conditions at the boundaries that include refractive index effects. Two effects are now discussed as required for a detailed heat transfer analysis.

18-6.1 Effect of Refractive Index on Intensity Crossing an Interface

Consider radiation with intensity $i'_{\lambda,1}$ in an ideal dielectric medium of refractive index n_1. Let the radiation in solid angle $d\omega_1$ pass into an ideal dielectric medium of refractive index n_2 as in Fig. 18-16. As a result of the differing indices of refraction, the rays change direction as they pass into medium 2. The radiation

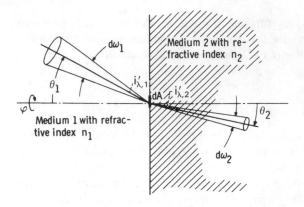

Figure 18-16 Beam with intensity $i'_{\lambda,1}$ crossing interface between two ideal dielectric media having unequal refractive indices.

in solid angle $d\omega_1$ at incidence angle θ_1 passes into solid angle $d\omega_2$ at angle of refraction θ_2. For simplicity in examining this angular effect within the medium, the amount of reflection of $i'_{\lambda,1}$ at the interface is temporarily neglected. This will be included in detail in Sec. 18-6.3. Briefly, to include the surface reflection, the $i'_{\lambda,1}(\theta_1)$ used here is replaced by $\rho'_\lambda(\theta_1)i'_{\lambda,i}(\theta_1)$, where $\rho'_\lambda(\theta_1)$ is the directional-hemispherical reflectivity of the interface and $i'_{\lambda,i}(\theta_1)$ is the incoming intensity. Thus the $i'_{\lambda,1}$ used here can be regarded as the portion of the incoming intensity that actually crosses the interface and passes into the material.

After allowing for reflection, the energy of the radiation is conserved in crossing the interface. From the definition of intensity this energy conservation is given by

$$i'_{\lambda,1} \cos\theta_1 \, dA \, d\omega_1 \, d\lambda_1 = i'_{\lambda,2} \cos\theta_2 \, dA \, d\omega_2 \, d\lambda_2 \tag{18-39}$$

where dA is an area element in the plane of the interface. Using the relation for solid angle, $d\omega = \sin\theta \, d\theta \, d\varphi$, results in (18-39) becoming (noting that the increment of circumferential angle $d\varphi$ is not changed in crossing the interface)

$$i'_{\lambda,1} \sin\theta_1 \cos\theta_1 \, d\theta_1 \, d\lambda_1 = i'_{\lambda,2} \sin\theta_2 \cos\theta_2 \, d\theta_2 \, d\lambda_2 \tag{18-40}$$

From (4-34) Snell's law relates the indices of refraction to the angles of incidence and refraction by

$$\frac{n_1}{n_2} = \frac{\sin\theta_2}{\sin\theta_1} \tag{18-41}$$

Then by differentiation

$$n_1 \cos\theta_1 \, d\theta_1 = n_2 \cos\theta_2 \, d\theta_2 \tag{18-42}$$

Substituting (18-41) and (18-42) into (18-40) gives

$$\frac{i'_{\lambda,1} d\lambda_1}{n_1^2} = \frac{i'_{\lambda,2} d\lambda_2}{n_2^2} \tag{18-43}$$

Although (18-43) was derived for radiation crossing the interface of two media, the equation also holds for intensity at any point inside a medium with variable refractive index as long as the local properties of the medium are independent of direction (isotropic). Thus in a partially transparent isotropic medium, for either the spectral intensity or, by integrating over all wavelengths, the total intensity in a medium with spectrally independent refractive index, there is the relation

$$\frac{i'_\lambda \, d\lambda}{n^2} = \text{const} \qquad \frac{i'}{n^2} = \text{const} \tag{18-44}$$

In terms of frequency, which does not change with n, we can write either

$$\frac{i'_\nu d\nu}{n^2} = \text{const} \quad \text{or} \quad \frac{i'_\nu}{n^2} = \text{const} \tag{18-45}$$

without restriction. Hence for spectral calculations with variable n, it is better to work with *frequency* than with wavelength.

18-6.2 The Effect of Angle for Total Reflection

Consider a volume element dV inside a region with refractive index n_2 as in Fig. 18-17. Suppose that diffuse radiation of intensity i'_1 is incident upon the boundary of this region from a region having refractive index n_1, where $n_1 < n_2$. Radiation incident at grazing angles to the interface ($\theta_1 \approx 90°$) will be refracted into medium 2 at a maximum value of θ_2 given by

$$\sin \theta_{2,\max} = \frac{n_1}{n_2} \sin 90° = \frac{n_1}{n_2} \tag{18-46}$$

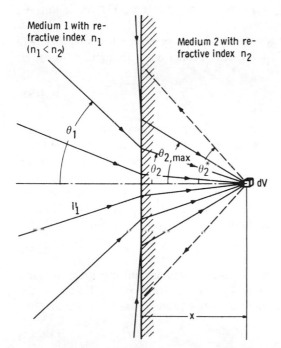

Medium 1 with re-
fractive index n_1
($n_1 < n_2$)

Medium 2 with re-
fractive index n_2

Figure 18-17 Effect of refraction on radiation transport in media with nonunity refractive index.

Hence the volume element in medium 2 will receive direct radiation from medium 1 only at angular directions within the range

$$0 \leqslant \theta_2 \leqslant \theta_{2,\text{max}} \quad \left(= \sin^{-1}\frac{n_1}{n_2}\right) \tag{18-47}$$

Now consider emission from dV. The portion of this emission that enters region 1 will be along paths found by reversing the arrows on the solid lines in Fig. 18-17. However, there is also radiation emitted from dV along paths such as those shown by the dashed lines in Fig. 18-17 that are incident on the interface at angles θ_2^*, where

$$\sin \theta_2^* > \frac{n_1}{n_2} \tag{18-48}$$

From (18-41), this means that such a ray would enter medium 1 at an angle given by

$$\sin \theta_1 = \frac{n_2}{n_1} \sin \theta_2^* > \frac{n_2 n_1}{n_1 n_2} = 1 \tag{18-49}$$

But $\sin \theta_1$ *cannot* be greater than unity for real values of θ_1. This means that any ray incident on the interface from medium 2 at an angle greater than

$$\theta_{2,\text{max}} = \sin^{-1}\frac{n_1}{n_2} \tag{18-50}$$

cannot enter medium 1 and is totally reflected at the interface. The angle defined by (18-50) is the *angle for total reflection*.

From Sec. 2-5.12 the blackbody spectral intensity emitted locally *inside* a medium with a nonunity refractive index that is a function of frequency, $n(\nu)$, is,

$$i_{\nu b,m}'(\nu) \, d\nu = \frac{2n^2(\nu)C_1 \nu^3}{c_o^4(e^{C_2\nu/c_oT} - 1)} \, d\nu \tag{18-51}$$

If spectral variations of $n(\nu)$ are important, an integration over frequency should be performed during the solution to obtain total energy quantities, provided that data for $n(\nu)$ as a function of frequency are available. It is best to use frequency as the spectral variable, as frequency does not change as energy travels within a medium with variable refractive index, while wavelength and wavenumber do change. If the refractive index is constant with frequency, integrating (18-51) over all ν yields the local total emitted blackbody intensity inside a medium,

$$i_{b,m}' = n^2 i_b' \tag{18-52}$$

where i_b' is for $n = 1$. Consequently, for an absorbing-emitting gray medium (refractive index not a function of frequency) with absorption coefficient a, the total energy emitted by a volume element at temperature T has an n^2 factor

$$dQ_e = 4n^2 a\sigma T^4 \, dV \tag{18-53}$$

From (18-52) it might appear that because $n > 1$, the intensity radiated from a dielectric medium into air could be larger than blackbody radiation i'_b. This is not the case, as some of the energy emitted within the medium is reflected back into the emitting body at the inside surface of the medium-air interface. Consider a thick ideal dielectric medium ($\kappa = 0$) at uniform temperature and with refractive index n. The maximum intensity received at an element dA on the interface from all directions within the medium is $n^2 i'_b$. Only the energy received within a cone having a vertex angle θ_{max} relative to the normal of dA can penetrate through the interface; for incidence angles larger than θ_{max}, the energy is totally reflected into the medium. Hence, the maximum amount of energy received at dA that can leave the medium is $\int_{\theta=0}^{\theta_{max}} n^2 i'_b \, dA \cos \theta \, 2\pi \sin \theta \, d\theta = 2\pi n^2 i'_b \, dA \, (\sin^2 \theta_{max}/2)$. From (18-46), with $n_1 = 1$ and $n_2 = n$ in this case, $\sin \theta_{max} = 1/n$ so the total hemispherical emissive power leaving the interface is $2\pi n^2 i'_b/2n^2 = \pi i'_b$. Dividing by π gives i'_b as the maximum diffuse intensity that can leave the interface, which is the expected blackbody radiation intensity. For a real interface there is partial internal reflection for $0 \le \theta \le \theta_{max}$, and the intensity leaving the interface is less than i'_b.

18-6.3 Interface Conditions for Analysis of Radiation within a Plane Layer

In Sec. 14-5 a plane layer of semitransparent medium was analyzed, and the solution depended on boundary conditions (14-50) giving the intensities *within the medium* at the boundaries. Later results of Chap. 14 were for a layer bounded by diffuse gray or black walls, so the boundary intensities were $1/\pi$ times the outgoing diffuse fluxes from the walls. Here we consider a plane layer within a medium that has a different refractive index, such as a glass plate in air or in water. Radiation is incident from the surrounding medium, and some of it will cross the boundaries and enter the plane layer. Expressions are obtained for the intensities *inside* the plane layer at the boundaries. The analysis of Chap. 14 can then be applied using these *internal* boundary conditions.

The geometry is shown in Fig. 18-18, which is a more general situation than in Fig. 14-9. As mentioned in connection with (18-45), it is convenient to use frequency as the spectral variable in situations where radiation crosses an interface between media having different n. Both the plane layer and surrounding medium are assumed to be ideal dielectrics with regard to interface behavior. A superscript s (for "surroundings") is used to designate conditions *outside* the layer. The angles θ and φ give the direction within the medium. The θ_i and φ_i fix incident directions at the boundaries. The $dq_{\nu,s}$ are the spectral fluxes *incident* on the layer from within the surrounding medium; these fluxes are assumed uniform over the layer boundary.

The intensity $i_\nu^+(0, \theta, \varphi)$ leaving boundary 1 *inside* the medium is composed of a transmitted portion from $i'_{\nu,s}(0, \theta_i, \varphi_i)$ and the reflected portion of $i_\nu^-(0, \theta_i, \varphi_i)$. The bidirectional spectral reflectivity (3-20) relates the reflected and incident intensities as

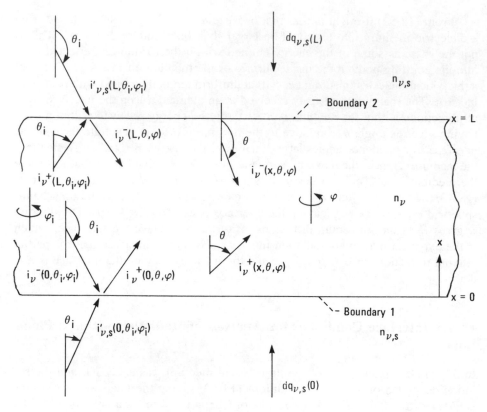

Figure 18-18 Intensities in a plane layer surrounded by a medium with a different refractive index.

$$\rho_\nu''(\theta, \varphi, \theta_i, \varphi_i) = \frac{i_{\nu,r}''(\theta, \varphi, \theta_i, \varphi_i)}{i_{\nu,i}'(\theta_i, \varphi_i) \cos \theta_i \, d\omega} \tag{18-54}$$

Similarly, a bidirectional transmissivity of the interface can be defined as

$$\tau_\nu''(\theta, \varphi, \theta_i, \varphi_i) = \frac{i_{\nu,\tau}''(\theta, \varphi, \theta_i, \varphi_i)}{i_{\nu,i}'(\theta_i, \varphi_i) \cos \theta_i \, d\omega} \tag{18-55}$$

where $i_{\nu,\tau}''$ is the intensity in the direction θ, φ obtained by transmission. The τ_ν'' contains a $(n_\nu/n_{\nu,s})^2$ factor to account for the effect in Sec. 18-6.1. By integrating over all incident solid angles at boundary 1, the intensity inside the layer leaving the boundary is:

$$i_\nu^+(0, \theta, \varphi) = \int_{\varphi_i=0}^{2\pi} \int_{\theta_i=0}^{\pi/2} \tau_{\nu,1}''(\theta, \varphi, \theta_i, \varphi_i) i_{\nu,s}'(0, \theta_i, \varphi_i) \cos \theta_i \sin \theta_i \, d\theta_i \, d\varphi_i$$

$$+ \int_{\varphi_i=0}^{2\pi} \int_{\theta_i=\pi/2}^{\pi} \rho_{\nu,1}''(\theta, \varphi, \theta_i, \varphi_i) i_\nu^-(0, \theta_i, \varphi_i) \cos \theta_i \sin \theta_i \, d\theta_i \, d\varphi_i \tag{18-56}$$

Similarly, inside the medium at boundary 2,

$$i_\nu^-(L, \theta, \varphi) = \int_{\varphi_i=0}^{2\pi} \int_{\theta_i=\pi/2}^{\pi} \tau_{\nu,2}''(\theta, \varphi, \theta_i, \varphi_i) i_{\nu,s}'(L, \theta_i, \varphi_i) \cos \theta_i \, \sin \theta_i \, d\theta_i \, d\varphi_i$$

$$+ \int_{\varphi_i=0}^{2\pi} \int_{\theta_i=0}^{\pi/2} \rho_{\nu,2}''(\theta, \varphi, \theta_i, \varphi_i) i_\nu^+(L, \theta_i, \varphi_i) \cos \theta_i \sin \theta_i \, d\theta_i \, d\varphi_i \quad (18\text{-}57)$$

As in Chap. 14, it is convenient to use the variable $\mu = \cos \theta$ to obtain

$$i_\nu^+(0, \mu, \varphi) = \int_{\varphi_i=0}^{2\pi} \left\{ \int_{\mu_i=0}^{1} [\tau_{\nu,1}''(\mu, \varphi, \mu_i, \varphi_i) i_{\nu,s}'(0, \mu_i, \varphi_i) \right.$$

$$\left. - \rho_{\nu,1}''(\mu, \varphi, -\mu_i, \varphi_i) i_\nu^-(0, -\mu_i, \varphi_i)] \mu_i \, d\mu_i \right\} d\varphi_i \quad (18\text{-}58)$$

$$i_\nu^-(L, \mu, \varphi) = \int_{\varphi_i=0}^{2\pi} \left\{ \int_{\mu_i=0}^{1} [-\tau_{\nu,2}''(\mu, \varphi, -\mu_i, \varphi_i) i_{\nu,s}'(L, -\mu_i, \varphi_i) \right.$$

$$\left. + \rho_{\nu,2}''(\mu, \varphi, \mu_i, \varphi_i) i_\nu^+(L, \mu_i, \varphi_i)] \mu_i \, d\mu_i \right\} d\varphi_i \quad (18\text{-}59)$$

As a special case, let the layer have optically smooth surfaces and let the incident fluxes $dq_{\nu,s}$ at 0 and L be diffuse. As discussed in Chap. 4, the reflection at an interface depends on the component of polarization. The analysis should consider the portion of radiation in each polarization component and then add the two energies to obtain the total quantity. For simplicity, this effect is neglected at present, and average values of the surface properties are used. The two components of polarization can be included by using Eqs. (4-57a,b), as is illustrated in Example 18-2.

The internal reflections from the optically smooth interfaces are specular; hence $\theta = \pi - \theta_i$ and $\varphi = \varphi_i + \pi$, and from (4-58a) at $x = L$,

$$\frac{i_\nu^-(L, \theta, \varphi)}{i_\nu^+(L, \theta_i, \varphi_i)} = \frac{1}{2} \frac{\sin^2(\theta_i - \chi)}{\sin^2(\theta_i + \chi)} \left[1 + \frac{\cos^2(\theta_i + \chi)}{\cos^2(\theta_i - \chi)} \right] \quad (18\text{-}60)$$

and similarly for $i_\nu^+(0, \theta, \varphi)/i_\nu^-(0, \theta_i, \varphi_i)$ where $\pi - \theta_i$ is used on the right in place of θ_i. The χ is determined from Snell's law, so that for the internal reflections,

$$\frac{\sin \chi}{\sin(\pi - \theta_i)} = \frac{\sin \chi}{\sin \theta_i} = \frac{n_\nu}{n_{\nu,s}} \qquad \text{at } x = 0 \text{ (boundary 1)}$$

$$\frac{\sin \chi}{\sin \theta_i} = \frac{n_\nu}{n_{\nu,s}} \qquad \text{at } x = L \text{ (boundary 2)}$$

where the subscript s corresponds to the surroundings outside the layer. Note that for some conditions and directions there is total internal reflection so the intensity reflected from the interface has the same magnitude as the incident intensity. This occurs when the index of refraction inside the layer is greater than that outside, $n_\nu > n_{\nu,s}$. There is total reflection when $\theta_i > \theta_{max}$ where $\theta_{max} = \sin^{-1}(n_{\nu,s}/n_\nu)$.

For the transmitted intensity the factor $(n_\nu/n_{\nu,s})^2$ is included to account for the effect discussed in Sec. 18-6.1, and incidence is from the surrounding medium onto the layer. Then, using (4-58a) for either boundary 1 or 2 gives

$$\frac{i_\nu^+(0,\theta,\varphi)}{i_{\nu,s}'(0,\theta_i,\varphi_i)} = \frac{i_\nu^-(L,\theta,\varphi)}{i_{\nu,s}'(L,\theta_i,\varphi_i)} = \left\{1 - \frac{1}{2}\frac{\sin^2(\theta_i-\theta)}{\sin^2(\theta_i+\theta)}\left[1+\frac{\cos^2(\theta_i+\theta)}{\cos^2(\theta_i-\theta)}\right]\right\}\left(\frac{n_\nu}{n_{\nu,s}}\right)^2 \quad (18\text{-}61)$$

where θ and θ_i are related by:

$$\sin\theta/\sin\theta_i = n_{\nu,s}/n_\nu$$

For diffuse incident fluxes the externally incident intensities are given by

$$i_{\nu,s}'(0,\theta_i,\varphi_i) = dq_{\nu,s}(0)/\pi\,d\nu \qquad (18\text{-}62a)$$

$$i_{\nu,s}'(L,\theta_i,\varphi_i) = dq_{\nu,s}(L)/\pi\,d\nu \qquad (18\text{-}62b)$$

This detailed analysis is not carried further, as it goes beyond the intended scope of this treatment. A development is in [2], where the literature is also reviewed. A few noteworthy references are those by Gardon [23–26], who treated problems of thermal radiation in glass, where effects of the refractive index are significant. Reference [23] includes some analysis of perpendicular and parallel polarization contributions; in [24, 25] a comprehensive analysis of the heat treatment of glass is given. This analysis of heat treatment includes the effects of conduction within the glass and convection at the surface. In [26] a review is given of radiant heat transfer as studied by researchers in the glass industry, and a digest of much of the literature on the subject up to 1961 is presented. Condon [27] gives a more recent review of radiation problems in the glass industry. As has been pointed out in a number of references discussing heat transfer by combined radiation and conduction in semitransparent media (see [26, 28], for example), care must be taken in an analysis to use true molecular conductivity values for the thermal conductivity k. High-temperature measurements of k in semitransparent materials must be corrected to eliminate the radiative contribution to the heat flux, which produces an apparent increase in k. In [29–33], the emittance is given from layers of material with nonunity refractive index, both semi-infinite in extent [31, 33] and bounded by a substrate [29, 30, 32]. The radiation and conduction heat transfer through a high-temperature semi-transparent layer of slag is analyzed in [34]. Index of refraction effects were analyzed in [35] for plane layers exposed to external radiation and convection. Internal reflections tended to make the temperature distributions more uniform in the central portions of the layer as compared with a layer with $n = 1$. This is also shown in [35a] for a radiating layer with absorption and scattering; the heat flow and temperature results for $n > 1$ are related to those for $n = 1$. Unconfined layers subjected to external radiation were analyzed in [36]. In [37] the response to an external pulse of radiation was analyzed. Inhomogeneous films may provide useful absorption properties for solar energy collection [38, 39]. An interesting example of ray tracing in a nonplanar geometry with nonunity refractive index is for radiation absorption by spheres (water droplets) [40]. The emittance of a sphere of scattering medium with Fresnel conditions at the boundary is analyzed in [41].

EXAMPLE 18-2 A volume element dV is located inside glass at $x = 3$ cm from an optically smooth planar interface with air as in Fig. 18-19. A diffuse-

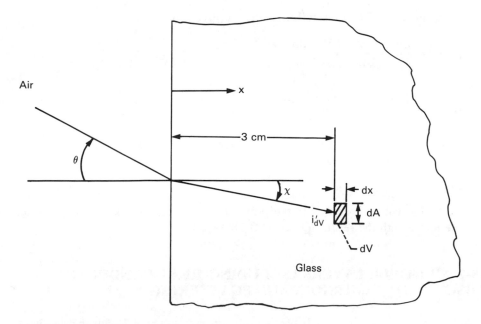

Figure 18-19 Coordinates for Example 18-2.

gray radiation flux $q_i = 40$ W/cm^2 in the air is incident on the glass surface. The absorption coefficient of the glass is assumed constant at 0.08 cm^{-1} and the refractive index of the glass is $n = 1.52$. Scattering is neglected. Determine the energy absorption rate per unit volume in dV.

Since the incident energy is diffuse and unpolarized, the incident intensity in each component of polarization is $q_i/2\pi$. The fraction of the intensity transmitted through the interface depends on the angle of incidence and is $1 - \rho'(\theta)$ where $\rho'(\theta)$ is given for each component of polarization by Eqs. (4-57a,b). From Eq. (18-43) the intensity inside the medium for each component of polarization is obtained by multiplying by n^2. The intensity in the medium in direction χ then becomes $(q_i/2\pi)[1 - \rho(\theta(\chi))]n^2$ where χ is the angle of refraction. The path length traveled from the interface to the volume element is $x/\cos \chi$. Using Bouguer's law, the fraction reaching dV is $\exp(-ax/\cos \chi)$. The fraction $a\,dx/\cos \chi$ of this incident energy at dV, $i'_{dV}\,dA \cos \chi$, is then absorbed in the volume element $dA\,dx$. The energy absorption for all directions of the arriving energy is found by integrating over $0 \le \chi \le \chi_{max}$ where χ_{max} is given by Snell's law as $\sin^{-1}(1/n)$. The integration over all incident solid angles introduces the factor for solid angle, $2\pi \sin \chi\,d\chi$. The energy absorbed per unit volume in dV is then obtained by doing this integration for each of the components of polarization and summing the results. This yields

$$\frac{dQ}{dV} = aq_i n^2 \int_0^{\chi_{max}} [2 - \rho_\parallel(\chi) - \rho_\perp(\chi)] \exp\left(-\frac{ax}{\cos \chi}\right) \sin \chi\,d\chi$$

From Eqs. (4-57a,b) the reflectivities are given by

$$\rho'_\parallel(\chi) = \left[\frac{n^2 \cos\theta - (n^2 - \sin^2\theta)^{1/2}}{n^2 \cos\theta + (n^2 - \sin^2\theta)^{1/2}} \right]^2$$

$$\rho'_\perp(\chi) = \left[\frac{(n^2 - \sin^2\theta)^{1/2} - \cos\theta}{(n^2 - \sin^2\theta)^{1/2} + \cos\theta} \right]^2$$

where the $\theta = \theta(\chi)$ on the right-hand side of the $\rho(\chi)$ equations is given in terms of χ by Snell's law as

$$\theta(\chi) = \sin^{-1}(n \sin\chi)$$

Using the values specified in the problem, the integration is carried out numerically with the result $dQ/dV = 3.38$ W/cm^3.

18-7 MULTIPLE LAYERS INCLUDING HEAT CONDUCTION, ABSORPTION, EMISSION, AND SCATTERING

The previous sections in this chapter were primarily concerned with the amount of energy *transmitted* through single- or multiple-layered windows and the effect on transmission of multiple reflections at the interfaces. If *temperature distributions* are needed within the multiple layers while including heat conduction, emission of energy, scattering, and internal heat generation, it is necessary to use the energy and radiative transfer equations as in Chap. 14 with the n^2 index of refraction factor included in the internal blackbody emission. The boundary conditions must include continuity of temperature and conduction heat flux across each interface (see, for example, discussions in [42] and [43]). As described in the previous section, the detailed directional reflection and transmission effects at an interface can become complicated ([44, 44a] are recent formulations including Fresnel reflection at a boundary). Simplifications are often made, such as assuming that the interfaces act as diffuse surfaces. Some examples using this approximation are in [45–50].

To illustrate the analysis for a multilayer situation, consider the transient thermal behavior of a three-layer system between two opaque diffuse-gray walls as in Fig. 18-20. All of the interfaces are assumed diffuse, and each layer is gray with isotropic scattering and constant properties. The transient energy equation is written for each layer by use of Eqs. (14-26) and (14-45). The square of the index of refraction appears in the emission term within the medium when $n \neq 1$. The wall emissivity is for emission into a medium, and this may differ from that into air because the index of refraction of the surrounding medium is different from unity (see Fig. 4-6, where the emissivity depends on the ratio of refractive indices of the emitting material and the surrounding medium). Note that for a black boundary the total emission into a surrounding medium with index of refraction n is $n^2 \sigma T_{\text{wall}}^4$ (see Sec. 2-5.12). For the first layer (layer I),

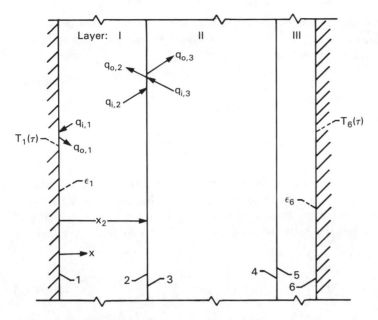

Figure 18-20 Multiple semitransparent layers between opaque parallel walls.

$$\rho c \frac{\partial T}{\partial \tau} = k \frac{\partial^2 T}{\partial x^2} + 4 \frac{a}{\Omega} [\pi I'(\kappa) - n^2 \sigma T^4(\kappa)] \tag{18-63}$$

where κ is the optical coordinate measured from the left boundary of the layer and all properties are for layer I. The integral equation for $I'(\kappa)$ is given by Eq. (14-69) as

$$I'(\kappa) = (1 - \Omega)n^2 \frac{\sigma T^4(\kappa)}{\pi} + \frac{\Omega}{2} \left[\frac{q_{o,1}}{\pi} E_2(\kappa) \right.$$

$$\left. + \frac{q_{o,2}}{\pi} E_2(\kappa_D - \kappa) + \int_0^{\kappa_D} I'(\kappa^*) E_1(|\kappa^* - \kappa|) \, d\kappa^* \right] \tag{18-64}$$

Since Eq. (18-63) has a second derivative in x, two spatial boundary conditions are needed. At each interface there is continuity of both temperature (contact resistance is assumed zero) and conduction heat flux so that at interface 2–3 between layers I and II (see [51] for additional discussion)

$$T_2 = T_3 \quad \text{and} \quad k_I \frac{\partial T}{\partial x}\bigg|_2 - k_{II} \frac{\partial T}{\partial x}\bigg|_3 \tag{18-65}$$

The wall temperature T_1 is a specified function of time. Similar equations and boundary conditions are written for the other layers and interfaces. The first derivative in time in (18-63) requires that the initial temperature distribution be specified.

Expressions for the q_o values are needed at the walls and interfaces. It is assumed that at each interface there is no absorption of energy, only reflection and transmission. Then at wall 1 and interfaces 2 and 3

$$q_{o,1} = \epsilon_1 n_I^2 \sigma T_1^4 + (1 - \epsilon_1) q_{i,1} \tag{18-66a}$$

$$q_{o,2} = q_{i,3}(1 - \rho_3) + q_{i,2}\rho_2 \tag{18-66b}$$

$$q_{o,3} = q_{i,2}(1 - \rho_2) + q_{i,3}\rho_3 \tag{18-66c}$$

and similarly at interfaces 4 and 5 and wall 6. These relations have introduced the incident fluxes q_i, each of which can be found from the flux leaving the other interface of the layer and from the source function within the layer (emitted and scattered energy). From the relations used in Eq. (14-74) there is obtained for layer I

$$q_{i,1} = 2q_{o,2}E_3(\kappa_D) + 2\pi \int_0^{\kappa_D} I'(\kappa)E_2(\kappa)\, d\kappa \tag{18-67a}$$

$$q_{i,2} = 2q_{o,1}E_3(\kappa_D) + 2\pi \int_0^{\kappa_D} I'(\kappa)E_2(\kappa_D - \kappa)\, d\kappa \tag{18-67b}$$

where κ_D is the optical thickness $(a_I + \sigma_{sI})x_2$ of layer I. The q_i are written in a similar fashion for the other layers using the optical distances in the individual layers. By use of Eqs. (18-67) and (18-66) the q_o can be eliminated from (18-64). The resulting energy and source function equations (18-63) and (18-64) are solved simultaneously by numerical procedures as described in Chap. 17 and [35, 45, 49].

REFERENCES

1. Hodgman, Charles D. (ed.): "Handbook of Chemistry and Physics," 60th ed., Chemical Rubber Publishing Company, Cleveland, Ohio, 1979.
2. Viskanta, R. and E. E. Anderson: Heat Transfer in Semi-Transparent Solids, in J. P. Hartnett and T. F. Irvine, Jr. (eds.), "Advances in Heat Transfer," vol. 11, pp. 317–441, Academic Press, New York, 1975.
3. Siegel, R.: Net Radiation Method for Transmission through Partially Transparent Plates, *Sol. Energy,* vol. 15, pp. 273–276, 1973.
4. Edwards, D. K.: Solar Absorption by Each Element in an Absorber-Coverglass Array, *Sol. Energy,* vol. 19, pp. 401–402, 1977.
5. Shurcliff, William A.: Transmittance and Reflection Loss of Multi-Plate Planar Window of a Solar-Radiation Collector: Formulas and Tabulations of Results for the Case of $n = 1.5$, *Sol. Energy,* vol. 16, pp. 149–154, 1974.
6. Duffie, J. A., and William A. Beckman: "Solar Energy Thermal Processes," Wiley, New York, 1974.
7. Wijeysundera, N. E.: A Net Radiation Method for the Transmittance and Absorptivity of a Series of Parallel Regions, *Sol. Energy,* vol. 17, pp. 75–77, 1975.

8. Siegel, R.: Net Radiation Method for Enclosure Systems Involving Partially Transparent Walls, NASA TN D-7384, August 1973.

9. Viskanta, R., D. L. Siebers, and R. P. Taylor: Radiation Characteristics of Multiple-Plate Glass Systems, *Int. J. Heat Mass Transfer,* vol. 21, pp. 815–818, 1978.

10. Mitts, S. J., and T. F. Smith: Solar Energy Transfer through Semitransparent Plate Systems, *J. Thermophys. Heat Transfer,* vol. 1, no. 4, pp. 307–312, 1987.

11. Thornton, B. S. and Q. M. Tran: Optimum Design of Wideband Selective Absorbers with Provisions for Specified Included Layers, *Sol. Energy,* vol. 20, no. 5, pp. 371–378, 1978.

12. Zeller, A. F.: High T_c Superconductors as Thermal Radiation Shields, *Cryogenics,* vol. 30, pp. 545–546, June 1990.

13. Snail, K. A.: Analytical Solutions for the Reflectivity of Homogeneous and Graded Selective Absorbers, *Sol. Energy Mater.,* vol. 12, pp. 411–424, 1985.

14. Hsieh, C. K., and R. Q. Coldewey: Study of Thermal Radiative Properties of Antireflection Glass for Flat-Plate Solar Collector Covers, *Sol. Energy,* vol. 16, pp. 63–72, 1974.

15. Musset, A., and A. Thelen: Multilayer Antireflection Coatings, in "Progress in Optics," vol. 8, pp. 203–237, American Elsevier, New York, 1970.

16. Forsberg, C. H., and G. A. Domoto: Thermal-Radiation Properties of Thin Metallic Films on Dielectrics, *J. Heat Transfer,* vol. 94, no. 4, pp. 467–472, 1972.

17. Taylor, R. P., and R. Viskanta: Spectral and Directional Radiation Characteristics of Thin-Film Coated Isothermal Semitransparent Plates, *Waerme Stoffuebertrag.,* vol. 8, pp. 219–227, 1975.

18. Heavens, O. S.: "Optical Properties of Thin Solid Films," Dover, New York, 1965.

19. Roux, J. A., A. M. Smith, and F. Shahrokhi: Effect of Boundary Conditions on the Radiative Reflectance of Dielectric Coatings, pp. 131–144, in "Thermophysics and Spacecraft Thermal Control," *Progress in Astronautics and Aeronautics,* vol. 35, MIT Press, Cambridge, Mass., 1974.

20. Palik, E. D., N. Ginsburg, H. B. Rosenstock, and R. T. Holm: Transmittance and Reflectance of a Thin Absorbing Film on a Thick Substrate, *Appl. Opt.,* vol. 17, no. 21, pp. 3345–3347, 1978.

21. Palik, E. D. (ed.): "Handbook of Optical Constants of Solids," Academic Press, New York, 1985.

22. Chen, G., and C.L. Tien: Partial Coherence Theory of Thin Film Radiative Properties. Presented at the 1991 ASME Winter Annual Meeting, ASME HTD vol. 84, pp. 9–20, 1991.

22a. Phelan, P. E., G. Chen, and C. L. Tien: Thickness-Dependent Radiative Properties of Y–Ba–Cu–O Thin Films, *J. Heat Transfer,* vol. 114, no. 1, pp. 227–233, 1992.

23. Gardon, Robert: The Emissivity of Transparent Materials, *J. Am. Ceram. Soc.,* vol. 39, no. 8, pp. 278–287, 1956.

24. Gardon, Robert: Calculation of Temperature Distributions in Glass Plates Undergoing Heat-Treatment, *J. Am. Ceram. Soc.,* vol. 41, no. 6, pp. 200–209, 1958.

25. Gardon, Robert: Appendix to Calculation of Temperature Distributions in Glass Plates Undergoing Heat Treatment, Ref. [24], Mellon Institute, Pittsburgh, 1958.

26. Gardon, Robert: A Review of Radiant Heat Transfer in Glass, *J. Am. Ceram. Soc.,* vol. 44, no. 7, pp. 305–312, 1961.

27. Condon, Edward U.: Radiative Transport in Hot Glass, *J. Quant. Spectrosc. Radiat. Transfer,* vol. 8, no. 1, pp. 369–385, 1968.

28. Araki, N.: Effect of Radiation Heat Transfer on Thermal Diffusivity Measurements, *Int. J. Thermophys.,* vol. 11, no. 2, pp. 329–337, 1990.

29. Caren, R. P., and C. K. Liu: Effect of Inhomogeneous Thin Films on the Emittance of a Metal Substrate, *J. Heat Transfer,* vol. 93, no. 4, pp. 466–468, 1971.

30. Baba, H., and A. Kanayama: Directional Monospectral Emittance of Dielectric Coating on a Flat Metal Substrate, paper 75-664, AIAA, May 1975.

31. Armaly, B. F., T. T. Lam, and A. L. Crosbie: Emittance of Semi-Infinite Absorbing and Isotropically Scattering Medium with Refractive Index Greater than Unity, *AIAA J.,* vol. 11, no. 11, pp. 1498–1502, 1973.

32. Anderson, E. E.: Estimating the Effective Emissivity of Nonisothermal Diatherminous Coatings, *J. Heat Transfer,* vol. 97, pp. 480–482, 1975.

33. Isard, J. O.: Surface Reflectivity of Strongly Absorbing Media and Calculation of the Infrared Emissivity of Glasses, *Infrared Phys.,* vol. 20, pp. 249–256, 1980.

34. Viskanta, R., and D. M. Kim: Heat Transfer through Irradiated, Semi-Transparent Layers at High Temperature, *J. Heat Transfer,* vol. 102, no. 1, pp. 182–184, 1980.

35. Spuckler, C. M. and R. Siegel: Refractive Index Effects on Radiative Behavior of a Heat Absorbing-Emitting Layer, *J. Thermophys. Heat Transfer,* 1992 (in press).

35a. Siegel, R., and C. M. Spuckler: Effect of Index of Refraction on Radiation Characteristics in a Heated Absorbing, Emitting, and Scattering Layer, *J. Heat Transfer,* 1992 (in press).

36. Schwander, D., G. Flamant, and G. Olalde, Effects of Boundary Properties on Transient Temperature Distributions in Condensed Semitransparent Media, *Int. J. Heat Mass Transfer,* vol. 33, no. 8, pp. 1685–1695, 1990.

37. Heping, T., B. Maestre, and M. Lallemand: Transient and Steady-State Combined Heat Transfer in Semi-Transparent Materials Subjected to a Pulse or a Step Irradiation, *J. Heat Transfer,* vol. 113, no. 1, pp. 166–173, 1991.

38. Heavens, O. S.: Optical Properties of Thin Films — Where To?, *Thin Solid Films,* vol. 50, pp. 157–161, 1978.

39. Fan, J. C. C.: Selective-Black Absorbers Using Sputtered Cermet Films, *Thin Solid Films,* vol. 54, pp. 139–148, 1978.

40. Harpole, G. M.: Radiative Absorption by Evaporating Droplets, *Int. J. Heat Mass Transfer,* vol. 23, no. 1, pp. 17–26, 1980.

41. Wu, C.-Y., and C.-J. Wang: Emittance of a Finite Spherical Scattering Medium with Fresnel Boundary, *J. Thermophys. Heat Transfer,* vol. 4, no. 2, pp. 250–252, 1990.

42. Amlin, D. W., and S. A. Korpela: Influence of Thermal Radiation on the Temperature Distribution in a Semi-Transparent Solid, *J. Heat Transfer,* vol. 101, no. 1, pp. 76–80, February 1979.

43. Chan, S. H., D. H. Cho, and G. Kocamustafaogullari: Melting and Solidification with Internal Radiative Transfer — A Generalized Phase Change Model, *Int. J. Heat Mass Transfer,* vol. 26, no. 4, pp. 621–633, 1983.

44. Rokhsaz, F., and R. L. Dougherty: Radiative Transfer within a Finite Plane-Parallel Medium Exhibiting Fresnel Reflection at a Boundary, in ASME HTD vol. 106, "Heat Transfer Phenomena in Radiation, Combustion and Fires," ASME-AIChE National Heat Transfer Conference, pp. 1–8, 1989.

44a. Reguigui, N. M., and R. L. Dougherty: Two-Dimensional Radiative Transfer in a Cylindrical Layered Medium with Reflective Interfaces, *J. Thermophys. Heat Transfer,* vol. 6, no. 2, pp. 232–241, 1992.

45. Ho, C.-H., and M. N. Özisik: Simultaneous Conduction and Radiation in a Two-Layer Planar Medium, *J. Thermophys. Heat Transfer,* vol. 1, no. 2, pp. 154–161, 1987.

46. Özisik, M. N., and S. M. Shouman: Radiative Transfer in an Isotropically Scattering Two-Region Slab with Reflecting Boundaries, *J. Quant. Spectrosc. Radiat. Transfer,* vol. 26, pp. 1–9, 1981.

47. Ho, C.-H., and M. N. Özisik: Combined Conduction and Radiation in a Two-Layer Planar Medium with Flux Boundary Condition, *Numer. Heat Transfer,* vol. 11, pp. 321–340, 1987.

48. Timoshenko, V. P., and M. G. Trenev: A Method for Evaluating Heat Transfer in Multilayered Semitransparent Materials, *Heat Transfer — Sov. Res.,* vol. 18, no. 5, pp. 44–57, 1986.

49. Tsai, C.-F., and G. Nixon: Transient Temperature Distribution of a Multilayer Composite Wall with Effects of Internal Thermal Radiation and Conduction, *Numer. Heat Transfer,* vol. 10, pp. 95–101, 1986.

50. Tarshis, L. A., S. O'Hara, and R. Viskanta: Heat Transfer by Simultaneous Conduction and Radiation for Two Absorbing Media in Intimate Contact, *Int. J. Heat Mass Transfer,* vol. 12, no. 3, pp. 333–347, 1969.

51. Song, B., and R. Viskanta: Deicing of Solids Using Radiant Heating, *J. Thermophys. Heat Transfer,* vol. 4, no. 3, pp. 311–317, 1990.

PROBLEMS

18-1 A horizontal glass plate 0.3 cm thick is covered by a plane layer of water 0.6 cm thick. A beam of radiation is incident from air onto the upper surface of the water at an incidence angle of 45°. What is the path length of radiation through the glass? (Take $n_{H_2O} = 1.33$, and $n_{glass} = 1.53$.)

Answer: 0.338 cm

18-2 Do Example 15-6 modified to include surface reflection and refraction effects.

Answer: 750 W/(m² · sr)

18-3 A radiation flux q_s is normally incident on a series of two glass sheets, each 0.35 cm thick, in air. What is the fraction of T that is transmitted? ($n_{glass} = 1.53$, and $a_{glass} = 12$ m⁻¹.) What is the fraction transmitted for a single plate 1 cm thick in air?

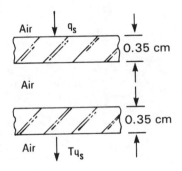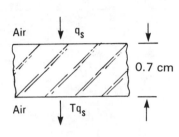

Answer: 2 plates, $T = 0.776$; 1 plate, $T = 0.842$

18-4 Prove that for an absorbing-emitting layer with $L > \lambda$, $T + A + R = 1$.

18-5 Derive an analytical expression that shows whether or not the total absorption in a system of n plates and m plates, $A_{(m+n)}$, is independent of whether radiation is first incident on the n or the m plates.

18-6 As a result of surface treatment, a partially transparent plate has a different reflectivity at each surface. For radiation incident on surface 1 in a single direction, obtain an expression for the overall reflectance and transmittance in terms of ρ_1, ρ_2, and τ.

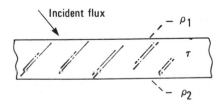

Answer: $R = \dfrac{\rho_1 + \rho_2(1 - 2\rho_1)\tau^2}{1 - \rho_1\rho_2\tau^2}$, $\quad T = \dfrac{(1 - \rho_1)(1 - \rho_2)\tau}{1 - \rho_1\rho_2\tau^2}$

18-7 Two parallel, partially transparent plates have different τ values and a different ρ at each surface. Obtain an expression for the overall transmittance of the two-plate system for radiation incident from a single direction.

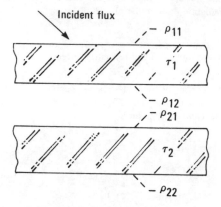

Incident flux

ρ_{11}

τ_1

ρ_{12}
ρ_{21}

τ_2

ρ_{22}

Answer: $T = \dfrac{T_1 T_2}{1 - R_{12} R_{21}}$

where $T_n = \dfrac{(1 - \rho_{n1})(1 - \rho_{n2})\tau_n}{1 - \rho_{n1}\rho_{n2}\tau_n^2}$

and $R_{nm} = \dfrac{\rho_{nm} + \rho_{nn}(1 - 2\rho_{nm})\tau_n^2}{1 - \rho_{nm}\rho_{nn}\tau_n^2}$

18-8 In a solar still, a thin layer of condensed water is flowing down a glass plate. The plate has a transmittance τ, and the water layer is assumed nonabsorbing. Obtain an expression for the overall transmittance T of the system for radiation incident on the glass in a single direction.

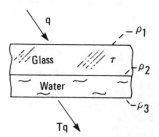

q

ρ_1

Glass τ

ρ_2

Water

ρ_3

Tq

Answer: $T = \dfrac{(1 - \rho_1)(1 - \rho_2)(1 - \rho_3)\tau}{(1 - \rho_2\rho_3)(1 - \rho_1\rho_2\tau^2) - \rho_1\rho_3(1 - \rho_2)^2\tau^2}$

18-9 The two glass plates in Prob. 18-3 are followed by a collector surface with absorptivity $\alpha_c = 0.95$. What is the fraction A_c absorbed by the collector surface for normally incident radiation?

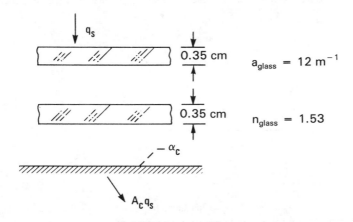

Answer: $A_c = 0.743$

18-10 Using the net radiation method, derive Eq. (18-19) for a thick dielectric film on a dielectric substrate.

Answer: $R = \dfrac{\rho_1 + \rho_2(1 - 2\rho_1)\tau^2}{1 - \rho_1\rho_2\tau^2}$

18-11 Extend Prob. 18-10 to a system of two differing dielectric layers coated onto a dielectric substrate.

Answer: $R = \rho_1 + \left[\dfrac{(1 - \rho_1)^2\tau_1^2(1 - \rho_2\rho_3\tau_2^2)}{(1 - \rho_1\rho_2\tau_1^2)(1 - \rho_2\rho_3\tau_2^2) - (1 - \rho_2)^2\rho_1\rho_3\tau_1^2\tau_2^2} \right]\left[\dfrac{\rho_2 + (1 - 2\rho_2)\rho_3\tau_2^2}{1 - \rho_2\rho_3\tau_2^2} \right]$

18-12 Two opaque gray plates have a nonabsorbing transparent plate between them. The transparent plate has surface reflectivities ρ. Derive a relation for the heat transfer from plate 1 to plate 2. Neglect conduction in the transparent plate, and neglect the fact that ρ depends on the angle of incidence.

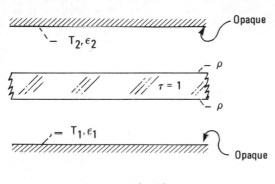

Answer: $q = \dfrac{\sigma(T_1^4 - T_2^4)}{1/\epsilon_1 + 1/\epsilon_2 - 1 + 2\rho/(1 - \rho)}$

18-13 Water is flowing between two identical glass plates adjacent to an opaque plate. Derive an expression for the fraction of radiation incident from a single direction that is absorbed by the water in terms of the ρ and τ values of the interfaces and layers. (Include only radiative energy transfer.)

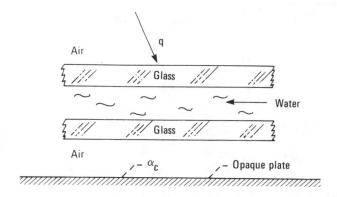

18-14 Do Prob. 15-10 with the index of refraction for glass equal to 1.53 and for the liquid equal to 1.35.

18-15 The sun shines on a flat-plate solar collector that has a single 0.5-cm-thick glass cover (use the data in Fig. 5-34 and assume that properties are constant below 1 μm) over a silicon oxide–coated aluminum absorber plate (use the data from Fig. 5-29 and assume that properties are constant below 0.4 μm). Determine the effective solar absorptivity of the collector for normal incidence (i.e., determine the fraction of incident solar energy that is actually absorbed by the collector plate).

18-16 A diffuse radiation flux q_{air} is incident on a dielectric medium having an absorption coefficient a and a simple index of refraction n. The air–medium interface is slightly rough so that it acts like a diffuse surface. It has a directional-hemispherical reflectivity of ρ'. Derive a relation to determine the amount of incident energy directly absorbed per unit volume, dQ/dV, in the medium as a function of depth x from the surface. For $n = 1.52$, as in Example 18-2, obtain an estimate for the surface reflectivity for diffuse incident radiation by using Fig. 4-7. Using this value and the conditions in Example 18-2, compute dQ/dV. How does the result compare with that in Example 18-2 and what is the reason for the difference?

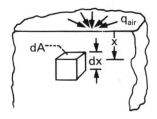

Answer: 3.08 W/cm³

18-17 A semi-transparent layer of fused silica has a complex index of refraction of $n - i\kappa = 1.42 - i1.45 \times 10^{-4}$ at the wavelength $\lambda = 3.5 \ \mu$m. Nonpolarized diffuse radiation in air at this wavelength is incident on the layer. What is the value of the overall transmittance of the layer for this diffuse incident radiation?

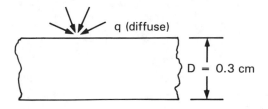

Answer: $T = 0.1311$

18-18 A series of three parallel glass plates is being used to absorb incident infrared radiation. The incident radiation is at wavelength $\lambda = 4 \ \mu$m. The complex index of refraction of the glass at this wavelength is $n - i\kappa = 1.40 - i5.8 \times 10^{-5}$. The plates have optically smooth surfaces. The radiation is incident from air in a direction normal to the plates. What fraction of the incident radiation is absorbed by the center plate?

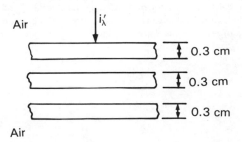

Answer: 0.233

18-19 A plane layer of semitransparent emitting and absorbing hot material is at uniform temperature. The surfaces of the layer are optically smooth, and the layer material is nonscattering. The simple index of refraction of the layer is n and the absorption coefficient is a (the layer is assumed gray). Obtain an expression for the emittance of the layer into vacuum.

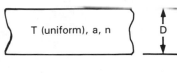

18-20 Two very thick dielectric regions with small absorption coefficients are in perfect contact at an optically smooth interface A_i. Prior to being placed in contact, region 1 was heated to a uniform temperature T_1 while region 2 was kept very cold at $T_2 \ll T_1$. The indices of refraction are such that $n_1 > n_2 > 1$. Derive an expression for the heat flux emitted across A_1 from region 1 into region 2. Evaluate the result for $n_1 = 2.2$, $n_2 = 1.5$, $T_1 = 625$ K, and $T_2 \approx 0$.

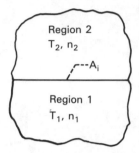

Answer: 17,780 W/m^2

18-21 Two very thick dielectric regions with small absorption coefficients are separated by a plane layer of a third dielectric material that is of thickness $D = 2$ cm. The two interfaces are optically smooth. The plane layer between the two thick regions is perfectly transparent. This plane layer is at low temperature, having been inserted suddenly between the two dielectric regions. For the temperatures shown, calculate the net heat flux being transferred by radiation from medium 1 into medium 2. Include the effects of interface reflections. The effect of heat conduction is neglected. The indices of refraction are given in the figure.

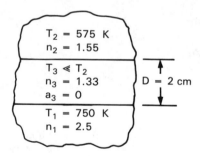

CONVERSION FACTORS, RADIATION CONSTANTS, AND BLACKBODY FUNCTIONS

Table A-1 lists some fundamental constants of radiation physics. Tables of conversion factors between the International System of Units (SI) and other systems of units are in Tables A-2 and A-3. In Table A-4 values of various radiation constants are given in both SI and U.S. conventional system (USCS) engineering units. Table A-5 lists blackbody emission properties as functions of the variable λT, again in both SI and USCS units. All values in Tables A-1, A-4, and A-5 are based on values of the physical constants given in [1].

With regard to Table A-5, the $F_{0-\lambda T}$ values were calculated using the series obtained by integrating the blackbody radiation function using integration by parts as discussed in Chap. 2. This series form was used to evaluate the blackbody integrals in [2] and is also given in [3]. The relation is repeated here for convenience:

$$F_{0-\lambda T} = \frac{15}{\pi^4} \sum_{n=1}^{\infty} \left[\frac{e^{-n\zeta}}{n} \left(\zeta^3 + \frac{3\zeta^2}{n^2} + \frac{6\zeta}{n^2} + \frac{6}{n^3} \right) \right]$$

where $\zeta = C_2/\lambda T$. This series is quite useful for computer solutions to provide on-line calculation of the $F_{0-\lambda T}$ function. Some polynomial approximations are also given in [4] and [5].

Table A-1 Values of the fundamental physical constants [1]

Definition	Symbol and value
Bohr electron radius	$a_0 = (4\pi/\mu_0 c_0^2)(\hbar^2/m_e e^2)$ $= 0.529177249 \times 10^{-10}$ m
Speed of light in vacuum	$c_0 = 2.99792458 \times 10^8$ m/s
Electronic charge	$e = 1.60217733 \times 10^{-19}$ C
Planck's constant	$h = 6.6260755 \times 10^{-34}$ J·s $\hbar = h/2\pi = 1.05457266 \times 10^{-34}$ J·s
Boltzmann's constant	$k = 1.380658 \times 10^{-23}$ J/K
Electron rest mass	$m_e = 9.1093897 \times 10^{-31}$ kg
Avogadro's number	$N_a = 6.02252 \times 10^{26}$ particles/kg-mole
Classical electron radius	$r_0 = (\mu_0/4\pi)(e^2/m_e)$ $= 2.81794092 \times 10^{-15}$ m
Thomson cross section	$\sigma_T = 8\pi r_0^2/3 = 6.6524616 \times 10^{-29}$ m^2
Permeability of vacuum	$\mu_0 = 4\pi \times 10^{-7}$ N·s^2/C^2
Electron volt	1 eV $= 1.60217733 \times 10^{-19}$ J
Ionization potential of hydrogen atom	$(c_0^2\mu_0/4\pi)(e^2/2a_0)$ $= [c_0^4\mu_0^2/(4\pi)^2](e^4 m_e/2\hbar^2)$ $= 13.6056981$ eV

Table A-2 Conversion factors for lengths [6]

	Mile, mi	Kilometer, km	Meter, m	Foot, ft	Inch, in
1 mile=	1	1.609	1609	5280	6.336×10^4
1 kilometer=	0.6214	1	10^3	3.281×10^3	3.937×10^4
1 meter=	6.214×10^{-4}	10^{-3}	1	3.281	39.37
1 foot=	1.894×10^{-4}	3.048×10^{-4}	0.3048	1	12
1 inch=	1.578×10^{-5}	2.540×10^{-5}	2.540×10^{-2}	8.333×10^{-2}	1
1 centimeter=	6.214×10^{-6}	10^{-5}	10^{-2}	3.281×10^{-2}	0.3937
1 millimeter=	6.214×10^{-7}	10^{-6}	10^{-3}	3.281×10^{-3}	0.03937
1 micrometer=	6.214×10^{-10}	10^{-9}	10^{-6}	3.281×10^{-6}	3.937×10^{-5}
1 nanometer=	6.214×10^{-13}	10^{-12}	10^{-9}	3.281×10^{-9}	3.937×10^{-8}
1 angstrom=	6.214×10^{-14}	10^{-13}	10^{-10}	3.281×10^{-10}	3.937×10^{-9}

	Centimeter, cm	Millimeter, mm	Micrometer, μm	Nanometer, nm	Angstrom, Å
1 mile=	1.609×10^5	1.609×10^6	1.609×10^9	1.609×10^{12}	1.609×10^{13}
1 kilometer=	10^5	10^6	10^9	10^{12}	10^{13}
1 meter=	10^2	10^3	10^6	10^9	10^{10}
1 foot=	30.48	3.048×10^2	3.048×10^5	3.048×10^8	3.048×10^9
1 inch=	2.540	25.40	2.540×10^4	2.540×10^7	2.540×10^8
1 centimeter=	1	10	10^4	10^7	10^8
1 millimeter=	10^{-1}	1	10^3	10^6	10^7
1 micrometer=	10^{-4}	10^{-3}	1	10^3	10^4
1 nanometer=	10^{-7}	10^{-6}	10^{-3}	1	10
1 angstrom=	10^{-8}	10^{-7}	10^{-4}	10^{-1}	1

Table A-3 Useful conversion factors [6]

Area	Mass
$1\ \text{ft}^2 = 0.0929030\ \text{m}^2$ $1\ \text{in}^2 = 6.4516 \times 10^{-4}\ \text{m}^2$ $1\ \text{m}^2 = 10.7639\ \text{ft}^2$	$1\ \text{lbm} = 0.453592\ \text{kg}$ $1\ \text{kg} = 2.20462\ \text{lbm}$

Volume	Density
$1\ \text{ft}^3 = 0.028317\ \text{m}^3$ $1\ \text{m}^3 = 35.315\ \text{ft}^3$	$1\ \text{lbm/ft}^3 = 16.0185\ \text{kg/m}^3$ $1\ \text{kg/m}^3 = 0.062428\ \text{lbm/ft}^3$

Heat	Heat rate per unit area
$(1\ \text{kJ} = 1\ \text{kW·s})$ $1\ \text{kJ} = 0.94782\ \text{Btu}^a = 0.23885\ \text{kcal}^a$ $1\ \text{Btu} = 1.0551\ \text{kJ} = 0.25200\ \text{kcal}$ $1\ \text{kcal} = 4.1868\ \text{kJ} = 3.9683\ \text{Btu}$ $1\ \text{kW·h} = 3.60 \times 10^6\ \text{J}$	$1\ \text{W/m}^2 = 0.31700\ \text{Btu}^a/(\text{h·ft}^2)$ $\quad = 0.85985\ \text{kcal}^a/(\text{h·m}^2)$ $1\ \text{Btu}/(\text{h·ft}^2) = 3.1546\ \text{W/m}^2$ $\quad = 2.7125\ \text{kcal}/(\text{h·m}^2)$ $1\ \text{kcal}/(\text{h·m}^2) = 1.1630\ \text{W/m}^2$ $\quad = 0.36867\ \text{Btu}/(\text{h·ft}^2)$

Heat rate	Heat transfer coefficient
$1\ \text{W} = 3.4121\ \text{Btu}^a/\text{h} = 0.85985\ \text{kcal}^a/\text{h}$ $1\ \text{Btu/h} = 0.29307\ \text{W} = 0.25200\ \text{kcal/h}$ $1\ \text{kcal/h} = 1.1630\ \text{W} = 3.9683\ \text{Btu/h}$	$1\ \text{W}/(\text{m}^2\text{·K}) = 0.17611\ \text{Btu}^a/(\text{h·ft}^2\text{·°R})$ $\quad = 0.85985\ \text{kcal}^a/(\text{h·m}^2\text{·K})$ $1\ \text{Btu}/(\text{h·ft}^2\text{·°R}) = 5.6783\ \text{W}/(\text{m}^2\text{·K})$ $\quad = 4.8824\ \text{kcal}/(\text{h·m}^2\text{·K})$ $1\ \text{kcal}/(\text{h·m}^2\text{·K}) = 1.1630\ \text{W}/(\text{m}^2\text{·K})$ $\quad = 0.20482\ \text{Btu}/(\text{h·ft}^2\text{·°R})$

Thermal conductivity	Specific heat
$1\ \text{W}/(\text{m·K}) = 0.57779\ \text{Btu}/(\text{h·ft·°R})$ $\quad = 0.85985\ \text{kcal}/(\text{h·m·K})$ $1\ \text{Btu}/(\text{h·ft·°R}) = 1.7307\ \text{W}/(\text{m·K})$ $\quad = 1.4882\ \text{kcal}/(\text{h·m·K})$ $1\ \text{kcal}/(\text{h·m·K}) = 1.1630\ \text{W}/(\text{m·K})$ $\quad = 0.67197\ \text{Btu}/(\text{h·ft·°R})$	$1\ \text{kJ}/(\text{kg·K}) = 0.23885\ \text{Btu}/(\text{lb·°R})$ $\quad = 0.23885\ \text{kcal}/(\text{kg·K})$ $1\ \text{Btu}/(\text{lb·°R}) = 4.1868\ \text{kJ}/(\text{kg·K})$ $\quad = 1.0000\ \text{kcal}/(\text{kg·K})$ $1\ \text{kcal}/(\text{kg·K}) = 4.1868\ \text{kJ}/(\text{kg·K})$ $\quad = 1.0000\ \text{Btu}/(\text{lb·°R})$

Energy per unit mass	Temperature
$1\ \text{kJ/kg} = 0.42992\ \text{Btu/lb} = 0.23885\ \text{kcal/kg}$ $1\ \text{Btu/lb} = 2.3260\ \text{kJ/kg} = 0.55556\ \text{kcal/kg}$ $1\ \text{kcal/kg} = 4.1868\ \text{kJ/kg} = 1.8000\ \text{Btu/lb}$	$K = \frac{5}{9}°R = \frac{5}{9}(°F + 459.67) = °C + 273.15$ $°R = \frac{9}{5}K = \frac{9}{5}(°C + 273.15) = °F + 459.67$ $°F = \frac{9}{5}°C + 32$ $°C = \frac{5}{9}(°F - 32)$

^{a}International Steam Table (for all Btu and kcal)

Table A-4 Radiation constants [1]

Symbol	Definition	Value
C_1	Constant in Planck's spectral energy (or intensity) distribution	0.18878×10^8 Btu·μm^4/(h·ft^2·sr) 0.59552197×10^8 W·μm^4/(m^2·sr) $0.59552197 \times 10^{-16}$ W·m^2/sr
C_2	Constant in Planck's spectral energy (or intensity) distribution	$25{,}897.84$ μm·°R $14{,}387.69$ μm·K 0.01438769 m·K
C_3	Constant in Wien's displacement law	5215.961 μm·°R 2897.756 μm·K 0.002897756 m·K
C_4	Constant in equation for maximum blackbody intensity	6.8712×10^{-14} Btu/(h·ft^2·μm·°R^5·sr) 4.095790×10^{-12} W/(m^2·μm·K^5·sr)
σ	Stefan-Boltzmann constant	0.17123×10^{-8} Btu/(h·ft^2·°R^4) 5.67051×10^{-8} W/(m^2·K^4)
q_{solar}	Solar constant	429 ± 7 Btu/(h·ft^2) 1353 ± 21 W/m^2
T_{solar}	Effective surface radiating temperature of the sun	5780 K, 10,400° R

Table A-5 Blackbody functions

Wavelength-temperature product λT		Blackbody hemispherical spectral emissive power divided by fifth power of temperature $e_{\lambda b}/T^5$		Black-body fraction	Difference between successive $F_{0-\lambda T}$ values
$\mu m \cdot K$	$\mu m \cdot °R$	$W/(m^2 \cdot \mu m \cdot K^5)$	$Btu/(h \cdot ft^2 \cdot \mu m \cdot °R^5)$	$F_{0-\lambda T}$	ΔF
500	900	381.28E-20	639.64E-22	1.30E-9	0.00000
550	990	323.87E-19	543.33E-21	1.35E-8	0.00000
600	1,080	185.42E-18	311.07E-20	9.29E-8	0.00000
650	1,170	786.04E-18	131.87E-19	4.67E-7	0.00000
700	1,260	263.74E-17	442.45E-19	1.84E-6	0.00000
750	1,350	735.27E-17	123.35E-18	5.95E-6	0.00000
800	1,440	176.61E-16	296.28E-18	1.64E-5	0.00001
850	1,530	375.68E-16	630.25E-18	3.99E-5	0.00002
900	1,620	722.93E-16	121.28E-17	8.70E-5	0.00005
950	1,710	127.97E-15	214.68E-17	1.74E-4	0.00008
1000	1,800	211.15E-15	354.23E-17	3.21E-4	0.00015
1050	1,890	328.24E-15	550.66E-17	5.56E-4	0.00024
1100	1,980	484.92E-15	813.51E-17	9.11E-4	0.00036
1150	2,070	685.67E-15	115.03E-16	0.00142	0.00051
1200	2,160	933.46E-15	156.60E-16	0.00213	0.00071
1250	2,250	122.95E-14	206.27E-16	0.00308	0.00095
1300	2,340	157.34E-14	263.96E-16	0.00432	0.00123
1350	2,430	196.29E-14	329.31E-16	0.00587	0.00155
1400	2,520	239.47E-14	401.74E-16	0.00779	0.00192
1450	2,610	286.39E-14	480.45E-16	0.01011	0.00232
1500	2,700	336.50E-14	564.53E-16	0.01285	0.00274
1550	2,790	389.20E-14	652.94E-16	0.01605	0.00320
1600	2,880	443.84E-14	744.60E-16	0.01972	0.00367
1650	2,970	499.76E-14	838.42E-16	0.02388	0.00416
1700	3,060	556.34E-14	933.33E-16	0.02854	0.00466
1750	3,150	612.97E-14	102.83E-15	0.03369	0.00516
1800	3,240	669.08E-14	112.25E-15	0.03934	0.00565
1850	3,330	724.17E-14	121.49E-15	0.04549	0.00614
1900	3,420	777.77E-14	130.48E-15	0.05211	0.00662
1950	3,510	829.51E-14	139.16E-15	0.05920	0.00709
2000	3,600	879.04E-14	147.47E-15	0.06673	0.00753
2025	3,645	902.89E-14	151.47E-15	0.07066	0.00393
2050	3,690	926.09E-14	155.36E-15	0.07469	0.00403
2075	3,735	948.61E-14	159.14E-15	0.07882	0.00413
2100	3,780	970.44E-14	162.80E-15	0.08306	0.00423
2125	3,825	991.55E-14	166.35E-15	0.08738	0.00433
2150	3,870	101.19E-13	169.76E-15	0.09180	0.00442
2175	3,915	103.16E-13	173.06E-15	0.09630	0.00450
2200	3,960	105.04E-13	176.22E-15	0.10089	0.00459
2225	4,005	106.85E-13	179.26E-15	0.10556	0.00467
2250	4,050	108.59E-13	182.17E-15	0.11031	0.00475
2275	4,095	110.25E-13	184.95E-15	0.11514	0.00482
2300	4,140	111.82E-13	187.60E-15	0.12003	0.00490
2325	4,185	113.33E-13	190.12E-15	0.12500	0.00496
2350	4,230	114.75E-13	192.51E-15	0.13002	0.00503
2375	4,275	116.10E-13	194.77E-15	0.13511	0.00509

(Table continues on next page)

Table A-5 Blackbody functions (*Continued*)

Wavelength-temperature product λT		Blackbody hemispherical spectral emissive power divided by fifth power of temperature $e_{\lambda b}/T^5$		Blackbody fraction $F_{0-\lambda T}$	Difference between successive $F_{0-\lambda T}$ values ΔF
$\mu m \cdot K$	$\mu m \cdot °R$	$W/(m^2 \cdot \mu m \cdot K^5)$	$Btu/(h \cdot ft^2 \cdot \mu m \cdot °R^5)$	$F_{0-\lambda T}$	ΔF
2400	4,320	117.37E-13	196.91E-15	0.14026	0.00515
2425	4,365	118.57E-13	198.92E-15	0.14546	0.00520
2450	4,410	119.69E-13	200.80E-15	0.15071	0.00525
2475	4,455	120.74E-13	202.56E-15	0.15601	0.00530
2500	4,500	121.72E-13	204.20E-15	0.16136	0.00535
2525	4,545	122.62E-13	205.72E-15	0.16675	0.00539
2550	4,590	123.46E-13	207.12E-15	0.17217	0.00542
2575	4,635	124.23E-13	208.41E-15	0.17763	0.00546
2600	4,680	124.93E-13	209.59E-15	0.18312	0.00549
2625	4,725	125.56E-13	210.65E-15	0.18865	0.00552
2650	4,770	126.13E-13	211.61E-15	0.19419	0.00555
2675	4,815	126.64E-13	212.46E-15	0.19977	0.00557
2700	4,860	127.09E-13	213.21E-15	0.20536	0.00559
2725	4,905	127.48E-13	213.86E-15	0.21097	0.00561
2750	4,950	127.81E-13	214.42E-15	0.21660	0.00563
2775	4,995	128.08E-13	214.88E-15	0.22224	0.00564
2800	5,040	128.30E-13	215.25E-15	0.22789	0.00565
2825	5,085	128.47E-13	215.53E-15	0.23355	0.00566
2850	5,130	128.59E-13	215.72E-15	0.23922	0.00567
2875	5,175	128.65E-13	215.83E-15	0.24489	0.00567
2898[a]	5,216	128.67E-13	215.87E-15	0.25000	0.00511
2900	5,220	128.67E-13	215.87E-15	0.25056	0.00056
2925	5,265	128.65E-13	215.82E-15	0.25624	0.00567
2950	5,310	128.57E-13	215.70E-15	0.26191	0.00567
2975	5,355	128.46E-13	215.51E-15	0.26757	0.00567
3000	5,400	128.31E-13	215.25E-15	0.27323	0.00566
3025	5,445	128.11E-13	214.92E-15	0.27889	0.00565
3050	5,490	127.88E-13	214.53E-15	0.28453	0.00564
3075	5,535	127.61E-13	214.08E-15	0.29016	0.00563
3100	5,580	127.30E-13	213.57E-15	0.29578	0.00562
3125	5,625	126.96E-13	213.00E-15	0.30139	0.00561
3150	5,670	126.59E-13	212.38E-15	0.30697	0.00559
3175	5,715	126.19E-13	211.70E-15	0.31255	0.00557
3200	5,760	125.76E-13	210.98E-15	0.31810	0.00555
3225	5,805	125.30E-13	210.21E-15	0.32364	0.00553
3250	5,850	124.82E-13	209.39E-15	0.32915	0.00551
3275	5,895	124.30E-13	208.54E-15	0.33464	0.00549
3300	5,940	123.77E-13	207.64E-15	0.34011	0.00547
3325	5,985	123.21E-13	206.70E-15	0.34555	0.00544
3350	6,030	122.63E-13	205.73E-15	0.35097	0.00542
3375	6,075	122.03E-13	204.72E-15	0.35637	0.00539
3400	6,120	121.41E-13	203.68E-15	0.36173	0.00537
3425	6,165	120.77E-13	202.61E-15	0.36707	0.00534
3450	6,210	120.11E-13	201.50E-15	0.37238	0.00531
3475	6,255	119.44E-13	200.37E-15	0.37766	0.00528

[a]$\lambda_{max}T$

Table A-5 (*Continued*)

Wavelength-temperature product λT		Blackbody hemispherical spectral emissive power divided by fifth power of temperature $e_{\lambda b}/T^5$		Blackbody fraction	Difference between successive $F_{0-\lambda T}$ values
$\mu\text{m·K}$	$\mu\text{m·°R}$	$\text{W}/(\text{m}^2\text{·}\mu\text{m·K}^5)$	$\text{Btu}/(\text{h·ft}^2\text{·}\mu\text{m·°R}^5)$	$F_{0-\lambda T}$	ΔF
3500	6,300	118.75E-13	199.22E-15	0.38291	0.00525
3525	6,345	118.05E-13	198.04E-15	0.38813	0.00522
3550	6,390	117.33E-13	196.84E-15	0.39332	0.00519
3575	6,435	116.60E-13	195.61E-15	0.39848	0.00516
3600	6,480	115.86E-13	194.37E-15	0.40360	0.00512
3625	6,525	115.11E-13	193.11E-15	0.40869	0.00509
3650	6,570	114.34E-13	191.83E-15	0.41375	0.00506
3675	6,615	113.57E-13	190.53E-15	0.41878	0.00502
3700	6,660	112.79E-13	189.22E-15	0.42377	0.00499
3725	6,705	112.00E-13	187.90E-15	0.42872	0.00496
3750	6,750	111.21E-13	186.56E-15	0.43364	0.00492
3775	6,795	110.40E-13	185.21E-15	0.43853	0.00489
3800	6,840	109.59E-13	183.86E-15	0.44338	0.00485
3825	6,885	108.78E-13	182.49E-15	0.44819	0.00481
3850	6,930	107.96E-13	181.12E-15	0.45297	0.00478
3875	6,975	107.14E-13	179.73E-15	0.45771	0.00474
3900	7,020	106.31E-13	178.35E-15	0.46241	0.00471
3925	7,065	105.48E-13	176.95E-15	0.46708	0.00467
3950	7,110	104.65E-13	175.56E-15	0.47172	0.00463
3975	7,155	103.81E-13	174.16E-15	0.47631	0.00460
4000	7,200	102.97E-13	172.75E-15	0.48087	0.00456
4050	7,290	101.30E-13	169.94E-15	0.48987	0.00901
4100	7,380	996.16E-14	167.12E-15	0.49873	0.00886
4150	7,470	979.37E-14	164.30E-15	0.50744	0.00871
4200	7,560	962.62E-14	161.49E-15	0.51600	0.00856
4250	7,650	945.94E-14	158.69E-15	0.52442	0.00841
4300	7,740	929.33E-14	155.91E-15	0.53269	0.00827
4350	7,830	912.84E-14	153.14E-15	0.54081	0.00812
4400	7,920	896.46E-14	150.39E-15	0.54878	0.00798
4450	8,010	880.22E-14	147.67E-15	0.55662	0.00783
4500	8,100	864.14E-14	144.97E-15	0.56431	0.00769
4550	8,190	848.23E-14	142.30E-15	0.57186	0.00755
4600	8,280	832.49E-14	139.66E-15	0.57927	0.00741
4650	8,370	816.94E-14	137.05E-15	0.58654	0.00727
4700	8,460	801.59E-14	134.48E-15	0.59367	0.00714
4750	8,550	786.44E-14	131.94E-15	0.60067	0.00700
4800	8,640	771.51E-14	129.43E-15	0.60754	0.00687
4850	8,730	756.79E-14	126.96E-15	0.61428	0.00674
4900	8,820	742.29E-14	124.53E-15	0.62089	0.00661
4950	8,910	728.02E-14	122.13E-15	0.62737	0.00648
5000	9,000	713.97E-14	119.78E-15	0.63373	0.00636
5050	9,090	700.16E-14	117.46E-15	0.63996	0.00623
5100	9,180	686.58E-14	115.18E-15	0.64608	0.00611
5150	9,270	673.23E-14	112.94E-15	0.65207	0.00600
5200	9,360	660.12E-14	110.74E-15	0.65795	0.00588
5250	9,450	647.24E-14	108.58E-15	0.66371	0.00576

(*Table continues on next page*)

Table A-5 Blackbody functions (*Continued*)

Wavelength-temperature product λT		Blackbody hemispherical spectral emissive power divided by fifth power of temperature $e_{\lambda b}/T^5$		Blackbody fraction $F_{0-\lambda T}$	Difference between successive $F_{0-\lambda T}$ values ΔF
$\mu m\cdot K$	$\mu m\cdot °R$	$W/(m^2\cdot\mu m\cdot K^5)$	$Btu/(h\cdot ft^2\cdot\mu m\cdot °R^5)$		
5300	9,540	634.59E-14	106.46E-15	0.66937	0.00565
5350	9,630	622.18E-14	104.38E-15	0.67491	0.00554
5400	9,720	610.00E-14	102.34E-15	0.68034	0.00543
5450	9,810	598.05E-14	100.33E-15	0.68566	0.00533
5500	9,900	586.33E-14	983.64E-16	0.69089	0.00522
5550	9,990	574.83E-14	964.36E-16	0.69600	0.00512
5600	10,080	563.56E-14	945.45E-16	0.70102	0.00502
5650	10,170	552.52E-14	926.92E-16	0.70594	0.00492
5700	10,260	541.69E-14	908.75E-16	0.71077	0.00482
5750	10,350	531.08E-14	890.95E-16	0.71550	0.00473
5800	10,440	520.68E-14	873.51E-16	0.72013	0.00463
5850	10,530	510.49E-14	856.42E-16	0.72468	0.00455
5900	10,620	500.51E-14	839.68E-16	0.72914	0.00446
5950	10,710	490.74E-14	823.28E-16	0.73351	0.00437
6000	10,800	481.17E-14	807.22E-16	0.73779	0.00428
6050	10,890	471.79E-14	791.50E-16	0.74199	0.00420
6100	10,980	462.62E-14	776.10E-16	0.74611	0.00412
6150	11,070	453.63E-14	761.02E-16	0.75015	0.00404
6200	11,160	444.83E-14	746.26E-16	0.75411	0.00396
6250	11,250	436.22E-14	731.81E-16	0.75800	0.00388
6300	11,340	427.79E-14	717.67E-16	0.76181	0.00381
6350	11,430	419.53E-14	703.82E-16	0.76554	0.00374
6400	11,520	411.45E-14	690.27E-16	0.76921	0.00366
6450	11,610	403.55E-14	677.00E-16	0.77280	0.00359
6500	11,700	395.81E-14	664.02E-16	0.77632	0.00352
6550	11,790	388.23E-14	651.31E-16	0.77978	0.00346
6600	11,880	380.82E-14	638.87E-16	0.78317	0.00339
6650	11,970	373.56E-14	626.70E-16	0.78650	0.00333
6700	12,060	366.46E-14	614.79E-16	0.78976	0.00326
6750	12,150	359.51E-14	603.13E-16	0.79296	0.00320
6800	12,240	352.71E-14	591.72E-16	0.79610	0.00314
6850	12,330	346.05E-14	580.55E-16	0.79918	0.00308
6900	12,420	339.54E-14	569.62E-16	0.80220	0.00302
6950	12,510	333.17E-14	558.93E-16	0.80517	0.00297
7000	12,600	326.93E-14	548.46E-16	0.80808	0.00291
7050	12,690	320.82E-14	538.22E-16	0.81093	0.00286
7100	12,780	314.85E-14	528.20E-16	0.81374	0.00280
7150	12,870	309.00E-14	518.39E-16	0.81649	0.00275
7200	12,960	303.28E-14	508.79E-16	0.81919	0.00270
7250	13,050	297.67E-14	499.39E-16	0.82183	0.00265
7300	13,140	292.19E-14	490.19E-16	0.82443	0.00260
7350	13,230	286.83E-14	481.19E-16	0.82699	0.00255
7400	13,320	281.57E-14	472.38E-16	0.82949	0.00251
7450	13,410	276.43E-14	463.75E-16	0.83195	0.00246
7500	13,500	271.40E-14	455.31E-16	0.83437	0.00242
7550	13,590	266.48E-14	447.05E-16	0.83674	0.00237
7600	13,680	261.65E-14	438.96E-16	0.83907	0.00233

Table A-5 (*Continued*)

Wavelength-temperature product λT		Blackbody hemispherical spectral emissive power divided by fifth power of temperature $e_{\lambda b}/T^5$		Black-body fraction $F_{0-\lambda T}$	Difference between successive $F_{0-\lambda T}$ values ΔF
μm·K	μm·°R	W/(m²·μm·K⁵)	Btu/(h·ft²·μm·°R⁵)		
7650	13,770	256.93E-14	431.04E-16	0.84135	0.00229
7700	13,860	252.31E-14	423.29E-16	0.84360	0.00225
7750	13,950	247.79E-14	415.70E-16	0.84580	0.00220
7800	14,040	243.36E-14	408.27E-16	0.84797	0.00217
7850	14,130	239.03E-14	401.00E-16	0.85010	0.00213
7900	14,220	234.78E-14	393.87E-16	0.85219	0.00209
7950	14,310	230.62E-14	386.90E-16	0.85424	0.00205
8000	14,400	226.55E-14	380.07E-16	0.85625	0.00202
8050	14,490	222.57E-14	373.38E-16	0.85823	0.00198
8100	14,580	218.66E-14	366.84E-16	0.86018	0.00195
8150	14,670	214.84E-14	360.42E-16	0.86209	0.00191
8200	14,760	211.10E-14	354.14E-16	0.86397	0.00188
8250	14,850	207.43E-14	347.99E-16	0.86581	0.00185
8300	14,940	203.84E-14	341.96E-16	0.86762	0.00181
8350	15,030	200.32E-14	336.06E-16	0.86941	0.00178
8400	15,120	196.87E-14	330.28E-16	0.87116	0.00175
8450	15,210	193.50E-14	324.62E-16	0.87288	0.00172
8500	15,300	190.19E-14	319.07E-16	0.87457	0.00169
8550	15,390	186.95E-14	313.64E-16	0.87623	0.00166
8600	15,480	183.78E-14	308.31E-16	0.87787	0.00163
8650	15,570	180.67E-14	303.09E-16	0.87947	0.00161
8700	15,660	177.62E-14	297.98E-16	0.88105	0.00158
8750	15,750	174.63E-14	292.97E-16	0.88261	0.00155
8800	15,840	171.71E-14	288.06E-16	0.88413	0.00153
8850	15,930	168.84E-14	283.25E-16	0.88563	0.00150
8900	16,020	166.03E-14	278.54E-16	0.88711	0.00148
8950	16,110	163.28E-14	273.92E-16	0.88856	0.00145
9000	16,200	160.58E-14	269.39E-16	0.88999	0.00143
9050	16,290	157.93E-14	264.95E-16	0.89140	0.00140
9100	16,380	155.33E-14	260.59E-16	0.89278	0.00138
9150	16,470	152.79E-14	256.33E-16	0.89413	0.00136
9200	16,560	150.30E-14	252.15E-16	0.89547	0.00134
9250	16,650	147.85E-14	248.05E-16	0.89679	0.00131
9300	16,740	145.46E-14	244.02E-16	0.89808	0.00129
9350	16,830	143.11E-14	240.08E-16	0.89935	0.00127
9400	16,920	140.80E-14	236.22E-16	0.90060	0.00125
9450	17,010	138.54E-14	232.43E-16	0.90183	0.00123
9500	17,100	136.33E-14	228.71E-16	0.90305	0.00121
9550	17,190	134.15E-14	225.06E-16	0.90424	0.00119
9600	17,280	132.02E-14	221.49E-16	0.90541	0.00117
9650	17,370	129.93E-14	217.98E-16	0.90657	0.00115
9700	17,460	127.88E-14	214.54E-16	0.90770	0.00114
9750	17,550	125.87E-14	211.16E-16	0.90882	0.00112
9800	17,640	123.90E-14	207.85E-16	0.90992	0.00110
9850	17,730	121.96E-14	204.60E-16	0.91101	0.00108
9900	17,820	120.06E-14	201.42E-16	0.91207	0.00107

(*Table continues on next page*)

Table A-5 Blackbody functions (*Continued*)

Wavelength-temperature product λT		Blackbody hemispherical spectral emissive power divided by fifth power of temperature $e_{\lambda b}/T^5$		Blackbody fraction $F_{0-\lambda T}$	Difference between successive $F_{0-\lambda T}$ values ΔF
μm·K	μm·°R	W/(m²·μm·K⁵)	Btu/(h·ft²·μm·°R⁵)		
9950	17,910	118.20E-14	198.29E-16	0.91312	0.00105
10000	18,000	116.37E-14	195.22E-16	0.91416	0.00103
10100	18,180	112.81E-14	189.26E-16	0.91618	0.00202
10200	18,360	109.39E-14	183.51E-16	0.91814	0.00196
10300	18,540	106.09E-14	177.97E-16	0.92004	0.00190
10400	18,720	102.91E-14	172.64E-16	0.92188	0.00184
10500	18,900	998.44E-15	167.50E-16	0.92367	0.00179
10600	19,080	968.91E-15	162.55E-16	0.92540	0.00173
10700	19,260	940.43E-15	157.77E-16	0.92709	0.00168
10800	19,440	912.97E-15	153.16E-16	0.92872	0.00163
10900	19,620	886.48E-15	148.72E-16	0.93031	0.00159
11000	19,800	860.93E-15	144.43E-16	0.93185	0.00154
11100	19,980	836.26E-15	140.29E-16	0.93334	0.00150
11200	20,160	812.46E-15	136.30E-16	0.93480	0.00145
11300	20,340	789.48E-15	132.45E-16	0.93621	0.00141
11400	20,520	767.29E-15	128.72E-16	0.93758	0.00137
11500	20,700	745.86E-15	125.13E-16	0.93892	0.00133
11600	20,880	725.16E-15	121.66E-16	0.94021	0.00130
11700	21,060	705.16E-15	118.30E-16	0.94147	0.00126
11800	21,240	685.83E-15	115.06E-16	0.94270	0.00123
11900	21,420	667.14E-15	111.92E-16	0.94389	0.00119
12000	21,600	649.08E-15	108.89E-16	0.94505	0.00116
12100	21,780	631.62E-15	105.96E-16	0.94618	0.00113
12200	21,960	614.72E-15	103.13E-16	0.94728	0.00110
12300	22,140	598.38E-15	100.39E-16	0.94835	0.00107
12400	22,320	582.57E-15	977.34E-17	0.94939	0.00104
12500	22,500	567.27E-15	951.68E-17	0.95041	0.00101
12600	22,680	552.47E-15	926.83E-17	0.95139	0.00099
12700	22,860	538.13E-15	902.78E-17	0.95236	0.00096
12800	23,040	524.25E-15	879.50E-17	0.95329	0.00094
12900	23,220	510.81E-15	856.95E-17	0.95420	0.00091
13000	23,400	497.79E-15	835.10E-17	0.95509	0.00089
13100	23,580	485.17E-15	813.94E-17	0.95596	0.00087
13200	23,760	472.95E-15	793.44E-17	0.95681	0.00084
13300	23,940	461.11E-15	773.57E-17	0.95763	0.00082
13400	24,120	449.62E-15	754.30E-17	0.95843	0.00080
13500	24,300	438.49E-15	735.63E-17	0.95922	0.00078
13600	24,480	427.70E-15	717.52E-17	0.95998	0.00076
13700	24,660	417.23E-15	699.97E-17	0.96072	0.00075
13800	24,840	407.08E-15	682.93E-17	0.96145	0.00073
13900	25,020	397.23E-15	666.41E-17	0.96216	0.00071
14000	25,200	387.68E-15	650.38E-17	0.96285	0.00069
14200	25,560	369.40E-15	619.72E-17	0.96419	0.00133
14400	25,920	352.17E-15	590.82E-17	0.96546	0.00127
14600	26,280	335.93E-15	563.57E-17	0.96667	0.00121
14800	26,640	320.61E-15	537.86E-17	0.96783	0.00116

Table A-5 (*Continued*)

Wavelength-temperature product λT		Blackbody hemispherical spectral emissive power divided by fifth power of temperature $e_{\lambda b}/T^5$		Blackbody fraction	Difference between successive $F_{0-\lambda T}$ values
μm·K	μm·°R	W/(m²·μm·K⁵)	Btu/(h·ft²·μm·°R⁵)	$F_{0-\lambda T}$	ΔF
15000	27,000	306.14E-15	513.59E-17	0.96893	0.00110
15200	27,360	292.47E-15	490.65E-17	0.96999	0.00106
15400	27,720	279.54E-15	468.97E-17	0.97100	0.00101
15600	28,080	267.32E-15	448.46E-17	0.97196	0.00096
15800	28,440	255.75E-15	429.06E-17	0.97289	0.00092
16000	28,800	244.80E-15	410.68E-17	0.97377	0.00088
16200	29,160	234.42E-15	393.27E-17	0.97461	0.00084
16400	29,520	224.58E-15	376.76E-17	0.97542	0.00081
16600	29,880	215.25E-15	361.11E-17	0.97620	0.00078
16800	30,240	206.39E-15	346.25E-17	0.97694	0.00074
17000	30,600	197.98E-15	332.14E-17	0.97765	0.00071
17200	30,960	189.99E-15	318.74E-17	0.97834	0.00068
17400	31,320	182.40E-15	306.01E-17	0.97899	0.00066
17600	31,680	175.18E-15	293.90E-17	0.97962	0.00063
17800	32,040	168.32E-15	282.37E-17	0.98023	0.00061
18000	32,400	161.78E-15	271.41E-17	0.98081	0.00058
18200	32,760	155.56E-15	260.97E-17	0.98137	0.00056
18400	33,120	149.63E-15	251.02E-17	0.98191	0.00054
18600	33,480	143.98E-15	241.54E-17	0.98243	0.00052
18800	33,840	138.59E-15	232.50E-17	0.98293	0.00050
19000	34,200	133.45E-15	223.88E-17	0.98341	0.00048
19200	34,560	128.54E-15	215.65E-17	0.98387	0.00046
19400	34,920	123.86E-15	207.79E-17	0.98431	0.00045
19600	35,280	119.39E-15	200.29E-17	0.98474	0.00043
19800	35,640	115.11E-15	193.12E-17	0.98516	0.00041
20000	36,000	111.03E-15	186.26E-17	0.98555	0.00040
20200	36,360	107.12E-15	179.71E-17	0.98594	0.00038
20400	36,720	103.38E-15	173.44E-17	0.98631	0.00037
20600	37,080	998.08E-16	167.44E-17	0.98667	0.00036
20800	37,440	963.84E-16	161.70E-17	0.98701	0.00035
21000	37,800	931.05E-16	156.20E-17	0.98735	0.00033
21200	38,160	899.64E-16	150.93E-17	0.98767	0.00032
21400	38,520	869.54E-16	145.88E-17	0.98798	0.00031
21600	38,880	840.68E-16	141.04E-17	0.98828	0.00030
21800	39,240	813.00E-16	136.39E-17	0.98858	0.00029
22000	39,600	786.45E-16	131.94E-17	0.98886	0.00028
23000	41,400	668.78E-16	112.20E-17	0.99014	0.00128
24000	43,200	572.24E-16	960.01E-18	0.99123	0.00109
25000	45,000	492.47E-16	826.18E-18	0.99217	0.00094
26000	46,800	426.09E-16	714.82E-18	0.99297	0.00081
27000	48,600	370.50E-16	621.57E-18	0.99368	0.00070
28000	50,400	323.67E-16	543.00E-18	0.99429	0.00061
29000	52,200	284.00E-16	476.45E-18	0.99482	0.00053
30000	54,000	250.21E-16	419.76E-18	0.99529	0.00047
31000	55,800	221.29E-16	371.25E-18	0.99571	0.00042

(Table continues on next page)

Table A-5 Blackbody functions (*Continued*)

Wavelength-temperature product λT		Blackbody hemispherical spectral emissive power divided by fifth power of temperature $e_{\lambda b}/T^5$		Blackbody fraction $F_{0-\lambda T}$	Difference between successive $F_{0-\lambda T}$ values ΔF
μm·K	μm·°R	W/(m²·μm·K⁵)	Btu/(h·ft²·μm·°R⁵)		
32000	57,600	196.43E-16	329.53E-18	0.99607	0.00037
33000	59,400	174.95E-16	293.51E-18	0.99640	0.00033
34000	61,200	156.33E-16	262.27E-18	0.99669	0.00029
35000	63,000	140.12E-16	235.07E-18	0.99695	0.00026
36000	64,800	125.95E-16	211.30E-18	0.99719	0.00023
37000	66,600	113.53E-16	190.46E-18	0.99740	0.00021
38000	68,400	102.60E-16	172.12E-18	0.99759	0.00019
39000	70,200	929.52E-17	155.94E-18	0.99776	0.00017
40000	72,000	844.12E-17	141.61E-18	0.99792	0.00016
41000	73,800	768.29E-17	128.89E-18	0.99806	0.00014
42000	75,600	700.78E-17	117.57E-18	0.99819	0.00013
43000	77,400	640.52E-17	107.46E-18	0.99831	0.00012
44000	79,200	586.59E-17	984.09E-19	0.99842	0.00011
45000	81,000	538.22E-17	902.94E-19	0.99851	0.00010
46000	82,800	494.73E-17	829.97E-19	0.99861	0.00009
47000	84,600	455.54E-17	764.23E-19	0.99869	0.00008
48000	86,400	420.15E-17	704.86E-19	0.99877	0.00008
49000	88,200	388.14E-17	651.15E-19	0.99884	0.00007
50000	90,000	359.11E-17	602.45E-19	0.99890	0.00007
52000	93,600	308.75E-17	517.97E-19	0.99902	0.00012
54000	97,200	266.91E-17	447.78E-19	0.99912	0.00010
56000	100,800	231.93E-17	389.09E-19	0.99921	0.00009
58000	104,400	202.49E-17	339.70E-19	0.99929	0.00008
60000	108,000	177.57E-17	297.90E-19	0.99935	0.00007
62000	111,600	156.37E-17	262.33E-19	0.99941	0.00006
64000	115,200	138.24E-17	231.92E-19	0.99946	0.00005
66000	118,800	122.66E-17	205.78E-19	0.99951	0.00005
68000	122,400	109.22E-17	183.23E-19	0.99955	0.00004
70000	126,000	975.66E-18	163.68E-19	0.99959	0.00004

REFERENCES

1. Cohen, E. R., and B. N. Taylor: The 1986 CODATA Recommended Values of the Fundamental Physical Constants, *J. Res. Nat. Bur. Stand.,* vol. 92 no. 2, pp. 85–95, 1987.
2. Lowan, A. N., Technical Director: "Planck's Radiation Functions and Electronic Functions," Federal Works Agency Work Projects Administration for the City of New York, under the sponsorship of the U.S. National Bureau of Standards Computation Laboratory, 1941.
3. Chang, S. L., and K. T. Rhee: Blackbody Radiation Functions, *Int. Commun. Heat Mass Transfer,* vol. 11, pp. 451–455, 1984.

4. Pivovonsky, Mark, and Max R. Nagel: "Tables of Blackbody Radiation Functions," Macmillan, New York, 1961.
5. Wiebelt, John A.: "Engineering Radiation Heat Transfer," Holt, Rinehart & Winston, New York, 1966.
6. Mechtly, E. A.: The International System of Units. Physical Constants and Conversion Factors, NASA SP-7012, 2d rev., 1973.

SOURCES OF DIFFUSE
CONFIGURATION FACTORS

This appendix contains tables of references to about 230 configuration factors that are available in the literature. The table is composed of three parts. Part A is for configuration factors between two elemental surfaces; part B gives references for factors between an elemental and a finite surface; part C is for factors between two finite areas. More than one reference is given for some factors, and in certain cases the reference in which a factor was originally derived is not given because of the difficulty in obtaining these earlier works.

The factors are arranged in the following manner: Factors involving only plane surfaces are given first, followed by those involving cylindrical bodies, conical bodies, spherical bodies, and more complex bodies. Within each category, progression is from simpler to more complex geometries.

Table B-1 Table of references for configuration factors

(A) Factors for two differential elements

Configu- ration number	Geometry	Configuration	Source
A-1	Two elemental areas in arbitrary configuration		Eq. (6-5)
A-2	Two elemental areas lying on parallel generating lines		Example 6-1
A-3	Elemental area to infinitely long strip of differential width lying on parallel generating line		Appendix C and Refs. [1,2]
A-4	Infinitely long strip of differential width to similar strip on parallel generating line		Appendix C and Ref. [1]
A-5	Strip of finite length and differential width to strip of same length on parallel generating line		Appendix C and Refs. [2,3]
A-6	Corner element of end of square channel to sectional wall element on channel		Example 6-19
A-7	Exterior element on tube surface to exterior element on adjacent parallel tube of same diameter		Refs. [2,4]
A-8	Exterior element on partitioned tube to similar element on adjacent parallel tube of same diameter		Refs. [2,4]

(*Table continues on next page*)

Table B-1 Table of references for configuration factors
(A) Factors for two differential elements (*Continued*)

Configuration number	Geometry	Configuration	Source
A-9	Two ring elements on interior of right circular cylinder		Appendix C and Refs. [2,5,6,7]
A-10	Band of differential length on inside of cylinder to differential ring on cylinder base		Refs. [2,8]
A-11	Elements inside a channel of varying cross section; includes circular to circular and circular to rectangular		Ref. [9]
A-12	Differential area on ring element to ring element on adjacent fin with view partially blocked by solid coaxial cylinder		Ref. [10]

Table B-1 Table of references for configuration factors
(A) (*Continued*)

Configu-ration number	Geometry	Configuration	Source
A-13	Ring element on fin to ring element on adjacent fin		Refs. [2,10,11]
A-14	Two elements on interior of right circular cone		Refs. [2,7,12,13]
A-15	Ring elements on two concentric axisymmetric bodies		Ref. [14]
A-16	Two differential elements on interior of spherical cavity		Appendix C and Refs. [1,2,6,13,15,16]

(*Table continues on next page*)

Table B-1 Table of references for configuration factors

(A) Factors for two differential elements (*Continued*)

Configu-ration number	Geometry	Configuration	Source
A-17	Band on outside of sphere to band on another sphere		Ref. [17] (equal radii); [18] (unequal radii)
A-18	Two differential elements on exterior of toroid		Ref. [19]
A-19	Element on exterior of toroid to ring element on exterior of toroid		Ref. [19]
A-20	Element on exterior of toroid to hoop element on exterior of toroid		Ref. [19]

Table B-1 Table of references for configuration factors

(B) Factors for exchange between differential element and finite area

Configu- ration number	Geometry	Configuration	Source
B-1	Plane element to plane extending to infinity and intersecting plane of element at angle Φ		Refs. [2,20–22]
B-2	Plane strip element of any length to plane of finite width and infinite length		Example 6-5
B-3	Plane element to infinitely long surface of arbitrary shape generated by line moving parallel to itself and plane of element		Appendix C and Refs. [2,20–24]
B-4	Strip element of finite length to rectangle in plane parallel to strip; strip is opposite to one edge of rectangle		Appendix C and Refs. [2,6,20–22]; arbitrary strip [25]
B-5	Strip element of finite length to plane rectangle that intercepts plane of strip at angle Φ and with one edge parallel to strip		Appendix C and Refs. [2,6] for Φ = 90° only; [20–22]

(Table continues on next page)

Table B-1 Table of references for configuration factors
(B) Factors for exchange between differential element and finite area (*Continued*)

Configuration number	Geometry	Configuration	Source
B-6	Plane element to plane rectangle; normal to element passes through corner of rectangle; surfaces are on parallel planes		Appendix C and Refs. [1,2,6,20, 23,26,27]
B-7	Area element to any parallel rectangle		Ref. [1]; arbitrary element [25]
B-8	Plane element to plane rectangle; planes containing two surfaces intersect at angle Φ		Appendix C and Refs. [1,2,6], for $\Phi = 90°$ only; [20–22]
B-9	Plane element to right triangle in plane parallel to plane of element; normal to element passes through vertex of triangle		Example 6-17
B-10	Plane element to a parallel isosceles triangle; normal to element passes through vertex		Appendix C and Refs. [2,28]
B-11	Plane element to parallel regular coaxial polygon; normal to element passes through polygon center		Appendix C and Ref. [28]

Table B-1 Table of references for configuration factors (B) (*Continued*)

Configu-ration number	Geometry	Configuration	Source
B-12	Plane element to plane area with added triangular area; element is on corner of rectangle with one side in common with plane at angle Φ		Refs. [2,20–22, 29]
B-13	Same geometry as preceding with triangle reversed relative to plane element		Refs. [2,20–22, 29]
B-14	Plane element to circular disk on plane parallel to that of element		Appendix C and Refs. [1,2,6,10, 20– 22,30]; nonparallel element [31]
B-15	Plane element to segment of parallel coaxial disk		Appendix C and Ref. [28]
B-16	Plane element to segment of disk in plane parallel to element		Refs. [2,6]

(*Table continues on next page*)

Table B-1 Table of references for configuration factors
(B) Factors for exchange between differential element and finite area (*Continued*)

Configuration number	Geometry	Configuration	Source
B-17	Plane element to circular disk; planes containing element and disk intersect at 90°, and centers of element and disk lie in plane perpendicular to those containing areas		Appendix C, Refs. [6,20–22,24,32], and Example 6-4
B-18	Strip element of finite length to perpendicular circular disk located at one end of strip		Refs. [24,32]
B-19	Plane element to a thin coaxial ring parallel to the element		Refs. [2,28]
B-20	Plane element to ring area in plane perpendicular to element		Example 6-7
B-21	Radial and wedge elements on circular disk to disk in parallel plane		Refs. [26,32]
B-22	Plane element to ring sector on circular disk parallel to element		Refs. [2,30]

Table B-1 Table of references for configuration factors

(B) (*Continued*)

Configu-ration number	Geometry	Configuration	Source
B-23	Plane element to annular ring on circular disk parallel to element with cylindrical blockage of view		Refs. [2,10]
B-24	Plane element to sector of circular disk parallel to element		Ref. [30]
B-25	Area element to parallel elliptical plate		Appendix C and Refs. [2,23,29]
B-26	Plane element to annular disk with conical blockage of view		Ref. [33]
B-27	Infinitely long cylinder to parallel infinitely long strip element, $a \geq r$		Appendix C and Refs. [2,20–22, 34][a]
B-28	Plane of infinite width and infinite length to infinitely long strip on the surface of a parallel cylinder		Appendix C and Refs [2,20,22]

[a]Factor incorrect in Refs. [20–22], see [34].

(*Table continues on next page*)

Table B-1 Table of references for configuration factors
(B) Factors for exchange between differential element and finite area (*Continued*)

Configuration number	Geometry	Configuration	Source
B-29	Plane element to right circular cylinder of finite length; normal to element passes through center of one end of cylinder and is perpendicular to cylinder axis		Appendix C and Refs. [2,6,20–22]
B-30	Plane vertical element to circular cylinder tilted toward the element		Ref. [35]
B-31	Plane element to interior of coaxial right circular cylinder		Refs. [2,28]
B-32	Element is at end of wall on inside of finite-length cylinder enclosing concentric cylinder of same length; factor is from element to inside surface of outer cylinder		Refs. [20–22,24, 32]
B-33	Elemental strip of finite length to parallel cylinder of same length, normals at ends of strip pass through cylinder axis		Refs. [2,20–22,24,32]

Table B-1 Table of references for configuration factors
(B) (*Continued*)

Configuration number	Geometry	Configuration	Source
B-34	Strip or element on plane parallel to cylinder axis to cylinder of finite length		Refs. [2,24,32]
B-35	Infinitely long strip of differential width to parallel semicylinder		Refs. [2,36]
B-36	Infinite strip on any side of any of three fins to tube or environment, and infinite strip on tube to fin or environment		Refs. [2,37]
B-37	Element and strip element on interior of finite cylinder to interior of cylindrical surface		Refs. [2,24,32]
B-38	Elemental strip on inner surface of outer concentric cylinder to interior surface of outer concentric cylinder		Refs. [20–22,24, 32]

(Table continues on next page)

Table B-1 Table of references for configuration factors

(B) Factors for exchange between differential element and finite area (*Continued*)

Configuration number	Geometry	Configuration	Source
B-39	Elemental strip on inner surface of outer concentric cylinder to either annular end		Refs. [20–22,24, 32]
B-40	Element on inside of outer finite concentric cylinder to inside cylinder or annular end		Refs. [2,24,32]
B-41	Strip element on exterior of inner finite-length concentric cylinder to inside of outer cylinder or to annular end		Refs. [24,32]
B-42	Strip on plane inside cylinder of finite length to inside of cylinder		Refs. [24,32]
B-43	Area element on interior of cylinder to base of second concentric cylinder; cylinders are one atop other		Refs. [24,32]

Table B-1 Table of references for configuration factors
(B) (*Continued*)

Configu- ration number	Geometry	Configuration	Source
B-44	Ring element on fin to tube		Ref. [11]
B-45	Ring element on interior of right circular cylinder to end of cylinder		Appendix C and Refs. [5,7]; [7] also for a cone
B-46	Exterior element on tube surface to finite area on adjacent parallel tube of same diameter		Refs. [2,4]
B-47	Exterior element on tube surface of partitioned tube to finite area on adjacent parallel tube of same diameter		Ref. [4]
B-48	Element on wall of right circular cone to base of cone		Ref. [38]

(*Table continues on next page*)

Table B-1 Table of references for configuration factors
(B) Factors for exchange between differential element and finite area (*Continued*)

Configu-ration number	Geometry	Configuration	Source
B-49	Plane element on a ring to an inverted cone; ring and cone have the same axis and plane of ring does not intersect cone		Ref. [39]
B-50	Plane element to truncated cone		Ref. [33]
B-51	Plane element on a ring to an inverted cone; ring and cone have the same axis and plane of ring intersects cone		Refs. [39–41]
B-52	Plane element to cone; plane of element may or may not intersect cone		Ref. [42]
B-53	Cone to coaxial differential conical ring		Ref. [42]

Table B-1 Table of references for configuration factors
(B) (*Continued*)

Configu-ration number	Geometry	Configuration	Source
B-54	Any infinitesimal element on interior of sphere to any finite element on interior of same sphere		Appendix C, Refs. [1,2], and Sec. 7-5.2
B-55	Spherical point source to rectangle; point source is on one corner of rectangle that intersects with receiving rectangle at angle Φ		Refs. [1,20–22, 29]
B-56	Plane element to sphere; normal to element passes through center of sphere		Appendix C and Refs. [2,28]
B-57	Plane element to sphere; tangent to element passes through center of sphere		Appendix C and Refs. [2,28]
B-58	Area element to sphere; in [46] the element can be on the surface of a sphere or cylinder		Refs. [21,43–46]

(Table continues on next page)

Table B-1 Table of references for configuration factors

(B) Factors for exchange between differential element and finite area (*Continued*)

Configu-ration number	Geometry	Configuration	Source
B-59	Plane element to spherical cap		Ref. [47]
B-60	Plane element to spherical sector		Ref. [47]
B-61	Sphere to ring element oriented normal to sphere axis		Refs. [2,34]
B-62	Differential spherical band to a finite coaxial cylindrical area		Ref. [48]
B-63	Differential spherical band to finite circular ring		Ref. [48]
B-64	Differential ring on cylinder base to finite cylindrical strip with spherical blocking		Ref. [48]

Table B-1 Table of references for configuration factors
(B) (*Continued*)

Configu-ration number	Geometry	Configuration	Source
B-65	Spherical element to sphere		Appendix C and Refs. [2,28]
B-66	Elemental area on sphere to finite area on second sphere		Ref. [18]
B-67	Area element to axisymmetric surface—paraboloid, cone, cylinder (formulation given, factors are not evaluated)		Ref. [49]
B-68	Element on interior (or exterior) of any axisymmetric body of revolution to band of finite length on interior (or exterior)		Refs. [50[b],51[b],52]
B-69	Element on exterior of toroid to toroidal segment of finite width		Ref. [19]
B-70	Element on exterior of toroid to toroidal band of finite width		Ref. [19]

[b]Kernels of integrals and limits are formulated in terms of appropriate variables, but integrations are not carried out explicitly.

(*Table continues on next page*)

Table B-1 Table of references for configuration factors

(B) Factors for exchange between differential element and finite area (*Continued*)

Configuration number	Geometry	Configuration	Source
B-71	Element and ring element on exterior of toroid to entire exterior of toroid		Refs. [19,53]
B-72	Slender torus to element on perpendicular axis		Ref. [23]

Table B-1 Table of references for configuration factors

(C) Factors for two finite areas

Configuration number	Geometry	Configuration	Source
C-1	Two infinitely long plates of equal finite width W and one common edge of included angle Φ		Appendix C and Ref. [2]
C-2	Two infinitely long plates of unequal width with one common edge and included angle $\Phi = 90°$		Appendix C and Refs. [2,21]
C-3	Finite rectangle to infinitely long rectangle of same width and with one common edge		Refs. [2,54]

Table B-1 Table of references for configuration factors (C) (*Continued*)

Configu-ration number	Geometry	Configuration	Source
C-4	Two finite rectangles of same width with common edge and included angle Φ		Appendix C and Refs. [1,2,26,27] for Φ = 90° only; Refs. [6,20–22, 24,54[c]]
C-5	Two rectangles with common edge and included angle Φ		Ref. [21]
C-6	Two rectangles with one edge of each parallel, and with one corner touching; planes containing rectangles intersect at angle Φ		Ref. [21]
C-7	Two rectangles of same width with two parallel edges; planes containing rectangles intersect at angle Φ		Ref. [6] for Φ = 90° only; Ref. [21]
C-8	Two rectangles with two parallel edges; planes containing rectangles intersect at angle Φ		Refs. [20,21,24]; [55] for Φ = 90°
C-9	Two infinitely long directly opposed parallel strips of same finite width		Appendix C and Refs. [2,21,26]

[c]Ref. [54] indicates that tabulated values for this case are incorrect in all other references. Corrected values are listed in Ref. [54].

(Table continues on next page)

Table B-1 Table of references for configuration factors
(C) Factors for two finite areas (*Continued*)

Configuration number	Geometry	Configuration	Source
C-10	Parallel, directly opposed rectangles of same width and length		Appendix C and Refs. [1,2,6,20–22,26,27,29,56]
C-11	Two rectangles in parallel planes with one rectangle directly opposite portion of other		Refs. [21,23,56]
C-12	Two unequal parallel squares that are coaxial and have parallel sides		Refs. [2,57]
C-13	Two rectangles of arbitrary size in parallel planes; all edges lie parallel to either x or y axis		Refs. [2,20,21, 24,55,56]
C-14	Rectangle to arbitrarily oriented rectangle of arbitrary size		Refs. [58,59][d]
C-15	Rectangle to right triangle		Ref. [25]

[d]Available as general computer program only.

Table B-1 Table of references for configuration factors
(C) (*Continued*)

Configuration number	Geometry	Configuration	Source
C-16	Plane polygon surfaces		Ref. [60]
C-17	Two flat plates of arbitrary shape and arbitrary orientation		Ref. [61][e]
C-18	Finite areas on interior of square channel		Ref. [24]
C-19	Factor between bases of right convex prism of regular triangular, square, pentagonal, hexagonal, or octagonal cross section		Refs. [2,54]
C-20	Factors between various sides, and sides and bases of regular hexagonal prism		Refs. [2,54]

[e]Kernels of integrals and limits are formulated in terms of appropriate variables, but integrations are not carried out explicitly.

(*Table continues on next page*)

Table B-1 Table of references for configuration factors
(C) Factors for two finite areas (*Continued*)

Configu-ration number	Geometry	Configuration	Source
C-21	Circular disk to arbitrarily placed rectangle in parallel plane (using configuration-factor algebra with configuration C-24)		Ref. [62]
C-22	Circle to arbitrarily placed rectangle in plane parallel to normal to circle (using configuration-factor algebra with configuration C-24)		Ref. [62]
C-23	Disk to arbitrarily oriented rectangle or disk of arbitrary size		Ref. [58][f]
C-24	Circular disk to parallel right triangle; normal from center of circle passes through one acute vertex		Ref. [62]
C-25	Parallel, directly opposed plane circular disks[g]		Appendix C and Refs. [1,2,6,13, 20–22,24,26,29, 32,63[g]]
C-26	Parallel noncoaxial disks		Ref. [64]

[f]Available as general computer program only.
[g]Reference [63] also has nonparallel disks that can both be inscribed in a sphere.

Table B-1 Table of references for configuration factors
(C) (*Continued*)

Configu-ration number	Geometry	Configuration	Source
C-27	Parallel disk and disk segment		Ref. [64]
C-28	Directly opposed ring and disk of arbitrary radius		Refs. [2,21,32]
C-29	Parallel, directly opposed plane ring areas		Refs. [20,32] and Example 6-8
C-30	Circular disk to segment of parallel circular disk with disk centers on the same axis		Ref. [65]
C-31	Sector of circular disk to sector of parallel circular disk		Ref. [30]
C-32	Entire inner wall of finite cylinder to ends		Refs. [66,67]

(*Table continues on next page*)

Table B-1 Table of references for configuration factors
(C) Factors for two finite areas (*Continued*)

Configu- ration number	Geometry	Configuration	Source
C-33	Internal surface of cylindrical cavity to cavity opening		Refs. [21 (Fig. 6-14), 68]
C-34	Inner surface of cylinder to annulus on one end		Refs. [2,29,32,67]
C-35	Inner surface of cylinder to disk at one end of cylinder		Refs. [2,29,32,67]
C-36	Portion of inner surface of cylinder to remainder of inner surface		Refs. [26,32,67]
C-37	Finite ring areas on interior of right circular cylinders to separate similar areas and to ends		Refs. [2,24,26,67]
C-38	Finite areas on interior of right circular cylinder		Ref. [24]
C-39	The interior cylindrical halves of a cylinder of finite length		Ref. [69]

Table B-1 Table of references for configuration factors
(C) (*Continued*)

Configu-ration number	Geometry	Configuration	Source
C-40	Infinite cylinder to parallel infinitely long plane of finite width		Appendix C and Refs. [6,20,21, 34][h]
C-41	Infinitely long plane of finite width to infinitely long cylinder		Refs. [32,34]
C-42	Rows of infinitely long parallel cylinders in square or equilateral-triangular arrays		Refs. [2,70]
C-43	Infinite plane to first, second, and first plus second rows of infinitely long parallel tubes of equal diameter		Refs. [1,2,26, 27,29,71] (randomly overlapping cylinders [72])
C-44	Finite length cylinder to rectangle with two edges parallel to cylinder axis and of length equal to cylinder		Refs. [6,29]
C-45	Finite cylinder to finite rectangle of same length		Ref. [73]

---

[h]This factor is given incorrectly in all references except [6,34] and Appendix C.

(Table continues on next page)

Table B-1 Table of references for configuration factors
(C) Factors for two finite areas (*Continued*)

Configuration number	Geometry	Configuration	Source
C-46	Cylinder to any rectangle in plane perpendicular to cylinder axis (using configuration-factor algebra with configuration C-50)		Ref. [62]
C-47	Cylinder to any rectangle in plane parallel to cylinder axis (using configuration-factor algebra with configuration C-50)		Ref. [62]
C-48	Finite area on exterior of cylinder to finite area on plane parallel to cylinder axis		Ref. [24]
C-49	Finite area on exterior of cylinder to finite area on skewed plane		Ref. [24]
C-50	Outside surface of cylinder to perpendicular right triangle; triangle is in plane of cylinder base with one vertex of triangle at center of base		Ref. [62]

Table B-1 Table of references for configuration factors
(C) (*Continued*)

Configuration number	Geometry	Configuration	Source
C-51	Cylinder and plane of equal length parallel to cylinder axis; plane inside cylinder; all factors between plane and inner surface of cylinder		Refs. [24,32]
C-52	Inner surface of cylinder to disk of same radius		Refs. [24,32]
C-53	Interior surface of circular cylinder of radius R to disk of radius r where $r < R$; disk is perpendicular to axis of cylinder, and axis passes through center of disk (using configuration-factor algebra with configuration C-25)		Example 6-9
C-54	Outer surface of cylinder to annular disk at end of cylinder		Refs. [74–76]
C-55	Annular ring to similar annular ring, each at end of cylinder		Refs. [11,24,32]

(*Table continues on next page*)

Table B-1 Table of references for configuration factors
(C) Factors for two finite areas (*Continued*)

Configuration number	Geometry	Configuration	Source
C-56	Annular ring to annular ring of different outer diameter, each at end of cylinder		Refs. [10,77]
C-57	Factors for interchange between fins and tube (given in algebraic form, untabulated)		Refs. [11,76]
C-58	Finite area on exterior of cylinder to finite area on exterior of parallel cylinder		Ref. [24]
C-59	Cylinder of arbitrary length and radius to rectangle, disk, or cylinder of arbitrary size and orientation		Refs. [58,59][i]
C-60	Cylinder and plate with arbitrary orientation		Ref. [61][j]
C-61	Concentric cylinders of infinite length; inner to outer cylinder; outer to inner cylinder; outer cylinder to itself		Appendix C and Ref. [20]
C-62	Nonconcentric infinitely long circular cylinders, inner cylinder to area on outer cylinder		Ref. [29]

[i]Available as general computer program only.
[j]Kernels of integrals and limits are formulated in terms of appropriate variables, but integrations are not carried out explicitly.

Table B-1 Table of references for configuration factors (C) (*Continued*)

Configu-ration number	Geometry	Configuration	Source
C-63	Inside surface of outer concentric cylinder of finite length to inner cylinder of same length		Appendix C and Refs. [6,20,26, 29,32,75,78]
C-64	Inside surface of outer concentric cylinder to itself		Appendix C and Refs. [6,11,20, 29,32,78]
C-65	Inside surface of outer concentric cylinder to either end of annulus		Refs. [6,20,29, 32,76,78]
C-66	Concentric cylinders of different finite lengths— portion of inner cylinder to entire outer cylinder		Refs. [32,75]
C-67	Concentric cylinders of different finite lengths— portion of inside of outer to outside of entire inner cylinder		Refs. [24,32,61[k]]
C-68	Factors between various areas on concentric cylinders of unequal length		Ref. [79]

[k]Kernels of integrals and limits are formulated in terms of appropriate variables, but integrations are not carried out explicitly.

(*Table continues on next page*)

Table B-1 Table of references for configuration factors
(C) Factors for two finite areas *(continued)*

Configuration number	Geometry	Configuration	Source
C-69	Two parallel cylinders with unequal radii and of the same length		Ref. [80]
C-70	Parallel cylinders of different radii and length— any portions of outer curved surfaces		Ref. [61][1]
C-71	Two cylinders in an oblique orientation		Ref. [81]
C-72	Concentric cylinders of different radii, one atop the other; factors between inside of upper cylinder and inside or base of lower cylinder		Refs. [24,32]
C-73	Cylinder to a coaxial paraboloid		Ref. [82]

[1]Kernels of integrals and limits are formulated in terms of appropriate variables, but integrations are not carried out explicitly.

Table B-1 Table of references for configuration factors (C) (*Continued*)

Configuration number	Geometry	Configuration	Source
C-74	Inside of right circular cylinder to outside of right circular cylinder of the same height and on the same axis		Ref. [83]
C-75	Infinitely long parallel semicylinders of same diameter		Example 6-14 and Refs. [2,6]
C-76	Finite area on exterior of inner cylinder to finite area on interior of concentric outer cylinder		Ref. [24]
C-77	Two tubes connected with fin of finite thickness; length can be finite or infinite; all factors between finite surfaces formulated in terms of integrations between differential strips		Ref. [3]
C-78	Two tubes connected with tapered fins of finite thickness; tube length can be finite or infinite; all factors between finite surfaces formulated in terms of integrations between differential strips		Ref. [3]

(*Table continues on next page*)

Table B-1 Table of references for configuration factors
(C) Factors for two finite areas (*Continued*)

Configu-ration number	Geometry	Configuration	Source
C-79	Sandwich tube and fin structure of infinite or finite length; all factors between finite surfaces formulated in terms of integrations between differential strips		Ref. [3]
C-80	Concentric cylinders connected by fin of finite thickness; length finite or infinite; all factors between finite surfaces formulated in terms of integrations between differential strips		Ref. [3]
C-81	Exterior of infinitely long cylinder to interior of concentric semicylinder		Example 7-5
C-82	Interior of infinitely long semicylinder 1 to interior of semicylinder 2 when concentric parallel cylinder 3 is present		Example 7-5
C-83	Between axisymmetrical sections of right circular cone		Refs. [2,24,67]
C-84	Between axisymmetrical sections of right circular cone and base or ring or disk on base		Refs. [26,67]

Table B-1 Table of references for configuration factors (C) (*Continued*)

Configuration number	Geometry	Configuration	Source
C-85	Internal surface of conical cavity to cavity opening		Example 6-12 and Ref. [67]; Refs. [21 (Fig. 6-14), 68][m]
C-86	Entire inner surface of frustum of cone to ends		Refs. [66,67]
C-87	Circular disk and right circular cone with axis normal to the disk and with vertex at center of disk		Ref. [84]; [74] for disk not touching cone
C-88	Two nested right circular cones having common axis and vertex, and closed by a ring in the base		Ref. [84]
C-89	Between any two of the ten surfaces 1, 2, 3,...,7, 1 + 2, 3 + 4, 5 + 6, inside frustum of right circular cone		Ref. [85]

---

[m]Results are not the configuration factor, and abscissa of Fig. 6-14 of [21] is ten times too large.

(Table continues on next page)

Table B-1 Table of references for configuration factors
(C) Factors for two finite areas (*Continued*)

Configuration number	Geometry	Configuration	Source
C-90	Right circular cone of arbitrary size to rectangle, disk, cylinder, or cone of arbitrary size and orientation		Refs. [58,59][n]
C-91	Cone to arbitrarily skewed plate		Ref. [61][o]
C-92	Annular ring to truncated cone		Ref. [74]
C-93	Disk to coaxial ellipsoid		Ref. [74]
C-94	Disk to coaxial paraboloid		Ref. [74]

[n]Available as general computer program only.

[o]Kernels of integrals and limits are formulated in terms of appropriate variables, but integrations are not carried out explicitly.

Table B-1 Table of references for configuration factors (C) (*Continued*)

Configuration number	Geometry	Configuration	Source
C-95	Two coaxial cones		Ref. [42]
C-96	Cone to coaxial paraboloid that may or may not intersect the cone		Ref. [42]
C-97	Cone to coaxial ellipsoid		Ref. [42]
C-98	Internal surface of spherical cavity to cavity opening		Example 6-12 and Refs. [21 (Fig. 6-14), 68][p]
C-99	Any finite area on interior of sphere to any other finite area on interior		Appendix C and Ref. [1]
C-100	Sphere to coaxial disk		Appendix C and Refs. [2,29,34]

[p]Results are not the configuration factor, and abscissa of Fig. 6-14 of [21] is 10 times too large.

(*Table continues on next page*)

Table B-1 Table of references for configuration factors
(C) Factors for two finite areas (*Continued*)

Configuration number	Geometry	Configuration	Source
C-101	Sphere to segment of coaxial disk		Appendix C and Refs. [29,34]
C-102	Sphere to sector of coaxial disk		Appendix C and Refs. [29,34]
C-103	Sphere to rectangle		Ref. [62]
C-104	Sphere to rectangle normal to sphere axis		Ref. [34]
C-105	Sphere to arbitrary rectangle (using configuration-factor algebra and configuration C-103)		Refs. [58[q],62,86]
C-106	Sphere to regular polygon normal to sphere axis		Ref. [34]

[q]Available as general computer program only.

Table B-1 Table of references for configuration factors
(C) (*Continued*)

Configuration number	Geometry	Configuration	Source
C-107	Sphere to disk not on axis		Ref. [34]
C-108	Sphere to sector of circular disk		Ref. [86]
C-109	Sphere to segment of circular disk		Ref. [86]
C-110	Sphere to ellipse		Ref. [86]
C-111	Sphere of arbitrary diameter to disk or cone of arbitrary size and orientation		Refs. [58,59]'

'Available as general computer program only.

(*Table continues on next page*)

Table B-1 Table of references for configuration factors
(C) Factors for two finite areas (*Continued*)

Configuration number	Geometry	Configuration	Source
C-112	Sphere to arbitrarily skewed plate		Ref. [61][s]
C-113	Sphere to cylinder		Refs. [58',59',87]
C-114	Sphere, cylinder, or cone and disk in intersecting plane		Ref. [64]
C-115	Cone to sphere having same diameter as base of cone; axis of cone passes through center of sphere		Refs. [17,58', 59',88]
C-116	Concentric spheres; inner to outer sphere; outer to inner sphere; outer sphere to itself		Appendix C, Refs. [20,21], and Example 6-11
C-117	Area on surface of sphere to rectangle in plane perpendicular to axis of sphere		Ref. [24]
C-118	Sphere to sphere		Refs. [2,17 (equal radii), 58',59', 87,89–92]

---

[s]Kernels of integrals and limits are formulated in terms of appropriate variables, but integrations are not carried out explicitly.

[t]Available as general computer program only.

Table B-1 Table of references for configuration factors
(C) (*Continued*)

Configu-ration number	Geometry	Configuration	Source
C-119	Sphere to coaxial cylinder of finite length		Ref. [93]
C-120	Sphere to coaxial ellipsoid or paraboloid		Ref. [88]
C-121	Area on sphere to cap on another sphere		Ref. [18]
C-122	Cap on sphere to cap on another sphere		Ref. [18]; sphere to spherical cap, Ref. [88]
C-123	Area on sphere to area on another sphere		Ref. [18]
C-124	Cap on sphere to band on another sphere		Ref. [18]
C-125	Band on one sphere to band on another sphere		Ref. [18]

(*Table continues on next page*)

Table B-1 Table of references for configuration factors

(C) Factors for two finite areas (*Continued*)

Configuration number	Geometry	Configuration	Source
C-126	Infinite plane to adjacent bed of randomly overlapping spheres all of the same radius		Ref. [72]
C-127	Internal surface of hemispherical cavity to cavity opening		Example 6-12 and Ref. [67]; Refs. [21 (Fig. 6-14), 68]"
C-128	Between axisymmetrical section of hemisphere and base or ring or disk on base		Refs. [29,67]
C-129	Between axisymmetrical sections of hemisphere		Refs. [24,67]
C-130	Annular ring around base of hemisphere to hemisphere		Ref. [94]
C-131	Annular ring around base of hemisphere to section of hemisphere		Ref. [94]
C-132	Hemisphere to coaxial hemisphere in contact		Ref. [95]
C-133	Sphere to hemisphere		Ref. [87]

"Results are not the configuration factor, and abscissa of Fig. 6-14 of [21] is 10 times too large.

Table B-1 Table of references for configuration factors
(C) (*Continued*)

Configuration number	Geometry	Configuration	Source
C-134	Sphere to ellipsoid		Refs. [58ᵛ,87]
C-135	Ellipsoid of arbitrary major and minor axes to rectangle, disk, cylinder, cone, or ellipsoid of arbitrary size and orientation		Ref. [58]ᵛ
C-136	From Moebius strip to itself		Ref. [96]
C-137	Exterior of toroid to itself		Refs. [19,53]
C-138	Segment of finite width on toroid to exterior of toroid		Ref. [19]
C-139	Toroidal band of finite width to exterior of toroid		Ref. [19]
C-140	Toroid of arbitrary size to rectangle, disk, cylinder, sphere, cone, ellipsoid, or toroid of arbitrary size and orientation		Ref. [58]ᵛ

ᵛAvailable as general computer program only.

(*Table continues on next page*)

Table B-1 Table of references for configuration factors
(C) Factors for two finite areas (*Continued*)

Configuration number	Geometry	Configuration	Source
C-141	Arbitrary polynomial of revolution to rectangle, disk, cylinder, sphere, cone, ellipsoid, toroid, or other arbitrary polynomial of revolution of arbitrary size and orientation (polynomials of fifth order or less)		Ref. [58]^w
C-142	General plane polygon to any general plane polygon or two or more intersecting or adjoining polygons		Ref. [97]^w

^wAvailable as general computer program only.

REFERENCES

1. Jakob, Max: "Heat Transfer," vol. 2, Wiley, New York, 1957.
2. Howell, J. R.: "A Catalog of Radiation Configuration Factors," McGraw-Hill, New York, 1982.
3. Sotos, Carol J., and Norbert O. Stockman: Radiant-Interchange View Factors and Limits of Visibility for Differential Cylindrical Surfaces with Parallel Generating Lines, NASA TN D-2556, 1964.
4. Sparrow, E. M., and V. K. Jonsson: Angle-Factors for Radiant Interchange between Parallel-Oriented Tubes, *J. Heat Transfer,* vol. 85, no. 4, pp. 382–384, 1963.
5. Usiskin, C. M., and R. Siegel: Thermal Radiation from a Cylindrical Enclosure with Specified Wall Heat Flux, *J. Heat Transfer,* vol. 82, no. 4, pp. 369–374, 1960.
6. Sparrow, E. M., and R. C. Cess: "Radiation Heat Transfer," Augmented edition, Hemisphere, Washington, D.C., 1978.
7. Buraczewski, C., and J. Stasiek: Application of Generalized Pythagoras Theorem to Calculation of Configuration Factors between Surfaces of Channels of Revolution, *Int. J. Heat Fluid Flow,* vol. 4, no. 3, pp. 157–183, 1983.
8. Sparrow, E. M., L. U. Albers, and E. R. G. Eckert: Thermal Radiation Characteristics of Cylindrical Enclosures, *J. Heat Transfer,* vol. 84, no. 1, pp. 73–81, 1962.
9. Eddy, T. L., and G. E. Nielsson: Radiation Shape Factors for Channels with Varying Cross Section, *J. Heat Transfer,* vol. 110, no. 1, pp. 264–266, 1988.

10. Minning, C. P.: Shape Factors between Coaxial Annular Discs Separated by a Solid Cylinder, *AIAA J.,* vol. 17, no. 3, pp. 318–320, March 1979.

11. Sparrow, E. M., G. B. Miller, and V. K. Jonsson: Radiative Effectiveness of Annular-Finned Space Radiators, Including Mutual Irradiation between Radiator Elements, *J. Aerospace Sci.,* vol. 29, no. 11, pp. 1291–1299, 1962.

12. Sparrow, E. M., and V. K. Jonsson: Radiant Emission Characteristics of Diffuse Conical Cavities, *J. Opt. Soc. Am.,* vol. 53, no. 7, pp. 816–821, 1963.

13. Kezios, Stothe P., and Wolfgang Wulff: Radiative Heat Transfer through Openings of Variable Cross Sections, *Proc. Third Int. Heat Transfer Conf. AIChE,* vol. 5, pp. 207–218, 1966.

14. Modest, M. F.: Radiative Shape Factors between Differential Ring Elements on Concentric Axisymmetric Bodies, *J. Thermophys. Heat Transfer,* vol. 2, no. 1, pp. 86–88, 1988.

15. Sparrow, E. M., and V. K. Jonsson: Absorption and Emission Characteristics of Diffuse Spherical Enclosures, NASA TN D-1289, 1962.

16. Nichols, Lester D.: Surface-Temperature Distribution on Thin-Walled Bodies Subjected to Solar Radiation in Interplanetary Space, NASA TN D-584, 1961.

17. Campbell, James P., and Dudley G. McConnell: Radiant-Interchange Configuration Factors for Spherical and Conical Surfaces to Spheres, NASA TN D-4457, 1968.

18. Grier, Norman T.: Tabulations of Configuration Factors between any Two Spheres and Their Parts, NASA SP-3050, 1969.

19. Grier, Norman T., and Ralph D. Sommers: View Factors for Toroids and Their Parts, NASA TN D-5006, 1969.

20. Hamilton, D. C., and W. R. Morgan: Radiant-Interchange Configuration Factors, NASA TN 2836, 1952.

21. Kreith, Frank: "Radiation Heat Transfer for Spacecraft and Solar Power Plant Design," International Textbook Company, Scranton, Pa., 1962.

22. Wiebelt, John A.: "Engineering Radiation Heat Transfer," Holt, Rinehart & Winston, New York, 1966.

23. Moon, Parry: "The Scientific Basis of Illuminating Engineering," Dover, New York, 1961.

24. Stevenson, J. A., and J. C. Grafton: Radiation Heat Transfer Analysis for Space Vehicles, Rept. SID-61-91, North American Aviation (AFASD TR 61-119, pt 1), Sept. 9, 1961.

25. Chung, B. T. F., and M. M. Kermani: Radiation View Factors from a Finite Rectangular Plate, *J. Heat Transfer,* vol. 111, no. 4, pp. 1115–1117, 1989.

26. Hottel, H. C., and A. F. Sarofim: "Radiation Transfer," McGraw-Hill, New York, 1967.

27. Hottel, H. C.: Radiant-Heat Transmission, in William H. McAdams (ed.), "Heat Transmission," 3d ed., pp. 55–125, McGraw-Hill, New York, 1954.

28. Chung, B. T. F., and P. S. Sumitra: Radiation Shape Factors from Plane Point Sources, *J. Heat Transfer,* vol. 94, no. 3, pp. 328–330, 1972.

29. Wong, H. Y.: "Handbook of Essential Formulae and Data on Heat Transfer for Engineers" Longman Group, London, 1977.

30. Charles P. Minning: Calculation of Shape Factors between Parallel Ring Sectors Sharing a Common Centerline, *AIAA J.,* vol. 14, no. 6, pp. 813–815, 1976.

31. Naraghi, M. H. N.: Radiation View Factors from Differential Plane Sources to Disks—A General Formulation, *J. Thermophys. Heat Transfer,* vol. 2, no. 3, pp. 271–274, 1988.

32. Leuenberger, H., and R. A. Person: Compilation of Radiation Shape Factors for Cylindrical Assemblies, paper no. 56-A-144, ASME, November 1956.

33. Holchendler, J., and W. F. Laverty: Configuration Factors for Radiant Heat Exchange in Cavities Bounded at the Ends by Parallel Disks and Having Conical Centerbodies, *J. Heat Transfer,* vol. 96, no. 2, pp. 254–257, 1974.

34. Feingold, A., and K. G. Gupta: New Analytical Approach to the Evaluation of Configuration Factors in Radiation from Spheres and Infinitely Long Cylinders, *J. Heat Transfer,* vol. 92, no. 1, pp. 69–76, 1970.

35. Rein, R. G., Jr., C. M. Sliepcevich, and J. R. Welker: Radiation View Factors for Tilted Cylinders, *J. Fire Flammability,* vol. 1, pp. 140–153, 1970.

36. Sparrow, E. M., and E. R. G. Eckert: Radiant Interaction between Fin and Base Surfaces, *J. Heat Transfer,* vol. 84, no. 1, pp. 12–18, 1962.

37. Holcomb, R. S., and F. E. Lynch: Thermal Radiation Performance of a Finned Tube with a Reflector, Rept. ORNL-TM-1613, Oak Ridge National Laboratory, April 1967.
38. Joerg, Pierre, and B. L. McFarland: Radiation Effects in Rocket Nozzles, Rept. S62-245, Aero-jet-General Corporation, 1962.
39. Minning, C. P.: Calculation of Shape Factors between Rings and Inverted Cones Sharing a Common Axis, *J. Heat Transfer,* vol. 99, no. 3, pp. 492–494, 1977. (See also discussion in *J. Heat Transfer,* vol. 101, no. 1, pp. 189–190, 1979).
40. Edwards, D. K.: Comment on "Radiation from Conical Surfaces with Nonuniform Radiosity," *AIAA J.* vol. 7, no. 8, pp. 1656–1659, 1969.
41. Bobco, R. P.: Radiation from Conical Surfaces with Nonuniform Radiosity, *AIAA J.,* vol. 4, no. 3, pp. 544–546, 1966.
42. Chung, B. T. F., M. M. Kermani, and M. H. N. Naraghi: A Formulation of Radiation View Factors from Conical Surfaces, *AIAA J.,* vol. 22, no. 3, pp. 429–436, 1984.
43. Cunningham, F. G.: Power Input to a Small Flat Plate from a Diffusely Radiating Sphere, with Application to Earth Satellites, NASA TN D-710, 1961.
44. Liebert, Curt H., and Robert R. Hibbard: Theoretical Temperatures of Thin-Film Solar Cells in Earth Orbit, NASA TN D-4331, 1968.
45. Goetze, Dieter, and Charles B. Grosch: Earth-Emitted Infrared Radiation Incident upon a Satellite, *J. Aerospace Sci.,* vol. 29, no. 5, pp. 521–524, 1962.
46. Juul, N. H.: Diffuse Radiation View Factors from Differential Plane Sources to Spheres, *J. Heat Transfer,* vol. 101, no. 3, pp. 558–560, 1979.
47. Naraghi, M. H. N.: Radiation View Factors from Spherical Segments to Planar Surfaces, *J. Thermophys. Heat Transfer,* vol. 2, no. 4, pp. 373–375, 1988.
48. Mahbod, B., and R. L. Adams: Radiation View Factors between Axisymmetric Subsurfaces within a Cylinder with Spherical Centerbody, *J. Heat Transfer,* vol. 106, no. 1, pp. 244–248, 1984.
49. Morizumi, S. J.: Analytical Determination of Shape Factors from a Surface Element to an Axisymmetric Surface, *AIAA J.,* vol. 2, no. 11, pp. 2028–2030, 1964.
50. Robbins, William H., and Carroll A. Todd: Analysis, Feasibility, and Wall-Temperature Distribution of a Radiation-Cooled Nuclear-Rocket Nozzle, NASA TN D-878, 1962.
51. Robbins, William H.: An Analysis of Thermal Radiation Heat Transfer in a Nuclear-Rocket Nozzle, NASA TN D-586, 1961.
52. Bernard, Jean-Joseph, and Jeanne Genot: "Diagrams for Computing the Radiation of Axisymmetric Surfaces (Propulsive Nozzles)," Office National d'Etudes et de Recherches Aerospatiales, Paris, France, ONERA-NT-185, 1971 (in French).
53. Sommers, Ralph D., and Norman T. Grier: Radiation View Factors for a Toroid: Comparison of Eckert's Technique and Direct Computation, *J. Heat Transfer,* vol. 91, no. 3, pp. 459–461, 1969.
54. Feingold, A.: Radiant-Interchange Configuration Factors between Various Selected Plane Surfaces, *Proc. R. Soc. London,* ser. A, vol. 292, no. 1428, pp. 51–60, 1966.
55. Chekhovskii, I. R., V. V. Sirotkin, Yu. V. Chu-Dun-Chu, and V. A. Chebanov: Determination of Radiative View Factors for Rectangles of Different Sizes, *High Temp.,* July 1979, Translation of Russian original, vol. 17, no. 1, Jan.–Feb., 1979.
56. Hsu, Chia-Jung: Shape Factor Equations for Radiant Heat Transfer between Two Arbitrary Sizes of Rectangular Planes, *Can. J. Chem. Eng.,* vol. 45, no. 1, pp. 58–60, 1967.
57. Crawford, Martin: Configuration Factor between Two Unequal, Parallel, Coaxial Squares, paper no. 72-WA/HT-16, ASME, November 1972.
58. Dummer, R. S., and W. T. Breckenridge, Jr.: Radiation Configuration Factors Program, Rept. ERR-AN-224, General Dynamics/Astronautics, February 1963.
59. Lovin, J. K., and A. W. Lubkowitz: User's Manual for "RAVFAC," a Radiation View Factor Digital Computer Program, Rept. HREC-0154-1; Lockheed Missiles and Space Co., Huntsville Research Park, Huntsville, Alabama; LMSC/HREC D148620 (Contract NAS8-30154), November, 1969.
60. Mathiak, F. U.: Berechnung von Konfigurationsfaktoren Polygonal Berandeter Ebener Gebiete

(Calculation of Form-Factors for Plane Areas with Polygonal Boundaries), *Waerme Stoffueber-trag.*, vol. 19, pp. 273–278, 1985.

61. Plamondon, Joseph A.: Numerical Determination of Radiation Configuration Factors for Some Common Geometrical Situations, Tech Rept. 32-127, Jet Propulsion Laboratory, California Institute of Technology, July 7, 1961.

62. Tripp, W., C. Hwang, and R. E. Crank: Radiation Shape Factors for Plane Surfaces and Spheres, Circles or Cylinders (Spec. Rept. 16), *Kansas State University Bulletin*, vol. 46, no. 4, 1962.

63. Feingold, A.: A New Look at Radiation Configuration Factors between Disks, *J. Heat Transfer*, vol. 100, no. 4, pp. 742–744, 1978.

64. Naraghi, M. H. N., and J. P. Warna: Radiation Configuration Factors from Axisymmetric Bodies to Plane Surfaces, *Int. J. Heat Mass Transfer*, vol. 31, no. 7, pp. 1537–1539, 1988.

65. Lin, S., P.-M. Lee, J. C. Y. Wang, W.-L. Dai, and Y.-S. Lou: Radiant-Interchange Configuration Factors between Disk and a Segment of Parallel Concentric Disk, *Int. J. Heat Mass Transfer*, vol. 29, no. 3, pp. 501–503, 1986.

66. Bien, Darl D.: Configuration Factors for Thermal Radiation from Isothermal Inner Walls of Cones and Cylinders, *J. Spacecr. Rockets*, vol. 3, no. 1, pp. 155–156, 1966.

67. Buschman, Albert J., Jr., and Claud M. Pittman: Configuration Factors for Exchange of Radiant Energy between Axisymmetrical Sections of Cylinders, Cones, and Hemispheres and Their Bases, NASA TN D-944, 1961.

68. Stephens, Charles, W., and Alan M. Haire: Internal Design Considerations for Cavity-Type Solar Absorbers, *ARS J.*, vol. 31, no. 7, pp. 896–901, 1961.

69. Hahne, E., and M. K. Bassiouni: The Angle Factor for Radiant Interchange within a Constant Radius Cylindrical Enclosure, *Lett. Heat Mass Transfer*, vol. 7, pp. 303–309, 1980.

70. Cox, Richard L.: Radiative Heat Transfer in Arrays of Parallel Cylinders, Ph.D. thesis, Tennessee University, Knoxville, 1976.

71. Kuroda, Z., and T. Munakata: Mathematical Evaluation of the Configuration Factors between a Plane and One or Two Rows of Tubes, *Kagaku Sooti* (Chemical Apparatus, Japan) pp. 54–58, Nov. 1979 (in Japanese).

72. Tseng, J. W. C., and W. Strieder: View Factors for Wall to Random Dispersed Solid Bed Transport, *J. Heat Transfer*, vol. 112, no. 3, pp. 816–819, 1990.

73. Wiebelt, J. A., and S. Y. Ruo: Radiant-Interchange Configuration Factors for Finite Right Circular Cylinder to Rectangular Planes, *Int. J. Heat Mass Transfer*, vol. 6, no. 2, pp. 143–146, 1963.

74. Naraghi, M. H. N., and B. T. F. Chung: Radiation Configuration Factors between Disks and a Class of Axisymmetric Bodies, *J. Heat Transfer*, vol. 104, no. 3, pp. 426–431, 1982.

75. Rea, Samuel N.: Rapid Method for Determining Concentric Cylinder Radiation View Factors, *AIAA J.*, vol. 13, no. 8, pp. 1122–1123, 1975.

76. Masuda, H.: Radiant Heat Transfer on Circular-Finned Cylinders, Report Inst. High Speed Mech., Tohoku University, vol. 27, no. 225, pp. 67–89, 1973.

77. Bornside, D. E., and R. A. Brown: View Factor between Differing-Diameter, Coaxial Disks Blocked by a Coaxial Cylinder, *J. Thermophys. Heat Transfer*, vol. 4, no. 3, pp. 414–416, 1990.

78. Aleksandrov, V. T.: Determination of the Angular Radiation Coefficients for a System of Two Coaxial Cylindrical Bodies, *Inzh. Fiz. Zh.*, vol. 8, no. 5, pp. 609–612, 1965.

79. Shukla, K. N., and D. Ghosh: Radiation Configuration Factors for Concentric Cylinder Bodies, *Indian J. Technol.*, vol. 23, pp. 244–246, 1985.

80. Juul, N. H.: View Factors in Radiation between Two Parallel Oriented Cylinders of Finite Length, *J. Heat Transfer*, vol. 104, no. 2, pp. 384–388, 1982.

81. Ameri, A., and J. D. Felske: Radiation Configuration Factors for Obliquely Oriented Finite Length Circular Cylinders, *Int. J. Heat Mass Transfer*, vol. 25, no. 5, pp. 728–736, 1982.

82. Saltiel, C., and M. H. N. Naraghi: Radiative Configuration Factors from Cylinders to Coaxial Axisymmetric Bodies, *Int. J. Heat Mass Transfer*, vol. 33, no. 1, pp. 215–218, 1990.

83. Reid, R. L., and J. S. Tennant: Annular Ring View Factors, *AIAA J.*, vol. 11, no. 10, pp. 1446–1448, 1973.

84. Kobyshev, A. A., I. N. Mastiaeva, Iu. A. Surinov, and Iu. P. Iakovlev: Investigation of the Field of Radiation Established by Conical Radiators, *Aviats. Tekh.* vol. 19, no. 3, pp. 43–49, 1976 (in Russian).
85. Wang, J. C. Y., S. Lin, P.-M. Lee, W.-L. Dai, and Y.-S. Lou: Radiant-Interchange Configuration Factors inside Segments of Frustum Enclosures of Right Circular Cones," *Int. Commun. Heat Mass Transfer,* vol. 13, pp. 423–432, 1986.
86. Sabet, M., and B. T. F. Chung: Radiation View Factors from a Sphere to Nonintersecting Planar Surfaces, *J. Thermophys. Heat Transfer,* vol. 2, no. 3, pp. 286–288, 1988.
87. Watts, R. G.: Radiant Heat Transfer to Earth Satellites, *J. Heat Transfer,* vol. 87, no. 3, pp. 369–373, 1965.
88. Chung, B. T. F., and M. H. N. Naraghi: A Simpler Formulation for Radiative View Factors from Spheres to a Class of Axisymmetric Bodies, *J. Heat Transfer,* vol. 104, no. 1, pp. 201–204, 1982.
89. Jones, L. R.: Diffuse Radiation View Factors between Two Spheres, *J. Heat Transfer,* vol. 87, no. 3, pp. 421–422, 1965.
90. Juul, N. H.: Diffuse Radiation Configuration View Factors between Two Spheres and Their Limits, *Lett. Heat Mass Transfer,* vol. 3, no. 3, pp. 205–211, 1976.
91. Juul, N. H.: Investigation of Approximate Methods for Calculation of the Diffuse Radiation Configuration View Factor between Two Spheres, *Lett. Heat Mass Transfer,* vol. 3, pp. 513–522, 1976.
92. J.D. Felske: "Approximate Radiation Shape Factors between Two Spheres," *J. Heat Transfer,* vol. 100, no. 3, pp. 547–548, 1978.
93. Chung, B. T. F., and M. H. N. Naraghi: Some Exact Solutions for Radiation View Factors from Spheres, *AIAA J.,* vol. 19, pp. 1077–1081, 1981.
94. Ballance, J. O., and J. Donovan: Radiation Configuration Factors for Annular Rings and Hemispherical Sectors, *J. Heat Transfer,* vol. 95, no. 2, pp. 275–276, 1973.
95. Wakao, Noriaki, Koichi Kato, and Nobuo Furuya: View Factor between Two Hemispheres in Contact and Radiation Heat-Transfer Coefficient in Packed Beds, *Int. J. Heat Mass Transfer,* vol. 12, pp. 118–120, 1969.
96. Stasenko, A. L.: Self-irradiation Coefficient of a Moebius Strip of Given Shape, *Akad. Nauk SSSR, Izv. Energetika Transport,* pp. 104–107, July–August 1967.
97. Toups, K. A.: A General Computer Program for the Determination of Radiant Interchange Configuration and Form Factors: Confac-1, Rept. SID-65-1043-1, North American Aviation, Inc. (NASA CR-65256), October 1965.

CATALOG OF SELECTED CONFIGURATION FACTORS

1	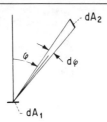	Area dA_1 of differential width and any length, to infinitely long strip dA_2 of differential width and with parallel generating line to dA_1.
		$$dF_{d1-d2} = \frac{\cos \varphi}{2} d\psi = \tfrac{1}{2} d(\sin \varphi)$$
2	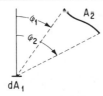	Area dA_1 of differential width and any length to any cylindrical surface A_2 generated by a line of infinite length moving parallel to itself and parallel to the plane of dA_1.
		$$F_{d1-2} = \tfrac{1}{2}(\sin \varphi_2 - \sin \varphi_1)$$

3

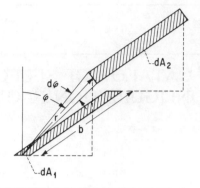

Strip of finite length b and of differential width, to differential strip of same length on parallel generating line.

$$dF_{d1-d2} = \frac{\cos \varphi}{\pi} \, d\varphi \, \tan^{-1} \frac{b}{r}$$

4

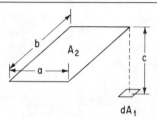

Plane element dA_1 to plane parallel rectangle; normal to element passes through corner of rectangle.

$$X = \frac{a}{c} \qquad Y = \frac{b}{c}$$

$$F_{d1-2} = \frac{1}{2\pi} \left(\frac{X}{\sqrt{1 + X^2}} \tan^{-1} \frac{Y}{\sqrt{1 + X^2}} + \frac{Y}{\sqrt{1 + Y^2}} \tan^{-1} \frac{X}{\sqrt{1 + Y^2}} \right)$$

5

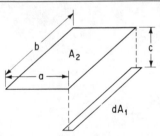

Strip element to rectangle in plane parallel to strip; strip is opposite one edge of rectangle.

$$X = \frac{a}{c} \qquad Y = \frac{b}{c}$$

$$F_{d1-2} = \frac{1}{\pi Y} \left(\sqrt{1 + Y^2} \tan^{-1} \frac{X}{\sqrt{1 + Y^2}} - \tan^{-1} X + \frac{XY}{\sqrt{1 + X^2}} \tan^{-1} \frac{Y}{\sqrt{1 + X^2}} \right)$$

6

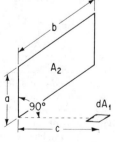

Plane element dA_1 to rectangle in plane 90° to plane of element.

$$X = \frac{a}{b} \qquad Y = \frac{c}{b}$$

$$F_{d1-2} = \frac{1}{2\pi} \left(\tan^{-1} \frac{1}{Y} - \frac{Y}{\sqrt{X^2 + Y^2}} \tan^{-1} \frac{1}{\sqrt{X^2 + Y^2}} \right)$$

7

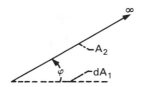

Element dA_1 of differential width and any length to semi-infinite plane. Plane containing dA_1 and semi-infinite plane intersect at angle φ at edge of semi-infinite plane.

$$F_{d1-2} = \tfrac{1}{2}(1 + \cos \varphi)$$

8

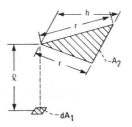

Plane element dA_1 to parallel isosceles triangle of height h and sides r; normal to element passes through vertex of triangle.

$$H = \frac{h}{l} \qquad R = \frac{r}{l}$$

$$F_{d1-2} = \frac{H}{\pi\sqrt{1 + H^2}} \tan^{-1} \sqrt{\frac{R^2 - H^2}{1 + H^2}}$$

9

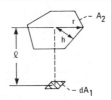

Plane element dA_1 to parallel polygon having n equal sides; normal to element passes through center of polygon.

$$H = \frac{h}{l} \qquad R = \frac{r}{l}$$

$$F_{d1-2} = \frac{nH}{\pi\sqrt{1 + H^2}} \tan^{-1} \sqrt{\frac{R^2 - H^2}{1 + H^2}}$$

10

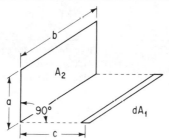

Strip element dA_1 to rectangle in plane at $90°$ to plane of strip.

$$X = \frac{a}{b} \qquad Y = \frac{c}{b}$$

$$F_{d1-2} = \frac{1}{\pi}\left[\tan^{-1}\frac{1}{Y} + \frac{Y}{2}\ln\frac{Y^2(X^2 + Y^2 + 1)}{(Y^2 + 1)(X^2 + Y^2)} - \frac{Y}{\sqrt{X^2 + Y^2}}\tan^{-1}\frac{1}{\sqrt{X^2 + Y^2}}\right]$$

11

Two infinitely long, directly opposed parallel plates of the same finite width.

$$H = \frac{h}{w}$$

$$F_{1-2} = F_{2-1} = \sqrt{1 + H^2} - H$$

12

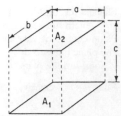

Identical, parallel, directly opposed rectangles.

$$X = \frac{a}{c} \qquad Y = \frac{b}{c}$$

$$F_{1\text{-}2} = \frac{2}{\pi XY} \left\{ \ln \left[\frac{(1 + X^2)(1 + Y^2)}{1 + X^2 + Y^2} \right]^{1/2} + X \sqrt{1 + Y^2} \, \tan^{-1} \frac{X}{\sqrt{1 + Y^2}} \right.$$

$$\left. + Y \sqrt{1 + X^2} \, \tan^{-1} \frac{Y}{\sqrt{1 + X^2}} - X \tan^{-1} X - Y \tan^{-1} Y \right\}$$

13

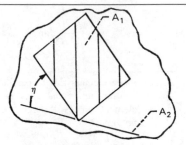

Finite rectangle A_1 of any size, tilted at angle η relative to an infinite plane A_2.

$$F_{1\text{-}2} = \tfrac{1}{2}(1 - \cos \eta)$$

14

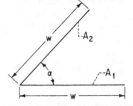

Two infinitely long plates of equal finite width w, having one common edge and having an included angle α to each other.

$$F_{1\text{-}2} = F_{2\text{-}1} = 1 - \sin \frac{\alpha}{2}$$

15

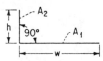

Two infinitely long plates of unequal widths h and w, having one common edge and having an angle of 90° to each other.

$$H = \frac{h}{w}$$

$$F_{1\text{-}2} = \tfrac{1}{2}(1 + H - \sqrt{1 + H^2})$$

16

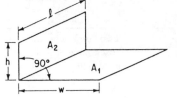

Two finite rectangles of same length, having one common edge and having an angle of 90° to each other.

$$H = \frac{h}{l} \qquad W = \frac{w}{l}$$

$$F_{1\text{-}2} = \frac{1}{\pi W} \left(W \tan^{-1} \frac{1}{W} + H \tan^{-1} \frac{1}{H} - \sqrt{H^2 + W^2} \, \tan^{-1} \frac{1}{\sqrt{H^2 + W^2}} \right.$$

$$\left. + \tfrac{1}{4} \ln \left\{ \frac{(1 + W^2)(1 + H^2)}{1 + W^2 + H^2} \left[\frac{W^2(1 + W^2 + H^2)}{(1 + W^2)(W^2 + H^2)} \right]^{W^2} \left[\frac{H^2(1 + H^2 + W^2)}{(1 + H^2)(H^2 + W^2)} \right]^{H^2} \right\} \right)$$

17

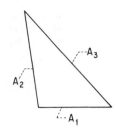

Infinitely long enclosure formed by three plane areas.

$$F_{1-2} = \frac{A_1 + A_2 - A_3}{2A_1}$$

18

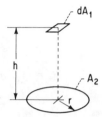

Plane element dA_1 to circular disk in plane parallel to element; normal to element passes through center of disk.

$$F_{d1-2} = \frac{r^2}{h^2 + r^2}$$

19

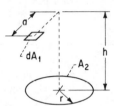

Plane element dA_1 to circular disk in plane parallel to element.

$$H = \frac{h}{a} \qquad R = \frac{r}{a}$$

$$Z = 1 + H^2 + R^2$$

$$F_{d1-2} = \frac{1}{2}\left(1 - \frac{1 + H^2 - R^2}{\sqrt{Z^2 - 4R^2}}\right)$$

20

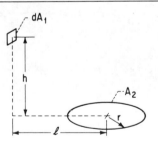

Plane element dA_1 to circular disk; planes containing element and disk intersect at 90°; $l \geq r$.

$$H = \frac{h}{l} \qquad R = \frac{r}{l}$$

$$Z = 1 + H^2 + R^2$$

$$F_{d1-2} = \frac{H}{2}\left(\frac{Z}{\sqrt{Z^2 - 4R^2}} - 1\right)$$

21

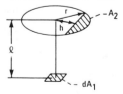

Plane element dA_1 to parallel circular segment; normal to element passes through center of disk containing segment.

$$H = \frac{h}{l} \qquad R = \frac{r}{l}$$

$$F_{d1-2} = \frac{1}{\pi}\left(\frac{R^2}{1 + R^2}\cos^{-1}\frac{H}{R} - \frac{H}{\sqrt{1 + H^2}}\tan^{-1}\sqrt{\frac{R^2 - H^2}{1 + H^2}}\right)$$

22

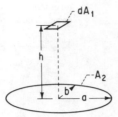

Plane element dA_1 to elliptical plate in plane parallel to element; normal to element passes through center of plate.

$$F_{d1-2} = \frac{ab}{\sqrt{(h^2 + a^2)(h^2 + b^2)}}$$

23

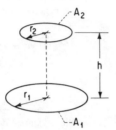

Parallel circular disks with centers along the same normal.

$$R_1 = \frac{r_1}{h} \qquad R_2 = \frac{r_2}{h}$$

$$X = 1 + \frac{1 + R_2^2}{R_1^2}$$

$$F_{1-2} = \frac{1}{2}\left[X - \sqrt{X^2 - 4\left(\frac{R_2}{R_1}\right)^2}\right]$$

24

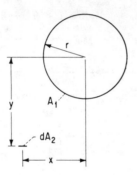

Strip element dA_2 of any length to infinitely long cylinder; $y \geq r$.

$$X = \frac{x}{r} \qquad Y = \frac{y}{r}$$

$$F_{d2-1} = \frac{Y}{X^2 + Y^2} \qquad (Y \geq 1)$$

25

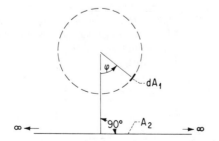

Element of any length on cylinder to plane of infinite length and width.

$$F_{d1-2} = \tfrac{1}{2}(1 + \cos \varphi)$$

26

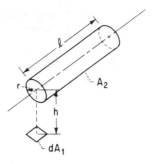

Plane element dA_1 to right circular cylinder of finite length l and radius r; normal to element passes through one end of cylinder and is perpendicular to cylinder axis.

$$L = \frac{l}{r} \qquad H = \frac{h}{r}$$

$$X = (1 + H)^2 + L^2$$
$$Y = (1 - H)^2 + L^2$$

$$F_{d1-2} = \frac{1}{\pi H} \tan^{-1} \frac{L}{\sqrt{H^2 - 1}} + \frac{L}{\pi} \left[\frac{X - 2H}{H\sqrt{XY}} \tan^{-1} \sqrt{\frac{X(H - 1)}{Y(H + 1)}} - \frac{1}{H} \tan^{-1} \sqrt{\frac{H - 1}{H + 1}} \right]$$

27

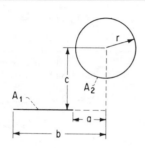

Infinitely long plane of finite width to parallel infinitely long cylinder.

$$F_{1-2} = \frac{r}{b - a} \left(\tan^{-1} \frac{b}{c} - \tan^{-1} \frac{a}{c} \right)$$

28

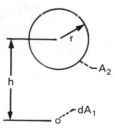

Infinitely long cylindrical line source dA_1 to a parallel infinitely long cylinder A_2.

$$F_{d1-2} = \frac{1}{\pi} \sin^{-1} \left(\frac{r}{h} \right)$$

29

Infinitely long cylindrical line source dA_1 to any cylindrical surface A_2 generated by a line of infinite length moving parallel to itself and parallel to the line source.

$$F_{d1-2} = \frac{1}{2\pi} (\varphi_2 - \varphi_1)$$

30

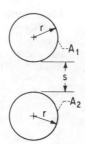

Infinitely long parallel cylinders of the same diameter.

$$X = 1 + \frac{s}{2r}$$

$$F_{1-2} = F_{2-1} = \frac{1}{\pi}\left(\sqrt{X^2 - 1} + \sin^{-1}\frac{1}{X} - X\right)$$

31

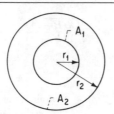

Concentric cylinders of infinite length.

$$F_{1-2} = 1$$

$$F_{2-1} = \frac{r_1}{r_2}$$

$$F_{2-2} = 1 - \frac{r_1}{r_2}$$

32

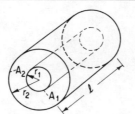

Two concentric cylinders of same finite length.

$$R = \frac{r_2}{r_1} \qquad L = \frac{l}{r_1}$$

$$A = L^2 + R^2 - 1$$

$$B = L^2 - R^2 + 1$$

$$F_{2-1} = \frac{1}{R} - \frac{1}{\pi R}\left\{\cos^{-1}\frac{B}{A} - \frac{1}{2L}\left[\sqrt{(A+2)^2 - (2R)^2}\cos^{-1}\frac{B}{RA} + B\sin^{-1}\frac{1}{R} - \frac{\pi A}{2}\right]\right\}$$

$$F_{2-2} = 1 - \frac{1}{R} + \frac{2}{\pi R}\tan^{-1}\frac{2\sqrt{R^2-1}}{L} - \frac{L}{2\pi R}\left[\frac{\sqrt{4R^2+L^2}}{L}\sin^{-1}\frac{4(R^2-1)+(L^2/R^2)(R^2-2)}{L^2 + 4(R^2-1)}\right.$$

$$\left. - \sin^{-1}\frac{R^2-2}{R^2} + \frac{\pi}{2}\left(\frac{\sqrt{4R^2+L^2}}{L} - 1\right)\right]$$

where for any argument ξ:

$$-\frac{\pi}{2} \leqslant \sin^{-1}\xi \leqslant \frac{\pi}{2}$$

$$0 \leqslant \cos^{-1}\xi \leqslant \pi$$

33

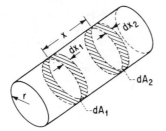

Two ring elements on the interior of a right circular cylinder.

$$X = \frac{x}{2r}$$

$$dF_{d1-d2} = \left[1 - \frac{2X^3 + 3X}{2(X^2+1)^{\frac{3}{2}}}\right]dX_2$$

34

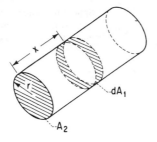

Ring element dA_1 on interior of right circular cylinder to circular disk A_2 at end of cylinder.

$$X = \frac{x}{2r}$$

$$F_{d1-2} = \frac{X^2 + \frac{1}{2}}{\sqrt{X^2 + 1}} - X$$

35

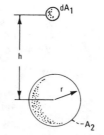

Spherical point source to a sphere of radius r.

$$R = \frac{r}{h}$$

$$F_{d1-2} = \frac{1}{2}(1 - \sqrt{1 - R^2})$$

36

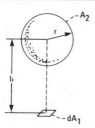

Plane element dA_1 to sphere of radius r; normal to center of element passes through center of sphere.

$$F_{d1-2} = \left(\frac{r}{h}\right)^2$$

37

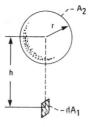

Plane element dA_1 to sphere of radius r; tangent to element passes through center of sphere.

$$H = \frac{h}{r}$$

$$F_{d1-2} = \frac{1}{\pi}\left(\tan^{-1}\frac{1}{\sqrt{H^2 - 1}} - \frac{\sqrt{H^2 - 1}}{H^2}\right)$$

38

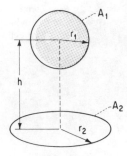

Sphere of radius r_1 to disk of radius r_2; normal to center of disk passes through center of sphere.

$$R_2 = \frac{r_2}{h}$$

$$F_{1-2} = \frac{1}{2}\left(1 - \frac{1}{\sqrt{1 + R_2{}^2}}\right)$$

39

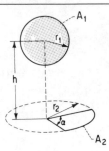

Sphere to sector of disk; normal to center of disk passes through center of sphere.

$$R_2 = \frac{r_2}{h}$$

$$F_{1-2} = \frac{\alpha}{4\pi}\left(1 - \frac{1}{\sqrt{1 + R_2{}^2}}\right)$$

40

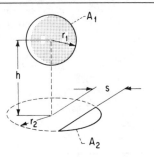

Sphere to segment of disk.

$$R_2 = \frac{r_2}{h} \qquad S = \frac{s}{h}$$

$$F_{1-2} = \frac{1}{8} - \frac{\cos^{-1}(S/R_2)}{2\pi\sqrt{1 + R_2{}^2}} + \frac{1}{4\pi}\sin^{-1}\frac{(1 - S^2)R_2{}^2 - 2S^2}{(1 + S^2)R_2{}^2}$$

41

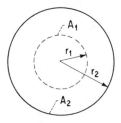

Concentric spheres.

$$F_{1-2} = 1$$

$$F_{2-1} = \left(\frac{r_1}{r_2}\right)^2$$

$$F_{2-2} = 1 - \left(\frac{r_1}{r_2}\right)^2$$

42

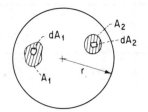

Differential or finite areas on the inside of a spherical cavity.

$$dF_{a1-a2} = dF_{1-d2} = \frac{dA_2}{4\pi r^2}$$

$$F_{d1-2} = F_{1-2} = \frac{A_2}{4\pi r^2}$$

RADIATIVE PROPERTIES

Tables of normal-total emissivities and normal-total absorptivities for incident solar radiation are provided here for convenience in working problems and to give the reader an indication of the magnitudes to be expected. As discussed in Chap. 5, many factors such as roughness and oxidation can strongly affect the radiative properties. No attempt is made here to describe in detail the condition of the material sample: hence the values given here are only reasonable approximations in some instances. For detailed information on radiative properties including sample descriptions and results from many sources, the reader is referred to the collections in [1–3]. Reference [3] in three volumes is very extensive. Some additional information is in [4]. Reference [5] has information on spacecraft materials and absorption for the solar spectrum. As will be seen from these references, for the same material there can sometimes be considerable differences in the property values measured by different investigators.

Normal-total emissivity

Metal	Surface temperature,[a] °F (K)	ϵ_n'
Aluminum:		
highly polished plate	400–1100 (480–870)	0.038–0.06
bright foil	70 (295)	0.04
polished plate	212 (373)	0.095
heavily oxidized	200–1000 (370–810)	0.20–0.33
Antimony, polished	100–500 (310–530)	0.28–0.31
Bismuth, bright	176 (350)	0.34
Brass:		
highly polished	500–700 (530–640)	0.028–0.031
polished	200 (370)	0.09
dull	120–660 (320–620)	0.22
oxidized	400–1000 (480–810)	0.60
Cadmium	77 (298)	0.02
Chromium, polished	100–2000 (310–1370)	0.08–0.40
Copper:		
highly polished	100 (310)	0.02
polished	100–500 (310–530)	0.04–0.05
scraped, shiny	100 (310)	0.07
slightly polished	100 (310)	0.15
black oxidized	100 (310)	0.78
Dow metal	0–600 (255–590)	0.15
Gold:		
highly polished	200–1100 (370–870)	0.018–0.035
polished	266 (400)	0.018
Haynes alloy X, oxidized	600–2000 (590–1370)	0.85–0.88
Iron:		
highly polished, electrolytic	100–500 (310–530)	0.05–0.07
polished	800–900 (700–760)	0.14–0.38
freshly rubbed with emery	100 (310)	0.24
wrought iron, polished	100–500 (310–530)	0.28
cast iron, freshly turned	100 (310)	0.44
iron plate, pickled, then rusted red	68 (293)	0.61
cast iron, oxidized at 1100°F	400–1100 (480–870)	0.64–0.78
cast iron, rough, strongly oxidized	100–500 (310–530)	0.95
Lead:		
polished	100–500 (310–530)	0.06–0.08
rough unoxidized	100 (310)	0.43
oxidized at 1100°F	100 (310)	0.63
Magnesium, polished	100–500 (310–530)	0.07–0.13
Mercury, unoxidized	40–200 (280–370)	0.09–0.12

[a]When temperatures and emissivities both have ranges, linear interpolation can be used over these values.

(Table continues on next page)

Normal-total emissivity (*Continued*)

Metal	Surface temperature,[a] °F (K)	ϵ_n'
Molybdenum:		
polished	100–500 (310–530)	0.05–0.08
polished	1000–2500 (810–1640)	0.10–0.18
polished	5000 (3030)	0.29
Monel:		
polished	100 (310)	0.17
oxidized at 1100°F	1000 (810)	0.45
Nickel:		
electrolytic	100–500 (310–530)	0.04–0.06
technically pure, polished	440–710 (500–650)	0.07–0.087
electroplated on iron, not polished	68 (293)	0.11
plate oxidized at 1100°F	390–1110 (470–870)	0.37–0.48
nickel oxide	1200–2300 (920–1530)	0.59–0.86
Platinum:		
electrolytic	500–1000 (530–810)	0.06–0.10
polished plate	440–1160 (500–900)	0.054–0.104
Silver, polished	100–1000 (310–810)	0.01–0.03
Stainless steel:		
type 304 foil (1 mil)	80 (300)	0.05
Inconel X foil (1 mil)	80 (300)	0.10
Inconel X, polished	−300–900 (90–760)	0.19–0.20
Inconel B, polished	−300–900 (90–760)	0.19–0.22
type 301, polished	75 (297)	0.16
type 310, smooth	1500 (1090)	0.39
type 316, polished	400–1900 (480–1310)	0.24–0.31
Steel:		
polished sheet	−300–0 (90–273)	0.07–0.08
polished sheet	0–300 (273–420)	0.08–0.14
mild steel, polished	500–1200 (530–920)	0.27–0.31
sheet with skin due to rolling	70 (295)	0.66
sheet with rough oxide layer	70 (295)	0.81
Tantalum	2500–5000 (1640–3030)	0.2–0.3
foil	80 (300)	0.05
Tin:		
polished sheet	93 (310)	0.05
bright tinned iron	76 (298)	0.043–0.064
Tungsten:		
polished	80 (300)	0.03
clean	100–1000 (310–810)	0.03–0.08
filament	80 (300)	0.032
filament	6000 (3590)	0.39
Zinc:		
polished	100–1000 (310–810)	0.02–0.05
galvanized sheet, fairly bright	100 (310)	0.23
gray oxidized	70 (295)	0.23–0.28

[a]When temperatures and emissivities both have ranges, linear interpolation can be used over these values.

Normal-total emissivity (*Continued*)

Dielectric	Surface temperature,[a] °F (K)	ϵ_n'
Alumina on Inconel	1000–2000 (810–1370)	0.65–0.45
Asbestos:		
cloth	199 (365)	0.90
paper	100 (310)	0.93
board	100 (310)	0.96
Asphalt pavement	100 (310)	0.93
Brick:		
white refractory	2000 (1370)	0.29
fireclay	1800 (1260)	0.75
rough red	100 (310)	0.93
Carbon, lampsoot	100 (310)	0.95
Ceramic, glazed earthenware	70 (295)	0.90
Clay, fired	158 (340)	0.91
Concrete, rough	100 (310)	0.94
Corundum, emery rough	200 (370)	0.86
Cotton cloth	68 (293)	0.77
Granite	70 (295)	0.45
Ice:		
smooth	32 (273)	0.966
rough crystals	32 (273)	0.985
Magnesium oxide, refractory	300–900 (420–760)	0.69–0.55
Marble, white	100 (310)	0.95
Mica	100 (310)	0.75
Paint:		
oil, all colors	212 (373)	0.92–0.96
red lead	200 (370)	0.93
lacquer, flat black	100–200 (310–370)	0.96–0.98
Paper:		
roofing	100 (310)	0.91
white	100 (310)	0.95
Plaster	100 (310)	0.91
Porcelain, glazed	70 (295)	0.92
Rokide A on molybdenum	600–1500 (590–1090)	0.79–0.60
Rubber, hard	68 (293)	0.92
Sand	68 (293)	0.76
Sandstone	100–500 (310–530)	0.83–0.90
Silicon carbide	300–1200 (420–920)	0.83–0.96
Silk cloth	68 (293)	0.78
Slate	100 (310)	0.67–0.80
Snow	20 (270)	0.82

[a]When temperatures and emissivities both have ranges, linear interpolation can be used over these values.

(*Table continues on next page*)

Normal-total emissivity (*Continued*)

Dielectric	Surface temperature,[a] °F (K)	ϵ_n'
Soil:		
black loam	68 (293)	0.66
plowed field	68 (293)	0.38
Soot, candle	200–500 (370–530)	0.95
Water, deep	32–212 (273–373)	0.96
Wood:		
sawdust	100 (310)	0.75
oak, planed	70 (295)	0.90
beech	158 (340)	0.94

[a]When temperatures and emissivities both have ranges, linear interpolation can be used over these values.

Normal-total absorptivity for incident solar radiation
(Receiving material at 300 K)

Metal	α_n'
Aluminum:	
highly polished	0.10
polished	0.20
Chromium, electroplated	0.40
Copper:	
highly polished	0.18
clean	0.25
tarnished	0.64
oxidized	0.70
Galvanized iron	0.38
Gold, bright foil	0.29
Iron:	
ground with fine grit	0.36
blued	0.55
sandblasted	0.75
Magnesium, polished	0.19
Nickel:	
highly polished	0.15
polished	0.36
electrolytic	0.40

Normal-total absorptivity
for incident solar radiation (*Continued*)
(Receiving material at 300 K)

Metal	α'_n
Platinum, bright	0.31
Silver:	
highly polished	0.07
polished	0.13
commercial sheet	0.30
Stainless steel #301, polished	0.37
Tungsten, highly polished	0.37

Dielectric	α'_n
Aluminum oxide (Al_2O_3)	0.06–0.23
Asphalt pavement, dust-free	0.93
Brick, red	0.75
Clay	0.39
Concrete roofing tile:	
uncolored	0.73
brown	0.91
black	0.91
Earth, plowed field	0.75
Felt, black	0.82
Graphite	0.88
Grass	0.75–0.80
Gravel	0.29
Leaves, green	0.71–0.79
Magnesium oxide (MgO)	0.15
Marble, white	0.46
Paint:	
aluminum	0.55
oil, zinc white	0.30
oil, light green	0.50
oil, light gray	0.75
oil, black on galvanized iron	0.90
Paper, white	0.28
Slate, blue gray	0.88
Snow, clean	0.2–0.35
Soot, coal	0.95
Titanium dioxide (TiO_2)	0.12
Zinc oxide	0.15
Zinc sulfide (ZnS)	0.21

(*Table continues on next page*)

Normal-total absorptivity
for incident solar radiation (*Continued*)
(Receiving material at 300 K)

Coating	α_n'
Black coatings:	
anodize black	0.88
carbon black paint NS-7	0.96
Ebanol C black	0.97
Martin black velvet paint	0.91
3M black velvet paint	0.97
Tedlar black plastic	0.94
Velestat black plastic	0.96
White coatings:	
barium sulfate with polyvinyl alcohol	0.06
Catalac white paint	0.23
Dow Corning white paint DC-007	0.19
magnesium oxide white paint	0.09
potassium fluorotitanate white paint	0.15
Tedlar white plastic	0.39
titanium oxide white paint with methyl silicone	0.20
zinc oxide with sodium silicate	0.15
Conversion coatings (values can vary significantly with coating thickness):	
Alzac A-2	0.16
black chrome	0.96
black copper	0.98
black irridite	0.62
black nickel	0.91
Vapor-deposited coatings on glass substrates:	
aluminum	0.08
chromium	0.56
gold	0.19
nickel	0.38
silver	0.04
titanium	0.52
tungsten	0.60
Plastic films with metal backing:	
Mylar film, 3-mil aluminum backing	0.17
Tedlar film, 1-mil gold backing	0.26
Teflon film, 2-mil aluminum backing	0.08
Teflon film, 1-mil gold backing	0.22
Teflon film, 2-mil silver backing	0.08

REFERENCES

1. Gubareff, G. G., J. E. Janssen, and R. H. Torborg: "Thermal Radiation Properties Survey," 2d ed., Honeywell Research Center, Minneapolis-Honeywell Regulator Co., Minneapolis, 1960.
2. Wood, W. D., H. W. Deem, and C. F. Lucks: "Thermal Radiative Properties," Plenum Press, New York, 1964.

3. Touloukian, Y. S., et al.: "Thermal Radiative Properties," vol. 7, "Metallic Elements and Alloys," vol. 8, "Nonmetallic Solids," vol. 9, "Coatings," Thermophysical Properties Research Center of Purdue University, Data Series, Plenum Publishing, 1970.

4. Svet, Darii Yakovlevich: "Thermal Radiation, Metals, Semiconductors, Ceramics, Partly Transparent Bodies, and Films," Consultants Bureau, Plenum Publishing, New York, 1965.

5. Henninger, J. H.: "Solar Absorptance and Thermal Emittance of Some Common Spacecraft Thermal-Control Coatings," NASA Reference Publication 1121, 1984.

EXPONENTIAL INTEGRAL RELATIONS AND TWO-DIMENSIONAL RADIATION FUNCTIONS

E-1 EXPONENTIAL INTEGRAL RELATIONS

A summary of some useful exponential integral relations is presented here. Additional relations are in [1–3].

For positive real arguments, the nth exponential integral is defined as

$$E_n(x) \equiv \int_0^1 \mu^{n-2} \exp\left(\frac{-x}{\mu}\right) d\mu \tag{E-1}$$

and only positive integral values of n will be considered here. An alternative form is

$$E_n(x) = \int_1^\infty \frac{1}{t^n} \exp(-xt) \, dt \tag{E-2}$$

By differentiating (E-1) under the integral sign, the recurrence relation is obtained:

$$\frac{d}{dx} E_n(x) = -E_{n-1}(x) \qquad n \geq 2$$

$$\frac{d}{dx} E_1(x) = -\frac{1}{x} \exp(-x) \tag{E-3}$$

Another recurrence relation obtained by integration is

$$nE_{n+1}(x) = \exp(-x) - xE_n(x) = \exp(-x) + x \frac{d}{dx} E_{n+1}(x) \qquad n \geq 1 \tag{E-4}$$

Also, integration results in

$$\int E_n(x)\, dx = -E_{n+1}(x) \tag{E-5}$$

By use of Eq. (E-4), all exponential integrals can be reduced to the first exponential integral given by

$$E_1(x) = \int_0^1 \mu^{-1} \exp\left(\frac{-x}{\mu}\right) d\mu \tag{E-6}$$

Alternative forms of $E_1(x)$ are

$$E_1(x) = \int_1^\infty t^{-1} \exp\left(-xt\right) dt = \int_x^\infty t^{-1} \exp\left(-t\right) dt \tag{E-7}$$

For $x = 0$ the exponential integrals are equal to

$$E_n(0) = \frac{1}{n-1} \qquad n \geq 2$$

$$E_1(0) = +\infty \tag{E-8}$$

For large values of x there is the asymptotic expansion

$$E_n(x) = \frac{\exp\left(-x\right)}{x} \left[1 - \frac{n}{x} + \frac{n(n+1)}{x^2} - \frac{n(n+1)(n+2)}{x^3} + \cdots \right] \tag{E-9}$$

Therefore, as $x \to \infty$, $E_n(x) \to \exp(-x)/x \to 0$.
   Series expansions are of the form

$$E_1(x) = -\gamma - \ln x + x - \frac{x^2}{2 \times 2!} + \frac{x^3}{3 \times 3!} - \cdots$$

$$= -\gamma - \ln x - \sum_{n=1}^\infty (-1)^n \frac{x^n}{n \times n!} \tag{E-10}$$

$$E_2(x) = 1 + (\gamma - 1 + \ln x)x - \frac{x^2}{1 \times 2!} + \frac{x^3}{2 \times 3!} - \cdots$$

$$E_3(x) = \tfrac{1}{2} - x + \tfrac{1}{2}(-\gamma + \tfrac{3}{2} - \ln x)x^2 + \frac{x^3}{1 \times 3!} - \cdots$$

where $\gamma = 0.577216$ is Euler's constant. The general series expansion given in [3] is

$$E_n(x) = \frac{(-x)^{n-1}}{(n-1)!} [-\ln x + \psi(n)] - \sum_{\substack{m=0 \\ (m \neq n-1)}}^\infty \frac{(-x)^m}{(m-n+1)m!} \tag{E-11}$$

where $\psi(1) = -\gamma$ and $\psi(n) = -\gamma + \sum_{m=1}^{n-1} \frac{1}{m} \qquad n \geq 2$

Some rough approximations are as follows (from Section 13-5.4):

$E_3(x) \approx \frac{1}{2} \exp(-1.8x)$

Using (E-3) gives

$E_2(x) \approx 0.9 \exp(-1.8x)$

In [4], the approximations are used,

$E_2(x) \approx \frac{3}{4} \exp(-\frac{3}{2}x)$

$E_3(x) \approx \frac{1}{2} \exp(-\frac{3}{2}x)$

Table E-1 Values of exponential integrals $E_n(x)$ [2]

x	$E_1(x)$	$E_2(x)$	$E_3(x)$	$E_4(x)$	$E_5(x)$
0	∞	1.00000	0.50000	0.33333	0.25000
0.01	4.03793	0.94967	0.49028	0.32838	0.24669
0.02	3.35471	0.91311	0.48097	0.32353	0.24343
0.03	2.95912	0.88167	0.47200	0.31876	0.24022
0.04	2.68126	0.85354	0.46332	0.31409	0.23706
0.05	2.46790	0.82784	0.45492	0.30949	0.23394
0.06	2.29531	0.80405	0.44676	0.30499	0.23087
0.07	2.15084	0.78184	0.43883	0.30056	0.22784
0.08	2.02694	0.76096	0.43112	0.29621	0.22486
0.09	1.91874	0.74124	0.42361	0.29194	0.22191
0.10	1.82292	0.72255	0.41629	0.28774	0.21902
0.15	1.46446	0.64104	0.38228	0.26779	0.20514
0.20	1.22265	0.57420	0.35195	0.24945	0.19221
0.25	1.04428	0.51773	0.32468	0.23254	0.18017
0.30	0.90568	0.46912	0.30004	0.21694	0.16893
0.35	0.79422	0.42671	0.27767	0.20250	0.15845
0.40	0.70238	0.38937	0.25729	0.18914	0.14867
0.50	0.55977	0.32664	0.22160	0.16524	0.13098
0.60	0.45438	0.27618	0.19155	0.14463	0.11551
0.70	0.37377	0.23495	0.16606	0.12678	0.10196
0.80	0.31060	0.20085	0.14432	0.11129	0.09007
0.90	0.26018	0.17240	0.12570	0.09781	0.07963
1.00	0.21938	0.14850	0.10969	0.08606	0.07045
1.20	0.15841	0.11110	0.08393	0.06682	0.05525
1.40	0.11622	0.08389	0.06458	0.05206	0.04343
1.60	0.08631	0.06380	0.04991	0.04068	0.03420
1.80	0.06471	0.04882	0.03872	0.03187	0.02698
2.00	0.04890	0.03753	0.03013	0.02502	0.02132
2.25	0.03476	0.02718	0.02212	0.01855	0.01592
2.50	0.02491	0.01980	0.01630	0.01378	0.01191
2.75	0.01798	0.01449	0.01205	0.01027	0.00892
3.00	0.01305	0.01064	0.00893	0.00767	0.00670
3.25	0.00952	0.00785	0.00664	0.00573	0.00504
3.50	0.00697	0.00580	0.00495	0.00430	0.00379

Tabulations of $E_n(x)$ are in [2, 3]. An abridged listing is given in Table E-1 for convenience. Reference [5] presents forms of a generalized exponential integral function that are convenient for numerical computation. The generalized forms include as a special case the $E_n(x)$ discussed here.

E-2 TWO-DIMENSIONAL RADIATION FUNCTIONS

For two-dimensional problems the exponential integral functions are replaced by another set of integral functions. These are the S_n functions and can be defined in a number of equivalent ways. An expression that is easily evaluated by numerical integration is

$$S_n(x) = \frac{2}{\pi} \int_0^{\pi/2} \exp\left(-\frac{x}{\cos \theta}\right) \cos^{n-1} \theta \, d\theta \tag{E-12}$$

An alternative form is

$$S_n(x) = \frac{2}{\pi} \int_1^{\infty} \frac{\exp(-xt) \, dt}{t^n(t^2 - 1)^{1/2}}, \qquad x \geq 0, \, n = 0, 1, 2, \ldots \tag{E-13}$$

Another form that has been used is

$$S_n(a\tau) = \frac{\tau^n}{\pi} \int_{-\infty}^{\infty} \frac{\exp[-a(x^2 + \tau^2)^{1/2}]}{(x^2 + \tau^2)(n + 1)/2} \, dx, \qquad a \geq 0, \, \tau \geq 0, \, n = 0, 1, 2, \ldots \tag{E-14}$$

From the definition of $S_n(x)$, the values at $x = 0$ can be found from ($n > 0$),

$$S_n(0) = \frac{1}{\pi^{1/2}} \frac{\Gamma(n/2)}{\Gamma[(n + 1)/2]} \tag{E-15}$$

This yields $S_1(0) = 1$, $S_2(0) = 2/\pi$, $S_3(0) = 1/2$, and $S_4(0) = 4/(3\pi)$. The $S_0(x)$ is related to the modified Bessel function by

$$S_0(x) = \frac{2}{\pi} K_0(x) \tag{E-16}$$

The derivative of $S_n(x)$ is given by

$$\frac{dS_n(x)}{dx} = -S_{n-1}(x), \qquad n \geq 1 \tag{E-17}$$

Hence the integral of S_n is

$$\int S_n(x) \, dx = -S_{n+1}(x), \qquad n \geq 0 \tag{E-18}$$

Table E-2 Values of the two-dimensional radiation function $S_n(x)$

x	$S_0(x)$	$S_1(x)$	$S_2(x)$	$S_3(x)$	$S_4(x)$
0	∞	1.00000	0.63662	0.50000	0.42441
0.002	4.03015	0.99067	0.63463	0.49873	0.42341
0.005	3.44684	0.97958	0.63167	0.49683	0.42192
0.02	2.56460	0.93598	0.61732	0.48746	0.41454
0.04	2.12411	0.88960	0.59908	0.47530	0.40491
0.07	1.76969	0.83169	0.57329	0.45772	0.39092
0.10	1.54512	0.78217	0.54910	0.44089	0.37744
0.15	1.29236	0.71168	0.51180	0.41438	0.35607
0.20	1.11581	0.65169	0.47776	0.38965	0.33597
0.25	0.98135	0.59940	0.44651	0.36656	0.31708
0.30	0.87374	0.55312	0.41772	0.34496	0.29929
0.35	0.78477	0.51172	0.39111	0.32475	0.28256
0.40	0.70953	0.47441	0.36648	0.30582	0.26680
0.45	0.64484	0.44059	0.34362	0.28807	0.25195
0.5	0.58850	0.40979	0.32237	0.27143	0.23797
0.6	0.49499	0.35580	0.28417	0.24115	0.21237
0.7	0.42050	0.31016	0.25093	0.21443	0.18962
0.8	0.35991	0.27124	0.22191	0.19082	0.16939
0.9	0.30986	0.23783	0.19650	0.16993	0.15137
1.0	0.26803	0.20899	0.17419	0.15142	0.13532
1.2	0.20277	0.16227	0.13728	0.12042	0.10826
1.5	0.13611	0.11218	0.09661	0.08571	0.07764
2.0	0.07251	0.06183	0.05442	0.04900	0.04484
3.0	0.02212	0.01964	0.01778	0.01633	0.01516
4.0	0.00710	0.00646	0.00595	0.00554	0.00520
5.0	0.00235	0.00217	0.00202	0.00190	0.00180

Values of the first five S_n functions were computed numerically from (E-12) and are in Table E-2. Additional information is in [6].

REFERENCES

1. Chandrasekhar, Subrahmanyan: "Radiative Transfer," Dover, New York, 1960.
2. Kourganoff, Vladimir: "Basic Methods in Transfer Problems," Dover, New York, 1963.
3. Abramowitz, Milton, and Irene A. Stegun (eds.): "Handbook of Mathematical Functions with Formulas, Graphs, and Mathematical Tables," *Appl. Math. Ser.* 55, National Bureau of Standards, 1964.
4. Cess, R. D., and S. N. Tiwari: "Infrared Radiative Energy Transfer in Gases," Advances in Heat Transfer, vol. 8, Academic Press, New York, 1972.
5. Breig, W. F., and A. L., Crosbie: Numerical Computation of a Generalized Exponential Integral Function, *Math. Comp.*, vol. 28, no. 126, pp. 575–579, 1974.
6. Yuen, W. W., and L. W. Wong: Numerical Computation of an Important Integral Function in Two-Dimensional Radiative Transfer, *J. Quant. Spectrosc. Radiat. Transfer*, vol. 29, no. 2, pp. 145–149, 1983.

INDEX TO INFORMATION IN TABLES AND FIGURES

(Table continues on next page)

Subject	Table or figure number
Energy and temperature relations for gray medium between gray surfaces:	
diffusion theory	Table 15-2
exchange-factor relations	Table 15-4
P-1 approximation	Table 16-3
Energy transfer between:	
concentric cylinders	Table 9-1
concentric spheres	Table 9-1
parallel plates	Table 9-1
Energy transfer for gray medium between black parallel plates	Table 14-3, Fig. 14-12
Equation of transfer approximations	Table 15-1
Exchange areas	Table 13-8
Exponential integral values:	
one-dimensional	Table E-1
two-dimensional	Table E-2
Fundamental numerical values	Table A-1
Flame temperatures	Table 13-9
Gas emittance:	
carbon dioxide	Figs. 12-10, 13-13, 13-18, 13-20
water vapor	Figs. 12-10b, 13-15, 13-19, 13-21, 13-22
Gray gas coefficients:	
carbon dioxide	Tables 13-5, 13-7
water vapor	Tables 13-6, 13-7
Kirchhoff's-law relations	Table 3-2
Mean beam-length relations	Table 13-4
Monte Carlo random-number relations:	
surface emission	Table 11-3
radiation in a medium	Table 15-5
Property prediction equations by electromagnetic theory, summary	Table 4-4
Radiation constants	Table A-4
Reflectivity reciprocity relations	Table 3-3
Refractive indices	Tables 4-2, 4-3, 18-1
Scattering behavior:	
cross sections	Table 12-1
independent and dependent regimes	Fig. 12-18
polarizability	Table 12-6
Soot:	
optical constants	Table 13-10
emittance	Figs. 13-30, 13-31
Surface properties:	
definitions	Table 3-1
emissivity values	Appendix D
solar-absorptivity values	Appendix D
Water properties:	
absorption coefficient	Table 5-2
solar transmission	Table 5-3

GLOSSARY

Absolute temperature—Temperature measured on one of the thermodynamic scales, either Kelvin or Rankine.

Absorbing media—Media that absorb electromagnetic radiation and convert it into internal energy.

Absorptance—The property of a medium that determines the fraction of radiant energy traveling along a path that will be absorbed within a given distance.

Absorption coefficient—The property of a medium that describes the amount of absorption of thermal radiation per unit path length within the medium.

Absorptivity—The property of a material that gives the fraction of energy incident on the material that is absorbed.

Adiabatic—A boundary or material possessing the quality of a perfect insulator, so that no heat is transferred through it.

Albedo for scattering—The ratio of the scattering coefficient to the extinction coefficient.

Angle for total reflection—The angle from the normal of the interface between two materials of differing refractive index at which total reflection of incident intensity occurs.

Band—A spectral interval containing many closely spaced spectral lines, so that radiation within that spectral interval is significantly absorbed.

Bandwidth—The spectral interval containing the major absorption capability of an absorbing band.

Black—Having the property of complete absorption of incident radiation of all wavelengths and from all directions.

Blackbody—Any object or material that completely absorbs all incident radiation. A blackbody emits energy in a manner described by the Stefan-Boltzmann law, Wien's displacement law, and the Planck spectral distribution of energy.

Bouguer's law—The mathematical relation describing the exponential reduction in intensity of radiation as it travels along a path of finite length within a medium.

Brewster's angle—The incidence angle at which there is zero reflection for the parallel polarized component of the radiation incident on an interface.

Broadening—The increase in the spectral width of an absorption line because of any of various mechanisms.

Cavity—Any surface concavity that acts to increase the apparent absorptivity of a surface by increasing multiple reflections (and therefore the absorption) of incident radiation.

Cold-medium approximation—An approximate form of the equation of radiation transfer derived by omitting the emission terms of the complete equation. The approximation is justified if the medium is at a low temperature so that its emission can be neglected relative to radiation from the boundaries.

Configuration factor—The fraction of uniform diffuse radiant energy leaving one surface that is incident upon a second surface.

Configuration-factor algebra—Mathematical relations between configuration factors.

Cosine law—The mathematical relation describing the variation of emissive power from a diffuse surface as varying with the cosine of the angle measured from the normal of the surface.

Crossed-string method—A method for easy calculation of the configuration factor between objects having one infinite dimension and a nonvarying cross-sectional geometry.

Cross section—The apparent projected area of a particle or object relative to its ability to absorb or scatter thermal radiation.

Curtis-Godson approximation—A method for computing radiative transfer through gases that makes use of the limiting absorption characteristics of weakly and strongly absorbing bands to compute absorption in all spectral regions.

Cutoff wavelength—The wavelength at which the absorptive properties of a surface or partially transparent medium undergo a large change, thus imparting spectral selectivity.

Dielectric—An electrically insulating material.

Differential approximation—An approximation to the equation of transfer derived by approximating the complete equation with a finite set of moment equations.

Diffuse surface—Surface that emits and/or reflects equal radiation intensity into all directions.

Diffusion approximation—An approximation to the equation of transfer derived by assuming the medium to be optically thick, so that the mean free path for radiation propagation is small.

Draper point—The temperature at which visible radiation emitted by a heated blackbody in darkened surroundings becomes visible to the human eye.

Drude theory—A theoretical approach to predicting the radiative properties of materials, based on electromagnetic theory.

Elsasser model—A model of the absorption properties of an absorption band based on the assumption that the individual lines in the band have a Lorentz shape and are equally spaced and of the same shape.

Emissive power—The rate of radiative energy emission per unit area from a surface.

Emissivity—The property of a body that describes its ability to emit radiation as compared with the emission from a blackbody at the same temperature.

Emittance—The property of an isothermal material that describes the ability of a given thickness of the material to emit energy as compared to emission by a blackbody at the same temperature.

Enclosure—A concept to account for the incident radiation from all directions of the surrounding space.

Equation of transfer—The mathematical relation describing the variation along a path of the intensity of radiation in an absorbing, emitting, and scattering medium.

Exchange areas—Factors accounting for the geometric and absorption effects on radiation between volumes or surfaces separated by attenuating media.

Exchange factor—For radiation exchange between surfaces in a system including specular surfaces, the fraction of diffuse energy leaving one surface that is incident on a second surface by direct paths and by all possible paths of specular reflection.

Exchange-factor approximation—An approximate method for calculating energy transfer in absorbing-emitting media with parallel radiation, convection, and/or conduction. The approximation is based on the assumption that the exchange factors for radiative equilibrium are unchanged by the presence of other heat transfer modes.

Exponential wide-band model—A model of an absorption band constructed of equally spaced Lorentz lines arranged with exponentially decreasing line strengths away from the band center. This model allows accurate description of the band absorption properties at long path lengths where there is increased absorption in the wings of the band.

Extinction coefficient—The property of a medium describing its ability to attenuate intensity per unit of path length. It is composed of the sum of the absorption and scattering coefficients.

Fin efficiency—The ratio of the actual energy lost from a fin to the amount that would be lost if the fin were at a uniform temperature equal to its root temperature.

Fourier's conduction law—The relation between energy flux by conduction heat transfer and the negative of the temperature gradient in the direction of the energy flux.

Frequency—Propagation velocity divided by wavelength; has the advantage that it does not change when a wave goes from one material into another with a different refractive index.

Fresnel equations—The equations describing the reflection characteristics of electromagnetic radiation at an optically smooth interface between media with differing refractive indices.

Gray—Having radiative properties that do not vary with wavelength.

Greenhouse effect—The trapping effect for radiation by substances that are transparent to radiation in the visible portion of the spectrum, such as solar energy, but are relatively opaque to radiation in the infrared portion of the spectrum, such as that emitted by low-temperature surfaces.

Hagen-Rubens relations—Relations between the optical, radiative, and electrical properties of materials developed from electromagnetic theory.

Half-width—One-half the spectral width of an absorption line at the point of half the maximum line intensity.

Hemispherical properties—Radiative properties of a surface element that are averaged over all solid angles passing through a hemisphere centered over the surface element.

Hohlraum—A heated cavity constructed in such a way that its opening closely approximates the absorption and emission properties of a blackbody.

Index of refraction—The property of a medium equal to the ratio of the speed of electromagnetic radiation in vacuum to the speed in the material. In an attenuating material, a complex refractive index is defined.

Induced emission—That portion of thermal radiation emission caused by the presence of a radiation field.

Infrared—The part of the electromagnetic spectrum in the wavelength region 0.7 to 1000 μm.

Intensity—Radiative energy passing through an area per unit solid angle, per unit of the area projected normal to the direction of passage, and per unit time.

Isotropic—Having no dependence on direction or angle.

Jump boundary condition—A boundary condition accounting for the discontinuity between the temperature of a bounding surface and the temperature of an absorbing, emitting, and scattering medium adjacent to the surface in the absence of heat conduction.

Kernel—That portion of the integrand that contains the dependent variable in an integral equation. If the dependent and independent variables can be interchanged without affecting the value of the kernel, then the kernel is symmetric. Symmetric kernels are necessary for some standard solution methods for certain integral equations.

Kirchhoff's law—The equality of the directional spectral emissivity and directional spectral absorptivity of a material. Various averaged emissivities and absorptivities are also equal if certain restrictions are met.

Lambert's cosine law—See *Cosine law*.

Line—A very narrow spectral region that absorbs radiation in a medium. The line shape and spectral location are determined by the quantum-mechanical properties of the medium.

Local thermodynamic equilibrium—Having sufficiently approached thermodynamic equilibrium at each location that thermodynamic properties such as temperature can be used to describe the state of each local volume element.

Luminescence—Emission of radiant energy where excitation is by means other than thermal agitation.

Luminous—Emitting radiation in the visible spectrum, usually because of the presence of incandescent particles.

Markov chain—A chain of events, the probability of each event in the chain being independent of all prior events.

Mean absorption coefficient—A spectral absorption coefficient weighted by a spectral energy distribution (for example, the Planck distribution) and averaged over the spectrum.

Mean beam length—An average path length traveled by radiation within a volume of absorbing medium.

Mie scattering—The general scattering theory derived from consideration of the interaction of an electromagnetic wave with a spherical particle, and applying over the entire range of particle diameters relative to wavelength.

Milne-Eddington approximation—An approximation to the differential form of the equation of transfer that assumes in a one-dimensional geometry that the radiation traveling with positive direction components is isotropic, while that with negative components is isotropic with a different value.

Monochromatic—Refers to radiation at one wavelength; spectral.

Monte Carlo method—A numerical technique based on a probabilistic model of a physical process. For radiation problems, a physical model that simulates the transfer behavior of radiative energy "bundles" is used to calculate energy transmission.

Natural broadening—The increased width of a spectral line due to uncertainty-principle effects on the energy states for the transition process producing the line.

Net-radiation method—A general method for writing a closed set of equations that can be solved for the radiative transfer between surfaces in an enclosure.

Newton-Raphson method—A technique for numerical solution of a set of simultaneous nonlinear algebraic equations.

Opacity—A measure of the ability of a medium to attenuate energy. The opacity is the extinction coefficient integrated over the path length; also called the *optical thickness*.

Optically smooth—See *Optical roughness*.

Optical roughness—The roughness of a surface relative to the wavelength of incident radiation. Surfaces with roughness much smaller than the wavelength of the incident radiation are optically smooth; those with roughness greater than the wavelength are optically rough.

Optical thickness—The extinction coefficient integrated over the physical thickness; for constant properties it is the product of extinction coefficient and thickness.

Optically thin—Having a small opacity that causes only slight attenuation of radiation along a given path.

Phase function—The mathematical description of a scattering medium that gives the relative amount of intensity that is scattered into various scattering angles from a given angle of incidence.

Planck distribution—The distribution of energy as a function of wavelength that is emitted at a given temperature by a blackbody.

Polarization—The concept of the amplitude of an electromagnetic wave having components that are perpendicular to one another and thus that can be treated separately and then added vectorially.

Poljak net-radiation method—See *Net-radiation method*.

Prevost's law—The concept that a body emits electromagnetic radiation even when it is in thermal equilibrium with its surroundings.

Radiative equilibrium—A system in the steady state, with no energy transfer by conduction or convection, is said to be in radiative equilibrium. Every element of such a system must be emitting radiation at the same rate as it absorbs radiation (no internal heat generation).

Radiosity—The rate at which radiant energy leaves a surface by combined emission and reflection of radiation.

Random number—A number chosen at random from a large set of numbers evenly distributed within an interval. The interval is usually 0 to 1.

Rayleigh-Jeans formula—A distribution of radiant energy with respect to wavelength for a blackbody at a given temperature; derived from the concept of equipartition of energy and valid only at very large values of wavelength times temperature.

Rayleigh scattering—The type of scattering of radiation that occurs when the wavelength of the incident radiation is much larger than the size of the scattering particles. Atmospheric scattering of solar radiation is generally of this type.

Reciprocity—The mathematical relation between the configuration factor for radiation from surface a to b and that from surface b to a.

Reflectivity—The property of a surface that describes the fraction of incident energy that is reflected from the surface.

Refraction—The change in direction of radiation upon crossing an interface between materials with differing indices of refraction.

Refractive index—See *Index of refraction*.

Rosseland diffusion equation—The relation between energy and the gradient of emissive power in an optically thick medium as derived in the diffusion approximation.

Rosseland mean absorption coefficient—The spectral absorption coefficient of a medium averaged over the spectrum as weighted by the derivative of the Planck distribution with respect to total emissive power. Applies to optically thick media.

Scattering coefficient—The property of a medium that describes the amount of scattering of thermal radiation per unit path length for propagation in the medium.

Scattering cross section—The apparent projected area of a particle relative to its ability to scatter radiation.

Schuster-Schwarzchild approximation—An approximation to the differential form of the equation of transfer for one-dimensional planar geometries, based on the assumption that intensities with positive components of direction have one value at all angles, while intensities with negative components have another value. The resulting equations are a simplified form of those derived from P-N and discrete-ordinate methods.

Selective surface—A surface with spectral radiation properties that vary widely with wavelength, causing emphasis or suppression of absorption and emission in different spectral regions.

Semigray approximation—A solution technique that assumes that radiant interchange can be treated in two separate spectral regions (usually the solar-dominated and infrared regions) so that the effects of spectrally selective surfaces and transmitting media can be accounted for.

Slip boundary condition—See *Jump boundary condition.*

Snell's law—The mathematical relation between the angles of incidence and refraction at an interface between two media and the refractive indices of the media.

Solid angle—Area intercepted on a unit sphere by a conical angle originating at the sphere center.

Source function—The mathematical relation describing the gain in intensity at a location because of both emission and scattering into the direction of the intensity.

Spectral—Having a dependence on wavelength or frequency; radiation within a very narrow region of wavelength or frequency.

Specular—Mirrorlike in reflection behavior.

Spontaneous emission—Emission of radiation because of a spontaneous change in energy level of a substance.

Stefan-Boltzmann constant—The proportionality constant σ between the blackbody hemispherical total emissive power e_b and the fourth power of the absolute temperature ($e_b = \sigma T^4$).

Stefan-Boltzmann law—The relation between blackbody hemispherical total emissive power and the fourth power of the absolute temperature.

Stimulated emission—See *Induced emission.*

Thermal radiation—Radiation detected as heat or light ($\lambda \approx 0.2$–1000 μm).

Total—Integrated or summed over all wavelengths or frequencies.

Transmittance—The property of a material that determines the fraction of energy at the origin of a path that will be transmitted through a given thickness.

Transmittance factor—The fraction of energy leaving one surface that is incident upon a second surface after absorption of some of the energy by an absorbing medium between the surfaces.

Transparent-gas approximation—An approximation to the equation of transfer that contains the assumption that no attenuation of energy occurs between the point of origin of the energy and the point of absorption.

True absorption coefficient—The absorption coefficient of a medium that would be measured in the absence of induced emission.

Unit-sphere method—A graphical method for the determination of configuration factors, based on the fact that if an image of a surface is formed on a unit hemispherical surface and then projected onto the base of the hemisphere, the resulting projected area is simply related to the configuration factor between the original surface and an area element at the center of the hemisphere.

Wavenumber—The reciprocal of the wavelength; number of waves per unit length of propagation.

Wien's displacement law—The mathematical relation showing that the product of the wavelength at the peak of a blackbody spectral energy distribution and the blackbody temperature is a constant.

Wien's formula—A spectral distribution of energy emission from an ideal blackbody that is derived from classical thermodynamics. It is quite accurate for most wavelength-temperature ranges of engineering interest.

Zone method—A method of computing radiant exchange in enclosures containing absorbing-emitting media that is based on dividing the boundary into finite areas and the medium into finite volumes. The areas and volumes are assumed to be individually uniform, and energy exchange among all elements and volumes is then calculated.

INDEX